Year	Name	Country	Contribution
1628	William Harvey	Britain	Demonstrates that the blood circulates and the heart is a pump.
1648	Jan B. van Helmont	Belgium	Shows that plants derive little substance from the soil.
1665	Robert Hooke	Britain	Uses the word "cell" to describe compartments he sees in cork under the microscope.
1668	Francesco Redi	Italy	Shows that decaying meat protected from flies does not spontaneously produce maggots.
1672	Marcello Malpighi	Italy	Microscopic studies allow him to discover that capillaries link arteries to veins.
1673	Anton van Leeuwenhoek	Holland	Uses microscope to view living microorganisms.
1700	John Ray	Britain	Describes many plants, and classifies flowering plants as either monocots or dicots.
1735	Carolus Linnaeus	Sweden	Initiates the binomial system of naming organisms.
1772	Joseph Priestley	Britain	Demonstrates that plants give off a gas required by animals.
1779	Jan Ingenhousz	Holland	Declares that sunlight is required for the green parts of plants to photosynthesize and purify the air.
1786	Luigi Galvani	Italy	Discovers that nerves can be electrically stimulated and this will lead to muscle contraction.
1796	Edward Jenner	Britain	Shows that vaccination with cowpox protects individuals from smallpox.
1809	Jean B. Lamarck	France	Supports the idea of evolution but thinks there is inheritance of acquired characteristics.
1825	Georges Cuvier	France	Founds the science of paleontology and shows that fossils are related to living forms.
1828	Karl E. von Baer	Germany	Establishes the germ layer theory of development.
1833	William Beaumont	United States	Presents evidence that digestion is a chemical process.
1837	René Dutrochet	France	Realizes that green pigment, chlorophyll, is necessary to photosynthesis.
1838	Matthias Schleiden	Germany	States that plants are multicellular organisms.
1839	Theodor Schwann	Germany	States that animals are multicellular organisms.
1851	Claude Bernard	France	Concludes that a relatively constant internal environment allows organisms to survive under varying conditions.
1858	Rudolf Virchow	Germany	States that cells come only from preexisting cells.
1858	Charles Darwin	Britain	Presents evidence that natural selection guides the evolutionary process.
1858	Alfred R. Wallace	Britain	Independently comes to same conclusions as Darwin.
1865	Louis Pasteur	France	Disproves the theory of spontaneous generation for bacteria; shows that infections are caused by bacteria and develops vaccines against rabies and anthrax.
1866	Gregor Mendel	Austria	Proposes basic laws of genetics based on his experiments with garden peas.
1869	Friedrich Miescher	Switzerland	Discovers that the nucleus contains a chemical he called nuclein, now termed DNA.
1878	Joseph Lister	Britain	Devises a method of sterilizing the operating room to prevent infection in surgical patients.

Anton van Leeuwenhoek

Charles Darwin

Louis Pasteur

INQUIRY
INTO LIFE

For my children

INQUIRY
INTO LIFE

Sylvia S. Mader

Sixth Edition

WCB

Wm. C. Brown Publishers

Book Team

Editor *Kevin Kane*
Developmental Editor *Carol Mills*
Production Editor *Sherry Padden*
Visuals/Design Consultant *Marilyn Phelps*
Designer *Mark Elliot Christianson*
Art Editor *Donna Slade*
Photo Editor *Mary Roussel*
Permissions Editor *Vicki Krug*
Visuals Processor *Joyce E. Watters*

 Wm. C. Brown Publishers

President *G. Franklin Lewis*
Vice President, Publisher *George Wm. Bergquist*
Vice President, Publisher *Thomas E. Doran*
Vice President, Operations and Production *Beverly Kolz*
National Sales Manager *Virginia S. Moffat*
Advertising Manager *Ann M. Knepper*
Marketing Manager *Craig S. Marty*
Editor in Chief *Edward G. Jaffe*
Managing Editor, Production *Colleen A. Yonda*
Production Editorial Manager *Julie A. Kennedy*
Production Editorial Manager *Ann Fuerste*
Publishing Services Manager *Karen J. Slaght*
Manager of Visuals and Design *Faye M. Schilling*

Cover photos Front & Back © Teiji Saga/Allstock

The credits section for this book begins on page C-1, and is considered an extension of the copyright page.

Publisher's Note to the Instructor

Recycled Paper and New Binding Options

In mid-Spring of 1991, the sixth edition of *Inquiry into Life* will be made available in a variety of binding options on both standard and ***recycled paper stock.*** Our goal in offering each new binding option is to increase the number of ways in which you may successfully use the text. Our goal in offering each option on **recycled paper** is to take an important first step toward minimizing the environmental impact of our products.

But it's important to realize that each binding option will still be made available on standard paper. And it's important to understand why. While the quality of recycled paper is improving almost daily, it still presents publishers with a number of technical challenges. It's difficult to find a recycled stock that combines true environmental "friendliness" with the durability and graphic sensitivity (the capacity to reproduce graphic images precisely) of its standard counterparts. So as "a best possible compromise," a recycled stock has been chosen that is attractive and durable, but with one slight, unavoidable drawback. While its graphic sensitivity is good, the images on our recycled stock may not seem as crisp and precise as the ones on standard stock. That's why we're making both options available. You should feel free to decide now whether the recycled or standard paper versions better suit your needs. But our hope for the future is to offer books of the highest possible graphic quality on recycled stock only.

Also, because of the slight differences in standard and recycled paper, you should be aware of how the sixth edition of *Inquiry into Life* will be previewed. In order to present you with optimal graphics first, the sample you receive in the mail will be printed on standard, not recycled stock. Any of the other options (on either standard or recycled stock) will be sent to you upon special request, or will be previewed for you by your local Wm. C. Brown sales representative. But regardless of the options you evaluate, those printed on recycled stock will always be labeled as such on the text's cover. The price for the standard and recycled versions of each option will be the same (so no additional cost will be passed on to students).

Finally, only the text options described below and the Laboratory Manual to accompany the sixth edition of *Inquiry into Life* will be offered on **both** standard and recycled paper. All other ancillaries (excluding the Customized Lab Manual)—as well as all advertising pieces for *Inquiry into Life*—will be printed **exclusively on recycled stock,** subject to market availability.

What follows is a list of the binding options for the *Inquiry into Life* textbook, as well as its accompanying laboratory manual, with a brief description of each, when appropriate. If you have any questions about recycled paper use or any of the materials available with Mader's *Inquiry into Life,* please feel free to call either of us at 1-800-331-2111.

Kevin Kane
Senior Editor for Biology

Craig Marty
Marketing Manager for Biology

Binding Option	Description	Standard Paper Version	Recycled Paper Version
Inquiry into Life, sixth edition—**casebound**		0-697-10200-9	0-697-13747-3
Inquiry into Life, sixth edition—**paperbound**	The full text, paperback covered and available at a significantly reduced price when compared with the hardcover price.	0-697-13280-3	0-697-13748-1
Inquiry into Life, sixth edition Volume One: Cell, Plant, and Human Biology—**paperbound**	Volume One features the first four units or twenty-six chapters of the text, covering the fundamentals of cell biology, photosynthesis and plant biology, as well as human anatomy and physiology, development, and genetics. This paperback option is available at a significantly reduced price when compared with both the full-length casebound or full-length paperback prices.	0-697-13281-1	0-697-13749-X
Inquiry into Life, sixth edition Volume Two: Evolution, Diversity, Behavior, and Ecology—**paperbound**	Volume Two features the last two units or chapters 26-35 of the text, addressing the issues of evolution, diversity, behavior, and ecology. Paperback covered, it is priced at even less than Volume One.	0-697-13282-X	0-697-13750-3
Inquiry into Life, sixth edition Volumes One and Two Set—**paperbound "splits"**	The entire text offered in a package of two paperback "splits". It is available at the same price as the full-length casebound text.	0-697-13574-8	0-697-13751-1
The Laboratory Manual to accompany *Inquiry into Life,* sixth edition		0-697-10201-7	0-697-13752-X
The Customized Laboratory Manual to accompany *Inquiry into Life,* sixth edition	Individual one-color "Lab Separates" from the original Laboratory Manual are available on standard paper only, custom-bound according to your selection.	Contact your local Wm. C. Brown salesperson or call Beth Kundert, Wm. C. Brown Customized Publishing Service at 1-800-331-2111 for more details.	

Brief Contents

Volume One

Cell, Plant, and Human Biology

Introduction 2

Part One

The Cell, the Smallest Unit of Life 15

1 Chemistry and Life 16
2 Cell Structure and Function 43
3 Cell Membrane and Cell Wall Function 64
4 Cell Division 79
5 Cellular Metabolism 97

Part Two

Plant Biology 121

6 Photosynthesis 122
7 Plant Organization and Growth 141
8 Plant Physiology and Reproduction 154

Part Three

Human Anatomy and Physiology 171

9 Human Organization 172
10 Digestion 189
11 Circulation 218
12 Blood 238
13 The Lymphatic System and Immunity 257
14 Respiration 278
15 Excretion 298
16 The Nervous System 319
17 The Musculoskeletal System 346
18 Senses 366
19 Hormones 385

Part Four

Human Reproduction, Development, and Inheritance 409

20 The Reproductive System 410
21 Development 435
22 Patterns of Gene Inheritance 468
23 Patterns of Chromosome Inheritance 489
24 The Molecular Basis of Inheritance 507
25 Recombinant DNA and Biotechnology 533

Volume Two

Evolution, Diversity, Behavior, and Ecology

Part Five

Evolution and Diversity 551

26 The Origin of Life 552
27 Evolution 562
28 Viruses and Kingdoms Monera, Protista, and Fungi 583
29 The Plant Kingdom 609
30 The Animal Kingdom 629
31 Human Evolution 669

Part Six

Behavior and Ecology 685

32 Behavior Patterns 686
33 The Biosphere 700
34 Ecosystems 718
35 Human Population Concerns 735

The Inquiry into Life *Learning System* xiv
Preface xviii
Acknowledgments xxiv

Introduction 2

Introduction Concepts 2
Introduction Outline 2
Characteristics of Life 4
The Classification of Living Things 6
Ecosystems 7
    The Human Population 8
The Process of Science 8
    The Scientific Method 8
    Controlled Experiments 10
    Theories and Principles 11
Science and Social Responsibility 12
Summary 13
Study Questions 13
Objective Questions 13

Part One

The Cell, the Smallest Unit of Life 15

1 Chemistry and Life 16

Chapter Concepts 16
Chapter Outline 16
Atoms 17
    Isotopes 19
Reactions between Atoms 19
    Ionic Reactions 19
    Covalent Reactions 21
    Oxidation-Reduction
      Reactions 23
Inorganic Chemistry versus Organic
    Chemistry 24
Some Important Inorganic
    Molecules 24
    Water 24
      Acids and Bases 25
      Salts 28
Some Important Organic
    Molecules 29
    Proteins 29
    Carbohydrate 33
    Lipids 35
    Nucleic Acids 40
Summary 41
Study Questions 41
Objective Questions 42
Selected Key Terms 42

2 Cell Structure and Function 43

Chapter Concepts 43
Chapter Outline 43
The Cell Theory 44
    Microscopy 44
    Types of Cells 44
Eukaryotic Cell Organelles 47
    The Nucleus 48
    Membranous Canals and
      Vacuoles 50
    Energy-Related Organelles 53
    The Cytoskeleton 55
    Centrioles and Related
      Organelles 57
Cellular Comparisons 59
    Prokaryotic Cells versus
      Eukaryotic Cells 59
    Plant Cells versus Animal
      Cells 60
Summary 60
Study Questions 62
Objective Questions 62
Selected Key Terms 63

3 Cell Membrane and Cell Wall Function 64

Chapter Concepts 64
Chapter Outline 64
Outer Boundaries of Cells 65
    The Cell Membrane 65
    The Plant Cell Wall 67
    Permeability 68
Osmosis 69
    Tonicity 71
    The Importance of Osmosis 72

Contents

Transport by Carriers 72
  Facilitated Diffusion 73
  Active Transport 73
Endocytosis and Exocytosis 75
  Endocytosis 75
  Exocytosis 77
Summary 77
Study Questions 77
Objective Questions 77
Selected Key Terms 78

4 Cell Division 79

Chapter Concepts 79
Chapter Outline 79
The Life Cycle of Animals 81
Mitosis 83
  The Cell Cycle 83
  Overview of Animal Mitosis 84
  Stages of Mitosis 84
  Mitosis in Plants 88
Meiosis 89
  Overview of Animal Meiosis 89
  Stages of Meiosis 89
  Spermatogenesis and
    Oogenesis 92
  Meiosis in Plants 93
Comparison of Mitosis and Meiosis 93
  Occurrence 93
  Process 95
  Daughter Cells 95
Summary 95
Study Questions 95
Objective Questions 96
Selected Key Terms 96

5 Cellular Metabolism 97

Chapter Concepts 97
Chapter Outline 97
Metabolism 98
  Metabolic Pathways 98
  Enzymes 99
  Conditions Affecting Enzymatic
    Reactions 101
  Coenzymes 103
  Energy 104
Aerobic Cellular Respiration
    (Simplified) 105
  Subpathways 107
  Mitochondria 107
  Metabolites 108

Aerobic Cellular Respiration
    (Detailed) 110
  Glycolysis 111
  The Transition Reaction 111
  The Krebs Cycle 112
  The Respiratory Chain 112
Fermentation 114
  The Results of Fermentation 115
  The Usefulness of
    Fermentation 115
Summary 117
Study Questions 117
Objective Questions 117
Selected Key Terms 117
Further Readings for Part One 118

Part Two

Plant Biology 121

6 Photosynthesis 122

Chapter Concepts 122
Chapter Outline 122
Radiant Energy 123
  Sunlight 124
Chloroplasts 125
  Anatomy 125
  Overall Equations for
    Photosynthesis 125
Capturing the Energy of Sunlight 127
  Two Electron Pathways 128
Reducing Carbon Dioxide 131
  C_3 Photosynthesis 132
  C_4 Photosynthesis 135
  Carbohydrate Utilization 136
Autotrophic Bacteria 137
  Bacterial Photosynthesis 137
  Chemosynthesis 137

Comparison of Cellular Respiration
    and Photosynthesis 138
  Differences 138
  Similarities 139
Summary 139
Study Questions 139
Objective Questions 140
Selected Key Terms 140

7 Plant Organization and Growth 141

Chapter Concepts 141
Chapter Outline 141
The Flowering Plant 142
  Monocots and Dicots 142
  Tissue and Cell Types 143
The Root System 145
  Dicot Roots 145
  Monocot Roots 147
The Shoot System 147
  Primary Growth of Stems 147
  Secondary Growth of Stems 148
Leaves 150
  Leaf Veins 151
Summary 152
Study Questions 153
Objective Questions 153
Selected Key Terms 153

8 Plant Physiology and Reproduction 154

Chapter Concepts 154
Chapter Outline 154
Transport in a Plant 155
  Water Transport 155
  Organic Nutrient Transport 157
Plant Responses to
    Environmental Stimuli 159
  Flowering 159
Reproduction in Plants 163
  Asexual Reproduction 164
  Sexual Reproducton 164
  The Seed 167
Summary 168
Study Questions 169
Objective Questions 169
Selected Key Terms 169
Further Readings for Part Two 169

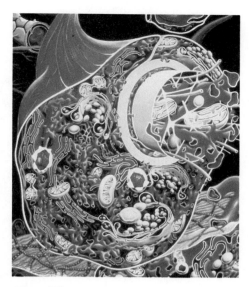

Part Three

Human Anatomy and Physiology 171

9 Human Organization 172

Chapter Concepts 172
Chapter Outline 172
Types of Tissues 173
    Epithelial Tissue 173
    Connective Tissue 176
    Muscular Tissue 179
    Nervous Tissue 180
Organs and Organ Systems 180
    The Skin 180
    Organ Systems 184
Homeostasis 185
    Body Temperature Control 186
Summary 187
Study Questions 187
Objective Questions 187
Selected Key Terms 187

10 Digestion 189

Chapter Concepts 189
Chapter Outline 189
The Digestive System 190
    The Mouth 190
    The Pharynx 192
    The Esophagus 193
    The Stomach 194
    The Small Intestine 195
    The Large Intestine 195
    Accessory Organs 197
Digestive Enzymes 200
    Conditions for Digestion 202

Nutrition 202
    Proteins 204
    Carbohydrates 204
    Lipids 206
    Vitamins and Minerals 209
Dieting 211
    Daily Energy Requirement 212
    Fad Diets versus Behavior
        Modification 213
    Eating Disorders 215
Summary 216
Study Questions 216
Objective Questions 217
Selected Key Terms 217

11 Circulation 218

Chapter Concepts 218
Chapter Outline 218
The Circulatory System 219
    Blood Vessels 219
    The Heart 221
    The Heartbeat 223
    Vascular Pathways 226
Features of the Circulatory
    System 228
    The Pulse 228
    Blood Pressure 229
    The Velocity of Blood Flow 231
Circulatory Disorders 231
    Hypertension 231
    Atherosclerosis 233
    Varicose Veins and Phlebitis 236
Summary 236
Study Questions 237
Objective Questions 237
Selected Key Terms 237

12 Blood 238

Chapter Concepts 238
Chapter Outline 238
The Transport Function of Blood 240
    Plasma 240
    Red Blood Cells
        (Erythrocytes) 241
    Capillary Exchange within the
        Tissues 246
The Blood-Clotting Function of
    Blood 248
The Infection-Fighting Function of
    Blood 249
    White Blood Cells
        (Leukocytes) 249
    Antibodies 251
    The Inflammatory Reaction 252

Blood Typing 253
    The ABO System 253
    The Rh System 254
Summary 255
Study Questions 256
Objective Questions 256
Selected Key Terms 256

13 The Lymphatic System and Immunity 257

Chapter Concepts 257
Chapter Outline 257
The Lymphatic System 258
    Lymphatic Vessels 258
    Lymphoid Organs 259
Immunity 261
    General Defense 261
    Specific Defense 263
Immunotherapy 270
    Induced Immunity 270
    Lymphokines 271
    Monoclonal Antibodies 272
Immunological Side Effects and
    Illnesses 273
    Allergies 274
    Tissue Rejection 275
    Autoimmune Diseases 276
Summary 276
Study Questions 277
Objective Questions 277
Selected Key Terms 277

14 Respiration 278

Chapter Concepts 278
Chapter Outline 278
Breathing 280
    The Passage of Air 280
The Mechanism of Breathing 284
    Inspiration 285
    Expiration 287
    Lung Capacities 287
External and Internal Respiration 288
    External Respiration 288
    Internal Respiration 291
Respiration and Health 292
    Common Respiratory
        Infections 292
    Lung Disorders 293
Summary 296
Study Questions 296
Objective Questions 297
Selected Key Terms 297

15 Excretion 298

Chapter Concepts 298
Chapter Outline 298
Excretory Substances and Organs 299
    Nitrogenous End Products 300
    Other Excretory Substances 301
    Organs of Excretion 302
The Urinary System 303
    The Path of Urine 303
    Kidneys 304
Urine Formation 307
    Pressure Filtration 308
    Selective Reabsorption 308
    Tubular Excretion 310
    Reabsorption of Water 310
Regulatory Functions of the
    Kidneys 312
    Adjustment of Blood pH and Ion
    Balance 312
    Blood Volume 313
Problems with Kidney Function 315
    Kidney Replacement 315
Summary 316
Study Questions 317
Objective Questions 317
Selected Key Terms 318

16 The Nervous System 319

Chapter Concepts 319
Chapter Outline 319
Neurons 320
    The Structure of a Neuron 320
    The Nerve Impulse 322
Transmission across a Synapse 325
    Summation and Integration 327
    Neurotransmitter Substances 328
The Peripheral Nervous System 328
    The Somatic Nervous
    System 331
    The Autonomic Nervous
    System 332
The Central Nervous System 334
    The Spinal Cord 334
    The Brain 335
Drug Abuse 340
    Drug Action 340
    Alcohol 341
    Marijuana 342
    Cocaine 342
    Heroin 343
    Designer Drugs 344
Summary 344
Study Questions 344
Objective Questions 345
Selected Key Terms 345

17 The Musculoskeletal System 346

Chapter Concepts 346
Chapter Outline 346
The Skeleton 347
    Functions 347
    Structure 348
Skeletal Muscles: Macroscopic
    View 352
    The Anatomy of Whole
    Muscle 352
    The Physiology of Whole
    Muscle 354
    The Effect of Exercise 355
Skeletal Muscles: Microscopic
    View 358
    The Anatomy of a Muscle
    Fiber 358
    The Physiology of a Muscle
    Fiber 360
    Innervation 362
Summary 364
Study Questions 365
Objective Questions 365
Selected Key Terms 365

18 Senses 366

Chapter Concepts 366
Chapter Outline 366
General Receptors 367
    The Skin 367
    Muscles and Joints 369
Special Senses 369
    Chemoreceptors 369
    The Photoreceptor—the Eye 371
    A Mechanoreceptor—the
    Ear 378
Summary 383
Study Questions 383
Objective Questions 384
Selected Key Terms 384

19 Hormones 385

Chapter Concepts 385
Chapter Outline 385
Endocrine Glands 386
    The Mechanism of Hormonal
    Action 387
The Hypothalamus and the
    Pituitary Gland 388

    The Posterior Pituitary 388
    The Anterior Pituitary 391
Thyroid and Parathyroid Glands 394
    The Thyroid Gland 394
    Parathyroid Glands 395
Adrenal Glands 395
    The Adrenal Medulla 396
    The Adrenal Cortex 396
The Pancreas 398
    Diabetes Mellitus 398
Other Endocrine Glands 400
    Gonads 400
    The Thymus 400
    The Pineal Gland 400
    Still Other Glands 403
Environmental Signals 404
    Redefinition of a Hormone 405
Summary 405
Study Questions 406
Objective Questions 406
Selected Key Terms 406
Further Readings for Part Three 407

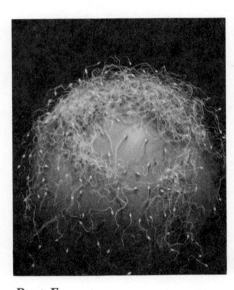

Part Four

Human Reproduction, Development, and Inheritance 409

20 The Reproductive System 410

Chapter Concepts 410
Chapter Outline 410
The Male Reproductive System 411
    Testes 412
    The Genital Tract 412
    Orgasm in Males 414
    The Regulation of the Male
    Hormone Levels 415

The Female Reproductive System 416
    Ovaries 417
    The Genital Tract 418
    Orgasm in Females 419
    The Regulation of Female
      Hormone Levels 419
The Control of Reproduction 425
    Birth Control 425
    Infertility 426
Sexually Transmitted Diseases 429
    AIDS 429
    Genital Herpes 430
    Genital Warts 431
    Gonorrhea 432
    Chlamydia 432
    Syphilis 433
Summary 433
Study Questions 433
Objective Questions 434
Selected Key Terms 434

21 Development 435

Chapter Concepts 435
Chapter Outline 435
Early Developmental Stages 436
    The Morula 437
    The Blastula 437
    The Gastrula 437
    The Neurula 439
Differentiation and
    Morphogenesis 441
    Differentiation 441
    Morphogenesis 443
Human Embryonic and
    Fetal Development 445
    Embryonic Development 447
    Fetal Development 456
    Birth 458
Human Development after Birth 460
    Adulthood and Aging 461
Summary 465
Study Questions 466
Objective Questions 466
Selected Key Terms 467

22 Patterns of Gene Inheritance 468

Chapter Concepts 468
Chapter Outline 468
Mendel's Laws 469
    The Inheritance of a Single
      Trait 471
    The Inheritance of
      Multitraits 474
Genetic Disorders 479
    Autosomal Recessive
      Disorders 480
    Autosomal Dominant
      Disorders 482
Beyond Mendel's Laws 483
    Polygenic Inheritance 483
    Multiple Alleles 484
    Degrees of Dominance 485
Summary 486
Study Questions 487
Objective Questions 487
Additional Genetic Problems 487
Selected Key Terms 488

23 Patterns of Chromosome Inheritance 489

Chapter Concepts 489
Chapter Outline 489
Chromosome Inheritance 490
    Normal Inheritance 490
    Abnormal Autosomal Chromosome
      Inheritance 491
    Abnormal Sex Chromosome
      Inheritance 495
Sex-Linked Inheritance 496
    X-Linked Genetics
      Problems 496
    Recessive X-Linked
      Disorders 497
    Sex-Influenced Traits 500
Mapping the Human
    Chromosomes 501
    Linkage Data 501
    Human-Mouse Cell Data 502
    Genetic Marker Data 503
    DNA Probe Data 504
Summary 504
Study Questions 505
Objective Questions 505
Additional Genetics Problems 505
Selected Key Terms 506

24 The Molecular Basis of Inheritance 507

Chapter Concepts 507
Chapter Outline 507
DNA 509
    The Structure of DNA 509
    Functions of DNA 511
Protein Synthesis 516
    The Code of Heredity 517
    Transcription 517
    Translation 520
    Summary of Protein
      Synthesis 522
Control of Gene Expression 524
    The Control of Gene Expression in
      Prokaryotes 524
    The Control of Gene Expression in
      Eukaryotes 526
Gene Mutations 527
    Substitutions, Alterations, and
      Deletions of Bases 527
    Transposons 528
Cancer, a Failure in Genetic
    Control 528
    Causes of Cancer 529
Summary 531
Study Questions 531
Objective Questions 531
Selected Key Terms 532

25 Recombinant DNA and Biotechnology 533

Chapter Concepts 533
Chapter Outline 533
Basic Biotechnology Laboratory
    Techniques 534
    Recombinant DNA 535
    Other Biotechnology
      Techniques 536
Biotechnology Products 537
    Hormones and Similar Types of
      Proteins 538
    DNA Probes 539
    Vaccines 539
Genetically Engineered Bacteria 540
    Uses for Bacteria in Agriculture
      and Industry 540
    Ecological Considerations 540
Transgenic Organisms 541
    Transgenic Plants 541
    Transgenic Animals 544
Gene Therapy in Humans 545
    Modifying Stem Cells 545
    Developing Delivery Systems 546
    The Human Genome Project 548
Summary 548
Study Questions 548
Objective Questions 549
Selected Key Terms 549
Further Readings for Part Four 549

Part Five

Evolution and Diversity   551

26 The Origin of Life   552

Chapter Concepts   552
Chapter Outline   552
Chemical Evolution   553
   The Primitive Atmosphere   553
   Simple Organic Molecules   553
   Macromolecules   555
Biological Evolution   556
   Protocells   556
   True Cells   557
Summary   560
Study Questions   560
Objective Questions   560
Selected Key Terms   561

27 Evolution   562

Chapter Concepts   562
Chapter Outline   562
Evidences for Evolution   563
   The Fossil Record   563
   Comparative Anatomy   566
   Comparative Embryology   568
   Vestigial Structures   568
   Comparative Biochemistry   570
   Biogeography   570
The Evolutionary Process   571
   Population Genetics   571
   The Synthetic Theory   573

   Examples of Natural
      Selection   577
Speciation   580
   The Process of Speciation   580
   Adaptive Radiation   581
Summary   581
Study Questions   582
Objective Questions   582
Selected Key Terms   582

28 Viruses and Kingdoms Monera, Protista, and Fungi   583

Chapter Concepts   583
Chapter Outline   583
Viruses   584
   Life Cycles   585
   Viroids and Prions   588
   Viral Infections   588
Kingdom Monera   589
   The Structure of Bacteria   589
   The Reproduction of
      Bacteria   590
   The Metabolism of Bacteria   591
   Types of Bacteria   592
Kingdom Protista   593
   Protozoans   593
   Algae   596
   Slime Molds and Water
      Molds   603
Kingdom Fungi   603
   Black Bread Molds   603
   Sac Fungi   604
   Club Fungi   606
   Other Types of Fungi   607
Summary   607
Study Questions   607
Objective Questions   607
Selected Key Terms   608

29 The Plant Kingdom   609

Chapter Concepts   609
Chapter Outline   609
Characteristics of Plants   610
   The Alternation of Generations
      Life Cycle   611
Nonvascular Plants   612
   Bryophytes   612
Vascular Plants   614
   Vascular Plants without
      Seeds   615
   Vascular Plants with Seeds   617

Seed Plants   618
   Gymnosperms   619
   Angiosperms   622
Comparisons between Plants   626
Summary   627
Study Questions   628
Objective Questions   628
Selected Key Terms   628

30 The Animal Kingdom   629

Chapter Concepts   629
Chapter Outline   629
Evolution and Classification   630
   The Evolution of Animals   630
   The Classification of
      Animals   630
Primitive Invertebrates   633
   Sponges   633
   Cnidarians   634
   Flatworms   637
   Roundworms   640
Advanced Invertebrates   642
   Mollusks   643
   Annelids   647
   Arthropods   648
   Echinoderms   654
Chordates   656
   Protochordates   656
   Vertebrate Chordates   656
   Comparisons between
      Vertebrates   665
Summary   667
Study Questions   667
Objective Questions   667
Selected Key Terms   668

31 Human Evolution   669

Chapter Concepts   669
Chapter Outline   669
The Evolution of Humans   670
   Prosimians   670
   Anthropoids   670
   Hominoids   673
   Hominids   673
   Australopithecines   674
Humans   676
   Homo habilis   676
   Homo erectus   676
   Homo sapiens   678
Summary   682
Study Questions   682
Objective Questions   683
Selected Key Terms   683
Further Readings for Part Five   683

Part Six

Behavior Ecology 685

32 Behavior Patterns 686

Chapter Concepts 686
Chapter Outline 686
Competition 687
    An Example of Competition 687
    The Exclusion Principle 687
Predation 688
    Predation Adaptations 689
    Prey Defense Adaptations 692
Symbiosis 693
    Parasitism 693
    Commensalism 695
    Mutualism 696
Summary 698
Study Questions 698
Objective Questions 699
Selected Key Terms 699

33 The Biosphere 700

Chapter Concepts 700
Chapter Outline 700
Terrestrial Biomes 701
    Treeless Biomes 702
    Forests 706
    Latitude versus Altitude 710
Aquatic Communities 710
    Freshwater Communities 710
    Saltwater Communities 712
Summary 716
Study Questions 716
Objective Questions 716
Selected Key Terms 717

34 Ecosystems 718

Chapter Concepts 718
Chapter Outline 718
The Nature of Ecosystems 720
    Habitat and Niche 720
    Ecosystem Composition 720
Energy Flow 721
    Food Chains and Food Webs 722
    Ecological Pyramids 723
Chemical Cycling 725
    The Carbon Cycle 725
    The Nitrogen Cycle 727
The Human Ecosystem 729
    The Country 730
    The City 731
    The Solution 732
Summary 733
Study Questions 734
Objective Questions 734
Selected Key Terms 734

35 Human Population Concerns 735

Chapter Concepts 735
Chapter Outline 735
Exponential Population Growth 736
    The Growth Rate 738
    The Doubling Time 738
    The Carrying Capacity 739
Human Population Growth 740
    Developed Countries 740
    Less-Developed Countries 741
Human Population and Pollution 743
    Land Degradation 743
    Water Pollution 747
    Air Pollution 750
A Sustainable World 754
    The Steady State 755
Summary 756
Study Questions 756
Objective Questions 757
Selected Key Terms 757
Further Readings for Part Six 757

Appendix A: Periodic Table of the
    Elements A-1
Appendix B: Drugs of Abuse A-3
Appendix C: The Metric System A-6
Appendix D: Classification of Organisms
    A-9
Appendix E: Answers to the Critical
    Thinking Questions A-11
Glossary G-1
Credits C-1
Index I-1

The Inquiry Into Life Learning System

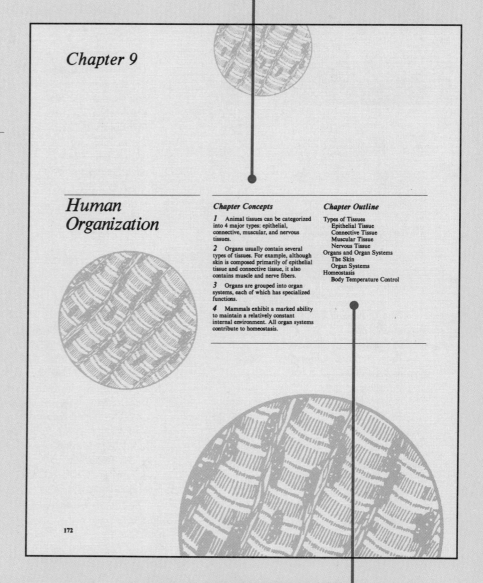

Chapter 9

Human Organization

Chapter Concepts

1 Animal tissues can be categorized into 4 major types: epithelial, connective, muscular, and nervous tissues.

2 Organs usually contain several types of tissues. For example, although skin is composed primarily of epithelial tissue and connective tissue, it also contains muscle and nerve fibers.

3 Organs are grouped into organ systems, each of which has specialized functions.

4 Mammals exhibit a marked ability to maintain a relatively constant internal environment. All organ systems contribute to homeostasis.

Chapter Outline

Types of Tissues
    Epithelial Tissue
    Connective Tissue
    Muscular Tissue
    Nervous Tissue
Organs and Organ Systems
    The Skin
    Organ Systems
Homeostasis
    Body Temperature Control

172

This book is about living things (fig. I.1); therefore, it is appropriate to first define life. Unfortunately, this is not done so easily—life cannot be given a simple, one-line definition. Since this is the case, it is customary to discuss life in terms of its characteristics. The following 5 characteristics commonly are attributed to living things.

1. Living things are organized and are made up of cells.
2. Living things grow and maintain their organization by taking chemicals and energy from the environment.
3. Living things respond to the external environment.
4. Living things reproduce and pass on their organization to their offspring.
5. Living things evolve, or change, and adapt to the environment.

Introduction

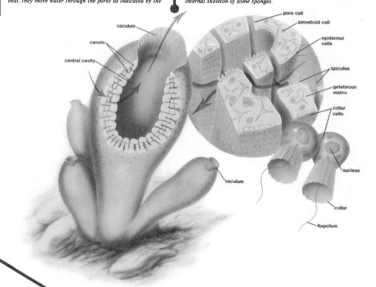

Figure 30.6
Generalized sponge anatomy. Epidermal cells form an outer layer of cells, and collar cells line the central cavity and the canals. The collar cells (enlarged) have flagella, and as they beat, they move water through the pores as indicated by the arrows. Food particles in the water are trapped by the collar cells and are digested within their food vacuoles. Amoeboid cells transport nutrients from cell to cell; spicules comprise an internal skeleton of some sponges.

process produces whole colonies that can become quite large. Like many less-specialized organisms, sponges are capable of **regeneration,** or growth of a whole from a small part.

The cellular organization of sponges is different from that of other animals, and the main opening of a sponge is used only as an exit, not an entrance. Further, movement is limited to the beating of the flagella, constriction of the osculum, and larval-stage swimming. Sponges are classified according to type of spicule.

Cnidarians

The body of a **cnidarian** (phylum Cnidaria) is a hollow, 2-layered sac, which accounts for the former name of these organisms—*coelenterate* means hollow sac. The outer layer of the sac, the ectoderm, is separated from the inner layer, the endoderm, by a jellylike material called **mesoglea.** All cnidarians have specialized stinging cells now termed cnidocytes, from a Greek word meaning sea nettles. Within these cells are the **nematocysts,** which are long, spirally

634   *Evolution and Diversity*

XV

Critical Thinking Questions

New to this edition are critical thinking boxes, which contain questions that require the student to form a hypothesis, come to a conclusion, or apply information in a new and different way. The critical thinking questions are placed in the margin next to pertinent text material. Answers to the critical thinking questions appear in appendix E.

Tables

Numerous strategically-placed tables list and summarize important information, making it readily accessible for efficient study.

6.3 Critical Thinking

The structure of chloroplasts and mitochondria can be compared.

1. What part of a mitochondrion compares to the stroma of a chloroplast?
2. Are the biochemical processes in each of these parts the same or the opposite? Explain.
3. What part of a mitochondrion compares to the thylakoid membrane of a chloroplast? to the thylakoid space of a chloroplast?
4. Are the biochemical processes in each of these parts the same or the opposite?

Table 6.3 Cellular Respiration versus Photosynthesis

Cellular Respiration	Photosynthesis
Mitochondrion	Chloroplast
Oxidation	Reduction
Releases energy	Requires energy
Requires O_2	Releases O_2
Releases CO_2	Requires CO_2

teria supply the animals with carbohydrate. It was a surprise to find communities of organisms living so deep in the ocean, where light never penetrates. Unlike photosynthesis, chemosynthesis does not require light energy.

Except for cyanobacteria, photosynthetic bacteria do not evolve oxygen as do plants because they utilize hydrogen from a source other than water. Chemosynthetic bacteria produce their own food by oxidizing inorganic compounds and therefore do not require light at all.

Comparison of Cellular Respiration and Photosynthesis

Differences

Both plant and animal cells carry on cellular respiration but only plant cells photosynthesize. The cellular organelle for cellular respiration is the mitochondrion, while the cellular organelle for photosynthesis is the chloroplast.

The overall equation for aerobic cellular respiration is the opposite of that for photosynthesis:

$$\text{energy} + 6CO_2 + 6H_2O \rightleftharpoons C_6H_{12}O_6 + 6O_2$$

The reaction in the forward direction represents photosynthesis, and the energy is the energy of the sun. The reaction in the opposite direction represents cellular respiration, and the energy is ATP.

Obviously, photosynthesis is the building up of glucose, while cellular respiration is the breaking down of glucose. See table 6.3 for a summarized list of differences between these processes.

Figure 6.12
Diagram illustrating similarities and differences of photosynthesis, which takes place in chloroplasts, and cellular respiration, which takes place in mitochondria. Both have a cytochrome system, located within membranes, where ATP is produced. Both have enzyme-catalyzed reactions in solution. The coenzyme NAD(P) operates in the membrane and the solution. Photosynthesis releases oxygen (O_2) and reduces carbon dioxide (CO_2) into carbohydrate; cellular respiration reduces oxygen and releases carbon dioxide.

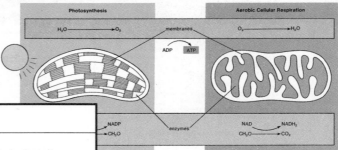

Plant Biology

There are risks to having a blood transfusion, such as an immune reaction and acquiring an infection. A cross-matching test between the donor's blood and the recipient's blood usually detects if the recipient has antibodies in the plasma that will react against antigens on the membrane of the donor's red blood cells and vice versa. Blood also is screened for the presence of 2 viruses that are especially troublesome. They are the hepatitis B virus and the AIDS virus. Blood donors are questioned carefully, and their blood is

Artificial Blood

tested for the presence of these viruses. Despite the care that is taken to avoid immune reactions and the transference of disease, it would be advantageous to develop an artificial blood that has neither of these risks.

Investigation is proceeding in 3 directions. Enrico Bucci, a blood-substitute specialist at the University of Maryland School of Medicine, is working with modifying hemoglobin itself. He links several hemoglobin molecules so that they do not become "lost" from the circulatory system and uses chemicals to modify the complexes' oxygen affinity so that they are more likely to give up the oxygen when needed. Many other problems remain, however. The hemoglobin is taken from whole blood, and it alone may contain infectious material. Modified hemoglobin also seems to cause a generalized constriction of the body's blood vessels, making oxygenation of tissues more difficult.

Anthony Hunt and colleagues at the University of California at San Francisco are working with artificial red blood cells called neohemocytes (NHCs). To make the cells, purified human hemoglobin taken from outdated donor blood is encapsulated in a lipid bilayer membrane. The artificial cells are much smaller than normal human red blood cells, and they do not contain as much hemoglobin. However, when tested in rats, the animals survived until they were sacrificed for gross toxicity studies. The investigators believe that the tests are successful enough to warrant further study. They want to improve the stability and the vascular retention time of the cells since they are removed from the bloodstream and are broken down at a faster rate than normal cells.

George Groveman is director of new-products marketing at Alpha Therapeutic, a subsidiary of a Japanese firm that is based in Los Angeles. His firm is working on a third possibility, a substance called perfluorocarbon oil (PFC) emulsion that can be transfused and can carry oxygen much like hemoglobin does. This substance has served as a blood substitute for humans in emergency situations in which only the oxygen-carrying function of blood was required. FDA approval has been sought by a Japanese corporation to market PFC under the trade name of Fluosol-DA, but thus far the FDA has denied permission on the grounds that the clinical trials were not successful enough. There is some hope, though, that PFC will be approved for localized use. For example, it may be helpful to administer Fluosol-DA when a person is suffering a heart attack or is undergoing thrombolytic therapy (p. 234).

Although researchers have been working for 20 years to produce a blood substitute for general use, the prospects are still in the future. Anthony Hunt says, "Physiological systems always turn out to be more complicated than we thought."

View inside a capillary in which blood has been replaced with a 25% suspension of hemoglobin-containing synthetic neohemocytes (NHCs). A normal red blood cell is shown for scale. Unlike living red blood cells, neohemocytes are nearly spherical and can have one, 2, or 3 chambers.

Red blood cells that are not engaged in oxygen transport assist in transporting carbon dioxide (CO_2). First, hemoglobin (Hb) combines with carbon dioxide to form *carbaminohemoglobin*:

$$Hb + CO_2 \underset{\text{lungs}}{\overset{\text{tissues}}{\rightleftharpoons}} HbCO_2$$

However, this combination with hemoglobin actually represents only a small portion of the carbon dioxide in the blood. Most of the carbon dioxide is transported as the *bicarbonate ion* (HCO_3^-). This ion forms after carbon dioxide

Text Line Art

Graphic diagrams placed immediately after or within textual passages help clarify difficult concepts and enhance learning.

Readings

Throughout **Inquiry into Life,** selected readings reinforce major concepts in the book. Most readings are written by the author, but a few are excerpted from popular magazines. A reading may provide insight into the process of science or show how a particular kind of scientific knowledge is applicable to the students' everyday lives.

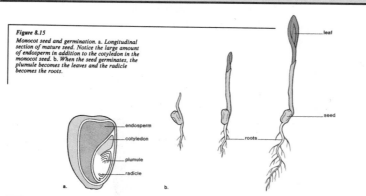

Figure 8.15
Monocot seed and germination. a. Longitudinal section of mature seed. Notice the large amount of endosperm in addition to the cotyledon in the monocot seed. b. When the seed germinates, the plumule becomes the leaves and the radicle becomes the roots.

leaf

seed

endosperm

cotyledon

plumule

radicle

a.   b.   roots

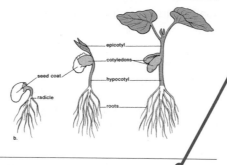

Figure 8.16
Dicot seed and germination. a. Longitudinal section of dicot seed shows 2 large cotyledons, one on either side of the embryo. b. Germination of a dicot seed makes it easier to detect that the epicotyl gives rise to the leaves; the hypocotyl becomes a portion of the stem, and the radicle becomes the roots.

epicotyl

cotyledons

hypocotyl

plumule

radicle

seed coat

radicle

roots

cotyledons

a.   b.

Summary

Transport of water occurs within xylem, and transport of organic nutrients occurs within phloem. The cohesion-tension theory of xylem transport states that transpiration creates a tension that pulls water upward in xylem. This transport works only because water molecules are cohesive. Most of the water taken in by a plant is lost through stomata by transpiration. Stomata open when guard cells take up water (H₂O), stretching their thin side walls. Water enters the guard cells after potassium (K⁺) ions have entered.

Both stimulatory and inhibitory hormones help to control certain plant growth patterns. There are hormones that stimulate growth (auxins, gibberellins, cytokinins) and hormones that inhibit growth (ethylene and abscisic acid). Plant hormones most likely control flowering. Short-day plants flower when the nights are longer than a critical length, and long-day plants flower when the nights are shorter than a critical length. Some plants are day-length neutral. Phytochrome, a plant pigment that can respond to day length, is believed to communicate with a biological clock that in some unknown way brings about flowering.

Flowering plants exhibit an alternation of generations life cycle that includes a separate male and female gametophyte. The pollen grain is the male gametophyte. Pollen grains are produced within the stamens of a flower. The female gametophyte is produced within the ovule of a flower. Following pollina[...] matures [...] become [...] contain [...] plumule [...] (endosp [...] seed ge [...] the sho [...]

168   *Plant Biology*

Chapter Summaries

Chapter summaries offer a concise review of material in each chapter. Students may read them before beginning the chapter to preview the topics of importance, and they may also use them to refresh their memories after they have a firm grasp of the concepts presented in each chapter.

Study Questions

These questions appear at the end of each chapter. They call for specific, short essay answers that truly challenge students' mastery of the chapter's basic concepts.

Study Questions

1. Explain the cohesion-tension theory of water transport. (p. 156)
2. What events precede the opening and the closing of stomata by guard cells? (p. 156)
3. Explain the pressure-flow theory of phloem transport. (p. 158)
4. Name 5 plant hormones, and state their function. (p. 160)
5. Define photoperiodism, and discuss its relationship to flowering in certain plants. (p. 161)
6. What is the phytochrome conversion cycle, and what are some possible functions of phytochrome in plants? (p. 162)
7. How do plants reproduce asexually? sexually? (p. 164)
8. Describe the development of a female gametophyte from the megaspore mother cell to the production of an egg. (p. 165)
9. Describe the development of the male gametophyte from the microspore mother cell to the production of sperm. (p. 165)
10. Contrast the monocot seed and seedling with the dicot seed and seedling. (pp. 167–68)

Objective Questions

1. The transport of water and minerals is dependent upon _____ , which occurs whenever the stomata are open.
2. Stomata open when _____ followed by _____ enter guard cells.
3. The _____ theory explains the transport of solutes in sieve-tube cells.
4. Short-day plants (will, will not) _____ flower when a longer-than-critical-length night is interrupted by a flash of light.
5. _____ is the pigment that is believed to signal a biological clock in plants that exhibit photoperiodism.
6. Plants have a life cycle called

7. The female gametophyte develops within the _____ of a flower, and the male gametophyte develops within the

8. Monocots have seeds with one _____ , while dicots have a seed with 2.

Answers to Objective Questions
1. transpiration 2. potassium ions, water 3. pressure-flow 4. will not 5. Phytochrome 6. alternation of generations 7. ovule, anther 8. cotyledon

Label this Diagram.
See figure 8.12 (p. 164) in text.

Selected Key Terms

cotyledon (kot″ĭ-le′don) the seed leaf of the embryo of a plant. *167*
epicotyl (ep″ĭ-kot′il) the plant embryo portion above the cotyledons that contributes to stem development. *167*
hypocotyl (hi″po-kot′il) the plant embryo portion below the cotyledons that contributes to stem development. *167*

photoperiodism (fo″to-pe′re-od-izm) a response to light and dark; particularly in reference to flowering in plants. *161*
phytochrome (fi′to-krōm) a plant pigment that is involved in photoperiodism in plants. *162*
plumule (ploo′mūl) the shoot tip and the first 2 leaves of a plant. *167*

pollination (pol″ĭ-na′shun) the delivery of pollen by wind or animals to the stigma of a pistil in flowering plants and fertilization. *166*
radicle (rad′ik′l) the embryonic root of a plant. *167*
spore (spōr) the haploid structure produced by the sporophyte that develops into a gametophyte. *165*

Further Readings for Part Two

Albersheim, P., and A. G. Darvill. September 1985. Oligosaccharins. *Scientific American.*
Alberts, B., et al. 1983. *Molecular biology of the cell.* New York: Garland Publishing.
Barrett, S. C. H. September 1987. Mimicry in plants. *Scientific American.*
Bold, H. C. 1980. *Morphology of plants and fungi.* 4th ed. New York: Harper & Row, Publishers, Inc.
Brill, W. J. March 1977. Biological nitrogen fixation. *Scientific American.*
Childress, J. J., et al. May 1987. Symbiosis in the deep sea. *Scientific American.*
Cronquist, A. 1982. *Basic botany.* 2d ed. New York: Harper & Row, Publishers, Inc.

Dickerson, E. March 1980. Cytochrome and the evolution of energy metabolism. *Scientific American.*
Epel, D. November 1977. The program of fertilization. *Scientific American.*
Miller, K. R. October 1979. The photosynthetic membrane. *Scientific American.*
Nassau, Kurt. April 1980. The causes of color. *Scientific American.*
Niklas, K. J. July 1987. Aerodynamics of wind pollination. *Scientific American.*
Raven, H., et al. 1986. *Biology of plants.* 4th ed. New York: Worth Publishers, Inc.
Rayle, D., and H. L. Wedberg. 1980. *Botany: A human concern.* Boston: Houghton Mifflin.

Rost, R., et al. 1984. *Botany: A brief introduction to plant biology.* 2d ed. New York: John Wiley and Sons.
Salisbury, F. B., and C. W. Ross. 1985. *Plant physiology.* 3d ed. Belmont, Calif.: Wadsworth.
Shepard, J. F. May 1982. The regeneration of potato plants from leaf-cell protoplasts. *Scientific American.*
Stryer, L. 1988. *Biochemistry.* 3d ed. San Francisco: W. H. Freeman.
Yougan, D. C., and B. I. Mars. June 1987. Molecular mechanisms of photosynthesis. *Scientific American.*
Zimmerman, M. H. March 1963. How sap moves in trees. *Scientific American.*

Plant Physiology and Reproduction   169

Objective Questions

Located at the end of each chapter, these questions require multiple-choice answers that test basic recall of the chapter's key points. Answers to these questions appear on the same page. New to this edition is the frequent use of labeling exercises to the objective questions section.

Selected Key Terms

A selected list of boldface, key terms from the chapter appears at the end of each chapter. Each term is accompanied by its phonetic spelling, definition, and a page number indicating where the term is introduced and defined in the chapter.

Further Readings

A list of readings at the end of each part suggests references that can be used for further study of topics covered in the chapters of that part. The items listed in this section were carefully chosen for readability and accessibility.

Preface

Inquiry into Life is written for the introductory-level student who would like to develop a working knowledge of biology. Educational theory tells us that students are most interested in knowledge of immediate practical application. This text attempts to remain true to this idea. Its basic theme is knowledge about and understanding the human animal and its place within the material world. Plants and other animals are included, because humans cannot understand themselves in perspective unless they understand other living things. All organisms share an evolutionary history, in which they have adapted to life in a particular way. Humans can better understand themselves when they appreciate this unity of life, while at the same time seeing its diversity. Concerned citizens need to realize that humans are not the pivot point, nor even the culmination of life. They are a part of a great, interrelated network.

While this text covers the whole field of basic biology, it emphasizes the application of this knowledge to human concerns. Concurrent with this approach is an emphasis on concepts and principles, rather than on detailed, high-level scientific data and terminology. The latter can always be added by the instructor in more advanced study, after a firm base of knowledge has been established.

The sixth edition of *Inquiry into Life* has the same organization and style as its previous five editions. New chapters have been added on plant reproduction, human genetics, and biotechnology, but they have been placed at logical points within the same chapter sequence.

With so much happening in so many areas of biological research, it is difficult to keep succeeding editions up to date without allowing them to become lengthy and erudite. But these undesirable tendencies have been vigorously avoided. With the addition of new material in botany and genetics, a corresponding amount of information has been judiciously pared from the coverage of animal diversity, behavior, and ecology. And all chapters remain succinct and easy to understand.

The sixth edition of *Inquiry into Life* also has a greatly improved art program. Instructors will find the same illustrations they have come to trust, but most have been newly rendered. Also, every page, indeed every word, has been carefully examined to make sure the text is as helpful as possible to students. Also new to this edition are *Critical Thinking* questions, which appear in the margin next to appropriate narrative material. These questions require students to interpret data, come to a conclusion, or to apply conceptual information to new and different situations. In short, they challenge students to think beyond simple memorization.

Organization of Text

As stated in The Publisher's Note earlier, *Inquiry into Life* will now be made available in a number of different binding options for convenient use in a number of different class settings. Because each chapter of the text is written to "stand alone" with a minimal amount of cross referencing, the text may be successfully used in any number of class settings within the context of any binding option. Chapter readings from the text may be assigned in virtually any order, according to individual instructors' needs.

Volume One: Cells, Plants, and Human Biology

Introduction

That part of the introductory chapter which reviews the characteristics of life is designed to show that humans are but one kind of living thing and all types or organisms share the same characteristics.

The process of science is explained with reference to the discovery of the cause for Lyme disease. The new section on controlled laboratory experiments expands our coverage of the scientific method.

Part One, The Cell, the Smallest Unit of Life

The first part introduces basic biological principles and serves as a foundation for the parts that follow. In order to understand how the multicellular organism functions, it is necessary to understand how the cell functions. Since cells are made up of chemicals, we begin by considering some basic chemistry essential to the cell. Our study of cells also includes how they grow and reproduce, two vital functions in the life of the cell.

Part Two, Plant Biology

Flowering plants serve as our basis for the study of plant biology in this part. Plant cells carry on photosynthesis, the process by which they make their own organic food after capturing energy from the sun.

There are now two chapters devoted to plant structure and function. The first outlines the anatomy of the root, stems, and leaves and the second discusses the physiology and reproduction of plants.

Part Three, Human Anatomy and Physiology

Humans serve as our basis for discussing animal anatomy and physiology. The contribution of each system to homeostasis is our theme for this part.

Nutrition and dieting have been completely rewritten in a way that will make this information more serviceable to students. The Nervous System chapter now includes a section on drugs of abuse. There is a new chapter called The Lymphatic System and Immunity, which covers both general and specific defense mechanisms. The other chapters have been reorganized and portions rewritten to improve their content.

Part Four, Human Reproduction, Development, and Inheritance

Human reproduction, development, and inheritance are the topics for this part. The development chapter has been rewritten to include the most recent findings regarding differentiation and morphogenesis. It also includes a new section on aging. There are now four genetics chapters. The first two chapters, which cover Mendelian genetics, include the very latest information regarding human genetic disorders. The other chapters concern biochemical genetics. A new chapter called Recombinant DNA and Biotechnology introduces students to techniques that are making important contributions to society.

Volume Two: Evolution, Diversity, Behavior, and Ecology

Part Five, Evolution and Diversity

After an introductory chapter that presents the principles of evolution, there is a survey of living things from the origin of life to human evolution. Knowledge of the diversity of life illustrates our relationship with other living things and makes us aware of the need to preserve and protect all forms of life.

The very newest 5 kingdom classification system is used; therefore, all algae are now placed in the Kingdom Protista.

Part Six, Behavior and Ecology

In this part there is a single chapter devoted to behavior before biomes, ecosystems, and human population concerns are discussed. The future existence of human beings is dependent on our preserving the natural world, and the goal of this part is to make students aware of this dependence and what should be done to protect the balance of nature. The last chapter considers modern ecological concerns such as global warming, tropical rain forest destruction, and the growing ozone hole.

Aids to the Reader

Inquiry into Life includes a number of aids that have helped students study biology successfully and enjoyably.

History of Biology End Sheets

The front and back cover list major contributions to the field of biology in a concise, chronological manner. Students may refer to these whenever it is appropriate.

Text Introduction

The introductory chapter surveys the field of biology as a whole and prepares the student for study of its individual portions. In particular it discusses the characteristics of life and the scientific method.

Part Introductions

An introduction for each part highlights the central ideas of that part and specifically tells the reader how the topics within each part contribute to biological knowledge.

Chapter Concepts

Each chapter begins with a list of concepts stressed in the chapter. This listing introduces the student to the chapter by organizing its content into a few meaningful sentences. The concepts provide a framework for the content of each chapter.

Chapter Outlines

In addition to the chapter concepts, each chapter has a chapter outline. These will allow students to tell at a glance how the chapter is organized and what major topics have been included in the chapter. The chapter outlines include the first and second level heads for the chapter.

Readings

Two types of readings are included in the text. Readings chosen from popular magazines illustrate the applications of concepts to modern concerns. These spark interest by illustrating that biology is an important part of everyday life. The second type of reading, usually written by the author, is designed to expand, in an interesting way, on the core information presented in each chapter. Topics such as coronary heart disease and endometriosis are addressed in these readings.

Critical Thinking Questions

New to this edition are critical thinking questions that require the student to form a hypothesis, come to a conclusion, or apply information in a new and different way. The critical thinking questions are placed in the margin next to pertinent text material. Answers to the critical thinking questions appear in appendix E.

The critical thinking questions are also designed to increase the active participation of students in the learning process.

Tables and Illustrations

Numerous tables and illustrations appear in each chapter and are placed near their related textual discussion. The tables clarify complex ideas and summarize sections of the narrative. Once students have achieved an understanding of the subject matter by examining the chapter concepts and the text, these tables can be used as an important review tool. The photographs and drawings have been selectively chosen and designed to help students visualize structures and processes.

Boldfaced Words

New terms appear in boldface print as they are introduced within the text and are immediately defined in context. If any of these terms are reintroduced in later chapters, they are italicized. Key terms are defined in the end-of-chapter glossary and all boldface terms are in the text glossary, where a phonetic pronunciation is given along with appropriate page references.

Internal Summary Statements

Summary statements are placed at strategic locations throughout the chapter. These immediately reinforce the concept that has just been discussed. The summary statements will aid student retention of the chapter's main points.

Chapter Summaries

Chapter summaries offer a concise review of material in each chapter. Students may read them before beginning the chapter to preview the topics of importance, and they may also use them to refresh their memories after they have a firm grasp of the concepts presented in each chapter.

Chapter Questions

Study questions and objective questions are at the close of each chapter. The study questions allow students to test their understanding of the information in the chapter. The objective questions allow students to quiz themselves with short fill-in-the-blank questions. New to this edition is the frequent use of labeling exercises to the objective questions section.

Selected Key Term Lists

Major boldfaced terms within the chapter are defined at the end of each chapter for more convenient review. Selected key terms are listed with their phonetic pronunciation, carefully defined, and page referenced. All boldface terms are still listed alphabetically with their pronunciations, definitions, and page references in the text glossary at the end of the book.

Further Readings

For those students who would like more information about a particular topic or are seeking references for a research paper, each part ends with a listing of articles and books to help them get started. Usually the entries are *Scientific American* articles and specialty books that expand on the topics covered in the chapter.

Appendix and Glossary

The appendix contains optional information for student referral. It includes Periodic Table of the Elements, and a review of the metric system. An important part of the appendix is the Classification System of Organisms used in the text.

The text glossary defines the terms most necessary for making the study of biology successful. By using this tool, students can review the definitions of the most frequently used terms.

Index

The text also includes an index in the back matter of the book. By consulting the index it is possible to determine on what page or pages various topics are discussed.

Additional Aids

Instructor's Manual/Test Item File

The Instructor's Manual/Test Item File revised by Les Wiemerslage is designed to assist instructors as they plan and prepare for classes using *Inquiry into Life*. Possible course organizations for semester and quarter systems are suggested, along with alternate suggestions for sequencing the chapters. A general discussion and an extended lecture outline are provided for each chapter; together these give a brief overview, and a complete set of lecture notes for each chapter. For previous users of the text, sixth edition changes are noted for convenient comparison of the sixth edition to the fifth edition. Approximately 50 objective test questions and several essay questions are provided for each chapter. A list of suggested audiovisuals for the various topics and a list of suppliers are included at the end of the Instructor's Manual.

Student Study Guide

The Student Study Guide that accompanies the text was revised by Les Wiemerslage. For each text chapter, there is a corresponding Study Guide chapter that includes a chapter outline and numerous study questions and objective questions. The study questions are designed to help students carefully digest the content of the chapter while the objective questions allow students to practice for an examination. Answers are listed at the end of each chapter.

Extended Lecture Outline Software

The highly detailed outlines for each text chapter that appear in the Instructor's Manual are also available on IBM, Apple, and Macintosh diskettes for flexibility and convenience to the instructor.

Laboratory Manual

The author also wrote the *Laboratory Manual* that accompanies *Inquiry into Life*. With few exceptions, each chapter in the text has an accompanying laboratory exercise in the manual (some chapters have more than one accompanying exercise). In this way, instructors will be better able to emphasize particular portions of the curriculum if they wish. New to this edition is a Metric Measurements laboratory. The thirty-three laboratory sessions in the manual were designed to further help students appreciate the scientific method and learn the fundamental concepts of biology and the specific content of each chapter. All exercises have been tested for student interest, preparation time, and feasibility.

Customized Laboratory Manual

The Laboratory Manual's 32 exercises are now available as individual "lab separates," so instructors can custom-tailor the manual to their particular course needs. The separates, which are published in one-color at a greatly reduced price, will be collated and bound by WCB on request.

Laboratory Resource Guide

More extensive information regarding preparation can be found in the *Laboratory Resource Guide*. This guide, developed by the author and Dr. Trudy McKee, will assist the instructor in making the laboratory experience a meaningful one for the student. The guide includes suggested sources for materials and supplies, directions for making up solutions and otherwise setting up the laboratory, expected results for the exercises, and suggested answers to all questions in the laboratory manual.

Transparencies and Lecture Enrichment Kit

A set of 250 transparency acetates also accompanies the text. These feature key illustrations from the text in both two and full color. They are accompanied by a Lecture Enrichment Kit, which is a set of lecture notes featuring a summary of all textual information on the process or element depicted in each transparency, as well as additional high-interest information not presented in the text.

Slides

Often instructors prefer to use slides rather than transparencies. These are available upon special request. It is possible to acquire slides of all transparency art pieces.

Slides of Photomicrographs and Electron Micrographs

This addition to the ancillary program features 50 slides of high interest photomicrographs and electron micrographs, most of which are scanning electron micrographs.

Critical Thinking Case Study Workbook

Written by Robert Allen, this ancillary includes thirty critical thinking case studies. These case studies are designed to immerse students in the "process of science" and challenge them to solve problems in the same way biologists do. The case studies are divided into three levels of difficulty (introductory, intermediate, and advanced) to afford instructors greater choice and flexibility.

Critical Thinking Case Study Workbook Answer Key

The answers to each Critical Thinking Case Study are presented in this Key available to instructors.

Videodisk with Hypercard and Linkway

BioSci II, a general biology videodisc program with Hypercard and Linkway interface from VideoDiscovery is now available with this text. The videodisk includes photos and art found in *Inquiry into Life* as well as a broad assortment of illustrations, photos, charts, diagrams, and motion pictures with narration for use as lecture support, individual student study, or student group activities.

Videodisc Correlation Direction

This manual keys appropriate photos, illustrations, film and video clips from the BioSci II videodisc to their corresponding chapter of coverage in *Inquiry into Life,* 6/e.

Comparative Anatomy & Physiology Supplement

Instructors who wish to supplement the human anatomy and physiology coverage in *Inquiry into Life,* 6/e with comparative, evolutionary coverage of animal biology can use this free supplement, which is page-referenced to corresponding topics in the text.

Animal Behavior Supplement

Additional material on animal behavior within species is available to instructors who wish to supplement the chapter on Behavior Patterns.

WCB Testpak with Enhanced QuizPak and GradePak

WCB TestPak, a computerized testing service, provides instructions with either a mail-in/call-in testing program or the complete test item file on diskette for use with the IBM PC, Apple, or Macintosh computer. WCB TestPak requires no programming experience.

WCB QuizPak, a part of TestPak, provides students with true/false, multiple choice, and matching questions for each chapter in the text. Using this portion of the program will help students to prepare for examinations. Also included with the **WCB** QuizPak is an on-line testing option to allow professors to prepare tests for students to take on the computer. The computer will automatically grade the test and update a gradebook file.

WCB GradePak, also a part of TestPak, is a computerized grade management system for instructors. This program tracks student performance on examinations and assignments. It will compute each student's percentage and corresponding letter grade, as well as the class average. Printouts can be made utilizing both text and graphics.

Visuals Testbank

A set of 50 transparency masters are available for use by instructors. These feature line art from the text wth labels deleted for student quizzing or for student practice.

Instructor's Binder

An instructor's binder is available with *Inquiry into Life*. The binder can be used to house the text, Instructor's manual with test items, transparencies, and Lecture Enrichment units collated by chapter. This will facilitate preparation of lectures and help instructors utilize the many additional aids available with *Inquiry into Life*.

Acknowledgments

The personnel at Wm. C. Brown Publishers have always lent their talents to the success of *Inquiry into Life*. My editor, Kevin Kane, directed the efforts of all. Carol Mills, my developmental editor, served as a liaison between the editor, myself, and those in production. Sherry Padden was the production editor, Donna Slade the art editor, Mary Roussel the photo researcher, Mark Christianson the designer, Vicki Krug the permissions editor, and Joyce Watters the visuals processor. My thanks to each of them for a job well done.

A special word of thanks goes to Carlyn Iverson who was the primary artist for the book. Her beautiful renderings will be enjoyed by all. Chris Creek and Mark Lefkowitz also provided several outstanding illustrations.

Many instructors have contributed not only to this edition of *Inquiry into Life* but to previous editions. The author is extremely thankful to each one, for we have all worked diligently to remain true to our calling and to provide a product that will be the most useful to our students.

In particular, it is appropriate to acknowledge the help of the following individuals.

For the sixth edition:

D. Daryl Adams
Mankato State University

Steve Adams
Lake City Community College

Richard Blazier
Parkland College

Kathleen Burt-Utley
University of New Orleans

Lyle F. Chichester
Central Connecticut State University

Janet Dettloff
Wayne County Community College

Judith P. Downing
Bloomsburg University

Marilyn Gallup Gotkin
Nassau Community College

Ivan Huber
Fairleigh Dickinson University

D. K. Pearce
Northern Kentucky University

Hayden Williams
Golden West College

For previous editions:

Dr. Quamar A. Abbasi
Olive-Harvey College

Ken Abbott
Yavapai College

Steve Adams
Lake City Community College

Gordon Ashcroft
San Juan College

James Averett
Nassau Community College

William Barnes
Clarion University

Karen Belcher
Lake City Community College

Michael Bell
Richland College

Barbara Berkley
St. Paul's College

John Biehl
Riverside City College

Paul Biersuck
Nassau Community College

Dick Birkholz
Sheridan College

Clyde Bottrell
Tarrant County Junior College, South Campus

Hessel Bouma III
Calvin College

Sheila Brown
University of S. Mississippi

Kathy Burt-Utley
University of New Orleans

Donald Butler
Duquesne University

Joseph Cancannon
St. Johns University

William Carden
Grossmont College

Eldon Carins
Auburn University at Montgomery

William Carr
Alexander City State Junior College

Marie Casciano
Cace University

John M. Chapin
St. Petersburgh Junior College

David Cherney
Azusa Pacific College

S. Chidambaram
Canisius College

Al Chiscon
Purdue University

Barbara Clarke
American University

Roy B. Clarkson
West Virginia University

Joseph W. Cliborn
Pearl River Junior College

Charles Cottingham
South Carolina State College

Jerry A. Clonts
Anderson College

Arthur A. Cohen
Massachusetts Bay Community College

Don Collier
John C. Calhoun State Community College

James V. A. Conkey
Truckee Meadows Community College

Lesta J. Cooper-Freytag
University of Cincinnati-Raymond Walters College

John Cowlishaw
Oakland University

Shirley Crawford
SUNY Agricultural & Technical College

Neil Crenshaw
Indian River Community College

Robert Dahm
William Wright College

Patrick Daley
Lewis & Clark Community College

Ron Daniel
Cal Poly, Pomona

James M. Davenport
Washtenaw Community College

Lester Davis
Sandhills Community College

Russell Davis
University of Arizona

H. Douglas Dean
Pepperdine University

George DeHullu
Northern Essex Community College

Joyce Denham
University of Minnesota

John Dixon
University of Wisconsin at Eau Claire

Glen Drews
Chaffey College

Carlos Estol
New York University

Kathy Fergen
Miami-Dade Community College

Steven A. Fink
West Los Angeles College

Fred First
Wytheville Community College

Marie Fitzgerald
Holyoke Community College

James E. Forbes
Hampton University

Bernard L. Frye
University of Texas-Arlington

Gary Fungle
Santa Barbara City College

Carol Gerding
Cuyahoga Community College, West Campus

Donald C. Giersch
Triton College

Matthew R. Gilligan
Savannah State College

Sushil K. Gilotra
Delgado College

Albert Gordon
Southwest Missouri State University

Jack P. Gottschang
University of Cincinnati

Lewis Grey
Truman City College-Chicago

Malkiat S. Guram
Voorhees College

Martin Hahn
William Paterson College

Laszlo Hanzely
Northern Illinois University

Laird Hartman
University of South Dakota-Springfield

William R. Hawkins
Mt. St. Antonio College

Jerry Henderson
St. Louis Community College

John Hoagland
Danville Junior College

Harry R. Holloway, Jr.
University of North Dakota

Sadako Houghten
Los Angeles Pierce College

Ivan Huber
Farleigh Dickenson University-Madison

Frank Johnson
Massasoit Community College

Neil Johnson
N. Dakota State School of Science

Isidore A. Julien
Roxbury Community College

James Kane
Muskegon Community College

Vincent A. Kissel
CUNY-Bronx Community College

George Klee
Kent State University-Stark

Dean H. Kruse
Rancho Santiago College

Mark Levinthal
Purdue University

Mary Lou Longo
American International College

Benjamin Lowenhaupt
Edinboro State College

Edward Lucier
Becker Junior College

John F. Lyon
University of Lowell

Madhu Narayan Mahadeva
University of Wisconsin-Oshkosh

George Mason
East Central Junior College

Priscilla Mattson
University of Lowell

Joyce Maxwell
California State University at Northridge

Betty McAtee
Prince George's Community College

Jack Dennis McCullough
Stephen F. Austin State University

Harry S. McDonald
Stephen F. Austin State University

W. Donald McGavock
Tusculum College

Ellen McGloflin
Sanford University

James McIver
Idaho State College

Heather McKean
Eastern Washington University

Roger McPherson
Clarion University of Pennsylvania

Jim Milek
Casper College

Raymond A. Miller
University of Kentucky-Somerset Community College

Thomas H. Milton
Richard Bland College

Jean Monier
College of Notre Dame

John Munford
Wabash College

Steven N. Murray
California State University at Fullerton

Malcolm Nason
North Shore Community College

Milton Nathanson
CUNY Queens College

Hossam A. Negm
Grambling State University

Robert Neher
University of LaVerne

William Neilson
Harrisburg Area Community College

Diane R. Nelson
East Tennessee State University

Edwin Lane Netherland
Cameron University

Jerry Nielsen
Western Oklahoma State College

Carl Oney
Ononoga Community College

Lynne Osborn
Middlesex Community College

Ann Robinson
Carnegie-Mellon University

David C. Robinson
Albany State College

J. Rosko
St. Thomas Aquinas College

Jim Royce
Iowa Central Community College

Kenneth S. Saladin
Georgia College

James Sandoval
LA City College

Robert Schodorf
Lake Michigan College

A. Floyd Scott
Austin Peay State University

John Sharp
John Tyler Community College

Ted Sherill
Eastfield College

Jean Shuemaker
Gallaudet College

Robert C. Simpson
Lord Fairfax Community College

Morton Sinclair
Erie Community College, City Campus

Ethel Sloane
University of Wisconsin at Milwaukee

Judith Slon
Hilbert College

James Smith
Lawson State Community College

Jeffrey L. Smith
Dillard University

Warren Smith
Central State University

Dixie Stone
Washington Technical College

David Sulter
College of the Desert

R. Bruce Sundrud
Harrisburg Area Community College

Maurice Sweatt
Canada College

Virginia Teagarden
Lees-McRae College

Wesley Thompson
Rio Hondo College

Doris Tingle
North Greenville College

Ben Van Wagner
Bethany Bible College

Alexander Varkey
Liberty University

Mario A. Vecchiarelli
Housatonic Regional Community College

Miryam Z. Wahrman
William Paterson College

George Washington
Jackson State University

Stephen Wheeler
Alvin Junior College

William Whittaker
Greenville Technical College

Billy Williams
Dyersburg State Community College

Michael Willig
Texas Tech University

Bernard L. Woodhouse
Savannah State College

Gary Wrinkle
West Los Angeles College

Tommy Wynn
North Carolina State University

W. Brooke Yeager
Luzerne Community College

INQUIRY
INTO LIFE

Introduction

Introduction Concepts

1   Although life is difficult to define, it can be recognized by certain common characteristics.

2   The information in this book has been gathered by using the scientific method, a process based on certain identifiable procedures.

3   There are various approaches to the study of biology. (The approach chosen in this text stresses human biology.)

Introduction Outline

Characteristics of Life
The Classification of Living Things
Ecosystems
    The Human Population
The Process of Science
    The Scientific Method
    Controlled Experiments
    Theories and Principles
Science and Social Responsibility

 This book is about living things (fig. I.1); therefore, it is appropriate to first define life. Unfortunately, this is not done so easily—life cannot be given a simple, one-line definition. Since this is the case, it is customary to discuss life in terms of its characteristics. The following 5 characteristics commonly are attributed to living things.

1. Living things are organized and are made up of cells.
2. Living things grow and maintain their organization by taking chemicals and energy from the environment.
3. Living things respond to the external environment.
4. Living things reproduce and pass on their organization to their offspring.
5. Living things evolve, or change, and adapt to the environment.

Characteristics of Life

Living things are organized and are made up of cells (fig. I.2). Plants and animals are organisms in which organs are composed of tissues that contain many cells, the smallest units of life. Cells are made up of molecules that contain atoms, the smallest units of matter nondivisible by chemical means. It is of interest that a large proportion of a cell (over 50% by weight) is the molecule water, and life as we know it is dependent upon the presence of water.

Living things grow and maintain their organization by taking chemicals and energy from the environment (figs. I.3 and I.4). Only plants and plantlike organisms are capable of utilizing inorganic chemicals and the energy of the sun in the process of making their own organic food. Other organisms, such as animals, must take in preformed food as a source of chemicals and energy. The chemicals and the energy obtained from the environment are used in part to maintain the organization of living organisms. Each living organism has its own particular organization that is maintained only because energy can be acquired from the environment.

Living things respond to the external environment (fig. I.4). This characteristic commonly is acknowledged by people who recognize a living thing by its ability to move. A multicellular animal can move because it possesses a nervous system, but all living things, including plants, possess a variety of mechanisms by which they respond to the physical environment and the biological environment. Responses to the environment constitute the organism's behavior.

Figure I.2

Organization of plants and animals. Cells are composed of molecules; tissues are made up of cells; organs are composed of tissues; and organisms contain organs.

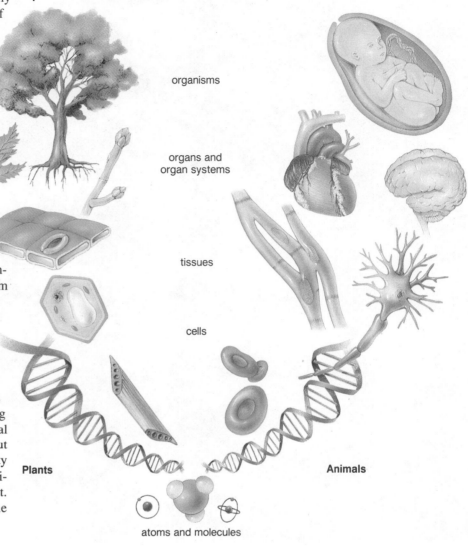

organisms

organs and organ systems

tissues

cells

Plants   **Animals**

atoms and molecules

Figure I.3

Early stages in the growth of a bean seedling. People who do not realize that growth is a characteristic of life do not realize that plants are alive.

Living things reproduce and pass on their organization to their off-spring (figs. I.4 and I.5). A unicellular organism reproduces asexually simply by dividing into 2 new organisms. Because the offspring have the same hereditary factors, or genes, as the single parent, they tend to have the same structure and function as this parent. In contrast, multicellular organisms often reproduce by means of sexual reproduction. Because male and female parents each contribute one-half the total number of genes to an offspring, it will have characteristics of both parents but will not resemble either one exactly.

Living things evolve, or change, and adapt to the environment (fig. I.5). Evolution begins when certain members of a **species** (a group of similarly constructed organisms that share common genes) happen to inherit a genetic change that causes them to be better suited, or adapted, to a particular environment. These organisms tend to survive and to have more offspring than the other members of the species that are not as well suited to the environment. Each successive generation has members that are better adapted to the environment. In this way, evolution produces adaptation to the environment.

Figure I.6

Classification of organisms. In this text, organisms are classified into the 5 kingdoms described here.

Kingdom	Representative Organisms	Description
Monera (monerans)	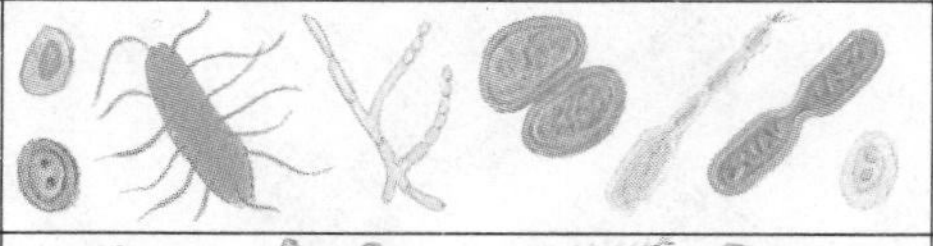	bacteria, including cyanobacteria
Protista (protists)	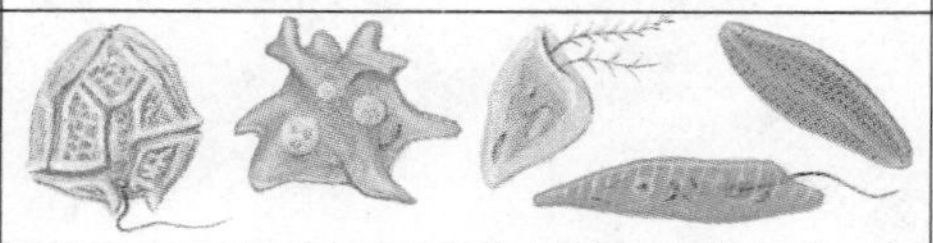	protozoans, all algae, and slime molds
Fungi (fungi)	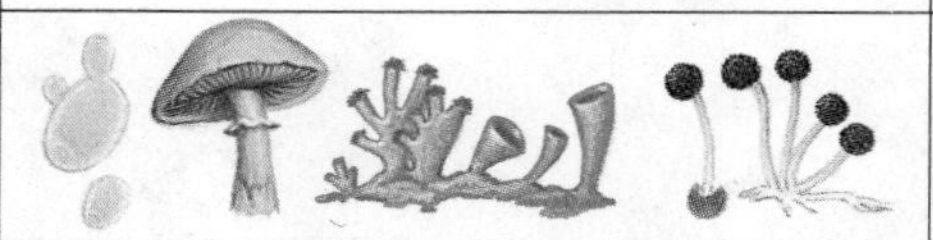	molds and mushrooms
Plantae (plants)		mosses, ferns, various trees, and flowering plants
Animalia (animals)	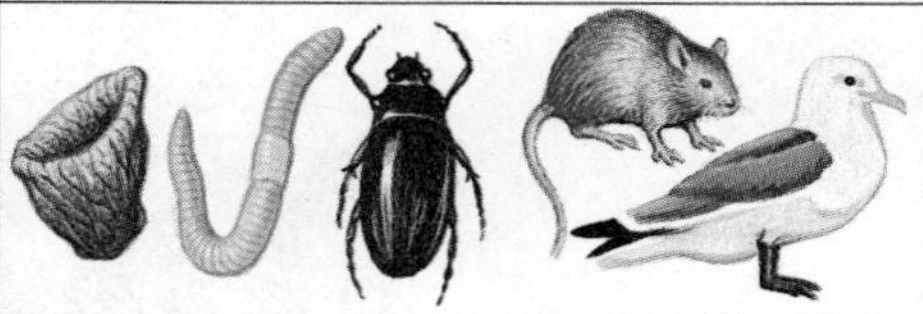	sponges, worms, insects, fishes, amphibians, reptiles, birds, and mammals

The Classification of Living Things

Taxonomists, biologists who classify living things, give each type of organism a scientific name in Latin. The scientific name is a binomial (*bi*—two; *nomen*—name). For example, the name for humans is *Homo sapiens,* and for corn, it is *Zea mays*. The first word is the genus name, and the second word is the specific name that tells which species in a genus is being discussed. Note that the genus name is capitalized, but the species name is not.

Taxonomists group species into larger and larger groups according to their shared characteristics:

Categories	Human Being	Corn
Kingdom	Animalia	Plantae
Phylum (Division)	Chordata	Anthophyta
Class	Mammalia	Monocotyledons
Order	Primates	Commelinales
Family	Hominidae	Poaceae
Genus	*Homo*	*Zea*
Species	*Homo sapiens*	*Zea mays*

In between genus and kingdom, each classification category contains more species that have fewer specific characteristics in common, until finally the species that are in the same kingdom have only general characteristics in common. For example, all species in the genus *Zea* look pretty much the same, that is, like corn plants, while those species in the plant kingdom are quite different, as when we compare grass to trees.

Taxonomists disagree about the number of kingdoms there are. Many today, however, recognize the 5 kingdoms that are listed in figure I.6.

Known organisms are given a scientific name and are classified into taxonomic categories by scientists.

Figure I.7
*Natural area versus one developed by humans.
a. Not only do other species of life thrive in a
natural area, they benefit the lives of humans
also. For example, they absorb pollutants if not
overwhelmed. b. Developed areas are restrictive
to other types of organisms, and they often
produce the pollution that is harmful to all
forms of life.*

Ecosystems

The organization of life goes beyond separate and individual organisms. All
living things on earth are a part of the **biosphere,** an interconnected system
spanning the surface of the earth that reaches up into the atmosphere and
down into the seas, wherever organisms exist. In any portion of the biosphere,
such as a particular forest or pond, all members of one species belong to a
population. The various populations interact with one another and with the
physical environment. We call the total of all populations in one natural setting, together with their physical environment, an **ecosystem** (fig. I.7a).

The interactions of the populations in an ecosystem, such as a forest or
a pond, tend to keep the ecosystem in a dynamic balance that allows it to
perpetuate itself. As in the case of the living organism, it is the organization
of an ecosystem that usually allows it to perpetuate itself.

*Organisms are members of populations within ecosystems, units of the biosphere.
Here, the populations interact in such a way that the system is kept in dynamic
balance.*

The Human Population

The human population tends to modify existing ecosystems for its own purposes. For example, humans clear forests or grasslands in order to grow crops; later, they build houses on what was once farmland; and then, finally, they convert small towns into cities. With each progressive step, fewer and fewer of the original organisms remain, until the ecosystem is completely altered. In the end, there are largely only humans and their domesticated plants and animals where once there were many diverse populations.

More and more ecosystems undergo modifications as the human population increases in size. Presently, there is much concern because the tropical rain forests of Brazil are being destroyed by large numbers of persons who are starting to live and to farm there. This comes at a time when we are beginning to realize how dependent we are on intact ecosystems and the services they perform for us. For example, the tropical rain forests act like a giant sponge that absorbs carbon dioxide (CO_2), a pollutant we pour into the atmosphere from the burning of fossil fuels, like oil and coal. An increased amount of carbon dioxide in the atmosphere is expected to have many adverse effects on the biosphere.

As more and more of the biosphere is converted to towns and cities (fig. I.7b), natural processes cannot function adequately, and in the end, the human population itself is threatened. The recognition that the biosphere needs to be preserved is one of the most important developments of our new ecological awareness. It makes us realize that an ever-increasing human population size is a possible threat to the continued existence of humanity.

The human population tends to modify existing ecosystems to suit its own needs. We now realize we are dependent upon the normal functioning of the biosphere, and it should be preserved as much as possible.

The Process of Science

Science helps human beings understand the natural world. It is concerned solely with information gained by observing and testing that world. Science aims to be objective rather than subjective. It is very difficult to make objective observations and to come to objective conclusions because human beings often are influenced by their particular prejudices. Still, anything less than a completely objective observation or conclusion is not considered scientific. Finally, scientific conclusions are subject to change whenever new findings so dictate. Quite often in science, new studies, which might utilize new techniques and equipment, indicate that previous conclusions need to be modified or changed entirely.

Scientists ask questions and carry on investigations that pertain to the natural world. The conclusions of these investigations are tentative and subject to change.

The Scientific Method

Scientists, including biologists, employ an approach to gathering information known as the **scientific method.** Although this approach is as varied as scientists themselves, there are still certain processes that can be identified as typical of the scientific method. Figure I.8 outlines the essential steps in the scientific method. First, accumulated data are used to formulate a hypothesis, which becomes the basis for new observations and/or experimentation. The collected data help a scientist come to a conclusion that either supports or does not support the hypothesis. When many other observations and experiments come to the conclusion consistent with a particular hypothesis, the hypothesis is called a theory. A theory that is considered proven by many scientists sometimes is called a principle or a law.

Introduction

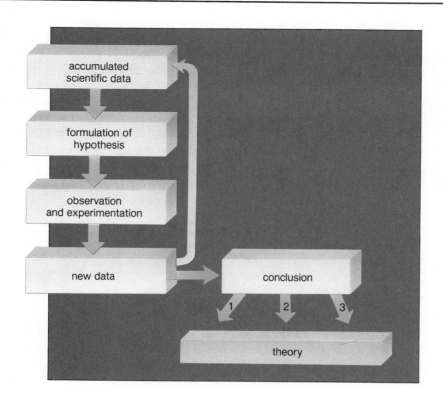

Figure I.8

Figure I.8
The scientific method. Inductive reasoning is used to formulate the hypothesis from accumulated scientific data, and deductive reasoning is used to decide which further observations and experiments are appropriate in order to test the hypothesis. The new data allow the researcher to come to a general conclusion about the phenomenon being studied. Several such conclusions (labeled 1, 2, 3) enable scientists to develop a comprehensive theory. For example, studies in comparative embryology, comparative anatomy, and paleontology all support the theory of evolution.

Let us now discuss the scientific method in more detail by taking an explicit example.

The Formulation of the Hypothesis

New observations as well as previous observations (called accumulated data in figure I.8) permit scientists to form **hypotheses,** tentative explanations of observed phenomena. To arrive at a hypothesis, scientists use various methods of reasoning, especially inductive reasoning. *Inductive reasoning* allows you to arrive at a generalization after observing specific facts. For example, in 1975 a number of adults and children in the city of Lyme, Connecticut were diagnosed as having rheumatoid arthritis. Since children rarely get rheumatoid arthritis, health officials called on Allen C. Steere, an authority on rheumatology to investigate the matter. He found that (1) most of the victims lived in heavily wooded areas, (2) the disease was not contagious, (3) symptoms first occurred in summer, and (4) several victims remembered a strange bull's-eye rash occurring several weeks before the onset of arthritis-like symptoms. On the basis of his findings, Steere formulated the hypothesis that he was dealing with a disease caused by an agent (possibly a virus) transmitted by an arthropod (the group to which insects, spiders, and ticks belong). He named the disease Lyme disease for the town where it was first observed.

Experimentation and Conclusion

Once the hypothesis has been stated, then deductive reasoning comes into play. *Deductive reasoning* begins with an "if . . . then" statement: If Lyme disease is caused by a virus, then it should be possible to find evidence of the virus in the blood of Lyme disease victims. Deductive reasoning allows scientists to decide which **data** (factual information) to collect. For example, Steere decided to test the blood of Lyme disease victims for the presence of every known arthropod-transmitted virus (fig. I.9). Not a single test result was positive. Finally in 1977, one victim knew he had been bitten by a tick and had saved

Figure I.9
This laboratory technician is testing blood samples for the presence of an infectious agent.

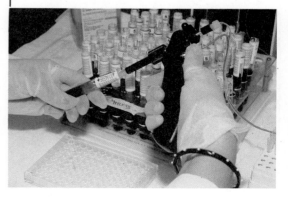

it. He gave it to Steere and it was identified as *I. dammini,* called the deer tick because adult ticks mate while feeding on deer. Steere decided to see if the natural distribution of the deer tick corresponded to the outbreak of Lyme disease. When this proved to be the case, Steere's original hypothesis had been supported. Later, a spirochete bacterium was isolated from deer ticks by Willy Burgdorfer, an authority on tickborne diseases, and the blood of Lyme disease victims tested positive for this bacterium. The new spirochete was named *Borrelia burgdorfei* after Burgdorfer.

Because hypotheses always are subject to modification, they actually never can be proven true; however, they can be proven false. When the data do not support the hypothesis, the hypothesis has to be rejected. Therefore, some think of science as what is left after alternative hypotheses have been rejected.

Even though the scientific method is quite variable, it is possible to point out certain steps that characterize it: making observations, formulating a hypothesis, testing, and coming to a conclusion.

Reporting the Data

It is customary to report findings in a scientific journal so that the design and the results of the experiment are available to all. For example, data about tickborne diseases are often reported in the *Clinical Microbiology Review* journal. It is necessary to give other researchers details on how the experiment was done because experimental results must be repeatable; that is, other scientists using the same procedures must get the same results. If they do not, the original data cannot be considered supportive of the hypothesis.

Often, too, the authors of a report suggest other experiments that could be done to clarify or to broaden our understanding of the matter under study. People reading the report also may think of other experiments to do. For example, other experimenters showed that the bull's-eye rash was due to the Lyme disease spirochete.

Observations and the results of experiments are published in a journal, where they can be examined by all. These results are expected to be repeatable; that is, they can be obtained by anyone following the exact same procedure.

Controlled Experiments

When scientists are studying a phenomenon, they often perform controlled experiments in a laboratory. There are 2 major variables in a controlled experiment—the experimental variable and the dependent variable. The *experimental variable* is the element being tested, and the *dependent variable* is the result or change that is observed.

A controlled experiment contains a **control sample,** which goes through all the steps of the experiment except the one being tested. In other words, the control sample is not subjected to the experimental variable.

The Design of the Experiment

Suppose, for example, physiologists want to determine if sweetener S is a safe food additive (fig. I.10). On the basis of available information, they hypothesize that sweetener S has no effect on health at a low concentration, but it causes bladder cancer at a high concentration. They then might decide to feed sweetener S to groups of mice at ever-greater percentages of the total dietary intake of food.

I.1 Critical Thinking

1. A variable is an element that changes. Why is sweetener S called the experimental variable in the described experiment?
2. With reference to figure I.10, explain this definition: A control group goes through all the steps of an experiment except the one being tested.
3. Why is bladder cancer the dependent variable in the described experiment?
4. Does the experiment have elements that are constant and not variable? What are they?
5. What is the value of including a control group in an experiment?

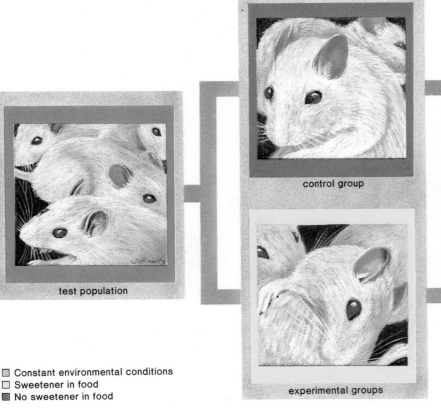

test population

control group

experimental groups

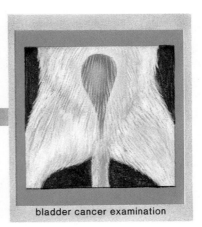

bladder cancer examination

Figure I.10

Design of a controlled experiment. From left *to* right: *genetically identical mice are divided randomly into the control group and the experimental groups. All groups are exposed to the same environmental conditions, such as housing, temperature, and water supply. Only the experimental groups are subjected to the test (experimental variable); in this case, the presence of sweetener S in the food is the variable. At the end of the experiment, all mice are examined for bladder cancer (dependent variable).*

☐ Constant environmental conditions
☐ Sweetener in food
■ No sweetener in food

Group 1: diet contains no sweetener S (the control group)
Group 2: 5% of diet is sweetener S
Group 3: 10% of diet is sweetener S
 ↓
Group 11: 50% of diet is sweetener S

The researchers first would place a certain number of randomly chosen inbred (genetically identical) mice into the various groups—say 10 mice per group. If any of the mice are different from the others, it is hoped random selection will distribute them evenly among the groups. The researchers also would make sure that all conditions, such as availability of water, cage setup, and temperature of the surroundings, are the same for all groups. The food for each group would be exactly the same except for the amount of sweetener S.

Results of the Experiment

The data from this experiment no doubt would be presented in the form of a table or a graph (fig. I.11). Researchers might run a statistical test to determine if the difference in the number of cases of bladder cancer between the various groups is significant. After all, if a significant number of mice in the control group develop cancer, the results are invalid. On the basis of the results, the experimenters try to develop a recommendation concerning the safety

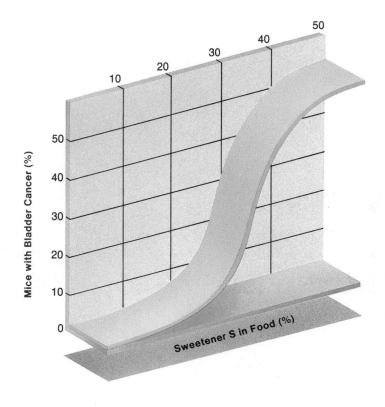

Figure I.11

Presenting the data. Scientists often acquire mathematical data that they report in the form of tables or graphs. Mathematical data are more decisive and objective than visual observations. The data in this instance suggest that there is an ever-greater chance of bladder cancer if the food is more than 10% sweetener S. Similar experiments will be repeated many times to test these results, and statistical analyses will be done to see if the results are significant or simply due to chance alone.

Table I.1 *Unifying Theories of Biology*

Name of Theory	Explanation
Cell	All organisms are composed of cells.
Biogenesis	Life comes only from life.
Evolution	All living things have a common ancestor and are adapted to a particular way of life.
Gene	Organisms contain coded information that dictates their form, function, and behavior.

I.2 Critical Thinking

1. Scientific hypotheses must be falsifiable. Why is the hypothesis "Every human being has a guardian angel" not falsifiable?
2. Why is the hypothesis "Biotin is required for good health" falsifiable?
3. In what way are religious beliefs different from scientific beliefs?

of sweetener S in the food of humans. They might determine that any amount of sweetener S over 10% of food intake is expected to cause an ever-greater incidence of bladder cancer, for example.

Many scientists work in laboratories, where they carry out controlled experiments.

Theories and Principles

The ultimate goal of science is to understand the natural world in terms of **theories,** interpretations based on the conclusions of many experiments and many observations (fig. I.8). In a movie, a detective might claim to have a theory about the crime, or you might say that you have a theory about the win-lose record of your favorite baseball team, but in science, the word *theory* is reserved for a conceptual scheme that is supported by a large number of observations and has not been found lacking yet.

Table I.1 lists some of the unifying theories of biology. You can see that, in general, the theories pertain to the characteristics of life listed earlier in this chapter. Further, they apply to various aspects of living things. For example, the theory of evolution enables scientists to understand the history of life, the variety of living things, and the anatomy, the physiology, and the development of organisms—even their behavior. Because the theory of evolution has been supported by so many observations and experiments for over a hundred years, some biologists refer to the principle of evolution. As mentioned earlier, they believe this is the appropriate terminology for theories that are generally accepted as valid by an overwhelming number of scientists.

Science and Social Responsibility

There are many ways in which science has improved our life. The most obvious examples are in the field of medicine. The discovery of antibiotics, such as penicillin, and the vaccines for polio, measles, and mumps have increased our life span by decades. Cell biology research is helping us to understand the mechanisms that cause cancer. Genetic research has produced new strains of agricultural plants that have eased the burden of feeding our burgeoning world population.

Science also has produced conclusions that we find disturbing and has fostered technologies that have proven to be ecologically disastrous if not controlled properly. Too often we blame science for this and think that scientists are duty bound to pursue only those avenues of research that will not conflict with our system of values and/or result in environmental degradation, particularly when applied in an irresponsible manner. Yet science, by its very nature, is impartial and simply attempts to study natural phenomena. Science does not make ethical or moral decisions. Instead, all men and women have a responsibility to decide how best to use scientific knowledge so that it benefits the human species and all living things.

The information presented in this text has been gathered by using the scientific method. The text, while covering all aspects of biology, focuses on human biology. It is hoped thereby that your study of biology will allow you to make wise decisions regarding your own individual well-being and also the well-being of the species.

Summary

All living things display certain characteristics: they are made up of cells; they maintain their organization by taking chemicals and energy from the environment; they respond to external stimuli; and they reproduce. As species, living things evolve and change. Evolution accounts for the diversity of life we see about us.

When studying the world of living things, biologists and other scientists use the scientific method. First, inductive reasoning based on previous and current observations and data is used to formulate a hypothesis. Then deductive reasoning is used to decide which new experiments and observations should be done to test the hypothesis. The data may support a hypothesis or may prove it false. Hypotheses cannot be proven true. All conclusions are subject to revision whenever new findings dictate. Still, there are some concepts, such as the theory of evolution, that have been supported for so long and by so many observations and experiments that they are generally accepted as true.

Science does not answer ethical questions; we must do this for ourselves. Knowledge provided by science, such as the contents of this text, can assist you in making decisions that will be beneficial to human beings and to other living things.

Study Questions

1. Name the 5 characteristics of life, and discuss each one. (pp. 3–5)
2. Support the statement "All living things are organized." (p. 4)
3. Food provides which 2 necessities for living things? (p. 4)
4. Explain the process by which living things become increasingly adapted to the environment. (p. 5)
5. Explain the scientific name of an organism. (p. 6)
6. Name the categories of classification from genus to kingdom. Which category contains more types of organisms having general characteristics in common? (p. 6)
7. What are ecosystems, and how are they modified by humans? (p. 7)
8. List a series of steps to explain the scientific method. (p. 8)
9. Explain the difference between inductive reasoning and deductive reasoning. (p. 9)
10. Give an example of a controlled experiment. Name the experimental and dependent variable. (p. 10)
11. What is the ultimate goal of science? Give an example that supports your answer. (p. 11)

Objective Questions

1. All living things are composed of _____ , the smallest units of a life.
2. All living things need a source of _____ and _____ to maintain themselves.
3. When a plant turns toward the light, it is _____ to an external stimulus.
4. When living things _____ , the offspring resemble the parents.
5. Living things are suited, that is, _____ , to their environment.
6. A type of reasoning called _____ reasoning helps scientists to formulate hypotheses.
7. Very often, the next step after formulation of the hypothesis is _____ , a type of testing that usually includes a control sample.
8. Scientists try to be objective; therefore, they prefer _____ data.
9. In science, the word _____ is often used to stand for a broad conceptual scheme that shows how various types of information are related.

Answers to Objective Questions

1. cells 2. chemicals, energy 3. responding 4. reproduce 5. adapted 6. inductive 7. experimentation 8. mathematical 9. theory

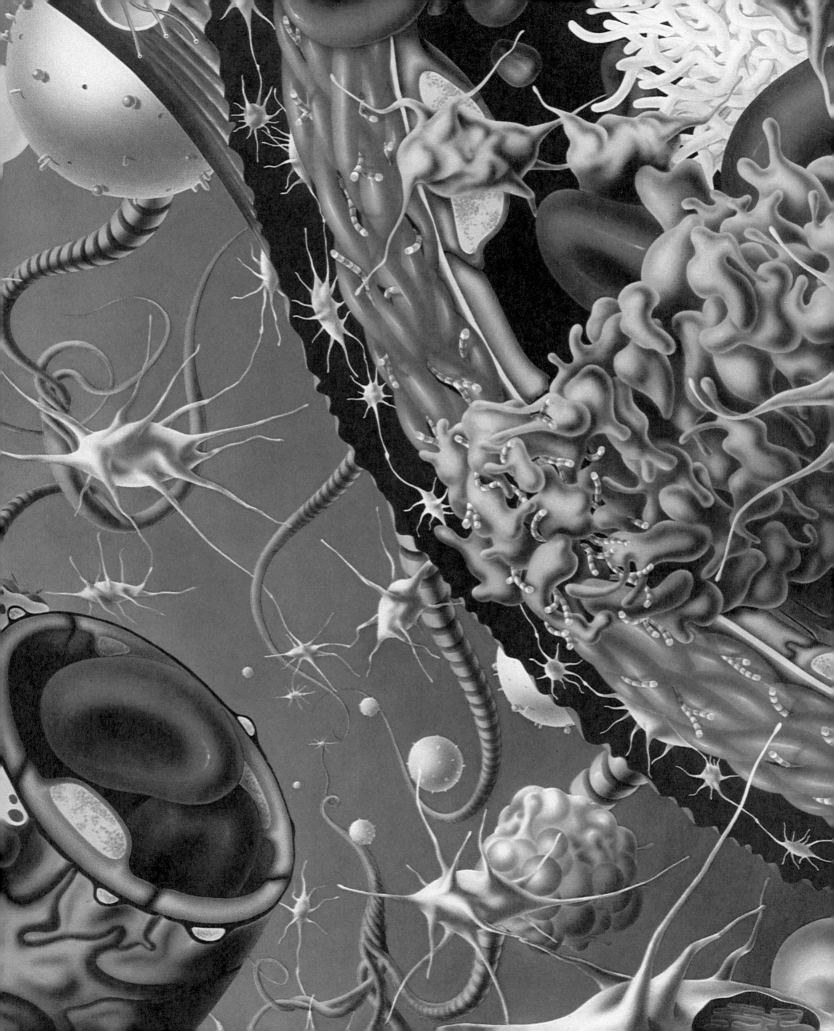

Part One

The Cell, the Smallest Unit of Life

The cell is the smallest living thing, and all the characteristics of life are found here. An understanding of cell structure, physiology, and biochemistry serves as a foundation to an understanding of multicellular forms. Part one studies each aspect of cellular biology in detail and thereby covers the fundamental concepts of biology.

Principles of inorganic and organic chemistry are discussed before a study of cell structure is undertaken. The cell is bounded by a membrane and contains organelles, many of which are also membranous. It is membrane that regulates the entrance and exit of molecules and determines how cellular organelles carry out their functions.

Cell reproduction is dependent upon mitotic cell division, whereas animal and plant reproduction are dependent upon meiotic cell division. Both types of cell division are introduced in part one. Cells require energy for growth and reproduction, and the biochemical means by which this energy is provided is considered in chapter 5.

Ruptured blood vessels are releasing red blood cells, white blood cells, and platelets into tissue fluid. Some platelets are forming a plug preventing further loss of blood. In this artist's representation, axons are represented as coiled cords, and protective macrophages look like balls of cotton.

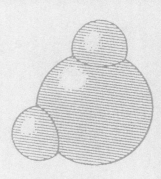

Chapter 1

Chemistry and Life

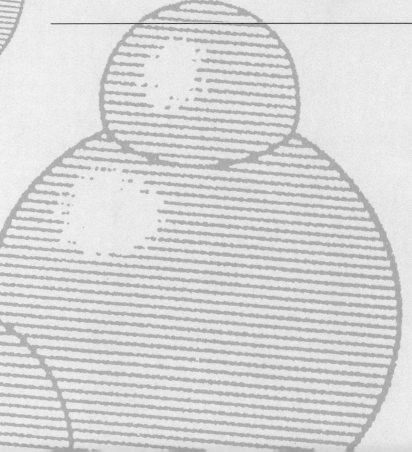

Chapter Concepts

1 Atoms, the smallest units of matter, react with one another to form molecules.

2 All living organisms are composed only of inorganic and organic molecules.

3 Some important inorganic molecules in living organisms are water, acids, bases, and salts.

4 Some important organic molecules in living organisms are proteins, carbohydrates, lipids, and nucleic acids, each of which forms when smaller molecules join.

Chapter Outline

Atoms
   Isotopes
Reactions between Atoms
   Ionic Reactions
   Covalent Reactions
   Oxidation-Reduction Reactions
Inorganic Chemistry versus Organic
   Chemistry
Some Important Inorganic Molecules
   Water
   Acids and Bases
   Salts
Some Important Organic Molecules
   Proteins
   Carbohydrates
   Lipids
   Nucleic Acids

Figure 1.1
Like nonliving things, all living things are
composed only of chemicals.

Figure 1.2

Representation of an atom. The nucleus contains
protons and neutrons; the shells contain
electrons. The first shell is complete with 2
electrons, and every shell thereafter may contain
as many as 8 electrons.

p = protons
n = neutrons
● = electrons

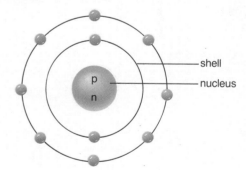

It is not always easy to understand that living things, like nonliving things, are composed of chemicals. After all, it is not possible to see the chemicals that make up an organism's body (fig. 1.1). However, a few minutes' reflection regarding the dietary needs of the body usually convinces us that humans indeed are made of chemicals. For example, calcium is needed to maintain the bones, iron is necessary to prevent anemia, and adequate amino acid intake is required to build muscles.

Because living things, including humans, are composed only of chemicals, it is absolutely essential for a biology student to have a basic understanding of chemistry.

Atoms

An **atom** is the smallest unit of matter nondivisible by chemical means. While it is possible to split an atom by physical means, an atom is the smallest unit to enter into chemical reactions. For our purposes, it is permissible to think of an atom as having a central **nucleus,** where subatomic particles called **protons** and **neutrons** are located, and *shells,* where **electrons** orbit about the nucleus (fig. 1.2). Electrons have varying amounts of energy. Those with the greatest amount of energy are located in the shells farthest from the nucleus. Other important features of protons, neutrons, and electrons are their charge and weight, which are indicated in table 1.1.

Table 1.1 Subatomic Particles

Name	Charge	Weight
Electron	One negative unit	Almost no weight
Proton	One positive unit	One atomic unit
Neutron	No charge	One atomic unit

Figure 1.3

Periodic table of the elements (simplified). See appendix A for the complete table. Each element has an atomic number, an atomic symbol, and an atomic weight. The elements in a darker color are the most common, and those in a lighter color are also common in living things. The dark line separates metals (to the left) from nonmetals (to the right).

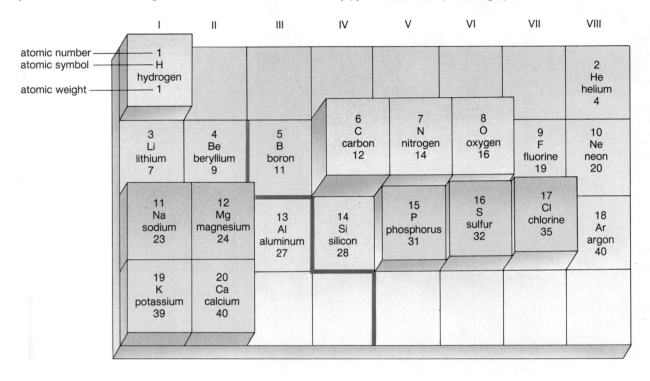

The periodic table of the elements in appendix A shows all the atoms that presently are known. An **element** is any substance that contains just one type of atom. Figure 1.3 gives a simplified table that highlights the elements most common to living things. Notice that in the table, each atom has a symbol; for example, C = carbon and N = nitrogen. Each type of atom also has an **atomic number;** for example, carbon is number 6 and nitrogen is number 7. *The atomic number equals the number of protons.* Each type of atom also has an **atomic weight,**[1] or mass. In the table given here, carbon has an atomic weight, or mass, of 12 and nitrogen has an atomic weight of 14. *The atomic weight equals the number of protons plus the number of neutrons.*

Now, it is possible to diagram a specific electrically neutral atom (fig. 1.4). In an *electrically neutral atom,* the number of protons (+) is equal to the number of electrons (−). The first shell of an atom can contain up to 2 electrons; thereafter, each shell of those atoms in the simplified table (fig. 1.3) can contain up to 8 electrons.[2]

It is readily apparent that the elements in the periodic table are horizontally arranged in order of increasing atomic number and weight, but they also are vertically arranged according to similar chemical properties. For example, the atoms in the first column all have one electron in the outermost shell, and we will see that they give up this electron in chemical reactions. The Roman numerals at the top of each column indicate how many electrons there are in the outer shell of the elements listed in that column. An exception is helium, whose outer shell has only 2 electrons.

Figure 1.4

Carbon atom. The diagram of the atom shows that the number of protons (the atomic number) equals the number of electrons when the atom is electrically neutral. Carbon also can be written in the manner shown below the diagram. The subscript is the atomic number, and the superscript is the atomic weight.

p = protons
n = neutrons
= electrons

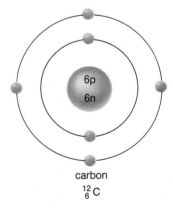

carbon
$^{12}_{6}C$

---

[1]Atomic weights are relative weights. The most common isotope of carbon has been assigned a weight of 12, and the other atoms are either lighter or heavier than carbon.

[2]The maximum number of electrons for any shell except the outer shell is $2n^2$, where *n* is the shell number. However, we will consider only atoms 1–20, in which all shells can have 8 electrons except the first shell, which has 2 electrons.

Isotopes

The atomic weights given in the simplified periodic table (fig. 1.3) have been rounded off; for example, the actual atomic weight of carbon (C) is 12.011. This is because the atomic weight is the average weight of each type of atom, and individual atoms can vary in weight. When they do vary, they are called **isotopes** of one another. Isotopes of carbon can be written in the following manner, where the subscript stands for the atomic number and the superscript stands for the weight:

$$^{12}_{6}\text{C} \qquad ^{13}_{6}\text{C} \qquad ^{14}_{6}\text{C*}$$

*radioactive

The number of protons (6) in these isotopes does not vary, but the weight does. This indicates that the number of neutrons must be responsible for the weight difference since electrons have almost no weight.

Certain isotopes, called **radioactive isotopes,** are unstable and emit radiation, which can be detected photographically or with a special counter. Among those isotopes of carbon listed, only carbon 14 is radioactive, as the asterisk indicates. Radioactive isotopes are used widely in biological research and medical diagnostic procedures. For example, because the thyroid gland uses iodine, it is possible to administer radioactive iodine and then to view a scan of the thyroid gland sometime later (fig. 1.5).

All matter is composed of atoms that are arranged in the periodic table of the elements according to increasing weight. The weight of an atom is dependent on the number of protons and neutrons in the nucleus, while the chemical properties are dependent on the electrons in the shells.

Reactions between Atoms

Usually, reactions between atoms involve the electrons in their outer shell. The *octet rule,* based on chemical findings, states that atoms react with one another in order to achieve 8 electrons in their outer shell. Hydrogen (H) is an exception to this rule. In hydrogen, the first shell, which is complete with 2 electrons only, is the outer shell.

When 2 or more atoms of different elements combine, a **compound** results. A **molecule** is the smallest part of a compound that still has the properties of that compound. Molecules also can form when 2 or more atoms of the same element react with one another.

Ionic Reactions

In one type of reaction, atoms give up or take on electrons in order to achieve a completed outer shell. Such atoms, which thereafter carry a charge, are called **ions,** and the reaction is called an **ionic reaction.** In ionic reactions, atoms lose or gain electrons to produce a compound that contains ions in a fixed ratio to one another. The formula for the compound shows the proper ratio of atoms in the compound.

For example, figure 1.6 depicts a reaction between sodium (Na) and chlorine (Cl) in which chlorine takes an electron from sodium. The resulting ions in the compound sodium chloride (Na^+Cl^-) have 8 electrons each in the outer shell. Notice that when sodium gives up an electron, the second shell, with 8 electrons, becomes the completed outer shell. Also, the sodium (Na^+) ion carries a positive charge because it has given up an electron. Chlorine, on the other hand, receives an electron for a total of 8 electrons in the outer shell. This causes the chloride (Cl^-) ion to carry a negative charge.

Figure 1.5
Use of radioactive iodine. a. *Drawing of the anatomical shape and location of the thyroid gland.* b. *A scan of the thyroid gland 24 hours after the patient was administered radioactive iodine.*

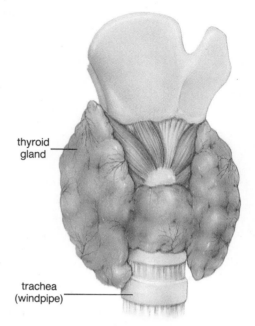

thyroid gland

trachea (windpipe)

a.

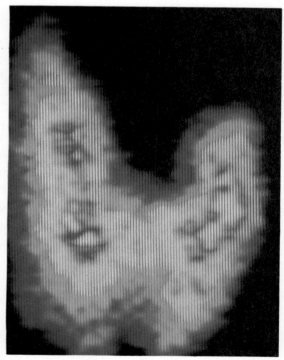

b.

Ionic reaction. a. *When the neutral atom sodium (Na) becomes an ion, it loses an outer electron and then has 8 electrons in the outer shell. The sodium (Na⁺) ion carries a positive charge because it has one more proton than electrons.* b. *When the neutral atom chlorine (Cl) becomes an ion, it receives an outer electron and then has 8 electrons in the outer shell. The chloride* (Cl⁻) *ion carries a negative charge because it has one less proton than electrons.* c. *When sodium reacts with chlorine, sodium gives an electron to chlorine, and the compound sodium chloride (Na⁺Cl⁻) results. The 2 ions are held together by their opposite charge.*

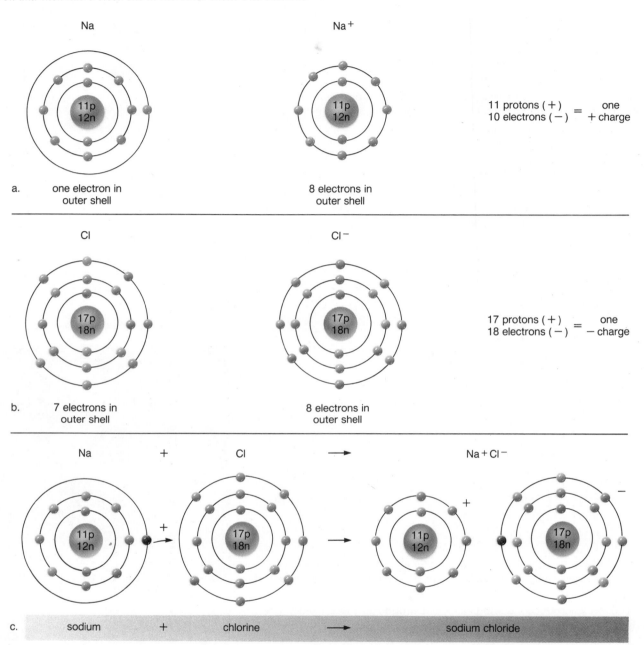

Na   Na⁺

11p 12n   11p 12n

11 protons (+)   one
10 electrons (−)   = + charge

a.   one electron in outer shell   8 electrons in outer shell

Cl   Cl⁻

17p 18n   17p 18n

17 protons (+)   one
18 electrons (−)   = − charge

b.   7 electrons in outer shell   8 electrons in outer shell

Na + Cl ⟶ Na⁺Cl⁻

11p 12n + 17p 18n ⟶ 11p 12n   17p 18n

c.   sodium + chlorine ⟶ sodium chloride

The attraction between oppositely charged sodium ions and chloride ions is called an **ionic bond,** and it is such bonds that are found within ionic compounds. We are quite familiar with the ionic compound sodium chloride (Na⁺Cl⁻) because it is table salt, which we use to enliven the taste of foods.

Ionic-bond formation occurs when a **metal** reacts with a nonmetal. Metals are those atoms that appear to the left of the dark line in the simplified periodic table (fig. 1.3), and nonmetals are those that appear to the right of this dark line. Metals lose electrons and become positively charged. In contrast, nonmetals gain electrons and become negatively charged. This principle is illustrated in figure 1.7.

Figure 1.7

Ionic reactions. When a metal reacts with a nonmetal, an ionic compound results. The metal becomes positively charged, and the nonmetal becomes negatively charged. a. The metal magnesium (Mg) gives up 2 electrons to oxygen (O), a nonmetal. In the compound magnesium oxide ($Mg^{++}O^{--}$), the magnesium (Mg^{++}) ion carries 2 positive charges (why?), while the oxygen (O^{--}) ion carries 2 negative charges (why?). b. The metal calcium (Ca) gives up one electron to each of 2 nonmetal chlorines (Cl). In the compound calcium chloride ($Ca^{++}Cl_2^{-}$), the calcium (Ca^{++}) ion carries 2 positive charges, and each chloride (Cl^{-}) ion carries one negative charge.

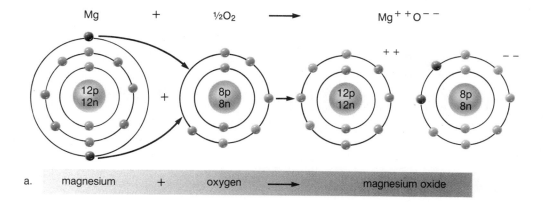

a. magnesium + oxygen → magnesium oxide

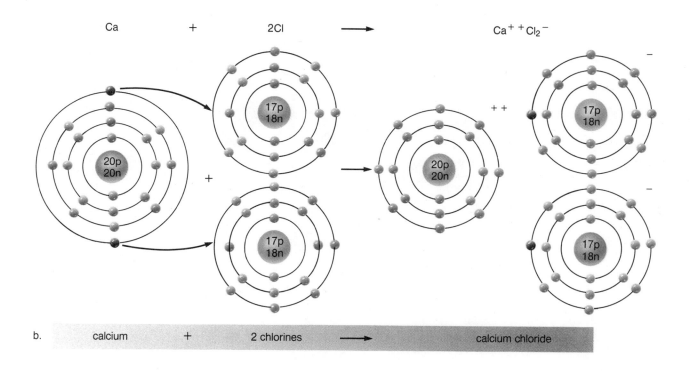

b. calcium + 2 chlorines → calcium chloride

With the exception of hydrogen, atoms react with each other in order to achieve 8 electrons in the outer shell. In an ionic reaction, positively and negatively charged ions are formed when electrons are transferred from metals to nonmetals. The attraction between ions is called an ionic bond.

Covalent Reactions

When nonmetals join with nonmetals, a **covalent reaction** occurs (fig. 1.8). In a **covalent bond,** the atoms share electrons instead of losing or gaining them. The overlapping outer shells in figure 1.8 indicate that the atoms are sharing electrons. Sharing is usually equal; each atom contributes one electron to each

Figure 1.8
Covalent reactions. After a covalent reaction, the atoms share electrons, and each atom has 8 electrons in the outer shell. To show this, it is necessary to count the shared electrons as belonging to both bonded atoms. a. Formation of chlorine gas. b. Formation of nitrogen gas. c. Formation of water.

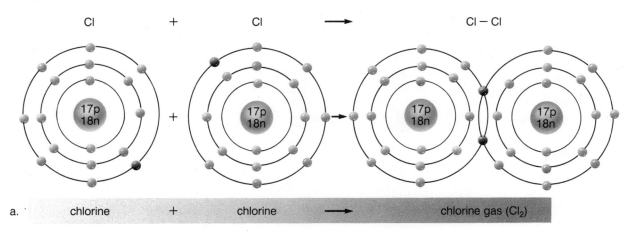

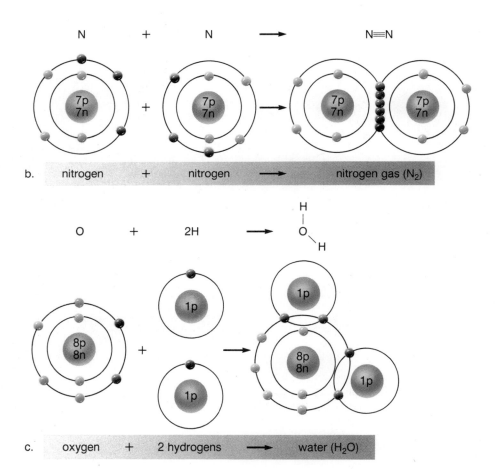

pair that is shared. These electrons spend part of their time in the outer shell of each atom; therefore, they may be counted as belonging to both atoms. When this is done, each atom has 8 electrons in the outermost shell.

Instead of drawing complete diagrams of molecules, electron-dot diagrams sometimes are used. For example, in reference to figure 1.8a, each chlorine atom can be represented by its symbol, and the electrons in the outer shell can be designated by dots. The shared electrons are placed between the 2 sharing atoms, as shown here:

$$:\ddot{C}l\cdot + \cdot\ddot{C}l: \longrightarrow :\ddot{C}l:\ddot{C}l:$$

Electron-dot diagrams are a bit cumbersome, and covalent bonds often are indicated simply by straight-line structural formulas.[3] At other times, even the lines are omitted, and molecular formulas that indicate only the number of each type of atom are given:

$$Cl-Cl \quad \text{or} \quad Cl_2$$

Even if the molecule is written as Cl_2, it is easy to tell that the 2 chlorines are sharing electrons because (1) they are both nonmetals and (2) no charge is indicated. Additional examples of electron-dot, structural, and molecular formulas are shown in figure 1.9.

Double Bonds

Besides a single bond, like that between the 2 chlorine (Cl) atoms, a double bond or a triple bond can form in order for 2 atoms to complete their octet. In a triple bond, 2 atoms share 3 pairs of electrons between them. For example, in figure 1.8b, the reaction between 2 nitrogen (N) atoms results in a triple bond because each nitrogen atom requires 3 electrons to achieve a total of 8 electrons in the outermost shell. Notice that in the diagrammatic representation, 6 electrons are placed in the outer overlapping shells and that 3 straight lines are indicated in the structural formula for nitrogen gas.

In a covalent reaction, nonmetals share electrons so that all atoms involved can have completed outer shells. Each shared pair of electrons is a covalent bond; double bonds and even triple bonds are possible.

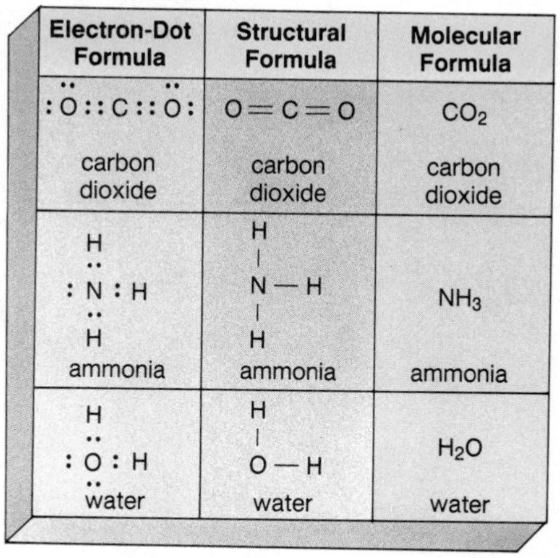

Oxidation-Reduction Reactions

When oxygen combines with a metal, oxygen receives electrons and becomes negatively charged; the metal loses electrons and becomes positively charged. For example, consider the reaction that is illustrated in figure 1.7a.

$$Mg + \tfrac{1}{2}O_2 \longrightarrow Mg^{++}O^{--}$$

In such cases, it is obviously appropriate to say that the metal has been oxidized and that because of oxidation, the metal has lost electrons. Then we need only admit that the oxygen has been reduced because it has gained electrons, or negative charges.

Today, the terms **oxidation** and **reduction** are applied to many ionic reactions, whether or not oxygen is involved. Very simply, *oxidation refers to the loss of electrons, and reduction refers to the gain of electrons.* In this ionic reaction, $Na + Cl \longrightarrow Na^+ Cl^-$, the sodium atom has been oxidized (loss of electron) and the chlorine atom has been reduced (gain of electron).

The terms *oxidation* and *reduction* also are applied to certain covalent reactions. In this case, however, oxidation is the loss of hydrogen (H) atoms,

---

[3]Structural formulas show the orientation of the atoms to one another and try to reflect how the atoms are arranged in space.

Figure 1.9

Electron-dot, structural, and molecular formulas. In the electron-dot formula, only the atoms in the outermost shell are designated. In the structural formula, the lines represent a pair of electrons that are being shared between 2 atoms. The molecular formula indicates only the number of each type of atom found within a molecule.

Electron-Dot Formula	Structural Formula	Molecular Formula
$:\ddot{O}::C::\ddot{O}:$ carbon dioxide	$O=C=O$ carbon dioxide	CO_2 carbon dioxide
H $:N:H$ H ammonia	H $\|$ N—H $\|$ H ammonia	NH_3 ammonia
H $:\ddot{O}:H$ water	H $\|$ O—H water	H_2O water

Table 1.2 *Inorganic Chemistry versus Organic Chemistry*

Inorganic Compounds	Organic Compounds
Usually contain metals and nonmetals	Always contain carbon and hydrogen
Usually have ionic bonding	Always have covalent bonding
Always contain a small number of atoms	May be quite large with many atoms
Often associated with nonliving materials	Often associated with living organisms

Figure 1.10

Hydrogen bonding between water molecules. Water molecules are polar: each hydrogen (H) atom carries a partial positive charge, and each oxygen (O) atom carries a partial negative charge. The polarity of the water molecules brings about hydrogen bonding between the molecules in the manner shown. The dotted lines represent hydrogen bonds.

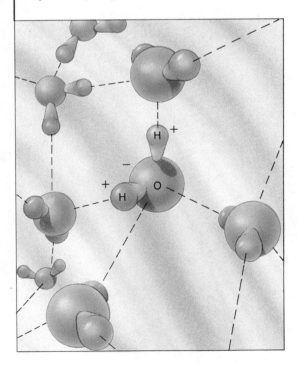

and reduction is the gain of hydrogen atoms. A hydrogen atom contains one proton and one electron; therefore, when a molecule loses a hydrogen atom, it has lost an electron, and when a molecule gains a hydrogen atom, it has gained an electron. We will have occasion to refer to this form of oxidation-reduction reaction again in chapter 5 because it is important in many metabolic reactions occurring in cells.

When oxidation occurs, an atom becomes oxidized (loses electrons). When reduction occurs, an atom becomes reduced (gains electrons). These 2 processes occur concurrently in oxidation-reduction reactions.

Inorganic Chemistry versus Organic Chemistry

There are 2 types of chemistry pertinent to our study, **inorganic chemistry** and **organic chemistry.** Ionic reactions are common to inorganic chemistry, while covalent reactions are common to organic chemistry. Table 1.2 lists the important differences between inorganic and organic compounds. As will be apparent in the following discussion, both types of compounds are involved in the proper functioning of the living organism.

Some Important Inorganic Molecules

Water

Water (H_2O) is not an organic molecule because it does not contain carbon (C), but as figure 1.8*c* shows, water is covalently bonded. Often, the atoms in a covalently bonded molecule share electrons evenly, but in water, the electrons spend more time circulating the larger oxygen (O) atom than the smaller hydrogen (H) atoms. Therefore, there is a slight positive charge on the hydrogen atoms and a slight negative charge on the oxygen atom. Because it is charged, water is called a **polar molecule,** and hydrogen bonding occurs between water molecules (fig. 1.10). A **hydrogen bond** occurs whenever a covalently bonded hydrogen atom is attracted to a negatively charged atom some distance away. The hydrogen bond is represented by a dotted line in figure 1.10, because it is a weak bond that is easily broken.

Characteristics of Water

Hydrogen bonds are relatively weak, but they still cause water molecules to be cohesive and to cling together. Without hydrogen bonding between molecules, water would boil at −80° C rather than 100° C and freeze at −100° C rather than 0° C, making life impossible. Instead, water is a liquid at body temperature. It absorbs a great deal of heat before it becomes warm and evaporates, and it gives off this heat as it cools down and freezes. This property allows great bodies of water, such as the oceans, to maintain a relatively constant temperature. It also helps to keep an animal's body temperature within normal limits and even accounts for the cooling effect of sweating.

The cohesiveness of water allows it to fill tubular vessels, and water is an excellent transport medium for distributing substances and heat throughout the body. Its cohesive property is obvious whenever we observe the surface tension of bodies or containers of water.

Because of hydrogen bonding, liquid water is denser than ice. Therefore, ice floats on liquid water, and bodies of water always freeze from the top down, making ice-skating possible (fig. 1.11). Furthermore, the layer of ice protects the organisms below and helps them to survive the winter.

Water, being a polar molecule, acts as a solvent and dissolves various chemical substances, particularly other polar molecules. This property of water greatly facilitates chemical reactions in cells.

The Cell, the Smallest Unit of Life

Dissociation Polarity also causes water molecules to tend to **dissociate,** or ionize, in this manner:

$$H—O—H \longrightarrow H^+ + OH^-$$

The hydrogen (H^+) ion has lost an electron; the hydroxide (OH^-) ion has gained the electron. Very few molecules actually dissociate; therefore, few hydrogen ions and hydroxide ions result (fig. 1.12).

Acids and Bases

Acids are compounds that dissociate in water and release hydrogen (H^+) ions (or protons[4]). For example, an important inorganic acid is hydrochloric acid (HCl), which dissociates in this manner:

$$HCl \longrightarrow H^+ + Cl^-$$

Dissociation is almost complete, and this acid is called a strong acid. If hydrochloric acid is added to a beaker of water (fig. 1.13), the number of hydrogen (H^+) ions increases.

Bases are compounds that dissociate in water and release hydroxide (OH^-) ions. For example, an important inorganic base is sodium hydroxide (NaOH), which dissociates in this manner:

$$NaOH \longrightarrow Na^+ + OH^-$$

Dissociation is complete, and sodium hydroxide is called a strong base. If sodium hydroxide is added to a beaker of water (fig. 1.14), the number of hydroxide ions increases.

---

[4]A hydrogen atom contains one electron and one proton. A hydrogen ion has only one proton and is often called a proton.

Figure 1.11

Ice. Hydrogen bonding causes ice to be less dense than liquid water; therefore, lakes freeze from the top down. This protects the organisms that live in the water beneath the ice and it also permits humans to enjoy ice-skating.

Figure 1.12

Dissociation of water molecules. In water, there are always a few molecules that have dissociated. Dissociation produces an equal number of hydrogen (H^+) ions and hydroxide (OH^-) ions. (These figures are for illustration and are not mathematically accurate.)

Figure 1.13

Hydrochloric acid (HCl). Hydrochloric acid is an acid that releases hydrogen (H^+) ions as it dissociates in water. Notice that the addition of hydrochloric acid to this beaker has caused it to have more hydrogen ions than hydroxide (OH^-) ions.

Figure 1.14

Sodium hydroxide (NaOH), as a base. Sodium hydroxide releases hydroxide (OH^-) ions as it dissociates. Notice that the addition of sodium hydroxide to this beaker has caused it to have more hydroxide ions than hydrogen (H^+) ions.

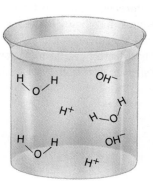

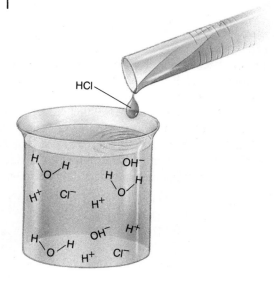

Normally, rainwater has a pH of about 5.6 because the carbon dioxide in the air combines with water to give a weak solution of carbonic acid. Rain falling in northeastern United States and southeastern Canada has a pH between 5.0 and 4.0. One has to remember that a pH of 4 is 10 times more acidic than a pH of 5 to appreciate the increase in acidity this represents.

There is very strong evidence now that this observed increase in rainwater acidity is a result of the burning of fossil fuels like coal and oil, as well as the gasoline derived from oil. When these substances are

Acid Deposition

burned, sulfur dioxide and nitrogen oxides are produced and combine with water vapor in the atmosphere to form acids. These acids fall out of the atmosphere as rain or snow, in a process properly called wet deposition but more often called acid rain. Also, dry particles of sulfate and nitrate salts can fall out of the atmosphere, and this is called dry deposition. The net result of the burning of fossil fuels and related products is the presence of acids in the air, and these return to the earth sometime later.

These air pollutants can be carried in the wind to places far distant from their origin. Acid deposition in Canada and northeastern United States is due to the burning of fossil fuels in factories and power plants located in the Midwest. Similarly, the Scandinavian countries are the recipients of air pollutants from England and the rest of northern Europe. Unfortunately, regulations that re-quire the use of tall smokestacks to reduce local air pollution only cause the pollutants to be carried further away (see figure).

Acid deposition has created a very serious situation in certain areas of the world. In the United States, vulnerable areas include not only the Northeast, but also the Great Smoky Mountains, the lakes of Wisconsin and Minnesota, the Pacific Northwest, and the Colorado Rockies. The soil in these areas is thin and lacks lime-stone (calcium carbonate, $CaCO_3$), which can buffer acid deposition. The first cry of alarm came from Sweden, which reported as early as the 1960s that its lakes were dying. Today, dying lakes are found across northeastern North America and eastern and western Europe. The acid deposition leaches aluminum from the soil and carries this element into the lakes. The acid also allows the conversion of mercury deposits in lake bottom sediments to soluble methyl mercury. Not only do the lakes become more acidic, they also show accumulation of substances that are toxic to living things. Sweden now reports 15,000 of its lakes are too acidic to support any higher aquatic life.

In the early 1980s, evidence began to accumulate that acid precipitation was also damaging to forests. As of mid-1986, some 19 countries in Europe reported damage to their woodlands ranging from roughly 5 to 15% of the forested area in Yugoslavia and Sweden to 50% or more in the Netherlands, Switzerland, and West Germany. More than one-fifth of Europe's forests are now damaged.

These aren't the only effects of acid deposition. Other effects include reduction of agricultural yields, damage to marble and limestone monuments and buildings, and even illnesses in humans. Acid deposition has been implicated in the increased incidence of lung cancer and possibly colon cancer in residents of the East Coast. Tom McMillan, Canadian Minister of the Environment, says that acid rain is "destroying our lakes, killing our fish, undermining our tourism, retarding our forests, harming our agriculture, devastating our heritage and threatening our health."

There are of course, things that can be done. We could

a. whenever possible use alternative energy sources, such as solar, wind, hydropower, and geothermal energy.
b. use low-sulfur coal or remove the sulfur impurities from coal before it is burned.
c. require the use of scrubbers to remove sulfur from factory and power plant emissions.
d. require people to use mass transit rather than drive their own automobiles.
e. reduce our energy needs through other means of energy conservation.

These measures and possibly others could be taken immediately. It is only necessary for us to determine that they are worthwhile.

Sources: This reading is based on these references: G. Tyler Miller, 1985, *Living in the Environment*. Wadsworth Publishing Co., Belmont, CA and Lester R. Brown, et al., 1988, *State of the World*. W.W. Norton and Co., New York, NY.

1.1 Critical Thinking

1. Water fulfills which of the criteria for an inorganic compound? (See table 1.2.)
2. Water fulfills which of the criteria for an organic compound?
3. Which of water's characteristics (pp. 24–25) help the body to function? How?
4. Since it is helpful in living things, why is water not termed an organic compound?

pH

The pH[5] scale is used to indicate the acidity and basicity (alkalinity) of a solution. A pH of exactly 7 is neutral pH. Pure water has an equal number of hydrogen (H^+) and hydroxide (OH^-) ions, and therefore, one of each is formed when water dissociate. The fraction of water molecules that dissociate is 10^{-7} (or 0.0000001), which is the source of the pH value for neutral solutions. The pH scale was devised to simplify discussion of the hydrogen ion concentration $[H^+]$ and consequently, also of the hydroxide ion concentration $[OH^-]$; it eliminates cumbersome numbers. For example,

a. 1×10^{-6} $[H^+]$ = pH 6
b. 1×10^{-7} $[H^+]$ = pH 7
c. 1×10^{-8} $[H^+]$ = pH 8

[5]pH is defined as the negative logarithm of the hydrogen ion concentration.

The Cell, the Smallest Unit of Life

The cause and effect of acid deposition.

Acidic Solutions In order to understand the relationship between hydrogen ion concentration [H$^+$] and pH, consider the following question. Of the 2 values, *a* or *c*, listed on page 26, which indicates a higher hydrogen ion concentation than pH 7 (neutral pH) and therefore refers to an acidic solution? A number with a smaller negative exponent indicates a greater quantity of hydrogen (H$^+$) ions than one with a larger negative exponent. Therefore, *a* is an acidic solution.

Basic Solutions Bases add hydroxide ions to solutions and increase the hydroxide ion concentration [OH$^-$] of water. Basic solutions, then, have fewer hydrogen ions compared to hydroxide ions. Which of the values—*a* or *c*—in the previous list refers to a basic solution? Item *c* refers to a basic solution because it indicates a lower hydrogen ion concentration [H$^+$] (greater hydroxide ion concentration) than pH 7.

Figure 1.15

The pH scale. The proportionate amount of hydrogen ions to hydroxide ions is indicated by the diagonal line. Any pH above 7 is basic, while any pH below 7 is acidic.

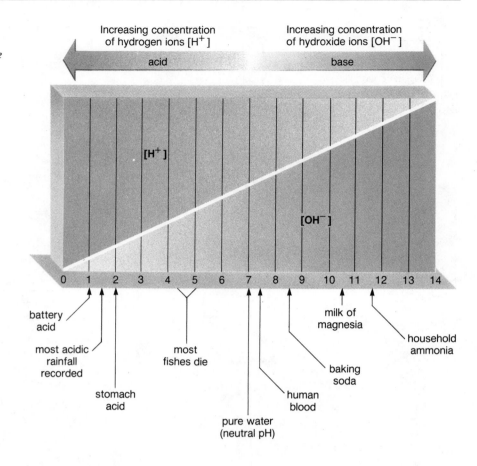

Increasing concentration of hydrogen ions [H⁺]

Increasing concentration of hydroxide ions [OH⁻]

acid

base

$[H^+]$

$[OH^-]$

0 1 2 3 4 5 6 7 8 9 10 11 12 13 14

battery acid

most acidic rainfall recorded

stomach acid

most fishes die

pure water (neutral pH)

human blood

baking soda

milk of magnesia

household ammonia

The pH Scale   The **pH scale** (fig. 1.15) ranges from 0 to 14. It uses whole numbers instead of negative exponents of the number 10 to indicate the hydrogen ion concentration [H⁺]. As you move down the pH scale, each unit has 10 times the acidity of the previous unit, and as you move up the scale, each unit has 10 times the basicity of the previous unit. A pH of 7 has an equal concentration of hydrogen (H⁺) ions and hydroxide (OH⁻) ions. Above pH 7 there are more hydroxide ions than hydrogen ions, and below pH 7 there are more hydrogen ions than hydroxide ions.

Buffers   Usually, the pH in living things is maintained within a narrow range. For example, when healthy, the pH of our blood is always about 7.4. Buffers help to keep the pH constant. A **buffer** is a chemical or a combination of chemicals that can take up excess hydrogen (H⁺) ions or excess hydroxide (OH⁻) ions. When an acid is added to a solution, a buffer takes up excess hydrogen ions, and when a base is added to a solution, a buffer takes up excess hydroxide ions. Therefore, a buffer resists changes in pH when either hydrogen ions or hydroxide ions are added. Eventually, however, a buffer can be overwhelmed, and the pH will change.

Salts

When an acid reacts with a base, a salt molecule and a water molecule result. In the case of a strong acid and a strong base, the reaction is complete:

$$HCl + NaOH \longrightarrow Na^+Cl^- + HOH$$

salt   water

Further, if an equal amount of both take part in this reaction, neutralization occurs; the resulting solution is neither acidic nor basic. In the neutralization

The Cell, the Smallest Unit of Life

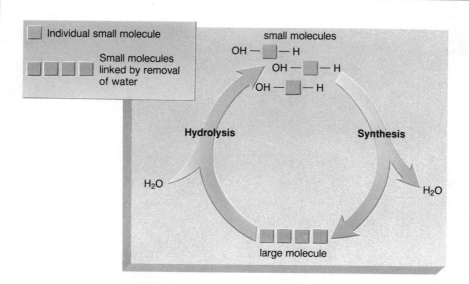

Individual small molecule

Small molecules linked by removal of water

small molecules

OH — ▢ — H

OH — ▢ — H

OH — ▢ — H

Hydrolysis

Synthesis

H_2O

H_2O

large molecule

Figure 1.16
Synthesis and hydrolysis of an organic polymer. When small molecules, or monomers, join to form a polymer (synthesis), water is released, and then when a polymer is broken down (hydrolysis), water is added.

process, the hydrogen (H^+) ions from the acid and the hydroxide (OH^-) ions from the base combine to form water. The salt consists of the positive ion of the base and the negative ion of the acid.

Acids have a pH that is less than 7, and bases have a pH that is greater than 7. Buffers help to keep the pH of internal body fluids near pH 7 (neutral) because a buffer can absorb both hydrogen ions and hydroxide ions.

Some Important Organic Molecules

The chemistry of carbon accounts for the formation of the very large number of organic compounds we associate with living organisms. Carbon is a non-metal with 4 electrons in the outer shell. In order to achieve 8 electrons in the outer shell, it shares with other nonmetals. Carbon shares electrons with as many as 4 other atoms. Many times, carbon atoms share with each other to form a ring or a chain of carbon atoms.

Usually, the large organic compounds characterisic of living things—proteins, carbohydrates, fats, and nucleic acids—are formed when smaller organic molecules join (fig. 1.16). Each of the smaller molecules is a unit of the larger molecule. The bonds that join any 2 of these units form after the removal of hydrogen (H^+) ions from one molecule and hydroxide (OH^-) ions from the next molecule. As water forms, **dehydration synthesis** occurs.

Large organic molecules built in this manner sometimes are called macromolecules or polymers. Each of the smaller molecules is called a monomer. Polymers can be broken down in a manner opposite to synthesis; the addition of water leads to the disruption of the bonds linking the monomers. During this process, called **hydrolysis,** one molecule takes on hydrogen ions and the next takes on hydroxide ions.

Proteins

Proteins are large macromolecules that sometimes have a structural function. For example, in humans, keratin is a protein that makes up hair and nails, while collagen is a protein found in all types of connective tissue, including ligaments, cartilage, bones, and tendons. The muscles (fig. 1.17) contain proteins and this accounts for their ability to contract.

Some proteins function as **enzymes,** necessary contributors to the chemical workings of the cell and therefore of the body. Enzymes are organic catalysts that speed up chemical reactions. They work so quickly that when

Figure 1.17
The well-developed muscle cells of an athlete. These cells contain many protein molecules. A major concern today is steroid use to promote the buildup of muscles. This practice can be detrimental to health.

Figure 1.18

Representative amino acids. Amino acids differ by their R group; the simplest R group is a single hydrogen. Those that contain carbon vary as shown. Notice that each amino acid is shown twice and that the second drawing is the simplified structural formula.

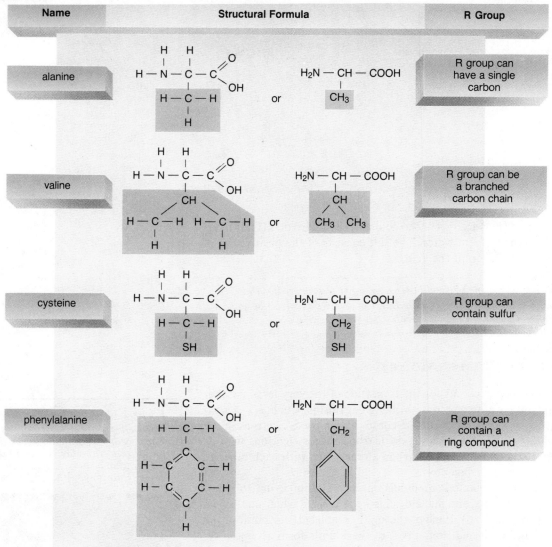

Name	Structural Formula	R Group
alanine		R group can have a single carbon
valine		R group can be a branched carbon chain
cysteine		R group can contain sulfur
phenylalanine		R group can contain a ring compound

present, a reaction that normally takes several hours or days takes only a fraction of a second. Specific enzymes in the body carry out synthetic reactions to build up macromolecules; others carry out hydrolytic reactions that break down macromolecules (fig. 1.16).

Amino Acids

Proteins form when a very large number of **amino acids** join. Most amino acids have this structural formula:

amino   acid   amino   acid

Figure 1.19

Formation of a peptide. On the left-hand side of the equation, there are 2 different amino acids, as signified by the difference in the R group notations. As the peptide bond forms, water is given off—the water molecule on the right-hand side of the equation is derived from components removed from the amino acids on the left-hand side. During hydrolysis, water is added as the peptide bond is broken.

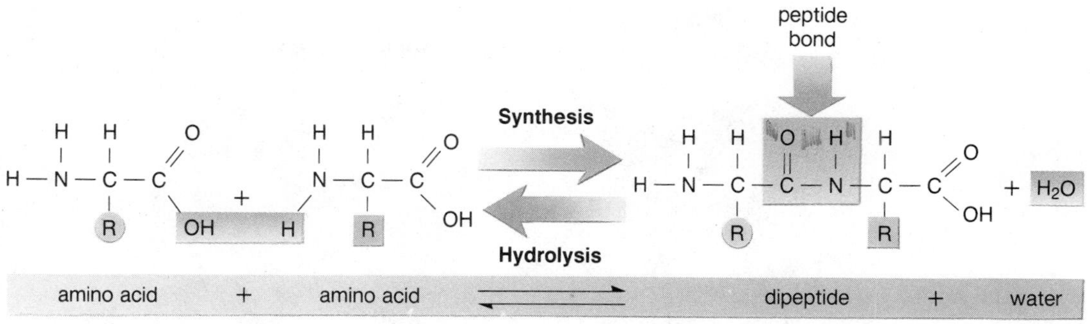

The name *amino acid* refers to the fact that the molecule has 2 functional groups: an **amino group** ($-NH_2$) and an **acid** or carboxyl **group** (COOH). Amino acids differ from one another by their *R group,* the *R*emainder of the molecule. In amino acids, the R group varies from a single hydrogen atom to complicated rings (fig. 1.18). There are about 20 different common amino acids found in the proteins of living things. Each one has a different R group.

Peptides

The bond that joins 2 amino acids is called a **peptide bond.** As you can see in figure 1.19, when dehydration synthesis occurs, the acid group of one amino acid reacts with the amino group of another amino acid, and water is given off. A **dipeptide** contains only 2 amino acids, but when up to 10 or 20 amino acids have joined, the resulting chain is called a **polypeptide.** A very long polypeptide of approximately 75 amino acids or more may be called a **protein.**

The atoms associated with the resulting peptide bond—oxygen (O), carbon (C), nitrogen (N), and hydrogen (H)—share the electrons in such a way that the oxygen atom carries a partial negative charge and the hydrogen atom carries a partial positive charge:

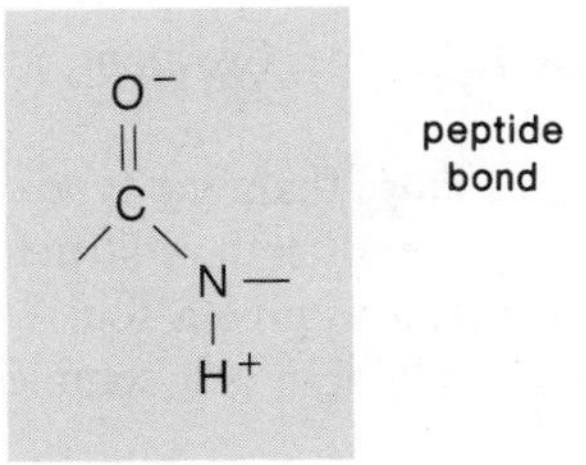

peptide bond

Therefore, the peptide bond is a polar bond, and hydrogen bonding, represented by the dotted lines in figure 1.20, is a frequent occurrence in polypeptides and proteins.

Levels of Structure

Proteins commonly have 3 levels of organization in their structure (fig. 1.20), although some have a fourth level also. The primary structure is the linear sequence of the amino acids joined by peptide bonds. Any number of the 20 different amino acids can be joined in any sequence, just as if you were making a necklace from associated beads. Any given protein has a characteristic sequence of amino acids.

The secondary structure of a protein comes about when the polypeptide chain takes a particular orientation in space. One common arrangement of the

Chemistry of Life

Figure 1.20

Levels of organization in the structure of a protein. Primary structure of a protein is the order of the amino acids; secondary structure is often an alpha (α) helix in which hydrogen bonding occurs along the length of a polypeptide, as indicated by the dotted lines; and in globular proteins, the tertiary structure is the twisting and the turning of the helix that takes place because of bonding between the R groups. Enzymes are globular proteins.

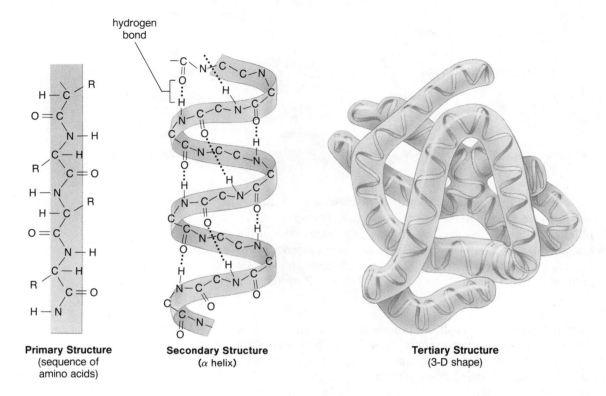

Primary Structure
(sequence of amino acids)

Secondary Structure
(α helix)

Tertiary Structure
(3-D shape)

chain is the alpha helix, or right-handed coil, with 3.6 amino acids per turn. Hydrogen bonding between amino acids, in particular, stabilizes the helix.

The tertiary structure of a protein is its final three-dimensional shape. In a structural protein like collagen, the helical chains lie parallel to one another. Enzymes are globular proteins in which the helix bends and twists in different ways. The tertiary shape of a protein is maintained by various types of bonding between the R groups. Covalent, ionic, and hydrogen bonding are all seen.

Some proteins have more than one type of polypeptide chain, each with its own primary, secondary, and tertiary structures. Within the protein, these separate chains are arranged to give a fourth level of structure, termed the quaternary structure. Hemoglobin is a complex protein having a quaternary structure.

The final shape of a protein is very important to its function, as is emphasized again in the discussion of enzyme activity (chap. 5). When proteins are exposed to extreme heat and pH, they undergo an irreversible change in shape called denaturation. For example, we are all aware that the addition of acid to milk causes curdling and that heating causes egg white, a protein called albumin, to coagulate. Denaturation occurs because the normal bonding between the R groups has been disturbed. Once a protein loses its normal shape, it no longer is able to perform its usual function.

Proteins, which contain joined amino acids, have both structural and metabolic[6] functions in cells. All enzymes, molecules that speed up metabolic reactions, are proteins.

---

[6]Metabolism is all the chemical reactions that occur in a cell.

The Cell, the Smallest Unit of Life

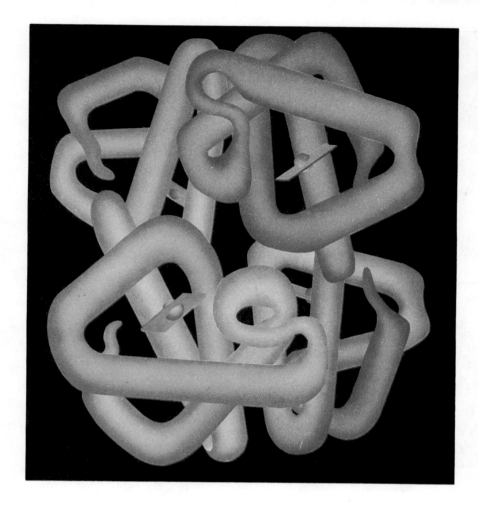

Hemoglobin molecule. Hemoglobin has a quaternary structure because it has 4 polypeptide chains arranged as shown. Each polypeptide has a tightly bound nonprotein heme group (represented here as a plane) that contains an oxygen-carrying iron atom (represented as a sphere).

Carbohydrates

A carbohydrate is characterized by the presence of the atomic grouping H—C—OH, in which the ratio of hydrogen (H) atoms to oxygen (O) atoms is approximately 2:1. Because water has this same ratio of hydrogen atoms to oxygen atoms, the name carbohydrate, which means *hydrates of carbon,* was given them. If the number of carbon atoms in a compound is low (from 3 to 7), then the carbohydrate is a simple sugar, or monosaccharide. Thereafter, larger carbohydrates are created by joining monosaccharides in the manner described in figure 1.16 for the synthesis of organic compounds.

Monosaccharides

As their name implies, **monosaccharides** are simple sugars having only one unit (fig. 1.21). These compounds often are designated by the number of carbons they contain; for example, **pentose** sugars, such as ribose, have 5 carbons and **hexose** sugars, such as glucose, have 6 carbons. Glucose is the primary energy source of the body, and most carbohydrate polymers can be broken down into monosaccharides that either are or can be converted to glucose. Other common monosaccharides are fructose, found in fruits, and galactose, a constituent of milk. These 3 monosaccharides all have the molecular formula $C_6H_{12}O_6$, but they differ in the shape of the ring and/or in the arrangement of the hydrogen (H) atoms and the hydroxide (OH) groups attached to the ring.

Disaccharides

The term **disaccharide** tells us that there are 2 monosaccharide units joined in the compound. When 2 glucose molecules are joined, **maltose** (fig. 1.22) is formed. The chemical equation for this reaction indicates that the forward direction is a dehydration synthesis and the backward reaction is a hydrolysis.

Figure 1.21

Common monosaccharides. a. Ribose, a pentose—a 5-carbon sugar. Deoxyribose has one less oxygen (O) atom attached to the second carbon (C) atom compared to ribose. b. Glucose, a hexose—a 6-carbon sugar. The small numbers count the carbon atoms.

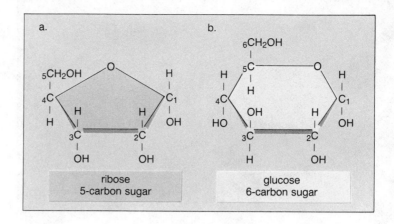

Figure 1.22

Synthesis and hydrolysis of maltose, a disaccharide containing 2 glucose units. During synthesis, a bond forms between the 2 glucose molecules as the components of water are removed. During hydrolysis, the components of water are added as the bond is broken.

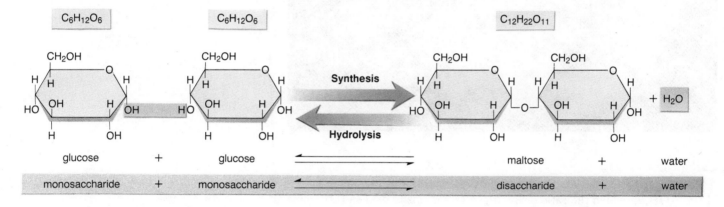

You also may be interested in knowing that when *glucose* and *fructose* are joined, the disaccharide **sucrose** is formed. Sucrose is derived from plants and is commonly known as table sugar.

Polysaccharides

A **polysaccharide** is a carbohydrate that contains a large number of monosaccharide molecules. There are 3 polysaccharides that are common in organisms: starch, glycogen, and cellulose. All of these are polymers, or chains, of glucose, just as a necklace might be made up of only one type of bead. Even though all 3 polysaccharides contain only glucose, they are distinguishable from one another.

As figure 1.23*a* shows, **starch** has few side branches, or chains of glucose that branch off from the main chain. Starch is the storage form of glucose in plants. Just as we store orange juice as a concentrate, plants store starch as a concentrate of glucose. This analogy is appropriate because, like the synthetic reaction described in figure 1.16, water is removed when glucose molecules are joined to form starch. The following equation also represents the synthesis of starch:

$$\text{glucose} \underset{\text{hydrolysis}}{\overset{\text{synthesis}}{\rightleftharpoons}} \text{starch} + H_2O$$

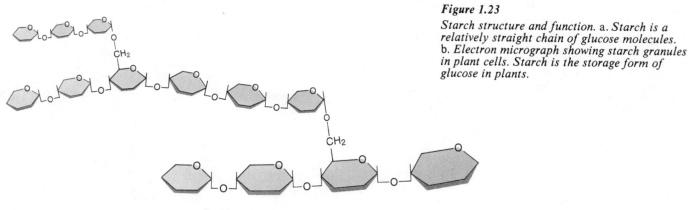

a. Starch molecule
(somewhat branched)

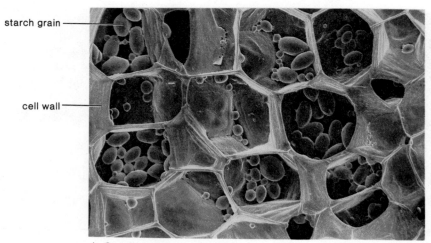

starch grain —

cell wall —

b. Starch granules in plant cells

Glycogen is characterized by the presence of side chains of glucose (fig. 1.24). Glycogen is the storage form of glucose in animals. After an animal eats, the liver stores glucose as glycogen; in between eating, the liver releases glucose so that the blood concentration of glucose is always about 0.1%.

The polymer **cellulose** is found in plant cell walls and accounts in part for the strong nature of these walls. In fact, it may be said that cellulose is the primary structural component of plants. In cellulose (fig. 1.25*a*), the glucose units are joined by a slightly different type of linkage compared to that of starch and glycogen. While this might seem to be a technicality, actually it is important because we are unable to digest foods containing this type of linkage; therefore, cellulose passes through our digestive tract as bran or roughage. Recently, it has been suggested that roughage in the diet is necessary to good health and may even help to prevent colon cancer.

A carbohydrate is a hydrate of carbon. The monosaccharide glucose frequently is used as an energy source in cells. The polysaccharides starch and glycogen are storage compounds in plant and animal cells, respectively, and the polysaccharide cellulose is found in plant cell walls.

Lipids

Lipids are water-insoluble substances obtainable from plant and animal sources. The familiar lipids are the fats (e.g., lard and butter) and oils (e.g., corn oil

Figure 1.24

Glycogen structure and function.
a. Glycogen is a highly branched
polymer of glucose molecules. The
branching allows breakdown to proceed
at several points simultaneously.
b. Electron micrograph showing
glycogen granules in liver
cells. Glycogen is the
storage form of glucose
in animals.

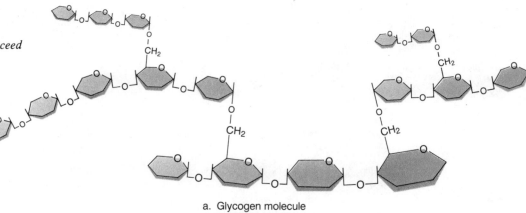

a. Glycogen molecule
(highly branched)

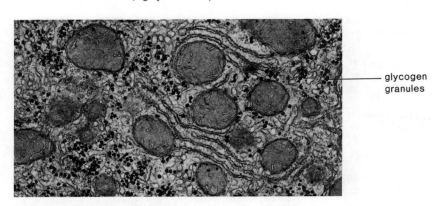

glycogen
granules

b. Glycogen granules in liver cell

Figure 1.25

Cellulose structure and function. a. Cellulose
contains a slightly different type of linkage
between glucose molecules compared to starch
and glycogen. b. Plant cells have walls made up
of cellulose. The rigidity of the cell walls permits
nonwoody plants to stand upright as long as they
receive an adequate supply of water.

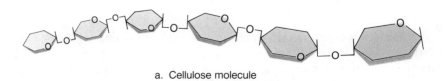

a. Cellulose molecule
(different bonding pattern)

b. Cellulose makes up plant cell walls

Figure 1.26

Synthesis and hydrolysis of a neutral fat. Three fatty acids plus glycerol react to produce a fat molecule and 3 water molecules. A fat molecule plus 3 water molecules react to produce 3 fatty acids and glycerol.

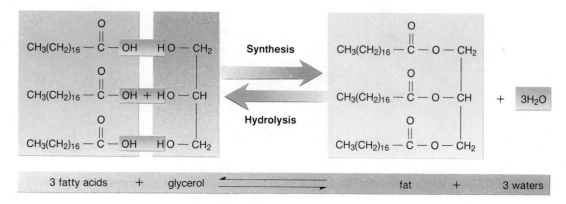

and soybean oil), which are distinguishable in that fats are solids at room temperature while oils are liquids at room temperature. In the body, fats serve as long-term energy sources. Adipose tissue is composed of cells that contain many molecules of fat.

Fats and Oils

A fat (or an oil) is formed when one molecule of glycerol reacts with 3 fatty acids. **Glycerol** is a compound with 3 hydrates of carbon. Notice in figure 1.26 that glycerol has 3 —OH groups and that it therefore can react with 3 fatty acids to form a fat molecule and 3 water molecules. Again, the larger fat molecule is formed by dehydration synthesis in the forward direction. The backward direction shows how fat can be hydrolyzed to its components. A fat is sometimes called a *triglyceride* because of its makeup. Sometimes the term *neutral fat* is used because the molecule has no groups that can ionize and become charged. Therefore, a fat is nonpolar.

Fatty Acids

Each **fatty acid** has a long chain of carbon atoms with hydrogens attached, and it ends in an acid group (fig. 1.27). Most of the fatty acids in cells contain 16 or 18 carbon (C) atoms per molecule, although smaller ones also are found. Fatty acids are either saturated or unsaturated. Saturated fatty acids have no

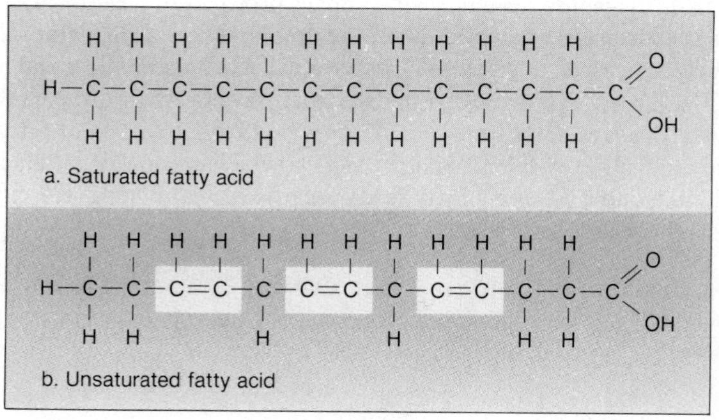

Figure 1.27

Saturated versus unsaturated fatty acids. Fatty acids are either saturated (have no double bonds) or unsaturated (have double bonds). a. In a saturated fatty acid, the carbons carry all the hydrogen atoms possible. b. In this unsaturated fatty acid, there is a double bond at the third from last carbon and at other carbons.

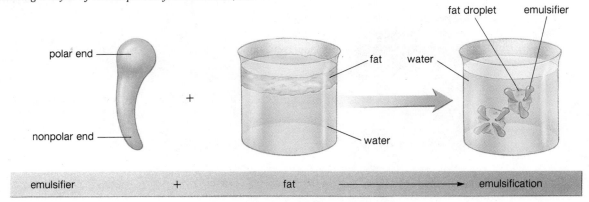

double bonds between the carbon atoms. The carbon chain is saturated, so to speak, with all the hydrogen (H) atoms that can be held. Butter is called a saturated fat because it contains saturated fatty acids.

Unsaturated fatty acids have double bonds in the carbon chain wherever the number of hydrogen atoms is less than 2 per carbon atom. Vegetable oils are called unsaturated fats because they contain unsaturated fatty acids. The presence of unsaturated fatty acids accounts for the liquid nature of vegetable oils. Vegetable oils are hydrogenated to make margarine. Polyunsaturated margarine still contains a large number of unsaturated or double bonds.

Soaps   Strictly speaking, soaps are not lipids, but they are considered here as a matter of convenience. A soap is a salt formed by a fatty acid and an inorganic base. For example,

$$NaOH + RCOOH \longrightarrow RCOO^- Na^+$$

sodium   fatty   soap
hydroxide acid

Fats do not mix with water because they are nonpolar; a soap has a non-polar end (the carbon chain represented by R) and a polar end (the charged end). Therefore, a soap does mix with water. When soaps are added to oils, the oils, too, mix with water. Figure 1.28 shows how a soap positions itself about an oil droplet so that the nonpolar ends project into the fat droplet, while the polar ends project outward. Now, the droplet disperses in water. This process of causing an oil to disperse in water is called **emulsification,** and it is said that an emulsion has formed. Emulsification occurs when dirty clothes are washed with soaps and detergents. Also, prior to the digestion of fatty foods, fats are emulsified by bile. Usually, a person who has had the gallbladder removed has trouble digesting fatty foods because the gallbladder stores bile for use during the digestive process.

Phospholipids

Phospholipids, as their name implies, contain a phosphate group:

$$O^- - \overset{\overset{\displaystyle O}{\|}}{\underset{\underset{\displaystyle O^-}{|}}{P}} - O^-$$

1.2 Critical Thinking

1. Actin and myosin are the proteins found in muscles. How do you predict they differ?
2. Starch and glycogen both are composed of glucose. How do they differ?
3. Oleic acid and linoleic acid are both unsaturated fatty acids. How do you predict they differ?

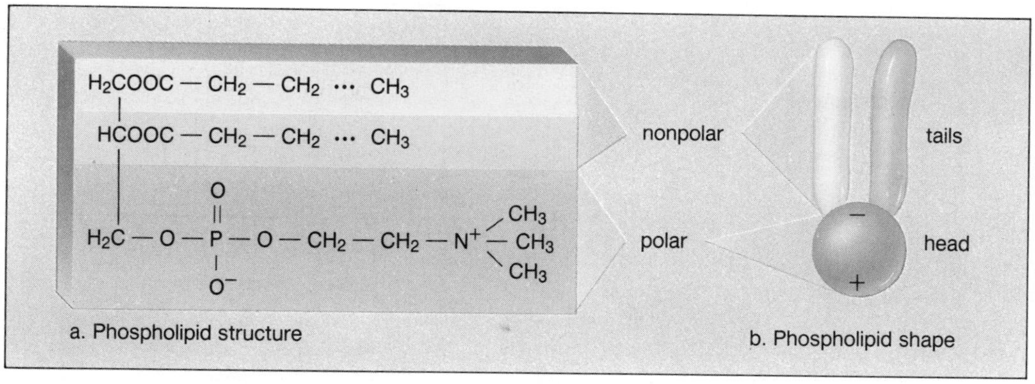

Figure 1.29
Phospholipid structure. Phospholipids are constructed similarly to fats except that they contain a phosphate group.
a. Lecithin, shown here, has a side chain that contains both a phosphate group and a nitrogen-containing group.
b. The polar portion of the molecule is soluble in water, whereas the 2 hydrocarbon chains are not soluble in water. This causes the molecule to arrange itself as shown.

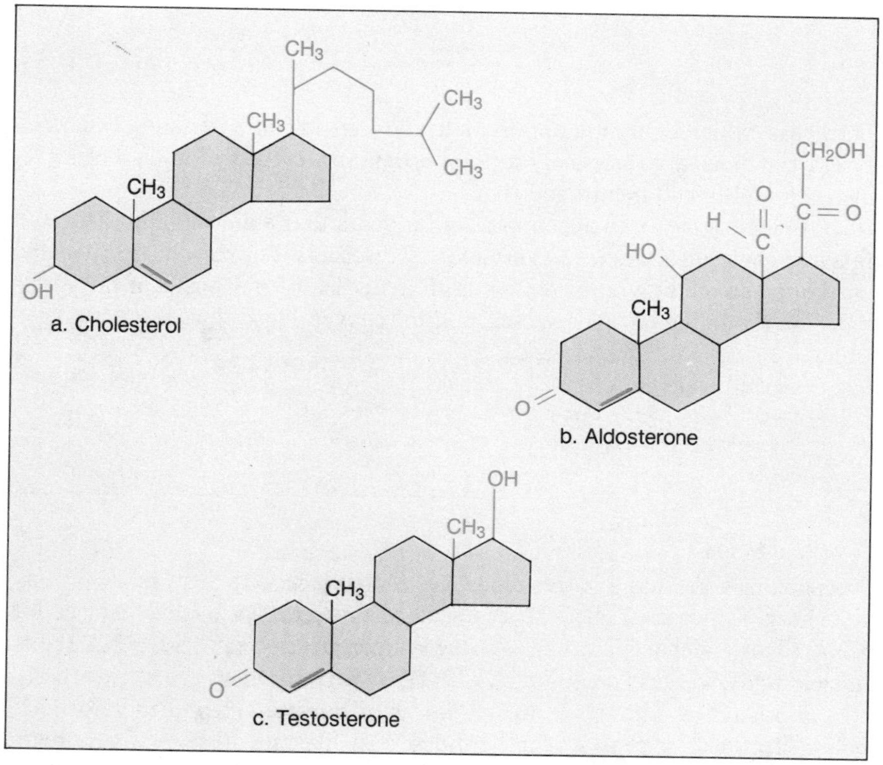

a. Phospholipid structure

b. Phospholipid shape

Figure 1.30
Steroid diversity. Like cholesterol in (a), steroid molecules have 4 adjacent rings, but their effects on the body largely depend on the type of chain attached at the location indicated. The chain in (b) is found in aldosterone, which is involved in the regulation of sodium blood levels, while the chain in (c) is found in testosterone, the male sex hormone.

a. Cholesterol

b. Aldosterone

c. Testosterone

Essentially, phospholipids are constructed as fats are, except that in place of the third fatty acid, there is a phosphate group or a grouping that contains both phosphate and nitrogen (fig. 1.29). These molecules are not electrically neutral as are the fats because the phosphate group can ionize. Notice, then, that the phospholipids have a nonpolar region that is not soluble in water and a polar region that is soluble in water. This property has important consequences for the structure of cell membranes, as we will see in the next chapter.

Steroids

Steroids are lipids that have entirely different structures than fats. Each steroid has a backbone of 4 fused carbon rings and varies from the others primarily according to the type of functional group attached to it (fig. 1.30). *Cholesterol* is the precursor of several other steroids, such as aldosterone, a hormone that helps to regulate the sodium content of the blood, and the sex

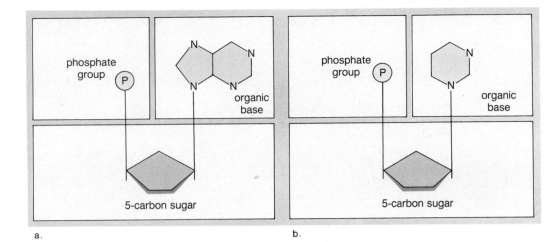

Figure 1.31

Generalized nucleotides. All nucleotides contain a phosphate group, a pentose sugar, and a nitrogen-containing organic base. The 2 types of nucleotides differ as to whether the base has (a) 2 rings or (b) one ring.

Figure 1.32

Generalized nucleic acid strand. Nucleic acid polymers contain a chain of nucleotides. Each strand has a backbone made of sugar and phosphate molecules. The bases project to the side.

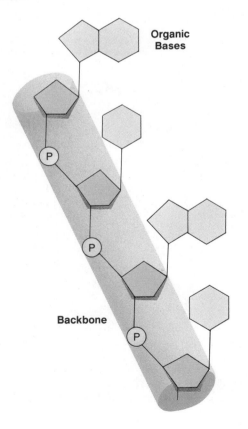

hormones, which help to maintain male and female characteristics. Such different functions are due solely to the particular attached group. Cholesterol also is found in cell membrane (p. 65).

Evidence has been accumulating for years that a diet high in saturated fats and cholesterol can lead to circulatory disorders. For reasons that are discussed in the reading on page 234, such a diet leads to deposits of fatty material on the lining of blood vessels and to reduced blood flow.

Lipids include nonpolar fats, long-term energy-storage molecules that form from glycerol and 3 fatty acids, and the related phospholipids, which have a charged group. The steroids have an entirely different structure, similar to that of cholesterol.

Nucleic Acids

Nucleic acids are huge macromolecular compounds with very specific functions in cells; for example, the genes are composed of a nucleic acid called **DNA** (deoxyribonucleic acid). Another important nucleic acid, **RNA** (ribonucleic acid), works in conjunction with DNA to bring about protein synthesis.

Both DNA and RNA are *polymers of nucleotides* and therefore are polynucleotides, or chains of nucleotides. Just like the other synthetic reactions we have studied in this section, these units are joined to form nucleic acids by the removal of water molecules:

$$\text{nucleotides} \underset{\text{hydrolysis}}{\overset{\text{synthesis}}{\rightleftharpoons}} \text{nucleic acid} + H_2O$$

Nucleotides

Every **nucleotide** is a molecular complex of 3 types of molecules: a phosphoric acid (phosphate) molecule, a pentose sugar, and a nitrogen-containing, organic base. DNA is composed of nucleotides that contain the sugar deoxyribose, while RNA has nucleotides that have the sugar ribose. The bases in both DNA and RNA have either a single ring or a double ring. Figure 1.31 shows generalized nucleotides because the specific type of base is not designated; the phosphate is simply represented as (P). When nucleotides join, they form a polymer in which the backbone is made up of phosphate-sugar-phosphate-sugar, with the bases projecting to one side of the backbone (fig 1.32). Such

The Cell, the Smallest Unit of Life

a polymer is called a strand. RNA is single stranded, and DNA is double stranded. The strands in DNA are held together by hydrogen bonding between the bases (fig. 24.3). This is only a brief description of the structure and the function of DNA and RNA; they are considered again in greater detail in chapter 24.

Nucleic acids are of 2 types, DNA and RNA. Both of these are polymers of nucleotides; DNA is composed of nucleotides having the sugar deoxyribose, while RNA contains the sugar ribose. DNA makes up the genes and, along with RNA, controls the metabolism of the cell.

Adenosine Triphosphate (ATP) ATP is a nucleotide that is used as a carrier of energy in cells. The structure of ATP is similar to that shown in figure 1.31*a*. Adenine is the base, the sugar is ribose, and there are 3 phosphate groups instead of one. It is customary to draw the molecule as shown in figure 1.33, with the 3 phosphate groups shown on the right. ATP is known as the energy molecule because the triphosphate unit contains 2 high-energy bonds, represented in figure 1.33 by wavy lines.

Figure 1.33
Structure of ATP. ATP is a nucleotide with 3 phosphate units; 2 of the phosphate bonds are high-energy bonds, indicated by wavy lines.

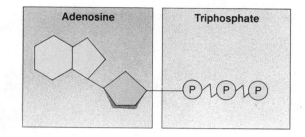

Summary

All matter is made up of atoms, each having a weight that is dependent on the number of protons and neutrons in the nucleus and chemical properties that are dependent on the number of electrons in the outermost shell. Atoms react with one another in order to acquire 8 electrons in the outermost shell. In ionic reactions, metals give electrons to nonmetals, and in covalent reactions, nonmetals share electrons.

Water, acids, bases, and salts are important inorganic compounds. Pure water has a neutral pH; acids decrease and bases increase the pH of water. The organic molecules of interest are proteins, carbohydrates, lipids, and nucleic acids. Polymers (macromolecules) arise when dehydration synthesis joins monomers, and polymers break up upon hydrolysis. Table 1.3 gives the makeup and the function of significant polymers.

Table 1.3 *Organic Compounds Associated with Living Things*

Polymer	Monomer	Function
Proteins	Amino acids	Enzymes speed up chemical reactions; structural components (e.g., muscle proteins)
Carbohydrates		
Starch	Glucose	Energy storage in plants
Glycogen	Glucose	Energy storage in animals
Cellulose	Glucose	Plant cell walls
Lipids		
Fats and Oils	Glycerol, 3 fatty acids	Long-term energy storage
Phospholipids	Glycerol, 2 fatty acids, phosphate group	Cell membrane structure
Nucleic Acids		
DNA	Nucleotides with deoxyribose sugar	Genetic material
RNA MESSENGER, RIBOSOMAL	Nucleotides with ribose sugar	Protein synthesis

TRANSCRIPTION TRANSLATION

Study Questions

1. Name the subatomic particles of an atom; describe their charge, weight, and location in the atom. (p. 17)
2. Draw the atomic diagram for calcium. (p. 18)
3. State the octet rule, and explain how it relates to chemical reactions. (p. 19)
4. Give an example of an ionic reaction, and explain it. Mention in your explanation compound, ion, formula, and ionic bond. (p. 19)

5. Give an example of a covalent reaction, and explain it. (p. 21)
6. Explain an oxidation-reduction reaction in terms of loss or gain of electrons. (p. 23)
7. Name 4 general differences between inorganic and organic compounds. (p. 24)
8. On the pH scale, which numbers indicate a basic solution? an acidic solution? Why? (p. 28)

9. What are buffers, and why are they important to life? (p. 28)
10. Explain dehydration synthesis of organic compounds and hydrolytic breakdown of organic compounds. (p. 29)
11. What are some functions of proteins? What is the monomer in proteins? What are a peptide bond, a dipeptide, and a polypeptide? (pp. 29–31)

12. Discuss the primary, secondary, and tertiary structures of proteins. (p. 31)
13. Name some monosaccharides, disaccharides, and polysaccharides, and state appropriate functions. What is the most common monomer for these? (pp. 33–35)
14. Name some important lipids, and state their function. What is a saturated fatty acid? an unsaturated fatty acid? How is a fat formed? (pp. 35–39)
15. What are the 2 types of nucleic acids? What is the monomer of both? (p. 40)

Objective Questions

1. The atomic number is equal to the number of _____ in an atom.
2. An ion is negatively charged when it has more _____ than _____ .
3. In a covalent bond, the atoms _____ electrons.
4. _____ take up either hydrogen (H^+) ions or hydroxide (OH^-) ions and therefore act to stabilize the pH.
5. When an acid is added to water, the number of hydrogen ions _____ and the pH _____ .
6. _____ are organic catalysts that speed up chemical reactions.
7. The bond that joins 2 amino acids is called a _____ bond.
8. The sequence, or order, of amino acids in a protein is termed its _____ structure.

9. The simple sugar _____ is the primary energy source of the body.
10. The polymer _____ is found in plant cell walls.
11. Fatty acids having no double bonds are said to be _____ .
12. Fats are composed of _____ and 3 fatty acids.
13. Nucleic acids are composed of _____ .
14. DNA is _____ stranded, while RNA is single stranded.

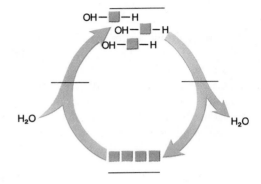

Label this Diagram.

See figure 1.16 (p. 29) in text.

Selected Key Terms

acid (as′id) a solution in which pH is less than 7; a substance that contributes or liberates hydrogen ions (protons) in a solution. *25*

amino acid (ah-me′no as′id) a unit of protein that takes its name from the fact that it contains an amino group (NH_2) and an acid group (COOH). *30*

atom (at′om) smallest unit of matter nondivisible by chemical means. *17*

ATP adenosine triphosphate; a compound containing adenine, ribose, and 3 phosphates, 2 of which are high-energy phosphates. It is the "common currency" of energy for most cellular processes. *41*

base (bās) a solution in which pH is more than 7; a substance that contributes or liberates hydroxide ions in a solution; alkaline; opposite of acidic. Also, a term commonly applied to one of the components of a nucleotide. *25*

buffer (buf′er) a substance or compound that prevents large changes in the pH of a solution. *28*

DNA deoxyribonucleic acid; a nucleic acid, the genetic material that directs protein synthesis in cells. *40*

electron (e-lek′tron) a subatomic particle that has almost no weight and carries a negative charge; travels in a shell about the nucleus of an atom. *17*

emulsification (e-mul″si-fi-ka′shun) the act of dispersing one liquid in another. *38*

enzyme (en′zīm) a protein catalyst that speeds up a specific reaction or a specific type of reaction. *29*

hydrogen bond (hi′dro-jen bond) a weak attraction between a hydrogen atom carrying a partial positive charge and another atom carrying a partial negative charge. *24*

ion (i′on) an atom or group of atoms carrying a positive or a negative charge. *19*

isotope (i′so-tōp) one of two or more atoms with the same number of protons and electrons but differing in the number of neutrons and, therefore, in weight. *19*

lipid (lip′id) a group of organic compounds that are insoluble in water; notably fats, oils, and steroids. *35*

neutron (nu′tron) a subatomic particle that has a weight of one atomic mass unit, carries no charge, and is found in the nucleus of an atom. *17*

nucleic acid (nu-kle′ik as′id) a large organic molecule made up of joined nucleotides; for example, DNA and RNA. *40*

pentose (pen′tōs) a 5-carbon sugar; deoxyribose is a pentose found in DNA; ribose is a pentose found in RNA. *33*

peptide bond (pep′tīd bond) the bond that joins 2 amino acids. *31*

pH a measure of the hydrogen ion concentration [H^+]; any pH below 7 is acidic, and any pH above 7 is basic. *28*

polysaccharide (pol″e-sak′ah-rīd) a macromolecule composed of many units of sugar. *34*

protein (pro′te-in) a macromolecule composed of one or several long polypeptides. *31*

proton (pro′ton) a subatomic particle found in the nucleus of an atom that has a weight of one atomic mass unit and carries a positive charge; a hydrogen ion. *17*

RNA ribonucleic acid; a nucleic acid that assists DNA in the production of proteins within the cell. *40*

The Cell, the Smallest Unit of Life

Chapter 2

Cell Structure and Function

Chapter Concepts

1 A cell is highly organized and contains organelles that carry out specific functions.

2 The nucleus controls the metabolic functioning and the structural characteristics of the cell.

3 The endoplasmic reticulum (ER), the Golgi apparatus, vacuoles, and lysosomes are all membranous structures concerned with production, digestion, or transport of molecules.

4 Mitochondria are concerned with conversion of glucose energy into ATP energy. Chloroplasts absorb the energy of the sun in order to produce glucose.

5 Centrioles and related structures are concerned with the shape and/or the movement of the cell.

Chapter Outline

The Cell Theory
   Microscopy
   Types of Cells
Eukaryotic Cell Organelles
   The Nucleus
   Membranous Canals and Vacuoles
   Energy-Related Organelles
   The Cytoskeleton
   Centrioles and Related Organelles
Cellular Comparisons
   Prokaryotic Cells versus Eukaryotic Cells
   Plant Cells versus Animal Cells

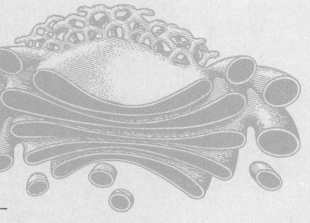

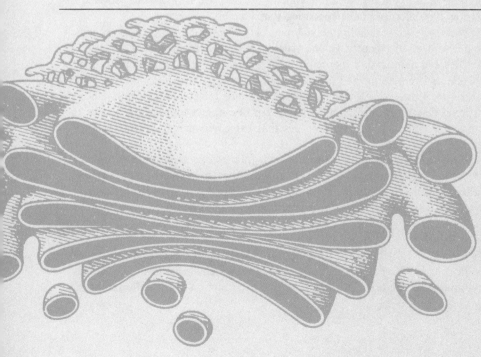

The Cell Theory

 All living things are made up of **cells.** Cells come in many different shapes and sizes, but no matter what the shape or the size, each carries on the functions associated with life—interacting with the environment, obtaining chemicals and energy, growing, and reproducing.

The cell marks the boundary between the nonliving and the living. The molecules that serve as food for a cell and the organic molecules that make up a cell are not alive, and yet the cell is alive. The answer to what life is will have to be found within the cell because the smallest living organisms are single cells, while larger organisms are **multicellular**—composed of many cells. The cell theory states that all living things are composed of cells and new cells arise only from preexisting cells.

Microscopy

With some exceptions, such as various types of eggs, cells readily are not visible to the eye; therefore, a microscope is needed to view them. Three types of microscopes are used most commonly. *Light microscopes* utilize light to view the object, and electron microscopes utilize electrons. A *transmission electron microscope* gives us an image of the interior of the object, while a *scanning electron microscope* provides a three-dimensional view of the surface of an object.

Transmission electron microscopes have a much greater resolving power than light microscopes. **Resolving power** is the capacity to distinguish between 2 points. If 2 points are seen as separate, then the image appears more detailed than if the 2 points are seen as one point. The resolving power of microscopes improves as the wavelength of the illumination shortens, and an electron beam has a much shorter wavelength than a visible light ray. At the very best, a light microscope can distinguish 2 points separated by 200 nm (nanometer = 1×10^{-6} mm), but the transmission electron microscope can distinguish 2 points separated by about 0.2 nm. The greater the resolving power the more detailed the image and the higher the useful magnification possible. The useful magnification of a light microscope is only $\times 1,000$, while that of an electron microscope is in excess of $\times 100,000$.

For scanning electron microscopy, a narrow beam of electrons is scanned over the surface of the specimen, which has been coated with a thin metal layer. The metal gives off secondary electrons that are collected to produce a television-type picture of the specimen's surface on a screen.

New types of scanning microscopes, called scanning-probe microscopes, have been invented recently. In one type, laser light is focused on a tiny probe that is pressed against an organic polymer. Reflected light tells the position of the probe as it moves up and down the polymer, and these data allow a computer to give a three-dimensional picture of a single biological molecule. Unlike traditional electron microscopes, scanning-probe microscopes are sensitive to electron movements and atomic forces that exist between the probe and the material.

A picture obtained by using the light microscope sometimes is called a photomicrograph, and a picture resulting from the use of electron microscopes is called a transmission electron micrograph (TEM) or a scanning electron micrograph (SCM), depending on the type of microscope used (fig. 2.1).

Types of Cells

Cells usually are divided into 2 main groups: **prokaryotic cells** and **eukaryotic cells.** Both types of cells have a cell membrane, which surrounds the contents, or **cytoplasm,** of the cell. The structure and the function of the cell membrane

The Cell, the Smallest Unit of Life

Figure 2.1

Comparison of micrographs of red blood cells within a blood vessel. a. Light micrograph. b. Transmission electron micrograph. c. Scanning electron micrograph with color added. d. Technician operating a transmission electron microscope.

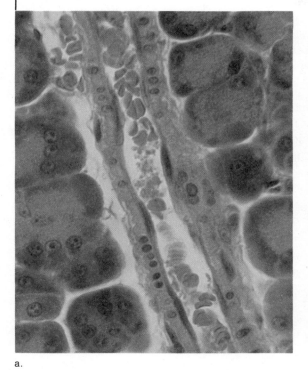

a.

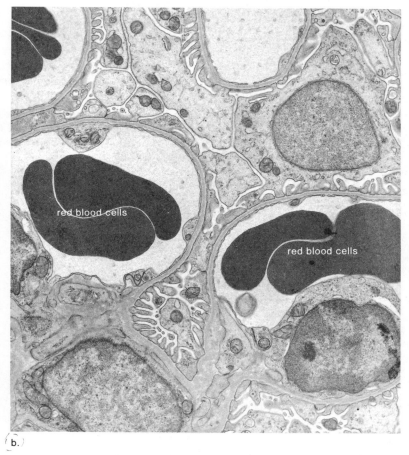

red blood cells

red blood cells

b.

c.

d.

Figure 2.2

Fluid-mosaic model of intracellular membrane. Proteins float within a phospholipid bilayer and are also located at the inner side. Cholesterol is present, particularly in animal cell membrane.

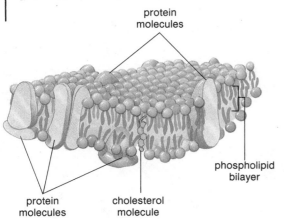

protein molecules

protein molecules

cholesterol molecule

phospholipid bilayer

are discussed at greater length in chapter 3. For now, we will mention that the cell membrane regulates the passage of molecules into and out of the cell.

The 2 main types of cells can be distinguished: eukaryotic cells have a true nucleus (*eu* means true, *karyon* means nucleus), and prokaryotic cells lack a true nucleus (*pro* means before). The nucleus in eukaryotic cells is membrane bound, and these cells contain several other types of small membranous bodies known as **organelles.**

All organisms are composed of cells. The internal structure of these cells has been revealed by electron microscopy. All eukaryotic cells contain a nucleus and various other organelles, each having a detailed structure and function.

The Structure of Membrane

The **fluid-mosaic model** of membrane structure is supported by both electron microscopy and biochemical analysis. Protein molecules form a pattern (mosaic) within a phospholipid bilayer, which is in a liquid (fluid) state (having

Figure 2.3

Animal and plant cells. These generalized representations are based on electron micrographs. a. Animal cell. b. Plant cell (p. 47).

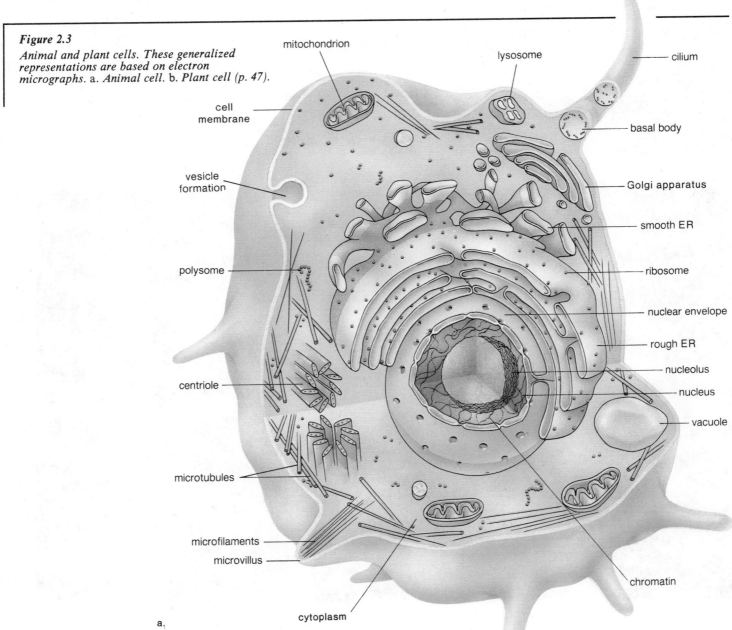

mitochondrion

lysosome

cilium

cell membrane

basal body

Golgi apparatus

vesicle formation

smooth ER

ribosome

polysome

nuclear envelope

rough ER

nucleolus

centriole

nucleus

vacuole

microtubules

microfilaments

microvillus

chromatin

cytoplasm

a.

the consistency of light oil). Notice the manner in which the phospholipid molecules arrange themselves (fig. 2.2). Their structure, discussed in chapter 1, causes each molecule to have a polar head and 2 nonpolar tails. Within the phospholipid bilayer, the tails face inward and the heads face outward, where they are likely to encounter a watery environment. Membranes also contain cholesterol, another type of lipid, which is arranged as depicted in figure 2.2.

The protein component of the membrane is more variable than the phospholipid bilayer. Depending on the membrane, the protein molecules reside above or below the phospholipids, extend from top to bottom of the membrane, or simply penetrate a short distance into the membrane.

Eukaryotic Cell Organelles

Figure 2.3 depicts and table 2.1 lists the structures, or organelles, found in plant and animal cells, the eukaryotic cells discussed in this chapter. Plant and animal cells differ somewhat, and for now, you should note that plant cells,

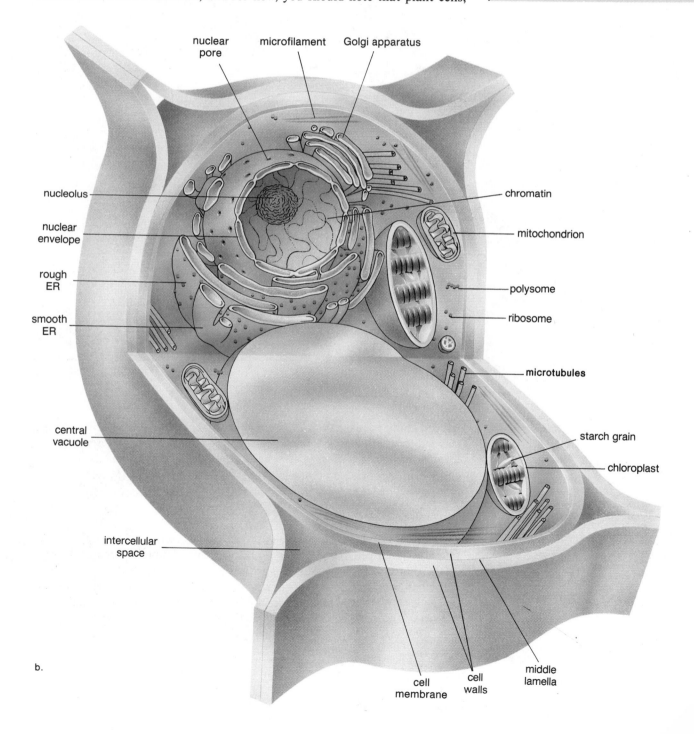

b.

Table 2.1 Eukaryotic Structures in Animal Cells and Plant Cells

Name	Composition	Function
Cell wall*	Contains cellulose fibrils	Support and protection
Cell membrane	Phospholipid bilayer with embedded proteins	Passage of molecules into and out of cell
Nucleus	Nuclear envelope surrounding the nucleoplasm, chromosomes, and nucleoli	Cellular reproduction and control of protein synthesis
Nucleolus	Concentrated area of chromatin, RNA, and proteins	Ribosome formation
Ribosome	Protein and RNA in 2 subunits	Protein synthesis
Endoplasmic reticulum (ER)	Membranous flattened channels and tubular canals	Synthesis of proteins and other substances and transport by vesicle formation
Rough	Studded with ribosomes	Protein synthesis
Smooth	No ribosomes	Various; lipid synthesis in some cells
Golgi apparatus	Stack of membranous saccules	Processing, packaging, and secretion of proteins
Vacuole and vesicle	Membranous sacs	Storage of substances
Lysosome**	Membranous vesicle containing digestive enzymes	Intracellular digestion
Mitochondrion	Inner membrane (cristae) within outer membrane	Cellular respiration
Chloroplast*	Grana within inner and outer membranes	Photosynthesis
Cytoskeleton	Microtubules and microfilaments	Shape of cell and movement of its parts
Cilia and flagella**	9 + 2 pattern of microlubules	Movement of cell
Centriole**	9 + 0 pattern of microtubules	Microtubule organization; forms basal bodies

*Plant cells
**Animal cells

but not animal cells, have a cell wall. The various cellular organelles are specialized for particular functions. For the sake of discussion, the organelles are divided into 5 categories: (1) the nucleus, (2) membranous canals and vacuoles, (3) energy-related organelles, (4) the cytoskeleton, and (5) centrioles and related organelles.

The Nucleus

The **nucleus** (fig. 2.4), which is about 5 μm in diameter, is the largest organelle found within a cell. The nucleus is enclosed by a double-layered **nuclear envelope** that is actually continuous with the endoplasmic reticulum discussed in the following section. There are pores in this envelope through which certain molecules are believed to pass from the cytoplasm to the nucleoplasm, the fluid portion of the nucleus, or from the nucleoplasm to the cytoplasm.

The Cell, the Smallest Unit of Life

Figure 2.4

The nucleus. a. Electron micrograph of a nucleus (N) with a clearly defined nucleolus (Nu) and irregular patches of chromatin scattered throughout the nucleoplasm. The nuclear envelope (NE) contains pores indicated by the arrows. b. A drawing of a nucleus surrounded by endoplasmic reticulum (ER). Mitochondria (M) are also shown in (a) and (b).

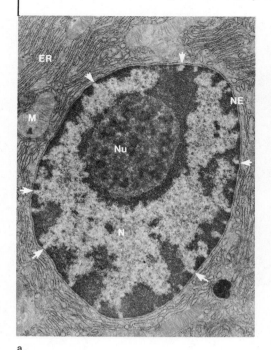

a.

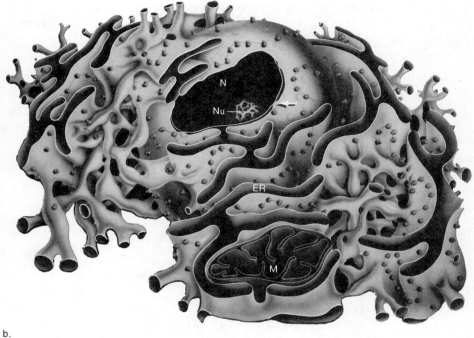

b.

The nucleus is of primary importance in the cell because it is the control center that oversees the metabolic functioning of the cell and ultimately determines the cell's characteristics. Within the nucleus, there are masses of threads called **chromatin,** so called because they take up stains and become colored. Chromatin is indistinct in the nondividing cell, but it condenses to rodlike structures called **chromosomes** at the time of cell division. Biochemical analysis shows that chromatin and therefore chromosomes contain DNA (deoxyribonucleic acid) along with certain proteins and some RNA. This is not surprising because we already know that chromosomes contain the genes and that the genes are composed of DNA. DNA, with the help of 2 types of RNA called messenger RNA (mRNA) and transfer RNA (tRNA), controls protein (enzyme) synthesis within the cytoplasm. It is this function that allows DNA to control the cell.

Nucleoli

One or more **nucleoli** (singular, nucleolus) are present in the nucleus. These dark-staining bodies are actually specialized parts of chromatin in which another type of RNA, called ribosomal RNA (rRNA), is produced. Ribosomes are organelles in the cytoplasm that contain not only rRNA but also proteins. These proteins, like all others, are made in the cytoplasm. They migrate into the nucleus and join with rRNA, forming subunits that pass out into the cytoplasm. When subunits combine in a certain way, ribosomes are formed.

The nucleus contains chromatin that condenses into chromosomes just prior to cell division. Chromosomes contain DNA that, with the help of RNA, directs protein synthesis in the cytoplasm. Another type of RNA, rRNA, is made within the nucleolus before migrating to the cytoplasm, where it is incorporated into ribosomes.

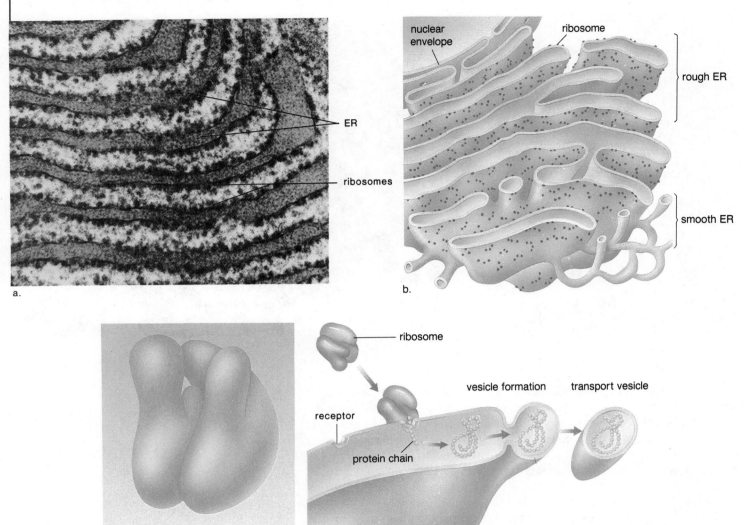

Figure 2.5
Endoplasmic reticulum. a. Electron micrograph of rough ER shows a cross section of many flattened vesicles with ribosomes attached to the sides that abut the cytoplasm. b. Drawing that shows the 3 dimensions of the ER. c. Model of a single ribosome illustrates that each is actually composed of 2 subunits. d. Method by which the ER acts as a transport system.

a.

nuclear envelope

ribosome

rough ER

smooth ER

b.

ribosome

vesicle formation   transport vesicle

receptor

protein chain

c. Ribosome

d.

Membranous Canals and Vacuoles

The endoplasmic reticulum (ER), the Golgi apparatus, vacuoles, and lysosomes (fig. 2.3*a*) are structurally and functionally related membranous structures. Ribosomes are not composed of membrane, but they are included in this category because they often are associated intimately with the ER.

The Endoplasmic Reticulum

The **endoplasmic reticulum** (fig. 2.5) forms a membranous system of tubular canals that is continuous with the nuclear envelope and branches throughout the cytoplasm. Ribosomes can be attached to the ER, and if so, the reticulum is called **rough ER;** if ribosomes are not present, the ER is called **smooth ER** (fig. 2.5*b*). Rough ER specializes in protein synthesis, but smooth ER has different functions in different cells. Sometimes it specializes in the production of lipids. For example, smooth ER is abundant in cells of the testes and the adrenal cortex, both of which produce steroid hormones. In muscle cells, smooth ER acts as a storage area for calcium ions that are released when contraction

Figure 2.6

Golgi apparatus in an animal cell. The Golgi apparatus receives protein-containing vesicles from the smooth ER at its inner face. These proteins are processed as they pass from saccule to saccule of the central region. Finally, they are repackaged in vesicles that leave the outer face. Some of these vesicles fuse with the cell membrane, allowing secretion to occur, and others are lysosomes that contain digestive enzymes. Electron micrograph (a) and drawing (b) of a Golgi apparatus in cross section.

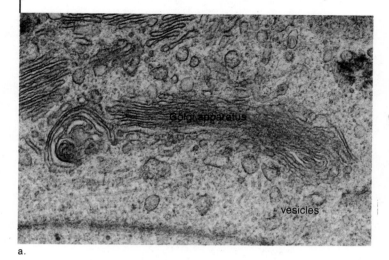

a.

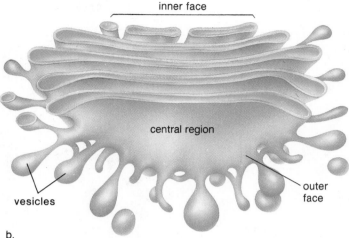

b.

occurs. In the liver, it is involved in the detoxification of drugs, including alcohol. It is quite possible that drugs are detoxified within structures called **peroxisomes,** membrane-bound vacuoles often attached to smooth ER that contain enzymes capable of carrying out oxidation of various substances.

Ribosomes   **Ribosomes** look like small, dense granules in low-power electron micrographs (fig. 2.5*a*), but higher-resolution micrographs show that each contains 2 subunits (fig. 2.5*c*). Each of these subunits has a particular mix of rRNA and proteins. We already have mentioned that rRNA is joined with proteins within the nucleus, but the 2 subunits are not assembled into one ribosome until they reach the cytoplasm.

Ribosomes can lie free within the cytoplasm, where they also are involved in protein synthesis. In these instances, several ribosomes, each of which is producing the same type of protein, are arranged in a functional group called a *polysome.* Most likely, these proteins are for use inside the cell.

The ribosomes attached to rough ER are making proteins for export from the cell. The proteins enter the lumen (interior space) of the rough ER (fig. 2.5*d*) and proceed to the lumen of the smooth ER. A small vesicle then pinches off from the smooth ER. Most vesicles formed in this way move through the cytoplasm to the Golgi apparatus, where the proteins are received and are processed further. This is how the ER serves as a transport system.

The Golgi Apparatus

The **Golgi apparatus** (fig. 2.6) is named for the person who first discovered its presence in cells. It is composed of a stack of about a half-dozen or more saccules that look like hollow pancakes. In animal cells, one side of the stack (the inner face) is directed toward the nucleus, and the other side of the stack (the outer face) is directed toward the cell membrane. Vesicles occur at the edges of the saccules.

Biochemical analyses suggest that the Golgi apparatus receives protein-filled vesicles from the ER at its inner face. After this, the proteins move from one saccule to the next via newly formed vesicles. In the meantime, the proteins are modified by the enzymes present within the saccules. Vesicles containing the modified proteins move to different locations in the cell (fig. 2.7).

2.2 Critical Thinking

1. It is possible to bathe a cell in radioactively tagged amino acids and then later to detect photographically the location of radiation and therefore the amino acids in the cell. Why would you suggest using radioactive sulfur (i.e., the amino acids cysteine and methionine contain sulfur) rather than radioactive carbon?

2. An investigator uses the procedure outlined in question 1 to support the belief that proteins move from the cytoplasm into the nucleus. If so, where will radiation first appear, and where will it subsequently appear in the cell?

3. An investigator uses this same procedure to support the belief that proteins move from the rough ER to secretory vesicles. Where will the radiation first appear, and where will it subsequently appear in the cell?

Figure 2.7

Golgi apparatus function. The Golgi apparatus receives vesicles from the smooth ER and thereafter forms at least 2 types of vesicles, lysosomes and secretory vesicles. Lysosomes contain hydrolytic enzymes that can break down large molecules. Vesicles bringing large molecules into a cell sometimes join with lysosomes, forming structures called secondary lysosomes. Thereafter, the molecules are digested. The secretory vesicles formed at the Golgi apparatus also discharge their contents at the cell membrane.

Figure 2.8

Lysosomes in a kidney cell of a mammal. The small dark body is a primary lysosome that has not yet fused with an incoming vesicle. The larger bodies are secondary lysosomes that have fused with such vesicles. The dark appearance of the lysosomes results from staining for a particular enzyme, acid phosphatase, the presence of which is the test for this organelle.

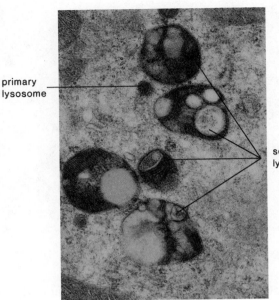

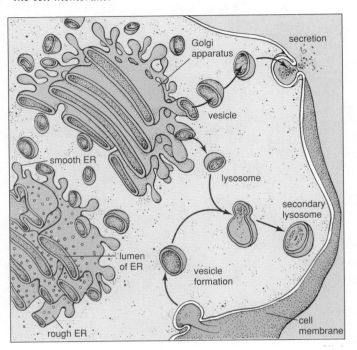

Some of the vesicles formed by the Golgi apparatus move to the cell membrane, where they discharge their protein contents. Because this is called *secretion,* it often is said that the Golgi apparatus is involved in processing, packaging, and secretion.

Vacuoles

A **vacuole** is a large membranous sac; a vesicle is smaller. Animal cells have vacuoles, but they are much more prominent in plant cells. Typically, plant cells have one or 2 large vacuoles filled with a watery fluid that gives added support to the cell. Most of the central area of the plant cell is occupied by vacuole, and the other contents of the cell are pushed to the sides (fig. 2.3*b*).

Vacuoles are most often storage areas. Plant vacuoles contain not only water, sugars, and salts but also pigments and toxic substances. The pigments are responsible for many of the red, blue, or purple colors of flowers and some leaves. The toxic substances help to protect a plant from predaceous animals. (As long as the substance is contained within a vacuole, it is not harmful to the plant.)

Lysosomes

Lysosomes (fig. 2.8) are membrane-bound vesicles formed by the Golgi apparatus that contain *hydrolytic enzymes*. Hydrolytic enzymes digest macromolecules in the manner described in figure 1.16, page 29. Macromolecules sometimes are brought into a cell by vesicle formation at the cell membrane

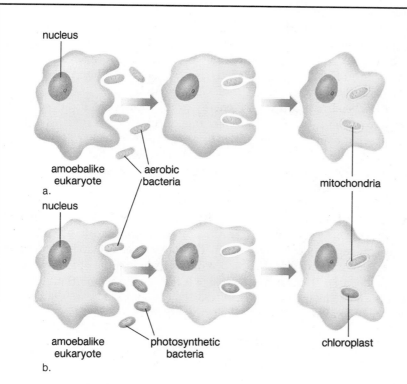

Figure 2.9
The endosymbiotic theory explains the origin of mitochondria and chloroplasts in eukaryotic cells. a. Mitochondria may be derived from aerobic bacteria that were engulfed by a larger amoebalike eukaryote. b. Similarly, chloroplasts may be derived from photosynthetic bacteria that were engulfed by a larger amoebalike eukaryote. The evidence of the endosymbiotic theory is as follows.
- *Mitochondria and chloroplasts are similar to bacteria in size and structure.*
- *Both organelles are bounded by a double membrane—the outer membrane may be derived from the engulfing vesicle, and the inner one may be derived from the cell membrane of the original prokaryote.*
- *Mitochondria and chloroplasts contain a limited amount of genetic material and are capable of self-reproduction. Their DNA is a circular strand that resembles the bacterial chromosome.*
- *Although most of the proteins within mitochondria and chloroplasts are now produced by the eukaryotic host, they do possess their own ribosomes, and they do produce some proteins. Their ribosomes resemble those of bacteria.*

(fig. 2.7). A lysosome can fuse with such a vesicle and can digest its contents into simpler molecules that then enter the cytoplasm. Some white blood cells defend the body by engulfing bacteria, a process that involves vesicle formation at the cell membrane. When lysosomes fuse with these vesicles, the bacteria are digested. It should come as no surprise, then, that even parts of a cell are digested by its own lysosomes (called autodigestion because auto = self). Normal cell rejuvenation most likely takes place in this way, but autodigestion is also important during development. For example, when a tadpole becomes a frog, the enzymes within lysosomes are utilized to digest the cells of the tail. The fingers of a human embryo are at first webbed, but they are freed from one another following lysosomal action.

Occasionally, a child is born with a metabolic disorder involving a missing or inactive lysosomal enzyme. In these cases, the lysosomes fill to capacity with macromolecules that cannot be broken down. The cells become so full of these lysosomes that the child dies. The best-known of these lysosomal storage disorders is Tay-Sachs disease, discussed on page 481.

The ER is a membranous system of tubular canals that can be smooth or rough. Proteins synthesized at the rough ER are processed and are packaged in vesicles by the Golgi apparatus. Some vesicles discharge their contents at the cell membrane, and some are lysosomes that digest any material enclosed therein.

Energy-Related Organelles

The energy-related organelles, chloroplasts and mitochondria, are transformers, changing one form of energy into another. While chloroplasts are unique to plant cells, mitochondria are found in both plant cells and animal cells.

The endosymbiotic theory (fig. 2.9), which is accepted widely today, states that chloroplasts and mitochondria were originally prokaryotes that came to reside inside eukaryotic cells.

Mitochondria

Most mitochondria (singular, mitochondrion) are between 0.5 μm and 1.0 μm in diameter and 7 μm in length, although the size and the shape can vary. Mitochondria are bounded by a double membrane. The inner membrane is folded to form little shelves, called *cristae,* which project into the *matrix,* an inner space filled with a gel-like fluid (fig. 2.10).

Mitochondria produce ATP energy in the form of ATP (adenosine triphosphate). Every cell uses a certain amount of ATP energy to synthesize molecules, but many cells use ATP to carry out their specialized function. For example, muscle cells use ATP for muscle contraction that produces movement, and nerve cells use it for the conduction of nerve impulses so we are aware of our environment.

Mitochondria often are called the powerhouses of the cell: just as a powerhouse burns fuel to produce electricity, the mitochondria use glucose products to produce ATP molecules. In the process, mitochondria use up oxygen and give off carbon dioxide and water. The oxygen you breathe in enters cells and then the mitochondria; the carbon dioxide you breathe out is released by the mitochondria. Because gas exchange is involved, it is said that mitochondria carry on **cellular respiration.** A shorthand way to indicate the chemical transformation associated with cellular respiration is

$$\text{carbohydrate } + \text{ oxygen} \longrightarrow \text{carbon dioxide } + \text{ water } + \text{ energy}$$

The energy in this equation stands for ATP molecules. As we will see in chapter 5, the matrix of a mitochondrion contains enzymes for breaking down glucose products. Then ATP production occurs at the cristae. The molecules that aid in the production of energy are located in an assembly-line fashion on these membranous shelves.

Chloroplasts

Chloroplasts range from about 4 μm to 6 μm in diameter and from 1 μm to 5 μm in length. They belong to a group of plant organelles known as plastids. Among the plastids are the leucoplasts, which store starch, and the chromoplasts, which contain red and orange pigments.

A chloroplast is bounded by a double membrane. Inside the structure there is even more membrane organized into flattened sacs called *thylakoids.* The thylakoids are piled up like stacks of coins, and each stack is called a *granum* (plural, grana). There are membranous connections between the grana called lamellae (fig. 2.11). The fluid-filled space about the grana is called the *stroma.*

Chloroplasts carry on **photosynthesis,** during which light energy (photo) is used to produce food molecules like glucose (synthesis). Chloroplasts take in carbon dioxide, water, and radiant energy from the sun in order to produce glucose. They give off oxygen, which exits from the leaves as a gas. Again, we can use the shorthand method to describe what has been said:

$$\text{energy } + \text{ carbon dioxide } + \text{ water} \longrightarrow \text{carbohydrate } + \text{ oxygen}$$

The equation for photosynthesis is the opposite of cellular respiration, as you can see by comparing the shorthand statements for each.

Figure 2.10

Mitochondria structure. a. *Electron micrograph.* b. *Generalized drawing in which the outer membrane and portions of the inner membrane have been cut away to reveal the cristae.*

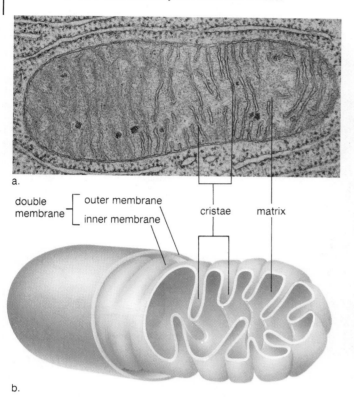

Figure 2.11

Chloroplast structure. a. *Electron micrograph.* b. *Generalized drawing in which the outer and inner membranes have been cut away to reveal the grana.*

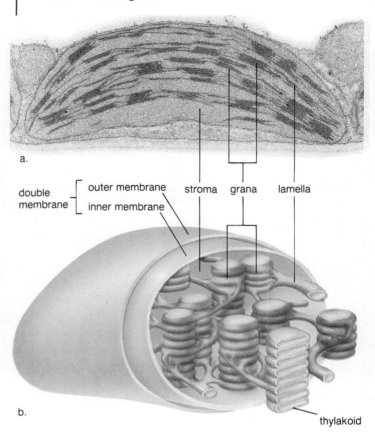

Like mitochondria, chloroplasts are highly organized. The green pigment **chlorophyll** is found within the grana, and this makes chloroplasts and leaves green. Chlorophyll absorbs the energy of the sun, and chloroplasts transform this energy into ATP molecules, the type of energy used by enzymes within the stroma to make glucose.

Mitochondria are the powerhouses of the cell because they convert carbohydrate energy to ATP energy. Chloroplasts contain chlorophyll and carry out photosynthesis.

The Cytoskeleton

Within the cytoplasm, several types of filamentous protein structures form a **cytoskeleton** (fig. 2.12) that helps to maintain the cell's shape, anchors the organelles, and allows the cell or its contents to move. Cells typically have a particular shape; in animals, muscle cells are tubular and nerve cells have long, skinny processes. Sometimes cells can change their shape or even move about by means of cilia or flagella.

Microfilaments are long, extremely thin fibers (approximately 7 nm in diameter) that usually occur in bundles. In most cells, microfilaments are composed of a protein called actin. Actin microfilaments are well known for their role in muscle contraction. During contraction, they interact with other

Figure 2.12

The cytoskeleton. a. Light micrograph of a cell treated to reveal microfilaments (red) and microtubules (green). Notice how the microfilaments extend into the processes, and the microtubules seem to emanate from a region near the nucleus (oval area). b. Drawing based on electron micrographs shows how the cytoskeleton anchors the organelles. It also can allow these organelles to move about. c. A microtubule is a cylinder composed of tubulin protein molecules. d. A microfilament most often is composed of actin protein molecules.

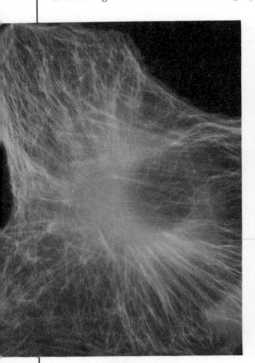

a. Cytoskeleton (micrograph)

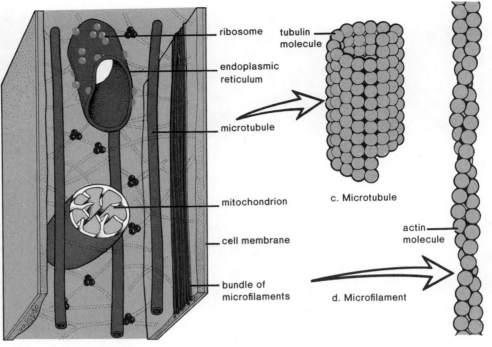

b. Cytoskeleton (drawing)

protein microfilaments composed of myosin. Actin microfilaments are seen in various types of cells, especially where movement occurs. For example, microfilaments seen in microvilli (singular, microvillus) that project from certain cells most likely account for the ability of microvilli to alternately shorten and extend.

Microtubules are shaped like thin cylinders and are several times larger than microfilaments (about 25 nm in diameter). Each cylinder contains 13 rows of tubulin, a globular protein, arranged in a helical fashion. Aside from existing independently in the cytoplasm, microtubules also are found in certain organelles, such as cilia, flagella, and centioles. These structures are associated with animal cells.

Remarkably, both microfilaments and microtubules assemble and disassemble within the cell. When they are assembled, the protein molecules are bonded, and when they are disassembled, the protein molecules are not attached to one another. When microfilaments and microtubules are assembled, the cell has a particular shape, and when they disassemble, the cell can change shape.

The cytoskeleton contains microfilaments (usually composed of the protein actin) and microtubules (composed of the protein tubulin). The cytoskeleton maintains the shape of the cell and also directs the movement of its parts.

The Cell, the Smallest Unit of Life

Figure 2.13

Centrioles. a. Drawing of centrioles showing their 9 + 0 arrangement of microtubule triplets. (The zero in this equation means that there are no microtubules in the center of the organelle.) b. Electron micrograph of centrioles verifies that they *lie at right angles to one another. Notice the large number of microtubules near the centrioles, which lie within the microtubule organizing region.*

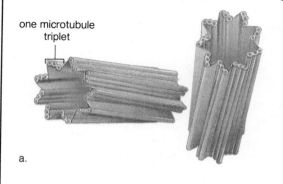

one microtubule triplet

a.

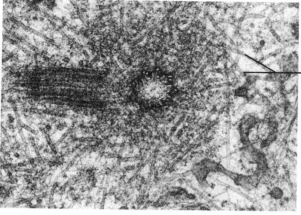

microtubules

b.

Centrioles and Related Organelles

Centrioles

There is a pair of **centrioles** lying at right angles to one another in a typical animal cell, but not in a typical plant cell (fig. 2.3). Before an animal cell divides, the centrioles duplicate, and the members of each new pair also lie at right angles to one another. Duplication ensures that each daughter cell will receive a pair of centrioles.

Centrioles (fig. 2.13) have a 9 + 0 pattern of microtubule triplets. You can verify this by counting the number of triplets at the perimeter of a centriole—there are 9—and noting that there are no microtubules in the center of the centriole.

For quite some time, investigators have tried to determine the function of centrioles. Although the centrioles duplicate just before cell division, they are not necessary to the process of cell division. Plant cells that lack centrioles still divide normally, and if centrioles are removed from embryonic animal cells, these cells divide normally also. It is of significance, however, that centrioles are composed of microtubules, and it might very well be that they play a role in the organization of microtubules within the cell. Centrioles are actually a part of a microtubule organizing region, which includes other proteins and substances lying just outside the nucleus. Microtubules begin to assemble in this region, and then they grow outward, extending through the entire cytoplasm.

In motile cells, centrioles give rise to **basal bodies** that direct the organization of microtubules within cilia and flagella. In other words, a basal body does for a cilium (or flagellum) what the microtubule organizing region does for the cell.

Cilia and Flagella

Cilia and **flagella** are hairlike projections of cells that can move either in an undulating fashion, like a whip, or stiffly, like an oar. Cells that have these organelles are capable of movement. For example, single-celled paramecia move by means of cilia; sperm cells, carrying genetic material to the egg, move

2.3 Critical Thinking

1. A microtubular spindle apparatus appears in plant and animal cells at the time of cell division. Why does this suggest that centrioles are not necessary to the formation of the spindle apparatus?
2. What evidence is there to suggest centrioles are necessary to microtubule organization in animal cells?
3. In animal cells, each newly formed cell receives a pair of centrioles. Why might centrioles be necessary to animal cells but not to most plant cells?

Figure 2.14

Anatomy of cilia and flagella. a. Drawing showing that the 9 + 2 pattern of microtubules within a cilium or a flagellum is derived (in some unknown way) from the 9 + 0 pattern in a basal body. b. Cross-section drawing of the 9 + 2 pattern shows the exact arrangement of microtubules. Notice the clawlike arms of the outer doublets and the spokes that connect them to the central microtubules. c. Electron micrograph of the cross section of a basal body. Basal bodies are derived from centrioles.

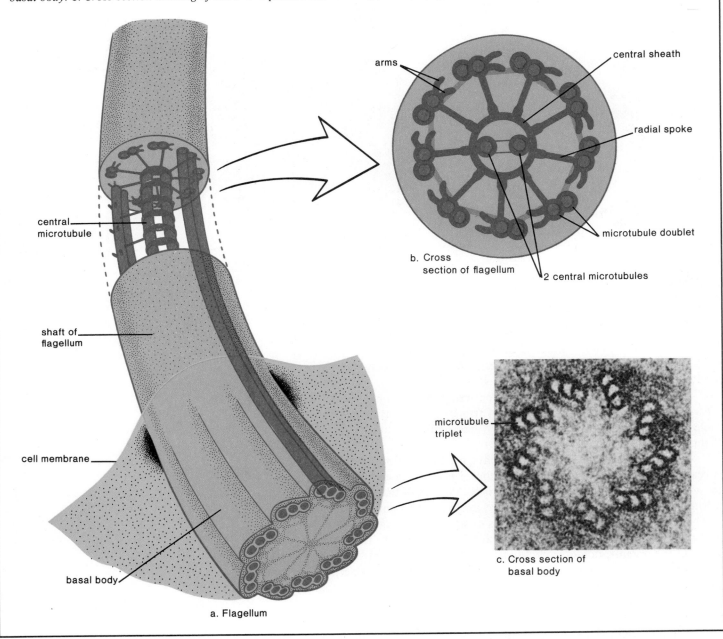

arms

central sheath

radial spoke

central microtubule

microtubule doublet

b. Cross section of flagellum

2 central microtubules

shaft of flagellum

cell membrane

basal body

microtubule triplet

c. Cross section of basal body

a. Flagellum

by means of flagella. The cells that line our upper respiratory tract are ciliated. These cilia sweep debris trapped within mucus back up into the throat, an action that helps to keep the lungs clean.

Cilia are much shorter than flagella, but even so they both are constructed similarly (fig. 2.14*a*). They both are membrane-bound cylinders enclosing a matrix area. In the matrix are microtubule doublets arranged in a circle around 2 central microtubules (fig. 2.14*b*). This is called the 9 + 2 pattern of microtubules. Each doublet also has a pair of arms projecting toward

The Cell, the Smallest Unit of Life

Table 2.2 *Comparison of Prokaryotic Cells and Eukaryotic Cells*

	Prokaryotic (1–10 μm in diameter)	Eukaryotic Cells (10–100 μm in diameter)	
		Animal	Plant
Cell membrane	yes	yes	yes
Cell wall	yes (not cellulose)	no	yes (cellulose)
Nuclear envelope	no	yes	yes
Nucleolus	no	yes	yes
Mitochondria	no	yes	yes
Chloroplasts	no	no	yes
Endoplasmic reticulum (ER)	no	yes	yes
Ribosomes	yes (smaller)	yes	yes
Vacuoles	some	yes (small)	yes (usually large, single vacuole)
Golgi apparatus	no	yes	yes
Lysosomes	no	always	often
Cytoskeleton	no	yes	yes
Contrioles	no	yes	no
9 + 2 cilia or flagella	no	often	seldom

a neighboring doublet and radial spokes extending toward the central pair of microtubules. Recent evidence indicates that cilia and flagella move when the microtubule doublets slide along one another. The clawlike arms and the spokes seem to be involved in causing this sliding action, which requires ATP energy.

Centrioles are small cylinders that contain microtubules. They are a part of a microtubule organizing region lying near the nucleus of animal cells. They also give rise to basal bodies, which organize the arrangement of microtubules within cilia and flagella.

Cellular Comparisons

Prokaryotic Cells versus Eukaryotic Cells

We mentioned previously that prokaryotic cells lack a true nucleus. Table 2.2 compares prokaryotic cells to the 2 types of eukaryotic cells (plant and animal cells) studied in this chapter. You will notice that prokaryotic cells, represented only by bacteria, including cyanobacteria (formerly called blue-green algae), also lack most of the other types of organelles we have been discussing (fig. 2.15). This does not mean, however, that these cells do not carry on the functions performed by organelles in eukaryotes. The functions simply occur within the cytoplasm of these much smaller cells. For example, prokaryotes have a chromosome, but it is not enclosed within a nucleus. The bacterial chromosome is composed of a single circular DNA macromolecule located within an area called the *nucleoid region*. Similarly, they have respiratory enzymes,

Figure 2.15
Prokaryotic cells. Notice the lack of discrete organelles.
a. Generalized nonphotosynthetic bacterium. b. Generalized
cyanobacterium, a photosynthetic bacterium.

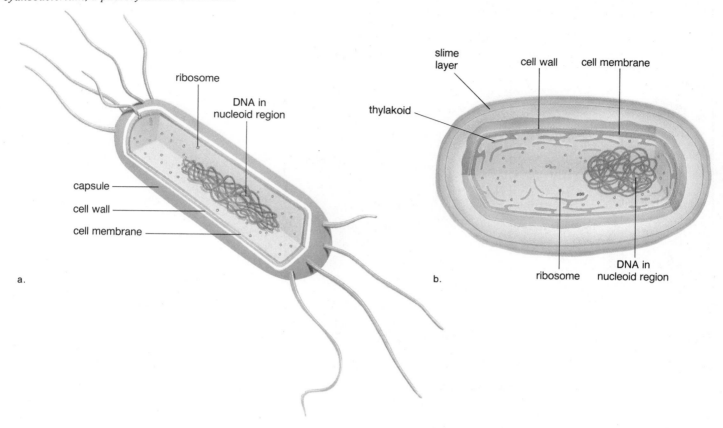

but not mitochondria; instead, the respiratory enzymes are free within the cytoplasm or they are associated with the cell membrane. When prokaryotes have chlorophyll, there are no chloroplasts. Within cyanobacteria, chlorophyll is associated with individual thylakoids.

In addition to a cell membrane, prokaryotes have a cell wall, and if motile, most possess flagella. However, the structures of these cell features differ from those found in eukaryotic cells. Outside the cell wall, there may be a capsule or a slime layer. Both of these are polysaccharides, but a capsule is denser than a slime layer.

Plant Cells versus Animal Cells

Table 2.2 also compares plant and animal cells. It is wise to keep in mind that while both plant and animal cells have a cell membrane, plant cells also have a cell wall. Both types of cells have mitochondria, but plant cells also have chloroplasts. Animal cells are more likely than plant cells to have centrioles and basal bodies, leading to the formation of cilia and flagella.

Summary

Table 2.2 lists the organelles we have studied in the chapter. The nucleus is a large organelle of primary importance because it controls the rest of the cell. Within the nucleus lies the chromatin, which condenses to become chromosomes during cell division.

Proteins are made at the rough ER before being packaged at the Golgi apparatus. Golgi-derived lysosomes fuse with incoming vacuoles to digest any material enclosed within. Lysosomes also carry out autodigestion of old parts of cells.

Mitochondria and chloroplasts are the energy-related organelles. During the process of cellular respiration, mitochondria convert carbohydrate energy to ATP energy, and during photosynthesis, chloroplasts form carbohydrates.

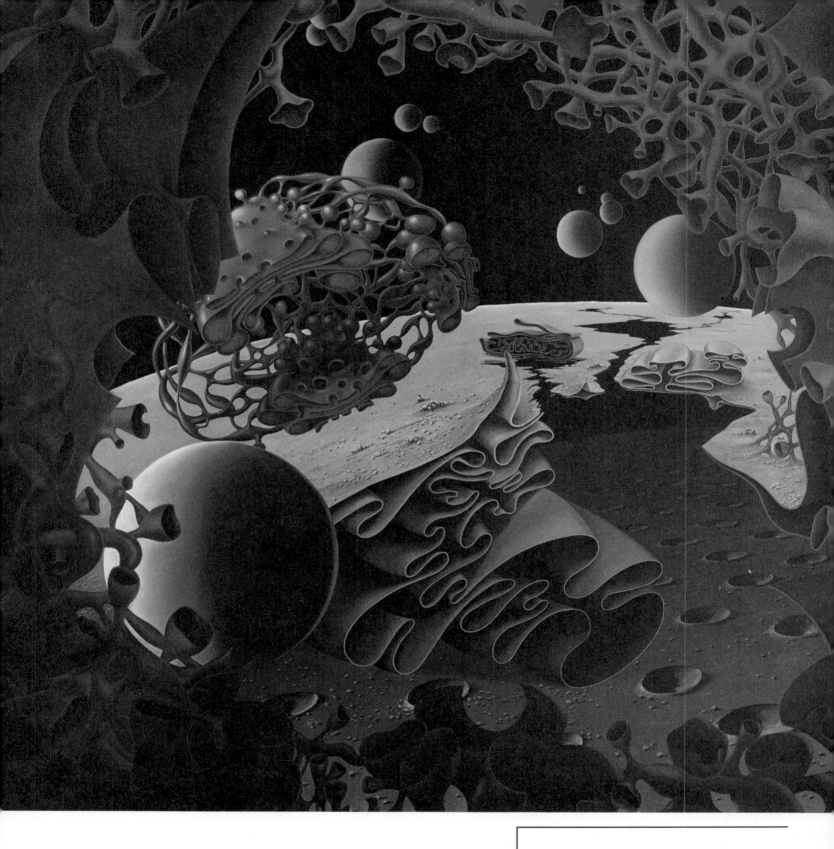

Artist's representation of the interior of an animal cell in which the vacuoles appear to float among the other organelles. Clearly seen are the folds of the endoplasmic reticulum and a weblike nuclear envelope with the Golgi apparatus attached. Parallel rows of circular craters mark the location of vesicles in the cell membrane.

Microfilaments and microtubules are found within a cytoskeleton that maintains the cell shape and permits movement of cell parts. Centrioles are found within a microtubule organizing region that lies just outside the nucleus of animal cells, and they also produce basal bodies that give rise to cilia and flagella.

Prokaryotic cells lack the organelles typically found in eukaryotic cells, such as plant and animal cells. Nevertheless, they carry on all the same functions as eukaryotic cells.

Study Questions

1. Briefly define the cell theory. (p. 44)
2. Describe the structure and the biochemical makeup of membrane. (p. 46)
3. Describe the nucleus and its contents, including the terms *DNA* and *RNA* in your description. (p. 48)
4. What is the nucleolus, and what function does it perform for the cell? (p. 49)
5. Describe the structure and the function of ER. Include the terms *rough ER, smooth ER,* and *ribosomes* in your description. (p. 50)
6. Describe the structure and the function of the Golgi apparatus. Mention vacuoles and lysosomes in your description. (pp. 51–53)
7. Describe the structure and the function of mitochondria and chloroplasts. (p. 54)
8. Describe the structure and the function of microfilaments, microtubules, centrioles, cilia, and flagella. (pp. 55–58)
9. What are the 2 main types of cells, and how do they differ structurally? (p. 59)
10. What are the structural differences between animal cells and plant cells? (p. 60)

Objective Questions

1. Electron microscopes have a greater _____ power than light microscopes.
2. The fluid-mosaic model of membrane structure says that _____ molecules drift about within a _____ bilayer.
3. Chromosomes are located within the _____ , an organelle that controls the rest of the cell.
4. Rough ER has _____ , but smooth ER does not.
5. Lysosomes contain _____ enzymes.
6. Vesicles derived from ER make their way to the _____ , an organelle that functions in packaging and secretion.
7. Both plant and animal cells have _____ , where glucose products provide energy for ATP formation.
8. Photosynthesis takes place within _____ .
9. Microfilaments and microtubules are a part of the _____ , the framework of the cell that provides its shape and regulates movement of its organelles.
10. Basal bodies that organize the microtubules within cilia and flagella are derived from _____ .

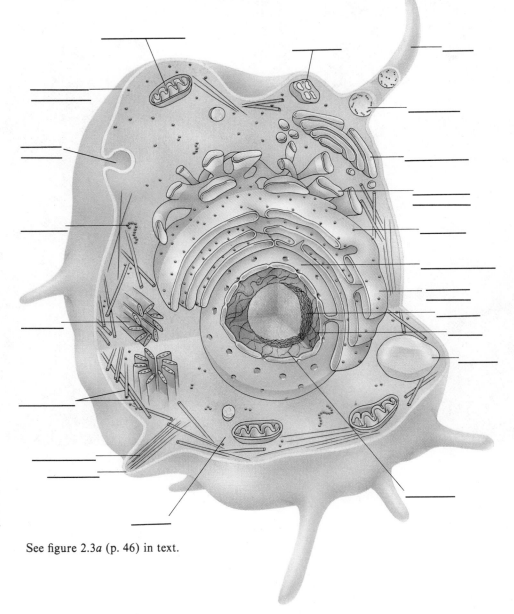

See figure 2.3*a* (p. 46) in text.

The Cell, the Smallest Unit of Life

Selected Key Terms

cell (sel) the structural and functional unit of an organism; the smallest structure capable of peforming all the functions necessary for life. *44*

centriole (sen'tre-ōl) a short, cylindrical organelle in animal cells that contains microtubules in a 9 + 0 pattern and that is associated with the formation of basal bodies. *57*

chloroplast (klo'ro-plast) an organelle that contains chlorophyll and where photosynthesis takes place. *54*

chromosome (kro'mo-sōm) rod-shaped body in the nucleus, particularly during cell division, that contains the hereditary units, or genes. *49*

cytoplasm (si'to-plazm'') the ground substance of cells located between the nucleus and the cell membrane. *44*

cytoskeleton (si''to-skel'ĕ-ton) filamentous protein structure found throughout the cytoplasm that helps maintain the shape of the cell. *55*

endoplasmic reticulum (ER) (en-do-plaz'mic rĕ-tik'u-lum) a complex system of tubules, vesicles, and sacs in cells, sometimes having attached ribosomes. Rough ER has ribosomes; smooth ER does not. *50*

eukaryotic (u''kar-e-ot'ik) possessing the membranous organelles characteristic of complex cells. *44*

Golgi apparatus (gol'ge ap''ah-ra'tus) an organelle consisting of concentrically folded membranes that functions in the packaging and secretion of cellular products. *51*

lysosome (li'so-sōm) an organelle in which digestion takes place due to the action of hydrolytic enzymes. *52*

microfilament (mi''kro-fil'ah-ment) an extremely thin fiber found within the cytoplasm that is involved in the maintenance of cell shape and the movement of cell contents. *55*

microtubule (mi''kro-tu'būl) an organelle composed of 13 rows of globular proteins; found in multiple units in several other organelles, such as the centriole, the cilia, and the flagella. *56*

mitochondrion (mi''to-kon'dre-on) an organelle in which cellular respiration produces the energy molecule ATP. *54*

nucleolus (nu-kle'o-lus) an organelle found inside the nucleus; composed largely of rRNA for ribosome formation. *49*

nucleus (nu'kle-us) a large organelle containing the chromosomes and acting as a control center for the cell. *48*

organelle (or''gan-el') specialized structure within cells, such as the nucleus, the mitochondria, and the endoplasmic reticulum. *46*

prokaryotic (pro''kar-e-ot'ik) lacking the organelles found in complex cells; bacteria including cyanobacteria. *44*

ribosome (ri'bo-som) minute particle found attached to the ER or loose in the cytoplasm that is the site of protein synthesis. *51*

Chapter 3

Cell Membrane and Cell Wall Function

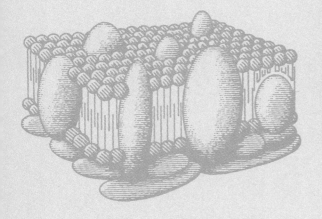

Chapter Concepts

1   The cell membrane is differentially permeable, allowing some substances to pass through freely and restricting the passage of other substances.

2   Small molecules can pass through a cell membrane along a concentration gradient either by diffusion or facilitated diffusion, the latter requiring protein carriers.

3   Active transport, the passage of molecules against a concentration gradient, requires protein carriers and an expenditure of cellular energy.

4   Vesicle formation takes large molecules into the cell and vesicle fusion discharges substances from the cell.

Chapter Outline

Outer Boundaries of Cells
    The Cell Membrane
    The Plant Cell Wall
    Permeability
Osmosis
    Tonicity
    The Importance of Osmosis
Transport by Carriers
    Facilitated Diffusion
    Active Transport
Endocytosis and Exocytosis
    Endocytosis
    Exocytosis

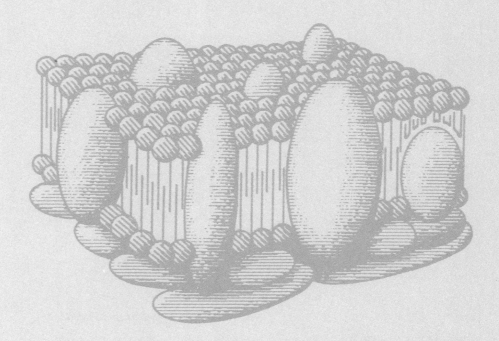

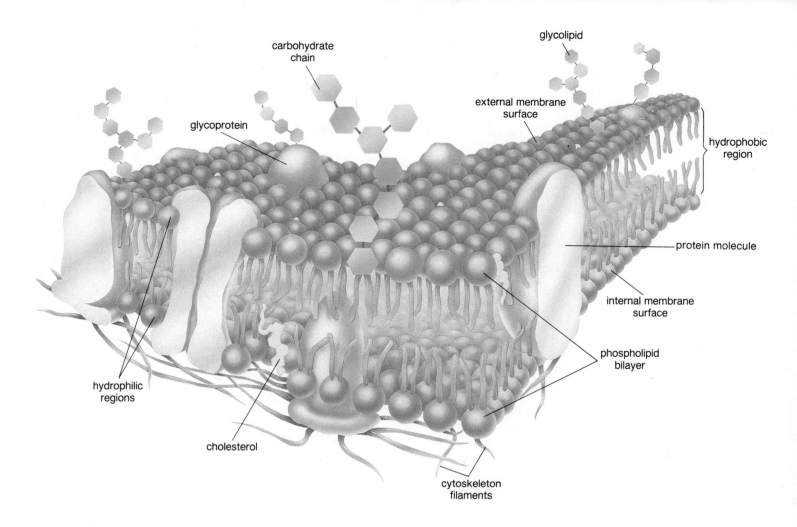

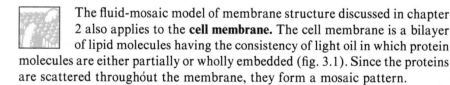

Outer Boundaries of Cells

The Cell Membrane

The fluid-mosaic model of membrane structure discussed in chapter 2 also applies to the **cell membrane.** The cell membrane is a bilayer of lipid molecules having the consistency of light oil in which protein molecules are either partially or wholly embedded (fig. 3.1). Since the proteins are scattered throughout the membrane, they form a mosaic pattern.

The Lipid Bilayer

Most of the lipids in the cell membrane are phospholipids. Each **phospholipid** molecule has a polar (charged) head and 2 nonpolar tails (see fig. 1.29). When surrounded by water, phospholipid molecules form a bilayer naturally. The heads, being polar, are attracted to the water, which is also polar; therefore, the heads face outward. The polar heads are said to be hydrophilic (water loving). The nonpolar tails are not attracted to the water. They face inside, away from the water, and are said to be hydrophobic (water hating).

Some of the lipids in the cell membrane are *glycolipids*. Glycolipids are constructed similarly to phospholipids except the polar head consists of a chain of sugar molecules. Glycolipids only occur in the outer half of the bilayer, where they function in cell-to-cell recognition. Different types of cells have different glycolipids.

Figure 3.1
Fluid-mosaic model of the cell membrane. The membrane is composed of a phospholipid bilayer with embedded proteins. The hydrophilic heads of the phospholipids are at the surfaces of the membrane, and the hydrophobic tails make up the interior of the membrane. The carbohydrate chains of glycolipids and glycoproteins are involved in cell-to-cell recognition. Proteins also serve other functions, such as receptors for chemical messengers, conducting molecules through the membrane, and enzymes in metabolic reactions.

Figure 3.2

Membrane protein diversity. These are some of the functions performed by proteins found in the cell membrane.

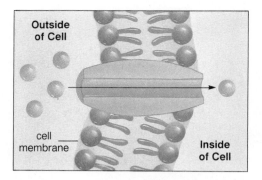

Channel protein

A protein that allows a particular molecule or ion to cross the cell membrane freely as it enters or exits the cell. Recently, it has been shown that cystic fibrosis, an inherited disorder, is caused by a faulty chloride (Cl⁻) channel. When the channel is not functioning normally, a thick mucus collects in airways and in pancreatic and liver ducts.

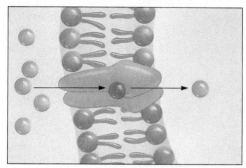

Carrier protein

A protein that selectively interacts with a specific molecule or ion so that it can cross the cell membrane to enter or exit the cell. The carrier protein that transports sodium (Na⁺) ions and potassium (K⁺) ions across the cell membrane requires ATP energy. The inability of some persons to use up energy for sodium-potassium transport has been suggested as the cause of their obesity.

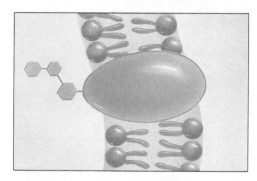

Cell recognition protein

A glycoprotein that identifies the cell. For example, the MHC (major histocompatibility complex) glycoproteins are different for each person so organ transplants are difficult to achieve. Cells with foreign MHC glycoproteins are attacked by blood cells responsible for immunity.

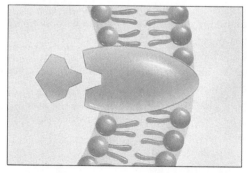

Receptor protein

A protein that is shaped in such a way that a specific molecule can bind to it. Recently, it has been shown that Pygmies are short not because they do not produce enough growth hormone, but because their cell membrane growth hormone receptors are faulty and cannot interact with growth hormone.

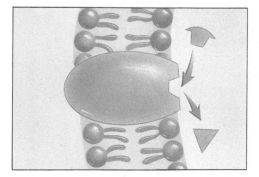

Enzymatic protein

A protein that catalyzes a specific reaction. For example, there is a cell membrane protein called adenylate cyclase that is involved in ATP metabolism. Polluted water can contain cholera bacteria, which release a toxin that interferes with the proper functioning of adenylate cyclase. So many sodium ions and so much water leave intestinal cells that the individual dies from severe diarrhea.

Animal cell membranes, in particular, also contain a substantial number of *cholesterol* molecules. These molecules lend stability to the lipid bilayer and prevent a drastic decrease in fluidity at low temperatures.

Proteins

The basic structure of the cell membrane is determined by the lipid bilayer, but the functions of the membrane are carried out largely by the proteins. Some protein molecules span the membrane and have attached carbohydrate chains. In this case, they are called glycoproteins. Other proteins are located on the inner surface of the membrane, where they are held in place by cytoskeleton filaments.

The proteins of the membrane carry out various functions. *Glycoproteins,* like glycolipids, make cell-to-cell recognition possible. Together, the glycoproteins and the glycolipids make up the "fingerprints" of the cell. As we will discuss in more detail, certain proteins are involved in the passage of molecules through the membrane. Some of these have a *channel* through which a substance simply can move across the membrane; others are *carriers* that combine with a substance and help it to move across the membrane. Still other proteins are *receptors;* each type of receptor has a specific shape that allows a specific molecule to bind to it. The binding of a molecule, such as a hormone, can influence the metabolism of the cell. Viruses often must attach to receptors before they enter a cell. Some proteins have an *enzymatic function* and carry out metabolic reactions. Figure 3.2 depicts the various functions of membrane proteins.

The Plant Cell Wall

In addition to a cell membrane, plant cells are surrounded by a **cell wall** (fig. 3.3) that varies in thickness depending on the function of the cell. All plant cells have a primary cell wall, having as its main constituent *cellulose* molecules united into threadlike microfibrils. Several microfibrils, in turn, are found in fibrils, and within the wall there are layers of fibrils lying at right angles to one another for added strength. Some cells in woody plants have a secondary cell wall that forms inside the primary cell wall. Secondary cells contain lignin, a substance that makes secondary cell walls even stronger than primary cell walls. Plant support cells have secondary walls; the cell dies and the strong wall remains as support material.

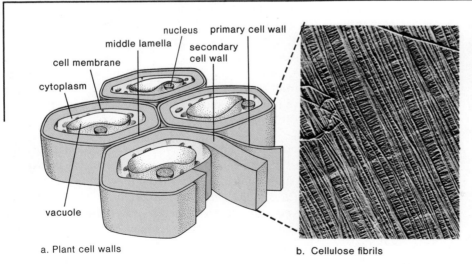

a. Plant cell walls

b. Cellulose fibrils

Figure 3.3

Plant cell wall. a. All plant cells have a primary cell wall, and some have a secondary cell wall. The middle lamella is a region between cells that contains a sticky substance. b. Electron micrograph of alternating layers of cellulose fibrils that make up the cell wall.

Table 3.1 Passage of Molecules into and out of Cells

Name	Direction	Requirements	Examples
Simple Diffusion	Toward lesser concentration	Concentration gradient	Lipid-soluble molecules, water, and gases
Facilitated Diffusion	Toward lesser concentration	Carrier and concentration gradient	Sugars and amino acids
Active Transport	Toward greater concentration	Carrier plus energy	Sugars, amino acids, and ions
Endocytosis Phagocytosis	Toward inside	Vesicle formation	Cells and subcellular material
Pinocytosis	Toward inside	Vesicle formation	Macromolecules
Exocytosis	Toward outside	Vesicle fuses with cell membrane	Macromolecules

Humans make use of the cellulose from plant cell walls. Cotton, rayon, flax, hemp, paper, and even wood are derived from plant cell walls. Because lignin causes paper to turn yellow with age, it usually is removed before paper is made from wood pulp. Today, the lignin is used in the production of various synthetic products, such as rubber, plastics, pigments, and adhesives.

Animal and plant cells are surrounded by a cell membrane. The cell membrane is composed of a lipid bilayer in which proteins having various functions are found. Plant cells have a cell wall in addition to a cell membrane.

Permeability

The plant cell wall is freely permeable, but the cell membrane in both animal and plant cells is not. Instead, it regulates the entrance and the exit of molecules to and from the cell.

Sometimes the cell membrane is said to be semipermeable because certain small molecules can pass through the membrane but large molecules cannot pass through the membrane. When you consider that some small molecules pass through the cell membrane quickly while others have difficulty passing through, however, it is probably better to call the cell membrane selectively or *differentially permeable.*

As listed in table 3.1, there are 3 general means by which substances can enter and exit cells from the surrounding medium: (1) diffusion (simple and facilitated), (2) active transport, and (3) endocytosis and exocytosis.

Diffusion

Diffusion is a physical process that can be observed with any type of molecule. Diffusion is such a universal phenomenon that there is a physical law, called the law of diffusion, that states *molecules move from the area of greater concentration to the area of lesser concentration until equally distributed.* For example, when a few crystals of dye are placed in water (fig. 3.4), the dye molecules move in various directions, but their net movement is in a direction away from the concentrated source. The net movement of the water molecules is in the opposite direction. Eventually, the dye is dissolved evenly in the water, resulting in a colored solution. The dye molecules and the water molecules continue to move about, but there is no net movement in any one direction.

The Cell, the Smallest Unit of Life

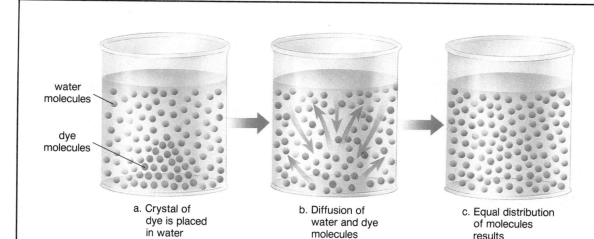

Figure 3.4

Process of diffusion. Diffusion is spontaneous, and no energy is required to bring it about. a. When dye crystals are placed in water, they are concentrated in one area. b. The dye dissolves in the water, and there is a net movement of dye molecules away from the area of concentration. There is a net movement of water molecules in the opposite direction. c. Eventually, the water and the dye molecules are distributed equally throughout the container.

a. Crystal of dye is placed in water

b. Diffusion of water and dye molecules

c. Equal distribution of molecules results

The movement of molecules by diffusion alone is a slow process. The rate of diffusion is affected by the **concentration gradient** (the difference in concentration of the diffusing molecules between the 2 regions), the size and the shape of the molecule, and the temperature. Also, diffusion in a liquid medium is slower than in a gaseous medium; however, distribution of molecules in cytoplasm is sped up by the even-constant flow of the cytoplasm that is called cytoplasmic streaming.

The chemical and physical properties of the cell membrane allow just a few types of molecules to enter and to exit by means of simple diffusion. Lipid-soluble molecules, such as alcohols, can diffuse through the membrane because lipids are the membrane's main structural components.

Gases also can diffuse through the lipid bilayer; this is the mechanism by which oxygen enters cells and carbon dioxide exits cells. As an example, consider the movement of oxygen from the air sacs (alveoli) of the lungs to the blood in the lung capillaries (fig. 3.5). After inhalation (breathing in), the concentration of oxygen in the alveoli is greater than the concentration of oxygen in the blood; therefore, oxygen diffuses into the blood.

Water passes into and out of cells with relative ease. It now appears that some membrane proteins form *channels* (fig. 3.2) with a pore large enough to allow the passage of water and some ions. Other molecules cannot utilize these channels because they are either too large or they carry a charge that prevents passage. (The membrane is usually positively charged outside and negatively charged inside. Therefore, negatively charged ions tend to move from inside to outside, and positively charged ions tend to move in the opposite direction.) The fact that water can penetrate a membrane has important biological consequences, as described in the discussion that follows.

The cell membrane is differentially permeable. A few types of small molecules can simply diffuse through the cell membrane from the area of greater concentration to the area of lesser concentration.

Osmosis

The diffusion of water across a differentially permeable membrane has been given a special term; it is called osmosis. **Osmosis** is the net movement of water molecules from the area of greater concentration of water to the area of lesser concentration of water across a differentially permeable membrane.

Figure 3.5

Gas exchange in lungs. Oxygen (dots) diffuses into the capillaries of the lungs because there is a greater concentration of oxygen in the air sacs (alveoli) than in the capillaries.

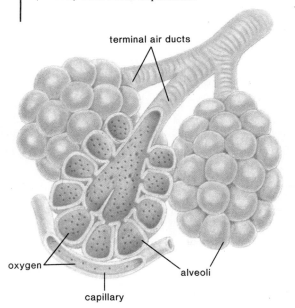

terminal air ducts

oxygen

alveoli

capillary

Figure 3.6

Osmosis demonstration. a. *A thistle tube, covered at the broad end by a membrane, contains a sugar solution. The beaker contains only solvent (water).* b. *The solute (green circles) is unable to pass through the membrane, but the water passes through in both directions. There is a net movement of water* toward the inside of the thistle tube, where the solute concentration is higher and the water concentration is lower. c. *In the end, the level of the solution rises in the thistle tube until a pressure equivalent to osmotic pressure builds.*

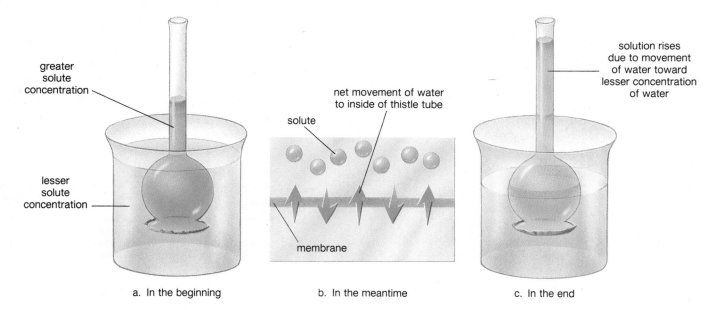

greater solute concentration

lesser solute concentration

net movement of water to inside of thistle tube

solute

membrane

solution rises due to movement of water toward lesser concentration of water

a. In the beginning

b. In the meantime

c. In the end

To illustrate osmosis, a thistle tube covered at one end by a differentially permeable membrane is placed in a beaker of distilled water. The tube contains a solution of both a **solute** (e.g., sugar molecules) and a **solvent** (water)(fig. 3.6). Obviously, the lesser solute and greater water concentration is outside the tube, so there will be a net movement of water from outside the tube to inside through the membrane. The solute (sugar molecules) is unable to pass through the membrane; therefore, the level of the solution within the tube rises (fig. 3.6c). The eventual height of the solution in the tube indicates the degree of **osmotic pressure** caused by the flow of water from the area of higher water concentration to the area of lower water concentration.

Notice that in this illustration of osmosis,

1. a differentially permeable membrane separates a solution from pure water;
2. a difference in solute and water concentrations exists on the 2 sides of the membrane;
3. the membrane does not permit passage of the solute particles;
4. the membrane permits passage of water, and water moves from the area of lesser solute (greater water) concentration to the area of greater solute (lesser water) concentration;
5. an osmotic pressure is present; the amount of liquid increases on the side of the membrane with the greater solute concentration.

These considerations will be important as we discuss osmosis in relation to cells placed in different solutions. The cell membrane does allow solutes, such as sugars and salts, to pass through, but the difference in permeability between water and these solutes is so great that cells in sugar and salt solutions do have to cope with the osmotic movement of water.

Osmosis is the diffusion of water across a differentially permeable membrane. The presence of osmotic pressure is evident when there is an increased amount of water on the side of the membrane that has the greater solute concentration.

Figure 3.7

Osmosis in animal and plant cells. The arrows indicate the net movement of water. In an isotonic solution, a cell neither gains nor loses water; in a hypotonic solution, a cell gains water; and in a hypertonic solution, a cell loses water.

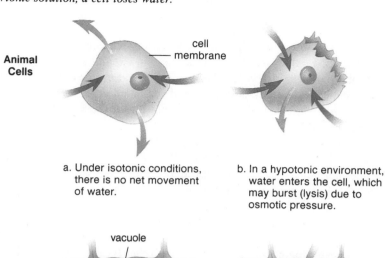

Animal Cells

a. Under isotonic conditions, there is no net movement of water.

b. In a hypotonic environment, water enters the cell, which may burst (lysis) due to osmotic pressure.

c. In a hypertonic environment, water leaves the cell, which shrivels (crenation).

Plant Cells

d. Under isotonic conditions, there is no net movement of water.

e. In a hypotonic environment, vacuoles fill with water, turgor pressure develops, and chloroplasts are seen next to the cell wall.

f. In a hypertonic environment, vacuoles lose water, the cytoplasm shrinks (plasmolysis), and chloroplasts are seen in the center of the cell.

Tonicity

Cells can be placed in solutions that have the same concentration solute, a greater solute concentration, or a lesser solute molecules concentration than the cell. These solutions are called isotonic, hypertonic, and hypotonic, respectively. Figure 3.7 depicts and describes the effects of these solutions on cells.

Isotonic Solutions

In the laboratory, cells normally are placed in solutions that cause them neither to gain nor to lose water. Such solutions are said to be **isotonic solutions;** that is, the solute concentration is the same on both sides of the membrane, and therefore there is no net gain or loss of water. The term *iso* means the same as and the term *tonicity* refers to the strength of the solution. It is possible to determine, for example, that a 0.9% solution[1] of the salt sodium chloride (Na^+Cl^-) is isotonic to red blood cells because the cells neither swell nor shrink when placed in such a solution (fig. 3.7*a*).

[1]Percent solutions are grams of solute per 100 ml of solvent. Therefore, a 0.9% solution is 0.9 (gm) sodium chloride in 100 ml of water.

Hypotonic Solutions

Solutions that cause cells to swell or even to burst due to an intake of water are said to be **hypotonic solutions.** The prefix *hypo* means less than and refers to a solution with a *lesser concentration of solute* (greater concentration of water) than the cell. If a cell is placed in a hypotonic solution, water enters the cell; the net movement of water is from the outside to the inside of the cell.

Any salt solution less that 0.9% is a hypotonic solution to red blood cells. Red blood cells placed in such a solution expand and even burst due to the buildup of osmotic pressure (fig. 3.7b). The term *lysis* is used to refer to disrupted cells. Hemolysis is disrupted red blood cells.

A hypotonic solution brings about **turgor pressure** due to swelling of the cell. When a plant cell is placed in a hypotonic solution, we can observe expansion of the cytoplasm because the large central vacuole gains water, and the cell membrane pushes against the rigid cell wall (fig. 3.7e). The plant cell does not burst because the cell wall does not give way. Turgor pressure in plant cells is extremely important to maintenance of the erect position of the plant.

Hypertonic Solutions

Solutions that cause cells to shrink or to shrivel due to a loss of water are said to be **hypertonic solutions.** The prefix *hyper* means greater than and refers to a solution with a *greater concentration of solute* (lesser concentration of water) than the cell. If a cell is placed in a hypertonic solution, water leaves the cell; the net movement of water is from the inside to the outside of the cell.

A 10% solution of sodium chloride (Na^+Cl^-) is hypertonic to red blood cells. In fact, any solution with a concentration higher than 0.9% sodium chloride is hypertonic to red blood cells. If red blood cells are placed in this solution, they shrink (fig. 3.7c). The term *crenation* refers to red blood cells in this condition.

A hypertonic solution brings about **plasmolysis,** shrinking of the cytoplasm due to osmosis. When a plant cell is placed in a hypertonic solution, we can observe the cytoplasm shrinking because the large central vacuole loses water, and the cell membrane pulls away from the cell wall (fig. 3.7f).

The Importance of Osmosis

Osmosis occurs constantly in living organisms. For example, due to osmosis, water is absorbed from the human large intestine, is retained by the kidneys, and is taken up by the blood. Since living things contain a very high percentage of water, osmosis is an extremely important physical process for their continued good health.

When a cell is placed in an isotonic solution, it neither gains nor loses water. When a cell is placed in a hypotonic solution (lesser solute concentration than an isotonic solution) the cell gains water. When a cell is placed in a hypertonic solution (greater solute concentration than an isotonic solution), the cell loses water and the cytoplasm shrinks. See table 3.2.

Transport by Carriers

The presence of the cell membrane impedes the passage of all but a few substances. Yet, biologically useful molecules do enter and exit the cell at a rapid rate because there are carrier proteins in the membrane. **Carrier proteins** are specific; each can combine with only a certain type of molecule, which is then

3.2 Critical Thinking

1. Contrast the manner in which alcohol and water enter a cell.
2. Contrast the manner in which sodium (Na^+) ions and chloride (Cl^-) ions exit a cell (see fig. 3.2).
3. Contrast the manner in which amino acids and proteins enter a cell.
4. How might the proteins from question 3 be digested?

The Cell, the Smallest Unit of Life

Table 3.2 *Effect of Osmosis on the Cell*

Tonicity of Solution	Concentrations		Net Movement of Water	Effect on Cell
	Solute	Water		
Isotonic	Same as cell	Same as cell	None	None
Hypotonic	Less than cell	More than cell	Cell gains water	Swells, turgor pressure
Hypertonic	More than cell	Less than cell	Cell loses water	Shrinks, plasmolysis

transported through the membrane. It is not completely understood how carrier proteins function, but after they combine with a molecule, they are believed to undergo a change in shape that causes the molecule to be moved across the membrane. Carrier proteins are required for facilitated diffusion and active transport (table 3.1).

Some of the proteins in the cell membrane are carriers that transport biologically useful molecules into and out of the cell.

Facilitated Diffusion

Facilitated diffusion explains the passage of molecules, such as glucose and amino acids, across the cell membrane even though they are lipid insoluble. The passage of these molecules is facilitated by their reversible combination with carrier proteins that in some manner transport them through the cell membrane. These carrier proteins are specific. For example, various sugar molecules of identical size might be present inside or outside the cell, but certain ones cross the membrane hundreds of times faster than others. As stated earlier, this is the reason that the membrane can be called differentially permeable.

A model for facilitated diffusion (fig. 3.8) shows that after a carrier has assisted the movement of a molecule to the other side of the membrane, it is free to assist the passage of other similar molecules. Neither simple diffusion, explained previously, nor facilitated diffusion by means of carrier proteins requires an expenditure of energy, because the molecules are moving down a concentration gradient in the same direction they would tend to move anyway.

Active Transport

Due to **active transport,** molecules or ions move through the cell membrane, accumulating either inside or outside the cell in the region of *greater* concentration. For example, iodine collects in the cells of the thyroid gland; sugar is absorbed completely from the gut by the cells lining the digestive tract; and sodium is sometimes almost completely withdrawn from urine by cells lining the kidney tubules.

Both carrier proteins and an expenditure of energy (fig. 3.9) are needed to transport substances from an area of lesser concentration to an area of greater concentration. In this case, energy (ATP molecules) is required for the carrier to combine with the substance to be transported. Therefore, it is not surprising that cells primarily involved in active transport, such as kidney cells, have a large number of mitochondria near the membrane where active transport is occurring (fig. 15.10).

Figure 3.8

Facilitated diffusion. A carrier protein speeds the rate at which a solute crosses a membrane in the direction of lesser concentration. Facilitated diffusion occurs only when there is a concentration gradient across the membrane; therefore, energy is not required.

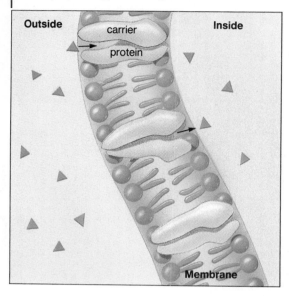

Figure 3.9

Active transport through a cell membrane. Active transport is apparent when a molecule crosses the cell membrane toward the area of greater concentration. An expenditure of ATP energy is required, presumably to allow a carrier protein to transport a molecule across the cell membrane contrary to its concentration gradient.

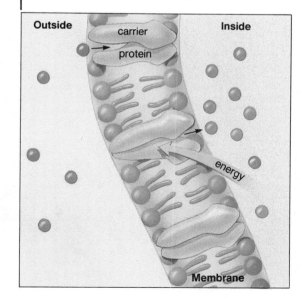

Figure 3.10

The sodium-potassium pump. The same protein carrier actively moves sodium (Na⁺) ions to outside the cell and potassium (K⁺) ions to inside the cell by means of the same type of carrier protein. This establishes not only a concentration gradient for each of these ions, but also an electrical gradient. The inside of a cell is negatively charged, while the outside is positively charged. Note that ATP energy is required.

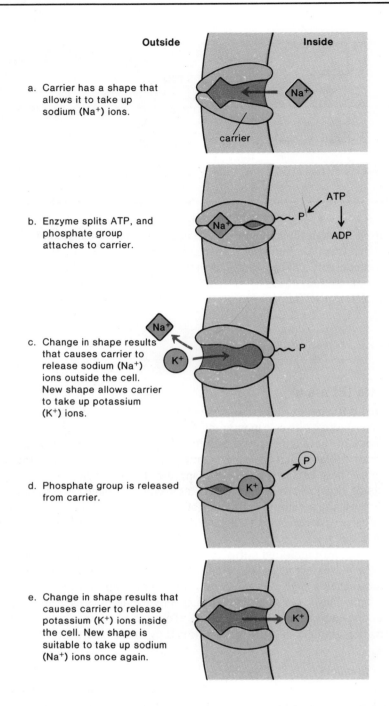

a. Carrier has a shape that allows it to take up sodium (Na⁺) ions.

b. Enzyme splits ATP, and phosphate group attaches to carrier.

c. Change in shape results that causes carrier to release sodium (Na⁺) ions outside the cell. New shape allows carrier to take up potassium (K⁺) ions.

d. Phosphate group is released from carrier.

e. Change in shape results that causes carrier to release potassium (K⁺) ions inside the cell. New shape is suitable to take up sodium (Na⁺) ions once again.

Proteins involved in active transport often are called *pumps* because just as a water pump uses energy to move water against the force of gravity, proteins use energy to move a substance against its concentration gradient. One type of pump that is active in all cells, but is especially associated with nerve and muscle cells, is the one that moves sodium (Na⁺) ions to the outside of the cell and potassium (K⁺) ions to the inside of the cell. These 2 events are presumed to be linked, and the carrier protein is called a *sodium-potassium pump*. A change in shape brought about by the attachment of a phosphate group to the carrier allows it to alternately combine with sodium ions and potassium ions (fig. 3.10). The phosphate group is donated by ATP, which is broken down enzymatically by the carrier.

The Cell, the Smallest Unit of Life

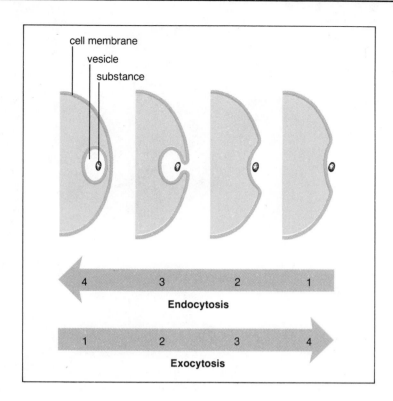

cell membrane
vesicle
substance

4 3 2 1

Endocytosis

1 2 3 4

Exocytosis

Figure 3.11
Endocytosis and exocytosis. During endocytosis, the cell membrane forms a vesicle around the substance to be taken into the cell. During exocytosis, a vesicle fuses with the membrane, and this allows an enclosed substance to leave the cell. Exocytosis occurs during secretion.

The passage of salt (Na^+Cl^-) across a cell membrane is of primary importance in cells. The chloride (Cl^-) ion does not cross the cell membrane unless it is attracted by positive charges. Once sodium ions have been pumped across a membrane, chloride ions simply diffuse through channels that allow its passage. It is now known that the chloride channels malfunction in persons with cystic fibrosis, and this leads to the symptoms of this inherited (genetic) disorder.

During facilitated diffusion (no energy required), small molecules follow their concentration gradient. During active transport (energy required), small molecules go against their concentration gradient.

Endocytosis and Exocytosis

Endocytosis and exocytosis are opposite processes (fig. 3.11 and table 3.1). During endocytosis, a vesicle forms at the cell membrane, and during exocytosis, a vesicle joins with the cell membrane.

Endocytosis

At times, molecules or larger substances are incorporated into cells by the process of **endocytosis,** which requires the formation of a vesicle or a vacuole. Endocytosis, even when not moving substances toward a greater concentration, requires energy.

When the material taken in by the process of endocytosis is quite large, the process is called **phagocytosis** (cell eating). Phagocytosis is common in amoeboid-type cells such as macrophages, large cells found in humans. These cells phagocytize bacteria and worn-out red blood cells, for example.

3.3 Critical Thinking

1. Exocytic vesicles add cell membrane to the cell, and endocytic vesicles remove cell membrane. In a cell in which the amount of cell membrane stays constant, how many exocytic vesicles per endocytic vesicles would you expect?
2. Imagine a cell that is moving from left to right. If vesicle formation is facilitating movement, where would you expect exocytosis to be occurring? Where would you expect endocytosis to be occurring?
3. Receptor-mediated endocytosis is a process by which a substance combines with a receptor before endocytosis brings the entire complex into the cell. Some investigators have proposed that the AIDS virus enters a cell in this manner. If so, what additional step is needed for the virus to enter the cell proper?

Macrophage. These "big eaters" are the body's scavengers. They take in (phagocytize) all sorts of debris, including worn-out red blood cells, as shown here.

The Cell, the Smallest Unit of Life

Vesicles also form around large-sized molecules such as proteins; this is called **pinocytosis** (cell drinking). Whereas phagocytosis can be seen with the light microscope, pinocytosis requires the use of the electron microscope.

Once formed, vesicles or vacuoles contain a substance enclosed by membrane. Digestion is required for this substance to be broken down and incorporated into the cytoplasm. Therefore, it is believed that lysosomes fuse with these bodies, allowing digestive enzymes to break down the molecules they contain.

Exocytosis

Exocytosis is the reverse of endocytosis and requires that a vesicle fuse with the membrane, thereby discharging its contents. As we saw in chapter 2, vesicles formed at the Golgi apparatus transport cell products out of the cell; this entire process is called *secretion*.

Summary

All cells are surrounded by a cell membrane. The cell membrane is a lipid bilayer with embedded proteins and also some proteins attached to its inner surface. The proteins have various functions. In addition to the cell membrane, plant cells have a cell wall. Substances cross cell membranes by diffusion, transport by carriers, and vesicle formation. The diffusion of water across a membrane is called osmosis. When a cell is placed in an isotonic solution, there is no net gain or loss of water. In a hypertonic solution, cells shrink, and in a hypotonic solution, cells swell. The presence of a cell wall makes it possible to observe plasmolysis and turgor pressure when plant cells are placed in a hypertonic and hypotonic solution, respectively.

Facilitated diffusion, which moves molecules toward a lesser concentration, and active transport, an energy-requiring process, which moves molecules toward a greater concentration, require a protein carrier. Vesicle formation during endocytosis permits molecules (pinocytosis) and debris (phagocytosis) to be taken into cells. Exocytosis describes the process of cell product secretion.

Study Questions

1. Describe the structure of the cell membrane, including the lipid bilayer and the various types of proteins. (pp. 65–67)
2. Why can a cell membrane be called semipermeable? differentially permeable? (p. 68)
3. What are the 3 mechanisms by which substances enter and exit cells? (p. 68)
4. Define diffusion, and give an example. (p. 68)
5. Define osmosis. Define isotonic, hypertonic, and hypotonic solutions, and give examples of these concentrations for red blood cells. (pp. 69–72)
6. Draw a simplified diagram of a red blood cell before and after being placed in these solutions. What terms are used to refer to the condition of the red blood cell in a hypertonic solution and in a hypotonic solution? (p. 70)
7. Draw a simplified diagram of a plant cell before and after being placed in these solutions. Describe the cell contents under these conditions. (p. 71)
8. How does facilitated diffusion differ from simple diffusion across the cell membrane? (p. 73)
9. How does active transport differ from facilitated diffusion? Give an example. (p. 73)
10. Draw diagrams that show endocytosis and exocytosis. Give an example for each of these. (p. 75)

Objective Questions

1. Both plant and animal cells have a cell membrane, but in addition, plant cells have a cell _____ .
2. Molecules diffuse from the area of _____ concentration to the area of _____ concentration.
3. In a hypertonic solution, cells _____ water and the cell contents _____ .
4. When plant cells are placed in a hypotonic solution, _____ is obvious because cell contents _____ against the cell wall.
5. During _____ diffusion, carriers move glucose and amino acids from the area of greater concentration to the area of lesser concentration.
6. During active transport, a carrier moves molecules from the area of _____ concentration to the area of _____ concentration.
7. Sodium ions and potassium ions move across the cell membrane in opposite directions due to the action of the _____ .
8. During endocytosis, _____ formation takes substances into the cell.

Answers to Objective Questions
1. wall 2. greater, lesser 3. lose, shrink 4. turgor pressure, press 5. facilitated 6. lesser, greater 7. sodium-potassium pump 8. vacuole

Label this Diagram.

See figure 3.1 (p. 65) in text.

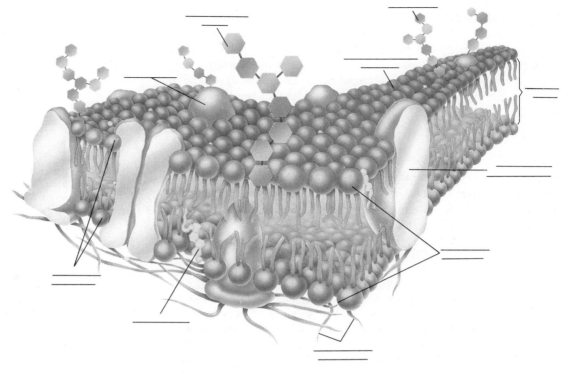

Selected Key Terms

active transport (ak'tiv trans'port)
transfer of a substance into or out of a
cell from lesser concentration to greater
concentration by a process that requires
a carrier and an expenditure of
energy. *73*

carrier protein (kar'e-er pro'te-in) a
protein molecule that combines with a
substance and transports it through the
cell membrane. *72*

cell membrane (sel mem'brān) a
membrane that surrounds the cytoplasm
of cells and regulates the passage of
molecules into and out of the cell. *65*

diffusion (dĭ-fu'zhun) the movement of
molecules from an area of greater
concentration to an area of lesser
concentration. *68*

endocytosis (en''do-si-to'sis) a process in
which a vesicle is formed at the cell
membrane to bring a substance into the
cell. *75*

exocytosis (eks''o-si-to'sis) a process in
which an intracellular vesicle fuses with
the cell membrane so that the vesicle's
contents are released outside the
cell. *77*

facilitated diffusion (fah-sil'ĭ-tāt-ed dĭ-
fu'zhun) passive transfer of a substance
into or out of a cell along a concentration
gradient by a process that requires a
carrier. *73*

hypertonic solution (hi''per-ton'ik so-
lu'shun) one that has a greater
concentration of solute and a lesser
concentration of water than the cell. *72*

hypotonic solution (hi''po-ton'ik so-
lu'shun) one that has a lesser
concentration of solute and a greater
concentration of water than the cell. *72*

isotonic solution (i''so-ton'ik so-lu'shun)
one that contains the same concentration
of solute and water as the cell. *71*

osmosis (oz-mo'sis) the movement of
water from an area of greater
concentration of water to an area of
lesser concentration of water across a
differentially permeable membrane. *69*

osmotic pressure (oz-mot'ik presh'ur)
pressure generated by and due to the
osmotic flow of water. *70*

phagocytosis (fag''o-si-to'sis) the taking
in of bacteria and/or debris by engulfing;
cell eating. *75*

pinocytosis (pin''o-si-to'sis) the taking in
of fluid along with dissolved solutes by
engulfing; cell drinking. *77*

plasmolysis (plas-mol'ĭ-sis) contraction
of the cell contents due to the loss of
water. *72*

solute (sol'ūt) a substance dissolved in a
solvent to form a solution. *70*

solvent (sol'vent) a fluid, such as water,
that dissolves solutes. *70*

turgor pressure (tur'gor presh'ur)
osmotic pressure that adds to the
strength of the cell. *72*

Chapter 4

Cell Division

Chapter Concepts

1   Cell division includes the division of both the cytoplasm and the nucleus.

2   The nucleus contains the gene-bearing chromosomes. Ordinary cell division ensures that each new cell receives a full complement of chromosomes and genes.

3   In animals, reduction division is required to produce the sex cells, which contain one-half the full number of chromosomes.

4   In plants, reduction division produces spores, structures that precede the formation of the sex cells in the life cycle of these organisms.

Chapter Outline

The Life Cycle of Animals
Mitosis
    The Cell Cycle
    Overview of Animal Mitosis
    Stages of Mitosis
    Mitosis in Plants
Meiosis
    Overview of Animal Meiosis
    Stages of Meiosis
    Spermatogenesis and Oogenesis
    Meiosis in Plants
Comparison of Mitosis and Meiosis
    Occurrence
    Process
    Daughter Cells

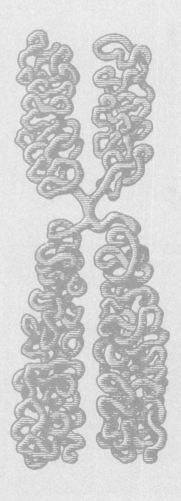

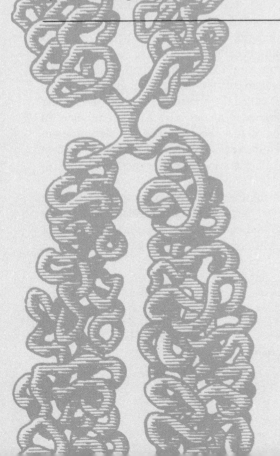

Figure 4.1

Human embryo at 6 weeks. Due to the process of cell division, this human embryo is made up of millions of cells. Each cell contains the same number and kinds of chromosomes, copies of the very ones that were inherited from its parents.

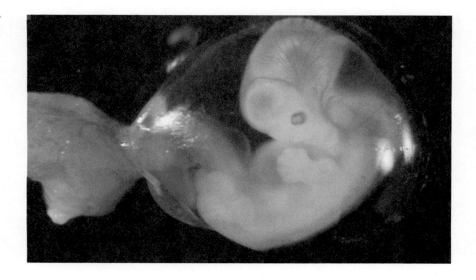

Cell division is necessary for growth (fig. 4.1) and repair of multi-cellular organisms and for reproduction of all organisms. Cell division requires division of both the nucleus and the cytoplasm. We learned in chapter 2 that the nucleus contains **chromatin,** long threads made up of the DNA double helix and associated proteins. At the time of cell division, chromatin becomes coiled, folded, and twisted to give a highly compacted structure. Individual **chromosomes** can be seen just as cell division is about to occur. The reading on page 82 describes the transition between chromatin and chromosomes in greater detail.

An examination of the chromosomes of multicellular organisms shows that each type of organism has a characteristic number—corn plants have 20 chromosomes, houseflies have 12, and humans have 46. The particular number has nothing to do with the complexity of the organism. For example, hydras (fig. 30.9), which are very simple microscopic organisms, have 32 chromosomes, many more than houseflies with 12. In order to view the chromosomes for counting, cells are treated, photographed, and enlarged. In cells about to divide, the chromosomes can be cut out of the photograph and arranged by pairs. (Pairs of chromosomes have the same size and general appearance.) The resulting display of chromosome pairs is called a **karyotype.**

A male karyotype is shown in figure 4.2. Although both males and females have 23 pairs of chromosomes, one pair is of unequal length in the male. The larger chromosome of this pair is called the X chromosome, and the smaller chromosome is called the Y chromosome. Females have 2 X chromosomes in their karyotype. The X and Y chromosomes are called the **sex chromosomes** because they carry genes that determine sex. The other chromosomes, known as **autosomes,** include all the pairs of chromosomes except the X and Y chromosomes.

Notice, as the enlargement in figure 4.2 shows, prior to division each chromosome is composed of 2 identical parts, called **chromatids.** These 2 sister chromatids are genetically identical and contain the same genes, the units of heredity that control the cell. The chromatids are held together at a region called the **centromere.**

Each organism has a characteristic number of chromosomes; humans have 46. A human karyotype shows 22 pairs of autosomes and 1 pair of sex chromosomes. The sex pair is an X chromosome and a Y chromosome in males and 2 X chromosomes in females. Each chromosome in a karyotype is composed of 2 sister chromatids held together at the centromere.

Figure 4.2

Human karyotype preparation. As illustrated here, the stain used can result in a banded appearance of the chromosomes. The bands help researchers to identify and to analyze the chromosomes. The enlargement of a chromosome on the far left shows that the chromosomes in a karyotype consist of 2 chromatids held together by a centromere. This is the appearance of chromosomes just before they divide.

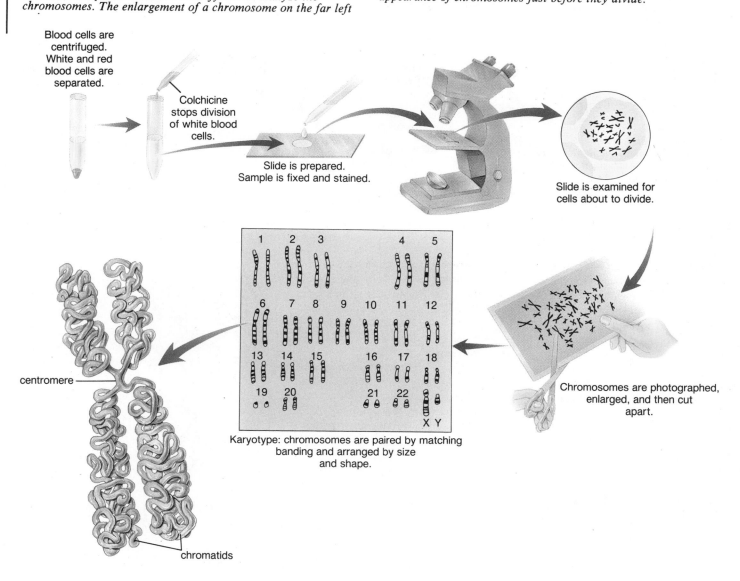

Blood cells are centrifuged. White and red blood cells are separated.

Colchicine stops division of white blood cells.

Slide is prepared. Sample is fixed and stained.

Slide is examined for cells about to divide.

Chromosomes are photographed, enlarged, and then cut apart.

Karyotype: chromosomes are paired by matching banding and arranged by size and shape.

centromere

chromatids

The Life Cycle of Animals

Advanced multicellular animals, including humans, typically have a life cycle (fig. 4.3) that requires 2 types of cell division: meiosis and mitosis.

Meiosis occurs in the sex organs of males and females. In males, it produces the cells that become sperm; in females, it produces the cells that become eggs. The sperm and the egg are the sex cells, or gametes. The **gametes** contain the N or **haploid** number of chromosomes. The "N" symbolizes a set of chromosomes. The haploid number has one chromosome from each of the pairs of chromosomes and therefore one chromosome of each kind. The haploid number in humans is 23 chromosomes.

A new individual comes into existence when the sperm of the male fertilizes the egg of the female. Therefore, the resulting fertilized egg, or **zygote,** has the **2N** or **diploid** number of chromosomes. In humans, the diploid number is 46 chromosomes because there are 23 pairs of chromosomes. Each parent

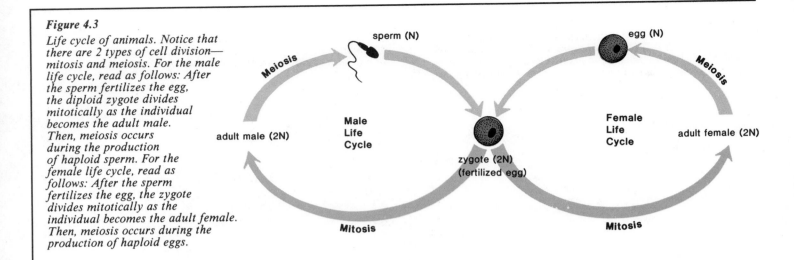

Figure 4.3

Life cycle of animals. Notice that there are 2 types of cell division—mitosis and meiosis. For the male life cycle, read as follows: After the sperm fertilizes the egg, the diploid zygote divides mitotically as the individual becomes the adult male. Then, meiosis occurs during the production of haploid sperm. For the female life cycle, read as follows: After the sperm fertilizes the egg, the zygote divides mitotically as the individual becomes the adult female. Then, meiosis occurs during the production of haploid eggs.

sperm (N)

egg (N)

Meiosis

Meiosis

Male
Life
Cycle

Female
Life
Cycle

adult male (2N)

adult female (2N)

zygote (2N)
(fertilized egg)

Mitosis

Mitosis

What's in a Chromosome? When early investigators decided that the genes are on the chromosomes, they had no idea of chromosome composition. By the mid-1900s, it was known that chromosomes are made up of both DNA and protein. Only in recent years, however, have investigators been able to produce models suggesting how chromosomes are organized.

A eukaryotic chromosome is more than 50% protein. Many of these proteins are concerned with DNA and RNA synthesis, but a large proportion, termed histones,

What's in a Chromosome?

seem to play primarily a structural role. A human cell contains 46 chromosomes, and the length of the DNA in each is about 5 cm. Therefore, a human cell contains at least 2 m of DNA. Yet, all of this DNA is packed into a nucleus that is about 5 µm in

diameter. The histones seem to be responsible for packaging the DNA so that it can fit into such a small space. The packing unit, termed a nucleosome, gives chromatin a beaded appearance in certain electron micrographs.

The accompanying drawing shows that the DNA double helix is wound around a core of histone molecules in a nucleosome. Notice how DNA stretches between the nucleosomes at the location of H_1 histone molecules. Whenever the H_1 molecules make contact, the chromatin shortens. At the time of cell division, the entire structure coils to give a more compacted structure. Then it folds and coils again to give the highly condensed form of the chromosome seen during the stage of cell division called metaphase. No doubt, compacted chromosomes are easier to move about than extended chromatin.

Chromosome versus chromatin structure.

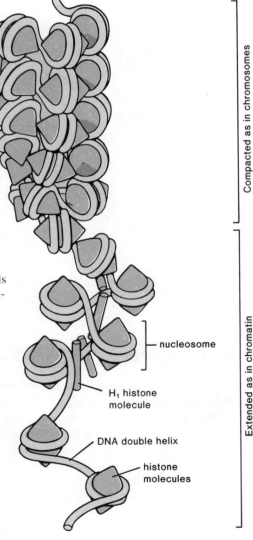

Compacted as in chromosomes

Extended as in chromatin

nucleosome

H_1 histone molecule

DNA double helix

histone molecules

Table 4.1 *Mitosis versus Meiosis in Animals*

Location	Cell Division	Description	Necessary for
Body cells	Mitosis	2N (diploid) → 2N(diploid)	Growth and repair
Sex organs	Meiosis	2N (diploid) → N(haploid)	Gamete production

contributes one chromosome of each of the pairs of chromosomes possessed by the new individual. As the individual grows and matures, **mitosis** occurs, and each cell in the body retains 46 chromosomes.

Table 4.1 summarizes the major differences between mitosis and meiosis in multicellular animals.

The life cycle of humans requires 2 types of cell divisions: mitosis and meiosis. Mitosis is responsible for growth and repair, while meiosis is required for gamete production.

Mitosis

Because of mitosis, each cell in an animal's body has the same number and kinds of chromosomes. Mitosis is important to the growth and the repair of multicellular organisms. When a baby develops in the mother's womb, mitosis occurs as a component of growth. As a wound heals, mitosis occurs, and the damage is repaired.

Mitosis also occurs during the process of asexual reproduction. In protists, for example, one division results in 2 organisms where before there was only one organism.

The Cell Cycle

The **cell cycle** consists of mitosis and interphase (fig. 4.4). The length of time required for the entire cell cycle varies according to the organism and even the type of cell within the organism, but 18–24 hours is typical for animal cells. Mitosis lasts for less than an hour to slightly more than 2 hours; for the rest of the cycle, the cell is in interphase.

During **interphase,** an animal cell resembles that shown in figure 2.3*a* (p. 46). The nuclear envelope and the nucleoli are visible. The chromosomes, however, are not visible because the chromosome material is "decompacted." Only indistinct chromatin is seen.

It used to be said that interphase was a resting stage, but we now know that this is not the case. The organelles are metabolically active and are carrying on their normal function. When and if the cell is going to divide, 2 duplication events occur during interphase. The DNA replicates, and the centrioles duplicate. Following DNA replication, each chromosome is duplicated and consists of sister chromatids. A nondividing cell has one pair of centrioles, but in a cell that is about to divide, this pair duplicates, and there are 2 pairs of centrioles outside the nucleus.

The cell cycle includes mitosis and interphase. During interphase, centriole duplication occurs and DNA replication causes each chromosome to have sister chromatids.

However, there is usually a limit to the number of times an animal cell enters the cell cycle and divides before degenerative changes lead to the death of the cell. Normal cells divide only about 50 times. Usually, they leave the

4.2 Critical Thinking

1. As cells grow, there is more cytoplasm per amount of cell membrane. Cell division restores the original cell membrane-to-cytoplasm ratio. Why is this useful to the cell?
2. What event during interphase provides chromosomes for daughter cells?
3. During the last growth portion of the cell cycle, cellular organelles increase in number. Why is this useful?
4. Specialized cells "break out" of the cell cycle and do not divide anymore. Give an example of a specialized cell you would not expect to divide again.

Figure 4.4

The cell cycle. Immature cells go through a cycle that consists of mitosis and interphase. During interphase, there is growth before and after DNA replication. During replication, DNA is copied. Eventually, some daughter cells "break out" of the cell cycle and become specialized cells performing a specific function.

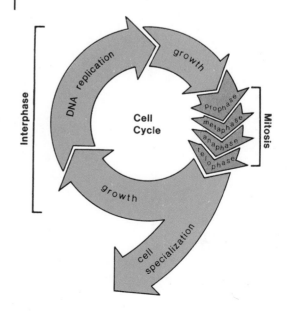

Figure 4.5

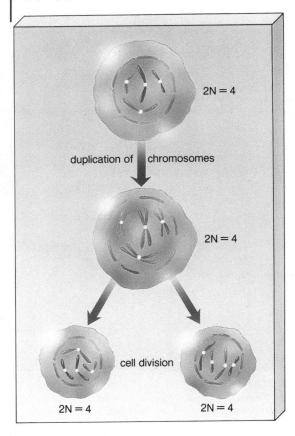

Mitosis overview. Following duplication, each chromosome in the mother cell contains 2 sister chromatids. During mitotic division, the sister chromatids separate so that daughter cells have the same number and kinds of chromosomes as the mother cell. (The blue chromosomes in the mother cell were inherited from one parent and the red chromosomes were inherited from the other parent.)

cell cycle and become specialized even before this number of divisions is reached. This can be contrasted to cancer cells, which can continue to divide indefinitely. Cancer cells have abnormal chromosomes and/or other irregularities of cell structure that can be associated with their ability to keep entering the cell cycle.

Control of the cell cycle is an active area of research today. Researchers have found that early stages release molecules that control later stages, and later stages release molecules that feed back to control early stages. A critical regulatory event occurs during the first stage—if the nutrient and control signals are positive, the cell always completes the rest of the cycle.

Overview of Animal Mitosis

Mitosis is cell division in which *the daughter cells retain the same number and kinds of chromosomes as the mother cell.*[1] Therefore, the mother cell and the daughter cells are genetically identical. The **mother cell** is the cell that divides, and the **daughter cells** are the resulting cells.

Figure 4.5 gives an overview of mitosis; for simplicity, only 4 chromosomes are depicted. Notice that each chromosome in the diagram goes through a cycle in which it is at first a singled chromosome, then a duplicated chromosome, and then a singled chromosome again:

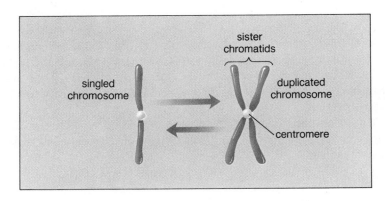

Just before cell division, the singled chromosomes duplicate as DNA replication occurs. Because of duplication, each chromosome in the mother cell contains 2 sister chromatids. During mitosis, sister chromatids separate and go to the newly forming cells, ensuring that each cell receives a copy of each kind of chromosome. This means that the chromosomes in newly formed daughter cells are singled chromosomes once more.

Stages of Mitosis

As an aid in describing the events of mitosis, the process is divided into 4 phases: prophase, metaphase, anaphase, and telophase (fig. 4.6). Although it is necessary to depict the stages of mitosis as if they could be separated, they are actually continuous and flow from one stage to another with no noticeable interruption.

Prophase

It is apparent during **prophase** that cell division is about to occur. There are now 2 pairs of centrioles outside the nucleus, and the pairs begin moving away from each other toward opposite ends of the nucleus. **Spindle fibers** appear between the separating centriole pairs. As the spindle appears, the nuclear envelope begins to fragment and the nucleolus begins to disappear.

[1]The term *mitosis* technically refers only to nuclear division but for convenience is used here to refer to division of the entire cell.

The Cell, the Smallest Unit of Life

Figure 4.6

Mitotic cell division. Mitosis has 4 stages, excluding interphase and daughter cells. Notice that these drawings are for mitosis in animal cells and that because centrioles duplicate in cells about to undergo mitosis, there are 2 pairs of centrioles in the late interphase cell at the start of the process.

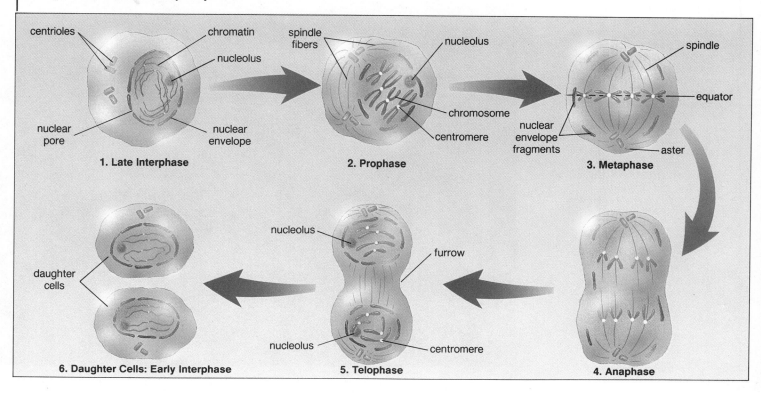

1. Late Interphase — centrioles, chromatin, nucleolus, nuclear envelope, nuclear pore

2. Prophase — spindle fibers, nucleolus, chromosome, centromere

3. Metaphase — spindle, equator, nuclear envelope fragments, aster

4. Anaphase

5. Telophase — nucleolus, furrow, centromere, nucleolus

6. Daughter Cells: Early Interphase — daughter cells

The chromosomes are now visible. Each is composed of sister chromatids held together at a centromere. As the chromosomes continue to shorten and to thicken, they move toward the equator of the spindle. Figure 4.7 is a micrograph of an animal cell at the time of late prophase. The chromosomes are randomly placed and have not aligned yet at the equator of the spindle.

The Function of Centrioles The entire spindle apparatus is shown in figure 4.8. It consists of asters, spindle fibers, and centrioles. Both the short **asters** radiating from the centrioles and the long spindle fibers are composed of microtubules. It is known that microtubules are capable of assembling and disassembling, which accounts for the appearance and the disappearance of asters and spindle fibers. It is possible that the centrioles are a part of organizing centers for spindle formation, but it also could be that their location at the poles simply ensures that each daughter cell receives a pair of centrioles.

Metaphase

During **metaphase,** the nuclear envelope is fragmented and the spindle occupies the region formerly occupied by the nucleus. Each chromosome is attached to the spindle and moves to the equator (center) of the spindle. Metaphase is characterized by a fully formed spindle, with the chromosomes, each composed of 2 chromatids, aligned at the equator (fig. 4.9). At the close of metaphase, the centromeres uniting the chromatids split.

Figure 4.7

Micrograph of late prophase. The chromosomes are moving toward the equator of the spindle. The equator is located midway between the asters found at the poles of the spindle.

cell membrane   chromosomes

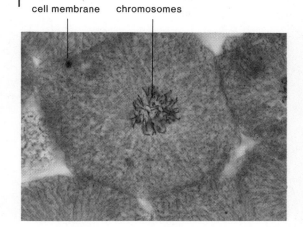

Figure 4.8

Spindle apparatus anatomy and function. a. *Artist's representation of the spindle apparatus from an animal cell. The poles (yellow) contain the centrioles within the asters. Polar fibers reach from the poles to the equator, where they overlap. Centromeric fibers are attached to the chromosomes. The chromosomes (blue) are moving toward the poles.* b. *Chromosome movement. Spindle fibers are composed of microtubules that can assemble and disassemble. It is believed that the polar fibers lengthen at the equator and that they slide past one another in the region of overlap. This pushes the chromosomes apart. The centromeric fibers disassemble at the poles, and this pulls the chromosomes toward the poles.*

a.

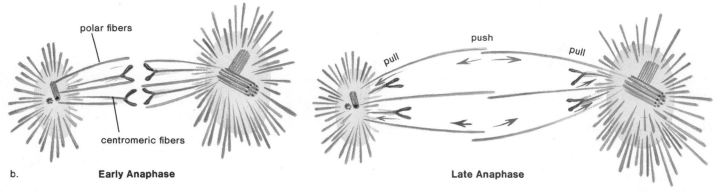

b.   **Early Anaphase**   **Late Anaphase**

Anaphase

During **anaphase,** the chromatids separate (fig. 4.10). *Once separated, the chromatids are called chromosomes.* The daughter chromosomes now move up (*ana* means up) to the opposite poles of the spindle. Separation of the sister chromatids ensures that each cell receives a copy of each type of chromosome and thereby has a full complement of genes.

The Movement of Chromosomes   The mechanism of chromosome movement during anaphase is the subject of much investigation and speculation. Two types of spindle fibers have been identified. The so-called polar fibers extend from the poles to the equator of the spindle; there they overlap one another (fig. 4.8). These fibers increase in length (by assembling new protein sub-

Figure 4.9

Micrograph of metaphase. The chromosomes are now lined up along the equator of the spindle.

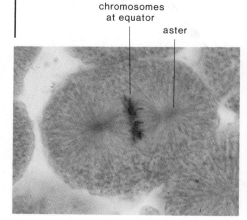

chromosomes
at equator

aster

Figure 4.10

Micrograph of anaphase. Separation of sister chromatids results in chromosomes that are "pushed and pulled" by spindle fibers to opposite poles of the spindle.

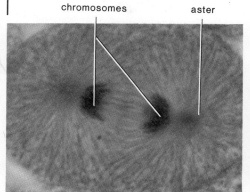

chromosomes

aster

Figure 4.11

Micrograph of late telophase. Furrowing has resulted in 2 daughter cells separated by membrane. Remnants of the spindle still are seen, but these will disappear as the daughter nuclei reform.

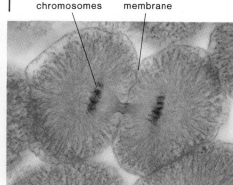

chromosomes

new
cell
membrane

Table 4.2 *Stages of Mitosis*

Stage	Events
Prophase	Replication has occurred and each chromosome is composed of 2 sister chromatids.
Metaphase	Duplicated chromosomes are at the equator (center) of the cell.
Anaphase	Sister chromatids separate, and each is now termed a chromosome.
Telophase	At each pole, there is a diploid number of chromosomes, the same number and kinds of chromosomes as the mother cell.

Figure 4.12

Cytokinesis in an animal cell. The single cell becomes 2 cells by a furrowing process that gradually separates the 2 daughter cells.

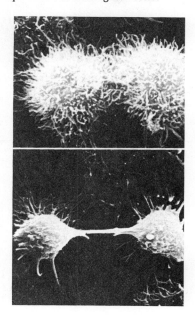

units), and then they slide past one another. This not only causes the spindle apparatus to increase in length, it also seems to provide a push that moves the chromosomes toward the poles.

The chromosomes themselves are not attached to the polar fibers. Each is attached to the second type of spindle fibers, the *centromeric fibers,* that extend from the region of the centromere to each of the poles. These fibers get shorter and shorter as the chromosomes move toward the poles, and eventually they disappear. Undoubtedly, disassembly is the mechanism by which these fibers shorten, providing a *pull* that moves the chromosomes toward the poles.

Telophase

Telophase begins when the chromosomes arrive at the poles. During telophase (fig. 4.11), the spindle disappears, possibly due to disassembly of the microtubules making up the spindle fibers. As the nuclear envelopes form and the nucleoli appear in each cell, the chromosomes untwist, and eventually only indistinct chromatin is visible again. Following nuclear division, cytoplasmic division, called **cytokinesis,** usually occurs. In animal cells, **furrowing,** or an indentation of the cell membrane between the 2 newly forming cells, divides the cytoplasm (fig. 4.12). Furrowing is complete when each cell is enclosed in a membrane. Microfilaments are believed to take part in the furrowing process since they are always in the vicinity.

Table 4.2 is a summary of the stages of mitosis.

Figure 4.13

Plant cell mitosis. Note the absence of centrioles and asters and the presence of cell wall. In telophase, a cell plate develops in between the 2 daughter cells. The cell plate marks the boundary of the new daughter cells, where new cell membrane and new cell wall will form for each cell. a. Longitudinal section of an onion root tip. Many of the cells are undergoing cell division because this portion of the onion contributes to root growth. b. Diagrammatic representation of the stages of mitosis.

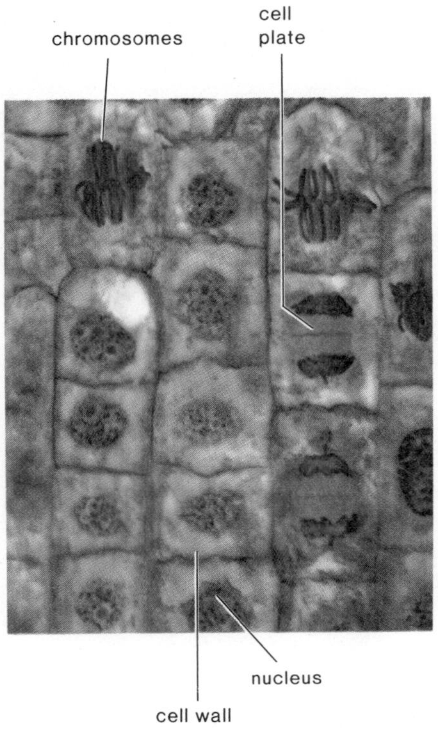

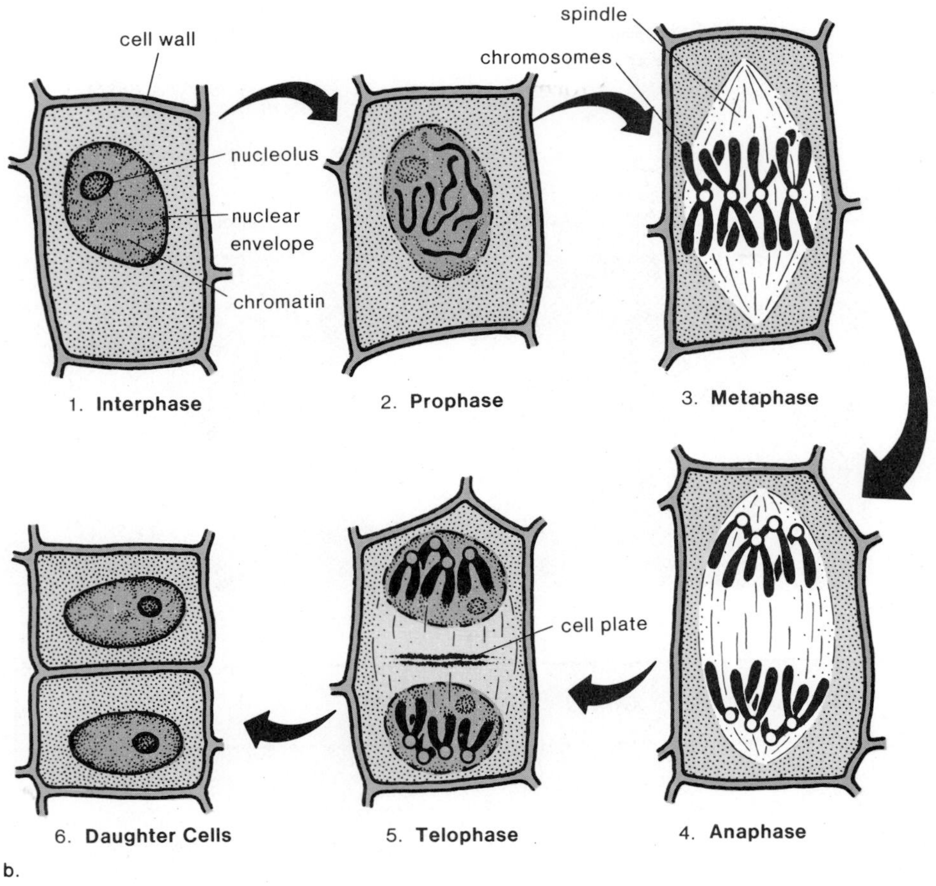

a.

b.

1. **Interphase** 2. **Prophase** 3. **Metaphase** 4. **Anaphase** 5. **Telophase** 6. **Daughter Cells**

Mitosis in Plants

There are 2 main differences between plant cell mitosis and animal cell mitosis. First of all, in higher plant cells, centrioles and asters are not seen during mitosis. However, spindle fibers do appear (fig. 4.13). It is interesting to note that animal cells deprived of centrioles will also form a spindle. It may be, therefore, that centrioles do *not* contribute to spindle formation.

The second difference pertains to cytokinesis. The rigid cell wall that surrounds plant cells does not permit division of the cytoplasm by means of furrowing. Instead, vesicles largely derived from the Golgi apparatus travel down the polar spindle fibers to the region of the equator. These vesicles fuse to form a membranous structure, the early **cell plate,** that spreads to the sides and marks the boundary of the 2 daughter cells. A daughter cell is complete once it forms a new cell membrane and a new cell wall next to the cell plate. No furrowing is observed in plant cells.

Mitosis ensures that each cell will have the same number of chromosomes. In a dividing cell, each chromosome consists of 2 chromatids. When the sister chromatids separate during anaphase, each newly forming cell receives the same number and kinds of chromosomes as the original cell.

The Cell, the Smallest Unit of Life

Meiosis

Meiosis is an important part of sexual reproduction in both animals and plants. In animals, it occurs during the production of the egg and the sperm. In plants, it occurs when spores are formed.

Meiosis keeps the chromosome number constant generation after generation; it also ensures that the next generation has a different genetic makeup than the previous generation. For example, in animals, the egg carries one-half the chromosomes and genes from the female parent and the sperm carries one-half the chromosomes and genes from the male parent. When the sperm fertilizes the egg, the zygote has the diploid number of chromosomes and a different combination of genes than either parent.

Overview of Animal Meiosis

Meiosis, which requires 2 cell divisions, results in *4 daughter cells, each having one of each kind of chromosome and therefore half the number of chromosomes as the mother cell.*[2] The mother cell has the diploid number of chromosomes, while the daughter cells have the haploid number of chromosomes. Therefore, meiosis often is called reduction division.

Recall that when a cell is 2N or diploid, the chromosomes occur in pairs. For example, the 46 chromosomes of humans occur in 23 pairs of chromosomes. These pairs are called **homologous chromosomes.** During meiosis, homologous chromosomes separate. Because of this, each daughter cell receives one of each kind of chromosome and half the total number of chromosomes. Each human sperm and egg has only 23 chromosomes.

Figure 4.14 presents an overview of meiosis, indicating the 2 cell divisions, **meiosis I** and **meiosis II.** Prior to meiosis I, duplication has occurred and the chromosomes consist of sister chromatids. Therefore, each chromosome can be called a **dyad.** During meiosis I, the homologous chromosomes come together and line up side by side due to a means of attraction still unknown. This so-called **synapsis** results in **tetrads,** associations of 4 chromatids that stay in close proximity during the first 2 phases of meiosis I. During synapsis, nonsister chromatids exchange genetic material as illustrated in figure 4.15. The exchange of genetic material between chromatids is called **crossing-over** and is an additional means by which new combinations of genes occur so that an offspring has a different genetic makeup than either of its parents.

Following synapsis during meiosis I, the homologous chromosomes separate. This separation means that one chromosome of every homologous pair reaches each gamete. There are no restrictions on the separation process; either chromosome of a homologous pair can occur in a gamete with either chromosome of any other pair. Therefore, any possible combination of chromosomes and genes can occur within the gametes.

Notice that at the completion of meiosis I (fig. 4.14), the chromosomes are still dyads. No replication of DNA nor duplication of chromosomes occurs between meiosis I and meiosis II. During meiosis II, the sister chromatids separate, resulting in 4 daughter cells. Each daughter cell has the haploid number of chromosomes. These chromosomes are singled chromosomes. You can count the number of centromeres to verify that the mother cell has the diploid number of chromosomes and each of the daughter cells has the haploid number.

Stages of Meiosis

Meiosis is divided into meiosis I and meiosis II. The same 4 stages seen in mitosis—prophase, metaphase, anaphase, and telophase—occur during both of these divisions. The complete designation indicates the stage and the division; for example, prophase I is the first stage of meiosis I.

[2]The term *meiosis* technically refers only to nuclear division but for convenience is used here to refer to the division of the entire cell.

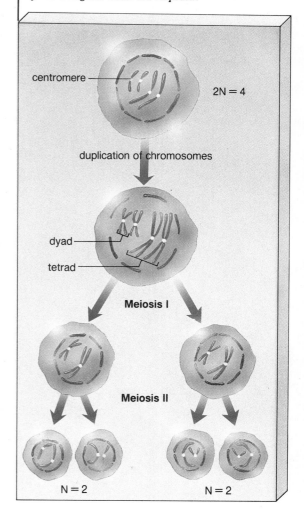

Figure 4.14
Overview of meiosis. Following duplication of chromosomes, the mother cell undergoes 2 divisions, meiosis I and meiosis II. During meiosis I, homologous chromosomes separate, and during meiosis II, chromatids separate. The final daughter cells are haploid.

centromere

2N = 4

duplication of chromosomes

dyad

tetrad

Meiosis I

Meiosis II

N = 2   N = 2

4.3 Critical Thinking

1. The drug colchicine prevents cell division from finishing (fig. 4.2). What part of a dividing cell do you suppose it disrupts?
2. Asexual reproduction ordinarily does not produce genetic variation. Why not?
3. Sexual reproduction does produce genetic variation. Why?
4. Would you expect sexual reproduction to aid the evolutionary process? Why?

Figure 4.15
Crossing-over. When homologous chromosomes are in synapsis, the nonsister chromatids exchange genetic material. Following crossing-over, there is a different combination of genes on the chromatids.

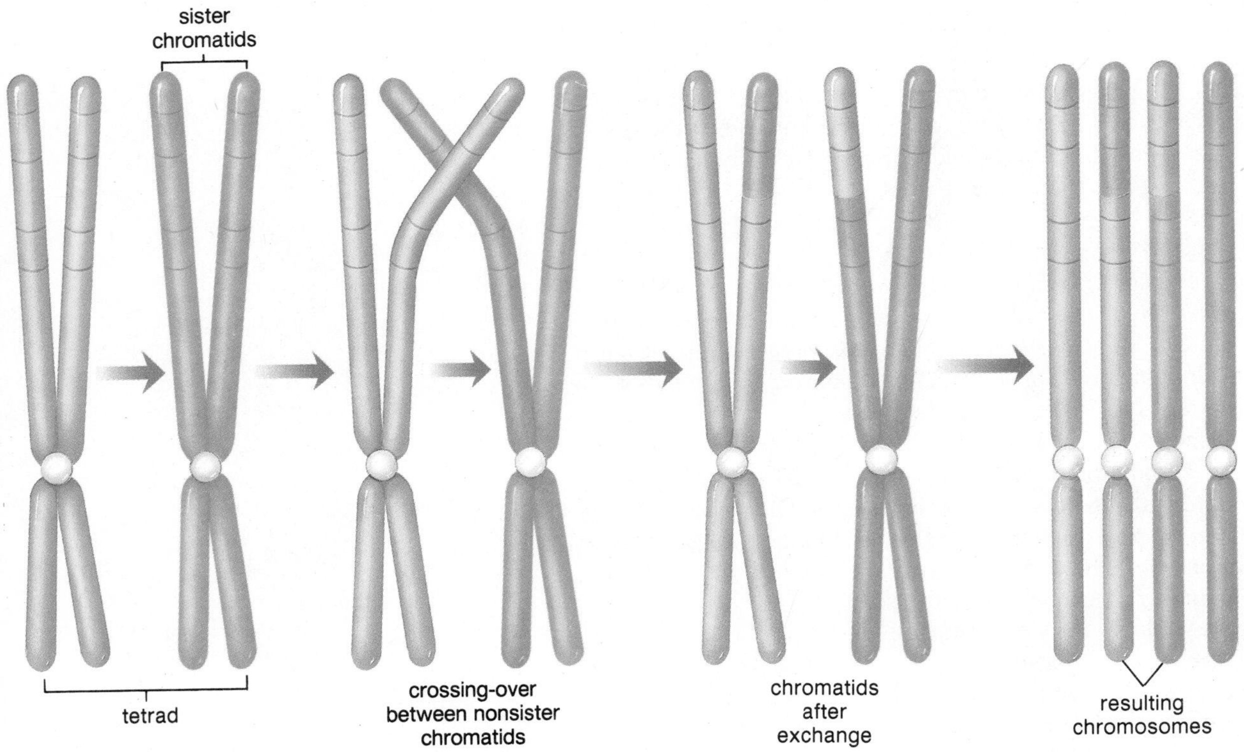

sister
chromatids

tetrad

crossing-over
between nonsister
chromatids

chromatids
after
exchange

resulting
chromosomes

The First Division

The stages of meiosis I as they appear in an animal cell are diagrammed in figure 4.16*a*. During *prophase I,* the spindle appears, while the nuclear envelope fragments and the nucleolus disappears. Homologous pairs of chromosomes undergo synapsis, forming tetrads. Crossing-over occurs now, but for simplicity, this event has been omitted from figure 4.16*a*. In *metaphase I,* tetrads line up at the equator of the spindle. During *anaphase I,* homologous chromosomes separate because the centromeres of each dyad are "pushed and pulled" to opposite poles. At the start of *telophase I,* each pole has received the haploid number of chromosomes. The nuclear envelope now reforms, and the nucleolus reappears as the spindle disappears. In certain species, the cell membrane furrows to give 2 cells, and in others, the second division begins without benefit of complete furrowing. Regardless, each daughter cell is now haploid and contains only one of each of the pairs of homologous chromosomes. The chromosomes are still dyads, and no duplication of chromosomes occurs during interphase.

The Second Division

The stages of meiosis II for an animal cell are diagrammed in figure 4.16*b*. At the beginning of *prophase II,* a spindle appears, while the nuclear envelope fragments and the nucleolus disappears. Dyads (one dyad for each pair of homologous chromosomes) are present, and each attaches to the spindle independently. During *metaphase II,* the chromosomes are lined up at the

The Cell, the Smallest Unit of Life

Figure 4.16

Meiosis. a. *Stages of meiosis I. During meiosis I, homologous chromosomes separate so that each daughter cell has only one chromosome from each original homologous pair. For simplicity, the results of crossing-over have not been depicted.* b. *Stages of meiosis II. During meiosis II, chromatids separate, and each daughter cell has the haploid number of chromosomes in single copy.*

a. **Meiosis I**

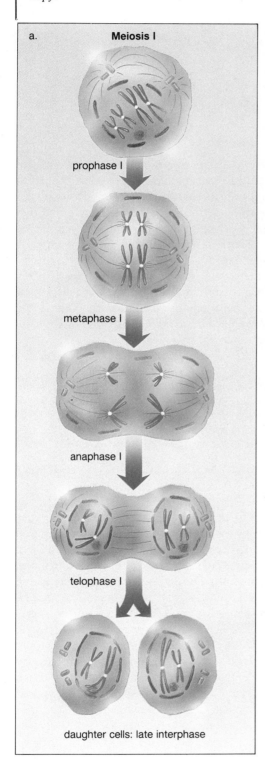

prophase I

metaphase I

anaphase I

telophase I

daughter cells: late interphase

b. **Meiosis II**

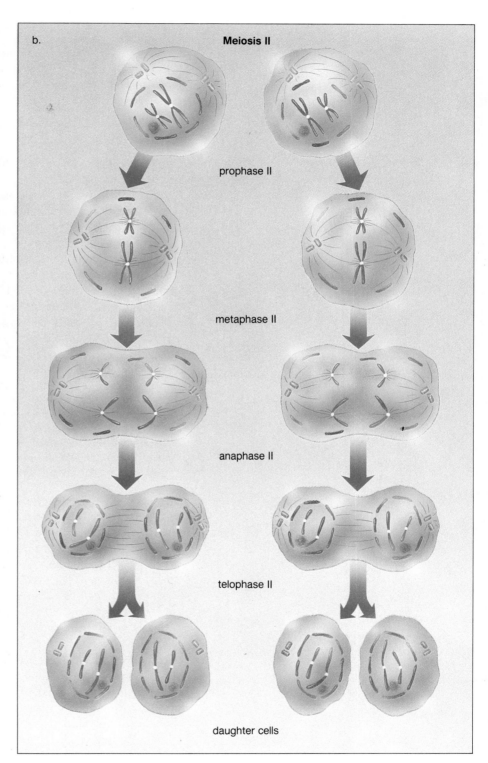

prophase II

metaphase II

anaphase II

telophase II

daughter cells

Table 4.3 Stages of Meiosis I and Meiosis II

Stages	Meiosis I	Meiosis II
Prophase	Homologous chromosomes, each composed of chromatids, synapse, forming tetrads.	Dyads are present; one dyad from each pair of homologous chromosomes.
Metaphase	Tetrads are at the equator.	Dyads are at the equator.
Anaphase	Homologous chromosomes separate, pulled to opposite poles by their centromere.	Sister chromatids separate and each is now termed a chromosome.
Telophase	At each pole there is one dyad from each pair of homologous chromosomes.	At each pole there is the haploid number and one of each kind of chromosome.

equator. At the close of metaphase, the centromeres split. During *anaphase II*, the sister chromatids of each dyad separate and move toward the poles. Each pole receives the same number of chromosomes. In *telophase II*, the spindle disappears as the nuclear envelopes reform. The cell membrane furrows to give 2 complete cells, each of which has the haploid or N number of chromosomes in single copy. Since each cell from meiosis I undergoes meiosis II, there are 4 daughter cells altogether.

Table 4.3 is a summary of the stages of meiosis I and meiosis II. This summary is appropriate for both plants and animals. The preceding discussion refers to animal cell meiosis because centrioles are present in the figures, the cells have no cell walls, and furrowing occurs to divide the cells. In animals, meiosis is involved in either spermatogenesis or oogenesis.

Meiosis involves 2 cell divisions. During meiosis I, homologous pairs of chromosomes come to lie side by side during synapsis. The nonsister chromatids of the resulting tetrad exchange chromosome pieces; this is called crossing-over. When the homologous pairs of chromosomes separate during meiosis I, each daughter cell receives one chromosome from each homologous pair. Separation of sister chromatids during meiosis II then produces a total of 4 daughter cells, each with the haploid number of chromosomes.

Spermatogenesis and Oogenesis

Spermatogenesis and oogenesis occur in the sex organs, the testes in males and the ovaries in females. During **spermatogenesis,** sperm are produced, and during **oogenesis,** eggs are produced. The gametes appear differently in the 2 sexes (fig. 4.17), and meiosis is different, too. The process of meiosis in males always results in 4 cells that become sperm. During oogenesis, the first meiotic division produces 2 cells, but one cell is much larger than the other. The smaller, nonfunctional cell is called a **polar body.** The second division also results in 2 cells, and again, the smaller cell is a nonfunctional polar body. Therefore, meiosis in females produces only one functional egg that contains a rich supply of cytoplasmic components. The polar bodies are a way to discard unnecessary chromosomes while retaining much of the cytoplasm that will serve as a source of nutrients for the developing embryo. Figure 4.17 shows how the sperm and the egg are adapted to their function. The sperm is a tiny flagellated cell that contributes only chromosomes to the new individual, while the large egg contributes most of the cytoplasm.

The Cell, the Smallest Unit of Life

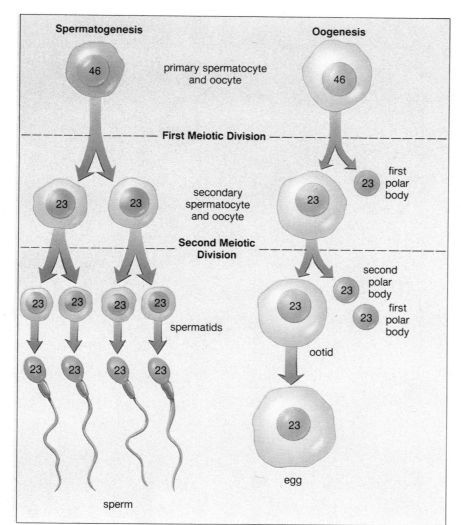

Figure 4.17
Spermatogenesis and oogenesis. Spermatogenesis produces 4 viable sperm, whereas oogenesis produces one egg and at least 2 polar bodies. In humans, both sperm and egg have 23 chromosomes each; therefore, following fertilization, the zygote has 46 chromosomes.

There is another difference between spermatogenesis and oogenesis. Spermatogenesis, once started, continues to completion, and mature sperm result. In contrast, oogenesis does not necessarily go to completion. Only if a sperm fertilizes the maturing egg does it undergo meiosis II; otherwise, it simply disintegrates. Regardless of this complication, however, both the sperm and the egg contribute the haploid number of chromosomes to the new individual. In humans, each contributes 23 chromosomes.

Meiosis in Plants

While the events of meiosis are essentially the same in plants as in animals, the cells in plants are called spores, not gametes. When spores germinate, a haploid generation occurs in the plant life cycle (fig. 8.13).

Comparison of Mitosis and Meiosis

Figure 4.18 compares mitosis to meiosis. The differences between these cellular divisions can be categorized according to occurrence and process.

Occurrence

Meiosis occurs only in cells that eventually give rise to the gametes. Mitosis occurs in all types of cells in the body.

Figure 4.18

Mitosis compared to meiosis.

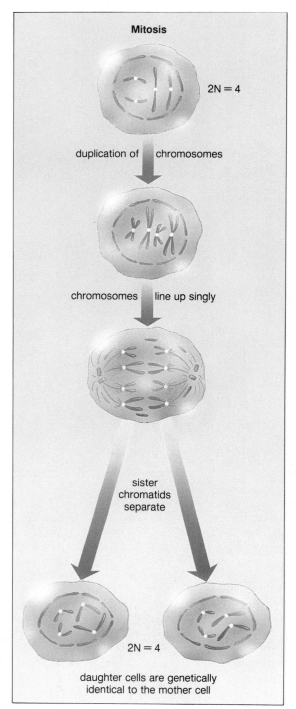

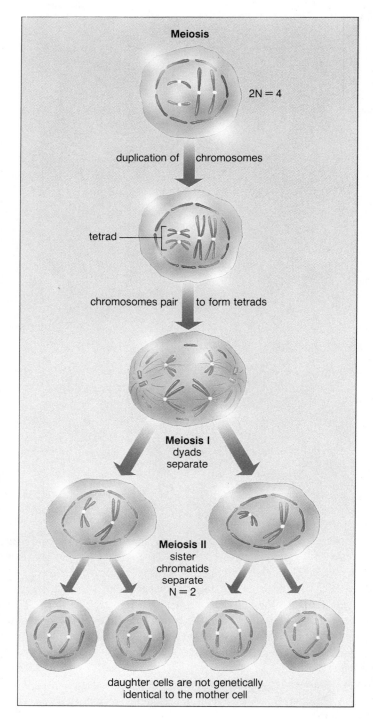

Process

Mitosis can be compared to meiosis I in this manner:

Mitosis	Meiosis I
Prophase	
No pairing of chromosomes	Pairing of homologous chromosomes
Metaphase	
Duplicated chromosomes (dyads) at equator	Tetrads at equator
Anaphase	
Sister chromatids separate	Homologous chromosomes separate
Telophase	
Daughter cells are diploid	Daughter cells are haploid

Mitosis can be compared to meiosis II in this manner:

Mitosis	Meiosis II
Metaphase	
Diploid number of dyads at equator	Haploid number of dyads at equator
Anaphase	
Sister chromatids separate	Sister chromatids separate
Telophase	
2 daughter cells	4 daughter cells

Daughter Cells

The genetic consequence of mitosis and meiosis are quite different.

1. Mitosis results in 2 daughter cells; meiosis results in 4 daughter cells.
2. The 2 daughter cells from mitosis are diploid; the 4 daughter cells from meiosis are haploid.
3. The mitotic daughter cells are genetically identical to each other and to the mother cell. The daughter cells from meiosis are not genetically identical to the mother cell. (The occurrence of crossing-over ensures that no daughter cell is identical to another also.)

Summary

The life cycle of higher organisms requires 2 types of cell division, mitosis and meiosis. Mitosis is required for growth and repair of body parts in both animals and plants. It ensures that all cells in the body have the diploid number and same kinds of chromosomes. Cells that are undergoing the cell cycle spend most of their time in interphase. During interphase, centrioles and chromosomes duplicate before mitosis begins. Following duplication, the chromosomes are made up of 2 chromatids held together at a centromere. Mitosis has 4 stages: prophase, (chromosomes move toward equator of spindle), metaphase (chromosomes are at equator), anaphase (sister chromatids separate), and telophase (nuclei and nucleolus reappear). Separation of sister chromatids ensures that each newly forming cell receives the same number and kinds of chromosomes as the original cell. The cytoplasm is partitioned by furrowing in animals and by cell-plate formation in plants.

Meiosis involves 2 cell divisions. During meiosis I, the homologous pairs of chromosomes (following crossing-over between nonsister chromatids) separate, and during meiosis II the sister chromatids separate. The result of these divisions is 4 cells with the haploid number of chromosomes. Meiosis is a part of gamete formation in animals and spore formation in plants.

Mitosis can be contrasted with meiosis as depicted in figure 4.18.

Study Questions

1. Describe the normal karyotype of a human being. What is the difference between a male karyotype and a female karyotype? (p. 80)
2. Relate the terms *diploid (2N)* and *haploid (N)* to mitosis in somatic cells and to meiosis in sex organs. (p. 81)
3. Give several instances of when mitosis occurs in humans. (p. 83)
4. Explain the makeup of a chromosome just prior to cell division. (p. 84)
5. Describe the stages of animal mitosis, including in your description the terms *centriole, nucleolus, spindle,* and *furrowing.* (pp. 84–87)
6. Name 2 differences between plant cell mitosis and animal cell mitosis. (p. 88)
7. What is the importance of meiosis in the life cycle of any organism? (p. 89)
8. Describe the stages of meiosis I, including in your description the term *tetrad.* (p. 90)
9. Describe the stages of meiosis II, including in your description the term *dyad.* (p. 90)
10. Explain the fact that oogenesis produces one mature egg, but spermatogenesis results in 4 sperm. (p. 93)
11. Compare mitosis to both meiosis I and meiosis II. (p. 93)

Objective Questions

1. During interphase, the chromosomes are not visible because they are extended into fine threads called _____ .

2. If an organism has 12 chromosomes, it has _____ homologous pairs.

3. Just prior to division, every chromosome is composed of 2 identical parts called _____ .

4. If the mother cell has 24 chromosomes, following mitosis the daughter cells have _____ chromosomes.

5. As the organelles called _____ separate and move to the poles, the spindle fibers appear.

6. _____ is the stage of mitosis during which the chromatids separate and become chromosomes.

7. Cytokinesis in an animal cell occurs by a _____ process, while in a plant cell it involves the formation of a _____ .

8. Whereas mitosis results in 2 daughter cells, meiosis produces _____ daughter cells.

9. During anaphase I of meiosis, the _____ separate. This means that eventually the gametes will have the haploid number of chromosomes.

10. Meiosis ensures that the zygote will have a _____ combination of genes than either parent.

Label this Diagram.

See figure 4.6 (p. 85) in text.

_____

Answers to Objective Questions

1. chromatin 2. 6 3. chromatids 4. 24 5. centrioles 6. Anaphase 7. furrowing, cell plate 8. 4 9. homologous chromosome pairs 10. different.

Selected Key Terms

aster (as'ter) short ray of microtubule that appears at the end of the spindle apparatus in animal cells during cell division. *85*

autosome (aw'to-sōm) chromosome other than sex chromosome. *80*

centromere (sen'tro-mēr) a region of attachment of a chromosome to spindle fibers that is generally seen as a constricted area. *80*

chromatid (kro'mah-tid) one of the 2 identical parts of a chromosome following replication of DNA. *80*

crossing-over (kros'ing o'ver) the exchange of corresponding segments of genetic material between nonsister chromatids of homologous chromosomes during synapsis of meiosis I. *89*

cytokinesis (si''to-ki-ne'sis) division of the cytoplasm of a cell. *87*

diploid (dip'loid) the 2N number of chromosomes; twice the number of chromosomes found in gametes. *81*

gamete (gam'et) one of 2 reproductive cells that join in fertilization to form a zygote; most often an egg or a sperm. *81*

haploid (hap'loid) the N number of chromosomes; half the diploid number; the number characteristic of gametes that contain only one set of chromosomes. *81*

homologous chromosome (ho-mol'o-gus kro'mo-sōm) similarly constructed; homologous chromosomes have the same shape and contain genes for the same traits. *89*

karyotype (kar'e-o-tīp) the arrangement of all the chromosomes within a cell by pairs in a fixed order. *80*

meiosis (mi-o'sis) a type of cell division that occurs during the production of gametes or spores by means of which the 4 daughter cells receive half the haploid number of chromosomes. *81*

mitosis (mi-to'sis) type of cell division in which daughter cells receive the exact chromosome and genetic makeup of the mother cell; occurs during growth and repair. *83*

oogenesis (o''o-jen'ĕ-sis) production of an egg in females by the process of meiosis and maturation. *92*

sex chromosome (seks kro'mo-sōm) chromosome responsible for the development of characteristics associated with maleness or femaleness; an X or Y chromosome. *80*

spermatogenesis (sper''mah-to-jen'ĕ-sis) production of sperm in males by the process of meiosis and maturation. *92*

spindle fiber (spin'd'l fi'ber) microtubule bundle in eukaryotic cells that are involved in the movement of chromosomes during mitosis and meiosis. *84*

synapsis (si-nap'sis) the attracting and pairing of homologous chromosomes during prophase I of meiosis. *89*

tetrad (tet'rad) a set of 4 chromatids resulting from the pairing of homologous chromosomes during prophase I of meiosis. *89*

zygote (zi'gōt) diploid cell formed by the union of 2 gametes; the product of fertilization. *81*

Chapter 5

Cellular Metabolism

Chapter Concepts

1 A metabolic pathway is a series of reactions controlled by enzymes.

2 Enzymes are protein molecules that speed up chemical reactions and only function properly when they retain their normal shape.

3 Cells require the energy molecule ATP to drive forward synthetic reactions and for various other functions.

4 Cellular respiration, a metabolic pathway involving mitochondria, provides the necessary energy to form ATP molecules.

5 During cellular respiration, glucose products are oxidized and concurrently ATP is formed from ADP + Ⓟ molecules.

Chapter Outline

Metabolism
    Metabolic Pathways
    Enzymes
    Conditions Affecting Enzymatic
        Reactions
    Coenzymes
    Energy
Aerobic Cellular Respiration
    (Simplified)
    Subpathways
    Mitochondria
    Metabolites
Aerobic Cellular Respiration (Detailed)
    Glycolysis
    The Transition Reaction
    The Krebs Cycle
    The Respiratory Chain
Fermentation
    The Results of Fermentation
    The Usefulness of Fermentation

Figure 5.1
Multicellular organisms contain millions of cells that are actively carrying out metabolic reactions. Among these, cellular respiration provides the energy needed to jog or to perform any manner of physical activities.

Metabolism

Cells are not static; they are dynamic. Drawings of cells and even microscopic slides of cells give us the impression that cells are inactive, when actually, cells are constantly active. Pinocytotic and phagocytotic vesicles are being formed constantly, organelles are moving about, and division may be taking place. A vital part of this activity is constantly occurring chemical reactions, which collectively are termed the metabolism of the cell. **Metabolism** is the sum of all chemical activities occurring inside a living cell.

Metabolic Pathways

Reactions do not occur haphazardly in cells; they are usually a part of a metabolic pathway. *Metabolic pathways* begin with a particular reactant and terminate with an end product. While it is possible to write an overall equation for a pathway as if the beginning *reactant* went to the end *product* in one step, actually there are many minute steps in between. In the pathway, one reaction leads to the next reaction, which leads to the next reaction, and so forth in an organized, highly structured manner. This arrangement makes it possible for one pathway to lead to several others, especially since various pathways have several molecules in common. Also, metabolic energy is captured and utilized more easily if it is released in small increments rather than all at once.

The Cell, the Smallest Unit of Life

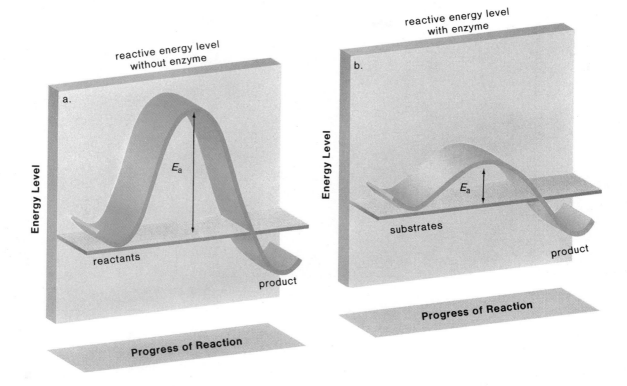

Figure 5.2
Energy of activation without and with enzyme. Enzymes speed up the rate of chemical reactions because they lower the required energy of activation. a. Required energy of activation (E_a) *is much higher when an enzyme is not available.* b. *Required energy of activation* (E_a) *is much lower when an enzyme is available.*

Metabolic pathways can be represented by the following diagram, as long as we realize that side branches can occur at any juncture:

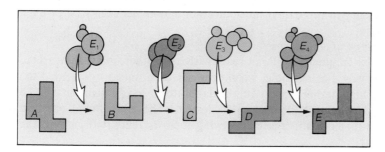

In this pathway, the letters are products of the previous reaction and reactants of the next reaction. The numbers in the pathway identify different enzymes. The reactant(s) in an enzymatic reaction is (are) called the **substrate(s)** for that enzyme. *A* is the beginning substrate, and *E* is the end product of the pathway.

A metabolic pathway is a series of reactions that proceed in an orderly, step-by-step manner. Each reactant is a substrate for a particular enzyme.

Enzymes

Every reaction in a cell requires a specific enzyme. **Enzymes** are organic catalysts, globular protein molecules that speed up chemical reactions (fig. 5.2). An enzyme is very specific in its action and can speed up only one particular reaction or one type of reaction. Table 5.1 indicates that the name of an enzyme

Cellular Metabolism   **99**

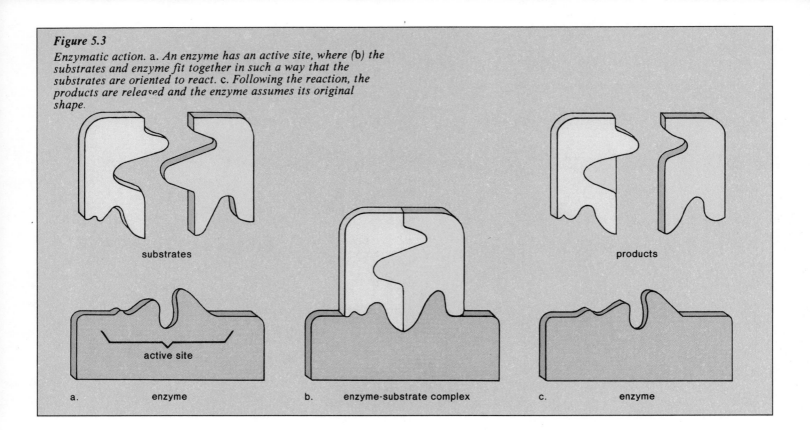

Figure 5.3

Enzymatic action. a. An enzyme has an active site, where (b) the substrates and enzyme fit together in such a way that the substrates are oriented to react. c. Following the reaction, the products are released and the enzyme assumes its original shape.

substrates

products

active site

a.   enzyme   b.   enzyme-substrate complex   c.   enzyme

Table 5.1 *Enzymes Named for Their Substrate*

Substrate	Enzyme
Lipid	Lipase
Urea	Urease
Maltose	Maltase
Ribonucleic acid	Ribonuclease
Lactose	Lactase

often is formed by adding *ase* to the name of its substrate. Some enzymes are named for the action they perform; for example, a dehydrogenase is an enzyme that removes hydrogen atoms from its substrate.

No reaction can occur in a cell unless the reaction's own enzyme is present and active. For example, if E_2 in the preceding diagram is missing or not functioning, the pathway will shut down and stop at *B*. Since enzymes are so necessary in cells, their mechanism of action has been studied extensively.

The Energy of Activation

Molecules frequently will not react with each other unless they are activated in some way. For example, wood does not burn unless it is heated to a moderately high temperature. In the laboratory, too, activation very often is achieved by heating the reaction flask so that the number of effective collisions between molecules increases, allowing them to react with one another more frequently.

The energy that must be supplied for molecules to react with one another is called the **energy of activation.** Figure 5.2 compares the reactive energy level required in the absence of an enzyme with the reactive energy level required if an enzyme is present. The difference between the nonreactive energy level and the reactive energy level is the necessary energy of activation (E_a). For example, the hydrolysis of casein (the protein found in milk) uses an energy of activation of 20,600 Kcal/mole[1] in the absence of an enzyme, but only 12,000 Kcal/mole are used if the appropriate enzyme is present. This example tells us that less energy and therefore less heat is used to bring about a reaction when an enzyme is present. Enzymes lower the energy of activation by binding with their substrate in such a way that a reaction can occur.

[1]Using Kcal (Kcalorie) is a common way to measure heat, and a mole is the molecular weight of a substance expressed in grams.

The Enzyme-Substrate Complex

The equation that is pictorially shown in figure 5.3 often is used to indicate that an enzyme forms a complex with its substrate:

$$E + S \rightarrow ES \rightarrow E + P$$

In this equation, E = enzyme, S = substrate, ES = enzyme-substrate complex, and P = product.

Notice in figure 5.3 that an enzyme is unaltered by the reaction. Only a small amount of enzyme actually is needed in a cell because enzymes are used over and over again. When the enzyme-substrate complex forms, the substrates seemingly are attracted to the enzyme because their shapes fit together like a *key fits a lock*. However, it is now thought that the enzyme very well may undergo a slight change in shape in order to more perfectly accommodate the substrates. This is called the induced-fit model because as binding occurs, the enzyme is induced (undergoes a slight alteration) to achieve maximum fit. The region of substrate attachment is called the **active site,** and it is here that the reaction takes place (fig. 5.4). After the reaction is completed, the product is released, and the active site returns to its original state.

The Product

Only certain products can be produced by any particular reactant, and an enzyme cannot bring about any other products. However, the presence of an active enzyme can determine whether or not a reaction takes place. For example, if substance A can react to form either substance B or substance C, then the enzyme that is active—enzyme 1 or enzyme 2—determines which product (substance D or substance F) is produced:

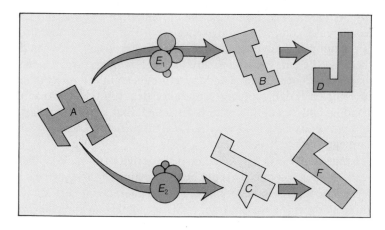

Enzymes are proteins that speed up chemical reactions by lowering the energy of activation. They do this by forming an enzyme-substrate complex.

Conditions Affecting Enzymatic Reactions

Enzymes normally allow reactions to proceed quite rapidly. For example, the breakdown of hydrogen peroxide into water (H_2O) and oxygen (O_2) can occur 600,000 times a second when the enzyme catalase is present. How quickly an enzyme works, however, is affected by certain conditions.

Substrate Concentration

Generally, enzyme activity increases as substrate concentration increases because there are more collisions between substrate molecules and the enzyme.

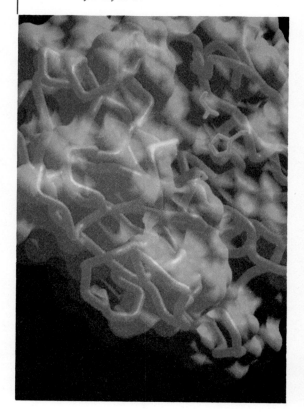

Figure 5.4
Computer-generated model of the functional backbone of the enzyme trypsin. Trypsin breaks peptide bonds, and it has been found that the 3 R groups shown in green are in the active site and interact with the members of the peptide bond in order to hydrolyze it.

Figure 5.5

Rate of an enzymatic reaction as a function of temperature and pH. a. At first, as with most chemical reactions, the rate of enzymatic reaction doubles with every 10° C rise in temperature. In this graph, which is typical of a mammalian enzyme, the rate is maximum at about 40° C and then it decreases until it stops

altogether, indicating that the enzyme now is denatured. b. Pepsin, an enzyme found in the stomach, acts best at a pH of about 2, while trypsin, an enzyme found in the small intestine, prefers a pH of about 8. Other pHs do not maintain properly the shape that enables these proteins to bind with their substrates.

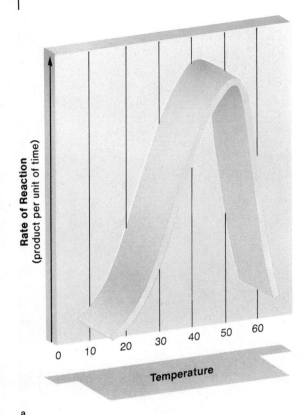

a.

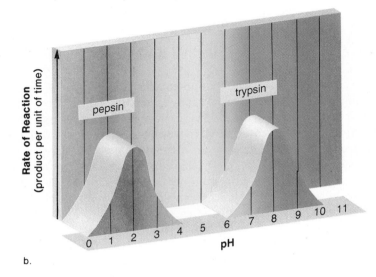

b.

As more substrate molecules fill active sites, more product results per unit time. But when the enzyme's active sites are filled almost continuously with substrate, the enzyme's rate of activity cannot increase anymore. Maximum velocity has been reached.

Temperature and pH

A higher temperature generally results in an increase in enzyme activity (fig. 5.5a). As the temperature rises, the movement of enzyme and substrate molecules increases, and more active sites are filled. If the temperature rises beyond a certain point, however, the reaction time levels off and then declines rapidly because the enzyme is denatured at high temperatures. A denatured protein has lost its normal shape and therefore its ability to form an enzyme-substrate complex. High temperatures cause denaturation because the hydrogen bonding between amino acids is disrupted; that is, the secondary structure of the protein is affected (fig. 1.20). Denaturation is observed when the white of an egg (albumin) is cooked.

A change in pH also can affect enzyme activity (fig. 5.5b). There is an optimal pH for each enzyme. This pH helps to maintain the enzyme's tertiary structure because it helps to maintain the normal interactions between R groups of the amino acids within the enzyme. A change in pH can alter the ionization of R groups and disrupt the normal interactions, and a change in shape, that is, denaturation, eventually occurs. Without its normal shape, the enzyme is unable to combine efficiently with its substrate.

Inhibition

In *competitive inhibition,* another molecule is so close in shape to an enzyme's substrate that this molecule can compete with the true substrate for the active site of the enzyme. Such a molecule is designated as *I* for inhibitor in these reactions:

irreversible: $I + E \longrightarrow EI \longrightarrow$ no further reaction
reversible: $\quad I + E \longrightarrow EI \longrightarrow E + I$

5.1 Critical Thinking

1. Pepsin is an enzyme that breaks down protein. A student has a test tube that contains pepsin, egg white, and water. What conditions would you recommend to ensure digestion of the egg white?
2. If all the conditions are perfect, how could you increase the yield (i.e., amount of product—amino acids—per unit of time)?
3. The instructor adds an inhibitor to the test tube. How could the student tell if reversible or irreversible inhibition is taking place?

The Cell, the Smallest Unit of Life

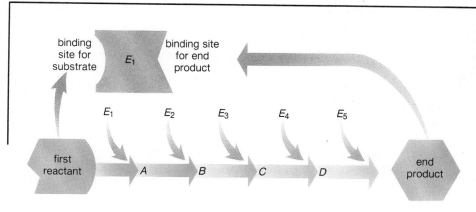

Figure 5.6
Noncompetitive feedback inhibition. The end product of this metabolic pathway can bind to the enzyme E_1, and when it does, the pathway is shut down because the binding of the end product changes the enzyme's shape so that the substrate cannot bind to the active site. In this way, noncompetitive feedback inhibition helps to regulate the output of metabolic pathways.

Any molecule that binds with an enzyme other than its substrate is an inhibitor of the reaction because only the binding of the substrate allows a product to be produced. The first reaction given represents irreversible inhibition, and the second is reversible inhibition. Irreversible inhibition of an enzyme is not common and usually is caused by a poison. For example, penicillin causes the death of bacteria due to the irreversible inhibition of an enzyme needed to form the bacterial cell wall. In humans, hydrogen cyanide is an inhibitor of a very important enzyme (cytochrome oxidase) present in all cells, and this accounts for its lethal effect.

Reversible inhibition is common in cells, and an enzyme can be inhibited competitively by its own product if it closely resembles the enzyme's substrate. One interesting example of reversible inhibition, though, concerns 2 foreign substances in cells. The enzyme alcohol dehydrogenase combines with either alcohol or ethylene glycol (automobile antifreeze). The breakdown product of the latter damages the kidneys so the medical remedy for accidental ingestion of antifreeze is administration of alcohol to the point of intoxication. Most of the ethylene glycol then is excreted harmlessly.

One type of reversible inhibition is called *noncompetitive inhibition* because a molecule binds to an enzyme at a site other than the active site. The binding of the molecule causes the enzyme to assume a shape that prevents it from binding with its substrate. Noncompetitive reversible inhibition is the normal way by which metabolic pathways are regulated in cells. Consider, for example, that it would be possible to slow down a metabolic pathway if the end product inhibited the first enzyme (E_1) noncompetitively (fig. 5.6). When there is an adequate quantity of end product, the pathway is shut down, and when more end product is needed, the pathway is active. This type of *feedback inhibition* is quite common in cells.

An enzyme has a shape appropriate to its substrate. Any environmental factor that affects the shape of a protein also affects the ability of an enzyme to speed up its reaction.

Coenzymes

Many enzymes require a nonprotein cofactor to assist them in carrying out their function. Some cofactors are ions: for example, magnesium (Mg^{++}) ions, potassium (K^+) ions, and calcium (Ca^{++}) ions very often are involved in enzymatic reactions.

Some other cofactors, called **coenzymes,** are organic molecules that bind to enzymes and serve as carriers for chemical groups of electrons. In this case, the protein portion of the enzyme accounts for is specificity, that is, the ability of the enzyme to form an enzyme-substrate complex and to speed up only one particular reaction. The coenzyme portion of the enzyme participates in the reaction.

Coenzyme	Enzyme
nonprotein	protein
helper	specificity

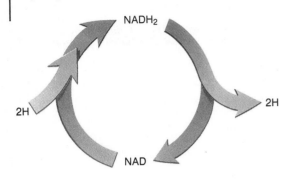

A coenzyme is generally a large molecule that the body is incapable of synthesizing without the ingestion of a vitamin. Vitamins are organic dietary requirements needed in small amounts only. Niacin (or nicotinate), thiamine (or vitamin B_1), riboflavin, folate, and biotin are just a few examples of well-known vitamins that are part of coenzymes.

The NAD Cycle

Presently, we want to consider a coenzyme known as **NAD**[2] that contains the vitamin niacin. **Dehydrogenase enzymes** often utilize the coenzyme NAD, which removes and passes hydrogen (H) atoms from one substrate to another. Therefore, dehydrogenases are agents of oxidation (removal of hydrogen atoms) and reduction (addition of hydrogen atoms). Oxidation usually is defined as the removal of electrons, but note that when hydrogen atoms are removed from a substrate, electrons are also removed since a hydrogen atom consists of $e^- + H^+$.

Only a small amount of a coenzyme like NAD is present in a cell because the same coenzyme molecule is used over and over again, just as an enzyme is used over and over again. Figure 5.7 illustrates this point. After NAD accepts hydrogen atoms and is reduced to $NADH_2$, $NADH_2$ is apt to pass the hydrogen atoms to another carrier, becoming oxidized to NAD again.

Coenzymes are nonprotein organic molecules that participate in a reaction speeded up by an enzyme. NAD is a coenzyme for enzymes that speed up oxidation-reduction reactions.

Energy

When cells require energy, a certain kind of energy must be available. Electricity is the type of energy we use to light our home and to run our electric appliances. Cells, through the process of evolution, have come to depend on the molecule *ATP* (adenosine triphosphate) whenever they require energy, just as our society has come to depend on electricity.

The ATP Reaction

ATP (fig. 5.8) is a nucleotide composed of the base adenine and the sugar ribose (together called adenosine) and 3 phosphate groups. The wavy lines in the formula for ATP indicate high-energy phosphate bonds; when these bonds are broken, an unusually large amount of energy is released. Because of this property, ATP is the energy currency of cells. When cells "need" something, they "spend" ATP.

[2]Nicotinamide adenine dinucleotide.

Figure 5.8
ATP reaction. ATP, the energy molecule in cells, has 2 high-energy phosphate bonds (indicated in the figure by wavy lines). When cells require energy, the last phosphate bond is broken, and a phosphate molecule is released.

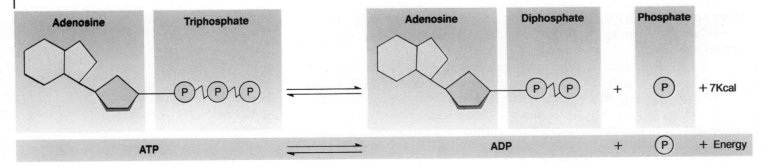

ATP is used in body cells for synthetic reactions, active transport, nervous conduction, and muscle contraction. When energy is required for these processes, the end phosphate group is removed from ATP, breaking down the molecule to ADP (adenosine diphosphate) and phosphate ((P)) (fig. 5.8).

The ATP Cycle

The reaction shown in figure 5.9 occurs in both directions; ATP is both broken down and is built up from ADP and (P) molecules. Since ATP breakdown is occurring constantly, there is always a ready supply of ADP and (P) to rebuild ATP.

Figure 5.9 illustrates the ATP cycle in a diagrammatic way. Notice that when ATP is broken down, energy is released and when it is built up, energy is required. We will see that the breakdown of organic molecules, such as glucose, releases the energy used for ATP buildup.

ATP is the energy molecule in cells because it contains high-energy phosphate bonds. ATP breaks down to ADP + (P) + energy, and it can be rebuilt when these are available.

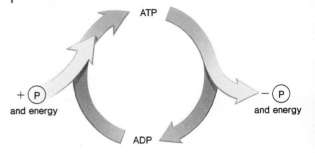

Figure 5.9
The ATP cycle. When ADP joins with a P molecule, energy is required; but when ATP breaks down to ADP and a ℗ molecule, energy is given off.

Aerobic Cellular Respiration (Simplified)

Both plants and animals, whether they reside in the water or on the land, carry on aerobic cellular respiration (fig. 5.10). The term **aerobic** means that oxygen is required for the process. An overall equation for aerobic cellular respiration is given in figure 5.10. As the equation suggests, cellular respiration most often begins with glucose, but as we will see later, other molecules also can be used. Most importantly, we want to realize that as glucose is broken down, ATP molecules are produced, and this is the reason the ATP reaction is drawn using a curved arrow above the glucose reaction arrow.

While an overall equation for aerobic cellular respiration is useful, it does not indicate how glucose or other metabolites actually are broken down. During cellular respiration, organic molecules are oxidized by the removal of hydrogen (H) atoms. Oxidation releases the energy needed for ATP buildup:

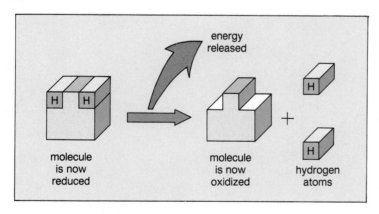

Most often, these hydrogen atoms are accepted by NAD, which then becomes NADH₂. Eventually, the hydrogen atoms are delivered to oxygen, which then is reduced to water.

The breakdown of glucose requires many individual reactions. They are a part of 3 individual subpathways: *glycolysis,* the *Krebs cycle,* and the *respiratory chain.* The *transition reaction* acts like a bridge to connect glycolysis with the Krebs cycle, as illustrated in figure 5.11.

In figure 5.11, each arrow represents a different enzyme and the letters represent the product of the previous reaction and the substrate for the next.

Figure 5.10

Aerobic cellular respiration. a. Almost all organisms, whether they reside in the water or on the land, carry on aerobic cellular respiration. b. The overall equation for aerobic cellular respiration shows that glucose is broken down completely to carbon dioxide (CO_2) and water (H_2O). ATP buildup is indicated by a curved arrow above the reaction arrow. The curved arrow indicates that as glucose is oxidized, ATP is produced. In other words, the energy of oxidation is used to form ATP molecules. (36 ATP rather than 38 ATP are commonly produced in eukaryotic cells for reasons explained on page 114.)

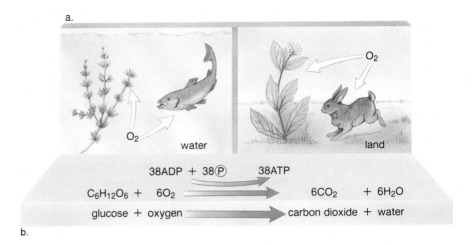

Figure 5.11

The subpathways of aerobic cellular respiration. The overall equation (fig. 5.10) does not indicate that the process requires these 3 subpathways plus the transition reaction.

Glycolysis: Glucose is broken down to 2 molecules of pyruvate (PYR) for a net gain of 2 ATP molecules. The hydrogen (H) atoms removed are taken to the respiratory chain.

Transition Reaction: Carbon dioxide (CO_2) and hydrogen (H) atoms are removed from pyruvate (PYR). Active acetate (AA) results. The hydrogen (H) atoms are taken to the respiratory chain.

Krebs Cycle: Active acetate (AA) enters the cycle; carbon dioxide (CO_2) and hydrogen (H) atoms exit the cycle. There is a gain of 2 ATP molecules per glucose molecule.

Respiratory Chain: The chain receives hydrogen (H) atoms, and as they pass down the chain from carrier to carrier (cytochromes are carriers), 34 ATP molecules are built up. Oxygen (O_2) acts as the final acceptor for hydrogen (H) atoms, and water (H_2O) results.

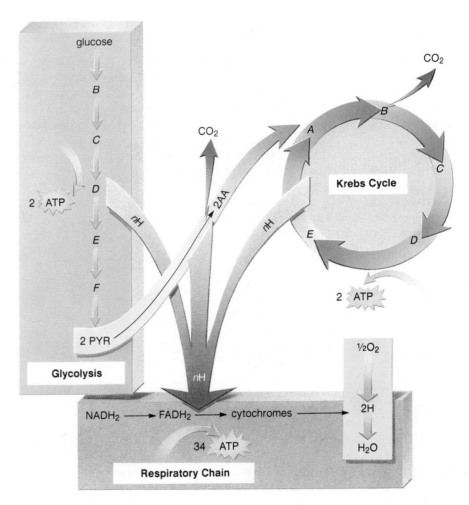

Notice how each pathway resembles a conveyor belt in which a beginning substrate continuously enters at the start, and after a series of reactions, end products leave at the termination of the belt. It is important to realize, too, that all 3 pathways are going on at the same time. They can be compared to the inner workings of a watch in which all parts are synchronized.

The Cell, the Smallest Unit of Life

Subpathways

It is possible to relate the reactants and the products of the overall equation for glucose breakdown with the subpathways (fig. 5.11).

- Glucose ($C_6H_{12}O_6$) is associated with **glycolysis,** the breakdown of glucose to 2 molecules of **pyruvate** (PYR). Oxidation by removal of hydrogen (H) atoms provides enough energy for the buildup of 2 ATP molecules. Glycolysis takes place outside the mitochondria and does not utilize oxygen (O_2). The other subpathways take place inside the mitochondria, where oxygen is utilized.

- Carbon dioxide (CO_2) is associated with the transition reaction and the Krebs cycle. During the **transition reaction,** PYR is oxidized to **active acetate** (AA). AA enters the **Krebs cycle,** a cyclical series of oxidation reactions that give off carbon dioxide and produce one ATP molecule. Notice that since glycolysis ends with 2 molecules of PYR, the transition reaction and the Krebs cycle occur twice per glucose molecule. Altogether, then, the Krebs cycle accounts for 2 ATP molecules per glucose molecule.

- Oxygen and water (H_2O) are associated with the respiratory chain. The coenzyme $NADH_2$ carries hydrogen atoms removed during glycolysis, the transition reaction, and the Krebs cycle to the respiratory chain. The **respiratory chain** (often called the electron transport system or the cytochrome system) is a series of carriers that pass hydrogen atoms (or only the electrons) from one to the other until they finally are received by oxygen, which is then reduced to water. As electrons pass from one molecule to the next, oxidation occurs, and this releases the energy used for ATP buildup.

- ATP is associated with glycolysis, the Krebs cycle, and the respiratory chain. Most ATP, however, is produced by the respiratory chain. The chain can produce as many as 34 ATP molecules per glucose molecule.

Table 5.2 summarizes our discussion of aerobic cellular respiration. This chart assumes that aerobic cellular respiration produces 38 ATP per glucose molecule.

Mitochondria

Mitochondria are called the powerhouses of the cell because they produce ATP molecules. It is interesting to think about how our body provides the reactants for complete glucose breakdown and how it disposes of the products (fig. 5.12a). The air we breathe contains *oxygen,* and the food we eat contains *glucose.* These enter the bloodstream, which carries them about the body, and they diffuse into each and every cell. Glycolysis occurs in the cytoplasm, and the end product pyruvate (PYR) enters the mitochondria. In mitochondria, pyruvate is broken down to carbon dioxide (CO_2) and water (H_2O) as ATP is produced. All 3 of these diffuse out of the mitochondria into the cytoplasm. The *ATP* is utilized inside the cytoplasm for energy-requiring processes. Carbon dioxide diffuses out of the cell and enters the bloodstream. The bloodstream takes the *carbon dioxide* to the lungs, where it is exhaled. The *water* molecules produced, called metabolic water, only become important if they are the organism's only supply of water. In these cases, metabolic water can help to prevent dehydration of the organism.

A mitochondrion has a double membrane (fig. 5.12b). The inner membrane forms **cristae,** shelflike projections that jut out into the **matrix,** the innermost compartment that is filled with a gel-like substance. The transition

Table 5.2 *Overview of Aerobic Cellular Respiration*

Name of Pathway	Result
Glycolysis	Removal of H from substrates Produces 2 ATP*
Transition Reaction	Removal of H from substrates Releases CO_2
Krebs Cycle	Removal of H from substrates Releases CO_2 Produces 2 ATP after 2 turns
Respiratory Chain	Accepts H from other pathways and passes them on to O_2, producing H_2O Produces 34 ATP†

*This is a net gain of 2 ATP for reasons that are given in the detailed discussion.
†Usually less than 34 ATP are produced for reasons that are given in the detailed discussion.

Figure 5.12

Mitochondrion structure and function. a. *Glycolysis occurs in the cytoplasm of a cell, and pyruvate (PYR) enters mitochondria along with oxygen (O_2) and ADP. ATP, carbon dioxide (CO_2), and water (H_2O) exit mitochondria. ATP is the energy currency used in the cytoplasm of the cell.* b. *Mitochondria have a double membrane with a space between. The inner membrane forms cristae that project into the matrix, the innermost compartment.*

c. *The transition reaction and the Krebs cycle are located in the matrix. The respiratory chain is located on the cristae. The chemiosmotic theory states there is a hydrogen (H^+) ion buildup in the intermembrane space, and when these hydrogen ions flow down their electrochemical gradient through F_1 particles, ATP is produced.*

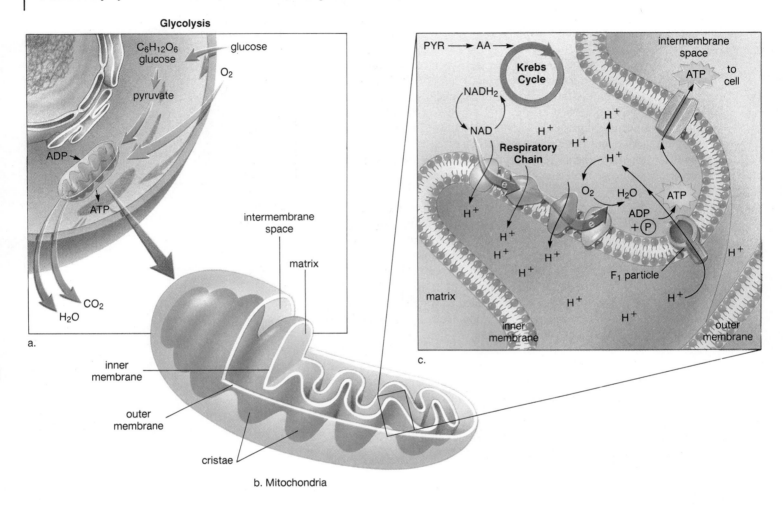

b. Mitochondria

reaction and the Krebs cycle occur within the matrix; the carriers of the respiratory chain are located on the cristae. $NADH_2$ brings hydrogen atoms ($e^- + H^+$) to the chain. The respiratory chain also is called the electron transport system because some carriers accept only electrons (e^-). The released hydrogen (H^+) ions build up within the intermembrane space, creating an electrochemical gradient. When the hydrogen ions flow down this gradient through channel proteins (sometimes called F_1 particles) located in the inner membrane, ATP is produced by ATP synthetase enzymes associated with the channel proteins (fig. 5.12c). This is called chemiosmotic ATP synthesis.

Mitochondria are the powerhouses of the cell; they produce most of the ATP molecules. The transition reaction and the Krebs cycle are located in the matrix, and the respiratory chain is located on the cristae.

Metabolites

We already have seen that glucose can be utilized for aerobic cellular respiration. Other molecules also can be utilized. When a fat is used as an energy source, it breaks down to glycerol and 3 fatty acids. As figure 5.13 indicates,

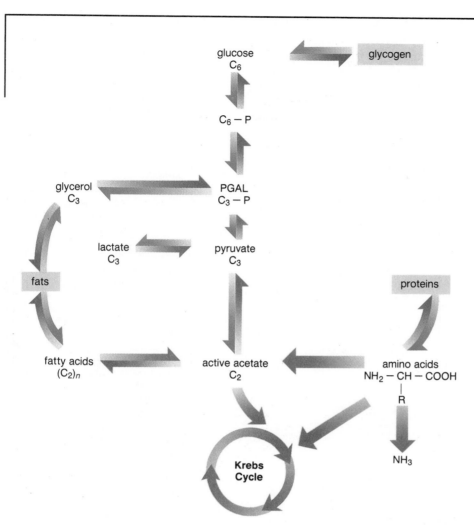

Figure 5.13
Interrelationship of metabolic pathways.
Glycolysis and the Krebs cycle can serve both to
break down or to build up the molecules noted.

glycerol is converted easily to PGAL, a metabolite in the glycolytic pathway. The fatty acids are converted to active acetate, which enters the Krebs cycle. A fatty acid that contains 18 carbons results in 9 active acetate (AA) molecules. Calculation shows that respiration of these can produce a total of 108 ATP molecules. For this reason, fats are an efficient form of stored energy—there are 3 long fatty acid chains per fat molecule.

The carbon skeleton of amino acids can enter the Krebs cycle (fig. 5.13). Before the skeleton enters, the amino acid must undergo **deamination,** or the removal of the amino group. Later, this group is converted to ammonia (NH_3), an excretory product of cells. Just where the carbon skeleton enters the Krebs cycle is dependent on the length of the R group, since this determines the number of carbons left following deamination.

Synthetic Reactions

The substrates making up the pathways in figure 5.13 also can be used as starting materials for synthetic reactions. In other words, compounds that enter the pathways are oxidized to substrates that can be used for biosynthesis. This is the cell's **metabolic pool,** in which one type of molecule can be converted to another. In this way, carbohydrate intake can result in the formation of fat. PGAL molecules can be converted to glycerol molecules, and acetyl groups can be joined to form fatty acids. Fat synthesis follows. This explains why you can get fat from eating too much candy, ice cream, and cake.

Some metabolites of the Krebs cycle can be converted to amino acids. Plants are able to synthesize all of the amino acids they need. Animals, however, lack some of the enzymes necessary for synthesis of all amino acids.

Figure 5.14

Glycolysis, a metabolic pathway. The pathway begins with glucose and ends with pyruvate. Net gain of 2 ATP molecules can be calculated by subtracting those expended from those produced. Print in the boxes explains each reaction.

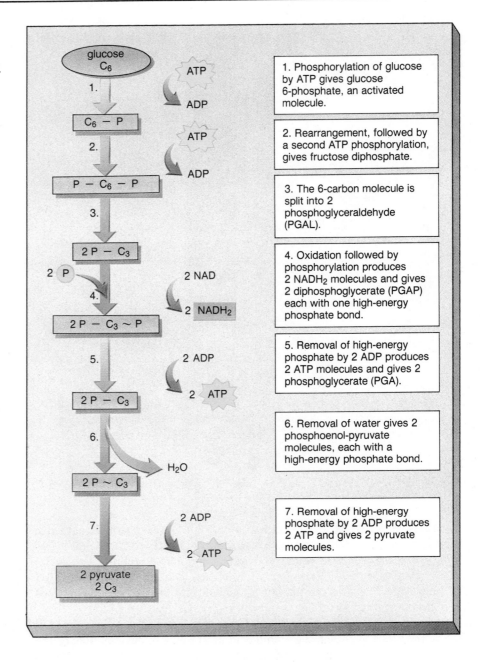

1. Phosphorylation of glucose by ATP gives glucose 6-phosphate, an activated molecule.

2. Rearrangement, followed by a second ATP phosphorylation, gives fructose diphosphate.

3. The 6-carbon molecule is split into 2 phosphoglyceraldehyde (PGAL).

4. Oxidation followed by phosphorylation produces 2 NADH$_2$ molecules and gives 2 diphosphoglycerate (PGAP) each with one high-energy phosphate bond.

5. Removal of high-energy phosphate by 2 ADP produces 2 ATP molecules and gives 2 phosphoglycerate (PGA).

6. Removal of water gives 2 phosphoenol-pyruvate molecules, each with a high-energy phosphate bond.

7. Removal of high-energy phosphate by 2 ADP produces 2 ATP and gives 2 pyruvate molecules.

Humans (and white rats), for example, can synthesize 9 of the common amino acids, but they cannot synthesize the other 11. The amino acids that cannot be synthesized must be supplied by the diet; they are called the essential amino acids. The nonessential amino acids can be synthesized. It is quite possible for animals to suffer from protein deficiency if their diet does not contain adequate quantities of all the essential amino acids.

All the reactions involved in cellular respiration are a part of a metabolic pool; the metabolites from the pool can be used either as energy sources or as substrates for various synthetic reactions.

Aerobic Cellular Respiration (Detailed)

Now that we have taken a look at some of the cell's main metabolic pathways, it is possible to define aerobic cellular respiration with more precision. Aerobic cellular respiration occurs when oxygen (O_2) is the final acceptor for the hydrogen (H) atoms delivered to the respiratory chain by $NADH_2$ or $FADH_2$. Glucose ($C_6H_{12}O_6$) is most often the beginning substrate for aerobic cellular respiration, a process that results in ATP buildup. The complete breakdown of glucose requires glycolysis, the transition reaction, the Krebs cycle, and the respiratory chain.

Glycolysis

Glycolysis takes place within the cytoplasm (fig. 5.12a) and outside the mitochondria. Glycolysis (fig. 5.14) is the breakdown of glucose to the end product pyruvate (PYR). As 2 molecules of ATP (steps 1 and 2) are hydrolyzed, the substrates are phosphorylated and are made ready to react. Later, hydrogen (H) atoms are removed from the substrates of the pathway and are picked up by NAD. Altogether, the glycolytic pathway produces 2 $NADH_2$ molecules. Removal of hydrogen atoms means that oxidation has occurred. Oxidation releases energy, and some of this energy is captured within the substrate molecules and then is used to form 4 ATP molecules. Subtracting the 2 molecules that were used to get started, there is a net gain of 2 ATP molecules for the glycolytic pathway.

Figure 5.14 gives the individual reactions of the glycolytic pathway. The reactants are represented by the number of carbon atoms they contain (each molecule also contains hydrogen and oxygen atoms). Also, table 5.3 summarizes the results of the reactions.

When glycolysis is a part of aerobic cellular respiration, the end product pyruvate enters the mitochondria, where oxygen (O_2) is utilized. However, glycolysis is a pathway that can occur even when oxygen is not present in cells; glycolysis does not need to be a part of aerobic cellular respiration. As we will see on page 114, glycolysis is also a part of fermentation. Fermentation is an anaerobic process; it does not require oxygen.

Glycolysis takes place within the cytoplasm. During glycolysis, glucose is broken down to pyruvate with a net gain of 2 ATP molecules and 2 NADH₂ molecules.

The Transition Reaction

As shown in figure 5.12c, the transition reaction and the Krebs cycle are located within the matrix of mitochondria. In the transition reaction, pyruvate (PYR) is converted to active acetate (AA), and carbon dioxide (CO_2) is given off in the process:

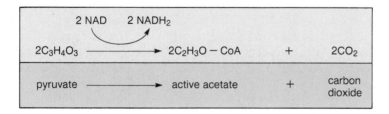

This is an oxidation reaction in which hydrogen (H) atoms are removed from pyruvate by a dehydrogenase that uses NAD as the coenzyme. Also, note that active acetate is an acetyl group (C_2H_3O—) attached to a coenzyme called **coenzyme A** (CoA). This coenzyme activates the acetyl group; hence the name *active acetate.* Notice that since glycolysis results in 2 pyruvate molecules, the transition reaction occurs twice per glucose molecule. Altogether, the reaction produces 2 molecules of carbon dioxide, 2 molecules of active acetate, and 2 molecules of $NADH_2$ per glucose molecule.

Table 5.3 *Summary of Glycolysis*

Reaction	NADH₂	ATP
1		−1 ATP
2		−1 ATP
3		
4	2 NADH₂	
5		+2 ATP
6		
7		+2 ATP
Net gain	2 NADH₂	+2 ATP

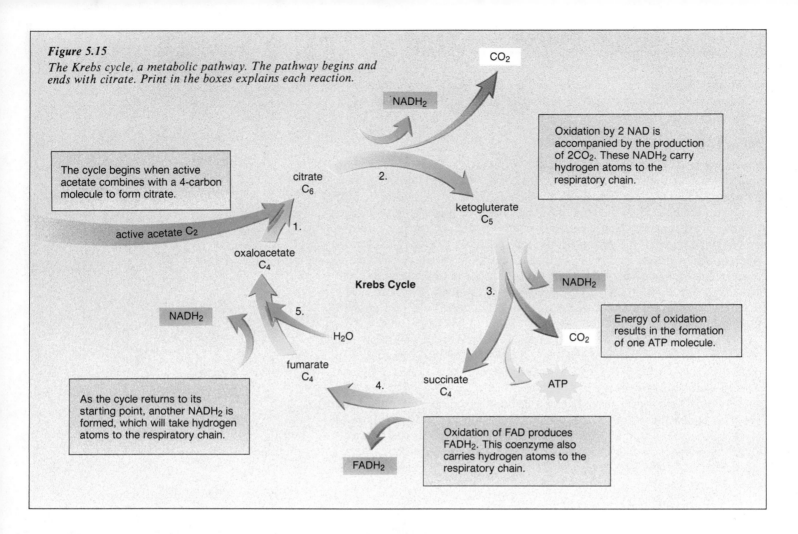

Figure 5.15
The Krebs cycle, a metabolic pathway. The pathway begins and ends with citrate. Print in the boxes explains each reaction.

The cycle begins when active acetate combines with a 4-carbon molecule to form citrate.

Oxidation by 2 NAD is accompanied by the production of 2CO₂. These NADH₂ carry hydrogen atoms to the respiratory chain.

Energy of oxidation results in the formation of one ATP molecule.

As the cycle returns to its starting point, another NADH₂ is formed, which will take hydrogen atoms to the respiratory chain.

Oxidation of FAD produces FADH₂. This coenzyme also carries hydrogen atoms to the respiratory chain.

Table 5.4 Summary of the Krebs Cycle				
Step	**CO₂**	**NADH₂**	**FADH₂**	**ATP**
1				
2	CO₂	NADH₂		
3	CO₂	NADH₂		ATP
4			FADH₂	
5		NADH₂		
	2 CO₂	3 NADH₂	FADH₂	ATP

Note: Summary is per turn. Cycle turns twice per glucose molecule.

The Krebs Cycle

The Krebs cycle, represented in detail in figure 5.15, is named for the person who discovered it. It is called a cycle because it is a series of reactions, or steps, that begin and end with **citrate** (citric acid). Sometimes the cycle is called the *citric acid cycle*. During this series of reactions that occur in the matrix of mitochondria (fig. 5.12c), *oxidative decarboxylation* occurs. **Oxidative decarboxylation** means that hydrogen (H) atoms and carbon dioxide (CO_2) are removed from the substrates at the same time. Oxidation results in energy that is partially captured by the molecules within the cycle, and this energy is used to form one ATP molecule per turn. The carbon dioxide given off is a metabolic waste and is excreted by cells. Most of the hydrogen atoms are picked up by NAD, but a few are taken by FAD. **FAD** is another coenzyme of oxidation-reduction. FAD is used infrequently compared to NAD. Altogether, the Krebs cycle turns twice per glucose molecule and produces 4 CO_2, 6 NADH₂, 2 FADH₂, and 2 ATP molecules. Table 5.4 summarizes the reactions of the Krebs cycle.

The Respiratory Chain

The respiratory chain is located on the cristae of the mitochondria (fig. 5.12c). It is a series of carriers that pass electrons from one carrier to another; therefore, the respiratory chain often is called the electron transport system. The electrons are at first a part of the hydrogen atoms ($e^- + H^+$) attached to NAD or to FAD. These are the same hydrogen atoms that were removed from the metabolites of glycolysis, the transition reaction, and the Krebs cycle. When the electrons enter the respiratory chain (fig. 5.16), they are at a high-energy

Figure 5.16

Electron transport system. $NADH_2$ delivers 2 hydrogen (H) atoms to a carrier at the top of the chain. Each pair of electrons delivered directly by $NADH_2$ ultimately is responsible for the formation of 3 ATP molecules. Those delivered by $FADH_2$ are responsible for the formation of 2 ATP molecules. ATP is the result of chemiosmotic phosphorylation: the carriers in the chain deposit hydrogen (H^+) ions in the intermembrane space of mitochondria, creating a hydrogen ion reservoir. The energy released as these hydrogen ions flow down their electrochemical gradient is used to form ATP from ADP and P. Oxygen (O_2) is the final acceptor for the electrons, and it is reduced to water (H_2O). (red. = reduction, ox. = oxidation)

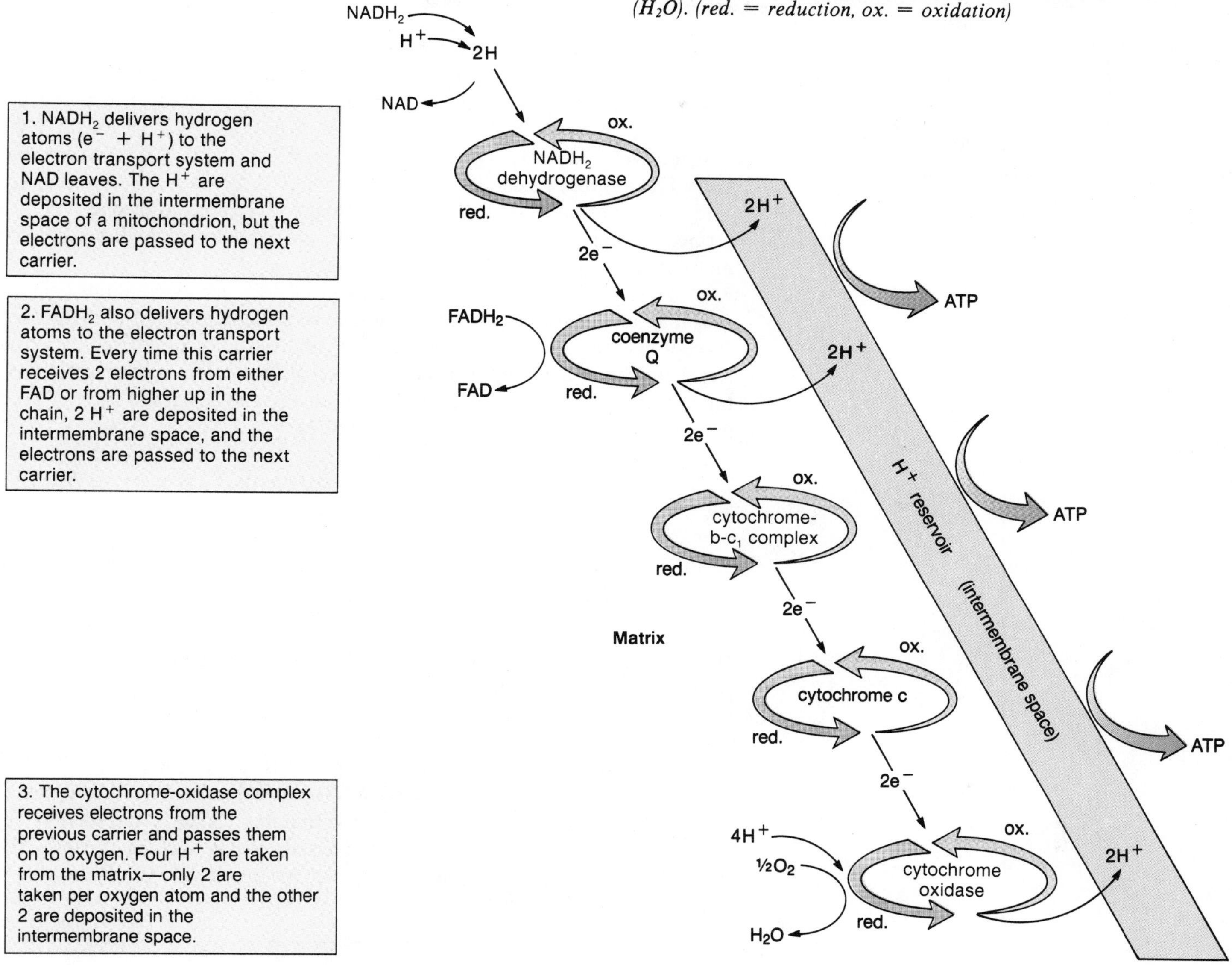

1. $NADH_2$ delivers hydrogen atoms (e^- + H^+) to the electron transport system and NAD leaves. The H^+ are deposited in the intermembrane space of a mitochondrion, but the electrons are passed to the next carrier.

2. $FADH_2$ also delivers hydrogen atoms to the electron transport system. Every time this carrier receives 2 electrons from either FAD or from higher up in the chain, 2 H^+ are deposited in the intermembrane space, and the electrons are passed to the next carrier.

3. The cytochrome-oxidase complex receives electrons from the previous carrier and passes them on to oxygen. Four H^+ are taken from the matrix—only 2 are taken per oxygen atom and the other 2 are deposited in the intermembrane space.

level, but as they are passed downhill from one carrier to the next, they lose energy as oxidation occurs. Some of the carriers in the respiratory chain are cytochrome molecules. Because of this, still another name for the respiratory chain is the cytochrome system.

Although $NADH_2$ and $FADH_2$ bring hydrogen atoms to the chain, certain carriers (e.g., cytochrome molecules) accept only electrons, and the hydrogen (H^+) ions are deposited in the intermembrane space of the mitochondrion. This establishes an electrochemical gradient that leads to ATP production. As the hydrogen ions pass from the area of greater concentration to the area of lesser concentration through a channel protein (called an F_1 particle), the released energy is used to make ATP (fig. 5.12c). This is termed chemiosmotic ATP synthesis. Chemiosmotic ATP synthesis is the way the energy of oxidation is used to make ATP molecules.

Table 5.5 *ATP Produced per Glucose Molecule*

	Direct	By Way of Respiratory Chain	
Glycolysis	2 ATP	2 NADH$_2$ =	6 ATP*
Transition reaction		2 NADH$_2$ =	6 ATP
Krebs cycle	2 ATP	6 NADH$_2$ =	18 ATP
		2 FADH$_2$ =	4 ATP
Subtotal	4 ATP		34 ATP
Grand total		38 ATP*	

*The numbers in this column and the total number of ATP are usually less because the chain does not always produce the maximum possible number of ATP per NADH$_2$.

Oxygen (O$_2$) is the final acceptor for electrons at the end of the respiratory chain. After accepting electrons, oxygen combines with hydrogen ions to form water:

$$\tfrac{1}{2}O_2 + 2e^- + 2H^+ \longrightarrow H_2O$$

Energy Yield per Glucose Molecule

In order to calculate the number of ATP molecules per glucose molecule, it is customary to first consider the number of ATP that are made directly during glycolysis and the Krebs cycle (first column, table 5.5). Then, we must realize that for every NADH$_2$ that enters the respiratory chain, 3 ATP result. For every FADH$_2$ that enters the respiratory chain, 2 ATP result. (Examine figure 5.15 and note that FAD delivers its hydrogen (H) atoms after one ATP has formed.) Using this method of calculation, we arrive at 38 ATP per glucose molecule (table 5.5).

Thirty-eight is the maximum number of ATP per glucose molecule. In actuality, NADH$_2$ from glycolysis cannot cross the mitochondrial membrane (except in certain cells), and its hydrogen atoms usually are ferried by a carrier that delivers them to FAD. This means that for every cytoplasmic NADH$_2$ formed, there usually are only 2 ATP produced, and the grand total of ATP per glucose molecule is often 36 instead of 38.

The transition reaction, the Krebs cycle, and the respiratory chain are located in mitochondria. These reactions account for 30 of the total number of ATP molecules produced per glucose molecule.

Fermentation

If oxygen (O$_2$) is not available to cells, the respiratory chain soon becomes inoperative. Electrons plug up the system when the final acceptor, oxygen, is not present. In this case, cells begin to ferment so that some ATP energy still can be produced.

The glycolytic pathway runs as long as it is supplied with "free" NAD, that is, NAD that can pick up hydrogen (H) atoms. Normally, NADH$_2$ passes hydrogen atoms to the respiratory chain within mitochondria and thereby becomes "free" of hydrogen atoms. If the system is not working because of a lack of oxygen, NADH$_2$ can pass hydrogen atoms to pyruvate (PYR), as shown in the following reactions:

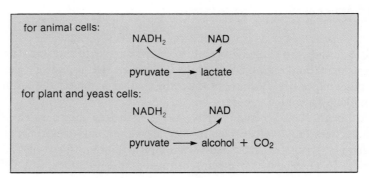

This NAD now is capable of picking up more hydrogen atoms so that the glycolytic pathway can keep operating. **Fermentation** is an anaerobic process; that is, it does not involve oxygen.

The Cell, the Smallest Unit of Life

Wine, beer, and whiskey production all require yeast fermentation. To produce wine, grape juice is allowed to ferment. After the grapes are picked, they are crushed in order to collect the juice. To make wine in the old days, wine makers simply relied on spontaneous fermentation by yeasts that were on the grape skins, but now many add specially selected cultures of yeast. It also is common practice to maintain the temperature at about 20°C for white wines and 28°C for red wines. Fermentation ceases after most of the sugar has been converted to alcohol. Various

Alcoholic Beverages

methods are used to clarify the wine, that is, to remove any suspended materials. Also, many fine wines improve when they are allowed to "age" during barrel or bottle storage.

Brewing beer is more complicated than wine production. Usually grains of barley first are malted; that is, they are allowed to germinate for a short time so that amylase enzymes that will break down the starch content of the grain are produced. After the germinated grains have been crushed and mixed with water, the malt wort is separated from the spent grains and traditionally is boiled with hops (an herb derived from the hop plant) to give flavor to the beer. Then the hop wort is seeded with a strain of yeast that converts the sugars in the wort to alcohol and carbon dioxide (CO_2). At the end of fermentation, the yeast is separated from the beer, which is then allowed to mature for an appropriate period. After filtration and pasteurization, the beer is packaged.

The production of whiskey (from grains), brandy (from grapes), and rum (from molasses) differs from wine and beer production chiefly in that the alcohol is removed from the fermented substance by distillation. Most often in the United States, corn or rye is used in the production of whiskey. These grains are ground up and mashed to release their starch content. Then amylase enzymes are added to convert the starch to fermentable sugars. Yeast then is added so that fermentation can occur. Following fermentation, the alcohol is concentrated by distillation. A warm temperature causes the alcohol to become gaseous and to rise in a column, where it condenses to a liquid before entering a collecting vessel. The alcohol content of the collecting vessel is much higher following this distillation process. The distillate usually is stored, quite often in an oak barrel, to improve the aroma and the taste of the final product.

Barley (top), grapes (middle), and corn (bottom). *Fermentation of these produces beer, wine, and whiskey, respectively.*

The Results of Fermentation

Fermentation (fig. 5.17) consists of the glycolytic pathway plus one additional reaction in which pyruvate (PYR) accepts hydrogen (H) atoms and is reduced to either lactate (animals) or alcohol and carbon dioxide (yeast). Therefore, fermentation represents an incomplete breakdown of glucose. The energy yield for this anaerobic process is only 2 ATP molecules per glucose molecule. The end products (lactate or alcohol) are toxic to cells and can lead to the death of the organism.

The Usefulness of Fermentation

Despite its low yield of ATP, fermentation does have its place because it can provide a *rapid burst* of ATP energy. In our own body, it is the muscle cells that are most likely to carry on fermentation. When the muscles are working vigorously over a short period of time, as when we run, fermentation provides a way to produce ATP energy even though oxygen is temporarily in limited supply. At first, the blood carries away all the lactate formed in the muscles, but eventually lactate begins to build up in the muscles, changing the pH and causing muscle fatigue and discomfort. When we stop running, our body has

Figure 5.17

Fermentation, an anaerobic process. Fermentation consists of the glycolytic pathway plus one additional reaction in which the end product of the glycolytic pathway, pyruvate, accepts hydrogen (H) atoms and is reduced. This "frees" NAD so that it can return to pick up more hydrogen atoms.

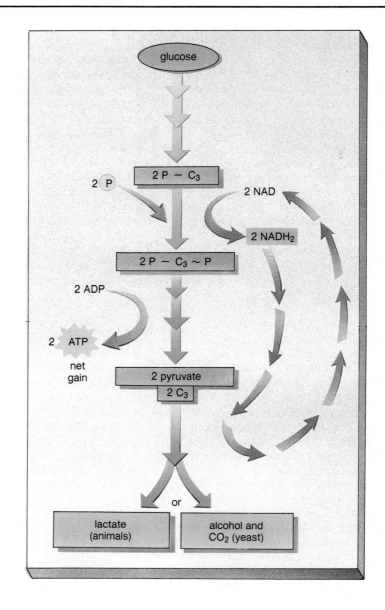

a chance to catch up while the lactate is reconverted to pyruvate by the removal of hydrogen (H) atoms, which can now be transported to the electron transport system. In the meantime, we are said to be in **oxygen debt,** as signified by the fact that we keep on breathing very heavily for a time.

Fermentation allows yeast cells to grow and to divide anaerobically for a time. Eventually, though, they are killed by the very alcohol they produce if the initial glucose level is high. Presumably, human beings were delighted to discover this form of fermentation, as the ethyl alcohol produced has been consumed in great quantity for thousands of years. In addition, we use the carbon dioxide (CO_2) to make bread rise.

Fermentation, which involves the reduction of pyruvate to lactate or to carbon dioxide and alcohol, gives a net yield of only 2 ATP. Although this is a very low yield, it provides a way for a cell to produce ATP molecules even when oxygen is not available.

The Cell, the Smallest Unit of Life

Summary

Metabolic pathways are a series of reactions that proceed in an orderly, step-by-step manner. Each of these reactions requires a specific enzyme. Any environmental factor that affects the shape of a protein also affects the ability of an enzyme to do its job. Sometimes enzymes require coenzymes, nonprotein portions that participate in the reaction. NAD is a coenzyme.

Aerobic cellular respiration (the breakdown of glucose to carbon dioxide and water) requires 3 subpathways: glycolysis, the Krebs cycle, and the respiratory chain. As electrons (usually contributed by $NADH_2$) pass down the chain, 34 ATP (maximum) are generated (in addition to 4 ATP produced directly by glycolysis and the Krebs cycle). Other types of molecules, such as amino acids and neutral fats, also can be respired in cells.

If oxygen is not available in cells, the respiratory chain is inoperative, and fermentation (anaerobic cellular respiration) occurs. Pyruvate from glycolysis is reduced by $NADH_2$ to either lactate (animal cells) or alcohol and carbon dioxide (yeast), substances harmful to these cells. Now that NAD is available once again, glycolysis can reoccur. Fermentation only gives a net gain of 2 ATP, and it eventually causes the death of the animal and the yeast cells.

Study Questions

1. Discuss and draw a diagram to describe a metabolic pathway. (p. 99)
2. Discuss and give a reaction to describe the specificity of enzymatic actions. (p. 101)
3. Name several factors that affect the yield of enzymatic reactions, and explain the effect of these factors. (p. 101)
4. Define coenzyme, and give several examples. What is the NAD cycle? (p. 103)
5. Which molecule is the common energy currency in cells? Why is this molecule appropriate for the task? (p. 104)
6. Give the overall equation for aerobic cellular respiration, and discuss the equation in general. (pp. 105–106)
7. Name and describe the events within the 3 subpathways and the transition reaction that make up aerobic cellular respiration. (pp. 106–107)
8. How do our body cells obtain glucose and oxygen? What happens to the carbon dioxide (CO_2) given off by these cells? (pp. 107–108)
9. Explain why glycolysis directly produces a net gain of 2 ATP. How many ATP are produced as a *result* of glycolysis? (p. 111)
10. Explain the term *oxidative decarboxylation*, which occurs in the Krebs cycle. How many ATP are produced as a *result* of the Krebs cycle? (p. 112)
11. The respiratory chain is composed of what type of molecules? How does the chain produce ATP? (pp. 112–13)
12. Calculate the number of ATP that are actually produced as hydrogen (H) atoms pass down the respiratory chain. (p. 114)
13. What is the benefit and the drawbacks of fermentation? (p. 115)

Objective Questions

1. Every reaction that occurs in a cell requires a(n) _____ .
2. If more substrate or more enzyme is added, you would expect an enzymatic reaction to _____ .
3. NAD is a coenzyme of
   _____ .
4. Glycolysis is the breakdown of glucose to 2 molecules of _____ .
5. The Krebs cycle begins and ends with
   _____ .
6. The respiratory chain is located on the _____ of mitochondria.
7. The final acceptor for hydrogen (H) atoms at the end of the respiratory chain is _____ .
8. Carbon dioxide (CO_2) formation should be associated with the _____ cycle and the _____ reaction.
9. The immediate acceptor of hydrogen (H) atoms during fermentation is
   _____ .
10. Fermentation only results in a net gain of _____ ATP molecules.
11. Fatty acids are broken down to _____ molecules, which enter the Krebs cycle.

Selected Key Terms

active acetate (ak′tiv as′ĕ-tāt) an acetyl group attached to coenzyme A; a complex that forms during the transition reaction that links glycolysis to the Krebs cycle. *107*

active site (ak′tiv sīt) the region on the surface of an enzyme where the substrate binds and where the reaction occurs. *101*

aerobic (a″er-ōb′ik) growing or metabolizing only in the presence of oxygen, as in aerobic cellular respiration. *105*

coenzyme (ko-en′zīm) a nonprotein molecule that aids the action of an enzyme, to which it is loosely bound. *103*

coenzyme A (ko-en′zīm) a coenzyme that participates in the transition reaction and carries the organic product to the Krebs cycle. *111*

deamination (de-am″ĭ-na′shun) removal of an amino group ($-NH_2$) from an amino acid or other organic compound. *109*

dehydrogenase enzyme (de-hi′dro-jen-ās en′zīm) an enzyme that accepts hydrogen (H) atoms, speeding up the process of dehydrogenation. *104*

FAD a coenzyme of oxidation; a dehydrogenase that participates in hydrogen (electron) transport within the mitochondria. *112*

fermentation (fer″men-ta′shun) anaerobic breakdown of carbohydrates that results in organic end products, such as alcohol and lactate. *115*

glycolysis (gli-kol'ĭ-sis)   the metabolic pathway that converts sugars to simpler compounds and ends with pyruvate.   *107*

Krebs cycle (krebz si'kl)   a series of reactions found within the matrix of mitochondria that give off carbon dioxide. Also called the citric acid cycle because the reactions begin and end with citrate.   *107*

metabolism (mē-tab'o-lizm)   all of the chemical changes that occur within a cell.   *98*

NAD   a coenzyme of oxidation; a dehydrogenase that frequently participates in hydrogen transport.   *104*

oxidative decarboxylation (ok″sĭ-da'tiv de″kar-bok″sĭ'-la'shun)   a reaction that involves the release of carbon dioxide as oxidation occurs.   *112*

pyruvate (pi'roo-vāt)   the end product of glycolysis; pyruvic acid.   *107*

respiratory chain (re-spi'rah-to″re chān)   a series of carriers within the inner mitochondrial membrane that pass electrons one to the other from a higher energy level to a lower energy level; the energy released is used to build ATP energy; also called the electron transport system or the cytochrome system.   *107*

transition reaction (tran-zish'un re-ak'shun)   a reaction within aerobic cellular respiration during which hydrogen and carbon dioxide are removed from pyruvate; connects glycolysis to Krebs cycle.   *107*

Further Readings for Part One

Alberts, B., et al. 1983. *Molecular biology of the cell.* New York: Garland Publishing.

Allen, R. D. February 1987. The microtubule as an intracellular engine. *Scientific American.*

Avers, C. J. 1981. *Cell biology.* 2d ed. New York: D. Van Nostrand.

Baker, J. J. W., and G. E. Allen. 1981. *Matter, energy, and life.* 4th ed. Reading, Mass.: Addison-Wesley.

Berns, M. W. 1983. *Cells.* 3d ed. New York: Holt, Rinehart & Winston.

Cronquist, A. 1982. *Basic botany.* 2d ed. New York: Harper & Row Publishers, Inc.

Dautry-Varsat, A., and H. F. Lodish. May 1984. How receptors bring proteins and particles into cells. *Scientific American.*

de Duve, C. May 1983. Microbodies in the living cell. *Scientific American.*

Dickerson, E. March 1980. Cytochrome and the evolution of energy metabolism. *Scientific American.*

Dustin, P. August 1980. Microtubules. *Scientific American.*

Grivell, L. A. March 1983. Mitochondrial DNA. *Scientific American.*

Karplus, M., and J. A. McCammon. April 1986. The dynamics of proteins. *Scientific American.*

Lake, J. A. July 1981. The ribosome. *Scientific American.*

McIntosh, J. R., and K. L. McDonald. October 1989. The mitotic spindle. *Scientific American.*

Miller, K. R. October 1979. The photosynthetic membrane. *Scientific American.*

Ostro, M. J. January 1987. Lipsosomes. *Scientific American.*

Porter, K. R., and M. A. Bonneville. 1973. *Fine structure of cells and tissues.* 4th ed. Philadelphia: Lea and Febiger.

Porter, K. R., and J. B. Tucker. March 1981. The ground substance of the living cell. *Scientific American* (offprint 1494).

Rothman, J. September 1985. The compartmental organization of the Golgi apparatus. *Scientific American.*

Scientific American. October 1985. The molecules of life.

Sharon, N. November 1980. Carbohydrates. *Scientific American.*

Sheeler, P., and D. E. Bianchi. 1980. *Cell biology: Structure, biochemistry, and function.* New York: John Wiley and Sons.

Stryer, L. 1981. *Biochemistry.* 2d ed. San Francisco: W. H. Freeman.

Unwin, N. February 1984. The structure of proteins in biological membranes. *Scientific American.*

Wolfe, S. L. 1980. *Biology of the cell.* Belmont, Calif.: Wadsworth.

Wickramasinghe, H. K. October 1989. Scanning-probe microscopes. *Scientific American.*

Part Two

Plant Biology

Plants capture radiant energy from the sun and use this energy to convert inorganic chemicals to organic chemicals. Photosynthesis is therefore a 2-step process. The first step, which requires light energy, produces ATP and $NADPH_2$. The second step uses these molecules to reduce carbon dioxide (CO_2) molecules to carbohydrate molecules.

Plant form and function can be related to a plant's ability to carry out photosynthesis. Flowering plants, representative of plants in general, are composed of a root system and a shoot system; the latter includes the stems, leaves, and flowers. Photosynthesis occurs primarily in leaves containing the green pigment chlorophyll, which is capable of absorbing the energy of the sun.

Photosynthesis is essential to the continued existence of all living things. Its by-product, oxygen (O_2), makes cellular respiration possible, and its primary product, carbohydrate, is food for the entire biosphere.

A dandelion inflorescence has produced many single-seeded fruits. Dandelion fruits are adapted to dispersal by wind. Each fruit has a tiny "parachute" that keeps it afloat in the breezes.

Chapter 6

Photosynthesis

Chapter Concepts

1 Photosynthetic organisms produce the organic molecules that are used as a source of food and chemical energy by all living things.

2 Photosynthesis takes place primarily within chloroplasts.

3 Photosynthesis has 2 subpathways. The first drives the second by providing the energy and the hydrogen (H) atoms needed by the second to reduce carbon dioxide (CO_2) to carbohydrate.

4 Photosynthesis can be compared to cellular respiration; both similarities and differences exist.

Chapter Outline

Radiant Energy
    Sunlight
Chloroplasts
    Anatomy
    Overall Equations for Photosynthesis
Capturing the Energy of Sunlight
    Two Electron Pathways
Reducing Carbon Dioxide
    C_3 Photosynthesis
    C_4 Photosynthesis
    Carbohydrate Utilization
Autotrophic Bacteria
    Bacterial Photosynthesis
    Chemosynthesis
Comparison of Cellular Respiration and
    Photosynthesis
    Differences
    Similarities

a. Autotrophs   b. Herbivores   c. Carnivores   d. Omnivores

 Only plants, algae, and a few bacteria carry on photosynthesis. As the name implies, **photosynthesis** refers to the ability of these organisms to make their own food in the presence of sunlight.

Radiant Energy

The food produced through photosynthesis eventually becomes the food for the rest of the living world. For example, plants use the organic food they produce as a source of metabolic energy and as building blocks for growth. Animals, including humans, either eat plants directly or eat animals that have eaten plants, and so forth. Another way to express this idea is to say that **autotrophs,** which are able to synthesize organic molecules from inorganic raw materials, not only feed themselves but also **heterotrophs,** which must take in preformed organic molecules (fig. 6.1). After food is eaten and digested by heterotrophs, they use the resulting small molecules either as building blocks for growth or as a source of metabolic energy.

Plants also supply energy in another sense—their bodies became the fossil fuel coal that is an energy source today. For the following reasons, then, it is correct to say that all life is ultimately dependent on the energy of the sun.

1. Sunlight supplies the energy needed for photosynthesis.
2. The food made by photosynthetic organisms becomes the food for the biosphere. This food is used not only for growth but also for metabolic energy.
3. The bodies of plants became the fossil fuel coal, upon which we are still dependent today.
4. Sunlight can be an energy source, called solar energy, for private and industrial use.

Figure 6.1

Methods of acquiring organic food.
a. Autotrophs, represented by green plants, algae, and a few types of bacteria (the latter 2 presented in circles) produce food for themselves and for (b) herbivores, which feed directly on plants or plant products, (c) carnivores, which feed on herbivores or other carnivores, and (d) omnivores, which feed on all of these.

Figure 6.2

Solar energy. a. Energy from the sun is categorized into the types listed. Gamma rays have the shortest wavelength and the greatest energy content. Radio waves have the longest wavelength and the least energy content. Visible (white) light actually contains many colors of light, as is detected when the light is passed through a prism. b. Absorption spectrum of chlorophyll a and chlorophyll b shows that these 2 pigments best absorb violet-blue and red-orange light. c. Leaves appear green to us because the green light is transmitted or reflected by chlorophyll.

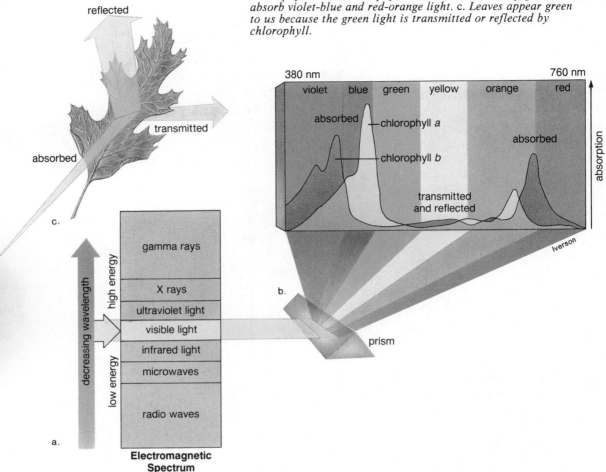

Sunlight

Radiant energy from the sun can be described in terms of its wavelength and its energy content. Figure 6.2 lists the different types of radiant energy, from the shortest wavelength, gamma rays, to the longest, radio waves. The shorter wavelengths contain more energy than the longer ones. *White* or *visible light* is only a small portion of this spectrum. Visible light itself contains various wavelengths of light, as can be proven by passing it through a prism; then we see all the different colors that make up visible light. (Actually, of course, it is our eyes that interpret these wavelengths as colors.) The colors in visible light range from violet (the shortest wavelength) to blue, green, yellow, orange, and red (the longest wavelength). The energy content is highest for violet light and lowest for red light.

The pigments found within photosynthesizing cells are capable of absorbing various portions of visible light. The absorption spectrum for chlorophyll *a* and chlorophyll *b* is shown in figure 6.2*b*. Both chlorophyll *a* and chlorophyll *b* absorb violet, blue, and red light better than the light of other colors. Because green light is only minimally absorbed and is primarily transmitted or reflected (fig. 6.2*c*), leaves appear green to us.

Photosynthesis is absolutely essential for the continuance of life because it supplies food and energy for the biosphere.

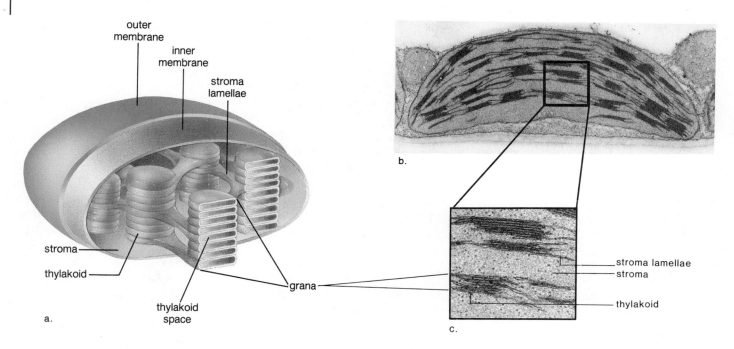

Figure 6.3

Chloroplast anatomy. a. *A chloroplast is bounded by a double membrane. Within, the grana contain pigments (e.g., chlorophyll) that absorb the energy of sunlight. The stroma contains enzymes that reduce carbon dioxide (CO_2).* b. *Electron micrograph of a chloroplast cross section showing the grana, their connecting stroma lamellae, and the stroma.* c. *A granum is a stack of membranous sacs called thylakoids. Each thylakoid has a thylakoid space.*

Chloroplasts

Chloroplasts are the organelles found in a plant cell (fig. 6.3). Photosynthesis takes place within chloroplasts.

Anatomy

The chloroplast is membranous. A double membrane, or envelope, surrounds a large central space called the **stroma.** The stroma contains many different enzymes that function to reduce carbon dioxide (CO_2) and to incorporate it into organic compounds. Here, too, membrane forms the **grana.** Each granum is a stack of flattened sacs, or disks, called **thylakoids.** Grana are connected to each other by **stroma lamellae. Chlorophyll** is found within the membrane of the grana, making them the energy-generating system of the chloroplast. Each thylakoid, being saclike, contains a space. Movement of hydrogen (H^+) ions from the thylakoid space across the thylakoid membrane is believed to be important in the photosynthetic process.

Overall Equations for Photosynthesis

Sometimes the overall equation for photosynthesis is written in this manner:

$$\text{energy} + CO_2 + 2H_2O \longrightarrow (CH_2O) + H_2O + O_2$$

In this equation, (CH_2O) represents a generalized carbohydrate, the organic food produced by photosynthesis. The parentheses indicate that CH_2O is not a specific molecule. Water (H_2O) appears on both sides of the equation because water is both utilized and produced during photosynthesis. This equation also has the advantage of keeping the chemical arithmetic correct in that it is known that the *oxygen (O_2) given off by photosynthesis comes from the water*. This was proven experimentally by exposing plants first to carbon dioxide (CO_2) and then to water that contained an isotope of oxygen called heavy oxygen (O^{18}). Only when heavy oxygen was a part of water (indicated by the color red in the preceding equation) did this isotope appear in oxygen given off by the plant. Therefore, oxygen released by chloroplasts comes from water and not from carbon dioxide.

Another overall equation for photosynthesis is this:

$$\text{light energy} + 6CO_2 + 6H_2O \longrightarrow C_6H_{12}O_6 + 6O_2$$

This equation has the advantage of being the exact opposite of the equation for cellular respiration. It shows the sugar glucose as the organic food produced by photosynthesis. When we compare cellular respiration to photosynthesis at the end of this chapter, we make use of this equation. In the meantime, remember that cellular respiration takes place in mitochondria and photosynthesis takes place in chloroplasts. Plant cells have both of these organelles.

Sometimes the overall equation for photosynthesis is written like this:

$$\text{light energy} + CO_2 + H_2O \longrightarrow (CH_2O) + O_2$$

This is the simplest overall equation that can be written for photosynthesis. Like the first equation, it represents carbohydrate simply as (CH_2O). Simpler molecules than glucose first are formed during photosynthesis, and these later are joined to give larger carbohydrates.

Notice that photosynthesis is an energy-requiring synthetic reaction. As with certain other synthetic reactions, photosynthesis involves reduction: hydrogen atoms ($e^- + H^+$) are added to the carbon-oxygen combination when a carbohydrate is formed. This addition of hydrogen requires the formation of new bonds and therefore requires energy. It is not surprising to learn that carbon dioxide and water are low-energy molecules and carbohydrate is a high-energy molecule.

Chloroplasts carry on photosynthesis, an energy-requiring reaction that results in carbohydrate synthesis. First the energy of sunlight must be captured, then carbon dioxide must be reduced to carbohydrate.

Two Subpathways

An overall equation for photosynthesis does not indicate that the process actually involves 2 subpathways (fig. 6.4). The first subpathway takes place in the thylakoids, where chlorophyll is located, and the second takes place in the stroma, where many enzymes are located. The first subpathway uses light energy. As water is split and oxygen (O_2) is released, $NADPH_2$[1] and ATP are made. The second subpathway uses the $NADPH_2$ and ATP produced in the first subpathway to reduce carbon dioxide ($R{-}CO_2$) to carbohydrate.

It was discovered some years ago that the process of photosynthesis involves 2 subreactions. At that time, names were proposed for these subreactions: light reaction was suggested for the first subreaction because it requires light, and dark reaction was suggested for the second because it does not require light. The subreactions were detected by noting that when light is being

---

[1]**NADP** is similar in structure to NAD except that it carries an extra phosphate group, as indicated by the additional P.

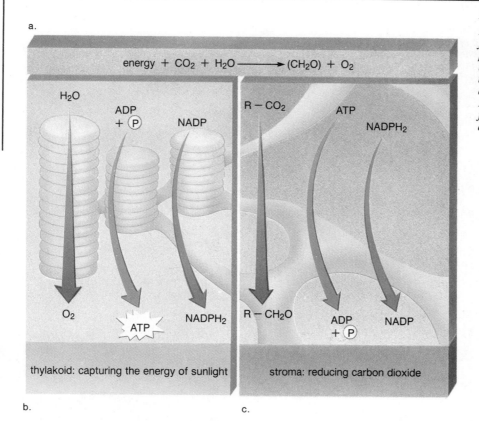

a.

$$\text{energy} + CO_2 + H_2O \longrightarrow (CH_2O) + O_2$$

b. thylakoid: capturing the energy of sunlight

c. stroma: reducing carbon dioxide

absorbed maximally by a photosynthetic system, a rise in temperature still
increases the rate of photosynthesis. This indicates that the first subpathway
is dependent primarily on light, while the second subpathway is dependent
primarily on temperature. Today, the terms *light reaction* and *dark reaction*
are being phased out because they do not adequately point out other differences between the 2 subpathways.

*The overall equation for photosynthesis does not indicate that the process involves
2 subpathways. The first subpathway captures the energy of sunlight needed by the
second to reduce carbon dioxide to carbohydrate.*

Capturing the Energy of Sunlight

The first subpathway of photosynthesis requires the participation of 2 photosystems, called **Photosystem I** and **Photosystem II.** Electron micrographs of
the thylakoid membrane show 2 different kinds of bound particles that correspond to these 2 photosystems (fig. 6.5). Each photosystem particle contains
(a) many pigment molecules that are a part of a **light-harvesting antenna** and
(b) other molecules that are electron acceptors. Like television antennae aimed
to pick up signals, the leaves of a plant turn so their antennae of the photosystems can collect as much solar energy as possible. The pigment molecules
in an antenna are either chlorophyll molecules or carotenoid molecules. One
chlorophyll *a* molecule is special; it is the reaction-center chlorophyll molecule.

Photosynthesis begins when the antenna of each photosystem absorbs
light energy and funnels it to their respective reaction center (fig. 6.6). The
reaction-center chlorophyll *a* molecule of the Photosystem I antenna has an

Figure 6.5
Grana anatomy. a. Location of grana within a chloroplast. Each granum is composed of thylakoids. b. Drawing of the particles embedded in the thylakoid membrane. These particles are believed to be structural components of the 2 photosystems involved in capturing solar energy. c. Electron micrograph of these particles.

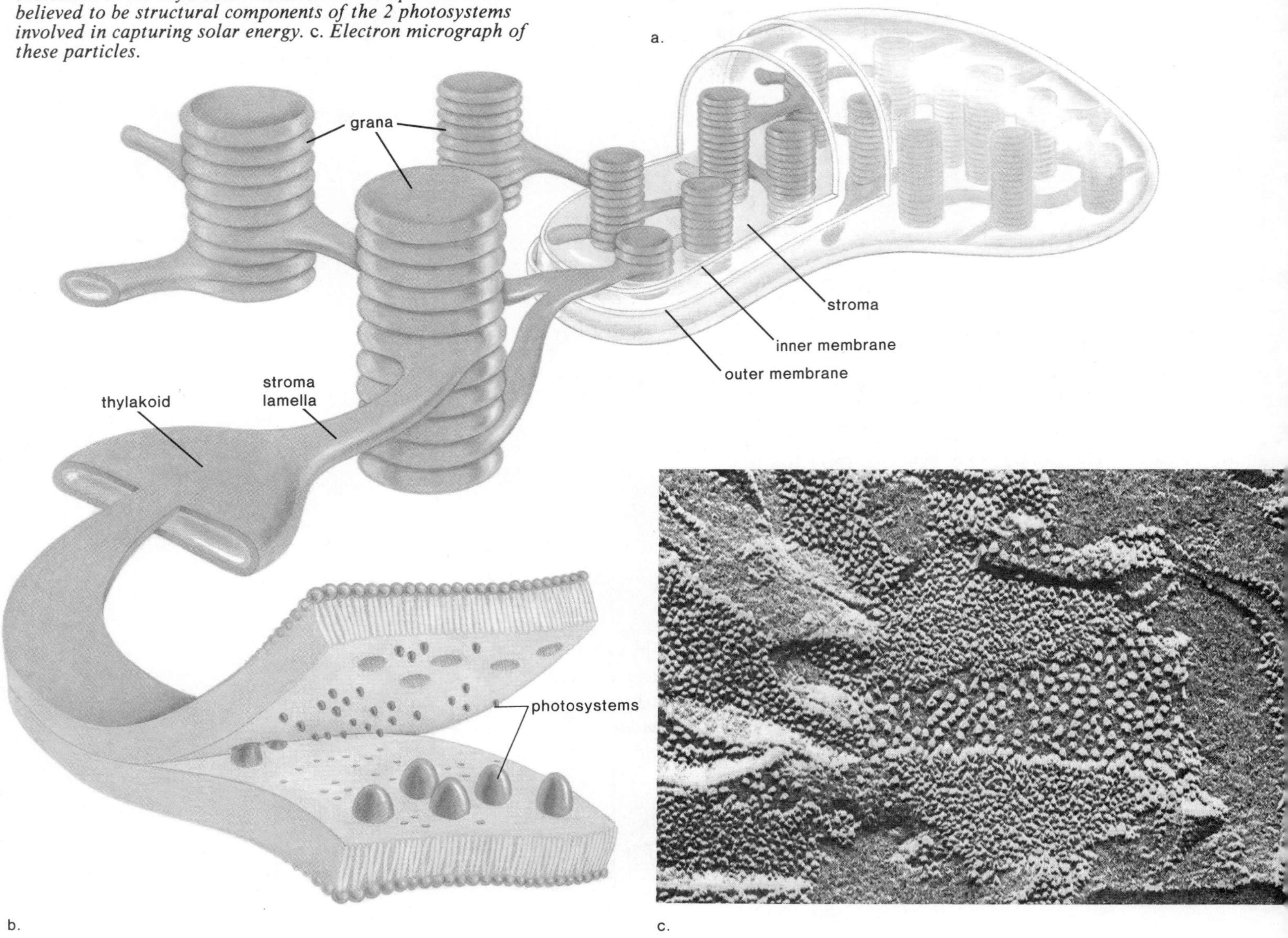

a.

grana

stroma

inner membrane

outer membrane

thylakoid

stroma lamella

photosystems

b.

c.

6.1 Critical Thinking

1. A plant exchanges materials with its environment (surroundings). What substances does it take from the environment, and what substances does it give to the environment?
2. What part of a chloroplast gives off (O_2)? What is the source of this oxygen? What part of a chloroplast takes up (CO_2)? What happens to the carbon dioxide?
3. How do plants and animals use the oxygen given off by a plant? What is the source of carbon dioxide used by a plant?
4. In what ways does a plant use newly formed glucose?

absorption spectrum that peaks at a wavelength of around 700 nm and therefore is called P700 (P stands for pigment). The reaction-center chlorophyll *a* molecule of the Photosystem II antenna has an absorption spectrum that peaks at a slightly shorter wavelength; it is called P680. The light energy received at a reaction center energizes electrons within the chlorophyll *a* molecule, and these electrons then pass to an acceptor molecule.

In a photosystem, the light-harvesting antenna absorbs solar energy and funnels it to a reaction-center chlorophyll a *molecule, which then sends energized electrons to an acceptor molecule.*

Two Electron Pathways

There are 2 possible pathways that electrons can take during the first phase of photosynthesis, but only one pathway, the **noncyclic electron pathway,** produces both ATP and NADPH₂. The other, the **cyclic electron pathway,** generates only ATP. Both pathways are diagrammed in figure 6.7.

The subpathway of photosynthesis located in the thylakoid membrane contains a noncyclic electron pathway (producing both ATP and NADPH₂) and a cyclic electron pathway (producing only ATP).

The Noncyclic Electron Pathway

From figure 6.7, you can see that each photosystem absorbs sunlight at the same time. However, it is easier to describe the events as if they occur in a sequential manner and as if they begin with Photosystem II.

1. As P680 absorbs solar energy (lower left), electrons (e^-) become so highly charged that they can leave the thylakoid-bound chlorophyll *a* molecule. The "hole" left in the molecule is filled by electrons that come from the splitting of water:

$$H_2O \longrightarrow 2H^+ + 2e^- + \tfrac{1}{2}O_2$$

The freed oxygen (O_2) is the oxygen gas given off during photosynthesis.

2. The electrons that leave P680 are received by an acceptor molecule that sends them down an electron transport system consisting of a series of thylakoid-bound carriers, some of which are cytochrome molecules. For this reason, this electron transport system often is called a **cytochrome system.** As the electrons pass downhill from one cytochrome molecule to another, energy is made available for ATP formation:

$$ADP + \textcircled{P} + energy \longrightarrow ATP$$

3. Electrons leaving the cytochrome system are picked up by P700, the reaction-center chlorophyll *a* molecule of Photosystem I. As this antenna receives solar energy, the electrons are boosted to their highest energy level yet, and they move from the reaction center to an acceptor that passes them on to NADP, which also combines with hydrogen (H^+) ions from the stroma. In the end, NADPH₂ results:

$$NADP + 2e^- + 2H^+ \longrightarrow NADPH_2$$

The preceding events (paragraphs 1–3) make up the process of **noncyclic photophosphorylation** because it is possible to trace electrons in a one-way (noncyclic) direction from water (H_2O) to NADPH₂ and because the energy from sunlight is used to produce ATP (photophosphorylation).

The noncyclic electron pathway produces both ATP and NADPH₂, which are used in the stroma to reduce carbon dioxide.

The Cyclic Electron Pathway

The other electron pathway in chloroplasts is the cyclic electron pathway. In this pathway, electrons leave P700 and eventually return to it (dotted line in figure 6.7) instead of reacting with NADP. Before they return to P700, they

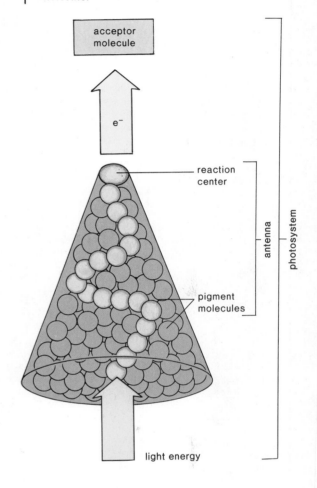

Figure 6.6

Components of a photosystem. An antenna consists of several hundred chlorophyll and carotenoid molecules plus a reaction-center chlorophyll a molecule. Light energy is absorbed and then passed from molecule to molecule (shown in yellow) until it reaches a reaction-center chlorophyll a molecule. Thereafter, energized electrons are passed to an acceptor molecule.

Figure 6.7

Photophosphorylation (ATP formation) occurs in thylakoid membranes. In noncyclic photophosphorylation, electrons move from water to P680 to an acceptor molecule that passes them down a transport system to P700, which sends them to another acceptor molecule before they are finally sent to NADP. Then

NADP combines with hydrogen (H⁺) ions from stroma and becomes $NADPH_2$. In cyclic photophosphorylation (dotted lines), electrons pass from P700 to an acceptor molecule that sends them down the electron transport system before they return to P700.

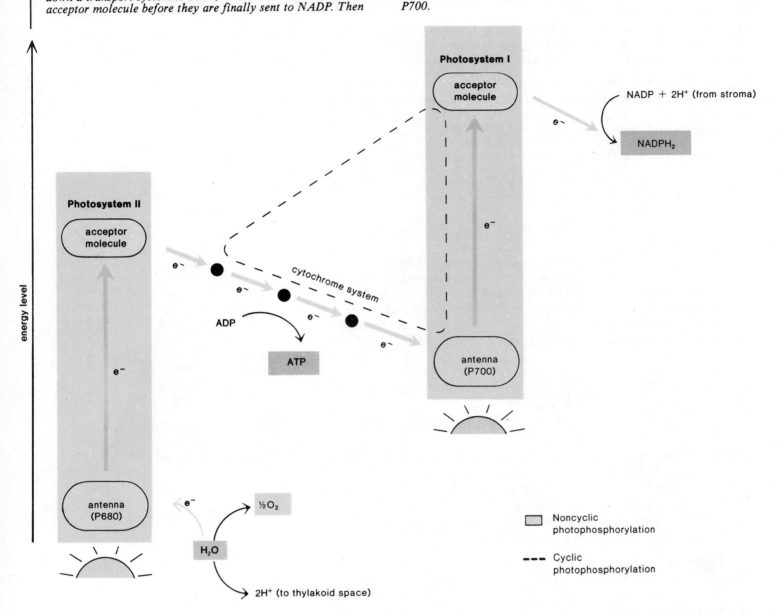

pass down the cytochrome system, and ATP is produced as previously described. This pathway sometimes is called **cyclic photophosphorylation** because it is possible to trace electrons in a cycle from P700 to P700 and because the energy from sunlight is used to produce ATP (photophosphorylation). It is believed that in plants, the cyclic pathway is utilized whenever carbon dioxide (CO_2) is in such limited supply that carbohydrate is not being produced. At these times, all available NADP is already $NADPH_2$ and there is no more NADP to receive electrons. The energized electrons from Photosystem I have to go somewhere, and under these circumstances, the cyclic electron pathway acts as a safety valve.

Even though cyclic photophosphorylation does not produce $NADPH_2$, it does provide an independent means by which ATP can be generated. Perhaps this form of photophosphorylation evolved before the noncyclic form, simply as a means to make ATP. There are some photosynthetic bacteria today that utilize the cyclic electron pathway only.

Table 6.1 Capturing the Energy of Sunlight

Noncyclic Electron Pathway

Participants	*Function*	*Results*
Water	Splits to give O_2, H^+ Electrons	Becomes O_2 in atmosphere Collects in thylakoid space Passes to Photosystem II
Photosystem II	Absorbs solar energy	Gives energized electrons to cytochrome system
Cytochrome system	Electron transport	Produces ATP
Photosystem I	Absorbs solar energy	Gives energized electrons to NADP
NADP	Final H^+ and acceptor for electrons	Becomes $NADPH_2$

Cyclic Electron Pathway

Participants	*Function*	*Results*
Photosystem I	Absorbs solar energy	Gives energized electrons to cytochrome system
Cytochrome system	Electron transport	Produces ATP

The cyclic electron pathway utilizes only Photosystem I and produces only ATP. It probably evolved before Photosystem II.

The characteristics of the first subpathway in photosynthesis are summarized in table 6.1.

Chemiosmotic ATP Synthesis

The thylakoid space acts as a reservoir for hydrogen (H^+) ions. First, for every water molecule that is split at the beginning of noncyclic photophosphorylation, 2 hydrogen ions stay behind in the thylakoid space. Second, some of the thylakoid-bound carriers not only carry electrons, they also take hydrogen ions from the stroma and deposit them in the thylakoid space. (The hydrogen ions taken up by NADP come from the stroma.)

Because of the large number of hydrogen ions in the thylakoid space compared to the stroma, an extreme electrochemical gradient is present. When these hydrogen ions flow out of the thylakoid space and down this gradient through special channel proteins (called CF_1 particles in chloroplasts), ATP is produced by an ATP synthetase enzyme associated with the channel protein (fig. 6.8). This is called chemiosmotic ATP synthesis.

Reducing Carbon Dioxide

The reduction of carbon dioxide (CO_2) occurs in the stroma of the chloroplast by means of a series of reactions known as the **Calvin-Benson cycle.** Figure 6.9 is a simplified diagram of the Calvin-Benson cycle, and figure 6.10 gives these reactions in greater detail. Usually, the Calvin-Benson cycle takes up carbon dioxide directly from the atmosphere. When it does, **C₃ photosynthesis** results.

Figure 6.8

Chloroplast function. a. Chloroplasts are organelles found in plant cells that carry on photosynthesis. b. Photosynthesis consists of 2 subpathways. The first occurs in the grana, where sunlight energy is captured, where water (H_2O) is split and gives off oxygen (O_2), and where ATP and $NADPH_2$ are produced. The second subpathway occurs in the stroma, where carbon dioxide (CO_2) is fixed and reduced after being incorporated into the Calvin-Benson cycle. Reduction uses the ATP and the $NADPH_2$ from the first subpathway. c. Chemiosmotic ATP synthesis. When water is split, the hydrogen (H^+) ions remain in the thylakoid space. Also, some of the cytochrome system carriers deposit hydrogen (H^+) ions inside the thylakoid space. When these flow out of the thylakoid space and down their electrochemical gradient through a special channel protein called CF_1 particle, ATP is produced.

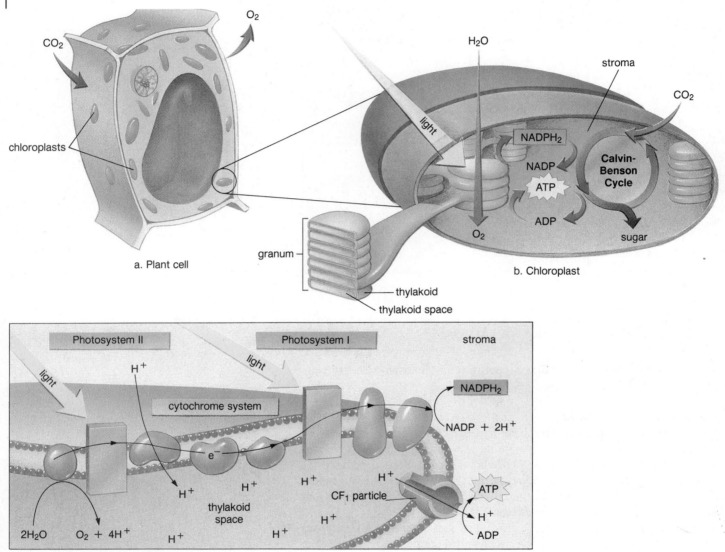

a. Plant cell

b. Chloroplast

c. Chemiosmotic ATP synthesis

C₃ Photosynthesis

The Calvin-Benson cycle, which was named for the men who discovered it, can be divided into 3 stages: carbon dioxide (CO_2) fixation, carbon dioxide reduction, and **ribulose bisphosphate (RuBP)** regeneration.

Carbon Dioxide Fixation

During the first reaction of the Calvin-Benson cycle, RuBP, a 5-carbon molecule, combines with carbon dioxide (CO_2). The resulting 6-carbon molecule immediately breaks down to form 2 molecules of PGA (phosphoglycerate), a

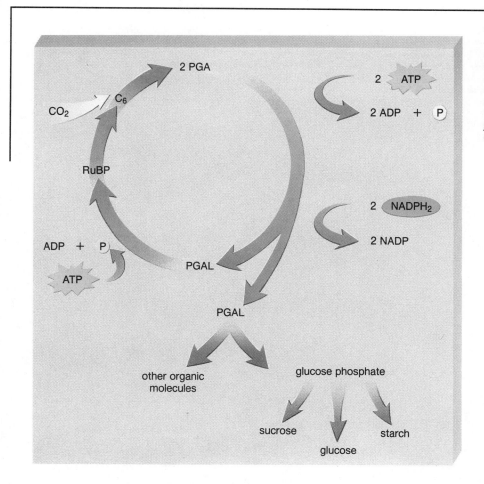

Figure 6.9
The Calvin-Benson cycle (simplified). RuBP accepts carbon dioxide (CO_2), forming a 6-carbon molecule (C_6), which immediately breaks down to 2 PGA molecules that are reduced to 2 PGAL molecules, one of which represents the net gain of the cycle. One PGAL molecule can be combined with another PGAL molecule to give glucose-6 phosphate, which can be metabolized to other organic molecules.

3-carbon molecule. When a C_3 molecule is the first molecule detected following carbon dioxide fixation (the attachment of carbon dioxide to an organic molecule[2]), it is called C_3 photosynthesis.

Reduction of Carbon Dioxide
Each molecule of PGA now undergoes reduction to PGAL (phosphoglyceraldehyde) in 2 steps:

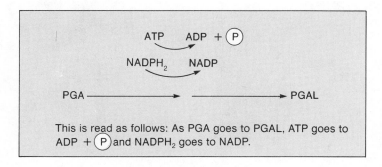

This is read as follows: As PGA goes to PGAL, ATP goes to ADP + (P) and NADPH₂ goes to NADP.

The reactions that reduce PGA to PGAL use the $NADPH_2$ and the ATP produced by noncyclic photophosphorylation. Look again at the overall equation for photosynthesis on page 126. Notice that carbohydrate (CH_2O) cotains hydrogen (H) atoms, whereas carbon dioxide (CO_2) does not. When carbon (C) binds to hydrogen atoms, it has accepted electrons plus hydrogen ions. Therefore, carbon has been reduced. The reduction process is a building-up or a

---

[2]Not to be confused with *carbon fixation*, which refers to both the uptake and the reduction of carbon dioxide to a carbohydrate.

Figure 6.10

The Calvin-Benson cycle (detailed). The molecules in the cycle have been multiplied by 3 because after 5 molecules of PGAL (15 carbons) are converted to 3 molecules of RuBP (15 carbons), one molecule of PGAL remains. Since it takes 2 PGAL molecules to produce one 6-carbon sugar, the numbers shown must be remultiplied by 2 in order for a monosaccharide to be the end product. Compare steps 3 and 4 to steps 5 and 4 in figure 5.14.

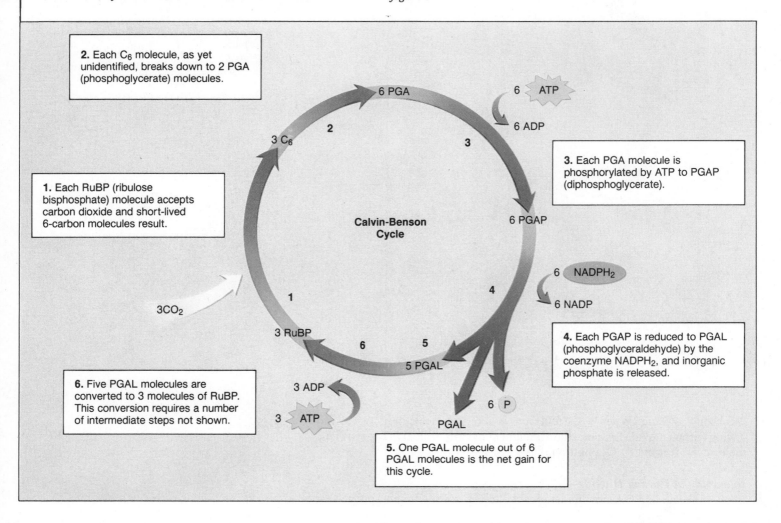

2. Each C_6 molecule, as yet unidentified, breaks down to 2 PGA (phosphoglycerate) molecules.

1. Each RuBP (ribulose bisphosphate) molecule accepts carbon dioxide and short-lived 6-carbon molecules result.

3. Each PGA molecule is phosphorylated by ATP to PGAP (diphosphoglycerate).

4. Each PGAP is reduced to PGAL (phosphoglyceraldehyde) by the coenzyme $NADPH_2$, and inorganic phosphate is released.

6. Five PGAL molecules are converted to 3 molecules of RuBP. This conversion requires a number of intermediate steps not shown.

5. One PGAL molecule out of 6 PGAL molecules is the net gain for this cycle.

Calvin-Benson Cycle

6 PGA
3 C_6
6 ATP
6 ADP
6 PGAP
6 $NADPH_2$
6 NADP
6 P
$3CO_2$
3 RuBP
5 PGAL
PGAL
3 ADP
3 ATP

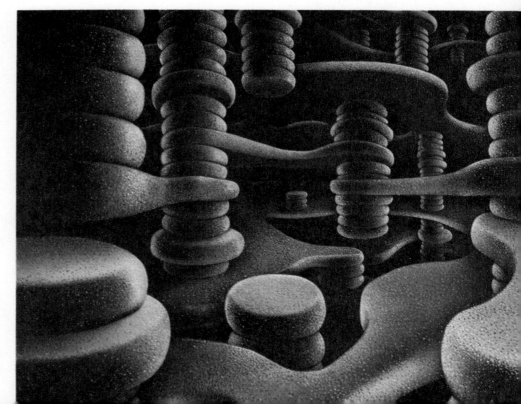

The interior of a chloroplast. The stroma of a chloroplast is a fluid-filled space surrounding the grana. In this drawing the grana held together by the stroma lamellae look like majestic pillars built from thylakoids placed one on top the other. The grana make ATP and $NADPH_2$ so that the enzymes of the stroma can reduce carbon dioxide to a carbohydrate.

Table 6.2 *Reducing Carbon Dioxide*

Participant	Function	Results
RuBP	Takes up CO_2	CO_2 fixation
CO_2	Provides carbon atoms	Reduced to CH_2O
ATP	Provides energy for reduction of CO_2 and generation of RuBP	Broken down to ADP + (P)
$NADPH_2$	Provides hydrogen atoms for reduction	Oxidized NADP
PGAL	End product of photosynthesis	Glucose phosphate from 2 PGAL

synthetic process because it requires the formation of new bonds. Hydrogen atoms and energy are needed for reduction synthesis, and these are supplied by $NADPH_2$ and ATP.

PGAL is the immediate photosynthetic product of the Calvin-Benson cycle. For every 6 turns of the Calvin-Benson cycle, there is a net gain of one PGAL molecule. Therefore, *PGAL sometimes is said to be the end product of photosynthesis*. However, since 2 PGAL molecules can combine to form glucose phosphate (or simply glucose), often is called the end product of photosynthesis. Within the leaves of plants, glucose molecules are converted to the disaccharide sucrose for transport to other parts of the plant.

Regeneration of RuBP

For every 6 turns of the Calvin-Benson cycle, 5 molecules of PGAL are used to reform 3 molecules of RuBP:

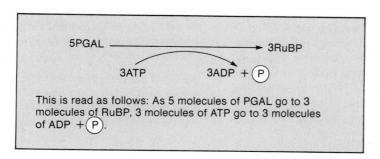

This is read as follows: As 5 molecules of PGAL go to 3 molecules of RuBP, 3 molecules of ATP go to 3 molecules of ADP + (P).

This reaction also utilizes some of the ATP produced by photophosphorylation.

During photosynthesis, ATP and $NADPH_2$ are utilized to reduce carbon dioxide to PGAL within the stroma. Two PGAL molecules then join to form glucose phosphate.

Table 6.2 summarizes the reactions that occur during reduction of carbon dioxide (CO_2).

C_4 Photosynthesis

In some plants, the first detected product of carbon dioxide (CO_2) fixation is not a C_3 molecule as it is when the 5-carbon RuBP of the Calvin-Benson cycle takes up carbon dioxide from the atmosphere. Instead, a C_4 molecule is the first detected product because carbon dioxide first is taken up by a 3-carbon organic molecule before it eventually is passed to the Calvin-Benson cycle. It

6.2 Critical Thinking

1. Energy is needed to maintain cellular and organismal organization. The oxidation of glucose releases 686 Kcal/mole. Specifically, what do cells do with the energy of glucose breakdown?
2. How much energy do you suppose it takes to reduce carbon dioxide (CO_2) to glucose? From where does this energy ultimately come?
3. Why would it be correct to say that photosynthesis drives cellular respiration?
4. In general, what role is played by glucose in all organisms?

The world's population continues to increase, and as it does, it encroaches on agricultural land. In the United States, where there were once fields of grain, there are now towns, suburbs, and shopping malls. More food must be produced on less land; therefore, agricultural yields must be increased constantly. Despite advanced technology and the development of mutant plants, the average yield per acre has not significantly increased in recent years. New ideas are needed, and one may have been discovered.

Increasing Crop Yields: C_3 versus C_4 Photosynthesis

Observation has shown that temperate-zone plants, such as wheat, alfalfa, and potatoes, do not take up carbon dioxide (CO_2) as efficiently as plants such as corn, which are adapted to climates of high-light intensity and high temperature. The fault lies with the first enzyme of the Calvin-Benson cycle. This enzyme can catalyze one of 2 reactions:

1. $CO_2 + RuBP \longrightarrow \longrightarrow C_3$
2. $O_2 + RuBP \longrightarrow \longrightarrow CO_2 + H_2O$

The first reaction is a normal part of **photosynthesis**. The second reaction is called photorespiration because oxygen (O_2) is taken up and carbon dioxide is given off to the environment. Photorespiration, which competes with the photosynthetic reaction, accounts for the lower yield of temperate-zone plants because it obviously does not lead to carbohydrate synthesis.

How is photorespiration avoided in plants that are adapted to a hot, dry climate? First, chloroplasts occur in bundle sheath cells, so called because they encircle the vascular tissue of the leaf. The other cells of

the leaf specialize in capturing carbon dioxide and passing it to the bundle sheath cells. Carbon dioxide is captured by a reaction that is catalyzed by an enzyme called pepco. This reaction occurs even when there is a low concentration of carbon dioxide in the leaves:

3. $CO_2 + PEP \longrightarrow C_4$

These C4 molecules are transported into the bundle sheath cells, where carbon dioxide is released (see figure).

Notice that it is possible to make a distinction between the 2 groups of plants according to the first molecule detected following carbon dioxide uptake. In temperate-zone plants, C_3 is always the first molecule detected following carbon dioxide uptake (see reaction number 1), and in the tropical-zone plants under discussion, C_4 is the first molecule detected (see reaction number 3). Therefore, it is now customary to speak of C_3 versus C_4 photosynthesis—C_4 being much more efficient. Scientists now are exploring the possiblity of transforming temperate-zone plants into C_4 photosynthesizers. If this can be accomplished, agricultural yields will be greatly increased.

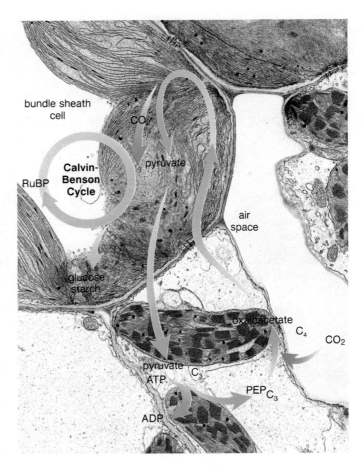

C_4 photosynthesis. The C_4 pathway is superimposed on an electron micrograph of the cells of the mesophyll and the bundle sheath in corn. During the C_4 pathway, pepco fixes carbon dioxide (CO_2) to PEP, forming a C_4 molecule that carries carbon dioxide to the Calvin-Benson cycle in bundle sheath cells. Pyruvate, a C_3 molecule, then returns to the mesophyll, where ATP is used to reform PEP.

has been discovered that plants carrying on C_4 **photosynthesis** are adapted to a hot, dry climate and are able to fix carbon dioxide even when the concentration of carbon dioxide in the leaf is low. The reasons for this are discussed in the reading above.

Carbohydrate Utilization

Carbohydrate (PGAL, glucose phosphate, sucrose, and so forth), synthesized by photosynthesis in the leaves, is transported to other parts of the plant, usually in the form of sucrose. Although carbohydrate can be stored temporarily in the form of starch, much of this carbohydrate eventually is broken down by cellular respiration (just as in animals) to produce the ATP needed for cell metabolism.

Plants not only have the enzymatic capability to produce sugars from carbon dioxide (CO_2) and water (H_2O), they also have the ability, by way of the metabolic pathways diagrammed in figure 5.13 and by even more complex pathways, to produce all the organic molecules they require:

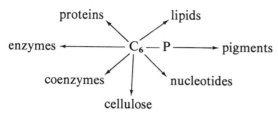

In comparison to the animal cell, algal and plant cells have enormous biochemical capabilities. Biochemically speaking, plant cells are due our utmost respect because they are capable of making all necessary organic molecules using only inorganic molecules as nutrients.

Autotrophic Bacteria

Bacterial Photosynthesis

Although the cyanobacteria carry on photosynthesis much like plants, some other photosynthetic bacteria do not release oxygen (O_2) when they photosynthesize because they do not use water (H_2O) as a hydrogen (H) donor. The green sulfur bacteria and the purple sulfur bacteria use hydrogen gas (H_2) and hydrogen sulfide (H_2S) as hydrogen donors. These bacteria usually live in anaerobic conditions, like the muddy bottoms of marshes, and cannot photosynthesize in the presence of oxygen.

Chemosynthesis

Some bacteria can oxidize inorganic compounds, such as ammonia, nitrates, and sulfides, and can trap the small amount of energy released to synthesize carbohydrate. This is called **chemosynthesis.** While this capability is not believed to contribute greatly to support of life on land, it recently has been found sufficient for support of an entire community 3 km below sea level. Using deep-diving research submarines, scientists have been examining the midoceanic ridge system, where hot minerals spew out from the inner earth. Here, bacteria in the vent water oxidize hydrogen sulfide (H_2S) and carry on chemosynthesis. Chemosynthetic bacteria also live within the cells of giant tubeworms (fig. 6.11) and colonize the gills of giant clams. The blood of these animals is specialized to carry hydrogen sulfide to the bacteria, and in return, the bac-

Figure 6.11

Deep-sea ecosystem. At midoceanic ridges, there are dense clusters of tubeworms and clams plus other types of animals. Both tubeworms and clams are home to chemosynthetic bacteria that derive energy by oxidizing hydrogen sulfide (H_2S). The bacteria then produce organic molecules that are utilized by their hosts.

6.3 Critical Thinking

The structure of chloroplasts and mitochondria can be compared.

1. What part of a mitochondrion compares to the stroma of a chloroplast?
2. Are the biochemical processes in each of these parts the same or the opposite? Explain.
3. What part of a mitochondrion compares to the thylakoid membrane of a chloroplast? to the thylakoid space of a chloroplast?
4. Are the biochemical processes in each of these parts the same or the opposite?

Except for cyanobacteria, photosynthetic bacteria do not evolve oxygen as do plants because they utilize hydrogen from a source other than water. Chemosynthetic bacteria produce their own food by oxidizing inorganic compounds and therefore do not require light at all.

Comparison of Cellular Respiration and Photosynthesis

Differences

Both plant and animal cells carry on cellular respiration but only plant cells photosynthesize. The cellular organelle for cellular respiration is the mitochondrion, while the cellular organelle for photosynthesis is the chloroplast.

The overall equation for aerobic cellular respiration is the opposite of that for photosynthesis:

$$energy + 6CO_2 + 6H_2O \rightleftharpoons C_6H_{12}O_6 + 6O_2$$

The reaction in the forward direction represents photosynthesis, and the energy is the energy of the sun. The reaction in the opposite direction represents cellular respiration, and the energy is ATP.

Obviously, photosynthesis is the building up of glucose, while cellular respiration is the breaking down of glucose. See table 6.3 for a summarized list of differences between these processes.

Table 6.3 *Cellular Respiration versus Photosynthesis*

Cellular Respiration	Photosynthesis
Mitochondrion	Chloroplast
Oxidation	Reduction
Releases energy	Requires energy
Requires O_2	Releases O_2
Releases CO_2	Requires CO_2

Figure 6.12

Diagram illustrating similarities and differences of photosynthesis, which takes place in chloroplasts, and cellular respiration, which takes place in mitochondria. Both have a cytochrome system, located within membranes, where ATP is produced. Both have enzyme-catalyzed reactions in solution. The coenzyme NAD(P) operates in the membrane and the solution. Photosynthesis releases oxygen (O_2) and reduces carbon dioxide (CO_2) into carbohydrate; cellular respiration reduces oxygen and releases carbon dioxide.

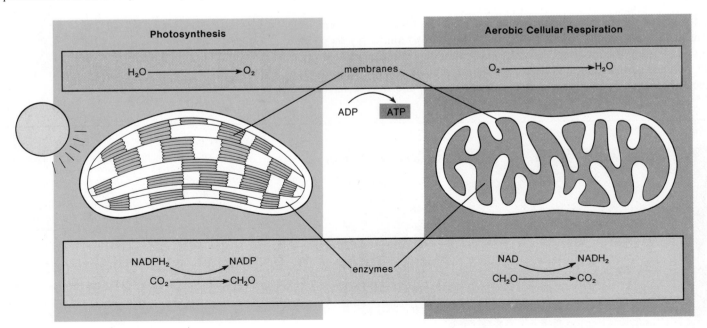

Similarities

Both photosynthesis and cellular respiration are metabolic pathways within cells and therefore consist of a series of reactions that the overall reaction does not indicate. Both pathways make use of a cytochrome system located in membrane (fig. 6.12) to generate a supply of ATP, and both make use of a hydrogen carrier—cellular respiration uses NAD, and photosynthesis uses NADP.

Both pathways utilize this overall reaction, but in opposite directions. For photosynthesis, read from left to right in the following diagram, and for cellular respiration, read from right to left.

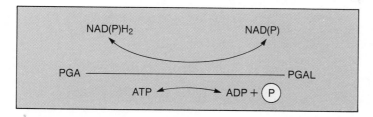

Both photosynthesis and cellular respiration occur in plant cells. While both of these occur during the daylight hours, only cellular respiration occurs at night. During daylight hours, the rate of photosynthesis exceeds the rate of cellular respiration, resulting in a net increase and storage of glucose. The stored glucose is used to support cellular metabolism, which continues during the night.

Summary

Photosynthesis, carried out largely by plants and algae, is absolutely essential for the continuance of life because it supplies the biosphere with food and energy. Photosynthesis takes place in chloroplasts and requires 2 subpathways: photophosphorylation occurs within thylakoid membranes of the grana, and the Calvin-Benson cycle occurs within the stroma. Noncyclic photophosphorylation produces the ATP and NADPH$_2$ needed to reduce carbon dioxide CO$_2$. During C$_3$ photosynthesis, carbon dioxide enters the Calvin-Benson cycle directly from the atmosphere and PGAL molecules exit from it. Two PGAL molecules join to form glucose phosphate, a molecule that can be metabolized to all the substances a plant needs. Table 6.4 summarizes the roles played by the various participants in photosynthesis.

Table 6.4 *Participants in Photosynthesis*

	Participant	Role
Thylakoid membrane	Sunlight	Provides energy
	Chlorophyll	Absorbs energy
	Water	Donates electrons and releases oxygen
	ADP + ℗	Forms ATP
	NADP	Accepts electrons and H$^+$ and becomes NADPH$_2$
Stroma	RuBP	Takes up CO$_2$
	CO$_2$	Reduced to PGAL
	ATP	Provides energy for reduction
	NADPH$_2$	Provides electrons for reduction
	2 PGAL	Becomes glucose phosphate

Like plants, cyanobacteria carry out photosynthesis, but other types of bacteria do not release oxygen when they photosynthesize. Chemosynthetic bacteria remove hydrogen (H) atoms from a number of substances in order to capture energy. A surprising find is deep-ocean communities supported by chemosynthetic bacteria.

Study Questions

1. Why are almost all living things dependent upon the process of photosynthesis and the energy of the sun? (p. 123)
2. Which rays of light are most important for photosynthesis? Why? (p. 124)
3. Give an overall equation for photosynthesis and the equations for the 2 subpathways involved in the process. (pp. 125–27)
4. Describe the structure of a chloroplast, and indicate where photophosphorylation and the Calvin-Benson cycle occur. In what way are these 2 processes connected? (pp. 125, 127, 131)
5. Trace the path of electrons during noncyclic photophosphorylation and during cyclic photophosphorylation. (pp. 129–30)
6. Give the primary steps of the Calvin-Benson cycle, indicating the reaction that represents the reduction of carbon dioxide (CO$_2$). (pp. 131–35)

7. Describe the role of each participant in both subpathways of photosynthesis. (pp. 131, 135)

8. Why is it correct to say that a plant cell is more biochemically competent than an animal cell? (p. 137)

9. Discuss bacterial photosynthesis. (p. 137)

10. Contrast cellular respiration and photosynthesis in at least 5 ways. How are the 2 cellular processes similar? (pp. 138–39)

Objective Questions

1. Life requires a continual supply of energy. Ultimately, this energy comes from _____ .

2. An organism that makes its own food is called an _____ .

3. The large fluid-filled central space within a chloroplast is called the _____ .

4. The photosystems are located within the _____ membranes of the grana.

5. Chlorophyll molecules are located within the _____ of the photosystems.

6. Lying between Photosystem I and Photosystem II, the _____ system converts ADP to ATP.

7. In addition to ATP, noncyclic photophosphorylation produces the _____ needed by the Calvin-Benson cycle.

8. During C_3 photosynthesis, RuBP of the Calvin-Benson cycle fixes _____ .

9. The Calvin-Benson cycle is found in the _____ of the chloroplast.

10. The thylakoid membranes within chloroplasts can be compared to the _____ within mitochondria because they both contain a cytochrome system.

Answers to Objective Questions

1. sunlight 2. autotroph 3. stroma 4. thylakoid 5. antennae 6. cytochrome 7. NADPH₂ 8. carbon dioxide 9. stroma 10. cristae

Label this Diagram.

See figure 6.3*a* (p. 125) in text.

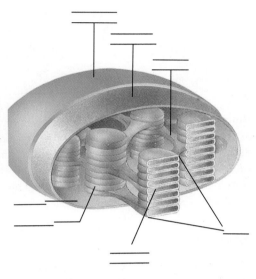

Selected Key Terms

autotroph (aw′to-trōf) an organism that is capable of making its food (organic molecules) from inorganic molecules. *123*

C₃ photosynthesis (se thre fo″to-sin′thĕ-sis) photosynthesis that utilizes the Calvin-Benson cycle to take up and reduce carbon dioxide; so named because the first molecule detected after CO₂ uptake is a C₃ molecule. *131*

C₄ photosynthesis (se for fo″to-sin′thĕ-sis) photosynthesis in which the first detected molecule following CO₂ uptake is a C₄ molecule. Later, this same CO₂ is made available to the Calvin-Benson cycle. *136*

Calvin-Benson cycle (kal′vin ben′sun si′kl) a circular series of reactions by which CO₂ fixation occurs within chloroplasts. *131*

chemosynthesis (ke″mo-sin′the-sis) the process of making food by using energy derived from the oxidation of reduced molecules in the environment. *137*

chlorophyll (klo′ro-fil) the green pigment found in photosynthesizing organisms that is capable of absorbing energy from the sun's rays. *125*

cyclic photophosphorylation (sik′lik fo″to-fos″for-i-la′shun) the synthesis of ATP by a cytochrome system within thylakoid membranes utilizing electrons received from and returning to P700 chlorophyll molecules within Photosystem I. *130*

cytochrome system (si′to-krōm sis′tem) a series of cytochrome molecules present in thylakoid membranes and inner mitochondrial membranes that pass electrons one to the other from a higher energy level to a lower energy level; the energy released is used to build ATP. *129*

granum (gra′num) stack of flattened membranous vesicles in a chloroplast where chlorophyll is located and photosynthesis begins. *125*

heterotroph (het′er-o-trof″) an organism that cannot synthesize organic compounds from inorganic substances and therefore must acquire food from external sources. *123*

NADP a coenzyme of reduction; a hydrogenase that frequently donates hydrogen atoms to metabolites. *126*

noncyclic photophosphorylation (non-sik′lik fo″to-fos″for-i-la′shun) the passage of solar-energized electrons within chloroplasts from water to NADP; results in the generation of ATP and NADPH₂. *129*

photosynthesis (fo″to-sin′the-sis) in plants, the process of making carbohydrate from carbon dioxide and water by using the energy of the sun. *123*

Photosystem I and **Photosystem II** (fo″to-sis′tem) molecular units located within the membrane of a thylakoid that capture solar energy, making photophosphorylation possible. *127*

ribulose bisphosphate (ri′bu-lōs bis-fos′fāt)(RuBP) the molecule that acts as the acceptor for carbon dioxide within the Calvin-Benson cycle. *132*

stroma (stro′mah) the interior portion of a chloroplast. *125*

stroma lamella (stro′mah lah-mel′e) membranous connection between adjacent thylakoids of the grana. *125*

thylakoid (thi′lah-koid) an individual flattened vesicle found within a granum (pl. grana). *125*

Chapter 7

Chapter Concepts

1 The vegetative organs of a plant are the roots, the stems, and the leaves.

2 Shoot tips and root tips continually grow and produce new cells that become specialized for particular functions.

3 Both roots and stems in woody plants increase not only in length but also in girth.

4 Leaves contain cells that are specialized to carry on photosynthesis.

Chapter Outline

The Flowering Plant
    Monocots and Dicots
    Tissue and Cell Types
The Root System
    Dicot Roots
    Monocot Roots
The Shoot System
    Primary Growth of Stems
    Secondary Growth of Stems
Leaves
    Leaf Veins

Plant Organization and Growth

The Flowering Plant

The body of a flowering plant has 2 divisions: the root system and the shoot system (fig. 7.1). The root system anchors the plant in the soil, and the stem holds the leaves aloft to catch the rays of the sun. The roots absorb water (H_2O) and minerals; these nutrients then are transported in the stem to the leaves, which receive carbon dioxide (CO_2) from the air and carry on photosynthesis. A carbohydrate (CH_2O) produced by photosynthesis first is transported in the stem to the rest of the plant until finally it reaches the root, where it is stored as starch.

Monocots and Dicots

Flowering plants are divided into the **monocots** and the **dicots** depending on the type of cotyledon(s) or seed leaf (leaves) the embryonic plant has. **Cotyledons** provide nutrient molecules for growing embryos before the true leaves begin photosynthesizing. Adult monocots and dicots have several other structural differences, as illustrated in figure 7.2.

Figure 7.1

Root system and shoot system of a plant. A plant has 2 main divisions: the root system below ground and the shoot system containing the stems and the leaves above ground. Vascular tissue transports materials from the roots to the leaves and vice versa. A plant grows lengthwise by the production of new cells at the terminal bud (shoot tip) and the root tip.

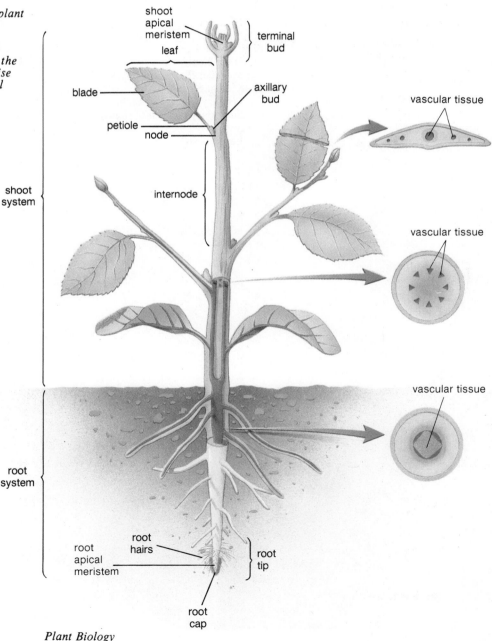

Figure 7.2

Monocots versus dicots. Four features are used to distinguish monocots from dicots: the number of cotyledons, the arrangement of vascular bundles, the pattern of leaf veins, and the number of flower parts.

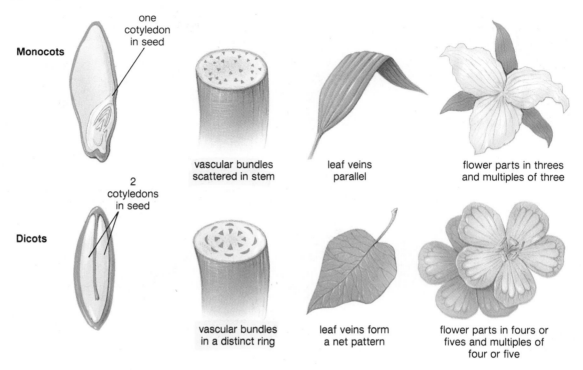

Monocots

one cotyledon in seed

vascular bundles scattered in stem

leaf veins parallel

flower parts in threes and multiples of three

Dicots

2 cotyledons in seed

vascular bundles in a distinct ring

leaf veins form a net pattern

flower parts in fours or fives and multiples of four or five

Table 7.1 Vegetative Organs and Major Tissues

	Roots	Stems	Leaves
Function	Absorb water and minerals Anchor plant Store materials	Transport water and nutrients Support leaves Help store materials	Carry on photosynthesis
Tissue Epidermis*	Root hairs absorb water and minerals	Protect inner tissues	Stomata carry on gas exchange
Cortex†	Store water and products of photosynthesis	Carry on photosynthesis, if green	————
Endodermis†	Regulate passage of minerals into vascular tissue		
Vascular‡	Transport water and nutrients	Transport water and nutrients	Transport water and nutrients
Pith†	Store water and products of photosynthesis	Store products of photosynthesis	
Mesophyll†	————	————	Carry on gas exchange and photosynthesis

MERISTEM TISSUE

Note: Plant tissues belong to one of 3 tissue systems:
*Dermal tissue system
†Ground tissue system
‡Vascular tissue system

Tissue and Cell Types

This chapter considers the basic anatomy of roots, stems, and leaves of a flowering plant. These are vegetative organs, meaning they are not concerned with reproduction. Some of the tissues found within each organ are listed in table 7.1. In addition to the tissue types listed, plants also contain embryonic tissue called **meristem tissue,** which is unspecialized and is capable of continual cell

Figure 7.3

Plant cell types. A meristem cell is embryonic, and by the process of maturation, it may become one of the specialized cells shown. The epidermal cell of the dermal tissue system is found in epidermis, the outermost tissue in all organs of the plant. Parenchymal and sclerenchymal cells of the ground tissue system are found in cortex and pith, for example. Vessel elements and tracheids of the vascular tissue system are found in xylem (fig. 8.2). The sieve-tube cells and their companion cells of the vascular tissue system are found in the phloem (fig. 8.5).

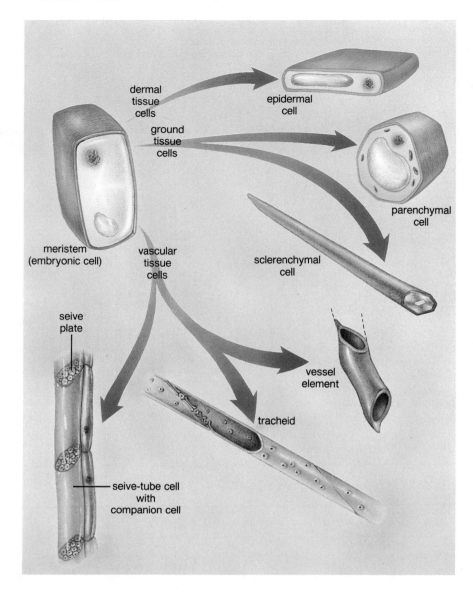

division. Cell division is followed by differentiation into the cell types depicted in figure 7.3. **Parenchyma** and **sclerenchyma** are found in most tissues. Parenchymal cells are relatively unspecialized and correspond best to the generalized cell of a plant (fig. 2.3*b*). Sclerenchymal cells are hollow, nonliving cells with extremely strong walls that support plant tissues and organs. Two tissue types, epidermis and endodermis, do not have parenchymal and sclerenchymal cells. **Epidermis,** composed of epidermal cells, covers the entire body of nonwoody and young woody plants. Epidermis protects inner body parts and prevents the plant from drying out. In addition, the epidermis has specialized structures and functions that are discussed when each vegetative organ is considered. In contrast to epidermis, **endodermis,** which nearly universally is found in roots, contains only endodermal cells. The endodermis is involved in regulating the flow of ions and molecules in the root.

The other cell types shown in figure 7.3 are found in the vascular (transport) tissue, xylem and phloem. **Xylem** transports water (H_2O) and minerals from the roots to the leaves, and **phloem** transports organic nutrients, usually from the leaves to the roots. Xylem contains 2 types of conducting cells: tracheids and vessel elements. Both of these types of conducting cells are hollow and nonliving.

Plants need only inorganic nutrients in order to produce all the organic molecules that make up their body. Aside from carbon (C), hydrogen (H), and oxygen (O_2) (obtained from carbon dioxide (CO_2) and water (H_2O)), the mineral elements listed in table 7.A are required nutrients for plants. The mineral elements are classified as **macronutrients** when they are used by plants in great amounts and as **micronutrients** when they are needed by plants in very small amounts. Both types of minerals are found in the soil but in low concentrations; not only must a plant be able to take them up, it must also be able to concentrate them. Fortunately, the root system of a plant is designed for just this purpose. As the root system grows, it branches and branches again so that the roots are exposed to a tremendous amount of soil. It has been estimated that a rye plant has roots totaling about 900 km (more than 650 mi) in length.

Further, because of the extensive number of root hairs, the total surface area is about 635 m², or more than 7,000 square feet! Water, and possibly minerals too, enters root hairs by diffusion, but eventually active transport is used to concentrate the minerals within the organs of a plant. A plant uses a great deal of ATP for active transport.

We are lucky that plants can concentrate minerals, for we often are dependent on them for our basic supply of such ions as sodium to maintain blood pressure, calcium to build bones and teeth, and iron to help carry oxygen to our cells. Once plants have taken the minerals up, they are incorporated into proteins, fats, and vitamins. When we eat plants, we are supplied with minerals and all types of organic molecules, some of which become building blocks for our own cells and some of which are used as an energy source.

Table 7.B lists examples of plant foods that humans consume. Each type of food is associated with a particular organ of a flowering plant: root, stem, leaf, or flower.

Plants as Food

Table 7.A *Inorganic Nutrients Necessary for Plant Life*

Compound	Element Supplied
Macronutrients	
KNO_3	K;N
$CaNO_3$	Ca;N
$NH_4H_2PO_4$	N;P
$MgSO_4$	Mg;S
Micronutrients	
KCl	Cl
H_3BO_3	B
$MnSO_4$	Mn
$ZnSO_4$	Zn
$CuSO_4$	Cu
H_2MoO_4	Mo
Fe-EDTA	Fe

From E. Epstein, *Mineral Nutrition of Plants: Principles and Perspectives.* Copyright © 1972 John Wiley & Sons Inc., New York, NY. Reprinted by permission of John Wiley & Sons, Inc.

Table 7.B *Food from Plants*

Plant Part	Foods
Roots	Sweet potato, beet, radish, carrot, turnip, parsnip
Stems	White potato, sugar cane, asparagus
Leaves	Cabbage, kale, spinach, lettuce, tea leaves
Petioles*	Celery, rhubarb
Seeds †	Pea, navy bean, lima bean, nuts, coffee bean
Fruits †	Wheat, rice, corn, oat, rye, string bean, apple, orange, peach, tomato, squash

* Part of a leaf
† Derived from flower parts

The conducting cells of phloem are sieve-tube cells, each of which has a companion cell. Sieve-tube cells, which have perforated end walls called sieve plates, contain cytoplasm but no nuclei. Strands of cytoplasm, called plasmodesmata, extend from one cell to the other through the sieve plates.

The body of a plant is composed of tissues differentiated to perform various functions and containing specialized cells. One tissue, the meristem, remains undifferentiated and continually capable of dividing and producing new cells.

The Root System

The root system functions to anchor a plant in the soil and to absorb water and minerals from the soil. The reading above discusses this vital absorption activity and how it relates to the extensive formation of plant roots. Roots also store the products of photosynthesis they receive from the leaves.

Dicot Roots

A longitudinal section of a dicot root tip (fig. 7.4*a*) reveals zones where cells are in various stages of maturation. Below the *zone of cell division* (root apical meristem), cells are being added continuously to the **root cap,** a thimble-shaped mass of parenchymal cells that is a protective covering for the root tip. Cells in the root cap have to be replaced constantly because they are ground off as the root pushes through rough soil particles. Outer root-cap cells contain a slimy substance that assists movement through the soil. Above the zone of cell

Plant Organization and Growth

Figure 7.4

Dicot root tip. a. Root tip is divided into 4 regions, best seen in a longitudinal section such as this. b. Vascular cylinder of a dicot root contains the vascular tissue. Xylem is typically star shaped, and phloem lies between the points of the star. c. Casparian strip. Because of the Casparian strip (waxy substance), water and solutes must pass through the endodermal cells. In this way, endodermal cells regulate the passage of materials into the vascular cylinder.

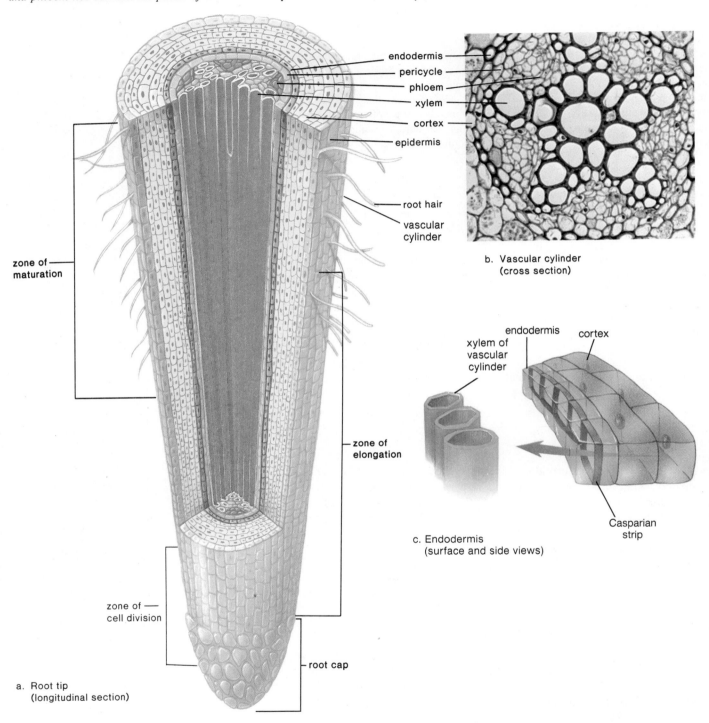

endodermis
pericycle
phloem
xylem
cortex
epidermis

root hair
vascular cylinder

b. Vascular cylinder (cross section)

zone of maturation

xylem of vascular cylinder
endodermis
cortex

Casparian strip

c. Endodermis (surface and side views)

zone of elongation

zone of cell division

root cap

a. Root tip (longitudinal section)

division, cells also are being added continuously. Here, they elongate as they become specialized. Therefore, this region of a root is called the *zone of elongation*. Finally, there is a region in which cells are mature and specialized. This region is called the *zone of maturation*. The zone of maturation is recognizable even in a whole root because root hairs are borne by many of the epidermal cells. Root hairs add tremendously to the total absorptive surface area of roots.

The absorbed water and minerals pass through the cortex, a tissue composed of parenchymal cells. The water and minerals must pass through the endodermis before entering the **vascular cylinder** (fig. 7.4b). The single layer of endodermal cells fit snugly together and are bordered on 4 of their 6 sides by a strip of waxy material known as the **Casparian strip.** This strip does not permit water and solutes to pass *between* adjacent endodermal cells; therefore, the only access to the vascular cylinder is *through* the endodermal cells themselves, as shown by the arrow in figure 7.4c.

Within the vascular cylinder, water and minerals are transported upward by way of the xylem and the products of photosynthesis most often are transported downward by way of the phloem for storage in the cortex. Lying between the endodermis and the vascular tissue is the **pericycle,** composed of parenchymal cells, that retains the ability to undergo cell division and on occasion produces branch roots. The pericycle also contributes to the formation of vascular cambium, which is meristematic tissue lying between xylem and phloem that is capable of producing new vascular tissue.

Monocot Roots

Monocot roots often have *pith,* which is centrally located ground tissue. In a monocot root, pith is surrounded by a ring of alternating xylem and phloem bundles (fig. 7.5). They also have pericycle, endodermis, cortex, and epidermis.

The root system of a plant absorbs water and minerals that cross the epidermis and cortex before entering the endodermis, the tissue that regulates the entrance of molecules into the vascular cylinder.

The Shoot System

The shoot system includes the stem and the leaves. First, we will consider the growth of the stem, and in the next section, we will examine the leaf. A stem supports leaves, flowers, and fruits; it conducts material to and from the roots and the leaves; and it helps store water and the products of photosynthesis.

Even highly modified stems can be recognized by the presence of nodes, internodes, buds, and leaves (fig. 7.1). A *node* is the point on a stem at which leaves or buds are attached, and an *internode* is the segment of a stem between nodes.

Primary Growth of Stems

The terminal bud of each growing stem contains shoot meristem (fig. 7.6). Whereas root apical meristem is protected by a root cap, shoot apical meristem is protected by newly formed leaves within a bud. Shoot apical meristem produces cells that will become the stem and the leaves. At first, the nodes are very close together; then, internode growth occurs so that the nodes are distant from one another. This pattern of growth complicates the growth of stems, and it is not possible to divide the entire stem into zones of cell division, elongation, and maturation as with the root.

Axillary buds, which are usually dormant but may develop into branch shoots, are seen in the axes of mature leaves. Inactive buds are covered by protective bud scales. In the temperate zone, the terminal bud stops growing in the winter and is then protected by bud scales also. In the spring, when growth resumes, these scales fall off and leave a scar. You can tell the age of a stem by counting these bud-scale scars.

Herbaceous Stems

Mature nonwoody stems, also called **herbaceous stems,** exhibit only primary growth.

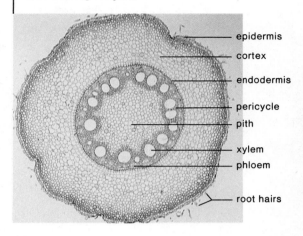

Figure 7.5

Cross section of the root of a corn plant (a monocot), showing the vascular cylinder surrounding the pith.

- epidermis
- cortex
- endodermis
- pericycle
- pith
- xylem
- phloem
- root hairs

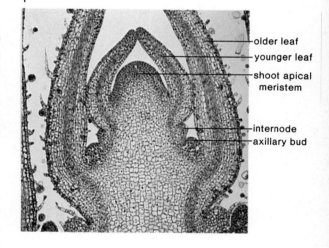

Figure 7.6

Terminal bud anatomy. The shoot apical meristem is surrounded by 2 sets of leaves, a younger pair and an older pair. Axillary buds are seen at the axes of the older pair of leaves. Axillary buds can give rise to side branches when they are active.

- older leaf
- younger leaf
- shoot apical meristem
- internode
- axillary bud

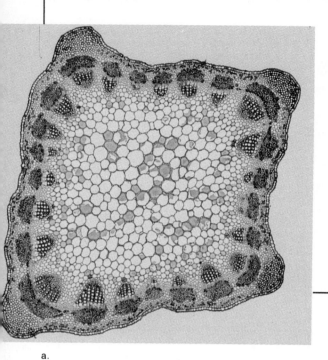

a.

Figure 7.7

Dicot herbaceous stem anatomy. a. Cross section of alfalfa stem shows that the vascular bundles occur in a ring. b. Drawing of a section of the stem, with tissues in the bundle and in the stem identified.

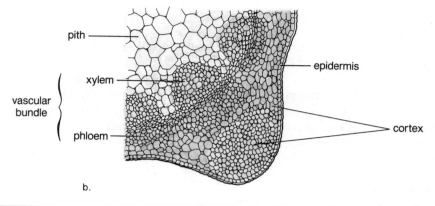

b.

The outermost tissue of herbaceous stems is the epidermis, which is covered by a waxy cuticle to prevent water loss. These stems have distinctive **vascular bundles**, where xylem and phloem are found. In each bundle, xylem is typically found toward the inside of the stem and phloem is found toward the outside.

In the dicot herbaceous stem (fig. 7.7), the bundles are arranged in a distinct ring that separates the cortex from the central pith. The cortex is sometimes green and carries on photosynthesis, and the pith may function as a storage site for the products of photosynthesis. In the monocot herbaceous stem (fig. 7.8), the vascular bundles are scattered throughout the stem, and there is no well-defined cortex nor well-defined pith.

Secondary Growth of Stems

Secondary growth of stems is seen primarily in woody plants, such as trees that live for many years. Almost all trees are dicots. Primary growth in woody plants occurs for a short distance beneath the apical meristem, at which point secondary growth begins. Secondary growth occurs because of meristematic tissue called vascular cambium and cork cambium.

Vascular cambium begins as meristematic cells between the xylem and the phloem of each vascular bundle. Then, these cells join to form a ring of meristematic tissue. The division of this meristematic tissue takes place mostly in a plane parallel to the surface of the plant. The secondary tissues produced by the vascular cambium, called secondary xylem and secondary phloem, therefore add to the girth of the stem instead of to its length (fig. 7.9).

Cork cambium is located beneath the epidermis. When it begins to divide, it produces tissue that disrupts and replaces the epidermis with cork cells. Cork cells are impregnated with suberin, a fatty substance that makes them waterproof. Dead cork allows gas exchange only in pockets of loosely arranged cells, called **lenticels**, that are not impregnated with suberin.

Cross Section of Woody Stem

As a result of secondary growth, a woody stem (fig. 7.10) has an entirely different type of organization than a dicot herbaceous stem. After secondary growth has continued for a time, it is no longer possible to make out individual vascular bundles. Instead, a woody stem has 3 distinct areas: the bark, the wood, and the pith. The bark contains cork, cork cambium, cortex, and phloem. Although secondary phloem is produced each year by vascular cambium, it

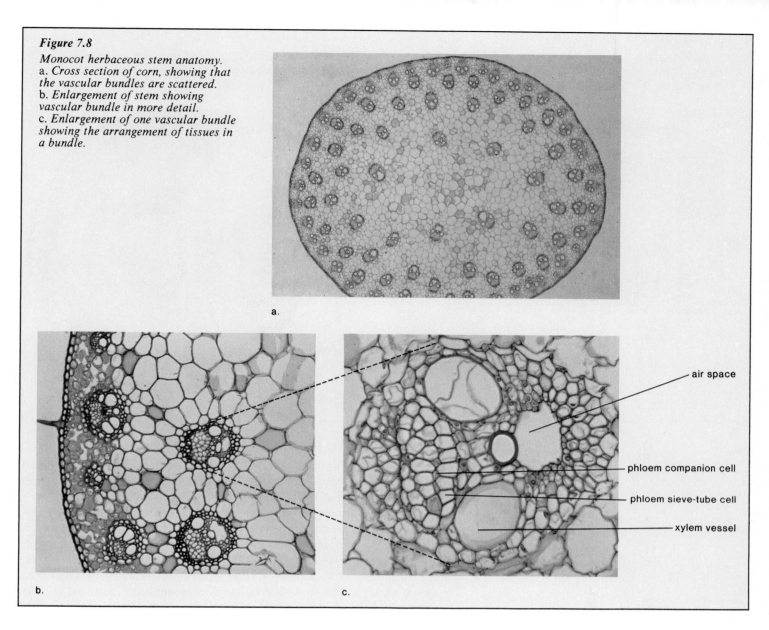

Figure 7.8

Monocot herbaceous stem anatomy.
a. Cross section of corn, showing that the vascular bundles are scattered.
b. Enlargement of stem showing vascular bundle in more detail.
c. Enlargement of one vascular bundle showing the arrangement of tissues in a bundle.

a.

b.

c.

air space

phloem companion cell

phloem sieve-tube cell

xylem vessel

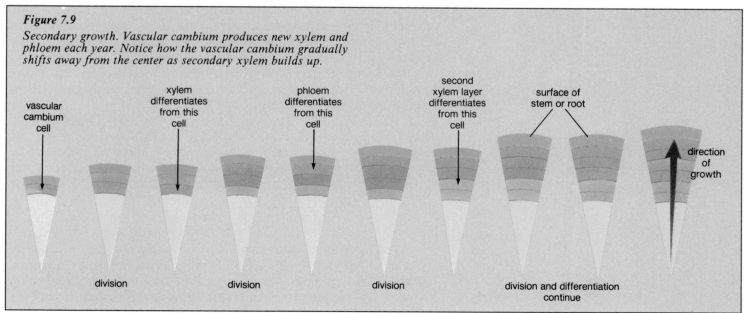

Figure 7.9

Secondary growth. Vascular cambium produces new xylem and phloem each year. Notice how the vascular cambium gradually shifts away from the center as secondary xylem builds up.

vascular cambium cell

xylem differentiates from this cell

phloem differentiates from this cell

second xylem layer differentiates from this cell

surface of stem or root

direction of growth

division

division

division

division and differentiation continue

Figure 7.10

Dicot woody stem. a. *A drawing indicating the location of cork, phloem, vascular cambium, and xylem that accumulates to give annual rings.* b. *A photomicrograph of a cross section shows the 3 parts of a woody stem: bark, wood, and pith. Carolina Biological Supply Company.*

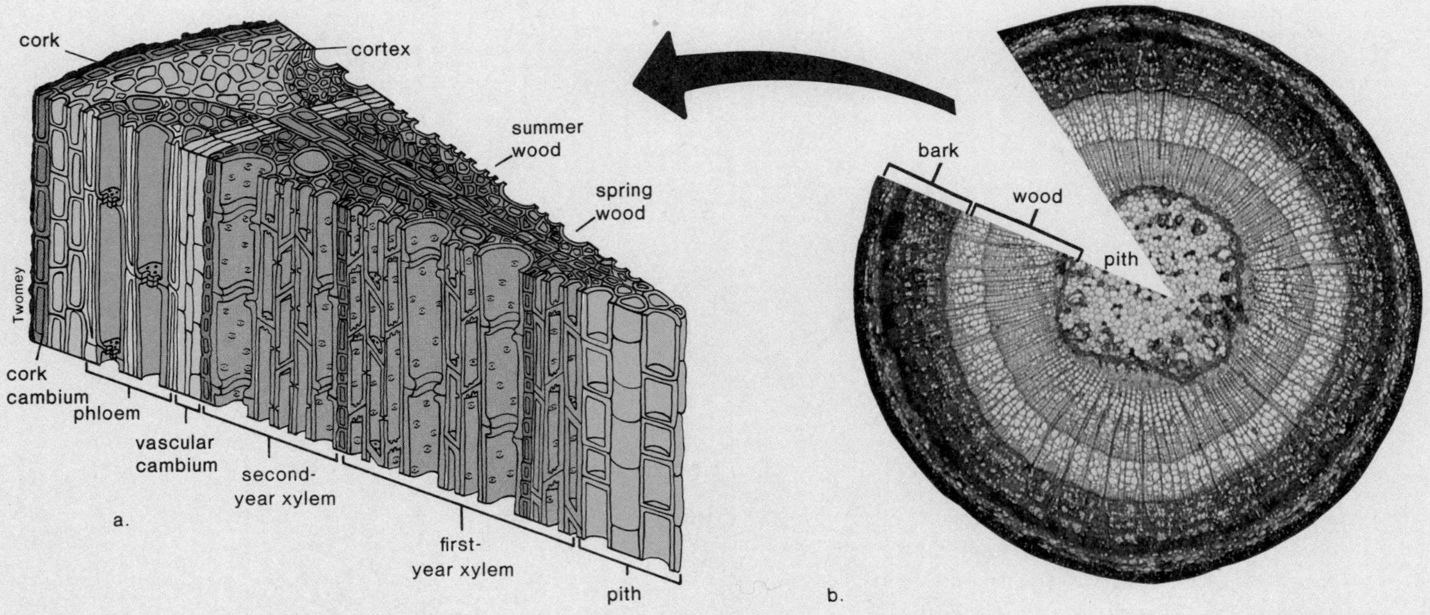

does not build up. The outer layers of phloem are crushed as they are pushed against the bark. Because the newly formed phloem is in the bark of a tree, even partial removal of the bark can seriously damage a tree.

The wood of trees that grow in seasonal climates contains rings of secondary xylem that are called growth rings because there is one for each season of growth. In temperate regions with one growing season per year, these are called annual rings. It is easy to tell where one ring begins and another ends. In the spring, when moisture is plentiful, the xylem cells are much larger (i.e., in spring wood) than later in the summer, when moisture is scarcer (i.e., in summer wood).

In large trees, only the more recently formed layer of xylem, the sapwood, functions in water transport. The older inner part, called the heartwood, becomes plugged with deposits, such as resins, gums, and other substances. Heartwood may help support a tree, although some trees stand erect and live for many years after the heartwood has rotted away.

Like stems, the roots of woody plants also show secondary growth. Figure 7.11 illustrates that the secondary growth of roots arises and progresses in the same manner as that of the stem.

Within the shoot system, the stem transports water and nutrients between the leaves and the roots. Stem anatomy differs according to whether the plant is a monocot or a dicot and whether the plant is herbaceous or woody.

Leaves

Leaves are the organs of photosynthesis in vascular plants. A leaf usually consists of a flattened **blade** and a **petiole,** which connects the blade to the stem. The blade may be single or it may be composed of several leaflets. Externally, it is possible to see the pattern of the **leaf veins** that bring water to the leaf

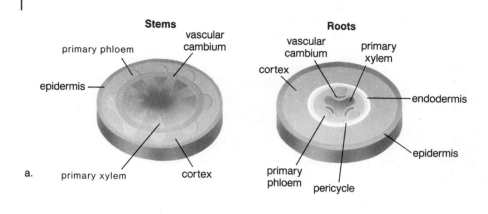

Stems

primary phloem

vascular cambium

epidermis

a.

primary xylem

cortex

Roots

vascular cambium

primary xylem

cortex

endodermis

epidermis

primary phloem

pericycle

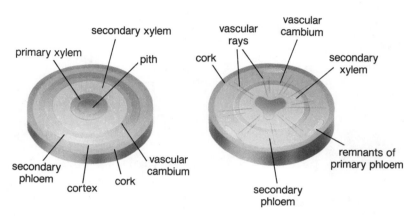

secondary xylem

primary xylem

pith

secondary phloem

cortex

cork

vascular cambium

b.

vascular rays

vascular cambium

cork

secondary xylem

remnants of primary phloem

secondary phloem

Figure 7.11

Cross sections of woody roots and stems. Both stems and roots experience secondary growth. a. Stems. Above is a young stem in which the vascular cambium is just forming a ring. Below is an older stem in which there is secondary xylem and secondary phloem. The primary xylem remains but the primary phloem has disappeared. b. Roots. Above is a young dicot root in which the vascular cambium is located between the primary xylem and phloem. Below is an older root in which there is secondary xylem and secondary phloem. The primary xylem remains but there are only remnants of the primary phloem.

and are the final extensions of vascular tissue. Leaf veins have a net pattern in dicot leaves and a parallel pattern in monocot leaves (fig. 7.2).

The cross section of a typical dicot leaf of a temperate-zone plant is shown in figure 7.12a and b. At the top and the bottom of the leaf is a layer of epidermal tissue that often bears protective hairs and glands that produce irritating substances. These features may prevent the leaf from being eaten by insects. The *epidermis* is covered by a waxy *cuticle* that keeps the leaf from drying out. Unfortunately, it also prevents gas exchange because the cuticle is not gas permeable. However, the epidermis, particularly the lower epidermis, contains openings called **stomata** (singular, stoma) that allow gases to move into and out of the leaf. Each stoma has 2 **guard cells** that regulate its opening and closing.

The body of a leaf is composed of **mesophyll tissue,** which has 2 types of layers: **palisade mesophyll** is a layer of cells that contains elongated cells, and **spongy mesophyll** is a layer of cells that contains irregular cells bounded by air spaces. The parenchymal cells of the palisade and spongy layers have many chloroplasts and carry on most of the photosynthesis for the plant. The loosely packed arrangement of the cells in the spongy layer increases the surface area for gas exchange.

Leaf Veins

A cross section of a leaf shows that leaf veins consist of a strand of xylem and a strand of phloem surrounded by a bundle sheath. The bundle sheath differs in C_3 and C_4 plants. As discussed on page 135, these terms refer to the number of carbon atoms in the first-detected molecule after carbon dioxide (CO_2) uptake. The bundle sheath cells in C_3 plants do not contain chloroplasts, and mesophyll tissue is divided into the palisade layer and the spongy layer, as just mentioned. The bundle sheath cells in the leaves of C_4 plants characteristically

7.1 Critical Thinking

1. What is the function of leaf epidermis, and how does the structure suit the function?
2. What is the function of the spongy layer of mesophyll tissue, and how does the structure suit the function?
3. What is the function of leaf veins, and how does the structure suit the function?
4. What is the difference in structure between C_3 and C_4 leaves, and how does the structure suit the function?

Figure 7.12

Leaf anatomy. a. *Drawing of a leaf from a C₃ (photosynthesis) plant that is adapted to a temperate-zone climate. In a C₃ leaf, the mesophyll is divided into a palisade and a spongy layer. The bundle sheath cells lack chloroplasts.* b. *A scanning electron micrograph of the mesophyll.* c. *Drawing of a leaf vein from a C₄ (photosynthesis) plant that is adapted to a hot, dry climate. In a C₄ leaf, the mesophyll is arranged around the bundle sheath, and the bundle sheath cells have chloroplasts.*

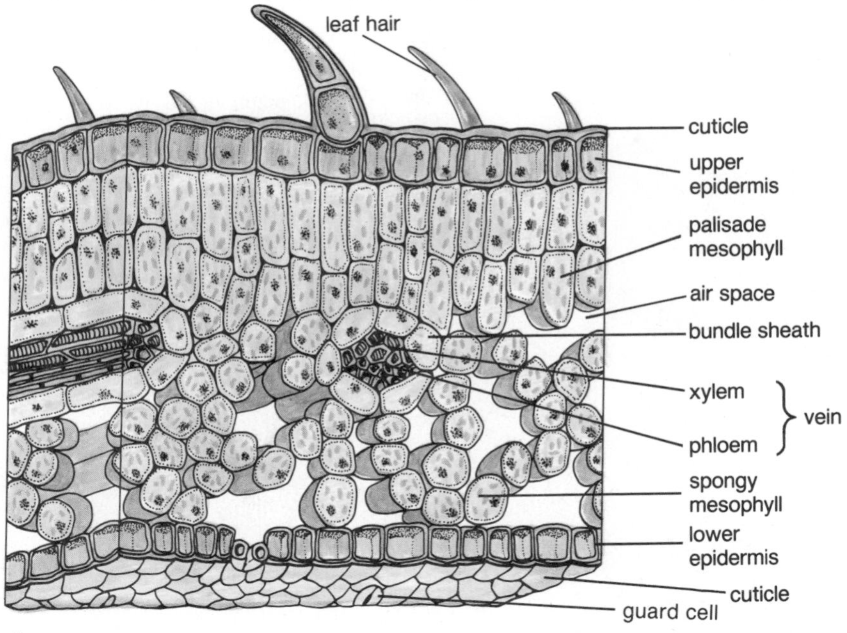

have chloroplasts, and some of the mesophyll cells are packed closely around the bundle sheath in a radial fashion (fig. 7.12c). These mesophyll cells are specialized to pass carbon dioxide to the bundle sheath cells after it has been fixed.

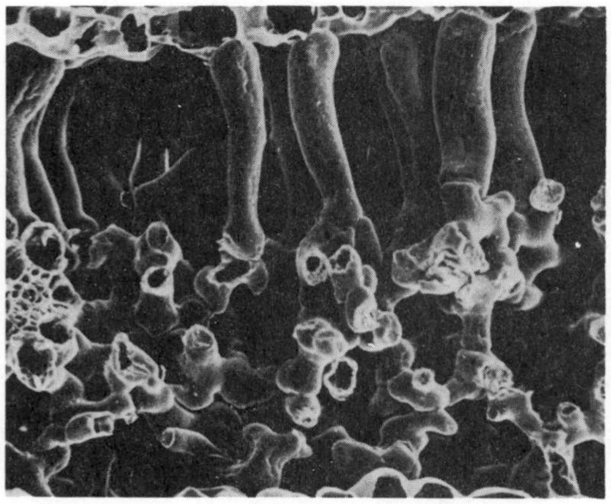

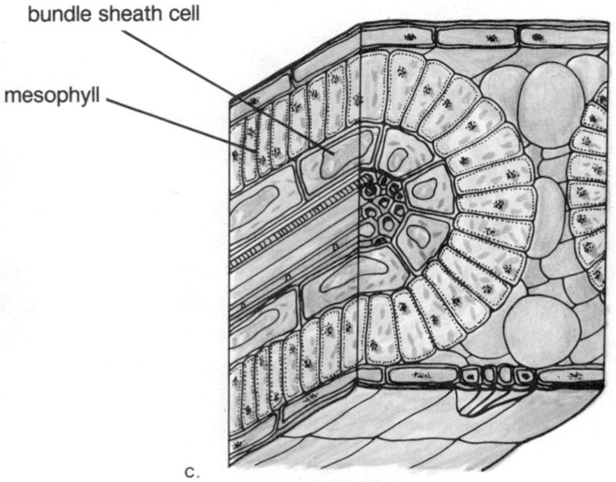

Within the shoot system, the leaves carry on photosynthesis. The leaf is covered by epidermis, and mesophyll tissue and leaf veins are within. In C₃ plants, mesophyll contains a palisade layer and a spongy layer of cells.

Summary

The body of a flowering plant is divided into the root system and the shoot system, which contains the stems and the leaves. Water and minerals enter root hairs at the zone of maturation of a root and cross (in dicot roots) the epidermis, the cortex, and the endodermis before entering the vascular cylinder. Thereafter, water and minerals pass up the shoot system of a plant. Dicot herbaceous stems have vascular bundles arranged in a ring, and monocots have scattered vascular bundles. Woody dicot roots and stems experience secondary growth. Xylem and phloem are separated by a vascular cambium that produces secondary xylem and phloem each year. In the stems of temperate-zone plants, it is possible to see annual rings composed of secondary xylem.

Finally, water and minerals arrive at the leaf veins within leaves, where photosynthesis occurs. A cross section of a leaf shows the epidermis, with stomata mostly on the underside. In C₃ leaves, mesophyll tissue within the leaf has a palisade layer and a spongy layer. In C₄ leaves, some mesophyll cells surround the bundle sheath cells, which contain chloroplasts.

Study Questions

1. Contrast the root system and the shoot system of a plant. (p. 142)
2. Contrast monocots with dicots in 4 ways. (p. 143)
3. Name the cell types found in flowering plants. (p. 144)
4. Name and discuss the zones of a root tip. (pp. 145–46)
5. Describe the anatomy of a dicot root tip, both longitudinal and in cross section. How does the anatomy of a monocot root differ from this? (p. 147)
6. Describe the anatomy of a dicot herbaceous stem and of a dicot woody stem in cross section. How does the anatomy of a monocot stem differ from this? (p. 147)
7. Contrast the primary growth with the secondary growth of stems and roots. (pp. 148–50)
8. Describe the anatomy of a C_3 leaf in cross section. How does the anatomy of a C_4 leaf differ from this? (p. 151)

Objective Questions

1. The _____ tissue is unspecialized and is capable of continual cell division.
2. In the roots, epidermal cells have _____ , and in the leaves, the epidermis contains _____ .
3. The conducting cells in xylem are called _____ and _____ .
4. In a dicot root, mature cells are found in the zone of _____ .
5. In a dicot root, the _____ regulates the passage of ions into the vascular cylinder.
6. In a monocot stem, the vascular bundles are said to be _____ .
7. In a woody stem, the secondary xylem builds up and forms the _____ , which can be counted to tell the age of a tree.
8. The mesophyll contains 2 layers, the _____ layer and the spongy layer.

Answers to Objective Questions

1. meristem 2. root hairs, stomata 3. vessel elements, tracheids 4. maturation 5. endodermis 6. scattered 7. annual rings 8. palisade

Selected Key Terms

Casparian strip (kas-par'e-an strip) band of waxy material found on endodermal cells of plants; prevents passage of molecules outside of cells. *147*

cotyledon (kot''i-le'don) the seed leaf of the embryo of a plant. *142*

dicot (di'kot) dicotyledon; a type of angiosperm distinguished particularly by the presence of 2 cotyledons in the seed. *142*

endodermis (en''do-der'mis) a plant tissue consisting of a single layer of cells that surrounds and regulates the entrance of materials, particularly into the vascular cylinder of roots. *144*

epidermis (ep''i-der'mis) the outer layer of cells of organisms, including plants. *144*

herbaceous stem (her-ba'shus stem) nonwoody stem. *147*

lenticel (len'ti-sel) a pocket of loosely arranged cells in cork that permit gas exchange. *148*

meristem tissue (mer'i-stem tish'u) plant tissue that always remains undifferentiated and capable of dividing to produce new cells. *143*

mesophyll tissue (mes'o-fil tish'u) the middle tissue of a leaf made up of parenchymal cells that carries on photosynthesis and gas exchange. *151*

monocot (mon'o-kot) monocotyledon; a type of angiosperm in which the seed has only one cotyledon, such as corn and lily. *142*

palisade mesophyll (pal'i-sād mes'o-fil) the upper layer of the mesophyll of a leaf. *151*

phloem (flo'em) the vascular tissue in plants that transports organic nutrients. *144*

spongy mesophyll (spun'je mes'o-fil) the lower layer of the mesophyll of a leaf. *151*

stoma (sto'mah) opening in the leaves of plants through which gas exchange takes place (pl. stomata). *151*

vascular bundle (vas'ku-lar bun'd'l) tissues that include xylem and phloem enclosed by a sheath and typically found in herbaceous plant stems. *148*

vascular cambium (vas'ku-lar kam'be-um) a cylindrical sheath of meristematic tissue that produces secondary xylem and phloem. *148*

vascular cylinder (vas'ku-lar sil'in-der) a central region of roots that contains vascular and other tissues. *147*

xylem (zi'lem) the vascular tissue in plants that transports water and minerals. *144*

Chapter 8

Plant Physiology and Reproduction

Chapter Concepts

1 Theories have been formulated to explain the transport of water and minerals and the transport of organic nutrients in plants.

2 Plants respond to outside stimuli with changes in their growth pattern.

3 Flowering is controlled by the length of daylight (photoperiod) in some plants.

4 The sex organs of a flowering plant are located in the flower.

5 Seeds within fruits are the products of sexual reproduction in flowering plants. Germination of a seed results in another plant.

Chapter Outline

Transport in a Plant
    Water Transport
    Organic Nutrient Transport
Plant Responses to Environmental
    Stimuli
    Flowering
Reproduction in Plants
    Asexual Reproduction
    Sexual Reproduction
    The Seed

Transport in a Plant

 In order to carry on photosynthesis, plants require only water (H_2O) and carbon dioxide (CO_2). After water is absorbed by the root system, it is transported to the leaves in xylem, a vascular tissue that is continuous throughout the body of the plant (fig. 8.1). Following photosynthesis, organic nutrients are transported throughout the plant by another vascular tissue, phloem.

Water Transport

As discussed in chapter 7, xylem contains 2 types of conducting cells; vessel cells and tracheids, each of which is specialized for transport (fig. 8.2). The nonliving vessel cells are called **vessel elements** because piled on top of one another, they form a continuous pipeline that stretches from the root system to the leaves.

This is an open pipeline because the elements have no end walls separating one from the other. The elongated **tracheids** are also dead at maturity, but they have tapered end walls that are not perforated. Both types of cells have secondary walls that contain lignin, an organic substance that makes the walls tough and hard. Even so, water can move between the lateral walls of both types of cells and the end walls of tracheids because of pits, depressions where the secondary wall does not form.

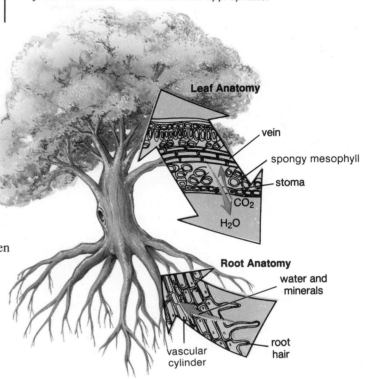

Figure 8.1

Function of xylem and phloem. The upward arrow on the left indicates that water is transported in xylem from the roots to the leaves, where it exits at the stomata. The double-headed arrow on the right indicates that organic nutrients (e.g., sugar) can be transported by phloem either from the leaves to the roots or from the roots to the leaves as is appropriate.

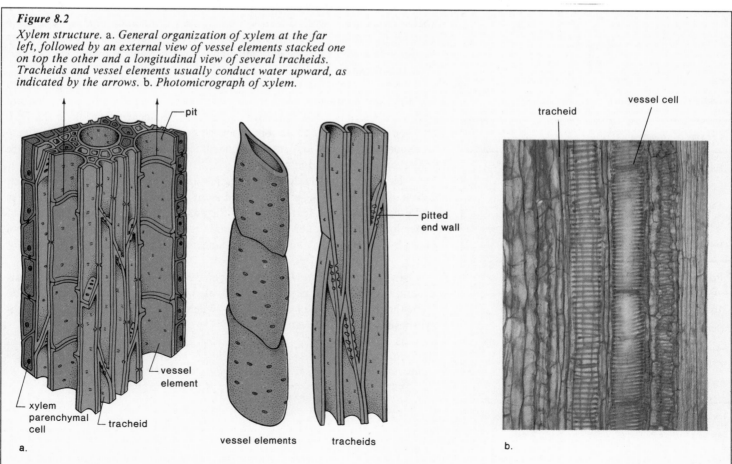

Figure 8.2

Xylem structure. a. *General organization of xylem at the far left, followed by an external view of vessel elements stacked one on top the other and a longitudinal view of several tracheids. Tracheids and vessel elements usually conduct water upward, as indicated by the arrows.* b. *Photomicrograph of xylem.*

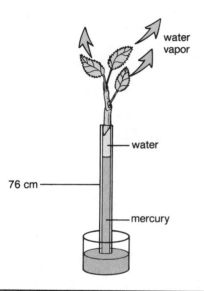

water vapor

water

76 cm

mercury

Table 8.1 Transpiration Rates		
Per Day Midsummer		
Ragweed		6–7 liters
3.5 m apple tree		10–20 liters
4 m cactus		0.02 liters
Coconut palm		70–80 liters
Date palm		400–500 liters
Per Growing Season		
Tomato	100 days	113 liters
Sunflower	90 days	450 liters
Apple tree	188 days	6,800 liters
Coconut	365 days	15,900 liters
Date palm	365 days	132,100 liters

Table 8.2 from *Botany: A Human Concern,* Second Edition, by David L. Rayle and Hale L. Wedberg, copyright © 1980 by Saunders College Publishing, a division of Holt, Rinehart and Winston, Inc., reprinted by permission.

The Cohesion-Tension Theory of Water Transport

How is it possible for water (H_2O) absorbed by root hairs to be transported from the roots to the leaves within the xylem of even very tall trees? The 2 factors necessary to this process already have been mentioned:

1. The vessel elements (and tracheids) within xylem form a continuous pipeline from the roots to the leaves (fig. 8.3). Water fills this pipeline. Water enters a root primarily at the root hairs and makes its way across the cortex until it enters xylem within the vascular cylinder. Polar water molecules adhere to the walls of the vessel elements, and because of hydrogen bonding, they cling together. *Cohesion* of water molecules within the xylem pipeline is absolutely necessary for water transport in plants.
2. **Transpiration,** the evaporation of water from a leaf, occurs at the stomata. Other water molecules continuously take the place of water molecules that evaporate. In this way, transpiration exerts a pull—creates a tension—that draws a column of water up the vessel elements from the roots to the leaves.

The tension created by transpiration would not be effective if it were not for the cohesive property of water. Therefore, the theory of water transport in xylem is called the **cohesion-tension theory.**

Transpiration is dependent on whether the stomata are open or closed.

Water is transported from the roots to the leaves in xylem. Transpiration at the leaves pulls up a column of water molecules, which exhibit cohesion.

Stomatal Opening and Closing

Notice in figure 8.4 that each stoma is surrounded by 2 guard cells. The guard cell walls are thinner at the ends and the sides away from the stomatal opening. When water (H_2O) enters guard cells and turgor pressure increases (p. 72), their thicker, inner walls do not move, but their thinner, outer side walls begin to stretch. The guard cells bend inward, causing the stoma to open. You can simulate this effect by placing your hands together and allowing your fingers to curve outward.

Water enters the guard cells by osmosis because of a high potassium (K^+) ion concentration. Stomata are open when a plant is photosynthesizing and when there is a low carbon dioxide (CO_2) concentration inside the leaf. In some unknown way, these conditions trigger a potassium pump that actively transports potassium ions into the guard cells, creating an osmotic pressure that eventually opens the stomata.

However, stomata do not open unless there is a plentiful supply of water. Much of the water that is transported from the roots to the leaves evaporates and escapes from the leaf by way of the stomata (table 8.1). The amount of water lost in this manner is phenomenal, but evaporation keeps the leaf cool enough to function in the bright sun and is also critical to the transport of water.

In most plants, the stomata are open during the day when photosynthesis occurs; water is transported to the leaves, which also receive carbon dioxide from the air. Stomata close at night—when water is not utilized for photosynthesis—to prevent water loss.

Stomata tend to open when a plant is photosynthesizing. Stomata open when potassium ions are pumped into the guard cells, which then take up water.

Figure 8.3

Cohesion-tension theory of water transport. Water (H$_2$O) enters a plant at the root hairs and evaporates (transpires) at the leaves. Xylem elements form a continuous pipeline from the roots to the leaves, and this pipeline is completely full of water. Therefore, transpiration exerts a pull on the water column that causes it to move upward.

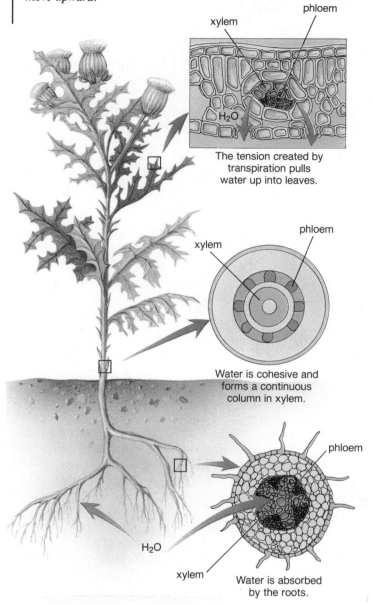

The tension created by transpiration pulls water up into leaves.

Water is cohesive and forms a continuous column in xylem.

Water is absorbed by the roots.

Figure 8.4

Opening of stomata. a. Scanning electron micrograph of an open stoma. b. When a stoma opens first potassium (K$^+$) ions and then water enter guard cells. This increases the turgor pressure of guard cells so that the stoma opens. c. When a stoma closes, first potassium ions and then water exit guard cells. This decreases the turgor pressure of guard cells so that the stoma closes.

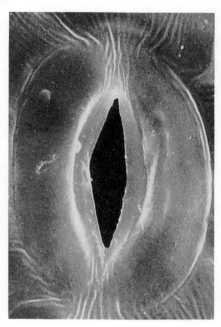

a.

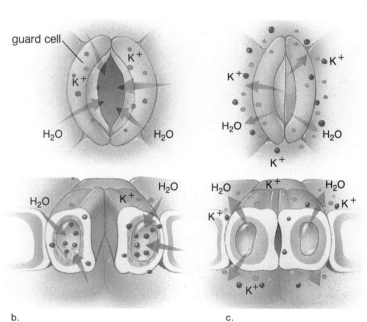

b.

c.

Organic Nutrient Transport

The conducting cells in phloem are sieve-tube cells, each of which typically has a companion cell (fig. 8.5). **Sieve-tube cells** contain cytoplasm but no nucleus; as their name implies, these cells have pores in their end walls that make these walls resemble a sieve. Through these pores, strands of cytoplasm, called plasmodesmata, extend from one cell to the other. The smaller **companion cells** are more generalized cells and have nuclei. It is speculated that the companion-cell nucleus controls and maintains the lives of both cells.

Figure 8.5

Phloem structure. a. General organization of phloem at the far left, followed by an external view of 2 sieve-tube cells and their companion cells. b. Photomicrograph of phloem from Tilia americana *sieve tubes.*

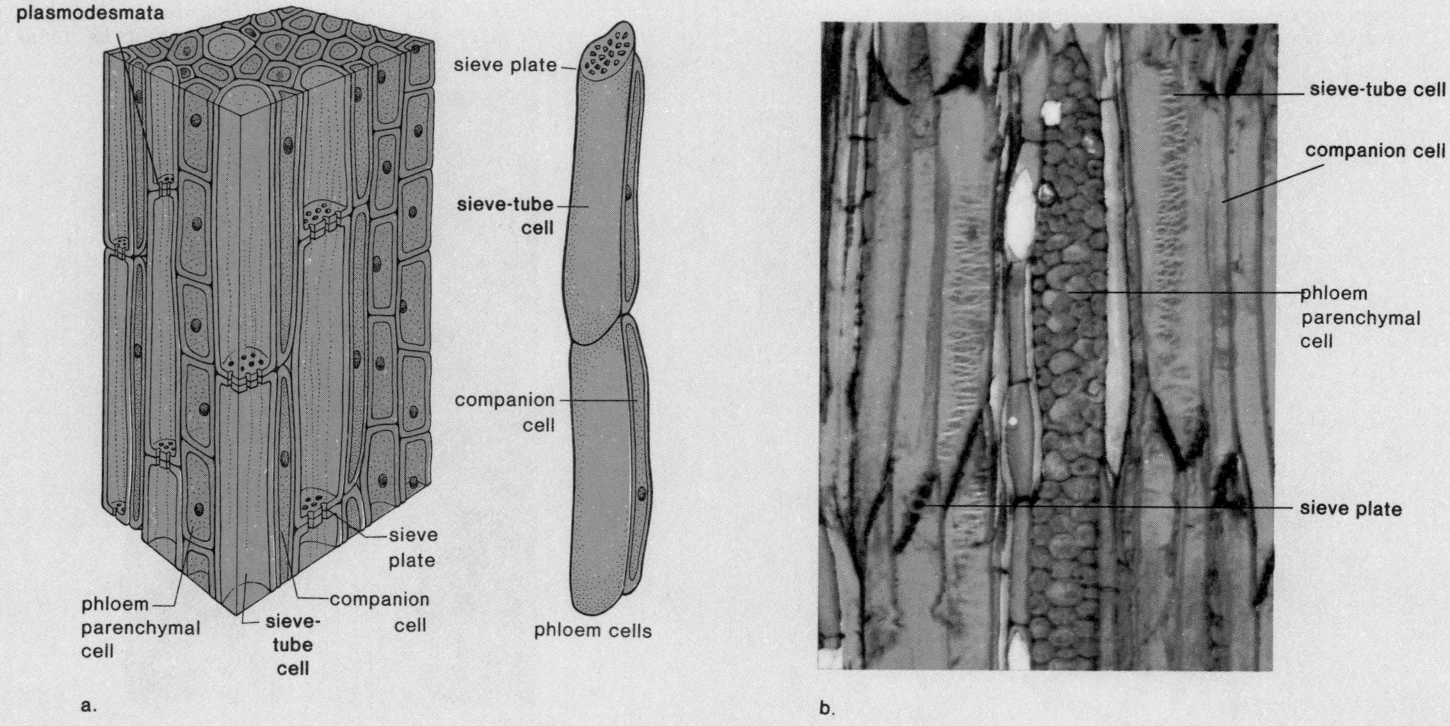

Chemical analysis of phloem sap shows that it is composed chiefly of sugar and that the concentration of nutrients is 10–13% by volume. Samples for chemical analysis most often are obtained using aphids (fig. 8.6), small insects that are phloem feeders. The aphid drives its stylet, a short mouthpart that functions like a hypodermic needle, between the epidermal cells and withdraws sap from a sieve-tube cell. If the aphid is anesthetized by ether, the body can be cut away carefully, leaving the stylet, which exudes the phloem contents for collection and analysis.

The Pressure-Flow Theory of Phloem Transport

As discussed, phloem is found from the leaves to the roots of a plant (fig. 8.7). The sieve-tube cells are connected by plasmodesmata that extend through the adjoining sieve-tube plates. In order to understand the movement of solute within phloem, it is important to envision this continuous stream of cytoplasm from the leaves to the roots. Also, remember that there is a companion cell for each sieve-tube cell.

1. During the growing season, the leaves are photosynthesizing and are producing carbohydrate. Carbohydrate, in the form of sucrose, a sugar, is actively transported into sieve-tube cells, and water follows passively by osmosis. Active transport and osmosis are possible because sieve-tube cells have a living cell membrane. The energy needed for sugar transport is provided by the companion cells. The buildup of water within the sieve-tube cells at the leaves creates a *pressure.*
2. At the roots (or at other places in the plant), sugar is actively transported out of the sieve-tube cells, and water follows passively by osmosis. The exit of sugar and water at the roots means that the

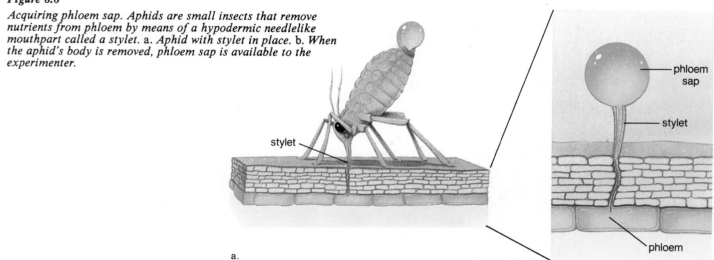

Figure 8.6

Acquiring phloem sap. Aphids are small insects that remove nutrients from phloem by means of a hypodermic needlelike mouthpart called a stylet. a. Aphid with stylet in place. b. When the aphid's body is removed, phloem sap is available to the experimenter.

stylet

phloem sap

stylet

phloem

a.

b.

pressure created at the leaves causes a *flow* of water from the leaves (where pressure is high) to the roots (where pressure is low). As the water flows along, it brings sugar with it (fig. 8.7).

The **pressure-flow theory** can account for the observed reversal of flow in phloem, for example, in the spring before the leaves are out. At that time, when the plant is not photosynthesizing, the roots serve as a *source* of sugar, and the other parts of the plant serve as a *sink*. Water enters sieve-tube cells at the source and flows toward the sink, carrying sugar with it.

Phloem transports organic nutrients in a plant. Typically, sugar and then water enter sieve-tube cells in the leaves. This creates a pressure that causes water to flow to the roots, carrying sugar with it.

Plant Responses to Environmental Stimuli

Plants respond to environmental stimuli, such as light, day length, gravity, and temperature, usually by changing their growth pattern. Among the principal internal factors that regulate such responses are plant hormones. A **hormone** is a chemical messenger produced in small amounts by one part of the body that is active in a different part of the body. Generally, plant hormones are produced by the meristematic regions of a plant and are transported in vascular tissue. Responses can be observed in almost every part of the plant's body. Table 8.2 lists the major types of plant hormones and their function. However, you should realize that these hormones often interact to control physiological responses.

Each naturally occurring hormone has a specific chemical structure. Other chemicals, some of which differ only slightly from the natural hormones, also affect the growth of plants. These and the naturally occurring hormones sometimes are grouped and are called plant growth regulators. The reading on page 160 discusses the various uses of plant growth regulators.

Flowering

The effect of photoperiod (length of daylight compared to the length of darkness) on plants is particularly obvious in the temperate zone. In the spring, plants respond to increasing day length by initiating growth, and in the fall, they respond to decreasing day length by ceasing growth processes. Among its effects, day length also regulates flowering in some plants; for example, violets and tulips flower in the spring, asters and goldenrods flower in the fall.

Figure 8.7

Pressure-flow theory of phloem transport. Sugar and water enter sieve-tube cells at a source. This creates a pressure that causes sap to flow in phloem. Sieve-tube cells form a continuous pipeline from a source to a sink, where sucrose and water exit sieve-tube cells.

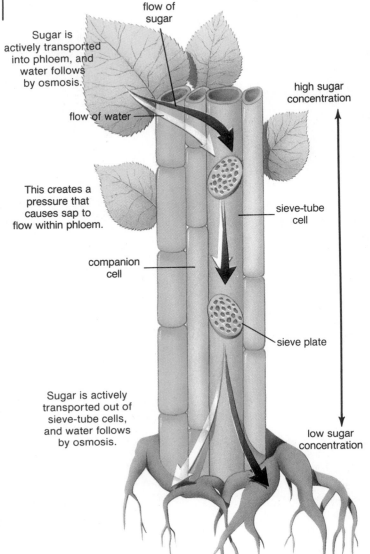

Sugar is actively transported into phloem, and water follows by osmosis.

flow of sugar

flow of water

high sugar concentration

This creates a pressure that causes sap to flow within phloem.

sieve-tube cell

companion cell

sieve plate

Sugar is actively transported out of sieve-tube cells, and water follows by osmosis.

low sugar concentration

Plant growth involves production of cells by means of cell division, enlargement of these cells, and finally, differentiation as the cells take on specific functions. Three types of hormones are known to promote plant growth: the cytokinins stimulate cell division, the auxins bring about cell division and enlargement of plant cells; and the gibberellins promote enlargement of cells and to a lesser extent cell division. A fourth class of plant hormones, termed inhibitors, retard or prevent growth in general. Plant growth regulators include natural hormones and related synthetic hormones. Today, plant growth regulators are used to bring about an increase in crop yields just as fertilizers, irrigation, and pesticides have done in the past.

Plants bend toward light, and experiments with oat seedlings have shown that bending occurs because auxin is transported to the

Plant Growth Regulators

shady side of the shoot. This can be proven by removing the tip of a shoot and placing an auxin-containing agar block on one side of the stump. The cells on this side elongate, causing bending to occur.

Since the time these experiments first were performed, many commercial uses for auxins have been discovered. Auxins can cause the base of a shoot to form new roots so that new plants can be started from cuttings. When sprayed on trees, auxins can prevent fruit from dropping too soon. Auxins also inhibit the growth of axillary buds; potatoes sprayed with auxin will not sprout and thus can be stored longer. In high concentrations, auxins are used as herbicides that prevent the growth of broad-leaved plants. The synthetic auxins known as 2,4D and 2,4,5T were used as defoliants during the Vietnam War.

Gibberellins cause the entire plant to grow larger. Before World War II, the Japanese studied a disease they called "foolish seedling disease" because the young plants grew rapidly, became spindly, and fell over. They found that this disease was caused by gibberellins secreted by a fungus that had infected the plants. Since then, it has been discovered that the application of gibberellins can cause

Table 8.2 Plant Hormones

Type	Primary Example	Notable Function
Growth Promoters		
Auxins	Indolacetic acid (IAA)	Cell elongation
Gibberellins	Gibberellic acid (GA)	Stem elongation
Cytokinins	Zeatin	Cell division
Growth Inhibitors		
Abscisic acid	Abscisic acid (ABA)	Dormancy
Ethylene	Ethylene	Leaf and fruit drop

The effect of gibberellic acid (GA₃) on Thompson seedless grapes (Vitis vinifera). *Control grapes (left). GA₃ sprayed at bloom and at fruit set (right). Almost all grapes sold in stores now are treated with gibberellic acid.*

seeds to germinate and plants, such as cabbages, to bolt (meaning rapid stem elongation) and flower. Gibberellins are used commercially to increase the size of plants. Treatment of sugarcane with as little as 56 grams per acre increases the yield of cane by more than 5 metric tons.

Cytokinins were discovered when mature carrot and tobacco plant cells began to divide when grown in coconut milk. Testing revealed the presence of cytokinins in the milk. Later, scientists were able to grow entire plants from single cells in test tubes when various plant hormones were present in correct proportions.

Nurseries now culture all sorts of plants with assembly-line efficiency. Plant breeders are extremely interested in utilizing a modification of the tissue culture technique in which most often leaf cells are treated to produce *protoplasts*, cells that chemically have been stripped of their outer wall. (A single protoplast will give rise to a new plant identical in nature to the original plant.) It is faster and easier to test protoplasts instead of entire plants for desired characteristics, such as resistance to bacteria and fungi, high temperatures, and drought. Also selected protoplasts have been

genetically engineered as described in chapter 25 and the resulting plants have shown the characteristics dictated by the inserted genes. Perhaps protoplast technology someday will allow botanists to alter the genetic makeup of a variety of plants.

The hormone ethylene, which is classified as an inhibitor, causes fruit to ripen. Fruits commonly are kept in cold storage to prevent the release of ethylene. Many synthetic inhibitors simply oppose the action of the natural stimulatory hormones (auxins, gibberellins, and cytokinins). The application of synthetic inhibitors can cause leaf and fruit drop. Removal of the leaves of cotton plants by chemical means aids harvesting; thinning the fruit of young fruit trees produces larger fruit from the trees as they mature; and retarding the growth of some plants increases their hardiness. For example, an inhibitor has been used to reduce stem length in wheat plants so that they do not fall over in heavy winds and rain. Other synthetic inhibitors mimic the action of ethylene and cause ripening of fruit and other crops. Fields and orchards now are sprayed with synthetic growth regulators just as they are sprayed with pesticides.

Investigators who first studied the effect of **photoperiodism** on flowering came to the conclusion that plants could be divided into the following 3 groups (fig. 8.8):

Short-day (long-night) plants—flower when the photoperiod is shorter than a critical length. (Good examples are cocklebur, poinsettia, and chrysanthemum.)

Long-day (short-night) plants—flower when the photoperiod is longer than a critical length. (Good examples are wheat, barley, clover, and spinach.)

Day-neutral plants—flowering is not dependent on a photoperiod. (Good examples are tomato and cucumber.)

Later, experiments were done with artificial light and dark time periods that did not necessarily correspond to a normal 24-hour day. It was discovered that the cocklebur, a short-day plant, would flower as long as the dark period was continuous for at least 8½ hours, regardless of the length of the light period.

Figure 8.8

Day length (night length) effect on 2 types of plants. a. Short-day (long-night) plant. 1. When the day is shorter (the night longer) than a critical length, this type of plant flowers. 2. It does not flower when the day is longer (night is shorter) than the critical length. 3. It also does not flower if the longer-than-critical-length night is interrupted by a flash of light. b. Long-day (short-night) plant. 1. When the day is shorter (the night longer) than a critical length, this type plant does not flower. 2. It flowers when the day is longer (the night shorter) than a critical length. 3. It flowers if the slightly longer-than-critical-length night is interrupted by a flash of light.

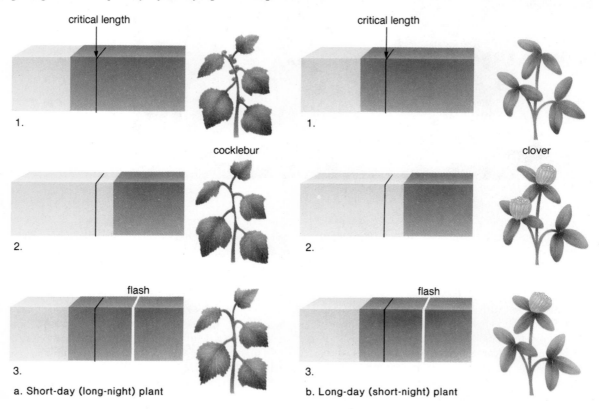

a. Short-day (long-night) plant

b. Long-day (short-night) plant

Further, if this critical-length dark period was interrupted by a brief flash of light, the cocklebur would not flower. (Interrupting the light period with darkness had no effect.)

Similar results have also been found for long-day plants. These require a shorter dark period than a critical length regardless of the length of the light period. However, if a longer-than-critical-length night is interrupted by a brief flash of light, long-day plants will flower. We can conclude, then, that it is a critical length of darkness (specific for each plant) that controls flowering, not the length of the light period. Of course, in nature, shorter days always go with longer nights and vice versa.

Short-day plants require a period of darkness that is longer than a critical length and long-day plants require a period of darkness that is shorter than a critical length in order to flower.

Phytochrome

If flowering is dependent on the length of day and night, plants must have some way to detect these periods. Many years of research led to the discovery of the detector—a plant pigment called phytochrome. **Phytochrome** is a blue-green leaf pigment that alternately exists in 2 forms (fig. 8.9):

P_r (phytochrome red) absorbs red light (of 660 nm wavelength) and is converted to P_{fr}.

P_{fr} (phytochrome far-red) absorbs far-red light (of 730 nm wavelength) and is converted to P_r.

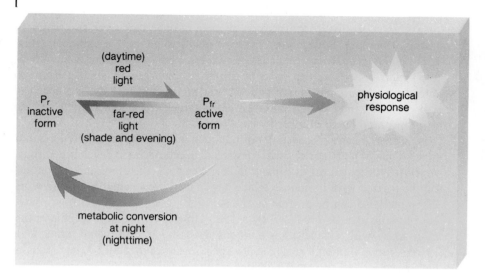

Figure 8.9

The $P_r \rightleftarrows P_{fr}$ conversion cycle. The inactive form of phytochrome, P_r, is prevalent during the night. At sunset or in the shade when there is more far-red light, P_{fr} is converted to P_r. Also during the night, metabolic processes cause P_{fr} to be replaced by P_r. P_{fr}, the active form of phytochrome, is prevalent during the day because at that time there is more red light than far-red light.

Sunlight contains more red light than far-red light; therefore, P_{fr} is apt to be present in plant leaves during the day. In the shade and at sunset, there is more far-red light than red light; therefore, P_{fr} is converted to P_r as night approaches. There is also a slow metabolic replacement of P_{fr} by P_r during the night. It was thought for some time that this slow reversion provided a means for the plant to measure the length of the night. However, it is now assumed that the active form of phytochrome, P_{fr}, signals a biological clock, an internal system that controls the timing of certain physiological and behavioral responses—in this case, flowering. (Biological clocks are discussed in more detail in the reading for chapter 19.)

Just how this biological clock is signaled is unknown. Hormones are probably involved, but this has not been proven. At one time, researchers thought they would find a special flowering hormone called florigen, but such a hormone never has been discovered.

Phytochrome is a leaf pigment that alternates between 2 forms (P_{fr} during the day and P_r during the night). The $P_r \rightleftarrows P_{fr}$ conversion cycle apparently allows a plant to detect photoperiod changes that can result in flowering.

Other Functions of Phytochrome The $P_r \rightleftarrows P_{fr}$ conversion cycle now is known to control other growth functions in plants. It promotes seed germination and inhibits stem elongation, for example. The presence of P_{fr} indicates to some seeds that sunlight is present and conditions are favorable for germination. This is why some seeds must be only partly covered with soil when planted. Germination of other seeds is inhibited by light so they must be planted deeper. Following germination, the presence of P_r indicates that stem elongation may be needed to reach sunlight. Seedlings that are grown in the dark *etiolate:* that is, the stem increases in length and the leaves remain small (fig. 8.10). Once the seedling is exposed to sunlight and P_r is converted to P_{fr}, the seedling begins to grow normally—the leaves expand and the stem branches.

Figure 8.10

Accessory functions of phytochrome. Phytochrome has other functions besides regulating flowering. For example, if P_{fr} is prevalent, as it is in the shade, the seedling etiolates—the stem elongates and the leaves remain small (left). However, if P_r is prevalent, as it is in bright sunlight, the stem does not elongate and the leaves expand (right). These effects are due to phytochrome.

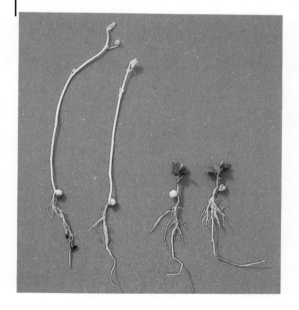

Reproduction in Plants

Plants can reproduce both asexually (without gametes) or sexually (with gametes).

Figure 8.11

Asexual reproduction of trees. Each dish shows several shoots of Douglas fir growing from a single cotyledon. Such cultures are part of a research project for cloning genetically improved trees. Subsequently, the shoots will be cut, rooted, and planted in the forest. Tissue culture propagation is expected to play an important part in bringing forest yields toward their theoretical maximum.

Asexual Reproduction

Asexual reproduction, also known as vegetative propagation, is common in plants. In vegetative propagation, a portion of one plant gives rise to a completely new plant. Both plants now have identical genes. For example, strawberry plants grow from the nodes of runners, aboveground horizontal stems, and violets grow from the nodes of rhizomes, underground stems. White potatoes are actually portions of underground stems, and each eye is a node that will produce a new potato plant. Sweet potatoes are modified roots and can be propagated by planting sections of the root.

Other means of asexual reproduction have great commercial importance. Once a plant variety with desired characteristics is developed through vegetative propagation, cuttings can be taken from the plant and the cut end can be treated with hormones to encourage it to grow roots. A cutting or axillary bud also can be grafted to the stem of a plant that has a root. Today, entire plants can be produced by tissue culture (fig. 8.11), a technique that most likely will replace other methods. Usually, an embryonic tissue is removed from a plant and is placed in a special culture medium. After the tissue has grown for awhile, it is subdivided, and many identical plants are produced from a very small number of starting cells.

Plants reproduce both asexually and sexually. Asexual reproduction occurs when a portion of one plant gives rise to an entirely new plant that is genetically identical to the original plant. Grafting and tissue culture propagation have commercial importance today.

Sexual Reproduction

Plants also reproduce sexually. This may come as a surprise to those who never thought of plants as being male and female. Sexual reproduction is defined properly as reproduction requiring gametes, often an egg and a sperm. In flowering plants, the sex organs are located in the flower.

Flower Anatomy

Figure 8.12 shows the parts of a typical flower. The **sepals,** most often green, form a whorl about the **petals,** the color of which accounts for the attractiveness of many flowers. In the center of the flower is a small vaselike structure, the **pistil,** which usually has 3 parts: the **stigma,** an enlarged sticky knob; the **style,** a slender stalk; and the **ovary,** an enlarged base. The ovary contains a number of ovules that play a significant role in reproduction. Grouped about

Figure 8.12
Parts of a typical flower, a sporophyte structure.

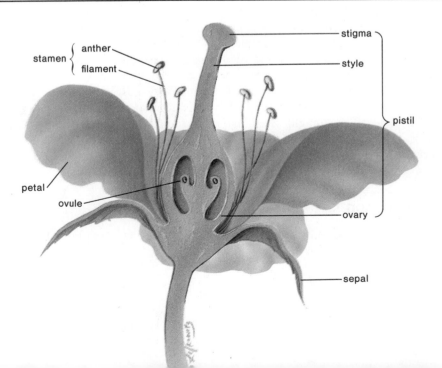

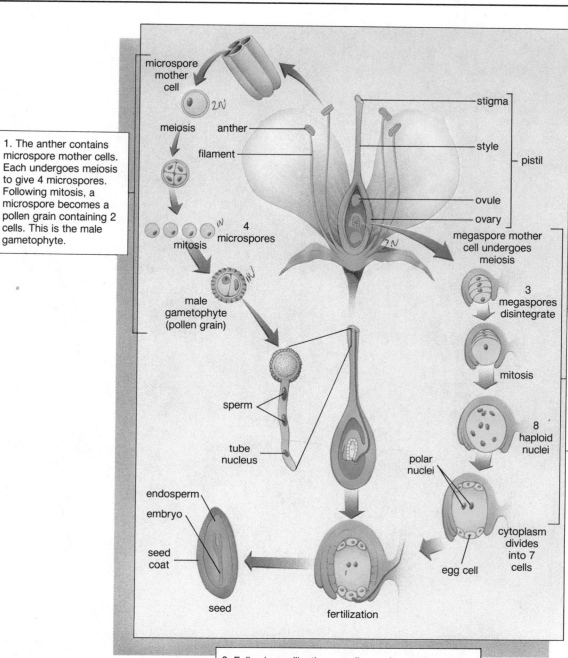

Figure 8.13
*Life cycle of a
flowering plant. The
life cycle involves
production of eggs and
sperm by gametophyte
generations and
development of an
embryo-containing
seed.*

microspore
mother
cell

2N

meiosis   anther

filament

4
microspores

mitosis

male
gametophyte
(pollen grain)

sperm

tube
nucleus

endosperm

embryo

seed
coat

seed   fertilization

stigma

style

pistil

ovule

ovary

megaspore mother
cell undergoes
meiosis

3
megaspores
disintegrate

mitosis

8
haploid
nuclei

polar
nuclei

egg cell

cytoplasm
divides
into 7
cells

1. The anther contains
microspore mother cells.
Each undergoes meiosis
to give 4 microspores.
Following mitosis, a
microspore becomes a
pollen grain containing 2
cells. This is the male
gametophyte.

2. There is a megaspore
mother cell in each ovule
found in the ovary. The
megaspore mother cell
undergoes meiosis to
give one functional
megaspore, which
divides mitotically. The
result is a female
gametophyte that
contains an egg cell.

3. Following pollination, a pollen grain germinates to
give a pollen tube, which contains 2 sperm that travel
down the tube to the ovule. During double
fertilization, one sperm joins with the egg and the
other joins with 2 polar nuclei. The ovule now matures
and becomes a seed.

the pistil are a number of **stamens,** each of which has 2 parts: the **anther,** a
saclike container, and the **filament,** a slender stalk.

Flowering plants have a life cycle called **alternation of generations** (see
chapter 29) because it contains 2 generations: the sporophyte and the gametophyte. The sporophyte is a diploid (2N) generation that produces haploid
spores by meiosis. **Spores** give rise to the gametophyte, which is the haploid
(N) generation. The gametophyte produces gametes that join to give rise to
the sporophyte once again.

A plant that flowers is actually the **sporophyte** that produces spores.
Within the ovary, each **ovule** contains a megaspore (*mega* means large) mother
cell (fig. 8.13) that undergoes meiosis to produce 4 haploid megaspores. Three

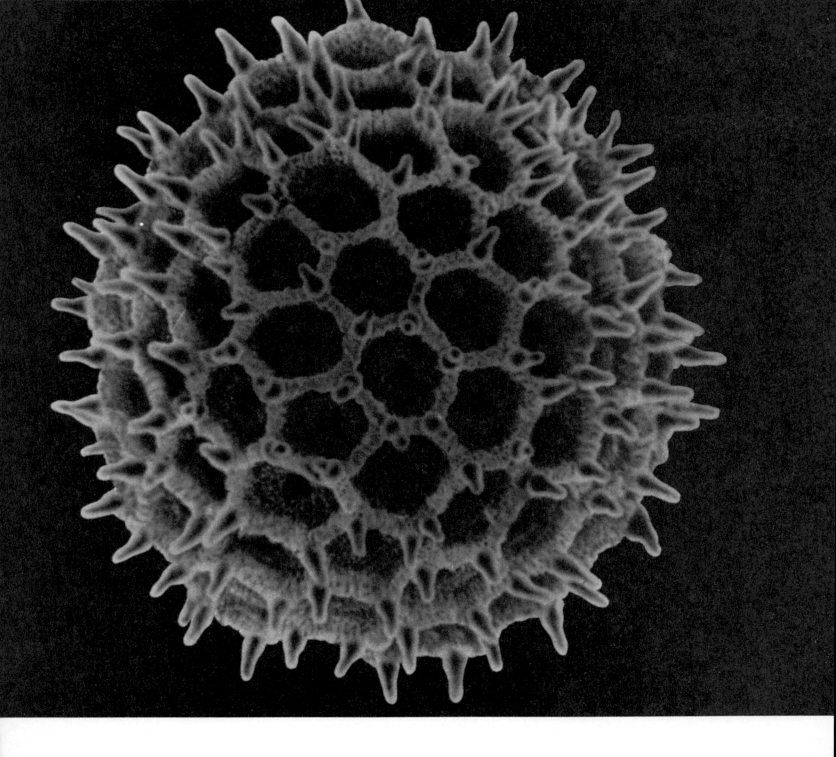

of these megaspores disintegrate, leaving one functional megaspore that divides mitotically. The result is the *female gametophyte,* which typically consists of 8 nuclei embedded in a mass of cytoplasm that is partly differentiated into cells. One of these cells is an egg.

The anther contains numerous microspore (*micro* means small) mother cells, each of which undergoes meiosis to produce 4 haploid cells called microspores. The microspores usually separate, and each one becomes a **pollen grain** (fig. 8.13), or male gametophyte. At this point, the young male gametophyte contains 2 nuclei, the *generative nucleus* and the *tube nucleus.*

Pollination occurs when pollen is windblown or carried by insects, birds, or bats to the stigma of the same type of plant. Only then does a pollen grain germinate and produce a long pollen tube. This pollen tube grows within the style until it reaches an ovule in the ovary. Before fertilization occurs, the generative nucleus divides, producing 2 sperm that have no flagellae. On reaching the ovule, the pollen tube discharges the sperm. One of the 2 sperm migrates to and fertilizes the egg, forming a zygote; the other sperm migrates to and unites with the polar nuclei, producing a 3N (triploid) endosperm nucleus. The endosperm nucleus divides to form endosperm, food for the developing plant. Note that flowering plants have *double fertilization.* One fertilization produces the zygote; the other produces endosperm.

In a flower, the ovary, a part of the pistil, contains eggs within ovules, and the pollen produced by the anther contains sperm. Upon pollination, the pollen is transported to the pistil and germinates. The sperm pass down the pollen tube, and one fertilizes the egg. The other fuses with the polar nuclei. The endosperm nucleus divides to give endosperm. The ovule now matures to produce a seed, and the ovary becomes a fruit.

The Seed

Following fertilization, each ovule becomes a mature seed containing an embryo that has at least 3 parts: the **cotyledon(s),** or seed leaf (leaves); the **epicotyl,** that portion of the embryo above the attachment of the cotyledon(s); and the **hypocotyl,** which lies below the attachment of the cotyledons and becomes a portion of the stem. The epicotyl contains the apical meristem of the shoot and sometimes bears young leaves, in which case it is called a **plumule.** A **radicle,** the embryonic root, may be at the lower end of the hypocotyl.

Monocots have seeds with one cotyledon and dicots have seeds with 2 cotyledons (fig. 7.2). Cotyledons typically provide nutrient molecules for the growing embryo. In monocot embryos, the cotyledon rarely stores food; rather, it absorbs food molecules from the endosperm and passes them to the embryo. In many dicot embryos, the cotyledons replace the endosperm, which typically already has transferred its nutrients to the cotyledons.

A seed contains an embryo and stored food, and is covered by a *seed coat.* In flowering plants, all seeds also are enclosed within a fruit that develops from the ovary (fig. 8.14) and, at times, from other accessory parts. Although peas, beans, tomatoes, and cucumbers commonly are called vegetables by laypeople, botanists categorize them as fruits. Fruits protect seeds and also sometimes aid in their dispersal. For example, winged dry fruits, like that of a maple tree, are adapted to distribution by the wind, while fleshy fruits, like that of a cherry, are eaten by birds and the seeds are deposited some distance away.

The mature seed typically contains an embryo consisting of the cotyledon(s), the epicotyl, and the hypocotyl. Upon germination, a radicle (root) appears, and the epicotyl gives rise to the leaves.

Seed Germination A seed normally will not germinate unless environmental factors are favorable. Water (H_2O) and oxygen (O_2) are essential to the completion of germination in nearly all seeds, but light is not necessarily a requirement because some seeds prefer darkness. Most seeds need a temperature that is above freezing but below 45° C. The first event normally observed in a germinating seed is the emergence of the radicle, the first indication of the roots (figs. 8.15 and 8.16). This is followed shortly by the appearance and the expansion of the seedling shoot formed from the elongating epicotyl. In dicots, the cotyledons degenerate after the nourishment they provide is consumed by the developing plant. As the seedling emerges from the soil, the shoot may be hook-shaped, protecting the delicate leaves, but once the seed is above ground, the stem straightens out and the leaves expand as photosynthesis begins. The plant continues to grow as long as it lives because of the meristematic tissue.

Figure 8.14
Development of fruit from a flower. Once seed formation begins, the ovary and the accessory parts begin to enlarge and to grow larger, until only remnants of the other flower parts remain and finally disappear entirely, leaving only the mature fruit. a. Fruit formation has begun. b. Fruit is enlarging. c. Fruit is mature.

a.

b.

c.

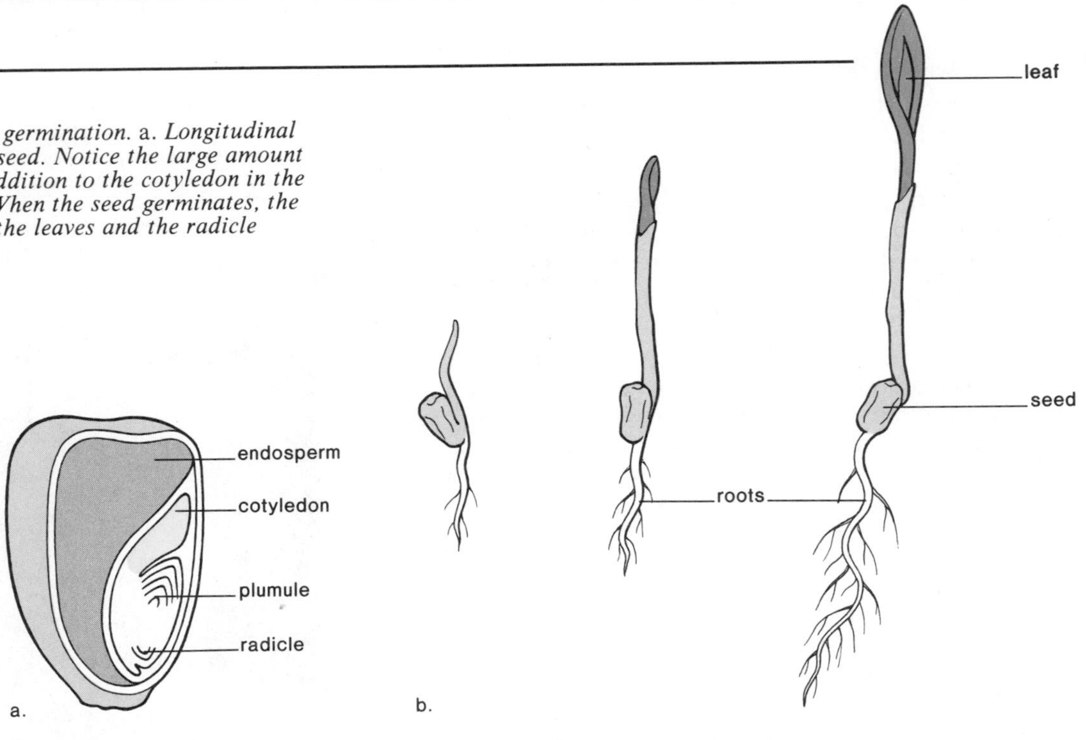

Figure 8.15

Monocot seed and germination. a. Longitudinal section of mature seed. Notice the large amount of endosperm in addition to the cotyledon in the monocot seed. b. When the seed germinates, the plumule becomes the leaves and the radicle becomes the roots.

a.

b.

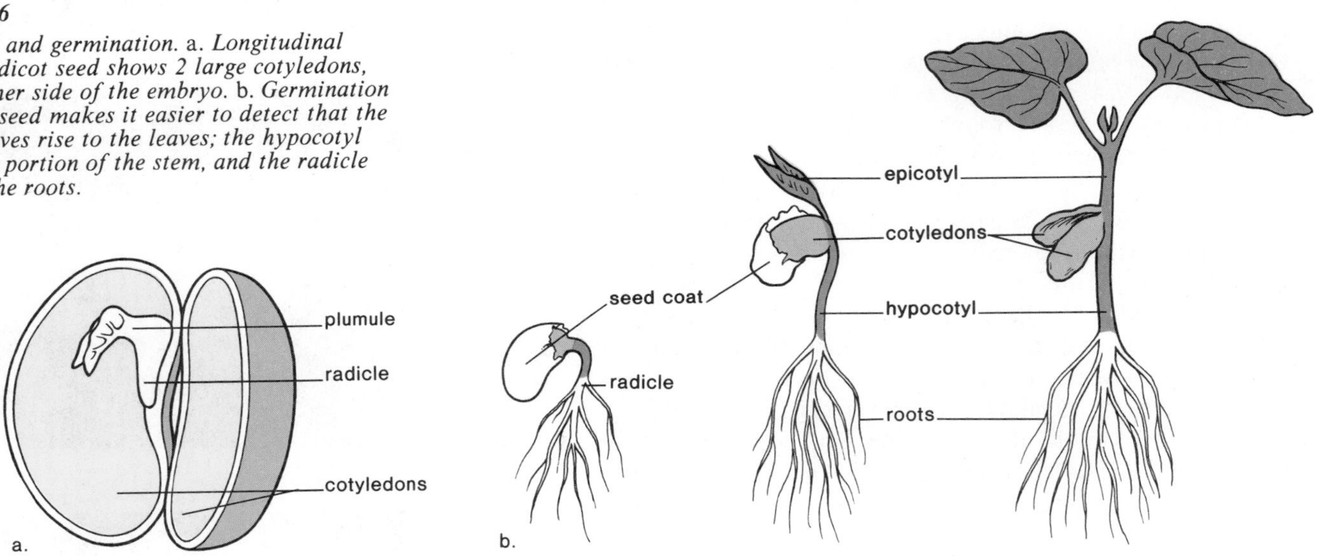

Figure 8.16

Dicot seed and germination. a. Longitudinal section of dicot seed shows 2 large cotyledons, one on either side of the embryo. b. Germination of a dicot seed makes it easier to detect that the epicotyl gives rise to the leaves; the hypocotyl becomes a portion of the stem, and the radicle becomes the roots.

a.

b.

Summary

Transport of water occurs within xylem, and transport of organic nutrients occurs within phloem. The cohesion-tension theory of xylem transport states that transpiration creates a tension that pulls water upward in xylem. This transport works only because water molecules are cohesive. Most of the water taken in by a plant is lost through stomata by transpiration. Stomata open when guard cells take up water (H_2O), stretching their thin side walls. Water enters the guard cells after potassium (K^+) ions have entered.

Both stimulatory and inhibitory hormones help to control certain plant growth patterns. There are hormones that stimulate growth (auxins, gibberellins, cytokinins) and hormones that inhibit growth (ethylene and abscisic acid). Plant hormones most likely control flowering. Short-day plants flower when the nights are longer than a critical length, and long-day plants flower when the nights are shorter than a critical length. Some plants are day-length neutral. Phytochrome, a plant pigment that can respond to day length, is believed to communicate with a biological clock that in some unknown way brings about flowering.

Flowering plants exhibit an alternation of generations life cycle that includes a separate male and female gametophyte. The pollen grain is the male gametophyte. Pollen grains are produced within the stamens of a flower. The female gametophyte is produced within the ovule of a flower. Following pollination and fertilization, the ovule matures to become the seed and the ovary becomes the fruit. The enclosed seeds contain the embryo (hypocotyl, epicotyl, plumule, radicle) and stored food (endosperm and/or cotyledons). When a seed germinates, the root appears below and the shoot appears above.

Study Questions

1. Explain the cohesion-tension theory of water transport. (p. 156)
2. What events precede the opening and the closing of stomata by guard cells? (p. 156)
3. Explain the pressure-flow theory of phloem transport. (p. 158)
4. Name 5 plant hormones, and state their function. (p. 160)
5. Define photoperiodism, and discuss its relationship to flowering in certain plants. (p. 161)
6. What is the phytochrome conversion cycle, and what are some possible functions of phytochrome in plants? (p. 162)
7. How do plants reproduce asexually? sexually? (p. 164)
8. Describe the development of a female gametophyte from the megaspore mother cell to the production of an egg. (p. 165)
9. Describe the development of the male gametophyte from the microspore mother cell to the production of sperm. (p. 165)
10. Contrast the monocot seed and seedling with the dicot seed and seedling. (pp. 167–68)

Objective Questions

1. The transport of water and minerals is dependent upon _____ , which occurs whenever the stomata are open.
2. Stomata open when _____ followed by _____ enter guard cells.
3. The _____ theory explains the transport of solutes in sieve-tube cells.
4. Short-day plants (will, will not) _____ flower when a longer-than-critical-length night is interrupted by a flash of light.
5. _____ is the pigment that is believed to signal a biological clock in plants that exhibit photoperiodism.
6. Plants have a life cycle called _____ .

7. The female gametophyte develops within the _____ of a flower, and the male gametophyte develops within the _____ .
8. Monocots have seeds with one _____ , while dicots have a seed with 2.

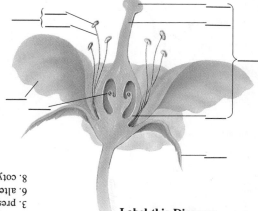

Label this Diagram.
See figure 8.12 (p. 164) in text.

Answers to Objective Questions

1. transpiration 2. potassium ions, water 3. pressure-flow 4. will not 5. Phytochrome 6. alternation of generations 7. ovule, anther 8. cotyledon

Selected Key Terms

cotyledon (kot″ĭ-le′don) the seed leaf of the embryo of a plant. *167*

epicotyl (ep″ĭ-kot′il) the plant embryo portion above the cotyledons that contributes to stem development. *167*

hypocotyl (hi″po-kot′il) the plant embryo portion below the cotyledons that contributes to stem development. *167*

photoperiodism (fo″to-pe′re-od-izm) a response to light and dark; particularly in reference to flowering in plants. *161*

phytochrome (fi′to-krōm) a plant pigment that is involved in photoperiodism in plants. *162*

plumule (ploo′mūl) the shoot tip and the first 2 leaves of a plant. *167*

pollination (pol″ĭ-na′shun) the delivery of pollen by wind or animals to the stigma of a pistil in flowering plants and fertilization. *166*

radicle (rad′ik′l) the embryonic root of a plant. *167*

spore (spōr) a haploid reproductive cell, produced by the diploid sporophyte of a plant, which asexually gives rise to the haploid gametophyte. *165*

Further Readings for Part Two

Albersheim, P., and A. G. Darvill. September 1985. Oligosaccharins. *Scientific American.*

Alberts, B., et al. 1983. *Molecular biology of the cell.* New York: Garland Publishing.

Barrett, S. C. H. September 1987. Mimicry in plants. *Scientific American.*

Bold, H. C. 1980. *Morphology of plants and fungi.* 4th ed. New York: Harper & Row, Publishers, Inc.

Brill, W. J. March 1977. Biological nitrogen fixation. *Scientific American.*

Childress, J. J., et al. May 1987. Symbiosis in the deep sea. *Scientific American.*

Cronquist, A. 1982. *Basic botany.* 2d ed. New York: Harper & Row, Publishers, Inc.

Dickerson, E. March 1980. Cytochrome and the evolution of energy metabolism. *Scientific American.*

Epel, D. November 1977. The program of fertilization. *Scientific American.*

Miller, K. R. October 1979. The photosynthetic membrane. *Scientific American.*

Nassau, Kurt. April 1980. The causes of color. *Scientific American.*

Niklas, K. J. July 1987. Aerodynamics of wind pollination. *Scientific American.*

Raven, H., et al. 1986. *Biology of plants.* 4th ed. New York: Worth Publishers, Inc.

Rayle, D., and H. L. Wedberg. 1980. *Botany: A human concern.* Boston: Houghton Mifflin.

Rost, R., et al. 1984. *Botany: A brief introduction to plant biology.* 2d ed. New York: John Wiley and Sons.

Salisbury, F. B., and C. W. Ross. 1985. *Plant physiology.* 3d ed. Belmont, Calif.: Wadsworth.

Shepard, J. F. May 1982. The regeneration of potato plants from leaf-cell protoplasts. *Scientific American.*

Stryer, L. 1988. *Biochemistry.* 3d ed. San Francisco: W. H. Freeman.

Yougan, D. C., and B. I. Mars. June 1987. Molecular mechanisms of photosynthesis. *Scientific American.*

Zimmerman, M. H. March 1963. How sap moves in trees. *Scientific American.*

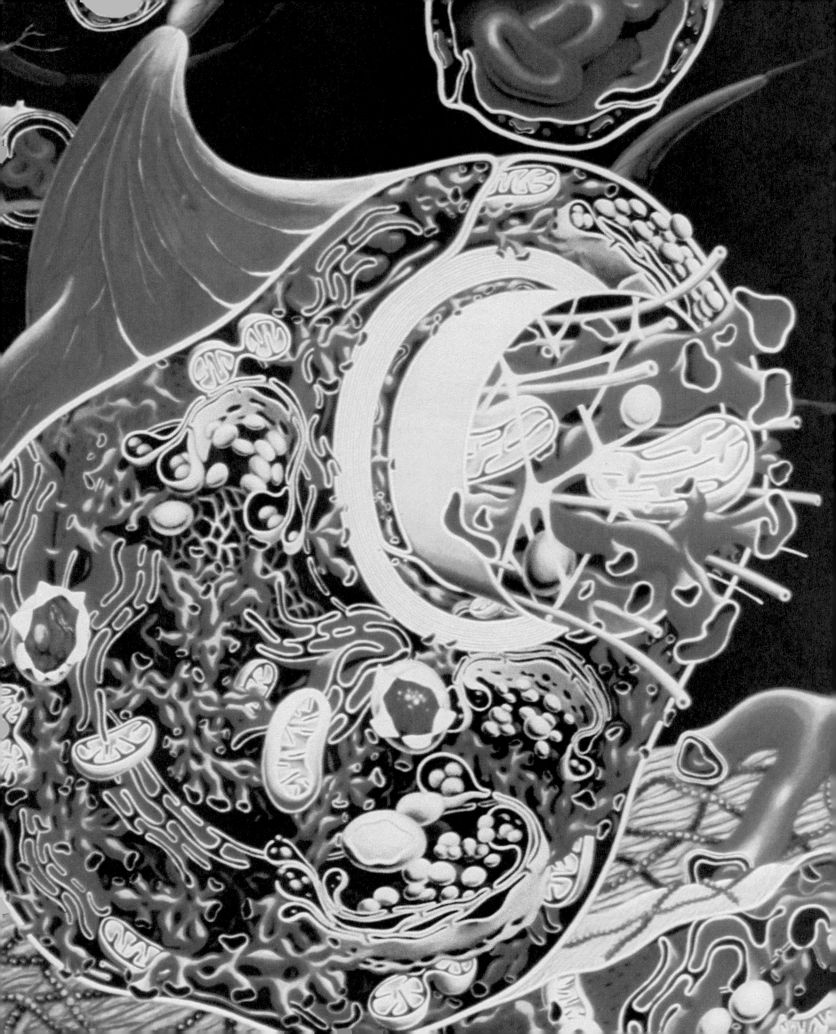

Part Three

Human Anatomy and Physiology

 The study of human anatomy and physiology serves as a guide to an understanding of the vertebrate body. A limited number of tissues make up organs, which form systems to carry out the functions assumed by the cell in less complex animals.

All body systems help to maintain a relatively constant internal environment. The digestive system provides nutrients, and the excretory system rids the body of metabolic wastes. The respiratory system supplies oxygen (O_2), but it also eliminates carbon dioxide (CO_2). The circulatory system carries nutrients and oxygen to and wastes from the cells so that tissue fluid composition remains constant. The immune system helps to protect the body from disease. The nervous and hormonal systems control body functions. The nervous system directs body movements, allowing the organism to manipulate the external environment, an important life-sustaining function.

Artist's representation of blood vessels and neurons sliced open to reveal their structure. Endothelial cells line the capillaries that contain many red blood cells. Inside the neurons you can make out the endoplasmic reticulum, mitochondria, and the Golgi apparatus.

171

Chapter 9

Human Organization

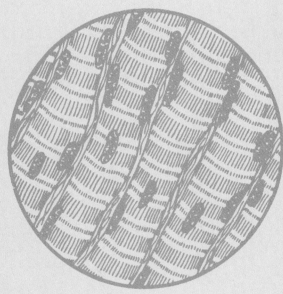

Chapter Concepts

1 Animal tissues can be categorized into 4 major types: epithelial, connective, muscular, and nervous tissues.

2 Organs usually contain several types of tissues. For example, although skin is composed primarily of epithelial tissue and connective tissue, it also contains muscle and nerve fibers.

3 Organs are grouped into organ systems, each of which has specialized functions.

4 Mammals exhibit a marked ability to maintain a relatively constant internal environment. All organ systems contribute to homeostasis.

Chapter Outline

Types of Tissues
 Epithelial Tissue
 Connective Tissue
 Muscular Tissue
 Nervous Tissue
Organs and Organ Systems
 The Skin
 Organ Systems
Homeostasis
 Body Temperature Control

In the chapters to follow, human anatomy and physiology are studied as representatives of vertebrate anatomy and physiology. Our study will be more meaningful if we first review human organization. Figure I.2 shows that the human body, like that of other organisms, has levels of organization. Cells of the same type are joined to form a tissue. Different tissues are found in an organ, and various types of organs are arranged into an organ system. Finally, the organ systems make up the organism.

As you study this chapter, note that the structure and the function of an organ system are dependent upon the structure and the function of the organ, tissue, and cell type contained therein. For example, the structure and the function of the skeletal muscle system are the same as that of the skeletal muscles, the muscular tissue, and the muscle cells.

Types of Tissues

The tissue of the human body can be categorized into 4 major types: *epithelial tissue* that covers body surfaces and lines body cavities; *connective tissue* that binds and supports body parts; *muscular tissue* that causes body parts to move; and *nervous tissue* that responds to stimuli and transmits impulses from one body part to another (fig. 9.1).

Epithelial Tissue

Epithelial tissue, also called epithelium, forms a continuous layer, or sheet, over the entire body surface and most of the body's inner cavities. On the external surface, it forms a covering that, like the epidermis in plants, protects the animal from injury and drying out. On internal surfaces, epithelial tissue may be specialized for other functions in addition to protection; for example, it secretes mucus along the digestive tract; it sweeps up impurities from the lungs by means of hairlike extensions called cilia; and it efficiently absorbs molecules from kidney tubules because of fine cellular extensions called microvilli.

There are 3 types of epithelial tissue. **Squamous epithelium** (fig. 9.2a) is composed of flat cells and is found lining the lungs and the blood vessels. **Cuboidal epithelium** (fig. 9.2b) contains cube-shaped cells and is found lining the kidney tubules. In **columnar epithelium** (fig 9.2c), the cells resemble pillars or columns, and nuclei usually are located near the bottom of each cell. This epithelium is found lining the digestive tract. An epithelium can have microvilli or cilia as appropriate for its particular function. For example, the oviducts are lined by ciliated columnar cells that beat to propel the egg and the embryo toward the uterus, or the womb.

An epithelium can be simple or stratified. Simple means that the tissue has a single layer of cells, and **stratified** means that the tissue has layers piled one on top of the other. Table 9.1 gives the body locations for squamous and stratified squamous epithelium, for example. One type of epithelium is **pseudostratified**—it appears to be layered, but actually true layers do not exist because each cell touches a baseline. The lining of the windpipe, or trachea, is *pseudostratified ciliated columnar epithelium* (fig. 9.3).

An epithelium sometimes secretes a product, in which case it is described as glandular. A gland can be a single epithelial cell, as in the case of the mucus-secreting goblet cells found within the columnar epithelium lining the digestive tract (fig. 9.2c), or a gland can contain numerous cells. Glands that secrete their product into ducts are called **exocrine glands,** and those that secrete their product directly into the bloodstream are called **endocrine glands.**

Epithelial tissue is classified according to the shape of the cell. There can be one or many layers of cells, and the layer lining a cavity can be ciliated and/or secretory.

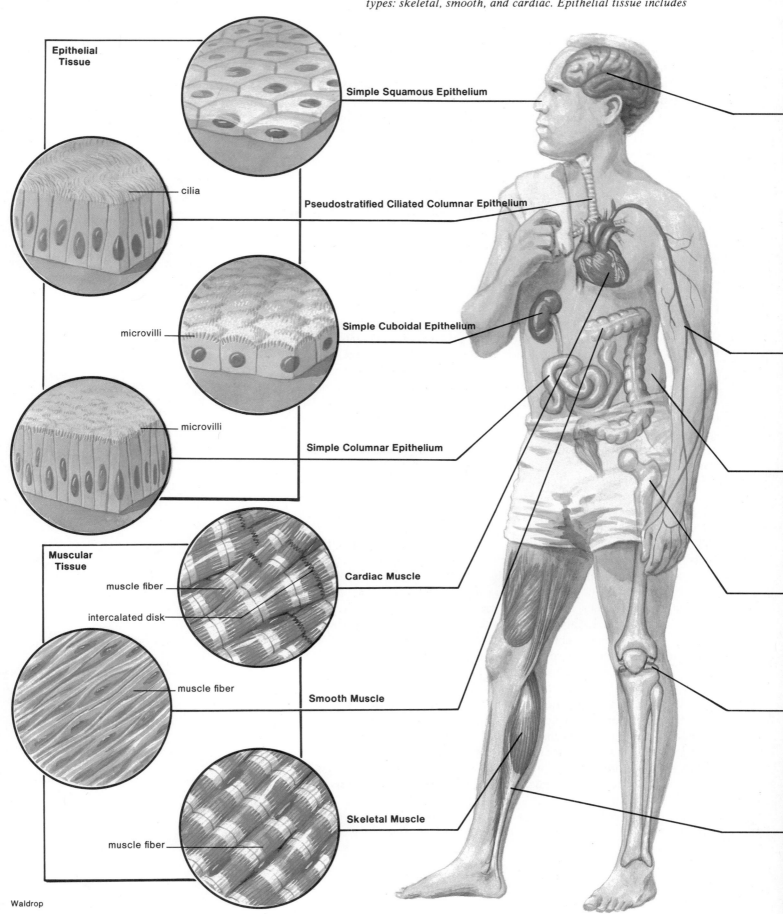

Figure 9.1

The major tissues in the human body. Reading clockwise, observe that the nervous tissue contains specialized cells called neurons. Connective tissue includes blood, adipose tissue, bone, cartilage, and fibrous connective tissue. Muscular tissue is of 3 types: skeletal, smooth, and cardiac. Epithelial tissue includes

Epithelial Tissue

Simple Squamous Epithelium

cilia

Pseudostratified Ciliated Columnar Epithelium

microvilli

Simple Cuboidal Epithelium

microvilli

Simple Columnar Epithelium

Muscular Tissue

muscle fiber

intercalated disk

Cardiac Muscle

muscle fiber

Smooth Muscle

muscle fiber

Skeletal Muscle

Waldrop

columnar, cuboidal, ciliated columnar, and squamous epithelium. Each type can be stratified or ciliated. Not all types of connective and epithelial tissue are shown.

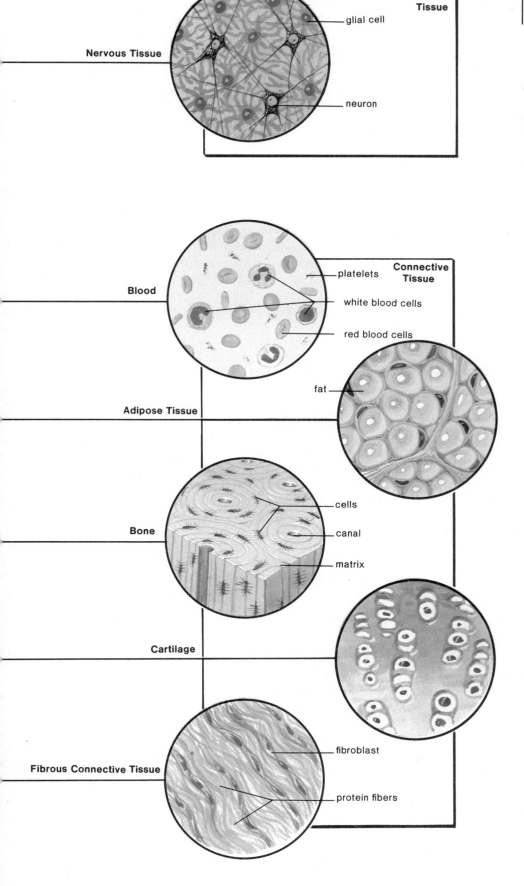

Nervous Tissue

Nervous Tissue
- glial cell
- neuron

Blood
Connective Tissue
- platelets
- white blood cells
- red blood cells

Adipose Tissue
- fat

Bone
- cells
- canal
- matrix

Cartilage

Fibrous Connective Tissue
- fibroblast
- protein fibers

Figure 9.2
Simple epithelial tissue. a. Simple squamous epithelium consists of a single layer of thin cells. b. Simple cuboidal epithelium is composed of cells that look like cubes. c. Simple columnar epithelium contains cells that resemble columns because they are elongated. Epithelial tissue lines cavities; lumen-cavity.

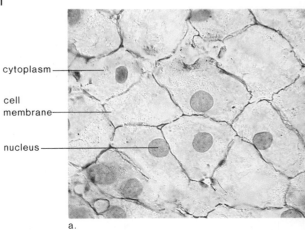

- cytoplasm
- cell membrane
- nucleus

a.

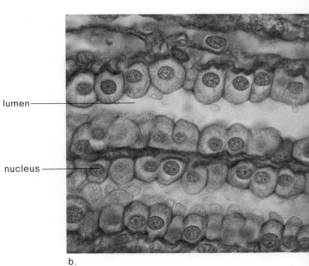

- lumen
- nucleus

b.

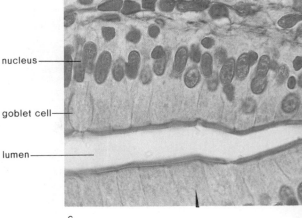

- nucleus
- goblet cell
- lumen

c.

Figure 9.3

Pseudostratified ciliated columnar epithelium forms the lining of the windpipe. a. Note that all cells touch the baseline and there are goblet cells. When you cough, material trapped in the mucus secreted by goblet cells is moved upward to the throat, where it can be swallowed.
b. Photomicrograph of pseudostratified ciliated columnar epithelium.

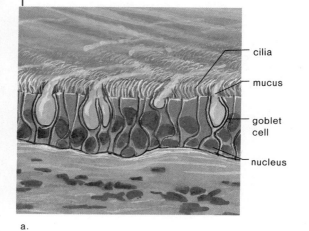

cilia

mucus

goblet cell

nucleus

a.

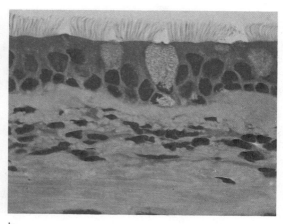

b.

Junctions between Cells

Epithelial tissue cells are packed tightly and are joined to one another in one of 3 ways: spot desmosomes, tight junctions, and gap junctions (fig. 9.4). In a **spot desmosome**, internal cytoplasmic plaques, firmly attached to the cytoskeleton within each cell, are joined by intercellular filaments. The cells are joined more closely in a **tight junction** because adjacent cell membrane (tight junction) proteins actually attach to each other, producing a zipperlike fastening. A **gap junction** is formed when 2 identical cell membrane channels join. This lends strength, but it also allows substances to pass between the 2 cells.

An epithelium often is joined to underlying connective tissue by a so-called *basement membrane*. We now know that the basement membrane is glycoprotein reinforced by fibers supplied by the connective tissue.

Connective Tissue

Connective tissue (table 9.2) binds structures together, provides support and protection, fills spaces, stores fat, and forms blood cells. As a rule, connective tissue cells are separated widely by a **matrix,** noncellular material found between cells. The matrix may have fibers of 2 types. White fibers contain collagen, a substance that gives them flexibility and strength. Yellow fibers contain elastin, a substance that is not as strong as collagen but is more elastic.

Loose Connective Tissue

Loose connective tissue binds structures (fig. 9.5). The cells of this tissue, which are mainly **fibroblasts,** are located some distance from one another and are separated by a jellylike matrix that contains many white collagen fibers and yellow elastic fibers. The collagen fibers occur in bundles and are strong and flexible. The elastic fibers form networks that when stretched return to their original length. As discussed previously, loose connective tissue commonly lies beneath epithelial layers. In certain instances, epithelium and its underlying connective tissue form body membranes (p. 184). In addition, adipose tissue (fig. 9.6) is a type of loose connective tissue in which the fibroblasts enlarge and store fat and in which the intercellular matrix is reduced.

Fibrous Connective Tissue

Fibrous connective tissue contains many collagenous fibers that are packed closely together. This type of tissue has more specific functions than loose connective tissue. For example, fibrous connective tissue is found in **tendons,** which

Figure 9.4

Junction between epithelial cells. Epithelial tissue cells are held tightly together by (a) spot desmosomes and (b) tight junctions. c. Gap junctions allow materials to pass from cell to cell.

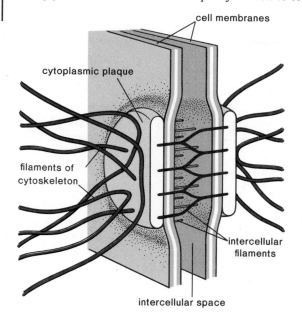

a. Spot desmosome

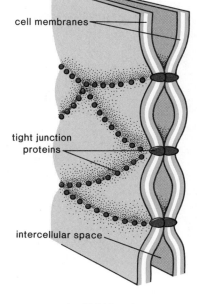

b. Tight junction

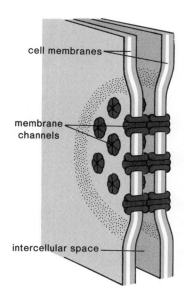

c. Gap junction

Table 9.2 Connective Tissue		
Type	**Function**	**Location**
Loose connective tissue	Binds organs	Beneath the skin; beneath most epithelial layers
Adipose tissue	Insulates; stores fat	Beneath the skin; around the kidneys
Fibrous connective tissue	Binds organs	Tendons; ligaments
Cartilage		
Hyaline cartilage	Supports; protects	Ends of bones; nose; rings in walls of respiratory passages
Elastic cartilage	Supports; protects	External ear; part of the larynx
Fibrocartilage	Supports; protects	Between bony parts of backbone and knee
Bone	Supports; protects	Bones of skeleton
Blood	Transports gases, nutrients, and wastes about body; infection fighting; blood clotting	Blood vessels

From John W. Hole, Jr., *Human Anatomy and Physiology,* 5th ed. Copyright © 1990 Wm. C. Brown Publishers, Dubuque, Iowa. All Rights Reserved. Reprinted by permission.

connect muscles to bones, and in **ligaments**, which connect bones to other bones at joints. Tendons and ligaments take a long time to heal following an injury because their blood supply is relatively poor.

Loose connective tissue and fibrous connective tissue, which bind body parts, differ according to the type and the abundance of fibers in the matrix.

Figure 9.5

Loose connective tissue. Loose connective tissue has plenty of space between components. This type of tissue is found surrounding and between the organs.

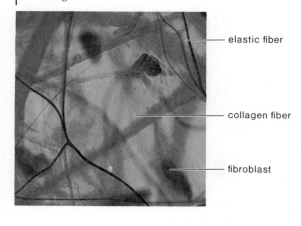

elastic fiber

collagen fiber

fibroblast

Figure 9.6

Adipose tissue. Adipose cells look like white ghosts because the fat has been washed out during preparation of the tissue. The nucleus of one cell is indicated by the arrow.

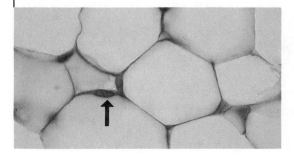

Figure 9.7

Hyaline cartilage. Hyaline cartilage cells, located in lacunae, are separated by a flexible matrix rich in protein and fibers. This type of cartilage forms the embryonic skeleton, which later is replaced by bone.

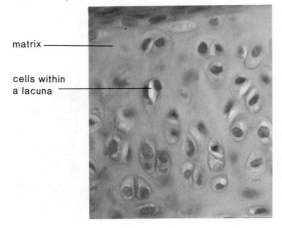

matrix ———

cells within
a lacuna ———

Figure 9.8

Compact bone. Compact bone is highly organized. The cells are arranged in circles about a central (Haversian) canal that contains a nutrient-bearing blood vessel.

osteocyte
within a
lacuna ———

canaliculi ———

Haversian
canal ———

Table 9.3 Blood Plasma

Water	92% of total
Inorganic ions (salts)	Na^+, Ca^{++}, K^+, Mg^{++}; Cl^-, HCO_3^-, HPO_4^-, SO_4^-
Gases	O_2, CO_2
Plasma proteins	Albumin, globulins, fibrinogen
Organic nutrients	Glucose, fats, phospholipids, amino acids, etc.
Nitrogenous waste products	Urea, ammonia, uric acid
Regulatory substances	Hormones, enzymes

Cartilage

In **cartilage,** the cells lie in small chambers called **lacunae** (singular, lacuna), separated by a matrix that is solid yet flexible. Unfortunately, because this tissue lacks a direct blood supply, it heals very slowly. There are 3 types of cartilage, distinguished by the type of fiber in the matrix.

Hyaline cartilage (fig. 9.7), the most common type of cartilage, contains only very fine collagenous fibers. The matrix has a milk-glass appearance. Hyaline cartilage is found in the nose, at the ends of the long bones and the ribs, and in the supporting rings of the windpipe. The fetal skeleton also is made of this type of cartilage. Later, the cartilaginous fetal skeleton is replaced by bone.

Elastic cartilage has more elastic fibers than hyaline cartilage. For this reason, it is more flexible and is found, for example, in the framework of the outer ear.

Fibrocartilage has a matrix containing strong collagenous fibers. Fibrocartilage is found in structures that withstand tension and pressure, such as the pads between the vertebrae in the backbone and the wedges found in the knee joint.

Bone

Bone is the most rigid connective tissue. It consists of an extremely hard matrix of calcium salts deposited around protein fibers. The minerals give bone rigidity, and the protein fibers provide elasticity and strength, much as steel rods do in reinforced concrete.

The shaft of a long bone is compact bone (fig. 9.8). In **compact bone,** bone cells (osteocytes) are located in lacunae that are arranged in concentric circles around tiny tubes called Haversian canals. Nerve fibers and blood vessels are in these canals. The latter bring the nutrients that allow bone to renew itself. The nutrients can reach all of the cells because there are minute canals (canaliculi) containing thin processes of the osteocytes that connect them with one another and with the Haversian canals.

The ends of a long bone contain spongy bone (fig. 17.5), which has an entirely different structure. Spongy bone contains numerous bony bars and plates separated by irregular spaces. Although lighter than compact bone, **spongy bone** still is designed for strength. Just as braces are used for support in buildings, the solid portions of spongy bone follow lines of stress.

Cartilage and bone are support tissues. Cartilage is more flexible than bone because the matrix is rich in protein and not calcium salts like that of bone.

Blood

Blood (fig. 9.9) is a connective tissue in which the cells are separated by a liquid called plasma, the contents of which are listed in table 9.3. Blood cells are of 2 types: **erythrocytes (red),** which carry oxygen, and **leukocytes (white),** which aid in fighting infection. Also present in plasma are *platelets,* which are important to the initiation of blood clotting. Platelets are not complete cells; rather, they are fragments of giant cells found in the bone marrow.

Blood is a connective tissue in which the matrix is plasma.

Blood is unlike other types of connective tissue in that the intercellular matrix (i.e., plasma) is not made by the cells. Plasma (table 9.3) is a mixture of different types of molecules that enter the blood at various locations. Some people do not classify blood as connective tissue; instead, they suggest a separate tissue category for blood called vascular tissue.

Figure 9.9

Blood, a liquid tissue. Blood is classified as connective tissue because the cells are separated by a matrix—plasma. Plasma, the liquid portion of blood, contains several types of cells (red blood cells, white blood cells, and platelets).

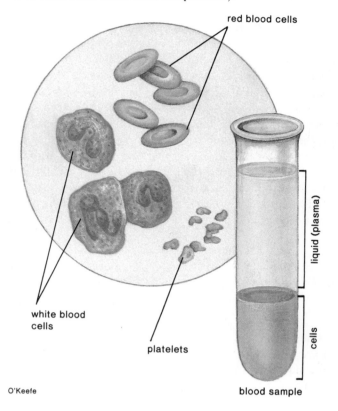

red blood cells

white blood cells

platelets

O'Keefe

liquid (plasma)

cells

blood sample

Figure 9.10

Muscular tissue. How do you distinguish a plant from an animal? One way is to detect rapid motion—only animals have contractile fibers that permit movement. a. Skeletal muscle is found within the muscles attached to the skeleton. Note the striations and the peripheral location of the nuclei in the multinucleate fibers. b. Smooth muscle is found in the walls of internal organs. Note the single nucleus and the lack of striations. c. Cardiac muscle pumps the heart. Note the branching of the fibers, the central position of the nuclei, and the presence of intercalated disks, which are folded cell membranes between adjacent individual cells.

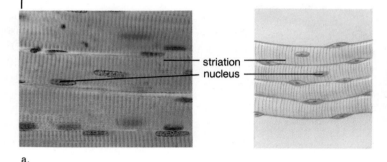

striation
nucleus

a.

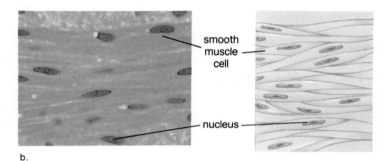

smooth muscle cell

nucleus

b.

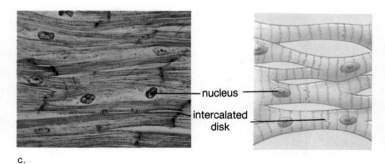

nucleus

intercalated disk

c.

Table 9.4 Muscular Tissue

Type	Fiber Appearance	Location	Control
Skeletal	Striated	Attached to skeleton	Voluntary
Smooth	Spindle shaped	Internal organs	Involuntary
Cardiac	Striated and branched	Heart	Involuntary

Muscular Tissue

Muscular (contractile) tissue is composed of cells that are called *muscle fibers.* Muscle fibers contain actin filaments and myosin filaments, whose interaction accounts for the movements we associate with animals. There are 3 types of vertebrate muscles: *skeletal, smooth,* and *cardiac* (table 9.4).

Skeletal muscle (fig. 9.10*a*) is attached to the bones of the skeleton; it moves body parts. It is under our voluntary control and contracts faster than all the other muscle types. Skeletal muscle cells are cylindrical and quite long— they run the length of the muscle. They arise during development when several cells fuse, giving one multinucleated cell. The nuclei are placed at the periphery of the cell, just inside the cell membrane.

Skeletal muscle cells are **striated.** There are light and dark bands perpendicular to the length of the cell. These bands are due to the placement of actin filaments and myosin filaments in the cell.

Smooth muscle is so named because the cells lack striations. The spindle-shaped cells form layers in which the thick middle portion of one cell is opposite the thin ends of adjacent cells. Consequently, the nuclei form an irregular pattern in the tissue (fig. 9.10b). Smooth muscle is not under voluntary control and therefore is said to be involuntary. Smooth muscle, found in walls of viscera (intestine, stomach, and other internal organs) and blood vessels, contracts more slowly than skeletal muscle but can remain contracted for a longer time. When the smooth muscle of the intestine contracts, it moves the food along, and when the smooth muscle of the blood vessels contracts, it constricts the blood vessels, helping to raise the blood pressure.

Cardiac muscle (fig. 9.10c), which is found only in the heart, is responsible for the heartbeat. Cardiac muscle seems to combine features of both smooth muscle and skeletal muscle. It has striations like skeletal muscle, but the contraction of the heart is involuntary for the most part. Cardiac muscle cells also differ from skeletal muscle cells in that they have a single, centrally placed nucleus. The cells are branched and seemingly fused one with the other, and the heart appears to be composed of one large interconnecting mass of muscle cells. Actually, cardiac muscle cells are separate and individual, but they are bound end to end at intercalated disks, areas of folded cell membrane between the cells.

All muscular tissue contains actin filaments and myosin filaments; these form a striated pattern in skeletal and cardiac muscle, but not in smooth muscle.

Nervous Tissue

The brain and the nerve cord (also called the spinal cord) contain conducting cells termed neurons. A **neuron** (fig. 9.11) is a specialized cell that has 3 parts: (1) *dendrites* that conduct impulses (send a message) to the cell body; (2) the *cell body* that contains most of the cytoplasm and the nucleus of the neuron; and (3) the *axon* that conducts impulses away from the cell body.

When axons and dendrites are long, they are called *nerve fibers*. Outside the brain and the spinal cord, nerve fibers are bound by connective tissue to form **nerves.** Nerves conduct impulses from sense organs to the spinal cord and the brain, where the phenomenon called sensation occurs. They also conduct nerve impulses away from the spinal cord and the brain to the muscles, causing them to contract.

In addition to neurons, nervous tissue contains **glial cells.** These cells maintain the tissue by supporting and protecting neurons. They also provide nutrients to neurons and help to keep the tissue free of debris.

Organs and Organ Systems

We tend to think that a particular organ contains one type of tissue. For example, we associate muscular tissue with muscles and nervous tissue with the brain. However, these organs also contain other types of tissue; for example, they contain loose connective tissue and blood. An **organ** is a structure that is composed of 2 or more tissues. An **organ system** contains many different organs.

We are going to consider the skin as an example of an organ. Some people even like to call the skin the *integumentary system,* especially since it cannot be placed in one of the other organ systems. However, the skin does not really have distinct organs.

The Skin

The skin (fig. 9.12) covers the body, protecting underlying parts from physical trauma, microbial invasion, and water loss. Skin also helps to regulate body temperature (p. 186), and because it contains sense organs, the skin helps us to be aware of our surroundings and to communicate with others.

Figure 9.11

Photo of neuron. Conduction of the nerve impulse is dependent on neurons, each of which has the 3 parts indicated. A dendrite takes nerve impulses to the cell body, and an axon takes them away from the cell body.

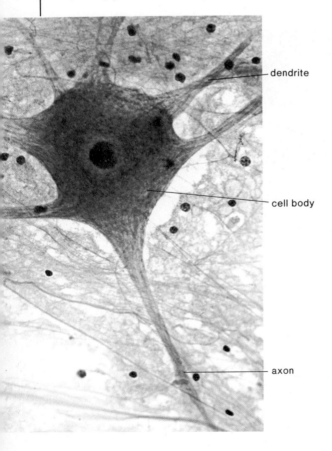

dendrite

cell body

axon

Figure 9.12

Human skin anatomy. Skin contains 3 layers: epidermis, dermis, and subcutaneous.

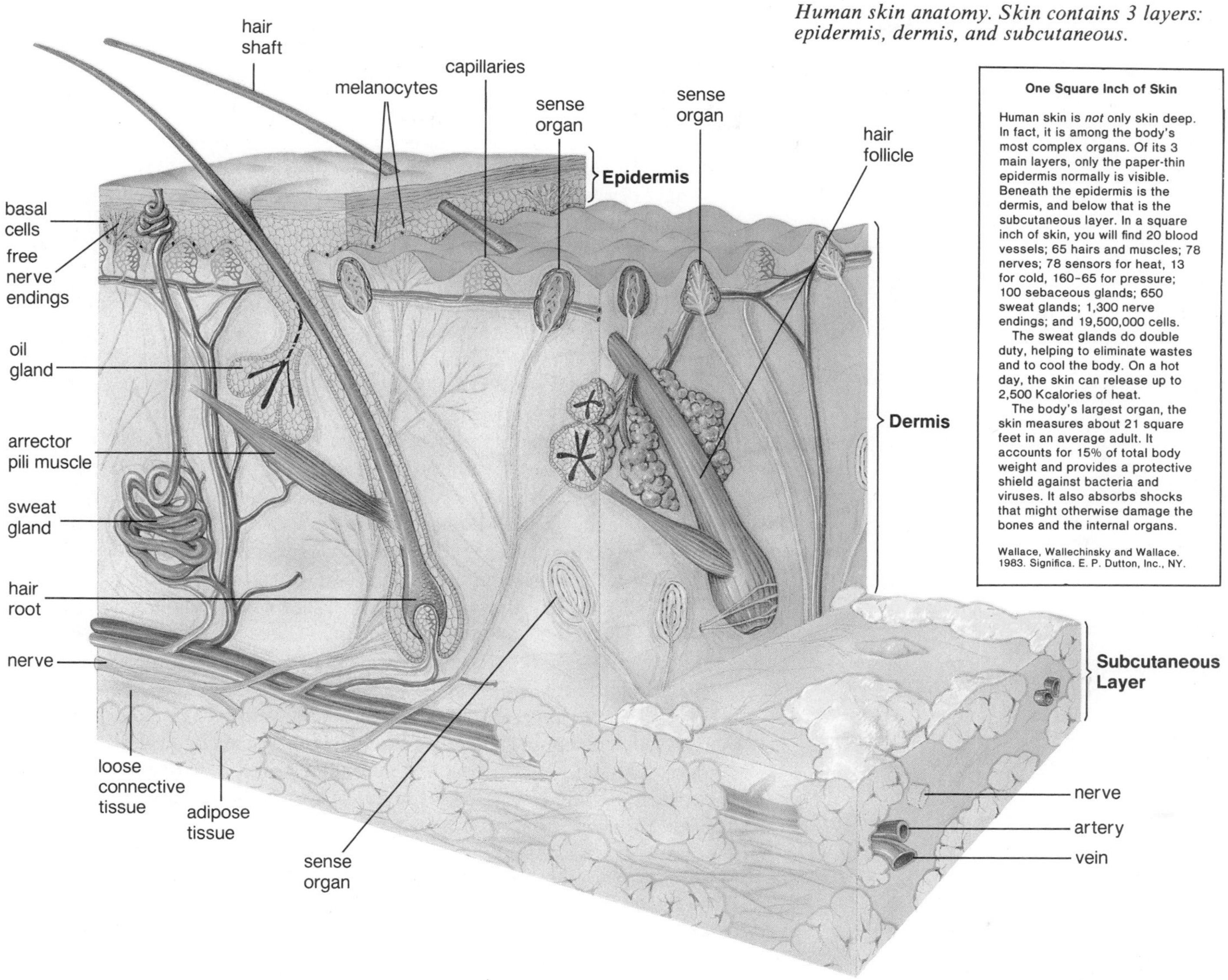

One Square Inch of Skin

Human skin is *not* only skin deep. In fact, it is among the body's most complex organs. Of its 3 main layers, only the paper-thin epidermis normally is visible. Beneath the epidermis is the dermis, and below that is the subcutaneous layer. In a square inch of skin, you will find 20 blood vessels; 65 hairs and muscles; 78 nerves; 78 sensors for heat, 13 for cold, 160–65 for pressure; 100 sebaceous glands; 650 sweat glands; 1,300 nerve endings; and 19,500,000 cells.

The sweat glands do double duty, helping to eliminate wastes and to cool the body. On a hot day, the skin can release up to 2,500 Kcalories of heat.

The body's largest organ, the skin measures about 21 square feet in an average adult. It accounts for 15% of total body weight and provides a protective shield against bacteria and viruses. It also absorbs shocks that might otherwise damage the bones and the internal organs.

Wallace, Wallechinsky and Wallace. 1983. Significa. E. P. Dutton, Inc., NY.

Skin has an outer epidermal layer (the epidermis) and an inner dermal layer (the dermis). Beneath the dermis, there is a subcutaneous layer that binds the skin to the underlying organs.

Skin Layers

The **epidermis** is the outer, thinner layer of the skin. It is made up of stratified squamous epithelium, which is produced continually by a bottom layer of cells termed basal cells. As newly formed cells are pushed to the surface, they gradually flatten and harden. Eventually, they die and are sloughed off. Hardening is caused by cellular production of a waterproof protein called *keratin*. Over much of the body, keratinization is minimal, but the palm of the hand and the sole of the foot have a particularly thick outer layer of dead keratinized cells arranged in spiral and concentric patterns. We call these patterns fingerprints and footprints.

Specialized cells in the epidermis called *melanocytes* produce melanin, the pigment responsible for skin color in dark-skinned persons. When you sunbathe, the melanocytes become more active, producing melanin in an attempt to protect the skin from the damaging effects of the ultraviolet (UV) radiation in sunlight.

Human Organization 181

The epidermis, the outer layer of skin, is made up of stratified squamous epithelial cells. New cells that are produced continually in the innermost layer of the epidermis push outward, become keratinized, die, and are sloughed off.

The **dermis** is a layer of fibrous connective tissue that is deeper and thicker than the epidermis. It contains elastic fibers and collagen fibers. The collagen fibers form bundles that interlace with each other and run, for the most part, parallel to the skin surface. As a person ages and is exposed to the sun, the number of fibers decreases, and those remaining have characteristics that make the skin less supple and cause wrinkling.

There are several types of structures in the dermis. A hair, except for the root, is formed of dead, hardened epidermal cells; the root is alive and resides in a *hair follicle* found in the dermis. Each follicle has one or more *oil* (sebaceous) *glands* that secrete sebum, an oily substance that lubricates the hair and the skin. Particularly on the nose and the cheeks, the sebaceous glands may fail to discharge, and the secretions collect and form "whiteheads" or "blackheads." The color of blackheads is due to oxidized sebum. If pus-inducing bacteria also are present, a boil or a pimple may result.

A smooth muscle called the *arrector pili* muscle is attached to the hair follicle in such a way that when contracted, the muscle causes the hair to "stand on end." When you have had a scare or are cold, goose bumps develop due to the contraction of these muscles.

Sweat (sudoriferous) *glands* are quite numerous and are present in all regions of the skin. A sweat gland begins as a coiled tubule within the dermis, but then it straightens out near its opening. Some sweat glands open into hair follicles, and others open onto the surface of the skin.

Small *sense organs* are present in the dermis. There are different sense organs for touch, pressure, pain, and temperature. The fingertips contain the most touch receptors, and these add to our ability to use our fingers for delicate tasks. The dermis also contains nerve fibers and blood vessels. When blood rushes into these vessels, a person blushes, and when blood is reduced in them, a person turns "blue."

The dermis is composed of fibrous connective tissue and lies beneath the epidermis. It contains hair follicles, sebaceous glands, and sweat glands. It also contains sense organs, blood vessels, and nerve fibers.

The **subcutaneous layer,** which lies below the dermis, is composed of loose connective tissue, including adipose tissue. Adipose tissue helps to insulate the body from either gaining heat from the outside or losing heat from the inside. A well-developed subcutaneous layer gives a rounded appearance to the body. Excessive development of this layer accompanies obesity.

Disorders

The skin is subject to many disorders, some of which are not serious. For example, when you have *dandruff,* the rate of keratinization is 2 or 3 times the normal rate in certain areas of the scalp. Other disorders of the skin are more serious.

Cancer of the Skin The most dangerous type of skin cancer is *malignant melanoma, death-causing cancer.* A melanoma is a darkly pigmented spot that resembles a nonmalignant mole. Although melanomas tend to occur on such sun-exposed areas as the chest of men and the legs of women, their relationship to the sun is unclear.

The 2 more common types of skin cancer, basal-cell carcinoma and squamous-cell carcinoma, definitely are related to exposure to sunlight. Ultraviolet radiation from the sun causes dividing skin cells to become cancerous. During the development of *squamous-cell carcinoma,* precancerous dark

Are tanning machines safe? Most dermatologists feel they are not. Tanning booths almost certainly are capable of causing skin to age, to degenerate, or to develop cancer.

On the whole, people who use tanning machines probably get away with it, but they are taking a chance. Tanning machines expose their patrons to intense ultraviolet light, which is capable of producing the acute and chronic side effects of exposure to sunlight. Tanning machines use mainly

That Oh, So Nice, Tan

UVA light, which is somewhat less potent in producing biological changes than the slightly shorter UVB waves, but sufficient doses of UVA can be deleterious. Also, extra UVA may enhance the carcinogenic potential of exposure to natural sunlight. People taking medications that induce photosensitivity may have severe reactions to the light from tanning machines. Certain diseases, notably lupus erythematosus, which are exacerbated by sunlight, also are made worse by tanning machines.

The light in these booths is intense so as to achieve in a few minutes the skin reaction that would otherwise take a longer period of baking in natural sunlight. People who tan in booths may do so without any clothing; thus they expose areas of skin that lack protective pigmentation built up from long-term exposure to

sunlight. Ultraviolet light has the potential to injure the retina; people who patronize tanning parlors always should wear protective goggles.

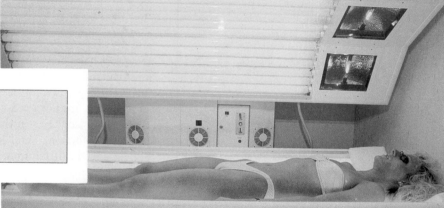

Tanning booth. *Tanning the skin most likely contributes to skin aging, skin degeneration, and skin cancer.*

The American Academy of Dermatology has formed a task force on photobiology. Its chairman, Dr. Leonard C. Harber, also is chairman of the Department of Dermatology at Columbia University College of Physicians and Surgeons. He says, "We want health warnings to appear in tanning parlors as they do on cigarette packs."

Excerpted from "That Oh, So Nice, Tan" from the March 1988 issue of *The Harvard Medical School Health Letter* ©1988 President and Fellows of Harvard College.

---

patches called actinic keratosis precede the rough scaly patches of skin cancer. In *basal-cell carcinoma,* newly produced cells no longer rise to the surface and become keratinized. Instead, these cells invade the dermis, and ulcers develop. Both of these types of cancer usually can be removed surgically.

In recent years, there has been a great increase in the number of persons with skin cancer, and physicians believe this is due to sunbathing or even to the use of tanning machines, as discussed in the reading above. These professionals strongly recommend that everyone stay out of the sun and refrain from using tanning machines. If persons must be in the sun, then sunscreens should be used. For example, a number-15 sunscreen lotion provides 15 times the natural sunburn protection. In other words, 15 minutes in the sun with protection from this strength of sunscreen is equivalent to one minute without protection. Even higher strengths are available.

Skin cancer is associated with ultraviolet radiation and appears in 3 forms. Malignant melanoma is the most dangerous skin cancer. Basal-cell carcinoma and squamous-cell carcinoma usually can be removed surgically.

Burns   Two factors affect the severity of a burn: the depth of the burn and the extent of the burned area. In *first-degree burns,* only the epidermis is affected. The burn site is painful, but there are no blisters or swelling. A classic example of a first-degree burn is moderate sunburn. The pain subsides within 48–72 hours, and the injury heals without further complications or scarring. The damaged skin peels off in about one week.

Table 9.5 Human Organ Systems

Name	Function
Digestive	Converts food particles to nutrient molecules
Circulatory	Transports molecules to and from cells
Immune	Defends against disease
Respiratory	Exchanges gases with the environment
Excretory	Eliminates metabolic wastes
Nervous and sensory	Regulates systems and response to environment
Musculoskeletal	Supports and moves organism
Hormonal	Regulates internal environment
Reproductive	Produces offspring

Figure 9.13

Organization of the human body. Like other mammals, humans have a dorsal nervous system and a well-developed coelom that contains the internal organs. The coelom is divided by the diaphragm into the thoracic cavity and the abdominal cavity.

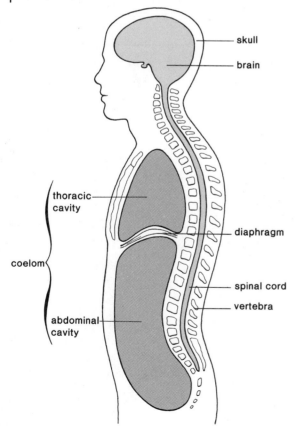

A *second-degree burn* extends through the entire epidermis and into part of the dermis. There is not only redness and pain, but blisters also develop in the region of the damaged tissue. The deeper the burn, the more prevalent the blisters, which increase in size during the hours after injury. Unless these blisters become infected, most second-degree burns heal without complication and with little scarring in 10–14 days.

More severe burns result in more permanent damage. *Third-degree burns* destroy the entire thickness of the skin. The surface of the burn is leathery and may be brown, tan, black, white, or red. There is no pain because the pain receptors have been destroyed. Blood vessels, sweat glands, sebaceous glands, and hair follicles all are destroyed. There are even *fourth-degree burns* that involve tissue down to the bone. Obviously, the chances of a person surviving such a burn are not good unless a very limited area of the body is affected.

Major concerns in the case of severe burns are fluid loss, heat loss, and bacterial infection. Fluid loss is counteracted by intravenous administration of a balanced salt solution. Heat loss is minimized by placing the burn patient in a warm environment. Bacterial infection is treated by the application of an antibacterial dressing.

As soon as possible, the damaged tissue is removed, and skin grafting is begun. Usually, the skin used for grafting is taken from other parts of the patient's body because there is little chance of graft rejection. However, experimentation with artificial skin and with new skin grown in the laboratory from only a small piece taken from the patient is underway. In one type of artificial skin, the inner layer is a lattice made from shark cartilage and collagen fibers from cowhide. The outer layer is rubberlike silicone plastic. Skin that has been grown in the laboratory functions in some respects like the patient's own skin.

Organ Systems

In this text, we will study the organ systems listed in table 9.5. Each of these systems has a specific location within the body. The central nervous system is located dorsally (toward the back); the brain is protected by the skull; and the spinal cord, which gives off spinal nerves, is protected by the vertebrae (fig. 9.13). The repeating units of vertebrae and spinal nerves show that humans are segmented animals, meaning that body parts reoccur at regular intervals.

Within the musculoskeletal system, the skeleton provides the surface area for attachment of the well-developed and powerful striated muscles. The musculoskeletal system makes up most of the body weight and is specialized for locomotion.

The other internal organs are found within a body cavity called the **coelom.** In humans and other mammals, the coelom is divided by a muscular diaphragm that assists breathing. The heart, a pump for the closed circulatory system, and the lungs are located in the upper (thoracic or chest) cavity. The major portion of the digestive system, the entire excretory system, and much of the reproductive system are located in the lower (abdominal) cavity. The major organs of the excretory system are the paired kidneys, and the accessory organs of the digestive system are the liver and the pancreas. Each sex has characteristic sex organs.

The preceding attributes are vertebrate characteristics, and in this text, human physiology is studied as representative of vertebrates in general.

Body Membranes

The term *membrane* at the organ level generally refers to a thin lining or covering composed of an epithelium overlying a layer of loose connective tissue. For example, mucous membrane lines the organs of the respiratory and digestive systems. This type of membrane, as its name implies, secretes mucus.

Serous membrane lines enclosed cavities and covers the organs that lie within these cavities, such as the heart, the lungs, and the kidneys. This type of membrane secretes a watery lubricating fluid.

The body is divided into cavities within which the organs are found. These cavities usually are lined with membrane.

Homeostasis

Homeostasis means that the internal environment remains relatively constant, regardless of the conditions in the external environment. In humans, for example:

1. Blood glucose concentration remains at about 0.1%.
2. The pH of the blood is always near 7.4.
3. Blood pressure in the brachial artery averages near 120/80.
4. Blood temperature averages around 37° C (98.6° F).

The ability of the body to keep the internal environment within a certain range allows humans to live in a variety of habitats, such as the arctic regions, the deserts, or the tropics.

This internal environment includes a tissue fluid that bathes all the tissues of the body. Tissue fluid is created when water, oxygen (O_2), and nutrient molecules leave a capillary (the smallest of the blood vessels), and it is purified when water, carbon dioxide (CO_2), and other waste molecules enter a capillary from the fluid (fig. 9.14). Tissue fluid remains constant only as long as blood composition remains constant. Although we are accustomed to using the word *environment* to mean the external environment of the body, it is important to realize that it is the internal environment of tissues that is ultimately responsible for our health and well-being.

The internal environment of the body consists of blood and tissue fluid, which bathes the cells.

Most systems of the body contribute to maintenance of a constant internal environment. The digestive system takes in and digests food, providing nutrient molecules that enter the blood and replace the nutrients that are constantly being used by the body cells. The respiratory system adds oxygen to the blood and removes carbon dioxide. The amount of oxygen taken in and carbon dioxide given off can be increased to meet body needs. The chief regulators of blood composition, however, are the liver and the kidneys. They monitor the chemical composition of plasma (table 9.3) and alter it as required. Immediately after glucose enters the blood, it can be removed by the liver for storage as glycogen. Later, the glycogen can be broken down to replace the glucose used by the body cells; in this way, the glucose composition of the blood remains constant. The hormone insulin, secreted by the pancreas, regulates glycogen storage. The liver also removes toxic chemicals, such as ingested alcohol and drugs and nitrogenous wastes given off by the cells. These are converted to molecules that can be excreted by the kidneys. The kidneys are also under hormonal control as they excrete wastes and salts, substances that can affect the pH level of the blood.

All the systems of the body contribute to homeostasis, that is, maintaining the relative constancy of the internal environment.

Although homeostasis is, to a degree, controlled by hormones, it is ultimately controlled by the nervous system. The brain contains centers that regulate such factors as temperature and blood pressure. Maintaining proper

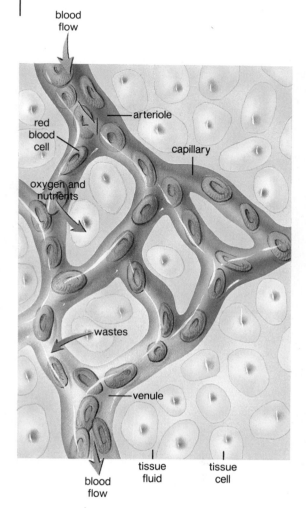

Figure 9.14
Formation of tissue fluid from blood. The internal environment of the body is the blood and the tissue fluid. Tissue cells are surrounded by tissue fluid, which is refreshed continually because nutrient molecules constantly exit and waste molecules continually enter the bloodstream as shown.

blood flow

red blood cell

arteriole

capillary

oxygen and nutrients

wastes

venule

blood flow

tissue fluid

tissue cell

9.2 Critical Thinking

1. Normal body temperature is said to be 37° C. Is body temperature always exactly 37° C, or does it fluctuate?
2. Why does the text use the phrase "*relative constancy* of the internal environment" when referring to homeostasis?
3. What type of stimuli activate the receptor and the regulator center shown in figure 9.16?
4. Does this account for fluctuation of body temperature above and below a certain temperature?

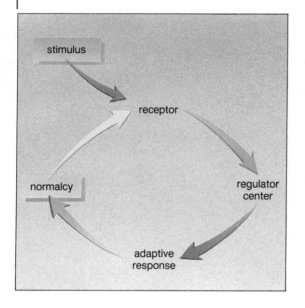

temperature and blood pressure levels requires receptors (sense organs) that detect unacceptable levels and signal a regulator center. If a correction is required, the center then directs an adaptive response (fig. 9.15). Once normalcy is obtained, the receptor no longer signals the center. This is called control by **negative feedback** because the regulator center shuts down until it is stimulated to activity once again. This type of homeostatic regulation results in fluctuation between 2 levels, as is illustrated in the following example of temperature control (fig. 9.16). Notice that feedback control is a self-regulatory mechanism.

Body Temperature Control

The receptor and the regulator center for body temperature are located in the hypothalamus. The receptor is sensitive to the temperature of the blood, and when the temperature falls below normal, the regulator center directs (via nerve impulses) the blood vessels of the skin to constrict. This conserves heat. Also, the arrector pili muscles pull hairs erect, and a layer of insulating air is trapped next to the skin. If body temperature falls even lower, the regulator center sends nerve impulses to the skeletal muscles, and shivering occurs. Shivering generates heat, and gradually body temperature rises to 37° C and perhaps higher. During the period of time the body temperature is normal, the receptor and the regulator center are not active, but once body temperature is higher than normal, they are reactivated. Now the regulator center directs the blood vessels of the skin to dilate. This allows more blood to flow near the surface of the body, where heat can be lost to the environment. The

Figure 9.16

Temperature control. When the body temperature rises, the regulator center directs the blood vessels to dilate and the sweat glands to be active. Now, the body temperature lowers. Then, the regulator center directs the blood vessels to constrict, hairs to stand on end, and even shivering to occur if needed. Now, the body temperature rises again. Because the regulator center is activated only by extremes, the body temperature fluctuates above and below normal.

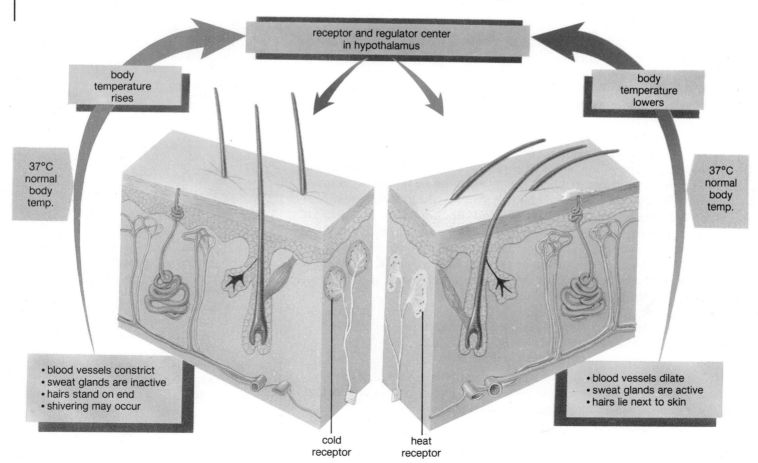

regulator center also activates the sweat glands because the evaporation of sweat also helps to lower body temperature. Gradually, body temperature decreases to 37° C and perhaps lower. Once body temperature is below normal, the cycle begins again.

Homeostasis of internal conditions is a self-regulatory mechanism that results in slight fluctuations above and below a mean. For example, body temperature rises above and drops below a normal temperature of 37° C.

Summary

Human tissues are categorized into 4 groups. Epithelial tissue covers the body and lines its cavities. Connective tissue often binds body parts. Contraction of muscular tissue permits movement of the body and its parts. Nerve impulses conducted by neurons within nervous tissue help to bring about coordination of body parts.

Different types of tissues are joined to form organs, each one having a specific function. Organs are grouped into organ systems. In vertebrates, the brain and the spinal cord are dorsally located, and the internal organs are located in the coelom composed of both the thoracic and abdominal cavities.

All organ systems contribute to the constancy of the internal environment. The nervous and hormonal systems regulate the other systems. Both of these are controlled by a feedback mechanism, which results in fluctuation above and below a mean.

Study Questions

1. Name the 4 major types of tissues. (p. 173)
2. What are the functions of epithelial tissue? Name the different kinds, and give a location for each. (pp. 173–75)
3. What are the functions of connective tissue? Name the different kinds, and give a location for each. (pp. 176–78)
4. What are the functions of muscular tissue? Name the different kinds, and give a location for each. (pp. 179–80)
5. Nervous tissue contains what type of cell? Which organs in the body are made up of nervous tissue? (p. 180)
6. Describe the structure of skin, and state at least 2 functions of this organ. (pp. 180–82)
7. In general terms, describe the location of the human organ systems. (p. 184)
8. List at least 4 vertebrate characteristics of humans. (p. 184)
9. Distinguish between cell membrane and body membrane. (p. 184)
10. What is homeostasis, and how is it achieved in the human body? (pp. 185–86)

Objective Questions

1. Most organs contain several different types of _____ .
2. Kidney tubules are lined by cube-shaped cells called _____ epithelium.
3. Pseudostratified ciliated columnar epithelium contains cells that appear to be _____ , have projections called _____ , and are _____ in shape.
4. Both cartilage and blood are classified as _____ tissue.
5. Cardiac muscle is _____ but involuntary.
6. Nerve cells are called _____ .
7. Skin has 3 layers: epidermis, _____ , and the subcutaneous layer.
8. Outer skin cells are filled with _____ , a waterproof protein that strengthens them.
9. Mucous membrane contains _____ tissue overlying _____ tissue.
10. Homeostasis is maintenance of the relative _____ of the internal environment, that is, the blood and _____ fluid.

Selected Key Terms

bone (bōn) connective tissue having a hard matrix of calcium salts deposited around protein fibers. *178*

cartilage (kar'ti-lij) a connective tissue in which the cells lie within lacunae embedded in a flexible matrix. *178*

coelom (se'lom) a body cavity of higher animals that contains internal organs, such as those of the digestive system. *184*

compact bone (kom'pakt bōn) hard bone consisting of Haversian systems cemented together. *178*

connective tissue (kŏ-nek'tiv tish'u) a type of tissue characterized by cells separated by a matrix that often contains fibers. *176*

dermis (der'mis) the thick skin layer that lies beneath the epidermis. *182*

epidermis (ep"ĭ-der'mis) the outer skin layer composed of stratified squamous epithelium. *181*

epithelial tissue (ep"ĭ-the'le-al tish'u) a type of tissue that lines cavities and covers the external surface of the body. *173*

homeostasis (ho"me-o-sta'sis) the relative maintenance of conditions, particularly the internal environment of birds and mammals, such as maintenance of temperature, blood pressure, and pH. *185*

hyaline cartilage (hi′ah-līn kar′tĭ-lij) cartilage composed of very fine collagenous fibers and a matrix of a milk-glass appearance. *178*

lacuna (lah-ku′nah) a small pit or hollow cavity, as in bone or cartilage, where a cell or cells are located. *178*

ligament (lig′ah-ment) dense connective tissue that joins bone to bone. *177*

muscular tissue (mus′ky-lar tish′u) a type of tissue that contains cells capable of contracting; skeletal muscles are attached to the skeleton, smooth muscle is found within walls of internal organs, and cardiac muscle comprises the heart. *179*

negative feedback (neg′ah-tiv fēd′bak) a self-regulatory mechanism that is activated by an imbalance and results in a fluctuation about a mean. *186*

neuron (nu′ron) nerve cell that characteristically has 3 parts: dendrites, cell body, axon. *180*

pseudostratified (su″-do strat′e-fīd) the appearance of layering in some epithelial cells when actually each cell touches a baseline and true layers do not exist. *173*

spongy bone (spun′je bōn) porous bone found at the ends of long bones. *178*

stratified (strat′ĭ-fīd) layered, as in stratified epithelium, which contains several layers of cells. *173*

striated (stri′āt-ed) having bands; cardiac and skeletal muscle are striated with bands of light and dark. *179*

subcutaneous layer (sub″ku-ta′ne-us la′er) a tissue layer found in vertebrate skin that lies just beneath the dermis and tends to contain adipose tissue. *182*

tendon (ten′don) dense connective tissue that joins muscle to bone. *176*

Chapter 10

Digestion

Chapter Concepts

1 Small molecules, such as amino acids, glucose, and fatty acids, that can cross cell membranes are the products of digestion that nourish the body.

2 Regions of the digestive tract are specialized to carry on specific functions; for example, the mouth is specialized to receive and to chew food, and the small intestine is specialized to absorb the products of digestion.

3 Digestive enzymes are specific hydrolytic enzymes and have a preferred temperature and pH.

4 Proper nutrition requires that the energy needs of the body be met and that the diet be balanced so that all vitamins, essential amino acids, and fatty acids are included.

Chapter Outline

The Digestive System
    The Mouth
    The Pharynx
    The Esophagus
    The Stomach
    The Small Intestine
    The Large Intestine
    Accessory Organs
Digestive Enzymes
    Conditions for Digestion
Nutrition
    Proteins
    Carbohydrates
    Lipids
    Vitamins and Minerals
Dieting
    Daily Energy Requirement
    Fad Diets versus Behavior
        Modification
    Eating Disorders

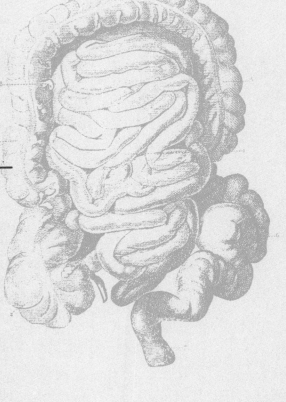

Table 10.1 Path of Food

Organ	Function	Special Feature	Function
Mouth	Receives food; digestion of starch	Teeth Tongue	Chewing of food Formation of bolus
Esophagus	Passageway		
Stomach	Storage of food; acidity kills bacteria; digestion of protein	Gastric glands	Release gastric juices
Small intestine	Digestion of all foods; absorption of nutrients	Intestinal glands Villi	Release intestinal juices Absorb nutrients
Large intestine	Absorption of water; storage of nondigestible remains		
Anus	Defecation		

Digestion takes place within a tube, often called the gut, that begins with the mouth and ends with the anus (table 10.1 and fig. 10.1). Digestion of food in humans is an extracellular process. Digestive enzymes are secreted into the gut by glands that reside in the gut lining or lie nearby. Food never is found within these *accessory glands,* only within the gut itself.

While the term *digestion,* strictly speaking, means the breakdown of food by enzymatic action, in this text the term is expanded to include both physical and chemical processes that reduce food to small soluble molecules. Only small molecules can cross cell membranes and be absorbed by the gut lining. Too often we are inclined to think that since we eat meat (protein), potatoes (carbohydrate), and butter (fat), these are the substances that nourish our bodies. Instead, it is the amino acids from the protein, the sugars from the carbohydrate, and the glycerol and fatty acids from the fat that actually enter the blood and are transported throughout the body to nourish our cells. Any component of food, such as cellulose, that is incapable of being digested to small molecules leaves the gut as waste material.

Digestion of food requires a cooperative effort between different parts of the body. We will see that the production of hormones and the performance of the nervous system achieve the cooperation of body parts.

The Digestive System

The functions of the digestive system are to ingest the food, to digest it to small molecules that can cross cell membranes, to absorb these nutrient molecules, and to eliminate nondigestible wastes.

The Mouth

The mouth receives the food in humans. Most people enjoy eating because of the combined sensations of smelling and tasting food. The olfactory receptors, located in the nose, are responsible for smelling; tasting is a function of the taste buds, located primarily on the tongue. (See chapter 18 for a description of these sense organs.)

The teeth chew the food into pieces convenient to swallow. During the first 2 years of life, the 20 deciduous or baby teeth appear. These will be replaced eventually by the adult teeth. Normally, adults have 32 teeth (fig. 10.2). One-half of each jaw has teeth of 4 different types: 2 chisel-shaped *incisors* for biting; one pointed *canine* for tearing; 2 fairly flat *premolars* for grinding; and 3 *molars,* more flattened for crushing. The last molars, called the wisdom teeth, may fail to erupt, or if they do, they are sometimes crooked and useless. Oftentimes, the extraction of the wisdom teeth is recommended.

Human Anatomy and Physiology

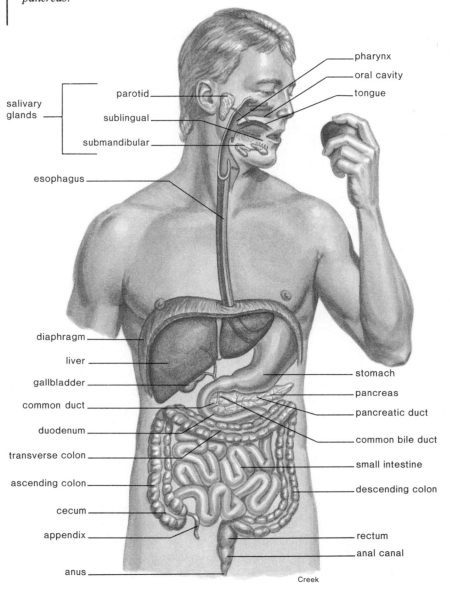

Figure 10.1
Trace the path of food from the mouth to the anus, and note the placement of the accessory organs of digestion, the liver and the pancreas.

salivary glands
parotid
sublingual
submandibular

esophagus

pharynx
oral cavity
tongue

diaphragm
liver
gallbladder
common duct
duodenum
transverse colon
ascending colon
cecum
appendix
anus

stomach
pancreas
pancreatic duct
common bile duct
small intestine
descending colon
rectum
anal canal

Creek

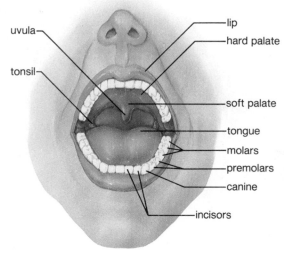

Figure 10.2
Diagram of the mouth showing the adult teeth. The sizes and shapes of the incisors, canine, premolars, and molars correlate with their function.

uvula
tonsil

lip
hard palate
soft palate
tongue
molars
premolars
canine
incisors

Each tooth (fig. 10.3) has a crown and a root. The crown has a layer of enamel, an extremely hard outer covering of calcium compounds; dentin, a thick layer of bonelike material; and an inner pulp that contains the nerves and the blood vessels. Dentin and pulp are also in the root. Tooth decay, or *caries,* commonly called a cavity, occurs when the bacteria within the mouth metabolize sugar and give off acids that corrode the tooth. Two measures can prevent tooth decay: eating a limited amount of sweets and daily brushing and flossing of teeth. It also has been found that fluoride treatments, particularly in children, can make the enamel stronger and more resistant to decay. Gum disease is more apt to occur with aging. Inflammation of the gums (gingivitis) can spread to the periodontal membrane (fig. 10.3) that lines the tooth socket. The individual then has **periodontitis,** characterized by a loss of bone and loosening of the teeth so that extensive dental work may be required. Stimulation of the gums in a manner advised by dentists is helpful in controlling this condition.

Figure 10.3

Longitudinal section of a canine tooth. Nerves and blood vessels are found within the pulp of a tooth.

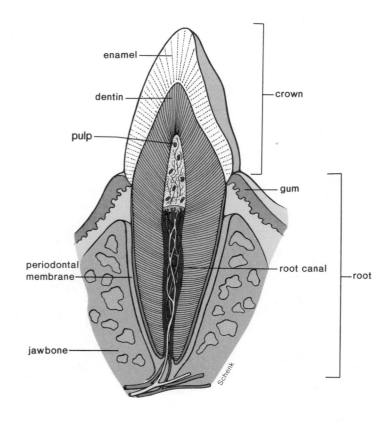

enamel

dentin

pulp

crown

gum

periodontal membrane

root canal

root

jawbone

Schenk

Figure 10.4

Swallowing. When food is swallowed, the soft palate covers the nasopharyngeal openings and the epiglottis covers the glottis so that the bolus must pass down the esophagus. Therefore, you do not breath when swallowing.

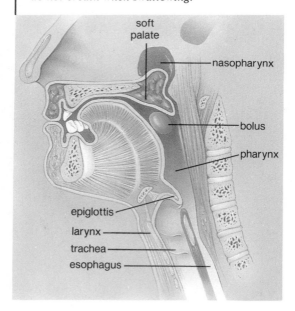

soft palate

nasopharynx

bolus

pharynx

epiglottis

larynx

trachea

esophagus

In humans, the roof of the mouth separates the air passages from the mouth cavity. The roof has 2 parts: an anterior **hard palate** and a posterior **soft palate** (fig. 10.2). The hard palate contains several bones, but the soft palate is only muscular. The soft palate ends in the *uvula,* a suspended process often mistaken by the layperson for the tonsils. In fact, the tonsils are at the sides of the oral cavity (fig. 10.2), at the base of the tongue, and in the nose (called adenoids). The tonsils play a minor role in protecting the body from disease-causing organisms, as is discussed in chapter 13.

There are 3 pairs of **salivary glands** that send their juices (saliva) by way of ducts to the mouth. The parotid glands lie at the sides of the face immediately below and in front of the ears. They become swollen when a person has the mumps, a viral infection most often seen in children. Each *parotid gland* has a duct that opens on the inner surface of the cheek at the location of the second upper molar. The *sublingual glands* lie beneath the tongue, and the *submandibular glands* lie beneath the lower jaw. The ducts from these glands open into the mouth under the tongue. You can locate all these openings if you use your tongue to feel for small flaps on the inside of your cheek and under your tongue. An enzyme within saliva begins the process of digesting food. Specifically, this enzyme acts on starch.

The tongue, which is composed of striated muscle with an outer layer of mucous membrane, mixes the chewed food with saliva. It then forms this mixture into a mass called a bolus in preparation for swallowing.

The salivary glands send saliva into the mouth, where the teeth chew the food and the tongue forms it into a bolus for swallowing.

The Pharynx

Swallowing (fig. 10.4) occurs in the **pharynx,** a region between the mouth and the esophagus, which is a long muscular tube leading to the stomach. Swallowing is a *reflex action,* which means the action usually is performed automatically and does not require conscious thought. Normally, during swallowing,

Figure 10.5

Wall of the esophagus. Like the rest of the digestive tract, several different types of tissues are found in the wall of the esophagus. Note the placement of circular muscle inside longitudinal muscle. This arrangement, found throughout the animal kingdom wherever a tube has 2 muscle layers, ensures that the action of the circular muscles does not interfere with that of the longitudinal muscle. a. Diagrammatic drawing. b. Scanning electron micrograph.
(Lu = central lumen; Mu = mucous membrane; Su = submucosa; Me = muscular layer; and Ad = adventitia, or serous layer.)

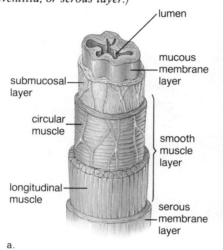

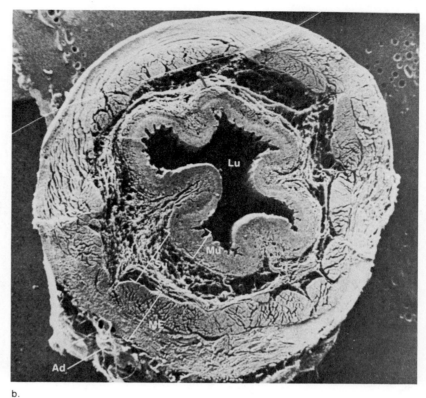

a.   b.

food enters the esophagus because the air passages are blocked. Unfortunately, we have all had the unpleasant experience of having food "go down the wrong way." The wrong way may be either into the nose or into the trachea (windpipe). If it is the latter, coughing usually forces the food up out of the trachea into the pharynx again. Food usually goes into the esophagus because the openings to the nose, called the *nasopharyngeal openings,* are covered when the soft palate moves back. The opening to the larynx (voice box) at the top of the trachea, called the **glottis,** is covered when the trachea moves up under a flap of tissue called the **epiglottis.** This is easy to observe in the up-and-down movement of the *Adam's apple,* a part of the larynx, when a person eats. Notice that breathing does not occur during swallowing because air passages are closed off.

The air passage and the food passage cross in the pharynx. When you swallow, the air passage usually is blocked off, and food must enter the esophagus.

The Esophagus

After swallowing occurs, the **esophagus** conducts the bolus through the thoracic cavity. The wall (fig. 10.5) of the esophagus is representative of the gut in general. A *mucous membrane layer* lines the **lumen** (space within the tube); this is followed by a *submucosal layer* of connective tissue that contains nerve and blood vessels, a *smooth muscle layer* having both longitudinal and circular muscles, and finally, a *serous membrane layer.*

A rhythmic contraction of the esophageal wall, called **peristalsis** (fig. 10.6), pushes the food along. Occasionally, peristalsis begins even though there is no food in the esophagus. This produces the sensation of a lump in the throat.

The esophagus extends from the back of the pharynx to just below the diaphragm, where it meets the stomach at an angle. The entrance of the esophagus into the stomach is marked by the presence of a constrictor, called the lower esophogeal sphincter, although muscle development is less than in a true

Figure 10.6

Peristalsis in the digestive tract. Rhythmic waves of muscle contraction move material along the digestive tract. The 3 drawings show how a peristaltic wave moves through a single section of gut over time.

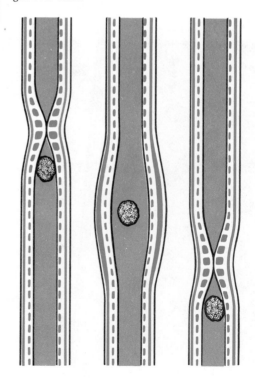

Digestion   **193**

Figure 10.7

Photomicrograph of the mucous membrane layer of the stomach. Gastric glands produce gastric juice rich in pepsin, an enzyme that digests protein to peptides.

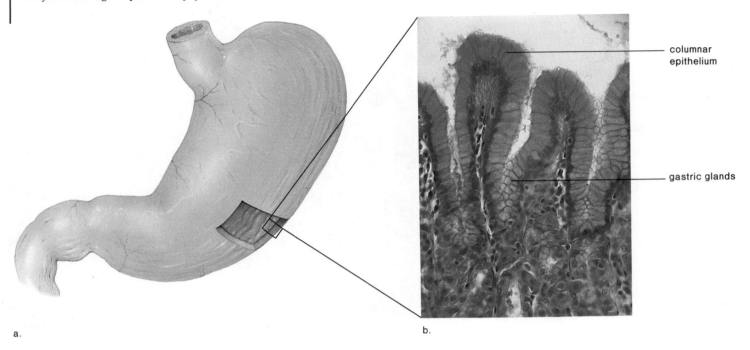

columnar
epithelium

gastric glands

a.

b.

Figure 10.8

Peptic ulcer in the stomach. Normally, the digestive tract produces enough mucus to protect itself from digestive juices. An ulcer often begins when an excessive amount of gastric juices is produced due to an increased amount of nervous stimulation.

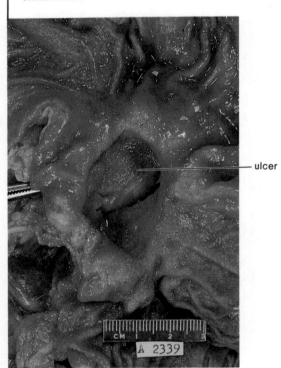

ulcer

sphincter. **Sphincters** are muscles that encircle tubes and act as valves; tubes close when sphincters contract, and they open when sphincters relax. When food is swallowed, the sphincter relaxes, allowing the bolus to pass into the stomach. Normally, this sphincter prevents the acidic contents of the stomach from entering the esophagus. Heartburn, which feels like a burning pain rising up into the throat, occurs when some of the stomach contents escape into the esophagus. When vomiting occurs, a reverse peristaltic wave causes the sphincter to relax, and the contents of the stomach are propelled upward through the esophagus.

The Stomach

The **stomach** (fig. 10.7) is a thick-walled, J-shaped organ that lies on the left side of the body beneath the diaphragm. The stomach is continuous with the esophagus above and the duodenum of the small intestine below. The stomach stores food. The wall of the stomach has 3 layers of muscle and also contains deep folds that disappear as the stomach fills. The muscular wall of the stomach churns, mixing the food with gastric secretions. When food leaves the stomach, it is a pasty material called *acid chyme.*

The columnar epithelium lining the stomach contains millions of microscopic digestive glands called **gastric glands** (the term gastric always refers to the stomach). The gastric glands produce gastric juice. Gastric juice contains hydrochloric acid (HC1) and a digestive enzyme that digests protein. The high acidity of the stomach (about pH 2) is beneficial because it kills most bacteria present in food. Although hydrochloric acid does not digest food, it does break down the connective tissue of meat and activates the digestive enzyme in gastric juice. Normally, the wall of the stomach is protected by a thick layer of mucus, but if by chance hydrochloric acid does penetrate this mucus, autodigestion of the wall can begin, and an ulcer results (fig. 10.8). An **ulcer** is an open sore in the wall caused by the gradual disintegration of tissue. It is believed that the most frequent cause of an ulcer is oversecretion of gastric juice due to too much nervous stimulation; persons under stress tend to have a greater incidence of ulcers. However, there is now evidence that a bacterial (*Campylobacter pyloridis*) infection may impair the ability of cells to produce mucus, which protects the wall from ulcer formation.

Normally, the stomach empties in about 2–6 hours. Acid chyme leaves the stomach and enters the small intestine by way of the pyloric sphincter. The pyloric sphincter repeatedly opens and closes, allowing acid chyme to enter the small intestine in small squirts only. This means that digestion in the small intestine will proceed at a rate slow enough to ensure thoroughness.

The stomach expands and stores food. While food is in the stomach, it churns, mixing food with acidic gastric juice.

The Small Intestine

The **small intestine** gets its name from its small diameter (compared to that of the large intestine), but perhaps it should be called the long intestine because it averages about 6.0 m (20 ft) in length compared to the large intestine, which is about 1.5 m (5 ft) long. The small intestine receives secretions from the liver and the pancreas, chemically and mechanically breaks down acid chyme, absorbs nutrient molecules, and transports undigested material to the large intestine.

The first 25 cm (10 in) of the small intestine is called the **duodenum.** Duodenal ulcers sometimes occur because gastric juice within acid chyme can digest the intestinal wall in this region. Ducts from the gallbladder and the pancreas join to form the common duct, which enters into the duodenum (see fig. 10.1).

The wall of the small intestine contains fingerlike projections called **villi** (fig. 10.9), and the villi themselves have microvilli that are visible microscopically. Because they are so numerous, the villi give the intestinal wall a soft, velvety appearance. Each villus has an outer layer of columnar epithelium and contains blood vessels and a small lymphatic vessel called a **lacteal.** The lymphatic system is an adjunct to the circulatory system and returns fluid to the veins. Villi cells produce an intestinal digestive juice that contains enzymes for finishing the digestion of acid chyme to small molecules that can cross cell membranes. They also greatly increase the surface area of the small intestine for absorption of nutrients.

Absorption of nutrient molecules across the wall of each villus continues until all small molecules have been absorbed. Therefore, absorption is an active process involving active transport of molecules across cell membranes and requiring an expenditure of cellular energy. Sugars and amino acids cross the columnar epithelial cells to enter the blood, and the components of fats rejoin before entering the lacteals.

The small intestine is specialized to absorb the products of digestion. It is quite long (3.0 m) and has fingerlike projections called villi, where nutrient molecules are absorbed into the circulatory and lymphatic systems.

The Large Intestine

The **large intestine,** which includes the cecum, the colon, the rectum, and the anal canal, is larger in diameter than the small intestine (6.5 cm compared to 2.5 cm). The large intestine absorbs water and salts. It also prepares and stores nondigestible material until it is defecated at the anus.

The *cecum,* which lies below the entrance of the small intestine, has a small projection called the veriform **appendix** (veriform means wormlike)(fig. 10.10). In humans, the appendix, like the tonsils, may play a role in immunity. This organ is subject to inflammation, a condition called appendicitis. It is better to have the appendix removed before the fluid content rises to the point that the appendix bursts because this can lead to generalized infection of the serous membranes of the abdominal cavity.

Figure 10.9

Anatomy of intestinal lining. a. The products of digestion are absorbed by villi, fingerlike projections of the intestinal wall, (b) each of which contains blood vessels and a lacteal. c. The scanning electron micrograph shows that the villi themselves are covered with microvilli (Mv). d. A transmission electron micrograph shows that the microvilli contain microfilaments (Mf). These allow limited motion of the microvilli. They extend to the terminal web (TW). Adjacent epithelial cells are joined by tight junctions termed zonula occludens (ZO).

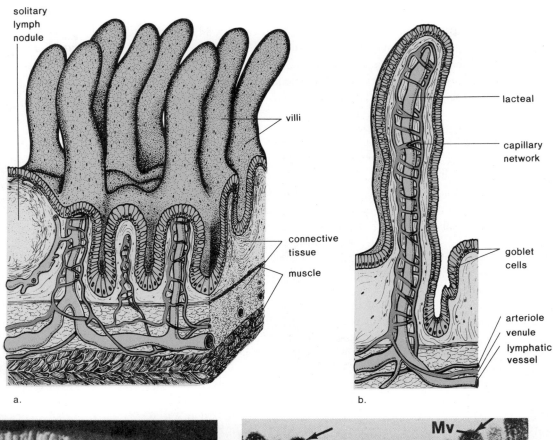

a.

b.

solitary lymph nodule

villi

connective tissue

muscle

lacteal

capillary network

goblet cells

arteriole
venule
lymphatic vessel

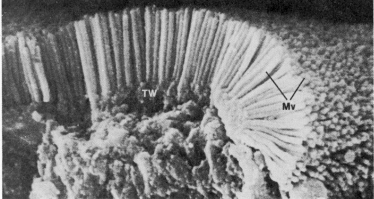

c.

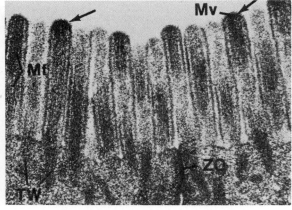

d.

The **colon** has 3 parts: the *ascending colon* goes up the right side of the body to the level of the liver; the *transverse colon* crosses the abdominal cavity just below the liver and the stomach; and the *descending colon* passes down the left side of the body to the rectum, the last 20 cm of the large intestine. The colon is subject to the development of *polyps,* small growths arising from the epithelial lining. Polyps, whether they are benign or cancerous, can be removed individually. If colon cancer is detected while it is still confined to a polyp, the outcome is expected to be a complete cure. The cause of colon cancer is not known yet, but a low-fat, high-fiber (roughage) diet that promotes regularity has been recommended as a protective measure.

The *rectum* ends in the anal canal, which opens at the **anus.** The stimulus to defecate arises from stimuli generated in the colon and the rectum (fig. 10.11). In addition to nondigestible remains, feces also contain certain excretory substances, such as bile pigments and heavy metals, and large quantities of the bacterium *Escherichia coli* and other bacteria.

The large intestine normally contains a large number of normally non-infectious bacteria that live off any substances that were not digested earlier. When they break this material down, they give off odorous molecules that cause the characteristic odor of feces. Some of the vitamins, amino acids, and other growth factors produced by these bacteria are absorbed by the gut lining. In this way, the bacteria perform a service for us.

Water is considered unsafe for swimming when the coliform bacterial count reaches a certain level. A high count is an indication of the amount of fecal material that has entered the water. The more fecal material present, the greater the possibility that pathogenic or disease-causing organisms are also present.

Diarrhea and Constipation

Two common everyday complaints associated with the large intestine are diarrhea and constipation.

The major causes of *diarrhea* are infection of the lower tract and nervous stimulation. In the case of infection, such as food poisoning caused by eating contaminated food, the intestinal wall becomes irritated, and peristalsis increases. Water is not absorbed as a protective measure, and the diarrhea that results serves to rid the body of the infectious organisms. In nervous diarrhea, the nervous system stimulates the intestinal wall, and diarrhea results. Prolonged diarrhea can lead to dehydration because of water loss and to disturbances in the heart's contraction due to an imbalance of salts in the blood.

When a person is constipated, the feces are dry and hard. One cause of this condition is that socialized persons have learned to inhibit defecation to the point that the desire to defecate is ignored. Two components of the diet can help to prevent constipation: water and fiber (roughage). Water intake prevents drying out of the feces, and fiber provides the bulk needed for elimination. It is possible that regularity helps to prevent colon cancer because feces are in contact with the membrane of the colon for less time. Even so, the frequent use of laxatives is discouraged, but if it should be necessary to take a laxative, a bulk laxative is the most natural because, like fiber, it produces a soft mass of cellulose in the colon. Lubricants, like mineral oil, make the colon slippery, and saline laxatives, like milk of magnesia, act osmotically—they prevent water from being absorbed and may even cause water to enter the colon, depending on the dosage. Some laxatives are irritants; they increase peristalsis to the degree that the contents of the colon are expelled.

Chronic constipation is associated with the development of hemorrhoids, a condition that is discussed on page 236.

The large intestine does not produce digestive enzymes; it does absorb water and salts. In diarrhea, too little water has been absorbed; in constipation, too much water has been absorbed.

Accessory Organs

The pancreas and the liver are the accessory organs of digestion. Figure 10.1 shows how ducts conduct pancreatic juice from the pancreas and bile from the liver to the duodenum.

The Pancreas

The **pancreas** lies deep in the abdominal cavity, resting on the posterior abdominal wall. It is an elongated and somewhat flattened organ that has both an endocrine function and an exocrine function. We now are interested in its exocrine function—most of its cells produce pancreatic juice that contains digestive enzymes for carbohydrate, protein, and fat. In other words, the pancreas secretes enzymes for the digestion of all types of food. The enzymes travel by way of the pancreatic duct and the common duct to the duodenum of the small intestine (fig. 10.1). Regulation of pancreatic secretion is discussed in the reading on page 198.

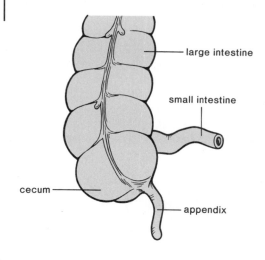

Figure 10.10
The anatomical relationship between the small intestine and the colon. The cecum is the blind end of the ascending colon. The appendix is attached to the cecum.

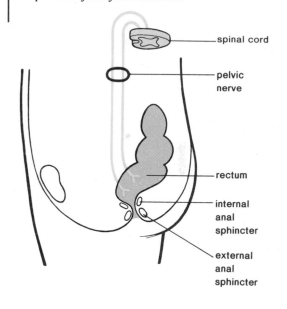

Figure 10.11
Defecation reflex. The accumulation of feces in the rectum causes it to stretch, which initiates a reflex action resulting in rectal contraction and expulsion of the fecal material.

The study of the control of digestive gland secretion began in the late 1800s. At that time, Ivan Pavlov showed that dogs would begin to salivate at the ringing of a bell because they had learned to associate the sound of the bell with being fed. Pavlov's experiments demonstrated that even the thought of food can cause the nervous system to order the secretion of digestive juices. If food is present in the mouth, the stomach, and the small intestine, digestive secretion occurs because of simple reflex action. The presence of food sets off nerve

Control of Digestive Gland Secretion

impulses that travel to the brain. Thereafter, the brain stimulates the digestive glands to secrete.

In this century, investigators have discovered that specific control of digestive secretions is achieved by hormones. A *hormone* is a substance that is produced by one set of cells but affects a different set of cells, the so-called target cells. Hormones are transported by the bloodstream. For

example, when a person has eaten a meal particularly rich in protein, the hormone **gastrin**, produced by the lower part of the stomach, enters the bloodstream and soon reaches the upper part of the stomach, where it causes the gastric glands to secrete more gastric juice.

Experimental evidence also has shown that the duodenal wall produces hormones, the most important of which are **secretin** and **CCK** (cholecystokinin). Acid, especially hydrochloric acid (HCl), present in acid chyme stimulates the release of secretin, while partially digested protein and fat stimulate the release of CCK. These hormones enter the bloodstream and signal the pancreas and the gallbladder to send secretions to the duodenum.

Another hormone recently has been discovered. (gastric inhibitory peptide), produced by the small intestine, apparently works opposite to gastrin because it inhibits gastric acid secretion. This is not surprising since very often the body has hormones with opposite effects.

Hormonal control of digestive gland secretions. Especially after eating a protein-rich meal, gastrin produced by the lower part of the stomach enters the bloodstream and thereafter stimulates the upper part of the stomach to produce more digestive juices. Acid chyme from the stomach causes the duodenum to release secretin and CCK. Both of these hormones stimulate the pancreas to secrete its digestive juices, and CCK alone stimulates the gallbladder to release bile.

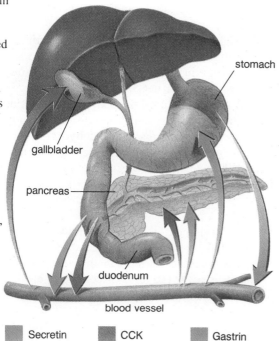

Secretin   CCK   Gastrin

The Liver
The **liver,** which is the largest gland in the body, lies mainly in the upper right of the abdominal cavity, under the diaphragm.

Bile Production   **Bile** is a yellowish green fluid because it contains the bile pigments bilirubin and biliverdin, which come from the breakdown of hemoglobin, the pigment found in red blood cells. Bile also contains bile salts (derived from cholesterol), which emulsify fat in the duodenum of the small intestine. When fat is emulsified, it breaks up into droplets that can be acted upon by a digestive enzyme from the pancreas (p. 201).

Up to 1,500 ml of bile are produced by the liver each day. This bile is sent by way of bile ducts to the gallbladder, where it is stored. The **gallbladder** is a pear-shaped, muscular sac attached to the undersurface of the liver. Here, water is absorbed so that bile becomes a thick, mucuslike material. Bile leaves the gallbladder by the common bile duct and proceeds to the duodenum by the common duct (fig. 10.1).

Other Functions of the Liver   In some ways, the liver acts as the gatekeeper to the blood. Once nutrient molecules have been absorbed by the small intestine, they enter the **hepatic portal vein,** pass through the blood vessels of the liver, and then enter the hepatic vein. The arrangement (fig. 10.12) is like this:

small intestine—hepatic portal vein—liver—hepatic vein

As blood passes through the liver, it removes poisonous substances and works to keep the contents of the blood constant. For example, excess glucose present in the hepatic portal vein is removed and is stored by the liver as glycogen:

$$glucose \longrightarrow glycogen + H_2O$$

Between eating periods, when the glucose level of the blood falls below 0.1%, glycogen is broken down to glucose, which enters the hepatic vein. In this way, the glucose content of the blood remains near 0.1%. It is interesting to note that glycogen sometimes is called animal starch because both starch and glycogen are made up of glucose molecules (p. 35).

If, by chance, the supply of glycogen or glucose runs short, the liver converts amino acids to glucose molecules:

$$amino\ acids \longrightarrow glucose + amino\ groups$$

Recall that amino acids contain nitrogen in the form of amino groups, whereas glucose contains only carbon, oxygen, and hydrogen. Therefore, before amino acids can be converted to glucose molecules, **deamination,** or the removal of amino groups from the amino acids, must take place. By an involved metabolic pathway, the liver converts these amino groups to urea:

$$H_2N - \overset{\overset{\displaystyle O}{\displaystyle \|}}{C} - NH_2$$

Urea is the usual nitrogenous waste product of humans; after its formation in the liver, it is transported to the kidneys for excretion.

The liver also makes blood proteins from amino acids. These proteins are not used as food for cells; rather, they serve important functions within the blood itself.

Altogether, we have mentioned the following functions of the liver:

1. Destroys old red blood cells and converts hemoglobin to the breakdown products (bilirubin and biliverdin) excreted along with bile salts in bile.
2. Produces bile, which is stored in the gallbladder before entering the small intestine, where it emulsifies fats.
3. Stores glucose as glycogen after eating and breaks down glycogen to glucose to maintain the glucose concentration of the blood between eating.
4. Produces urea from the breakdown of amino acids.
5. Makes blood proteins.
6. Detoxifies the blood by removing and metabolizing poisonous substances.

There are 2 accessory organs of digestion that send secretions to the duodenum via ducts. The pancreas produces pancreatic juice, which contains digestive enzymes for carbohydrate, protein, and fat. The liver produces bile, which is stored in the gallbladder.

Liver Disorders   When a person is *jaundiced,* there is a yellowish tint to the skin due to an abnormally large amount of bilirubin in the blood. In one type of jaundice, called *hemolytic jaundice,* red blood cells are broken down in such quantity that the liver cannot excrete the bilirubin fast enough, and the excess spills over into the bloodstream. In *obstructive jaundice,* there is an obstruction of the bile duct or damage to the liver cells, which causes an increased amount of bilirubin to enter the bloodstream. Obstructive jaundice often occurs when crystals of cholesterol precipitate out of bile and form **gallstones,** which on occasion also contain calcium carbonate. The stones may be so numerous that passage of bile along the bile duct is blocked, and the gallbladder must be removed (fig. 10.13). In the meantime, the bile leaves the liver by way of the blood, and a jaundiced appearance results.

Figure 10.12
Hepatic portal system. The hepatic portal vein takes the products of digestion from the digestive system to the liver, where they are processed before entering the circulatory system proper.

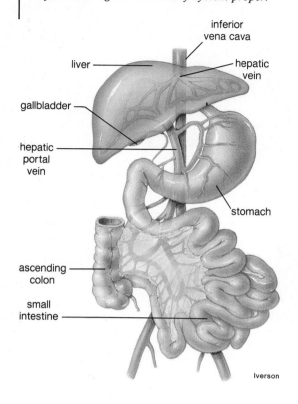

inferior
vena cava

liver

hepatic
vein

gallbladder

hepatic
portal
vein

stomach

ascending
colon

small
intestine

Iverson

Figure 10.13
Gallstones. After removal, this gallbladder was cut open to show its contents—numerous gallstones. The dime was added later to indicate the size of the stones.

Jaundice also can result from hepatitis, an inflammation of the liver sometimes due to viral infection. *Hepatitis A* (infectious hepatitis caused by HAV) usually is contracted after consuming food, drink, or shellfish containing HAV. The virus often causes only grippelike symptoms although abdominal discomfort, weakness, and even jaundice can develop. *Hepatitis B* (serum hepatitis caused by HBV) is a viral disease transmitted via blood transfusions, unsterile needles (of drug addicts), or any body secretion, such as saliva, semen, or milk. Although symptoms are somewhat the same as those for hepatitis A, they can last much longer. To recover from a severe attack, a long recuperation period commonly is required, during which time the patient is in a very weakened condition. *Hepatitis C* (non-A, non-B caused by HCV), like hepatitis B, is transmitted by virus-contaminated blood, but the symptoms are more likely mild. Because the HBV and HBC remain in the blood even for years, carriers (people who can pass on the disease even though they have no symptoms) often are seen. Evidence suggests that all 3 types of hepatitis can be spread by sexual contact.

Cirrhosis is a chronic disease of the liver in which the organ first becomes fatty. Liver tissue then is replaced by inactive fibrous scar tissue. In alcoholics, who often get cirrhosis of the liver, the condition most likely is caused by the excessive amounts of alcohol the liver is forced to break down. When alcohol, a 2-carbon compound, is metabolized, active acetate (AA) results, and these molecules can be synthesized to fatty acids. To accomplish this synthesis, smooth ER increases dramatically in the liver. This may be the first step toward cirrhosis.

The liver is a very critical organ. Any malfunction is a matter of considerable concern.

Digestive Enzymes

The digestive enzymes are **hydrolytic enzymes** that catalyze degradation by the introduction of water at specific bonds (fig. 1.16). Digestive enzymes are no different from any other enzyme of the body. For example, they are proteins having a particular shape that fits their substrate. They have a preferred pH. This pH maintains their shape, enabling them to speed up their specific reaction.

The various digestive enzymes are present in the digestive juices mentioned previously. We now consider each of the enzymes listed in table 10.2 as we discuss the digestion of starch, protein, and fat, the major components of food.

In the mouth, saliva from the salivary glands has a neutral pH and contains **salivary amylase,** an enzyme that acts on starch:

$$\text{starch} + \text{H}_2\text{O} \xrightarrow{\text{salivary amylase}} \text{maltose}$$

In this equation, salivary amylase is written above the arrow to indicate that it is neither a reactant nor a product in the reaction. It merely speeds up the reaction in which its substrate, starch, is digested to many molecules of maltose. Maltose is not one of the small molecules that can be absorbed by the gut lining. However, additional digestive action in the small intestine converts maltose to glucose.

In the stomach, gastric juice secreted by gastric glands has a very low pH—about 2—because it contains hydrochloric acid (HCl). Pepsinogen, a precursor that is converted to the enzyme **pepsin** when exposed to hydrochloric acid, is also present in gastric juice. Pepsin acts on protein to produce peptides:

$$\text{protein} + \text{H}_2\text{O} \xrightarrow{\text{pepsin}} \text{peptides}$$

Human Anatomy and Physiology

Table 10.2 *Comparison of Enzymes*

Enzyme	Source	Optimum pH	Type of Food Digested	Product
Salivary amylase	Saliva	Neutral	Starch	Maltose
Pepsin	Stomach	Acidic	Protein	Peptides
Pancreatic amylase	Pancreas	Basic	Starch	Maltose
Trypsin	Pancreas	Basic	Protein	Peptides
Lipase	Pancreas	Basic	Fat	Glycerol; fatty acids
Peptidases	Intestine	Basic	Peptides	Amino acids
Maltase	Intestine	Basic	Maltose	Glucose

Peptides vary in length, but they always consist of a number of linked amino acids. Peptides are too large to be absorbed by the gut lining. However, they later are broken down to amino acids in the small intestine.

Pancreatic juice, which enters the duodenum, is basic because it contains sodium bicarbonate ($NaHCO_3$). It also contains digestive enzymes for all types of food. One pancreatic enzyme, called **pancreatic amylase,** digests starch:

$$\text{starch} + H_2O \xrightarrow{\text{pancreatic amylase}} \text{maltose}$$

Trypsin is a pancreatic enzyme that digests protein:

$$\text{protein} + H_2O \xrightarrow{\text{trypsin}} \text{peptides}$$

Trypsin is secreted as trypsinogen, which is converted to trypsin in the duodenum.

Lipase digests fat droplets after they have been emulsified by bile salts:

$$\text{fat} \xrightarrow{\text{bile salts}} \text{fat droplets}$$

$$\text{fat droplets} + H_2O \xrightarrow{\text{lipase}} \text{glycerol} + \text{fatty acids}$$

The end products of lipase digestion, glycerol and fatty acids, are small enough to cross the cells of the intestinal villi, where absorption takes place. As mentioned previously, glycerol and fatty acids enter the cells of the villi, and within these cells, they rejoin to give fat, which enters the lacteals (fig. 10.9).

Peptidases and maltase produced by the epithelial cells of the intestinal villi complete the digestion of protein and starch to small molecules that cross into the cells of the villi. Peptides, which result from the first step in protein digestion, are digested to amino acids by peptidases:

$$\text{peptides} + H_2O \xrightarrow{\text{peptidases}} \text{amino acids}$$

Maltose, which results from the first step in starch digestion, is digested to glucose by maltase:

$$\text{maltose} + H_2O \xrightarrow{\text{maltase}} \text{glucose}$$

Other disaccharides, each of which has its own enzyme, are digested in the small intestine. The absence of any one of these enzymes can cause illness. For example, many people, including as many as 75% of American blacks, cannot digest lactose, the sugar found in milk, because they do not produce

Table 10.3 Digestive Enzymes

Reaction	Enzyme	Gland	Site of Action
Starch + H_2O ———→maltose	Salivary amylase Pancreatic amylase	Salivary Pancreas	Mouth Small intestine
Maltose + H_2O ———→glucose*	Maltase	Intestinal	Small intestine
Protein + H_2O ———→peptides	Pepsin Trypsin	Gastric Pancreas	Stomach Small intestine
Peptides + H_2O ———→amino acids*	Peptidases	Intestinal	Small intestine
Fat + H_2O ———→glycerol + fatty acids*	Lipase	Pancreas	Small intestine

*Absorbed by villi.
Note: Food is made up largely of carbohydrate (starch), protein, and fat. These very large macromolecules are broken down by digestive enzymes to small molecules that can be absorbed by intestinal villi. This table indicates the steps needed for carbohydrate digestion (starch and maltose), protein digestion (protein and peptides), and fat digestion (fat) and shows that they are all hydrolytic reactions.

10.2 Critical Thinking

1. Does the experiment described in figure 10.14 show that the enzyme pepsin prefers an acidic rather than a basic pH? Describe an experiment that would show this.
2. Does the experiment described in figure 10.14 show that pepsin will digest only protein and not starch, for example? Describe an experiment that would show this.
3. Does the experiment described in figure 10.14 show that better digestion occurs at a warm temperature rather than a cold temperature? Describe an experiment that would show this.
4. Describe the contents of a test tube that would give the best digestion of fat. Use test tube 4 in figure 10.14 as a guide but change the contents to suit the digestion of fat.

lactase, an enzyme that converts lactose to its components, glucose and galactose. Drinking untreated milk often gives these individuals the symptoms of *lactose intolerance* (diarrhea, gas, cramps) caused by a large quantity of undigested lactose in the intestine. In most areas, it is possible to purchase milk made lactose-free by the addition of lactase.

Table 10.3 lists the enzymes needed for the digestion of each of the major components of food. Two enzymes are needed for the digestion of starch to maltose, and 2 enzymes are needed for the digestion of protein to amino acids, but only one enzyme is required for the digestion of fat. (Not all digestive enzymes are listed; for example, there are intestinal nucleases that digest RNA and DNA to nucleotides.)

Conditions for Digestion

Laboratory experiments can show the necessary conditions for digestion (fig. 10.14). For example, the following 4 test tubes can be prepared and observed for the digestion of egg white, or the protein albumin.

1. H_2O + a small sliver of egg white (protein)
2. Pepsin + H_2O + a small sliver of egg white
3. HCl + H_2O + a small sliver of egg white
4. Pepsin + HCl + H_2O + a small sliver of egg white

All tubes now are placed in an incubator at body temperature for at least one hour. At the end of this time, we can predict that tube number 4 will show the best digestive action because the environmental conditions are appropriate. Tube number 3 does not contain the enzyme (pepsin) and tube number 2 has too high a pH (hydrochloric acid is lacking) so these 2 tubes are expected to show little or no digestion. Tube number 1 is a control tube, and no digestion is expected to occur in this tube. This experiment shows that for digestion to occur, the enzyme as well as the substrate and hydrochloric acid must be present.

Nutrition

The body requires many different types of organic molecules and a smaller number of various types of inorganic ions and compounds from the diet each day. Nutrition involves an interaction between food and the living organism, and a **nutrient** is a substance in food that is used by the body for the maintenance of health. In order to be sure the diet contains all the essential nutrients, it is recommended that we eat a balanced diet. A *balanced diet* is ensured by eating a variety of food from the 4 food groups pictured in figure 10.15.

Human Anatomy and Physiology

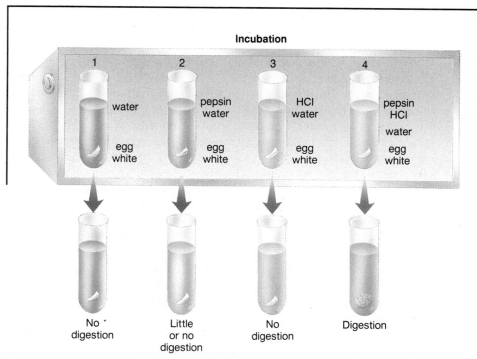

Incubation

1. water / egg white
2. pepsin water / egg white
3. HCl water / egg white
4. pepsin HCl water / egg white

No digestion

Little or no digestion

No digestion

Digestion

Figure 10.14

An experiment to demonstrate that enzymes digest food when the environmental conditions are correct. Tube number 1 lacks the enzyme pepsin, and no digestion occurs; tube number 2 has too high a pH, and little or no digestion occurs; tube number 3 has the proper pH because of the presence of hydrochloric acid, but still no digestion occurs because the enzyme is missing; tube number 4 contains the enzyme, and the environmental conditions are correct for digestion.

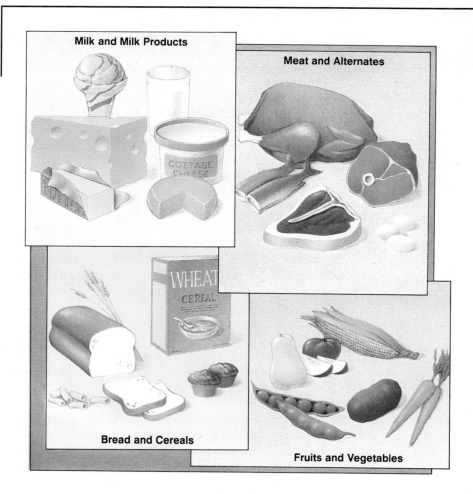

Milk and Milk Products

Meat and Alternates

Bread and Cereals

Fruits and Vegetables

Figure 10.15

The 4 food groups. A variety of foods from each group should be eaten daily.

In order to get a daily supply of essential nutrients, it is necessary to have a balanced diet.

Food consists largely of proteins, carbohydrates, and lipids (fats and cholesterol). Therefore, we begin by considering these substances.

Proteins

Foods rich in protein include red meat, fish, poultry, dairy products, legumes, nuts, and cereals. Following digestion of protein, amino acids enter the bloodstream and are transported to the tissues. Most of these amino acids are incorporated into structural proteins found in muscles, skin, hair, and nails. Others are used to synthesize such proteins as hemoglobin, plasma proteins, enzymes, and hormones.

Protein formation requires 20 different types of amino acids. Of these, 9 are required from the diet because the body is unable to produce them. These are termed the **essential amino acids.** The body produces the other 11 amino acids by simply transforming one type into another type. Some protein sources, such as meat, are *complete;* they provide all types of amino acids. Vegetables and grains supply us with amino acids, but each vegetable or grain alone is an *incomplete* protein source because at least one of the essential amino acids is absent. However, it is possible to combine foods in order to acquire all the essential amino acids. For example, the combinations of cereal with milk, or beans with rice provide all the essential amino acids.

A complete source of protein is absolutely necessary to ensure a sufficient supply of the essential amino acids.

Even though in this country we emphasize protein intake, it does not take very much protein to meet the daily requirement. In the United States, the *required dietary (daily) allowances (RDAs)* are determined by the National Research Council, a part of the National Academy of Sciences. The RDA for the reference woman (120 lb) is 44 g of protein a day. For the reference man (154 lb), it is 56 g of protein a day. A single serving of roast beef (3 oz) provides 25 g of protein, and a cup of milk provides 8 g.

Unfortunately, in this country, the manner in which we meet our daily requirement for protein is not always the most healthful way. The foods that are richest in protein are apt to be richest in fat (fig. 10.16).

Foods richest in protein also tend to be richest in fat.

While it is very important to meet the RDA for protein, consuming more actually can be detrimental. Calcium loss in the urine has been noted when dietary protein intake is over twice the RDA. Everything considered, it is probably a good idea to depend on protein from plant origins (e.g., whole-grain cereals, dark breads, legumes) to a greater extent than is the custom in this country.

Carbohydrates

The quickest, most readily available source of energy for the body is carbohydrates, which can be complex as in breads and cereals or simple as in candy, ice cream, and soft drinks. As mentioned previously, starches are digested to glucose, which is stored by the liver in the form of glycogen. Between eating, the blood glucose is maintained at about 0.1% by the breakdown of glycogen or by the conversion of amino acids to glucose. If necessary, these amino acids are taken from the muscles, even from heart muscle. To avoid this situation, it is suggested that the daily diet contain at least 100 g of carbohydrate. As a point of reference, a slice of bread contains approximately 14 g of carbohydrate.

10.3 Critical Thinking

1. Essential amino acids cannot be made by the body. From your knowledge of metabolic pathways, what is lacking in the cell?
2. Genes (DNA) control the production of enzymes. What happened to a gene if it no longer can code for a functioning enzyme?
3. Plants are able to make all types of amino acids. What would happen to a plant that was unable to make a particular amino acid? Why?
4. Why can animals survive even though they cannot make all types of amino acids?

Human Anatomy and Physiology

Figure 10.16
Protein-rich foods are frequently high in fat. Courtesy of National Dairy Council.

Figure 10.17
Complex carbohydrates. To meet our energy needs, dieticians recommend complex carbohydrates like those shown here rather than simple carbohydrates like candy and ice cream. The latter are more likely to cause weight gain than the recommended complex ones displayed.

Carbohydrates are needed in the diet to maintain the blood glucose level.

Actually, the dietary guidelines produced jointly by the U.S. Department of Agriculture and the Department of Health and Human Services recommend that we increase the proportion of carbohydrates per total energy content of the diet:

	Typical Diet (%)	Recommended Diet (%)
Proteins	12	12
Carbohydrates	46	58
Fats	42	30

Further, it is assumed that these carbohydrates are complex and not simple (fig. 10.17). Simple carbohydrates (e.g., sugars) are labeled "empty calories" by some dieticians because they contribute to energy needs and weight gain and are not a part of foods that supply other nutritional requirements. Table 10.4 gives suggestions on how to cut down on your consumption of dietary sugars (simple carbohydrates).

Table 10.4 Reducing Dietary Sugar

To reduce dietary sugar, the following suggestions are recommended.
1. Eat fewer sweets, such as candy, soft drinks, ice cream, and pastry.
2. Eat fresh fruits or fruits canned without heavy syrup.
3. Use less sugar—white, brown, or raw—and less honey and syrups.
4. Avoid sweetened breakfast cereals.
5. Eat less jelly.
6. Drink pure fruit juices, not imitations.
7. When cooking, use spices like cinnamon to flavor foods instead of using sugar.
8. Do not put sugar in tea or coffee.

In contrast to simple sugars, complex carbohydrates are likely to be accompanied by a wide range of other nutrients and by fiber, which is nondigestible plant material. Insoluble fiber, such as that found in wheat bran, has a laxative effect and therefore may reduce the risk of colon cancer. Soluble fiber, such as that found in oat bran, may possibly reduce cholesterol in the blood because it combines with cholesterol in the gut and prevents it from being absorbed.

While the diet should have an adequate amount of fiber, a high-fiber diet can be detrimental. Some evidence suggests that the absorption of iron, zinc, and calcium is impaired by a high-fiber diet.

Complex carbohydrates along with fiber are considered beneficial to health.

Carbohydrates usually provide most of the dietary Kcalories, even though they have fewer Kcalories per gram than fats:

Food Component	Kcal/g
Carbohydrate	4.1
Protein	4.1
Fat	9.3

A Kcalorie is a measurement unit used to indicate the energy content of food. All types of foods can be used as energy sources (fig. 5.13) in the body. Because carbohydrate makes up the bulk of the diet, it provides most of the Kcalories in the diet.

Lipids

Our discussion of lipids is divided into 2 parts: fats and cholesterol.

Fats

Fats are present not only in butter, margarine, and oils, but also in foods high in protein (fig. 10.16). After being absorbed, the products of fat digestion are transported by the lymph and the blood to the tissues. The liver can alter ingested fats to suit the body's needs, except it is unable to produce the fatty acid linoleic acid. Since this is required for phospholipid production, linoleic acid is considered an essential fatty acid.

Fats have the highest Kcaloric content, but they should not be avoided entirely because they contain the essential fatty acid linoleic acid.

While we need to be sure to ingest some fat in order to satisfy our need for linoleic acid, recent dietary guidelines (p. 205) suggest that we should reduce the amount of fat per total energy content of the diet from 40 to 30%. Dietary fat has been implicated in cancer of the colon, the pancreas, the prostate, and the breast (fig. 10.18). Many animal studies have shown that a high-fat diet stimulates the development of mammary tumors, while a low-fat diet does not. It also has been found that women who have a high-fat diet are more likely to develop breast cancer. Surprisingly, it has been discovered that a reduction in the amount of linoleic acid in the diet helps to prevent breast cancer. Linoleic acid is found in corn, safflower, sunflower, and other common plant oils but is not abundant in olive oil or in fatty fishes and marine animals.

There is very strong evidence that women who have a diet high in fat are more apt to develop breast cancer.

Fat is the component of food that has the highest energy content (9.3 Kcal/g compared to 4.1 Kcal/g for carbohydrate). Raw potatoes, which contain roughage, have about 0.9 Kcalories per gram, but when they are cooked in fat, the number of Kcalories jumps to 6 Kcalories per gram. Another problem

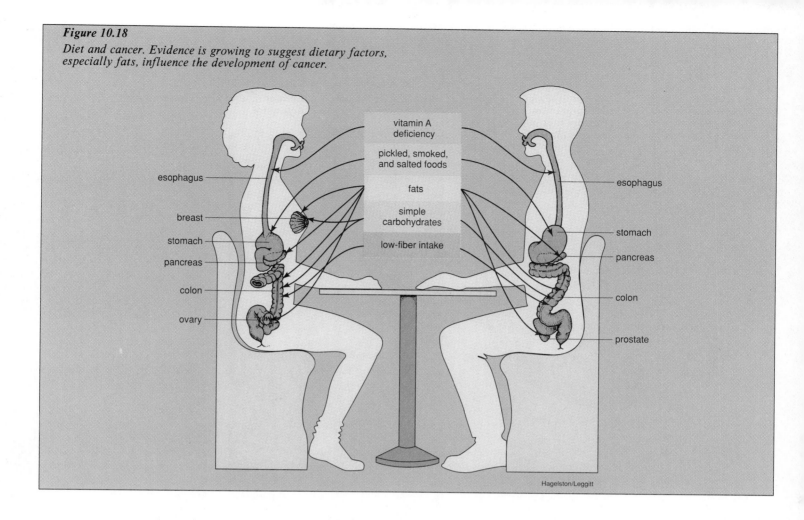

Figure 10.18
Diet and cancer. Evidence is growing to suggest dietary factors, especially fats, influence the development of cancer.

vitamin A deficiency

pickled, smoked, and salted foods

fats

simple carbohydrates

low-fiber intake

esophagus

breast

stomach

pancreas

colon

ovary

esophagus

stomach

pancreas

colon

prostate

Hagelston/Leggitt

for those trying to limit their Kcaloric intake is that fat is not always highly visible; butter melts on toast or potatoes. Table 10.5 gives suggestions for cutting down on the amount of fat in the diet.

As a nation, we have increased our consumption of fat from plant sources and have decreased our consumption from animal sources, such as red meat and butter (fig. 10.19). Most likely, this is due to recent studies linking diets high in saturated fats and cholesterol to hypertension and heart attack.

Cholesterol

The risk of cardiovascular disease includes many factors that are discussed on page 234. One of these factors, according to the National Heart, Lung, and Blood Institute, is a cholesterol blood level of 240 mg/100 ml or higher. If the cholesterol level is this high, additional testing can determine how much of each of 2 important subtypes of cholesterol is in the blood. Cholesterol is carried from the liver to the cells (including the endothelium of the arteries) by plasma proteins called *low-density lipoprotein (LDL)*, and is carried away from the cells to the liver by *high-density lipoprotein (HDL)*. Therefore, LDL is the type of lipoprotein that apparently contributes to formation of plaque (p. 234), which can clog the arteries, while HDL protects against the development of clogged arteries.

A diet low in saturated fat and cholesterol decreases the blood cholesterol level (LDL level) in some individuals. In order to achieve this type of diet, the suggestions in table 10.5 are helpful, as are certain protein sources. White fish, poultry, and shellfish are helpful; cheese, egg yolks, and liver are not helpful. It is recommended that egg whites substitute for egg yolks in both cooking and eating.

Table 10.5 Reducing Dietary Fat

To reduce dietary fat, the following suggestions are recommended.
1. Choose lean red meat, poultry, fish, or dry beans and peas as a protein source.
2. Trim fat off meat and remove skin from poultry before cooking.
3. Cook meat or poultry on a rack so that fat will drain off.
4. Broil, boil, or bake, rather than fry.
5. Limit your intake of butter, cream, hydrogenated oils, shortenings, and coconut and palm oil.*
6. Use herbs and spices to season vegetables instead of butter, margarine, or sauces. Use lemon juice instead of salad dressing.
7. Drink skim milk instead of whole milk, and use skim milk in cooking and baking.

*Although coconut and palm oils are from plant sources, they are saturated fats.

Figure 10.19

Fat content in the diet. a. *Americans acquire 43% of fat intake from animal fats, like butter, and from plant oils, like cooking oils; 36% comes from red meat, poultry, and fish; and lesser amounts come from the sources shown.* b. *The amount of fat acquired from vegetable sources is now larger than it was in the early 1900s.*

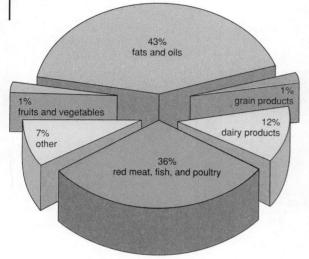

a. **Sources of Fat in the U.S. Diet, 1980**

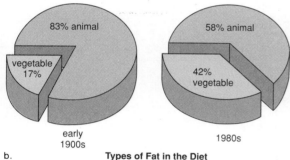

b. **Types of Fat in the Diet**

Figure 10.20

Illnesses due to vitamin deficiency. a. *Bowing of bones (rickets) due to vitamin D deficiency.* b. *Bleeding of gums (scurvy) due to vitamin C deficiency.* c. *Fissures of lips (cheilosis) due to riboflavin deficiency.* d. *Dermatitis of areas exposed to light (pellagra) due to niacin deficiency.*

a.

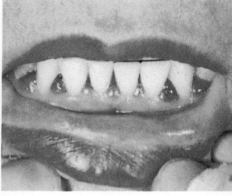

b.

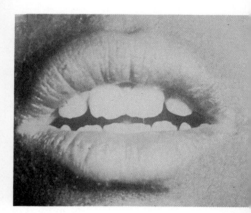

c.

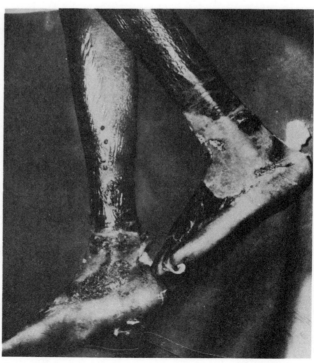

d.

Table 10.6 *Vitamins, Their Functions and Sources*

Vitamins	Role in body	Good food sources
Fat soluble		
vitamin A	assists in maintenance of healthy skin, hair, and mucous membranes; aids in the ability to see in dim light (night vision); essential for proper bone growth, tooth development, and reproduction	deep yellow/orange and dark green vegetables and fruits (carrots, broccoli, spinach, cantaloupe, sweet potatoes); cheese, milk, and fortified margarines
vitamin D	aids in the formation and maintenance of bones and teeth; assists in the absorption and use of calcium and phosphorus	milks fortified with vitamin D; tuna, salmon, or cod liver oil; also made in the skin when exposed to sunlight
vitamin E	protects vitamin A and essential fatty acids from oxidation; prevents cell membrane damage	vegetable oils and margarine, nuts, wheat germ and whole grain breads and cereals, green leafy vegetables
vitamin K	aids in synthesis of substances needed for clotting of blood; helps maintain normal bone metabolism	green leafy vegetables, cabbage, and cauliflower; also made by bacteria in intestines of humans, except for newborns
Water soluble		
vitamin C	important in forming collagen; helps maintain capillaries, bones, and teeth; aids in absorption of iron; helps protect other vitamins from oxidation	citrus fruits, berries, melons, dark green vegetables, tomatoes, green peppers, cabbage, and potatoes
thiamin	helps in release of energy from carbohydrates; promotes normal functioning of nervous system	whole-grain products, dried beans and peas, sunflower seeds, nuts
riboflavin	helps body transform carbohydrate, protein, and fat into energy	nuts, milk products, whole-grains, poultry, leafy green vegetables
niacin	helps body transform carbohydrate, protein, and fat into energy	nuts, poultry, fish, whole-grain products, leafy greens, beans
vitamin B_6	aids in the fat and amino acids use; aids protein formation	bananas, beans, poultry, nuts, leafy green vegetables
folic acid	aids in the formation of hemoglobin in red blood cells; aids in the formation of genetic material	dark green leafy vegetables, nuts, beans, whole-grain products, fruit juices
pantothenic acid	aids in the formation of hormones and certain nerve-regulating substances; helps in metabolism	nuts, beans, seeds, dark green leafy vegetables, poultry, dried fruit, milk
biotin	aids fatty acid formation; helps release carbohydrate energy	occurs widely in foods, especially eggs
vitamin B_{12}	aids in the formation of red blood cells and genetic material; helps the functioning of the nervous system	milk, yogurt, cheese, fish, poultry, and eggs; not found in plant foods unless fortified (such as in some breakfast cereals)

The intake of soluble fiber is believed to possibly reduce the cholesterol blood level by combining with cholesterol in the gut and carrying it out of the body. Foods high in soluble fiber are oat bran, oatmeal, beans, corn, and certain fruits, such as apples, citrus fruits, and cranberries.

Vitamins and Minerals

Vitamins

Vitamins are organic compounds (other than carbohydrate, fat, and protein) that the body is unable to produce but uses for metabolic purposes. Therefore, vitamins must be present in the diet; if they are lacking, various symptoms develop (fig. 10.20). There are many substances that are advertised as vitamins, but in reality there are only 13 vitamins (table 10.6). Table 10.6 also gives the RDA vitamin values for the reference female and the reference male. In general, carrots, squash, turnip greens, and collards are good sources of vitamin A. Citrus fruits and other fresh fruits and vegetables are natural sources of vitamin C. Sunshine and irradiated milk are primary sources of vitamin D, and whole grains are good sources of B vitamins.

It is not difficult to acquire the RDAs for vitamins if your diet is balanced because each vitamin is needed in small amounts only. Many vitamins are portions of coenzymes, or enzyme helpers. For example, niacin is part of the coenzyme NAD (p. 104), and riboflavin is part of another dehydrogenase, FAD. Coenzymes are needed in only small amounts because each can be used over and over again.

Table 10.7 *Minerals, Their Functions and Sources*

Minerals	Role in body	Good food sources
Major minerals		
calcium	used for building bones and teeth and maintaining bone strength; also involved in muscle contraction and blood clotting	all dairy products, dark green leafy vegetables, beans, nuts, sunflower seeds, dried fruit, molasses, canned fish
phosphorus	used to build bones and teeth; release energy from nutrients; and form DNA, cell membranes, and many enzymes	beans, sunflower seeds, milk, cheese, nuts, poultry, fish, lean meats
magnesium	used to build bones, produce proteins, release energy from muscle carbohydrate stores (glycogen), and regulate body temperature	sunflower and pumpkin seeds, nuts, whole-grain products, beans, dark green vegetables; dried fruit, lean meats
sodium	regulates body-fluid volume and blood acidity; aids in transmission of nerve impulses	most of the sodium in the American diet is added to food as salt (sodium chloride) in cooking, at the table, or in commercial processing. Animal products contain some natural sodium
chloride	is a component of gastric juice and aids in acid-base balance	table salt, seafood, milk, eggs, meats
potassium	assists in muscle contraction, the maintenance of fluid and electrolyte balance in the cells, and the transmission of nerve impulses; also aids in the release of energy from nutrients	widely distributed in foods, especially fruits and vegetables, beans, nuts, seeds, and lean meats
Minor minerals		
iron	involved in the formation of hemoglobin in the red blood cells of the blood and myoglobin in muscles; also a part of several proteins	molasses, seeds, whole-grain products, fortified breakfast cereals, nuts, dried fruits, beans, poultry, fish, lean meats
zinc	involved in the formation of protein (growth of all tissues), wound healing, and prevention of anemia; a component of many enzymes	whole-grain products, seeds, nuts, poultry, fish, beans, lean meats
iodine	integral component of thyroid hormones	table salt (fortified), dairy products, shellfish, and fish
copper	vital to enzyme systems and in manufacturing red blood cells. Needed for utilization of iron	nuts, oysters, seeds, crab, wheat germ, dried fruit, whole grains, legumes
manganese	needed for normal bone structure, reproduction, and the normal functioning of the central nervous system; is a component of many enzyme systems	whole grains, nuts, seeds, pineapple, berries, legumes, dark green vegetables, tea

From David C. Nieman, et al., *Nutrition.* Copyright © 1990 Wm. C. Brown Publishers, Dubuque, Iowa. All Rights Reserved. Reprinted by permission.

The National Academy of Sciences suggests that we eat more fruits and vegetables in order to acquire a good supply of vitamins C and A because these 2 vitamins may help to guard against the development of cancer. Nevertheless, they discourage the intake of excess vitamins by way of pills because this practice possibly can lead to illness. For example, excess vitamin C can cause kidney stones, and this excess is converted to oxalic acid, a molecule that is toxic to the body. Vitamin A taken in excess over long periods can cause hair loss, bone and joint pains, and loss of appetite. Excess vitamin D can cause an overload of calcium in the blood, which in children, leads to loss of appetite and retarded growth. Megavitamin therapy always should be supervised by a physician.

A properly balanced diet includes all the vitamins and the minerals needed by most individuals to maintain health.

Minerals

In addition to vitamins, various **minerals** are required by the body (table 10.7). Minerals are divided into macrominerals, which are recommended in amounts more than 100 mg per day, and microminerals (trace elements), which are recommended in amounts less than 20 mg per day. The macrominerals sodium, magnesium, phosphorus, chlorine, potassium, and calcium serve as constituents of cells and body fluids and as structural components of tissues. For example, calcium is needed for the construction of bones and teeth and for nerve conduction and muscle contraction.

The microminerals seem to have very specific functions. For example, iron is needed for the production of hemoglobin, and iodine is used in the production of thyroxin, a hormone produced by the thyroid gland. As research

Human Anatomy and Physiology

continues, more and more elements are added to the list of microminerals considered to be essential. During the past 3 decades, molybdenum, selenium, chromium, nickel, vanadium, silicon, and even arsenic have been found to be essential to good health in very small amounts.

Occasionally, it has been found that individuals do not receive enough iron (in women), calcium, magnesium, or zinc in their diet. Adult females need more iron in the diet than males (RDA of 18 mg compared to 10 mg) because they lose hemoglobin each month during menstruation. Stress can bring on a magnesium deficiency, and due to its high-fiber content, a vegetarian diet may make zinc less available to the body. However, a varied and complete diet usually supplies the mineral RDAs.

Calcium There has been much interest in the dietary addition of calcium supplements (fig. 10.21) to counteract osteoporosis, a degenerative bone disease that afflicts an estimated one-fourth of older men and one-half of older women in the United States. These individuals have porous bones that break easily because they lack sufficient calcium. In 1984, a National Institutes of Health conference on osteoporosis advised postmenopausal women to increase their daily intake of calcium to 1,500 mg and all others to 1,000 mg (compared with the RDA of 800 mg).

However, recent studies have shown that calcium supplements cannot prevent osteoporosis after menopause even when the dosage is 3,000 mg a day. The body becomes less able to take in calcium after about age 35. The most-effective defense against osteoporosis in older women now is believed to be estrogen replacement and exercise. In one survey, women aged 35–65 who took a 50-minute aerobics class 3 times a week lost only 2.5% of the density in their forearm bones, compared with 9.5% in women who did not exercise.

Young women can guard against osteoporosis by forming strong, dense bones before menopause. Eighteen-year-old women are apt to get only 679 mg of calcium a day when the RDA is 800 mg (or 1,000 mg according to the NIH). They should consume more calcium-rich foods, such as milk and dairy products. Taking calcium supplements may not be as effective; a cup of milk supplies 270 mg of calcium, while a 500-mg tablet of calcium carbonate provides only 200 mg. The excess supplemental calcium is not taken up by the body; it is not in a form that is *bioavailable*. However, an excess of bioavailable calcium can lead to kidney stones.

Dietary calcium and exercise, plus estrogen therapy if needed, are the best safeguards against osteoporosis.

Sodium The recommended amount of sodium intake per day is 400–3,300 mg, and the average American takes in 4,000–4,700 mg. In recent years, this imbalance has caused concern because high sodium intake has been linked to hypertension in some people. About one-third of the sodium we consume occurs naturally in foods; another one-third is added during commercial processing; and we add the last one-third either during home cooking or at the table in the form of table salt.

Clearly, it is possible for us to cut down on the amount of sodium in the diet. Table 10.8 gives recommendations for doing so.

Excess sodium in the diet can lead to hypertension; therefore, excess sodium intake should be avoided.

Dieting

When persons go on a diet, they usually begin by determining how many Kcalories they intend to consume each day.

Figure 10.21
Calcium in the diet. Many over-the-counter calcium supplements now are available to boost the amount in the diet.

Table 10.8 *Reducing Dietary Sodium*

To reduce dietary sodium, the following suggestions are recommended.
1. Use spices instead of salt to flavor foods.
2. Add little or no salt to foods at the table, and add only small amounts of salt when you cook.
3. Eat unsalted crackers, pretzels, potato chips, nuts, and popcorn.
4. Avoid frankfurters, ham, bacon, luncheon meat, smoked salmon, sardines, and anchovies.
5. Avoid processed cheese and canned or dehydrated soups.
6. Avoid prepared catsup and horseradish.
7. Avoid brine-soaked foods, such as pickles and olives.

Table 10.9 *Kcaloric Expenditures for Some Physical Activities*

Activity	Kcalories Per Minute	Kcalories Per Hour
Watching television	1.2	72
Eating	1.4	85
Classwork	1.7	101
Driving	3.0	180
Walking, 2 mph	3.6	213
4½ mph	6.7	401
going upstairs	17.5	1,052
Baseball—infield-outfield	4.7	284
pitching	6.0	362
Bicycling—slow	5.0	300
strenuous	10.8	650
Swimming—leisurely	5.0	300
rapidly	10.8	650
Tennis—doubles	5.8	350
singles	7.5	450
Badminton and volleyball—moderate	5.8	345
vigorous	9.8	590
Basketball—moderate	7.1	426
vigorous	8.5	512
Skiing—downhill	9.8	585
cross country, 5 mph	11.8	710
Jogging	10.0	600

Reproduced with permission from Jordan, H. A., *Finding Your Way to Slimming;* Figure 16, p. 14. Carolina Biology Reader Series, No. 141. Copyright 1983, Carolina Biological Supply Co., Burlington, North Carolina, U.S.A.

Figure 10.22

Diagram illustrating the relationship between caloric intake and weight gain or loss. In each instance, energy needs are divided between basal metabolism (basal metabolic rate) and physical activity. a. Energy content of food is greater than energy needs of the body—weight gain occurs. b. Energy content of food is less than energy needs of the body—weight loss occurs. c. Energy content of food equals energy needs of the body—no weight change occurs.

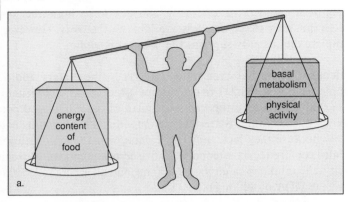

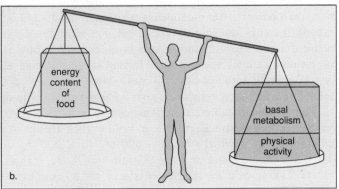

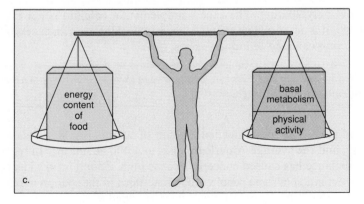

Daily Energy Requirement

Humans need energy primarily for basal metabolism and for physical activities. *Basal metabolism* involves breathing, circulating blood, and maintaining body temperature and muscle tone. The basal metabolic rate (*BMR*) is best measured 14 hours after the last meal, with the subject lying down at complete physical and mental rest. The BMR is usually lower for women than for men and, in general, is affected by size, shape, weight, age, activity of the endocrine glands, and similar individual characteristics. The BMR for the reference female of 120 pounds is 1,320 Kcalories, and for the reference male weighing 154 pounds, the BMR is 1,848 Kcalories.

Figure 10.23
*Behavior and food intake. Social situations
sometimes cause us to indulge in fattening foods.*

The energy used for various voluntary activities can be determined by consulting tables such as table 10.9, which lists the number of Kcalories per hour that are required for all sorts of activities. From such tables, it is possible to estimate the number of Kcalories needed each day for voluntary activities.

The averge daily Kcaloric requirement for our reference female and reference male is expected to be as follows:

	Female	Male
Sedentary	1,700	2,500
Moderately active	2,100	2,900
Very active	2,500	3,300

Figure 10.22 indicates that weight loss occurs if the Kcaloric intake is reduced, while the same level of activity is maintained. The net Kcaloric restriction needed for losing 1 pound of body weight is 3,500. So, if you want to lose a half-pound in one week, you have to reduce the Kcaloric intake by 250 calories a day.[1] If you are a sedentary male weighing 154 pounds, this means you are striving for a total intake of 2,250 Kcalories per day.

Reducing the Kcaloric intake and/or increasing the amount of exercise eventually results in weight loss.

In order to determine whether you are meeting your Kcaloric goal, it is necessary to keep track of the Kcalories in the food you are eating. Table 10.10 offers a shortcut method for doing this. If you want to be more precise, you will have to consult larger and more complicated tables.

Fad Diets versus Behavior Modification

Nutritionists tell us that the best way to lose weight is to modify our behavior and to lose the weight slowly. Behavior modification requires that we examine our eating behaviors (fig. 10.23), identify the situations that cause us to snack unnecessarily, and work at changing these. For example, if you are used to

---

[1]The recommended body weight for men and women can be determined in the following way.
　　(Add or subtract 10 pounds depending on frame.)
Women: 100 pounds for first 5 feet of height; 5 pounds per inch over 5 feet.
Men: 140 pounds for first 5 feet of height; 5 pounds per inch over 5 feet.

Table 10.10 *Nutrient and Energy Content of Foods*

Breads
1 slice of bread
¾ cup ready-to-eat cereal
⅓ cup corn
1 small potato
(1 bread = 15 g carbohydrate, 20 g protein, and 70 Kcal)

Milk
1 cup skim milk   (2% milk—add 5 g fat)
1 cup skim-milk yogurt, plain   (Whole milk—add 8 g fat)
1 cup buttermilk
½ cup evaporated skim milk or milk dessert
(1 milk = 12 g carbohydrate, 8 g protein, and 80 Kcal)

Vegetables
½ cup greens
½ cup carrots
½ cup beets
(1 vegetable = 5 g carbohydrate, 1 g protein, and 25 Kcal)

Fruits
½ small banana
1 small apple
½ cup orange juice
½ grapefruit
(1 fruit = 10 g carbohydrate and 40 Kcal)

Meats (lean)
1 oz lean red meat
1 oz chicken meat without the skin
1 oz any fish
¼ cup canned tuna or 1 oz low-fat cheese
(1 oz low-fat meat = 7 g protein, 3 g fat, and 55 Kcal)

Meats (medium fat)
1 oz pork loin
1 egg
¼ cup creamed cottage cheese
(1 medium-fat meat = 7 g protein, 5½ g fat, and about 80 Kcal)

Meats (high fat)
1 oz country-style ham
1 oz cheddar cheese
1 small hot dog (frankfurter)
(1 high-fat meat = 7 g protein, 8 g fat, and 100 Kcal)

Peanut butter
Peanut butter is like a meat in terms of its protein content but is very high in fat. It is estimated that 2 Tbs peanut butter = 7 g protein, 15½ g fat, and about 170 Kcal.

Fats
1 tsp butter or margarine
1 tsp any oil
1 Tbs salad dressing
1 strip crisp bacon
5 small olives
10 whole Virginia peanuts
(1 fat = 5 g fat and 45 Kcal)

Legumes (beans and peas)
Legumes are like meats because they are rich in protein and iron, but they are lower in fat than meat. They contain much starch.
It is estimated that ½ cup legumes = 15 g carbohydrate, 9 g protein, 3 g fat, and 125 Kcal.

Miscellaneous Foods

	Protein	Grams Fat	Carbohydrate	Kcal		Protein	Grams Fat	Carbohydrate	Kcal
Ice cream (1 cup)	5	14	32	274	Beer (1 can light)	1	0	14	110
Cake (1 piece)	3	1	32	149	Soft drink	0	0	37	148
Doughnut (1)	3	11	22	199	Soup (1 cup)	2	3	15	95
Pie (1 piece)	3	15	51	351	Coffee and tea	0	0	0	0
Caramel candy (1 oz)	1	3	22	119					

Note: For caloric values not included in this report, consult Whitney and Hamilton, 1984. *Understanding nutrition.* 3d ed. New York: West Publishing Co. Reprinted by permission from B. L. Frye and R. L. Neill, "A Laboratory Exercise in Nutrition" in *American Biology Teacher,* September 1984, p. 372. Copyright © 1984 National Association of Biology Teachers, Reston, VA.

having cake and ice cream for dessert, substitute fruit. If you pass an ice cream shop every day and tend to stop in, then change your route and do not go by this shop.

Physical activity is a very important addition to your daily routine. Aside from helping to firm up the body after weight loss, exercise also puts you in the mood to keep the weight off, and it burns off calories.

The reading on page 215 lists and discusses some of the types of fad diets that are popular. Several are dangerous to your health and do not work because the weight loss is simply regained once the diet ceases. Our body is not adapted to rapid weight loss. First, if Kcalories are severely and suddenly restricted, the body burns protein rather than fat. This protein is removed from the muscles, even the heart muscle. Second, the body apparently has a "set point" for its usual amount of fat. If the amount of stored fat falls below this point, the fat cells are believed to signal the brain by the release of a chemical substance. In response, the metabolic rate is lowered so that fewer Kcalories

Many people end up willing to try any gimmick that is supposed to lead to permanent weight loss. The diets that have been promoted in recent years range from the merely ineffective to the outright dangerous, with a scattering of reasonably sound ones. Most weight-loss aids, schemes, and plans fall into one of the following categories.

Pills. Some of the widely marketed over-the-counter drugs sold for dieting are supposed to suppress the appetite; others claim to "burn" fat. Appetite suppressants may work at first, but weight-control re-

body's prime source of energy. Despite common opinion, starches are not fattening—rather, fat is fattening.

Fasting. Some years ago, a popular diet had people foregoing food entirely and living on liquid protein drinks and vitamins. A few people on this regime died, probably because their bodies were forced to digest so much muscle that their heart muscles failed.

Liquid diets. While some liquid diets provide sufficient protein and vitamins, they may restrict the dieter to 400 Kcalo-

High-carbohydrate, low-fat diets. Most of the sound diets fall into this category. High-carbohydrate, high-fiber foods are good sources of energy and nutrients, and most are low in fat. Combined with exercise, this kind of diet promotes gradual loss of body fat. Still, the number of Kcalories allotted per day must be sufficient to prevent the body from consuming muscle tissue. A few high-carbohydrate diet plans on the market cut Kcalories too severely.

"Crash" Diets

searchers say that to lead to permanent weight loss, a pill would have to be taken for life, like blood-pressure pills. Doctors say most diet pills are useless, and some—in particular, the amphetamines and similar prescription drugs—are addictive. So-called starch-blocker pills, which bind to starch and make it indigestible, are now illegal in the United States but still are sold on the black market. They cause a number of adverse side effects, including abdominal cramps.

Low-carbohydrate diets. Severely cutting back on carbohydrate upsets the body's chemical balance in such a way that fluids are depleted. While this gives the illusion of weight loss, fat is not lost, and the water weight eventually will be regained. Besides, carbohydrate is the

ries or less a day. Because the body cannot burn fat quickly enough to compensate for so few calories, however, muscle also is digested. Most doctors do not recommend cutting Kcalories to fewer than 1,200 a day on any diet.

Single-category diets. These programs call for a diet restricted entirely to one kind of food, such as fruits or vegetables or rice alone. However, no single category of food provides enough nutrients to maintain healthy body tissue. Some dieters in recent years had a dramatic revelation of the inadequacy of such diets—their hair and fingernails fell out!

Nutrition and weight loss. *Nutritionists advise that it is not necessary to purchase diet foods. Simply eat a variety of wholesome foods in small quantities and exercise. Dieting should be considered a lifelong project rather than a short-term one.*

Reprinted from the 1988 *Medical and Health Annual,* copyright 1987, with the permission of Encyclopaedia Britannica, Inc., Chicago, Illinois.

are needed to stay at the same weight. This set-point hypothesis also explains why people tend to immediately regain any weight that has been lost through dieting. There is some evidence that exercise and avoidance of fatty foods lowers the set point for body fat.

The overall conclusion, then, is that in order to keep unwanted pounds off, a long-term program is needed. This program should include regular exercise and a balanced diet that avoids fatty foods.

Fad diets can be dangerous to your health. A balanced diet containing a reduced number of Kcalories along with an exercise program offer long-term weight control.

Eating Disorders

Authorities recognize 3 primary eating disorders: obesity, bulimia, and anorexia nervosa. Although they exist in a continuum as far as body weight is concerned, there is much overlap between them.

Obesity is defined as a body weight of more than 20% above the ideal weight. It most likely is caused by a combination of factors, including endocrinal, metabolic, and social factors. The social factors include the eating habits of other family members. Obese individuals need to consult a physician if they want to bring their body weight down to normal and to keep it there permanently.

Obesity has many complex causes that possibly can be detected by a physician.

Bulimia can coexist with either obesity or anorexia nervosa. People, usually young women, who are afflicted have the habit of eating to excess and then purging themselves by some artificial means, such as vomiting or laxatives. These individuals usually are depressed, but whether the depression causes or is caused by the bulimia cannot be determined. While individual psychological help does not seem to be effective, there is some indication that group therapy helps. The possibility of a hormonal disorder has not been ruled out, however.

Anorexia nervosa is diagnosed when an individual is extremely thin but still claims to "feel fat" and continues to diet. It is possible these individuals have an incorrect body image that makes them think they are fat. It is also possible they have various psychological problems, including a desire to suppress their sexuality. Menstruation ceases in very thin women.

Both bulimia and anorexia nervosa are serious disorders that require the assistance of competent medical personnel.

Summary

In the mouth, food is chewed and starch is acted upon by salivary amylase. After swallowing, peristaltic action moves the food along the esophagus to the stomach. Here, pepsin, in the presence of hydrochloric acid (HCl), acts on protein. In contrast, the small intestine has a basic pH environment. Here, fat is emulsified by bile salts to fat droplets before being acted upon by pancreatic lipase. Protein is digested by pancreatic trypsin, and starch is digested by pancreatic amylase. The cells that line the intestinal wall produce intestinal enzymes to finish the digestion of protein and carbohydrate. Only nondigestible material passes from the small intestine to the large intestine. The large intestine absorbs water from this material. It also contains a large population of bacteria that can use the material as food. In the process, the bacteria produce vitamins that can be absorbed and used by our body.

The walls of the small intestine have fingerlike projections called villi within which are blood capillaries and a lymphatic lacteal. Amino acids and glucose enter the blood; glycerol and fatty acids re-form to give fat before entering the lacteal. The blood from the small intestine moves into the hepatic portal vein, which goes to the liver, an organ that monitors and contributes to blood composition.

A balanced diet is required for good health. Food should provide us with all necessary vitamins, amino acids, fatty acids, and an adequate amount of energy. If the Kcaloric value of food consumed is greater than that needed for body functions and activity, weight gain occurs.

Study Questions

1. List the parts of the digestive tract, anatomically describe them, and state the contribution of each to the digestive process. (pp. 190–97)
2. Discuss the absorption of the products of digestion into the circulatory system. (p. 195)
3. What is the common intestinal bacterium? What do these bacteria do for us? (p. 197)
4. List the accessory glands, and describe the part they play in the digestion of food. (pp. 197–200)
5. What are gastrin, secretin, and CCK? Where are they produced? What are their functions? (p. 198)
6. List 6 functions of the liver. How does the liver maintain a constant glucose level in the blood? (p. 199)
7. What is jaundice? cirrhosis of the liver? (p. 199)
8. Discuss the digestion of starch, protein, and fat, listing all the steps that occur to bring about digestion of each of these. (pp. 200–202)
9. Give reasons why carbohydrates, fats, proteins, vitamins, and minerals are all necessary to good nutrition. (pp. 202–11)
10. What factors determine how many Kcalories should be ingested? (pp. 211–13)

Objective Questions

1. In the mouth, salivary _____ digests starch to _____ .
2. When swallowing, the _____ covers the opening to the larynx.
3. The _____ takes food to the stomach, where _____ is primarily digested.
4. The gallbladder stores _____ , a substance that _____ fat.
5. The pancreas sends digestive juices to the _____ , the first part of the small intestine.
6. Pancreatic juice contains _____ for digesting protein, _____ for digesting starch, and _____ for digesting fat.
7. Whereas pepsin prefers a _____ pH, the enzymes found in pancreatic juice prefer a _____ pH.
8. The products of digestion are absorbed into the cells of the _____ , fingerlike projections of the intestinal wall.
9. After eating, the liver stores glucose as _____ .
10. The diet should include a complete protein source, one that includes all the _____ .

Answers to Objective Questions

1. amylase, maltose 2. epiglottis 3. esophagus, protein 4. bile, emulsifies 5. duodenum 6. trypsin, pancreatic amylase, lipase 7. strongly acidic, slightly basic 8. villi 9. glycogen 10. essential amino acids

Label this Diagram.

See figure 10.1 (p. 191) in text.

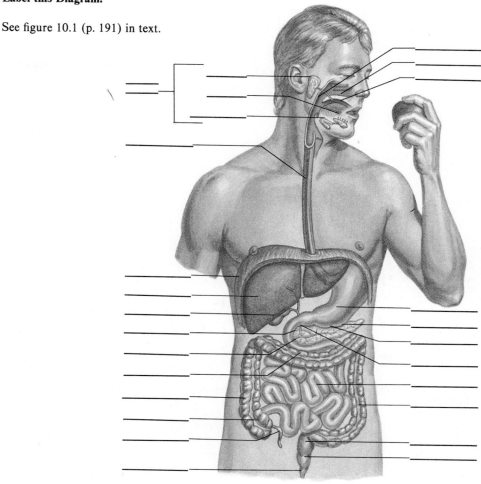

Selected Key Terms

amylase (am'i-lās) a starch-digesting enzyme secreted by the salivary glands (salivary amylase) and the pancreas (pancreatic amylase). *200*

CCK cholecystokinin (ko''le-sis''to-ki'nin) hormone produced by the duodenum that stimulates release of bile from the gallbladder. *198*

colon (ko'lon) the large intestine of vertebrates. *196*

epiglottis (ep''i-glot'is) a structure that covers the glottis during the process of swallowing. *193*

esophagus (ĕ-sof'ah-gus) a tube that transports food from the mouth to the stomach. *193*

gastric gland (gas'trik gland) gland within the stomach wall that secretes gastric juice. *194*

gastrin (gas'trin) a hormone secreted by cells of the stomach wall to regulate the release of pepsin. *198*

glottis (glot'is) slitlike opening between the vocal cords. *193*

hard palate (hard pal'at) anterior portion of the roof of the mouth that contains several bones. *192*

hydrolytic enzyme (hi-dro–lit'ik en'zīm) an enzyme that catalyzes a reaction in which the substrate is broken down with the addition of water. *200*

lipase (li'-pās) a fat-digesting enzyme secreted by the pancreas. *201*

lumen (lu'men) the cavity inside any tubular structure, such as the lumen of the gut. *193*

pepsin (pep'sin) a protein-digesting enzyme secreted by gastric glands. *200*

peristalsis (per''i-stal'sis) a rhythmic contraction that serves to move the contents along in tubular organs, such as the digestive tract. *193*

pharynx (far'ingks) a common passageway (throat) for both food intake and air movement. *192*

salivary gland (sal'i-ver-e gland) a gland associated with the mouth that secretes saliva. *192*

secretin (se-kre'tin) hormone secreted by the small intestine that stimulates the release of pancreatic juice. *198*

soft palate (soft pal'at) entirely muscular posterior portion of the roof of the mouth. *192*

sphincter (sfingk'ter) a muscle that surrounds a tube and closes or opens the tube by contracting and relaxing. *194*

trypsin (trip'sin) a protein-digesting enzyme secreted by the pancreas. *201*

villus (vil'us) fingerlike projection that lines the small intestine and functions in absorption. *195*

vitamin (vi'tah-min) essential requirement in the diet, needed in small amounts, that is often a part of coenzymes. *209*

Chapter 11

Circulation

Chapter Concepts

1 In human beings, blood, kept in motion by the pumping of the heart, circulates through a series of vessels.

2 The heart is actually a double pump: the right side pumps blood to the lungs, and the left side pumps blood to the rest of the body.

3 The purpose of circulation is to deliver blood to the capillaries, where exchange of molecules takes place.

4 Although the circulatory system is very efficient, it is still subject to various degenerative disorders.

Chapter Outline

The Circulatory System
  Blood Vessels
  The Heart
  The Heartbeat
  Vascular Pathways
Features of the Circulatory System
  The Pulse
  Blood Pressure
  The Velocity of Blood Flow
Circulatory Disorders
  Hypertension
  Atherosclerosis
  Varicose Veins and Phlebitis

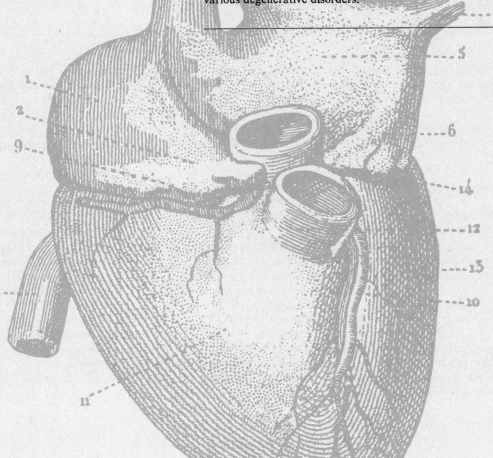

Figure 11.1

Blood vessels. a. Blood leaving the heart moves from an artery to arterioles to capillaries to venules and then returns to the heart by way of a vein. b. Arteries have well-developed walls with a thick middle layer of elastic tissue and smooth muscle.

c. Capillary walls are one cell thick. d. Veins have flabby walls, particularly because the middle layer is not as thick. Veins have valves that point toward the heart.

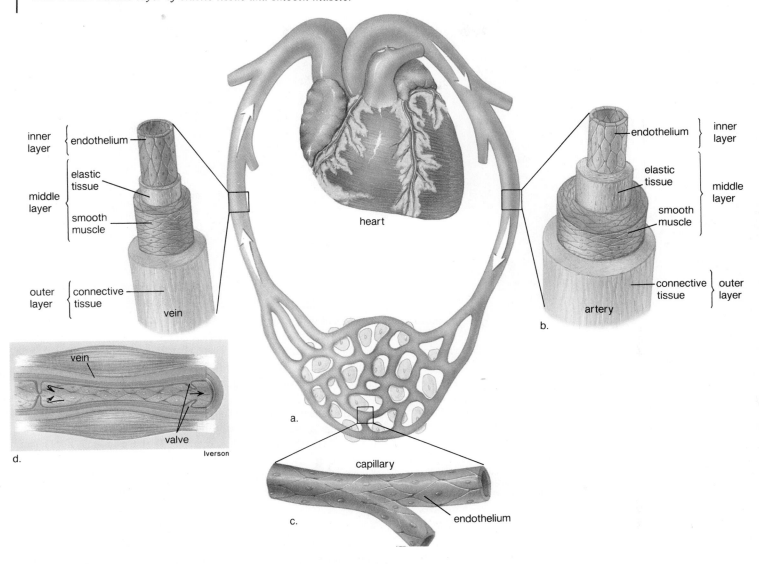

Single-celled organisms do not need a circulatory system. Their watery environment brings them their food and removes their wastes. But most of our 60–100 trillion cells are far removed from the external environment and need to be serviced. The circulatory system brings them their daily supply of nutrients, such as amino acids and glucose, and takes away their wastes, such as carbon dioxide and ammonia. At the center of the system is the heart (fig. 11.1), which keeps the blood moving along its predetermined circular path. Circulation of the blood is so important that if the heart discontinues beating for only a few minutes, death will result.

The Circulatory System

Blood Vessels

The blood vessels are arranged so that they continually carry blood from the heart to the tissues and then return it from the tissues to the heart. Blood vessels are of 3 types: the **arteries** (and **arterioles**) carry blood away from the heart; the **capillaries** exchange material with the tissues; and the **veins** (and **venules**) return blood to the heart.

Figure 11.2

Anatomy of a capillary bed. Capillary beds form a maze of vessels that lie between an arteriole and a venule. Blood can move directly between the arteriole and the venule by way of a shunt. When sphincter muscles are closed, blood flows through the shunt. When sphincter muscles are open, the capillary bed is open, and blood flows through the capillaries. As blood passes through a capillary, it gives up its oxygen (O_2). Therefore, blood goes from oxygenated in the arteriole (red color) to deoxygenated (blue color) in the vein.

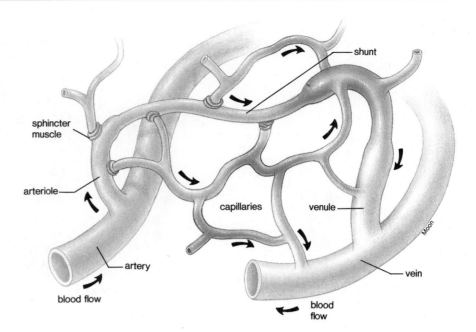

Arteries and Arterioles

Arteries have thick walls (fig. 11.1*b*) because in addition to an inner endothelial layer and an outer fibrous connective tissue layer, they have a thick middle layer that contains elastic and muscular tissues. The elastic tissue is loose connective tissue that is rich in elastic fibers, and the muscular tissue is smooth muscle. The elastic fibers enable an artery to expand and to accommodate the sudden increase in blood volume that results after each heartbeat. Arterial walls are so thick that the walls themselves are supplied with blood vessels.

Arterioles are small arteries just visible to the naked eye. The middle layer of arterioles has some elastic tissue but is composed mostly of smooth muscle, the fibers of which encircle the arteriole. If these muscle fibers contract, the bore of the arteriole gets smaller; if the fibers relax, the bore of the arteriole enlarges. Whether arterioles are constricted or dilated affects blood pressure. The greater the number of vessels dilated, the lower the blood pressure.

Capillaries

Arterioles branch into small vessels called capillaries. Each capillary is an extremely narrow, microscopic tube with a wall composed of only one layer of endothelial cells (fig. 11.1*c*). *Capillary beds* (networks of many capillaries) are present in all regions of the body; consequently, a cut to any body tissue draws blood. The capillaries are the most important part of a closed circulatory system because an exchange of nutrient and waste molecules takes place across their thin walls. Oxygen and nutrients diffuse out of a capillary into the tissue fluid that surrounds cells, and carbon dioxide and other wastes diffuse into the capillary (see fig. 12.6). Some water also leaves a capillary; any excess is picked up by lymphatic vessels, which return it to the blood circulatory system. The lymphatic system is discussed in chapter 13.

Since the capillaries serve the cells, the heart and the other vessels of the circulatory system can be thought of as a means by which blood is conducted to and from the capillaries. Not all capillary beds (fig. 11.2) are open or in use at the same time. After eating, the capillary beds of the digestive tract are usually open; during muscular exercise, the capillary beds of the skeletal muscles are open. Most capillary beds have a shunt through which blood moves directly from arteriole to venule when the capillary bed is closed. Sphincter muscles encircle the entrance to each capillary. These are constricted, preventing blood from entering the capillaries, when the bed is closed,

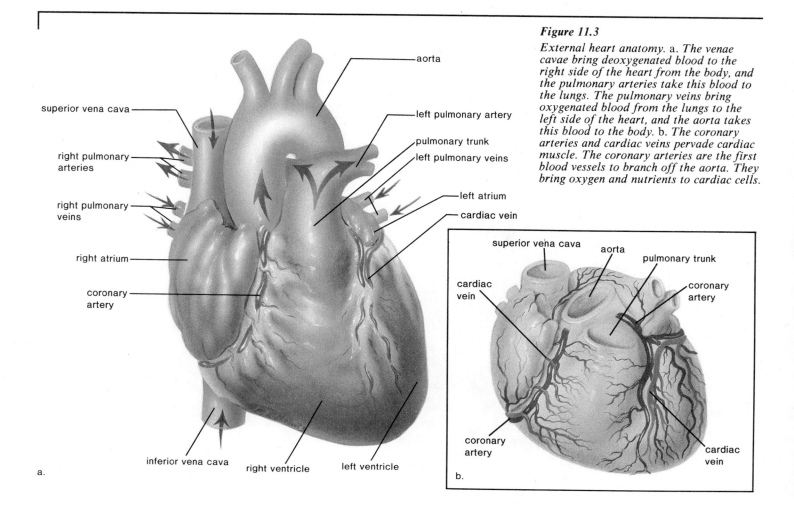

Figure 11.3

External heart anatomy. a. *The venae cavae bring deoxygenated blood to the right side of the heart from the body, and the pulmonary arteries take this blood to the lungs. The pulmonary veins bring oxygenated blood from the lungs to the left side of the heart, and the aorta takes this blood to the body.* b. *The coronary arteries and cardiac veins pervade cardiac muscle. The coronary arteries are the first blood vessels to branch off the aorta. They bring oxygen and nutrients to cardiac cells.*

and they are relaxed when the bed is open. As expected, the larger the number of capillary beds open, the lower the blood pressure.

Veins and Venules

Veins and venules take blood from the capillary beds to the heart. First, the venules drain the blood from the capillaries and then join to form a vein (fig. 11.1*d*). The wall of a venule has the same 3 layers as an artery, but the wall is much thinner than that of an artery because the middle layer of elastic and smooth muscle tissues is poorly developed. Within some veins, especially in the major veins of the arms and the legs, there are **valves** that allow blood to flow only toward the heart when they are open and prevent the backward flow of blood when they are closed.

At any given time, more than half of the total blood volume is in the veins and the venules. If a loss of blood occurs, for example due to hemorrhaging, nervous stimulation causes the veins to constrict, providing more blood to the rest of the body. In this way, the veins act as a blood reservoir.

Arteries and arterioles carry blood away from the heart; veins and venules carry blood to the heart; and capillaries join arterioles to venules.

The Heart

The **heart** is a cone-shaped (fig. 11.3), muscular organ about the size of a fist. It is located between the lungs directly behind the sternum and is tilted so that the apex is directed to the left. The major portion of the heart is called the **myocardium** and consists largely of cardiac muscle tissue. The muscle fibers

---

11.1 Critical Thinking

1. Draw a diagram of the heart in which the atria are above and the ventricles are below. Imagine that the vessels attached to the atria enter from above the heart and the vessels attached to the ventricles exit from below. Where would you place (a) superior vena cava, (b) aorta, (c) pulmonary vein, and (d) pulmonary artery in the diagram of your heart?

2. During fetal development, a blood vessel connects the pulmonary artery to the aorta. What effect does this have on the pulmonary circuit? Under these circumstances, would the aorta carry fully oxygenated blood?

Figure 11.4

Internal view of the heart. a. The right side of the heart contains deoxygenated blood. The venae cavae empty into the right atrium, and the pulmonary trunk and arteries leave the right ventricle. The left side of the heart contains oxygenated blood. The pulmonary veins enter the left atrium, and the aorta leaves from the left ventricle. b. A diagrammatic representation of the heart that allows you to trace the path of the blood. On the right: venae cavae, right atrium, right ventricle, pulmonary arteries to lungs. On the left: pulmonary veins, left atrium, left ventricle, aorta to body. Restate this and put in the name of the valves where appropriate.

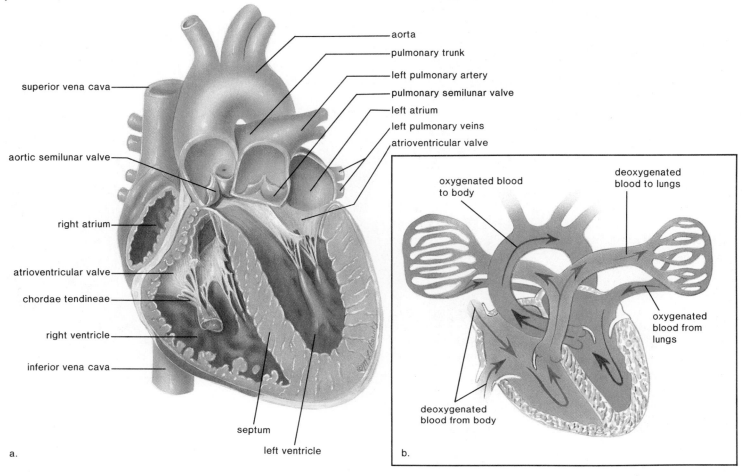

within the myocardium are branched and joined to one another so tightly that prior to studies with the electron microscope, it was thought that they formed one continuous muscle cell. Now it is known that there are individual fibers in the myocardium (p. 180). The inner surface of the heart is lined with endothelial tissue called *endocardium,* which resembles squamous epithelium. The outside of the heart is surrounded by a membrane called *pericardium,* and this forms a sac, called the pericardial sac, within which the heart is located. Normally, this sac contains a small quantity of liquid to lubricate the heart.

Internally (fig. 11.4), the heart has a right side and a left side separated by the **septum.** The heart has 4 chambers: 2 upper, thin-walled **atria** (singular, atrium), sometimes called auricles, and 2 lower, thick-walled **ventricles.** The atria are much smaller than the strong, muscular ventricles.

The heart also has valves that direct the flow of blood and prevent its backward flow. The valves that lie between the atria and the ventricles are called the **atrioventricular valves.** These valves are supported by strong fibrous strings called chordae tendineae. These cords, which are attached to muscular projections of the ventricular walls, support the valves and prevent them from inverting. The atrioventricular valve on the right side is called the tricuspid valve because it has 3 cusps, or flaps. The valve on the left side is called the bicuspid, or mitral, because it has 2 flaps. There are also **semilunar valves,** which resemble half moons, between the ventricles and their attached vessels.

Humans have a 4-chambered heart (2 atria and 2 ventricles), in which the right side is separated from the left side by a septum.

The Path of Blood in the Heart

We can trace the path of blood through the heart (fig. 11.4*b*) in the following manner:

- The superior (anterior) **vena cava** and the inferior (posterior) vena cava carrying deoxygenated blood (low in oxygen and high in carbon dioxide) enter the right atrium.
- The right atrium sends blood through an atrioventricular valve (the tricuspid valve) to the right ventricle.
- The right ventricle sends blood through the pulmonary semilunar valve into the pulmonary trunk and the pulmonary arteries to the lungs.
- The pulmonary veins carrying oxygenated blood (high in oxygen and low in carbon dioxide) from the lungs enter the left atrium.
- The left atrium sends blood through an atrioventricular valve (the bicuspid, or mitral, valve) to the left ventricle.
- The left ventricle sends blood through the aortic semilunar valve into the aorta to the body proper.

From this description, you can see that deoxygenated blood never mixes with oxygenated blood and that blood must pass through the lungs in order to pass from the right side to the left side of the heart. In fact, the heart is a *double pump* because the right side of the heart sends blood through the lungs, and the left side sends blood throughout the body (fig. 11.5).

There are 2 circular pathways (circuits) for blood in the body: one from the heart to the lungs and back to the heart, and another from the heart to the body and back to the heart. Since the left ventricle has the harder job of pumping blood to the entire body, its walls are thicker than those of the right ventricle, which pumps blood to the lungs.

The right side of the heart pumps blood to the lungs, and the left side of the heart pumps blood to the tissues.

The Heartbeat

The Cardiac Cycle

From this description of the path of blood through the heart, it might seem that the right and left sides of the heart beat independently of one another, but actually, they contract together. First, the 2 atria contract simultaneously; then the 2 ventricles contract at the same time. The word **systole** refers to contraction of heart muscle, and the word **diastole** refers to relaxation of heart muscle. The heart contracts, or beats, about 70 times a minute, and each heartbeat lasts about 0.85 second. Each heartbeat, or *cardiac cycle* (fig. 11.6), consists of the following elements:

Time	Atria	Ventricles
0.15 sec	Systole	Diastole
0.30 sec	Diastole	Systole
0.40 sec	Diastole	Diastole

This shows that while the atria contract, the ventricles relax, and vice versa, and that all chambers rest at the same time for 0.40 second. The short systole of the atria is appropriate since the atria send blood only into the ventricles.

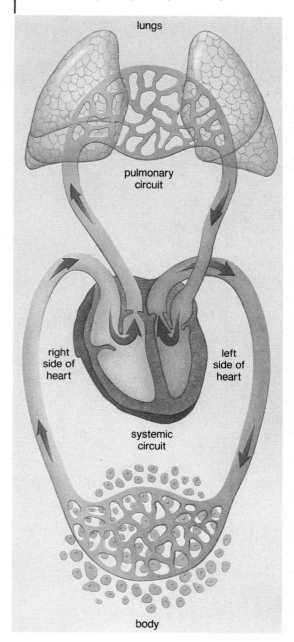

Figure 11.5
Diagram of pulmonary and systemic circuits. The blue-colored vessels carry deoxygenated blood, while the red-colored vessels carry oxygenated blood. Notice that the blood cannot move from the right side of the heart to the left side without passing through the lungs.

lungs

pulmonary circuit

right side of heart

left side of heart

systemic circuit

body

Figure 11.6

*Stages in the cardiac cycle. Note when the
semilunar valves and the atrioventricular valves
are open or closed.*

During the first phase of
the cardiac cycle (0.15 sec),
the atria are in systole and
pumping blood into the
ventricles.

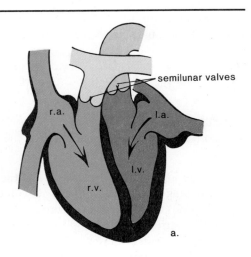

a.

During the second phase of
the cardiac cycle (0.30 sec),
the ventricles are in systole.
Deoxygenated blood goes to the
lungs by way of the pulmonary
arteries and oxygenated blood
goes to the body by way of the
aorta.

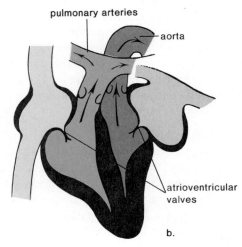

b.

During the third phase of the
cardiac cycle, all chambers
are in diastole (0.40 sec).
Deoxygenated blood enters the
right side of the heart from
the vena cavae. Oxygenated
blood enters the left side
of the heart from the
pulmonary veins.

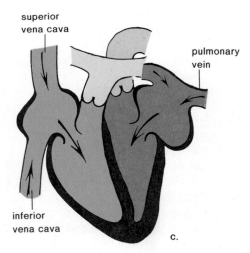

c.

It is the muscular ventricles that actually pump blood out into the circulatory
system proper. When the word *systole* is used alone, it usually refers to the
left ventricular systole.

*The heartbeat is divided into 3 phases. First, the atria contract, and then the
ventricles contract. (When the atria are in systole, the ventricles are in diastole, and
vice versa.) Finally, all chambers are in diastole.*

Heart Sounds

When the heart beats, the familiar lub-DUPP sound is heard as the valves of the heart close. The lub is caused by vibrations of the heart when the atrioventricular valves close, and the DUPP is heard when vibrations occur due to the closing of the semilunar valves (fig. 11.7). Heart murmurs, or a slight slush sound after the lub, are often due to ineffective valves that allow blood to pass back into the atria after the atrioventricular valves have closed. Rheumatic fever resulting from a bacterial infection is one cause of a faulty valve, particularly the bicuspid valve. If operative procedures are unable to open and/or restructure the valve, it can be replaced with an artificial valve.

The heart sounds are due to the closing of the heart valves.

The Cardiac Conduction System

The heart will beat independently of any nervous stimulation. In fact, it is possible to remove the heart of a small animal, such as a frog, and to watch it undergo contraction in a petri dish. The reason for this is a unique type of tissue called nodal tissue, having both muscular and nervous characteristics, located in 2 regions of the heart. The first of these, the **SA (sinoatrial) node,** is found in the upper dorsal wall of the right atrium; the other, the **AV (atrioventricular) node,** is found in the base of the right atrium very near the septum (fig. 11.8a). The SA node initiates the heartbeat and automatically sends out an excitation impulse every 0.85 second to cause the atria to contract. When the impulse reaches the AV node, it signals the ventricles to contract by way of specialized fibers called Purkinje fibers. The SA node is called the **pacemaker** because it usually keeps the heartbeat regular. If the SA node fails to work properly, the heart still beats, but irregularly. To correct this condition, it is possible to implant in the body an artificial pacemaker that automatically gives an electric shock to the heart every 0.85 second. This causes the heart to beat regularly again.

The SA node is the natural pacemaker that keeps the heart beating regularly.

Electrocardiogram (ECG)   With the contraction of any muscle, including the myocardium, ionic changes occur that can be detected by electrical recording devices. Therefore, it is possible to study the heartbeat by recording voltage changes that occur when the heart contracts. (Voltage, which in this case is measured in millivolts, is the difference in polarity between 2 electrodes attached to the body.) The record that results is called an **electrocardiogram** (fig. 11.8b), which clearly shows an atrial phase and a ventricular phase. The first wave in the electrocardiogram, called the P wave, represents the excitation and contraction of the atria. The second wave, or the QRS wave, occurs during ventricular excitation and contraction. The third, or T, wave is caused by the recovery of the ventricles. An examination of the electrocardiogram indicates whether the heartbeat has a normal or an irregular pattern.

The conduction system of the heart includes the SA node, the AV node, and the Purkinje fibers. With an ECG, it is possible to determine if the conduction system, and therefore the beat of the heart, is regular.

Nervous Control of the Heartbeat

The rate of the heartbeat is also under nervous control. A cardiac center in the medulla oblongata (p. 336) of the brain can alter the beat of the heart by way of the *autonomic nervous system* (p. 332). This system is made up of 2 divisions: the *parasympathetic system,* which promotes those functions we tend to associate with normal activities, and the *sympathetic system,* which brings

Figure 11.7

Heart valves. a. *Close-up view of closed semilunar valves.* b. *Drawing showing the relative positions of all the valves.*

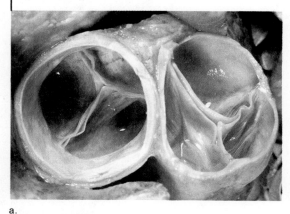

a.

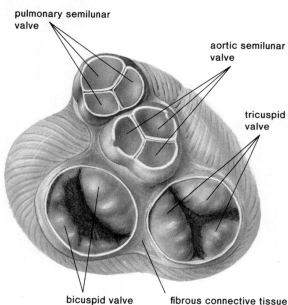

pulmonary semilunar valve

aortic semilunar valve

tricuspid valve

bicuspid valve

fibrous connective tissue

b.

Figure 11.8

Control of the heart cycle. a. The SA node sends out a stimulus that causes the atria to contract. When this stimulus reaches the AV node, it signals the ventricles to contract by way of the Purkinje fibers. b. A normal ECG indicates that the heart is functioning properly. The P wave occurs as the atria contract; the QRS wave occurs as the ventricles contract; and the T wave occurs when the ventricles are recovering from contraction. c. Abnormal ECGs: sinus tachycardia is an abnormally fast heartbeat due to a fast pacemaker; ventricular fibrillation is irregular heartbeat due to irregular stimulation of the ventricles; and mitral stenosis occurs because the bicuspid (mitral) valve is obstructed.

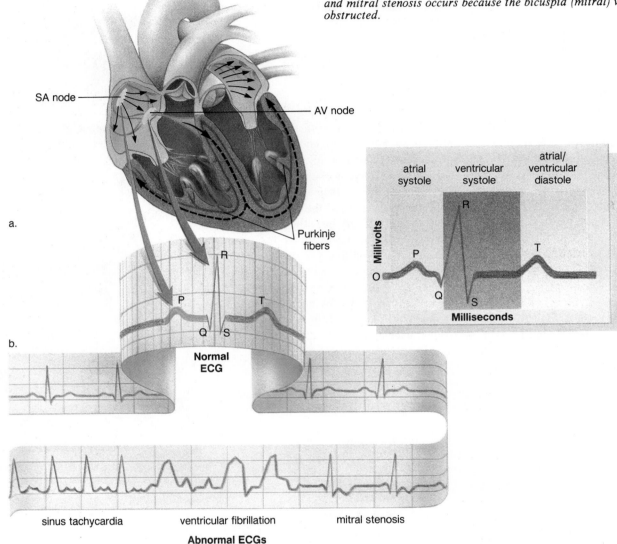

about those responses we associate with times of stress. For example, the parasympathetic system causes the heartbeat to slow down, and the sympathetic system increases the heartbeat. Various factors, such as the relative need for oxygen or blood pressure, determine which of these systems is activated.

The heart rate is regulated largely by the autonomic nervous system.

Vascular Pathways

The cardiovascular system, which is represented in figure 11.9, includes 2 circuits: the **pulmonary circuit,** which circulates blood through the lungs, and the **systemic circuit,** which serves the needs of body tissues.

The Pulmonary Circuit

The path of blood through the lungs can be traced as follows. Blood from all regions of the body first collects in the right atrium and then passes into the

Human Anatomy and Physiology

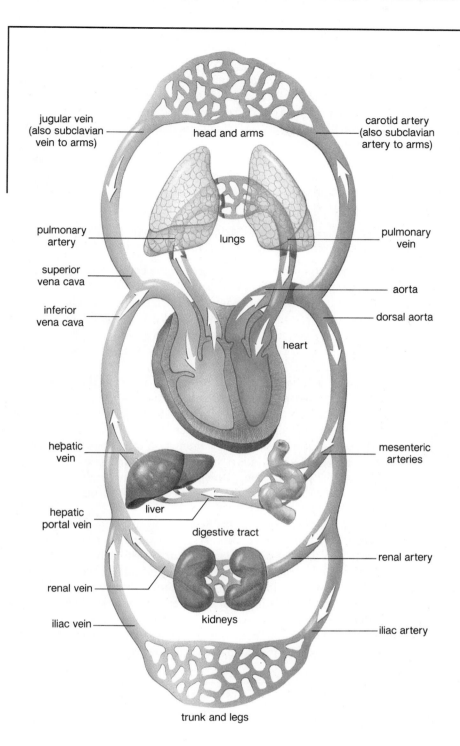

jugular vein
(also subclavian
vein to arms)

head and arms

carotid artery
(also subclavian
artery to arms)

pulmonary
artery

lungs

pulmonary
vein

superior
vena cava

aorta

inferior
vena cava

dorsal aorta

heart

hepatic
vein

mesenteric
arteries

hepatic
portal vein

liver

digestive tract

renal artery

renal vein

iliac vein

kidneys

iliac artery

trunk and legs

Figure 11.9

Blood vessels in the pulmonary and systemic circuits. The blue-colored vessels carry deoxygenated blood, and the red-colored vessels carry oxygenated blood; the arrows indicate the flow of blood. Compare this diagram, useful for learning to trace the path of blood, to figure 11.10 in order to realize that both arteries and veins go to all parts of the body. Also, there are capillaries in all parts of the body. No cell is located far from a capillary.

right ventricle, which pumps it into the pulmonary trunk. The pulmonary trunk divides into the *pulmonary arteries,* which divide into the arterioles of the lungs. The arterioles take blood to the pulmonary capillaries, where carbon dioxide and oxygen are exchanged. The blood then enters the pulmonary venules that lead back through the *pulmonary veins* to the left atrium. Since the blood in the pulmonary arteries is deoxygenated but the blood in the pulmonary veins is oxygenated, it is not correct to say that all arteries carry oxygenated blood and all veins carry deoxygenated blood. It is just the reverse in the pulmonary circuit.

The pulmonary arteries take deoxygenated blood to the lungs, and the pulmonary veins return oxygenated blood to the heart.

The Systemic Circuit

The systemic circuit includes all of the other arteries and veins shown in figure 11.9. The largest artery in the systemic circuit is the **aorta,** and the largest veins are the *superior* and *inferior* venae cavae. The superior vena cava collects blood from the head, the chest, and the arms, and the inferior vena cava collects blood from the lower body regions. Both enter the right atrium. The aorta and the venae cavae serve as the major pathways for blood in the systemic circuit.

The path of systemic blood to any organ in the body begins in the left ventricle, which pumps blood into the aorta. Branches from the aorta go to the major body regions and organs. For example, the path of blood to the kidneys can be traced as follows:

> left ventricle—aorta—renal artery—renal arterioles, capillaries, venules—renal vein—inferior vena cava—right atrium

To trace the path of blood to any organ in the body, you need only mention the aorta, the proper branch of the aorta, the organ, and the vein returning blood to the vena cava. In most instances, the artery and the vein that serve the same organ are given the same name (fig. 11.9). In the systemic circuit, unlike the pulmonary system, arteries contain oxygenated blood and have a bright red color, but veins contain deoxygenated blood and appear a purplish color.

The **coronary arteries** (fig. 11.3*b*), which are a part of the systemic circuit, are extremely important because they serve the heart muscle itself. (The heart is not nourished by the blood in its chambers.) The coronary arteries arise from the aorta just above the aortic semilunar valve. They lie on the exterior surface of the heart, where they branch off in various directions into the arterioles. The coronary capillary beds join to form venules. The venules converge to form the cardiac veins, which empty into the right atrium. The coronary arteries have a very small diameter and may become blocked, as discussed on page 234.

The body has a portal system, the hepatic portal system (see fig. 10.12), that is associated with the liver. A portal system begins and ends in capillaries; in this instance, the first set of capillaries occurs at the villi of the small intestine and the second occurs in the liver. Blood passes from the capillaries of the villi into venules that join to form the *hepatic portal vein,* a vessel that connects the villi of the intestine with the liver. The *hepatic vein* leaves the liver and enters the inferior vena cava.

While figure 11.9 is helpful in tracing the path of the blood, remember that all parts of the body receive both arteries and veins, as illustrated in figure 11.10.

The systemic circuit takes blood from the left ventricle of the heart to the right atrium of the heart. It serves the body proper.

Features of the Circulatory System

When the left ventricle contracts, the blood is sent into the aorta under pressure.

The Pulse

The surge of blood entering the arteries causes their elastic walls to swell, but then they almost immediately recoil. This alternating expansion and recoil of an arterial wall can be felt as a **pulse** in any artery that runs close to the body's surface. It is customary to feel the pulse by placing several fingers on the radial artery, which lies near the outer border of the palm side of the wrist. The carotid artery is another good location to feel the pulse (fig. 11.11). Normally, the pulse rate indicates the rate of the heartbeat because the arterial walls pulse whenever the left ventricle contracts.

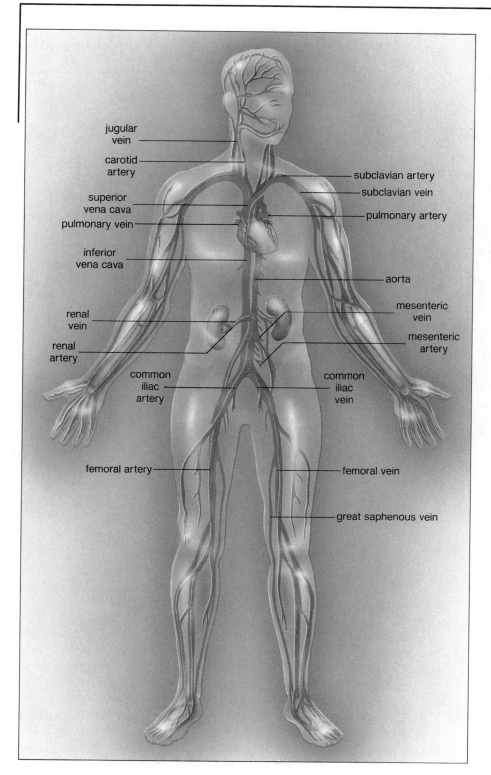

Figure 11.10
Human circulatory system. A more realistic representation of major blood vessels in the body shows that arteries and veins go to all parts of the body. The superior and inferior venae cavae take their names from their relationship to which organ?

jugular vein

carotid artery

superior vena cava

pulmonary vein

inferior vena cava

renal vein

renal artery

common iliac artery

femoral artery

subclavian artery

subclavian vein

pulmonary artery

aorta

mesenteric vein

mesenteric artery

common iliac vein

femoral vein

great saphenous vein

The pulse rate indicates the heartbeat rate.

Blood Pressure

Blood pressure is the pressure of the blood against the wall of a blood vessel.

Measurement of Blood Pressure

A sphygmomanometer is used to measure blood pressure (fig. 11.12). This instrument consists of a hollow cuff connected by tubing to a compressible

Figure 11.11

The common carotid artery. This artery is located in the neck region and can be used to take the pulse. The pulse indicates how rapidly the heart is beating.

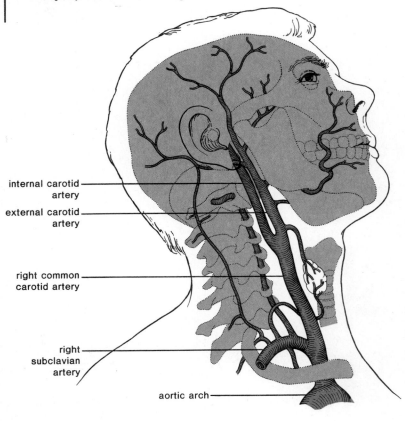

internal carotid artery

external carotid artery

right common carotid artery

right subclavian artery

aortic arch

Figure 11.12

Determination of blood pressure using a sphygmomanometer. The technician inflates the cuff with air and then as he or she gradually reduces the pressure, he or she listens by means of a stethoscope for the sounds that indicate the blood is moving past the cuff in an artery. A pressure gauge on the cuff is used to tell the systolic and diastolic blood pressures.

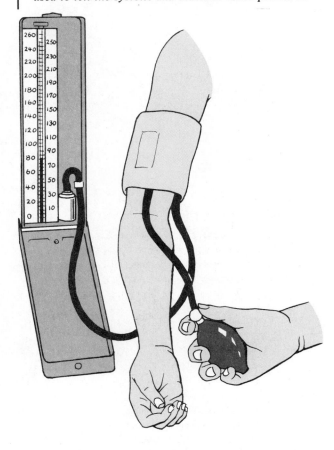

11.2 Critical Thinking

Smoking cigarettes increases blood pressure. Considering this, what do you predict that smoking does to the

a. heartbeat rate?
b. bore of the arteries?
c. capillary beds in the fingers and the toes?
d. resistance of blood flow through the lungs?

bulb and to a pressure gauge. The cuff is placed about the upper arm over the brachial artery and is inflated with air. Eventually, the brachial artery is squeezed shut, and when a stethoscope is placed in the crook of the elbow, no sounds are heard. Air is slowly released, and the cuff is deflated until a sharp sound is heard through the stethoscope. The examiner glances at the manometer, or pressure gauge, and notes the pressure at this point. This is the value assigned to **systolic blood pressure,** the highest arterial pressure reached during ejection of blood from the heart. The systolic pressure has overcome the pressure exerted by the cuff and has caused the blood to flow in the artery.

The cuff is further deflated while the examiner continues to listen. Tapping sounds become louder as the pressure is lowered. Finally, the sounds become abruptly dull and muffled just before there are no sounds at all. Now the examiner again notes the pressure. This is **diastolic blood pressure,** the lowest arterial pressure. Diastolic pressure occurs while the heart ventricles are relaxing.

Normal resting blood pressure for a young adult is said to be 120 mm of mercury (Hg) over 80 mm, or simply 120/80. The higher number is the systolic pressure, and the lower number is the diastolic pressure.

Blood Pressure throughout the Body

Actually, 120/80 is the expected blood pressure in the brachial artery of the arm; blood pressure decreases with distance from the left ventricle. Blood pressure is, therefore, higher in the arteries than in the arterioles. Further, there is a sharp drop in blood pressure when the arterioles reach the capillaries. The decrease can be correlated with the increase in the total cross-sectional area of the vessels as blood moves through arteries, arterioles, and then into capillaries. There are more arterioles than arteries and many more capillaries than arterioles (fig. 11.13).

Blood pressure steadily decreases from the aorta to the veins.

The Velocity of Blood Flow

The velocity of blood flow varies in different parts of the circulatory system (fig. 11.13). Blood pressure accounts for the velocity of the blood flow in the arterial system and therefore, as blood pressure decreases due to the increased cross-sectional area of the arterial system, so does velocity. The blood moves more slowly through the capillaries than it does through the aorta. This is important because the slow progress allows time for the exchange of molecules between the blood and the tissues.

Blood pressure cannot account for the movement of blood through the venules and the veins since they lie on the other side of the capillaries. Instead, movement of the blood through the venous system is due to skeletal muscle contraction. When the skeletal muscles contract, they press against the weak walls of the veins. This causes the blood to move past a *valve* (fig. 11.14). Once past the valve, the blood cannot return. The importance of muscle contraction in moving blood in the venous system can be demonstrated by forcing a person to stand rigidly still for a number of hours. Frequently, fainting occurs because the blood collects in the limbs, robbing the brain of oxygen. In this case, fainting is beneficial because the resulting horizontal position aids in getting blood to the head. Blood flow gradually increases in the venous system (fig. 11.13) due to a progressive reduction in the cross-sectional area as small venules join to form veins. The 2 venae cavae together have a cross-sectional area only about double that of the aorta. The blood pressure is lowered in the thoracic cavity whenever the thorax expands during inspiration. This also aids the flow of venous blood into the thoracic cavity because blood flows in the direction of reduced pressure.

Blood pressure accounts for the flow of blood in the arteries and the arterioles; skeletal muscle contraction accounts for the flow of blood in the venules and the veins.

Circulatory Disorders

During the past 30 years, the number of deaths due to cardiovascular disease has declined more than 30%. Even so, more than 50% of all deaths in the United States still are attributable to cardiovascular disease. The number of deaths due to hypertension, stroke, and heart attack is greater than the number due to cancer and accidents combined.

Cardiovascular disease is the number one killer in the United States.

Hypertension

It is estimated that about 20% of all Americans suffer from *hypertension*, high blood pressure indicated by a blood pressure reading. Women of any age are considered to have hypertension if their blood pressure reading is 160/95 or above. For a man under age 45, a reading above 130/90 is hypertensive, and beyond age 45, a reading above 140/95 is considered hypertensive. While both systolic and diastolic pressures are considered important, it is the diastolic pressure that is emphasized when medical treatment is being considered.

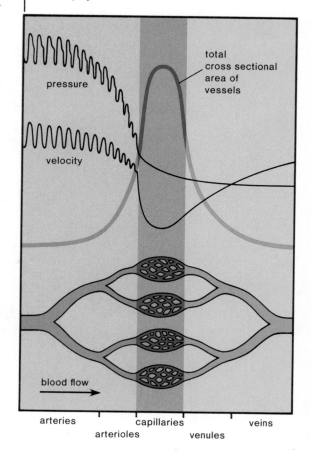

Figure 11.13
Diagram illustrating how velocity and blood pressure are related to the total cross-sectional area of blood vessels. Capillaries have the greatest cross-sectional area and the least pressure and velocity. Skeletal muscle contraction, not blood pressure, accounts for the velocity of blood in the veins.

pressure

total cross sectional area of vessels

velocity

blood flow

arteries
arterioles
capillaries
venules
veins

Figure 11.14

Skeletal muscle contraction moves blood in veins.
a. Muscle contraction exerts pressure against the
vein, and blood moves past the valve. b. Once
blood has moved past the valve, it cannot return.

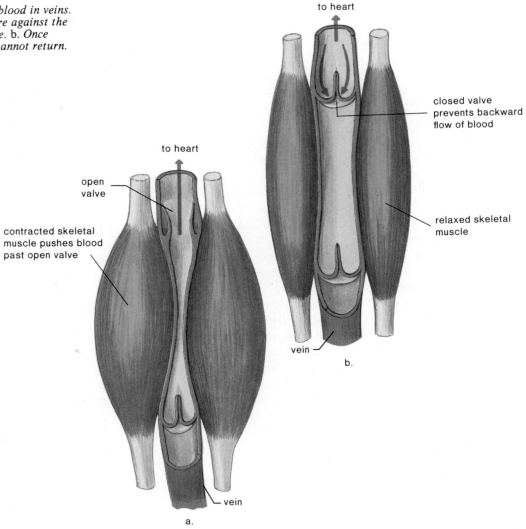

to heart

open
valve

contracted skeletal
muscle pushes blood
past open valve

vein

a.

to heart

closed valve
prevents backward
flow of blood

relaxed skeletal
muscle

vein

b.

The reasons for the development of hypertension are various. One possible scenario is described in figure 11.15. Blood pressure normally rises with excitement or alarm due to the involvement of the sympathetic nervous system (p. 332), which causes arterioles to constrict and the heart to beat faster. When arterioles are constricted, reduced blood flow to the kidneys causes these organs to release renin, a molecule that brings about vasoconstriction and sodium retention, as described in chapter 15. Vasoconstriction and sodium retention lead to high blood pressure. Excess sodium in the blood causes an increase in blood volume due to water retention.

Medical treatment can control hypertension with the following drugs. Sympathetic-blocking agents act at arrow 1 in figure 11.15 and prevent action of the sympathetic nervous system. Two types of drugs that act at arrow 2 are available. Calcium channel blockers are new drugs that prevent constriction of arteries because calcium is needed for muscle contraction (p. 361). Use of vasodilators results in relaxation of the muscles in arterial walls. Another new drug is an enzyme inhibitor that works at arrow 3 to prevent renin from producing its normal effects (p. 313). Finally, diruretics act at arrow 4 and cause the kidneys to excrete excess salts and fluids.

Since these drugs can have side effects that are at best unpleasant, it is wise to adopt a life-style that protects against the development of hypertension. The reading on page 234 discusses good health habits that lower the risk of cardiovascular disorders.

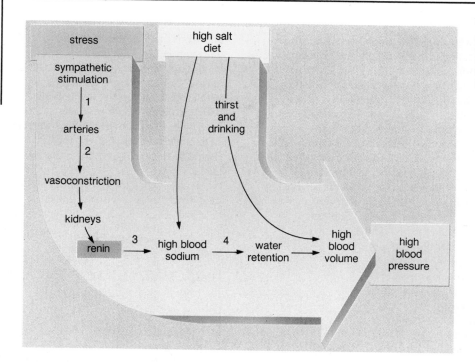

Figure 11.15
A scheme that explains the development of high blood pressure due to either stress or a high salt diet. The numbers (1–4) indicate the site of action of antihypertensive drugs, as discussed in the text.

Diet, stress, and kidney involvement are implicated in the development of hypertension in some persons.

Atherosclerosis

Hypertension also is seen in individuals who have *atherosclerosis* (formerly called arteriosclerosis), an accumulation of soft masses of fatty materials, particularly cholesterol, beneath the inner linings of arteries. Such deposits are called *plaque,* and as it develops, it tends to protrude into the vessel and interferes with the flow of blood. Atherosclerosis begins in early adulthood and develops progressively through middle age, but symptoms may not appear until an individual is 50 or older. To prevent its onset and development, a diet low in saturated fat and cholesterol is recommended by the American Heart Association and other organizations, as discussed in the reading on page 234.

Plaque can cause a clot to form on the irregular arterial wall. As long as the clot remains stationary, it is called a *thrombus,* but when and if it dislodges and moves along with the blood, it is called an *embolus*. If *thromboembolism* is not treated, complications can arise, as mentioned in the following section.

Development of atherosclerosis, which is associated with a high cholesterol blood level, can lead to thromboembolism.

Stroke and Heart Attack

Both strokes and heart attacks are associated with hypertension and atherosclerosis. A *stroke* occurs when a portion of the brain dies due to a lack of oxygen. A stroke, characterized by paralysis or death, often results when a small arteriole bursts or is blocked by an embolus. A person sometimes is forewarned of a stroke by a feeling of numbness in the hands or the face, difficulty in speaking, or temporary blindness in one eye.

Because sudden cardiac death happens once every 72 seconds in the United States, it is well to identify the factors that predispose an individual to cardiovascular disease. The risk factors for cardiovascular disease include the following:

Male sex
Family history of heart attack
    under age 55
Smoking more than 10 cigarettes a day
Severe obesity (30% or more
    overweight)
Hypertension
Unfavorable HDL and LDL cholesterol
    blood levels
Impaired circulation to the brain
    or the legs
Diabetes mellitus

Hypertension is well recognized as a major factor in cardiovascular disease, and 2 controllable behaviors contribute to hypertension. Smoking cigarettes, including filtered cigarettes, causes hypertension, as does obesity. It is best to never take up the habit of smoking cigarettes, but most of the detrimental effects can be reversed if you stop smoking. Since it is very difficult for obese individuals to lose weight, it is recommended that weight control be a lifelong endeavor.

Investigators have identified several behaviors that may help to reduce the possibility of heart attack and stroke. Exercise seems to be critical. Sedentary individuals have a risk of cardiovascular disease that is about double that of those who are very active. One physician, for example, recommends that his patients walk for one hour, 3 times a week. Stress reduction also is desirable. The same investigator recommends everyday meditation and yoga-like stretching and breathing exercises to reduce stress.

Another behavior that is much in the news of late is the adoption of a diet that is low in saturated fats and cholesterol (see p. 207) because such a diet is believed by many to protect against the development of cardiovascular disease. Cholesterol is ferried in the blood by 2 types of plasma proteins called LDL (low-density lipoprotein) and HDL (high-density lipoprotein). LDL (called "bad" lipoprotein) takes cholesterol to the tissues from the liver, and HDL (called "good" lipoprotein) transports cholesterol out of the tissues to the liver. When the LDL level in blood is abnormally high or the HDL level is abnormally low, cholesterol accumulates in the cells. When cholesterol-laden cells line the arteries, plaque develops, which interferes with circulation.

Cholesterol guidelines have been established by the National Heart, Lung, and Blood Institute. According to the institute, everyone should know his or her cholesterol blood level. Individuals with a borderline-high cholesterol blood level (200-239 mg/100 ml) should be tested further if they already have heart disease or if they have 2 known risk factors for cardiovascular disease (see list). Individuals with a high cholesterol blood level (240 mg/100 ml) always should be tested further. Persons with an LDL cholesterol level of over 130 mg/100 ml should be treated if they have other risk factors, and those with an LDL cholesterol level of 160 mg/100 ml should be treated even if this is the only risk factor.

Persons with a total-to-HDL cholesterol ratio higher than 4.5 also are considered to be at risk. Heart attack has occurred in

Cardiovascular Disease

individuals who have a normal total cholesterol level, but who also have an unfavorable total-to-HDL cholesterol ratio. For example, if a person's total cholesterol blood level is 200, but the HDL level is only 25 mg/100 ml, then the total-to-HDL cholesterol ratio is 8.0, and circulatory difficulties most likely will develop.

First and foremost, treatment for unfavorable cholesterol levels consists of adopting a diet that is low in saturated fat and cholesterol (see p. 207). Although the prescribed diet does not lower cholesterol blood level in all persons, it is expected to do so for most individuals. If diet alone does not bring down the cholesterol blood level, drugs can be prescribed. Some of the drugs

A *heart attack* occurs when a portion of the heart muscle dies because of a lack of oxygen. Due to atherosclerosis, the coronary artery may be partially blocked. The individual may then suffer from *angina pectoris,* characterized by a radiating pain in the left arm. When a coronary artery is completely blocked, perhaps because of thromboembolism, a heart attack occurs.

Stroke and heart attack are associated with both hypertension and atherosclerosis.

Medical and Surgical Treatment
Medical and surgical treatments now are available for blocked coronary arteries.

Thrombolytic Therapy   Medical treatment for thromboembolism includes 2 drugs that can be given intravenously to dissolve a clot: streptokinase and tPA. Both drugs convert plasminogen, a molecule found in blood, into plasmin, an enzyme that can dissolve a blood clot. In fact, tPA, which stands for tissue plasminogen activator, is the body's very own way of converting plasminogen to plasmin. Streptokinase and tPA are used particularly when it is known that a clot is present.

Human Anatomy and Physiology

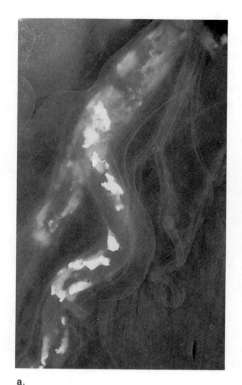

a.

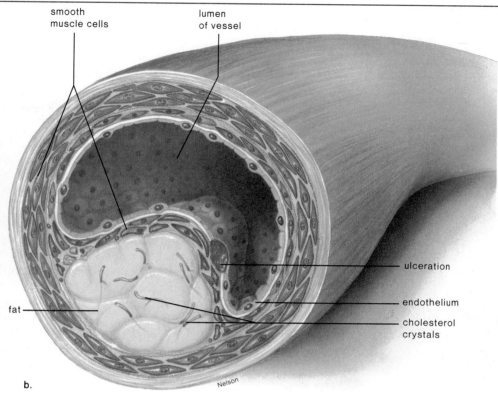

smooth
muscle cells

lumen
of vessel

ulceration

endothelium

fat

cholesterol
crystals

Nelson

b.

Plaque. a. *Plaque (yellow) in the coronary artery of a heart patient.* b. *Cross section of plaque shows its composition and indicates how it bulges out into the lumen of an artery, obstructing blood flow.*

act in the intestine to remove cholesterol, and others act in the body to prevent its production. These drugs reduce the blood level of cholesterol, but their long-term side effects are not completely known and may

be serious. Considering this, some investigators do not recommend these drugs to lower cholesterol blood levels.

A diet low in saturated fat and cholesterol can lower the total cholesterol blood level and the LDL level of some individuals, but this diet most likely will not raise the HDL level. Aside from certain drugs that apparently can raise HDL level, exercise is sometimes effective.

There is nothing that can be done about some of the cardiovascular risk factors, such as male gender and family history. However, other risk factors likely can be controlled if the individual believes it is worth the effort. It is clear that the 4 great admonitions for a healthy life—heart-healthful diet, regular exercise, proper weight maintenance, and refraining from smoking—all contribute to acceptable blood pressure and cholesterol blood levels.

If a person has symptoms of angina or a thrombolytic stroke, then an anticoagulant drug, such as aspirin, may be given. Aspirin works by inhibiting platelets, cell fragments that trigger clot formation. Aspirin reduces the stickiness of platelets and thus lowers the probability that a clot will form. There is evidence that aspirin protects against first heart attacks, but there is no clear support for taking aspirin every day to prevent strokes in symptom-free people. Physicians warn that long-term use of aspirin might have harmful effects, including bleeding in the brain.

Arterial Plaque Surgical procedures are available to clear clogged arteries. In one procedure, a plastic tube is threaded into an artery of an arm or a leg and is guided through a major blood vessel toward the heart. When the tube reaches the region of plaque in a coronary or carotid artery (fig. 11.11), a balloon attached to the end of the tube inflates, forcing the vessel open. The problem with this procedure is the vessel may not remain open, and worse, it may cause clots to form. As an alternative, it is possible to open the narrowed vessel and to remove the plaque. This requires more invasive surgery, however.

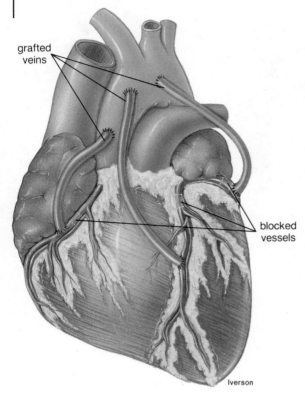

Figure 11.16

Coronary bypass operation. During this operation, the surgeon grafts segments of a leg vein between the aorta and the coronary vessels, bypassing areas of blockage. Patients who are ill enough to require surgery often receive 2 or 3 bypasses in a single operation.

grafted veins

blocked vessels

Iverson

Thousands of persons each year have *coronary bypass* surgery. During this operation, surgeons take a segment of another blood vessel, often from a large vein in the leg, and stitch one end to the aorta and the other end to a coronary artery past the point of obstruction (fig. 11.16).

Once the heart is exposed, some physicians use lasers to open up clogged coronary vessels. Presently, this technique is used in conjunction with coronary bypass operations, but eventually it may be possible to use lasers independently, without opening the thoracic cavity.

Donor Heart Transplants and Artificial Heart Implants   Persons with weakened hearts eventually may suffer from *congestive heart failure,* meaning the heart no longer is able to pump blood adequately. These individuals, depending on their age, are candidates for a donor heart transplant. The difficulties with a donor heart transplant are, first, one of availability and, second, the tendency of the body to reject foreign organs. Sometimes, it is possible to repair the heart instead of replacing it. For example, a back muscle can be wrapped around a heart weakened by the removal of myocardial tissue. An artificial implant pacemaker causes the muscle to contract regularly and helps to pump the blood.

On December 2, 1982, Barney Clark became the first person to receive an artificial heart. The heart's 2 polyurethane ventricles were attached to Clark's own atria and blood vessels by way of Dacron fittings. Two long tubes were stretched between the artificial heart and an external machine that periodically sent bursts of air into the ventricles, forcing the blood out into the aorta and the pulmonary trunk. As yet, no one has survived for longer than a few months after receiving an artificial heart due to complications, such as formation of blood clots, following the operation. For this reason, artificial hearts are not implanted except on a temporary basis in patients awaiting a transplant.

Varicose Veins and Phlebitis

Varicose veins are abnormal and irregular dilations in superficial (near the surface) veins, particularly those in the lower legs. Varicose veins in the rectum, however, are commonly called piles, or more properly, *hemorrhoids.* Varicose veins develop when the valves of the veins become weak and ineffective due to a backward pressure of the blood. The problem can be aggravated when venous blood flow is obstructed by crossing the legs or by sitting in a chair so that its edge presses against the back of the knees.

Phlebitis, or inflammation of a vein, is a more serious condition, particularly when a deep vein is involved. Blood in the inflamed vessel may clot, in which case thromboembolism occurs. An embolus that originates in a systemic vein eventually may come to rest in a pulmonary arteriole, blocking circulation through the lungs. This condition, termed *pulmonary embolism,* can result in death.

Summary

The movement of blood in the circulatory system is dependent on the beat of the heart. During the cardiac cycle, the SA node (pacemaker) initiates the beat and causes the atria to contract. The AV node picks up the stimulus and initiates contraction of the ventricles. The heart sounds, lub-DUPP, are due to the closing of the atrioventricular valves, followed by the closing of the semilunar valves.

The circulatory system is divided into the pulmonary and systemic circuits. In the pulmonary circuit, the pulmonary artery takes blood from the right ventricle to the lungs, and the pulmonary veins return it to the left atrium. To trace the path of blood in the systemic circuit, start with the aorta from the left ventricle. Follow its path until it branches to an artery going to a specific organ. It can be assumed that the artery divides into arterioles and capillaries, and that the capillaries lead to venules. The vein that takes blood to the vena cava most likely has the same name as the artery.

Blood pressure accounts for the flow of blood in the arteries, but because blood pressure drops off after the capillaries, it cannot cause blood flow in the veins.

Skeletal muscle contraction pushes blood past a venous valve, which then shuts, preventing backward flow. The velocity of blood flow is slowest in the capillaries, where exchange of nutrient and waste molecules takes place.

Hypertension and atherosclerosis are 2 circulatory disorders that lead to heart attack and to stroke. Medical and surgical procedures are available to control cardiovascular disease, but the best policy is prevention by following a heart-healthy diet, getting regular exercise, maintaining a proper weight, and not smoking cigarettes.

Study Questions

1. What types of blood vessels are there? Discuss their structure and function. (pp. 219–21)
2. Trace the path of blood in the heart, mentioning the vessels attached to and the valves within the heart. (p. 223)
3. Describe the cardiac cycle (using the terms systole and diastole), and explain the heart sounds. (p. 223)
4. Describe the cardiac conduction system and an ECG. Tell how an ECG is related to the cardiac cycle. (p. 225)
5. Trace the path of blood in the pulmonary circuit as it travels from and returns to the heart. (p. 226)
6. Trace the path of blood from the mesenteric arteries to the aorta, indicating which of the vessels are in the systemic circuit and which are in the pulmonary circuit. (p. 227)
7. What is blood pressure, and why is the average normal arterial blood pressure said to be 120/80? (p. 229)
8. In which type of vessel is blood pressure highest? lowest? Velocity is lowest in which type of vessel, and why is it lowest? Why is this beneficial? What factors assist venous return of the blood? (p. 230)
9. What is atherosclerosis? (p. 233) Name 2 illnesses associated with hypertension and thromboembolism. (p. 233) Discuss the medical and surgical treatment of cardiovascular disease. (pp. 234–36)

Objective Questions

1. Arteries are blood vessels that take blood _____ from the heart.
2. When the left ventricle contracts, blood enters the _____ .
3. The pulmonary veins carry blood _____ in oxygen.
4. The right side of the heart pumps blood to the _____ .
5. The _____ node is known as the pacemaker.
6. The blood vessels that serve the heart are the _____ arteries and veins.
7. The pressure of blood against the walls of a vessel is termed _____ .
8. Blood moves in arteries due to _____ and in veins due to _____ .
9. Reducing the amount of _____ and _____ in the diet reduces the chances of plaque buildup in arteries.
10. Varicose veins develop when _____ become weak and ineffective.

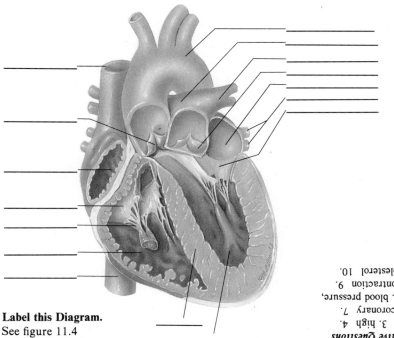

Label this Diagram.
See figure 11.4 (p. 222) in text.

Selected Key Terms

aorta (a-or′tah) major systemic artery that receives blood from the left ventricle. *228*

arteriole (ar-te′re-ōl) vessel that takes blood from arteries to capillaries. *219*

artery (ar′ter-ē) vessel that takes blood away from the heart; characteristically possessing thick elastic and muscular walls. *219*

atrium (a′tre-um) chamber; particularly an upper chamber of the heart that lies above the ventricles. *222*

AV node (a-ve nōd) a small region of neuromuscular tissue that transmits impulses received from the SA node to the ventricular walls. *225*

capillary (kap′ĭ-lar″ē) microscopic vessel connecting arterioles to venules through the thin walls of which molecules either exit or enter the blood. *219*

coronary artery (kor′ŏ-na-re ar′ter-ē) artery that supplies blood to the wall of the heart. *228*

diastole (di-as′to-le) relaxation of heart chamber. *223*

pulmonary circuit (pul′mo-ner″e ser′kit) that part of the circulatory system that takes deoxygenated blood to and oxygenated blood away from the lungs. *226*

SA node (es a nōd) small region of neuromuscular tissue that initiates the heartbeat. Also called the pacemaker. *225*

systemic circuit (sis-tem′ik ser′kit) that part of the circulatory system that serves body parts other than the gas-exchanging surfaces in the lungs. *226*

systole (sis′to-le) contraction of a heart chamber. *223*

valve (valv) membranous extension of a vessel or the heart wall that opens and closes, ensuring one-way flow. *221*

vein (vān) vessel that takes blood to the heart; characteristically having nonelastic walls. *219*

vena cava (ve′nah ka′vah) large systemic vein that returns blood to the right atrium of the heart. *223*

ventricle (ven′tri-k′l) cavity in an organ, such as a lower chamber of the heart. *222*

venule (ven′ūl) vessel that takes blood from capillaries to veins. *219*

Chapter 12

Blood

Chapter Concepts

1 Blood, which is composed of cells and a fluid containing many inorganic and organic molecules, has 3 primary functions: transport, clotting, and infection fighting.

2 Exchange of molecules between blood and tissue fluid takes place across capillary walls.

3 Blood is typed according to the antigens present on the red blood cells.

4 All of the functions of blood can be correlated with the ability of the body to maintain a constant environment.

Chapter Outline

The Transport Function of Blood
    Plasma
    Red Blood Cells (Erythrocytes)
    Capillary Exchange within the
      Tissues
The Blood-Clotting Function of Blood
The Infection-Fighting Function of
    Blood
    White Blood Cells (Leukocytes)
    Antibodies
    The Inflammatory Reaction
Blood Typing
    The ABO System
    The Rh System

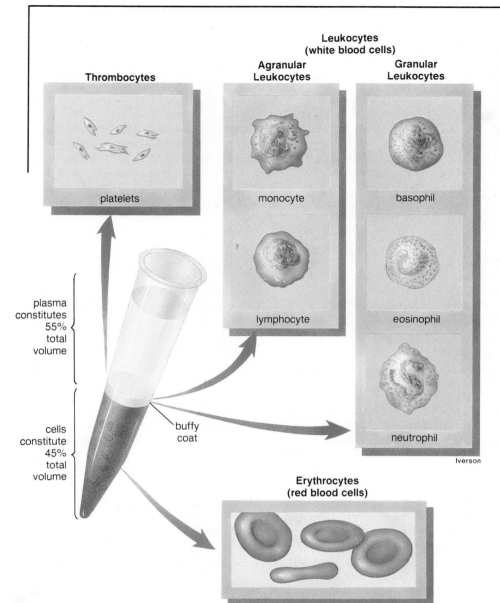

Thrombocytes

platelets

**Leukocytes
(white blood cells)**

**Agranular
Leukocytes**

monocyte

lymphocyte

**Granular
Leukocytes**

basophil

eosinophil

neutrophil

Iverson

plasma
constitutes
55%
total
volume

cells
constitute
45%
total
volume

buffy
coat

**Erythrocytes
(red blood cells)**

Figure 12.1

Composition of blood. When blood is transferred to a test tube and is centrifuged, it forms 2 layers. The transparent yellow top layer is plasma, the liquid portion of blood. The blood cells are in the bottom layer: the white blood cells (leukocytes) and the platelets (thrombocytes) are in a narrow buff-colored band above the red blood cells (erythrocytes). The white blood cells are either granular leukocytes or agranular leukocytes.

 It is a curious fact that more than half the body is water; the total quantity of water is around 70% of body weight. By far, most of this water is found within the cells. A much smaller amount lies outside the cells. The water outside the cells is found (1) in *tissue fluid* surrounding the cells, (2) in *lymph* contained with lymphatic vessels, and (3) in *blood* vessels.

Blood is a liquid tissue. Its functions are absolutely necessary to the continued existence of the organism. The functions of blood include the following:

1. transporting gases, nutrients, and wastes about the body
2. clotting to prevent loss of blood by bleeding
3. fighting infection or invasion of the body by microorganisms

If blood is transferred from a person's vein to a test tube and is prevented from clotting, it separates into 2 layers (fig. 12.1). The lower layer consists of red blood cells (erythrocytes), white blood cells (leukocytes), and blood platelets (thrombocytes). Collectively, these are called the **formed elements.** Formed elements make up about 45% of the total volume of whole blood. The upper layer, called plasma, contains a variety of inorganic and organic substances dissolved or suspended in water. Plasma accounts for about 55% of the total

Blood

Table 12.1 *Components of Blood*

Blood	Function	Source
Formed elements		
Red blood cells	Transport oxygen (O_2)	Bone marrow
Platelets	Clotting	Bone marrow
White blood cells	Fight infection	Bone marrow and lymphoid tissue
Plasma*		
Water	Maintain blood volume; transport molecules	Absorbed from intestine
Plasma proteins	Maintain blood osmotic pressure and pH	
Albumin	Transport	Liver
Fibrinogen	Clotting	Liver
Globulins		
Alpha and beta	Transport	Liver
Gamma	Fight infection	Lymphocytes
Gases		
Oxygen (O_2)	Cellular respiration	Lungs
Carbon dioxide (CO_2)	End product of metabolism	Tissues
Nutrients		
Fats, glucose, amino acids, etc.	Food for cells	Absorbed from intestinal villi
Salts	Maintain blood osmotic pressure and pH; aid metabolism	Absorbed from intestinal villi
Wastes		
Urea and ammonia (NH_3)	End products of metabolism	Tissues
Hormones, vitamins, etc.	Aid metabolism	Varied

*Plasma is 90–92% water, 7–8% plasma proteins, and not quite 1% salts; all other components are present in even smaller amounts.

volume of whole blood. Table 12.1 lists the components of blood, which are discussed in terms of the 3 functions mentioned earlier. All 3 of these functions can be related to blood's primary function of maintaining a constant internal environment, or *homeostasis*.

Blood is a liquid tissue. The liquid portion is termed plasma, and the solid portions are the formed elements.

The Transport Function of Blood

The transport function of blood helps to maintain the constancy of tissue fluid. Blood transports oxygen (O_2) from the lungs and the nutrients from the intestine to the capillaries of the body, where they enter tissue fluid. Here, blood also takes up carbon dioxide (CO_2) and any other waste molecules given off by the cells and transports them away. Carbon dioxide exits blood at the lungs and urea, a substance formed by the liver following deamination of amino acids, travels by way of the bloodstream to the kidneys and is excreted. Figure 12.2 diagrams the transport function of blood, indicating the manner in which this function helps to keep the internal environment relatively constant.

Homeostasis is only possible because blood brings nutrients to the cells and removes their wastes.

Plasma

Plasma (table 12.1) is the liquid portion of blood. Small organic molecules, such as glucose and urea, simply dissolve in plasma, but large organic molecules combine with proteins for transport.

Plasma Proteins

Plasma proteins make up 8–9% of plasma. These molecules assist in transporting large organic molecules in blood. For example, the molecule bilirubin,

12.1 Critical Thinking

1. The liver has sinusoids, special blood-filled regions where proteins can pass from liver cell to blood. Why would this be expected since the liver makes blood proteins?

2. In kidney disease, proteins are lost from blood by excretion. What do you predict happens to blood volume under these circumstances? Why?

3. Proteins help to buffer blood (the amino group accepts hydrogen ions and the carboxyl group gives them up). What would happen to the pH of blood if blood proteins were lost by excretion?

4. Cells do not get their proteins from blood. How do they get them?

a breakdown product of hemoglobin, is transported by **albumin,** and the **alpha** and **beta globulins**[1] transport hormones and fat-soluble molecules. The lipoproteins that transport cholesterol are globulins.

The movement of blood in the arteries is dependent upon blood pressure, but blood needs a certain *viscosity,* or thickness, in order to exert pressure. This property of blood is largely dependent on plasma proteins and red blood cells. Maintenance of blood viscosity is, therefore, another way that proteins contribute to the transport of molecules.

Blood also needs a certain *volume* in order to exert a pressure. Because plasma proteins are too large to pass through a capillary wall, the fluid within capillaries is always an area of lesser concentration of water compared to tissue fluid, and water, therefore, passes into capillaries. Maintenance of blood volume is associated particularly with albumin, the smallest and most plentiful plasma protein (table 12.2).

Certain plasma proteins have specific functions that are not duplicated by any other proteins. We will discuss how fibrinogen is necessary to blood clotting. We also will discuss the gamma globulins, which are antibodies that help to fight infection.

Plasma proteins assist in the transport function of blood. They serve as carriers for some molecules, and they help to maintain the viscosity and the volume of blood.

Table 12.2 Plasma Proteins

Name	Function
Albumin	Maintains blood volume and transports bilirubin
Globulins Alpha and beta	Transport hormones and fat-soluble molecules
Gamma	Fight infection
Fibrinogen	Causes blood clotting

Red Blood Cells (Erythrocytes)

Red blood cells also are involved in the transport function of blood. They contain the respiratory pigment **hemoglobin,** which carries oxygen (O_2) (fig. 12.3). Since hemoglobin is a red pigment, the cells appear red, and their color also makes the blood red. There are between 4 and 6 million red blood cells per mm^3 of whole blood, and each of these cells contains about 200 million hemoglobin molecules. If this much hemoglobin were suspended within the plasma rather than enclosed within the cells, blood would be so thick the heart would have difficulty pumping it.

Each hemoglobin molecule (fig. 12.3c) contains 4 polypeptide chains that make up the protein **globin,** and each chain is associated with **heme,** a complex iron-containing group. Iron combines loosely with oxygen, and in this way, oxygen is carried in blood.

Humans are active, warm-blooded animals; the brain and the muscles often require much oxygen within a short period of time. Plasma carries only about 0.3 ml of oxygen per 100 ml, but whole blood carries 20 ml of oxygen

---

[1]When globulins undergo electrophoresis (are put in an electrical field), they separate into major components called alpha globulin, beta globulin, and gamma globulin. Almost all circulating antibodies are found in the gamma globulin fraction.

Figure 12.2

Diagram illustrating the transport function of blood. Oxygen (O_2) is transported from the lungs to the tissues, and carbon dioxide (CO_2) is transported from the tissues to the lungs. Urea, a nitrogenous end product produced by the liver, is excreted by the kidneys. Glucose ($C_6H_{12}O_6$) is absorbed by the gut and may be stored temporarily in the liver as glycogen before it is transported to the tissues.

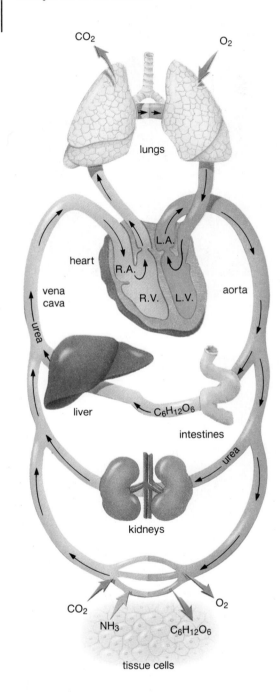

Figure 12.3

Physiology of red blood cells. a. *Red blood cells move single file through the capillaries.* b. *Each red blood cell is a biconcave disk containing many molecules of hemoglobin, the respiratory pigment.* c. *Hemoglobin contains 4 polypeptide chains, 2 of* which are alpha (α) chains and 2 of which are beta (β) chains. The plane in the center of each chain represents an iron-containing heme group. Oxygen combines loosely with iron when hemoglobin is oxygenated.

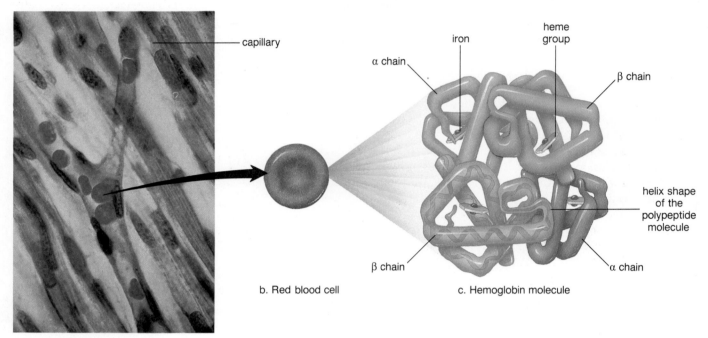

a. Blood capillary

b. Red blood cell

c. Hemoglobin molecule

per 100 ml. This shows that hemoglobin increases the carrying capacity of blood more than 60 times. Although the iron portion of hemoglobin carries oxygen, the equation for oxygenation of hemoglobin is usually written as

$$Hb + O_2 \underset{\text{tissues}}{\overset{\text{lungs}}{\rightleftharpoons}} HbO_2$$

The hemoglobin on the right, which is combined with oxygen, is called *oxyhemoglobin*. Oxyhemoglobin forms in the lungs and has a bright red color. The hemoglobin on the left, which has given up oxygen to tissue fluid, has a dark purple color.

Hemoglobin is an excellent oxygen carrier because it combines loosely with oxygen in the cool, neutral conditions of the lungs. It then readily gives up oxygen under the warm and more acidic conditions of the tissues.

Carbon monoxide (CO)[2], present in automobile exhaust, combines with hemoglobin more readily than oxygen, and it stays combined for several hours, regardless of the environmental conditions. Accidental death or suicide from carbon monoxide poisoning occurs because the hemoglobin of blood is not available for oxygen transport. This transport function of blood is so important that life can be sustained temporarily by giving a patient a hemoglobin substitute transfusion when whole blood is not available or cannot be given. The reading on page 243 discusses the possible benefits of this "artificial blood."

Oxygen is transported to the tissues in combination with hemoglobin, a pigment found in red blood cells.

[2]Carbon monoxide is an unusual molecule because it has this electron dot formula:

$$:C::O:$$

There are risks to having a blood transfusion, such as an immune reaction and acquiring an infection. A cross-matching test between the donor's blood and the recipient's blood usually detects if the recipient has antibodies in the plasma that will react against antigens on the membrane of the donor's red blood cells and vice versa. Blood also is screened for the presence of 2 viruses that are especially troublesome. They are the hepatitis B virus and the AIDS virus. Blood donors are questioned carefully, and their blood is

Artificial Blood

tested for the presence of these viruses. Despite the care that is taken to avoid immune reactions and the transference of disease, it would be advantageous to develop an artificial blood that has neither of these risks.

Investigation is proceeding in 3 directions. Enrico Bucci, a blood-substitute specialist at the University of Maryland School of Medicine, is working with modifying hemoglobin itself. He links several hemoglobin molecules so that they do not become "lost" from the circulatory system and uses chemicals to modify the complexes' oxygen affinity so that they are more likely to give up the oxygen when needed. Many other problems remain, however. The hemoglobin is taken from whole blood, and it alone may contain infectious material. Modified hemoglobin also seems to cause a generalized constriction of the body's blood vessels, making oxygenation of tissues more difficult.

Anthony Hunt and colleagues at the University of California at San Francisco are working with artificial red blood cells called neohemocytes (NHCs). To make the cells, purified human hemoglobin taken from outdated donor blood is encapsulated in a lipid bilayer membrane. The artificial cells are much smaller than normal human red blood cells, and they do not contain as much hemoglobin. However, when tested in rats, the animals survived until they were sacrificed for gross toxicity studies. The investigators believe that the tests are successful enough to warrant further study. They want to improve the stability and the vascular retention time of the cells since they are removed from the bloodstream and are broken down at a faster rate than normal cells.

George Groveman is director of new-products marketing at Alpha Therapeutic, a subsidiary of a Japanese firm that is based in Los Angeles. His firm is working on a third possibility, a substance called perfluorocarbon oil (PFC) emulsion that can be transfused and can carry oxygen much like hemoglobin does. This substance has served as a blood substitute for humans in emergency situations in which only the oxygen-carrying function of blood was required. FDA approval has been sought by a Japanese corporation to market PFC under the trade name of Fluosol-DA, but thus far the FDA has denied permission on the grounds that the clinical trials were not successful enough. There is some hope, though, that PFC will be approved for localized use. For example, it may be helpful to administer Fluosol-DA when a person is suffering a heart attack or is undergoing thrombolytic therapy (p. 234).

Although researchers have been working for 20 years to produce a blood substitute for general use, the prospects are still in the future. Anthony Hunt says, "Physiological systems always turn out to be more complicated than we thought."

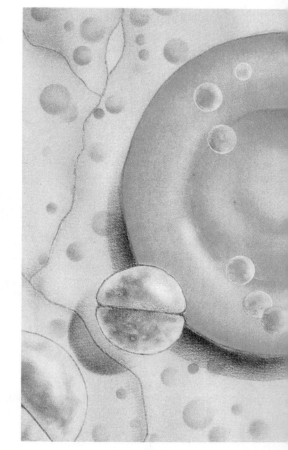

View inside a capillary in which blood has been replaced with a 25% suspension of hemoglobin-containing synthetic neohemocytes (NHCs). A normal red blood cell is shown for scale. Unlike living red blood cells, neohemocytes are nearly spherical and can have one, 2, or 3 chambers.

Red blood cells that are not engaged in oxygen transport assist in transporting carbon dioxide (CO_2). First, hemoglobin (Hb) combines with carbon dioxide to form *carbaminohemoglobin:*

$$Hb + CO_2 \underset{\text{lungs}}{\overset{\text{tissues}}{\rightleftharpoons}} HbCO_2$$

However, this combination with hemoglobin actually represents only a small portion of the carbon dioxide in the blood. Most of the carbon dioxide is transported as the *bicarbonate ion* (HCO_3^-). This ion forms after carbon dioxide

Table 12.3 Hemoglobin			
Components	**Structure**	**Function**	**Feature of Metabolism**
Heme	Nonprotein; contains iron	Carries O_2	Becomes bile pigments
Globin	Four polypeptide chains	Carries CO_2 Takes up H^+	————

Figure 12.4

Blood cell formation in red bone marrow. Multipotent stem cells give rise to specialized stem cells. The myeloid stem cell gives rise to still other cells that become red blood cells, platelets, and all the white blood cells except lymphocytes. The lymphoid stem cell gives rise to the lymphocytes.

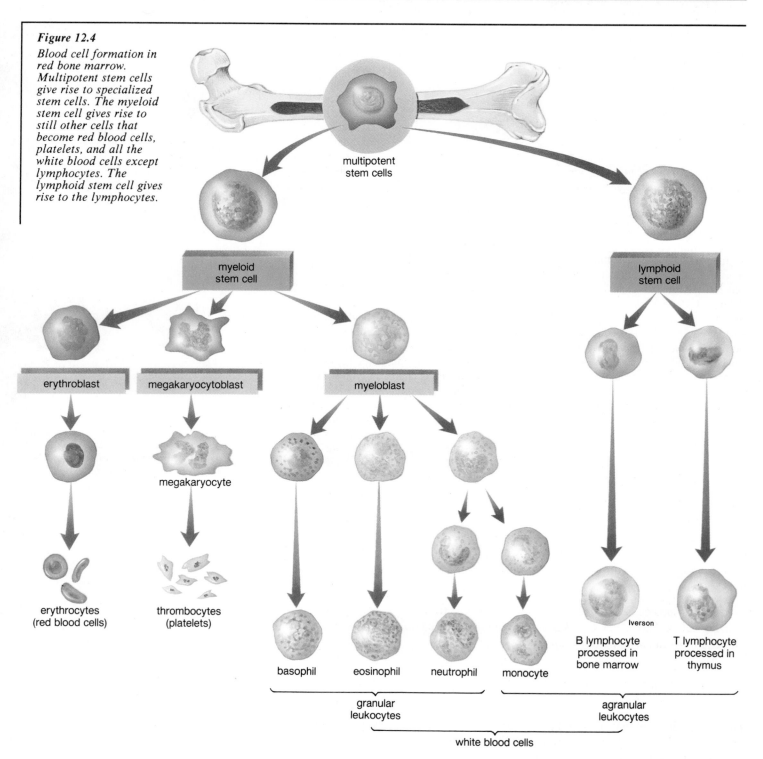

multipotent stem cells

myeloid stem cell

lymphoid stem cell

erythroblast

megakaryocytoblast

myeloblast

megakaryocyte

erythrocytes (red blood cells)

thrombocytes (platelets)

basophil

eosinophil

neutrophil

monocyte

Iverson

B lymphocyte processed in bone marrow

T lymphocyte processed in thymus

granular leukocytes

agranular leukocytes

white blood cells

Human Anatomy and Physiology

combines with water. Carbon dioxide combined with water forms carbonic acid (H_2CO_3). This dissociates (breaks down) to a hydrogen (H^+) ion and a bicarbonate ion:

$$CO_2 + H_2O \underset{lungs}{\overset{tissues}{\rightleftharpoons}} H_2CO_3 \underset{lungs}{\overset{tissues}{\rightleftharpoons}} H^+ + HCO_3^-$$

An enzyme within red blood cells, called *carbonic anhydrase,* speeds up this reaction. Most of the released hydrogen ions, are absorbed by the globin portions of hemoglobin, and the bicarbonate ions diffuse out of red blood cells to be carried in the plasma. Hemoglobin, which combines with a hydrogen ion, called reduced hemoglobin, can be symbolized as HHb. This combination of hydrogen and hemoglobin plays a vital role in maintaining the pH of blood.

Once systemic venous blood reaches the lungs, the reaction just described takes place in the reverse: the bicarbonate ion joins with a hydrogen ion to form carbonic acid, and this splits into carbon dioxide and water. The carbon dioxide diffuses out of blood into the lungs for expiration. Now, hemoglobin is ready again to transport oxygen. Table 12.3 summarizes the structure and function of hemoglobin.

Hemoglobin participates in the transport of carbon dioxide in blood and helps to buffer the blood.

The Life Cycle of Red Blood Cells

Red blood cells are manufactured continuously in the bone marrow of the skull, the ribs, the vertebrae, and the ends of the long bones. The number produced increases whenever arterial blood carries a reduced amount of oxygen (O_2), as happens when an individual first takes up residence at a high altitude. Under these circumstances, the kidneys (and probably other organs as well) produce a hormone called erythropoietin that stimulates cell division and differentiation of erythrocyte stem cells in the bone marrow (fig. 12.4).

Before they are released from the bone marrow into blood, red blood cells lose their nucleus and acquire hemoglobin (fig. 12.5). Possibly because they lack a nucleus, red blood cells only live about 120 days. They are destroyed chiefly in the *liver* and the *spleen,* where they are engulfed by large phagocytic cells. When red blood cells are broken down, the hemoglobin is released. The iron is recovered and is returned to the bone marrow for reuse. The heme portion of the molecule undergoes chemical degradation and is excreted by the liver in the bile as bile pigments. These bile pigments contribute to the color of feces.

Anemia

When there is an insufficient number of red blood cells or the cells do not have enough hemoglobin, the individual suffers from **anemia** and has a tired, run-down feeling. In iron-deficiency anemia, the hemoglobin blood level is low. It may be that the diet does not contain enough iron. Certain foods, such as raisins and liver, are rich in iron, and the inclusion of these in the diet can help to prevent this type of anemia.

In another type of anemia, called pernicious anemia, the digestive tract is unable to absorb enough vitamin B_{12}. This vitamin is essential to the proper formation of red blood cells; without it, immature red blood cells tend to accumulate in the bone marrow in large quantities. A special diet and administration of vitamin B_{12} by injection is an effective treatment for pernicious anemia.

Illness (anemia) results when the blood has too few red blood cells and/or not enough hemoglobin.

Figure 12.5

Maturation of red blood cells (erythrocytes). Red blood cells are made in the bone marrow, where stem cells continuously divide. During the maturational process, an erythrocyte loses its nucleus and gets much smaller.

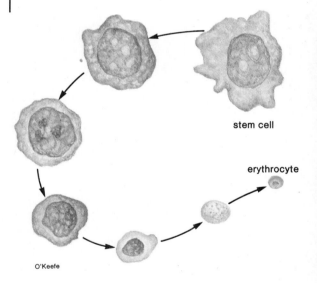

stem cell

erythrocyte

O'Keefe

Figure 12.6

Diagram of a capillary illustrating the exchanges that take place and the forces that aid the process. At the arterial end of a capillary, the blood pressure is higher than the osmotic pressure, and therefore water (H_2O), oxygen (O_2), amino acids and glucose ($C_6H_{12}O_6$) tend to leave the bloodstream. At the venous end of a
capillary, the osmotic pressure is higher than the blood pressure and therefore water (H_2O), carbon dioxide (CO_2), and other waste molecules tend to enter the bloodstream. Notice that the red blood cells and the plasma proteins are too large to exit from a capillary.

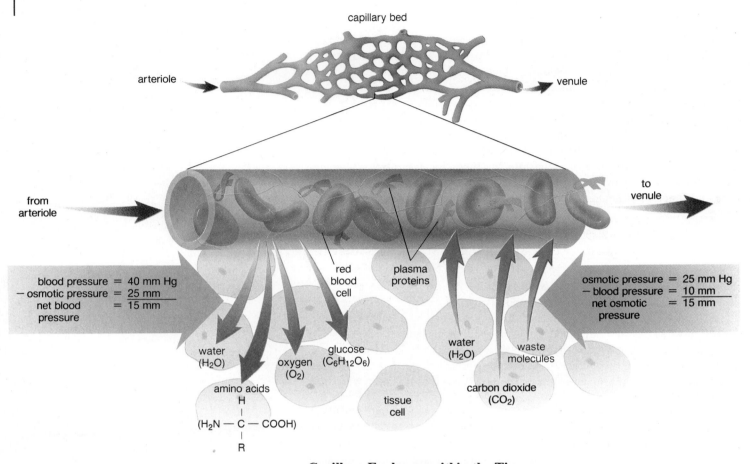

capillary bed

arteriole

venule

from arteriole

to venule

blood pressure = 40 mm Hg
− osmotic pressure = 25 mm
net blood = 15 mm
pressure

osmotic pressure = 25 mm Hg
− blood pressure = 10 mm
net osmotic = 15 mm
pressure

red blood cell

plasma proteins

water (H_2O)

oxygen (O_2)

glucose ($C_6H_{12}O_6$)

amino acids

H
|
(H_2N — C — COOH)
|
R

tissue cell

water (H_2O)

waste molecules

carbon dioxide (CO_2)

Capillary Exchange within the Tissues

The Arterial End of the Capillary

When arterial blood enters the tissue capillaries (fig. 12.6), it is bright red because red blood cells are carrying oxygen. It is also rich in nutrients that are dissolved in the plasma. At the arterial end of the capillary, blood pressure (40 mm Hg) is higher than the osmotic pressure of the blood (15 mm Hg). Blood pressure, you recall, is created by the pumping of the heart; the osmotic pressure is caused by the presence of salts and in particular by the plasma proteins that are too large to pass through the wall of the capillary. Since the blood pressure is higher than the osmotic pressure, fluid together with oxygen and nutrients (glucose and amino acids), exit from the capillary. This is a *filtration* process because large substances, such as red blood cells and plasma proteins, remain, but small substances, such as water and nutrient molecules, leave the capillaries. Tissue fluid, created by this process, consists of all the components of plasma except the proteins.

The Midsection

Along the length of the capillary, molecules follow their concentration gradient as diffusion occurs. Diffusion, you recall, is the movement of molecules from an area of greater concentration to an area of lesser concentration. The area of greater concentration for nutrients is always blood because after these molecules have passed into the tissue fluid, they are taken up and metabolized

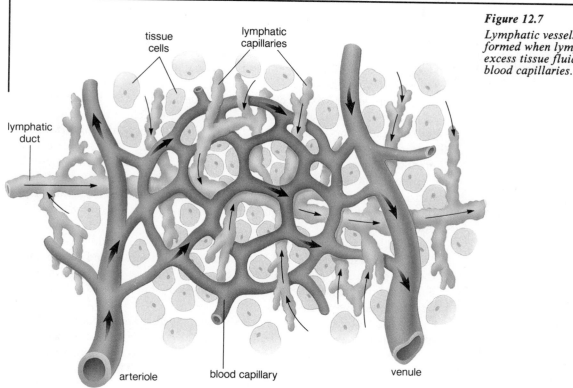

tissue cells

lymphatic capillaries

lymphatic duct

arteriole

blood capillary

venule

Figure 12.7
Lymphatic vessels. Arrows indicate that lymph is formed when lymphatic capillaries take up excess tissue fluid. Lymphatic capillaries lie near blood capillaries.

by the tissue cells. The cells use glucose ($C_6H_{12}O_6$) and oxygen (O_2) in the process of cellular respiration, and they use amino acids for protein synthesis. Following cellular respiration, the cells give off carbon dioxide (CO_2) and water (H_2O). Whenever the cells break down amino acids, they remove the amino group, which is released as ammonia (NH_3). Carbon dioxide and ammonia, waste products of metabolism, leave the cell by diffusion. Since tissue fluid is always the area of greater concentration for these waste materials, they diffuse into the capillary.

Oxygen and nutrient molecules (e.g., glucose and amino acids) exit a capillary near the arterial end; waste molecules (e.g., carbon dioxide and ammonia) enter a capillary near the venous end.

The Venous End of the Capillary

At the venous end of the capillary, blood pressure is much reduced (10 mm Hg), as can be verified by reviewing figure 11.13 (p. 231). However, there is no reduction in osmotic pressure (25 mm Hg), which tends to pull fluid into the capillary. As water enters a capillary, it brings with it carbon dioxide and other additional waste molecules. Blood that leaves the capillaries is deep purple in color because red blood cells contain reduced hemoglobin. Carbon dioxide (CO_2) becomes the bicarbonate ion (HCO_3^-), and this molecule is dissolved in the plasma.

Retrieving fluid by means of osmotic pressure is not completely effective. There is always some fluid that is not picked up at the venous end. This excess tissue fluid enters the lymphatic capillaries (fig. 12.7). Lymph is tissue fluid contained within lymphatic vessels. Lymph is returned to the systemic venous blood when the major lymphatic vessels enter the subclavian veins (p. 258).

Lymphatic capillaries lie in close proximity to blood capillaries, where they collect excess tissue fluid.

Figure 12.8

Serum formation. When blood clots, serum is squeezed out as a solid plug is formed. In a blood vessel, this plug can help to prevent further blood loss.

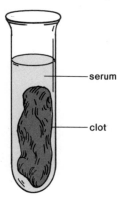

serum

clot

Table 12.4 Body Fluids

Name	Composition
Blood	Formed elements and plasma
Plasma	Liquid portion of blood
Serum	Plasma minus fibrinogen
Tissue fluid	Plasma minus proteins
Lymph	Tissue fluid within lymphatic vessels

The Blood-Clotting Function of Blood

When an injury to a blood vessel occurs, **clotting,** or coagulation, of blood takes place. This is obviously a protective mechanism to prevent excessive blood loss. As such, blood clotting is another mechanism by which blood components maintain homeostasis.

There are at least 12 clotting factors in the blood that participate in the formation of a blood clot. We will discuss the roles played by platelets, prothrombin, and fibrinogen. **Platelets** result from fragmentation of certain large cells, called megakaryocytes, in the bone marrow (fig. 12.4). These cells are produced at a rate of 200 billion a day, and the bloodstream carries more than a trillion. **Fibrinogen** and **prothrombin** are proteins manufactured and deposited in blood by the liver. Vitamin K is necessary to the production of prothrombin, and if by chance this vitamin is missing from the diet, hemorrhagic disorders develop.

If blood is allowed to clot in a test tube, a yellowish fluid develops above the clotted material (fig. 12.8). This fluid is called **serum,** and it contains all the components of plasma except fibrinogen. Because we now have used a number of different terms to refer to various body fluids related to blood, table 12.4 reviews these terms for you.

When a blood vessel in the body is damaged, platelets clump at the site of the puncture and partially seal the leak. They and the injured tissues release a clotting factor called prothrombin activator that converts prothrombin to thrombin. This reaction requires calcium (Ca^{++}) ions. **Thrombin,** in turn, acts as an enzyme that severs 2 short amino acid chains from each fibrinogen molecule. These activated fragments then join end to end, forming long threads of **fibrin.** Fibrin threads wind around the platelet plug in the damaged area of the blood vessel and provide the framework for the clot. Red blood cells also are trapped within the fibrin threads (fig. 12.9); these cells make a clot appear red.

The steps necessary for blood clotting upon injury are quite complex, but they can be summarized in this simplified manner:

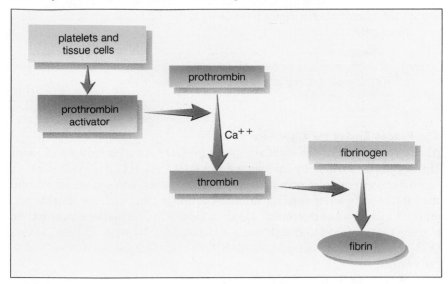

A blood clot consists of platelets and red blood cells entangled within fibrin threads.

A fibrin clot is only temporarily present. As soon as blood vessel repair is initiated, an enzyme called plasmin destroys the fibrin network and restores the fluidity of plasma. This is a protective measure because a blood clot can act as a thrombus or an embolus. In either case, it interferes with circulation and even can cause the death of tissues in the area.

Figure 12.9
Scanning electron micrograph showing a red blood cell (erythrocyte) caught in the fibrin threads of a clot. Fibrin threads form from activated fibrinogen, a normal component of blood plasma.

The Infection-Fighting Function of Blood

The body defends itself against microorganisms in various ways (see chapter 13). Two components of blood, white blood cells and antibodies, contribute to this defense. Their roles are explored briefly here.

Diseases and disorders that affect the quantity of white blood cells are particularly threatening. **Leukemia** is a form of cancer characterized by uncontrolled production of abnormal white blood cells. These cells accumulate in the bone marrow, the lymph nodes, the spleen, and the liver so that these organs are unable to function properly. Since the initiation of "total therapy," the combined use of chemotherapy with radiation therapy, a large number of leukemia patients are able to remain in remission for years.

White blood cells fight infection. They attack bacteria and viruses that have invaded the body.

White Blood Cells (Leukocytes)

White blood cells differ from red blood cells in that they are usually larger, have a nucleus, lack hemoglobin, and without staining, appear white in color. White blood cells are not as numerous as red blood cells, with only 7,000–8,000 cells per mm^3.

Table 12.5 lists the different types of white blood cells, and figure 12.10 shows 2 in detail. On the basis of structure, it is possible to divide white blood cells into the granular leukocytes and the agranular leukocytes. The **granular**

Table 12.5 White Blood Cells (Leukocytes)

Granular Leukocytes (Polymorphonuclear)			Function
	Size	Granules stain	
Neutrophils	9–12 μm	Lavender	Phagocytize primarily bacteria
Eosinophils	9–12 μm	Red	Phagocytize and destroy antigen-antibody complexes
Basophils	9–12 μm	Deep blue	Congregate in tissues; release histamine when stimulated
Agranular Leukocytes			
	Size	Type of nucleus	
Lymphocytes	8–10 μm	Indented	B type produce antibodies in blood and lymph
Monocytes	12–20 μm	Irregular	Become macrophages— phagocytize bacteria and viruses

leukocytes (neutrophils, eosinophils, and basophils) have granules in the cytoplasm that contain powerful digestive enzymes. Some have a many-lobed nucleus joined by nuclear threads; therefore, they also are called polymorphonuclear. Granular leukocytes are formed and mature in the bone marrow (fig. 12.4). The **agranular leukocytes** (monocytes and lymphocytes) do not have prominent granules in the cytoplasm and a nucleus that is mononuclear. Agranular leukocytes are also produced in the bone marrow but certain of the lymphocytes mature in the thymus. Agranular leukocytes are stored in the spleen, the lymph nodes, the tonsils, and other lymphoid organs.
and other lymphoid organs.

Infection fighting by white blood cells is dependent primarily on the neutrophils, which comprise 60–70% of all leukocytes, and the lymphocytes, which make up 25–30% of the leukocytes. Neutrophils are phagocytic; they destroy many bacteria by traveling to the site of invasion and engulfing the foe. Some lymphocytes produce immunoglobulins or **antibodies,** that combine with foreign substances to inactivate them. Neutrophils and lymphocytes can be compared in the following manner:

Neutrophils	*Lymphocytes*
Granules in cytoplasm	*No granules in cytoplasm*
Polymorphonuclear	*Mononuclear*
Produced in the bone marrow	*Produced in lymphoid tissue*
Phagocytic	*Make gamma globulins (antibodies)*

Ordinarily, the total number of white blood cells increases when there is infection. Sometimes, when only one type of white blood cell increases in number, it is possible to help to diagnose the illness by doing a differential white blood cell count, involving microscopic examination of a blood sample and counting the number of each type of white blood cell up to a total of 100 cells. For example, in *infectious mononucleosis,* the characteristic finding is a large number of lymphocytes of the B type that are atypical in appearance. This condition, caused by an Epstein-Barr viral infection, takes its name from the fact that lymphocytes are mononuclear. On the other hand, the *AIDS* (acquired immune deficiency syndrome) virus attacks the T type of lymphocyte,

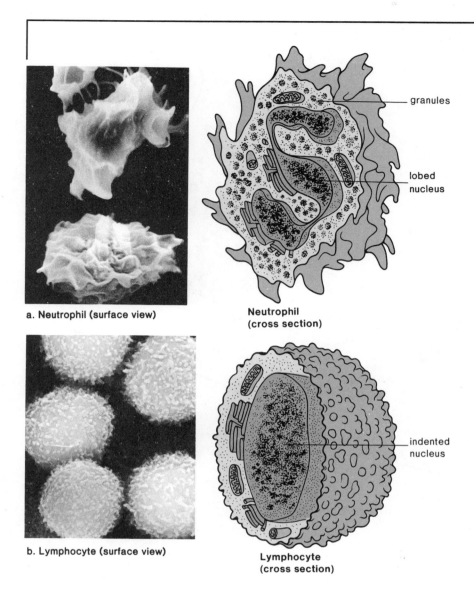

a. Neutrophil (surface view)

Neutrophil
(cross section)

granules

lobed
nucleus

b. Lymphocyte (surface view)

Lymphocyte
(cross section)

indented
nucleus

Figure 12.10
*Two leukocytes of interest. a. Scanning electron
micrograph and an artist's representation of a
neutrophil. b. Scanning electron micrograph and
an artist's representation of a lymphocyte.*

Figure 12.11
*Antigen-antibody reaction. The variable region of
antibodies (shown in purple) causes each type of
antibody to combine with only one type of
antigen. Quite often, the antigen-antibody
reaction produces an inactive complex.
Eosinophils are believed to phagocytize these
complexes.*

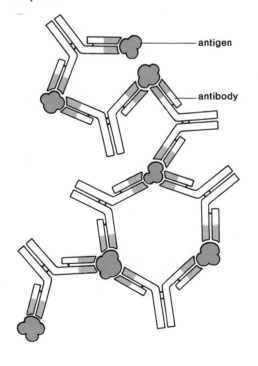

antigen

antibody

and the patient eventually has a reduced number of this type of lymphocyte in his or her blood (p. 266). Infection is diagnosed by testing for the presence of antibodies against the virus in blood.

Antibodies

Bacteria and their toxins cause lymphocytes of the B type to produce antibodies because they contain antigens. **Antigens** are protein, sometimes polysaccharide, molecules that are foreign to the individual. When activated, each B lymphocyte produces only one type of antibody that is specific for one type of antigen. *Antibodies* combine with their antigen (fig. 12.11) in such a way that the antigen is rendered harmless. Sometimes antibodies cause agglutination (clumping) of the antigen or simply prepare it for phagocytosis. In any case, keep in mind that the antigen is the foreigner, and the antibody is the molecule prepared by the body.

The antigen-antibody reaction is a lock-and-key reaction in which the molecules fit together like a lock and key. Specificity is possible because each type of antibody has variable regions, unique sequences of amino acids that result in receptor sites capable of combining with one type of antigen. In other words, this particular sequence of amino acids shapes a site where the antibody fits a specific antigen.

Figure 12.12

Inflammatory reaction. When a blood vessel is injured, release of bradykinin stimulates the pain nerve ending and the mast cells that secrete histamine. Histamine dilates blood vessels. Neutrophils congregate at the injured site. These amoeboid cells squeeze through the capillary wall and begin to phagocytize bacteria, especially those that have been attacked by antibodies.

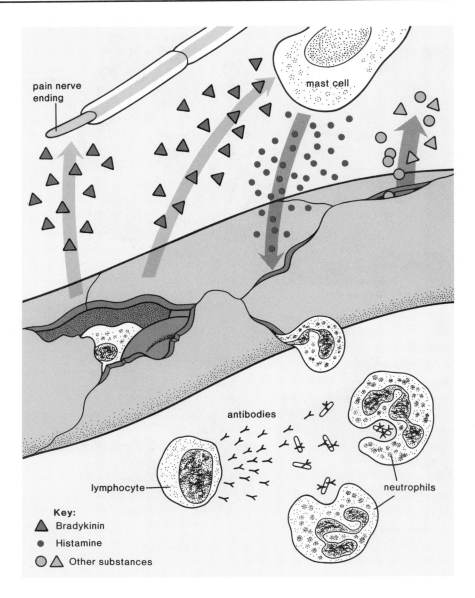

Immunity

An individual is actively immune when the body has antibodies that can react to a disease-causing antigen. Blood in these individuals contains lymphocytes that are capable of producing the necessary antibodies for a length of time. Exposure to the antigen, either naturally or by way of a vaccine, can cause active immunity to develop. Chapter 13 deals with immunity and explores the topic in detail.

Lymphocytes are responsible for immunity. B lymphocytes produce antibodies that specifically combine with disease-causing antigens.

The Inflammatory Reaction

Whenever the skin is broken due to a minor injury, a series of events occurs that is known as the **inflammatory reaction** because there is swelling and reddening at the site of the injury. Figure 12.12 illustrates the participants in the inflammatory reaction. One participant, the mast cells, possibly is derived from basophils, a type of white blood cell that takes up residence in the tissues.

When an injury occurs, a capillary and several tissue cells are apt to rupture and to release certain precursors that lead to the presence of **bradykinin,** a molecule that (1) initiates nerve impulses resulting in the sensation of

Human Anatomy and Physiology

pain, (2) stimulates mast cells to release **histamine,** another molecule that together with bradykinin (3) causes the capillary to dilate and to become more permeable. The enlarged capillary causes the skin to redden, and its increased permeability allows proteins and fluids to escape so that swelling results.

Any break in the skin allows bacteria and viruses to enter the body. In figure 12.12, a B lymphocyte is releasing antibodies that attack the bacteria, preparing them for **phagocytosis.** When a neutrophil phagocytizes a bacterium, an intracellular vacuole is formed. The engulfed bacterium is destroyed by hydrolytic enzymes when the vacuole combines with a granule.

Also present in tissues are monocyte-derived **macrophages,** large phagocytic cells that are able to devour a hundred invaders and still survive. Some tissues, particularly connective tissue, have resident macrophages that routinely act as scavengers, devouring old blood cells, bits of dead tissue, and other debris. Macrophages also are capable of bringing about an explosive increase in the number of leukocytes by liberating a substance that passes by way of the blood to the bone marrow, where it stimulates the production and the release of white blood cells, usually neutrophils.

As the infection is being overcome, some neutrophils die. These, along with dead tissue, cells, and bacteria and living white blood cells, form **pus,** a thick, yellowish fluid. Pus indicates that the body is trying to overcome the infection.

The inflammatory reaction is a "call to arms"—it marshalls phagocytic white blood cells to the site of invasion by bacteria.

Blood Typing

The ABO System

The cell membrane of red blood cells contains molecules (p. 65) that may differ from one individual to the next. When the blood of one individual is given to another, certain of these molecules can act as antigens in the recipient. These antigens are known as antigen A and antigen B, and blood is typed as described in table 12.6 according to whether or not these antigens are present

---

12.3 Critical Thinking

1. Type O blood used to be called the universal donor for blood transfusions. Why?
2. Type AB blood used to be called the universal recipient. Why?
3. Newborns already have a particular blood type. What does this tell you about blood type?
4. Other systems aside from the ABO system exist for typing blood. Why is this to be expected?

Table 12.6 Blood Groups

Type	Antigen on Red Blood Cells	Antibody in Plasma	U.S. Black (%)*	U.S. Caucasian (%)*
A	A	anti-B	25	41
B	B	anti-A	20	7
AB	A,B	none	4	2
O	none	anti-A, anti-B	51	50

*Blood type frequency for other races is not available.

on a person's red blood cells. For example, if a person has type A blood, the A antigen is on his or her red blood cells. This molecule is not an antigen to this individual, although it can be an antigen to a recipient of his or her blood.

In the simplified ABO system, there are 4 types of blood: A, B, AB, and O. Type O blood has neither the A antigen nor the B antigen on red blood cells; the other types of blood are designated by the antigen(s) present on red blood cells.

Within the plasma, there are antibodies to the antigens that are *not* present on the person's red blood cells. Therefore, for example, type A blood has an antibody called anti-B in the plasma. Type AB blood has neither anti-A nor anti-B antibodies because both antigens are on the red blood cells. This is reasonable because if the same antigen and antibody are present, **agglutination,** or clumping of red blood cells, occurs. Agglutination of red blood cells can cause blood to stop circulating in small blood vessels, and this leads to organ damage. It also is followed by hemolysis, which brings about the death of the individual.

For a recipient to receive blood from a donor, the recipient's plasma must not have an antibody that causes the donor's cells to agglutinate. For this reason, it is important to determine each person's blood type. Figure 12.13 demonstrates a way to use the antibodies derived from plasma to determine the blood type. If clumping occurs after a sample of blood is exposed to a particular antibody, the person has that type of blood.

The Rh System

Another important antigen in matching blood types is the **Rh factor.** Persons with this particular antigen on the red blood cells are Rh positive (Rh⁺); those without it are Rh negative (Rh⁻). Rh negative individuals normally do not have antibodies to the Rh factor, but they may make them when exposed to the Rh factor. It is possible to use anti-Rh antibodies for blood testing. When Rh positive blood is mixed with anti-Rh antibodies, agglutination occurs.

The designation of blood type usually also includes whether the person has the Rh factor (Rh positive) or does not have the Rh factor (Rh negative) on the red blood cells.

Figure 12.13

Blood typing. The standard test to determine ABO and Rh blood type consists of putting a drop of anti-A antibodies, anti-B antibodies, and anti-Rh antibodies on a slide. To each of these, a drop of the person's blood is added. a. If agglutination occurs, as seen in the lower photo, the person has this antigen on red blood cells. b. Several possible results.

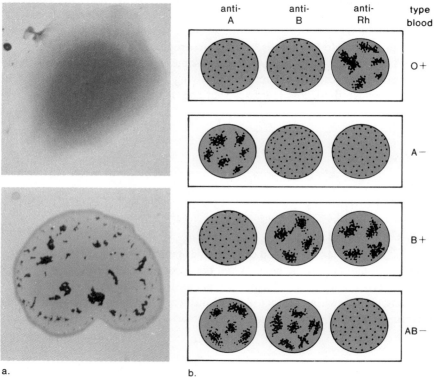

a.

b.

During pregnancy (fig. 12.14), if the mother is Rh negative and the father is Rh positive, the child may be Rh positive. The Rh positive red blood cells may begin leaking across the placenta into the mother's circulatory system, as placental tissues normally break down before and at birth. This causes the mother to produce anti-Rh antibodies. In this or a subsequent pregnancy with another Rh positive baby, anti-Rh antibodies (but not anti-A and anti-B antibodies discussed earlier) may cross the placenta and destroy the child's red blood cells. This is called hemolytic disease of the newborn (HDN).

The Rh problem has been solved by giving Rh negative women an Rh immunoglobulin injection either midway through the first pregnancy or no later than 72 hours after giving birth to any Rh positive child. This injection contains anti-Rh antibodies that attack any of the baby's red blood cells in the mother's blood before these cells can stimulate her immune system to produce her own antibodies. The injection is not beneficial if the woman has already begun to produce antibodies; therefore, the timing of the injection is most important.

The possibility of hemolytic disease of the newborn exists when the mother is Rh negative and the father is Rh positive.

Figure 12.14
Diagram describing the development of hemolytic disease of the newborn.

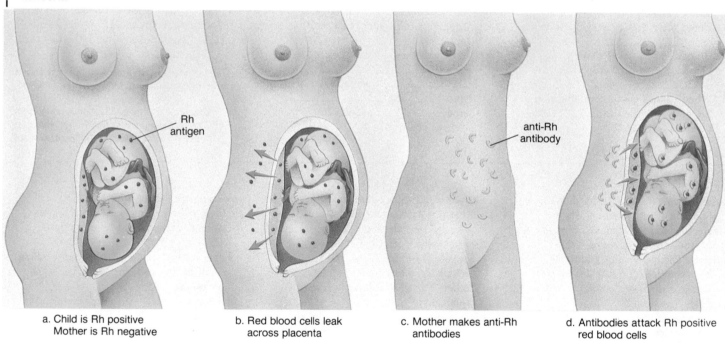

a. Child is Rh positive Mother is Rh negative

b. Red blood cells leak across placenta

c. Mother makes anti-Rh antibodies

d. Antibodies attack Rh positive red blood cells

Summary

Nutrients and wastes are transported in plasma, but oxygen is combined with hemoglobin within red blood cells. The end result of transport is capillary exchange with tissue fluid, regulated by blood pressure and osmotic pressure. Blood clotting requires a series of enzymatic reactions involving platelets, prothrombin, and fibrinogen. In the final reaction, fibrinogen becomes fibrin threads that entrap red blood cells.

White blood cells and gamma globulin proteins are required to fight infections. The 2 most prevalent white blood cells are the neutrophils and the lymphocytes. Neutrophils are involved in the inflammatory reaction, and lymphocytes are involved in immunity development.

Blood transfusions require compatible blood types. Of consideration are the antigens (A and B) on the red blood cells and the antibodies (anti-A and anti-B) in the plasma. Another important antigen is the Rh antigen, particularly because an Rh negative mother may produce anti-Rh antibodies that will attack the red blood cells of an Rh positive fetus.

Study Questions

1. Define blood, plasma, tissue fluid, lymph, and serum. (pp. 239, 248)
2. Name 3 functions of blood, and tell how they are related to the maintenance of homeostasis. (p. 240)
3. State the major components of plasma. Name the plasma proteins, and tell their common function as well as their specific functions. (p. 241)
4. Give the equation for the oxygenation of hemoglobin. Where does this reaction occur? Where does the reverse reaction occur? (p. 242)
5. Give an equation that indicates how carbon dioxide (CO_2) commonly is carried in blood. Indicate the direction of the reaction in the tissues and in the lungs. In what ways does hemoglobin aid the process of transporting carbon dioxide? (p. 245)
6. Discuss the life cycle of red blood cells. (p. 245)
7. What forces operate to facilitate exchange of molecules across the capillary wall? (pp. 246–47)
8. Name the steps that take place when blood clots. Which substances are present in blood at all times, and which appear during the clotting process? (p. 248)
9. Name and discuss 2 ways that blood fights infection. Associate each of these with a particular type of white blood cell. (pp. 249–52)
10. Describe the inflammatory reaction, and give a role for each type of cell and chemical that participates in the reaction. (pp. 252–53)
11. What are the 4 ABO blood types in humans? For each, state the antigen(s) on the red blood cells and the antibody(ies) in the plasma. (pp. 253–55)
12. Problems can arise during childbearing if the mother is which Rh type and the father is which Rh type? Explain why this is so. (p. 255)

Objective Questions

1. The liquid part of blood is called _____ .
2. Red blood cells carry _____ , and white blood cells _____
3. Hemoglobin that is carrying oxygen is called _____ .
4. Human red blood cells lack a _____ and only live about _____ days.
5. When a blood clot occurs, fibrinogen has been converted to _____ threads.
6. The most common granular leukocyte is the _____ , a phagocytic white blood cell.
7. Lymphocytes are made in _____ tissue and produce _____ that react with antigens.
8. At a capillary, _____ , _____ , and _____ leave the arterial end, and _____ and _____ enter the venous end.
9. Type AB blood has the antigens _____ and _____ on red blood cells and _____ antibodies in plasma.
10. Hemolytic disease of the newborn can occur when the mother is _____ and the father is _____ .

Answers to Objective Questions
1. plasma 2. oxygen, fight infection 3. oxyhemoglobin 4. nucleus, 120 5. fibrin 6. neutrophil 7. lymphoid, antibodies 8. oxygen, amino acids, glucose; carbon dioxide, other wastes 9. A, B, no 10. Rh−, Rh+

Selected Key Terms

agglutination (ah-gloo″tĭ-na′shun) clumping of cells, particularly in reference to red blood cells involved in an antigen-antibody reaction. *254*

agranular leukocyte (ah-gran′u-lar lu″ko-sīt) white blood cell that does not contain distinctive granules. *250*

antibody (an′tĭ-bod″e) a protein produced in response to the presence of some foreign substance in blood or the tissues. *250*

antigen (an′tĭ-jen) a foreign substance, usually a protein, that stimulates the immune system to produce antibodies. *251*

clotting (klot′ing) process of blood coagulation, usually when injury occurs. *248*

fibrinogen (fi-brin′o-jen) plasma protein that is converted into fibrin threads during blood clotting. *248*

formed element (form′d el′ĕ-ment) a constituent of blood that is either cellular (red blood cells and white blood cells) or at least cellular in origin (platelets). *239*

granular leukocyte (gran′u-lar lu″ko-sīt) white blood cell that contains distinctive granules. *249*

inflammatory reaction (in-flam′ah-to″re re-ak′shun) a tissue response to injury that is characterized by dilation of blood vessels and accumulation of fluid in the affected region. *252*

macrophage (mak′ro-fāj) a large cell derived from a monocyte that ingests foreign material and cellular debris. *253*

phagocytosis (fag″o-si-to′sis) the taking in of bacteria and/or debris by engulfing. *253*

platelet (plāt′let) a formed element that is necessary to blood clotting. *248*

prothrombin (pro-throm′bin) plasma protein that is converted to thrombin during the process of blood clotting. *248*

pus (pus) thick, yellowish fluid composed of dead phagocytes, dead tissue, and bacteria. *253*

serum (se′rum) light-yellow liquid left after clotting of blood. *248*

thrombin (throm′bin) an enzyme that converts fibrinogen to fibrin threads during blood clotting. *248*

Chapter 13

The Lymphatic System and Immunity

Chapter Concepts

1 The lymphatic vessels form a one-way system that transports lymph from the tissues and fat from the lacteals to certain cardiovascular veins.

2 Lymphocytes are produced and accumulate in the lymphoid organs.

3 The body has various general (nonspecific) ways to protect itself from disease.

4 Immunity is specific and requires 2 types of lymphocytes, B lymphocytes and T lymphocytes. Both of these are produced in the bone marrow.

5 While immunity preserves our existence, it also is responsible for certain undesirable effects, such as tissue rejection, allergies, and autoimmune diseases.

6 Immunotherapy involves the use of vaccines to achieve long-lasting immunity and the use of antibodies to provide temporary immunity.

Chapter Outline

The Lymphatic System
    Lymphatic Vessels
    Lymphoid Organs
Immunity
    General Defense
    Specific Defense
Immunotherapy
    Induced Immunity
    Lymphokines
    Monoclonal Antibodies
Immunological Side Effects and
    Illnesses
    Allergies
    Tissue Rejection
    Autoimmune Diseases

Figure 13.1

Lymphatic vessels. The lymphatic vessels drain excess fluid from the tissues and return it to the cardiovascular system. The thoracic duct and the right lymphatic duct are the major lymphatic vessels. The enlargement shows the lymphatic vessels, called lacteals, that are present in the intestinal villi.

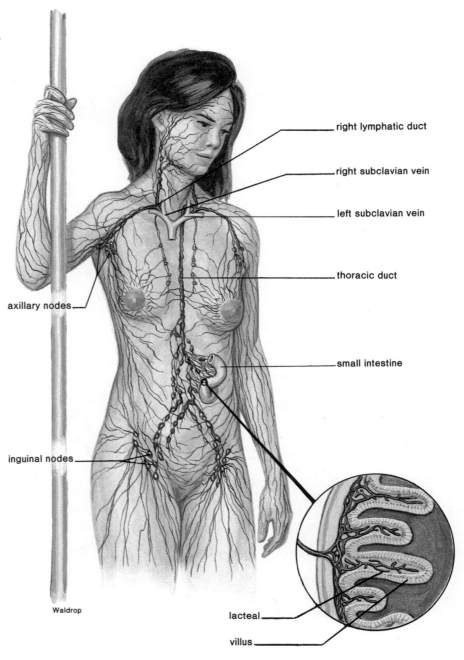

right lymphatic duct

right subclavian vein

left subclavian vein

thoracic duct

axillary nodes

small intestine

inguinal nodes

Waldrop

lacteal

villus

The Lymphatic System

The **lymphatic system** consists of lymphatic vessels (fig. 13.1) and the lymphoid organs. This system, which is closely associated with the cardiovascular system, has 3 main functions: (1) lymphatic vessels take up excess tissue fluid and return it to the bloodstream; (2) lymphatic capillaries absorb fats at the intestinal villi and transport them to the bloodstream (p. 195); and (3) the lymphatic system helps to defend the body against disease.

Lymphatic Vessels

Lymphatic vessels are quite extensive; every region of the body is supplied richly with lymphatic capillaries. The construction of the larger lymphatic vessels is similar to that of cardiovascular veins, including the presence of valves.

Also, the movement of lymph within these vessels is dependent upon skeletal muscle contraction. When the muscles contract, the lymph is squeezed past a valve that closes, preventing the lymph from flowing backwards.

The lymphatic system begins with lymphatic capillaries that lie near blood capillaries. These capillaries take up fluid that has diffused from and has not been reabsorbed by the blood capillaries (fig. 12.7). Once tissue fluid enters the lymphatic vessels, it is called **lymph.** The lymphatic capillaries join to form lymphatic vessels that merge before entering one of 2 ducts: the thoracic duct or the right lymphatic duct.

The *thoracic duct* is much larger than the right lymphatic duct. It serves the lower extremities, the abdomen, the left arm, and the left side of the head and the neck. In the thorax, the left thoracic duct enters the left subclavian vein. The *right lymphatic duct* serves only the right arm and the right side of the head and the neck. It enters the right subclavian vein.

The lymphatic system is a one-way system. Lymph flows from a capillary to ever-larger lymphatic vessels and finally to a lymphatic duct, which enters a subclavian vein.

Edema

Edema is localized swelling caused by the accumulation of tissue fluid. Tissue fluid accumulates if (1) too much of it is being made and/or (2) not enough of it is being drained away. There are several reasons either of these situations occurs. Too much tissue fluid is made whenever blood pressure is increased or osmotic pressure is decreased at a capillary (fig. 12.6). An obstruction, such as a thromboembolism, can prevent venous blood from leaving a capillary and increase blood pressure; an abnormally low number of plasma proteins can decrease osmotic pressure. The liver may not be making enough plasma proteins, or perhaps the plasma proteins are being excreted due to kidney disease. It is also possible that the capillaries are losing plasma proteins because they are "leaky" due to the inflammatory response (p. 252).

Pulmonary edema is a life-threatening condition that can complicate the recovery of a patient suffering from congestive heart failure. In congestive heart failure, the heart is not pumping and keeping the blood flowing adequately. Blood backs up in the pulmonary circuit, and the increase in blood pressure causes excess tissue fluid and an increase in interstitial pressure to the point that the walls of the air sacs in the lungs rupture. When fluid accumulates in the lungs, the patient can suffocate.

The malfunction of the lymphatic system causes tissue fluid to accumulate because it is not being drained away. During an operation for breast cancer, lymph nodes and lymphatic vessels sometimes are removed because they can be involved in the spread of cancerous cells. This procedure results in an inability of the body to collect tissue fluid; edema results. In the tropics, infection of lymphatic vessels by a parasitic worm can result in elephantiasis, a condition in which a limb swells and supposedly resembles the limb of an elephant (fig. 13.2).

Edema is characterized by localized swelling due to an abnormal accumulation of tissue fluid. Either too much tissue fluid is being made and/or not enough is being drained.

Lymphoid Organs

The lymphoid organs include the bone marrow, the lymph nodes, the spleen, and the thymus (fig. 13.3).

Bone Marrow

In the adult, bone marrow is present only in the bones of the skull, the sternum, the ribs, the clavical, the spinal column, and the ends of the femur and the

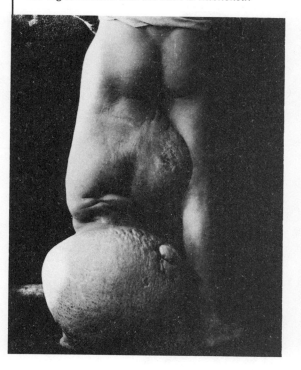

Figure 13.2
Elephantiasis. When the lymphatic vessels are blocked due to an infection by filarial worms, extreme edema results. Because tissue fluid is not being drained away by the lymphatic vessels, the leg is swollen and the skin is thickened.

13.1 Critical Thinking

1. Why would you expect edema when blood pressure rises, but not when it decreases?
2. Why would you expect edema when the osmotic pressure decreases, but not when it rises?
3. Suppose blood proteins leak into the tissues so that there is an equal amount on either side of the capillary wall. Under these circumstances, what happens to the osmotic pressure of the blood? What will happen to blood pressure? What will happen to tissue fluid formation?

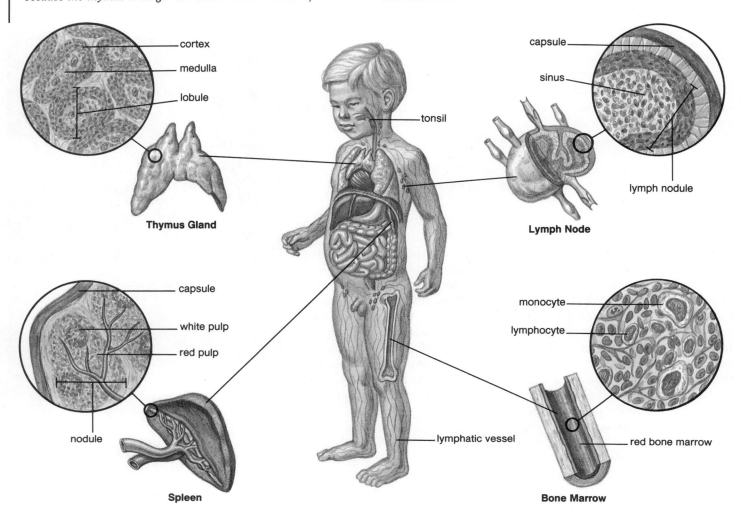

Figure 13.3

Lymphoid organs. The bone marrow is the site of lymphocyte and monocyte (macrophage) production and B cell maturation. The thymus is the site of T cell maturation. (A child is shown because the thymus is larger in children than in adults.) The

lymph node is the site of lymphocyte and macrophage accumulation (the tonsils are modified lymph nodes). The spleen is the site of lymphocyte, macrophage, and red blood cell accumulation.

cortex

medulla

lobule

Thymus Gland

tonsil

capsule

sinus

lymph nodule

Lymph Node

capsule

white pulp

red pulp

nodule

Spleen

monocyte

lymphocyte

red bone marrow

lymphatic vessel

Bone Marrow

humerus (p. 351). The bone marrow consists of a network of connective tissue fibers, called reticular fibers, that are produced by cells called reticular cells. These and the cells that are developing into the blood cells are packed about thin-walled venous sinuses. Differentiated blood cells enter the bloodstream at these sinuses.

Radioactive tracer studies have shown that the bone marrow is the site of origination for all the types of blood cells (fig. 12.4), including both the granular and agranular leukocytes. In other words, the bone marrow contains lymphoid tissue that produces lymphocytes. The B (for bone marrow) lymphocytes mature in the bone marrow, and the T (for thymus) lymphocytes mature in the thymus gland. The structure and the function of B and T lymphocytes are discussed in the following section.

The bone marrow also contains monocytes that have developed into resident macrophages. These large phagocytic cells help to cleanse the marrow and the adjacent blood sinuses.

Lymph Nodes
At certain points along lymphatic vessels, small (about 2.5 cm), ovoid or round structures called lymph nodes occur. A lymph node has a fibrous connective tissue capsule. Connective tissue also divides a node into nodules. Each nodule

contains a sinus (open space) filled with many lymphocytes and macrophages. As lymph passes through the sinuses, it is purified of infectious organisms and any other debris.

While nodules usually occur within lymph nodes, they also can occur singly or in groups. The *tonsils* are composed of partly encapsulated lymph nodules. There are also nodules called *Peyer's patches* within the intestinal wall.

The lymph nodes occur in groups in certain regions of the body. For example, the inguinal nodes are in the groin and the axillary nodes are in the arm pits.

Lymph nodes are divided into sinus-containing lobules, where the lymph is cleansed by phagocytes.

The Spleen

The spleen is located in the upper left abdominal cavity just beneath the diaphragm. The spleen is constructed the same as a lymph node. Outer connective tissue divides the organ into lobules that contain sinuses. In the spleen, however, the sinuses are filled with blood instead of lymph. Especially since the blood vessels of the spleen can expand, this organ serves as a blood reservoir and makes blood available in times of low pressure or when the body needs extra oxygen in the blood.

A spleen nodule contains red pulp and white pulp. Red pulp contains red blood cells, lymphocytes, and macrophages. The white pulp contains only lymphocytes and macrophages. Both types of pulp help to purify the blood that passes through the spleen. If the spleen ruptures due to injury, it can be removed. Although the functions of the spleen are duplicated by other organs, the individual is expected to be slightly more susceptible to infections and may have to take antibiotic therapy indefinitely.

The spleen is divided into sinus-containing lobules, where the blood is cleansed by phagocytes.

The Thymus

The thymus is located along the trachea behind the sternum in the upper thoracic cavity. This gland varies in size, but it is larger in children than in adults. The thymus also is divided into lobules by connective tissue. The T lymphocytes mature in these lobules. Those in the medulla are more mature than those in the cortex of the thymus.

The thymus secretes thymosin, a molecule that is believed to be an inducing factor; that is, it causes pre-T cells to become T cells. Thymosin also may have other functions in immunity.

The thymus is divided into lobules, where T lymphocytes mature.

Immunity

The body is prepared to protect itself from foreign substances and cells, including infectious microbes. The *first line of defense* is available immediately because it involves mechanisms that are nonspecific. The *second line of defense* takes a little longer to act because it is highly specific and contains mechanisms that are tailored to a particular threat.

General Defense

The environment contains many types of organisms that are able to invade and to infect the body. There are 3 general defense mechanisms that are useful against all types of organisms: barriers to entry, phagocytic white blood cells, and protective proteins.

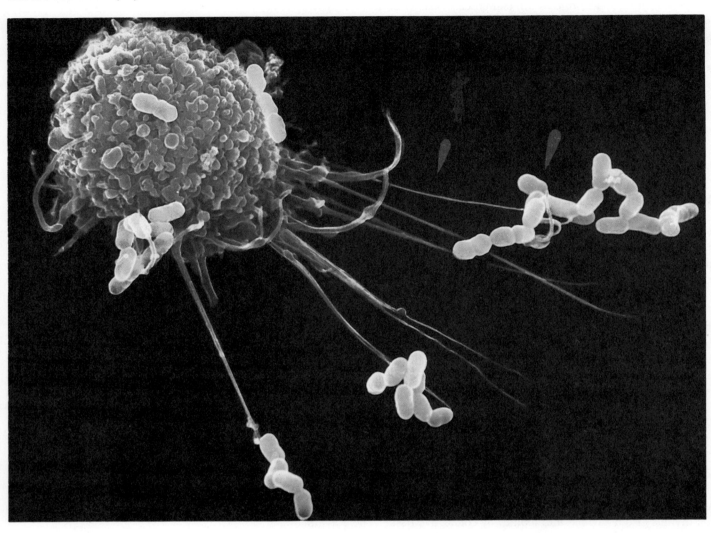

Barriers to Entry

The skin and the mucous membrane lining the respiratory and digestive tracts serve as mechanical barriers to entry by bacteria and viruses. The secretions of the oil glands in the skin contain chemicals that weaken or kill bacteria. The respiratory tract is lined by cells that sweep mucus and trapped particles up into the throat, where they can be swallowed. The stomach has an acidic pH that inhibits the growth of many types of bacteria. A mix of bacteria that normally reside in the intestine and other organs, such as the vagina, prevent pathogens from taking up residence.

Phagocytic White Blood Cells

If microbes do gain entry to the body, as described in the section on the inflammatory response on page 252, other nonspecific forces come into play. For example, neutrophils and monocyte-derived macrophages (fig. 13.4) are phagocytic white blood cells that engulf some bacteria upon contact. Infections may be accompanied by fever, which is a protective response because phagocytes function better at a higher-than-normal body temperature.

Figure 13.5

*Action of the complement system. The complement system is a
number of proteins always present in the plasma. When
activated, some of these form pores in bacterial cell wall
membranes, allowing fluids and salts to enter, until the cell
eventually bursts.*

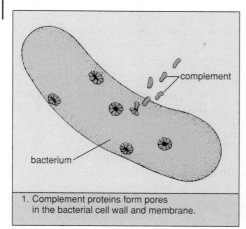

1. Complement proteins form pores
   in the bacterial cell wall and membrane.

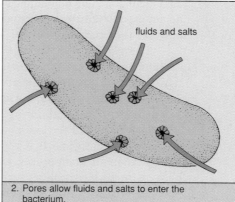

2. Pores allow fluids and salts to enter the
   bacterium.

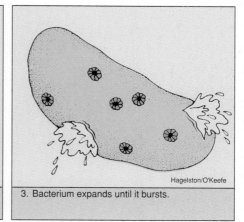

3. Bacterium expands until it bursts.

Protective Proteins

The **complement system** is a series of proteins produced by the liver that are
present in the plasma. When the first protein is activated, a cascade of reactions occurs. Every protein molecule in the series activates another in a predetermined sequence. In the end, certain proteins form pores in bacterial cell
walls and membranes. These pores allow fluids and salt to enter the bacterial
cell, until it bursts (fig. 13.5).

If complement is unable to destroy the microbes directly, some complement proteins still can coat them and other proteins attract phagocytes to the
scene. Although complement is a general defense mechanism, it also plays a
role in specific defense, as we will see later.

When most viruses infect a tissue cell, the infected cell produces and
secretes interferon. **Interferon** binds to receptors on noninfected cells, and this
action causes these cells to prepare for possible attack by producing substances
that interfere with viral replication.

A cell that has bound interferon is protected against any type of virus;
therefore, interferon should be very useful in preventing viral infection. However, interferon is specific to the species; only human interferon can be used
in humans. It used to be quite a problem to collect enough interferon for clinical and research purposes, but now interferon is made by recombinant DNA
technology.

---

*The first line of defense against disease is nonspecific. It consists of barriers to
entry, phagocytic white blood cells, and protective proteins.*

Specific Defense

Sometimes, we are threatened by an invasion by microorganisms that cannot
be counteracted successfully by the general defense mechanisms. In such cases,
it is necessary to develop immunity against a specific antigen. **Antigens** are
usually protein (or polysaccharide) molecules that specific lymphocytes recognize as foreign to the body. Antigens occur on bacteria and viruses, but they
also can be part of a foreign cell or a cancerous cell. Ordinarily, we do not
become immune to our body's own normal cells; therefore, it is said that the
immune system is able to distinguish self from nonself.

Immunity usually lasts for some time. For example, once we recover from the measles, we usually cannot be infected by the measles virus a second time. Immunity is primarily the result of the action of the B lymphocytes and the T lymphocytes, which have different functions. **B lymphocytes,** also called B cells, become plasma cells that produce **antibodies,** proteins that are capable of combining with and inactivating antigens. These antibodies are secreted into the blood and the lymph. In contrast, **T lymphocytes,** also called T cells, do not produce antibodies. Instead, certain T cells directly attack cells bearing antigens they recognize. Other T cells regulate the immune response.

Lymphocytes are capable of recognizing an antigen because they have receptor molecules on their surface. The shape of the receptors on any particular lymphocyte are complementary to a portion of one specific antigen. It is often said that the receptor and the antigen fit together like *a lock and a key.* It is estimated that during our lifetime, we encounter a million different antigens so we need the same number of different lymphocytes for protection against those antigens. It is remarkable that so much diversification occurs during the maturation process that in the end there is a different type of lymphocyte for each possible antigen. Despite this great diversity, none of the lymphocytes ordinarily attacks the body's own cells. It is believed that if by chance a lymphocyte arises that is equipped to respond to the body's own proteins, it normally is suppressed and develops no further.

There are 2 types of lymphocytes. B cells produce and secrete antibodies that combine with antigens. Certain T cells directly attack antigen-bearing cells, and others regulate the immune response.

The Action of B Cells

The receptor on a B cell is called a membrane-bound antibody because it is structured like an antibody. When a B cell encounters a bacterial cell or a toxin bearing an appropriate antigen, it is activated; that is, it has the potential to produce many **plasma cells** that will secrete antibodies against this antigen (fig. 13.6). (For the B cell to realize this potential, it must be stimulated by a helper T cell [p. 266]). All of the plasma cells derived from one parent lymphocyte are called a clone, and a clone produces the same type of antibody. Notice that a B cell does not clone until its antigen is present. The *clonal selection theory* states that the antigen selects which B cell will produce a clone of plasma cells.

Once antibody production is sufficient, the antigen disappears from the system, and the development of plasma cells ceases. However, some members of a clone do not participate in antibody production; instead, they remain in the bloodstream as **memory B cells.** Memory B cells are capable of producing the antibody specific to a particular antigen for some time. As long as these cells are present, the individual is said to be actively immune: future antibody production is possible because the memory cells can produce more plasma cells if the same antigen invades the system again.

Defense by B cells is called **antibody-mediated immunity** because B cells produce antibodies. It also is called humoral immunity because these antibodies are present in the bloodstream.

B cells are responsible for antibody-mediated immunity. After they recognize an antigen, they (if stimulated by a helper T cell) divide to produce both antibody-secreting plasma cells and memory B cells.

Antibodies   The most common type of antibody (IgG) is a Y-shaped protein molecule having 2 arms. Each arm has a long "heavy" chain and a short "light" chain of amino acids. These chains have *constant regions,* where the sequence of amino acids is set, and *variable regions,* where the sequence of amino acids

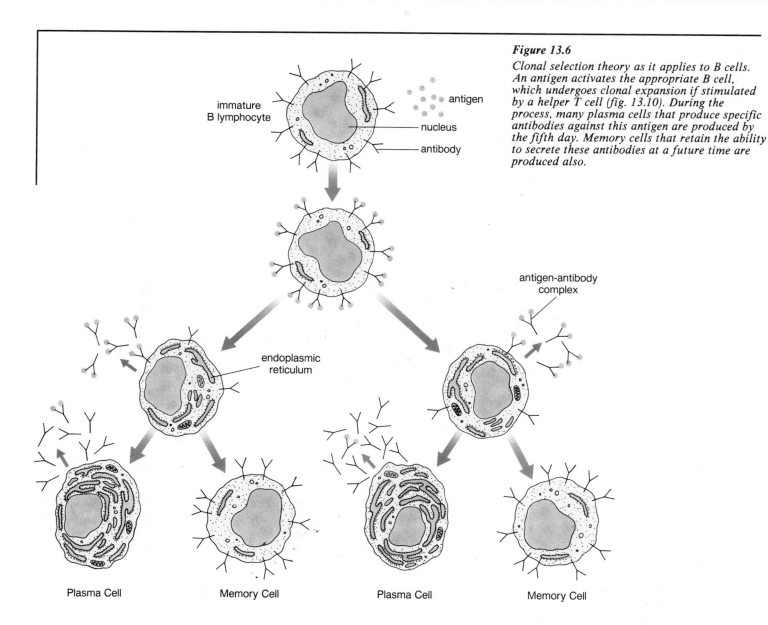

immature
B lymphocyte

antigen

nucleus

antibody

antigen-antibody
complex

endoplasmic
reticulum

Plasma Cell

Memory Cell

Plasma Cell

Memory Cell

Figure 13.6

Clonal selection theory as it applies to B cells. An antigen activates the appropriate B cell, which undergoes clonal expansion if stimulated by a helper T cell (fig. 13.10). During the process, many plasma cells that produce specific antibodies against this antigen are produced by the fifth day. Memory cells that retain the ability to secrete these antibodies at a future time are produced also.

varies (fig. 13.7). The antigen binds to the antibody at the variable regions of one arm in a lock-and-key manner. In other words, the variable regions form an antibody-binding site that is specific for a particular antigen.

The constant regions are not identical among all the antibodies. Instead, they are the same for different classes of antibodies. Most antibodies found in the blood belong to the class IgG (immunoglobulin G) (table 13.1).

The antigen-antibody reaction can take several forms, but quite often the antigen-antibody reaction produces complexes of antigens combined with antibodies (fig. 12.11). Such an antigen-antibody complex, sometimes called the immune complex, marks the antigen for destruction by other forces. For example, the complex may be engulfed by neutrophils or macrophages or it may activate a portion of blood serum called complement. As mentioned earlier, complement refers to a series of different proteins with various functions. Some make microbes more susceptible to phagocytosis. Others bring about lysis of bacterial cells—several of the proteins form a channel that allows water and salts to enter the cell until it bursts (fig. 13.5).

An antibody combines with its antigen in a lock-and-key manner. The antibody-antigen reaction can lead to complexes that contain several molecules of antibody and antigen.

Table 13.1 *Antibodies*

Classes	Description
IgG	Main antibody type in circulation; attacks microorganisms and their toxins
IgA	Main antibody type in secretions, such as saliva and milk; attacks microorganisms and their toxins
IgE	Antibody type on mast cells; responsible for allergic reactions
IgM	Antibody type found in circulation; largest antibody, with 5 subunits
IgD	Antibody type found primarily as a membrane-bound Ig

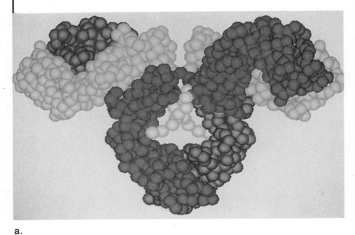

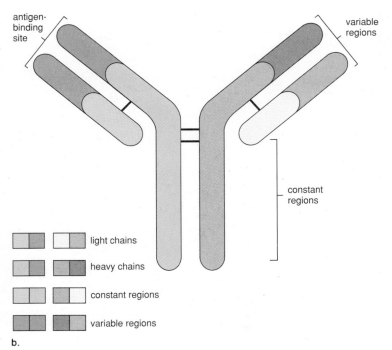

Figure 13.7

Structure of the most common antibody (IgG). a. Computer model of an antibody. b. Schematic drawing. Each arm of an antibody contains a light (short) chain and a heavy (long) chain. The amino acid chains make up constant regions and variable regions. In the variable regions, the amino acid sequence varies so that there is a specific antibody for any particular antigen. This antigen binds to the variable regions in a lock-and-key manner.

antigen-binding site

variable regions

constant regions

light chains

heavy chains

constant regions

variable regions

a.

b.

The Actions of T Cells

There are 4 different types of T cells: cytotoxic T cells, helper T cells, memory T cells, and suppressor T cells. All 4 types look alike but can be distinguished by their functions.

Cytotoxic T (T_C) cells sometimes are called killer T cells. In immune individuals, they attack and destroy cells bearing a foreign antigen, such as virus-infected or cancerous cells. These T cells have storage vacuoles that contain a chemical called perforin because it perforates cell membranes. The perforin molecules form a pore in the membrane that allows water and salts to enter. The cell under attack then swells and eventually bursts (fig. 13.8).

It often is said that T cells are responsible for **cell-mediated immunity,** characterized by destruction of antigen-bearing cells. Of all the T cells, only T_C cells are involved in this type of immunity.

Helper T (T_H) cells regulate immunity by enhancing the response of other immune cells. In response to an antigen, they enlarge and secrete lymphokines, including interferon and the interleukins. Lymphokines are stimulatory molecules that cause T_H cells to clone and other immune cells to perform their functions. For example, T_H cells stimulate macrophages to phagocytize and B cells to manufacture antibodies.[1] Because the AIDS virus attacks T_H cells, it inactivates the immune response. This is discussed in the reading on page 274.

When an activated T_H cell divides, the clone contains **suppressor T (T_S) cells** and **memory T (T_M) cells.** Once there is a sufficient number of T_S cells, the immune response ceases. Following suppression, a population of T_M cells persists, perhaps for life. These cells are able to secrete lymphokines and to stimulate macrophages and B cells whenever the same antigen enters the body once again.

T_C cells are responsible for cell-mediated immunity; T_H cells promote the immune response; T_S cells suppress the immune response; and T_M cells maintain immunity.

[1]The lymphokine secreted by helper T cells that stimulates B cells once was called B cell growth factor. Now it is called interleukin-4.

Figure 13.8

Cell-mediated immunity. a. *Scanning electron micrograph showing cytotoxic T cells attacking a cancer cell.* b. *The cancer cell now is destroyed.* c. *During the killing process, the vacuoles in a T cell fuse with the cell membrane and release units of the* protein perforin. *These units combine to form pores in the target cell membrane. Thereafter, fluid and salts enter so that the target cell eventually bursts.*

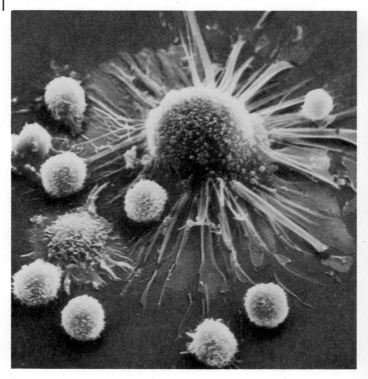

a.

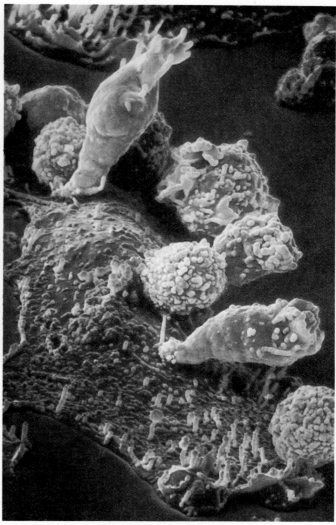

b.

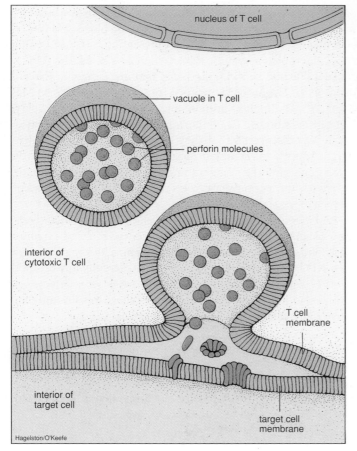

nucleus of T cell

vacuole in T cell

perforin molecules

interior of cytotoxic T cell

T cell membrane

interior of target cell

target cell membrane

Hagelston/O'Keefe

c.

Figure 13.9

T cell activation. a. *Either a macrophage or a B cell presents an antigen to a helper T cell. To accomplish this, the antigen has to be digested to peptides that are combined with an MHC protein. The complex is presented to the T cell. In return, the helper T cell produces and secretes lymphokines that stimulate T cells and other immune cells.* b. *Cells infected with a virus present one of the viral proteins along with an MHC protein to a cytotoxic T cell. This causes the cytotoxic T cell to attack and to destroy any cell infected with the same virus. (see fig. 13.8).*

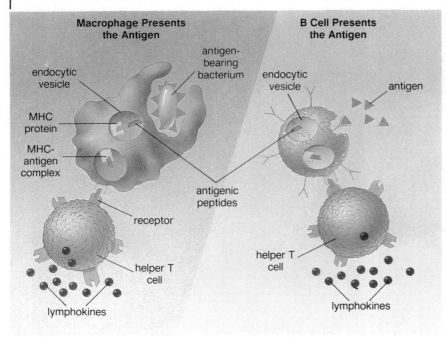

a.

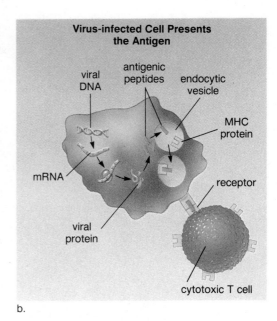

b.

13.2 Critical Thinking

1. A mouse is irradiated so that its bone marrow and thymus are destroyed. It then is resupplied only with bone marrow. The mouse is unable to form antibodies. Why?

2. A mixture of B cells is exposed to a specific radiolabeled antigen in vitro (within laboratory glassware). Would you expect all B cells to bind with the antigen and to be radiolabeled?

3. When B cells and T cells are incubated in vitro with a radiolabeled antigen, binding to certain B cells occurs but not to T cells. Why?

4. Human beings communicate by sight, sound, and touch. How do immune cells communicate with one another?

The Activation of Cytotoxic and Helper T Cells   T cells have receptors just as B cells do. Unlike B cells, however, T_C cells and T_H cells are unable to recognize an antigen that simply is present in lymph or blood. Instead, the antigen must be presented to them by an antigen-presenting cell (APC). When an APC, such as a macrophage, engulfs a bacterium or a virus, the APC enzymatically breaks it down to peptide fragments that are antigenic (have the properties of an antigen). The antigenic peptide fragment is linked to an **MHC** (major histocompatibility complex) **protein,** and together they are displayed to a T cell at the cell membrane. The importance of MHC proteins first was recognized when it was discovered that they contribute to the specificity of tissues and make it difficult to transplant tissue from one person to another. In other words, the donor and the recipient must be histo-(tissue) compatible (the same or nearly so) for a transplant to be successful without the administration of immunosuppressive drugs.

Figure 13.9*a* shows a macrophage and a B cell presenting an antigen to a T_H cell. Once a T_H cell recognizes an antigen, it undergoes clonal expansion, producing T_S cells and T_M cells that also can recognize this same antigen. While well-known APCs are macrophages and B cells, actually any cell in the body can be an APC. For example, figure 13.9*b* shows a virus-infected cell presenting an antigen to a T_C cell. Once a T_C cell recognizes an antigen, it attacks and destroys any cell that is infected with the same virus.

In order for T_C and T_H cells to recognize an antigen, the antigen, along with an MHC protein, must be presented to them by an APC.

Table 13.2 and figure 13.10 summarize our discussion of B cells and T cells.

Table 13.2 *Some Properties of B Cells and T Cells*

Property	B Cells	T Cells Cytotoxic	Helper
Type of immunity	Antibody mediated	Cell mediated	———
Antigen recognition	Direct recognition	Must be presented by APC	Must be presented by APC
Response	Become antibody-producing plasma cells	Search and destroy antigen-bearing cells	Secrete lymphokines; stimulate other immune cells
Final response	Memory cells	———	Suppressor and memory cells

Figure 13.10

Summary of B cell and T cell functions. Both cells are produced in the bone marrow, but only B cells mature there. T cells mature in the thymus. An antigen activates the appropriate B cell, which divides and differentiates to give a clone of antibody-producing plasma cells and memory cells. An antigen activates a T cell (either cytotoxic T or helper T) if it is presented with an MHC by an APC. Cytotoxic T cells then attack and destroy antigen-bearing cells; helper T cells divide to give a clone of helper T cells that produce lymphokines and also suppressor T and memory T cells.

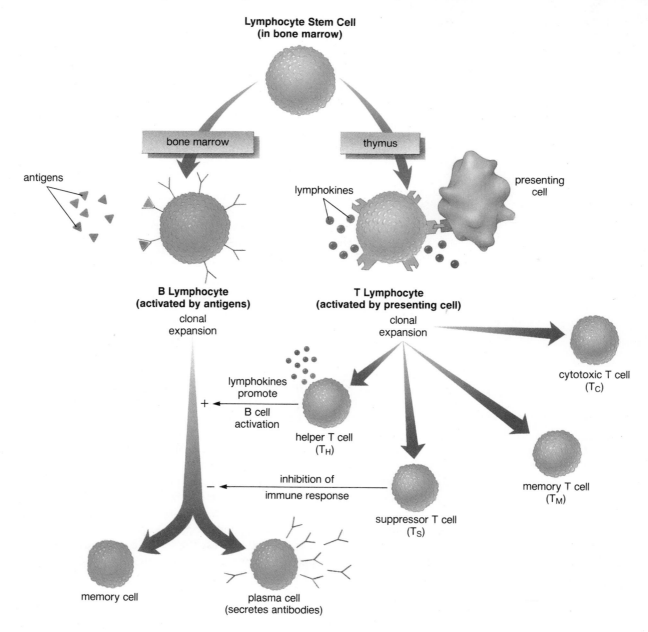

Recommended Immunization Schedule

Age	Vaccines
2 months	1st DTP[a]
	1st polio[b]
4 months	2nd DTP
	2nd polio
6 months	3rd DTP
9 months	Measles[c]
15 months	MMR[d]
	4th DTP
	3rd polio
18 months	HbCV[e]
4–6 years	5th DTP
	4th polio
14–16 years	TD[f]

[a]Diphtheria, tetanus, and pertussis (whooping cough)

[b]Trivalent oral containing poliovirus types 1, 2, and 3

[c]Measles only, vaccine should be administered in high-risk areas

[d]Measles, mumps, and rubella (German measles) virus

[e]Haemophilus influenzae b conjugated to a protein carrier

[f]Compared to DTP, contains same amount of tetanus but reduced dose of diphtheria, repeat every 10 years

Immunotherapy

The immune system can be manipulated to help people avoid or recover from diseases. Some of these techniques have been utilized for a long time, and some are relatively new.

Induced Immunity

Active immunity, which provides long-lasting protection against a disease-causing organism, develops after an individual is infected with a virus or a bacterium. In many instances today, however, it is not necessary to suffer an illness to become immune because it is possible to be artificially immunized against a disease. One recommended immunization schedule for children is given in figure 13.11. The importance of following a recommended immunization schedule is being demonstrated at this time. There have been outbreaks of childhood communicable diseases among college-age people because they were not immunized properly when they were younger.

Immunization requires the use of **vaccines,** which are traditionally bacteria and viruses (antigens) that have been treated so that they are no longer virulent (able to cause disease). New methods of producing vaccines are being developed. For example, it is possible to use the recombinant DNA technique to mass-produce a protein that can be used as a vaccine. This method is being used to prepare a vaccine against hepatitis B.

After a vaccine is given, it is possible to determine the amount of antibody present in a sample of serum—this is called the *antibody titer.* After the first exposure to an antigen, a primary response occurs. For a period of several days, no antibodies are present; then, there is a slow rise in the titer, followed by a gradual decline (fig. 13.12). After a second exposure, a secondary response may occur. If so, the titer rises rapidly to a level much greater than before. The second exposure in that case often is called the *"booster"* because it boosts the antibody titer to a high level. The antibody titer now may be high enough to prevent disease symptoms even if the individual is exposed to the disease. If so, the individual is now immune to that particular disease. A good secondary response can be related to the number of plasma cells and memory cells in the serum. Upon the second exposure, these cells already are present, and antibodies can be produced rapidly.

Vaccines can be used to make people actively immune.

Passive immunity occurs when an individual is given antibodies (immunoglobulins) to combat a disease. Since these antibodies are not produced by the individual's B cells, passive immunity is short-lived. For example, newborn infants possess passive immunity because antibodies have crossed the placenta from the mother's blood. These antibodies soon disappear, however, so that within a few months, infants become more susceptible to infections. Breast feeding (fig. 13.13) prolongs the passive immunity an infant receives from the mother because there are antibodies in the mother's milk.

Even though passive immunity does not last, it sometimes is used to prevent illness in a patient who has been unexpectedly exposed to an infectious disease. Usually, the person receives an injection of a serum containing antibodies. This may have been taken from donors who have recovered from the illness. In the past, horses were immunized, and serum was taken from them to provide the needed antibodies. Horses were used to produce antibodies against diphtheria, botulism, and tetanus. Occasionally, a patient who received these antibodies became ill because the serum contained proteins that the individual's immune system recognized as foreign. This was called serum sickness.

Passive immunity is short-lived because the antibodies are administered to and not made by the individual.

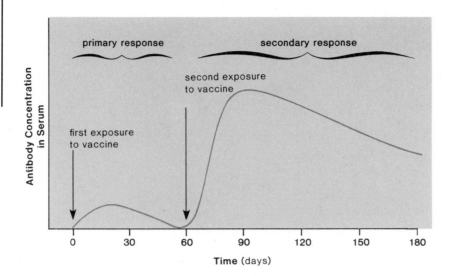

primary response   secondary response

second exposure
to vaccine

Antibody Concentration in Serum

first exposure
to vaccine

0   30   60   90   120   150   180

Time (days)

Figure 13.12

Development of active immunity due to immunization. The primary response after the first exposure of a vaccine is minimal, but the secondary response that may occur after the second exposure shows a dramatic rise in the amount of antibody present in serum.

Lymphokines

Lymphokines are being investigated as possible adjunct therapy for cancer and AIDS because they stimulate white blood cell formation and/or function.

Both interferon and various interleukins have been used as immunotherapeutical drugs, particularly to potentiate the ability of the individual's own T cells (and possibly B cells) to fight cancer.

Interferon is a substance produced by leukocytes, fibroblasts, and probably most cells in response to a viral infection. When it is produced by T cells, it is called a lymphokine. Interferon still is being investigated as a possible cancer drug, but thus far it has proven to be effective only in certain patients, and the exact reasons as yet cannot be discerned. For example, interferon has been found to be effective in up to 90% of patients with a type of leukemia known as hairy-cell leukemia (because of the hairy appearance of the malignant cells).

When and if cancerous cells carry an altered protein on their cell surface, by all rights they should be attacked and destroyed by T_C cells. Whenever cancer develops, it is possible that the T_C cells have not been activated. In that case, the use of lymphokines might awaken the immune system and lead to the destruction of the cancer. In one technique being investigated, researchers first withdraw T cells from the patient and activate the cells by culturing them in the presence of an interleukin. The cells then are reinjected into the patient, who is given doses of interleukin to maintain the killer activity of the T cells.

Those who are actively engaged in interleukin research believe that interleukins soon will be used as adjuncts for vaccines, for the treatment of chronic infectious diseases, and perhaps for the treatment of cancer. Interleukin antagonists also may prove helpful in preventing skin and organ rejection, autoimmune diseases, and allergies.

Other Lymphokines

Several growth factors that raise the blood cell count of particular white blood cells have been discovered. The best known of these is *GM-CSF* (granulocytemacrophage colony-stimulating factor), which enhances production of macrophages, neutrophils, eosinophils, and possibly red blood cells. This growth factor has been given to patients with AIDS, and it did raise their white blood cell count temporarily. Presently, the growth factor is being produced by recombinant DNA technology, and more extensive clinical trials are expected soon.

Figure 13.13

Example of passive immunity. Breast feeding is believed to provide a newborn with antibodies during the period of time when the child is not yet producing antibodies.

Figure 13.14

One possible method of producing human monoclonal antibodies. a. Blood sample is taken from a patient. b. Inactive B lymphocytes from the sample are exposed to an antigen. c. Activated lymphocytes are fused with myeloma cells. d. Scanning electron micrograph of cells fusing. e. Resulting hybridomas divide repeatedly, giving many cells that produce monoclonal antibodies.

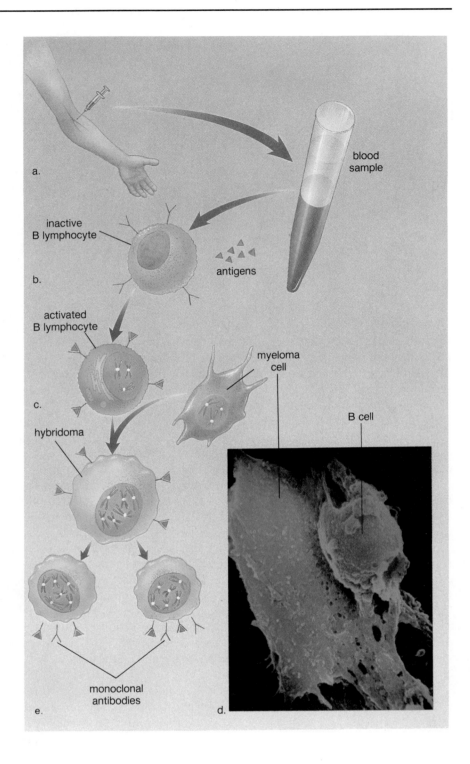

a.

blood sample

inactive B lymphocyte

antigens

b.

activated B lymphocyte

myeloma cell

c.

hybridoma

B cell

monoclonal antibodies

e.

d.

Lymphokines and other blood cell growth factors show some promise of potentiating the individual's own immune system.

Monoclonal Antibodies

Every plasma cell derived from the same B cell secretes antibodies against a specific antigen, as previously discussed. These are **monoclonal antibodies** because all of them are the same type (mono) and because they are produced by plasma cells derived from the same B cell (clone).

Human Anatomy and Physiology

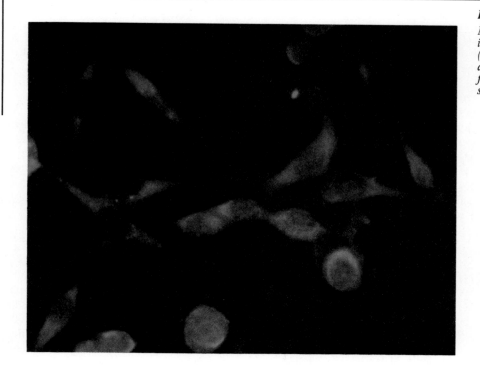

Figure 13.15
Monoclonal antibody use. These cells are infected with herpes simplex type 1 virus (HSV-1), which can be detected by a monoclonal antibody specific for the virus and tagged with a fluorescent dye. When the cells fluoresce, it shows that the virus is inside the cells.

One method of producing monoclonal antibodies in vitro (in laboratory glassware) is depicted in figure 13.14. B lymphocytes are removed from the body (today, usually a mouse) and are exposed to a particular antigen. Then they are fused with a myeloma cell (a malignant plasma cell) because these cells, unlike normal plasma cells, live and divide indefinitely. The fused cells are called hybridomas; *hybrid* because they result from the fusion of 2 different cells and *oma* because one of the cells is a cancer cell.

At present, monoclonal antibodies are being used for quick and certain diagnosis of various conditions. For example, a particular hormone is present in the urine of a pregnant woman. A monoclonal antibody can be used to detect the hormone and so indicate that the woman is pregnant. Monoclonal antibodies also are used for identifying infections (fig. 13.15). They are so accurate they even can sort out the different types of T cells in a blood sample. And because they can distinguish between cancerous and normal tissue cells, they are used to carry radioactive isotopes or toxic drugs only to tumors so these can be destroyed selectively.

Monoclonal antibodies are considered to be a biotechnology product because the production process makes use of a living system to mass-produce the product.

Monoclonal antibodies are produced in pure batches—they react specifically with just one type of molecule (antigen); therefore, they can distinguish one cell, or even one molecule, from another.

13.3 Critical Thinking

1. In this text, antigen originally was defined as a foreign substance in the body. Expand this definition by telling what an antigen does in the body.
2. It is possible to tag different types of monoclonal antibodies with different dyes so you can tell them apart. Knowing this, how would you produce and use monoclonal antibodies to distinguish T_H, T_M, T_S, and T_C cells in a blood sample?
3. How would you prove that a monoclonal antibody is specific to the herpes virus (HSV-2) that causes genital herpes, but not to the one (HSV-1) that causes cold sores?
4. In a manufacturing process called affinity purification, an impure mixture containing a desired substance is passed through a tube containing a large number of an appropriate antibody molecule fixed to a solid support. Why does this process result in purification of the product?

Immunological Side Effects and Illnesses

The immune system protects us from disease because it can tell self from nonself. Sometimes, however, the immune system is underprotective, as when an individual develops cancer, or is overprotective, as when an individual has allergies.

AIDS (acquired immune deficiency syndrome) is a sexually transmitted disease that results in the inability to achieve immunity. Therefore, the victim contracts various infectious diseases and often develops a virulent type of skin cancer (Kaposi's sarcoma). Many also show brain impairment, with ever-increasing neuromuscular and psychological consequences. Since the number of deaths from AIDS reported thus far in the United States is at least 128,000 and those diagnosed rarely live more than a few years, there is

AIDS: Its Treatment and Transmission

great interest in developing techniques to treat those who are infected and to prevent the spread of infection to others.

Investigators have identified several viruses that can cause AIDS, but the vast majority of infections in the United States are caused by HIV-1. This virus (see figure) primarily attacks helper T cells because it has an outer envelope protein (gp-120) that combines with a molecule (called CD4 antigen) in the cell membrane of these lymphocytes. Thereafter, the virus enters helper T cells. Monocytes have CD4 molecules in fewer number and therefore are less susceptible to viral infection. They

act as a reservoir for the virus and distribute it about the body, including the lungs and the brain.

The AIDS viruses are retroviruses, RNA viruses that are capable of incorporating a DNA copy of their genes into the host cell chromosome. While the HIV-1 virus is incorporated into the host chromosome, it is latent and does not multiply, and the individual does not display any symptoms of AIDS. When the virus is activated, it multiplies and bursts forth from helper T cells, destroying them. Soon the person may show the symptoms of AIDS. It is possible that infection with another disease-causing antigen activates the AIDS viruses. Certainly, AIDS is seen more often in individuals who have another sexually transmitted disease in addition to AIDS.

Since antibodies to the HIV-1 virus are found in the bloodstream of those infected, it would seem the body has an ability to mount an offensive to the invader. Eventually, however, the body succumbs. Various explanations have been given:

1. Infected lymphocytes are known to fuse with others, some of which are not infected. If this is the way the virus spreads, then it need never return to the bloodstream, where antibodies are located.

2. There is evidence that the HIV-1 virus alters the MHC proteins that appear on the surfaces of infected cells. This would make cytotoxic T cells incapable of recognizing and destroying infected cells.
3. The HIV-1 virus mutates so rapidly that new antibodies have to be made constantly, and therefore there are never enough of the right kind to be useful.

Development of a treatment for AIDS is taking 2 main avenues. There are drugs that can prolong the life of those infected, and vaccines are being developed to prevent infection possibly. The drug zidovudine (AZT) has been shown to be effective in those with full-blown AIDS and also seems to prevent the progression of the disease in infected persons who have fewer than 500 helper T cells per mm^3 but who exhibit no symptoms. A new drug called dideoxyinosine (DDI), which like AZT works by preventing viral replication in cells, is undergoing clinical trials and shows fewer side effects.

Two experimental vaccines for HIV-1 virus are being studied clinically. Both of these vaccines utilize only gp-120, the outer viral envelope protein labeled in the figure. One vaccine uses gp-120 directly. The other uses a vaccinia (cowpox) virus that has a gp-120 gene inserted into it. The hope is that after the vaccinia virus enters host cells, the gene will use the machinery of the cells to produce gp-120 proteins, which will then appear on the host cell membrane and

Allergies

Allergies are caused by an overactive immune system that forms antibodies to substances that usually are not recognized as foreign substances. Unfortunately, allergies usually are accompanied by coldlike symptoms, or even at times, by severe systemic reactions, such as anaphylactic shock, a sudden drop in blood pressure.

Of the 5 varieties (table 13.1) of antibodies—IgA, IgD, IgE, IgG, and IgM—IgE antibodies cause allergies. IgE antibodies are found in the bloodstream, but they, unlike the other types of antibodies, also reside in the membrane of *mast cells* found in the tissues. Some investigators contend that mast cells are basophils that have left the bloodstream and have taken up residence in the tissues. In any case, when the *allergen,* an antigen that provokes an allergic reaction, attaches to the IgE antibodies on mast cells, these cells release histamine and other substances that cause mucus secretion and airway constriction, resulting in the characteristic symptoms of allergy. On occasion, basophils and other white blood cells release these chemicals into the bloodstream. The increased capillary permeability that results from this can lead to fluid loss and shock.

will be recognized by the body's immune system. Thereafter, cytotoxic T cells are expected to destroy all cells bearing this antigen.

An entirely different approach is being taken by Jonas Salk, who invented the polio vaccine. His vaccine utilizes whole HIV-1 viruses killed by treatment with chemicals and radiation. So far, this vaccine has been found to be effective against the HIV-1 virus in chimpanzees.

It will be some time before an AIDS vaccine is available to the public; in the meantime, it is advisable to know how the HIV-1 virus is transmitted and to take precautions against infection. The following are excerpts from the Surgeon General's report on AIDS, which was sent to all U.S. households in 1986.

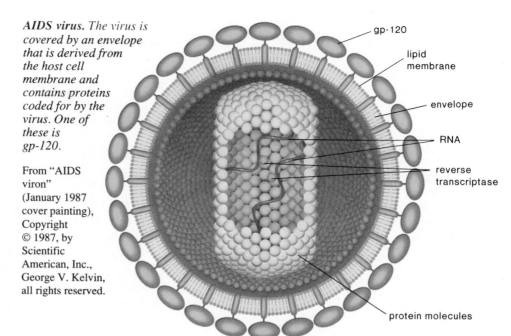

AIDS virus. The virus is covered by an envelope that is derived from the host cell membrane and contains proteins coded for by the virus. One of these is gp-120.

From "AIDS viron" (January 1987 cover painting), Copyright © 1987, by Scientific American, Inc., George V. Kelvin, all rights reserved.

gp-120
lipid membrane
envelope
RNA
reverse transcriptase
protein molecules

• Although the AIDS virus is found in several body fluids, a person acquires the virus during sexual contact with an infected person's blood or semen and possibly vaginal secretions. . . . Small (unseen by the naked eye) tears in the surface lining of the vagina or rectum may occur during insertion of the penis, fingers, or other objects, thus opening an avenue for entrance of the virus directly into the bloodstream. . . .

• Although the initial discovery was in the homosexual community . . . AIDS is found in heterosexual people as well. AIDS is not a black or white disease. AIDS is not just a male disease. AIDS is found in women; it is found in children. In the future AIDS will probably increase and spread among people who are not homosexual or intravenous drug abusers.

• Unless it is possible to know with absolute certainty that neither you nor your sexual partner is carrying the virus of AIDS, you must use protective behavior. Absolute certainty means not only that you and your partner have (had) a mutually faithful monogamous sexual relationship (for at least 5 years), and . . . that neither you nor your partner has used illegal intravenous drugs.

• If you suspect that (your partner) has been exposed by previous heterosexual or homosexual behavior or use of intravenous drugs with shared needles and syringes, a rubber (condom) should always be used during (start to finish) sexual intercourse (vagina or rectum).

• The risk of infection increases according to the number of sexual partners one has, male or female. The more partners you have, the greater the risk of becoming infected with the AIDS virus.

• No one should shoot up drugs . . . . However, many drug users are addicted to drugs and for one reason or another have not changed their behavior. For these people, the only way not to get AIDS is to use a clean, previously unused needle, syringe or any other implement [for injection].

• Shaking hands, hugging, social kissing, crying, coughing or sneezing will not transmit the AIDS virus . . . . You cannot get AIDS from toilets, doorknobs, telephones, office machinery, or household furniture. You cannot get AIDS from body massages, masturbation, or any non-sexual body contact.

Source: U.S. Surgeon General's Report.

Allergy shots sometimes prevent the onset of allergic symptoms. Injections of the allergen cause the body to build up high quantities of IgG antibodies, and these combine with allergens received from the environment before they have a chance to reach the IgE antibodies located in the membrane of mast cells.

Allergic symptoms are caused by the release of histamine and other substances from mast cells.

Tissue Rejection

Certain organs, such as skin, the heart, and the kidneys, could be transplanted easily from one person to another if the body did not attempt to *reject* them.

Rejection occurs because the transplanted organ is foreign to the individual. Both T_C cells and/or antibodies can bring about disintegration of the foreign tissue.

Organ rejection can be controlled in 2 ways: careful selection of the organ to be transplanted and the administration of immunosuppressive drugs. It is best if the transplanted organ has the same type of MHC proteins as those of the recipient, because T_C cells learn to recognize foreign MHC proteins. The immunosuppressive drug cyclosporine has been in use for some years. A new experimental drug, FK-506, eventually may replace cyclosporine as the drug of choice for transplant patients. In more than 100 patients taking FK-506, the rate of organ rejection was one-sixth that of cyclosporine.

When an organ is rejected, the immune system is attacking cells that bear different MHC proteins from those of the individual.

Autoimmune Diseases

Certain human illnesses are believed to be due to the production of antibodies that act against an individual's own tissues. In myasthenia gravis, autoantibodies attack the neuromuscular junctions so that the muscles do not obey nervous stimuli. Muscular weakness results. In MS (multiple sclerosis), antibodies attack the myelin of nerve fibers, causing various neuromuscular disorders. A person with SLE (systemic lupus erythematosus) forms various antibodies to different constituents of the body, including the DNA of the cell nucleus. The disease sometimes results in death, usually due to kidney damage. In rheumatoid arthritis, the joints are affected. When an autoimmune disease occurs, a viral infection of tissues often has set off an immune reaction to the body's own tissues. There is evidence to suggest that type I diabetes is the result of this sequence of events, as well as heart damage following rheumatic fever.

Autoimmune diseases seem to be preceded by a viral infection that fools the immune system into attacking the body's own tissues.

Summary

The lymphatic system consists of lymphatic vessels and lymphoid organs. The lymphatic vessels collect excess tissue fluid and fat molecules at lacteals and carry these to the cardiovascular system. Lymphocytes are produced and accumulate in the lymphoid organs (bone marrow, thymus, lymph nodes, and spleen).

The body is prepared to defend itself in both a generalized and a specific manner. Barriers to entry, phagocytic white blood cells, and protective chemicals react to any threat. The immune response is specific to a particular antigen and requires 2 types of lymphocytes, both of which are produced in the bone marrow. B cells mature in the bone marrow, and T cells mature in the thymus.

B cells directly recognize an antigen and give rise to antibody-secreting plasma cells if stimulated to do so by T_H cells. In order for the T cell to recognize an antigen, the antigen must be presented by an APC. There are 4 types of T cells. T_C cells kill cells on contact; T_H cells stimulate other immune cells; T_S cells suppress the immune response. There are also memory T and memory B cells that remain in the body and provide long-lasting immunity.

Immunity can be fostered by immunotherapy. Vaccines are available to promote active immunity, and antibodies sometimes are available to provide an individual with short-term passive immunity. Lymphokines, notably interferon and interleukins, plus other blood cell growth factors, are used to promote the body's ability to recover from cancer and AIDS.

Immunity has certain undesirable side effects. Allergies are due to an overactive immune system that forms antibodies to substances not normally recognized as foreign. T_C cells attack transplanted organs, although immunosuppressive drugs are available. Autoimmune illnesses occur when antibodies against the body's own cells form.

Study Questions

1. What is the lymphatic system, and what are its 3 functions? (p. 258)
2. Describe the structure and the function of the bone marrow, lymph nodes, the spleen, and the thymus. (pp. 259–61)
3. What are the general defense mechanisms of the body? (pp. 261–63)
4. B cells are responsible for which type of immunity? What is the clonal selection theory? (p. 264)
5. Describe the structure of an antibody, including the terms *variable regions* and *constant regions*. (p. 264)
6. Name the 4 types of T cells, and state their functions. (p. 266)
7. Explain the process by which a T cell is able to recognize an antigen. (p. 268)
8. How is active immunity achieved? How is passive immunity achieved? (p. 270)
9. What are lymphokines, and how are they used in immunotherapy? (p. 271)
10. How are monoclonal antibodies produced, and what are their applications? (pp. 272–73)
11. Discuss allergies, tissue rejection, and autoimmune diseases as they relate to the immune system. (pp. 274–76)

Objective Questions

1. Lymphatic vessels collect excess _____ and return it to the _____ veins.
2. The function of lymph nodes is to _____ the lymph.
3. T lymphocytes have passed through the _____ .
4. A stimulated B cell produces antibody-secreting _____ cells and _____ cells that are ready to produce the same type of antibody at a later time.
5. B cells are responsible for _____ -mediated immunity.
6. In order for a T cell to recognize an antigen, it must be presented by an _____ along with an MHC protein.
7. T cells produce _____ , which are stimulatory chemicals for all types of immune cells.
8. Cytotoxic T cells are responsible for _____ -mediated immunity.
9. Allergic reactions are associated with the release of _____ from mast cells.
10. The body recognizes foreign cells because they bear different _____ proteins than the body's cells.
11. Immunization with _____ brings about active immunity.
12. Hybridomas produce _____ antibodies.

Answers to Objective Questions

1. tissue fluid, subclavian 2. purify 3. thymus 4. plasma, memory 5. antibody 6. APC 7. lymphokines 8. cell 9. histamine 10. MHC 11. vaccines 12. monoclonal

Selected Key Terms

B lymphocyte (lim′fo-sīt) a lymphocyte that matures in the bone marrow and differentiates into antibody-producing plasma cells when stimulated by the presence of a specific antigen. *264*

complement system (kom′plĕ-ment sis′tem) a series of proteins in plasma that produce a variety of effects once an antigen-antibody reaction has occurred. *263*

helper T cell (hel′per te sel) T lymphocyte that stimulates other immune cells to perform their respective function. *266*

interferon (in″ter-fēr′on) a protein formed by a cell infected with a virus that can increase the resistance of other cells to the virus. *263*

cytotoxic T cell (si″to-tok′sik te sel) T lymphocyte that attacks cells bearing foreign peptides. *266*

lymph (limf) fluid having the same composition as tissue fluid and carried in lymphatic vessels. *259*

lymphatic system (lim-fat′ik sis′tem) vascular system that takes up excess tissue fluid and transports it to the bloodstream. *258*

lymphokine (lim′fo-kīn) molecule secreted by T lymphocytes that has the ability to affect the characteristics of lymphocytes and monocytes. *271*

memory B cell (mem′o-re be sel) persistant population of B cells that produce a specific antibody and account for the development of active immunity. *264*

MHC protein (pro′te-in) major histocompatibility protein; a surface molecule that serves as a genetic marker. *268*

monoclonal antibody (mon″-o-klōn′al an′tĭ-bod″ē) antibody of one type that is produced by cells that are derived from a lymphocyte that is fused with a cancerous cell. *272*

plasma cell (plaz′mah sel) cell derived from B cell lymphocyte that is specialized to mass-produce antibodies. *264*

suppressor T cell (su-pres′ or te sel) T lymphocyte that suppresses certain other T and B lymphocytes from continuing to divide and perform their respective function. *266*

T lymphocyte (lim′fo-sīt) a lymphocyte that matures in the thymus and occurs in 4 varieties, one of which kills antigen-bearing cells outright. *264*

vaccine (vak′sēn) antigens prepared in such a way that they can promote active immunity without causing disease. *270*

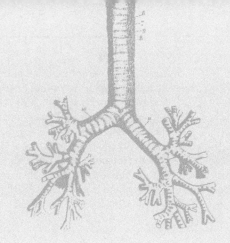

Chapter 14

Respiration

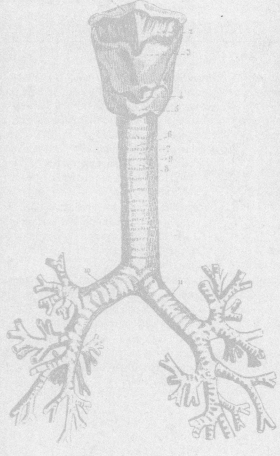

Chapter Concepts

1 As air passes along the respiratory tract, it is filtered, warmed, and saturated with water before gas exchange takes place across a very extensive moist surface.

2 Breathing brings in oxygen needed by the cells for cellular respiration and rids the body of carbon dioxide, a by-product of cellular respiration.

3 The respiratory pigment hemoglobin combines with oxygen in the lungs and releases oxygen in the tissues. It also aids in the transport of carbon dioxide, largely by its ability to buffer.

4 The respiratory tract is especially subject to disease because it serves as an entrance for infectious agents. Polluted air contributes to 2 major lung disorders—emphysema and cancer.

Chapter Outline

Breathing
    The Passage of Air
The Mechanism of Breathing
    Inspiration
    Expiration
    Lung Capacities
External and Internal Respiration
    External Respiration
    Internal Respiration
Respiration and Health
    Common Respiratory Infections
    Lung Disorders

Figure 14.1
Oxygen need of body. Exercising increases the body's need for oxygen (O_2) because cellular respiration is sped up in order to provide ATP for muscle contraction. The heart pumps faster to deliver oxygen to the tissues in a more timely manner.

Breathing is more eminently necessary than eating. While it is possible to stop eating altogether for several days, it is not possible to remain alive for longer than several minutes without breathing. Breathing supplies the body with the oxygen needed for cellular respiration, as indicated in the following equation.

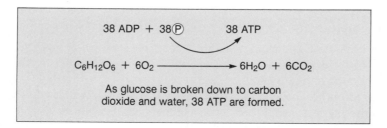

As glucose is broken down to carbon dioxide and water, 38 ATP are formed.

This equation indicates that the body requires oxygen to convert the energy within glucose ($C_6H_{12}O_6$) to phosphate-bond energy.[1] Therefore, the more energy expended, the greater the need for oxygen (fig. 14.1). The minimum amount of oxygen a person consumes at complete rest, without eating previously, is related to the basal metabolic rate (p. 212). The average young adult male utilizes about 250 ml of oxygen per minute in a basal or restful state. Exercise and digestion of food raise the oxygen need. The average amount of oxygen needed with mild exercise is 500 ml of oxygen per minute. The equation for cellular respiration also indicates that cells produce carbon dioxide (CO_2). This metabolic end product must be eliminated from the body by the breathing process.

Breathing is necessary to supply the body with oxygen so that ATP can be formed by cellular respiration.

[1]The body requires oxygen (O_2) for the respiration of fats and amino acids as well as for glucose. 36 ATP are commonly produced in most eukaryotic cells (p. 114).

Table 14.1 *Composition of Inspired and Expired Air*

Component of Air	Inspired Air (%/vol)	Expired Air (%/vol)
Nitrogen (N_2)	79.00	79.60
Oxygen (O_2)	20.96	16.02*
Carbon dioxide (CO_2)	0.04	4.38

*We exhale some oxygen, making mouth-to-mouth resuscitation possible.

Altogether, the term *respiration* can be used to refer to the complete process of getting oxygen to body cells for cellular respiration and the reverse process of ridding the body of carbon dioxide given off by cells. Respiration can be said to include the following components.

1. **Breathing:** entrance and exit of air into and out of the lungs
2. **External respiration:** exchange of the gases oxygen (O_2) and carbon dioxide (CO_2) between air and blood
3. **Internal respiration:** exchange of the gases (oxygen and carbon dioxide) between blood and the tissue fluid
4. **Cellular respiration:** production of ATP in cells

In this chapter, we study the first 3 components of the respiratory process. Cellular respiration was discussed in chapter 5.

Breathing

The normal breathing rate is about 14–20 times per minute. Breathing consists of taking air in, **inspiration** (inhalation), and forcing air out, **expiration** (exhalation). Expired air contains less oxygen (O_2) and more carbon dioxide (CO_2) than inspired air, indicating that the body takes in oxygen and gives off carbon dioxide (table 14.1).

The Passage of Air

During inspiration and expiration, air is conducted toward or away from the lungs by a series of cavities, tubes, and openings, listed in order in table 14.2 and illustrated in figure 14.2.

As air moves in along the air passages, it is filtered, warmed, and moistened. Filtering is accomplished by coarse hairs and cilia in the region of the nostrils and by cilia alone in the rest of the nose and the windpipe. In the nose, the hairs and the cilia act as a screening device. In the trachea, cilia beat upward, carrying mucus, dust, and occasional bits of food that "went down the wrong way" into the pharynx, where the accumulation can be swallowed or expectorated. The air is warmed by heat given off by the blood vessels lying close to the surface of the lining of the air passages, and it is moistened by the wet surface of these passages.

On the other hand, as air moves out during expiration, it cools and loses its moisture. As the gas cools, it deposits its moisture on the lining of the windpipe and the nose, and the nose may even drip as a result of this condensation. The air still retains so much moisture, however, that upon expiration on a cold day, it condenses and forms a small cloud.

Air is warmed, filtered, and moistened as it moves from the nose toward the lungs.

Each portion of the air passage has its own structure and function, as described in the sections that follow.

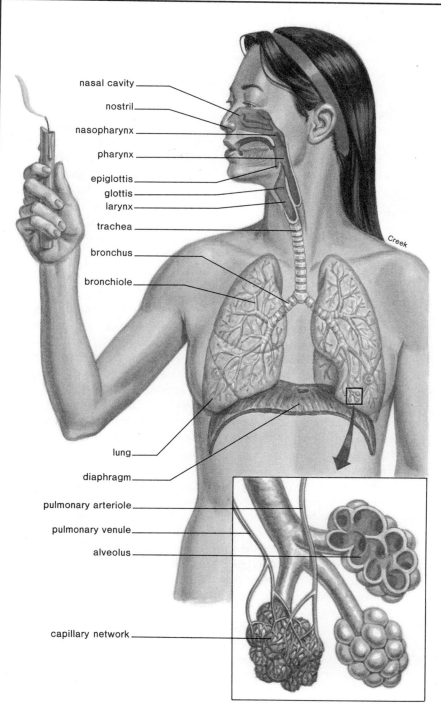

Figure 14.2
Diagram of the human respiratory tract, with internal structure of one lung revealed in an enlargement of a section of this lung. Gas exchange occurs in the alveoli, which are surrounded by a capillary network. Notice that the pulmonary arteriole carries deoxygenated blood (colored blue) and the pulmonary venule carries oxygenated blood (colored red). (See page 227.)

nasal cavity

nostril

nasopharynx

pharynx

epiglottis

glottis

larynx

trachea

bronchus

bronchiole

Creek

lung

diaphragm

pulmonary arteriole

pulmonary venule

alveolus

capillary network

Table 14.2 Path of Air	
Structure	**Function**
Nasal cavities	Filter, warm, and moisten air
Nasopharynx	Passage of air from nose to throat
Pharynx (throat)	Connection to surrounding regions
Glottis	Passage of air into larynx
Larynx (voice box)	Sound production
Trachea (windpipe)	Passage of air to bronchi
Bronchi	Passage of air to each lung
Bronchioles	Passage of air to each alveolus
Alveoli	Air sacs for gas exchange

The Nose

The nose contains 2 *nasal cavities*, narrow canals with convoluted lateral walls that are separated from one another by a median septum. In the narrow upper recesses of the nasal cavities are special ciliated cells (see fig. 18.4) that act as odor receptors. Nerves lead from these cells to the brain, where the impulses are interpreted as smell.

The tear (lacrimal) glands drain into the nasal cavities by way of tear ducts. For this reason, crying produces a runny nose.

The nasal cavities also communicate with the cranial sinuses, air-filled spaces in the skull that are lined with mucous membrane. If these membranes are inflamed due to a cold or an allergic reaction, mucus can accumulate in the sinuses and cause a sinus headache.

The nasal cavities empty into the nasopharynx, a chamber just beyond the soft palate. The *eustachian tubes* lead from the nasopharynx to the middle ears (see fig. 18.13).

The nasal cavities, which receive air, open into the nasopharynx.

The Pharynx

The **pharynx** is in the back of the throat; therefore, air taken in by either the nose or the mouth enters the pharynx. In the pharynx, the air passage and the food passage temporarily join. The trachea, which lies in front of the esophagus, is normally open, allowing the passage of air, but the esophagus normally is closed and opens only when swallowing occurs. The larynx lies at the top of the trachea.

Air from either the nose or the mouth enters the pharynx, as does food. The passage of air continues in the larynx and the trachea.

The Larynx

The **larynx** can be imagined as a triangular box whose apex, the Adam's apple, is located at the front of the neck. At the top of the larynx is a variable-sized opening called the **glottis.** When food is being swallowed, the glottis is covered by a flap of tissue called the **epiglottis** so that no food passes into the larynx. If, by chance, food or some other substance does gain entrance to the larynx, reflex coughing usually occurs to expel the substance. If this reflex is not sufficient, it may be necessary to resort to the Heimlich maneuver (fig. 14.3).

At the edges of the glottis and embedded in mucous membrane are elastic ligaments called the **vocal cords** (fig. 14.4). These cords, stretching from the back to the front of the larynx at the sides of the glottis, vibrate when air is expelled past them through the glottis. Vibration of the vocal cords produces sound. The high or low pitch of the voice depends upon the length, the thickness, and the degree of elasticity of the vocal cords and the tension at which they are held. The loudness, or intensity, of the voice depends upon the amplitude of the vibrations, or the degree to which vocal cords vibrate.

At the time of puberty, the growth of the larynx and the vocal cords is much more rapid and accentuated in the male than in the female, causing the male to have a more prominent Adam's apple and a deeper voice. The voice "breaks" in the young male due to his inability to control the longer vocal cords.

The larynx is the voice box because it contains the vocal cords at the sides of the glottis, an opening sometimes covered by the epiglottis.

The Trachea

The **trachea** is a tube held open by C-shaped cartilaginous rings. Ciliated mucous membrane (see fig. 9.3) lines the trachea, and normally these cilia keep the windpipe free of debris. Smoking is known to destroy the cilia, and consequently the soot in cigarette smoke collects in the lungs. Smoking is discussed more fully at the end of this chapter.

If the trachea is blocked because of illness or accidental swallowing of a foreign object, it is possible to insert a tube by way of an incision made in the trachea. This tube acts as an artificial air intake and exhaust duct. The operation is called a *tracheostomy.*

Bronchi

The trachea divides into 2 **bronchi** (singular, bronchus) that enter the right and left lungs and then branch into a great number of smaller passages called

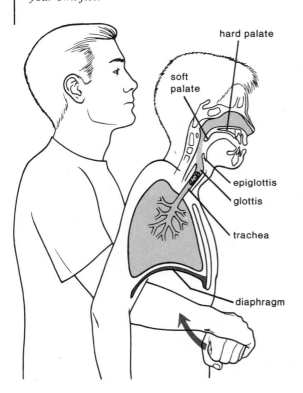

Figure 14.3

The Heimlich maneuver. More than 8 Americans choke to death each day on food lodged in the trachea. A simple process termed the abdominal thrust (Heimlich) maneuver can save the life of a person who is choking. The abdominal thrust maneuver is performed as follows. If the victim is standing or sitting: (1) Stand behind the victim or the victim's chair, and wrap your arms around his or her waist; (2) grasp your fist with your other hand, and place the fist against the victim's abdomen, slightly above the navel and below the rib cage; (3) press your fist into the victim's abdomen with a quick upward thrust; (4) repeat several times if necessary. If the victim is lying down: (1) Position the victim on his or her back; (2) face the victim, and kneel on his or her hips; (3) with one of your hands on top of the other, place the heel of your bottom hand on the abdomen, slightly above the navel and below the rib cage; (4) press into the victim's abdomen with a quick upward thrust; (5) repeat several times if necessary. If you are alone and choking, use anything that applies force just below your diaphragm. Press into a table or a sink, or use your own fist.

hard palate
soft palate
epiglottis
glottis
trachea
diaphragm

Human Anatomy and Physiology

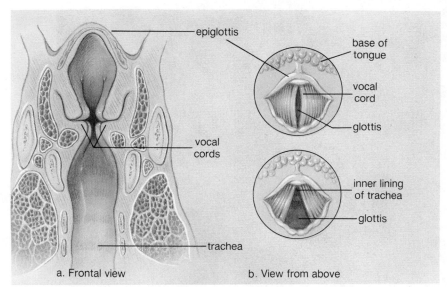

epiglottis

base of tongue

vocal cord

glottis

vocal cords

inner lining of trachea

glottis

trachea

a. Frontal view

b. View from above

Iverson

Figure 14.4
Operation of vocal cords. The vocal cords lie at the edges of the glottis. When air is expelled from the larynx, the cords vibrate, producing the voice. a. Frontal view. b. View from above.

bronchioles. The 2 bronchi resemble the trachea in structure, but as the bronchial tubes divide and subdivide, their walls become thinner and the small rings of cartilage are no longer present (fig. 14.5). Each bronchiole terminates in an elongated space enclosed by a multitude of air pockets, or sacs, called **alveoli** (singular, alveolus) (fig. 14.2), which make up the lungs.

Lungs

Within the lungs, each alveolar sac consists of only one layer of squamous epithelium surrounded by blood capillaries. Gas exchange occurs between air in the alveoli and blood in the capillaries (fig. 14.2).

A film of lipoprotein that lines the alveoli of mammalian lungs lowers the surface tension and prevents them from closing. The lungs collapse in some newborn babies, especially premature infants, who lack this film. This condition, called infant respiratory distress syndrome, often results in death.

There are approximately 300 million alveoli, having a total cross-sectional area of 50–70 m². This is about 40 times the surface area of the skin. Because of their many air spaces, the lungs are very light; normally, a piece of lung tissue dropped in a glass of water floats.

Air moves from the trachea and the 2 bronchi, which are held open by cartilaginous rings, into the lungs. The lungs are composed of air sacs called alveoli.

Externally, the lungs are cone-shaped organs that lie on both sides of the heart in the thoracic (chest) cavity. Each lung has a narrow and rounded apex that approaches the neck. The base of each lung is broad and concave in shape so as to fit upon the convex surface of the diaphragm. The other surfaces of the lungs follow the contours of the ribs and the organs in the thoracic cavity.

Figure 14.6 shows the relationship of the pulmonary vessels to the trachea and the bronchial tubes. The branches of the pulmonary artery, carrying deoxygenated blood, accompany the bronchial tubes and form a mass of capillaries around the alveoli. The 4 pulmonary veins, carrying oxygenated blood, collect blood from these capillaries. These veins then empty into the left atrium of the heart.

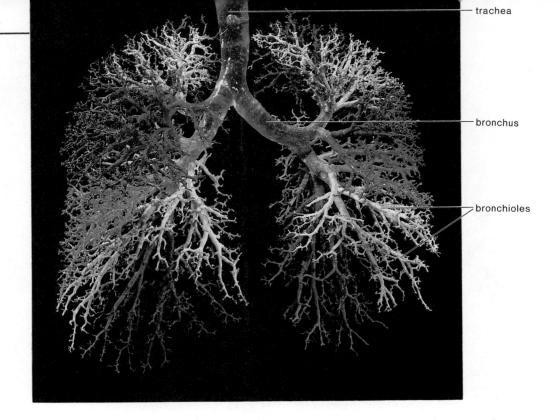

Figure 14.5

Cast of lungs showing the number of airways. The bronchi branch into the bronchioles that branch and rebranch until termination in the alveoli. Each lobe of the lung is in a different color in this cast.

trachea

bronchus

bronchioles

Figure 14.6

Posterior view of the heart and the lungs showing the relationship of the pulmonary vessels to the trachea and the bronchial tubes. Trace the path of air to the left lung, and trace the path of blood from the heart to the left lung and return. Why are the arteries blue and the veins red?

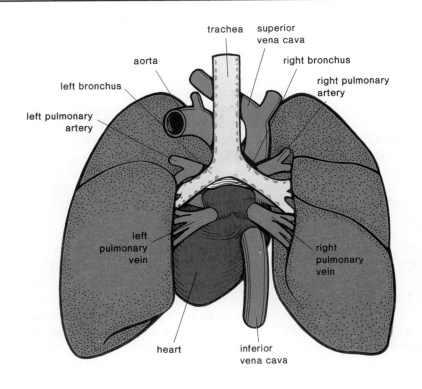

aorta

left bronchus

left pulmonary artery

left pulmonary vein

trachea

superior vena cava

right bronchus

right pulmonary artery

right pulmonary vein

heart

inferior vena cava

The Mechanism of Breathing

In order to understand **ventilation,** the manner in which air is drawn into and expelled out of the lungs, it is necessary to remember first that when you are breathing, there is a continuous column of air from the pharynx to the alveoli of the lungs; that is, the air passages are open.

Human Anatomy and Physiology

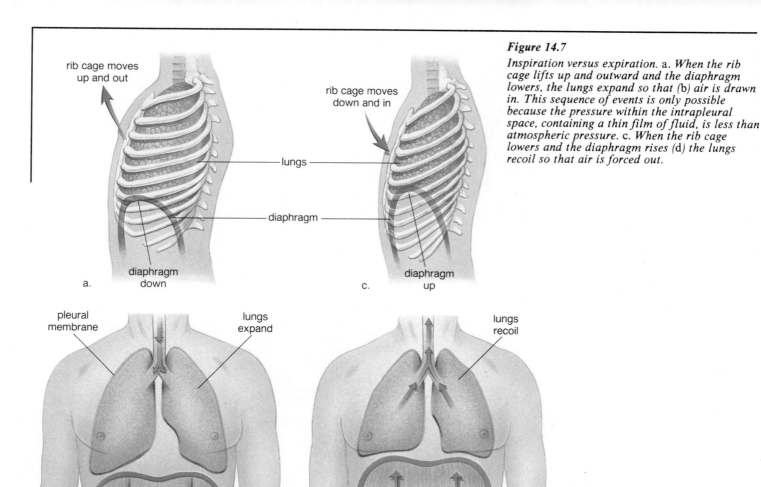

Figure 14.7

Inspiration versus expiration. a. When the rib cage lifts up and outward and the diaphragm lowers, the lungs expand so that (b) air is drawn in. This sequence of events is only possible because the pressure within the intrapleural space, containing a thin film of fluid, is less than atmospheric pressure. c. When the rib cage lowers and the diaphragm rises (d) the lungs recoil so that air is forced out.

Secondly, note that the lungs lie within the sealed-off thoracic cavity. The **ribs,** hinged to the vertebral column at the back and to the *sternum* (breastbone) at the front, along with the muscles that lie between them, make up the top and the sides of the thoracic cavity. The **diaphragm,** a dome-shaped horizontal muscle, forms the floor of the thoracic cavity. The lungs themselves are enclosed by the **pleural membranes** (fig. 14.7). The outer pleural membrane adheres closely to the thoracic cavity wall and the diaphragm, and the inner membrane is fused to the lungs. The 2 pleural layers lie very close to one another, separated only by a thin film of fluid. Normally, the intrapleural pressure is less than atmospheric pressure. The importance of this reduced pressure is demonstrated when, by design or accident, air enters the intrapleural space. The lungs collapse, and inspiration is impossible.

The lungs are completely enclosed and by way of the pleural membranes, adhere to the thoracic cavity walls.

Inspiration

Carbon dioxide (CO_2) and hydrogen (H^+) ions are the primary stimuli that cause us to breathe. When the concentration of carbon dioxide and subsequently the concentration of hydrogen ions (see following sections) reach a certain level in blood, the *respiratory center* in the medulla oblongata, the stem portion of the brain, is stimulated. This center is not affected by low oxygen (O_2) levels. There also are chemoreceptors in the *carotid bodies,* located in the carotid arteries, and in the *aortic bodies,* located in the aorta,

Figure 14.8

Nervous control of breathing. During inspiration, the respiratory center stimulates the intercostal (rib) muscles and the diaphragm to contract by way of the phrenic (efferent) nerve. Nerve impulses from the expanded alveoli traveling by way of the vagus (afferent) nerve then inhibit the respiratory center. Lack of stimulation causes the rib muscles and the diaphragm to relax, and expiration follows.

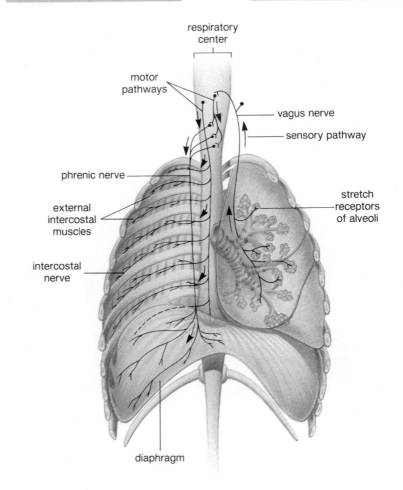

Inspiration requires that the phrenic nerve and these structures be active.

Expiration requires that the vagus nerve and these structures be active.

respiratory center

motor pathways

vagus nerve

sensory pathway

phrenic nerve

external intercostal muscles

stretch receptors of alveoli

intercostal nerve

diaphragm

that respond primarily to hydrogen ion concentration [H⁺] but also to the level of carbon dioxide and oxygen in blood. These bodies communicate with the respiratory center.

In humans, the breathing rate is regulated by the amount of carbon dioxide in blood.

When the breathing center is stimulated, a nerve impulse goes out by way of nerves to the diaphragm and the rib cage (fig. 14.8). In its relaxed state, the *diaphragm* is dome shaped, but upon stimulation, it contracts and lowers. When the rib muscles contract, the *rib cage* moves upward and outward. Both of these contractions serve to increase the size of the thoracic cavity. As the thoracic cavity increases in size, the lungs expand. When the lungs expand, air pressure within the enlarged alveoli lowers and is immediately rebalanced by air rushing in through the nose or the mouth.

Inspiration (fig. 14.7*a* and *b*) is the active phase of breathing. During this time, the diaphragm contracts, the rib muscles contract, the lungs are pulled open, and air comes rushing in. Note that air comes in because the lungs already have opened up; air does not force the lungs open. This is why it sometimes is said that *humans breathe by negative pressure*. The creation of a partial vacuum sucks air into the lungs.

Human Anatomy and Physiology

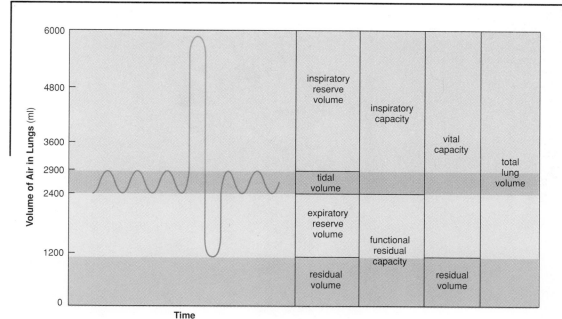

Figure 14.9
Vital capacity. A spirometer measures the maximum amount of air that can be inhaled and exhaled. When an individual inspires, a pen moves up, and when he expires, a pen moves down. The resulting pattern, such as the one shown here, is called a spirograph.

Stimulated by nervous impulses, the rib cage lifts up and out and the diaphragm lowers to expand the thoracic cavity and the lungs, allowing inspiration to occur.

Expiration

When the lungs are expanded, stretching of the alveoli stimulates special receptors in the alveolar walls, and these receptors initiate nerve impulses that travel from the inflated lungs to the breathing center. When the impulses arrive at the medulla oblongata, the center is inhibited and stops sending signals to the diaphragm and the rib cage. The *diaphragm* relaxes and resumes its dome shape (fig. 14.7c and d). The abdominal organs press up against the diaphragm. The *rib cage* moves down and inward. The elastic lungs recoil, and air is pushed out.

Table 14.3 summarizes the events that occur during inspiration and expiration. It is clear that while inspiration is an active phase of breathing, normally expiration is passive because the breathing muscles automatically relax following contraction. It is possible, in deeper and more rapid breathing, for both phases to be active because there is another set of rib muscles whose contraction can forcibly cause the thoracic wall to move downward and inward. Also, when the abdominal wall muscles are contracted, there is an increase in pressure that helps to expel air.

When nervous stimulation ceases, the rib cage lowers and the diaphragm rises, allowing the lungs to recoil and expiration to occur.

Table 14.3 Breathing Process

Inspiration	Expiration
Medulla sends stimulatory message to diaphragm and rib muscles.	Stretch receptors in lungs send inhibitory message to medulla.
Diaphragm contracts and flattens.	Diaphragm relaxes and resumes a dome position.
Rib cage moves up and out.	Rib cage moves down and in.
Lungs expand.	Lungs recoil.
Negative pressure builds in lungs.	Positive pressure builds in lungs.
Air is pulled in.	Air is forced out.

Lung Capacities

When we breathe, the amount of air moved in and out with each breath is called the **tidal volume.** Normally, the tidal volume is about 500 ml, but we can increase the amount inhaled and exhaled by deep breathing. The total volume of air that can be moved in and out during a single breath is called the **vital capacity** (fig. 14.9). First, we can increase inspiration by as much as 3,100 ml of air. This is called the *inspiratory reserve volume.* Similarly, we can increase expiration by contracting the thoracic muscles. This is called the *expiratory reserve volume* and measures approximately 1,400 ml of air. Vital capacity is the sum of tidal, inspiratory reserve, and expiratory reserve volumes.

Figure 14.10

Distribution of air in the lungs. The air in a *does not reach the alveoli immediately; therefore, this is called dead space. The air in* c *represents the amount of residual air that has not left the lungs. Only the air in* b *brings additional oxygen (O_2) for respiration.*

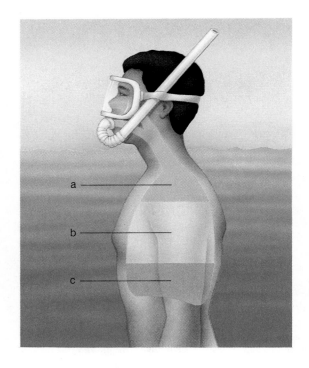

Note in figure 14.9 that even after very deep breathing, some air (about 1,000 ml) remains in the lungs; this is called the **residual volume.** This air is no longer useful for gas exchange purposes. In some lung diseases, such as emphysema (p. 294), the residual volume builds up because the individual has difficulty emptying the lungs. This means that the lungs tend to be filled with useless air, and as you can see from examining figure 14.9, the vital capacity is reduced.

Dead Space

Some of the inspired air never reaches the lungs; instead it fills the conducting airways (fig. 14.10). These passages are not used for gas exchange and therefore are said to contain *dead space*. To ventilate the lungs, then, it is better to breathe more slowly and deeply because it ensures that a greater percentage of the tidal volume reaches the lungs.

If we breathe through a tube, we increase the amount of dead space and increase the amount of air that never reaches the lungs. Any device that increases the amount of dead space beyond maximal inhaling capacity spells death for the individual because the air inhaled never reaches the alveoli.

The manner in which we breathe has physiological consequences.

External and Internal Respiration

External Respiration

The term *external respiration* refers to the exchange of gases between air in the alveoli and blood in the pulmonary capillaries (fig. 14.11). The wall of an alveolus consists of a thin, single layer of cells, and the wall of a blood capillary also consists of such a layer. Since neither wall offers resistance to the passage of gases, *diffusion* alone governs the exchange of oxygen (O_2) and carbon dioxide (CO_2) between alveolar air and blood. Active cellular absorption and secretion do not appear to play a role. Rather, the direction in which the gases move is determined by the pressure or tension gradients between blood and inspired air.

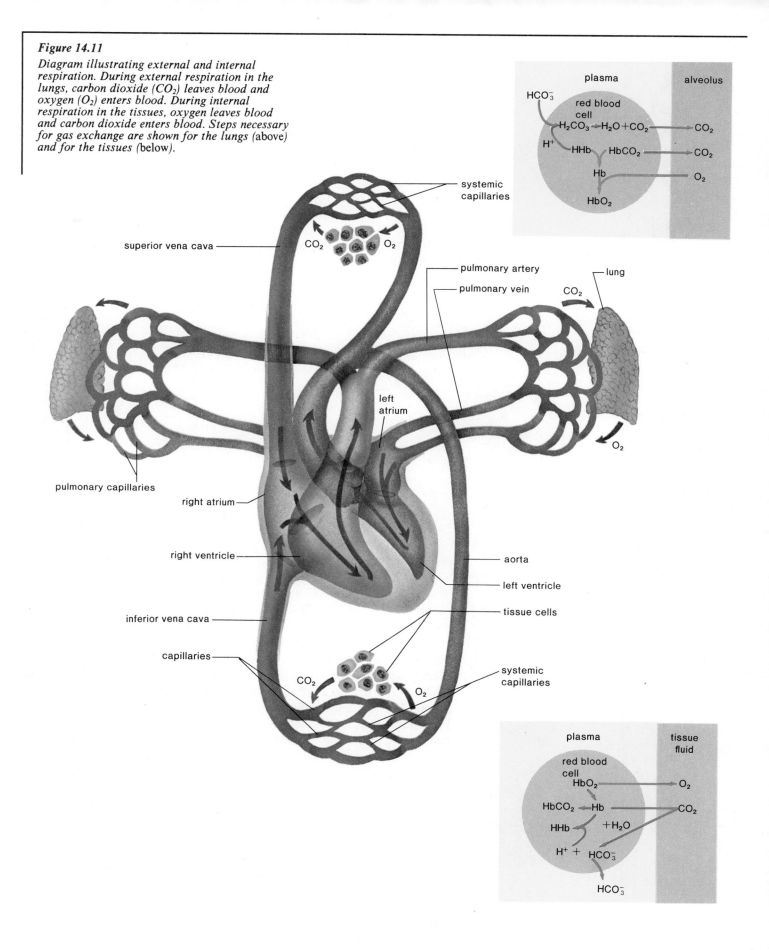

Figure 14.11

Diagram illustrating external and internal respiration. During external respiration in the lungs, carbon dioxide (CO_2) leaves blood and oxygen (O_2) enters blood. During internal respiration in the tissues, oxygen leaves blood and carbon dioxide enters blood. Steps necessary for gas exchange are shown for the lungs (above) and for the tissues (below).

systemic capillaries

superior vena cava

CO_2 O_2

pulmonary artery

pulmonary vein

lung

CO_2

left atrium

O_2

pulmonary capillaries

right atrium

right ventricle

aorta

left ventricle

inferior vena cava

tissue cells

capillaries

CO_2 O_2

systemic capillaries

plasma   alveolus

HCO_3^-

red blood cell

$H_2CO_3 \rightarrow H_2O + CO_2$ CO_2

H^+ HHb $HbCO_2$ CO_2

Hb O_2

HbO_2

plasma   tissue fluid

red blood cell

HbO_2 O_2

$HbCO_2 \leftarrow Hb$ CO_2

HHb $+H_2O$

$H^+ +$ HCO_3^-

HCO_3^-

Atmospheric air contains little carbon dioxide, but blood flowing into the lung capillaries is almost saturated with the gas. Therefore, *carbon dioxide diffuses out of blood into the alveoli.* The pressure pattern is the reverse for oxygen. Blood coming into the pulmonary capillaries is deoxygenated, and alveolar air is oxygenated; therefore, *oxygen diffuses into the capillary.* Breathing at high altitudes is less effective than at low altitudes because the air pressure is lower, making the concentration of oxygen (and other gases) lower than normal; therefore, less oxygen diffuses into the blood. Breathing problems do not occur in airplanes because the cabin is pressurized to maintain an appropriate pressure. Emergency oxygen is available in case the pressure is, for one reason or another, reduced.

As blood enters the pulmonary capillaries (fig. 14.11), most of the carbon dioxide is being carried as the bicarbonate ion (HCO_3^-). As the little remaining free carbon dioxide begins to diffuse out, the following reaction is driven to the right:

$$H^+ + HCO_3^- \longrightarrow H_2CO_3 \longrightarrow H_2O + CO_2 \uparrow$$

(bicarbonate ion)

"Up" arrow indicates carbon dioxide is leaving the body.

The enzyme carbonic anhydrase (p. 245), present in red blood cells, speeds up the reaction. As the reaction proceeds, the respiratory pigment **hemoglobin** gives up the hydrogen (H^+) ions it has been carrying; HHb becomes Hb.

Now, hemoglobin more readily takes up oxygen and becomes oxyhemoglobin.

$$Hb + O_2 \longrightarrow HbO_2 \downarrow$$

(oxyhemoglobin)

"Down" arrow indicates that oxygen is entering the body.

It is remarkable that at the partial pressure[2] of oxygen in the lungs (P_{O_2} = about 100 mm Hg), hemoglobin is about 95% saturated (fig. 14.12). Hemoglobin takes up oxygen in increasing amounts as the P_{O_2} increases and likewise gives it up as the P_{O_2} decreases. The curve begins to level off at about 90 mm Hg. This means that hemoglobin easily retains oxygen in the lungs but tends to release it in the tissues. This effect is potentiated by the fact that hemoglobin takes up oxygen more readily in the cool temperature (fig. 14.12*a*) and neutral pH (fig. 14.12*b*) of the lungs. On the other hand, it gives up oxygen more readily at the warmer temperature and more acidic pH of the tissues.[3]

External respiration, the exchange of oxygen for carbon dioxide between air within alveoli and blood in pulmonary capillaries, is dependent on the process of diffusion.

---

[2]Air exerts pressure, and the amount of pressure each gas exerts in air is called its partial pressure, symbolized by a capital P.

[3]

	pH	Temperature
Lungs	7.40	37° C (98.6° F)
Body	7.38	38° C (100.4° F)

Human Anatomy and Physiology

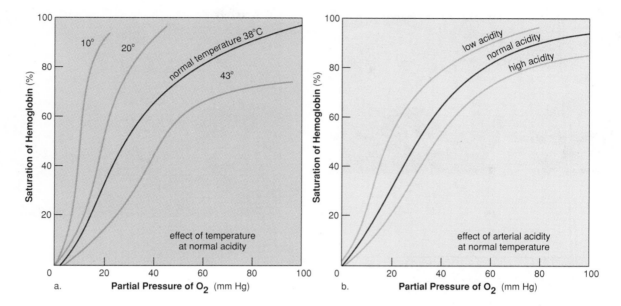

Figure 14.12

Hemoglobin saturation curves. Percent saturation of hemoglobin with oxygen (O_2) varies with the partial pressure of oxygen. Hemoglobin is more saturated in the lungs, where the temperature and acidity are lower, than in the tissues, where the temperature and acidity are higher. a. These data show that it takes a lower partial pressure of oxygen to saturate hemoglobin when the temperature is 10° C than when the temperature is 43° C. b. These data show that it takes a lower partial pressure of oxygen to saturate hemoglobin when the acidity is low than when the acidity is high.

Internal Respiration

The term *internal respiration* refers to the exchange of gases between blood in systemic capillaries and tissue fluid. When blood enters the systemic capillaries (fig. 14.11), oxyhemoglobin gives up oxygen which diffuses out of blood into the tissues:

$$HbO_2 \longrightarrow Hb + O_2$$

(oxyhemoglobin) (hemoglobin) (oxygen)

Diffusion of oxygen out of blood into the tissues occurs because the oxygen concentration of tissue fluid is low—the cells continuously use up oxygen in cellular respiration.

Diffusion of carbon dioxide into blood from the tissues occurs because carbon dioxide concentration of tissue fluid is high. Carbon dioxide is produced continuously by cells, and it collects in tissue fluid. Carbon dioxide enters the blood and then the red blood cells, where a small amount is taken up by hemoglobin, forming carbaminohemoglobin (p. 243). Most carbon dioxide combines with water to form carbonic acid, which dissociates to hydrogen ions and the bicarbonate ion. The enzyme carbonic anhydrase, present in red blood cells, speeds up the reaction:

$$CO_2 + H_2O \rightleftharpoons H_2CO_3 \rightleftharpoons H^+ + HCO_3^-$$

carbon dioxide   water   carbonic acid   hydrogen ion   bicarbonate ion

Figure 14.13

How colds are spread. a. Millions of viruses can be carried in the droplets of moisture in our breath, visible on a cold day. b. Computer graphics allows us to see the structure of a cold virus. These germs are spread through the air when a person with a cold sneezes or coughs.

a.

b.

14.2 Critical Thinking

1. Why is it better to give a person a mixture of oxygen (O_2) and carbon dioxide (CO_2) gases to stimulate breathing?
2. Why is it impossible for a person to commit suicide by holding his or her breath? (Hint—what gas builds up?)
3. Why is it workable for the carotid and aortic bodies to be sensitive to the hydrogen ion concentration [H^+] of the blood rather than O_2?
4. Why would you predict that the respiratory center is sensitive to the *presence* of carbon dioxide rather than to the *absence* of oxygen in the blood?

The globin portion of hemoglobin combines with excess hydrogen ions produced by the reaction, and Hb becomes HHb called reduced hemoglobin. In this way, the pH of blood remains fairly constant. The bicarbonate ion diffuses out of the red blood cells and is carried in the plasma.

Internal respiration, the exchange of oxygen for carbon dioxide between blood in the tissue capillaries and tissue fluid, is dependent on the process of diffusion.

Respiration and Health

We have seen that the full length of the respiratory tract is lined with a warm, wet mucous membrane lining that is constantly exposed to environmental air. The quality of this air, determined by the pollutants and the germs it contains, can affect our health.

Germs frequently spread from one individual to another by way of the respiratory tract. Droplets from one single sneeze can be loaded with billions of bacteria or viruses. The mucous membranes are protected by the production of mucus and by the constant beating of the cilia, but if the number of infective agents is large and/or our resistance is reduced, respiratory infection can result.

Common Respiratory Infections

Infections of the nasal cavities, the throat, the trachea, the bronchi, and the associated organs are fairly common.[4] Vaccines for these infections are not in wide use, and if they are viral in nature, antibiotics are not helpful. Since viruses take over the machinery of the cell when they reproduce, it is difficult to develop drugs that affect the virus without affecting the cell itself.

The Common Cold

A cold is a viral infection that usually begins as a scratchy sore throat, followed by a watery mucous discharge from the nasal cavities. There is rarely a fever, and symptoms are usually mild, requiring little or no medication. While colds have a short duration, immunity is also brief. Since it is estimated that there are over 150 cold-causing viruses (fig. 14.13), it is very difficult to avoid infection.

[4]Allergies are discussed on pages 274–75.

Human Anatomy and Physiology

Influenza

"Flu" is a viral infection of the respiratory tract, which is accompanied by aches and pains in the joints. There is usually fever, and the illness lasts for a longer length of time than a cold. Immunity is possible, but only the vaccine developed for the particular virus or viruses prevalent in a season is successful in protecting the individual during a flu epidemic. Flu viruses constantly mutate, which is why a new viral illness rapidly spreads from person to person and from place to place. Pandemics, in which a newly mutated flu virus spreads about the world, have occurred regularly about every 10 years.

Bronchitis

Viral infections can spread from the nasal cavities to the sinuses (sinusitis), to the middle ears (otitis media), to the larynx (laryngitis), and to the bronchi (bronchitis). Acute bronchitis usually is caused by a secondary bacterial infection of the bronchi, resulting in a heavy mucus discharge with much coughing. Acute bronchitis usually responds to antibiotic therapy. Chronic bronchitis is not necessarily due to infection. It often is caused by constant irritation of the lining of the bronchi, which as a result undergo degenerative changes, including the loss of cilia and their normal cleansing action. There is frequent coughing, and the individual is more susceptible to respiratory infections. Chronic bronchitis most often affects cigarette smokers.

Strep Throat

Strep throat is a very severe throat infection caused by the bacterium *Streptococcus pyogenes*. Swallowing may be difficult, and there is fever. Unlike a viral infection, strep throat should be treated with antibiotics. If not treated, it can lead to complications, such as rheumatic fever, in which the heart valves may be permanently affected.

Respiratory infections due to viral infections cannot be treated by antibiotics; bacterial infections can.

Lung Disorders

Pneumonia and tuberculosis are 2 serious infections of the lungs that ordinarily can be controlled by antibiotics. Two other illnesses discussed, emphysema and lung cancer, are not due to infections; in most instances, they are due to cigarette smoking.

Pneumonia

Most forms of pneumonia are caused by either a bacterium or a virus that has infected the lungs. The demise of AIDS patients is usually due to a particularly rare form of pneumonia caused by the protozoan *Pneumocystis carinii*. Sometimes, pneumonia is localized in specific lobes of the lungs. These lobes become inoperative as they fill with mucus and pus. Obviously, the more lobes involved, the more serious the infection.

Tuberculosis

Tuberculosis is caused by the tubercle bacillus. It is possible to tell if a person has ever been exposed to tuberculosis with a skin test in which a highly diluted extract of the bacilli is injected into the skin of the patient. A person who has never been in contact with the bacillus shows no reaction, but one who has developed immunity to the organism shows an area of inflammation that peaks in about 48 hours. If these bacilli invade the lung tissue, the cells build a protective capsule about the foreigners to isolate them from the rest of the body. This tiny capsule is called a *tubercle*. If the resistance of the body is high, the imprisoned organisms die, but if the resistance is low, the organisms eventually

Figure 14.14

Scanning electron micrograph of the lungs of a person with emphysema. There are large cavities in the lungs due to the breakdown of alveoli.

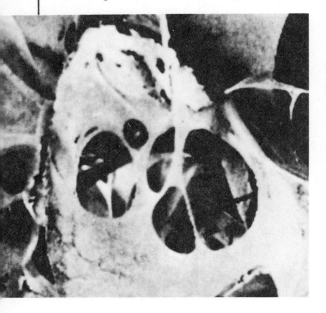

can be liberated. If a chest X ray detects tubercles, the individual is put on appropriate drug therapy to ensure the localization of the disease and the eventual destruction of any live bacterial organisms.

Emphysema

Emphysema refers to the destruction of lung tissue, with accompanying ballooning or inflation of the lungs due to trapped air. The trouble stems from the destruction and the collapse of the bronchioles. When this occurs, the alveoli are cut off from renewed oxygen supply and the air within them is trapped. The trapped air very often causes the alveolar walls to rupture (fig. 14.14) and a fibrous thickening of associated blood vessel walls. The victim is breathless and may have a cough. Since the surface area for gas exchange is reduced, not enough oxygen reaches the heart and the brain. Even so, the heart works furiously to force more blood through the lungs, which can lead to a heart condition. Lack of oxygen to the brain can make the person feel depressed, sluggish, and irritable.

Chronic bronchitis and emphysema are 2 conditions most often caused by smoking.

Pulmonary Fibrosis

Inhaling particles, such as silica (sand), coal dust, and asbestos (fig. 14.15), can lead to pulmonary fibrosis in which fibrous connective tissue builds up in the lungs. Breathing capacity can be seriously impaired, and the development of cancer is common. Since asbestos has been used so widely as a fireproofing and insulating agent, unwarranted exposure has occurred.

Lung Cancer

Lung cancer used to be more prevalent in men than in women, but recently lung cancer has surpassed breast cancer as a cause of death in women. This

Figure 14.15

Asbestos fibers. a. Polarized light photograph shows the long, thin, flexible, and strong fibers that make up asbestos. b. A scanning electron micrograph of macrophages reveals an "asbestos body" in the lung tissue of a person exposed to asbestos for some time. Asbestos bodies are fibers that have been coated with iron, plasma proteins, and other materials.

a.

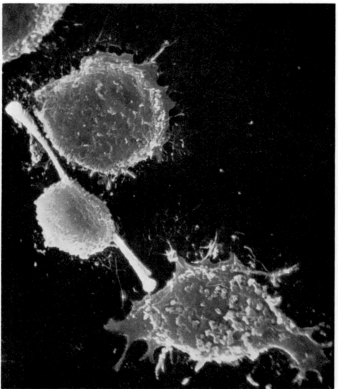

b.

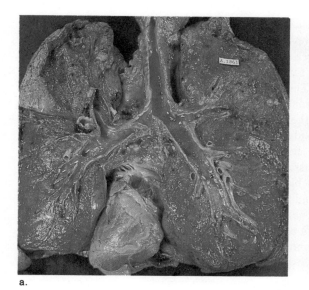

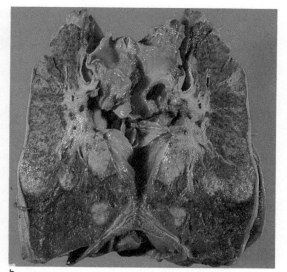

a.

b.

Normal lung versus cancerous lung. a. *Normal lung with heart in place. Notice the healthy red color.* b. *Lungs of a heavy smoker. Notice how black the lungs are except where cancerous tumors have formed.*

The Risks of Smoking versus the Benefits of Quitting

Based on available statistics, the American Cancer Society informs us of the risks of smoking versus the benefits of quitting.

Risks of Smoking

Shortened life expectancy. Twenty-five-year-old, 2-pack-a-day smokers have a life expectancy of 8.3 years shorter than nonsmoking contemporaries. Other smoking levels: proportional risk.

Lung cancer. Smoking cigarettes is "major cause in both men and women."

Larynx cancer. In all smokers (including pipe and cigar), it is 2.9-17.7 times that of nonsmokers.

Mouth cancer. Cigarette smokers have 3-10 times as many oral cancers as nonsmokers. Pipes, cigars, chewing tobacco are also major risk factors. Alcohol seems to be synergistic carcinogen with smoking.

Cancer of esophagus. Cigarettes, pipes, and cigars increase risk of dying of esophageal cancer about 2-9 times. Synergistic relationship exists between smoking and alcohol.

Cancer of bladder. Cigarette smokers have 7-10 times risk of bladder cancer as nonsmokers. Also synergistic with certain exposed occupations: dye-stuffs, etc.

Benefits of Quitting

Reduced risk of premature death after quitting. After 10-15 years, exsmokers' risk approaches that of those who have never smoked.

Gradual decrease in risk after quitting. *After 10-15 years, risk approaches that of those who never smoked.*

Gradual reduction of risk after smoking cessation. *Reaches normal after 10 years.*

Reducing or eliminating smoking/drinking reduces risk in first few years; *risk drops to level of nonsmokers in 10-15 years.*

Since risks are dose related, reducing or eliminating smoking/ drinking *should have risk-reducing effect.*

Risk decreases gradually to that of nonsmokers over 7 years.

Risks of Smoking

Cancer of pancreas. Cigarette smokers have 2-5 times risk of dying of pancreatic cancer as nonsmokers.

Coronary heart disease. Cigarette smoking is major factor; responsible for 120,000 excess U.S. deaths from coronary heart disease (CHD) each year.

Chronic bronchitis and pulmonary emphysema. Cigarette smokers have 4-25 times risk of death from these diseases as nonsmokers. Damage seen in lungs of even young smokers.

Stillbirth and low birthweight. Smoking mothers have more stillbirths and babies of low birthweight, who are more vulnerable to disease and death.

Children of smoking mothers are smaller, underdeveloped physically and socially, 7 years after birth.

Peptic ulcer. Cigarette smokers get more peptic ulcers and die more often of them; cure is more difficult in smokers.

Allergy and impairment of immune system.

Alters pharmacologic effects of many medicines, diagnostic tests, and greatly increases risk of thrombosis with oral contraceptives.

Benefits of Quitting

Since there is evidence of dose-related risk, reducing or eliminating smoking should have risk-reducing effect.

Sharply decreases risk after one year of not smoking. After 10 years, exsmokers' risk is same as that of those who never smoked.

Cough and sputum disappear during first few weeks after quitting. *Lung function may improve* and rate of deterioration may slow down.

Women who stop smoking before fourth month of pregnancy *eliminate risk of stillbirth and low birthweight* caused by smoking.

Since children of nonsmoking mothers are bigger and more advanced socially, inference is that *not smoking during pregnancy might avoid such underdeveloped children.*

Exsmokers get ulcers, but these are *more likely to heal rapidly and completely* than those of smokers.

Since these are direct, immediate effects of smoking, they are obviously *avoidable by not smoking.*

Majority of blood components elevated by smoking return to normal after cessation. Nonsmokers on the Pill have much lower risks of thrombosis.

Reproduced by permission of the American Cancer Society, Inc.

can be linked to an increase in the number of women who smoke today. Autopsies on smokers have revealed the progressive steps by which the most common form of lung cancer develops. The first event appears to be thickening or callusing of the cells that line the bronchi. (Callusing occurs whenever cells are exposed to irritants.) Then there is a loss of cilia so that it is impossible to prevent dust and dirt from settling in the lungs. Following this, cells with atypical nuclei appear in the callused lining. A disordered collection of cells with atypical nuclei is considered to be cancer in situ (at one location). A final step occurs when some of these cells break loose and penetrate the other tissues, a process called metastasis. Now the cancer has spread. The tumor may grow until the bronchus is blocked, cutting off the supply of air to that lung. The lung then collapses, and the secretions trapped in the lung spaces become infected, with the result pneumonia or the formation of a lung abscess. The only treatment that offers a possibility of cure, before secondary growths have had time to form, is to remove the lung completely. This operation is called *pneumonectomy.*

The incidence of lung cancer is much higher in individuals who smoke than in those who do not smoke.

A recent finding is that *involuntary smoking,* simply breathing in air filled with cigarette smoke, also can cause lung cancer and other illnesses associated with smoking. The reading on page 295 lists these various illnesses associated with smoking. If a person stops both voluntary and involuntary smoking and if the body tissues are not already cancerous, they return to normal.

Summary

Air enters and exits the lungs by way of the respiratory tract (table 14.2). Inspiration begins when the breathing center in the medulla oblongata sends excitatory nerve impulses to the diaphragm and the rib cage. As they contract, the diaphragm lowers and the rib cage moves up and out; the lungs expand, creating a partial vacuum that causes air to rush in. Nerves within the expanded lungs then send inhibitory impulses to the breathing center. As the diaphragm relaxes, it resumes its dome shape, and as the rib cage retracts, air is pushed out of the lungs during expiration.

External respiration occurs when carbon dioxide (CO_2) leaves blood and oxygen (O_2) enters blood at the alveoli. Oxygen is transported to the tissues in combination with hemoglobin. Internal respiration occurs when oxygen leaves blood and carbon dioxide enters blood at the tissues. Carbon dioxide is carried to the lungs in the form of the bicarbonate ion (HCO_3^-).

There are a number of illnesses associated with the respiratory tract. In addition to colds and flu, pneumonia and tuberculosis are serious lung infections. Two illnesses that have been attributed to breathing polluted air are emphysema and lung cancer.

Study Questions

1. Name and explain the 4 parts of respiration. (p. 280)
2. List the parts of the respiratory tract. What are the special functions of the nasal cavity, the larynx, and the alveoli? (pp. 281–84)
3. What are the steps in inspiration and expiration? How is breathing controlled? (pp. 285–87)
4. Why can we not breathe through a very long tube? (p. 288)
5. What physical process is believed to explain gas exchange? (p. 290)
6. What 2 equations are needed to explain external respiration? (p. 290)
7. How is hemoglobin remarkably suited to its job? (p. 291)
8. What 2 equations are needed to explain internal respiration? (p. 291)
9. Name and discuss some infections of the respiratory tract. (pp. 292–93)
10. What are emphysema and pulmonary fibrosis, and how do they affect a person's health? (p. 294)
11. By what steps is cancer believed to develop in the person who smokes? (pp. 294–96)

Objective Questions

1. In tracing the path of air, the
_____ immediately follows the
pharynx.
2. The lungs contain air sacs called
_____ .
3. The breathing rate is primarily
regulated by the amount of
_____ and _____ in
blood.
4. Air enters the lungs after they have
_____ .
5. Carbon dioxide (CO_2) is carried in
blood as the _____ ion.
6. The hydrogen (H^+) ions given off when
carbonic acid (H_2CO_3) dissociates are
carried by _____ .
7. Gas exchange is dependent on the
physical process of _____ .
8. Reduced hemoglobin becomes
oxyhemoglobin in the _____ .
9. The most likely cause of emphysema
and chronic bronchitis is
_____ .
10. Most cases of lung cancer actually
begin in the _____ .

Answers to Objective Questions
1. larynx 2. alveoli 3. CO_2, H^+ 4. expanded
5. bicarbonate 6. globin portion of hemoglobin
7. diffusion 8. lungs 9. cigarette smoking
10. bronchi

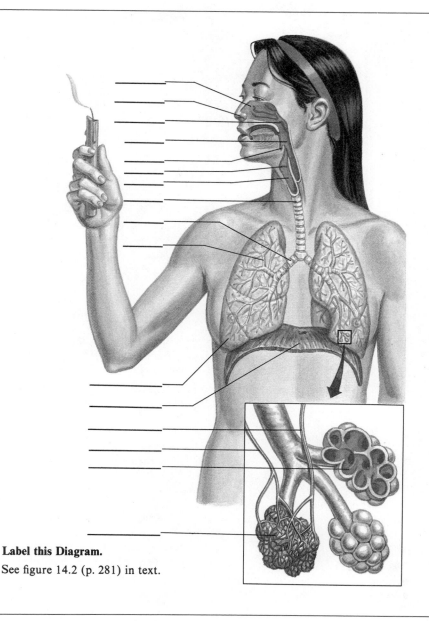

Label this Diagram.

See figure 14.2 (p. 281) in text.

Selected Key Terms

alveolus (al-ve′o-lus) (pl. alveoli)
saclike structure that is an air sac of a
lung. *283*

bronchus (brong′kus) (pl. bronchi) one
of the 2 major divisions of the trachea
leading to the lungs. *282*

bronchiole (brong′ke-ōl) the smaller air
passages in the lungs of mammals. *283*

diaphragm (di′ah-fram) a sheet of
muscle that separates the thoracic cavity
from the abdominal cavity. *285*

expiration (eks″pi-ra′shun) process of
expelling air from the lungs; exhalation.
280

external respiration (eks-ter′nal res″pi-
ra′shun) exchange between blood and
alveoli of carbon dioxide and oxygen.
280

hemoglobin (he″mo-glo′-bin) a red, iron-
containing pigment in blood that
combines with and transports oxygen.
290

inspiration (in″spi-ra′shun) the act of
breathing in. *280*

internal respiration (in-ter′nal res″pi-
ra′shun) exchange between blood and
tissue fluid of oxygen and carbon
dioxide. *280*

rib (rib) bone hinged to the vertebral
column and sternum that, with muscle,
defines the top and sides of the thoracic
cavity. *285*

trachea (tra′ke-ah) a tube that is
supported by C-shaped cartilagenous
rings that lies between the larynx and
the bronchi; the windpipe. *282*

ventilation (ven″tī-la′shun) breathing;
the process of moving air into and out of
the lungs. *284*

vocal cord (vo′kal kord) fold of tissue
within the larynx that creates vocal
sounds when it vibrates. *282*

Chapter 15

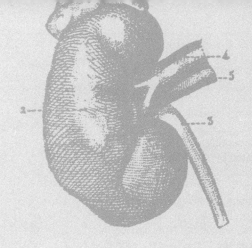

Excretion

Chapter Concepts

1 Excretion rids the body of unwanted substances, particularly the end products of metabolism.

2 Several organs assist in the process of excretion, but the kidneys, which are a part of the urinary system, are the primary organs of excretion.

3 The formation of urine by the more than 1 million nephrons present in each kidney serves not only to rid the body of nitrogenous wastes but also to regulate the salt/water content and the pH of blood.

4 The kidneys, the malfunction of which causes illness and perhaps death, are important organs of homeostasis.

Chapter Outline

Excretory Substances and Organs
    Nitrogenous End Products
    Other Excretory Substances
    Organs of Excretion
The Urinary System
    The Path of Urine
    Kidneys
Urine Formation
    Pressure Filtration
    Selective Reabsorption
    Tubular Excretion
    Reabsorption of Water
Regulatory Functions of the Kidneys
    Adjustment of Blood pH and Ion
      Balance
    Blood Volume
Problems with Kidney Function
    Kidney Replacement

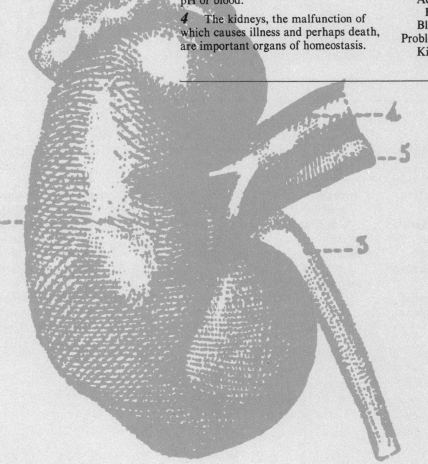

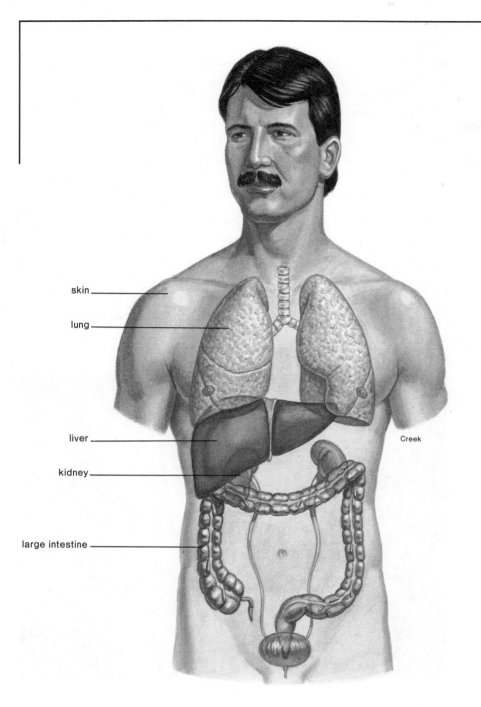

skin

lung

liver

kidney

large intestine

Creek

Figure 15.1

The organs of excretion. The lungs excrete carbon dioxide (CO_2); the liver excretes hemoglobin breakdown products in bile; the large intestine excretes certain heavy metals; the skin excretes perspiration; and the kidneys excrete urine. Excretion, ridding the body of metabolic wastes, should not be confused with defecation, ridding the body of nondigestable remains.

The composition of blood serving the tissues remains relatively constant due to both the continual addition of substances needed by cells and the continual removal of substances not needed by cells. In previous chapters, we discussed how the digestive tract and the lungs add nutrients and oxygen to blood. In this chapter, we discuss how the organs of excretion (fig. 15.1) remove substances from blood and thereby help to maintain homeostasis.

Excretory Substances and Organs

Excretion rids the body of metabolic wastes. Among these wastes are the toxic products listed in table 15.1. In addition, salts and water are constantly excreted.

Several of the end products excreted by humans are related to nitrogen metabolism since amino acids, nucleotides, and creatine all contain nitrogen.

Excretion

Table 15.1 Some Metabolic End Products

Name	End Product of	Primarily Excreted by
Nitrogenous Wastes		
Ammonia	Amino acid metabolism	Kidneys
Urea	Amino acid metabolism	Kidneys and skin
Uric acid	Nucleotide metabolism	Kidneys
Creatinine	Creatine phosphate metabolism	Kidneys
Other Wastes		
Bile pigments	Hemoglobin metabolism	Liver
Carbon dioxide	Cellular respiration	Lungs
Metals		
Iron	Hemoglobin metabolism	Large intestine
Calcium	Muscle and nerve metabolism	

Figure 15.2

Marine fishes. Most aquatic animals, including most bony fishes, excrete ammonia as their nitrogenous waste. Marine fishes, such as these, excrete excess salt by way of their gills.

Nitrogenous End Products

Ammonia (NH_3) arises from the deamination, or removal, of amino groups from amino acids. Deamination occurs in the liver. Ammonia is extremely toxic, and only animals living in water, who continually flush out their body with water, excrete ammonia (fig. 15.2). In our body, ammonia is converted to urea by the liver.

Urea is produced in the liver by a complicated series of reactions called the urea cycle. In this cycle, carrier molecules take up carbon dioxide (CO_2) and 2 molecules of ammonia to release finally a combined product, urea:

$$H_2N - \overset{\overset{\textstyle O}{\|}}{C} - NH_2$$

Uric acid is excreted as the general nitrogenous end product of many terrestrial animals that need to conserve water (fig. 15.3). In humans, uric acid only occurs when nucleotides are broken down metabolically; if uric acid is present in excess, it precipitates out of the plasma. Crystals of uric acid sometimes collect in the joints, producing a painful ailment called gout.

Creatinine is an end product of muscle metabolism. It results when creatine phosphate, a molecule that serves as a reservoir of high-energy phosphate, breaks down.

Other Excretory Substances

Other excretory substances are bile pigments, carbon dioxide (CO_2), ions (salts), and water (H_2O).

Bile Pigments

Bile pigments are derived from the heme portion of hemoglobin and are incorporated into bile within the liver (fig. 15.4). Although the liver produces bile, it is stored in the gallbladder before passing into the small intestine by way of ducts. If for any reason a bile duct is blocked, bile spills out into the blood, producing a condition called jaundice in which the skin is discolored (p. 199).

Carbon Dioxide

The lungs are the major organs of *carbon dioxide* (CO_2) excretion, although the kidneys are also important. The kidneys excrete bicarbonate ions (HCO_3^-), the form in which carbon dioxide is carried in the blood.

Ions

Various ions (salts) that have participated in metabolism are excreted. The blood level of these ions is important to the pH, the osmotic pressure, and the electrolyte balance of blood. The balance of potassium (K^+) ions and sodium (Na^+) ions is important to nerve conduction. The level of calcium (Ca^{++}) ions in blood affects muscle contraction; iron (Fe^{++}) ions take part in hemoglobin metabolism; and magnesium (Mg^{++}) ions help many enzymes to function properly.

Water

Water (H_2O) is an end product of metabolism; it also is taken into the body when food and liquids are consumed. The amount of fluid in the blood helps

Figure 15.3
Seabirds on cliff. Birds excrete uric acid, a solid material, as their nitrogenous waste. It is mixed with fecal material in a common repository for the urinary, digestive, and reproductive systems. Seabirds congregate in such numbers that their droppings build up to give a nitrogen-rich substance called guano. At one time, guano was harvested for natural fertilizer.

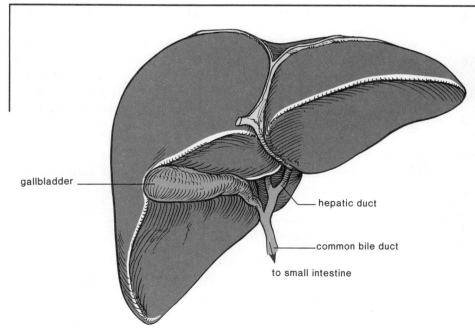

gallbladder

hepatic duct

common bile duct

to small intestine

Figure 15.4
The liver. This is an organ of excretion because it breaks down hemoglobin and these products become the bile pigments. The liver makes bile, which is stored in the gallbladder before being sent to the small intestine by way of ducts.

Excretion

to determine blood pressure. Treatment of hypertension sometimes includes the administration of a diuretic drug that increases the excretion of sodium and water by the kidneys.

Urea, salts, and water are the primary constituents of human urine. Carbon dioxide is excreted as a gas in the lungs and as the bicarbonate ion in the kidneys.

Organs of Excretion

The kidneys are the primary excretory organs, but there are other organs that also function in excretion (fig. 15.1), such as those described in the discussion that follows.

The Skin

The sweat glands in the skin (see fig. 9.12) excrete perspiration, which is a solution of water, salt, and some urea. A sweat gland is made up of a coiled tubule portion in the dermis and a narrow, straight duct that exits from the epidermis. Although perspiration is an excretion, we perspire not so much to rid the body of waste as to cool the body. The body cools because heat is lost as perspiration evaporates. Sweating keeps the body temperature within normal range during muscular exercise or when the outside temperature rises. In times of renal failure, more urea than usual may be excreted by the sweat glands, to the extent that a so-called urea frost is observed on the skin.

The Liver

The liver excretes bile pigments that are incorporated into bile, a substance stored in the gallbladder before it passes into the small intestine by way of ducts (fig. 15.4). The yellow pigment found in urine, called urochrome, also is derived from the breakdown of heme, but this pigment is deposited in blood and subsequently is excreted by the kidneys.

Lungs

The process of expiration (breathing out) not only removes carbon dioxide (CO_2) from the body, it also results in the loss of water (H_2O). The air we exhale contains moisture, as demonstrated by blowing onto a cool mirror.

The Large Intestine

Certain salts, such as those of iron and calcium, are excreted directly into the cavity of the large intestine by the epithelial cells lining it. These salts leave the body in the feces.

At this point, it is helpful to remember that the term *defecation,* and not *excretion,* is used to refer to the elimination of feces from the body. Substances that are excreted are waste products of metabolism. Undigested food and bacteria, which make up feces, have never been a part of the functioning of the body, but salts that are passed into the gut are excretory substances because they were once metabolites in the body.

Kidneys

The kidneys excrete urine, which ordinarily contains organic wastes and salts (table 15.2). The kidneys are a part of the urinary system.

There are various organs that excrete metabolic wastes, but only the kidneys consistently rid the body of urea.

Table 15.2 Composition of Urine	
Water	95%
Solids	5%
Organic Wastes	(per 1,500 ml of urine)
Urea	30 g
Creatinine	1–2 g
Ammonia	1–2 g
Uric acid	1 g
Ions (Salts)	25g
Positive Ions	*Negative Ions*
Sodium	Chlorides
Potassium	Sulfates
Magnesium	Phosphates
Calcium	

Human Anatomy and Physiology

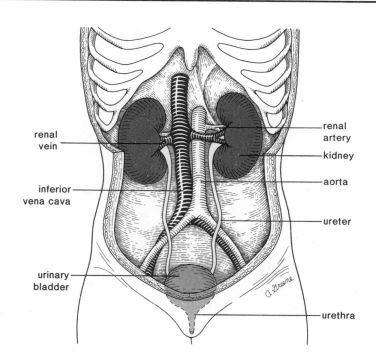

Figure 15.5

The urinary system. Urine is found only within the kidneys, the ureters, the urinary bladder, and the urethra.

The Urinary System

The urinary system includes the structures illustrated in figure 15.5 and listed in table 15.3. The organs are listed in order, according to the path of urine.

The Path of Urine

Urine is made by the **kidneys,** 2 bean-shaped, reddish-brown organs about the size of a fist. One kidney is found on either side of the vertebral column just below the diaphragm. The kidneys lie in depressions against the deep muscles of the back beneath the membranous lining of the abdominal cavity, where they also receive some protection from the lower rib cage. Each is covered by a tough fibrous capsule of connective tissue overlaid by adipose tissue.

The **ureters** are muscular tubes that convey the urine from the kidneys toward the bladder by peristaltic contractions. Urine enters the bladder by peristaltic contractions, in jets that occur at the rate of 5 per minute.

The **urinary bladder,** which can hold up to 600 ml of urine, is a hollow, muscular organ that gradually expands as urine enters. In the male, the urinary bladder lies ventral to the rectum, the seminal vesicles, and the vas deferens. In the female, the urinary bladder is ventral to the uterus and the upper vagina.

The **urethra,** which extends from the urinary bladder to an external opening, differs in length in the female and the male. In the female, the urethra lies ventral to the vagina and is only about 2.5 cm long. The short length of the female urethra invites bacterial invasion and explains why the female is more prone to bladder infections. In the male, the urethra averages 15 cm when the penis is relaxed. As the urethra leaves the urinary bladder, it is encircled by the prostate gland (see fig. 20.2). In older men, enlargement of the prostate gland can prevent urination, a condition that usually can be corrected surgically.

There is no connection between the genital (reproductive) and urinary systems in females (see fig. 20.6), but there is a connection in males. This double function does not alter the path of urine, and it is important to realize that urine is found only in those structures listed in table 15.3.

Table 15.3 Urinary System

Organ	Function
Kidneys	Produce urine
Ureters	Transport urine
Bladder	Stores urine
Urethra	Eliminates urine

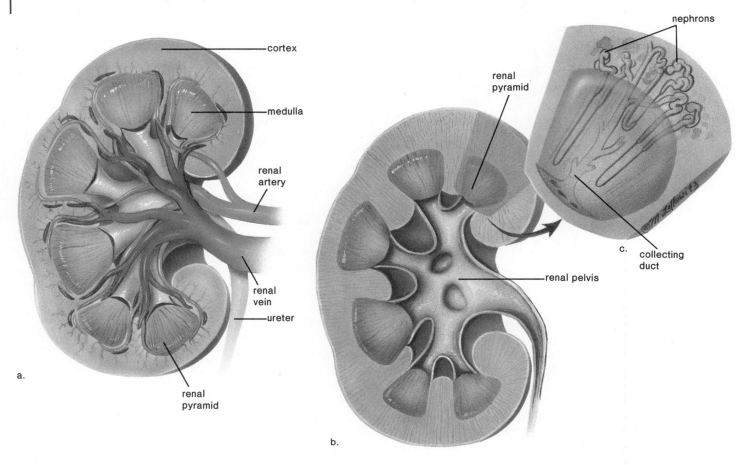

Figure 15.6

Gross anatomy of the kidney. a. *A longitudinal section of the kidney showing the blood supply. Note that the renal artery divides to give smaller arteries that frame the renal pyramids of the medulla. Smaller veins join to form the renal vein.* b. *Same section without the blood supply. Now it is easier to make out the cortex, the medulla, and the renal pelvis, which connects with the ureter.* c. *An enlargement of one renal pyramid showing the placement of the nephrons.*

Urination

When the urinary bladder fills with urine, stretch receptors send nerve impulses to the spinal cord; nerve impulses leaving the cord then cause the urinary bladder to contract and the sphincters to relax so that urination is possible. In older children and adults, it is possible for the brain to control this reflex, delaying urination until a suitable time.

Only the urinary system, consisting of the kidneys, the urinary bladder, the ureters, and the urethra, ever hold urine.

Kidneys

On the concave side of each kidney there is a depression where the renal blood vessels and the ureters enter (fig. 15.6a). When a kidney is sliced lengthwise, it is possible to make out 3 regions: (1) an outer granulated layer called the **cortex,** which dips down in between (2) a radially striated or lined layer called the **medulla,** and (3) an inner space, or cavity, called the renal **pelvis,** which is continuous with the ureter.

Upon closer examination, you can see that the medulla contains conical masses of tissue called renal pyramids. At the tip of each pyramid there is a tube that joins with others to form the renal pelvis.

Human Anatomy and Physiology

Figure 15.7

Nephron gross and microscopic anatomy. A nephron is made up of Bowman's capsule, the proximal convoluted tubule, the loop of Henle, the distal convoluted tubule, and the collecting duct. Bowman's capsule contains a capillary tuft called the glomerulus. Note the portions that are in the cortex and the portions that are in the medulla. The blowups show the types of the tissue at these different locations.

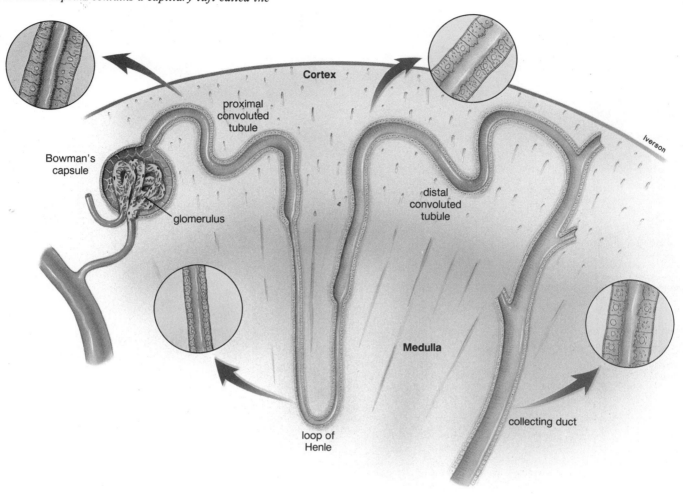

Nephrons

Microscopically, the kidney is composed of over 1 million **nephrons,** sometimes called renal or kidney tubules (fig. 15.7). Each nephron is made up of several parts. The blind end of the nephron is pushed in on itself to form a cuplike structure called **Bowman's capsule.** The outer layer of Bowman's capsule is composed of squamous epithelial cells; the inner layer is composed of specialized cells that allow easy passage of molecules. Next, there is a **proximal** (meaning near the Bowman's capsule) **convoluted tubule** in which the cells are cuboidal, with many mitochondria and an inner brush border (tightly packed microvilli). Then the cells become flat and the tube narrows and makes a U-turn to form the portion of the tubule called the **loop of Henle.** This leads to the **distal** (far from Bowman's capsule) **convoluted tubule,** where the cells are cuboidal, again with mitochondria but no brush border. The distal convoluted tubule enters the **collecting duct.**

Figure 15.6*b* shows that typically Bowman's capsules and convoluted tubules lie within the cortex and account for its granular appearance. Loops of Henle and collecting ducts lie within the triangular-shaped pyramids of the medulla. Because the loops and the ducts are longitudinal structures, they account for the striped appearance of the renal pyramids.

Figure 15.8

Steps in urine formation (simplified). a. The steps are noted at the location of the nephron where they occur. b. The molecules involved in the steps and the processes are listed.

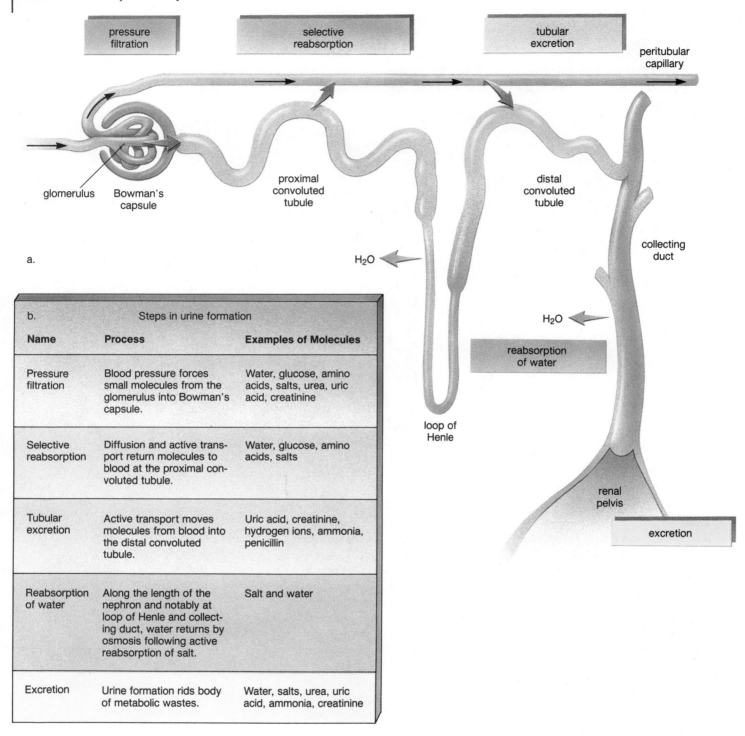

a.

pressure filtration

selective reabsorption

tubular excretion

peritubular capillary

glomerulus Bowman's capsule

proximal convoluted tubule

distal convoluted tubule

collecting duct

H_2O

H_2O

reabsorption of water

loop of Henle

renal pelvis

excretion

b.

Steps in urine formation		
Name	**Process**	**Examples of Molecules**
Pressure filtration	Blood pressure forces small molecules from the glomerulus into Bowman's capsule.	Water, glucose, amino acids, salts, urea, uric acid, creatinine
Selective reabsorption	Diffusion and active transport return molecules to blood at the proximal convoluted tubule.	Water, glucose, amino acids, salts
Tubular excretion	Active transport moves molecules from blood into the distal convoluted tubule.	Uric acid, creatinine, hydrogen ions, ammonia, penicillin
Reabsorption of water	Along the length of the nephron and notably at loop of Henle and collecting duct, water returns by osmosis following active reabsorption of salt.	Salt and water
Excretion	Urine formation rids body of metabolic wastes.	Water, salts, urea, uric acid, ammonia, creatinine

Figure 15.9

Nephron anatomy. a. *You can trace the path of blood about the nephron by following the arrows.* b. *Scanning electron micrograph of a glomerulus (the outer layer of Bowman's capsule has been removed.) The large cells adhering to the* glomerulus *are actually from the inner wall of Bowman's capsule. They are spaced wide apart; the spaces are called filtration slits.*

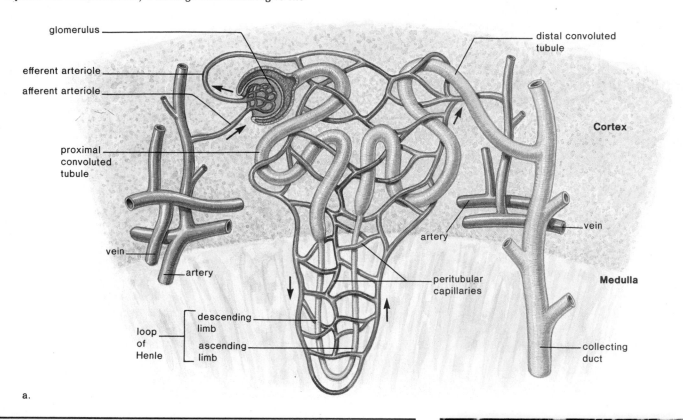

a.

b.

Table 15.4 Circulation about a Nephron	
Name of Structure	**Function**
Afferent arteriole	Brings arteriolar blood toward Bowman's capsule
Glomerulus	Capillary tuft enveloped by Bowman's capsule
Efferent arteriole	Takes arteriolar blood away from Bowman's capsule
Peritubular capillary network	Capillary bed that envelops the rest of the tubule
Venule	Takes venous blood away from the tubule

Note: Compare this table with figure 15.9.

Urine Formation

Figure 15.8 gives a simple overview of urine formation. Each nephron has its own blood supply, including 2 capillary regions. The **glomerulus** is a capillary tuft inside Bowman's capsule, and the **peritubular capillary** surrounds the rest of the nephron (table 15.4 and fig. 15.9). Urine formation requires the movement of molecules between these capillaries and the nephron. Three steps are involved: *pressure filtration, selective reabsorption,* and *tubular excretion.*

The pattern of blood flow about the nephron is critical to urine formation.

Table 15.5 Reabsorption from Nephron

Substance	Amount Filtered per Day	Amount Excreted per Day	Reabsorbtion (%)
Water (L)	180	1.8	99.0
Sodium (g)	630	3.2	99.5
Glucose (g)	180	0.0	100.0
Urea (g)	54	30.0	44.0

From A. J. Vander, et al., *Human Physiology,* 4th ed. Copyright © 1985 McGraw-Hill Publishing Company, New York, NY. Reprinted by permission of McGraw-Hill, Inc.

Pressure Filtration

Whole blood, of course, enters the afferent arteriole and the glomerulus (fig. 15.9b). Under the influence of glomerular blood pressure, which is usually about 60 mm Hg, small molecules move from the glomerulus to the inside of Bowman's capsule across the thin walls of each. This is a **pressure filtration** process because large molecules and formed elements are unable to pass through. In effect, then, blood that enters the glomerulus is divided into 2 portions: the filterable components and the nonfilterable components.

Filterable Blood Components	**Nonfilterable Blood Components**
Water	Formed elements (blood cells and
Nitrogenous wastes	platelets)
Nutrients	Proteins
Ions (salts)	

The filterable components form a filtrate called the **glomerular filtrate** that contains small dissolved molecules in approximately the same concentration as plasma. The filtrate stays inside of Bowman's capsule, and the nonfilterable components leave the glomerulus by way of the efferent arteriole.

A consideration of the preceding filterable substances leads us to conclude that if the composition of urine were the same as that of glomerular filtrate, the body would continually lose nutrients, water, and salts. Death from dehydration, starvation, and low blood pressure would quickly follow. Therefore, we can assume that the composition of the filtrate must be altered as this fluid passes through the remainder of the tubule.

During pressure filtration, water, salts, nutrient molecules, and waste molecules move from the glomerulus to the inside of Bowman's capsule. The filtered substances are called the glomerular filtrate.

Selective Reabsorption

Both passive and active reabsorption of molecules from the nephron to the blood of the peritubular capillary occur as the filtrate moves along the **proximal convoluted tubule.**

Because of *passive reabsorption,* even some urea is reabsorbed (table 15.5). However, we are particularly interested in the passive reabsorption of water (H_2O). Two factors aid this process. The nonfilterable proteins remain in blood, and salt is returned to blood. Following active reabsorption of sodium (Na^+) ions, chloride (Cl^-) ions follow passively, as does water. Therefore, water moves from the area of greater concentration in the filtrate to the area of lesser concentration in blood because of the osmolarity difference created largely by salt reabsorption. This process occurs along the length of the nephron, until eventually nearly all water and sodium ions have been reabsorbed (table 15.5).

Figure 15.10

The cells that line the lumen (inside) of the proximal convoluted tubule, where selective reabsorption takes place. a. This photo shows that the cells have a brushlike border composed of microvilli (mv) that greatly increases the surface area exposed to the lumen. The peritubular capillary surrounds the cells (nu = nucleus). b. Each cell has many mitochondria that supply the energy needed for active transport, the process that moves molecules from the lumen to the blood.

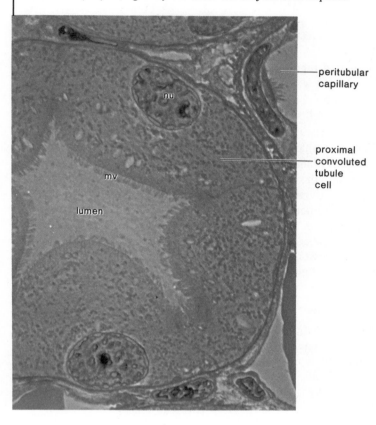

a.

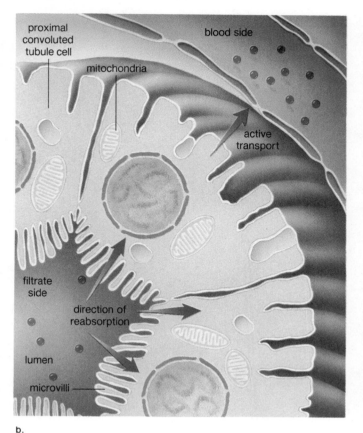

b.

The cells that line the proximal convoluted tubule are anatomically adapted for *active reabsorption* (fig. 15.10). These cells have numerous microvilli, each about 1 μm in length, that increase the surface area for reabsorption. In addition, the cells contain numerous mitochondria, which produce the energy necessary for active transport. Reabsorption by active transport is **selective reabsorption** because only molecules recognized by carrier molecules move across the membrane. After passing through the tubule cells, the molecules enter the blood of the peritubular capillary.

Glucose is an example of a molecule that ordinarily is reabsorbed completely (table 15.5). Such molecules are actually reabsorbed until their threshold level is obtained; thereafter, they appear in the urine. For example, the threshold level of glucose is about 180 mg of glucose per 100 ml of blood. After this amount is reabsorbed, any excess molecules present in the filtrate appear in the urine. In diabetes mellitus (sugar diabetes, p. 398), the filtrate contains excess glucose molecules because the liver fails to store glucose as glycogen.

We have seen that the filtrate that enters the proximal convoluted tubule is divided into 2 portions: the components that are reabsorbed from the tubule into blood and the components that are nonreabsorbed.

Reabsorbed Filtrate Components	**Nonreabsorbed Filtrate Components**
Most water	Some water
Nutrients	Much nitrogenous waste
Required ions (salts)	Excess ions (salts)

The substances that are not reabsorbed become the tubular fluid that enters the loop of Henle.

During selective reabsorption, nutrient and salt molecules are actively reabsorbed from the proximal convoluted tubule into the peritubular capillary, and water follows passively.

Tubular Excretion

The **distal convoluted tubule** continues the work of the proximal convoluted tubule in that salt and water are both reabsorbed. As before, sodium (Na^+) ions are actively reabsorbed into the blood capillary, and thereafter chloride (Cl^-) ions and water follow passively. In this region of the nephron also, substances are added to the urine by a process called tubular excretion, or augmentation. The cells that line this portion of the tubule have numerous mitochondria because **tubular excretion** is an active process just like selective reabsorption, although the molecules are moving in the opposite direction. Some molecules that are actively excreted are uric acid, creatinine, hydrogen (H^+) ions, ammonia, and penicillin.

During tubular excretion, certain molecules are actively secreted from the peritubular capillary into the fluid of the distal convoluted tubule. These molecules are found in urine.

Reabsorption of Water

Water is reabsorbed along the whole length of the nephron, but the excretion of a hypertonic urine (one that is more concentrated than blood) is dependent upon the action of the loop of Henle and the collecting duct.

The **loop of Henle,** which typically lies in the medulla (fig. 15.9a) is made up of a *descending* (going down) limb and an *ascending* (going up) limb. Salt (Na^+Cl^-) passively diffuses out of the lower portion of the ascending limb, but the upper thick portion of the limb actively transports salt out into the tissue of the outer medulla (fig. 15.11). Less and less salt is available for transport from the tubule as fluid moves up the thick portion of the ascending limb. In the end, there is an osmotic gradient within the tissues of the medulla: the concentration of salt is greater in the direction of the inner medulla. (Note that water cannot leave the ascending limb because the limb is impermeable to water.)

If you examine figure 15.11 carefully, you can see that the *innermost* portion of the inner medulla has the highest concentration of solutes. This cannot be due to salt because active transport of salt does not start until the thick portion of the ascending limb. Urea is believed to leave the lower portion of the collecting duct, and it is this molecule that contributes to the high solute concentration of the inner medulla. This urea passes into the loop of Henle and thereby eventually reaches the collecting duct once again.

Because of the solute concentration gradient of the medulla, water leaves the descending limb of the loop of Henle along its length. This is a *countercurrent mechanism*—the increasing concentration of solute encounters the decreasing number of water molecules in the descending limb, ensuring that water continues to leave the descending limb from the top to the bottom.

Fluid entering the **collecting duct** comes from the distal convoluted tubule. This fluid is isotonic to the cells of the cortex. This means that to this point, the net effect of reabsorption of water and salt is the production of a fluid that has the same tonicity as blood. Now, however, the collecting duct passes through the medulla, which is increasingly hypertonic, as previously explained (fig.

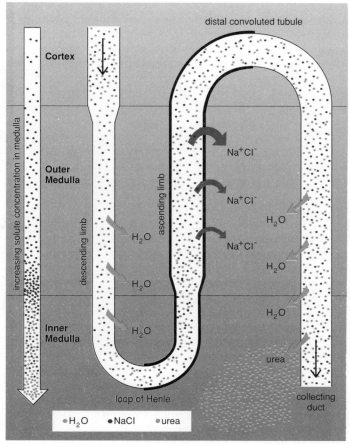

Figure 15.11
Reabsorption of water at the loop of Henle and the collecting duct. Salt (Na⁺Cl⁻) diffuses and is extruded by the ascending limb of the loop of Henle into the medulla; also, urea is believed to leave the collecting duct and to enter the tissues of the medulla. This creates a hypertonic environment that draws water out of the descending limb and the collecting duct. This water is returned to the circulatory system.

15.11). Therefore, water diffuses out of the collecting duct into the medulla, and the urine within the collecting duct becomes hypertonic to blood plasma.

Urine, the composition of which is listed in table 15.2, now passes out of the collecting duct into the pelvis of the kidney. Urine contains all the molecules that were not reabsorbed and the ones that underwent tubular excretion at the distal convoluted tubule.

It is interesting to note that human beings, along with other mammals, excrete a concentrated or hypertonic urine, not because water fails to enter the nephron, but because the water that enters is reabsorbed. A loop of Henle in *all* nephrons makes it possible for efficient reabsorption of water to occur. In lower vertebrates (fig. 15.12) up to and including the reptiles, the nephrons lack a loop of Henle and the kidney is incapable of producing a hypertonic urine. In birds, some nephrons have a loop of Henle, but the arrangement is not as efficient as in mammals. Reptiles and birds, unlike mammals, conserve water by excreting uric acid, a solid nitrogenous waste (fig. 15.3). This is possible because the urinary system empties into the cloaca, a common repository for the urinary, digestive, and reproductive systems.

Water diffuses from the descending limb of the loop of Henle and the collecting duct due to an increasingly hypertonic kidney medulla. Urine (table 15.2) formation is complete.

Figure 15.12

Presence or absence of loop of Henle in groups of vertebrate animals. Freshwater bony fishes and amphibians do not need to conserve water and there is no loop of Henle. Reptiles are terrestrial and conserve water by excreting a solid waste, uric acid, rather than by reabsorbing water. Only mammals are solely dependent on excreting a hypertonic urine to conserve water. They have a well-developed loop of Henle.

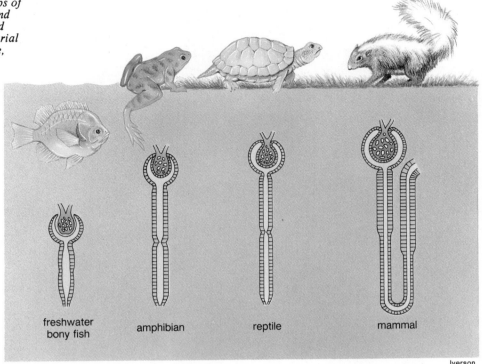

freshwater bony fish   amphibian   reptile   mammal

Iverson

Regulatory Functions of the Kidneys

Adjustment of Blood pH and Ion Balance

The kidneys help to maintain the pH level of blood within a narrow range, and the whole nephron takes part in this process. The excretion of hydrogen (H^+) ions and ammonia (NH_3), together with the reabsorption of sodium (Na^+) and bicarbonate ions (HCO_3^-), is adjusted to keep the pH within normal bounds. If blood is acidic, hydrogen ions are excreted in combination with ammonia, while sodium and bicarbonate ions are reabsorbed. This restores the pH because sodium ions promote the formation of hydroxide (OH^-) ions, while bicarbonate ions takes up hydrogen ions when carbonic acid (H_2CO_3) is formed. If blood is basic, fewer hydrogen ions are excreted and fewer sodium and bicarbonate ions are reabsorbed.

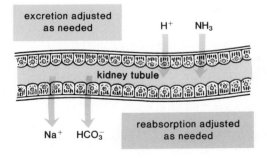

excretion adjusted as needed

H^+ NH_3

kidney tubule

Na^+ HCO_3^-

reabsorption adjusted as needed

These examples also show that the kidneys regulate the ion balance in blood by controlling the excretion and the reabsorption of various ions. Sodium (Na^+) is an important ion in plasma that must be regulated, but the kidneys also excrete or reabsorb other ions, such as the bicarbonate ion, potassium ions, and magnesium ions, as needed.

Human Anatomy and Physiology

Table 15.6 *Antidiuretic Hormone (ADH)*

Increase in ADH	Increased reabsorption of water	Less urine
Decrease in ADH	Decreased reabsorption of water	More urine

Blood Volume

Maintenance of blood volume and ion balance is under the control of hormones. **ADH (antidiuretic hormone)** is a hormone secreted by the posterior pituitary that primarily maintains blood volume. ADH increases the permeability of the collecting duct so that more water can be reabsorbed. In order to understand the function of this hormone, consider its name. *Diuresis* means increased amount of urine, and *antidiuresis* means decreased amount of urine. When ADH is present, more water is reabsorbed and a decreased amount of urine results. This hormone is secreted according to whether blood volume needs to be increased or decreased. When water is reabsorbed at the collecting duct, blood volume increases, and when water is not reabsorbed, blood volume decreases. In practical terms (table 15.6), if an individual does not drink much water on a certain day, the *posterior lobe of the pituitary* releases ADH, more water is reabsorbed, blood volume is maintained at a normal level, and, consequently, there is less urine. On the other hand, if an individual drinks a large amount of water and does not perspire much, the posterior lobe of the pituitary does not release ADH, more water is excreted, blood volume is maintained at a normal level, and a greater amount of urine is formed.

Drinking alcohol causes diuresis because it inhibits the secretion of ADH. The dehydration that follows is believed to contribute to the symptoms of a "hangover." Drugs called diuretics often are prescribed for high blood pressure. The drugs cause salts and water to be excreted; therefore, they reduce blood volume and blood pressure. Concomitantly, any *edema* (p. 259) present also is reduced.

Aldosterone, secreted by the adrenal cortex, is a hormone that primarily maintains sodium (Na^+) and potassium (K^+) ion balance. It causes the distal convoluted tubule to reabsorb sodium ions and to excrete potassium ions. The increase of sodium ions in the blood causes water to be reabsorbed, leading to an increase in blood volume and blood pressure.

Blood pressure is constantly monitored by the afferent arteriole cells within the juxtaglomerular apparatus. The juxtaglomerular apparatus (fig. 15.13) occurs at a region of contact between the afferent arteriole and the distal convoluted tubule. The afferent arteriole cells in the region secrete *renin* when blood pressure is insufficient to promote efficient filtration in the glomerulus. Renin is an enzyme that cleaves a large plasma protein, angiotensinogen. This process releases angiotensin I (fig. 15.13c). This molecule undergoes further cleavage by an enzyme called converting enzyme, which is present in the lining of the pulmonary capillaries of the lungs. Converting enzyme changes angiotensin I to angiotensin II, and angiotensin II stimulates the adrenal cortex to release aldosterone. Now, blood pressure rises.

A possible mechanism also has been suggested for the suppression of renin secretion. Perhaps the distal convoluted tubule cells in the juxtaglomerular apparatus are sensitive to sodium ion concentration in urine, and when there is a high sodium ion concentration in the distal convoluted tubule, they inhibit the afferent arteriole cells from secreting renin. If so, this mechanism may be faulty in certain individuals. The renin-angiotensin-aldosterone system

15.2 Critical Thinking

1. If 99% of water is reabsorbed (table 15.5), how can urine be 95% water (table 15.2)?
2. Carrier molecules work as fast as they can to return glucose to blood. Explain why excess glucose is not returned.
3. When CO_2 is excreted by the lungs, does blood become more acidic or more basic? If the HCO_3^- is excreted by the kidneys, does blood become more acidic or more basic?
4. The maintenance of normal blood pH is a very important function of the kidneys. What molecules in the cells are affected by pH changes?

Figure 15.13

Juxtaglomerular apparatus. a. This drawing shows how it is possible for the afferent arteriole and the distal convoluted tubule to lie next to one another. The juxtaglomerular apparatus occurs where they touch. b. Cross section shows exact location of the juxtaglomerular apparatus, which releases renin if the blood pressure in the afferent arteriole falls. c. Renin is an enzyme that cleaves angiotensinogen, a plasma protein made by the liver. This releases angiotensin I, which is changed to angiotensin II by a converting enzyme found in the lining of the pulmonary (lung) capillaries. Angiotensin II stimulates the adrenal cortex to release aldosterone into blood. Now, the blood pressure rises because sodium (Na^+) ions are reabsorbed to a greater extent.

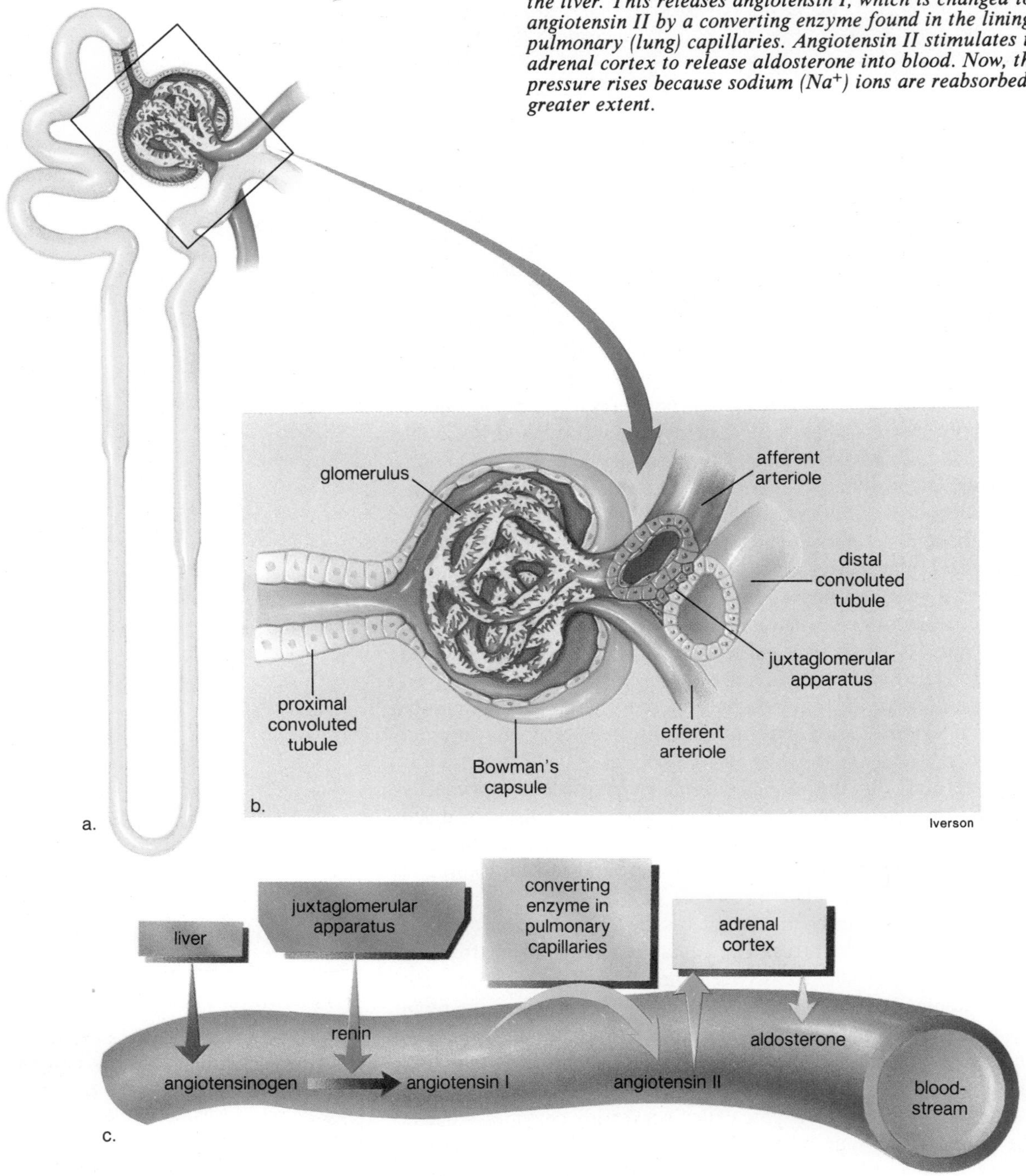

glomerulus

afferent arteriole

distal convoluted tubule

juxtaglomerular apparatus

proximal convoluted tubule

efferent arteriole

Bowman's capsule

a.

b.

Iverson

liver

juxtaglomerular apparatus

converting enzyme in pulmonary capillaries

adrenal cortex

renin

aldosterone

angiotensinogen → angiotensin I

angiotensin II

blood-stream

c.

always seems to be active in some patients with hypertension. In response to this possibility, there is a new medicine for hypertension that inhibits converting enzyme.

 The kidneys contribute to homeostasis by excreting urea. They also maintain both the pH and the ion balance of blood and regulate the volume of blood, 3 very important functions.

Problems with Kidney Function

Because of the great importance of the kidney to the maintenance of body fluid homeostasis, renal failure is a life-threatening event. There are many types of illnesses that cause progressive renal disease and renal failure.

Infections of the urinary tract themselves are a fairly common occurrence, particularly in the female since the urethra is considerably shorter than that of the male. If the infection is localized in the urethra, it is called *urethritis*. If it invades the urinary bladder, it is called *cystitis*. Finally, if the kidneys are affected, the infection is called *pyelonephritis*. Glomerular damage sometimes leads to blockage of the glomeruli so that no fluid moves into the tubules, or it can cause the glomeruli to become more permeable than usual. This is detected when a **urinalysis** is done. If the glomeruli are too permeable, albumin, white blood cells, or even red blood cells appear in the urine. Trace amounts of protein in the urine is not a matter of concern, however.

When glomerular damage is so extensive that more than two-thirds of the nephrons are incapacitated, waste substances accumulate in blood. This condition is called *uremia* because urea is one of these substances that accumulate. Although nitrogenous wastes can cause serious damage, the retention of water and salts is of even greater concern. The latter causes edema, fluid accumulation in the body tissues. Imbalance in the ionic composition of body fluids even can lead to loss of consciousness and to heart failure.

Kidney Replacement

Kidney Transplant

Patients with renal failure sometimes can undergo a kidney transplant operation, during which a functioning kidney from a donor is received. As with all organ transplants, there is the possibility of organ rejection. Receiving a kidney from a close relative has the highest chance of success. The current one-year survival rate is 97% if the kidney is received from a relative and 90% if it is received from a nonrelative.

Dialysis

If a satisfactory donor cannot be found for a kidney transplant, which is frequently the case, the patient can undergo dialysis treatments, utilizing either a kidney machine or continuous ambulatory peritoneal (abdominal) dialysis, or CAPD. Dialysis is defined as the diffusion of dissolved molecules through a semipermeable membrane. These molecules, of course, move across a membrane from the area of greater concentration to one of lesser concentration.

During hemodialysis (fig. 15.14), the patient's blood is passed through a semipermeable membranous tube that is in contact with a balanced salt (dialysis) solution. Substances more concentrated in blood diffuse into the dialysis solution, also called the dialysate. Conversely, substances more concentrated in the dialysate diffuse into blood. Accordingly, the artificial kidney can be utilized either to extract substances from blood, including waste products or toxic chemicals and drugs, or to add substances to blood, for example, bicarbonate ions (HCO_3^-) if blood is acidic. In the course of a 6-hour hemodialysis, from 50 to 250 grams of urea can be removed from a patient, which greatly exceeds the urea clearance of normal kidneys. Therefore, a patient need undergo treatment only about twice a week.

In the case of CAPD, a fresh amount of dialysate is introduced directly into the abdominal cavity from a bag attached to a permanently implanted plastic tube. Waste and water molecules pass into the dialysate from the surrounding organs before the fluid is collected 4 or 8 hours later. The individual can go about his or her normal activities during CAPD, unlike during hemodialysis.

15.3 Critical Thinking

1. Which of the steps given in figure 15.8 are also part of hemodialysis? Which are absent?
2. Why do patients not lose plasma proteins during hemodialysis?
3. Why is more urea excreted during hemodialysis than during urine formation?
4. What would you add to the dialysate to prevent the loss of glucose from blood?

Figure 15.14

Diagram of an artificial kidney. As the patient's blood circulates through dialysis tubing, it is exposed to a solution. Wastes exit from blood into the solution because of a preestablished concentration gradient. In this way, blood is not only cleansed, the pH also can be adjusted.

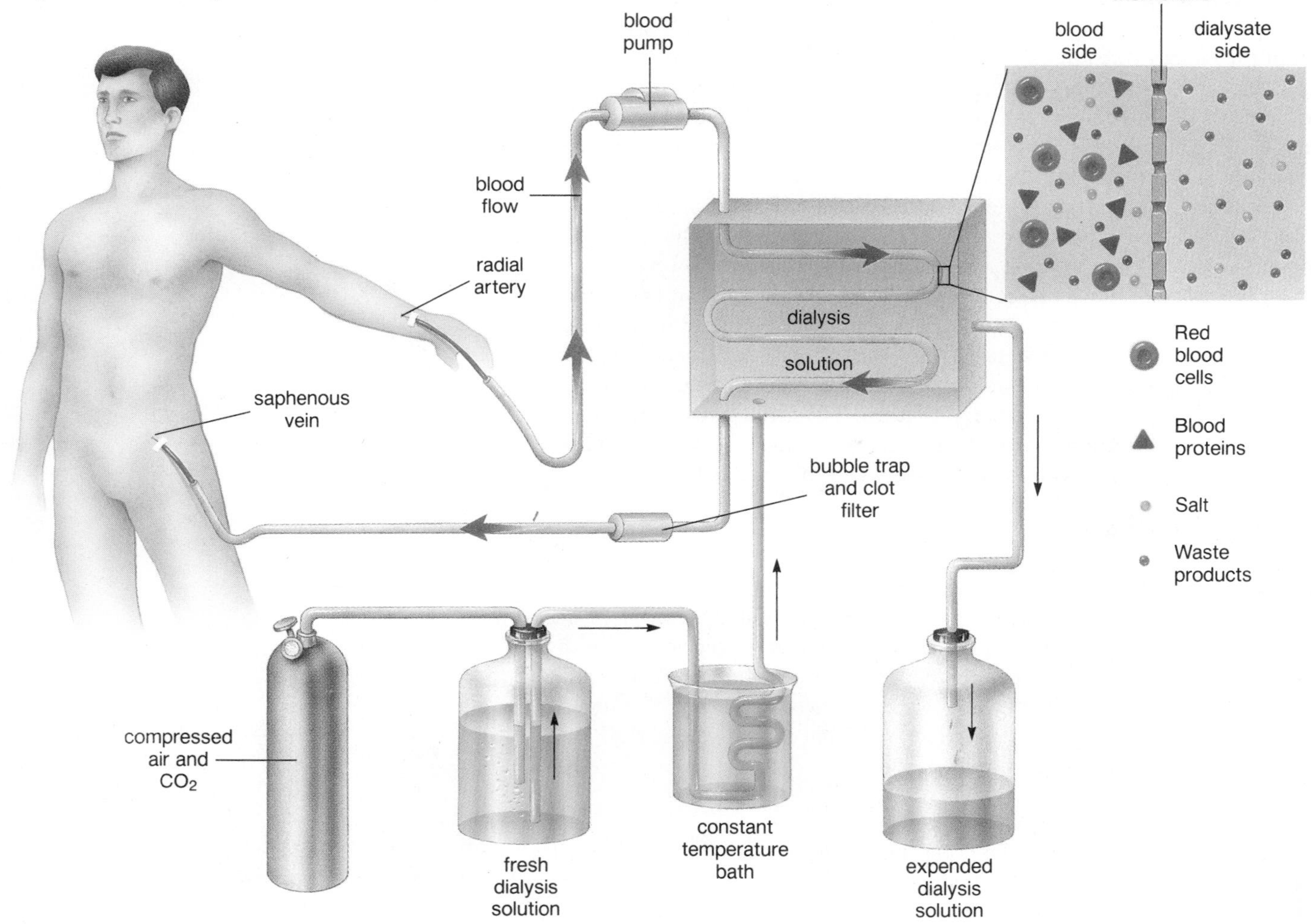

Kidney transplants and hemodialysis are available procedures for persons who have suffered renal failure.

Summary

The end products of metabolism are, for the most part, nitrogenous wastes, such as ammonia, urea, uric acid, and creatinine, all of which are excreted primarily by the urinary system. The urinary system contains the kidneys, whose macroscopic anatomy is dependent on nephrons. Urine formation requires 3 steps: during pressure filtration, small components of plasma pass into Bowman's capsule from the glomerulus due to blood pressure; during selective reabsorption, nutrients and sodium are actively reabsorbed from the proximal convoluted tubule back into blood; during tubular excretion, a few types of substances are actively secreted into the distal convoluted tubule from blood.

Water is reabsorbed along the length of the nephron, but it is the loop of Henle and the collecting duct that allow us to secrete a hypertonic urine. ADH, a hormone produced by the posterior pituitary, controls the reabsorption of water directly, and aldosterone from the adrenal cortex controls it indirectly by affecting sodium (Na^+) ion reabsorption. The whole nephron participates in maintaining the pH of blood by regulating the pH of urine. In practice, hydrogen (H^+) ions are excreted and sodium bicarbonate (HCO_3^-) ions are reabsorbed to maintain the pH.

Various problems can lead to kidney failure. In such cases, the person either can receive a kidney from a donor or undergo hemodialysis treatments by means of the kidney machine or CAPD.

Study Questions

1. Name 4 nitrogenous end products, and explain how each is formed in the body. (p. 300)
2. Name several excretory organs and the substances they excrete. (p. 302)
3. What is the composition of urine? (p. 302)
4. Give the path of urine. (p. 303)
5. Name the parts of a nephron. (p. 305)
6. Trace the path of blood about the nephron. (p. 306)
7. Describe how urine is made by telling what happens at each part of the tubule. (p. 307)
8. Explain these terms: *pressure filtration*, *selective reabsorption*, and *tubular excretion*. (pp. 308–10)
9. How does the nephron regulate the blood volume and the pH of blood? (p. 312)
10. Explain how the artificial kidney machine and CAPD work. (p. 315)

Objective Questions

1. The primary nitrogenous end product of humans is _____ .
2. The large intestine is an organ of excretion because it rids the body of _____ .
3. Urine leaves the urinary bladder in the _____ .
4. The capillary tuft inside Bowman's capsule is called the _____ .
5. _____ is a substance that is found in the filtrate, is reabsorbed, and is still in urine.
6. _____ is a substance that is found in the filtrate, is minimally reabsorbed, and is concentrated in the urine.
7. Tubular excretion takes place at the _____ , a portion of the nephron.
8. Reabsorption of water from the collecting duct is regulated by the hormone _____ .
9. In addition to excreting nitrogenous wastes, the kidneys adjust the _____ and _____ of blood.
10. Persons who have nonfunctioning kidneys often have their blood cleansed by _____ machines.

Answers to Objective Questions
1. urea 2. certain salts, e.g., calcium and iron 3. urethra 4. glomerulus 5. Water 6. Urea 7. distal convoluted tubule 8. ADH 9. volume, pH 10. hemodialysis

Label this Diagram.

See figure 15.9*a* (p. 308) in text.

antidiuretic hormone (an''ti-di''u-ret'ik hōr'mōn) ADH; sometimes called vasopressin, a hormone secreted by the posterior pituitary that controls the rate at which water is reabsorbed by the kidneys. *313*

Bowman's capsule (bo'manz kap'sūl) a double-walled cup that surrounds the glomerulus at the beginning of the nephron. *305*

collecting duct (kŏ-lekt'ing dukt) a tube that receives urine from several distal convoluted tubules. *305*

distal convoluted tubule (dis'tal kon'vo-lūt-ed tu'būl) highly coiled region of a nephron that is distant from Bowman's capsule. *305*

excretion (eks-kre'shun) removal of metabolic wastes. *299*

glomerular filtrate (fil'trāt) the filtered portion of blood that is contained within Bowman's capsule. *308*

glomerulus (glo-mer'u-lus) a cluster; for example, the cluster of capillaries surrounded by Bowman's capsule in a nephron. *307*

kidney (kid'nē) an organ in the urinary system that produces and excretes urine. *303*

nephron (nef'ron) the anatomical and functional unit of the vertebrate kidney; kidney tubule. *305*

pelvis (pel'vis) a hollow chamber in the kidney that lies inside the medulla and receives freshly prepared urine from the collecting ducts. *304*

peritubular capillary (per''i-tu'bu-lar kap'ĭ-lar''e) capillary that surrounds a nephron and functions in reabsorption during urine formation. *307*

pressure filtration (presh'ur fil-tra'shun) the movement of small molecules from the glomerulus into Bowman's capsule due to the action of blood pressure. *308*

proximal convoluted tubule (prok'sĭ-mal kon'vo-lūt-ed tu'būl) highly coiled region of a nephron near Bowman's capsule. *305*

selective reabsorption (sĕ-lek'tiv re''ab-sorp'shun) the movement of nutrient molecules, as opposed to waste molecules, from the contents of the nephron into blood at the proximal convoluted tubule. *309*

tubular excretion (tu'bu-lar eks-kre'shun) the movement of certain molecules from blood into the distal convoluted tubule so that they are added to urine. *310*

urea (u-re'ah) primary nitrogenous waste of mammals derived from amino acid breakdown. *300*

uric acid (u'rik as'id) waste product of nucleotide breakdown. *300*

ureter (u-re'ter) one of 2 tubes that take urine from the kidneys to the urinary bladder. *303*

urethra (u-re'thrah) tube that takes urine from the bladder to outside. *303*

urinary bladder (u'rĭ-ner''e blad'der) an organ where urine is stored before being discharged by way of the urethra. *303*

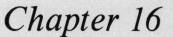

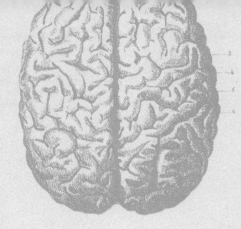

Chapter 16

The Nervous System

Chapter Concepts

1 The nervous system is made up of cells called neurons that are specialized to carry nerve impulses. A nerve impulse is an electrochemical change.

2 Transmission of impulses between neurons is accomplished by means of chemicals called neurotransmitter substances.

3 The nervous system consists of the central and peripheral nervous systems. The 2 systems are joined when a reflex occurs.

4 The central nervous system, made up of the spinal cord and the brain, is highly organized. Consciousness is a function only of the cerebrum, which is most highly developed in humans.

5 Drugs that affect the psychological state of the individual, such as alcohol, marijuana, cocaine, and heroin, are abused, sometimes to the detriment of the body.

Chapter Outline

Neurons
    The Structure of a Neuron
    The Nerve Impulse
Transmission across a Synapse
    Summation and Integration
    Neurotransmitter Substances
The Peripheral Nervous System
    The Somatic Nervous System
    The Autonomic Nervous System
The Central Nervous System
    The Spinal Cord
    The Brain
Drug Abuse
    Drug Action
    Alcohol
    Marijuana
    Cocaine
    Heroin
    Designer Drugs

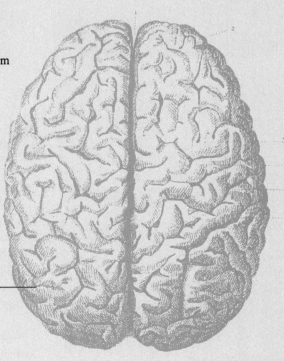

Figure 16.1

The nervous system. a. Nerve impulses flow along neurons (nerve cells) that are specialized for conducting and passing on nerve impulses from one to the other, much as a baton is passed from one runner to another in a relay race. b. The nervous system consists of the central nervous system (CNS) and the peripheral nervous system (PNS). Within the PNS, the nerves contain sensory neurons taking messages from sense organs to the CNS and motor neurons taking messages from the CNS to muscles and glands. Arrows show the direction of nerve impulses.

a.

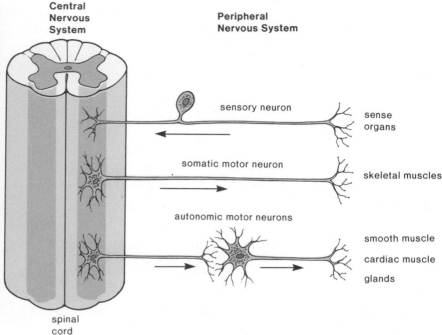

b.

The nervous system tells us that we exist and along with the muscles and sense organs accounts for our distinctly animal characteristic of quick reaction to environmental stimuli. The nerve cell is called a **neuron,** and it is neurons that carry nerve impulses (messages) from sense organs to a central receiving station (brain and spinal cord) before other impulses go to the muscles or the glands, which then react. As figure 16.1 indicates, the nervous system has 2 main divisions, the central and peripheral nervous systems. The **central nervous system (CNS),** which includes the brain and the spinal cord, lies in the midline of the body, while the **peripheral nervous system (PNS)** includes the nerves that project from the CNS. Somatic nerves include long fibers of sensory neurons and/or motor neurons; autonomic nerves include long fibers of motor neurons only.

Neurons

The Structure of a Neuron

All neurons (fig. 16.2) have 3 parts: dendrite(s), cell body, and axon. A **dendrite** conducts nerve impulses toward the **cell body,** the part of a neuron that contains the nucleus and other organelles. An **axon** conducts nerve impulses away from the cell body. There are 3 types of neurons: sensory neuron, motor neuron, and interneuron. A **sensory neuron** takes a message from a **receptor,** a sense organ, to the CNS and typically has a long dendrite and a short axon. A **motor neuron** takes a message away from the CNS to an **effector,** a muscle fiber or a gland, and has short dendrites and a long axon. Because motor neurons cause muscle fibers and glands to react, they are said to **innervate** these structures. Sometimes a sensory neuron is referred to as the *afferent neuron,* and the motor neuron is called the *efferent neuron.* These words, which are derived from Latin, mean running to and running away from, respectively. Obviously, they refer to the relationship of these neurons to the CNS.

Human Anatomy and Physiology

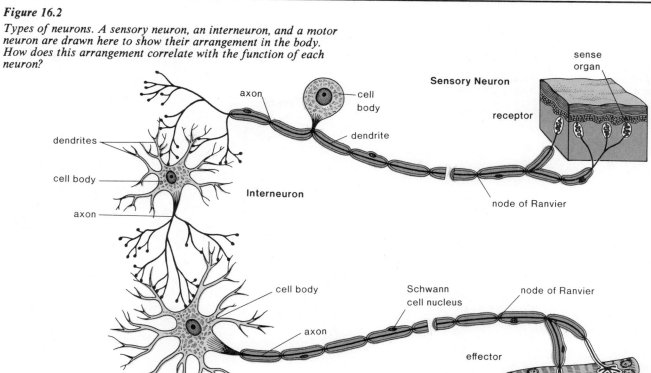

Figure 16.2

Types of neurons. A sensory neuron, an interneuron, and a motor neuron are drawn here to show their arrangement in the body. How does this arrangement correlate with the function of each neuron?

Table 16.1 Neurons		
Neuron	**Structure**	**Function**
Sensory neuron (afferent)	Long dendrite, short axon	Carry nerve impulses (messages) from periphery to the CNS*
Motor neuron (efferent)	Short dendrites, long axon	Carry nerve impulses (messages) from the CNS to periphery
Interneuron	Short dendrites, long or short axon	Carry nerve impulses (messages) within the CNS

*CNS = central nervous system

An **interneuron** (also called association neuron or connector neuron) always is found completely within the CNS and conveys messages between parts of the system. An interneuron has short dendrites and a long or a short axon. Table 16.1 summarizes the 3 types of neurons that also are illustrated in figure 16.2.

Although all neurons have the same 3 parts, each is specialized in structure and in function. Specialization is dependent on the location of the neuron in relation to the CNS.

The dendrites and the axons of neurons sometimes are called fibers, or processes. Most long fibers, whether dendrites or axons, are covered by tightly packed spirals of *Schwann cells* (fig. 16.3). Schwann cells first encircle an axon, and then as they wrap themselves around the axon many times, they lay

Figure 16.3

Neurilemma and myelin sheath. a. Axon of a motor neuron ending in a cross section of neurilemma and myelin sheath that encloses the long fibers of all neurons. The myelin sheath is composed of many layers of Schwann cell membrane and has a white, glistening appearance in the body. b. Electron micrograph of a cross section of an axon surrounded by neurilemma and myelin sheath.

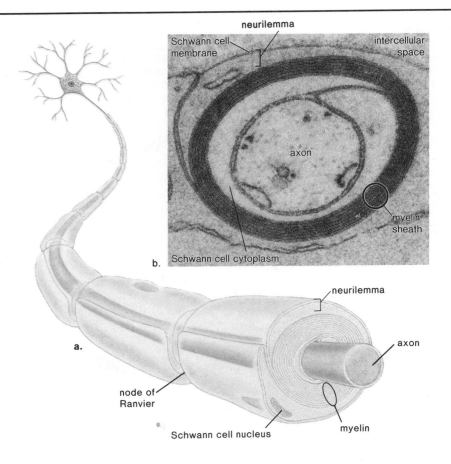

b.

Figure 16.4

Scientist working at an oscilloscope, the electrical recording device that measures changes in voltage wherever an electrode is placed on or inserted in a neuron.

down several layers of cellular membrane containing myelin, a lipid substance that is an excellent insulator. Myelin gives nerve fibers their white, glistening appearance. Because of the manner in which Schwann cells wrap themselves about nerve fibers, 2 sheaths are formed (fig. 16.3). The outermost sheath is called the **neurilemma,** or the cellular sheath, and the inner one is called the **myelin sheath.** The neurilemma plays an important role in nerve regeneration. If a nerve fiber is accidently severed, the part on the far side of the cell body degenerates except for the neurilemma, which can serve as a passageway for new growth stemming from the remaining portion of the nerve fiber. The myelin sheath is involved in nerve conduction and is discussed in the following section.

Schwann cells are one of several types of glial cells in the nervous system. Glial cells service the neurons—they have supportive and nutritive functions.

The Nerve Impulse

A neuron is specialized to conduct nerve impulses. The nature of a **nerve impulse** has been studied using giant axons from the squid and an instrument called a voltmeter. Voltage is a measure of the electrical potential difference between 2 points, which in this case are the inside and the outside of the axon. The change in voltage is displayed on an *oscilloscope,* an instrument with a screen that shows a trace, or pattern, indicating a change in voltage with time (fig. 16.4).

The Resting Potential

In the experimental setup shown in figure 16.5, an oscilloscope is wired to 2 electrodes, one inside and one outside a giant axon of the squid. The axon is essentially a membranous tube filled with cytoplasm axoplasm. When the axon is not conducting an impulse, the oscilloscope records a *membrane potential* (potential difference across a membrane) equal to about -65 mV (millivolts).

Human Anatomy and Physiology

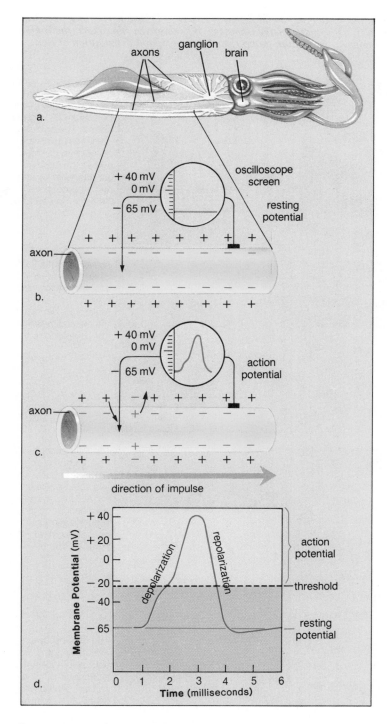

Figure 16.5
Nerve impulse. a. *The squid axons shown produce rapid muscular contraction so that the squid can move quickly.* b. *These axons are so large (about 1 mm in diameter) that a microelectrode can be inserted inside. When the axon is not conducting a nerve impulse, the electrode registers and the oscilloscope records a resting potential of −65 mV.* c. *When the axon is conducting a nerve impulse, the* threshold *for an action potential is achieved, and there is a rapid change in potential from −65 mV to +40 mV (called depolarization) followed by a return to −65 mV (called repolarization).* d. *Enlargement of action potential of nerve impulse.*

This reading indicates that the inside of the neuron is negative compared to the outside. This is called the **resting potential** because the axon is not conducting an impulse.

The existence of this polarity (charge difference) can be correlated with a difference in ion distribution on either side of the axomembrane (cell membrane of the axon). As figure 16.6a shows, there is a higher concentration of sodium (Na^+) ions outside the axon and a higher concentration of potassium (K^+) ions inside the axon. The unequal distribution of these ions is due to the action of the sodium-potassium pump. This is an active transport system in the cell membrane that pumps sodium ions out of and potassium ions into the axon (p. 74). The work of the pump maintains the unequal distribution of sodium and potassium ions across the axomembrane.

Figure 16.6

Action potential and resting potential. The action potential is the result of an exchange of sodium (Na^+) ions and potassium (K^+) ions, and it is recorded as a change in polarity by an oscilloscope (as shown on the right). So few ions are exchanged for each *action potential that it is possible for a nerve fiber to repeatedly conduct nerve impulses. Whenever the fiber rests, the sodium-potassium pump restores the original distribution of ions.*

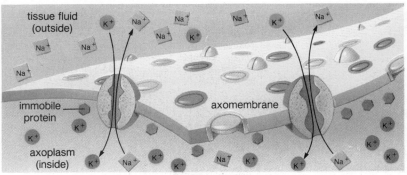

a. Resting potential

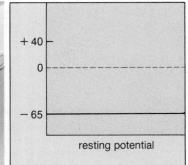

When a neuron is not conducting a nerve impulse, the sodium (Na^+) and potassium (K^+) gates are closed. The sodium-potassium pump maintains the uneven distribution of these ions across the axomembrane. The oscilloscope registers a resting potential of -65 mV inside compared to outside.

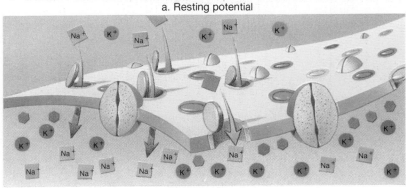

b. Sodium gates open

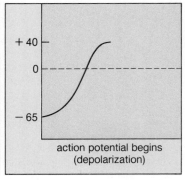

An action potential begins when the sodium gates open and sodium ions move to the inside. The oscilloscope registers a depolarization as the axoplasm reaches $+40$ mV compared to tissue fluid.

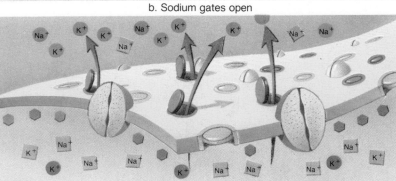

c. Potassium gates open

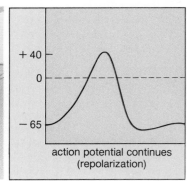

Recovery occurs as the sodium gates close and the potassium gates open, allowing potassium ions to move to the outside. The oscilloscope registers repolarization as the axoplasm again becomes -65 mV compared to tissue fluid.

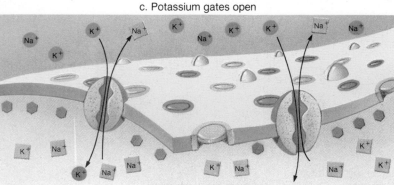

d. Sodium-potassium pump

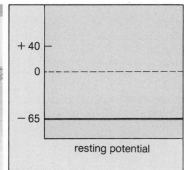

The oscilloscope registers -65 mV again, but the sodium-potassium pump is working to restore the original sodium and potassium ion distribution illustrated in *(a)*. The sodium and potassium gates are now closed but will open again in response to another stimulus.

The pump is always working because the membrane is somewhat permeable to these ions and they tend to diffuse toward their lesser concentration. Since the membrane is more permeable to potassium than to sodium, there are always more positive ions outside the axomembrane than inside; this accounts for the polarity recorded by the oscilloscope. There are also large, negatively charged proteins in the axoplasm, which are termed immobile in figure 16.6 because they are too large to cross the axomembrane.

The Action Potential

If the axon is stimulated to conduct a nerve impulse by an electric shock, by a sudden difference in pH, or by a pinch, a trace appears on the oscilloscope screen. This pattern, caused by rapid polarity changes and called the **action potential,** has an upswing and a downswing.

Sodium Gates Open   As the action potential swings up from -65 mV to $+40$ mV, sodium (Na^+) ions rapidly move across the axomembrane to the inside of the axon. The stimulation of the axon has caused the gates of the sodium channels to open temporarily, allowing sodium to flow into the axon. This sudden permeability of the axomembrane causes the oscilloscope to record a *depolarization:* the charge inside of the fiber changes from negative to positive as sodium ions enter (fig. 16.6b).

Potassium Gates Open   As the action potential swings down from $+40$ mV to at least -65 mV, potassium (K^+) ions rapidly move from the inside to the outside of the axon. The axomembrane suddenly has become permeable to potassium because the potassium gates of the potassium channels have opened temporarily, allowing potassium ions to flow out of the axon. The oscilloscope records a *repolarization* as the charge inside the axon becomes negative again (fig. 16.6c).

The Sodium-Potassium Pump   A fiber can conduct a volley of nerve impulses because only a small number of ions are exchanged with each impulse. When the fiber rests, however, there is a refractory period during which the sodium-potassium pump continues to return sodium (Na^+) ions to the outside and potassium (K^+) ions to the inside of the axon (fig. 16.6d). During the refractory period, a neuron is unable to conduct a nerve impulse.

All neurons, whether sensory or motor, transmit the same type of nerve impulse— an electrochemical change that is propagated along the nerve fiber(s).

The Speed of Conduction

The oscilloscope records changes at only one location in a nerve fiber. Actually, however, the action potential travels along the length of a fiber (fig. 16.7).

In invertebrates, nerve fibers do not have a myelin sheath as they do in vertebrates. The speed of conduction in invertebrate nerve fibers can reach 20 m per second, but the speed of conduction in myelinated vertebrate fibers can reach 200 m per second. Notice in figure 16.2 that the myelin sheath has gaps between Schwann cells called the **nodes of Ranvier.** The speed of conduction in myelinated fibers is much faster because the action potential jumps from one node of Ranvier to the next (fig. 16.7). This is called saltatory (saltatory means jumping) conduction.

Transmission across a Synapse

The mechanism by which an action potential passes from one neuron to another is not the same as the mechanism by which an action potential is conducted along a neuron. Each axon branches into many fine terminal branches, each of which is tipped by a small swelling, or terminal knob (fig. 16.8a). Each

Figure 16.7

Longitudinal section of a vertebrate axon illustrating the manner by which a nerve impulse travels down a long nerve fiber. The speed of the impulse is due to the fact that it jumps from one node of Ranvier to the next.

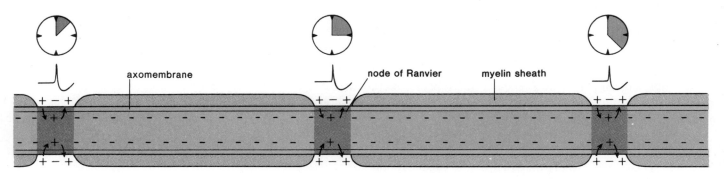

axomembrane node of Ranvier myelin sheath

Figure 16.8

Diagrammatic representation of a synapse at 3 different magnifications. a. Typically, a synapse is located wherever an axon is close to a dendrite or a cell body. Drawing based on a photomicrograph shows that there are several synaptic endings per axon because of terminal branching of the axon. b. Drawing based on low-power electron micrographs shows that a branch ends in a terminal knob having numerous synaptic vesicles, each filled with a neurotransmitter substance. This drawing makes it clear that a synapse contains a cleft (space) between the axon

and the dendrite or the cell body. c. Drawing based on high-power electron micrographs shows that a synapse consists of a presynaptic membrane, the synaptic cleft, and the postsynaptic membrane. When a nerve impulse reaches the synaptic vesicles, they move toward and fuse with the presynaptic membrane, discharging their contents. The neurotransmitter substance diffuses across the cleft and combines with a receptor. A nerve impulse may follow if enough sodium (Na$^+$) ion channels open.

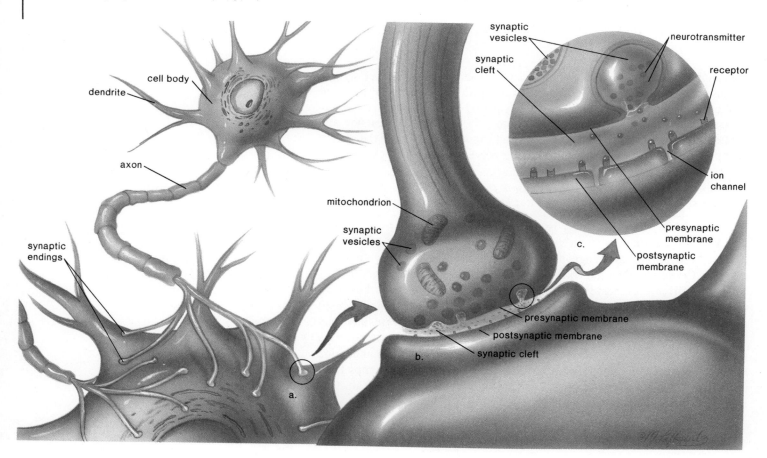

cell body
dendrite
axon
synaptic endings
synaptic vesicles
mitochondrion
presynaptic membrane
postsynaptic membrane
synaptic cleft
synaptic vesicles
synaptic cleft
neurotransmitter
receptor
ion channel
presynaptic membrane
postsynaptic membrane
a.
b.
c.

knob lies very close to the dendrite (or cell body) of another neuron. This region is called a **synapse,** and the knob is called a **synaptic ending.** The membrane of the knob is called the **presynaptic membrane,** and the membrane of the next neuron just beyond the knob is called the **postsynaptic membrane.** The small gap between is the **synaptic cleft.**

Transmission of nerve impulses across a synaptic cleft is carried out by **neurotransmitter substances,** which are stored in synaptic vesicles (fig. 16.8b) before their release. When nerve impulses traveling along an axon reach a synaptic ending, the axomembrane becomes permeable to calcium (Ca^{++}) ions. These ions then interact with microfilaments, causing them to pull the synaptic vesicles to the inner surface of the presynaptic membrane. When the vesicles merge with this membrane, a neurotransmitter substance is discharged into the synaptic cleft. The neurotransmitter molecules diffuse across the cleft to the postsynaptic membrane, where they bind with a receptor in a lock-and-key manner (fig. 16.8c).

Neurotransmitter substances can be excitatory or inhibitory. If the neurotransmitter substance is excitatory, the membrane potential of the postsynaptic membrane decreases, the sodium (Na^+) ion channels open at that locale, and the likelihood of the neuron firing (transmitting a nerve impulse) increases. If the neurotransmitter substance is inhibitory, the membrane potential of the postsynaptic membrane increases as the inside becomes more negative, and the likelihood of a nerve impulse decreases.

Summation and Integration

A dendrite or a cell body is on the receiving end of many synapses. Whether or not a neuron fires depends on **summation,** the net effect of all the excitatory and the inhibitory neurotransmitters received. If enough sodium (Na^+) ion channels open, excitation is sufficient to raise the membrane potential above threshold level (fig. 16.5d), and the neuron fires. Otherwise, it does not fire.

The CNS *integrates* (sums up) the information it receives from all over the body. Summation in a neuron is integration at the cellular level. Integration in the brain allows us to make decisions about the body in general.

Because a neuron either fires or does not fire, it is said to obey an all-or-none law. A **nerve** does not obey the all-or-none law because it contains many fibers, any number of which can be carrying nerve impulses. Therefore, a nerve can have degrees of performance.

16.1 Critical Thinking

1. Electricity is the flow of electrons within a wire. How is the nerve impulse different from this?
2. If a neurotransmitter substance is inhibitory, would you expect a higher or lower voltage reading compared to −65mV on the oscilloscope?
3. In the laboratory, an axon segment can conduct a nerve impulse in either direction. Why do nerve impulses go only from axon to dendrite or cell body across a synapse in the body?
4. Nerves cause muscles to contract. Assuming one nerve fiber per one muscle fiber of a muscle, explain how a nerve can bring about degrees of muscle contraction (see illustration below).

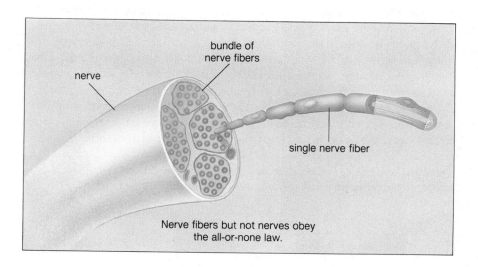

nerve

bundle of nerve fibers

single nerve fiber

Nerve fibers but not nerves obey the all-or-none law.

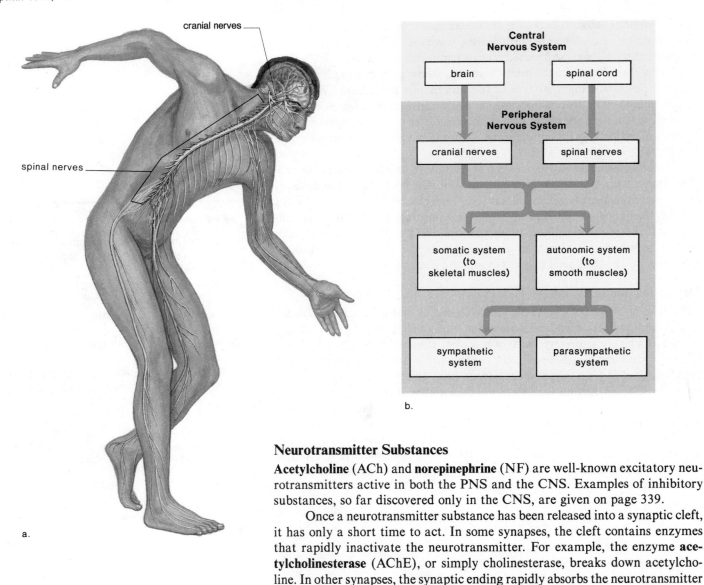

Figure 16.9
Peripheral nervous system (PNS) compared to central nervous system (CNS). a. The CNS lies in the center of the body, and the PNS lies to either side. b. The CNS contains the brain and the spinal cord, and the PNS contains the nerves. There are cranial and spinal nerves within the somatic and autonomic nervous systems. The nerves of the autonomic nervous system belong to either the sympathetic system or the parasympathetic system.

Neurotransmitter Substances

Acetylcholine (ACh) and **norepinephrine** (NF) are well-known excitatory neurotransmitters active in both the PNS and the CNS. Examples of inhibitory substances, so far discovered only in the CNS, are given on page 339.

Once a neurotransmitter substance has been released into a synaptic cleft, it has only a short time to act. In some synapses, the cleft contains enzymes that rapidly inactivate the neurotransmitter. For example, the enzyme **acetylcholinesterase** (AChE), or simply cholinesterase, breaks down acetylcholine. In other synapses, the synaptic ending rapidly absorbs the neurotransmitter substance, possibly for repackaging in synaptic vesicles or for chemical breakdown. The enzyme monoamine oxidase breaks down norepinephrine after it is absorbed. The short existence of neurotransmitters in the synapse prevents continuous stimulation (or inhibition) of postsynaptic membranes.

Transmission of nerve impulses across a synapse is dependent on a neurotransmitter substance that changes the permeability of the postsynaptic membrane.

The Peripheral Nervous System

The peripheral nervous system (PNS) is made up of nerves that are a part of either the somatic system or the autonomic system (fig. 16.9). The autonomic system is further divided into the sympathetic and parasympathetic systems. Nerves are structures that contain long dendrites and/or long axons. Each of these fibers is surrounded by myelin (fig. 16.3); therefore, nerves have a white,

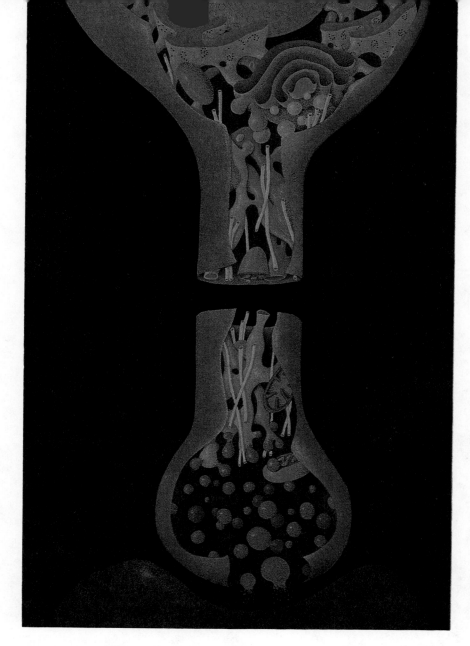

glistening appearance. There are no cell bodies in nerves; cell bodies are found only in the CNS or in the ganglia. **Ganglia** (singular, ganglion) are collections of cell bodies within the PNS.

Humans have 12 pairs of cranial nerves and 31 pairs of spinal nerves. **Cranial nerves** are either sensory nerves (having the long axons of sensory neurons only), motor nerves (having the long axons of motor neurons only), or mixed nerves (having long sensory dendrites and long motor axons) (table 16.2). All cranial nerves, except the vagus nerve, control the head, the face, the neck, and the shoulders. The *vagus nerve* controls the internal organs.

All **spinal nerves** (fig. 16.10) are mixed nerves that take impulses to and from the spinal cord. Their arrangement shows that humans are segmented animals: there is a pair of spinal nerves for each segment. Each spinal nerve emerges from the spinal cord as 2 short branches, the dorsal (toward back) and the ventral (toward front) roots. These roots join just before the nerve leaves the vertebral column, which is formed from the vertebrae that protect the cord. The spinal cord is part of the CNS and is discussed on page 334.

In the PNS, cranial nerves take impulses to and/or from the brain, and spinal nerves take impulses to and from the spinal cord.

Table 16.2 Nerves

Type of Nerve	Consists of	Function
Sensory nerve	Long axons of sensory neurons only	Carries message from receptors to CNS
Motor nerve	Long axons of motor neurons only	Carries message from CNS to effectors
Mixed nerve	Both long dendrites of sensory neurons and long axons of motor neurons	Carries message in dendrite to CNS and away from CNS in axons

Figure 16.10

The anatomy of the spinal cord. a. The CNS consists of the brain and the spinal cord. The brain is protected by the skull, and the spinal cord is protected by the vertebrae.
b. Cross section of the spine, showing spinal nerves. The human body has a total of 31 pairs of spinal nerves. c. This cross section of the spinal cord shows that a spinal nerve has a dorsal root and a ventral root. Also, the cord is protected by 3 layers of tissue called the meninges. Spinal meningitis is an infection of these layers.

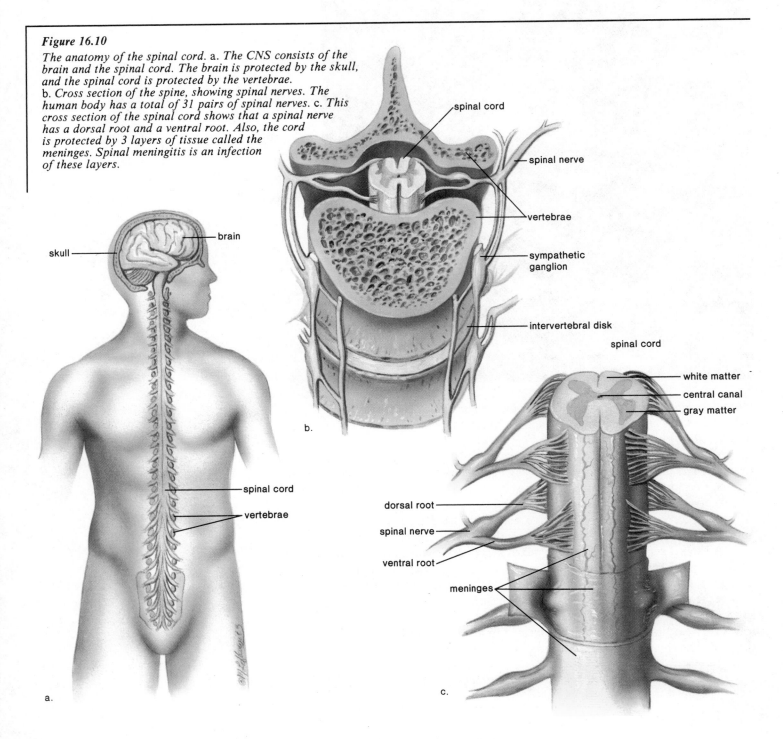

Human Anatomy and Physiology

Figure 16.11

Diagram of a reflex arc showing the detailed composition of a spinal nerve. When the receptors in the skin are stimulated, nerve impulses (see arrows) move along a sensory neuron to the spinal cord. (Note that the cell body of a sensory neuron is in a ganglion outside the cord.) The nerve impulses are picked up by an interneuron, which lies completely within the cord, and pass to the dendrites and the cell body of a motor neuron that lies ventrally within the cord. The nerve impulses then move along the axon of the motor neuron to an effector, such as a muscle fiber that contracts. The brain receives information concerning sensory stimuli by way of other interneurons, with long fibers in tracts that run up and down the cord within the white matter.

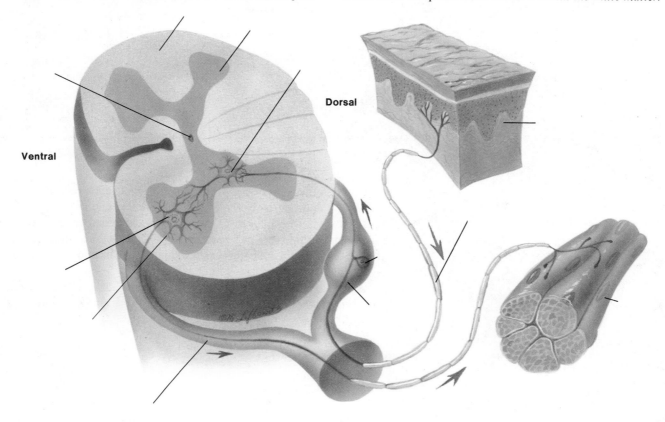

Dorsal

Ventral

The Somatic Nervous System

The **somatic nervous system** includes all nerves that serve the musculoskeletal system and the exterior sense organs, including those in the skin. Exterior sense organs are receptors that receive environmental stimuli and then initiate nerve impulses. Muscle fibers are effectors that bring about a reaction to the stimulus. Receptors are studied in chapter 18, and muscle effectors are studied in chapter 17.

The Reflex Arc

Reflexes are automatic, involuntary responses to changes occurring inside or outside the body. In the somatic nervous system, outside stimuli often initiate a reflex action. Some reflexes, such as blinking the eye, involve the brain, while others, such as withdrawing the hand from a hot object, do not necessarily involve the brain. Figure 16.11 illustrates the path of the second type of reflex action. Whenever a person touches a very hot object, a *receptor* in the skin generates nerve impulses that move along the dendrite of a *sensory neuron* toward the cell body and the CNS. The cell body of a sensory neuron is located in the **dorsal root ganglion** just outside the cord. From the cell body, the impulses travel along the axon of the sensory neuron and enter the cord by the dorsal root of a spinal nerve. The impulses then pass to many interneurons, one of which connects with a motor neuron. The short dendrites and the cell body of the *motor neuron* lead to the axon, which leaves the cord by way of the ventral root of a spinal nerve. The nerve impulses travel along the axon to *muscle fibers* that then contract so that the hand is withdrawn from the hot object. (See table 16.3 for a listing of these events.)

16.2 Critical Thinking

1. If you applied acid to the left leg of a frog, why might both legs respond?
2. If you severed just the dorsal root ganglion of the left sciatic nerve that serves the leg, would either leg be able to respond?
3. If you severed just the ventral root of the left sciatic nerve, would either leg be able to respond?
4. If you destroyed just the spinal cord, would either leg be able to respond?

Table 16.3 *Path of a Simple Reflex*

1. Receptor (formulates message)*	Generates nerve impulses
2. Sensory neuron (takes message to CNS)	Impulses move along dendrite (spinal nerve)† and proceed to cell body (dorsal root ganglia) and then go from cell body to axon (spinal cord)
3. Interneuron (passes message to motor neuron)	Impulses picked up by dendrites and pass through cell body to axon (spinal cord)
4. Motor neuron (takes message away from CNS)	Impulses travel through short dendrites and cell body (spinal cord) to axon (spinal nerve)
5. Effector (receives message)	Receives nerve impulses and reacts: glands secrete and muscles contract

*Phrases within parentheses state overall function.
†Words within parentheses indicate location of structure.

Various other reactions usually accompany a reflex response; the person may look in the direction of the object, jump back, and utter appropriate exclamations. This whole series of responses is explained by the fact that the sensory neuron stimulates several interneurons, which take impulses to all parts of the CNS, including the cerebrum, which in turn, makes the person conscious of the stimulus and his or her reaction to it.

The reflex arc is the main functional unit of the nervous system. It allows us to react to internal and external stimuli.

The Autonomic Nervous System

The autonomic nervous system, a part of the PNS, is made up of motor neurons that control the internal organs automatically and usually without need for conscious intervention. There are 2 divisions of the autonomic nervous system: the sympathetic and parasympathetic systems. Both of these (1) function automatically and usually subconsciously in an involuntary manner; (2) innervate all internal organs; and (3) utilize 2 motor neurons and one ganglion for each impulse. The first of these 2 neurons has a cell body within the CNS and a **preganglionic axon.** The second neuron has a cell body within the ganglion and a **postganglionic axon.**

The autonomic nervous system controls the functioning of internal organs without need of conscious control.

The Sympathetic System

The preganglionic fibers of the **sympathetic nervous system** arise from the middle or *thoracic-lumbar portion* of the spinal cord and almost immediately terminate in ganglia that lie near the cord. Therefore, in this system, the preganglionic fiber is short, but the postganglionic fiber that makes contact with an organ is long.

The sympathetic nervous system is especially important during emergency situations and is associated with "fight or flight." For example, it inhibits the digestive tract, but it dilates the pupil, accelerates the heartbeat, and increases the breathing rate. It is not surprising, then, that the neurotransmitter released by the postganglionic axon is norepinephrine, a chemical close in structure to adrenalin, a well-known heart stimulant.

The sympathetic nervous system brings about those responses we associate with "fight or flight."

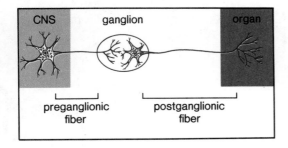

Figure 16.12

Structure and function of the autonomic nervous system. The sympathetic fibers arise from the thoracic and lumbar portion of the spinal cord; the parasympathetic fibers arise from the brain and the sacral portion of the cord. Each system innervates the same organs but have contrary effects. For example, the sympathetic system speeds up and the parasympathetic system slows down the beat of the heart.

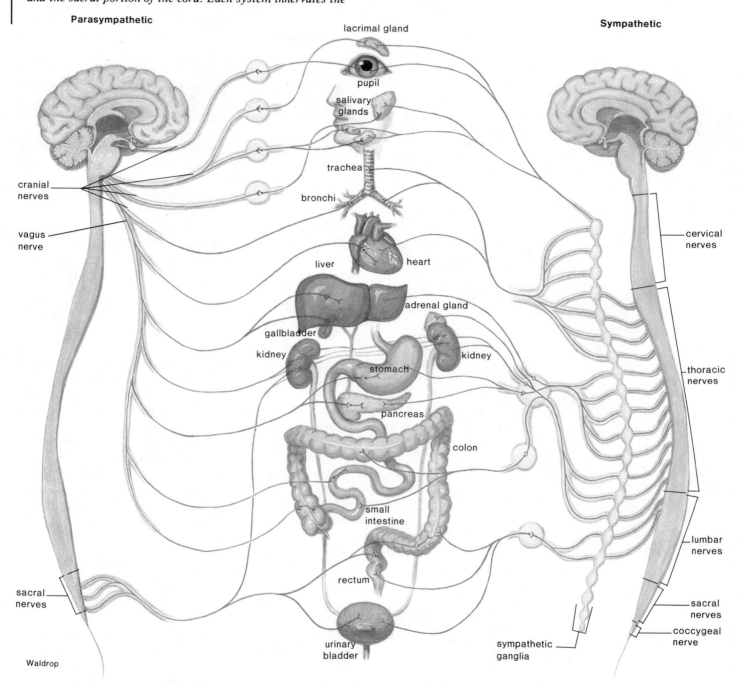

Parasympathetic

Sympathetic

lacrimal gland

pupil

salivary glands

trachea

bronchi

cranial nerves

vagus nerve

liver

heart

adrenal gland

gallbladder

kidney

stomach

kidney

pancreas

colon

small intestine

rectum

sacral nerves

urinary bladder

Waldrop

cervical nerves

thoracic nerves

lumbar nerves

sacral nerves

coccygeal nerve

sympathetic ganglia

The Parasympathetic System

A few cranial nerves, including the vagus nerve, together with fibers that arise from the bottom portion of the cord form the **parasympathetic nervous system** (fig. 16.12). Therefore, this system often is referred to as the *craniosacral portion* of the autonomic nervous system. In the parasympathetic nervous system, the preganglionic fiber is long and the postganglionic fiber is short because the ganglia lie near or within the organ.

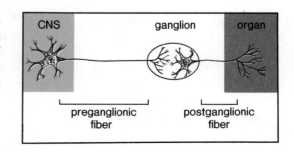

CNS   ganglion   organ

preganglionic fiber   postganglionic fiber

Table 16.4 *Sympathetic System versus Parasympathetic System*

Sympathetic System	Parasympathetic System
Fight or flight	Normal activity
Norepinephrine is neurotransmitter	Acetylcholine is neurotransmitter
Postganglionic fiber is longer than preganglionic fiber	Preganglionic fiber is longer than postganglionic fiber
Preganglionic fiber arises from middle portion of cord	Preganglionic fiber arises from brain and lower portion of cord

The parasympathetic system, sometimes called the "housekeeper system," promotes all the internal responses we associate with a relaxed state; for example, it causes the pupil of the eye to contract, it promotes digestion of food, and it retards the heartbeat. The neurotransmitter utilized by the parasympathetic system in acetylcholine.

Figure 16.12 contrasts the sympathetic and parasympathetic systems, and table 16.4 lists all the differences we have noted between these 2 systems.

The parasympathetic nervous system brings about the responses we associate with normally restful activities.

The Central Nervous System

The central nervous system (CNS) consists of the spinal cord and the brain. As figures 16.10 and 16.13 illustrate, the CNS is protected by bone: the brain is enclosed within the skull and the spinal cord is surrounded by vertebrae. Also, both the brain and the spinal cord are wrapped in 3 protective membranes known as **meninges;** meningitis is an infection of these coverings (fig. 16.10c). The spaces between the meninges are filled with **cerebrospinal fluid,** which cushions and protects the CNS. A small amount of this fluid sometimes is withdrawn for laboratory testing when a spinal tap (i.e., lumbar puncture) is done. Cerebrospinal fluid also is contained within the **central canal** of the spinal cord and the **ventricles** of the brain. The latter are interconnecting spaces that produce and serve as a reservoir for cerebrospinal fluid.

The Spinal Cord

The spinal cord lies along the middorsal line of the body (fig. 16.10). It has 2 main functions: (1) it is the center for many reflex actions, and (2) it provides a means of communication between the brain and the spinal nerves that leave the cord.

The path of a spinal reflex passes through the gray matter of the cord (fig. 16.11). It is gray because it contains cell bodies and short fibers that are unmyelinated. In cross section, the gray matter looks like a butterfly or the letter H. The axons of sensory neurons are found in the dorsal regions (horns) of the gray matter, and the dendrites and cell bodies of motor axons are found in the ventral regions (horns) of the gray matter. Short interneurons connect sensory to motor neurons on the same sides and opposite sides of the cord.

The white matter of the cord is found in between the regions of the gray matter (fig. 16.10c). The white matter of the cord is white because it contains myelinated long fibers of interneurons that run together in bundles called tracts. These tracts connect the cord to the brain. Dorsally, there are primarily ascending tracts that take information to the brain, and ventrally, there are primarily descending tracts that carry information down from the brain. Because

Figure 16.13

Figure 16.13
The human brain. Note how large the cerebrum is compared to the rest of the brain.

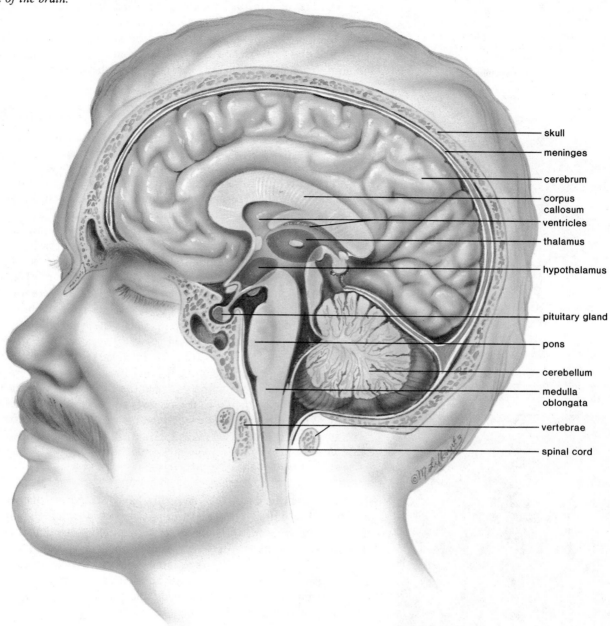

skull
meninges
cerebrum
corpus callosum
ventricles
thalamus
hypothalamus
pituitary gland
pons
cerebellum
medulla oblongata
vertebrae
spinal cord

the tracts at one point cross over, the left side of the brain controls the right side of the body and the right side of the brain controls the left side of the body.

The CNS lies in the midline of the body and consists of the brain and the spinal cord, where sensory information is received and motor control is initiated.

The Brain

The largest and most prominent portion of the human brain (fig. 16.13) is the cerebrum. Consciousness resides only in the cerebrum; the rest of the brain functions below the level of consciousness. In addition to the portions mentioned, remember that the unconscious brain contains many tracts that relay messages to and from the spinal cord.

Figure 16.14

The reticular activating system. The dotted arrows indicate that sensory impulses are sorted out by the system and the thalamus before being sent to the cerebrum. The solid arrows indicate that impulses are relayed to the cerebellum by the system in order to maintain muscle tone.

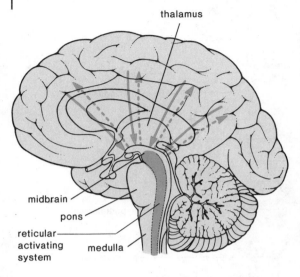

Figure 16.15

Skateboarding. This sport requires muscular coordination and balance that are controlled by the cerebellum.

The Unconscious Brain

The **medulla oblongata** lies closest to the spinal cord and contains centers for heartbeat, respiration, and vasoconstriction (blood pressure). It also contains reflex centers for vomiting, coughing, sneezing, hiccuping, and swallowing.

The **hypothalamus** is concerned with homeostasis, or the constancy of the internal environment, and contains centers for hunger, sleep, thirst, body temperature, water balance, and blood pressure. The hypothalamus controls the pituitary gland and thereby serves as a link between the nervous and endocrine systems.

The medulla oblongata and the hypothalamus both are concerned with control of the internal organs.

The *midbrain* and the *pons* contain tracts that connect the cerebrum with other parts of the brain. In addition, the pons functions with the medulla to regulate breathing rate, and the midbrain has reflex centers concerned with head movements in response to visual and auditory stimuli.

The **thalamus** is a central relay station for sensory impulses traveling upward from other parts of the cord and the brain to the cerebrum. It receives all sensory impulses (except those associated with the sense of smell) and channels them to appropriate regions of the cerebrum. In other words, it is the last portion of the brain for sensory input before the cerebrum.

The thalamus has connections to various parts of the brain by way of the diffuse thalamic projection system. This system is the upper part of the **RAS,** the **reticular activating system** (fig. 16.14), a complex network of cell bodies and fibers that extends from the medulla to the cerebrum. The RAS sorts out incoming stimuli, passing on only those that require immediate attention. The thalamus sometimes is called the gatekeeper to the cerebrum because it alerts the cerebrum to only certain sensory input. We are not aware of many of the sensory impulses received by the CNS.

The thalamus receives sensory impulses from other parts of the CNS and channels only certain of these to the cerebrum.

The **cerebellum,** a bilobed butterfly-shaped structure is the second largest portion of the brain. It functions in muscle coordination (fig. 16.15), integrating impulses received from higher centers to ensure that all the skeletal muscles work together to produce smooth and graceful motions. The cerebellum also is responsible for maintaining normal muscle tone and transmitting impulses that maintain posture. It receives information from the inner ear indicating the position of the body and sends impulses to those muscles whose contraction maintains or restores balance.

The cerebellum controls balance and complex muscular movements.

The Conscious Brain

The **cerebrum,** the only area of the brain responsible for consciousness, is the largest portion of the brain in humans. The outer layer of the cerebrum, called the cortex, is gray in color and contains cell bodies and short fibers. The cerebrum is divided into halves known as the right and left **cerebral hemispheres.** Each hemisphere contains 4 surface lobes: **frontal, parietal, temporal,** and **occipital** (fig. 16.16). Little is known about the functions of a fifth lobe, the insula, which lies inside beneath the surface.

Certain areas of the cerebral cortex have been "mapped" in great detail (table 16.5). For example, we know which portions of the frontal lobe control various parts of the body and which portions of the parietal lobe receive sensory information from these same parts. Each of the 4 lobes of the cerebral cortex contains association areas that receive information from the other lobes

Human Anatomy and Physiology

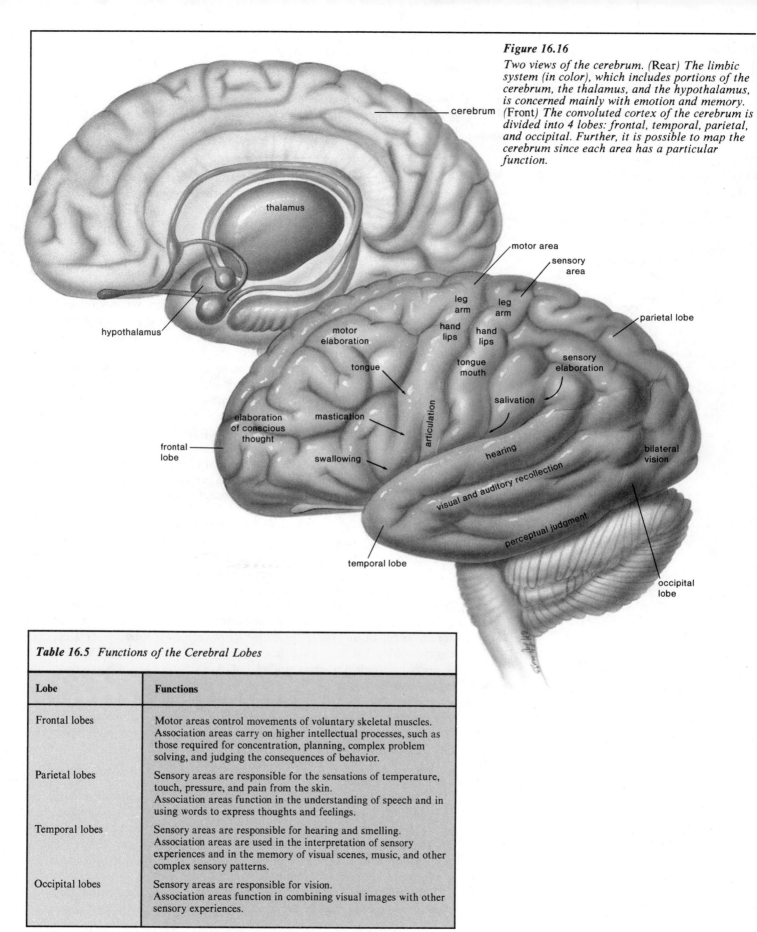

Figure 16.16

Two views of the cerebrum. (Rear) The limbic system (in color), which includes portions of the cerebrum, the thalamus, and the hypothalamus, is concerned mainly with emotion and memory. (Front) The convoluted cortex of the cerebrum is divided into 4 lobes: frontal, temporal, parietal, and occipital. Further, it is possible to map the cerebrum since each area has a particular function.

Table 16.5 Functions of the Cerebral Lobes	
Lobe	**Functions**
Frontal lobes	Motor areas control movements of voluntary skeletal muscles. Association areas carry on higher intellectual processes, such as those required for concentration, planning, complex problem solving, and judging the consequences of behavior.
Parietal lobes	Sensory areas are responsible for the sensations of temperature, touch, pressure, and pain from the skin. Association areas function in the understanding of speech and in using words to express thoughts and feelings.
Temporal lobes	Sensory areas are responsible for hearing and smelling. Association areas are used in the interpretation of sensory experiences and in the memory of visual scenes, music, and other complex sensory patterns.
Occipital lobes	Sensory areas are responsible for vision. Association areas function in combining visual images with other sensory experiences.

From John W. Hole, Jr., *Human Anatomy and Physiology,* 5th ed. Copyright © 1990 Wm. C. Brown Publishers, Dubuque, Iowa. All Rights Reserved. Reprinted by permission.

Figure 16.17

Encephalograms (EEGs), recordings of the electrical activity of the brain. The alpha waves, which appear when the subject is awake with eyes closed, are the most common. Second most common are the beta waves, recorded when the subject is awake with eyes open. Sleep has various stages, as indicated.

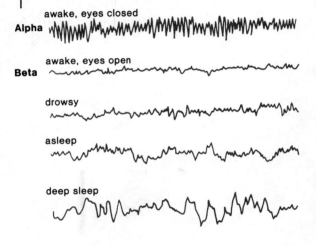

and integrate it into higher, more complex levels of consciousness. These areas are concerned with intellect, artistic and creative abilities, learning, and memory.

Consciousness is the province of the cerebrum, the most highly developed portion of the human brain. The cerebrum is responsible for higher mental processes, including the interpretation of sensory input and the initiation of voluntary muscular movements.

There has been a great deal of testing to determine whether the right and left halves of the cerebrum serve different functions. These studies tend to suggest that the left half of the brain is the verbal (word) half and the right half of the brain is the visual (spatial relation) and artistic half. However, other results indicate that such a strict dichotomy does not always exist between the 2 halves. In any case, the 2 cerebral hemispheres normally share information because they are connected by a horizontal tract called the **corpus callosum** (fig. 16.13).

Severing the corpus callosum can control severe epileptic seizures but results in a person with 2 brains, each with its own memories and thoughts. Today, use of the laser permits more precise treatment without this side effect. *Epilepsy* is caused by a disturbance of the normal communication between the RAS and the cortex. In a grand mal or epileptic seizure, the cerebrum is extremely excited. Due to a reverberation of signals within the RAS and the cerebrum, the individual loses consciousness, even while convulsions are occurring. Finally, the neurons fatigue and the signals cease. Following an attack, the brain is so fatigued the person must sleep for a while.

The EEG   The electrical activity of the brain can be recorded in the form of an **electroencephalogram** (**EEG**). Electrodes are taped to different parts of the scalp, and an instrument called the electroencephalograph records the so-called brain waves (fig. 16.17).

When the subject is awake, 2 types of waves are usual: *alpha waves,* with a frequency of about 6–13 per second and a potential of about 45 mV, predominate when the eyes are closed, and *beta waves,* with higher frequencies but lower voltage, appear when the eyes are open.

During an 8-hour sleep, there are usually 5 times when the brain waves become slower and larger than alpha waves. During each of these times, there are irregular flurries as the eyes move back and forth rapidly. When subjects are awakened during the latter, called **REM** (rapid eye movement) **sleep,** they always report that they were dreaming. The significance of REM sleep still is being debated, but some studies indicate that REM sleep is needed for memory to occur.

The EEG is a diagnostic tool; for example, an irregular pattern can signify epilepsy or a brain tumor. A flat EEG signifies lack of electrical activity of the brain, or brain death, and thus it can be used to determine the precise time of death.

The Limbic System

The **limbic system** involves portions of both the unconscious brain and the conscious brain. It lies just beneath the cerebral cortex (fig. 16.16) and contains neural pathways that connect portions of the frontal lobes, the temporal lobes, the thalamus, and the hypothalamus. Several masses of gray matter that lie deep within each hemisphere of the cerebrum, termed the *basal nuclei,* are also a part of the limbic system.

Stimulation of different areas of the limbic system causes the subject to experience rage, pain, pleasure, or sorrow. By causing pleasant or unpleasant feelings about experiences, the limbic system apparently guides the individual into behavior that is likely to increase the chance of survival.

Learning and Memory The limbic system also is involved in the processes of learning and memory. Learning requires memory, but just what permits memory to occur is not known definitely. Some investigators experiment with invertebrates, such as slugs and snails, because their nervous system is very simple and yet they can be conditioned to perform a particular behavior. In order to study this simple type of learning, it has been possible to insert electrodes into individual cells and to alter or record the electrochemical responses of these cells (fig. 16.18). This type of research has shown recently that learning is accompanied by an increase in the number of synapses, while forgetting involves a decrease in the number of synapses. In other words, the nerve-circuit patterns are constantly changing as learning, remembering, and forgetting occur. Within the individual neuron, learning involves a change in gene regulation, nerve protein synthesis, and an increased ability to secrete neurotransmitter substances.

Some investigators study learning and memory in monkeys. This work has led to the conclusion that the limbic system is absolutely essential to both short-term and long-term memory. An example of short-term memory in humans is the ability to recall a telephone number long enough to dial it; an example of long-term memory is the ability to recall the events of the day. It is believed that at first, impulses move only within the limbic circuit, but eventually the basal nuclei transmit the neurotransmitter acetylcholine (ACh) to the sensory areas where memories are stored. The involvement of the limbic system certainly explains why emotionally charged events result in our most vivid memories. The fact that the limbic system communicates with the sensory areas for touch, smell, vision, hearing and taste, accounts for the ability of any particular sensory stimulus to awaken a complex memory.

The limbic system is involved particularly in emotions and in memory and learning.

Neurotransmitters in the Brain

As discussed previously, neurotransmitters released at the end of axons affect the membrane potential of postsynaptic membranes. Some excitatory transmitters are amines, such as acetylcholine (ACh), norepinephrine (NE), serotonin, and dopamine. We already mentioned that acetylcholine and norepinephrine are neurotransmitters for the autonomic nervous system. *Serotonin* and *dopamine* are associated with behavioral states, such as mood, sleep, attention, learning, and memory. The inhibitory transmitters include the amino acids gamma-aminobutyrate (GABA) and glycine.

Both excitatory and inhibitory neurotransmitters are active in the brain.

In addition, a number of different types of peptides have been discovered in the CNS. The *endorphins,* molecules that are the body's natural opioids, are of particular interest. They are called opioids because the psychoactive drugs morphine and heroin, both of which are derived from opium, attach to the receptors for endorphin in the CNS. When endorphins are released, they, like opium, produce a feeling of elation and reduce the sensation of pain. For example, if endorphins are present, neurons do not release substance P, a neurotransmitter that brings about the sensation of pain. Exercise has been associated with the presence of endorphins, and this may account for the so-called "runner's high."

Neurotransmitter Disorders It has been discovered that several neurological illnesses, such as *Parkinson disease* and *Huntington disease,* are due to an imbalance of neurotransmitters. Parkinson disease is a condition characterized by a wide-eyed, unblinking expression, an involuntary tremor of the fingers and the thumbs, muscular rigidity, and a shuffling gait. All these symptoms

Figure 16.18
The snail Hermissenda. *Individual nerve cells are being stimulated by microelectrodes, simulating the signals that scientists had recorded previously when a snail learns to avoid light. When this snail is freed, it automatically avoids the light and does not need to be taught like other snails. To teach snails to avoid light, they are placed on a table that rotates every time they venture toward light.*

Figure 16.19

Drug action at synapses. a. Drug stimulates release of neurotransmitter. b. Drug blocks release of neurotransmitter. c. Drug combines with neurotransmitter, preventing its breakdown or reuptake. d. Drug mimics neurotransmitter. e. Drug blocks receptor so that neurotransmitter cannot be received.

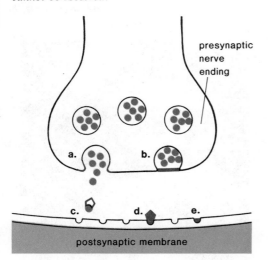

presynaptic nerve ending

postsynaptic membrane

Table 16.6 Drug Action		
Drug Action	**Neurotransmitter**	**Result**
Blocks	Excitatory	Depression
Enhances	Excitatory	Stimulation
Blocks	Inhibitory	Stimulation
Enhances	Inhibitory	Depression

are due to dopamine deficiencies. Huntington disease is characterized by a progressive deterioration of the individual's nervous system that eventually leads to constant thrashing and writhing movements and finally to insanity and death. The problem is believed to be the malfunction of the inhibitory neurotransmitter GABA. Most recently, it has been discovered that *Alzheimer disease,* a severe form of senility with marked memory loss found in 5–10% of all people over age 65, is due to deterioration in cells of the basal nuclei that use acetylcholine as a transmitter. It seems that a gene located on chromosome number 21, which normally is active only during development and which directs the production of a protein associated with neuron death, inexplicably has been turned on in these patients.

Some neurological illnesses are associated with the deficiency of a particular neurotransmitter in the brain.

Drug Abuse

A wide variety of drugs can be used to alter the mood and/or emotional state (see appendix B), but our discussion centers on the 4 most commonly abused drugs: alcohol, marijuana, cocaine, and heroin. *Drug abuse* is evident when a person takes a drug at a dose level and under circumstances that increase the potential for a harmful effect.

Drug abusers are apt to display a *physical dependence* on the drug (formerly called an addiction to the drug). Dependence has developed when the person (1) spends much time thinking about the drug or arranging to get it; (2) often takes more of the drug than was intended; (3) is *tolerant* to the drug—that is, must increase the amount of the drug to get the same effect; (4) has *withdrawal symptoms* when he or she stops taking the drug; and (5) has an ongoing desire to cut down on use.

Drug Action

Drugs that affect the nervous system have 2 general effects: (1) they affect the RAS (p. 336) and the limbic system, and (2) they either promote or decrease the action of a particular neurotransmitter. There are a number of different ways drugs can influence the transmission of neurotransmitters, some of which are described in figure 16.19. It is clear, as outlined in table 16.6, that stimulants can either enhance the action of an excitatory neurotransmitter or block the action of an inhibitory neurotransmitter. Depressants can either enhance the action of an inhibitory neurotransmitter or block the action of an excitatory neurotransmitter.

Drug abuse often results in physical dependence on the drug because the drug interferes with normal neurotransmitter function in the brain.

Table 16.7 Some Questions to Identify the Alcohol-Dependent Person
1. Do you occasionally drink heavily after a disappointment, a quarrel, or when the boss gives you a bad time?
2. When you are having trouble or feel under pressure, do you always drink more heavily than usual?
3. Have you noticed that you are able to handle more liquor than you did when you were first drinking?
4. Did you ever wake up the "morning after" and discover that you could not remember part of the evening before, even though your friends tell you that you did not "pass out"?
5. When drinking with other people, do you try to have a few extra drinks when others will not know it?
6. Are there certain occasions when you feel uncomfortable if alcohol is not available?
7. Have you recently noticed that when you begin drinking you are in more of a hurry to get the first drink than you used to be?
8. Do you sometimes feel a little guilty about your drinking?
9. Are you secretly irritated when your family or friends discuss your drinking?
10. Have you recently noticed an increase in the frequency of your memory "blackouts"?
11. When you are sober, do you often regret things you have done or said while drinking?
12. Have you often failed to keep the promises you have made to yourself about controlling or cutting down on your drinking?
13. Do more people seem to be treating you unfairly without good reason?
14. Do you eat very little or irregularly when you are drinking?
15. Do you get terribly frightened after you have been drinking heavily?

Source: National Council on Alcoholism.

Alcohol

The type of alcohol consumed is ethanol, the production of which is discussed on page 115. While it is possible to drink alcohol in moderation, the drug often is abused. Alcohol use becomes "abuse," or an illness, when alcohol ingestion impairs an individual's social relationships, health, job efficiency, or ability to avoid legal difficulties (fig. 16.20). Table 16.7 lists some of the questions that are used to identify the alcohol-dependent person.

Alcohol effects on the brain are biphasic: after consuming several drinks, blood alcohol concentration rises rapidly and the drinker reports feeling "high" and happy (euphoric). After 90 minutes and lasting until some 330–400 minutes after consumption, the drinker feels depressed and unhappy (dysphoric) (fig. 16.21). On the other hand, if the drinker continues to drink in order to maintain a high blood level of the alcohol, he or she will experience ever-increasing loss of control. Coma and death are even possible if a substantial amount of alcohol (1¼ pints of whiskey) is consumed within an hour.

The Mode of Action and Associated Illnesses

Recent research indicates that alcohol acts on the GABA (an inhibitory neurotransmitter) receptor and potentiates GABA's ability to increase chloride (Cl^-) ion uptake. Most likely, it does this by disordering membrane lipids.

Cirrhosis of the Liver The stomach and the liver contain the enzyme alcohol dehydrogenase, which begins the breakdown of alcohol to acetic acid. A new study reports that women have less of this enzyme in their stomach, and this may explain why women show a greater sensitivity to alcohol, including a greater chance of liver damage. Acetic acid can be used in the liver to produce energy, but the calories provided are termed "empty" because they contribute to energy needs and weight gain without supplying any other nutritional requirements. Worse still, the molecules (glucose and fatty acids) that the liver ordinarily uses as an energy source are converted to fats, and the liver cells become engorged with fat droplets. After a few years of being overtaxed, the liver cells begin to die, causing an inflammatory condition known as alcoholic hepatitis. Finally, scar tissue appears in the liver, and it no longer is able to

Figure 16.20

Drug abuse involvement in accidents. Those who abuse drugs, including alcohol, are more likely to be involved in automobile accidents. Unfortunately, others in addition to the abuser often suffer the consequences.

Figure 16.21

Typical blood alcohol curve for a normal drinker after intake of 1 ml of alcohol per kg of body weight. As blood alcohol concentration increases, the user often feels euphoric (happy), but as blood alcohol concentration declines, the user feels dysphoric (unhappy). In most states, a person is considered to be legally drunk when the blood alcohol content is 0.1%. This usually requires imbibing 3 mixed drinks within one and one-half hours.

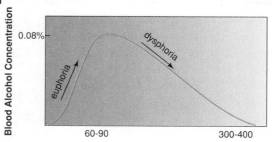

Blood Alcohol Concentration

0.08%

euphoria

dysphoria

60-90   300-400

Elapsed Time (minutes)

perform its vital functions. This condition is called cirrhosis of the liver, a frequent cause of death among drinkers. Brain impairment and generalized deterioration of other vital organs also are seen in heavy drinkers.

It should be stressed that the early signs of deterioration can be reversed if the habit of drinking to excess is given up.

Alcohol is the most abused drug in the United States. Its abuse often results in well-recognized illnesses and early death.

Marijuana

The dried flowering tops, leaves (fig. 16.22), and stems of the Indian hemp plant *Cannabis sativa* contain and are covered by a resin that is rich in THC (tetrahydrocannabinol). The names *cannabis* and *marijuana* can apply to the plant or to THC.

The effects of marijuana differ depending upon the strength and the amount consumed, the expertise of the user, and the setting in which it is taken. Usually, the user reports experiencing a mild euphoria along with alterations in vision and judgment that result in distortions of space and time. The inability to concentrate and to speak coherently and motor incoordination also can be involved.

Intermittent use of low-potency marijuana generally is not associated with obvious symptoms of toxicity, but heavy use can produce chronic intoxication. Intoxication is recognized by the presence of hallucinations, anxiety, depression, rapid flow of ideas, body image distortions, paranoid reactions, and similar psychotic symptoms. The terms *cannabis psychosis* and *cannabis delirium* refer to such reactions.

The Mode of Action

Marijuana is classified as an hallucinogen. It is possible that, like LSD (lysergic acid diethylamide), it has an effect on the action of serotonin, an excitatory neurotransmitter.

The use of marijuana does not seem to produce physical dependence, but a psychological dependence on the euphoric and sedative effects can develop. Craving or difficulty in stopping use also can occur as a part of regular heavy use.

Marijuana has been called a *gateway drug* because adolescents who have used marijuana also tend to try other drugs. For example, in a study of 100 cocaine abusers, 60% had smoked marijuana for more than 10 years.

Associated Illnesses

Usually marijuana is smoked in a cigarette form called a joint. Since this allows toxic substances, including carcinogens, to enter the lungs, chronic respiratory disease and lung cancer are considered dangers of long-term, heavy use. Some researchers claim that marijuana use leads to long-term brain impairment. Others report that males and females suffer reproductive dysfunctions. *Fetal cannabis syndrome,* which resembles fetal alcohol syndrome, has been reported.

Some psychologists are very concerned about the use of marijuana among adolescents because it can be used as a means to avoid coming to grips with the many personal problems that often develop during this maturational phase.

Although marijuana does not produce physical dependence, it does produce psychological dependence.

Cocaine

Cocaine is an alkaloid derived from the shrub *Erythroxylum cocoa. Cocaine* is sold in powder form and as *crack,* a more potent extract (fig. 16.23). Users often use the word *rush* to describe the feeling of euphoria that follows intake of the drug. Snorting (inhaling) produces this effect in a few minutes, injection, within 30 seconds, and smoking, in less than 10 seconds. Persons dependent upon the drug are, therefore, most likely to smoke cocaine. The rush only lasts a few seconds and then is replaced by a state of arousal that lasts from 5 to 30 minutes. Then the user begins to feel restless, irritable, and depressed. To overcome these symptoms, the user is apt to take more of the drug, repeating the cycle over and over again until there is no more drug left. A binge of this sort can go on for days, after which the individual suffers a *crash*. During the binge period, the user is hyperactive and has little desire for food or sleep, but has an increased sex drive. During the crash period, the user is fatigued, depressed, irritable, has memory and concentration problems, and displays no interest in sex. Indeed, men are often impotent. Other drugs, such as marijuana, alcohol, or heroin, often are taken to ease the symptoms of the crash.

The Mode of Action

Cocaine affects the concentration of dopamine, an excitatory neurotransmitter, in brain synapses. After release, dopamine ordinarily is withdrawn into the presynaptic cell for recycling and reuse. Cocaine prevents the reuptake of dopamine, and this causes an excess of dopamine in the synaptic cleft so that the user experiences the sensation of a rush. The adrenalin-like effects of dopamine account for the state of arousal that lasts for some minutes after the rush experience.

With continued cocaine use, the body begins to make less dopamine as a compensation for a seemingly excess supply. The user, therefore, now experiences *tolerance* (always needing more of the drug for the same effect), *withdrawal* (symptoms described previously when a drug is not taken), and an intense *craving* for cocaine. These are indications that the person is highly dependent upon the drug or, in other words, that cocaine is extremely addictive.

Associated Illnesses

Overdosing on cocaine is a real possibility. The number of deaths from cocaine and the number of emergency room admissions for drug reactions involving cocaine have increased greatly. High doses can cause seizures and cardiac and respiratory arrest.

Individuals who snort the drug can suffer damage to the nasal tissues and even perforation of the septum between the nostrils. Whether long-term cocaine abuse causes brain damage is not yet known, but this possibility is under investigation. It is known that babies born to addicts suffer withdrawal symptoms and may suffer neurological and developmental problems.

Heroin

Heroin is derived from morphine, an alkaloid of *opium*. Heroin usually is injected. After intravenous injection, the onset of action is noticeable within one minute and reaches its peak in 3–6 minutes. There is a feeling of euphoria along with relief of pain. Side effects can include nausea, vomiting, dysphoria, and respiratory and circulatory depression leading to death.

The Mode of Action

Heroin binds to receptors meant for the body's own opioids, the endorphins. As mentioned previously, the opioids are believed to alleviate pain by preventing the release of a neurotransmitter termed substance P from certain sensory neurons in the region of the spinal cord. When substance P is released, pain is felt, and when substance P is not released, pain is not felt. Evidence also indicates that there are opioid receptors in neurons that travel from the

Figure 16.23
Cocaine use. a. Crack, the ready-to-smoke form of cocaine that is a more potent and more deadly form than the powder. b. Users often smoke crack in a glass water pipe. The high produced consists of a "rush" lasting a few seconds, followed by a few minutes of euphoria. Continuous use makes the user extremely dependent on the drug.

a.

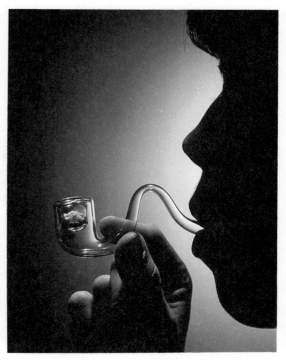

b.

spinal cord to the limbic system and that stimulation of these can cause a feeling of pleasure. This explains why opium and heroin not only kill pain but also produce a feeling of tranquility.

Individuals who inject heroin become physically dependent on the drug. With time, the body's production of endorphins decreases. Now *tolerance* develops so that the user needs to take more of the drug just to prevent *withdrawal* symptoms. The euphoria originally experienced upon injection no longer is felt.

Heroin withdrawal symptoms include perspiration, dilation of pupils, tremors, restlessness, abdominal cramps, gooseflesh, defecation, vomiting, and increases in systolic pressure and respiratory rate. Those who are excessively dependent may experience convulsions, respiratory failure, and death. Infants born to women who are physically dependent also experience these withdrawal symptoms.

Cocaine and heroin produce a very strong physical dependence. An overdose of these drugs can cause death.

Designer Drugs

Designer drugs are analogues; that is, they are chemical compounds of controlled substances slightly altered in molecular structure. One such drug is MPPP (1-methyl-4-phenylprionoxypiperidine), an analogue of the narcotic fentanyl. Even small doses of the drug are very toxic; MPPP already has caused many deaths on the West Coast.

Summary

The cell bodies of nerve cells are found in the CNS and the ganglia. Axons and dendrites make up nerves. The nerve impulse is a change in permeability of the axomembrane so that sodium (Na^+) ions move to the inside of a neuron and potassium (K^+) ions move to the outside. The nerve impulse is transmitted across the synapse by neurotransmitter substances.

During a spinal reflex, a sensory neuron transmits nerve impulses from a receptor to an interneuron, which in turn, transmits impulses to a motor neuron, which conducts them to an effector. Reflexes are automatic, and some do not require involvement of the brain.

Long fibers of sensory and/or motor neurons make up cranial and spinal nerves of the somatic and autonomic divisions of the PNS. While the somatic division controls skeletal muscle, the autonomic division controls smooth muscle and the internal organs.

The CNS consists of the spinal cord and the brain. Only the cerebrum is responsible for consciousness; the other portions of the brain have their own function. The cerebrum can be mapped, and each lobe also seems to have particular functions. Neurological drugs, although quite varied, have been found to affect the RAS and the limbic system by either promoting or preventing the action of neurotransmitters.

Study Questions

1. What are the 2 main divisions of the nervous system? Explain why these names are appropriate. (p. 320)
2. What are the 3 types of neurons? How are they similar, and how are they different? (p. 321)
3. What does the term *resting potential* mean, and how is it brought about? (p. 322) Describe the 2 parts of an action potential and the change that can be associated with each part. (pp. 324–25)
4. What is the sodium-potassium pump, and when is it active? (p. 325)
5. What is a neurotransmitter substance, where is it stored, how does it function, and how is it destroyed? (pp. 327–28) Name 2 well-known neurotransmitters. (p. 328)
6. What are the 3 types of nerves, and how are they anatomically different? functionally different? Distinguish between cranial and spinal nerves. (p. 330)
7. Trace the path of a reflex action after discussing the structure and the function of the spinal cord and the spinal nerve. (p. 331)
8. What is the autonomic nervous system, and what are its 2 major divisions? (p. 332) Give several similarities and differences between these divisions. (pp. 332–34)
9. Name the major parts of the brain, and give a function for each. (pp. 335–36)
10. Describe the EEG, and discuss its importance. (p. 338)
11. Describe the physiological effects and mode of action of alcohol, marijuana, cocaine, and heroin. (pp. 340–44)

Objective Questions

1. A(n) _____ carries nerve impulses away from the cell body.
2. During the upswing of the action potential, _____ ions are moving to the _____ of the nerve fiber.
3. The space between the axon of one neuron and the dendrite of another is called the _____ .
4. ACh is broken down by the enzyme _____ after it has altered the permeability of the postsynaptic membrane.
5. Motor nerves innervate _____ .
6. The vagus nerve is a(n) _____ nerve that controls the _____ .
7. In a reflex arc only the neuron called the _____ is completely within the CNS.
8. The brain and the spinal cord are covered by protective layers called _____ .
9. The _____ is that part of the brain that allows us to be conscious.
10. The _____ is the part of the brain responsible for coordination of body movements.

Label this Diagram.
See figure 16.2 (p. 321) in text.

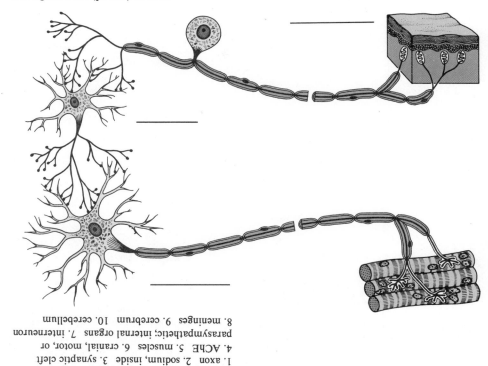

Answers to Objective Questions

1. axon 2. sodium, inside 3. synaptic cleft 4. AChE 5. muscles 6. cranial, motor, or parasympathetic; internal organs 7. interneuron 8. meninges 9. cerebrum 10. cerebellum

---

Selected Key Terms

axon (ak'son) process of a neuron that conducts nerve impulses away from the cell body. *320*

cell body (sel bod'e) portion of a nerve cell that includes a cytoplasmic mass and a nucleus and from which the nerve fibers extend. *320*

central nervous system (sen'tral ner'vus sis'tem) CNS; the brain and the spinal cord in vertebrate animals. *320*

cerebral hemisphere (ser'ĕ-bral hem'ĭ-sfēr) one of the large paired structures that together constitute the cerebrum of the brain. *336*

dendrite (den'drīt) process of a neuron, typically branched, that conducts nerve impulses toward the cell body. *320*

effector (ē-fek'tor) a structure, such as a muscle and a gland, that allows an organism to respond to environmental stimuli. *320*

ganglion (gang'gle-on) a collection of neuron cell bodies outside the CNS. *329*

innervate (in'er-vāt) to activate an organ, muscle, or gland by motor neuron stimulation. *320*

interneuron (in''ter-nu'ron) a neuron that is found within the CNS and that takes nerve impulses from one portion of the system to another. *321*

motor neuron (mo'tor nu'ron) a neuron that takes nerve impulses from the CNS to the effectors. *320*

myelin sheath (mi'ĕ-lin shēth) the Schwann cell membranes that cover long neuron fibers and give them a white glistening appearance. *322*

nerve impulse (nerv im'puls) an electrochemical change due to increased membrane permeability that is propagated along a neuron from the dendrite to the axon following excitation. *322*

neurotransmitter substance (nu''ro-trans'mit'er sub'stans) a chemical made at the ends of axons that is responsible for transmission across a synapse. *327*

parasympathetic nervous system (par''ah-sim''pah-thet'ik ner'vus sis'tem) that part of the autonomic nervous system that usually promotes those activities associated with a normal state. *333*

peripheral nervous system (pĕrif'er-al ner'vus sis'tem) PNS; nerves and ganglia that lie outside the CNS. *320*

receptor (re-sep'tor) a sense organ specialized to receive information from the environment. *320*

sensory neuron (sen'so-re nu'ron) a neuron that takes nerve impulses to the CNS afferent neuron. *320*

somatic nervous system (so-mat'ik ner'vus sis'tem) that portion of the PNS containing motor neurons that control skeletal muscles. *331*

sympathetic nervous system (sim'pah-thet'ik ner'vus sis'tem) that part of the autonomic nervous system that usually causes effects associated with emergency situations. *332*

synapse (sin'aps) the region between 2 nerve cells, where the nerve impulse is transmitted from one to the other, usually from axon to dendrite. *327*

Chapter 17

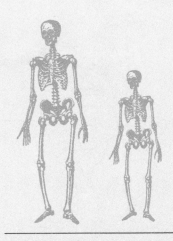

The Musculoskeletal System

Chapter Concepts

1 The skeleton, which contributes greatly to our general appearance, has various functions and is divided into the axial and appendicular skeletons.

2 Macroscopically, skeletal muscles work in antagonistic pairs and exhibit certain physiological characteristics.

3 Microscopically, muscle fiber contraction is dependent on actin filaments and myosin filaments and a ready supply of calcium (Ca^{++}) ions and ATP.

Chapter Outline

The Skeleton
    Functions
    Structure
Skeletal Muscles: Macroscopic View
    The Anatomy of Whole Muscle
    The Physiology of Whole Muscle
    The Effect of Exercise
Skeletal Muscles: Microscopic View
    The Anatomy of a Muscle Fiber
    The Physiology of a Muscle Fiber
    Innervation

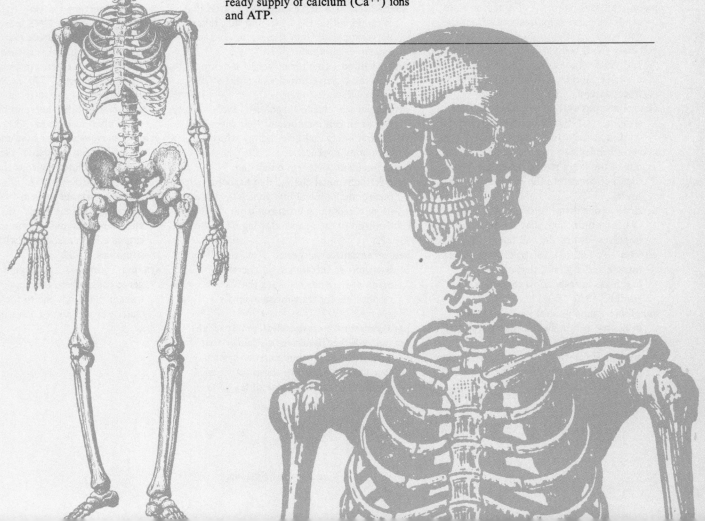

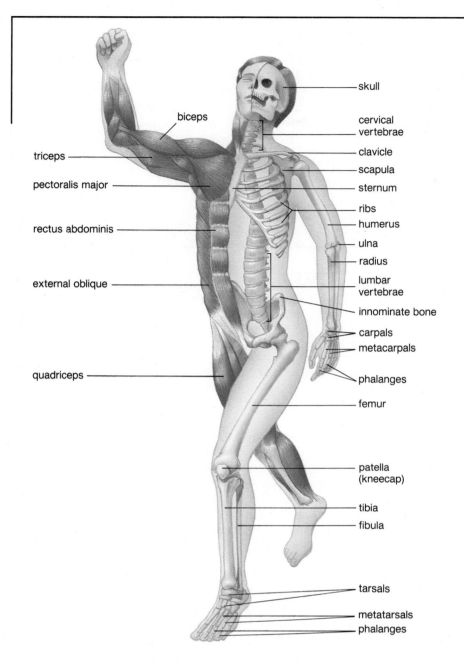

Figure 17.1
Major bones (right) and muscles (left) of the human body. The axial skeleton, composed of the skull, the vertebral column, the sternum, and the ribs, lies in the midline; the rest of the bones belong to the appendicular skeleton.

biceps

triceps

pectoralis major

rectus abdominis

external oblique

quadriceps

skull

cervical vertebrae

clavicle

scapula

sternum

ribs

humerus

ulna

radius

lumbar vertebrae

innominate bone

carpals

metacarpals

phalanges

femur

patella (kneecap)

tibia

fibula

tarsals

metatarsals

phalanges

Muscles and bones largely account for body weight and appearance. Working together, they allow us to perform many mechanical tasks, some of which require grace and agility.

The Skeleton

Functions

The skeleton (fig. 17.1), notably the large, heavy bones of the legs, supports the body against the pull of gravity. The skeleton also protects soft body parts. For example, the skull forms a protective encasement for the brain, as does the rib cage for the heart and the lungs. Flat bones, such as those of the skull, the ribs, and the breastbone, produce red blood cells in both adults and children. All bones are storage areas for inorganic calcium and phosphorus salts. Bones also provide sites for muscle attachment. The long bones, particularly those of the legs and the arms, permit flexible body movement.

Figure 17.2
The vertebral column. The vertebrae are named
according to their location in the column, which
is flexible due to the intervertebral disks. Note
the presence of the coccyx, the vestigial
"tailbone."

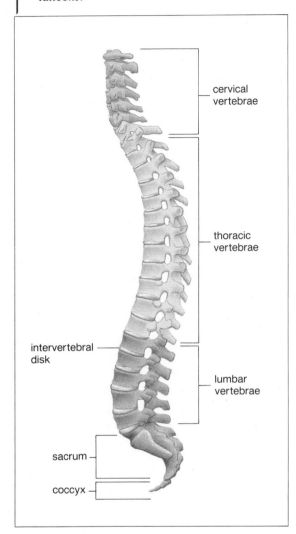

cervical
vertebrae

thoracic
vertebrae

intervertebral
disk

lumbar
vertebrae

sacrum

coccyx

Table 17.1 *Bones of the Skeleton*

Part	Bones
Axial skeleton	Skull, vertebral column, sternum, ribs
Appendicular skeleton	
Pectoral girdle	Clavicle, scapula
Arm	Humerus, ulna, radius
Hand	Carpals, metacarpals, phalanges
Pelvic girdle	Innominate bone
Leg	Femur, tibia, fibula
Foot	Tarsals, metatarsals, phalanges

The skeleton not only permits flexible movement, it also supports and protects the body, produces red blood cells, and serves as a storehouse for certain inorganic salts.

Structure

The skeleton can be divided into 2 parts: the **axial skeleton** and the **appendicular skeleton** (table 17.1).

The Axial Skeleton

The **skull,** or cranium, is composed of many bones fitted tightly together in adults. In newborns, certain bones are not completely formed and instead are joined by membranous regions called **fontanels,** all of which usually close by the age of 16 months. The bones of the skull contain the **sinuses,** air spaces lined by mucous membrane. Two of these, called the mastoid sinuses, drain into the middle ear. Mastoiditis, a condition that can lead to deafness, is an inflammation of these sinuses. Whereas the skull protects the brain, the several bones of the face join to support and to protect the special sense organs and to form the jawbones.

The **vertebral column** extends from the skull to the pelvis and forms a dorsal backbone that protects the spinal cord (see fig. 16.10). Normally, the vertebral column has 4 curvatures that provide more resiliency and strength than a straight column could. It is composed of many parts, called **vertebrae,** that are held together by bony facets, muscles, and strong ligaments. The vertebrae are named according to their location in the body (fig. 17.2).

Intervertebral disks between the vertebrae act as a kind of padding. They prevent the vertebrae from grinding against one another and also absorb shock caused by such movements as running, jumping, and even walking. Unfortunately, these disks weaken with age and can slip or even rupture. Pain results when the damaged disk presses up against the spinal cord and/or spinal nerves. The body may heal itself, or the disk can be removed surgically. If the latter occurs, the vertebrae can be fused, but this limits the flexibility of the body. Disks allow motion between the vertebrae so that we can bend forward, backward, and from side to side.

The vertebral column, directly or indirectly, serves as an anchor for all the other bones of the skeleton (fig. 17.1). All 12 pairs of **ribs** connect directly to the thoracic vertebrae in the back, and all but 2 pairs connect either directly or indirectly via shafts of cartilage to the **sternum** (breastbone) in the front. The lower 2 pairs of ribs are called "floating ribs" because they do not attach to the sternum.

The Appendicular Skeleton

The appendicular skeleton consists of the bones within the pectoral and pelvic girdles and the attached appendages. The pectoral (shoulder) girdle and appendages (arms and hands) are specialized for flexibility, but the pelvic girdle (hip bones) and appendages (legs and feet) are specialized for strength.

The components of the **pectoral girdle** (fig. 17.3) are linked loosely by ligaments rather than firm joints. Each **clavicle** (collar bone) connects with the sternum in front and the **scapula** (shoulder blade) behind, but the scapula is freely movable and is held in place by muscles. This allows it to freely follow the movements of the arm. The single long bone in the upper arm (fig. 17.3), the **humerus,** has a smoothly rounded head that fits into a socket of the scapula. The socket, however, is very shallow and much smaller than the head. Although this means that the arm can move in almost any direction, there is little stability. Therefore, this is the joint that is most apt to dislocate. The opposite end of the humerus meets the 2 bones of the lower arm, the **ulna** and the **radius,** at the elbow. (The prominent bone in the elbow is the topmost part of the ulna.) When the arm is held so that the palm is turned frontward, the radius and the ulna are about parallel to one another. When the arm is turned so that the palm is next to the body, the radius crosses in front of the ulna, a feature that contributes to the easy twisting motion of the lower arm.

The many bones of the hand increase its flexibility. The wrist has 8 **carpal** bones that look like small pebbles. From these, 5 **metacarpal** bones fan out to form a framework for the palm. The metacarpal bone that leads to the thumb is placed in such a way that the thumb can reach out and touch the other digits. (**Digits** is a term that refers to either fingers or toes.) Beyond the metacarpals are the **phalanges,** the bones of the fingers and the thumb. The phalanges of the hand are long, slender, and lightweight.

The **pelvic girdle** (fig. 17.4) consists of 2 heavy, large **innominate** hipbones. The innominate bones are anchored to the sacrum, and together these bones form a hollow cavity, the pelvis. The weight of the body is transmitted through the pelvis to the legs and then onto the ground. The largest bone in the body is the **femur,** or thigh bone. Although the femur is a strong bone, it is doubtful that the femurs of a fairy-tale giant could support the increase in weight. If a giant were 10 times taller than an ordinary human being, he or she would also be about 10 times wider and thicker, making him or her weigh about 1 thousand times as much. This amount of weight would break even giant-size femurs.

In the lower leg, the larger of the 2 bones, the **tibia** (fig. 17.4), has a ridge we call the shin. Both of the bones of the lower leg have a prominence that contributes to the ankle—the tibia on the inside of the ankle and the **fibula** on the outside of the ankle. Although there are 7 **tarsal** bones in the ankle, only one bone receives the weight of the body and passes it on to the heel and the ball of the foot. If you wear high-heel shoes, the weight is thrown even further toward the front of the foot. The **metatarsal** bones form the arches of the foot. There is a longitudinal arch from the heel to the toes and a transverse arch across the foot. These provide a stable, springy base for the body. If the tissues that bind the metatarsals together weaken, flatfeet are apt to result. The bones of the toes are called *phalanges,* just like those of the fingers, but in the foot, phalanges are stout and extremely sturdy.

The axial and appendicular skeletons contain the bones that are listed in table 17.1.

Long Bones

A long bone, such as the femur, illustrates principles of bone anatomy. When the bone is split open, as in figure 17.5, the longitudinal section shows that it

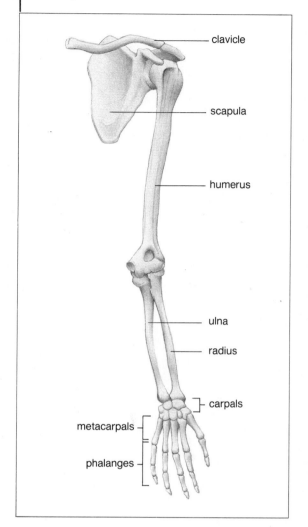

Figure 17.3

The bones of the pectoral girdle, the arm, and the hand. The humerus becomes the "funny bone" of the elbow.

clavicle

scapula

humerus

ulna

radius

carpals

metacarpals

phalanges

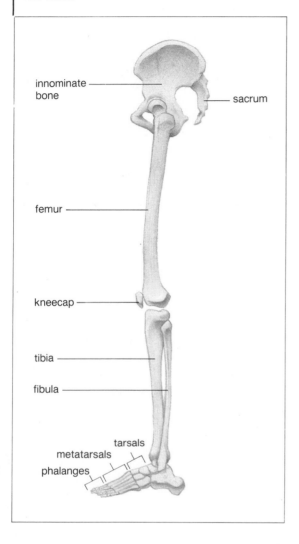

Figure 17.4

The bones of the pelvic girdle, the leg, and the foot. The femur is our strongest bone and withstands a pressure of 540 kg per 2.5 cm³ when we walk.

innominate bone

sacrum

femur

kneecap

tibia

fibula

tarsals

metatarsals

phalanges

is not solid but has a cavity, called the medullary cavity, bounded at the sides by compact bone and at the ends by spongy bone. Beyond the spongy bone, there is a thin shell of compact bone and finally a layer of cartilage.

Compact bone, as discussed on page 178, contains bone cells in tiny chambers called lacunae that are arranged in concentric circles around Haversian canals, which contain blood vessels and nerves. The lacunae are separated by a matrix that contains protein fibers of collagen and mineral deposits, primarily calcium and phosphorus salts.

Spongy bone contains numerous bony bars and plates separated by irregular spaces. Although lighter than compact bone, spongy bone is still designed for strength. Just as braces are used for support in buildings, the solid portions of spongy bone follow lines of stress. The spaces in spongy bone are often filled with **red bone marrow,** a specialized tissue that produces blood cells. The cavity of a long bone usually contains **yellow bone marrow,** which is a fat-storage tissue.

Bones are not inert. They are living tissue and contain cells that require nourishment.

The Growth and the Development of Bones   Most of the bones of the skeleton are cartilaginous during prenatal development. Later, bone-forming cells known as **osteoblasts** replace cartilage with bone. At first, there is only a primary ossification center at the middle of a long bone, but later, secondary centers form at the ends of the bones. A *cartilaginous disk* remains between the primary ossification center and each secondary center. The length of a bone is dependent on how long the cartilage cells within the disk continue to divide. Eventually, though, the disks disappear, and the bone stops growing as the individual attains adult height.

In the adult, bone is broken down and built up again continually. Bone-absorbing cells, called **osteoclasts,** are derived from cells carried in the bloodstream. As they break down bone, they remove worn cells and deposit calcium in the blood. Apparently, after about 3 weeks, osteoclasts disappear. The destruction caused by the osteoclasts is repaired by osteoblasts. As they form new bone, they take calcium from the blood. Eventually, some of these cells get caught in the matrix they secrete and are converted to **osteocytes,** the cells found within Haversian systems (fig. 17.5*b*).

Because of continual renewal and depending on the amount of physical activity or change in certain hormonal balances (see chap. 19), the thickness of bones can change. In most adults, the bones weaken due to a loss of mineral content. Strange as it may seem, adults seem to require more calcium in the diet than children in order to promote the work of osteoblasts. Due to a lack of estrogen, many older women suffer from *osteoporosis,* a condition in which weak and thin bones cause aches and pains and fracture easily. A tendency toward osteoporosis also may be augmented by lack of exercise and too little calcium in the diet.

Bone is living tissue, and it is always being rejuvenated.

Joints

Bones are linked at the joints, which often are classified according to the amount of movement they allow. Some bones, such as those that make up the cranium, are sutured together and are *immovable.* Other joints are *slightly movable,* such as the joints between the vertebrae. The vertebrae are separated by disks, described earlier, that increase their flexibility. Similarly, the 2 innominate bones are slightly movable where they are joined ventrally by cartilage. Owing to hormonal changes, this joint becomes more flexible during late pregnancy, which allows the pelvis to expand during childbirth.

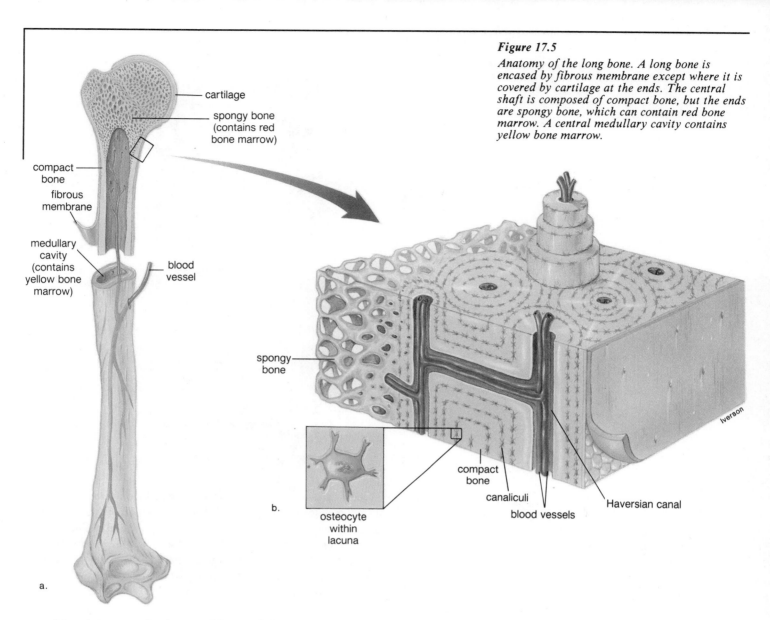

Figure 17.5

Anatomy of the long bone. A long bone is encased by fibrous membrane except where it is covered by cartilage at the ends. The central shaft is composed of compact bone, but the ends are spongy bone, which can contain red bone marrow. A central medullary cavity contains yellow bone marrow.

cartilage

spongy bone (contains red bone marrow)

compact bone

fibrous membrane

medullary cavity (contains yellow bone marrow)

blood vessel

spongy bone

osteocyte within lacuna

compact bone

canaliculi

blood vessels

Haversian canal

a.

b.

Iverson

Most joints are *freely movable* **synovial joints** in which the 2 bones are separated by a cavity. **Ligaments** are composed of fibrous connective tissue that binds the 2 bones to one another, holding them in place as they form a capsule. In a "double-jointed" individual, the ligaments are unusually loose. The joint capsule is lined by synovial membrane that produces *synovial fluid,* a lubricant for the joint.

The knee is an example of a synovial joint (fig. 17.6). As in other freely movable joints, the bones of the knee are capped by cartilage, but in the knee, there are also crescent-shape pieces of cartilage between the bones called **menisci** (singular, meniscus). These give added stability, helping to support the weight placed on the knee joint. Unfortunately, athletes often suffer injury of the menisci, known as torn cartilage. The knee joint also contains 13 fluid-filled sacs called bursae, which ease friction between tendons and ligaments and between tendons and bones. Inflammation of bursae is called bursitis. Tennis elbow is a form of bursitis.

There are different types of freely movable joints. The knee and elbow joints are *hinge joints* because like a hinged door, they permit movement largely in one direction only. More versatile are the ball-and-socket joints; for example, the ball of the femur fits into a socket on the innominate bone. *Ball-and-socket joints* allow movement in all planes and even a rotational movement.

Figure 17.6

The knee joint, an example of a freely movable synovial joint. Notice that there is a cavity between the bones that is encased by ligaments and lined by synovial membrane. The kneecap protects the joint.

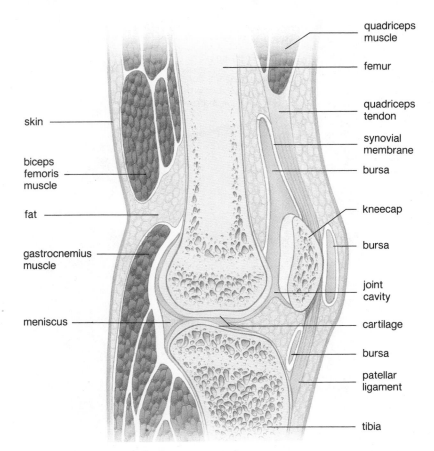

quadriceps muscle

femur

quadriceps tendon

synovial membrane

bursa

kneecap

bursa

joint cavity

cartilage

bursa

patellar ligament

tibia

skin

biceps femoris muscle

fat

gastrocnemius muscle

meniscus

Synovial joints are subject to *arthritis*. In rheumatoid arthritis, the synovial membrane becomes inflamed and thickens. Degenerative changes take place that make the joint almost immovable and painful to use. There is evidence that these effects are brought on by an autoimmune reaction. In old-age arthritis, or osteoarthritis, the cartilage at the ends of the bones disintegrates so that the 2 bones become rough and irregular. This type of arthritis is apt to affect the joints that have received the greatest use over the years.

Joints are classified according to the degree of movement. Some joints are immovable, some are slightly movable, and some are freely movable.

Skeletal Muscles: Macroscopic View

Muscles are effectors, which enable the organism to respond to a stimulus (p. 331). Skeletal muscles are attached to the skeleton, and their contraction accounts for voluntary movements. Involuntary muscles, both smooth and cardiac, were discussed on page 179.

The Anatomy of Whole Muscle

Muscles typically are attached to bone by **tendons** made of fibrous connective tissue. Tendons most often attach muscles to the far side of a joint so that the muscle extends across the joint (fig. 17.7). When the central portion of the muscle, called the belly, contracts, one bone remains fairly stationary and the other one moves. The **origin** of the muscle is on the stationary bone, and the **insertion** of the muscle is on the bone that moves.

When a muscle contracts, it shortens. Therefore, muscles can only pull; they cannot push. Because we need both to extend and to flex at a joint, muscles generally work in antagonistic pairs. For example, the biceps and the triceps are a pair of muscles that move the lower arm up and down (fig. 17.7).

Figure 17.7

Attachment of skeletal muscles as exemplified by the biceps and the triceps. The origin of a muscle is fairly stationary, while the insertion moves. These muscles are antagonistic. When the biceps contracts, the lower arm is raised, and when the triceps contracts, the lower arm is lowered.

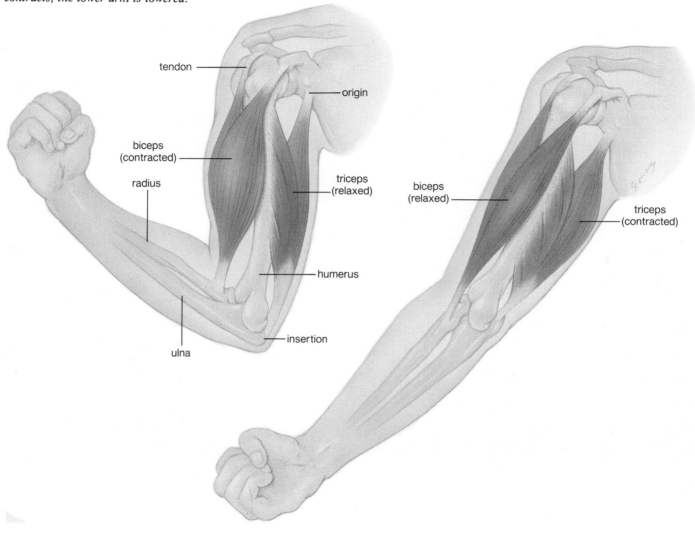

When the biceps contracts, the lower arm flexes, and when the triceps contracts, the lower arm extends.

Muscle Spindles

Muscle spindles are receptors that are widespread throughout skeletal muscles (fig. 17.8). A muscle spindle consists of several modified muscle fibers that have sensory nerve fibers wrapped around a short, specialized region somewhere near the middle of their length. A muscle spindle generates more nerve impulses when the muscle stretches and fewer when it contracts. These nerve impulses inform the central nervous system (CNS) of the state of this particular muscle so that contraction of this muscle can be coordinated with contraction of other muscles.

Information from muscle spindles also allows the CNS to maintain the **tone** of a skeletal muscle, a condition in which some fibers always are contracted. Muscle tone is particularly important in maintaining posture. If all the fibers within the neck, trunk, and leg muscles suddenly relaxed, the body would collapse.

17.2 Critical Thinking

1. Explain why the biceps causes the lower arm and not the upper arm (where it is located) to flex.
2. Why cannot the contraction of the biceps both flex and extend the lower arm?
3. What type of structure needed for a muscle to contract is missing from figure 17.7 but is present in 17.8?
4. Why do you suppose the muscles of the legs are larger than the muscles of the arms?

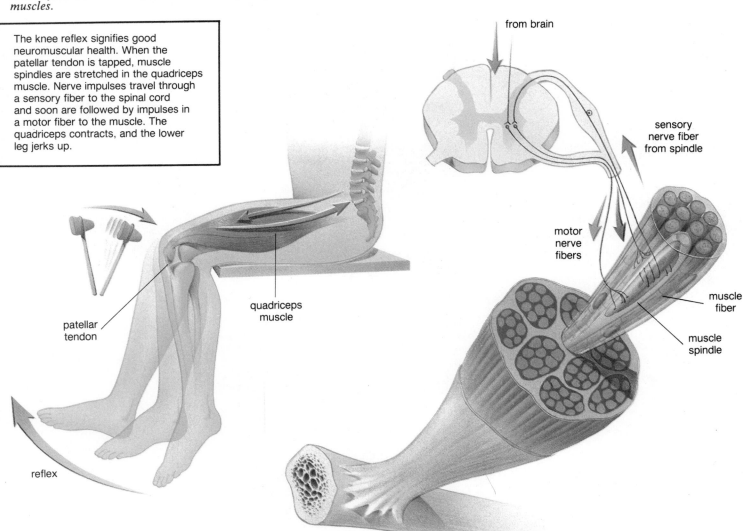

Figure 17.8

Muscle spindle structure and function. After associated muscle fibers stretch, spindle fibers aid in the coordination of muscular contraction by sending more nerve impulses to the CNS, which then adjusts contraction of this muscle and perhaps other muscles.

The knee reflex signifies good neuromuscular health. When the patellar tendon is tapped, muscle spindles are stretched in the quadriceps muscle. Nerve impulses travel through a sensory fiber to the spinal cord and soon are followed by impulses in a motor fiber to the muscle. The quadriceps contracts, and the lower leg jerks up.

from brain

sensory nerve fiber from spindle

motor nerve fibers

muscle fiber

muscle spindle

quadriceps muscle

patellar tendon

reflex

Whole skeletal muscles are attached to bones and work in antagonistic pairs. These muscles contain muscle spindles that regulate their contractile state.

The Physiology of Whole Muscle

In the laboratory, it is possible to study the contraction of individual whole muscles. Customarily, a calf muscle is removed from a frog and mounted so that one end is fixed and the other moveable. The mechanical force of contraction is transduced into an electrical current recorded by a **physiograph** (fig. 17.9*a*). The resulting visual pattern is called a **myogram.**

The All-or-None Response

A *single* muscle fiber (muscle cell, p. 179) either responds to a stimulus and contracts or it does not. At first, the stimulus may be so weak that no contraction occurs, but as soon as the strength of the stimulus reaches the *threshold stimulus,* the muscle fiber contracts completely. Therefore, a muscle fiber obeys the **all-or-none law.**

Contrary to that of an individual fiber, the strength of contraction of a whole muscle can increase according to the degree of stimulus beyond the threshold stimulus. A whole muscle contains many fibers, and the degree of contraction is dependent on the total number of fibers contracting. The *maximum stimulus* is the one beyond which the degree of contraction does not increase.

Muscle fibers obey the all-or-none law, but whole muscles do not obey this law.

The Muscle Twitch

If a muscle is placed in a physiograph and is given a maximum stimulus, it contracts and then relaxes. This action—a single contraction that lasts only a fraction of a second—is called a muscle **twitch.** Figure 17.9*b* is a myogram of a twitch, which is customarily divided into the *latent period,* or the period of time between stimulation and initiation of contraction, the *contraction period,* and the *relaxation period.*

If a muscle is exposed to 2 maximum stimuli in quick succession, it responds to the first but not to the second stimulus. This is because it takes an instant following a contraction for the muscle fibers to recover in order to respond to the next stimulus. The very brief moment following stimulation, during which a muscle is unresponsive, is called the refractory period.

Summation and Tetanus

If a muscle is given a rapid series of threshold stimuli, it can respond to the next stimulus without relaxing completely. In this way, muscle tension **summates** until maximal sustained tetanic contraction is achieved (fig. 17.9*c*). The myogram no longer shows individual twitches; rather, they are fused and blended completely into a straight line. **Tetanus** continues until the muscle fatigues due to depletion of energy reserves. **Fatigue** is apparent when a muscle relaxes even though stimulation continues.

Tetanic contractions occur whenever skeletal muscles are used actively. Ordinarily, however, only a portion of any particular muscle is involved—while some fibers are contracting, others are relaxing. Because of this, intact muscles rarely fatigue completely.

Muscle twitch, summation, and tetanus are related to the frequency with which a muscle is stimulated.

The Effect of Exercise

A regular exercise program, such as the one described in table 17.A in the reading on page 356, has many benefits. Increased endurance and strength of muscles are 2 possible benefits. Endurance is measured by the length of time the muscle can work before fatiguing, and strength is the force a muscle (or a group of muscles) can exert against a resistance.

A regular exercise program brings about physiological changes that build endurance, such as increased muscular stores of ATP and increased tolerance to lactate buildup. Muscle strength increases as muscle enlargement occurs due to exercise. When a muscle enlarges, the number of muscle fibers does not increase; however, the protein content of the muscle increases. This happens because the contractile elements in muscles—called myofibrils—that contain the protein filaments actin and myosin increase in number (fig. 17.10).

Aside from improved endurance and strength, an exercise program also helps many other organs of the body. Cardiac muscle enlarges, and the heart can work harder than before. The resting heart rate decreases. Lung and diffusion capacity increase. Body fat decreases, but bone density increases so that

Figure 17.9
Physiology of muscle contraction. a. A physiograph is an apparatus that can be used to record a myogram, a visual representation of the contraction of a muscle that has been dissected from an animal. b. Simple muscle twitch is composed of 3 periods: latent, contraction, and relaxation. c. Summation and tetanus. When a muscle is not allowed to relax completely between stimuli, the contractions increase in size. The muscle remains maximally contracted until it fatigues.

a.

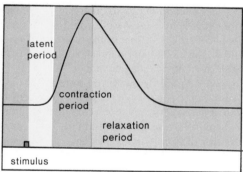
b.

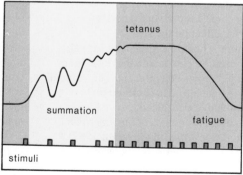
c.

. . . If a single researcher could be considered responsible for the fitness craze, it would be Dr. Ralph Paffenbarger, a specialist in heart disease and exercise with posts at both Stanford and Harvard universities. His landmark research, involving 17,000 Harvard alumni and 6,000 San Francisco longshoremen, showed that the men who exercised vigorously and burned at least 2,000 calories a week doing so cut their risk of dying from heart disease by half. The reduction was dramatic even if the men smoked or if their parents both had died of heart disease.

The problem was the public—and many health professionals—interpreted these

Exercise: Changing Perceptions

findings as meaning that anything less didn't do any good. If you weren't up for running 20 miles a week or an hour of tennis 5 times a week, or cross-country skiing for a half hour a day—well, you'd be just as well off sitting in front of the tube with a beer. The 2,000-calorie threshold was transformed into a "magic number"— above it, you were fit. Below it, you were a basket case.

Paffenbarger says he never believed that the 2,000-calorie figure was carved in stone. Neither did Dr. Arthur Leon, an epidemiologist at University of Minnesota's School of Public Health. Both were surprised, however, when Leon's research showed that a far more modest level of activity could exert a very powerful protective effect on the heart.

"People who don't want to do formal, sweaty exercise can be told that less can be beneficial," says Leon. "Just moving around more is a big help." He backs up his assertion with the results of his study, published in the *Journal of the American Medical Association* (in the fall of 1987). That study analyzed the relationship between heart attack and physical activity off the job in 12,138 men who participated in the Multiple Risk Factor Intervention Trial, a nationwide study conducted at 22 medical centers.

The trial found that men who were only moderately active—spending an average of 48 minutes a day on leisure-time physical activity—had one-third fewer heart attacks than their peers who moved around during leisure time an average of 16 minutes each day. And the moderately active group didn't spend all, or even most, of their exercise time huffing and puffing. Mostly, the report found, their activities were in the light-to-moderate range: lawn and garden work, bowling, ballroom dancing— activities we don't often even think of as exercise.

Indeed, some of the most dramatic gains are made by the sedentary folks whose initial efforts fall short of the magic 2,000 calories and whose activities never include much bouncing around. For example, after only 4 months of attending a twice-weekly, low-impact aerobics class at Northwestern University, men and women with rheumatoid arthritis—a chronic, severe form of joint disease—reported much less pain, swelling, fatigue, and depression than before they began exercising.

"The lower end of the spectrum gets the most benefit from its effort," says cardiologist Miller. "If you've just had a heart transplant, it takes very little—a walk around the gym—to get improvement. If you take a sedentary office worker, it doesn't take much more than adding a mile of walking a day to get improvement."

Leon and Paffenbarger point out that it's relatively easy to program greater amounts of activity into your normal day. Walk down to the accounting department at work instead of calling the accountant. Take stairs instead of elevators. Stroll during your lunchtime. A brisk walk—especially if you swing your arms—can get your heart rate up without the jouncing of jogging that many people dislike.

Leon's study showed that the benefits of exercise start to level off at a certain point—at least as far as fatal heart attack is concerned. The rate of fatal heart attack among the moderately active men was the same as the most active men in the study,

who devoted an average of 134 minutes a day to leisure-time physical activity.

Does this mean the most active group was no more fit than the moderately active group? Of course not. Fitness, stresses Bud Getchell, executive director of the nonprofit National Institute for Fitness and Sport in Indianapolis, includes flexibility, muscular strength, endurance, and body composition (lean mass vs. fat), as well as cardiovascular health. To achieve the type of fitness that would allow you to take on most vigorous activities with ease—from lifting a heavy box at work, to shoveling snow without hurting your back, to enjoying a game of pickup basketball—most fitness experts say you should spend about 3 hours (see table 17.A), spaced out during the week, doing activities that strengthen muscles and enhance flexibility, as well as challenge the heart.

Your body will reward your efforts. Although most people think the main benefit of fitness is reducing the risk of heart disease, there's mounting evidence that other tissues and organs benefit as well. Several studies at the Institute for Aerobics Research in Dallas, Harvard, and other institutions have found that more active people have lower rates of colon, brain, kidney, and reproductive cancers, as well as leukemia, than their more sedentary counterparts. And studies that looked at exercise along with other variables found this still held true when factors like age, diet, and socioeconomic background were taken into account. Researchers speculate this may be because activity promotes the delivery of more nutrients and oxygen to these organs and tissues.

Indeed, according to Dr. Everett L. Smith, director of the Biogerontology Laboratory at the University of Wisconsin-Madison, half of the functional decline between ages 30 and 70 could be prevented if we simply used our body more. Bone density, nerve function, and kidney efficiency, as well as overall strength and flexibility, largely can be preserved into our later years simply by keeping up an active life.

First appeared in *The Boston Globe* May 1, 1988. Copyright © 1988 Sy Montgomery. Reprinted by permission.

Table 17.A   A Checklist for Staying Fit

Children, 7-12	Teenagers, 13-18	Adults, 19-55	Seniors, 55 AND UP
Vigorous activity 1-2 hours daily	Vigorous activity 3-5 times a week	Vigorous activity for one-half hour, 3 times a week	Moderate exercise 3 times a week
Free play	Build muscle with calisthenics	Exercise to prevent lower back pain: aerobics, stretching, yoga	Plan a daily walk
Build motor skills through team sports, dance, swimming	Plan aerobic exercise to control buildup of fat cells	Take active vacations: hike, bicycle, cross-country ski	Daily stretching exercises
Encourage more exercise outside of physical education classes	Pursue tennis, swimming, riding— sports that can be enjoyed for a lifetime	Find exercise partners: join a running club, bicycle club, outing group	Learn a new sport: golf, fishing, ballroom dancing
Initiate family outings: bowling, boating, camping, hiking	Continue team sports, dancing, hiking, swimming		Try low-impact aerobics
			Before undertaking new exercises, consult your doctor

Fitness walkers. *The enclosed shopping mall, intended to be a boon for buyers, now also has become a haven for fitness walkers. Climate controlled and its passages unimpeded by curbs or stoplights, the mall provides a perfect environment for those determined to put in their daily mileage, rain or shine, while eliminating many of the outdoor hazards that may deter older pedestrians. Some mall walkers, like these Galleria Mall GoGetters in Glendale, California, have formed their own clubs. A few malls now issue special walking maps, while others open on holidays, even when the stores are closed, just to accommodate the local ramblers.*

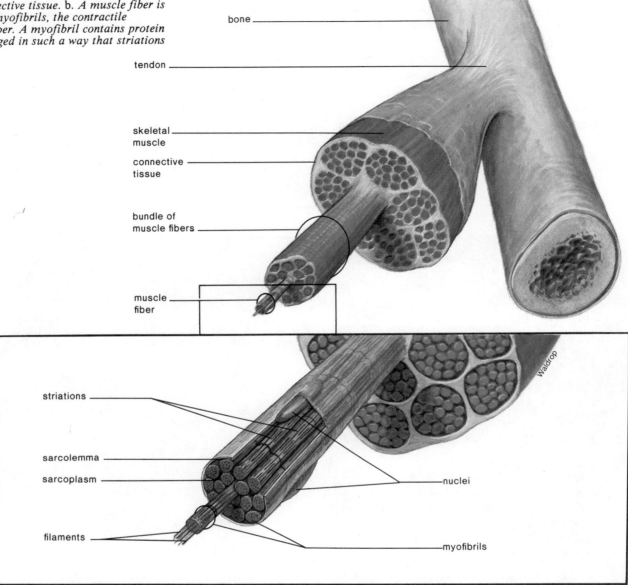

Figure 17.10

Anatomy of a skeletal muscle. a. Whole muscle contains several bundles of muscle fibers bound by fibrous connective tissue. b. A muscle fiber is packed full of myofibrils, the contractile elements in a fiber. A myofibril contains protein filaments arranged in such a way that striations are seen.

bone

tendon

skeletal muscle

connective tissue

bundle of muscle fibers

muscle fiber

striations

sarcolemma

sarcoplasm

nuclei

filaments

myofibrils

Waldrop

breakage is less apt to occur. Fat and cholesterol blood levels decrease along with blood pressure. The reading on page 356 particularly discusses studies showing that an exercise program lowers the risk of heart attack.

Skeletal Muscles: Microscopic View

A whole skeletal muscle (fig. 17.10) is composed of a number of bundles of **muscle fibers.**

The Anatomy of a Muscle Fiber

Each muscle fiber is a cell containing the usual cellular components, but special names have been assigned to some of these components, as indicated in table 17.2. A muscle fiber has some unique anatomical characteristics. For one thing, it has a T (for transverse) system (fig. 17.11). The **sarcolemma,** or cell membrane, forms these *tubules* that penetrate, or dip down, into the cell

Table 17.2 Muscle Cells

Component	Term
Cell membrane	Sarcolemma
Cytoplasm	Sarcoplasm
Endoplasmic reticulum	Sarcoplasmic reticulum

Human Anatomy and Physiology

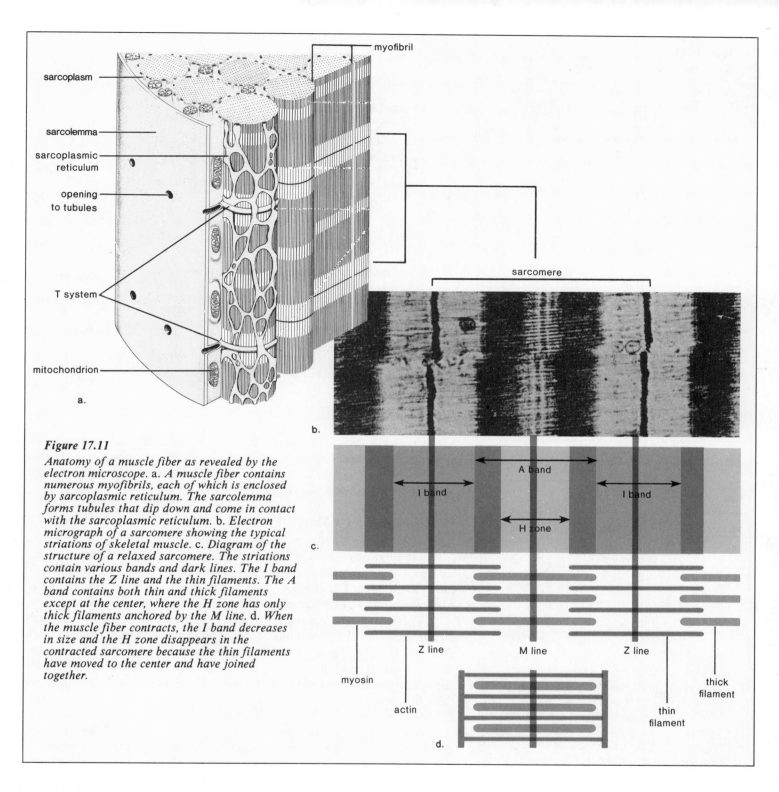

Figure 17.11

Anatomy of a muscle fiber as revealed by the electron microscope. a. A muscle fiber contains numerous myofibrils, each of which is enclosed by sarcoplasmic reticulum. The sarcolemma forms tubules that dip down and come in contact with the sarcoplasmic reticulum. b. Electron micrograph of a sarcomere showing the typical striations of skeletal muscle. c. Diagram of the structure of a relaxed sarcomere. The striations contain various bands and dark lines. The I band contains the Z line and the thin filaments. The A band contains both thin and thick filaments except at the center, where the H zone has only thick filaments anchored by the M line. d. When the muscle fiber contracts, the I band decreases in size and the H zone disappears in the contracted sarcomere because the thin filaments have moved to the center and have joined together.

so that they come into contact but do not fuse with expanded portions of modified ER, termed the **sarcoplasmic reticulum.** The expanded portions of the sarcoplasmic reticulum, called calcium-storage sacs, contain calcium (Ca^{++}) ions, which are essential for muscle contraction. The sarcoplasmic reticulum encases hundreds and sometimes even thousands of *myofibrils,* which are the contractile portions of the fibers.

Myofibrils

Myofibrils, the contractile elements in muscles, are cylindrical in shape and run the length of the muscle fiber. The light microscope shows that a myofibril

Figure 17.12

Sliding filament theory. a. *Relaxed sarcomere.*
b. *Contracted sarcomere. Note that during
contraction, the I band and the H zone decrease
in size. This indicates that the thin filaments
slide past the thick filaments. Even so, the thick
filaments do the work by pulling the thin
filaments by means of cross bridges.*

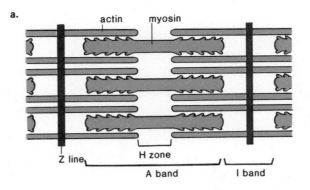

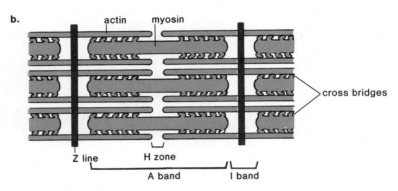

Table 17.3	Contractile Elements
Component	**Definition**
Myofibril	Muscle cell contractile subunit
Sarcomere	Functional unit of myofibril
Myosin	Thick filament
Actin	Thin filament

17.3 Critical Thinking

1. A muscle fiber obeys the all-or-none law.
   Therefore, do all myofibrils in a muscle
   fiber contract at the same time?
2. When a sarcomere contracts, does the Z
   line move? If so, in which direction?
3. A respiratory pigment in muscle called
   myoglobin receives oxygen (O_2) from
   hemoglobin. Which of these 2 respiratory
   pigments has the higher affinity for
   oxygen?
4. When we exercise, blood brings more
   oxygen to the muscles. What do the
   muscles specifically do with all this oxygen?

has light and dark bands called striations; it is their striations that cause skeletal muscle to appear striated (fig. 9.10a). The electron microscope shows that the striations of myofibrils are formed by the placement of protein filaments within contractile units called **sarcomeres** (fig. 17.11b).

The Physiology of a Muscle Fiber

A sarcomere extends between 2 dark lines called the *Z lines* (fig. 17.11c). The *I band* is a light-colored region that takes in the sides of 2 sarcomeres; therefore, an I band includes a Z line. The center dark region of a sarcomere is called the *A band*. The A band is interrupted by a light center portion, the *H zone,* and a fine, dark stripe called the *M line* cuts through the H zone.

These bands and zones relate to the placement of filaments within each sarcomere. A sarcomere contains 2 types of filaments, *thin filaments* and *thick filaments* (fig. 17.11c). The thin filaments are attached to a Z line, and the thick filaments are anchored by an M line. The I band is light because it contains only thin filaments; the A band is dark because it contains both thin and thick filaments, except at the center in the lighter H zone, where only thick filaments are found.

Sarcomere contraction is dependent on the action of 2 proteins, **actin** and **myosin,** that make up the thin filaments and the thick filaments, respectively (table 17.3). When a sarcomere contracts (fig. 17.11d), the actin filaments slide past the myosin filaments and approach one another. This causes the I band to shorten and the H zone to almost or completely disappear. The movement of actin filaments in relation to myosin filaments is called the **sliding filament theory** of muscle contraction. During the sliding process, the sarcomere shortens even though the filaments themselves remain the same length (fig. 17.12).

Table 17.4 Muscle Contraction

Name	Function
Actin filaments	Slide past myosin, causing contraction
Ca^{++}	Needed for myosin to bind to actin
Myosin filaments	a. Pull actin filaments by means of cross bridges b. Are enzymatic and split ATP
ATP	Supplies energy for bonding between myosin and actin

The overall formula for muscle contraction can be represented as follows:

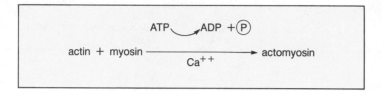

The participants in this reaction have the functions listed in table 17.4. Even though it is the actin filaments that slide past the myosin filaments, it is the myosin filaments that do the work. In the presence of calcium (Ca^{++}) ions and ATP, portions of a myosin filament called *cross bridges* (fig. 17.12) bend backward and attach to an actin filament. (Each cross bridge binds to an actin filament at a cross-bridge binding site; actomyosin represents this in the formula above.) After attaching, the cross bridges bend forward and the actin filament is pulled along. Now, ATP is broken down by myosin, and detachment occurs. Notice that myosin is not only a structural protein, it is also an ATPase enzyme. The cross bridges attach and detach some 50–100 times as the thin filaments are pulled to the center of a sarcomere.

The sliding filament theory states that actin filaments slide past myosin filaments because myosin has cross bridges that pull the actin filaments inward.

It is obvious from our discussion that ATP provides the energy for muscle contraction. In order to ensure a ready supply of ATP, muscle fibers contain **creatine phosphate** (phosphocreatine), a storage form of high-energy phosphate. Creatine phosphate does not participate directly in muscle contraction. Instead, it is used to regenerate ATP by the following reaction:

$$creatine \sim P + ADP \longrightarrow ATP + creatine$$

Oxygen Debt

When all of the creatine phosphate is depleted and no oxygen (O_2) is available for aerobic respiration, a muscle fiber can generate ATP by using fermentation, an anaerobic process (p. 114). Fermentation, which is apt to occur during strenuous exercise, can supply ATP for only a short time because lactate buildup produces muscular aching and fatigue that lasts a minute or so.

We all have had the experience of having to continue deep breathing following strenuous exercise. This continued intake of oxygen is required to complete the metabolism of lactate that has accumulated during exercise and

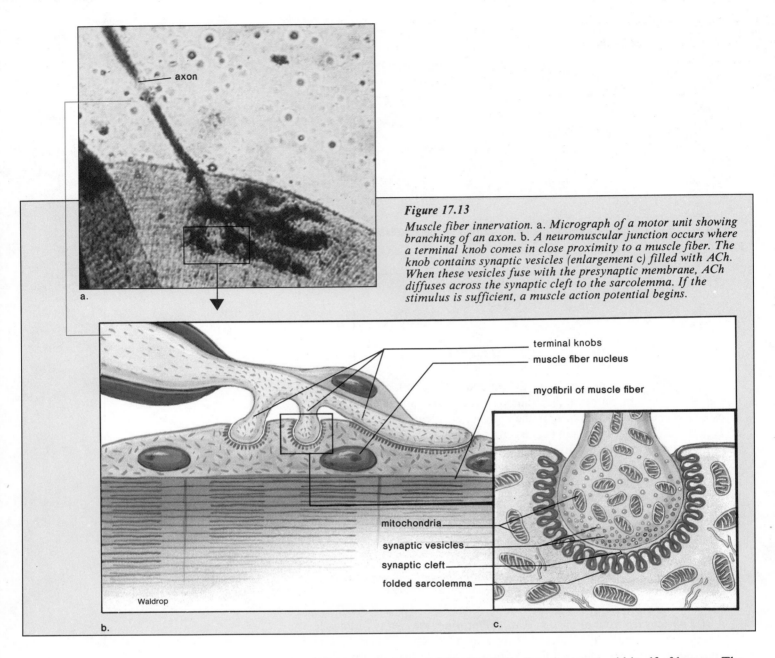

axon

Figure 17.13

Muscle fiber innervation. a. Micrograph of a motor unit showing branching of an axon. b. A neuromuscular junction occurs where a terminal knob comes in close proximity to a muscle fiber. The knob contains synaptic vesicles (enlargement c) filled with ACh. When these vesicles fuse with the presynaptic membrane, ACh diffuses across the synaptic cleft to the sarcolemma. If the stimulus is sufficient, a muscle action potential begins.

terminal knobs

muscle fiber nucleus

myofibril of muscle fiber

mitochondria

synaptic vesicles

synaptic cleft

folded sarcolemma

Waldrop

a.

b.

c.

represents an **oxygen debt** that the body must pay to rid itself of lactate. The lactate is transported to the liver, where one-fifth of it is completely broken down to carbon dioxide (CO_2) and water (H_2O) by means of the Krebs cycle and the respiratory chain (see chap. 5). The ATP gained by this respiration then is used to convert four-fifths of the lactate back to glucose.

Muscle contraction requires a ready supply of ATP. Creatine phosphate is used to generate ATP rapidly. If oxygen is in limited supply, fermentation produces ATP but results in oxygen debt.

Innervation

Muscles are innervated; that is, nerve impulses cause muscles to contract. Each motor axon of a nerve branches to several muscle fibers, and collectively, these muscle fibers are called a motor unit. Each branch has several terminal knobs, where there are synaptic vesicles filled with the neuromuscular transmitter acetylcholine (ACh). The region where a terminal knob lies in close proximity to the sarcolemma of a muscle fiber is called a neuromuscular junction. A

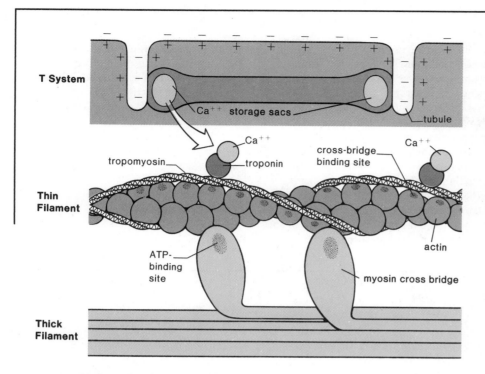

T System

Ca⁺⁺ storage sacs

tubule

tropomyosin

Ca⁺⁺

troponin

cross-bridge binding site

Ca⁺⁺

Thin Filament

ATP-binding site

actin

myosin cross bridge

Thick Filament

Figure 17.14

Detailed structure and function of sarcomere contraction. After calcium (Ca⁺⁺) ions are released from their storage sac, they combine with troponin, a protein that occurs periodically along tropomyosin threads. This causes the tropomyosin threads to shift their position so that cross-bridge binding sites are revealed along the actin filaments. The myosin filament extends its globular heads, forming cross bridges that bind to these sites. The breakdown of ATP by myosin causes the cross bridges to detach and to reattach farther along the actin. In this way, the actin filaments are pulled along past the myosin filaments.

neuromuscular junction (fig. 17.13) has the same components as a synapse: a presynaptic membrane, a synaptic cleft, and a postsynaptic membrane. Only in this case, the postsynaptic membrane is a portion of the sarcolemma of a muscle fiber.

Nerve impulses cause synaptic vesicles to merge with the presynaptic membrane and to release ACh into the synaptic cleft. When ACh reaches the sarcolemma, it is depolarized. The result is a **muscle action potential** that spreads over the sarcolemma and down the T system (fig. 17.11) to where calcium (Ca⁺⁺) ions are stored in calcium-storage sacs of the sarcoplasmic reticulum. When the action potential reaches a sac, calcium ions are released, and they diffuse into the sarcoplasm, where they participate in muscle contraction.

It is now necessary to consider the structure of a thin filament in more detail. Figure 17.14 shows the placement of 2 other proteins associated with a thin filament (the double row of twisted globular actin molecules). Threads of tropomyosin wind about a thin filament, and troponin occurs at intervals along the threads. After calcium ions are released from their storage sac, they combine with troponin. After binding occurs, the tropomyosin threads shift their position, and the cross-bridge binding sites are exposed.

The thick filament is a bundle of myosin molecules, each having a globular head. Each head is a cross bridge that has an ATP-binding site. After ATP attaches to ATP-binding sites, the cross bridges bend backward and attach to cross-bridge binding sites on the actin filaments. After attachment occurs, the cross bridges bend forward, pulling the actin filaments a short distance. Then the myosin heads break down ATP, and detachment of the cross bridges occurs. The actin filaments move nearer the center of the sarcomere each time the cycle is repeated.

The movement of the actin filaments causes muscle contraction. Contraction ceases when nerve impulses no longer stimulate the muscle fiber. With the cessation of a muscle action potential, calcium ions are pumped back into their storage sac by active transport. Relaxation now occurs.

A neuromuscular junction functions like a synapse except that a muscle action potential causes calcium ions to be released from calcium-storage sacs, and thereafter muscle contraction occurs.

Summary

The skeleton aids movement of the body while it also supports and protects the body. Bones serve as deposits for inorganic salts, and some bones are sites for blood-cell production. The skeleton is divided into 2 parts: (1) the axial skeleton, which is made up of the skull, the ribs, and the vertebrae; and (2) the appendicular skeleton, which is composed of the appendages and their girdles. Joints are regions where bones are linked.

Whole skeletal muscles work in antagonistic pairs and have degrees of contraction. Muscle fibers obey the all-or-none law; it is possible to study a single contraction (muscle twitch) and sustained contraction (summation and tetanus) by using a physiograph.

Muscle fibers are cells that contain myofibrils in addition to the usual components of cells. Longitudinally, myofibrils are divided into sarcomeres, where it is possible to note the arrangement of actin filaments and myosin filaments. When a sarcomere contracts, the actin filaments slide past the myosin filaments. Myosin has cross bridges that attach to and pull the actin filaments along. ATP breakdown by myosin is necessary for detachment to occur.

Innervation of a muscle fiber begins at a neuromuscular junction. Here, synaptic vesicles release ACh into the synaptic cleft. When the sarcolemma receives ACh, a muscle action potential moves down the T system to calcium-storage sacs. When calcium ions are released, contraction occurs. When calcium ions are actively transported back into the storage sacs, muscle relaxation occurs.

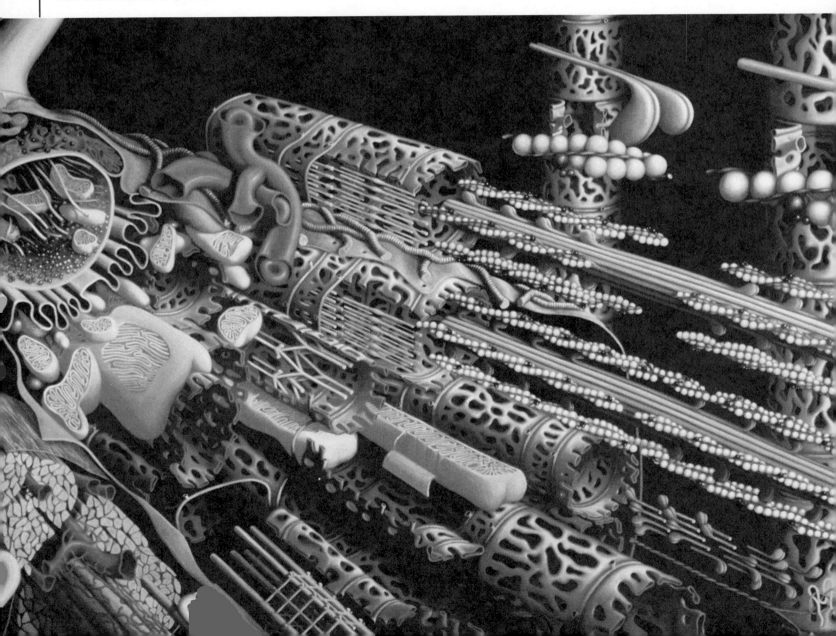

Skeletal muscle fiber anatomy. Looking from left *to* right, *we first see myofibrils packed as close together as they would be in a muscle fiber. Next, the enlarged but disrupted fiber gives a view of the sarcoplasmic reticulum of individual myofibrils. Then, the individual actin and myosin filaments of a sarcomere become apparent. Finally, on the* upper right, *note the globular heads of myosin about to attach to the twisted actin filaments. Once attachment occurs, muscle contraction will occur.*

Study Questions

1. Distinguish between the axial and appendicular skeletons. (pp. 348–49)
2. List the bones that form the pectoral and pelvic girdles. (p. 349)
3. Describe the anatomy of a long bone and of a freely movable joint. (p. 351)
4. Describe how muscles are attached to bones. Why do muscles act in antagonistic pairs? (p. 352)
5. Describe the significance of threshold and maximum stimuli, muscle twitch, summation, and tetanic contraction. (p. 355)
6. How is the tone of a muscle maintained, and how do muscle spindles contribute to the maintenance of tone? (pp. 353–54)
7. Discuss the microscopic anatomy of a muscle fiber and the structure of a sarcomere. What is the sliding filament theory? (pp. 358–60)
8. Give the function of each participant in the following reaction:

$$actin + myosin \xrightarrow[Ca^{++}]{ATP \quad ADP + \textcircled{P}} actomyosin$$

(p. 361)
9. Discuss the availability and the specific role of ATP during muscle contraction. What is oxygen debt, and how is it repaid? (p. 361)
10. What causes a muscle action potential? How does the muscle action potential bring about sarcomere and muscle fiber contraction? (pp. 362–63)

Objective Questions

1. The skull, the ribs, and the sternum are all in the _____ skeleton.
2. The vertebral column protects the _____ cord.
3. The 2 bones of the lower arm are the _____ and the _____ .
4. Most joints are freely movable _____ joints in which the 2 bones are separated by a cavity.
5. Muscles work in _____ pairs; the biceps flexes and the triceps extends the lower arm.
6. Maximal sustained contraction of a muscle is called _____ .
7. Actin and myosin filaments are found within cell inclusions called _____ , which are divided into units called _____ .
8. The molecule _____ serves as an immediate source of high-energy phosphate for ATP production in muscle cells.
9. The juncture between axon ending and muscle cell sarcolemma is called a _____ junction.
10. A muscle action potential causes _____ ions to be released from storage sacs, and this signals the muscle fiber to contract.

Answers to Objective Questions

1. axial 2. spinal 3. radius, ulna 4. synovial 5. antagonistic 6. tetanus 7. myofibrils, sarcomeres 8. creatine phosphate 9. neuromuscular 10. calcium

Selected Key Terms

actin (ak'tin) one of 2 major proteins of muscle; makes up thin filaments in myofibrils of muscle fibers. *See* myosin. *360*

appendicular skeleton (ap''en-dik'u-lar skel'ĕ-ton) portion of the skeleton forming the upper extremities, the pectoral girdle, the lower extremities, and the pelvic girdle. *348*

axial skeleton (ak'se-al skel'ĕ-ton) portion of the skeleton that supports and protects the organs of the head, the neck, and the trunk. *348*

compact bone (kom-pakt' bōn) bone in which cells, separated by a matrix of collagen and mineral deposits, are located within Haversian systems. *350*

creatine phosphate (kre'ah-tin fos'fāt) a compound unique to muscles that contains a high-energy phosphate bond. *361*

insertion (in-ser'shun) the end of a muscle that is attached to a movable bone. *352*

muscle action potential (mus'el ak'shun po-ten'shal) an electrochemical change due to increased sarcolemma permeability that is propagated down the T system and results in muscle contraction. *363*

myofibril (mi''o-fi'bril) the contractile portion of muscle fibers. *359*

myosin (mi'o-sin) one of 2 major proteins of muscle; makes up thick filaments in myofibrils and is capable of breaking down ATP. *See* actin. *360*

neuromuscular junction (nu''ro-mus'ku-lar jungk'shun) the point of contact between a nerve cell and a muscle fiber. *363*

origin (or'i-jin) end of a muscle that is attached to a relatively immovable bone. *352*

osteocyte (os'te-o-sīt) a mature bone cell. *350*

oxygen debt (ok'si-jen det) oxygen that is needed to metabolize lactate, a compound that accumulates during vigorous exercise. *362*

red bone marrow (red bon mar'o) tissue located in the cavity of bones that forms blood cells. *350*

synovial joint (si-no've-al joint) a freely movable joint. *351*

tetanus (tet'ah-nus) sustained muscle contraction without relaxation. *355*

tone (tōn) the continuous partial contraction of muscle; due to contraction of a small number of muscle fibers at all times. *353*

Chapter 18

Senses

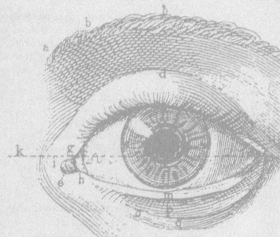

Chapter Concepts

1 Sense organs are sensitive to environmental stimuli and therefore are termed receptors. Each receptor responds to one type of stimulus, but they all initiate nerve impulses.

2 The sensation realized is the prerogative of the region of the cerebrum receiving nerve impulses.

3 There are sense receptors that respond to mechanical stimuli, chemical stimuli, and light energy. Our knowledge of the outside world is dependent on these stimuli.

Chapter Outline

General Receptors
    The Skin
    Muscles and Joints
Special Senses
    Chemoreceptors
    The Photoreceptor—the Eye
    A Mechanoreceptor—the Ear

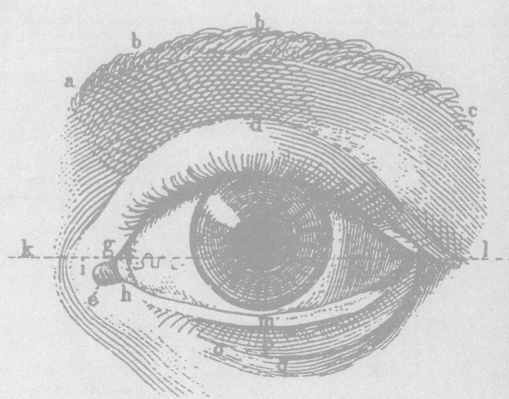

Table 18.1 *Receptors*

Receptors	Sense	Stimulus
General		
Temperature*	Hot-cold	Heat flow
Touch†	Touch	Mechanical displacement of tissue
Pressure†	Pressure	Mechanical displacement of tissue
Pain‡	Pain	Tissue damage
Proprioceptors†	Limb placement	Mechanical displacement
Special		
Eye*	Sight	Light
Ear†	Hearing	Sound waves
	Balance	Mechanical displacement
Taste buds‡	Taste	Chemicals
Olfactory cells‡	Smell	Chemicals

* Radioreceptors
† Mechanoreceptors
‡ Chemoreceptors

Figure 18.1
Sense perception. Do you see a vase in this picture or a man and a woman looking at each other? Our sense organs are dependent on the stimuli received, and they can be fooled. This vase was especially designed to commemorate the 25th year of the queen of England's reign. The man and the woman are Prince Philip and Queen Elizabeth.

 Sense organs receive external and internal stimuli (fig. 18.1); therefore, they are called **receptors.** Each type of receptor is sensitive to only one type of stimulus. Table 18.1 lists the receptors discussed in this chapter and the stimulus to which each reacts.

Receptors are the first components of the reflex arc described in chapter 16. When a receptor is stimulated, it generates nerve impulses that are transmitted to the spinal cord and/or the brain, but we are conscious of a sensation only if the impulses reach the cerebrum. The sensory portion of the cerebrum can be mapped according to the parts of the body and the type of sensation realized at different loci (see fig. 16.16).

General Receptors

Microscopic receptors (table 18.1) are present in the skin, in the visceral organs, and in the muscles and the joints. They are all specialized nerve endings for the detection of touch, pressure, pain, temperature (hot and cold), and proprioception. Proprioception refers to the sense of knowing the position of the limbs; for example, if you close your eyes and move your arm about slowly, you still have a sense of the location of your arm.

The Skin

The skin (fig. 18.2) contains receptors for touch, pressure, pain, and temperature. The skin is a mosaic of these tiny receptors, as you can determine by passing a metal probe slowly over the skin. At certain points, there is a feeling of pressure, at others, a feeling of hot or cold (depending on the temperature of the probe). Certain parts of the skin contain more receptors for a particular sensation; for example, the fingertips have an abundance of touch receptors.

A simple experiment suggests that temperature receptors are sensitive to the flow of heat. Fill 3 bowls with water—one cold, one warm, and one hot. Put your left hand in the cold water and your right in the hot water for a few moments. Your hands adjust, or adapt, to these temperatures so that when you put both hands in the warm water, each hand indicates a different water

Figure 18.2

Receptors in human skin. *The classical view is that each receptor has the function indicated. However, investigations in this century indicate that matters are not so clear cut. For example, microscopic examination of the skin of the ear shows only free nerve endings (pain receptors), and yet the skin of the ear is sensitive to all sensations. Therefore, it appears that the receptors of the skin are somewhat but not completely specialized.*

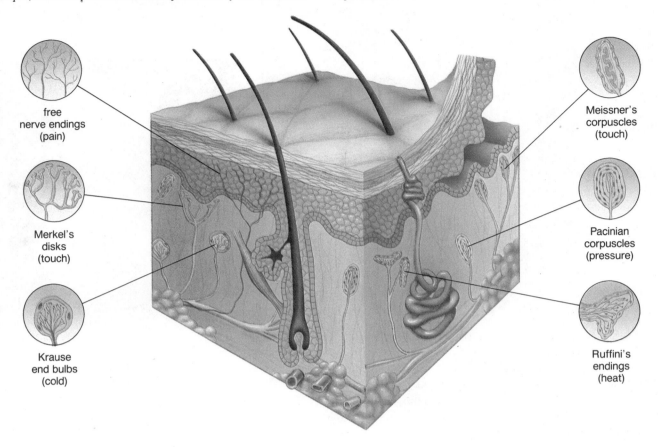

free nerve endings (pain)

Merkel's disks (touch)

Krause end bulbs (cold)

Meissner's corpuscles (touch)

Pacinian corpuscles (pressure)

Ruffini's endings (heat)

18.1 Critical Thinking

1. Why or why not should the term *receptor* be used for a sense organ?
2. Where would you expect a subject to report sensation if you stimulated each separate area of the sensory cortex for different portions of the body (see fig. 16.16)?
3. Does the biology of sensation support a mechanistic view of the body? Why or why not?

temperature. Therefore, it seems that when the outside temperature is higher than the temperature to which we have adjusted, we detect a sensation of warm or hot as heat flows into the skin. When the outside temperature is low enough that heat flows out from the skin, we detect coolness or cold.

Other skin receptors beside those for temperature also demonstrate adaptation. **Adaptation** occurs when the receptor becomes so accustomed to stimuli that it stops generating impulses, even though the stimulus is still present. The touch receptors adapt: they can quickly adapt to the clothing we put on so that we are not constantly aware of the feel of clothes against our skin.

The receptors of the skin can be used to illustrate that sensation actually occurs in the brain and not in the sense organ itself. If the nerve fiber from the sense organ is cut, there is no sensation. Also, since a nerve impulse is always the same electrochemical charge, the particular sensation realized does not have to do with the nerve impulse. The brain is responsible for the type of sensation felt and for the localization of the sensation. For example, if we connected a pain receptor in the foot to a nerve normally receiving impulses from a heat receptor in the hand and then proceeded to stick the pain receptor in the foot, the subject would report the feeling of warmth in the hand. The brain indicates the sensation and the localization. This realization is mildly disturbing because it makes us aware of how dependent we are on the anatomical wholeness of the body in order to be properly aware of our surroundings.

Figure 18.3

Taste buds. a. *Elevations on the tongue indicate the presence of taste buds. The location of those containing taste buds responsive to sweet, sour, salt, and bitter is indicated.* b. *Enlargement of elevations, called papillae.* c. *The taste buds occur along the walls of the papillae.* d. *Drawing shows the* various cells that make up a taste bud. Sensory cells in a bud end in microvilli that have receptors for the chemicals that exhibit the tastes noted in (a). When the chemicals combine with the receptors, nerve impulses are generated.

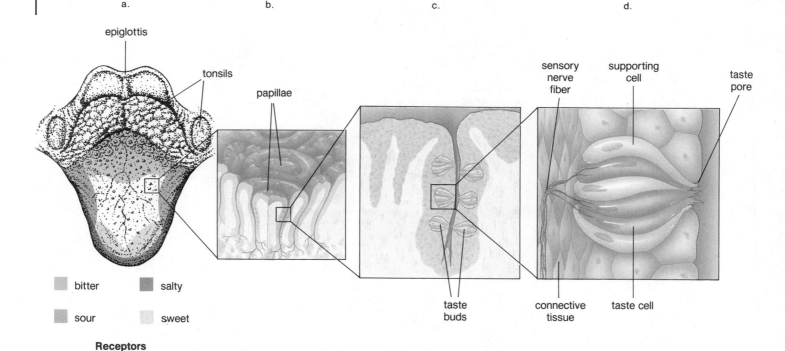

Muscles and Joints

The sense of position and movement of limbs (i.e., proprioception) is dependent upon receptors termed **proprioceptors.** Muscle spindles, discussed in chapter 17, sometimes are considered to be proprioceptors. Stretching of associated muscle fibers causes muscle spindles to increase the rate at which they fire, and for this reason, they are sometimes called **stretch receptors.** The *knee jerk* is a common example of the manner in which muscle spindles act as stretch receptors (fig. 17.8). When the legs are crossed at the knee and the tendon at the knee is tapped, both the tendon and the muscles in the thigh are stretched. Stimulated by the stretching, muscle spindles transmit impulses to the spinal cord, and thereafter the thigh muscles contract. This causes the lower leg to jerk upward in a kicking motion.

Proprioceptors are located in the joints and associated ligaments and tendons that respond to stretching, pressure, and pain. Nerve endings from these receptors are integrated with those received from other types of receptors so that we know the position of body parts.

Special Senses

The special senses include the chemoreceptors for taste and smell, the light receptors for sight, and the mechanoreceptors for hearing and balance.

Chemoreceptors

Taste and smell are called the *chemical senses* because these receptors are sensitive to certain chemical substances in the food we eat and the air we breathe.

Taste buds are located primarily on the tongue (fig. 18.3). Many lie along the walls of the papillae, the small elevations visible to the naked eye. Isolated ones also are present on the palate, the pharynx, and the epiglottis.

Figure 18.4

Olfactory cell location and anatomy. The olfactory area in humans is located high in the nasal cavity (right). Enlargement of the olfactory cells shows they are modified neurons located between supporting cells (left). The axons of these cells unite to form olfactory nerve fibers that terminate in the olfactory bulb. The dendrites of the olfactory cells have special structures known as olfactory cilia that project into the nasal cavity, where they are stimulated by chemicals in the air.

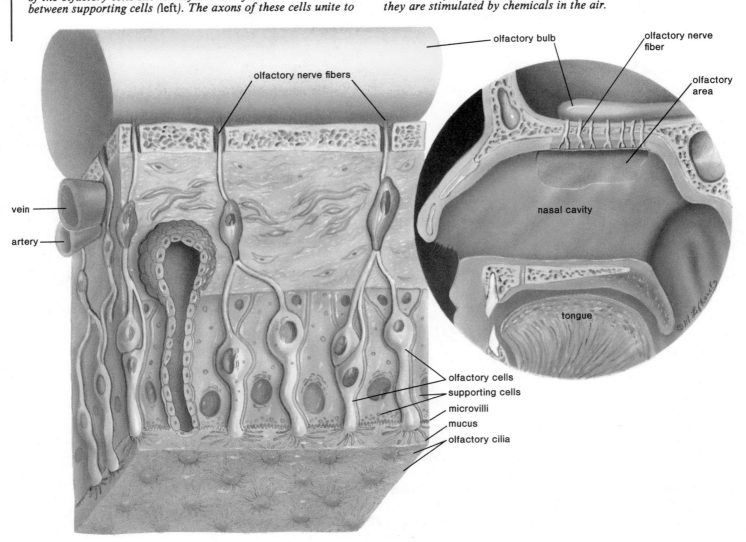

Taste buds are pockets of cells that extend through the tongue epithelium and open at a taste pore. Taste buds have supporting cells and a number of elongated cells that end in microvilli. These cells, which have associated nerve fibers, are sensitive to chemicals. Nerve impulses most probably are generated when the chemicals bind to receptor sites found on the microvilli.

The sense of taste has been shown to be inherited genetically, and foods taste differently to various people. This very well might account for the fact that some persons dislike a food that is preferred by others.

It is believed that there are 4 types of tastes (bitter, sour, salty, sweet) and that taste buds for each are concentrated on the tongue in particular regions (fig. 18.3a). Sweet receptors are most plentiful near the tip of the tongue. Sour receptors occur primarily along the margins of the tongue. Salt receptors are most common on the tip and the upper front portion of the tongue. Bitter receptors are located toward the back of the tongue.

The olfactory cells (fig. 18.4) are located high in the roof of the nasal cavity. Each cell ends in a tuft of about 5 cilia embedded in mucus that bear receptor sites for various chemicals. Research resulting in the stereochemical theory of smell suggests that different types of smell are related to the various shapes of molecules rather than to the atoms that make up the molecules. When chemicals combine with the receptor sites, nerve impulses are generated in the olfactory nerve fibers that project from the olfactory bulbs, which are paired masses of gray matter beneath the frontal lobes of the cerebrum. Within a bulb, an olfactory nerve takes sensory information to an olfactory area of the cerebral cortex. The olfactory receptors, like the touch and temperature receptors, adapt to outside stimuli. In other words, after a while, the presence of a particular chemical no longer causes the olfactory cells to generate nerve impulses, and we no longer are aware of a particular smell.

The sense of taste and the sense of smell supplement each other, creating a combined effect when interpreted by the cerebral cortex. For example, when we have a cold, we think that food has lost its taste, but actually we have lost the ability to sense its smell. This may work in reverse also. When we smell something, some of the molecules move from the nose down into the mouth region and stimulate the taste buds there. Therefore, part of what we refer to as smell actually may be taste.

The receptors for taste (taste buds) and the receptors for smell (olfactory cilia) work together to give us our sense of taste and our sense of smell.

The Photoreceptor—the Eye

The eyeball (fig. 18.5 and table 18.2), an elongated sphere about 2.5 cm in diameter, has 3 layers, or coats. The outer **sclera** is a white, fibrous layer except for the transparent cornea, the window of the eye. The middle, thin, dark brown layer, the **choroid,** contains many blood vessels and absorbs stray light rays. Toward the front, the choroid thickens and forms a ring-shaped structure, the ciliary body, containing the **ciliary muscle,** which controls the shape of the lens for near and far vision. Finally, the choroid becomes a thin, circular, muscular diaphragm, the **iris,** which regulates the size of a center hole, the **pupil,** through which light enters the eyeball. The **lens,** attached to the ciliary body by ligaments, divides the cavity of the eye into 2 chambers. A viscous, gelatinous material, the **vitreous humor,** fills the posterior cavity behind the lens. The anterior cavity between the cornea and the lens is filled with an alkaline, watery solution secreted by the ciliary body and called the **aqueous humor.**

A small amount of aqueous humor is produced continually each day. Normally, it leaves the anterior cavity by way of tiny ducts that are located where the iris meets the cornea. When a person has glaucoma, these drainage ducts are blocked, and aqueous humor builds up. The resulting pressure compresses arteries that serve the nerve fibers of the retina. The nerve fibers begin to die due to lack of nutrients, and the person becomes partially blind. Overtime, total blindness can result. The tendency toward glaucoma is inherited in some individuals.

The Retina

The inner layer of the eye, the **retina,** has 3 layers of cells (fig. 18.6). The layer closest to the choroid contains the sense receptors for sight, the **rods** and the **cones;** the middle layer contains bipolar cells; and the innermost layer contains ganglionic cells whose fibers become the **optic nerve.** Only the rods and the cones contain light-sensitive pigments, and therefore light must penetrate to the back of the retina before nerve impulses are generated. Nerve impulses initiated by the rods and the cones are passed to the bipolar cells, which in turn pass them to the ganglionic cells. The fibers of the ganglionic cells pass in front of the retina, forming the optic nerve, which turns to pierce the layers

Figure 18.5

Anatomy of the human eye. Notice that the sclera becomes the cornea; the choroid becomes the ciliary body and the iris. The retina contains the receptors for light, and vision is most acute due to the fovea centralis, where there are only cones. A blind spot occurs where the optic nerve leaves the retina and where there are no receptors for light.

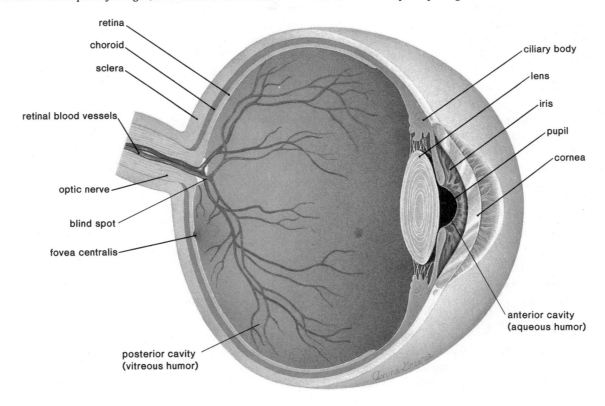

retina
choroid
sclera
retinal blood vessels
optic nerve
blind spot
fovea centralis
posterior cavity (vitreous humor)
ciliary body
lens
iris
pupil
cornea
anterior cavity (aqueous humor)

Table 18.2 Function of Parts of the Eye	
Part	**Function**
Lens	Refracts and focuses light rays
Iris	Regulates light entrance
Pupil	Admits light
Choroid	Absorbs stray light
Sclera	Protects eyeball
Cornea	Refracts light rays
Humors	Refracts light rays
Ciliary body	Holds lens in place, accommodation
Retina	Contains receptors for sight
Rods	Makes black-and-white vision possible
Cones	Makes color vision possible
Optic nerve	Transmits impulse
Fovea centralis	Contains the cones of the retina

of the eye. Notice in figure 18.6 that there are many more rods and cones than ganglionic cells. In fact, the retina has as many as 150 million rods but only 1 million ganglionic cells and optic nerve fibers. This means that there is considerable mixing of messages and a certain amount of integration before nerve impulses are sent to the occipital lobe of the brain. There are no rods or cones where the optic nerve passes through the retina; therefore, this is a **blind spot,** where vision is impossible.

The retina contains a very special region called the **fovea centralis** (fig. 18.5), an oval, yellowish area with a depression where there are only cone cells. Vision is most acute in the fovea centralis.

The eye has 3 layers: the outer sclera, the middle choroid, and the inner retina. Only the retina contains receptors for sight.

Physiology

Focusing   When we look at an object, light rays are **focused** on the retina (fig. 18.7). In this way, an *image* of the object appears on the retina. The image on the retina occurs when the rods and the cones in a particular region are excited. Obviously, the image is much smaller than the object. In order to produce this small image, light rays must be bent (refracted) and brought into focus. They are bent as they pass through the cornea. Further bending occurs as the rays pass through the lens and the humors.

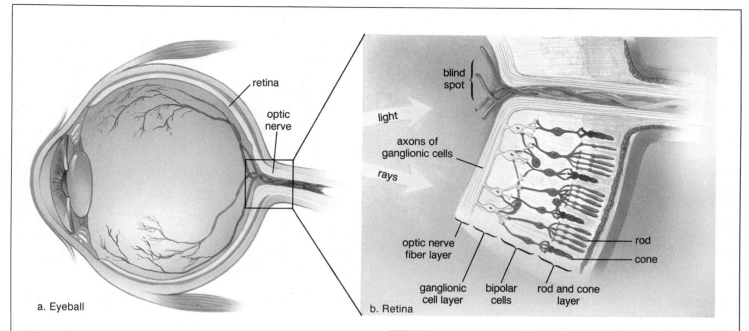

retina

optic nerve

a. Eyeball

blind spot

light

axons of ganglionic cells

rays

optic nerve fiber layer

ganglionic cell layer

bipolar cells

rod and cone layer

rod

cone

b. Retina

Figure 18.6

Anatomy of the retina. a. The retina is the inner layer of the eye. b. Rods and cones are located at the back of the retina, followed by the bipolar cells and the ganglionic cells, whose fibers become the optic nerve. Notice that rods share bipolar cells but cones do not. Cones, therefore, distinguish more detail. c. The photosensitive pigment is located in the membranous disks of the outer segment of rods and cones. d. Scanning electron micrograph of rods and cones. The cones are responsible for color vision, and the rods are responsible for night vision.

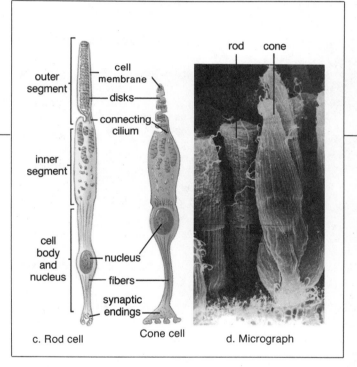

rod   cone

outer segment

cell membrane

disks

connecting cilium

inner segment

cell body and nucleus

nucleus

fibers

synaptic endings

c. Rod cell   Cone cell   d. Micrograph

Figure 18.7

Focusing. Light rays from each point on an object are bent by the cornea and the lens in such a way that they are directed to a single point after emerging from the lens. By this process, an inverted image of the object forms on the retina.

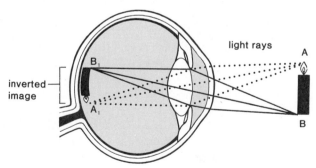

light rays

A

B₁

inverted image

A₁

B

Accommodation   Light rays are reflected from an object in all directions. If the eye is distant from an object, only nearly parallel rays enter the eye, and the cornea alone is needed for focusing. If the eye is close to the object, however, many of the rays are at sharp angles to one another, and additional focusing is required. When the lens provides this additional focusing power, **accommodation** occurs. Although the lens remains flat when we view distant objects, it rounds up when we view close objects. When rounded, the lens provides the additional refraction required to bring the divergent light rays to a sharp focus on the retina (fig. 18.8). The shape of the lens is controlled by the ciliary muscle within the ciliary body. When we view a distant object, the ciliary muscle is relaxed, causing the suspensory ligaments attached to the

Figure 18.8

Accommodation. a. *When the eye focuses on a far object, the lens is flat because the ciliary muscle is relaxed and the suspensory ligament is taut.* b. *When the eye focuses on a near object, the lens rounds up because the ciliary muscle contracts, causing the suspensory ligament to relax.*

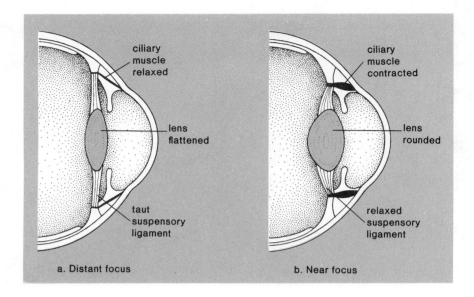

a. Distant focus

b. Near focus

ciliary body to be under tension; therefore, the lens remains relatively flat. When we view a close object, the ciliary muscle contracts, releasing the tension on the suspensory ligaments, and the lens rounds up due to its natural elasticity. Because close work requires contraction of the ciliary muscle, it very often causes eyestrain.

With aging, the lens loses some of its elasticity and is unable to accommodate. This usually necessitates the wearing of corrective lenses, as is discussed on page 377. The lens is also subject to *cataracts;* it can become opaque and unable to transmit rays of light. Special cells within the interior of the lens contain proteins called crystallin. Recent research suggests that cataracts develop when these proteins become oxidized, causing their three-dimensional shape to change. If so, researchers believe that they eventually may be able to find ways to restore the normal configuration of crystallin so that cataracts can be treated medically instead of surgically. For the present, however, surgery is the viable treatment. First, a surgeon opens the eye near the rim of the cornea. Zonulysin, an enzyme, may be used to digest away the ligaments holding the lens in place. Most surgeons then use a cryoprobe that freezes the lens for easy removal. An intraocular lens attached to the iris can be implanted in the eye so that the patient need not wear thick glasses or contact lenses.

The lens, assisted by the cornea and the humors, focuses images on the retina.

The Inverted Image   The image on the retina is upside down (fig. 18.7), and it is thought that perhaps this image is righted in the brain by experience. In one experiment, scientists wore glasses that inverted the field of vision. At first, they had difficulty adjusting to the placement of the objects, but they soon became accustomed to their inverted world. Experiments such as this suggest that if we see the world upside down, the brain learns to see it right side up.

Stereoscopic Vision   We can see well with either eye alone, but the 2 eyes functioning together provide us with **stereoscopic vision.** Normally, the 2 eyes are directed by the eye muscles toward the same object, and therefore the object is focused on corresponding points of the 2 retinas. Each eye, however, sends its own information to the brain about the placement of the object because each forms an image from a slightly different angle. These data are pooled to produce depth perception by a 2-step process. First, because the optic nerves cross at the optic chiasma (fig. 18.9), one-half of the brain receives

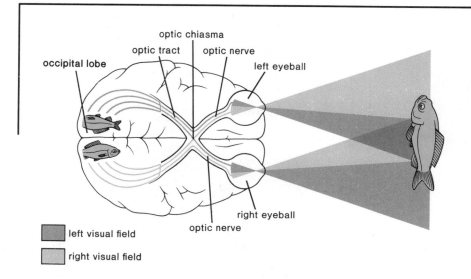

Figure 18.9
Optic chiasma. Both eyes "see" the entire object, but information from the right half of each retina goes to the right occipital lobe and information from the left half of the retina goes to the left occipital lobe because of the optic chiasma. When the information is pooled, the brain "sees" the entire object in depth.

optic chiasma

optic tract   optic nerve

occipital lobe   left eyeball

left visual field

right visual field

right eyeball

optic nerve

information from both eyes about the same part of an object. Later, the 2 halves of the brain communicate to arrive at a complete three-dimensional interpretation of the whole object.

The anatomy and the physiology of the brain allow us to see the world right side up and in 3 dimensions.

Biochemistry   In *dim light,* the pupils enlarge so that more rays of light can enter the eyes. As the rays of light enter, they strike the rods and the cones, but only the 150 million rods located in the periphery, or sides, of the eyes are sensitive enough to be stimulated by this faint light. The rods do not detect fine detail or color, so at night, for example, all objects appear to be blurred and to be a shade of gray. Rods do detect even the slightest motion, however, because of their abundance and position in the eyes.

Although it has been known for some time that vision generated by the rods is dependent on the presence of rhodopsin, the actual events only recently have been worked out in detail. Many molecules of rhodopsin, also called visual purple, are located within the membrane of the disks found in the outer segment (fig. 18.6c) of the rods. **Rhodopsin** is a complex molecule that contains a protein (*opsin*) and a pigment molecule called *retinal,* which is a derivative of vitamin A. When retinal absorbs light energy, it changes shape, and opsin is activated (fig. 18.10). The reactions that follow eventually end when many molecules of GMP (guanosine monophosphate) are converted to cyclic GMP.[1] Cyclic GMP, in turn, initiates an action potential in the cell membrane of the rod that travels along the membrane until it reaches the synaptic endings, where a neurotransmitter substance is released to excite a bipolar cell (fig. 18.6). Now, rhodopsin returns to its former configuration so that the series of events can reoccur and reoccur. Each stimulus generated lasts about one-tenth of a second. This is why we continue to see an image if we close our eyes immediately after looking at an object. It also allows us to see motion if still frames are presented at a rapid rate, as in "movies."

In *bright light,* the pupils get smaller so that less light enters the eyes. The cones, located primarily in the fovea, are active and detect the fine detail and color of an object. In order to perceive depth, as well as to see color, we turn our eyes so that reflected light from the object strikes the fovea centralis.

[1]GMP is a nucleotide that contains the base guanine, the sugar ribose, and one phosphate group. In cyclic GMP, the single phosphate is attached to ribose in 2 places.

Figure 18.10

Biochemistry of vision. Rhodopsin, a complex molecule containing the protein opsin and the pigment retinal, is present in the membranous disks within rod cells (fig. 18.6). When retinal absorbs light energy, it changes shape, and this activates opsin to begin a series of reactions that end when GMP is converted to cyclic GMP, a molecule that can trigger an action potential in the rod cell membrane.

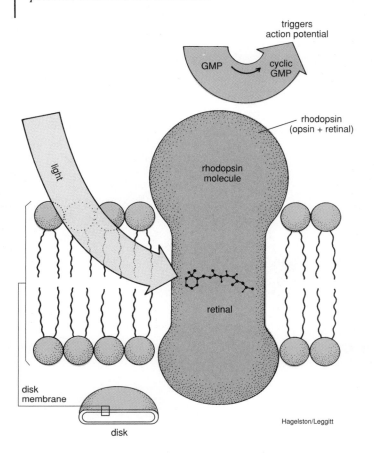

triggers
action potential

GMP → cyclic GMP

rhodopsin
(opsin + retinal)

rhodopsin
molecule

light

retinal

disk
membrane

disk

Hagelston/Leggitt

Figure 18.11

Test plates for color blindness. When looking at the plate top, the person with normal color vision sees the number 8, and when looking at the bottom plate, the person with normal color vision sees the number 12. The most common form of color blindness involves an inability to distinguish reds and greens.

These plates have been reproduced from Ishihara's Test for Colour Blindness published by Kanehara & Co., Ltd., Tokyo, Japan, but tests for color blindness cannot be conducted with this material. For accurate testing, the original plate should be used.

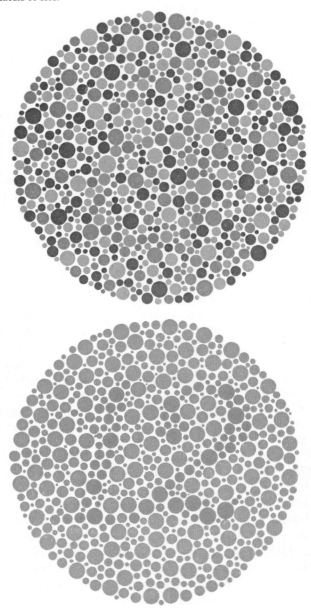

Color vision has been shown to depend on 3 kinds of cones that contain pigments sensitive to either blue, green, or red light. The nerve impulses generated from one type of cone not only stimulate certain cells in the visual cortex of the brain, they also inhibit the reception of impulses from other types of cones. For example, when we see red, certain cells in the brain are prohibited from receiving impulses from green cones. Similarly, impulses sent through blue cones tend to oppose the combination of signals sent by red and green cones—which together produce yellow. This process assists integration and enables the brain to tell the location of various colors in the environment (fig. 18.11). Complete color blindness is extremely rare. In most instances, a particular type of cone is lacking or deficient in number. The lack of red or green

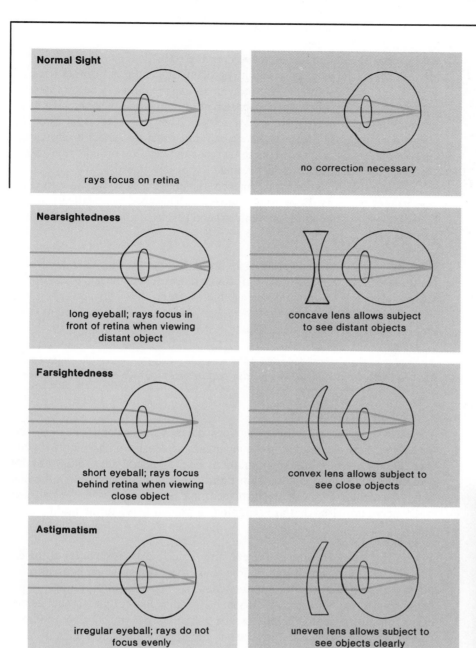

Normal Sight

rays focus on retina

no correction necessary

Nearsightedness

long eyeball; rays focus in front of retina when viewing distant object

concave lens allows subject to see distant objects

Farsightedness

short eyeball; rays focus behind retina when viewing close object

convex lens allows subject to see close objects

Astigmatism

irregular eyeball; rays do not focus evenly

uneven lens allows subject to see objects clearly

a.

b.

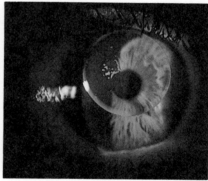

c.

Figure 18.12
Common abnormalities of the eye with possible corrective lenses. a. The cornea and the lens function in bringing light rays (lines) to focus, but sometimes they are unable to compensate for the shape of the eyeball or for an uneven cornea b. In these instances corrective lenses can allow the individual to see normally. c. Ophthalmologists can examine the fit of a contact lens by using a narrow beam of light from a "slit lamp" while looking through a biomicroscope.

cones is the most common, affecting about 5% of the American population. If the eye lacks red cones, the green colors are accentuated, and vice versa.

The sense receptors for sight are the rods and the cones. The rods are responsible for vision in dim light, and the cones are responsible for vision in bright light and for color vision. When either is stimulated, nerve impulses begin and are transmitted in the optic nerve to the brain.

Corrective Lenses

The majority of people can see what is designated as a size 20 letter 20 feet away and so are said to have 20/20 vision. Persons who can see close objects but cannot see the letters from this distance are said to be nearsighted. Nearsighted people can see near better than they can see far. These individuals often have an elongated eyeball, and when they attempt to look at a far object, the image is brought to focus in front of the retina (fig. 18.12). They can see near because they can adjust the lens to allow the image to focus on the retina, but to see far, these people must wear concave lenses that diverge the light rays so that the image can be focused on the retina.

Persons who can easily see the optometrist's chart but cannot see close objects well are farsighted; these individuals can see far away better than they can see near. They often have a shortened eyeball, and when they try to see near, the image is focused behind the retina. When the object is far away, the lens can compensate for the short eyeball, but when the object is close, these persons must wear a convex lens to increase the bending of light rays so that the image can be focused on the retina.

When the cornea or lens is uneven, the image is fuzzy because the light rays cannot be focused evenly on the retina. This condition, called astigmatism, can be corrected by an unevenly ground lens to compensate for the uneven cornea.

Bifocals   As mentioned earlier, with normal aging, the lens loses some of its ability to change shape in order to focus on close objects. Because nearsighted individuals still have difficulty seeing objects clearly in the distance, they must wear bifocals, which means that the upper part of the lens is for distant vision and the remainder is for near vision.

The shape of the eyeball determines the need for corrective lenses; the inability of the lens to accommodate as we age also requires corrective lenses for close vision.

A Mechanoreceptor—the Ear

The ear accomplishes 2 sensory functions: balance and hearing. The sense cells for both of these are located in the inner ear and consist of hair cells with cilia that respond to mechanical stimulation. Each hair cell has from 30 to 150 extensions called cilia, and contain tightly packed filaments. When the cilia of any particular hair cell are displaced in a certain direction, the cell generates nerve impulses that are sent along a cranial nerve to the brain.

Anatomy

Figure 18.13 is a drawing of the ear, and table 18.3 lists the parts of the ear. The ear has 3 divisions: outer, middle, and inner. The **outer ear** consists of the **pinna** (external flap) and the **auditory canal.** The opening of the auditory canal is lined with fine hairs and sweat glands. Modified sweat glands that secrete earwax, a substance that helps to guard the ear against the entrance of foreign materials, such as air pollutants, are in the upper wall of the canal.

The **middle ear** begins at the **tympanic membrane** (eardrum) and ends at a bony wall containing 2 small openings covered by membranes. These openings are called the **oval** and **round windows.** Three small bones are found between the tympanic membrane and the oval window. Collectively called the **ossicles,** individually they are the **hammer** (malleus), the **anvil** (incus), and the **stirrup** (stapes) because their shapes resemble these objects (fig. 18.13). The hammer adheres to the tympanic membrane, and the stirrup touches the oval window. The posterior wall has an opening that leads to sinuses within the mastoid portion of the temporal bone.

A **eustachian tube** extends from each middle ear to the nasopharynx and permit equalization of air pressure. Chewing gum, yawning, and swallowing in elevators and airplanes help to move air through the eustachian tubes upon ascent and descent.

Whereas the outer ear and the middle ear contain air, the inner ear is filled with fluid. The **inner ear** (fig. 18.14a), anatomically speaking, has 3 areas: the first 2, the semicircular canals and a vestibule, are concerned with balance; and the third, the cochlea, is concerned with hearing.

Human Anatomy and Physiology

Figure 18.13

Anatomy of the human ear. In the middle ear, the hammer, the anvil, and the stirrup amplify sound waves. Otosclerosis is a condition in which the stirrup becomes attached to the inner ear and is unable to carry out its normal function. It can be replaced by a plastic piston, and thereafter the individual hears normally because sound waves are transmitted as usual to the cochlea that contains the receptors for hearing.

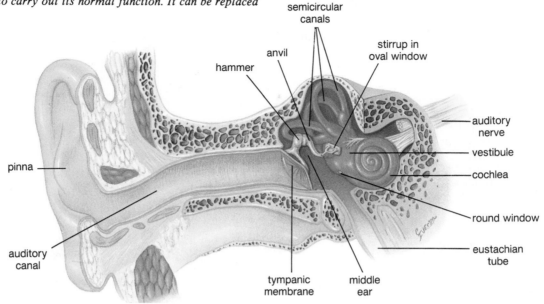

Table 18.3   The Ear

	Outer Ear	Middle Ear	Inner Ear	
			Cochlea	Sacs Plus Semicircular Canals
Function	Directs sound waves to tympanic membrane	Picks up and amplifies sound waves	Hearing	Maintains equilibrium
Anatomy	Pinna Auditory canal	Tympanic membrane Ossicles	Contains organ of Corti; auditory nerve starts here	Saccule and utricle Semicircular canals
Media	Air	Air (eustachian tube)	Fluid	Fluid

Path of vibration: Sound waves—vibration of tympanic membrane—vibration of hammer, anvil, and stirrup—vibration of oval window—fluid pressure waves in canals of inner ear lead to stimulation of hair cells—bulging of round window.

The **semicircular canals** are arranged so that there is one in each dimension of space. The base of each canal, called the **ampulla** (fig. 18.14*b*), is slightly enlarged. Within the ampullae are little hair cells whose cilia are inserted into a gelatinous material.

A vestibule, or a chamber, lies between the semicircular canals and the cochlea. It contains 2 small membranous sacs called the **utricle** and the **saccule** (fig. 18.14*c*). Within both of these are little hair cells whose cilia protrude into a gelatinous material. Resting on this substance are calcium carbonate granules, or **otoliths.**

The **cochlea** resembles the shell of a snail because it spirals. Within the tubular cochlea are 3 canals: the vestibular canal, the **cochlear canal,** and the tympanic canal. Along the length of the basilar membrane, which forms the lower wall of the cochlear canal, are little hair cells whose cilia come into contact with another membrane called the tectorial membrane. The hair cells of the cochlear canal plus the **tectorial membrane** are called the **organ of Corti** (fig. 18.14*d*). When this organ sends nerve impulses to the cerebral cortex, they are interpreted as sound.

Figure 18.14

Inner ear. a. *The inner ear contains the semicircular canals, a vestibule, and the cochlea. The cochlea has been cut to show the location of the organ of Corti.* b. *There is an ampulla at the base of each semicircular canal that contains the receptors (hair cells) for dynamic equilibrium.* c. *In a vestibule are the utricle and the saccule, small sacs that contain the receptors (hair cells) for static equilibrium.* d. *The sense organ for hearing, the organ of Corti, is in the cochlea. The organ of Corti also consists of hair cells.*

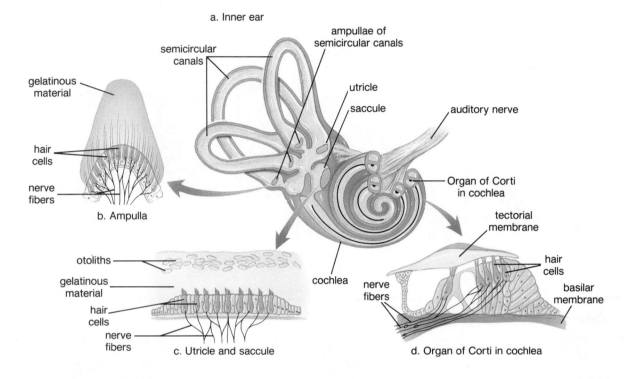

a. Inner ear

b. Ampulla

c. Utricle and saccule

d. Organ of Corti in cochlea

The outer ear, the middle ear, and the cochlea are necessary for hearing. The semicircular canals and the vestibule are concerned with the sense of balance.

Physiology

Balance (Equilibrium)   The sense of balance has been divided into 2 senses: *dynamic equilibrium,* requiring a knowledge of angular and/or rotational movement, and *static equilibrium,* requiring a knowledge of movement in one plane, either vertical or horizontal.

Dynamic equilibrium is required when the body is moving. At that time, the fluid within the semicircular canals flows over and displaces the gelatinous material within the ampullae (fig. 18.15a). This causes the cilia of the hair cells to bend and to initiate nerve impulses that travel to the brain. Continuous movement of the fluid in the semicircular canals causes one form of motion sickness.

When the body is still, the otoliths in the utricle and the saccule rest on the gelatinous material above the hair cells. Static equilibrium is required when the body moves horizontally or vertically. At that time, the otoliths are displaced and the gelatinous material sags, bending the cilia of the hair cells beneath (fig. 18.15b). Now the hair cells generate nerve impulses that travel to the brain.

Movement of fluid within the semicircular canals contributes to our sense of dynamic equilibrium. Movement of the otoliths within the utricle and the saccule is important for static equilibrium.

Figure 18.15

Sense of balance. The ampullae of the semicircular canals contain (a) hair cells with cilia embedded in a gelatinous material. b. When the head rotates, the material is displaced and the bending of the cilia initiates nerve impulses in sensory nerve fibers. This permits dynamic equilibrium. A vestibule contains the utricle and the saccule, sacs that (c) contain hair cells with cilia embedded in a gelatinous material. d. When the head bends, otoliths are displaced, causing the gelatinous material to sag and the cilia to bend. This initiates nerve impulses in sensory nerve fibers. This permits static equilibrium.

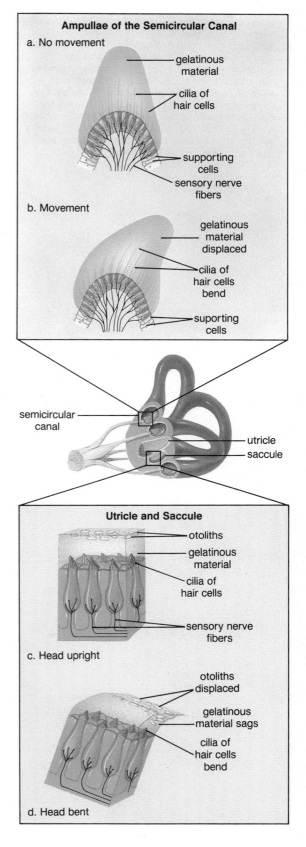

Ampullae of the Semicircular Canal

a. No movement — gelatinous material, cilia of hair cells, supporting cells, sensory nerve fibers

b. Movement — gelatinous material displaced, cilia of hair cells bend, suporting cells

semicircular canal — utricle, saccule

Utricle and Saccule

c. Head upright — otoliths, gelatinous material, cilia of hair cells, sensory nerve fibers

d. Head bent — otoliths displaced, gelatinous material sags, cilia of hair cells bend

Hearing   The process of hearing begins when sound waves enter the auditory canal. Just as ripples travel across the surface of a pond, sound travels by the successive vibrations of molecules. Ordinarily, sound waves do not carry much energy, but when a large number of waves strike the eardrum, it moves back and forth (vibrates) ever so slightly. The hammer then takes the pressure from the inner surface of the eardrum and passes it by way of the anvil to the stirrup in such a way that the pressure is multiplied about 20 times as it moves from the eardrum to the stirrup. The stirrup strikes the oval window, causing it to vibrate, and in this way, the pressure is passed to the fluid within the inner ear.

If the cochlea is unwound, as shown in figure 18.16b, you can see that the vestibular canal connects with the tympanic canal and that pressure waves move from one canal to the other toward the round window, a membrane that can bulge to absorb the pressure. As a result of the movement of the fluid within the cochlea, the basilar membrane moves up and down, and the cilia of the hair cells rub against the tectorial membrane. This bending of the cilia initiates nerve impulses that pass by way of the **auditory nerve** to the temporal lobe of the brain, where the impulses are interpreted as a sound.

The organ of Corti is narrow at its base but widens as it approaches the tip of the cochlear canal. Each part of the organ is sensitive to different wave frequencies, or pitch. Near the tip, the organ of Corti responds to low pitches, such as a tuba, and near the base, it responds to higher pitches, such as a bell or a whistle. The neurons from each region along the length of the cochlea lead to slightly different areas in the brain. The pitch sensation we experience depends upon which of these areas of the brain is stimulated.

Volume is a function of the amplitude of sound waves. Loud noises cause the fluid of the cochlea to oscillate to a greater degree, and this, in turn, causes the basilar membrane to move up and down to a great extent. The resulting increased stimulation is interpreted by the brain as loudness. It is believed that tone is an interpretation of the brain based on the distribution of hair cells stimulated.

The sense receptors for sound are hair cells on the basilar membrane (the organ of Corti). When the basilar membrane vibrates, the delicate hairs touch the tectorial membrane, initiating nerve impulses that are transmitted in the auditory nerve to the brain.

Deafness   There are 2 major types of deafness: *conduction deafness* and *nerve deafness.* Conduction deafness can be due to a congenital defect, as those that occur when a pregnant woman contracts German measles during the first trimester of pregnancy. (For this reason every female should be sure to be immunized against rubella before the childbearing years.) Conduction deafness also can be due to infections that have caused the ossicles to fuse, restricting the ability to magnify sound waves. Because respiratory infections can spread to the ear by way of the eustachian tubes every cold and ear infection should be taken seriously.

Figure 18.16

Sense of hearing. a. The organ of Corti is located within the cochlea. b. In the unwound cochlea, note that the organ of Corti consists of hair cells resting on the basilar membrane with the tectorial membrane above. The arrows represent the pressure waves that move from the oval window to the round window due to the motion of the stirrup. These pressure waves cause the basilar membrane to vibrate and the cilia of at least a portion of the 15,000 hair cells to bend against the tectorial membrane. The generated nerve impulses result in hearing.

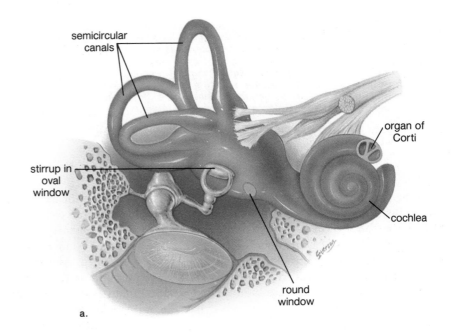

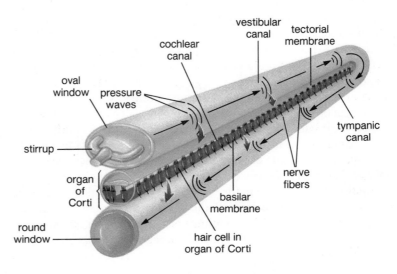

Nerve deafness most often occurs when cilia on the sense receptors within the cochlea have worn away. Since this can happen with normal aging, old people are more likely to have trouble hearing; however, nerve deafness also occurs when people listen to loud music amplified to 130 decibels. Because the usual types of hearing aids are not helpful for nerve deafness, it is wise to avoid subjecting the ears to any type of continuous loud noise. Costly cochlear implants that directly stimulate the auditory nerve are available, but those who have these electronic devices report that the speech they hear is like that of a robot.

"We have an idea of what noise does to the ear," David Lipscomb (of the University of Tennessee Noise Laboratory) says. "There's a pretty clear cause-effect relationship." And these scanning electron micrographs of the cochlea's tiny structures graphically document noise trauma to the inner ear.

Hair cells transmit the mechanical energy of sound waves into those neural impulses that the brain interprets as sound. Loud noise can damage or destroy hair cells, as these scanning electron micrographs illustrate.

Picturing the Effects of Noise

Hair cells come in 2 varieties: a single row of inner cells and a triple row of outer ones. "Outer cells degenerate before inner cells," notes Clifton Springs, New York, otolaryngologist Stephen Falk. The most subtle change wrought by noise is a development of vesicles, or blisterlike protrusions along the walls of the hair cells' cilia.

Continued assault by noise leads to the rupture of the vesicles and to damage. In addition, the "cuticular plate"—base tissue supporting the cilia—may soften, followed by swelling and ultimate degeneration of hair cells.

But sensory hair cells are not the only structures at risk. Adjacent inner ear cells. . . may undergo vacuolation—development of degenerative empty spaces in cells. Even nerve fibers synapsing at the hair cells' roots may die. In the final phase of noise-induced cochlear damage, the organ of Corti—of which hair cells and supporting cells are a part—is completely denuded of its natural components and is covered by a layer of scar tissue.

Reprinted with permission from SCIENCE NEWS, the weekly newsmagazine of science, copyright 1982 by Science Service, Inc.

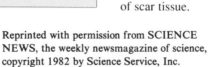

Damage to organ of Corti due to loud noise a. *Normal organ of Corti* b. *Organ of Corti after 24-hour exposure to noise level typical of rock music. Note scars where cilia have worn away.*

Summary

All receptors are the first part of a reflex arc, and they initiate nerve impulses that eventually reach the cerebrum, where sensation occurs. Among the general receptors are those located in the skin and proprioceptors located in the joints. The chemoreceptors for taste and smell are special receptors, as are the eyes and the ears.

Vision is dependent on the eye, the optic nerve, and the occipital lobe of the brain. The rods, receptors for vision in dim light, and the cones, receptors that depend on bright light and provide color and detailed vision, are located in the retina, the inner layer of the eyeball. The cornea, the humors, and especially the lens bring the light rays to focus on the retina. To see a close object, accommodation occurs as the lens rounds up. Due to the optic chiasma, both sides of the brain must function together to give us three-dimensional vision.

Hearing is a specialized sense dependent on the ear, the auditory nerve, and the temporal lobe of the brain. The outer and middle portions of the ear simply convey and magnify the sound waves that strike the oval window. Its vibrations set up pressure waves within the cochlea, which contains the organ of Corti, consisting of hair cells with the tectorial membrane above. When the hair cells strike this membrane, nerve impulses are initiated that finally result in hearing.

The ear also contains receptors for our sense of balance. Dynamic equilibrium is dependent on the stimulation of hair cells within the ampullae of the semicircular canals. Static equilibrium relies on the stimulation of hair cells by otoliths within the utricle and the saccule.

Study Questions

1. Name 3 factors that all receptors have in common. (p. 367)
2. What type of receptors are categorized as general, and what type are categorized as special receptors? (p. 367)
3. Discuss the receptors of the skin, the viscera, and the joints. (pp. 367–69)
4. Discuss the chemoreceptors. (pp. 369–71)
5. Describe the anatomy of the eye (pp. 371–72), and explain focusing and accommodation. (pp. 372–74)
6. Describe sight in dim light. What chemical reaction is responsible for vision in dim light? (p. 375) Discuss color vision. (p. 376)
7. Relate the need for corrective lenses to 3 possible shapes of the eye. (pp. 377–78) Discuss bifocals. (p. 378)
8. Describe the anatomy of the ear and how we hear. (pp. 378–79, 381)
9. Describe the role of the utricle, the saccule, and the semicircular canals in balance. (p. 380)
10. Discuss the 2 causes of deafness, including why young people frequently suffer loss of hearing. (pp. 381–82)

Objective Questions

1. The sense organs for position and movement are called _____ .

2. Taste buds and olfactory receptors are termed _____ because they are sensitive to chemicals in the air and food.

3. The receptors for sight, the _____ and the _____ , are located in the _____ , the inner layer of the eye.

4. The cones give us _____ vision and work best in _____ light.

5. The lens _____ for viewing close objects.

6. People who are nearsighted cannot see objects that are _____ . A _____ lens restores this ability.

7. The ossicles are the _____ , the _____ , and the _____ .

8. The semicircular canals are involved in our sense of dynamic _____ .

9. The organ of Corti is located in the _____ canal of the _____ .

10. Vision, hearing, taste, and smell do not occur unless nerve impulses reach the proper portion of the _____ .

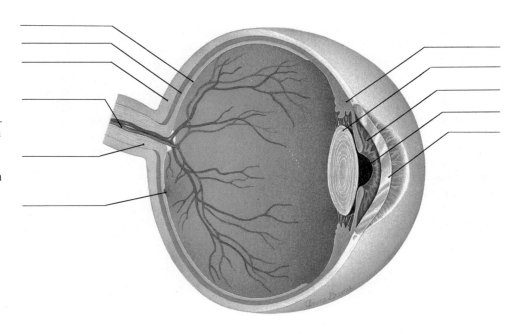

Label this Diagram.

See figure 18.5 (p. 372) in text.

Answers to Objective Questions

1. proprioceptors 2. chemoreceptors 3. rods, cones, retina 4. color, bright (day) 5. rounds up (accommodation) 6. distant, concave 7. hammer, anvil, stirrup 8. equilibrium 9. cochlear, cochlea 10. brain

---

Selected Key Terms

accommodation (ah-kom''o-da'shun) lens adjustment in order to see close objects. *373*

choroid (ko'roid) the vascular, pigmented middle layer of the eyeball. *371*

ciliary muscle (sil'e-er''e mus'el) a muscle that controls the curvature of the lens of the eye. *371*

cochlea (kok'le-ah) that portion of the inner ear that resembles a snail's shell and contains the organ of Corti, the sense organ for hearing. *379*

cone (kōn) bright-light receptor in the retina of the eye that detects color and provides visual acuity. *371*

fovea centralis (fo've-ah sen-tral'is) region of the retina consisting of densely packed cones that is responsible for the greatest visual acuity. *372*

lens (lenz) a clear membranelike structure found in the eye behind the iris. The lens brings objects into focus. *371*

organ of Corti (or'gan uv kor'ti) a portion of the inner ear that contains the receptors for hearing. *379*

otolith (o'to-lith) calcium carbonate granule associated with ciliated cells in the utricle and the saccule. *379*

proprioceptor (pro''pre-o-sep'tor) receptor that assists the brain in knowing the position of the limbs. *369*

retina (ret'ĭ-nah) the innermost layer of the eyeball that contains the rods and the cones. *371*

rhodopsin (ro-dop'sin) visual purple, a pigment found in the rods. *375*

rod (rod) dim-light receptor in the retina of the eye that detects motion but no color. *371*

saccule (sak'ūl) a saclike cavity that makes up part of the membranous labyrinth of the inner ear; contains receptors for static equilibrium. *379*

sclera (skle'rah) white fibrous outer layer of the eyeball. *371*

semicircular canal (sem''e-ser'ku-lar kah-nal') tubular structure within the inner ear that contains the receptors responsible for the sense of dynamic equilibrium. *379*

tympanic membrane (tim-pan'ik mem'brān) membrane located between the outer and middle ear that receives sound waves; the eardrum. *378*

utricle (u'tre-k'l) saclike cavity that makes up part of the membranous labyrinth of the inner ear; contains receptors for static equilibrium. *379*

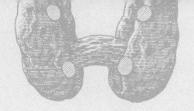

Chapter 19

Chapter Concepts

1 The endocrine system utilizes chemical messengers called hormones to bring about coordination of body parts.

2 Endocrine glands are usually ductless glands that secrete hormones directly into the bloodstream.

3 In general, secretion of hormones is controlled by a negative feedback mechanism.

4 Malfunctioning of endocrine glands can bring about a dramatic change in appearance and cause early death.

5 Chemical messengers are not unique to the endocrine system and most likely are present whenever one biological element influences the behavior or the chemistry of another element.

Chapter Outline

Endocrine Glands
    The Mechanism of Hormonal Action
The Hypothalamus and the Pituitary
    Gland
    The Posterior Pituitary
    The Anterior Pituitary
Thyroid and Parathyroid Glands
    The Thyroid Gland
    Parathyroid Glands
Adrenal Glands
    The Adrenal Medulla
    The Adrenal Cortex
The Pancreas
    Diabetes Mellitus
Other Endocrine Glands
    Gonads
    The Thymus
    The Pineal Gland
    Still Other Glands
Environmental Signals
Redefinition of a Hormone

Hormones

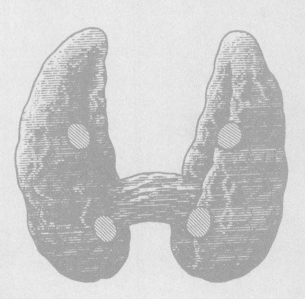

Figure 19.1

Anatomical location of major endocrine glands in the body. The hypothalamus produces the hormones secreted at the posterior pituitary and controls the anterior pituitary, which in turn controls the hormonal secretions of the thyroid, the adrenal cortex, and the sex organs (testes and ovaries). Female sex organs are shown in the box.

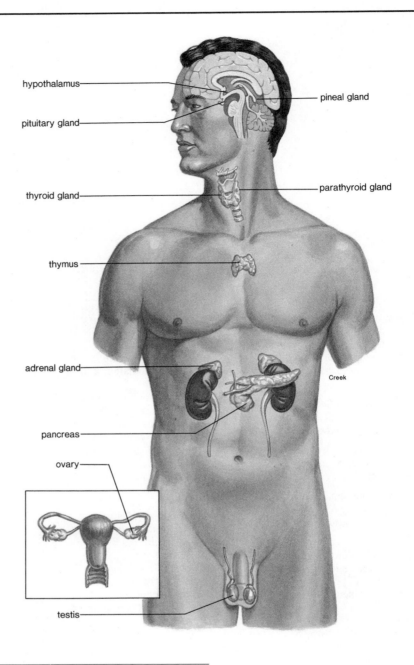

Endocrine Glands

Two major systems, the nervous system (chap. 16) and the **endocrine system,** coordinate the various activities of body parts. Both systems utilize chemical messengers (chemical signals) to fulfill their functions, and in chapter 16 we discussed the neurotransmitter substances that are released by one neuron and influence the excitability of another. In this chapter, we discuss the chemical messengers of the human endocrine system that are produced by the glands of internal secretion (fig. 19.1). These messengers are called **hormones** and are released directly into the bloodstream by these glands. Such glands, called **endocrine glands,** can be contrasted to exocrine glands. Exocrine glands have ducts, and they secrete their products into these ducts for transport into body cavities; for example, the salivary glands send saliva into the mouth by way of the salivary ducts. Endocrine glands are ductless; they secrete their hormones directly into the bloodstream for distribution throughout the body. The cells of these glands abut capillaries, whose thin walls allow easy entrance of hormones.

The nervous system reacts quickly to external and internal stimuli; you rapidly pull away your hand from a hot stove, for example. The endocrine system is slower because it takes time for a hormone to travel through the circulatory system to its *target organ.* You might think from the use of this terminology that the hormone is seeking out a particular organ, but quite the contrary, the organ is awaiting the arrival of the hormone. Those cells that can react to a hormone have specific receptors that combine with the hormone in a lock-and-key manner. Therefore, certain cells respond to one hormone and not to another depending on their receptors.

Sometimes, a hormone is defined as a chemical produced by one set of cells that affects a different set. This definition allows us to categorize all sorts of chemical messengers as hormones, even neurotransmitter substances! More is said about this later, but a certain amount of overlap between the nervous and endocrine systems is to be expected. The systems evolved together, no doubt making occasional use of the same chemical messengers and communicating not only with other systems but with each other as well. Later, several examples of such associations between the 2 systems are given.

Both the nervous and endocrine systems function to coordinate body parts and activities. Because they evolved at the same time, a certain amount of overlap between the chemical messengers utilized and the types of activities performed is to be expected.

The Mechanism of Hormonal Action

Most hormones are either *peptides, polypeptides, or proteins.* These molecules are coded for by genes and are synthesized at the ribosomes. Eventually, they are packaged into vesicles at the Golgi apparatus and are secreted at the cell membrane (fig. 2.7). There are some other hormones, such as norepinephrine, that are derived from the amino acid tyrosine; their production requires only a series of metabolic reactions within the cytoplasm.

For the purpose of discussing hormonal action, we can group the hormones that are peptides, polypeptides, and proteins with those derived from amino acids. We will call all of these peptide hormones.

Some hormones are steroids. Steroid hormones are produced by the adrenal cortex, the ovaries, and the testes (fig. 19.1). Steroids are derived from cholesterol (fig. 1.31) by a series of metabolic reactions. These hormones are stored in fat droplets in the cell cytoplasm until their release at the cell membrane.

Peptide Hormones

The receptors for peptide hormones are on the cell membrane (fig. 19.2*a*). When a hormone binds to a receptor, the resulting complex activates an enzyme that produces cyclic adenosine monophosphate (cyclic or cAMP). cAMP is a compound made from ATP, but it contains only one phosphate group that is attached to adenosine at 2 locations. cAMP now activates a particular enzyme in the cell; this enzyme in turn activates another, and so forth. In other words, a series of enzymatic reactions—an enzyme cascade—is set in motion. Because each enzyme can be used over and over again, at every step in the enzyme cascade, more reactions take place. The binding of a single hormone molecule eventually results in a thousandfold response. Notice that the peptide hormone never enters the cell. Therefore, these hormones sometimes are called the *first messenger,* while cAMP, which sets the metabolic machinery in motion, is called the *second messenger* (fig. 19.2*a*).

Steroid Hormones

Steroid hormones do not bind to cell surface receptors; they can enter the cell freely because they are lipids (fig. 19.2*b*). For example, radioactive scans of the reproductive organs clearly show that the hormones estrogen and progesterone enter the cells of the reproductive organs. Once inside, steroid hormones bind to receptors in the cytoplasm. The hormone-receptor complex then

Figure 19.2

Cellular activity of hormones. a. Peptide hormones combine with receptors located on the cell membrane. This promotes the production of cAMP, which in turn leads to activation of a particular enzyme. b. Steroid hormones pass through the cell membrane to combine with receptors; the complex activates certain genes, leading to protein synthesis.

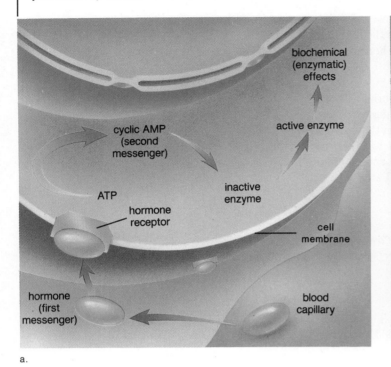

a.

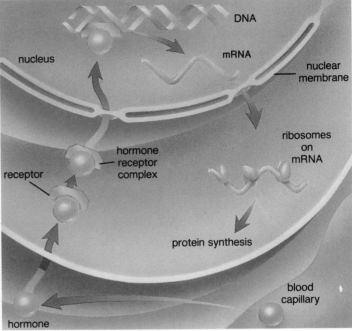

b.

enters the nucleus, where it binds with chromatin at a location that promotes activation of particular genes. Protein synthesis follows. In this manner, steroid hormones can lead to the synthesis of certain enzymes.

Steroids act more slowly than peptides because it takes more time to synthesize new proteins than to activate enzymes that are already present in the cell. Steroids have a more sustained effect on the metabolism of the cell, however, than peptide hormones.

Hormones are chemical messengers that influence the metabolism of the receiving cell. Peptide hormones activate existing enzymes in the cell, and steroid hormones bring about the synthesis of new enzymes.

The Hypothalamus and the Pituitary Gland

The hypothalamus is a portion of the brain that regulates the internal environment. For example, it helps to control heart rate, body temperature, and water balance, as well as secretions of the **pituitary gland.** The pituitary, a small gland about 1 cm in diameter, lies just below the hypothalamus (fig. 19.1) and is divided into 2 portions called the **posterior pituitary** and the **anterior pituitary.**

The Posterior Pituitary

The posterior pituitary is connected to the hypothalamus by means of a stalk-like structure. There are neurons in the hypothalamus that are called *neurosecretory cells* because they both respond to neurotransmitter substances and produce the hormones that are stored in and released from the posterior pituitary. The hormones pass from the hypothalamus through axons that terminate in the posterior pituitary (fig. 19.3).

The axon endings in the posterior pituitary store **antidiuretic hormone (ADH),** sometimes called vasopressin, and oxytocin. ADH, as discussed in chapter 15, promotes the reabsorption of water from the collecting duct, a

Figure 19.3

The hypothalamus produces 2 hormones, ADH and oxytocin, that are stored in and secreted by the posterior pituitary.

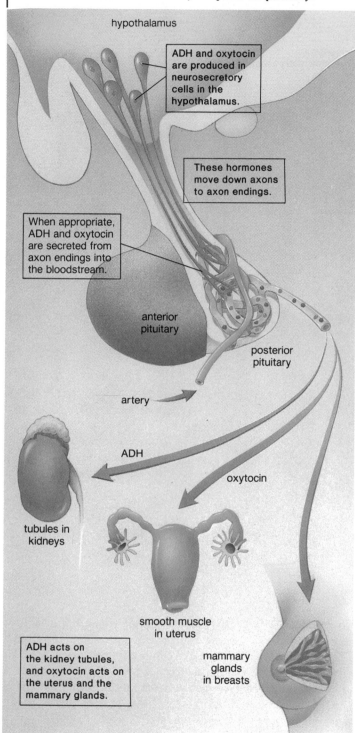

hypothalamus

ADH and oxytocin are produced in neurosecretory cells in the hypothalamus.

These hormones move down axons to axon endings.

When appropriate, ADH and oxytocin are secreted from axon endings into the bloodstream.

anterior pituitary

posterior pituitary

artery

ADH

oxytocin

tubules in kidneys

smooth muscle in uterus

mammary glands in breasts

ADH acts on the kidney tubules, and oxytocin acts on the uterus and the mammary glands.

Figure 19.4

Regulation of ADH secretion. Neurons in the hypothalamus are sensitive to the osmolarity of blood. When blood is concentrated, they send signals to the hypothalamus neurosecretory cells, which release ADH from their axon endings. ADH increases the permeability of the collecting ducts in the kidneys so that more water is reabsorbed. Once blood is diluted, ADH no longer is secreted. This is an example of control by negative feedback.

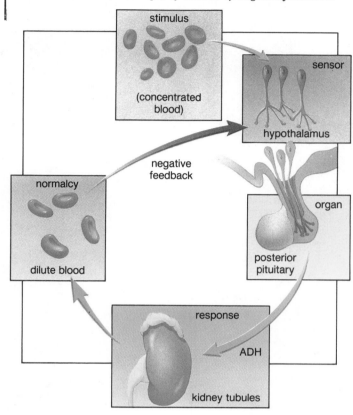

stimulus

(concentrated blood)

sensor

hypothalamus

negative feedback

normalcy

organ

dilute blood

posterior pituitary

response

ADH

kidney tubules

portion of the kidney tubule (nephron). The hypothalamus contains other nerve cells that are sensitive to the osmolarity of the blood. When these cells determine that the blood is too concentrated, ADH is released into the bloodstream from the axon endings in the posterior pituitary. As the blood becomes dilute, the hormone no longer is released. This is an example of control by negative feedback (fig. 19.4) because the effect of the hormone (diluted blood) acts to shut down the release of the hormone. Negative feedback mechanisms regulate the activities of most endocrine glands.

Figure 19.5
*Hypothalamus and anterior pituitary. The hypothalamus
controls the secretions of the anterior pituitary.*

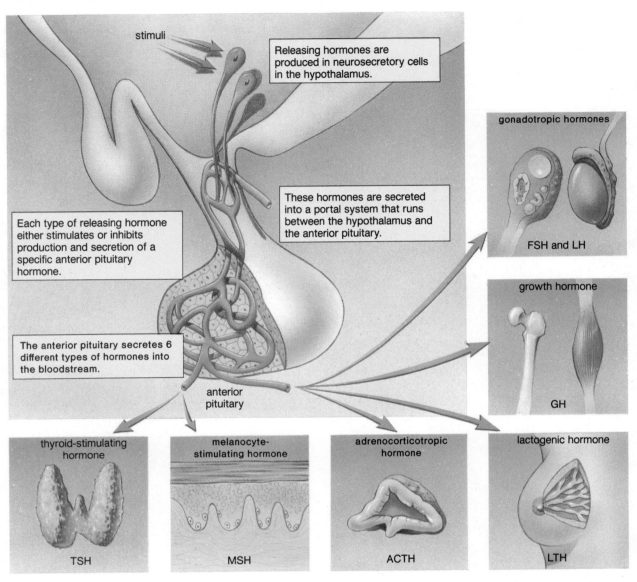

stimuli

Releasing hormones are
produced in neurosecretory cells
in the hypothalamus.

Each type of releasing hormone
either stimulates or inhibits
production and secretion of a
specific anterior pituitary
hormone.

These hormones are secreted
into a portal system that runs
between the hypothalamus and
the anterior pituitary.

The anterior pituitary secretes 6
different types of hormones into
the bloodstream.

anterior
pituitary

gonadotropic hormones

FSH and LH

growth hormone

GH

thyroid-stimulating
hormone

TSH

melanocyte-
stimulating hormone

MSH

adrenocorticotropic
hormone

ACTH

lactogenic hormone

LTH

Inability to produce ADH causes **diabetes insipidus** (watery urine), in
which a person produces copious amounts of urine with a resultant loss of salts
from the blood. This condition can be corrected by the administration of ADH.

Oxytocin is the other hormone that is made in the hypothalamus and is
stored in the posterior pituitary. Oxytocin causes the uterus to contract and is
used to artificially induce labor. It also stimulates the release of milk from the
mother's mammary glands when a baby is nursing.

It is appropriate to note that the neurosecretory cells in the hypothalamus provide an example of a way the nervous system and the endocrine system
are joined. This topic is discussed again later.

*The posterior pituitary stores 2 hormones, ADH and oxytocin, both of which are
produced by and released from neurosecretory cells in the hypothalamus.*

The Anterior Pituitary

A portal system composed of tiny blood vessels connects the anterior pituitary (fig. 19.5) to the hypothalamus. The hypothalamus controls the anterior pituitary by producing hypothalamic-releasing and release-inhibiting hormones that are transported to the anterior pituitary by blood within the portal system connecting the 2 organs. Each type of hypothalamic hormone, produced by neurosecretory cells, causes the anterior pituitary either to secrete or to stop secreting a type of hormone (table 19.1). There are 6 types of hormones produced by the anterior pituitary.

The anterior pituitary is controlled by the hypothalamic-releasing hormones. They are produced in neurosecretory cells and pass to the anterior pituitary by way of a portal system.

Three of the hormones produced by the anterior pituitary have a direct effect on the body. **Growth hormone (GH),** or somatotropin, dramatically affects physical appearance since it determines the height of the individual (fig. 19.6). If little or no GH is secreted by the anterior pituitary during childhood, the person could become a pituitary dwarf, although of perfect proportions, quite small in stature. If too much GH is secreted, the person could become a giant. Giants usually have poor health, primarily because GH has a secondary effect on blood sugar level, promoting an illness called diabetes (sugar) mellitus, discussed following.

GH promotes cell division, protein synthesis, and bone growth. It stimulates the transport of amino acids into cells and increases the activity of ribosomes, both of which are essential to protein synthesis. In bones, it promotes growth of the cartilaginous plates and causes osteoblasts to form bone (p. 350). Evidence suggests that the effects on cartilage and bone actually may be due to hormones called somatomedins, released by the liver. GH causes the liver to release somatomedins.

If the production of GH increases in an adult after full height has been attained, only certain bones respond. These are the bones of the jaw, the eyebrow ridges, the nose, the fingers, and the toes. When these begin to grow, the person takes on a slightly grotesque look with huge fingers and toes, a condition called **acromegaly** (fig. 19.7).

Lactogenic hormone (LTH), also called **prolactin,** is produced in quantity only after childbirth. It causes the mammary glands in the breasts to develop and to produce milk.

Melanocyte-stimulating hormone (MSH) causes skin color changes in lower vertebrates—but no one knows what it does in humans. However, it is derived from a molecule that is also the precursor for both adrenocorticotropic (ACTH) and the anterior pituitary endorphins. These endorphins are structurally and functionally similar to the endorphins produced in brain nerve cells.

GH and LTH are 2 hormones produced by the anterior pituitary. GH influences the height of children and overproduction brings about a condition called acromegaly in adults. LTH promotes milk production after childbirth.

Other Hormones Produced by the Anterior Pituitary

The anterior pituitary sometimes is called the *master gland* because it controls the secretion of other certain endocrine glands (fig. 19.5). As indicated in table 19.1, the anterior pituitary secretes the following hormones, which have an effect on other glands.

1. Thyroid-stimulating hormone (**TSH**)
2. Adrenocorticotropic hormone (**ACTH**), a hormone that stimulates the adrenal cortex
3. **Gonadotropic hormones** (FSH and LH) that stimulate the gonads, the testes in males and the ovaries in females

Figure 19.6
Giantism. Sandy Allen, one of the world's tallest women due to a higher than usual amount of GH produced by the anterior pituitary.

Table 19.1 The Principal Endocrine Glands and Their Hormones

Endocrine Gland	Hormone Released	Target Tissues/Organ
Hypothalamus	Hypothalamic-releasing hormones Hypothalamic release-inhibiting hormones	Anterior pituitary

Anterior pituitary	Thyroid-stimulating hormone (TSH, thyrotropic)
	Adrenocorticotropic hormone (ACTH)
	Gonadotropic hormones
	    Follicle-stimulating (FSH)
	    Luteinizing (LH)
	Lactogenic hormone (LTH, prolactin)
	Growth hormone (GH, somatotropin)
	
	Melanocyte-stimulating hormone (MSH)
		Thyroid
		Andrenal cortex
		Gonads
		
		Mammary glands
		Soft tissues, bones
		
		Melanocytes in skin

Posterior pituitary (storage of hypothalamic hormones)	Antidiuretic hormone (ADH, vasopressin) Oxytocin	Kidneys Uterus, mammary glands
Thyroid	Thyroxin Calcitonin	All tissues Bones, kidneys, gut
Parathyroids	Parathyroid hormone (PTH)	Bones, kidneys, gut
Adrenal cortex	Glucocorticoids (cortisol) Mineralocorticoids (aldosterone) Sex hormones	All tissues Kidneys Sex organs, skin, muscles, bones
Adrenal medulla	Epinephrine and norepinephrine	Cardiac and other muscles
Pancreas	Insulin Glucagon	Liver, muscles, adipose tissue Liver, muscles, adipose tissue

Gonads
  Testes
 	
  Ovaries
	
	Androgens (testosterone)
	
	Estrogen and progesterone
		
		Sex organs, skin, muscles, bones
		
		Sex organs, skin, muscles, bones

Thymus	Thymosins	T lymphocytes
Pineal gland	Melatonin	Circadian rhythms

*The word *diverse* in this table means that the symptoms have not been described as a syndrome in the medical literature.

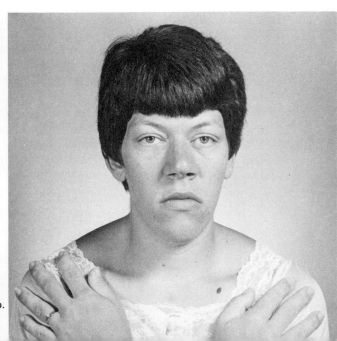

Figure 19.7

Acromegaly. The condition is caused by the overproduction of GH in the adult. It is characterized by an enlargement of the bones in the face and the fingers of an adult. a. At age 20, this individual was normal. b. At age 24, there is some enlargement of the nose, the jaw, and the fingers.

a.

b.

Chief Function of Hormone	Disorders Too Much/Too Little
Regulate anterior pituitary hormones	*See* anterior pituitary
Stimulates thyroid Stimulates adrenal cortex	*See* thyroid *See* adrenal cortex
Controls egg and sperm production Controls sex hormone production Stimulates milk production and secretion Stimulates cell division, protein synthesis, and bone growth Regulates skin color in lower vertebrates; unknown function in humans	*See* testis and ovary Giantism, acromegaly/dwarfism
Stimulates water reabsorption by kidneys	Diverse*/diabetes insipidus
Stimulates uterine muscle contraction and release of milk by mammary glands	
Increases metabolic rate; helps to regulate growth and development Lowers blood calcium level	Exophthalmic goiter/simple goiter myxedema, cretinism Tetany/weak bones
Raises blood calcium level	Weak bones/tetany
Raise blood glucose level; stimulate breakdown of protein Stimulate kidneys to reabsorb sodium and to excrete potassium Stimulate development of secondary sex characteristics (particularly in male)	Cushing's syndrome/Addison's disease
Stimulate fight or flight reactions; raise blood glucose level	
Lowers blood glucose level; promotes formation of glycogen, proteins, and fats Raises blood glucose level; promotes breakdown of glycogen, proteins, and fats	Shock/diabetes mellitus
Stimulate spermatogenesis; develop and maintain secondary male sex characteristics	Diverse/eunuch
Stimulate growth of uterine lining; develop and maintain secondary female sex characteristics	Diverse/masculinization
Stimulates maturation of T lymphocytes	
Involved in circadian and circannual rhythms; possibly involved in maturation of sex organs	

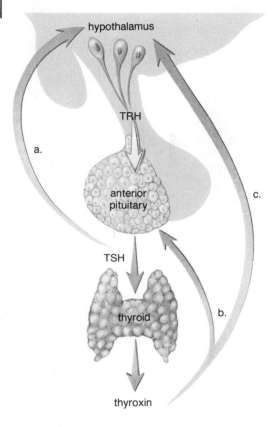

Figure 19.8

The hypothalamus–pituitary–thyroid control relationship. TRH (thyroid-releasing hormone) stimulates the anterior pituitary, and TSH (thyroid-stimulating hormone) stimulates the thyroid to secrete thyroxin. The level of thyroxin in the body is negatively controlled in 3 ways: (a) The level of TSH exerts feedback control over the hypothalamus; (b) the level of thyroxin exerts feedback control over the anterior pituitary; and (c) the level of thyroxin exerts feedback control over the hypothalamus. In this way, thyroxin controls its own secretion. Cortisol and sex hormone levels are controlled in similar ways.

TSH causes the thyroid to produce thyroxin; ACTH causes the adrenal cortex to produce cortisol; and gonadotropic hormones cause the gonads to secrete sex hormones. Notice that it is now possible to indicate a three-tiered relationship between the hypothalamus, the pituitary, and the other endocrine glands. The hypothalamus produces releasing hormones that control the anterior pituitary, and the anterior pituitary produces hormones that control the thyroid, the adrenal cortex, and the gonads. Figure 19.8 illustrates the feedback mechanism that controls the activity of all these glands.

The hypothalamus, the anterior pituitary, and the other endocrine glands controlled by the anterior pituitary all are involved in a self-regulating feedback loop.

Figure 19.9

Simple goiter. An enlarged thyroid gland often is caused by a lack of iodine in the diet. Without iodine, the thyroid is unable to produce thyroxin, and continued anterior pituitary stimulation causes the gland to enlarge.

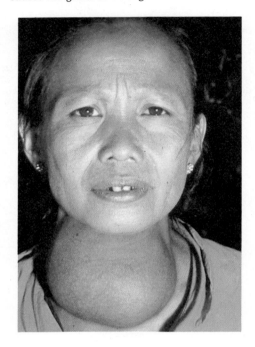

Thyroid and Parathyroid Glands

The Thyroid Gland

The thyroid gland (fig. 19.1) is located in the neck and is attached to the trachea just below the larynx. Internally, the gland is composed of a large number of follicles filled with thyroglobulin, the storage form of thyroxin. The production of both of these requires iodine. Iodine is actively transported into the thyroid gland, where the concentration can become as much as 25 times that of blood. If iodine is lacking in the diet, the thyroid gland enlarges, producing a goiter (fig. 19.9). (Salt in the United States is iodized for this reason.) The cause of goiter becomes clear if we refer to figure 19.8. When there is a low level of thyroxin in the blood, a condition called hypothyroidism, the anterior pituitary is stimulated to produce *TSH*. TSH causes the thyroid to increase in size so that enough **thyroxin** usually is produced. In this case, enlargement continues because enough thyroxin never is produced. An enlarged thyroid that produces some thyroxin is called a **simple goiter.**

Thyroxin

Thyroxin increases the metabolic rate. It does not have a target organ; instead, it stimulates most of the cells of the body to metabolize at a faster rate. The number of respiratory enzymes in the cell increases, as does oxygen (O_2) uptake.

If the thyroid fails to develop properly, a condition called **cretinism** results. Cretins (fig. 19.10) are short, stocky persons who have had extreme hypothyroidism since infancy and/or childhood. Thyroxin therapy can initiate growth, but unless treatment is begun within the first 2 months of life, mental retardation results. The occurrence of hypothyroidism in adults produces the condition known as **myxedema** (fig. 19.11), which is characterized by lethargy, weight gain, loss of hair, slowed pulse rate, decreased body temperature, and thickness and puffiness of the skin. The administration of adequate doses of thyroxin restores normal function and appearance.

Figure 19.10

Cretinism. Cretins are individuals who have suffered from thyroxin insufficiency since birth or early childhood. Skeletal growth is usually inhibited to a greater extent than soft tissue growth; therefore, the child appears short and stocky. Sometimes, the tongue becomes so large that it obstructs swallowing and breathing.

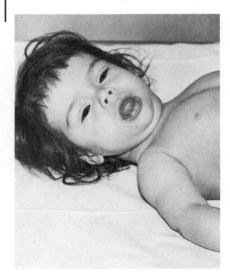

Figure 19.11

Myxedema. This condition is caused by thyroid insufficiency in the older adult. An unusual type of edema leads to swelling of the face and bagginess under the eyes.

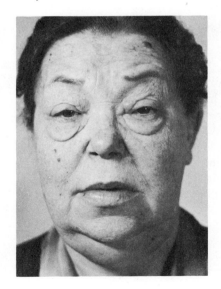

Figure 19.12

Exophthalmic goiter. Protruding eyes occur when an active thyroid gland enlarges.

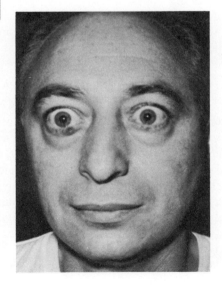

In the case of hyperthyroidism (too much thyroxin), the thyroid gland is enlarged and overactive, causing a goiter to form and the eyes to protrude because of edema in the tissues of the eye sockets and swelling of muscles that move the eyes. This type of goiter is called **exophthalmic goiter** (fig. 19.12). The patient usually becomes hyperactive, nervous, irritable, and suffers from insomnia. Removal or destruction of a portion of the thyroid by means of radioactive iodine sometimes is effective in curing the condition.

Calcitonin

In addition to thyroxin, the thyroid gland also produces the hormone **calcitonin.** This hormone helps to regulate the calcium level in the blood and opposes the action of parathyroid hormone. The interaction of these 2 hormones is discussed following.

The anterior pituitary produces TSH, a hormone that promotes the production of thyroxin by the thyroid, a gland subject to goiters. Thyroxin, which speeds up metabolism, can affect the body as a whole, as exemplified by cretinism and myxedema.

Parathyroid Glands

The parathyroid glands are embedded in the posterior surface of the thyroid gland, as shown in figure 19.13b. Many years ago, these 4 small glands sometimes were removed by mistake during thyroid surgery. Under the influence of **parathyroid hormone (PTH),** also called parathormone, the calcium level in blood increases and the phosphate level decreases. The hormone stimulates the absorption of calcium from the gut, the retention of calcium by the kidneys, and the demineralization of bone. In other words, PTH promotes the activity of osteoclasts, the bone-resorbing cells. Although this also raises the level of phosphate in the blood, PTH acts on the kidneys to excrete phosphate in the urine. When a woman stops producing the female sex hormone estrogen following menopause, she is more likely to suffer from osteoporosis, characterized by a thinning of the bones. Whether estrogen works counter to the action of PTH has not been determined yet.

If insufficient PTH is produced, the level of calcium in blood drops, resulting in **tetany.** In tetany, the body shakes from continuous muscle contraction. The effect really is brought about by increased excitability of the nerves, which fire spontaneously and without rest. Calcium plays an important role in both nervous conduction and muscle contraction.

The level of PTH secretion is controlled by a feedback mechanism involving calcium (fig. 19.13c). When the calcium level rises, PTH secretion is inhibited, and when the calcium level lowers, PTH secretion is stimulated.

As mentioned previously, the thyroid secretes calcitonin, which also influences blood calcium level. Although calcitonin has the opposite effect of PTH, particularly on the bones, its action is not believed to be as significant. Still, the 2 hormones function together to regulate the level of calcium in the blood.

PTH maintains a high blood level of calcium by promoting its absorption in the gut, its reabsorption by the kidneys, and demineralization of bone. These actions are opposed by calcitonin produced by the thyroid.

Adrenal Glands

The adrenal glands, as their name implies (*ad* = near; *renal* = kidneys), lie atop the kidneys (fig. 19.1). Each consists of an outer portion, called the *cortex,* and an inner portion, called the *medulla.* These portions, like the anterior pituitary and the posterior pituitary, have no functional connection with one another.

Figure 19.13
Thyroid and parathyroid glands. a. The thyroid is located in the neck in front of the trachea. b. The 4 parathyroid glands are embedded in the posterior surface of the thyroid gland. Yet, the parathyroid and thyroid glands have no anatomical or physiological connection with one another. c. Regulation of parathyroid hormone (PTH) secretion. A low blood level of calcium causes the parathyroids to secrete PTH, which causes the kidneys and the gut to retain calcium and osteoclasts to break down bone. The end result is an increased level of calcium in the blood. A high blood level of calcium inhibits secretion of PTH.

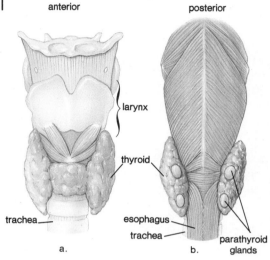

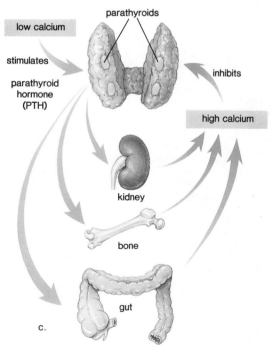

The Adrenal Medulla

The adrenal medulla secretes **norepinephrine** and **epinephrine** under conditions of stress. They bring about all those responses we associate with the "fight or flight" reaction: the blood glucose level and the metabolic rate increase, as do breathing and the heart rate. The blood vessels in the intestine constrict, and those in the muscles dilate. This increased circulation to the muscles causes them to have more stamina than usual. In times of emergency, the sympathetic nervous system *initiates* these responses, but they are maintained by secretions from the adrenal medulla.

The adrenal medulla is innervated by only one set of sympathetic nerve fibers. Recall that usually there are pre- and postganglionic nerve fibers for each organ stimulated. In this instance, what happened to the postganglionic neurons? It appears that the adrenal medulla may have evolved from a modification of the postganglionic neurons. Like the neurosecretory neurons in the hypothalamus, these neurons also secrete hormones into the bloodstream.

The adrenal medulla releases norepinephrine and epinephrine into the bloodstream. These hormones help us and other animals to cope with situations that threaten survival.

The Adrenal Cortex

Although the adrenal medulla can be removed with no ill effects, the adrenal cortex is absolutely necessary to life. The 2 major classes of hormones made by the adrenal cortex are the *glucocorticoids* and the *mineralocorticoids*. The cortex also secretes a small amount of male sex hormone and an even smaller amount of female sex hormone. All of these hormones are steroids.

Glucocorticoids

Of the various glucocorticoids, the hormone responsible for the greatest amount of activity is **cortisol.** Cortisol promotes the hydrolysis of muscle protein to amino acids that enter the blood. This leads to an increased level of glucose when the liver converts these amino acids to glucose. Cortisol also favors metabolism of fatty acids rather than carbohydrate. In opposition to insulin, therefore, cortisol raises the blood glucose level. Cortisol also counteracts the inflammatory response, which leads to the pain and the swelling of joints in arthritis and bursitis. The administration of cortisol aids these conditions because it reduces inflammation.

The secretion of cortisol by the adrenal cortex is under the control of the anterior pituitary hormone ACTH. Using the same kind of system shown in figure 19.8, the hypothalamus produces a releasing hormone (CRH) that stimulates the anterior pituitary to release ACTH. ACTH in turn stimulates the adrenal cortex to secrete cortisol, which regulates its own synthesis by negative feedback of both CRH and ACTH synthesis.

Mineralocorticoids

The secretion of mineralocorticoids, the most significant of which is **aldosterone,** is not under the control of the anterior pituitary. Aldosterone regulates the level of sodium and potassium in blood, its primary target organ being the kidney, where it promotes renal absorption of sodium and renal excretion of potassium (fig. 19.14). The level of sodium is particularly important to the maintenance of blood pressure because its concentration indirectly regulates the secretion of aldosterone. When the blood level of sodium is low, the kidneys secrete renin. Renin is an enzyme that converts the plasma protein angiotensinogen to angiotensin I, which becomes angiotensin II in the lungs. Angiotensin II stimulates the adrenal cortex to release aldosterone (see fig. 15.13). This is called the renin-angiotensin-aldosterone system. The effect of this

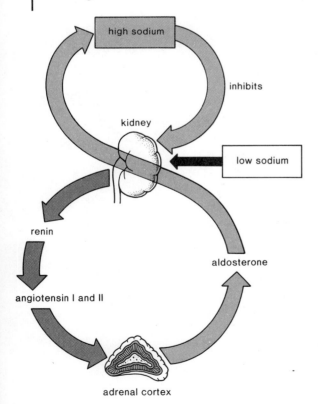

Figure 19.14

Renin–angiotensin–aldosterone system. If the blood level of sodium is low, the kidneys secrete renin. The increased renin acts via the increased production of angiotensin I and II to stimulate aldosterone secretion. Aldosterone promotes reabsorption of sodium by the kidneys; when the sodium level in the blood rises, the kidneys stop secreting renin.

system is to raise the blood pressure in 2 ways. First, angiotensin II constricts the arteries directly, and secondly, aldosterone causes the kidneys to reabsorb sodium. When the blood level of sodium is high, water is reabsorbed, and blood volume and pressure are maintained.

Lack of aldosterone would cause low blood sodium and blood volume; however, it recently has been found that there is another hormone that acts contrary to aldosterone. This hormone is called the atrial natriuretic hormone because (1) it is produced by the atria of the heart and (2) it causes natriuresis, the excretion of sodium. Once sodium is excreted so is water; therefore, blood volume and blood pressure decrease.

Cortisol, which raises the blood glucose level, and aldosterone, which raises the blood sodium level, are 2 hormones secreted by the adrenal cortex.

Sex Hormones

The adrenal cortex produces a small amount of both male and female sex hormones. In males, the cortex is a source of female sex hormones, and in females, it is a source of male hormones. A tumor in the adrenal cortex can cause the production of a large amount of sex hormones, which can lead to feminization in males and masculinization in females.

Disorders

When the level of adrenal cortex hormones is low, a person begins to suffer from Addison's disease. When the level of adrenal cortex hormones in the body is high, a person suffers from Cushing's syndrome:

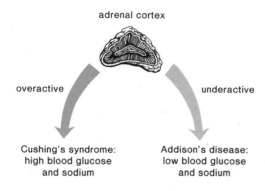

adrenal cortex

overactive   underactive

Cushing's syndrome: Addison's disease:
high blood glucose   low blood glucose
and sodium   and sodium

Addison's Disease   Because of the lack of cortisol, the Addison's disease patient is unable to maintain the glucose level of the blood, tissue repair is suppressed, and there is a high susceptibility to any kind of bodily stress. Even a mild infection can cause death. Due to the lack of aldosterone, the blood sodium level is low, and the person experiences low blood pressure along with acidosis and low pH. In addition, the patient has a peculiar bronzing of the skin (fig. 19.15).

Cushing's Syndrome   In Cushing's syndrome, a high level of cortisol causes a tendency toward diabetes mellitus, a decrease in muscular protein, and an increase in subcutaneous fat. Because of these effects, the person usually develops thin arms and legs and an enlarged trunk. Due to the high level of sodium in the blood, the blood is basic and the patient has hypertension and edema of the face, which gives the face a moon shape (fig. 19.16).

Addison's disease is due to adrenal cortex hyposecretion, and Cushing's syndrome is due to adrenal cortex hypersecretion.

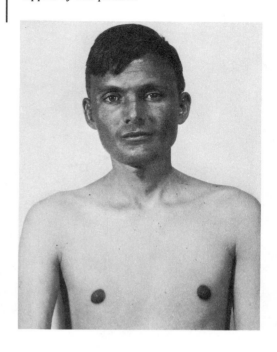

Figure 19.15
Addison's disease. This condition is characterized by a peculiar bronzing of the skin, as seen in the face and the thin skin of the nipples of this patient.

Figure 19.16
Cushing's syndrome. Persons with this condition tend to have an enlarged trunk and a moonlike face. Masculinization may occur in women due to the excessive male sex hormones in the body.

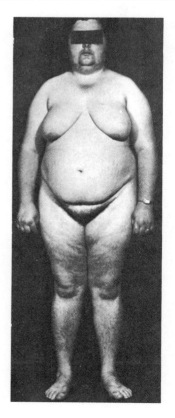

Figure 19.17

Gross and microscopic anatomy of the pancreas. The pancreas lies in the abdomen between the kidneys and near the duodenum. As an exocrine gland, it secretes digestive enzymes that enter the duodenum by the common duct. As an endocrine gland, it secretes insulin and glucagon into the bloodstream. The alpha cells of the islets of Langerhans produce glucagon, and the beta cells produce insulin (top right).

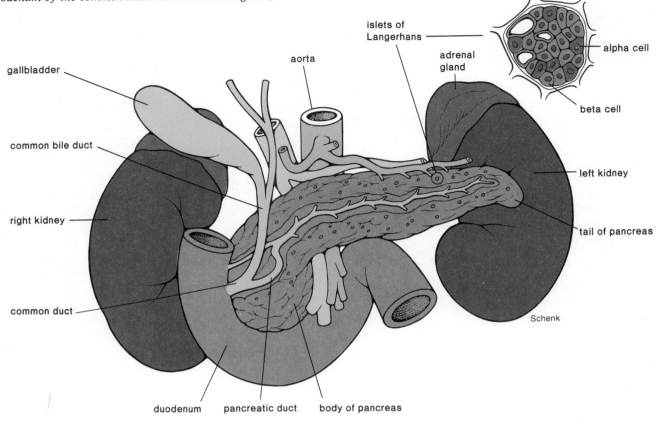

The Pancreas

The **pancreas** is a long organ that lies transversely in the abdomen (fig. 19.17) between the kidneys and near the duodenum of the small intestine. It is composed of 2 types of tissue—exocrine, which produces and secretes *digestive* juices that go by way of the pancreatic duct and the common duct to the small intestine, and endocrine, called the **islets of Langerhans,** which produces and secretes the hormones **insulin** and **glucagon** directly into the blood.

Insulin is secreted when there is a high level of glucose in blood, which usually occurs just after eating. Insulin has 3 different actions: (1) it stimulates liver, fat, and muscle cells to take up and metabolize glucose; (2) it stimulates the liver and the muscles to store glucose as glycogen; and (3) it promotes the buildup of fats and proteins and inhibits their use as an energy source. Therefore, insulin is a hormone that promotes storage of nutrients so that they are on hand during leaner times. It also helps to lower the blood glucose level.

Glucagon is secreted from the pancreas in between eating, and its effects are opposite to those of insulin. Glucagon stimulates the breakdown of stored nutrients and causes the blood sugar level to rise (fig. 19.18).

Diabetes Mellitus

The symptoms of **diabetes mellitus** (sugar diabetes) include the following:

Sugar in the urine
Frequent, copious urination
Abnormal thirst
Rapid loss of weight
General weakness

Drowsiness and fatigue
Itching of the genitals and the skin
Visual disturbances, blurring
Skin disorders, such as boils,
    carbuncles, and infection

Figure 19.18

Contrary effects of insulin and glucagon. When blood glucose level is high, the pancreas secretes insulin. Insulin promotes the storage of glucose as glycogen and the synthesis of proteins and fats as opposed to their use as energy sources. Therefore, insulin lowers the blood glucose level. When the blood glucose level is low, the pancreas secretes glucagon. Glucagon acts in opposition to insulin in all respects; therefore, glucagon raises the blood glucose level.

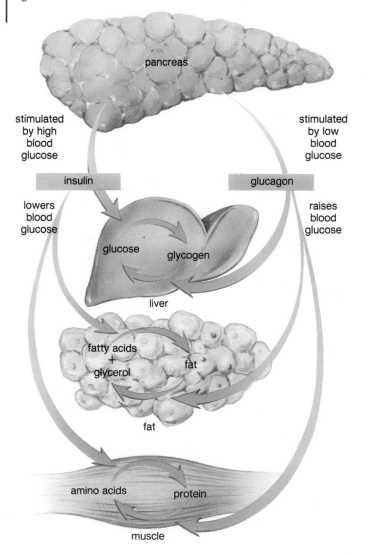

pancreas

stimulated by high blood glucose

stimulated by low blood glucose

insulin

glucagon

lowers blood glucose

raises blood glucose

glucose ⇄ glycogen

liver

fatty acids + glycerol ⇄ fat

fat

amino acids ⇄ protein

muscle

Table 19.2 Symptoms of Insulin Shock and Diabetic Coma

Insulin Shock	Diabetic Coma
Sudden onset	Slow, gradual onset
Perspiration, pale skin	Dry, hot skin
Dizziness	No dizziness
Heart palpitation	No palpitation
Hunger	No hunger
Normal urination	Excessive urination
Normal thirst	Excessive thirst
Shallow breathing	Deep, labored breathing
Normal breath odor	Fruity breath odor
Confusion, disorientation, strange behavior	Drowsiness and great lethargy leading to stupor
Urinary sugar absent or slight	Large amounts of urinary sugar
No acetone in urine	Acetone present in urine

From Henry Dolger and Bernard Seeman. *How to Live with Diabetes.* Copyright © 1972, 1965, 1958 W. W. Norton & Company, Inc., New York. Reprinted by permission.

19.1 Critical Thinking

1. If the pancreas is removed from an animal, what substance would you expect to find in the urine? Why?
2. Would your findings support the contention that the pancreas is the source of insulin? What do they prove?
3. How would you prove that insulin lowers blood sugar?
4. Once you have shown that insulin lowers blood sugar, do your findings in question 1 support the belief that the pancreas is the source of insulin? Why?
5. What could you do to prove that the pancreas is the source of insulin?

Many of these symptoms develop because sugar is not being metabolized by the cells. The liver fails to store glucose as glycogen, and all the cells fail to utilize glucose as an energy source. This means that the blood glucose level rises very high after eating, causing glucose to be excreted in the urine. More water than usual therefore is excreted so that the diabetic is extremely thirsty.

Since carbohydrate is not being metabolized, the body turns to the breakdown of protein and fat for energy. Unfortunately, the breakdown of these molecules leads to the buildup of acids in the blood (acidosis) and to respiratory distress. It is the latter that eventually can cause coma and death of the diabetic. The symptoms that lead to coma (table 19.2) develop slowly.

There are 2 types of diabetes. In *type I diabetes,* formerly called juvenile-onset diabetes, the pancreas is not producing insulin. Therefore, the patient must have daily insulin injections. These injections control the diabetic symptoms but still can cause inconveniences since either an overdose of insulin or the absence of regular eating can bring on the symptoms of insulin shock (table

19.2). These symptoms appear because the blood sugar level has fallen below normal levels. Since the brain requires a constant supply of sugar, unconsciousness can result. The cure is quite simple: an immediate source of sugar, such as a sugar cube or fruit juice, can counteract insulin shock immediately.

Obviously, insulin injections are not the same as a fully functioning pancreas that responds on demand to high glucose level by supplying insulin. For this reason, some doctors advocate an islet cell transplant.

Of the 12 million people who now have diabetes in the United States, at least 10 million have *type II diabetes,* formerly called maturity-onset diabetes. In this type of diabetes, now known to occur in obese people of any age, the pancreas is producing insulin, but the cells do not respond to it. At first, the cells lack the receptors necessary to detect the presence of insulin, and later, the cells are even incapable of taking up glucose. If type II diabetes is untreated, the results can be as serious as type I diabetes. Diabetics are prone to blindness, kidney disease, and circulatory disorders, including strokes. Pregnancy carries an increased risk of diabetic coma, and the child of a diabetic is somewhat more likely to be stillborn or to die shortly after birth. It is important, therefore, to prevent or to at least control type II diabetes. The best defense is a nonfattening diet and regular exercise. If that fails, there are oral drugs that make the cells more sensitive to the effects of insulin or that stimulate the pancreas to make more of it.

Diabetes mellitus is caused by a lack of insulin or insensitivity of cells to insulin. Insulin lowers blood glucose levels by causing the cells to take up glucose and the liver to convert it to glycogen.

Other Endocrine Glands

Gonads

The gonads are the endocrine glands that produce the hormones that determine sexual characteristics. As is discussed in detail in the following chapter, the *testes* produce the androgens—the most important of which is called testosterone—which are the male sex hormones, and the *ovaries* produce estrogen and progesterone, the female sex hormones. The secretion of these hormones is under the control of the gonadotropic hormones produced by the anterior pituitary.

The sex hormones bring about the secondary sex characteristics of males and females. Among other traits, males have greater muscular strength than females. As discussed in the reading on page 401, athletes and others sometimes take so-called **anabolic steroids,** which are synthetic steroids that mimic the action of testosterone, in order to improve their strength and physique. Unfortunately, this practice is accompanied by harmful side effects.

The Thymus

The **thymus** is a lobular gland that lies in the upper thorax (fig. 19.1). This organ reaches its largest size and is most active during childhood. With aging, the organ gets smaller and becomes fatty. Certain lymphocytes that originate in the bone marrow and then pass through the thymus are transformed into T cells (p. 260). The thymus produces various hormones called *thymosins,* which aid the differentiation of T cells and may stimulate immune cells in general. There is hope that these hormones can be used in conjunction with lymphokine therapy to restore or to stimulate T cell function in patients suffering from AIDS or cancer.

The Pineal Gland

The **pineal gland** produces the hormone called melatonin primarily at night. In fishes and amphibians, the pineal gland is located near the surface of the

Inside a glitzy health club in an affluent North Shore suburb of Boston, 19-year-old Matthew Creighton stands in front of a mirror that covers an entire wall and inspects his bulging biceps.

Creighton has been pumping iron for 6 years, going through the same routine 5 mornings a week. He began entering weightlifting tournaments last year and quickly realized that good training alone was not going to make him very competitive. On a September morning last year, Creighton met a man outside a health club and for $50 bought his first bottle of steroids. He has been a regular user since. . . .

The Growing Threat of Steroids

While much of the attention surrounding the nation's drug epidemic has focused on cocaine and heroin, muscle-enhancing drugs such as steroids are being used in steadily increasing numbers, and especially alarming is their growing popularity among youngsters.

Because steroids by and large have not been the target of federal law-enforcement authorities, there are no accurate figures on the number of steroid users. But the best estimate, according to federal officials, is that 1 million—3 million Americans use steroids, a figure that has increased steadily since the early 1970s.

The uproar surrounding the use of steroids by Canadian sprinter Ben Johnson in the Seoul Olympics has underscored an issue that has received little public attention in the past. For many years, steroids were used primarily by bodybuilders, weight-lifters, and other athletes, such as professional football players, to help them in competition. But in recent years, steroid use has expanded to fitness buffs, high school and college athletes, and skinny youngsters wanting to gain bulk quickly.

Anabolic steroids are synthetic hormones taken either orally or by injection. They were developed in the 1930s to prevent muscle atrophy in patients with debilitating illnesses. In some cases, they

also were given to burn victims and surgery patients to speed recovery. . . .

Steroids, which can be dispensed legally only through a prescription, also have been used to treat certain forms of rare anemia and breast cancer.

But those using steroids illegally to enhance their appearance take them in large quantities, sometimes as much as 40 or 50 times the recommended dosage, according to FDA officials. Prolonged use of such a large quantity can lead to stunted growth in youngsters and high blood cholesterol. In men, it can result in baldness, acne, shrunken testes, feminized breasts, and infertility.

In women, who normally produce very low levels of testosterone and therefore gain much more from steroids, such overuse can promote facial hair, deepening of the voice, and an enlarged clitoris.

While scientific research has not been exhaustive, some studies have linked steroid abuse to cancer of the liver, prostate, and testes, as well as to kidney diseases and atherosclerosis.

"No one has really conclusively determined the long-term effects of steroid use because most doctors don't want to appear to condone a practice many consider unhealthy," said Dr. Gloria Troendle, senior medical officer at the FDA (Federal Drug Administration).

But another FDA official in Washington, who asked not to be identified, said, "It seems the evidence is becoming clear that youngsters who use anabolic steroids in large doses for 2 or 3 months face the possibility of dying in their 30s or 40s."

Growing evidence also points to another conclusion that steroids, which some in the medical community say appear to be addictive, can damage the mind. In psychotic side effects, sometimes referred to as " 'roid mania," users of large quantities of steroids have experienced mania, wild aggression, and delusions, said Dr. Harrison Pope, a psychiatrist at McLean Hospital in Belmont who has studied the effects of steroids on the mind.

One patient Pope examined had a friend videotape him while he deliberately drove a car into a tree at 35 mph. "My hunch is that we are seeing only a small part of . . . the magnitude of the psychiatric effects of steroids," Pope said in a recent interview in *Currents*, a medical trade journal.

Steroids had their first nonmedical use in World War II when Nazi doctors gave them to soldiers in an attempt to make them more aggressive. Following the war, the Soviet Union and other Eastern European nations began dispensing steroids to their athletes. In the 1950s, an American doctor working with the U.S. weightlifting team learned that Soviet athletes were using steroids and markedly improving their performance. The doctor introduced steroids to lifters at the York (Pa.) Barbell Club, where they were an immediate hit.

In the 1960s, as their dangerous side effects started being documented, steroids fell into disfavor in the medical community. Since then, the FDA has banned most steroids. Currently, only about a dozen are approved for limited medical use.

The limited availability of steroids by prescription has helped trigger a flourishing black market. Smuggled into the United States from Mexico and to a lesser degree from Eastern Europe, the illicit trafficking has become a $100-million-a-year business, according to federal authorities.

Steroids are sold through the mail or in gyms and health clubs, federal authorities said. "You've got kids in the club making maybe $5 or $6 an hour, but if they're selling steroids, they can make as much as $900-$1,000 a week," said the operator of several health clubs in the Los Angeles area who did not want to give his name. . . .

Congress includes a provision raising the penalty for the distribution of steroids from a misdemeanor to a felony, with jail terms of 3 years for dispensing to adults and 6 years for dispensing to children. Still, most medical, sports, and law-enforcement officials agree that the nation's craving for steroids will not subside soon without more stringent law enforcement and better education about the dangers of the drugs. . .

Despite the health risks, steroids will continue to be popular among younger users who take them to "feel better about

Continued on next page

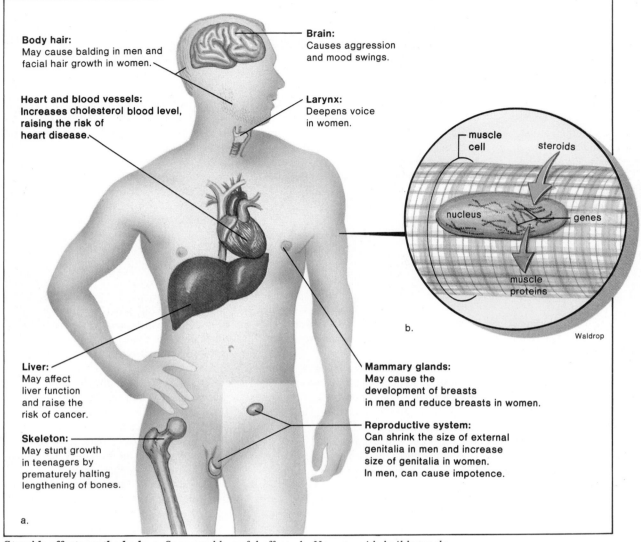

Body hair:
May cause balding in men and facial hair growth in women.

Heart and blood vessels:
Increases cholesterol blood level, raising the risk of heart disease.

Brain:
Causes aggression and mood swings.

Larynx:
Deepens voice in women.

muscle cell

steroids

nucleus

genes

muscle proteins

b.

Waldrop

Liver:
May affect liver function and raise the risk of cancer.

Skeleton:
May stunt growth in teenagers by prematurely halting lengthening of bones.

Mammary glands:
May cause the development of breasts in men and reduce breasts in women.

Reproductive system:
Can shrink the size of external genitalia in men and increase size of genitalia in women. In men, can cause impotence.

a.

Steroids effects on the body. a. *Suspected harmful effects.* b. *How steroids build muscle.*

themselves," said Dr. Jack Freinhar, a psychiatrist at Del Amo Hospital in Torrance, Calif., who treats adolescent steroid abusers.

"Physical risk doesn't matter when you're talking about saving the self," said Freinhar, who believes steroids can become an addiction he calls "reverse anorexia."

"With anorexia, one is never thin enough. In our culture, there's this push to be muscular. Most of the youngsters I treat have had a deprived childhood and they have holes in the self. They need acceptance. One way is to look good, and steroids provide a fast method. But they can

never get enough because each time they look in the mirror they still see themselves as too thin."

"The Growing Threat of Steroids," a three-part series by John Powers and Diego Rabadeneira, *The Boston Globe*, October 1988. Reprinted courtesy of *The Boston Globe*.

body and is a "third eye" that can receive light rays directly. In mammals, the pineal gland is located in the third ventricle of the brain and cannot receive direct light signals. However, it does receive nerve impulses from the eyes by way of the optic tract.

As discussed in the reading on page 403, most animals, including humans, go through daily cycles called circadian rhythms, and it is believed that the secretion of melatonin may be involved in regulating these cycles, particularly the sleep cycle. Physicians at the University of Texas are conducting an experiment with a melatonin pill. They want to see if taking the pill at 2 A.M. will set a person's body clock to local time and ease the symptoms of jet lag.

Certain behaviors of animals reoccur at regular intervals. Behaviors that occur on a daily basis are said to have a **circadian rhythm**. For example, some animals, like humans, usually are active during the day and sleep at night. Others, such as bats, sleep during the day and hunt at night. There are also behaviors that occur on a yearly basis that are called circannual behaviors. For example, birds migrate south in the fall, and the young of many animals are born in the spring.

Originally, it was assumed that environmental changes, such as the coming of

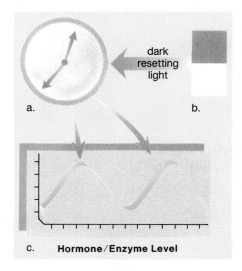

Biological clock system. *Biological clock systems have three components: (a) an internal timekeeper, (b) a means of detecting light/dark periods, (c) a method of communication that eventually brings about a behavioral response.*

Rhythmic or Cyclic Behavior

night or the length of the day, directly controlled cyclical behavior in animals, but now it is known that cyclical behaviors occur even when the associated stimulus (daylight or darkness) is lacking. For example, fiddler crabs are dark in color during the day and light in color at night, even when kept in a constant environment, but if the crabs are kept in the constant environment indefinitely, the timing of the daily change tends to drift and to become out of synchronization with the natural cycle. For this reason, it has been suggested that rhythmic behavior is under the control of an innate, internal biological clock that runs on its own but is reset by external

stimuli. Aside from keeping time, a biological clock also must be able to bring about the change in behavior. In the fiddler crab, for example, it must be able to stimulate the processes that cause the shell to change color. Therefore, a biological clock system needs (see also figure):

1. a time-keeping mechanism that *keeps time* independently of external stimuli (i.e., a minute is always a minute);
2. a receptor that is sensitive to light/dark periods and *can reset* the clock for circadian rhythms or indicate a change in the length of the day/night for circannual rhythms;
3. a communication mechanism by which the clock *induces* the appropriate behavior.

A review of the discussion concerning flowering in plants shows that the last 2 components of a biological clock system have been identified tentatively. Phytochrome is believed to be the receptor sensitive to light and dark periods, and plant hormones are believed to be the means by which flowering is induced. The time-keeping mechanism has not been identified, however.

In animals, we know that melatonin is produced at night by the pineal gland. It is possible that the level of melatonin controls both circadian and circannual behavior. Some animals reproduce in spring, when melatonin levels are high; therefore, the amount of melatonin produced begins to increase in the fall and to decrease in the spring. Melatonin is believed to inhibit development of the reproductive organs, which would account for why some animals reproduce in the spring. For animals in which the pineal gland is the "third eye," not only the clock but also the receptor are presumed to be in the pineal gland.

Many animals also go through yearly cycles. For example, the reproductive organs in hamsters shrink in size and weight as the amount of light decreases in fall and winter. This inhibits reproduction at a time of year when food is not plentiful. In humans, it has been noted that children with brain tumors that destroy the pineal gland experience early puberty. Therefore, it is thought that the pineal gland may be involved in regulating human sexual development.

Still Other Glands

Even organs that traditionally are not considered to be endocrine glands have been found to secrete hormones. For example, we previously discussed the hormones produced by the stomach and the small intestine. As mentioned on page 397, the heart produces *atrial natriuretic hormone,* which helps to regulate the sodium and water balance of the body. Not only does it promote renal excretion of sodium and water, it also inhibits the release of renin and the hormone aldosterone and ADH. Atrial natriuretic factor is a 28 amino acid peptide that is released not only by the atria but also by the aortic arch, ventricles, lungs, and pituitary gland in response to increases in blood pressure.

Figure 19.19

The 3 categories of environmental signals. Pheromones are chemical messengers that act at a distance between individuals. Endocrine hormones and neurosecretions typically are carried in the bloodstream and act at a distance within the body of a single organism. Some chemical messengers have local effects only; they pass between cells that are adjacent to one another. This, of course, includes neurotransmitter substances.

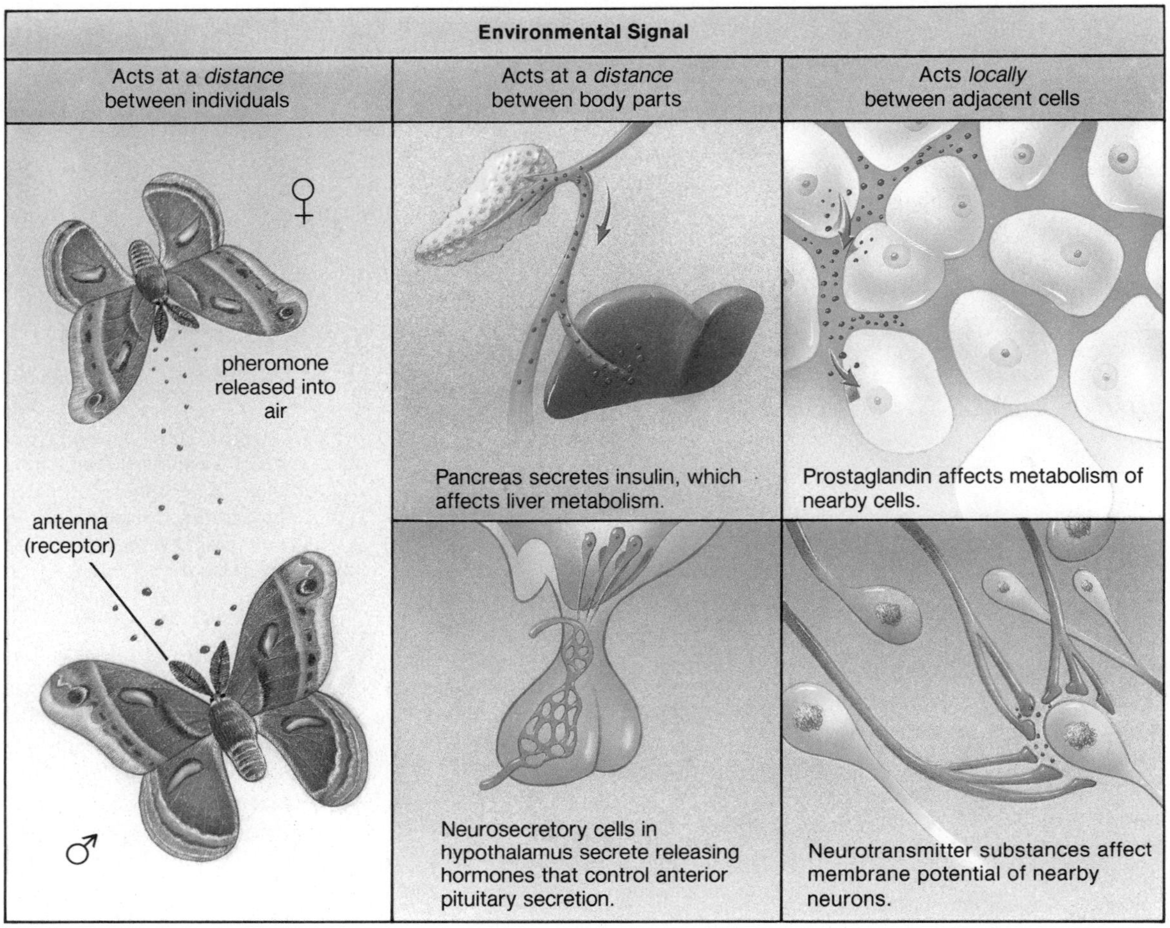

Environmental Signal

Acts at a *distance* between individuals	Acts at a *distance* between body parts	Acts *locally* between adjacent cells
♀ pheromone released into air antenna (receptor) ♂	Pancreas secretes insulin, which affects liver metabolism. Neurosecretory cells in hypothalamus secrete releasing hormones that control anterior pituitary secretion.	Prostaglandin affects metabolism of nearby cells. Neurotransmitter substances affect membrane potential of nearby neurons.

Environmental Signals

In this chapter, we concentrated on describing the functions of the human endocrine glands and their hormonal secretions. We already know that hormones are only one type of chemical messenger or environmental signal between cells. In fact, the concept of the environmental signal now has been broadened to include at least the following 3 different categories of messengers (fig. 19.19).

Environmental signals that act at a distance between individuals. Many organisms release chemical messengers, called **pheromones,** into the air or in externally deposited body fluids. These are intended to be messages for other members of the species. For example, ants lay down a pheromone trail to direct other ants to food, and the female silkworm moth releases bombykol, a sex attractant that is received by male moth antennae even several miles away. This chemical is so potent that it has been estimated that only 40 out of 40,000 receptors on the male antennae need to be activated in order for the male to respond. Mammals, too, release pheromones; the urine of dogs serves as a territorial marker, for example. Studies are being conducted to determine if humans also have pheromones. Certain studies have suggested that humans also respond to pheromones. For example, investigators have observed that women who live in close quarters tend to have menstrual cycles that coincide. They reason that this might be caused by a pheromone but they don't know what the pheromone might be or how it might exert its effect.

Environmental signals that act at a distance between body parts. This category includes the endocrine secretions, which traditionally have been called hormones. It also includes the secretions of the neurosecretory cells in the hypothalamus—the production and action of ADH and oxytocin illustrate the close relationship between the nervous system and the endocrine system. Neurosecretory cells produce these hormones, which are released when these cells receive nerve impulses. As another example of the overlap between the nervous and endocrine systems, consider that endorphins on occasion travel in the bloodstream, but they act on nerve cells to alter their membrane potential. Also, norepinephrine is secreted by the adrenal medulla but is also a neurotransmitter in the sympathetic nervous system.

Environmental signals that act locally between adjacent cells. Neurotransmitter substances belong in this category, as do substances that are sometimes called local hormones. For example, when the skin is cut, histamine is released by mast cells and promotes the inflammatory response. Nearby capillaries dilate, allowing more blood to reach the site, and they become leaky so that cells of the immune system are able to leave the capillaries and to attack incoming microorganisms.

Prostaglandins are local messengers derived from cell membrane phospholipids. There are many different types of prostaglandins produced by many different tissues. In the uterus, certain prostaglandins cause muscles to contract; therefore, they are implicated in the pain and discomfort of menstruation in some women. (Antiprostaglandin therapy is useful in these cases.) On the other hand, certain prostaglandins are being used to treat ulcers because they reduce gastric secretion, to treat hypertension because they lower blood pressure, and to prevent thrombosis because they inhibit platelet aggregation. Because the different prostaglandins can have contrary effects, however, it has been very difficult to standardize their use, and in most instances, prostaglandin therapy still is considered experimental.

Also, *growth factors* have a local effect that influences the development of adjacent cells. For example, nerve growth factor affects the development of the nervous system, and epidermal growth factor has effects on various types of cells.

Redefinition of a Hormone

Traditionally, a hormone was considered to be a secretion of an endocrine gland that was carried in the bloodstream to a target organ. In recent years, some scientists have broadened the definition of a hormone to include *all* types of chemical messengers. This change seemed necessary because those chemicals traditionally considered to be hormones now have been found in all sorts of tissues in the human body. For example, it is impossible for insulin produced by the pancreas to enter the brain because of the blood-brain barrier—a tight fusion of endothelial cells of the capillary walls that prevents passage of larger molecules like peptides. Yet, insulin has been found in the brain. It now appears that the brain cells themselves can produce insulin, which is used locally to influence the metabolism of adjacent cells. Also, some chemicals identical to the hormones of the endocrine system have been found in lower organisms, even in bacteria! A moment's thought about the evolutionary process helps to explain this; these regulatory chemicals may have been present in the earliest cells and only became specialized as hormones as evolution proceeded.

19.2 Critical Thinking

Environmental signals alter the behavior of target cells.

1. How would you modify this definition to be consistent with figure 19.19? Why?
2. What makes a target cell sensitive to a particular environmental signal?
3. Why is the term *behavior* a good one to use in the definition of an environmental signal? Give examples of change in behavior of the target cell, the organ, or the organism.
4. When is the environmental signal insulin present? When is an environmental signal like a neurotransmitter substance released?

Summary

Hormones are chemical messengers having a metabolic effect on cells. The hypothalamus produces the hormones ADH and oxytocin, released by the posterior pituitary. In addition, the hypothalamus produces releasing hormones that control the production of hormones by the anterior pituitary. In addition to GH and LTH, which affect the body directly, the anterior pituitary secretes hormones that control other endocrine glands: TSH stimulates the thyroid to release thyroxin; ACTH stimulates the adrenal cortex to release glucocorticoids; gonadotropins stimulate the gonads to release the sex hormones. The secretion of hormones is controlled by negative feedback. In the case of the hormones just mentioned, this mechanism involves the hypothalamus and the anterior pituitary, in addition to the hormonal gland in question. Other hormones are PTH, the mineralocorticoids, epinephrine, and insulin (table 19.1).

The most common illness due to hormonal imbalance is diabetes mellitus. This condition occurs when the islets of Langerhans within the pancreas fail to produce insulin. Insulin promotes the uptake of glucose by the cells and the conversion of glucose to glycogen, thereby lowering the blood glucose levels. Without the production of insulin, the blood sugar level rises, and some of it spills over into the urine. The real problem in diabetes mellitus, however, is acidosis, which may cause the death of the diabetic if therapy is not begun.

There are 3 categories of chemical messengers: those that act at a distance between individuals (pheromones); those that act at a distance within the individual (traditional endocrine hormones and secretions of neurosecretory cells); and local messengers (such as prostaglandins and neurotransmitter substances). Since there is great overlap between these categories, perhaps the definition of a hormone now should be expanded to include all of them.

Study Questions

1. Give the location in the human body of all the major endocrine glands. Name the hormones secreted by each gland, and describe their chief function. (p. 386–400)
2. Give a definition of endocrine hormones that includes their most likely source, how they are transported in the body, and how they are received. What does "target" organ mean? (p. 387)
3. Categorize endocrine hormones according to their chemical makeup. (p. 387)
4. Tell how the 2 major types of hormones influence the metabolism of the cell. (p. 387–88)
5. Explain the relationship of the hypothalamus to the posterior pituitary and to the anterior pituitary. (p. 388–91)
6. Explain the concept of negative feedback, and give an example involving ADH. (p. 389)
7. Explain why the anterior pituitary can be called the master gland. (p. 391) Give an example of the three-tiered relationship between the hypothalamus, the anterior pituitary, and other endocrine glands (p. 393).
8. Draw a diagram to explain the contrary actions of insulin and glucagon. Use your diagram to explain the symptoms of type I diabetes mellitus. (p. 399)
9. Categorize chemical messengers into 3 groups, and give examples of each group. (p. 403)
10. Give examples to show that there is an overlap between the mode of operation of the nervous system and that of the endocrine system. Explain why the traditional definition of a hormone may need to be expanded. (p. 405)

Objective Questions

1. The hypothalamus _____ the hormones _____ and _____ , released by the posterior pituitary.
2. The _____ secreted by the hypothalamus controls the anterior pituitary.
3. Generally, hormone production is self-regulated by a _____ mechanism.
4. Growth hormone is produced by the _____ pituitary.
5. Simple goiter occurs when the thyroid is producing _____ (too much or too little) _____ .
6. ACTH, produced by the anterior pituitary, stimulates the _____ of the adrenal glands.
7. An overproductive adrenal cortex results in the condition called _____ .
8. Parathyroid hormone increases the level of _____ in blood.
9. Type I diabetes mellitus is due to a malfunctioning _____ , while type II diabetes is due to limited uptake of insulin by _____ .
10. Prostaglandins are not carried in _____ as are hormones secreted by the endocrine glands.

Selected Key Terms

acromegaly (ak″ro-meg′ah-le) a condition resulting from an increase in GH production after adult height has been achieved. *391*

anterior pituitary (an-te′re-or pĭ-tu′ĭ-tār″e) a portion of the pituitary gland that produces 6 types of hormones and is controlled by hypothalamic-releasing hormones. *388*

cretinism (kre′tin-izm) a condition resulting from a lack of thyroid hormone in an infant. *394*

diabetes insipidus (di″ah-be′tēz in-sip′ĭ-dus) condition characterized by an abnormally large production of urine due to a deficiency of ADH. *390*

diabetes mellitus (di″ah-be′tēz mĕ-li′tus) condition characterized by a high blood glucose level and the appearance of glucose in the urine due to a deficiency of insulin production or uptake by cells. *398*

endocrine gland (en′do-krin gland) a gland that secretes hormones directly into blood or body fluids. *386*

exophthalmic goiter (ek″sof-thal′mik goi′ter) an enlargement of the thyroid gland accompanied by an abnormal protrusion of the eyes. *395*

islets of Langerhans (i′lets uv lahng′er-hanz) distinctive groups of cells within the pancreas that secrete insulin and glucagon. *398*

myxedema (mik″sĕ-de′mah) a condition resulting from a deficiency of thyroid hormone in an adult. *394*

pheromone (fer′o-mōn) a chemical substance secreted by one organism that influences the behavior of another. *404*

posterior pituitary (pos-tēr′e-or pĭ-tu′ĭ-tār″e) back lobe of the pituitary gland that secretes ADH and oxytocin produced by the hypothalamus. *388*

simple goiter (sim′p'l goi′ter) condition in which an enlarged thyroid produces low levels of thyroxin. *394*

Further Readings for Part Three

Aoki, C., and P. Siekevitz. December 1988. Plasticity in brain development. *Scientific American.*

Barr, M. L. 1983. *The human nervous system, an anatomical viewpoint.* 4th ed. New York: Harper & Row Publishers, Inc.

Bloom, F. E. October 1981. Neuropeptides. *Scientific American.*

Brown, M. S., and J. L. Goldstein. November 1984. How LDL receptors influence cholesterol and atherosclerosis. *Scientific American.*

Buisseret, P. D. August 1982. Allergy. *Scientific American.*

Doolittle, R. F. December 1981. Fibrinogen and fibrin. *Scientific American.*

Cantin, M., and J. Genest. February 1986. The heart as an endocrine gland. *Scientific American.*

Cohen, I. R. April 1988. The self, the world and autoimmunity. *Scientific American.*

Crapo, L. 1985. *Hormones, the messengers of life.* New York: W. H. Freeman.

Eckert, R., and D. Randall. 1983. *Animal physiology.* 2d ed. San Francisco: W. H. Freeman.

Feder, M. E., and W. W. Burggren. November 1985. Skin breathing in vertebrates. *Scientific American.*

Fine, A. August 1986. Transplantation in the central nervous system. *Scientific American.*

Goldstein, G. W., and A. L. Betz. September 1986. The blood-brain barrier. *Scientific American.*

Guyton, A. C. 1984. *Physiology of the human body.* 6th ed. Philadelphia: W. B. Saunders.

Hegarty, V. 1988. *Decisions in nutrition.* St. Louis, Mo.: Times Mirror/Mosby College Publishing.

Hole, J. W. 1987. *Human anatomy and physiology.* 4th ed. Dubuque, Ia.: Wm. C. Brown Publishers.

Hudspeth, A. J. January 1983. The hair cells of the inner ear. *Scientific American.*

Kennedy, R. C. July 1986. Anti-idiotypes and immunity. *Scientific American.*

Koretz, J. F., and G. H. Handelman. July 1988. How the human eye focuses. *Scientific American.*

Llinas, R. R. October 1982. Calcium in synaptic transmission. *Scientific American.*

Loeb, G. E. February 1985. The functional replacement of the ear. *Scientific American.*

Masland, R. H. December 1986. The functional architecture of the retina. *Scientific American.*

Mishkin, M., and T. Appenzeller. June 1987. The anatomy of memory. *Scientific American.*

Morell, P., and W. T. Norton. May 1980. Myelin. *Scientific American.*

Norman, D. A. 1982. *Learning and memory.* San Francisco: W. H. Freeman.

Orci, L., et al. September 1988. The insulin factory. *Scientific American.*

Robinson, T. F., et al. June 1986. The heart as a suction pump. *Scientific American.*

Rose, N. R. February 1981. Autoimmune diseases. *Scientific American.*

Rubenstein, E. March 1980. Diseases caused by impaired communication among cells. *Scientific American.*

Schnapt, J. L., and D. A. Baylor. April 1987. How photoreceptor cells respond to light. *Scientific American.*

Scientific American. October 1988. Entire issue is devoted to AIDS.

Snyder, S. H. October 1985. The molecular basis of communication between cells. *Scientific American.*

Stryer, L. July 1987. The molecules of visual excitation. *Scientific American.*

Tonegawa, S. October 1985. The molecules of the immune system. *Scientific American.*

Vander, A. J. 1985. *Human physiology: The mechanisms of body function.* 4th ed. New York: McGraw-Hill.

Wurtman, R. J. April 1982. Nutrients that modify brain function. *Scientific American.*

Young, J., and Z. Cohn. January 1988. How killer cells kill. *Scientific American.*

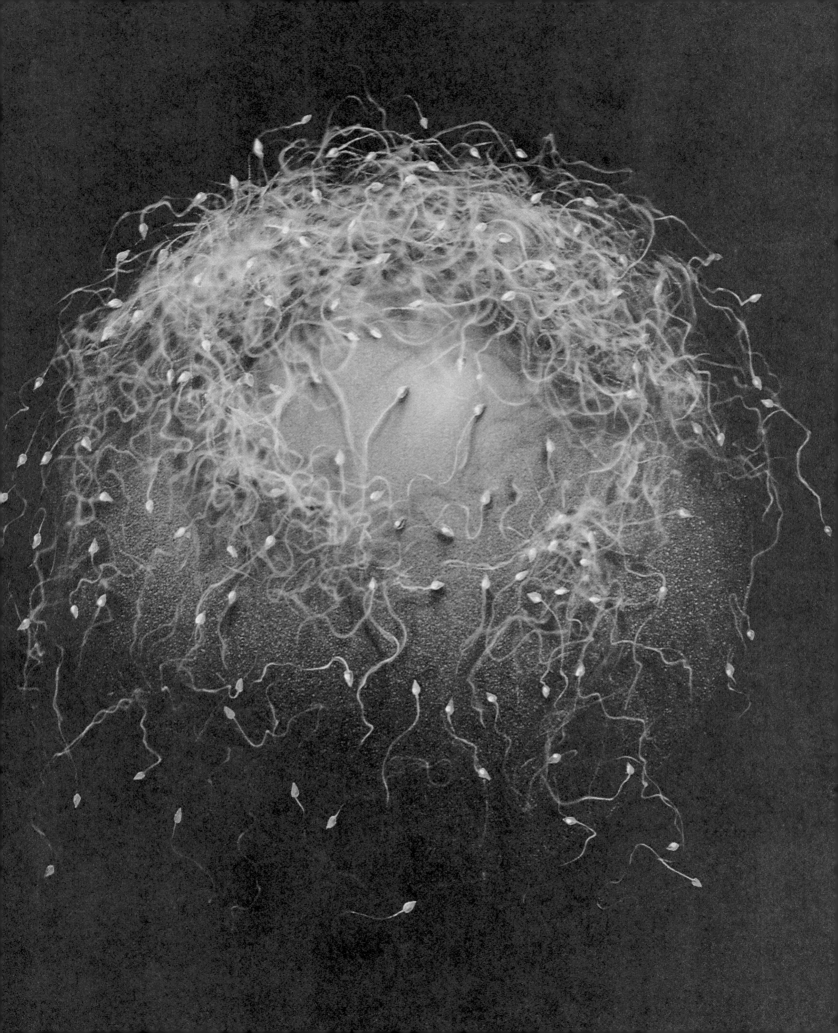

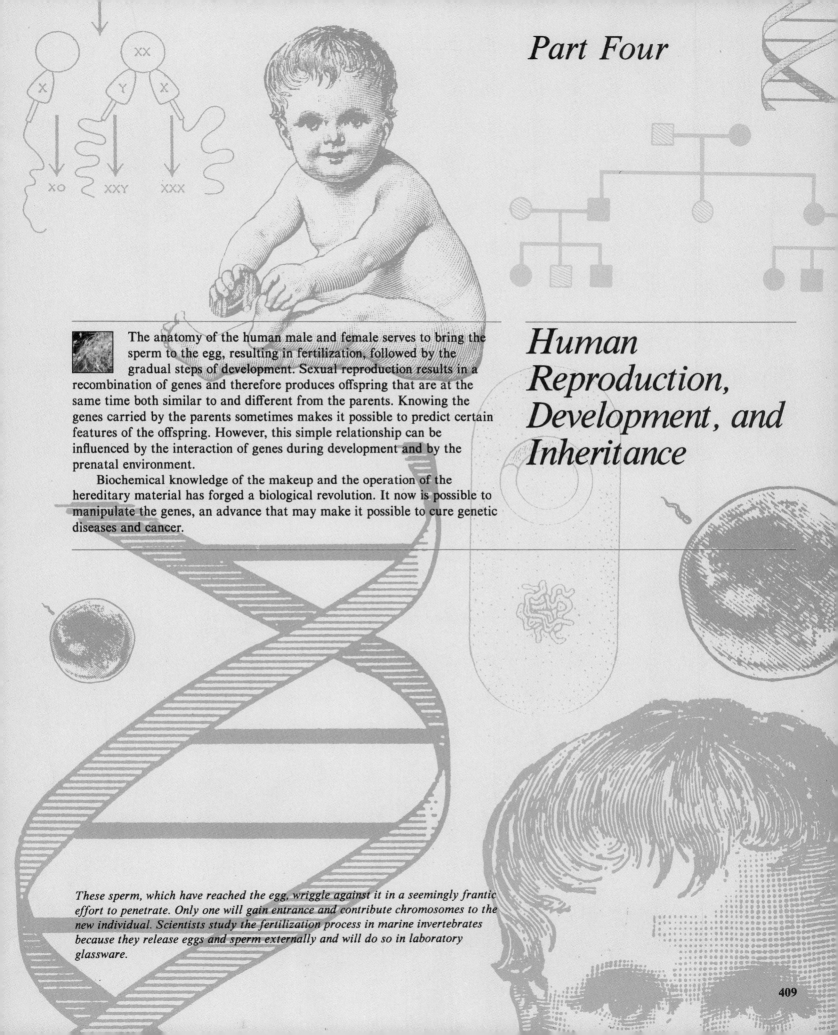

Part Four

Human Reproduction, Development, and Inheritance

The anatomy of the human male and female serves to bring the sperm to the egg, resulting in fertilization, followed by the gradual steps of development. Sexual reproduction results in a recombination of genes and therefore produces offspring that are at the same time both similar to and different from the parents. Knowing the genes carried by the parents sometimes makes it possible to predict certain features of the offspring. However, this simple relationship can be influenced by the interaction of genes during development and by the prenatal environment.

Biochemical knowledge of the makeup and the operation of the hereditary material has forged a biological revolution. It now is possible to manipulate the genes, an advance that may make it possible to cure genetic diseases and cancer.

These sperm, which have reached the egg, wriggle against it in a seemingly frantic effort to penetrate. Only one will gain entrance and contribute chromosomes to the new individual. Scientists study the fertilization process in marine invertebrates because they release eggs and sperm externally and will do so in laboratory glassware.

Chapter 20

The Reproductive System

Chapter Concepts

1 The male reproductive system is designed for the continuous production of a large number of sperm within a fluid medium.

2 The female reproductive system is designed for the monthly production of an egg and the preparation of the uterus for possible implantation of the fertilized egg.

3 Hormones control the reproductive process and the sex characteristics of the individual.

4 Birth control measures vary in effectiveness from those that are very effective to those that are minimally effective.

5 There are alternative methods of reproduction today, including in vitro fertilization followed by artificial implantation.

6 There are several serious and prevalent sexually transmitted diseases.

Chapter Outline

The Male Reproductive System
  Testes
  The Genital Tract
  Orgasm in Males
  The Regulation of Male Hormone
    Levels
The Female Reproductive System
  Ovaries
  The Genital Tract
  Orgasm in Females
  The Regulation of Female Hormone
    Levels
The Control of Reproduction
  Birth Control
  Infertility
Sexually Transmitted Diseases
  AIDS
  Genital Herpes
  Genital Warts
  Gonorrhea
  Chlamydia
  Syphilis

In advanced forms of sexual reproduction, there are 2 types of gametes (sex cells), both of which contribute the same number of chromosomes to the new individual (fig. 4.3, p. 82). The sperm are small and swim to the stationary egg, a much larger cell that contributes cytoplasm and organelles to the zygote (fig. 20.1). It seems reasonable that there are a large number of sperm to ensure that a few find the egg. In humans, after puberty the male continually produces sperm, which are stored temporarily before being released.

The Male Reproductive System

Figure 20.2 shows the reproductive system of the male, and table 20.1 lists the anatomical parts of this system.

Table 20.1 Male Reproductive System

Organ	Function
Testis	Produces sperm and sex hormones
Epididymis	Stores sperm as they mature
Vas deferens	Conducts and stores sperm
Seminal vesicle	Contributes to seminal fluid
Prostate gland	Contributes to seminal fluid
Urethra	Conducts sperm
Cowper's gland	Contributes to seminal fluid
Penis	Organ of copulation

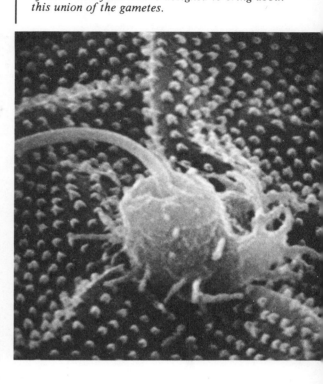

Figure 20.1

Fertilization. A single sperm enters the egg and then a new life begins. The reproductive systems of males and females are designed to bring about this union of the gametes.

Figure 20.2

Side view of the male reproductive system. Trace the path of the genital tract from a testis to the exterior. The seminal vesicles, the Cowper's gland, and the prostate gland produce seminal fluid and do not contain sperm. Notice that the penis in this drawing is not circumcised since the foreskin is present.

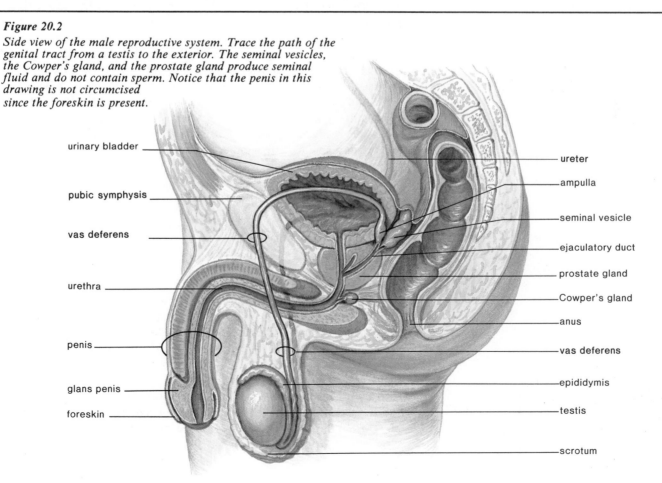

411

Testes

The **testes** lie outside the abdominal cavity of the male within the **scrotum.** The testes begin their development inside the abdominal cavity but descend into the scrotal sacs during the last 2 months of fetal development. If by chance the testes do not descend and the male is not treated or operated on to place the testes in the scrotum, sterility—the inability to produce offspring—usually follows. This is because the internal temperature of the body is too high to produce viable sperm.

Seminiferous Tubules

Fibrous connective tissue forms the wall of each testis and divides it into lobules (fig. 20.3). Each lobule contains 1–3 tightly coiled **seminiferous tubules** that have a combined length of approximately 250 m. A microscopic cross section through a tubule shows a lobule is packed with cells undergoing spermatogenesis (fig. 20.3c). These cells are derived from undifferentiated germ cells called spermatogonia (singular, spermatogonium) that lie just inside the outer wall and divide mitotically, always producing new spermatogonia. Some newly formed spermatogonia move away from the outer wall to increase in size and become primary spermatocytes that undergo meiosis, a type of cell division described in chapter 4. Although these cells have 46 chromosomes, they divide to give secondary spermatocytes, each with 23 chromosomes. Secondary spermatocytes divide to give spermatids, also with 23 chromosomes, but single stranded. Spermatids then differentiate into spermatozoa, or mature sperm. Also present in the tubules are the *sertoli* or nurse *cells,* that support, nourish, and regulate the spermatogenic cells.

Sperm   The mature sperm, or spermatozoan (fig. 20.3d), has 3 distinct parts: a head, a middle piece, and a tail. The *tail* contains the 9 + 2 pattern of microtubules typical of cilia and flagella (fig. 2.14, p. 58), and the *middle piece* contains energy-producing mitochondria. The *head* contains the 23 chromosomes within a nucleus. The tip of the nucleus is covered by a cap called the **acrosome,** which is believed to contain enzymes needed for fertilization. The human egg is surrounded by several layers of cells and a mucoprotein substance. The acrosome enzymes are believed to aid the sperm in reaching the surface of the egg and allowing a single sperm to penetrate the egg.

Each acrosome may contain such a minute amount of enzyme that it requires the action of many sperm to allow just one to actually penetrate the egg. This may explain why so many sperm are required for the process of fertilization. A normal human male usually produces several hundred million sperm per day, an adequate number for fertilization. Sperm are produced continually throughout a male's reproductive life.

In males, spermatogenesis occurs within the seminiferous tubules of the testes. Sperm have a head capped by an acrosome, where 23 chromosomes reside in the nucleus, a mitochondria-containing middle piece, and a tail with a 9 + 2 pattern of microtubules.

Interstitial Cells

The male sex hormones, the androgens, are secreted by cells that lie between the seminiferous tubules. Therefore, they are called **interstitial cells** (fig. 20.3b). The most important of the androgens is testosterone, whose functions are discussed on page 416.

The Genital Tract

Sperm are produced in the testes, but they mature in the **epididymis** (fig. 20.2), a tightly coiled tubule about 5–6 m (17 ft) in length that lies just outside each testis. During the 2–4-day maturational period, the sperm develop their characteristic swimming ability. Each epididymis joins with a **vas (ductus) deferens,**

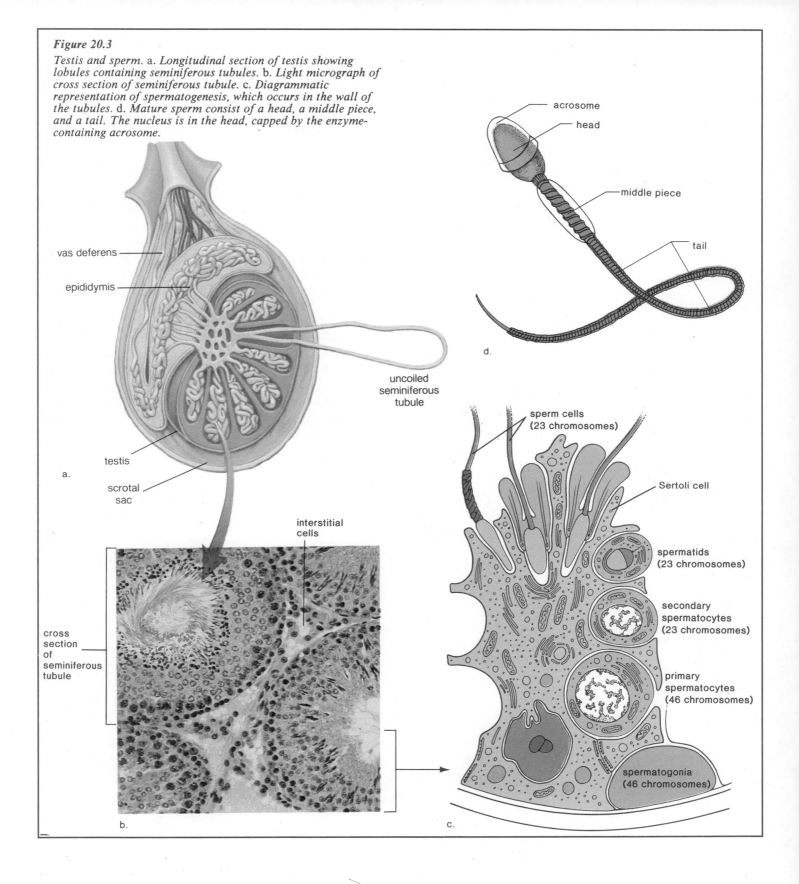

Figure 20.3

Testis and sperm. a. Longitudinal section of testis showing lobules containing seminiferous tubules. b. Light micrograph of cross section of seminiferous tubule. c. Diagrammatic representation of spermatogenesis, which occurs in the wall of the tubules. d. Mature sperm consist of a head, a middle piece, and a tail. The nucleus is in the head, capped by the enzyme-containing acrosome.

vas deferens

epididymis

testis

scrotal sac

a.

uncoiled seminiferous tubule

acrosome

head

middle piece

tail

d.

cross section of seminiferous tubule

interstitial cells

b.

sperm cells (23 chromosomes)

Sertoli cell

spermatids (23 chromosomes)

secondary spermatocytes (23 chromosomes)

primary spermatocytes (46 chromosomes)

spermatogonia (46 chromosomes)

c.

which ascends through a canal called the *inguinal canal* and enters the abdomen, where it curves around the bladder and empties into the urethra. Sperm are stored in the first part of a vas deferens. They pass from each vas deferens into the urethra only when ejaculation (p. 415) is imminent.

Spermatic Cords

The testes are suspended in the scrotum by the *spermatic cords,* each of which consists of fibrous connective tissue and muscle fibers that enclose the vas deferens, the blood vessels, and the nerves. The region of the inguinal canal, where the spermatic cord passes into the abdomen, remains a weak point in the abdominal wall. As such, it is frequently the site of hernias. A **hernia** is an opening or separation of some part of the abdominal wall through which a portion of an internal organ, usually the intestine, protrudes.

Seminal Fluid

At the time of ejaculation, sperm leave the penis in a fluid called **seminal fluid.** This fluid is produced by 3 types of glands—the seminal vesicles, the prostate gland, and the Cowper's glands. The **seminal vesicles** lie at the base of the bladder, and each has a duct that joins with a vas deferens. The **prostate gland** is a single doughnut-shaped gland that surrounds the upper portion of the urethra just below the bladder. In older men, the prostate can enlarge and squeeze off the urethra, making urination painful and difficult. This condition can be treated medically or surgically. **Cowper's glands** are pea-sized organs that lie posterior to the prostate on either side of the urethra.

Each component of seminal fluid seems to have a particular function. Sperm are more viable in a basic solution, and seminal fluid, which is milky in appearance, has a slightly basic pH (about 7.5). Swimming sperm require energy, and seminal fluid contains the sugar fructose, which presumably serves as an energy source. Seminal fluid also contains prostaglandins, chemicals that cause the uterus to contract. Some investigators now believe that uterine contraction is necessary to help propel the sperm toward the egg.

Orgasm in Males

The **penis** (fig. 20.4) is the copulatory organ of males. The penis has a long shaft and an enlarged tip called the glans penis. At birth, the glans penis (cone-shaped terminal portion of the penis) is covered by a layer of skin called the **foreskin,** or prepuce. Gradually, over a period of 5–10 years, the foreskin separates from the glans and may be retracted. Sometime near puberty, small glands located in the foreskin and glans begin to produce an oily secretion.

Figure 20.4

Penis anatomy. a. Beneath the skin and the connective tissue lies the urethra, surrounded by erectile tissue. This tissue expands to form the glans penis, which in uncircumcised males is partially covered by the foreskin (prepuce). b. Two other columns of erectile tissue in the penis are located dorsally.

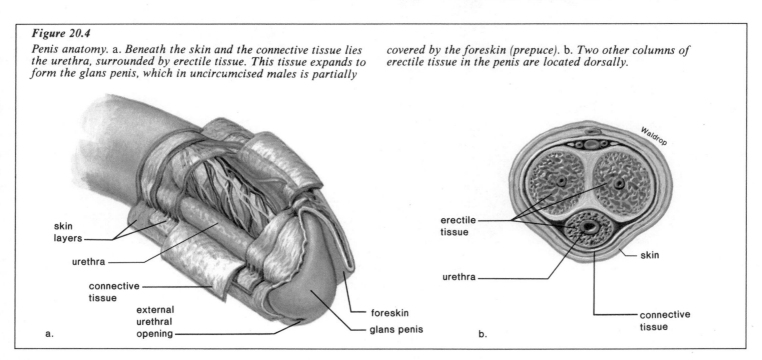

Human Reproduction, Development, and Inheritance

This secretion along with dead skin cells forms a cheesy substance known as smegma. In the child, no special cleansing method is needed to wash away smegma, but in the adult, the foreskin can be retracted to do so. **Circumcision** is the surgical removal of the foreskin usually soon after birth.

When the male is sexually aroused, the penis becomes erect and ready for intercourse. **Erection** is achieved because blood sinuses within the erectile tissue of the penis fill with blood. Parasympathetic impulses dilate the arteries of the penis, while the veins are compressed passively so that blood flows into the erectile tissue under pressure. If the penis fails to become erect, the condition is called **impotency.** There are medical and surgical remedies for impotency.

Ejaculation

As sexual stimulation intensifies, sperm enter the urethra from each vas deferens and the glands secrete seminal fluid. Sperm and seminal fluid together are called **semen.** Once semen is in the urethra, rhythmical muscle contractions cause it to be expelled from the penis in spurts. During ejaculation, a sphincter closes off the bladder so that no urine enters the urethra. (Notice that the urethra carries either urine or semen at different times.)

The contractions that expel semen from the penis are a part of male **orgasm,** the physiological and psychological sensations that occur at the climax of sexual stimulation. The psychological sensation of pleasure is centered in the brain, but the physiological reactions involve the genital (reproductive) organs and associated muscles, as well as the entire body. Marked muscular tension is followed by contraction and relaxation.

Following ejaculation and/or loss of sexual arousal, the penis returns to its normal flaccid state. After ejaculation, a male typically experiences a period of time, called the refractory period, during which stimulation does not bring about an erection. The length of the refractory period increases with age.

There may be in excess of 400 million sperm in the 3.5 ml of semen expelled during ejaculation. The sperm count can be much lower than this, however, and fertilization (fig. 20.1) still can take place.

Sperm mature in the epididymis and are stored in the vas deferens before entering the urethra just prior to ejaculation. The accessory glands (seminal vesicles, prostate gland, and Cowper's gland) produce seminal fluid. Semen, which contains sperm and seminal fluid, leaves the penis during ejaculation.

The Regulation of Male Hormone Levels

The hypothalamus has ultimate control of the testes' sexual functions because it secretes gonadotropic-releasing hormone (GnRH) that stimulates the anterior pituitary to produce the gonadotropic hormones. Two gonadotropic hormones, **FSH (follicle-stimulating hormone)** and **LH (luteinizing hormone),** are named for their function in females but exist in both sexes, stimulating the appropriate gonads in each. FSH promotes spermatogenesis in the seminiferous tubules, and LH promotes the production of testosterone in the interstitial cells. Sometimes, LH in males is given the name interstitial cell-stimulating hormone (ICSH).

The hormones mentioned are involved in a feedback process (fig. 20.5) that maintains the production of testosterone at a fairly constant level. For example, when the amount of testosterone in blood rises to a certain level, it causes the anterior pituitary to decrease its secretion of LH. As the level of testosterone begins to fall, the anterior pituitary increases its secretion of LH, and stimulation of the interstitial cells reoccurs. It should be emphasized that only minor fluctuations of testosterone level occur in the male and that the feedback mechanism in this case acts to maintain testosterone at a normal level. It long had been suspected that the seminiferous tubules produce a hormone that blocks FSH secretion. This substance, termed *inhibin,* recently has been isolated.

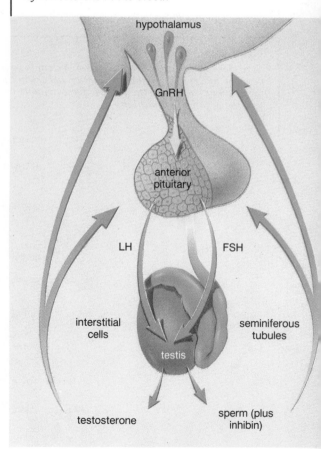

Figure 20.5

The hypothalamus–pituitary–testes control relationship. GnRH (gonadotropic-releasing hormone) stimulates the anterior pituitary to secrete the gonadotropic hormones FSH and LH. FSH stimulates the testes to produce sperm, and LH stimulates the testes to produce testosterone. Testosterone and inhibin exert negative feedback control over the hypothalamus and the anterior pituitary, and this ultimately regulates the level of testosterone in the blood.

hypothalamus

GnRH

anterior pituitary

LH

FSH

interstitial cells

seminiferous tubules

testis

testosterone

sperm (plus inhibin)

20.1 Critical Thinking

1. Using figure 20.5 as a guide, formulate a hypothesis to explain why anabolic steroids would shrink the size of the testes.
2. How might you test your hypothesis, for example in laboratory mice?
3. Would you predict that anabolic steroids raise or lower LDL blood levels (p. 234) leading to increased risk of heart disease.
4. How might you test your hypothesis, for example in laboratory mice?

Testosterone

The male sex hormone, **testosterone,** has many functions. It is essential for the normal development and functioning of the primary sex organs, those structures we just have discussed. It is also necessary for the maturation of sperm.

Greatly increased testosterone secretion at the time of puberty stimulates maturation of the penis and the testes. Testosterone also brings about and maintains the secondary sex characteristics in males that develop at the time of puberty. Testosterone causes growth of a beard, axillary (underarm) hair, and pubic hair. It prompts the larynx and the vocal cords to enlarge, causing the voice to change. It is responsible for the greater muscular strength of males, and this is the reason some athletes take supplemental amounts of *anabolic steroids,* which are either testosterone or related chemicals. The contraindications of taking anabolic steroids are discussed in the reading on page 401. Testosterone also causes oil and sweat glands in the skin to secrete; therefore, it is largely responsible for acne and body odor. Another side effect of testosterone activity is baldness. Genes for baldness probably are inherited by both sexes, but baldness is seen more often in males because of the presence of testosterone.

Testosterone is believed to be largely responsible for the sex drive. It may even contribute to the supposed aggressiveness of males.

In males, FSH promotes spermatogenesis and LH promotes testosterone production within the testes. Testosterone stimulates growth of the male genitals during puberty and is necessary for maturation of sperm and development of secondary sex characteristics.

Table 20.2 Female Reproductive System

Organ	Function
Ovary	Produces egg and sex hormones
Oviduct (fallopian tube)	Conducts egg toward uterus
Uterus (womb)	Houses developing fetus
Cervix	Contains opening to uterus
Vagina	Receives penis during copulation and serves as birth canal

The Female Reproductive System

Figure 20.6 illustrates the female reproductive system, and table 20.2 lists the anatomical parts of this system.

Figure 20.6

Side view of the female reproductive system. The ovaries produce one egg a month; fertilization occurs in the oviduct, and development occurs in the uterus. The vagina is the birth canal and the organ of copulation.

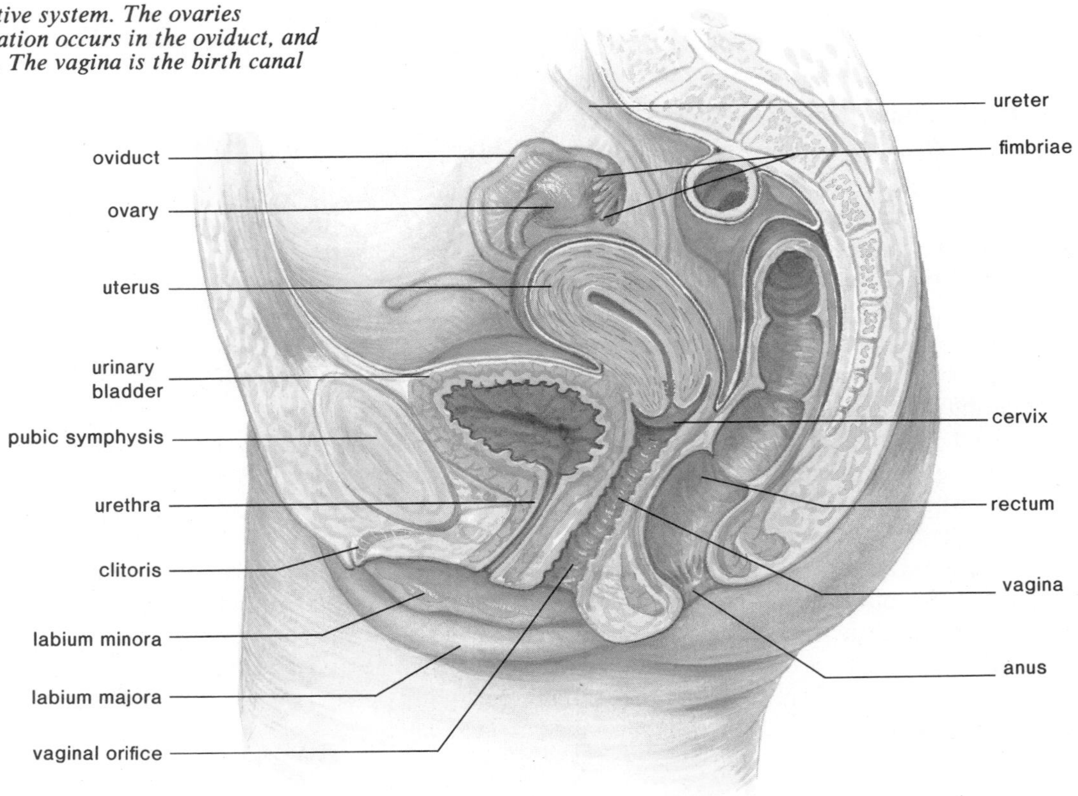

Ovaries

The **ovaries** lie in shallow depressions, one on each side of the upper pelvic cavity. A longitudinal section through an ovary shows that it is made up of an outer cortex and an inner medulla. There are many **follicles** in the cortex and each one contains an oocyte. A female is born with as many as 2 million follicles, but the number is reduced to 300,000–400,000 by the time of puberty. Only a small number of follicles (about 400) ever mature because a female usually produces only one egg per month during her reproductive years. Since oocytes are present at birth, they age as the woman ages. This is one possible reason why older women are more likely to produce children with genetic defects.

As the follicle undergoes maturation, it develops from a primary follicle to a secondary follicle to a **Graafian follicle** (fig. 20.7). In a primary follicle, the primary oocyte divides meiotically into 2 cells, each having 23 chromosomes (see fig. 4.17). One of these cells, termed the secondary oocyte, receives

Figure 20.7

Anatomy of ovary and follicle. a. As a follicle matures, the oocyte enlarges and is surrounded by a mantle of follicular cells and fluid. Eventually, ovulation occurs, the mature follicle ruptures, and the secondary oocyte is released. A single follicle actually goes through all stages in one place within the ovary. b. Scanning electron micrograph of a secondary follicle.

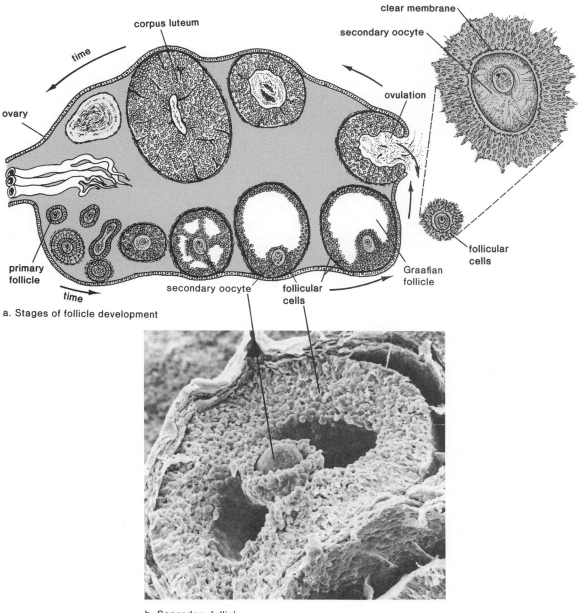

a. Stages of follicle development

b. Secondary follicle

almost all the cytoplasm. The other is a polar body that disintegrates. A secondary follicle contains the secondary oocyte pushed to one side of a fluid-filled cavity. In a Graafian follicle, the fluid-filled cavity increases to the point that the follicle wall balloons out on the surface of the ovary and bursts, releasing the secondary oocyte (often called an egg for convenience) surrounded by a clear membrane and follicular cells. This is referred to as **ovulation.** Once a follicle has lost its egg, it develops into a **corpus luteum,** a glandlike structure. If pregnancy does not occur, the corpus luteum begins to degenerate after about 10 days. If pregnancy does occur, the corpus luteum persists for 3–6 months. The follicle and the corpus luteum secrete the female sex hormones estrogen and progesterone, as discussed on page 420.

In females, oogenesis occurs within the ovaries, where one follicle reaches maturity each month. This follicle balloons out of the ovary and bursts to release the egg. The ruptured follicle develops into a corpus luteum. The follicle and the corpus luteum produce the female sex hormones estrogen and progesterone.

The Genital Tract

The female genital tract includes the oviducts, the uterus, and the vagina.

Oviducts

The oviducts, also called uterine or fallopian tubes, extend from the uterus to the ovaries. The oviducts are not attached to the ovaries; instead, they have fingerlike projections called **fimbriae** that sweep over the ovary at the time of ovulation. When the egg bursts (fig. 20.7) from the ovary during ovulation, it usually is swept up into an oviduct by the combined action of the fimbriae and the beating of cilia that line the oviducts.

Because the egg must traverse a small space before entering an oviduct, it is possible for the egg to get lost and instead to enter the abdominal cavity. Such eggs usually disintegrate, but in some rare cases, they have been fertilized in the abdominal cavity and have implanted themselves in the wall of an abdominal organ. Very rarely, such embryos have come to term, the child being delivered by surgery.

Once in the oviduct, the egg is propelled slowly by cilia movement and tubular muscle contraction toward the uterus. Fertilization, the completion of oogenesis, and zygote formation occurs in an oviduct. The developing embryo normally arrives at the uterus after several days and then embeds, or implants, itself in the uterine lining, which has been prepared to receive it. Occasionally, the embryo becomes embedded in the wall of an oviduct, where it begins to develop. Tubular pregnancies cannot succeed because the tubes are not anatomically capable of allowing full development to occur. An *ectopic pregnancy* is one that begins outside the uterus.

The Uterus

The **uterus** is a thick-walled, muscular organ about the size and the shape of an inverted pear. Normally, it lies above and is tipped over the urinary bladder. The oviducts join the uterus anteriorly, while posteriorly, the cervix enters into the vagina nearly at a right angle. A small opening in the cervix leads to the vaginal canal. Development of the embryo normally takes place in the uterus. This organ, sometimes called the womb, is approximately 5 cm wide in its usual state but is capable of stretching to over 30 cm to accommodate the growing baby. The lining of the uterus, called the **endometrium,** participates in the formation of the placenta (p. 453), which supplies nutrients needed for embryonic and fetal development. The endometrium has 2 layers: a basal layer and an inner functional layer. In the nonpregnant female, the functional layer of the endometrium varies in thickness according to a monthly reproductive cycle, called the uterine cycle (p. 420).

Cancer of the cervix is a common form of cancer in women. Early detection is possible by means of a **Pap test,** which requires that the removal of a few cells from the region of the cervix for microscopic examination. If the cells are cancerous, a hysterectomy may be recommended. A hysterectomy is the removal of the uterus. Removal of the ovaries in addition to the uterus is termed an ovariohysterectomy. Because the vagina remains, the woman still can engage in sexual intercourse.

The Vagina

The **vagina** is a tube that makes a 45-degree angle with the small of the back. The mucosal lining of the vagina lies in folds that extend as the fibromuscular wall stretches. This capacity to extend is especially important when the vagina serves as the birth canal, and it also can facilitate intercourse, when the vagina receives the penis during copulation.

External Genitalia

The external genital organs of the female (fig. 20.8) are known collectively as the **vulva.** The vulva includes 2 large, hair-covered folds of skin called the **labia majora.** They extend backward from the *mons pubis,* a fatty prominence underlying the pubic hair. The **labia minora** are 2 small folds lying just inside the labia majora. They extend forward from the vaginal opening to encircle and form a foreskin for the *clitoris,* an organ that is homologous to the penis. Although quite small, the clitoris has a shaft of erectile tissue and is capped by a pea-shaped glans. The glans clitoris also has sense receptors that allow it to function as a sexually sensitive organ.

The *vestibule,* a cleft between the labia minora, contains the openings of the urethra and the vagina. The vagina may be partially closed by a ring of tissue called the hymen. The hymen ordinarily is ruptured by initial sexual intercourse; however, it also can be disrupted by other types of physical activities. If the hymen persists after sexual intercourse, it can be surgically ruptured.

Notice that the urinary and reproductive systems in the female are entirely separate. For example, the urethra carries only urine, and the vagina serves only as the birth canal and the organ for sexual intercourse.

The egg enters the oviducts, which lead to the uterus followed by the vagina. The vagina opens into the vestibule, the location of female external genitalia.

Orgasm in Females

Sexual response in the female may be more subtle than in the male, but there are certain corollaries. The clitoris is believed to be an especially sensitive organ for initiating sexual sensations. It is possible for the clitoris to become ever so slightly erect as its erectile tissues become engorged with blood, but vasocongestion is more obvious in the labia minora, which expand and deepen in color. Erectile tissue within the vaginal wall also expands with blood, and the added pressure in these blood vessels causes small droplets of fluid to squeeze through the vessel walls and to lubricate the vagina.

Release from muscular tension occurs in females, especially in the region of the vulva and vagina but also throughout the entire body. Increased uterine motility may assist the transport of sperm toward the oviducts. Since female orgasm is not signaled by ejaculation, there is a wide range in normalcy of sexual response.

The Regulation of Female Hormone Levels

Hormonal regulation in the female is quite complex, so we begin with a simplified presentation and follow with a more in-depth presentation for those

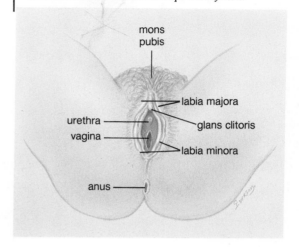

Figure 20.8
External genitalia of female. At birth, the opening of the vagina is partially occluded by a membrane called the hymen. Physical activities and sexual intercourse disrupt the hymen.

Table 20.3 Ovarian and Uterine Cycles (Simplified)

Ovarian Cycle	Events	Uterine Cycle	Events
Follicular phase Days 1–13	FSH	Menstruation Days 1–5	Endometrium breaks down
	Follicle maturation Estrogen	Proliferative phase Days 6–13	Endometrium rebuilds
Ovulation Day 14*			
Luteal phase Days 15–28	LH Corpus luteum Progesterone	Secretory phase Days 15–28	Endometrium thickens and glands are secretory

*Assuming a 28-day cycle

who wish to study the matter in greater detail. The following glands and hormones are involved in hormonal regulation.

Hypothalamus: secretes *GnRH* (gonadotropic-releasing hormone)
Anterior pituitary: secretes *FSH* (follicle-stimulating hormone) and *LH* (luteinizing hormone), the gonadotropic hormones
Ovaries: secrete estrogen and progesterone, the female sex hormones

Hormonal Regulation (Simplified)

The Ovarian Cycle The gonadotropic and sex hormones are not present in constant amounts in the female and instead are secreted at different rates during a monthly **ovarian cycle,** which lasts an average of 28 days but may vary widely in individuals. For simplicity's sake, it is convenient to emphasize that during the first half of a 28-day cycle (days 1–13, table 20.3), FSH from the anterior pituitary is promoting the development of a follicle in the ovary and that this follicle is secreting estrogen. As the estrogen blood level rises, it exerts feedback control over the anterior pituitary secretion of FSH so that this follicular phase comes to an end (fig. 20.9). The end of the follicular phase is marked by ovulation on the fourteenth day of the 28-day cycle. Similarly, it can be emphasized that during the last half of the ovarian cycle (days 15–28, table 20.3), anterior pituitary production of LH is promoting the development of a corpus luteum, which is secreting progesterone. As the progesterone blood level rises, it exerts feedback control over anterior pituitary secretion of LH so that the corpus luteum begins to degenerate. As the luteal phase comes to an end, menstruation occurs.

The Uterine Cycle The female sex hormones estrogen and progesterone have numerous functions, one of which is discussed here. The effect these hormones have on the endometrium of the uterus causes the uterus to undergo a cyclical series of events known as the **uterine cycle** (table 20.3). Cycles that last 28 days are divided as follows.

During *days 1–5*, there is a low level of female sex hormones in the body, causing the uterine lining to disintegrate and its blood vessels to rupture. A flow of blood, known as the *menses,* passes out of the vagina during a period of **menstruation,** also known as the menstrual period.

During *days 6–13*, increased production of estrogen by an ovarian follicle causes the endometrium to thicken and to become vascular and glandular. This is called the proliferative phase of the uterine cycle.

Ovulation usually occurs on the fourteenth day of the 28-day cycle.

Figure 20.9
The hypothalamus–pituitary–ovary control relationship.

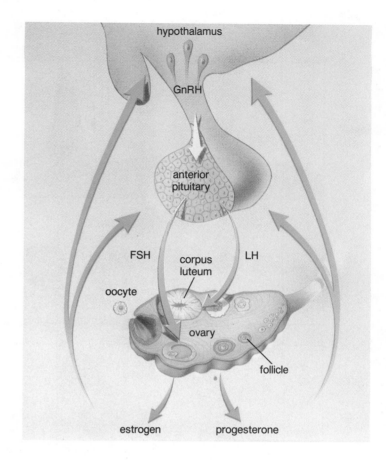

The hypothalamus produces GnRH (gonadotropic-releasing hormone)

GnRH stimulates the anterior pituitary to produce FSH (follicle-stimulating hormone) and LH (luteinizing hormone)

FSH stimulates the follicle to produce estrogen and LH stimulates the corpus luteum to produce progesterone

Estrogen and progesterone affect the sex organs (e.g., uterus) and the secondary sex characteristics, and exert feedback control over the hypothalamus and the anterior pituitary

During *days 15–28,* increased production of progesterone by the corpus luteum causes the endometrium to double in thickness and the uterine glands to mature, producing a thick mucoid secretion. This is called the secretory phase of the uterine cycle. The endometrium now is prepared to receive the developing embryo, but if pregnancy does not occur, the corpus luteum degenerates and the low level of sex hormones in the female body causes the uterine lining to break down. This is evident, due to the menstrual discharge that begins at this time. Even while menstruation is occurring, the anterior pituitary begins to increase its production of FSH and a new follicle begins to mature. Table 20.3 indicates how the ovarian cycle controls the uterine cycle.

Hormonal Regulation (Detailed)

Figure 20.10 shows the changes in blood concentration of all 4 hormones participating in the ovarian and uterine cycles. Notice that all 4 of these hormones (FSH, LH, estrogen, and progesterone) are present during the entire 28 days of the cycle. Therefore, in actuality, both FSH and LH *are* present during the follicular phase and both are needed for follicle development and egg maturation. The follicle secretes primarily estrogen and a very minimal amount of progesterone. Similarly, both LH and FSH are present in decreased amounts during the luteal phase. LH may be primarily responsible for corpus luteum formation, but the corpus luteum secretes both progesterone and estrogen. The effect that these hormones have on the endometrium has already been stated: Estrogen stimulates growth of the endometrium and readies it for reception of progesterone, which causes it to thicken and to become secretory.

20.2 Critical Thinking

1. Using figure 20.9 as a guide, formulate a hypothesis to explain why a pill that contains estrogen and progesterone could be used as a birth control pill.
2. How might you test your hypothesis, for example in laboratory mice?
3. In postmenopausal women (p. 424), there are usually increased levels of FSH and LH, but because the ovaries are unable to respond, there are decreased levels of estrogen and progesterone. How would you expect these levels to change in postmenopausal women who took birth control pills?
4. How might you test your prediction, for example in postmenopausal women?

Figure 20.10

Plasma hormonal levels associated with the ovarian and uterine cycles. During the follicular phase, FSH produced by the anterior pituitary promotes the maturation of a follicle in the ovary. The structure produces increasing levels of estrogen, which causes the endometrial lining of the uterus to thicken.

After ovulation and during the luteal phase, progesterone promotes the development of the corpus luteum. This structure produces increasing levels of progesterone, which causes the endometrial lining to become secretory. Menstruation begins when progesterone production declines to a low level.

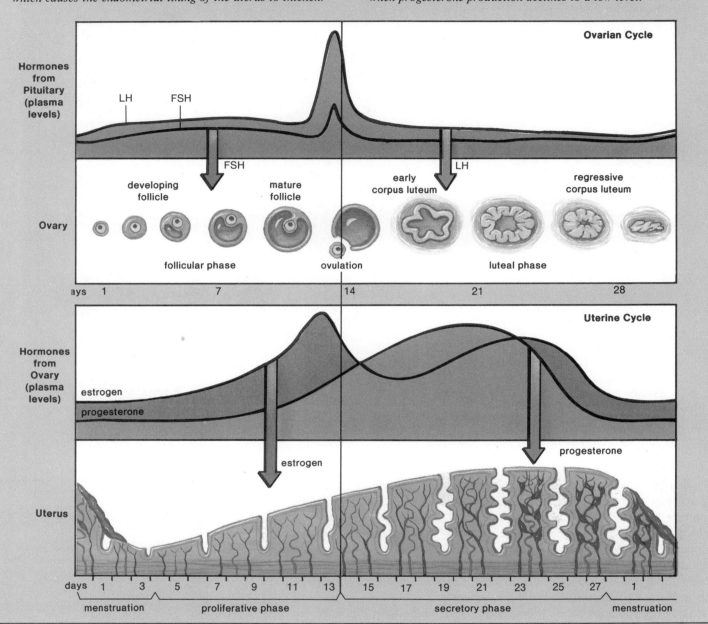

Feedback Control   As the estrogen level increases during the first part of the follicular phase, FSH secretion begins to decrease due to negative feedback. However, the high level of estrogen is believed to exert *positive feedback on the hypothalamus,* causing it to secrete GnRH, after which the pituitary momentarily produces an unusually large amount of FSH and LH. It is the surge of LH that is believed to promote ovulation. During the luteal phase, estrogen and progesterone bring about feedback inhibition as expected, and the levels of both LH and FSH decline steadily. In this way, all 4 hormones eventually reach their lowest levels, causing menstruation to occur. Therefore the corpus luteum degenerates unless pregnancy occurs. In some mammals, evidence suggests that prostaglandins (p. 405) are involved in degeneration, but this is not believed to be the case in humans.

Human Reproduction, Development, and Inheritance

During the first half of the ovarian cycle, FSH from the anterior pituitary causes maturation of a follicle, which secretes estrogen. After ovulation and during the second half of the cycle, LH from the anterior pituitary converts the follicle into the corpus luteum, which produces progesterone. Estrogen and progesterone regulate the uterine cycle in which the endometrium builds up and then is shed during menstruation.

Pregnancy

If pregnancy occurs, menstruation does not occur. Instead, the developing embryo embeds itself in the endometrial lining several days following fertilization. Once this process, called **implantation,** is complete, a female is *pregnant.* During implantation, an embryonic membrane surrounding the embryo produces a gonadotropic hormone called **h**uman **c**horionic **g**onadotropic hormone (HCG) that prevents degeneration of the corpus luteum and instead causes it to secrete even larger quantities of progesterone. The corpus luteum may be maintained for as long as 6 months, even after the placenta is fully developed.

The **placenta** (see fig. 21.18) originates from both maternal and fetal tissue and is the region of exchange of molecules between fetal and maternal blood, although there is no mixing of the 2 types of blood. After its formation, the placenta continues production of HCG and begins production of progesterone and estrogen. The latter hormones have 2 effects: they shut down the anterior pituitary so that no new follicles mature, and they maintain the lining of the uterus so that the corpus luteum is not needed. There is no menstruation during the 9 months of pregnancy.

Pregnancy Tests Pregnancy tests, which are readily available in hospitals, clinics, and now even drug and grocery stores, are based on the fact that HCG is present in the blood and the urine of a pregnant woman.

Before the advent of monoclonal antibodies, only a hospital blood test using radioactive material was available to detect pregnancy before the first missed menstrual period. Now there is a monoclonal antibody (p. 272) test for the detection of pregnancy 10 days after conception. This test can be done on a urine sample in a doctor's office, and the results are available within the hour.

The physical signs that oftentimes prompt a woman to have a pregnancy test are cessation of menstruation, increased frequency of urination, morning sickness, and increase in the size and the fullness of the breasts, as well as darkening of the areolae (fig. 20.11).

Female Sex Hormones

The female sex hormones estrogen and progesterone have many effects on the body. In particular, estrogen secreted at the time of puberty stimulates the growth of the uterus and the vagina. Estrogen is necessary for egg maturation and is largely responsible for the secondary sex characteristics in females. For example, it is responsible for the onset of the uterine cycle, as well as female body hair and fat distribution. In general, females have a more rounded appearance than males because of a greater accumulation of fat beneath the skin. Also, the pelvic girdle enlarges in females so that the pelvic cavity has a larger relative size compared to males; this means that females have wider hips. Both estrogen and progesterone also are required for breast development.

Breasts A female breast contains 15–25 lobules (fig. 20.11), each with its own mammary duct that begins at the nipple and divides into numerous other ducts that end in blind sacs called *alveoli.* In a nonlactating (nonmilk-producing) breast, the ducts far outnumber the alveoli because alveoli are made up of cells that can produce milk.

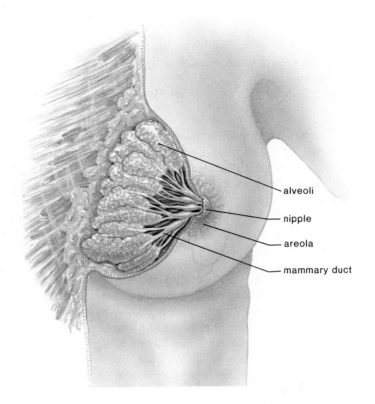

Figure 20.11

Female breast anatomy. The female breast contains lobules consisting of ducts and alveoli. The alveoli are lined by milk-producing cells in the lactating (milk-producing) breast.

alveoli

nipple

areola

mammary duct

Milk is not produced during pregnancy. *Lactogenic hormone* (prolactin) is needed for lactation (milk production) to begin, and the production of this hormone is suppressed because of the feedback inhibition estrogen and progesterone have on the pituitary during pregnancy. It takes a couple of days after delivery for milk production to begin, and in the meantime, the breasts produce a watery, yellowish white fluid called **colostrum,** which differs from milk in that it contains more protein and less fat. Colostrum is a source of passive immunity for the baby.

The continued production of milk requires continued production of lactogenic hormone, which occurs as long as the woman is breast-feeding. The hormone oxytocin is necessary to milk letdown. When a breast is suckled, the nerve endings in the areola are stimulated, and nerve impulses travel to the hypothalamus, which causes oxytocin to be released by the posterior pituitary. When this hormone arrives at the breasts, it causes contraction of the lobules so that milk flows into the ducts.

Menopause

Menopause, the period in a woman's life during which the ovarian and uterine cycles cease, is likely to occur between ages 45 and 55. The ovaries are no longer responsive to the gonadotropic hormones produced by the anterior pituitary, and the ovaries no longer secrete estrogen or progesterone. At the onset of menopause, the uterine cycle becomes irregular, but as long as menstruation occurs, it is still possible for a woman to conceive. Therefore, a woman usually is not considered to have completed menopause until there has been no menstruation for a year. The hormonal changes during menopause often produce physical symptoms, such as "hot flashes" that are caused by circulatory irregularities, dizziness, headaches, insomnia, sleepiness, and depression. Again, there is a great variation among women, and any of these symptoms may be absent altogether.

Women sometimes report an increased sex drive following menopause. It has been suggested that this may be due to androgen production by the adrenal cortex.

Table 20.4 Common Birth Control Methods

Name	Procedure	Methodology	Effectiveness*	Risk†
Vasectomy	Vas deferentia are cut and tied	No sperm in semen	Almost 100%	Irreversible sterility
Tubal ligation	Oviducts are cut and tied	No eggs in oviduct	Almost 100%	Irreversible sterility
Pill	Hormone medication must be taken daily	Shuts down pituitary	Almost 100%	Thromboembolism
IUD	Plastic coil must be inserted into uterus by physician	Prevents implantation	More than 90%	Infection
Diaphragm	Plastic cup inserted into vagina to cover cervix before intercourse	Blocks entrance of sperm into uterus	With jelly about 90%	—
Cervical cap	Rubber cup held by suction over cervix	Delivers spermicide near cervix	Almost 85%	Cancer of cervix?
Condom	Sheath that fits over erect penis at time of intercourse	Traps sperm	About 85%	—
Coitus interruptus (withdrawal)	Male withdraws penis before ejaculation	Prevents sperm from entering vagina	About 80%	—
Jellies, creams, foams	Contain spermicidal chemicals for insertion before intercourse	Kill a large number of sperm	About 75%	—
Rhythm method	Determine day of ovulation by record keeping; testing by various methods	Avoid day of ovulation	About 70%	—
Douche	Cleanses vagina and uterus after intercourse	Washes out sperm	Less than 70%	—

*Effectiveness is the average percentage of women who did not become pregnant in a population of 100 sexually active women using the technique for one year.
†Only condoms offer protection against AIDS.

Estrogen and to some extent progesterone affect the female genitals, promote development of the egg, and maintain the secondary sex characteristics. Lactogenic hormone causes the breasts to begin milk secretion after delivery, while another hormone, oxytocin, is responsible for milk letdown. When menopause occurs, FSH and LH are still produced by the anterior pituitary, but the ovaries no longer are able to respond.

The Control of Reproduction

Birth Control

Several means of birth control have been available for quite some time (table 20.4). The use of these contraceptive methods decreases the probability of pregnancy. A common way to discuss pregnancy rate is to indicate the number of pregnancies expected per 100 women per year. For example, it is expected that 80 out of 100 young women, or 80%, who are engaging regularly in unprotected intercourse will be pregnant within a year. Another way to discuss birth control methods is to indicate their effectiveness, in which case the emphasis is placed on the number of women who will not get pregnant. For example, with the least effective method given in table 20.4, we expect that within a year 70 out of 100, or 70%, of sexually active women will not get pregnant, while 30 women will get pregnant.

Future Means of Birth Control

There are 4 areas in which birth control investigations have been directed: morning-after medication, a long-lasting method, a medication that is specifically for males, and new barrier methods.

There is a new birth control pill (Ru 486) on the market in France consisting of a synthetic steroid that prevents progesterone from acting on the uterine lining because it has a high affinity for progesterone receptors. In clinical tests, the uterine lining sloughed off within 4 days in 85% of women who were less than a month pregnant. To improve the success rate, the drug is administered with a small dose of prostaglandin that causes contraction of the uterus to expel an embryo. The promoters of this treatment are using the term *contragestation* to describe its effects; however, it should be recognized that this medication, rather than preventing implantation, brings on an *abortion*, the loss of an implanted fetus. One day the medication might be used by many women who are experiencing delayed menstruation without knowing whether they are actually pregnant.

In this country, DES, a synthetic estrogen that affects the uterine lining making implantation difficult, sometimes is given following intercourse. Since large doses causing nausea and vomiting in the short run and possibly cancer in the long run are required, DES usually is given only for incest or rape.

Depo-Provera is an injectable contraceptive that is commercially available in many countries outside the United States. The injection contains crystals that gradually dissolve over a period of 3 months. The crystals contain a chemical related to progesterone; this chemical suppresses ovulation. The drug has not been approved for use in the United States because cancer developed in some test animals receiving the injections. More animal studies are now underway. An even more potent progesterone-like molecule for implantation under the skin is now close to being approved for sale in the United States. The *implant* consists of narrow tubes that slowly release the drug over a period of 5 years.

Various possibilities exist for a *"male pill."* Scientists have made analogues of gonadotropic-releasing hormones that interfere with the action of this hormone and prevent it from stimulating the pituitary. The seminiferous tubules produce a hormone termed *inhibin* that inhibits FSH production by the pituitary (p. 415). Testosterone and/or related chemicals can be used to inhibit spermatogenesis in males, but there are usually feminizing side effects because the excess is changed to estrogen by the body.

There has been a revival of interest in barrier methods of birth control, and a "female condom" now is being studied to determine its effectiveness against pregnancy and sexually transmitted diseases. The closed end of a large plastic tube is anchored by a plastic ring in the upper vagina, and the open end of the tube is held in place by a thinner ring that rests just outside the vagina.

There are numerous well-known birth control methods and devices available to those who wish to prevent pregnancy. Their effectiveness varies. In addition, new methods are expected to be developed.

Infertility

Sometimes, couples do not need to prevent pregnancy; conception or fertilization does not occur despite frequent intercourse. The American Medical Association estimates that 15% of all couples in this country are unable to have any children and therefore are properly termed *sterile;* another 10% have fewer children than they wish and therefore are termed *infertile*.

Infertility can be due to a number of factors. It is possible that fertilization occurs, but the embryo dies before implantation takes place. One area of concern is that radiation, chemical mutagens, and the use of psychoactive drugs contributes to sterility, possibly by causing chromosome mutations that prevent development from proceeding normally. The lack of progesterone also can prevent implantation, and therefore the proper administration of this hormone is sometimes helpful.

It also is possible that fertilization never takes place. There may be a congenital malformation of the reproductive tract or there may be an obstruction of the oviduct or vas deferens due to infection. Endometriosis, the spread of uterine tissue beyond the uterus, is also a cause of infertility, as discussed in the next reading (p. 428). Sometimes these physical defects can be corrected surgically. If no obstruction is apparent, it is possible to give females a substance rich in FSH and LH that is extracted from the urine of postmenopausal women. This treatment causes multiple ovulations and sometimes multiple pregnancies, however.

When reproduction does not occur in the usual manner, couples today are seeking alternative reproductive methods (fig. 20.12) that may include the following.

Figure 20.12
Mother and child. Sometimes couples utilize alternative methods of reproduction in order to experience the joys of parenthood.

Artificial Insemination by Donor (AID) Since the 1960s, there have been hundreds of thousands of births following artificial insemination, in which sperm are placed in the vagina by a physician. Sometimes a woman is artificially inseminated by her husband's sperm. This is especially helpful if the husband has a low sperm count—the sperm can be collected over a period of time and concentrated so that the sperm count is sufficient to result in fertilization. Often, however, a woman is inseminated by sperm acquired from a donor who is a complete stranger to her.

In Vitro Fertilization (IVF) Over a hundred babies have been conceived using IVF. First, a woman is given appropriate hormonal treatment. Then laparoscopy may be done. The laparoscope is a metal tube about the size of a pencil that is equipped with a tiny light and a telescopic lens. In this instance, it also is fitted with a tube for retrieving eggs. After insertion through a small incision near the woman's naval, the physician guides the laparoscope to the ovaries, where the eggs are sucked up into the tube. Alternately, it is possible to place a needle through the vaginal wall and to guide it by the use of ultrasound to the ovaries, where the needle is used to retrieve the eggs. This method is called transvaginal retrieval.

Concentrated sperm from the male is placed in a solution that approximates the conditions of the female genital tract. When the eggs are introduced, fertilization occurs. The resultant zygotes begin development, and after about 2–4 days, the embryos are inserted into the uterus of the woman, who is now in the secretory phase of her menstrual cycle. If implantation is successful, development is normal and continues to term.

Gamete Intrafallopian Transfer (GIFT) GIFT was devised as a means to overcome the low success rate (15–20%) of in vitro fertilization. The method is exactly the same as in vitro fertilization except the eggs and the sperm are immediately placed in the oviducts after they have been brought together. This procedure is helpful to couples whose eggs and sperm never make it to the oviducts; sometimes the egg gets lost between the ovary and the oviducts, and sometimes the sperm never reach the oviducts. GIFT has an advantage in that it is a one-step procedure for the woman—the eggs are removed and are reintroduced all in the same time period. For this reason, it is less expensive — $1,500 compared with $3,000 and up for in vitro fertilization.

Surrogate Mothers Over a hundred babies have been born to women paid to have them by other individuals who have contributed sperm (or egg) to the fertilization process.

If all the alternative methods discussed are considered, it is possible to imagine that a baby could have 5 parents: (1) sperm donor, (2) egg donor, (3) surrogate mother, and (4) and (5) adoptive mother and father.

Each month K.C. Esperance, 31, a San Francisco nurse practitioner, suffered menstrual cramps so agonizing that she would take to her bed, curl up and pray that she would live through the next couple of days. Doctor after doctor gave her the same ineffectual advice: rest, take some codeine and bear with it.

During her teens, Maria Menna Perper, 42, a New Jersey biochemist, suffered intestinal problems around the time of her period. By her late 30s, she felt "ecruciating, burning pain" in her colon every month "like clockwork." Eventually the pain became continuous, and it was impossible for her to work or even sit down.

Endometriosis

For Anne Hicks, 29, a Portland, Ore., real estate property manager, there were no obvious signs other than her inability to become pregnant.

Despite their differing complaints, each of the women eventually discovered that she suffered from the same insidious condition: endometriosis, an often unrecognized disease that afflicts anywhere from 4 million to 10 million American women and is a major cause of infertility. The condition is caused by the spread and growth of tissue from the lining of the uterus (or endometrium) beyond the uterine walls. These endometrial cells form bandlike patches and scars throughout the pelvis and around the ovaries and the Fallopian tubes, resulting in a variety of symptoms and degrees of discomfort. Because endometriosis has been associated with delayed childbearing, it is sometimes called the "career woman's disease." But recent studies have shown that the disorder strikes women of all socioeconomic groups and even teenagers, though those with heavier, longer, or more frequent periods may be especially susceptible. Says Dr. Donald Chatman of Chicago's Michael Reese Hospital, "Endometriosis is an equal-opportunity disease."

How the disease begins is something of a mystery. One theory ascribes it to "retrograde menstruation." Instead of flowing down through the cervix and the vagina, some menstrual blood and tissue back up through the Fallopian tubes and spill out into the pelvic cavity. Normally this errant flow is absorbed harmlessly, but in some cases the stray tissue implants itself outside the uterus and continues to grow. A second theory suggests that the disease arises from misplaced embryonic cells that have lain scattered around the abdominal cavity since birth. When the monthly hormonal cycles begin at puberty, says Dr. Howard Judd, director of gynecological endocrinology at UCLA Medical Center, "some of these cells get stirred up and could be a major cause of endometriosis."

If anything about endometriosis is clear, it is that once the disease has begun, it will probably get worse. Stimulated by the release of estrogen, the implanted tissue grows and spreads. Cells from the growths break away and are ferried by lymphatic fluid throughout the body, sometimes, although rarely, forming islands in the lungs, the kidneys, the bowel, or even the nasal passages. There they respond to the menstrual cycle, causing monthly bleeding from the rectum or wherever else they have settled.

The most common symptom of endometriosis is pain, which can occur during menstruation, urination, and sexual intercourse. Unfortunately, these warnings are often overlooked by women and their doctors. Cheri Bates, 31, of Seattle, describes the cramps she suffered as "outrageous," but she assumed they were "normal." By the time her condition was discovered, scar tissue covered her reproductive organs and parts of her bladder and intestines.

To confirm that a patient has endometriosis, doctors look for the telltale tissue by peering into the pelvic cavity with a fiber-optic instrument called a laparoscope. After

diagnosis, a number of treatments can be prescribed. One is pregnancy—if it is still feasible; the 9-month interruption of menstruation can help shrink misplaced endometrial tissue. Taking birth-control pills may also help, but more effective is a drug called danazol, a synthetic male hormone that stops ovulation and causes endometrial tissue to shrivel. But it also can produce acne, facial-hair growth, weight gain, and other side effects.

A new experimental treatment with perhaps fewer ill effects involves a synthetic substance called nafarelin, similar to gonadotropin-releasing hormone. Normally GnRH is released in bursts by the hypothalamus gland, eventually triggering the process of ovulation. But "if the GnRH stimulation is given continuously instead of in pulses," explains Dr. Robert Jaffe of the University of California, San Francisco, "the whole [ovulatory] system shuts off," and the endometrial implants "virtually melt away."

For severe cases of endometriosis, surgical removal of the ovaries and the uterus may be the only solution. But less extreme surgery can often help. At Atlanta's Northside Hospital, Dr. Camran Nezhat has had success with a high-tech procedure called videolaseroscopy, which employs a laparoscope rigged with a tiny video camera and a laser. The camera images, enlarged on a video screen, enable Nezhat to zero in on endometrial tissue and to vaporize it with the laser. In a study of 102 previously infertile patients, Nezhat found that 60.7% were able to conceive within 2 years of videolaseroscopy treatment.

Like many other doctors who see the unfortunate consequences of endometriosis, Nezhat is concerned that a "lot of women do not seek help for this problem." Any serious pain, he notes, needs investigating. Agrees Cheri Bates (a victim), "If a doctor tells you that suffering is a woman's lot in life, get another doctor."

Some couples are infertile. There may be a hormonal imbalance or a blockage of the oviducts. When corrective medical procedures fail, it is possible today to consider an alternative method of reproduction.

Sexually Transmitted Diseases

There are many diseases that are transmitted by sexual contact. Our discussion centers on 6 of the most prevalent: AIDS (acquired immune deficiency syndrome), genital herpes, genital warts, gonorrhea, chlamydia, and syphilis (fig. 20.13). AIDS, genital herpes, and genital warts are viral diseases; therefore, they are difficult to treat because the traditional antibiotics are not helpful. Other types of drugs have been developed to treat these, however. Although gonorrhea and chlamydia are treatable with appropriate antibiotic therapy, they are not always promptly diagnosed. Unfortunately, as yet there are no vaccines available for any of these infections.

AIDS

Acquired immune deficiency syndrome (AIDS) is caused by retroviruses that mainly infect helper T (T_H) cells and macrophages, white blood cells that are necessary to the normal functioning of the immune system (p. 274). The AIDS viruses are called human immunodeficiency virus type 1 and type 2, or HIV–1 and HIV–2. In the United States, the vast majority of infections are caused by HIV–1. The viruses usually stay hidden inside the host cell, but occasionally they do bud from T_H cells.

HIV–1 has various stages of infection. Often there are detectable antibodies to the virus in the bloodstream especially during the early stages. The presence of these antibodies is the diagnostic test for HIV–1 infection. During the stages referred to as AIDS-related complex (ARC), symptoms may include weight loss, swollen lymph nodes, night sweats, fatigue, fever, and diarrhea. Finally, the person may develop AIDS, characterized by the development of pneumonia, skin cancer, and neuromuscular and psychological disturbances. There may be 8–10 years between exposure to the virus and the final stage of AIDS. Most individuals die sometime during this period of time.

Transmission

HIV–1 is transmitted by infected blood cells or by body fluids (e.g., semen) contaminated by infected blood. In the United States, 2 groups initially were infected by HIV–1—homosexual men and intravenous drug abusers (IVDA) and their sexual partners. Today, the rate of infection is declining in the homosexual population, but it is rapidly increasing among IVDA and is spreading into the heterosexual population, especially among young people.

Certain portions of the country have been harder hit than others (fig. 20.14). Even in New York City, which reports the highest number of affected individuals, there are regions that have more cases than others. Here, the number of AIDS deaths among IVDA is higher than among homosexual men. The AIDS virus can cross the placenta, and in the part of New York City known as the Bronx, 1 in 43 babies is born with HIV antibodies in the blood. Some of these newborns may have the antibodies without having the virus, but 30–50% probably are infected.

Although IVDA can spread the disease to the general heterosexual population, today infection among the general population is still less than 4%. Health officials emphasize that unprotected intercourse with multiple partners or a single infected partner increases the chance of infection. The use of a condom reduces the risk, but the best preventive measure at this time is monogamy with a sexual partner who is not infected. Casual contact with someone who is infected, such as shaking hands, eating at the same table, or swimming in the same pool, does not transmit the virus.

Figure 20.13

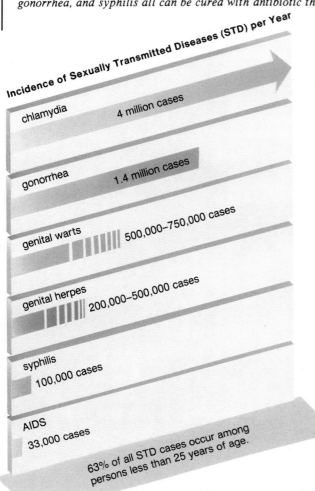

Figure 20.13
Statistics for the most common sexually transmitted diseases. These show that chlamydia, gonorrhea, and genital warts are all much more common than herpes and syphilis. Chlamydia, gonorrhea, and syphilis all can be cured with antibiotic therapy.

Incidence of Sexually Transmitted Diseases (STD) per Year

chlamydia — 4 million cases

gonorrhea — 1.4 million cases

genital warts — 500,000–750,000 cases

genital herpes — 200,000–500,000 cases

syphilis — 100,000 cases

AIDS — 33,000 cases

63% of all STD cases occur among persons less than 25 years of age.

Figure 20.14
Geography of AIDS. Number of AIDS cases in each state as reported to the Center for Disease Control, Inc., in 1990.

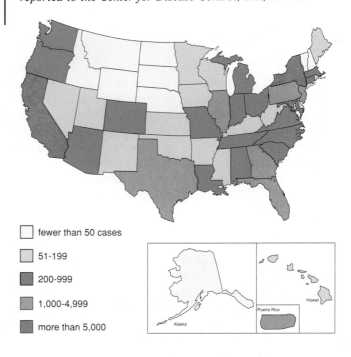

□ fewer than 50 cases

■ 51-199

■ 200-999

■ 1,000-4,999

■ more than 5,000

Alaska   Hawaii   Puerto Rico

Treatment

The drug zidovudine (AZT) has been shown to prolong the lives of those with full-blown AIDS and also seems to prevent the progression of the disease in HIV-infected persons having fewer than 500 T_H cells per mm^3 and who exhibit no symptoms. A new drug called dideoxyinosine (DDI), which like AZT works by preventing viral replication in cells, is undergoing clinical trials.

As discussed on page 274, researchers are trying to develop a vaccine for AIDS. The AIDS virus mutates frequently, but researchers have identified portions of the coat protein that they believe are relatively stable. When these are injected into the bloodstream, antibodies develop, but it is not yet known whether such antibodies offer protection against infection. Jonas Salk, who developed the polio virus vaccine, has developed a vaccine that contains whole AIDS viruses killed by treatment with chemicals and radiation. So far, this vaccine has been found to be effective in chimpanzees.

Genital Herpes

Genital herpes is caused by herpes simplex virus (fig. 20.15) of which there are 2 types: type 1 usually causes cold sores and fever blisters, while type 2 more often causes genital herpes.

Genital herpes is one of the more prevalent sexually transmitted diseases today (fig. 20.13); an estimated 40 million persons in the United States have it, with an estimated 500,000 new cases appearing each year. Immediately after infection, there are no symptoms, but the individual may experience a

Figure 20.15
Genital herpes virus. a. *Cell infected with herpes virus.*
b. *Enlarged model of herpes virus.*

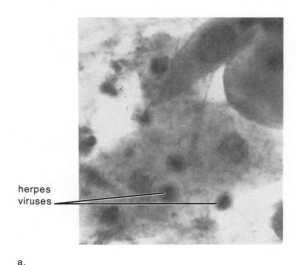

herpes
viruses

a.

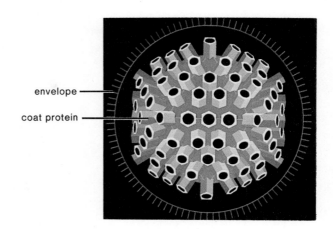

envelope

coat protein

b.

tingling or itching sensation before blisters appear at the infected site within
2–20 days. Once the blisters rupture, they leave painful ulcers that may take
as long as 3 weeks or as little as 5 days to heal. These symptoms may be accompanied by fever, pain upon urination, and swollen lymph nodes.

After the ulcers heal, the disease is only dormant, and blisters can reoccur
repeatedly at variable intervals. Sunlight, sex, menstruation, and stress seem
to cause the symptoms of genital herpes to reoccur. While the virus is latent,
it resides in nerve cells. Type 1 resides in a group of sensory ganglia located
near the brain, and type 2 resides in sacral ganglia that lie near the spinal
cord. Type 1 occasionally infects the eye, causing an eye infection that can
lead to blindness. Both type 1 and type 2 can cause CNS infections. Type 2
formerly was thought to cause a form of cervical cancer, but this no longer is
believed to be the case.

Infection of the newborn can occur if the child comes in contact with a
lesion in the birth canal. In 1–3 weeks, the infant is gravely ill and can become
blind, have neurological disorders including brain damage, or die. Birth by
cesarean section prevents these occurrences.

Genital Warts

Genital warts are caused by the human papillomaviruses (HPVs), which are
cubical DNA viruses that reproduce in the nuclei of cells. Plantar warts and
common warts also are caused by HPVs.

Some HPVs are sexually transmitted. Sometimes carriers do not have
any sign of warts, although flat lesions may be present. When present, the
warts commonly are seen on the penis and the foreskin of males and the vaginal opening in females. If the warts are removed, they may reoccur.

HPVs, rather than genital herpes, now are associated with cancer of the
cervix, as well as tumors of the vulva, the vagina, the anus, and the penis. Some
researchers believe that the viruses are involved in 90–95% of all cases of cancer
of the cervix. Physicians are disheartened that teenagers with multiple sex
partners seem to be particularly susceptible to HPV infections. More and more
cases of cancer of the cervix are being seen among this age group.

Presently, there is no cure for an HPV infection. A suitable medication
to treat genital warts before cancer occurs is being sought, and efforts also are
underway to develop a vaccine.

Figure 20.16
*Gonorrheal bacteria (Neisseria gonorrheae) in
male urethral discharge. If you look carefully,
you will notice that these round bacteria occur in
pairs; for this reason, they are called diplococci.*

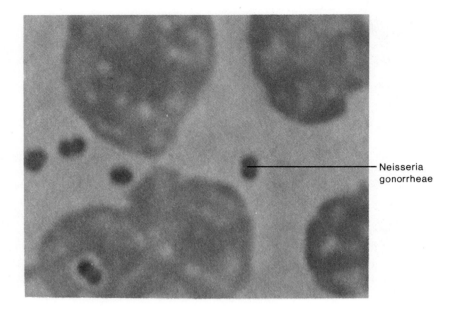

Neisseria
gonorrheae

Gonorrhea

Gonorrhea is caused by the bacterium *Neisseria gonorrheae,* which is a gram-negative, intracellular diplococcus, meaning that 2 cells generally stay together (fig. 20.16).

The diagnosis of gonorrhea in the male is not difficult as long as he displays typical symptoms (as many as 40% of males may be asymptomatic). The patient complains of pain on urination and has a thick, greenish yellow urethral discharge 3–5 days after contact. In the female, the bacteria may first settle within the urethra or near the cervix, from which they may spread to the oviducts, causing **pelvic inflammatory disease (PID).** As the inflamed tubes heal, they may become partially or completely blocked by scar tissue. As a result, the female is sterile or at best subject to ectopic pregnancy. Unfortunately, 60–80% of females are asymptomatic until they develop severe pains in the abdominal region due to PID. However, if the proper diagnosis is made, gonorrhea can be treated using antibiotics.

Homosexual males develop gonorrhea proctitis, or infection of the anus, with symptoms including pain in the anus and blood or pus in the feces. Oral sex can cause infection of the throat and the tonsils. Gonorrhea also can spread to other parts of the body, causing heart damage or arthritis. If, by chance, the person touches infected genitals and then his or her eyes, a severe eye infection can result.

Eye infection leading to blindness can occur as a baby passes through the birth canal. Because of this, all newborn infants receive eye drops containing antibacterial agents, such as silver nitrate, tetracycline, or penicillin, as a protective measure.

Chlamydia

Chlamydia is named for the tiny bacterium that causes it (*Chlamydia trachomatis*). New chlamydial infections occur at an even faster rate than gonorrheal infections (fig. 20.13). They are the most common cause of NGU, nongonococcal urethritis. About 8–21 days after exposure, men experience a mild burning sensation upon urination and a mucoid discharge. Women may have a vaginal discharge along with the symptoms of a urinary tract infection. Unfortunately, a physician mistakenly may diagnose a gonorrheal or urinary infection and prescribe the wrong type of antibiotic, or the person may never seek medical help. In either case, the infection eventually can cause PID and sterility or ectopic pregnancy.

If a newborn comes in contact with chlamydia during delivery, inflammation of the eyes or pneumonia can result. There are also those who believe that chlamydial infections increase the possibility of premature and stillborn births.

Syphilis

Syphilis is caused by a type of bacterium called *Treponema pallidum.* Syphilis has 3 stages, which can be separated by latent stages in which the bacteria are resting before multiplying again. During the *primary stage,* a hard chancre (ulcerated sore with hard edges) indicates the site of infection. The chancre can go unnoticed, especially since it usually heals spontaneously, leaving little scarring. During the *secondary stage,* proof that bacteria have invaded and spread throughout the body is evident when the victim breaks out in a rash. Curiously, the rash does not itch and is seen even on the palms of the hands and the soles of the feet. There can be hair loss and infectious gray patches on the mucous membranes, including the mouth. These symptoms disappear of their own accord.

During a *tertiary stage,* which lasts until the patient dies, syphilis may affect the cardiovascular system; weakened arterial walls (aneurysms) are seen, particularly in the aorta. In other instances, the disease may affect the nervous system; an infected person may show psychological disturbances, for example. Gummas, large destructive ulcers, may develop on the skin or within the internal organs in another variety of the tertiary stage.

Congenital syphilis is caused by syphilitic bacteria crossing the placenta. The child is born blind and/or with numerous anatomical malformations. Penicillin has been used as an effective antibiotic to cure syphilis.

The sexually transmitted diseases—AIDS, genital herpes, genital warts, gonorrhea, chlamydia, and syphilis—are prevalent at this time. AIDS often results in death within a few years; attacks of herpes reoccur throughout life; genital warts are associated with cervical cancer; and gonorrhea and chlamydia often lead to sterility. The best preventive measure against these diseases is monogamy with a partner who is free of them.

Summary

In males, spermatogenesis occurs within the seminiferous tubules of the testes, which also produce testosterone within the interstitial cells. Sperm mature in the epididymis and are stored in the vas deferens before entering the urethra, along with seminal fluid, prior to ejaculation. Hormonal regulation involving secretions from the hypothalamus, the anterior pituitary, and the testes in the male maintains testosterone at a fairly constant level.

In females, oogenesis occurs within the ovaries, where one follicle produces an egg each month. Fertilization, if it occurs, takes place in the oviducts, and the resulting embryo travels to the uterus, where it embeds itself in the uterine lining. In the nonpregnant female, hormonal regulation involves the ovarian and uterine cycles, dependent upon the hypothalamus, the anterior pituitary, and the female sex hormones estrogen and progesterone.

Numerous birth control methods and devices are available for those who wish to prevent pregnancy. Infertile couples increasingly are resorting to alternative methods of reproduction.

Sexually transmitted diseases are of concern to all. AIDS, genital herpes, genital warts, and chlamydia are presently of the greatest concern, but still prevalent are gonorrhea and syphilis.

Study Questions

1. Discuss the anatomy and the physiology of the testes. (p. 412) Describe the structure of sperm. (p. 412)
2. Give the path of sperm. (p. 412)
3. What glands produce seminal fluid? (p. 414)
4. Discuss the anatomy and the physiology of the penis. (p. 414) Describe ejaculation. (p. 415)
5. Discuss hormonal regulation in the male. Name 3 functions for testosterone. (p. 415)
6. Discuss the anatomy and the physiology of the ovaries. (p. 417) Describe ovulation. (p. 417)
7. Give the path of the egg. Where do fertilization and implantation occur? Name 2 functions of the vagina. (pp. 418–19)
8. Describe the external genitalia in females. (p. 419)
9. Compare male and female orgasm. (p. 414, p. 419)

10. Discuss hormonal regulation in the female, either simplified or detailed. (pp. 419–22) Give the events of the uterine cycle, and relate them to the ovarian cycle. (pp. 420–21) In what way is menstruation prevented if pregnancy occurs? (p. 423)

11. Name 4 functions of the female sex hormones. (p. 423) Describe the anatomy and the physiology of the breast. (pp. 423–24)

12. Discuss the various means of birth control and their relative effectiveness. (p. 425)

13. Describe the most common types of sexually transmitted diseases. (pp. 429–33)

Objective Questions

1. In tracing the path of sperm, the structure that follows the epididymis is the _____ .

2. The prostate gland, the Cowper's glands, and the _____ all contribute to seminal fluid.

3. The primary male sex hormone is _____ .

4. An erection is caused by the entrance of _____ into sinuses within the penis.

5. In the female reproductive system, the uterus lies between the oviducts and the _____ .

6. In the ovarian cycle, once each month a(n) _____ produces an egg. In the uterine cycle, the _____ lining of the uterus is prepared to receive the zygote.

7. The female sex hormones are _____ and _____ .

8. Pregnancy in the female is detected by the presence of _____ in blood or urine.

9. In vitro fertilization occurs in _____ .

10. Although a sexually transmitted disease, the AIDS virus mainly infects _____ cells.

11. Herpes simplex virus type 1 causes _____ , and type 2 causes _____ .

12. The most prevalent sexually transmitted disease today is _____ .

Label this Diagram.

See figure 20.2 (p. 411) in text.

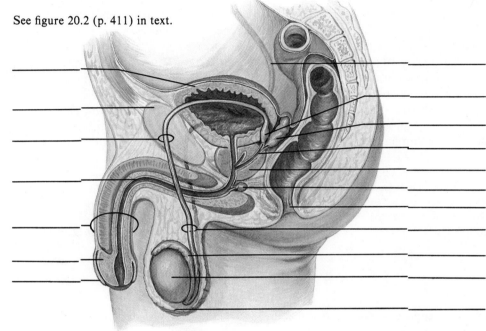

Answers to Objective Questions

1. vas deferens 2. seminal vesicles 3. testosterone 4. blood 5. vagina 6. follicle, endometrial 7. estrogen, progesterone 8. HCG 9. laboratory glassware 10. T, 11. cold sores, genital herpes 12. chlamydia

---

Selected Key Terms

endometrium (en″do-me′tre-um) the lining of the uterus that becomes thickened and vascular during the uterine cycle. *418*

erection (ĕ-rek′shun) referring to a structure, such as the penis, that is turgid and erect as opposed to being flaccid or lacking turgidity. *415*

Graafian follicle (graf′e-an fol′li-k'l) mature follicle within the ovaries that houses a developing egg. *417*

implantation (im″plan-ta′shun) the attachment and penetration of the embryo into the lining (endometrium) of the uterus. *423*

interstitial cell (in″ter-stish′al sel) hormone-secreting cell located between the seminiferous tubules of the testes. *412*

menopause (men′o-pawz) termination of the ovarian and uterine cycles in older women. *424*

menstruation (men″stroo a′shun) loss of blood and tissue from the uterus at the end of a uterine cycle. *420*

ovarian cycle (o-va′re-an si′k'l) monthly occurring changes in the ovary that affect the level of sex hormones in the blood. *420*

ovary (o′var-e) the female gonad, the organ that produces eggs, estrogen, and progesterone. *417*

semen (se′men) the sperm-containing secretion of males; seminal fluid plus sperm. *415*

seminiferous tubule (sem″i-nif′er-us tu′būl) highly coiled duct within the male testis that produces and transports sperm. *412*

testis (tes′tis) the male gonad, the organ that produces sperm and testosterone. *412*

uterine cycle (u′ter-in si′k'l) monthly occurring changes in the characteristics of the uterine lining. *420*

Chapter 21

Development

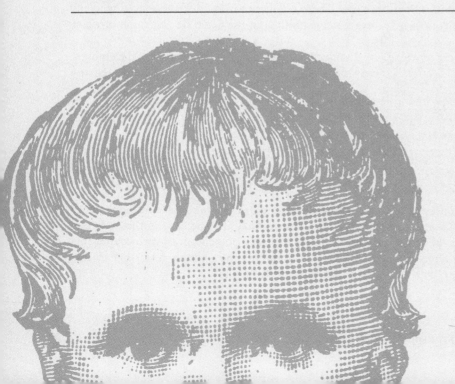

Chapter Concepts

1 Growth, differentiation, and morphogenesis are 3 aspects of development, a process that occurs throughout life.

2 The first stages of embryonic development in animals lead to the establishment of the embryonic germ layers.

3 Human embryos have the same extraembryonic membranes as those of reptiles and birds, but their function has been altered to suit internal development.

4 It is possible to outline precisely the steps in human embryonic and fetal development.

5 Investigation into aging shows hope of identifying underlying causes of degeneration and prolonging the health span of individuals.

Chapter Outline

Early Developmental Stages
    The Morula
    The Blastula
    The Gastrula
    The Neurula
Differentiation and Morphogenesis
    Differentiation
    Morphogenesis
Human Embryonic and Fetal
    Development
    Embryonic Development
    Fetal Development
    Birth
Human Development after Birth
    Adulthood and Aging

Figure 21.1

Human embryo and fetus. Human development is divided into the embryonic period (first 2 months) and fetal development (third through ninth month). a. Embryo is not recognizably human. b. Fetus is recognizably human.

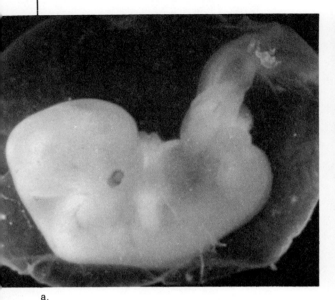

a.

b.

 The study of development concerns the events and the processes that occur as a single cell becomes a complex organism (fig. 21.1). These same processes also are seen as the newly born or hatched organism matures, as lost parts regenerate, as a wound heals, and even during aging. Therefore, today it is customary to stress that the study of development encompasses not only embryology (development of the embryo), but these other events as well.

Development requires growth, differentiation, and morphogenesis. When an organism increases in size, we say that it has grown. During **growth,** cells divide, get larger, and divide once again. **Differentiation** occurs when cells become specialized in structure and in function. A muscle cell looks and acts quite differently than a nerve cell, for example. **Morphogenesis** goes one step beyond growth and differentiation. It occurs when body parts are shaped and patterned into a certain form. There is a great deal of difference between your arm and your leg, for example, even though they contain the same types of tissues.

These processes are discussed as they apply to development of the embryo, but keep in mind that they also occur whenever an organism goes through any developmental change.

Growth, differentiation, and morphogenesis are 3 processes that are seen whenever a developmental change occurs.

Early Developmental Stages

All chordate embryos go through the same early stages of development, as listed in table 21.1. Chordates are animals that at some time in their life history have an elastic supporting rod known as a notochord. In vertebrates, this rod is replaced by the vertebral column. All the animals discussed in this chapter are vertebrates except the lancelet (p. 438).

Table 21.1 *Early Developmental Stages*

Stage	Requirement	Result
Morula	Cleavage	Compact ball of cells
Blastula	Morphogenesis and growth	Hollow ball of cells
Gastrula	Morphogenesis and growth	Embryo with 3 germ layers
Neurula	Differentiation by induction	Nervous system development

Human Reproduction, Development, and Inheritance

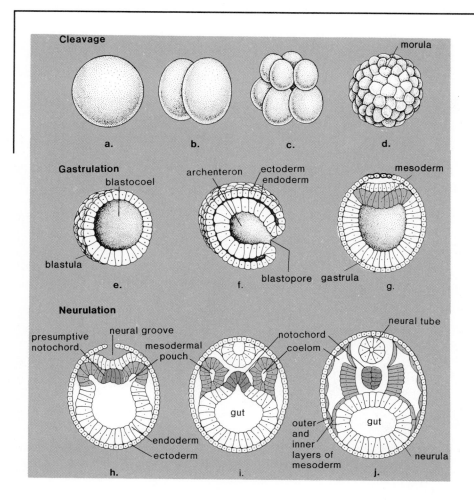

Figure 21.2
Lancelet development. In the first row, cleavage produces the morula, a ball of cells. In the second row, gastrulation by invagination produces the gastrula, an embryo with the 3 germ layers. In the third row, neurulation occurs; the embryo then has a neural tube and is called a neurula. Notice that the coelom develops by outpocketing from the primitive gut.

The Morula

Cleavage, the first event of development, is cell division without growth. It is best observed in an embryo such as the lancelet, which has little **yolk,** a rich nutrient material. (The yellow portion of a chick egg is the yolk.) Because a lancelet egg has little yolk, cell division is about equal, and the cells are fairly uniform in size (fig. 21.2 *a-d*). Cleavage continues until there is a solid ball of cells called the *morula.*

The Blastula

Following cleavage, the cells of the morula more or less position themselves to create a cavity. In the lancelet, a completely hollow ball called the *blastula* results, and the cavity within the ball is called the *blastocoel.* The human blastula is called the *blastocyst,* and therefore the cavity is called the blastocyst cavity. The blastocyst has a mass of cells—the *inner cell mass*—at one end. Figure 21.3 compares the appearance of a human embryo to that of a lancelet during the first stages of development.

In lancelets, cleavage results in a morula, which becomes the blastula when the blastocoel develops. In humans, the morula becomes the blastocyst, which contains an inner cell mass.

The Gastrula

During gastrulation in a lancelet, certain cells begin to push, or invaginate, into the blastocoel, creating a double layer of cells (fig. 21.2 *e-g*). The outer layer is called **ectoderm,** and the inner layer is called **endoderm.** The space

Figure 21.3

Comparison of lancelet development and human development. a. Comparative morula stages. Cleavage produces many cells in both. b. Comparative blastula stages. The observed cavity is the blastocoel in the lancelet, but the blastocyst cavity in the human. c. Comparative gastrula stages. The gastrula in the lancelet is spherical; the gastrula in the human is flattened. d. Comparative late gastrula stages. Outpocketing produces mesoderm in the lancelet, while invagination between ectoderm and endoderm produces mesoderm in the human. (l.s. = longitudinal section; c.s. = cross section)

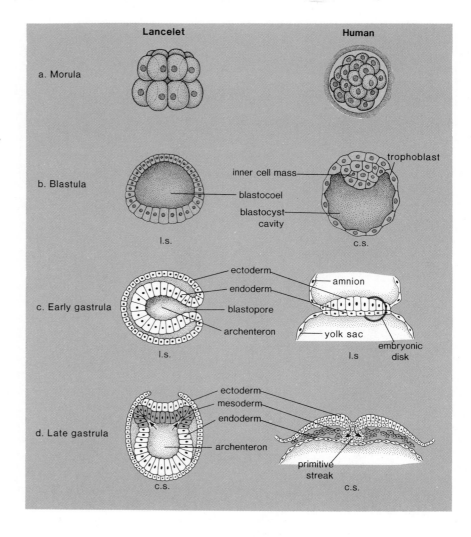

Figure 21.4

Human embryo at 16 days. The primitive node marks the extent of the primitive streak, where invagination occurs to establish the germ layers.

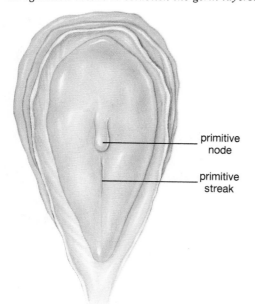

created by invagination becomes the gut and is called either the primitive gut or the **archenteron.** The pore, or hole, created by invagination is called the blastopore, and in a lancelet, as well as in vertebrates, this pore becomes the anus.

Gastrulation in lancelet is not complete until a third middle layer of cells, **mesoderm,** has formed. In the lancelet, this layer begins as outpocketings from the archenteron; these outpocketings grow in size until they meet and fuse. In effect, then, 2 layers of mesoderm are formed, and the space between them is called the coelom. A *coelom* is defined as a body cavity lined by mesoderm within which the internal organs form.

Figure 21.3 compares human gastrulation to that of the lancelet. In humans, a space called the amniotic cavity appears within the inner cell mass. The portion of the mass below this cavity is the embryonic disk, which elongates to form the primitive streak (fig. 21.4). Some of the upper cells within the primitive streak invaginate and spread out between the remaining cells of the upper layer, now called ectoderm, and the cells of the lower layer, now called endoderm. The invaginating cells are the mesoderm.

Because of the availability of chicken eggs, embryologists also have made a detailed study of this animal's development. In the chick, there is a primitive streak rather than a spherical gastrula because the yolk does not participate in the early stages of development. Human development resembles chick development, despite the fact that the human egg lacks yolk. The evolutionary history of these 2 animals can provide an answer to the amazing resemblance

Table 21.2 Organs Developed from the 3 Primary Germ Layers

Ectoderm	Mesoderm	Endoderm
Skin epidermis, including hair, nails, and sweat glands	All muscles	Lining of most of digestive tract, trachea, bronchi, lungs, gallbladder, and urethra
Nervous system, including brain, spinal cord, ganglia, and nerves	Dermis of skin	
	All connective tissue, including bone, cartilage, and blood	Liver
Retina, lens, and cornea of eye	Blood vessels	Pancreas
Inner ear	Kidneys	Thyroid, parathyroid, and thymus glands
Lining of nose, mouth, and anus	Reproductive organs	Urinary bladder
Tooth enamel		

of their early developmental stages. Both birds (e.g., chicks) and mammals (e.g., humans) are related to reptiles, and this evolutionary relationship manifests itself in the manner in which development proceeds.

Germ Layers

Ectoderm, mesoderm, and endoderm are called the primary *germ layers* of the embryo, and no matter how gastrulation takes place, the end result is the same: 3 germ layers are formed. Early development in most animals follows this same basic pattern. It is possible to relate the development of future organs to these germ layers, as is done for vertebrates in table 21.2.

During gastrulation, the 3 embryonic germ layers (ectoderm, mesoderm, and endoderm) arise. The development of organs can be related to these layers.

The Neurula

In chordate animals, newly formed mesoderm cells that lie along the main longitudinal axis of the animal coalesce to form the dorsal supporting rod called the notochord. The notochord persists in lancelets, but in humans, it is replaced later by the vertebral column.

Figure 21.5 shows an intact human embryo, allowing you to see the external appearance of the developing nervous system. At this point, the embryo is called a **neurula.** Later, the anterior end of the neural tube develops into the brain. The nervous system develops from ectoderm located just above the presumptive notochord. At first, a thickening of cells called the neural plate is seen along the dorsal surface of the embryo. Then, *neural folds* that develop on either side of a neural groove become the neural tube when they fuse. Figure 21.6 shows cross sections of frog development to illustrate the formation of the neural tube.

Midline mesoderm not contributing to the formation of the notochord now becomes 2 longitudinal masses of tissue. From these, blocklike portions called *somites* that give rise to segmental muscles develop. In vertebrates, the somites also produce the vertebral bones.

During neurulation, the neural tube develops just above the notochord.

The Vertebrate Cross Section

With the formation of the nervous system, it is possible to show a generalized diagram (fig. 21.7) of a vertebrate embryo to illustrate placement of parts.

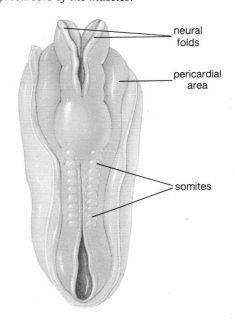

Figure 21.5
Human embryo at 21 days. The neural folds still need to close at the anterior and posterior ends of the embryo. The pericardial area contains the primitive heart, and the somites are the precursors of the muscles.

neural folds

pericardial area

somites

Figure 21.6
*Development of neural tube and coelom in a frog
embryo.*

Ectoderm cells that lie above the
presumptive notochord thicken to form a
neural plate.

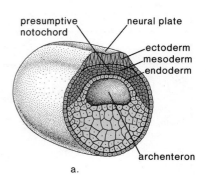

a.

The neural groove and folds are
noticeable as the neural tube begins to
form.

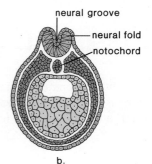

b.

Division of the mesoderm produces a
coelom completely lined by mesoderm.

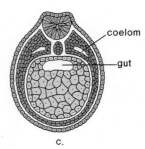

c.

Figure 21.7
*Typical cross section of a chordate embryo at the
neurula stage. Each of the germ layers indicated
by color (see key) can be associated with the later
development of particular parts (see table 21.2).
The somites give rise to the muscles of each
segment and the vertebrae that replace the
notochord.*

A neural tube and a coelom have now
developed.

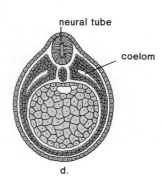

d.

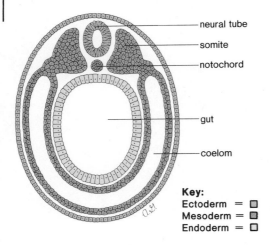

Key:
Ectoderm = ▢
Mesoderm = ▨
Endoderm = ▢

Correlation of figure 21.7 with table 21.2 helps you to relate the formation of
vertebrate structures and organs to the 3 embryonic layers of cells: ectoderm,
mesoderm, and endoderm. The skin and the nervous system develop from ectoderm; the muscles, the skeleton, the kidneys, the circulatory system, and the
gonads develop from mesoderm; and the lining of the digestive tract, the lungs,
the liver, and the pancreas develop from endoderm.

Figure 21.7 illustrates that embryonic vertebrates have a notochord and
a dorsal hollow nerve cord called the neural tube in the figure. Another characteristic of embryonic vertebrates is the presence of paired pharyngeal pouches

Figure 21.8

Human embryo at beginning of fifth week. a. Scanning electron micrograph. b. Drawing. The embryo is curled so that the head touches the heart, 2 organs whose development is further along than the rest of the body. The organs of the gastrointestinal tract are forming. The presence of the tail is an evolutionary remnant; its bones regress and become those of the coccyx. The arms and the legs develop from the bulges that are called limb buds.

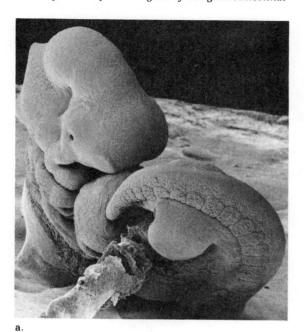

a.

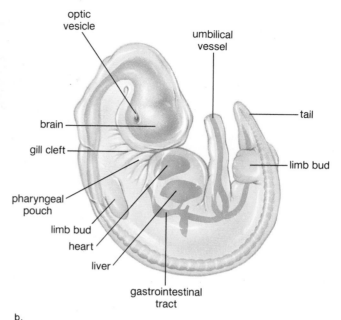

b.

bordering gill clefts (fig. 21.8). Only in primitive invertebrates (fishes and amphibian larvae) do the pouches and clefts become functioning gills. The fact that higher forms go through this embryonic stage of the lower forms indicates a relationship between them. The phrase *ontogeny* (development) *recapitulates* (repeats) *phylogeny* (evolutionary history) was coined some years ago as a dramatic way to suggest that all animals share the same embryonic stages. This theory has been modified today since embryos proceed only through those stages that are consistent with their later development. For example, in advanced vertebrates, although gills never form; the first pair of pharyngeal pouches becomes the auditory cavity of the middle ear and the eustachian tube. The second pair of pouches becomes the tonsils, while the third and fourth pairs become the thymus and the parathyroids. Therefore, pharyngeal pouches and gill clefts develop because they are necessary to later development.

All vertebrates at some time in their development portray a similar cross section that displays typical vertebrate embryonic characteristics: a dorsal hollow nerve cord, a notochord, and a coelom completely surrounded by mesoderm. Also, at some time in their embryonic history, vertebrates have pharyngeal pouches and clefts.

Differentiation and Morphogenesis

Differentiation and morphogenesis are 2 developmental processes that account for the specialization of tissues and the formation of organs that have an overall pattern of shape and form.

Differentiation

Differentiation is apparent when cells are specialized in structure and in function. The process of differentiation must start, though, long before we can recognize different types of cells. After all, the cells that make up ectoderm,

Figure 21.9

Totipotency experiment. The haploid nucleus of a frog's egg is destroyed by ultraviolet irradiation. Now it can receive a diploid nucleus taken from an intestinal cell of a tadpole. In some cases, the reconstituted cell develops into an adult frog.

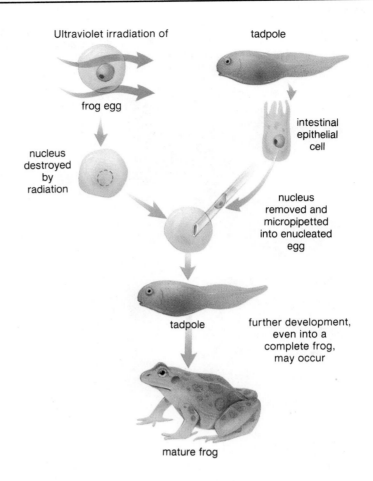

Ultraviolet irradiation of

frog egg

tadpole

nucleus destroyed by radiation

intestinal epithelial cell

nucleus removed and micropipetted into enucleated egg

tadpole

further development, even into a complete frog, may occur

mature frog

endoderm, and mesoderm in the gastrula look quite similar, but they are destined to develop into different organs. What causes differentiation to occur, and when does it begin?

The contents of the cytoplasm of an egg are not distributed uniformly. Following cleavage, therefore, the cytoplasmic content received by embryonic cells differs:

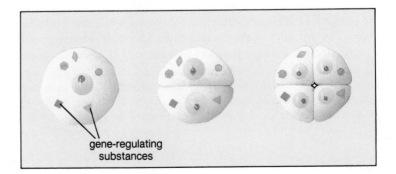

gene-regulating substances

These varying contents probably contain substances that play a role in regulating which genes are active in a particular cell. Figure 21.9 describes an experiment in which a nucleus from an intestinal cell of a tadpole is transplanted into an egg, the nucleus of which was destroyed. Development proceeds normally, showing that embryonic nuclei are totipotent—they contain all the genetic information required to bring about complete development of the organism. Therefore, differentiation cannot be due to the parceling of genes into the various embryonic cells. Instead, it must be due to the expression of

Figure 21.10

Importance of the gray crescent in frog development. a. The frog's egg is polar; there is an animal pole and a vegetal pole (contains the yolk). The position of the gray crescent can be correlated with the anterior/posterior and dorsal/ventral axes of the body. b. The first cleavage (1) normally divides the gray *crescent in half, and each daughter cell (2) is capable of developing into a complete tadpole, but if only one daughter cell receives the gray crescent, then only that cell can become a complete embryo.*

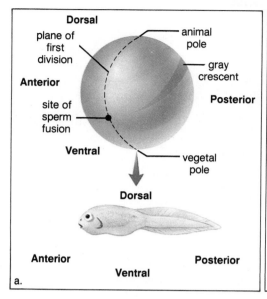

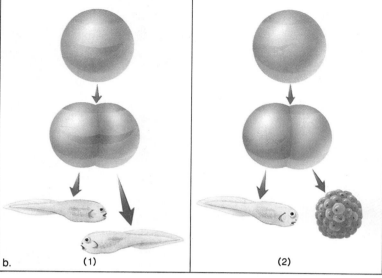

particular genes, controlled first by *ooplasmic segregation,* which is the distribution of maternal cytoplasmic contents to the cells of the morula.

The fact that the cytoplasm of the egg indeed does play a role in regulating development can be demonstrated by another experiment with a frog embryo. After the sperm fuses with the egg, some contents of the egg shift position, and a *gray crescent* appears on the egg opposite the point where the sperm entered (fig. 21.10). Normally, the first cleavage gives each daughter cell half of the gray crescent. In this case, each of the daughter cells has the potential to become a complete embryo. However, if the researcher causes the egg to divide so that only one daughter cell receives the gray crescent, only that cell can become a complete embryo.

Even though the first 2 cells of a frog embryo normally are able to develop into a complete embryo, this ability eventually disappears. At some point in all embryos—some earlier than others—the developmental potential of a cell is restricted to a particular fate. Only a transplant experiment that places the nucleus in the cytoplasmic contents of the complete egg shows that all the genes still are present in embryonic nuclei. This, too, indicates the importance of cellular cytoplasmic contents in early development.

Cytoplasmic substances unequally distributed in the egg are parceled out during cleavage. These substances initially influence which genes are active and how a cell differentiates.

Morphogenesis

As development proceeds, a cell's differentiation not only is influenced by its cytoplasmic contents, but also by signals given off by neighboring cells. Migration of cells occurs during gastrulation, and there is evidence that one set of cells can influence the migratory path taken by another set of cells. Some cells produce an extracellular matrix that contains fibrils, and in the laboratory, it can be shown that the orientation of these fibrils influences migratory

Figure 21.11
Experiments proving importance of presumptive notochord. In experiment A, presumptive nervous system tissue does not complete its development when moved from its location above the notochord. On the other hand, in experiment B, presumptive notochord can cause even presumptive belly ectoderm to develop into a nervous system.

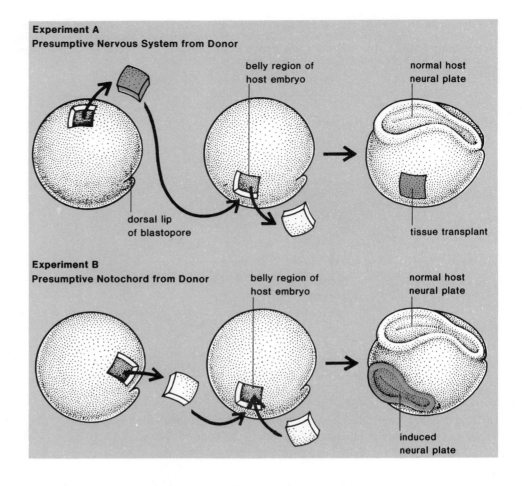

Experiment A
Presumptive Nervous System from Donor

dorsal lip of blastopore

belly region of host embryo

normal host neural plate

tissue transplant

Experiment B
Presumptive Notochord from Donor

belly region of host embryo

normal host neural plate

induced neural plate

cells. The cytoskeletons of the migrating cells are oriented in the same direction as the fibrils. Although this may not be an exact mechanism at work during gastrulation, it suggests that germ layer formation probably is influenced by environmental factors.

More specific information is known about neurulation. Experiments have shown that presumptive (potential) notochord tissue induces the formation of the nervous system (fig. 21.11). If presumptive nervous system tissue, located just above the notochord, is cut out and transplanted to the belly region of the embryo, it will not form a neural tube. On the other hand, if presumptive notochord tissue is cut out and transplanted beneath what would be belly ectoderm, this ectoderm differentiates into neural tissue.

There is another well-known example of **induction** (the ability of one tissue to influence the development of another tissue) during the development of the eye in frog embryos (fig. 21.12). The optic vesicles, which are lateral outgrowths from developing brain tissue, induce the overlying ectoderm to thicken and to become a lens. The developing lens in turn induces the optic vesicle to form the optic cup, where the retina develops.

Early investigators called the dorsal lip of the blastopore an organizer because it gives rise to notochord tissue, which in turn induces neural tube formation. They envisioned that morphogenesis is dependent upon a series of organizers that come into being sequentially and direct future development. Today, we believe the process of induction goes on continuously—neighboring cells are always influencing one another. Either direct contact or the produc-

Figure 21.12

The series of inductions accounting for the development of the vertebrate eye. a. The optic vesicles are lateral outgrowths from the rudimentary brain and induce the overlying ectoderm to thicken. b. and c. This presumptive lens (lens vesicle) in turn induces the optic vesicles to form the optic cup. d. The optic cup induces the formation of the lens and the cornea.

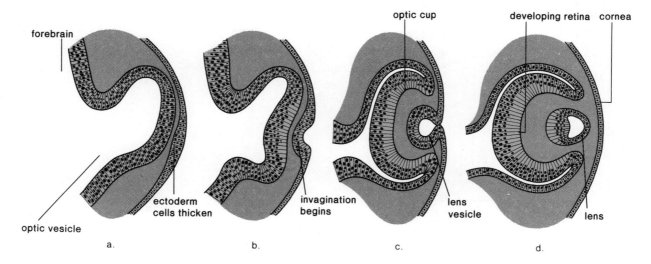

tion of a chemical acts as a signal that activates certain genes and brings about protein synthesis. This diagram shows how morphogenesis can be a sequential process:

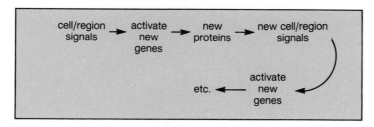

The diagram points out that there are both genetic and chemical aspects to morphogenesis. When the notochord induces the ectoderm to form the neural tube, a chemical signal is involved. The neural tube forms even if the 2 tissues are separated by a filter that allows only molecules to pass through. Presumably, this chemical messenger activates particular genes in the ectodermal tissue, and this is why one tissue forms the neural tube.

Morphogenesis is dependent upon signals (either contact or chemical) from neighboring cells. These signals are believed to activate particular genes.

Human Embryonic and Fetal Development

Human development is often divided into embryonic development (first 2 months) and fetal development (3–9 months) (fig. 21.1). The embryonic period consists of early formation of the major organs, and fetal development is a refinement of these structures.

Before we consider human development chronologically, we must understand the placement of **extraembryonic membranes.** Extraembryonic membranes are best understood by considering their function in reptiles and birds. In reptiles, these membranes made development on land first possible. If an embryo develops in the water, the water supplies oxygen for the embryo and

21.1 Critical Thinking

1. With the help of these questions, develop a scenario to explain why a particular type of cell gives off *particular* signals.
   a. What might be the effect on genes when a cell inherits a certain cytoplasmic composition?
   b. What do activated genes do?
   c. How might some of these proteins act?
2. With the help of these questions, develop a scenario to explain why embryonic development is so orderly.
   a. Tissue A just has become differentiated. What does it give off to affect tissue B?
   b. Having received a certain signal, what does tissue B do to affect tissue C?
   c. Having received a certain signal, what does tissue C do?

Table 21.3 *Extraembryonic Membranes in Chick and Human*

Name	Germ Layers	Location in Chick	Function in Chick	Location in Humans	Function in Humans
Chorion	Outer layer of ectoderm and inner layer of mesoderm	Lies next to shell	Gas exchange	Fetal half of placenta	Exchange with mother's blood
Amnion	Outer layer of mesoderm and inner layer of ectoderm	Surrounds embryo	Protection; prevention of desiccation	Same	Same
Allantois	Outer layer of mesoderm and inner layer of endoderm	Outgrowth of hindgut	Collection of nitrogenous waste	Same	Blood vessels become umbilical blood vessels
Yolk sac	Outer layer of mesoderm and inner layer of endoderm	Outgrowth of midgut; surrounds yolk	Provision of nourishment	Same, but contains no yolk	First site of blood cell formation

Figure 21.13

Extraembryonic membranes. The membranes, which are not part of the embryo, also are found during the development of chicks and humans, where each has a specific function (table 21.3).

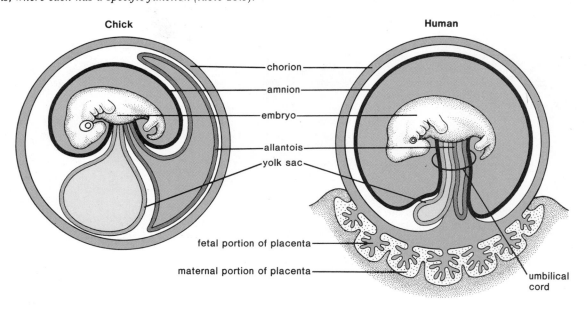

Chick

Human

chorion
amnion
embryo
allantois
yolk sac
fetal portion of placenta
maternal portion of placenta
umbilical cord

takes away waste products. The surrounding water prevents desiccation, or drying out, and provides a protective cushion. For an embryo that develops on land, all these functions are performed by the extraembryonic membranes.

In the chick, the extraembryonic membranes develop from extensions of the germ layers that spread out over the yolk. Each extraembryonic membrane consists of 2 germ layers (table 21.3). Figure 21.13 shows the chick within its hard shell surrounded by the membranes. The **chorion** lies next to the shell and carries on gas exchange. The **yolk sac** surrounds the remaining yolk. The **allantois** collects nitrogenous waste, and the **amnion** contains the amniotic fluid that bathes the developing embryo.

As figure 21.13 indicates, humans (and other mammals as well) also have these extraembryonic membranes. The chorion develops into the fetal half of the placenta; the yolk sac lacks yolk but is the first site of blood cell formation; the allantoic blood vessels become the umbilical blood vessels; and

Figure 21.14

Human development before implantation. Structures and events proceed counterclockwise. At ovulation, the secondary oocyte leaves the ovary. Fertilization occurs in the oviduct. As the zygote moves along the oviduct, it undergoes cleavage to produce a morula. The blastocyst forms and implants itself in the uterine lining.

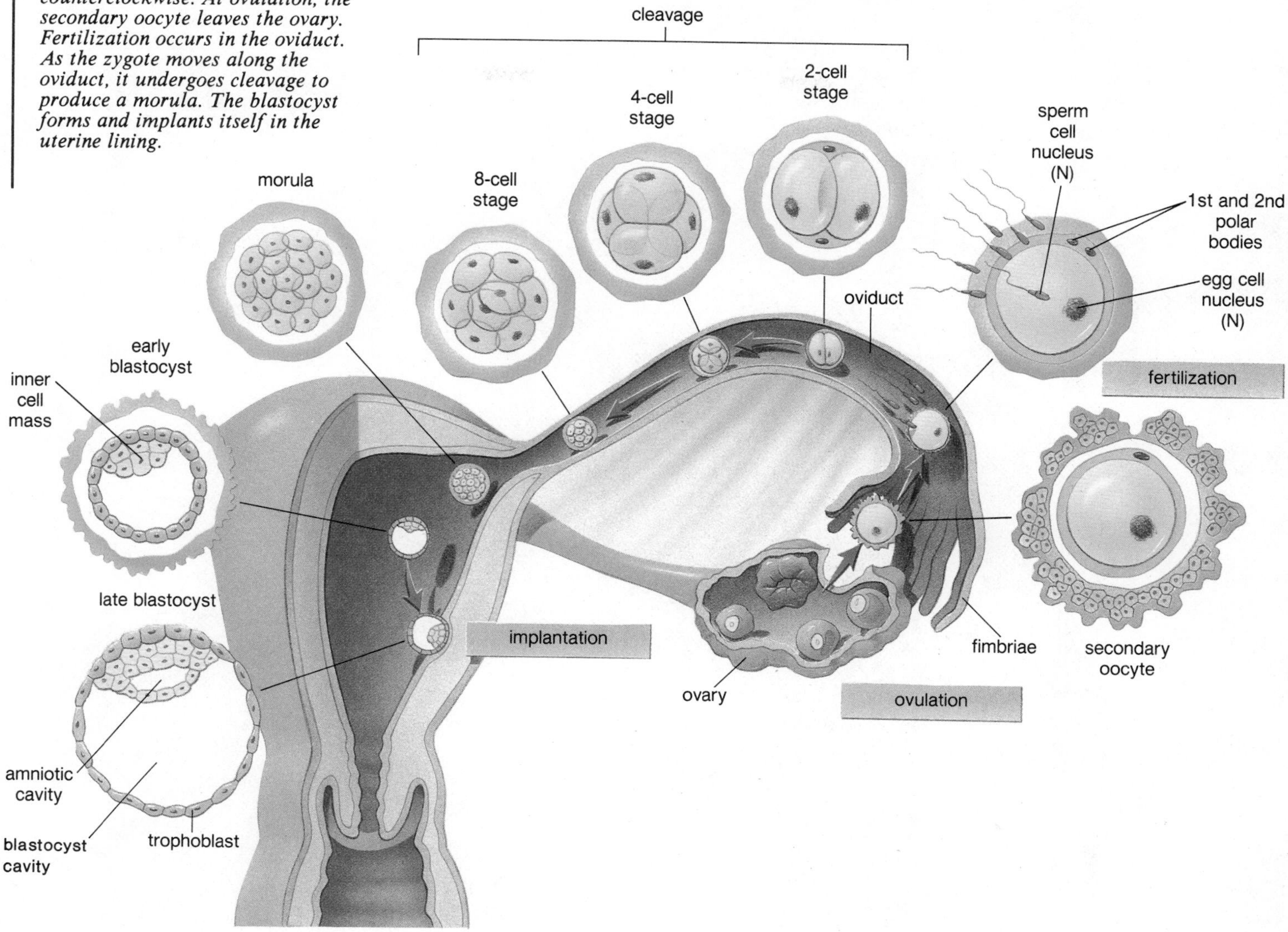

cleavage

2-cell stage

4-cell stage

sperm cell nucleus (N)

8-cell stage

1st and 2nd polar bodies

morula

egg cell nucleus (N)

oviduct

early blastocyst

fertilization

inner cell mass

late blastocyst

fimbriae

secondary oocyte

amniotic cavity

implantation

blastocyst cavity

trophoblast

ovary

ovulation

the amnion contains fluid to cushion and to protect the fetus. Thus, the function of the membranes has been modified to suit internal development, but their very presence indicates our relationship to birds and to reptiles. It is interesting to note that all animals develop in water, either directly or within amniotic fluid.

The presence of extraembryonic membranes in reptiles made development on land possible. Humans also have these membranes, but their function has been modified for internal development.

Embryonic Development

The First Week

Fertilization occurs in the upper third of an oviduct (fig. 21.14), and cleavage begins even as the embryo passes down this tube to the uterus. By the time the embryo reaches the uterus on the third day, it is a *morula*. The morula is not much larger than the zygote because although multiple cell divisions have occurred, there has been no growth of these newly formed cells. By about the

fifth day, the morula is transformed into the *blastocyst*. The blastocyst has a fluid-filled cavity, a single layer of outer cells called the **trophoblast,** and an inner cell mass. Later, the trophoblast, reinforced by a layer of mesoderm, gives rise to the *chorion,* one of the extraembryonic membranes (fig. 21.13). The *inner cell mass* eventually becomes the fetus. Each cell within the inner cell mass has the genetic capability of becoming a complete individual. Sometimes during human development, the inner cell mass splits, and 2 embryos start developing rather than one. These 2 embryos that share the same placenta (p. 453) are identical twins because they have inherited exactly the same chromosomes. Fraternal twins, who arise when 2 different eggs are fertilized by 2 different sperm, do not have identical chromosomes. It even has been known to happen that these "twins" have different fathers. There is a placenta for each fraternal twin during development.

During the first week, the human embryo undergoes cleavage. Then the morula becomes the blastocyst having 2 main parts, the outer trophoblast (becomes the chorion) and the inner cell mass (becomes the fetus).

Figure 21.15

Stages showing the early appearance of the extraembryonic membranes and the formation of the umbilical cord in the human embryo. a. At 2 weeks, the amniotic cavity appears. b. At 3 weeks, the chorion and the yolk sac are apparent. c. At 4 weeks, the body stalk and the allantois form. d. At 5 weeks, the embryo begins to take shape as the umbilical cord forms. e. Eventually, the umbilical cord is formed fully.

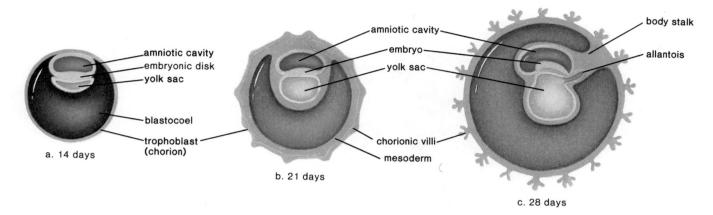

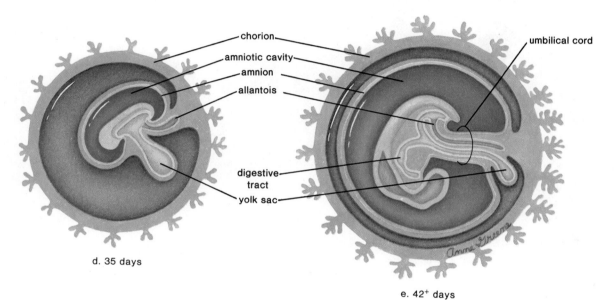

a. 14 days

b. 21 days

c. 28 days

d. 35 days

e. 42⁺ days

The Second Week

At the end of the first week, the embryo begins the process of *implanting* in the wall of the uterus. The trophoblast secretes enzymes to digest away some of the tissue and blood vessels of the uterine wall (fig. 21.14). The embryo is now about the size of the period at the end of this sentence. The trophoblast begins to secrete *HCG* (human chorionic gonadotropin), the hormone that is the basis for the pregnancy test and that serves to maintain the corpus luteum past the time it normally disintegrates. Because of this, the endometrium is maintained and menstruation does not occur.

As the week progresses, the inner cell mass detaches itself from the trophoblast, and 2 more extraembryonic membranes form (fig. 21.15). The *yolk sac,* which forms below the embryo, has no nutritive function, but it is the first site of blood cell formation. However, the *amnion* and its cavity are where the embryo (and then the fetus) develops. The amniotic fluid acts as an insulator against cold and heat and also absorbs any shock, such as a blow to the mother's abdomen.

Gastrulation occurs during the second week. The inner cell mass now has flattened into the *embryonic disk* composed of 2 layers of cells: *ectoderm* above and *endoderm* below. Once the embryonic disk elongates to become the

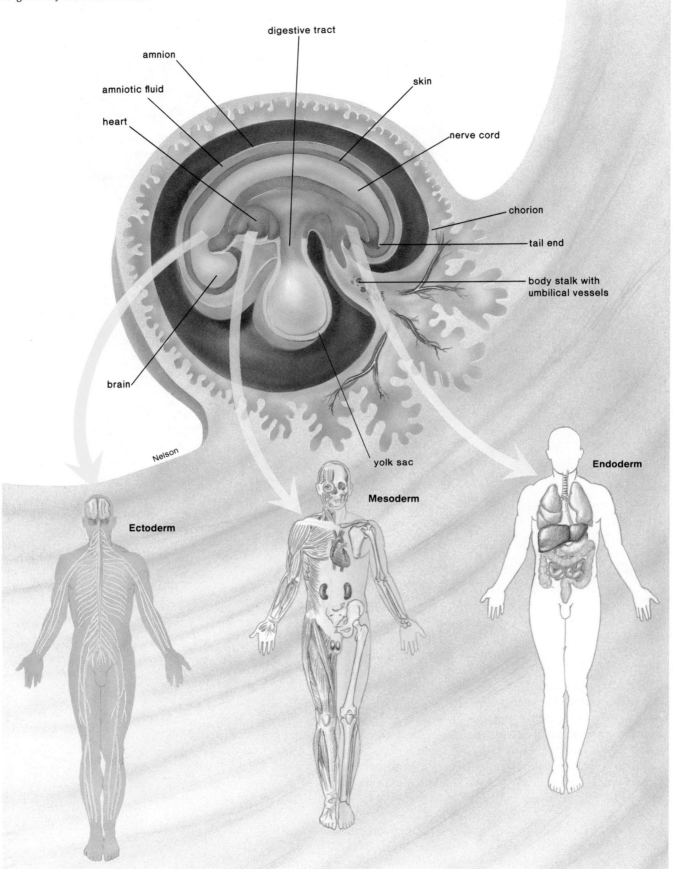

Figure 21.16
Germ layer theory. The organs of the body develop from one of the 3 germ layers as indicated.

digestive tract

amnion

amniotic fluid

skin

heart

nerve cord

chorion

tail end

body stalk with umbilical vessels

brain

Nelson

yolk sac

Endoderm

Mesoderm

Ectoderm

primitive streak (fig. 21.4), the third germ layer, mesoderm, forms by invagination of cells along the streak. The trophoblast is reinforced by mesoderm and becomes the chorion.

It is possible to relate the development of future organs to these germ layers (table 21.2). In general, ectoderm becomes the nervous system, the skin, the hair, and the nails; endoderm produces the inner linings of the digestive, respiratory, and urinary tracts; and mesoderm produces the muscles, the skeleton, and the circulatory systems (fig. 21.16).

The germ layers (ectoderm, endoderm, and mesoderm) are laid down during the third week of development. Development of the organs can be related to these germ layers.

The Third Week

Two important organ systems make their appearance during the third week. The nervous system is the first organ system to be visually evident. At first, a thickening appears along the entire dorsal length of the embryo, and then invagination occurs as neural folds appear. When the neural folds meet at the midline, the neural tube, which later develops into the brain and the nerve cord, is formed:

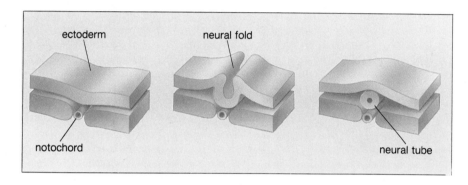

The notochord is replaced later by the vertebral column; the nerve cord then is called the spinal cord. Figure 21.16 shows a human embryo at the end of the third week, after neurulation has begun.

Development of the heart begins in the third week and continues into the fourth week. At first, there are right and left heart tubes; when these fuse, the heart begins pumping blood, even though the chambers of the heart are not formed fully yet. The veins enter posteriorly and the arteries exit anteriorly from this largely tubular heart, but later the heart twists so that all major blood vessels are located anteriorly.

During the third week, major organs, like the nerve cord and the heart, first make their appearance.

Fourth and Fifth Weeks

At four weeks, the embryo is barely larger than the height of this print. There is a bridge of mesoderm called the body stalk that connects the caudal end of the embryo with the chorion (fig. 21.15c). The fourth extraembryonic membrane, the *allantois* (fig. 21.13), is contained within this stalk, and its blood vessels become the umbilical blood vessels. Then the head and the tail lift up, and the body stalk moves anteriorly by constriction (fig. 21.15d). Once this process is complete, the **umbilical cord** that connects the developing embryo to the placenta is fully formed (fig. 21.15e).

Figure 21.17
Human embryo at the days indicated and at sizes noted.

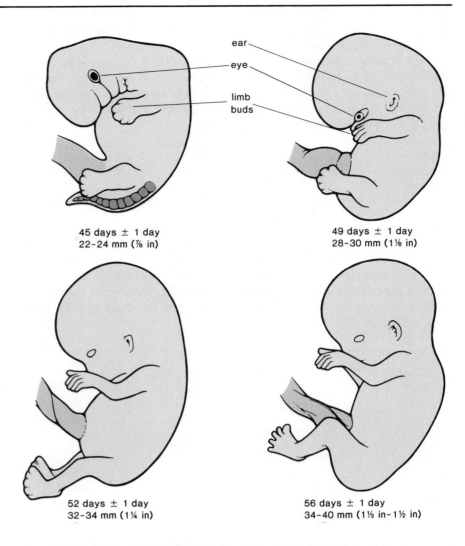

45 days ± 1 day
22–24 mm (⅞ in)

49 days ± 1 day
28–30 mm (1⅛ in)

52 days ± 1 day
32–34 mm (1¼ in)

56 days ± 1 day
34–40 mm (1⅓ in–1½ in)

Little flippers called limb buds appear (fig. 21.8); later, the arms and the legs develop from the limb buds, and even the hands and the feet become apparent. At the same time, during the fifth week, the head becomes larger than previously, and the sense organs become more prominent. It is possible to make out the developing eyes, ears, and even nose.

During the fourth and fifth weeks, human features, like the head, the arms, and the legs, begin to make their appearance.

Sixth through Eighth Weeks

There is a remarkable change in external appearance during the sixth through eighth weeks of development (fig. 21.17), from a form that is difficult to recognize as human to one that easily is recognizable as human. Concurrent with brain development, the head achieves its normal relationship with the body as a neck region develops. The nervous system is developed well enough to permit reflex actions, such as a startle response to being touched. At the end of this period, the embryo is about 38 mm (1½ in) long and weighs no more than an aspirin tablet, even though all organ systems are established.

The Placenta   The *placenta* begins formation once the embryo is implanted fully. Treelike extensions of the chorion called **chorionic villi** project into the

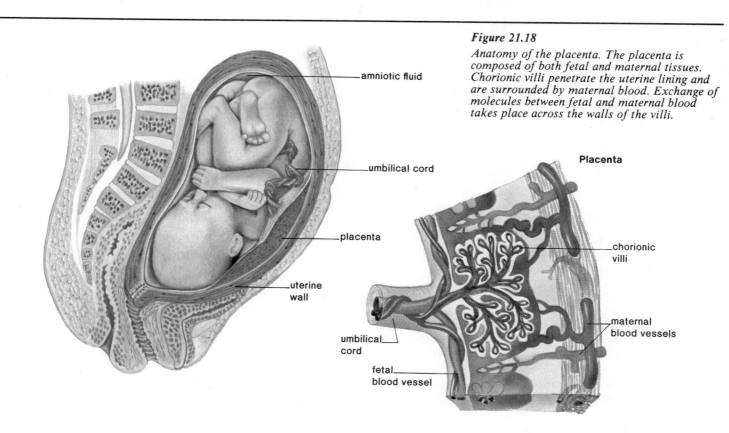

Figure 21.18

Anatomy of the placenta. The placenta is composed of both fetal and maternal tissues. Chorionic villi penetrate the uterine lining and are surrounded by maternal blood. Exchange of molecules between fetal and maternal blood takes place across the walls of the villi.

amniotic fluid

umbilical cord

placenta

uterine wall

Placenta

chorionic villi

maternal blood vessels

umbilical cord

fetal blood vessel

maternal tissues. Later, these disappear in all areas except where the placenta develops. By the tenth week, the placenta (fig. 21.18) is formed fully and begins to produce progesterone and estrogen (fig. 21.19). These hormones have 2 effects: due to their negative feedback effect on the hypothalamus and the anterior pituitary, they prevent any new follicles from maturing, and they maintain the lining of the uterus—now the corpus luteum is not needed. There is no menstruation during pregnancy.

The placenta has a fetal side contributed by the chorion and a maternal side consisting of uterine tissues. Notice in figure 21.18 how the chorionic villi are surrounded by maternal blood sinuses; yet, the blood of the mother and the fetus never mix since exchange always takes place across cell membranes. Carbon dioxide and other wastes move from the fetal side to the maternal side, and nutrients and oxygen move from the maternal side to the fetal side of the placenta. The umbilical cord stretches between the placenta and the fetus. Although it may seem that the umbilical cord travels from the placenta to the intestine, actually the umbilical cord simply is taking fetal blood to and from the placenta. The umbilical cord is the lifeline of the fetus because it contains the umbilical arteries and vein, which transport waste molecules (carbon dioxide and urea) to the placenta for disposal and take oxygen and nutrient molecules from the placenta to the rest of the fetal circulatory system.

Harmful chemicals also can cross the placenta, and this is of particular concern during the embryonic period, when various structures first are forming. Each organ or part seems to have a sensitive period during which a substance can alter its normal function. The reading on page 454 concerns the origination of birth defects.

At the end of the embryonic period, all organ systems are established and there is a mature and fully functioning placenta. The embryo is only about 38 mm (1½ in) long.

Figure 21.19

Hormones during pregnancy. Human chorionic gonadotropin is secreted by the trophoblast during the first 3 months of pregnancy. This maintains the corpus luteum, which continues to secrete estrogen and progesterone. At about 5 weeks of pregnancy, the placenta begins to secrete estrogen and progesterone in increasing amounts as the corpus luteum degenerates.

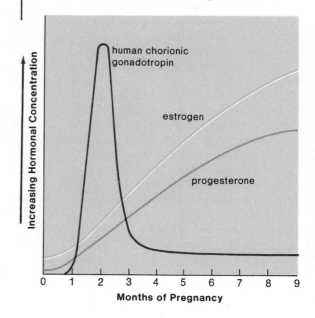

human chorionic gonadotropin

estrogen

progesterone

Increasing Hormonal Concentration

Months of Pregnancy

It is believed that at least 1 in 16 newborns has a birth defect, either minor or serious, and the actual percentage may be even higher. Most likely, only 20% of all birth defects are due to heredity. Those that are sometimes can be detected before birth. Amniocentesis allows the fetus to be tested for abnormalities of development. Chorionic villi sampling allows the embryo to be tested, and just recently, a method has been developed for screening eggs to be used for in vitro fertilization (see figure).

Treatment of the fetus in the womb is a rapidly developing area of medical expertise. Biochemical defects sometimes can be

Birth Defects

treated by giving the mother appropriate medicines. For example, if a baby is unable to synthesize vitamin B and/or is unable to use biotin efficiently, the mother can take these substances in doses large enough to prevent any untoward effects. Structural defects sometimes can be corrected by surgery. For example, if the fetus has water on the brain or is unable to pass urine, tubes that temporarily allow the fluid to pass out into the amniotic fluid can be inserted even while the fetus is still in the womb. Physicians are hopeful that eventually all sorts of structural defects can be corrected by lifting the fetus from the womb long enough for corrective surgery to be done.

It is recommended that all females take everyday precautions to protect any future and/or presently developing embryos and fetuses from defects that are not due to heredity. X-ray diagnostic therapy should be avoided during pregnancy because X rays are mutagenic to a developing embryo or fetus. Children born to women who received X-ray treatment are apt to have birth defects and/or to develop leukemia later on. Fetotoxic chemicals, such as pesticides and many organic industrial chemicals, are also mutagenic. Cigarette smoke not only contains carbon monoxide but also some of these very same fetotoxic chemicals. Babies born to smokers are often underweight and are subject to convulsions.

Pregnant Rh-negative women should receive a RhoGam injection to prevent the production of Rh antibodies. These antibodies can cause nervous system and heart defects.

Sometimes, birth defects are caused by microorganisms. Females can be immunized before the childbearing years for rubella (German measles), which causes birth defects such as deafness. Immunization for sexually transmitted diseases is not possible. The AIDS virus can cross the placenta, and over 1,500 babies who contracted AIDS while in their mother's womb are now mentally retarded. When a mother has herpes, gonorrhea, or chlamydia, newborns can become infected as they pass through the birth canal and might become blind or develop other mental and physical defects. Birth by cesarean section could prevent these occurrences.

Drugs of all types should be avoided. Certainly illegal drugs, like marijuana, cocaine, and heroin, should be completely avoided. "Cocaine babies" now make up 60% of drug-affected babies. Severe fluctuations in blood pressure accompany the use of cocaine and temporarily deprive the developing brain of oxygen. Cocaine babies have visual problems, lack coordination, and are mentally retarded. The drugs aspirin, caffeine (present in coffee, tea, and cola), and alcohol should be severely limited. It is not unusual for babies of drug addicts and alcoholics to display withdrawal symptoms and to have various abnormalities. Babies born to women who have about 45 drinks a month and as many as 5 drinks on one occasion are apt to have FAS (fetal alcohol syndrome). These babies have decreased weight, height, and head size, with malformation of the head and face. Mental retardation is common in FAS infants.

Medications can also sometimes cause problems. When the synthetic hormone DES was given to pregnant women to prevent miscarriage, their daughters showed various abnormalities of the reproductive organs and an increased tendency toward

cervical cancer. Other sex hormones, including birth control pills, possibly can cause abnormal fetal development, including abnormalities of the sex organs. The tranquilizer thalidomide is well known for having caused deformities of the arms and legs in children born to women who took the drug. Therefore, a woman has to be very careful about taking medications while pregnant.

Now that physicians and laypeople are aware of the various ways in which birth defects can be prevented, it is hoped that the incidence of birth defects will decrease in the future.

Three methods for genetic defect testing before birth. a. *Amniocentesis cannot be done until the sixteenth week of pregnancy. A long needle is passed through the abdominal wall to withdraw a small amount of amniotic fluid along with fetal cells. Since there are only a few cells in the amniotic fluid, testing must be delayed for 4 weeks until cell culture produces enough cells for testing purposes.*

b. *Chorionic villi sampling can be done as early as the fifth week of pregnancy. The doctor inserts a long, thin tube through the vagina into the uterus. With the help of ultrasound, which gives a picture of the uterine contents, the tube is placed between the lining of the uterus and the chorion. Then suction is used to remove a sampling of the chorionic villi cells. Chromosome analysis and biochemical tests for several different genetic defects can be done immediately on these cells.*

c. *Screening eggs for genetic defects is a new technique. Preovulation eggs are removed by aspiration after a telescope with a fiber-optic illuminator, called a laparoscope, is inserted into the abdominal cavity through a small incision in the region of the navel. The prior administration of FSH ensures that several eggs are available for screening. Only the chromosomes within the first polar body are tested because if the woman is heterozygous for a genetic defect and it is found in the polar body, then the egg must be normal. Normal eggs undergo in vitro fertilization and are placed in the prepared uterus. At present, only 1 in 10 attempts results in a birth, but it is known ahead of time that the child will be normal.*

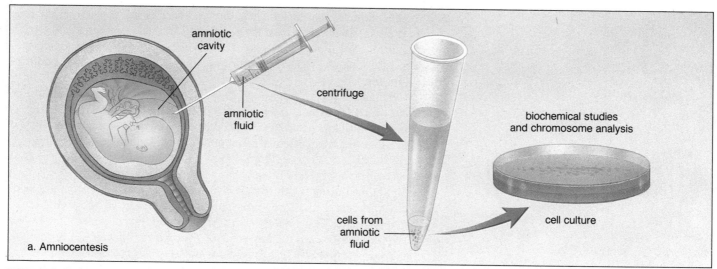

amniotic cavity

amniotic fluid

centrifuge

biochemical studies and chromosome analysis

cells from amniotic fluid

cell culture

a. Amniocentesis

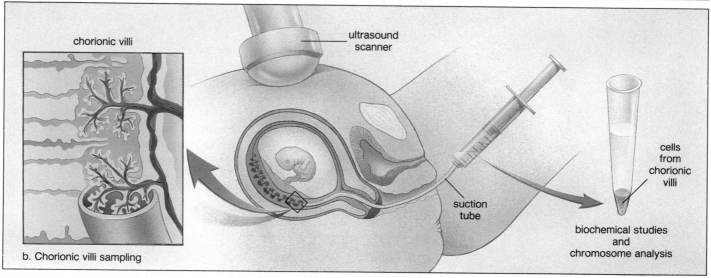

chorionic villi

ultrasound scanner

suction tube

cells from chorionic villi

biochemical studies and chromosome analysis

b. Chorionic villi sampling

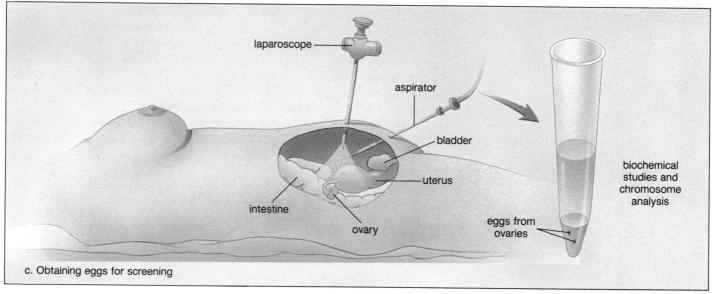

laparoscope

aspirator

bladder

uterus

intestine

ovary

eggs from ovaries

biochemical studies and chromosome analysis

c. Obtaining eggs for screening

Fetal Development

Third and Fourth Months

At the beginning of the third month, the fetal head is still very large, the nose is flat, the eyes are far apart, and the ears are distinctively present. Head growth now begins to slow down as the rest of the body increases in length. Epidermal refinements, such as eyelashes, eyebrows, hair on head, fingernails, and nipples, appear.

Cartilage is replaced by *bone* as ossification centers appear in most of the bones. Cartilage remains at the ends of the long bones, and ossification is not complete until age 18 or 20. The skull has 6 large membranous areas called *fontanels* that permit a certain amount of flexibility as the head passes through the birth canal and allow rapid growth of the brain during infancy. The fontanels disappear by 2 years of age.

Sometime during the third month, it is possible to distinguish males from females. Apparently, the Y chromosome has a gene called the testis determining factor gene (TDF) that triggers the differentiation of gonads into testes. Once the testes differentiate, they produce androgens, the male sex hormones. The androgens, especially testosterone, stimulate the growth of the male external genitalia. In the absence of androgens, female genitalia form. The ovaries do not produce estrogen because there is plenty of it circulating in the mother's bloodstream.

At this time, both testes and ovaries are located within the abdominal cavity, but later, in the last trimester of fetal development, the testes descend into the scrotal sacs (scrotum). Sometimes the testes fail to descend, and in that case, an operation may be done later to place them in their proper location.

During the fourth month, the fetal heartbeat is loud enough to be heard when a physician applies a stethoscope to the mother's abdomen. By the end of this month, the fetus is less than 140 mm (6 in) in length and weighs a little more than 200 g (½ lb).

During the third and fourth months, it is obvious that the skeleton is becoming ossified. The sex of the individual is now distinguishable.

Fifth through Seventh Months

During the fifth through seventh months, the mother begins to feel movement. At first, there is only a fluttering sensation, but as the fetal legs grow and develop, kicks and jabs are felt. The fetus, though, is in the fetal position with the head bent down and in contact with the flexed knees.

The wrinkled, translucent, pink-colored skin is covered by a fine down called **lanugo.** This in turn is coated with a white, greasy, cheeselike substance called **vernix caseosa,** which probably protects the delicate skin from the amniotic fluid. The eyelids now are open fully, however.

At the end of this period, weight has increased to almost 1,350 gm (3 lb) and the length to almost 300 mm (12 in). It is possible that if born now, the baby will survive.

Fetal Circulation   As figure 21.20 shows, the fetus has 4 circulatory features that are not present in adult circulation.

1. **Oval opening,** or *foramen ovale,* an opening between the 2 atria. This opening is covered by a flap of tissue that acts as a valve.
2. **Arterial duct,** or *ductus arteriosus,* a connection between the pulmonary artery and the aorta.
3. **Umbilical arteries** and **vein,** vessels that travel to and from the placenta, leaving waste and receiving nutrients.
4. **Venous duct,** or *ductus venosus,* a connection between the umbilical vein and the inferior vena cava.

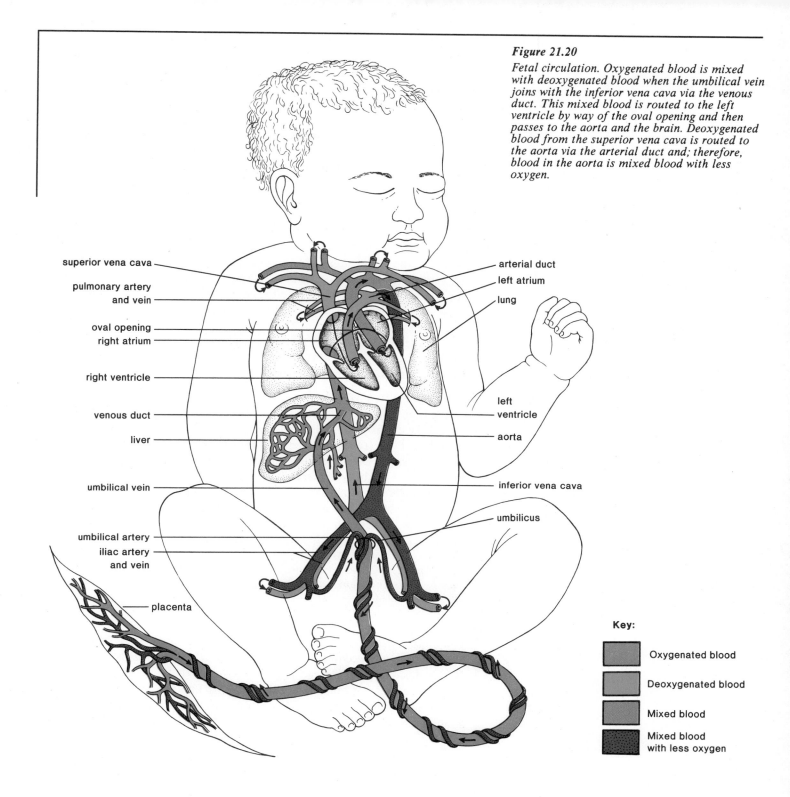

Figure 21.20
Fetal circulation. Oxygenated blood is mixed with deoxygenated blood when the umbilical vein joins with the inferior vena cava via the venous duct. This mixed blood is routed to the left ventricle by way of the oval opening and then passes to the aorta and the brain. Deoxygenated blood from the superior vena cava is routed to the aorta via the arterial duct and; therefore, blood in the aorta is mixed blood with less oxygen.

superior vena cava

pulmonary artery and vein

oval opening
right atrium

right ventricle

venous duct

liver

umbilical vein

umbilical artery
iliac artery and vein

placenta

arterial duct
left atrium

lung

left ventricle

aorta

inferior vena cava

umbilicus

Key:

Oxygenated blood

Deoxygenated blood

Mixed blood

Mixed blood with less oxygen

All of these features can be related to the fact that the fetus does not use its lungs for gas exchange since it receives oxygen and nutrients from the mother's blood by way of the placenta.

To trace the path of blood in the fetus, begin with the right atrium (fig. 21.20). From the right atrium, the blood may pass directly into the left atrium by way of the oval opening or it may pass through the atrioventricular valve into the right ventricle. From the right ventricle, the blood goes into the pulmonary artery, but because of the arterial duct, most of the blood then passes into the aorta. Therefore, by whatever route the blood takes, most of the blood reaches the aorta instead of the lungs.

Blood within the aorta travels to the various branches, including the iliac arteries that connect to the umbilical arteries leading to the placenta. Exchange between maternal blood and fetal blood takes place at the placenta. It is interesting to note that the blood in the umbilical arteries, which travels to the placenta, is low in oxygen, but the blood in the umbilical vein, which travels from the placenta, is high in oxygen. The umbilical vein enters the venous duct, which passes directly through the liver. The venous duct then joins with the inferior vena cava, a vessel that contains deoxygenated blood. The vena cava returns this "mixed blood" to the heart.

The most common of all cardiac defects in the newborn is the persistence of the oval opening. With the tying of the cord and the expansion of the lungs, blood enters the lungs in quantity. Return of this blood to the left side of the heart usually causes a flap to cover the opening. Incomplete closure occurs in nearly 1 out of 4 individuals, but even so, passage of the blood from the right atrium to the left atrium rarely occurs because either the opening is small or it closes when the atria contract. In a small number of cases, the passage of impure blood from the right side to the left side of the heart is sufficient to cause a "blue baby." Such a condition now can be corrected by open-heart surgery.

The arterial duct closes because endothelial cells divide and block off the duct. Remains of the arterial duct and parts of the umbilical arteries and vein later are transformed into connective tissue.

Eighth and Ninth Months

As the time of birth approaches, the fetus rotates so that the head is pointed toward the cervix (fig. 21.21a). If the fetus does not turn, then the likelihood of a breech birth (rump first) may call for a cesarean section. It is very difficult for the cervix to expand enough to accommodate this form of birth, and asphyxiation of the baby is more likely.

At the end of this time period, the fetus is about 530 mm (21 in) long and weighs about 3,400 gm (7½ lb). Weight gain is due largely to an accumulation of fat beneath the skin.

From the fifth to the ninth month, the fetus continues to grow and to gain weight. Babies born after 6 or 7 months may survive but are subject to various illnesses that can have lasting effects or cause an early death.

Birth

The uterus characteristically contracts throughout pregnancy. At first, light, often indiscernible contractions lasting about 20–30 seconds occur every 15–20 minutes, but near the end of pregnancy, they become stronger and more frequent so that the woman may think falsely that she is in labor. The onset of true labor is marked by uterine contractions that occur regularly every 15–20 minutes and last for 40 seconds or more. **Parturition,** which includes labor and expulsion of the fetus, usually is considered to have 3 stages.

The events that cause parturition still are not known entirely, but there is now evidence suggesting the involvement of prostaglandins. It may be, too, that the prostaglandins cause the release of oxytocin from the maternal posterior pituitary. Both prostaglandins and oxytocin cause the uterus to contract, and either hormone can be given to induce parturition.

Stages

During the *first stage* of parturition, the cervix dilates; during the *second,* the baby is born; and during the *third,* the afterbirth is expelled.

Stage 1   Prior to or concomitant with the first stage of parturition, there can be a "bloody show" caused by the expulsion of a mucus plug from the cervical canal. This plug prevents bacteria and sperm from entering the uterus during pregnancy.

Figure 21.21
Three stages of parturition. a. *Position of fetus just before birth begins.* b. *Dilation of cervix.* c. *Birth of baby.* d. *Expulsion of afterbirth.*

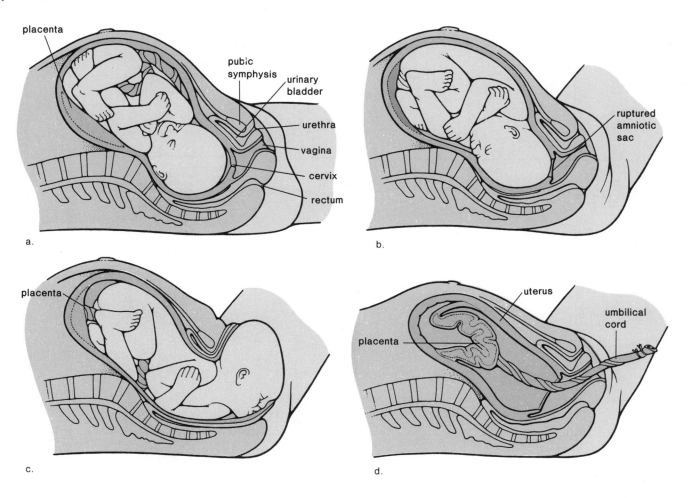

Uterine contractions during the first stage of labor occur in such a way that the cervical canal slowly disappears (fig. 21.21*b*) as the lower part of the uterus is pulled upward toward the baby's head. This process is called *effacement,* or "taking up the cervix." With further contractions, the baby's head acts as a wedge to assist cervical dilation. The baby's head usually has a diameter of about 10 cm; therefore, the cervix has to dilate to this diameter in order to allow the head to pass through. If it has not occurred already, the amniotic membrane is apt to rupture now, releasing the amniotic fluid, which escapes out the vagina. The first stage of labor ends once the cervix is dilated completely.

Stage 2 During the second stage of parturition, the uterine contractions occur every 1–2 minutes and last about one minute each. They are accompanied by a desire to push, or bear down. As the baby's head gradually descends into the vagina, the desire to push becomes greater. When the baby's head reaches the exterior, it turns so that the back of the head is uppermost (fig. 21.21*c*). Since the vagina may not expand enough to allow passage of the head without tearing, an *episiotomy* often is performed. This incision, which enlarges the opening, is stitched later and heals more perfectly than a tear. As soon as the head is delivered, the baby's shoulders rotate so that the baby faces either to the right or the left. The physician at this time may hold the head and guide it downward, while one shoulder and then the other emerges. The rest of the baby follows easily.

Figure 21.22

The average height of females (above) and males (below) from infancy through adolescence. Each has an adolescent growth spurt, but the male's growth spurt occurs about 2 years after the female's spurt.

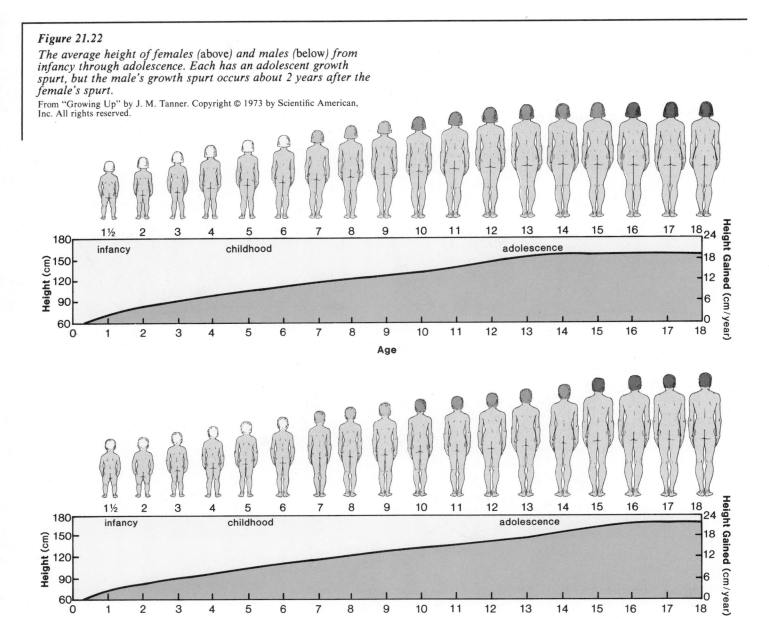

Once the baby is breathing normally, the umbilical cord is cut and tied, severing the child from the placenta. The stump of the cord shrivels and leaves a scar, which is the navel.

Stage 3 The placenta, or *afterbirth,* is delivered during the third stage of labor (fig. 21.21*d*). About 15 minutes after delivery of the baby, uterine muscular contractions shrink the uterus and dislodge the placenta. The placenta then is expelled into the vagina. As soon as the placenta and its membranes are delivered, the third stage of labor is complete.

During the first stage of birth, the cervix dilates; during the second, the child is born; and during the third, the afterbirth is expelled.

Human Development after Birth

Development does not cease once birth has occurred but continues throughout the stages of life: infancy, childhood, adolescence, and adulthood.

Infancy lasts until about 2 years of age. It is characterized by tremendous growth and sensorimotor development. During *childhood,* the individual grows, and the body proportions change (fig. 21.22). *Adolescence* begins with **puberty,** when the secondary sex characteristics appear and the sexual organs become functional. At this time, there is an acceleration of growth leading to changes in height, weight, fat distribution, and body proportions. Males commonly experience a growth spurt later than females; therefore, they grow for a longer period of time. Males are generally taller than females and have broader shoulders and longer legs relative to their trunk length.

Developmental changes keep occurring throughout infancy, childhood, adolescence, and adulthood.

Adulthood and Aging

Young adults are at their physical peak in muscle strength, reaction time, and sensory perception. The organ systems at this time are best able to respond to altered circumstances in a homeostatic manner. From now on, there is an almost imperceptible, gradual loss in certain of the body's abilities. **Aging** encompasses these progressive changes that contribute to an increased risk of infirmity, disease, and death (fig. 21.23).

Today, there is great interest in **gerontology,** the study of aging, because there are now more older individuals in our society than ever before and the number is expected to rise dramatically. In the next half-century, those over age 75 will rise from the present 13 million to 35–45 million, and those over age 80 will rise from 3 million to 6 million individuals. The human life span is judged to be a maximum of 110–115 years. The present goal of gerontology is not necessarily to increase the life span but to increase the health span, the number of years that an individual enjoys the full functions of all body parts and processes.

Theories of Aging

There are many theories about what causes aging. Three of these are considered here.

Genetic in Origin   Several lines of evidence indicate that aging has a genetic basis. (1) The children of long-lived parents tend to live longer than those of short-lived parents. Perhaps the genes are programmed to control aging and the time of death. The maximum life span of animals is species-specific; for humans it is about 110 years. (2) The number of times a cell divides is also species-specific. The maximum number of times human cells divide is around 50. Perhaps as we grow older, more and more cells are unable to divide any longer, and instead they undergo degenerative changes and die. (3) Some cell lines may become nonfunctional long before the maximum number of divisions has occurred. Whenever DNA replicates, mutations can occur, and this can lead to the production of nonfunctional proteins. Eventually, the number of inadequately functioning cells can build up, and this contributes to the aging process.

Whole Body Processes   A decline in the hormonal system can affect many different organs of the body. For example, type II diabetes is common in older individuals. The pancreas makes insulin, but the cells lack the receptors that enable them to respond. Menopause in women occurs for a similar reason. There is plenty of FSH in the bloodstream, but the ovaries do not respond. Perhaps aging results from the loss of hormonal activities and a decline in the functions they control.

The immune system, too, no longer performs as it once did, and this can affect the body as a whole. The thymus gland gradually decreases in size, and eventually most of it is replaced by fat and connective tissue. The incidence

Figure 21.23
Aging is a slow process during which the body undergoes changes that eventually bring about death even if no marked disease or disorder is present. Although the human life span probably cannot be expanded, it most likely is possible to expand the health span, the length of time the body functions normally.

of cancer increases among the elderly, which may signify that the immune system is no longer functioning as it should. This idea is substantiated, too, by the increased incidence of autoimmune diseases in older individuals.

It is possible, though, that aging is not due to the failure of a particular system that can affect the body as a whole, but to a specific type of tissue change that affects all organs and even the genes. It has been noticed for some time that proteins—such as collagen, which makes up the white fibers (p. 176) and is present in many support tissues—become increasingly cross-linked as people age. Undoubtedly, this cross-linking contributes to the stiffening and the loss of elasticity characteristic of aging tendons and ligaments. It also may account for the inability of organs, such as the blood vessels, the heart, and the lungs, to function as they once did. Some researchers now have found that glucose has the tendency to attach to any type of protein, which is the first step in a cross-linking process that ends with the formation of advanced glycosylation end products (AGEs).

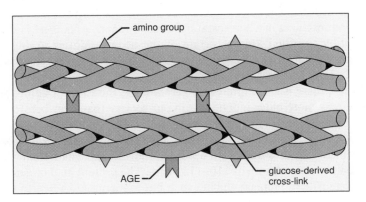

AGE-derived cross-links not only explain why cataracts develop, they also may contribute to the development of atherosclerosis and to the inefficiency of the kidneys in diabetics and older individuals. Even DNA-associated proteins seem capable of forming AGE-derived cross-links, and perhaps this increases the rate of mutations as we age. These researchers presently are experimenting with the drug aminoguanidine, which can prevent the development of AGEs.

Extrinsic Factors   The current data about the effects of aging often are based on comparisons of the characteristics of the elderly to younger age groups, but perhaps today's elderly were not as aware when they were younger of the importance of, for example, diet and exercise to general health. It is possible, then, that much of what we attribute to aging is instead due to years of poor health habits.

For example, osteoporosis is associated with a progressive decline in bone density in both males and females so that fractures are more likely to occur after only minimal trauma. Osteoporosis is common in the elderly—by age 65, one-third of women will have vertebral fractures, and by age 81, one-third of women and one-sixth of men will have suffered a hip fracture. While there is no denying that there is a decline in bone mass as a result of aging, certain extrinsic factors are also important. The occurrence of osteoporosis itself is associated with cigarette smoking, heavy alcohol intake, and perhaps inadequate calcium intake. Not only is it possible to eliminate these negative factors by personal choice, it also is possible to add a positive factor. A moderate exercise program has been found to slow down the progressive loss of bone mass.

Rather than collecting data on the average changes observed between different age groups, it might be more useful to note the differences within any particular age group. If this type of comparison is done, extrinsic factors that contribute to a decline and extrinsic factors that promote the health of an organ can be identified.

Figure 21.24

Percentage of function remaining of various organs in a 75–80-year-old person as measured against that of a 20-year-old person.

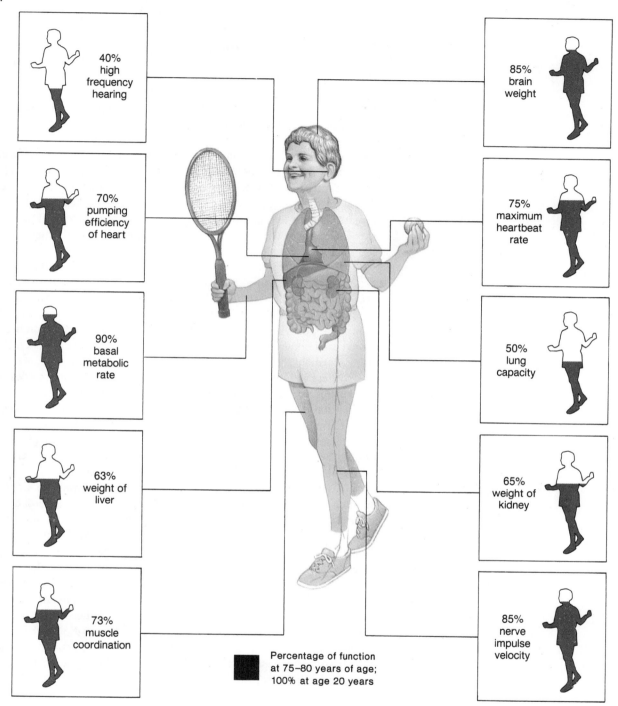

40%
high
frequency
hearing

70%
pumping
efficiency
of heart

90%
basal
metabolic
rate

63%
weight of
liver

73%
muscle
coordination

85%
brain
weight

75%
maximum
heartbeat
rate

50%
lung
capacity

65%
weight of
kidney

85%
nerve
impulse
velocity

Percentage of function
at 75–80 years of age;
100% at age 20 years

The Effect of Aging on Body Systems

Keeping in mind that we want to accept such data with reservations, we will still discuss in general the effects of aging on the various systems of the body. Figure 21.24 compares the percentage of function of various organs in a 75–80-year-old person to that of a 20-year-old person whose organs are assumed to function at 100% capacity. When making this comparison, we should keep in mind that the body has a vast functional reserve; it still can perform well even when not at 100% capacity.

The Skin As aging occurs, the skin becomes thinner and less elastic because the number of elastic fibers decreases and the collagen fibers undergo cross-linking as discussed previously. Also, there is less adipose tissue in the subcutaneous layer; therefore, older people are more likely to feel cold. The loss of thickness accounts for sagging and wrinkling of the skin.

Homeostatic adjustment to heat is also limited because there are fewer sweat glands for sweating to occur. There are fewer hair follicles so that the hair on the scalp and the extremities thins out. The number of sebaceous glands is reduced, and the skin tends to crack.

There is a decrease in the number of melanocytes, making the hair turn gray and the skin to pale. In contrast, some of the remaining pigment cells are larger, and pigmented blotches appear in the skin.

Processing and Transporting Cardiovascular disorders are the leading cause of death among the elderly. The heart shrinks because there is a reduction in cardiac muscle cell size. This leads to loss of cardiac muscle strength and reduced cardiac output. Still, it is observed that the heart, in the absence of disease, is able to meet the demands of increased activity. It can increase its rate to double or triple the amount of blood pumped each minute even though the maximum possible output declines.

Because the middle coat of arteries contains elastic fibers that most likely are subject to cross-linking, the arteries become more rigid with time, and their size is further reduced by plaque (p. 234). Therefore, blood pressure readings gradually rise. Such changes are common in individuals living in western industrialized countries but not in agricultural societies. As mentioned earlier, diet has been suggested as a way to control degenerative changes in the cardiovascular system (p. 207).

There is reduced blood flow to the liver, and this organ does not metabolize drugs as efficiently as before. This means that as a person gets older, less medication is needed to maintain the same level in the bloodstream.

Circulatory problems often are accompanied by respiratory disorders and vice versa. Growing inelasticity of lung tissue means that ventilation is reduced. Because we rarely use the entire vital capacity, these effects are not noticed unless there is increased demand for oxygen.

There is also reduced blood supply to the kidneys. The kidneys become smaller and less efficient at filtering wastes. Salt and water balance are difficult to maintain, and the elderly dehydrate faster than younger people. Difficulties involving urination include incontinence and the inability to urinate. In men, the prostate gland may enlarge and reduce the diameter of the urethra, making urination so difficult that surgery often is needed.

The loss of teeth, which often is seen in elderly people, is more apt to be the result of long-term neglect than a result of aging. The digestive tract loses tone and secretion of saliva and gastric juice is reduced, but there is no indication of reduced absorption. Therefore, an adequate diet, rather than vitamin and mineral supplements, is recommended. There are common complaints of constipation, increased amount of gas, and heartburn, but gastritis, ulcers, and cancer can also occur.

Integration and Coordination It often is mentioned that while most tissues of the body regularly replace their cells, some at a faster rate than others, the brain and the muscles do not. No new nerve or skeletal muscle cells are formed in the adult. However, contrary to previous opinion, recent studies show that few neural cells of the cortex are lost during the normal aging process. This means that cognitive skills remain unchanged even though there is characteristically a loss in short-term memory. Although the elderly learn more slowly than the young, they can acquire new material and remember it as well as the young. It is noted that when more time is given for the subject to respond, age differences in learning decrease.

Neurons are extremely sensitive to oxygen deficiency, and if neuron death does occur, it may not be due to aging itself but to reduced blood flow in narrowed blood vessels. Specific disorders, such as depression, Parkinson disease, and Alzheimer disease (p. 339), sometimes are seen, but they are not common. Reaction time, however, does slow, and more stimulation is needed for hearing, taste, and smell receptors to function as before. After age 50, there is a gradual reduction in the ability to hear tones at higher frequencies, and this can make it difficult to identify individual voices and to understand conversation in a group. The lens of the eye does not accommodate as well and also may develop a cataract. Glaucoma is more likely to develop because of a reduction in the size of the anterior chamber of the eye.

Loss of skeletal muscle mass is not uncommon, but it can be controlled by a regular exercise program. There is a reduced capacity to do heavy labor, but routine physical work should be no problem. A decrease in the strength of the respiratory muscles and inflexibility of the rib cage contribute to the inability of the lungs to expand as before, and reduced muscularity of the urinary bladder contributes to difficulties in urination.

As noted before, aging is accompanied by a decline in bone density. Osteoporosis, characterized by a loss of calcium and minerals from bone, is not uncommon, but there is evidence that proper health habits can prevent its occurrence. Arthritis, which restricts the motility of joints, also is seen. In arthritis, as the articular cartilage deteriorates, ossified spurs develop. These cause pain upon movement of the joint.

Weight gain occurs because the metabolic rate (p. 212) decreases and inactivity increases. Muscle mass is replaced by stored fat and retained water.

The Reproductive System Females undergo menopause, and thereafter the level of female sex hormones in blood falls markedly. The uterus and the cervix are reduced in size, and there is a thinning of the walls of the oviducts and the vagina. The external genitals become less pronounced. In males, the level of androgens falls gradually over the age span 50–90, but sperm production continues until death.

It is of interest that as a group, females live longer than males. Although their health habits may be poorer, it is also possible that the female sex hormone estrogen offers to women some protection against circulatory disorders when they are younger. Males suffer a marked increase in heart disease in their forties, but an increase is not noted in females until after menopause. Then women lead men in the incidence of stroke. Men are still more likely than women to have a heart attack, however.

Conclusion

We have listed many adverse effects due to aging, but it is important to emphasize that while such effects are seen, they are not a necessary occurrence (fig. 21.25). We must discover any extrinsic factors that precipitate these adverse effects and guard against them. Just as it is wise to make the proper preparations to remain financially independent when older, it is also wise to realize that biologically successful old age begins with the health habits developed when you are younger.

Figure 21.25
The aim of gerontology is to allow the elderly to enjoy living. This requires studying the debilities that can occur with aging and then making recommendations as to how best to forestall or prevent their occurrence.

Summary

Any developmental change requires 3 processes: growth, differentiation, and morphogenesis. Embryonic development has certain stages. During cleavage, there is division, but there is no overall growth. The result is a morula, which becomes the blastula when an internal cavity appears. During the gastrula stage, the germ layers (ectoderm, mesoderm, and endoderm) develop. At the neurula stage, the nervous system develops from midline mesoderm just above the notochord. At this point, it is possible to draw a typical cross section of a vertebrate embryo (fig. 21.7).

Differentiation begins with cleavage as the egg's cytoplasm is partitioned among the numerous cells. The cytoplasm is not uniform in content, and presumably each of the first few cells differ as to their cytoplasmic contents. Some probably contain substances that can influence gene activity—turning some genes on and others off. Morphogenesis involves the process of induction. Experiments show that the notochord induces formation of the nervous system, and the optic vesicles induce formation of the lens. Today, we envision induction as always present because cells are believed to constantly give off signals that influence the genetic activity of neighboring cells. This is the cause of morphogenesis.

Human development is divided into embryonic development and fetal development. The extraembryonic membranes appear early in human development. There are 4 membranes, as in the chick, but their function has been modified for internal development (table 21.3). During embryonic development, the body's various organs appear, and during fetal development, refinement of features occurs. The fetus is dependent upon the placenta for gas exchange and as a source of nutrient molecules. Birth has 3 stages: during the first stage, the cervix dilates; during the second, the child is born; and during the third, the afterbirth is expelled.

Development after birth consists of infancy, childhood, adolescence, and adulthood. Young adults are at their prime, and then the aging process begins. Aging encompasses progressive changes from about age 20 on that contribute to an increased risk of infirmity, disease, and death. Perhaps aging is genetic in origin, perhaps it is due to a change that affects the whole body, or perhaps it is due to extrinsic factors.

Study Questions

1. State, define, and give examples of the 3 processes that occur whenever a developmental change occurs. (p. 436)
2. Compare the process of cleavage and the formation of the blastula and the gastrula in the lancelet and the human. (pp. 437–38)
3. Name the germ layers, and state organs derived from each of the germ layers. (p. 439)
4. Draw a cross section of a typical chordate embryo at the neurula stage, and label your drawing. (p. 440)
5. Give reasons for suggesting that differentiation begins with the embryonic stage of cleavage. (p. 441)
6. Describe an experiment that helped investigators to conclude that the notochord induces formation of the neural tube. (p. 444) Give another well-known example of induction between tissues. (p. 444)
7. Give reasons for suggesting that morphogenesis is dependent upon signals given off by neighboring cells. What do the signals bring about in the receiving cells? (p. 445)
8. List the human extraembryonic membranes, give a function for each, and compare this function to that in the chick. (p. 446)
9. Describe in general what happens during embryonic and fetal development of the human. (pp. 447–58)
10. Trace the path of blood in the fetus from the umbilical vein to the aorta using 2 different routes. (p. 456)
11. Describe the 3 stages of parturition. (pp. 458–60)
12. Discuss 3 theories of aging. What are the major changes in body systems that have been observed as adults age? (pp. 461–62)

Objective Questions

1. When cells take on a specific structure and function, _____ occurs.
2. The morula becomes the _____ , a structure that contains the inner cell mass.
3. The _____ membranes include the chorion, the _____ , the yolk sac, and the allantois.
4. The blastocyst _____ itself in the uterine lining.
5. Formation of germ layers occurs during _____ .
6. The notochord _____ the formation of the nervous system.
7. During embryonic and fetal development, gas exchange occurs at the _____ .
8. During development, there is a connection between the pulmonary artery and the aorta called the _____ .
9. Fetal development begins with the _____ month.
10. If delivery is normal, the _____ appears before the rest of the body.

Label this Diagram.

See figure 21.13 (p. 446) in text.

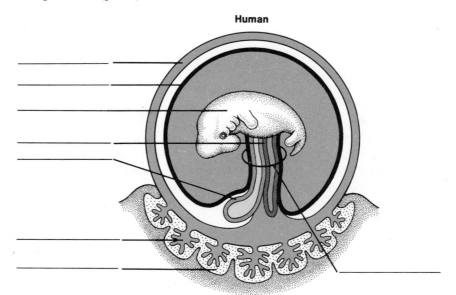

Human

Human Reproduction, Development, and Inheritance

allantois (ah-lan'to-is) one of the extraembryonic membranes; in reptiles and birds, it is a pouch serving as a repository for nitrogenous waste; in mammals, it is a source of blood vessels to and from the placenta. *446*

amnion (am'ne-on) an extraembryonic membrane; a sac around the embryo containing fluid. *446*

chorion (ko're-on) an extraembryonic membrane that forms an outer covering around the embryo; in reptiles and birds, it functions in gas exchange, in mammals, it contributes to the formation of the placenta. *446*

differentiation (dif''er-en''she-a'shun) the process and the developmental stages by which a cell becomes specialized for a particular function. *436*

ectoderm (ek'to-derm) the outer germ layer of the embryonic gastrula; it gives rise to the skin and the nervous system. *437*

endoderm (en'do-derm) an inner layer of cells that line the primitive gut of the gastrula. It becomes the lining of the digestive tract and associated organs. *437*

extraembryonic membrane (eks''trah-em''bre-on'ik mem'brān) membrane that is not a part of the embryo but is necessary to the continued existence and health of the embryo. *445*

induction (in-duk'shun) a process by which one tissue gives off signals that control the development of another, as when the embryonic notochord induces the formation of the neural tube. *444*

lanugo (lah-nu'go) downy hair on the body of a fetus; fetal hair. *456*

mesoderm (mes'o-derm) the middle germ layer of an animal embryo that gives rise to the muscles, the connective tissue, and the circulatory system. *438*

morphogenesis (mor''fo-jen'ĭ-sis) the movement of cells and tissues to establish the shape and the structure of an organism. *436*

parturition (par''tu-rish'un) the processes that lead to and include the birth of a human, and the expulsion of the extraembryonic membranes through the terminal portion of the female reproductive tract. *458*

trophoblast (trof'o-blast) the outer membrane that surrounds the human embryo, and when thickened by a layer of mesoderm, it becomes the chorion, an extraembryonic membrane. *448*

umbilical cord (um-bil'ĭ-kal kord) cord connecting the fetus to the placenta through which blood vessels pass. *451*

vernix caseosa (ver'niks ka''se-o'sah) cheeselike substance covering the skin of the fetus. *456*

yolk sac (yōk sak) one of the extraembryonic membranes within which yolk is found. *446*

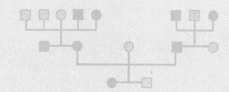

Chapter 22

Patterns of Gene Inheritance

Chapter Concepts

1 Genes, located on chromosomes, are passed from one generation to the next.

2 The Mendelian laws of genetics relate the genotype (inherited genes) to the phenotype (physical characteristics).

3 There are many exceptions to Mendel's laws, and these help to explain the wide variety in patterns of gene inheritance.

4 Humans are subject to various disorders due to inheritance of faulty genes.

Chapter Outline

Mendel's Laws
    The Inheritance of a Single Trait
    The Inheritance of Multitraits
Genetic Disorders
    Autosomal Recessive Disorders
    Autosomal Dominant Disorders
Beyond Mendel's Laws
    Polygenic Inheritance
    Multiple Alleles
    Degrees of Dominance

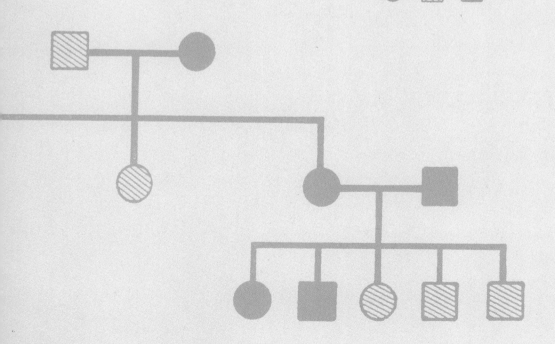

Figure 22.1

Figure 22.2

Diagrammatic representation of a homologous pair of chromosomes before and after replication. a. The letters represent alleles (alternate forms of a gene). Each allelic pair, such as Gg or Zz, is located on homologous chromosomes at a particular gene locus. b. Following replication, each sister chromatid carries the same alleles in the same order.

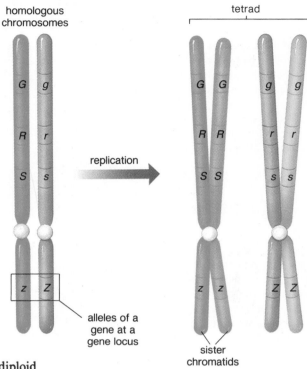

a.

b.

When a sperm fertilizes an egg, a new individual with the diploid number of chromosomes begins to develop. These chromosomes determine what the individual will be like; even if the zygote develops in a surrogate mother, the individual still will resemble the original parents (fig. 22.1).

The study of inheritance is an ongoing process that may have begun as soon as people noticed the resemblance between parent and offspring. The latest findings about heredity are covered in chapter 24, but first we consider the results of Gregor Mendel's investigations into heredity. Mendel's studies form the basis for the particulate model of heredity, in which it is assumed that genes are sections of chromosomes. For example, the letters on the homologous pair of chromosomes in figure 22.2 stand for genes that control particular traits. Genes, like the letters in the rectangles, are in a particular sequence and are at particular spots, or loci, on the chromosomes. Alternate forms of a gene having the same position on a pair of chromosomes and affecting the same trait are called **alleles**. In figure 22.2, *G* is an allele of *g*, and vice versa. Also, *R* is an allele of *r*, and vice versa. *G* could never be an allele for *R* because *G* and *R* are at different loci. According to the particulate model, each allelic pair controls a particular trait of the individual, such as color of hair, type of fingers, or length of nose.

Mendel's Laws

Gregor Mendel was a Catholic priest who in 1860 developed his laws of heredity after doing crosses between garden pea plants as described in the reading on page 470. Mendel did not use the terminology we just outlined. That came later, after biologists had more knowledge about chromosomes. Instead, Mendel said that every **trait**—for example, height—is controlled by 2 factors, or a pair

Mendel's use of pea plants as his experimental material was a good choice because pea plants are easy to cultivate, have a short generation time, and can be self-pollinated or cross-pollinated at will. Mendel selected certain traits for study and before beginning his experiments made sure his parental (P generation) plants bred true. He observed that when these plants self-pollinated, the offspring were like one another and like the parent plant. For example, a parent with yellow seeds always had offspring with yellow seeds; a plant with green seeds always had offspring with green seeds. Following that observation, Mendel cross-pollinated the plants by dusting the pollen

Mendel realized that these results were explainable, assuming (a) there are 2 factors for every trait; (b) one of the factors can be dominant over the other, which is recessive; and (c) the factors separate when the gametes are formed. He assigned letters to these factors and displayed his results similar to this:

P	yellow YY	x	green yy
F_1		all yellow	
F_1 x F_1	yellow Yy	x	yellow Yy
F_2		3 yellow : 1 green	

dominant characteristics, and therefore, he allowed the F_1 plants to self-pollinate. Among the F_2 generation he achieved an almost perfect ratio of 9:3:3:1 (table 22.A). For example, for every plant that had green, wrinkled seeds, he had approximately 9 that had yellow, round seeds, and so forth. Mendel saw that these results were explainable if pairs of factors separate independently from one another when the gametes form, allowing all possible combinations of factors to occur in the gametes. This would mean that the probability of achieving any 2 factors together in the F_2 offspring is the product of their chances of occurring separately. Thus, since the chance of yellow peas was 3/4 (in a one-trait cross) and the chance of round peas was 3/4 (in a one-trait cross), the chance of their occurring together was 9/16, and so forth.

Mendel achieved his success in genetics by studying large numbers of offspring, keeping careful records, and treating his data quantitatively. He showed that the application of mathematics to biology is extremely helpful in producing testable hypotheses.

Mendel's Results

of plants with yellow seeds on the stigma of plants with green seeds whose own anthers had been removed, and vice versa. Either way, the offspring (called F_1, or first filial generation) resembled the parents with yellow seeds. These results caused Mendel to allow F_1 plants to self-pollinate. Once he had obtained an F_2 generation, he observed the color of the peas produced. As table 22.A (p. 471). A indicates, he counted over 8,000 plants and found an approximate 3:1 ratio (about 3 plants with yellow seeds for every plant with green seeds) in the F_2 generation.

He believed that the F_2 plants with yellow seeds carried a dominant factor because his results could be related to the binomial equation $a^2 + 2ab + b^2$ in this manner: $a^2 = YY$; $2ab = 2Yy$, and $b^2 = yy$. Therefore, the plants with yellow seeds would be YY or Yy, and there would be 3 plants with yellow seeds for every plant with green seeds.

As a test to determine if the F_1 generation was indeed Yy, Mendel backcrossed it with the recessive parent, yy. His results of 1:1 indicated that he had reasoned correctly. Today, when a one-trait testcross is done, a suspected heterozygote is crossed with the recessive phenotype because this cross gives the best chance of producing the recessive phenotype.

Mendel performed a second series of experiments in which he crossed true-breeding plants that differed in 2 traits. For example, he crossed plants with yellow, round peas with plants with green, wrinkled peas. The F_1 generation always had both

Pea flower anatomy. *Self-pollination normally occurs within the flowers of the pea plant. In order to cross the dominant by the recessive, Mendel removed the anthers of the plant with the dominant trait and used pollen from the plant with the recessive trait to bring about cross-fertilization. He also performed the reverse cross in the same manner.*

Once the ovules developed into seeds (peas), they could be observed or planted in order to observe the results of a cross. The enlarged pod shows all possible shapes and colors of peas in a cross involving shape of the seed coat (smooth or wrinkled) and color of the seed coat (yellow or green).

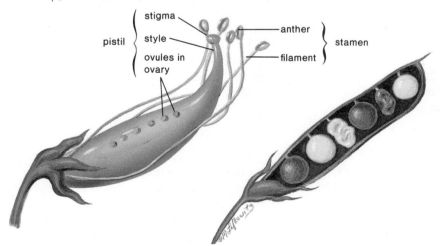

pistil { stigma
style
ovules in ovary

anther
filament } stamen

Table 22.A Mendel's Results			
Single-trait Cross	**F₁**	**F₂**	**Actual F₂ Ratio**
Yellow × green	All yellow	6,022 yellow 2,001 green	3.0:1.0
2-trait Cross	**F₁**	**F₂**	**Actual F₂ Ratio**
Yellow, round × green, wrinkled	All round, yellow	315 yellow, round 101 yellow, wrinkled 108 green, round 32 green, wrinkled	9.8:2.9:3.1:1.0

of factors. He observed that one of the factors controlling the same trait can be **dominant** over the other, which is **recessive.** For example, he found that a pea plant could show the dominant characteristic, say, tallness, while the recessive factor for shortness, although present, is not expressed.

Mendel's experiments led him to conclude that it is possible for a tall pea plant to pass on a factor for shortness. Therefore, he reasoned that while the individual plant has 2 factors for each trait, the gametes contained only one factor for each trait. This often is called Mendel's law of segregation.

Mendel's law of segregation: Each organism contains 2 factors for each trait, and the factors segregate during the formation of gametes. Each gamete then contains only one factor from each pair of factors. When fertilization occurs, the new organism has 2 factors for each trait, one from each parent.

The Inheritance of a Single Trait

Mendel suggested the use of letters to indicate factors so that *crosses* (gamete union resulting in offspring) could be described more easily. A capital letter indicates a *dominant* factor, and a lowercase letter indicates a *recessive* factor. The same procedure is used today, only the letters now are said to represent alleles. Also, Mendel's procedure and laws are applicable not only to peas but to all diploid individuals. Therefore, we now take as our example not peas but human beings.

Figure 22.3 illustrates the difference between a widow's peak and a continuous hairline. In doing a problem concerning hairline, the *key* is represented as

W = Widow's peak (dominant allele)
w = Continuous hairline (recessive allele)

The key tells us which letter of the alphabet to use for the gene in a particular problem. It also tells which allele is dominant, a capital letter signifying dominance.

The Genotype and the Phenotype

When we indicate the genes of a particular individual, 2 letters must be used for each trait mentioned. This is called the **genotype** of the individual. The genotype can be expressed not only using letters but also with a short descriptive phrase as table 22.1 shows. Therefore, the word **homozygous** means that the 2 members of the allelic pair in the zygote (*zygo*) are the same (*homo*); genotype *WW* is called *homozygous dominant* and *ww* is called *homozygous recessive.* The word **heterozygous** means that the members of the allelic pair are different (*hetero*); only *Ww* is heterozygous.

Figure 22.3
In humans, widow's peak (a) is dominant over continuous hairline (b).

a.

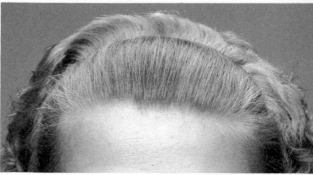

b.

Table 22.1 Genotype versus Phenotype

Genotype	Genotype	Phenotype
WW	Homozygous dominant	Widow's peak
Ww	Heterozygous	Widow's peak
ww	Homozygous recessive	Continuous hairline

As table 22.1 also indicates, the word **phenotype** refers to the physical characteristics of the individual—what the individual actually looks like. Also included in the phenotype are the microscopic and metabolic characteristics of the individual. Notice that both homozygous dominant (*WW*) and heterozygous (*Ww*) show the dominant phenotype.

Gamete Formation

Whereas the genotype has 2 alleles for each trait, the gametes have only one allele for each trait in accordance with Mendel's law of segregation. This, of course, is related to the process of meiosis. The alleles are present on a homologous pair of chromosomes, and these chromosomes separate during meiosis (see fig. 22.6). Therefore, the members of each allelic pair separate during meiosis, and there is only one allele for each trait in the gametes. When doing genetic problems, keep in mind that no 2 letters in a gamete should be the same. For this reason *Ww* represents a possible genotype, and the gametes for this individual could contain either a *W* or a *w*.

When doing genetics problems, the same letter is used for both the dominant and recessive alleles; a capital letter indicates the dominant and a lowercase letter indicates the recessive. A homozygous dominant individual is indicated by 2 capital letters, and a homozygous recessive individual is indicated by 2 lowercase letters. The genotype of a heterozygous individual is indicated by a capital letter and a lowercase letter. Unlike the situation in the individual, gametes have one letter of each type, either capital or lowercase, as appropriate. All possible combinations of letters indicate all possible gametes.

Practice Problems 1*

1. For each of the following genotypes, give all possible gametes.
   - a. *WW*
   - b. *WWSs*
   - c. *Tt*
   - d. *Ttgg*
   - e. *AaBb*
2. For each of the following, state whether a genotype or a gamete is represented.
   - a. *D*
   - b. *Ll*
   - c. *Pw*
   - d. *LlGg*

*Answers to problems are on page 488.

One-Trait Crosses

It is now possible for us to consider a particular cross. If a homozygous man with a widow's peak (fig. 22.3*a*) reproduces with a woman with a continuous hairline (fig. 22.3*b*), what kind of hairline will their children have?

In solving the problem, we use the key already established (p. 471) to indicate the genotype of each parent, we determine what the gametes are for each parent, we combine all possible gametes, and finally, we decide the genotypes and the phenotypes of all the offspring. In the format that follows, P

Human Reproduction, Development, and Inheritance

stands for the parental generation, and the letters in the P row are the genotypes of the parents. The second row shows that each parent has only one type of gamete in regard to hairline, and therefore all the children (F = filial generation) will have similar genotypes and phenotypes. The children are heterozygous (Ww) and will show the dominant characteristic, the widow's peak.

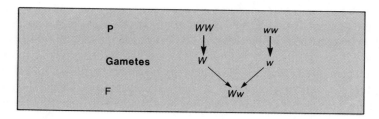

The children are **monohybrids;** that is, they are heterozygous for only one pair of alleles. If they reproduce with someone else of the same genotype, what type of hairline will the children have? In this problem ($Ww \times Ww$), each parent has 2 possible types of gametes (W or w), and we must ensure that all types of sperm have equal chance to fertilize all possible types of eggs. One way to do this is to use a **Punnett square** (fig. 22.4), in which all possible types of sperm are lined up vertically and all possible types of eggs are lined up horizontally (or vice versa), and every possible combination of gametes occurs within the squares.

After we determine the genotypes and the phenotypes of the offspring, we see that 3 have a widow's peak and 1 has a continuous hairline. This 3:1 ratio always is expected for a monohybrid cross. The exact ratio is more likely to be observed if a large number of matings take place and if a large number of offspring result. Only then do all possible sperm have an equal chance to fertilize all possible eggs. Naturally, we do not routinely observe hundreds of offspring from a single type of cross in humans. The best interpretation of figure 22.4 in humans is to say that each child has 3 chances out of 4 to have a widow's peak or 1 chance out of 4 to have a continuous hairline. It is important to realize that *chance has no memory;* for example, if 2 heterozygous parents already have 3 children with a widow's peak and are expecting a fourth child, this child still has a 75% chance of a widow's peak and a 25% chance of a continuous hairline.

Laws of Probability Another method of calculating the expected ratios uses the laws of probability. First, we must know that the probability (or chance) of 2 or more independent events occurring together is the product (multiplication) of their chance of occurring separately.

In the cross just considered ($Ww \times Ww$), what is the chance of obtaining either a W or a w from a parent?

The chance of $W = 1/2$
The chance of $w = 1/2$

Therefore, the probability of receiving these genotypes is as follows:

1. The chance of $WW = 1/2 \times 1/2 = 1/4$
2. The chance of $Ww = 1/2 \times 1/2 = 1/4$
3. The chance of $wW = 1/2 \times 1/2 = 1/4$
4. The chance of $ww = 1/2 \times 1/2 = 1/4$

Now we have to realize that the chance of an event that can occur in 2 or more independent ways is the sum (addition) of the individual chances. Therefore, the chance of offspring with a widow's peak (add chances of $WW, Ww,$ or wW from above) is 3/4. The chance of offspring with continuous hairline (only ww from above) is 1/4.

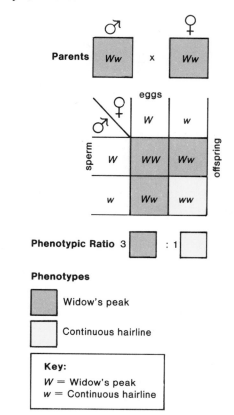

Figure 22.4
Monohybrid cross. In this cross, the parents are heterozygous for widow's peak. The chances of any child having a continuous hairline are 1 out of 4, or 25%.

When doing a genetics problem, it is assumed that all possible types of sperm fertilize all possible types of eggs. The results can be expressed as a probable phenotypic ratio; it is also possible to state the chances of an offspring showing a particular phenotype.

The One-Trait Testcross

If a plant, an animal, or a person has the dominant phenotype, it is not possible to tell by inspection if the organism is homozygous dominant or heterozygous. However, if the plant or the animal is crossed with a homozygous recessive, the results can indicate the original genotype. For example, figure 22.5 shows the different results if a man with a widow's peak reproduces with a woman who has a continuous hairline, is either homozygous dominant or heterozygous. (She must be homozygous recessive or she would not have a continuous hairline.) If the man is homozygous dominant, he can sire only children with widow's peaks. If he is heterozygous, the chances are 2:2 or 1:1 that a child will or will not have a widow's peak. Therefore, the cross of a possible heterozygote with an individual having the recessive phenotype gives the best chance of producing the recessive phenotype among the offspring. This type of cross is a one-trait **testcross.**

The results of a one-trait testcross can determine whether an individual who expresses the dominant phenotype is heterozygous or homozygous dominant. If any of the offspring of the testcross expresses the recessive phenotype, the parent with the dominant phenotype must be heterozygous.

Practice Problems 2*

1. Both a man and a woman are heterozygous for freckles. Freckles are dominant over no freckles. What are the chances that a child will have freckles?
2. A woman is homozygous dominant for short fingers. Short fingers are dominant over long fingers. Will any of her children have long fingers?
3. Both you and your sister or brother have attached earlobes, yet your parents have unattached earlobes. Unattached earlobes are dominant over attached earlobes. What are the genotypes of your parents?
4. A father has dimples, the mother does not have dimples, and all the children have dimples. Dimples are dominant over no dimples. Give the probable genotypes of all persons concerned.

*Answers to problems are on page 488.

The Inheritance of Multitraits

Although it is possible to consider the inheritance of just one trait, actually each individual passes on to his or her offspring 2 alleles for each of many traits. In order to arrive at a general understanding of multitrait inheritance, the inheritance of 2 traits is considered. The same principles apply to as many traits as we would want to consider.

Two Traits (Unlinked)

When Mendel performed two-trait crosses, he noticed that his results were attainable only if sperm with every possible combination of factors fertilized every possible egg. This caused him to formulate his second law, the law of independent assortment.

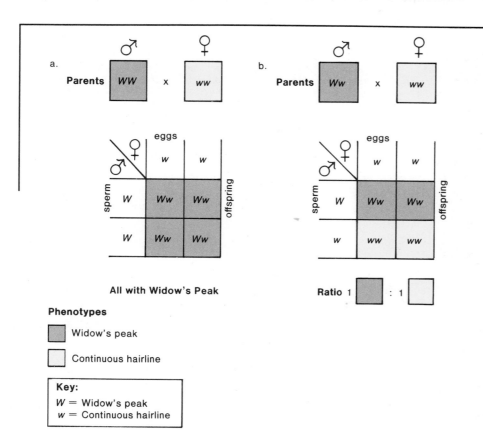

Phenotypes

☐ (dark) Widow's peak

☐ (light) Continuous hairline

Key:
W = Widow's peak
w = Continuous hairline

Figure 22.5

One-trait testcross. In this example, it is impossible to tell if the male parent is homozygous dominant or heterozygous by inspection. The results of reproduction with a homozygous recessive, however, can help to determine his genotype. If he is homozygous dominant (a), none of the offspring will show the recessive characteristic, but if he is heterozygous (b), there is a 50-50 chance that any offspring will show the recessive characteristic.

Mendel's law of independent assortment: Members of one pair of factors segregate (assort) independently of members of another pair of factors. Therefore, all possible combinations of factors can occur in the gametes.

Figure 22.6 illustrates that the law of segregation and the law of independent assortment hold because of the manner in which meiosis occurs. The law of segregation is dependent on the separation of members of homologous pairs of chromosomes. The law of independent assortment is dependent on the random arrangement of homologous pairs at the equator of the spindle during metaphase I. Because of this, the homologous pairs separate independently of one another.

Two-Trait Crosses

When doing a two-trait cross, the genotypes of the parents require 4 letters because there is an allelic pair for each trait. Also, the gametes of the parents will contain one letter of each kind in every possible combination in accordance with Mendel's law of independent assortment. Finally, in order to produce the probable ratio of phenotypes among the offspring, all possible matings are presumed to occur.

To give an example (fig. 22.7), let us cross a person homozygous for widow's peak and short fingers (*WWSS*) with a person who has a continuous hairline and long fingers (*wwss*). Because each parent has only one type of gamete, the F$_1$ offspring all will have the genotype *WwSs* and the same phenotype (widow's peak with short fingers). This genotype is called a **dihybrid** because the individual is heterozygous in 2 regards: hairline and fingers.

When a dihybrid reproduces with a dihybrid, each parent has 4 possible types of gametes, as shown in the Punnett square in figure 22.7. This Punnett square also shows the expected genotypes among 16 offspring if all possible sperm fertilize all possible eggs. An inspection of the various genotypes in the

22.1 Critical Thinking

1. Before Mendel formulated his law of segregation, what 2 alternative hypotheses might he have formulated about the kinds of gametes for a parent that is *Yy* (*Y* = yellow peas; *y* = green peas)?
2. How did his results of 3 yellow to 1 green support one of these hypotheses and not the other?
3. Before Mendel formulated his law of independent assortment, what 2 alternative hypotheses might he have formulated about the kinds of gametes for a parent in figure 22.6?
4. How did his results of 9:3:3:1 support one of these hypotheses and not the other?

Figure 22.6

Chromosome basis of Mendel's laws of segregation and independent assortment. The law of segregation states that homologous chromosomes segregate during meiosis I, therefore, there is only one chromosome (and one allele) of each kind in the gametes. The law of independent assortment states that homologous pairs segregate independently of other homologous pairs during meiosis I, and therefore all possible combinations of chromosomes (and alleles) are in the gametes. During meiosis II, the chromatids separate and therefore meiosis produces 4 daughter cells.

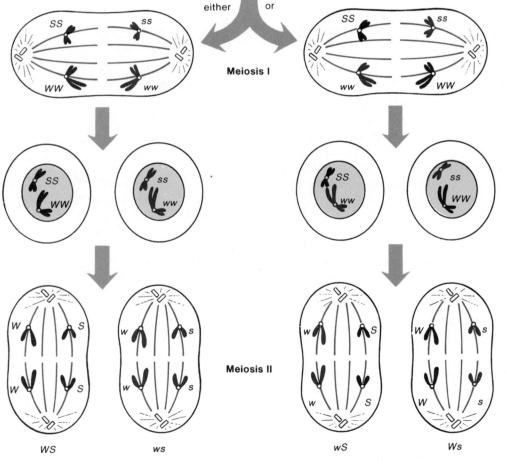

Tetrads are present in mother cell.

Alternative types of segregation are equally likely when homologous chromosomes separate independently of each other.

Each cell contains one of each kind of chromosome. All possible combinations of chromosomes and alleles are found in the gametes.

Human Reproduction, Development, and Inheritance

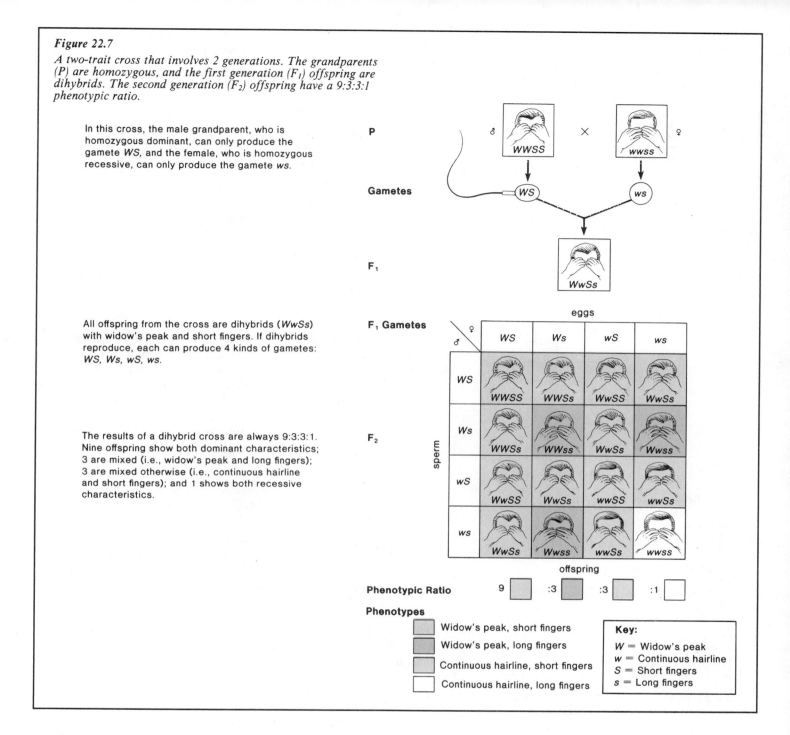

Figure 22.7

A two-trait cross that involves 2 generations. The grandparents (P) are homozygous, and the first generation (F₁) offspring are dihybrids. The second generation (F₂) offspring have a 9:3:3:1 phenotypic ratio.

In this cross, the male grandparent, who is homozygous dominant, can only produce the gamete *WS*, and the female, who is homozygous recessive, can only produce the gamete *ws*.

All offspring from the cross are dihybrids (*WwSs*) with widow's peak and short fingers. If dihybrids reproduce, each can produce 4 kinds of gametes: *WS, Ws, wS, ws.*

The results of a dihybrid cross are always 9:3:3:1. Nine offspring show both dominant characteristics; 3 are mixed (i.e., widow's peak and long fingers); 3 are mixed otherwise (i.e., continuous hairline and short fingers); and 1 shows both recessive characteristics.

Phenotypes

Widow's peak, short fingers

Widow's peak, long fingers

Continuous hairline, short fingers

Continuous hairline, long fingers

Key:

W = Widow's peak
w = Continuous hairline
S = Short fingers
s = Long fingers

square shows that the expected phenotypic ratio is 9 widow's peak and short fingers: 3 widow's peak and long fingers: 3 continuous hairline and short fingers: 1 continuous hairline and long fingers. This 9:3:3:1 phenotypic ratio is always expected for a dihybrid cross when simple dominance is present.

We can use this expected ratio to predict the chances of each child receiving a certain phenotype. For example, the chance of getting the 2 dominant phenotypes together is 9 out of 16, and the chance of getting the 2 recessive phenotypes together is 1 out of 16.

Probability   Instead of using a Punnett square to arrive at the chances of the different types of phenotypes in the cross under discussion, it is possible to use

Figure 22.8

Figure 22.8

Testcross. In this example, it is impossible to tell by inspection if the male parent is homozygous dominant or if he is heterozygous for both traits. However, reproduction with a female who is recessive for both traits is likely to show which he is. If he is heterozygous, there is a 25% chance that the offspring will show both recessive characteristics and a 50% chance that they will show one or the other of the recessive characteristics.

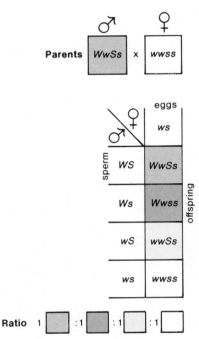

Ratio 1 ☐ : 1 ☐ : 1 ☐ : 1 ☐

Phenotypes

☐ Widow's peak, short fingers

☐ Widow's peak, long fingers

☐ Continuous hairline, short fingers

☐ Continuous hairline, long fingers

Key:
W = Widow's peak
w = Continuous hairline
S = Short fingers
s = Long fingers

---

the laws of probability we discussed before. For example, we already know the results for 2 separate monohybrid crosses are as follows:

1. Probability of widow's peak = 3/4
   Probability of short fingers = 3/4
2. Probability of continuous hairline = 1/4
   Probability of long fingers = 1/4

The probabilities for the dihybrid cross are, therefore,

Probability of widow's peak and short fingers = 3/4 × 3/4 = 9/16
Probability of widow's peak and long fingers = 3/4 × 1/4 = 3/16
Probability of continuous hairline and short fingers = 1/4 × 3/4 = 3/16
Probability of continuous hairline and long fingers = 1/4 × 1/4 = 1/16

The phenotypic ratio is 9:3:3:1.

Again, because all possible sperm must have an equal opportunity to fertilize all possible eggs to even approximate these results, a large number of offspring must be counted.

The Two-Trait Testcross

A plant or an animal that shows 2 dominant traits can be either heterozygous or homozygous for each one. Its genotype can be determined by a testcross with a homozygous recessive individual. For example, it is impossible to tell by inspection if an individual having a widow's peak and short fingers is homozygous dominant or heterozygous for each characteristic. A cross with the recessive phenotype has the best chance of determining the genotype of this individual. If the individual is heterozygous for each trait, 4 different types of gametes are possible. There is only one possible gamete for the recessive phenotype:

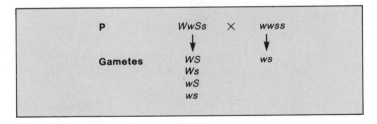

The Punnett square (fig. 22.8) shows that the expected ratio is 1 widow's peak with short fingers: 1 widow's peak with long fingers: 1 continuous hairline with short fingers: 1 continuous hairline with long fingers, or 1:1:1:1.

The results of a two-trait testcross can determine whether an individual who expresses the dominant phenotype is heterozygous or homozygous dominant for each trait. If any of the offspring of a two-trait testcross expresses the recessive phenotype, the parent with the dominant phenotype must be heterozygous for that trait.

Practice Problems 3*

Using the information provided in Practice Problems 2, solve these problems.

1. What is the genotype of the offspring if a man homozygous recessive for type of earlobes and homozygous dominant for type of hairline reproduces with a woman who is homozygous dominant for earlobes and homozygous recessive for hairline?

2. If the offspring of this cross reproduces with someone of the same genotype, then what are the chances that this couple will have a child with a continuous hairline and attached earlobes?
3. A person who has dimples and freckles reproduces with someone who does not. This couple has a child who does not have dimples or freckles. What is the genotype of all persons concerned?

*Answers to problems are on page 488.

We now have studied the pattern of simple Mendelian inheritance in regard to one- and two-trait crosses. Table 22.2 lists the crosses we have studied, along with their expected ratios. When doing genetics problems, it is not necessary to do a Punnett square if the expected ratio already is known.

Genetic Disorders

When studying human genetic disorders, biologists often construct pedigree charts that show the pattern of inheritance of a characteristic within a group of people. Let us contrast 2 possible patterns of inheritance in order to show how it is possible to determine whether the characteristic is an autosomal dominant or an autosomal recessive disorder. Autosomal disorders are caused by alleles on the autosomal chromosomes.

In both patterns, males are designated by squares and females are designated by circles. Shaded circles and squares indicate affected individuals. A line between a square and a circle represents a couple who have mated. A vertical line going downward leads, in these patterns, to a single child. (If there is more than one child, they are placed off a horizontal line.) Which pattern of inheritance do you suppose represents a dominant characteristic, and which represents a recessive characteristic?

In pattern I, the child is affected, but neither parent is; this can happen if the characteristic is recessively inherited. What are the chances that any offspring from this union will be affected? Because the parents are monohybrids, the chances are 1 in 4, or 25% (table 22.2). Notice that the parents also could be called **carriers** because they have a normal phenotype but are capable of having a child with a genetic disorder. See figure 22.9 for other ways to recognize an autosomal recessive pattern of inheritance.

Table 22.2 Phenotypic Ratios of Common Crosses	
Genotypes	**Phenotypes**
Monohybrid × monohybrid	3:1 (dominant to recessive)
Monohybrid × recessive*	1:1 (dominant to recessive)
Dihybrid × dihybrid	9:3:3:1 (9 both dominant, 3 one dominant, 3 other dominant, 1 both recessive)
Dihybrid × recessive*	1:1:1:1 (all possible combinations in equal number)

*Called a backcross because it is as if the offspring was mated back to the recessive parent. Also called a testcross because it can be used to test if the individual showing the dominant gene is homozygous dominant or heterozygous. For a definition of all terms, see the end-of-chapter key terms.

In pattern II, the child is affected, as is one of the parents. When a characteristic is dominant, an affected child usually has at least one affected parent. Of the 2 patterns, this one shows a dominant pattern of inheritance. What are the chances that any offspring from this union will be affected? Because this is a monohybrid by a recessive cross, the chances are 50% (table 22.2). See figure 22.10 for other ways to recognize an autosomal dominant pattern of inheritance.

22.2 Critical Thinking

1. Some individuals are albinos—they have no melanin in any of their skin cells. Melanin is a molecule produced by a biochemical pathway (p. 515). What fault have albinos inherited?
2. Considering your answer to question 1, why would you have predicted that albinism is a recessive rather than a dominant disorder?
3. What possible crosses would produce an offspring that is an albino? Are any of these individuals carriers?
4. Suppose you want to ensure that children born to an albino woman have normal pigmentation. Which of these—melanin, enzyme to produce melanin, or a normal gene—would you inject into the egg?

Autosomal Recessive Disorders

There are many autosomal recessive disorders that are inherited in a simple Mendelian manner, but only 3 of the better known are discussed. Others that are well known are albinism (lack of pigment); galactosemia (accumulation of galactose in the liver and mental retardation); thalassemia (production of abnormal type of hemoglobin); and xeroderma pigmentosum (inability to repair ultraviolet-induced damage). The homozygous recessive phenotype is more likely to occur among a group of people who tend to marry each other, which may explain why autosomal recessive disorders are sometimes more prevalent among members of a particular ethnic group.

Cystic Fibrosis

Cystic fibrosis is the most common lethal genetic disease among Caucasians in the United States. About 1 in 20 Caucasians is a carrier, and about 1 in 2,000 children born to this group has the disorder. In these children, the mucus in the lungs and the digestive tract is particularly thick and viscous. In the lungs, the mucus interferes with gas exchange. In the digestive tract, the thick

Figure 22.9

Sample pedigree chart for an autosomal recessive genetic disorder. Only those affected are shaded. Go through the chart and hatch mark all those who are carriers.

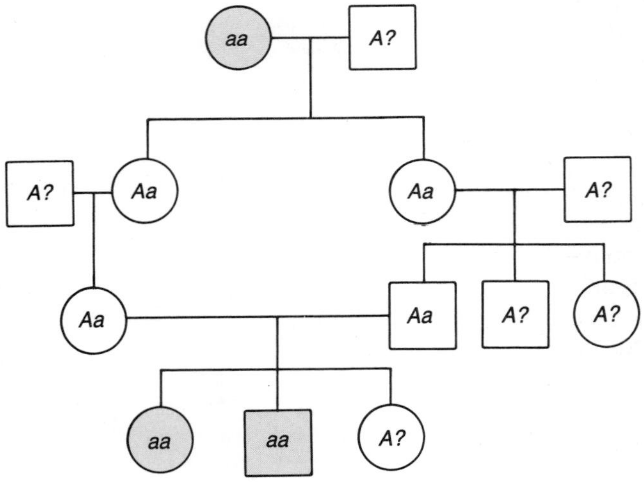

Autosomal Recessive Genetic Disorders

- Most affected children have normal parents.
- Heterozygotes have a normal phenotype.
- Two affected parents always will have affected children.
- Affected individuals who have noncarrier spouses will have normal children.
- Close relatives who marry are more likely to have affected children.
- Both males and females are affected with equal frequency.

Key:
aa = Affected
Aa = Carrier (appears normal)
AA = Normal

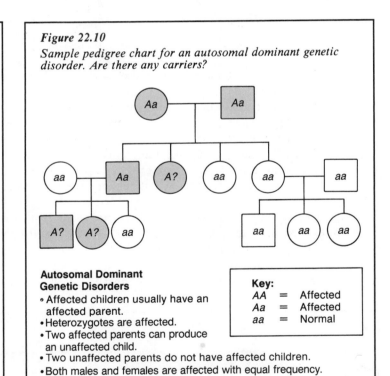

Figure 22.10

Sample pedigree chart for an autosomal dominant genetic disorder. Are there any carriers?

Autosomal Dominant Genetic Disorders

- Affected children usually have an affected parent.
- Heterozygotes are affected.
- Two affected parents can produce an unaffected child.
- Two unaffected parents do not have affected children.
- Both males and females are affected with equal frequency.

Key:
AA = Affected
Aa = Affected
aa = Normal

mucus impedes the secretion of pancreatic juices, and food cannot be properly digested; large, frequent, and foul-smelling stools occur. A few individuals have been known to survive childhood, but most die from recurrent lung infections.

In the past few years, much progress has been made in our understanding of cystic fibrosis. First of all, it was discovered that chloride (Cl⁻) ions fail to pass through cell membrane channels in these patients. Ordinarily, after chloride ions have passed through the membrane, water follows. It is believed that lack of water in the lungs causes the mucus to be so thick. Secondly, the cystic fibrosis gene, which is located on the number-7 chromosome, has been isolated, and soon there will be a test to identify carriers of the disease. The defective protein product has been identified, but its exact function is not yet known.

Tay-Sachs Disease

Tay-Sachs disease is the best-known genetic disease among U.S. Jewish people, most of whom are of central and eastern European descent. At first, it is not apparent that a baby has Tay-Sachs disease. However, development begins to slow down between 4 and 8 months of age as neurological impairment and psychomotor difficulties become apparent. Ophthalmologic examination reveals a characteristic red spot and yellowish accumulation in the region of the retina called the fovea centralis. The child gradually becomes blind and helpless, develops uncontrollable seizures, and eventually becomes paralyzed. There is no treatment or cure for Tay-Sachs disease, and most affected individuals die by the age of 3 or 4.

So-called late-onset Tay-Sachs disease occurs in adults. The symptoms are progressive mental and motor deterioration, depression, schizophrenia, and premature death. The gene for late-onset Tay-Sachs disease has been sequenced, and this form of the disorder apparently is due to one changed pair of bases in the DNA of the number-1 chromosome.

Tay-Sachs disease results from a lack of the enzyme hexosaminidase A (Hex A) and the subsequent storage of its substrate, a glycosphingolipid, in lysosomes. Although more and more lysosomes build up in many body cells (fig. 22.11), the primary sites of storage are the cells of the nervous system, which accounts for the onset and the progressive deterioration of psychomotor functions.

There is a test to detect carriers of Tay-Sachs. The test uses a sample of serum, white blood cells, or tears to determine whether Hex A activity is

---

Figure 22.11

Tay-Sachs disease. a. When Hex A is present, glycosphingolipids are broken down. b. When Hex A is absent, these lipids accumulate in lysosomes and lysosomes accumulate in the cell. c. Electron micrograph of cell crowded with lysosomes.

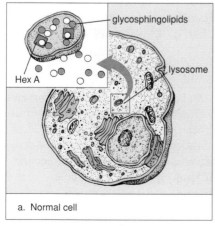

a. Normal cell

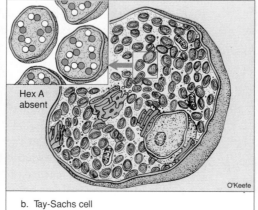

b. Tay-Sachs cell

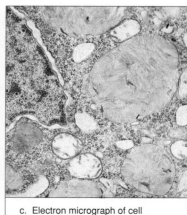

c. Electron micrograph of cell

present. Affected individuals have no detectable Hex A activity. Carriers have about half the level of Hex A activity found in normal individuals. Prenatal diagnosis of the disease also is possible following either amniocentesis or chorionic villi sampling.

Phenylketonuria (PKU)

Phenylketonuria (PKU) occurs in 1 in 20,000 births and so is not as frequent as the disorders previously discussed. When it does occur, the parents are very often close relatives. Affected individuals lack an enzyme that is needed for the normal metabolism of the amino acid phenylalanine, and an abnormal breakdown product, a phenylketone, accumulates in the urine. Newborns are routinely tested, and if they lack the necessary enzyme, they are placed on a diet low in phenylalanine. This diet must be continued until the brain is fully developed or else severe mental retardation develops. If a woman who is homozygous recessive for PKU wishes to have a normal child, she should resume her limited diet several months before getting pregnant; otherwise, she runs a high risk of having a microcephalic child—one with an abnormally small head.

Autosomal Dominant Disorders

There are many autosomal dominant disorders that are inherited in a simple Mendelian manner, but only 2 of the better known—neurofibromatosis and Huntington disease—are discussed here. Others that are well known are Marfan syndrome (connective tissue disorder); achondroplasia (dwarfism); brachydactyly (abnormally short fingers); porphyria (inability to metabolize porphyrins from hemoglobin breakdown); and hypercholesterolemia (elevated levels of cholesterol in blood).

Neurofibromatosis (NF)

Neurofibromatosis (NF), sometimes called von Recklinghausen disease, is one of the most common genetic disorders. It affects roughly 1 in 3,000 people, including an estimated 100,000 in the United States. It is seen equally in every racial and ethnic group throughout the world.

At birth or later, the affected individual may have 6 or more large tan spots on the skin. Such spots may increase in size and number and get darker. Small benign tumors (lumps) called neurofibromas may occur under the skin or in the muscles. Neurofibromas are made up of nerve cells and other cell types.

This genetic disorder shows *variable expressivity;* in most cases, symptoms are mild and patients live a normal life. In some cases, however, the effects are severe. Skeletal deformities, including a large head, are seen, and eye and ear tumors can lead to blindness and hearing loss. Many children with NF have learning disabilities and are overactive.

Only recently, researchers determined that the gene responsible for NF is located on the number-17 chromosome and developed a prenatal test for diagnosing the disorder. They believe the gene is rather large because of its varying effects and because about half of all NF cases are the result of new mutations in one of the parents.

Huntington Disease

As many as 1 in 10,000 persons in the United States have **Huntington disease (HD),** a neurological disorder that affects specific regions of the brain. Most individuals who inherit the allele appear normal until middle age. Then, minor disturbances in balance and coordination lead to progressively worse neurological disturbances. The victim becomes insane before death occurs.

Much has been learned about Huntington disease. The gene for the disease is located on the number-4 chromosome, and there is a test, of the type

Figure 22.12

Polygenic inheritance. a. *When you record the heights of a large group of young men, (b) the values follow a bell-shaped curve. Such distributions are seen when a trait is controlled by several sets of alleles.*

a.

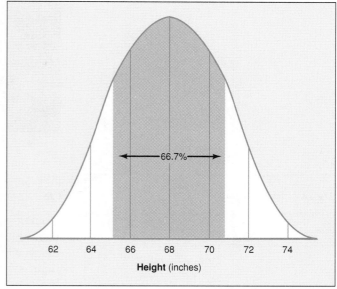

b.

described in figure 23.16, to determine if the dominant gene has been inherited. Because treatment is not available, however, few may want to have this information.

Research is being conducted, though, to determine the underlying cause of the disorder. It is known that the brain of a Huntington victim produces more than the usual amount of quinolinic acid, an excitotoxin that can overstimulate certain nerve cells. It is believed to lead to the death of these cells and to the subsequent symptoms of Huntington disease. Researchers are looking for chemicals that block quinolinic acid's action or inhibit quinolinic acid synthesis.

Some of the best-known genetic disorders in humans are either autosomal recessive or autosomal dominant disorders inherited in a simple Mendelian manner. Pedigree charts show the pattern of inheritance in a particular family.

Beyond Mendel's Laws

Certain traits, such as those just studied, follow the rules of simple Mendelian inheritance. There are, however, others that do not follow these rules.

Polygenic Inheritance

Two or more sets of alleles can affect the same trait, sometimes in an additive fashion. Polygenic inheritance can cause the distribution of human characteristics according to a bell-shaped curve, with most individuals exhibiting the average phenotype (fig. 22.12). The more genes that control the trait, the more continuous the distribution.

Skin Color

Just how many pairs of alleles control skin color is not known, but a range in colors can be explained on the basis of 2 pairs. When a black person has children by a white person, the children are intermediate, but 2 intermediates can

Figure 22.13

Inheritance of skin color. This white husband (aabb) and his intermediate wife (AaBb) had fraternal twins, one of whom is white and one of whom is intermediate.

produce children who range in skin color from black to white. If we assume that 2 pairs of alleles control skin color, then

Black = *AABB*
Dark = *AABb* or *AaBB*
Intermediate = *AaBb* or *AAbb* or *aaBB*
Light = *Aabb* or *aaBb*
White = *aabb*

If an intermediate reproduces with a white person, the very darkest individual possible is an intermediate, but a white child is also possible (fig. 22.13).

Polygenic Genetic Disorders

A number of serious genetic disorders, such as cleft lip or palate, clubfoot, congenital dislocation of the hip, and certain spinal conditions, traditionally are believed to be controlled by a combination of genes on autosomal chromosomes. This belief is being challenged by researchers who studied the inheritance of cleft palate in a large family in Iceland. These researchers reported the finding of a gene on the X chromosome that alone can cause cleft palate.

If a couple is concerned about the birth of a child with a neural tube defect, there is now a blood test that can be done. If the prospective mother has a high serum level of α-feto protein, further testing is advised. An analysis of the amniotic fluid following amniocentesis (p. 455) can reveal if there has been a leakage of neural tube substance into the fluid. If such a leakage has taken place, it usually is possible to diagnose the condition of the fetus.

Multiple Alleles

ABO Blood Type

Three alleles for the same gene control the inheritance of ABO blood types. These alleles determine the presence or absence of antigens on the red blood cells.

A = type A antigen on red blood cells
B = type B antigen on red blood cells
O = no antigens on the red blood cells

Each person has only 2 of the 3 possible alleles, and both *A* and *B* are dominant over *O*. Therefore, as table 22.3 shows, there are 2 possible genotypes for type A blood and 2 possible genotypes for type B blood. On the other hand, alleles *A* and *B* are fully expressed in the presence of the other. Therefore, if a person inherits one of each of these alleles, that person will have type AB blood. Type O blood can only result from the inheritance of 2 *O* alleles.

Table 22.3 Blood Groups

Phenotype	Genotype
A	*AA, AO*
B	*BB, BO*
AB	*AB*
O	*OO*

Human Reproduction, Development, and Inheritance

An examination of possible matings between different blood types sometimes produces surprising results; for example,

Parents: *AO* × *BO*
Children: *AB, OO, AO, BO*

Therefore, from this particular mating, every possible phenotype (types AB, O, A, B blood) is possible.

Blood typing sometimes can aid in paternity suits. However, a blood test of a supposed father only can suggest that he *might* be the father, not that he definitely *is* the father. For example, it is possible, but not definite, that a man with type A blood (having genotype *AO*) is the father of a child with type O blood. On the other hand, a blood test sometimes can prove definitely that a man is not the father. For example, a man with type AB blood cannot possibly be the father of a child with type O blood. Therefore, blood tests can be used legally only to exclude a man from possible paternity.

The Rh Blood Factor

The Rh blood factor is inherited separately from types A, B, AB, or O type blood. In each instance, it is possible to be Rh positive (Rh⁺) or Rh negative (Rh⁻). When you are Rh positive, there is a particular antigen on the red blood cells, and when you are Rh negative, it is absent. It can be assumed that the inheritance of this antigen is controlled by a single allelic pair in which simple dominance prevails: the Rh-positive allele is dominant over the Rh-negative allele. Complications arise when an Rh-negative woman reproduces with an Rh-positive man and the child in the womb is Rh positive. Under certain circumstances (p. 255), the woman may begin to produce antibodies that will attack the red blood cells of this baby or of a future Rh-positive baby.

Degrees of Dominance

The field of human genetics also has examples of incomplete dominance and codominance. For example, when a curly-haired Caucasian reproduces with a straight-haired Caucasian, their children will have wavy hair (fig. 22.14). We already mentioned that the multiple alleles controlling blood type are codominant. An individual with the genotype *AB* has type AB blood. Skin color, recall, is controlled by polygenes, and therefore it is possible to observe a range of skin colors among the members of a population in which Caucasians and blacks reproduce with one another.

Sickle-Cell Anemia

Sickle-cell anemia is an example of a human disorder that is controlled by incompletely dominant alleles. Individuals with the genotype $Hb^A Hb^A$ are normal, those with the $Hb^S Hb^S$ genotype have sickle-cell anemia, and those with the $Hb^A Hb^S$ genotype have **sickle-cell trait,** a condition in which the cells are sometimes sickle shaped. Two individuals with sickle-cell trait can produce children with all 3 phenotypes, as indicated in figure 22.15.

Among black Africans, the sickled cells seem to give protection against the malaria parasite, which uses red blood cells during its life cycle. Although infants with sickle-cell anemia often die, those with sickle-cell trait are protected from malaria, especially from ages 2 to 4. This means that in Africa, these children survive and grow to reproduce and to pass on the allele to their offspring. As many as 60% of blacks in malaria-infected regions of Africa have the allele. In the United States, about 10% of the black population carries the allele. A test can be done to detect the allele's presence; prenatal testing is also possible.

The red blood cells in persons with sickle-cell anemia cannot pass through small blood vessels easily. The sickle-shaped cells either break down or they clog blood vessels, and the individual suffers from poor circulation, anemia,

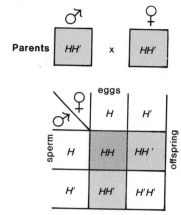

Figure 22.14
Incomplete dominance. Among Caucasians, neither straight nor curly hair is dominant. When 2 wavy-haired individuals reproduce, the offspring have a 25% chance of having either straight or curly hair and a 50% chance of having wavy hair, the intermediate phenotype.

Results 1 curly: 2 wavy: 1 straight

Genotypes   Phenotypes

HH = Curly hair

HH' = Wavy hair

$H'H'$ = Straight hair

Figure 22.15

Inheritance of sickle-cell anemia. a. In this example, both parents have the sickle-cell trait. Therefore, each child has a 25% chance of having sickle-cell anemia or of being perfectly normal and a 50% chance of having the sickle-cell trait. b. Sickled cells. Individuals with sickle-cell anemia have sickled red blood cells that tend to clump as illustrated here.

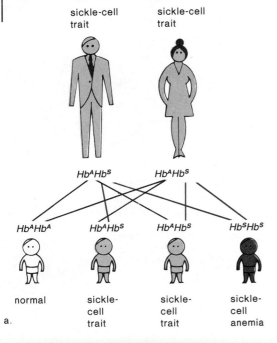

a.

b.

and sometimes internal hemorrhaging. Jaundice, episodic pain in the abdomen and joints, poor resistance to infection, and damage to internal organs are all symptoms of sickle-cell anemia.

Persons with sickle-cell trait do not usually have any difficulties unless they undergo dehydration or mild oxygen deprivation. At such times, the cells become sickle shaped, clogging blood vessels and leading to pain and even death. A study of the occurrence of sudden deaths during army basic training showed that a person with sickle-cell trait was 40 times more likely to die compared to normal recruits. This has caused the Army Medical Corps to advise drill instructors to train recruits more gradually, to give them enough to drink, and to make allowances for heat and humidity when planning their workouts.

There are many exceptions to Mendel's laws. These include polygenic inheritance, multiple alleles, and degrees of dominance.

Practice Problems 4*

1. What is the genotype of a person with straight hair? Could this individual ever have a child with curly hair?
2. What is the darkest child that could result from a mating between a light individual and a white individual?
3. What is the lightest child that could result from a mating between 2 intermediate individuals?
4. From the following blood types, determine which baby belongs to which parents:

Mrs. Doe	Type A
Mr. Doe	Type A
Mrs. Jones	Type A
Mr. Jones	Type AB
Baby 1	Type O
Baby 2	Type B

5. Prove that a child does not have to have the blood type of either parent by indicating what blood types *might* be possible when a person with type A blood reproduces with a person with type B blood.

*Answers to problems are on page 488.

Summary

In keeping with Mendel's laws of inheritance, it is customary to use letters to indicate the genotype and the alleles in the gametes of individuals. Homozygous dominant is indicated by 2 capital letters, and homozygous recessive is indicated by 2 lowercase letters. Heterozygous is indicated by a capital letter and a lowercase letter. Contrary to the individual, gametes have one letter of each type, either capital or lowercase as appropriate. All possible combinations of letters can occur in the gametes (except if the genes are linked as described in chapter 23). In doing an actual cross, it is assumed that all possible types of sperm fertilize all possible types of eggs. The results of some crosses can be determined by simple inspection, but certain others that commonly reoccur are given in table 22.2.

Studies of human genetics have shown that there are many autosomal genetic disorders that can be explained on the basis of simple Mendelian inheritance.

There are many exceptions to Mendel's laws, and these include polygenic inheritance (skin color), multiple alleles (ABO blood type), and degrees of dominance (curly hair). There are genetic disorders associated with these patterns of inheritance also.

Study Questions

1. What is Mendel's law of segregation? What do we call his factors today, and where are they located? (p. 471)
2. What is the difference between the genotype and the phenotype of an individual? For which phenotype in a one-trait problem are there 2 possible genotypes? (pp. 471–72)
3. What is Mendel's law of independent assortment? Relate Mendel's laws to one-trait and two-trait problems. (pp. 472–78)
4. What are the expected results from the following crosses? (p. 479)
   monohybrid × monohybrid
   monohybrid × recessive
   dihybrid × dihybrid
   dihybrid × recessive in both traits
5. Which of these crosses are called testcrosses? Why? (p. 474, p. 478)
6. What are the chances of the dominant phenotype(s) for each of these crosses? What does the phrase "chance has no memory" mean? (p. 473)
7. List ways it is possible to tell an autosomal recessive genetic disease and an autosomal dominant genetic disease by examining a pedigree chart. (p. 480)
8. Give an example of these patterns of inheritance: polygenic inheritance, multiple alleles, and degrees of dominance. (pp. 483–85)

Objective Questions

1. Whereas an individual has 2 genes for every trait, the gametes have _____ genes for every trait.
2. The recessive allele for the dominant gene W is _____ .
3. Mary has a widow's peak, and John has a continuous hairline. This is a description of their _____ .
4. W = widow's peak and w = continuous hairline; therefore, only the phenotype _____ could be heterozygous.
5. Two heterozygotes, each having a widow's peak, already have a child with a continuous hairline. The next child has what chance of having a continuous hairline? _____
6. In a testcross, an individual having the dominant phenotype is crossed with an individual having the _____ phenotype.
7. How many letters are required to designate the genotype of a dihybrid individual? _____
8. If a dihybrid is crossed with a dihybrid, how many offspring out of 16 are expected to have the dominant phenotype for both traits? _____
9. How many different phenotypes among the offspring are possible when a dihybrid is crossed with a dihybrid?

   _____
10. According to Mendel's law of independent assortment, a dihybrid can produce how many types of gametes having different combinations of genes?

   _____

Additional Genetic Problems

1. A woman heterozygous for polydactyly (dominant) reproduces with a normal man. What are the chances that their children will have 6 fingers and toes? (p. 472)
2. John cannot curl his tongue (recessive), but both his parents can curl their tongue. Give the genotype of all persons involved. (p. 472)
3. Parents who do not have Tay-Sachs disease (recessive) produce a child who has Tay-Sachs. What are the chances that each child born to this couple will have Tay-Sachs? (p. 472)
4. A man with widow's peak (dominant) who cannot curl his tongue (recessive) reproduces with a woman who has a continuous hairline and who can curl her tongue. They have a child who has a continuous hairline and cannot curl the tongue. Give the genotype of all persons involved. (p. 474)
5. Both Mr. and Mrs. Smith have freckles (dominant) and attached earlobes (recessive). Some of their children do not have freckles. What are the chances that their next child will have freckles and attached earlobes? (p. 474)
6. Mary has wavy hair (incomplete dominance) and reproduces with a man who has wavy hair. They have a child with straight hair. Give the genotype of all persons involved. (p. 485)
7. A man has type AB blood. What is his genotype? Could this man be the father of a child with type B blood? If so, what blood types could the child's mother have? (p. 484)
8. A woman with white skin has intermediate parents. If this woman marries a man with light skin, what is the darkest skin color possible for their children? the lightest? (p. 483)
9. Is the characteristic represented by the darkened individuals inherited as an autosomal dominant or an autosomal recessive? (p. 480)

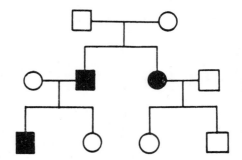

10. Fill in this pedigree chart to give the probable genotypes of the twins pictured in figure 22.13. (p. 484)

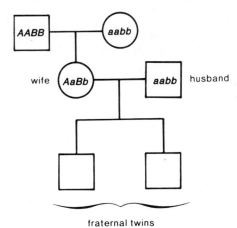

wife — AaBb
husband — aabb

AABB □ — aabb ○

fraternal twins

Selected Key Terms

allele (ah-lēl) an alternative form of a gene that occurs at a given chromosome site (locus). *469*

dihybrid (di-hi'brid) the offspring of parents who differ in 2 ways: shows the phenotype governed by the dominant alleles but carries the recessive alleles. *475*

dominant allele (dom'i-nant ah-lēl) hereditary factor that expresses itself, or a characteristic that is present when the genotype is heterozygous. *471*

genotype (jen'o-tīp) the genetic makeup of any individual. *471*

heterozygous (het''er-o-zi'gus) having 2 different alleles (as *Aa*) for a given trait. *471*

homozygous (ho''mo-zi'gus) having identical alleles (as *AA* or *aa*) for a given trait; pure breeding. *471*

monohybrid (mon''o-hi'brid) the offspring of parents who differ in one way only; shows the phenotype of the dominant allele but carries the recessive allele. *473*

phenotype (fe'no-tīp) the outward appearance of an organism caused by the genotype and environmental influences. *472*

Punnett square (pun'et skwār) a gridlike device that enables one to calculate the expected results of simple genetic crosses. *473*

recessive allele (re-ses'iv ah-lēl) hereditary factor that expresses itself, or a characteristic that is present only when the genotype is homozygous. *471*

testcross (test-kros) the backcross of a heterozygote with the recessive in order to determine its genotype. *474*

trait (trāt) specific term for a distinguishing feature studied in heredity. *469*

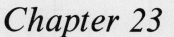

Chapter 23

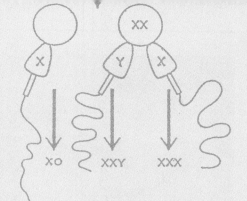

Chapter Concepts

1 Normally, humans inherit 22 pairs of autosomal chromosomes and 1 pair of sex chromosomes.

2 Abnormal chromosome inheritance occurs when there is either more or less than a pair of each autosomal chromosome or sex chromosome. It is also possible to inherit a mutated chromosome.

3 Certain genes, called X-linked genes, occur on the X chromosome and some control traits nonrelated to the sex of the individual.

4 Males always express inherited X-linked recessive disorders because they inherit only one X chromosome.

5 Various methods are available for determining the order of the genes (mapping) of the human chromosomes.

Chapter Outline

Chromosome Inheritance
  Normal Inheritance
  Abnormal Autosomal Chromosome
    Inheritance
  Abnormal Sex Chromosome
    Inheritance
Sex-Linked Inheritance
  X-Linked Genetics Problems
  Recessive X-Linked Disorders
  Sex-Influenced Traits
Mapping the Human Chromosomes
  Linkage Data
  Human-Mouse Cell Data
  Genetic Marker Data
  DNA Probe Data

Patterns of Chromosome Inheritance

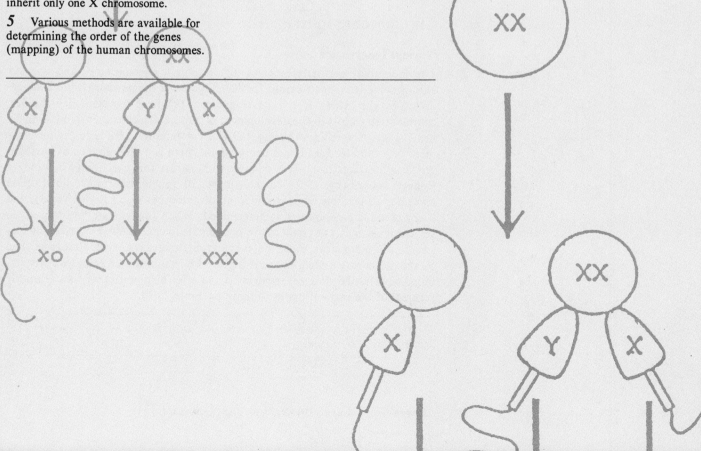

Figure 23.1

A girl! According to a somewhat controversial method advocated by Dr. Shettles of choosing the sex of your child, the X-bearing sperm is favored over the Y-bearing sperm if these requirements are met: (1) the vagina is acidic (a douche consisting of 2 tablespoons of white vinegar to a quart of water promotes this); (2) intercourse is frequent and penetration is shallow, but intercourse ceases 2 to 3 days before ovulation. On the other hand, the Y-bearing sperm is favored over the X-bearing sperm if the following requirements are met: (1) the vagina is alkaline (a douche consisting of 2 tablespoons of baking soda to a quart of water promotes this; let it stand 15 minutes before using); (2) abstinence is practiced until the day of ovulation, when penetration is deep.

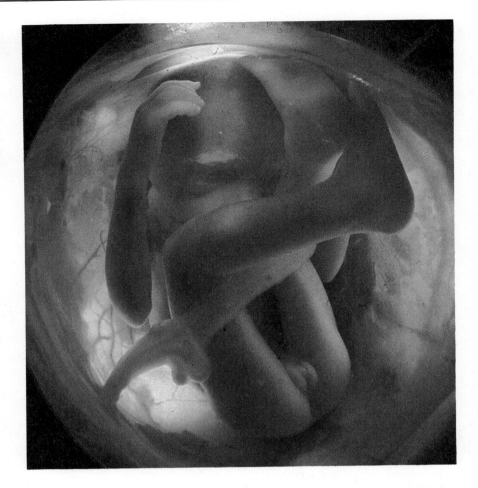

 In this chapter, we consider the chromosome inheritance of the individual, particularly as it relates to sex (fig. 23.1) and to the genes that are carried on the sex chromosomes.

Chromosome Inheritance

Normal Inheritance

The individual normally receives 22 autosomal chromosomes and one sex chromosome from each parent (p. 80). The sex of the newborn child is determined by the father. If a Y-bearing sperm fertilizes the egg, then the XY combination results in the development of a male. On the other hand, if an X-bearing sperm fertilizes the egg, the XX combination results in the development of a female. All factors being equal, there is a 50% chance of having a girl or a boy (fig. 23.1). It is possible to illustrate this probability by doing a **Punnett square** (fig. 23.2). In the square, all possible sperm are lined up on one side, all possible eggs are lined up on the other side (or vice versa), and every possible combination is determined. When this is done with regard to sex chromosomes, the results show one female to each male. However, for reasons that are not clear, more males than females are conceived, but from then on the death rate among males is higher. More males than females are spontaneously aborted, and this trend continues after birth until there is a dramatic reversal of the ratio of males to females (table 23.1).

The sex of a child is dependent on whether a Y-bearing or an X-bearing sperm fertilizes the X-bearing egg.

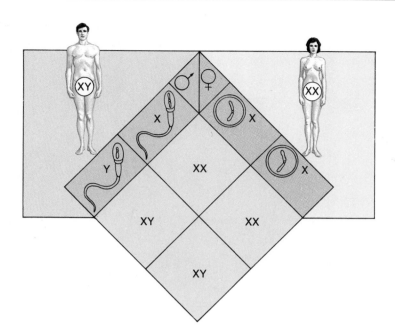

Figure 23.2
Inheritance of sex. In this Punnett square, the sperm and the eggs are shown as carrying only a sex chromosome. (Actually, they also carry 22 autosomes.) The offspring are either male or female, depending on whether an X or a Y chromosome is received from the male parent.

Male/Female Gene Differences

For some years, it has been proposed that a gene located on the Y chromosome brings about maleness. Embryos begin life with no evidence of a sex, but about the third month of development, males can be distinguished from females (p. 456). Investigators recently reported the finding of a gene they call the testis determining factor (TDF) gene on the Y chromosome. When this gene is lacking from the Y chromosome, the individual is a female, even though the chromosome inheritance is XY. On the other hand, if the gene is present in an XX individual, this person is a male.

Once males and females develop, there are certain cellular differences that can be detected, as discussed in the reading on page 493.

Abnormal Autosomal Chromosome Inheritance

Abnormal autosomal chromosome inheritance can occur due either to non-disjunction or to a chromosome mutation.

Nondisjunction

Sometimes individuals are born with either too many or too few autosomal chromosomes, most likely due to nondisjunction of chromosomes or sister chromatids during meiosis (fig. 23.3). **Nondisjunction** can occur during meiosis I if the homologous chromosomes fail to separate or during meiosis II if the sister chromatids fail to separate.

Down Syndrome   The most common autosomal abnormality is seen in individuals with **Down syndrome** (fig. 23.4). This syndrome is recognized easily. Its characteristics include a short stature; an oriental-like fold of the eyelids; stubby fingers; a wide gap between the first and second toes; a large, fissured tongue; a round head; a palm crease, the so-called simian line; and, unfortunately, mental retardation that sometimes can be severe.

Persons with Down syndrome usually have 3 number-21 chromosomes because the egg had 2 number-21 chromosomes instead of one. (In 23% of the cases studied, however, the sperm had the extra number-21 chromosome.) It appears that nondisjunction is most apt to occur in the older female since children with Down syndrome usually are born to women over age 40 (table 23.2). If a woman wishes to know whether or not her unborn child is affected by

Table 23.1 Sex Ratios in the United States	
Age	**Males:Females**
Birth	106:100
18 years	100:100
50 years	85:100
85 years	50:100
100 years	20:100

From John W. Hole, Jr., *Human Anatomy and Physiology,* 5th ed. Copyright © 1990 Wm. C. Brown Publishers, Dubuque, Iowa. All Rights Reserved. Reprinted by permission.

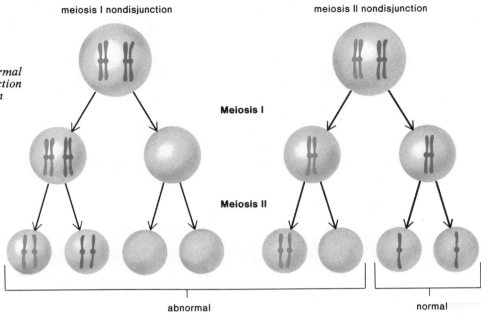

Figure 23.3

Nondisjunction of autosomal chromosomes during oogenesis. Nondisjunction can occur during meiosis I if a homologous pair of chromosomes fails to separate and during meiosis II if the sister chromatids fail to separate completely. In either case, the abnormal eggs carry an extra chromosome. Nondisjunction of the number-21 chromosome leads to Down syndrome.

meiosis I nondisjunction

meiosis II nondisjunction

Meiosis I

Meiosis II

abnormal

normal

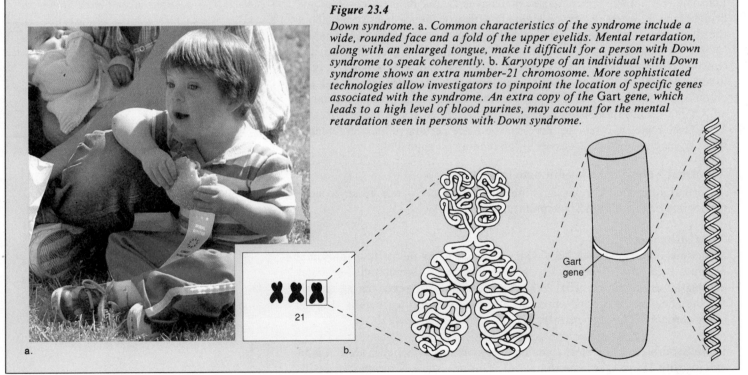

Figure 23.4

Down syndrome. a. Common characteristics of the syndrome include a wide, rounded face and a fold of the upper eyelids. Mental retardation, along with an enlarged tongue, make it difficult for a person with Down syndrome to speak coherently. b. Karyotype of an individual with Down syndrome shows an extra number-21 chromosome. More sophisticated technologies allow investigators to pinpoint the location of specific genes associated with the syndrome. An extra copy of the Gart gene, which leads to a high level of blood purines, may account for the mental retardation seen in persons with Down syndrome.

21

a.

b.

Gart gene

Down syndrome, she can elect to undergo chorionic villi testing or amniocentesis, 2 procedures discussed in the reading on page 454. Following this procedure, a karyotype can reveal whether the child has Down syndrome.

It is known that the genes that cause Down syndrome are located on the bottom third of the number-21 chromosome (fig. 23.4*b*), and there has been a lot of investigative work to discover the specific genes responsible for the characteristics of the syndrome. Thus far, investigators have discovered several genes that may account for various conditions seen in persons with Down syndrome. For example, they have located genes most likely responsible for

Human Reproduction, Development, and Inheritance

Table 23.2 *Frequency and Effects of the Most Common Chromosome Abnormalities in Humans*

Syndrome	Sex	Chromosomes	Frequency Miscarriages	Births	Fertility
Down General Mothers over 40	M or F	Extra 21	1/40	1/700 1/70	Infertile
Turner	F	XO	1/18	1/5,000	Sterile
Metafemale	F	XXX	0	1/700	Infertile
Klinefelter	M	XXY	0	1/2,000	Sterile
XYY	M	XYY	?	1/2,000	—

On occasion, such as at an athletic competition, it is important to be able to certify that an individual is a male or a female. Because physical examinations sometimes fail—as when a male has had a sex-change operation—officials often resort to examining the cells themselves.

You could, of course, do a karyotype, but there are easier methods. It so happens that XX females have small, darkly staining

Male/Female Cell Differences

masses of chromatin called Barr bodies (named after the person who first identified them) present in their nuclei (see figure). XY males have no comparable spots of chromatin in their nuclei. It turns out that a Barr body* is a condensed and at least to some degree inactive X chromosome as was proposed by Mary Lyon. The validity

of the *Lyon hypothesis* means that female cells function with a single X chromosome just as males do. Still, in some cells one X is condensed and in some cells the other X is condensed so that the female body is a mosaic of genetically different cells.

No doubt, you believe that observation of Barr bodies is not a guarantee of femaleness because we have already indicated that there are XX males, albeit rarely. We need another test, and there is one. There is an antigen, called the H-Y antigen, present in the cell membrane of males but not in females. It is called an antigen because females produce antibodies against it. To test for maleness it is possible to suspend a sample of white blood cells in a solution that contains some of these antibodies. If the cells carry the H-Y antigen, indicating that the person is a male,

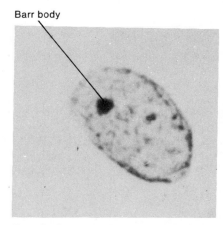
Barr body

Barr body. *Each female cell contains a Barr body, which is a condensed X chromosome.*

the antibodies bind with them. Now we can be certain who is a male and who is a female.

*How many Barr bodies does a person with Klinefelter syndrome have? a metafemale have?

the increased tendency toward leukemia, cataracts, accelerated rate of aging, and mental retardation. The gene for mental retardation, dubbed the *Gart* gene, causes an increased level of purines in the blood, a finding associated with mental retardation. It is hoped that someday it will be possible to find a way to control the expression of the *Gart* gene even before birth so that at least this symptom of Down syndrome does not appear.

Abnormal autosomal chromosome inheritance can be due to the inheritance of extra chromosomes. Down syndrome is caused by the inheritance of an extra number-21 chromosome.

Chromosome Mutations
It is possible that even though the correct number of chromosomes is inherited, one chromosome is defective in some way because of a chromosome mutation that occurred during meiosis.

Figure 23.5

Giant polytene chromosomes from salivary gland cells of Drosophila. *These chromosomes are composed of many hundreds of chromatids that lie side by side. Each chromosome has a characteristic banding pattern; the darker bands are places where DNA is more condensed. Sometimes it is possible to make out chromosome mutations in these chromosomes; for example, a duplication leads to the increased size of a particular band.*

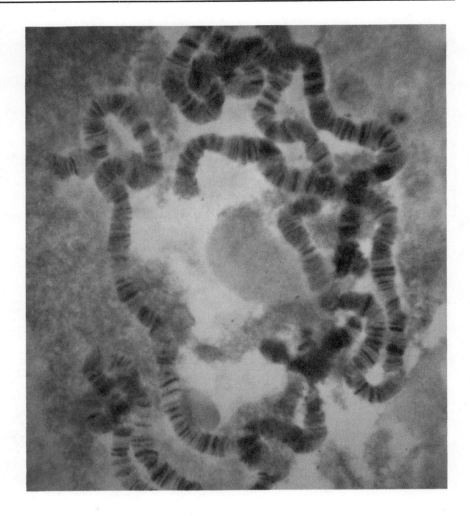

The process of crossing-over (p. 90) can lead to a chromosome mutation and so can various agents in the environment, such as radiation, certain organic chemicals, or even viruses, that can cause chromosomes to break apart. Ordinarily, when breaks occur in chromosomes, the 2 broken ends reunite to give the same sequence of genes. Sometimes, however, the broken ends of one or more chromosomes do not rejoin in the same pattern as before, and this results in a chromosome mutation. The existence of such mutations can be detected readily in chromosomes taken from salivary gland cells of a larval fruit fly (*Drosophila*) (fig. 23.5). In these cells, the chromosomes duplicate repeatedly, but instead of separating, they remain side by side to produce what are called giant polytene chromosomes. The large size of these chromosomes makes it possible to see some of the types of chromosome mutations discussed here.

An **inversion** (fig. 23.6) occurs when a segment of a chromosome is turned around 180 degrees. You might think this is not a problem because the same genes are present, but the new position can lead to altered gene activity. It also can lead to abnormal crossing-over during meiosis, resulting in duplications and deletions in the chromosomes of the gametes.

A **translocation** is the movement of a chromosome segment from one chromosome to another, nonhomologous chromosome. A **deletion** occurs when an end of a chromosome breaks off or when 2 simultaneous breaks lead to the loss of a segment. A **duplication,** the presence of a chromosome segment more than once in the same chromosome, can occur in 2 ways. A broken segment from one chromosome simply can attach to its homologue. The presence of a duplication in the middle of a chromosome is more likely due to unequal

Human Reproduction, Development, and Inheritance

crossing-over. This can occur when homologues are mispaired slightly. This simultaneously produces a gene duplication and a gene deletion during crossing-over:

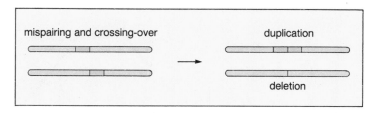

The production of mutated chromosomes during meiosis can cause reproductive problems in the individual. When genes are missing or when there are extra copies in the gametes, the zygote can fail to develop, or if it does develop, the resulting baby can have birth defects.

Cri du Chat Syndrome A chromosome deletion is responsible for **cri du chat** (cat's cry) **syndrome.** Affected individuals meow like a kitten when they cry, but more importantly, they tend to have a small head with malformations of the face and the body, and mental defectiveness usually causes retarded development. Chromosome analysis shows that a portion of one number-5 chromosome is missing (deleted), while the other number-5 chromosome is normal.

Abnormal autosomal chromosome inheritance can be due to the inheritance of a mutated chromosome. Cri du chat syndrome occurs when a portion of the number-5 chromosome is deleted.

Abnormal Sex Chromosome Inheritance

Abnormal sex chromosome constituencies (table 23.2) are also due to the occurrence of nondisjunction. Nondisjunction of the sex chromosomes during oogenesis can lead to an egg with either 2 X chromosomes or to no X chromosome. Nondisjunction of the sex chromosomes during spermatogenesis can result in a sperm that has no sex chromosome, both an X and a Y chromosome, 2 X chromosomes, or 2 Y chromosomes. These abnormal gametes sometimes result in miscarriages as noted in table 23.2, but other times, the zygote develops into an individual with one of the conditions listed in figure 23.7. The chromosome mutations noted in regard to autosomal chromosomes are not seen in the sex chromosomes, most likely because the zygote is not viable.

Sometimes a person inherits an abnormal combination of sex chromosomes due to nondisjunction of these chromosomes during meiosis.

Abnormalities

An X0 individual with **Turner syndrome** has only one sex chromosome, an X; the 0 signifies the absence of a second sex chromosome. Turner females are short, have a broad chest, and may have congenital heart defects. Because the ovaries never become functional, Turner females do not undergo puberty or menstruate, and there is a lack of breast development (fig. 23.8a). Although no overt mental retardation is reported, Turner females show reduced skills in interpreting spatial relationships.

A **metafemale** is an individual with more than 2 X chromosomes. It might be supposed that the XXX female is especially feminine, but this is not the case. Although in some cases there is a tendency toward learning disabilities, most metafemales have no apparent physical abnormalities except that they may have menstrual irregularities, including early onset of menopause.

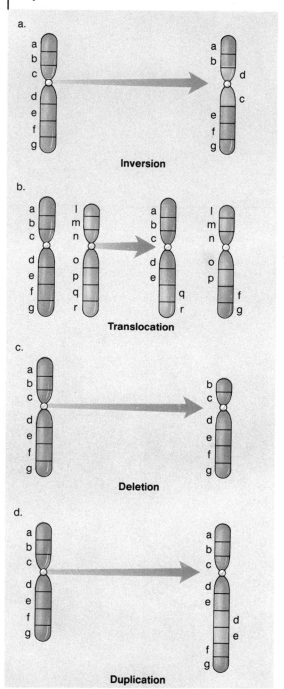

Figure 23.6
Types of chromosome mutations. a. Inversion occurs when a piece of a chromosome breaks loose and then rejoins in reversed direction. b. Translocation is the exchange of chromosome pieces between nonhomologous chromosomes. c. Deletion is the loss of a chromosome piece. d. Duplication occurs when the same piece is repeated within the chromosome.

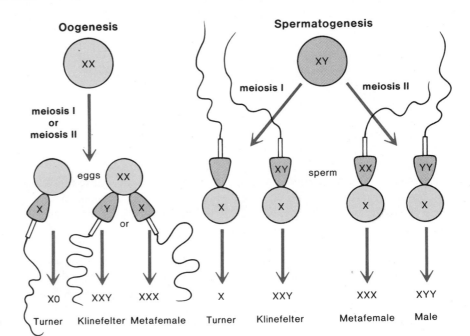

Figure 23.7

Nondisjunction of sex chromosomes. Nondisjunction of sex chromosomes during oogenesis followed by fertilization with normal sperm results in the conditions noted. Nondisjunction of sex chromosomes during spermatogenesis followed by fertilization of normal eggs results in the conditions noted.

Oogenesis

XX

meiosis I
or
meiosis II

eggs   XX

X   or   Y X

XO   XXY   XXX
Turner Klinefelter Metafemale

Spermatogenesis

XY

meiosis I   meiosis II

XY   sperm   XX   YY

X   X   X   X

X   XXY   XXX   XYY
Turner Klinefelter Metafemale Male

Figure 23.8

Abnormal sex chromosome inheritance. a. A male with Klinefelter (XXY) syndrome, which is marked by immature sex organs and development of the breasts. b. Female with Turner (XO) syndrome, which includes a bull neck, short stature, and immature sexual features.

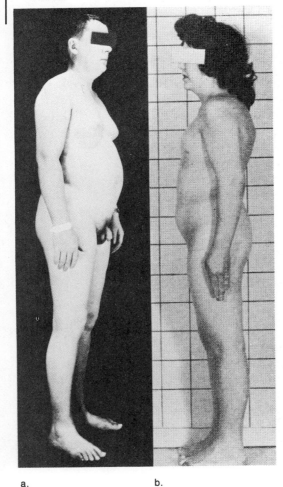

a.   b.

A male with **Klinefelter syndrome** has 2 or more X chromosomes in addition to a Y chromosome. Affected individuals are sterile males; the testes are underdeveloped and there may be some breast development (fig. 23.8b). These phenotypic abnormalities are not apparent until puberty, although some evidence of subnormal intelligence may be apparent before this time.

XYY also can result from nondisjunction during spermatogenesis. Affected males usually are taller than average, suffer from persistent acne, and tend to have barely normal intelligence. At one time, it was suggested that these men were likely to be criminally aggressive, but it has since been shown that the incidence of such behavior among them is no greater than among XY males.

Individuals sometimes are born with the sex chromosomes XO (Turner syndrome), XXX (metafemale), XXY (Klinefelter syndrome), and XYY. No matter how many X chromosomes there are, an individual with a Y chromosome is usually a male.

Sex-Linked Inheritance

The genes that determine the development of the sexual organs are on the sex chromosomes (p. 491). Even so, most of the genes on the sex chromosomes have nothing to do with sexual development and instead are concerned with other body traits. Genes on the sex chromosome are said to be **sex linked** because they are on the sex chromosomes. A few sex-linked alleles are presumed to be on the Y chromosome, but most of those discovered so far are only on the much larger X chromosome. These are called **X-linked genes.** The Y is blank for X-linked genes.

X-Linked Genetics Problems

Recall that when doing autosomal genetics problems, we represent the genotypes of males and females similarly, as shown in the following example for humans.

Key:
W = Widow's peak
w = Continuous hairline

Genotypes:
WW, Ww, or *ww*

Human Reproduction, Development, and Inheritance

However, as in the next example, when we set up the key for a sex-linked gene, males and females must be indicated by sex.

Key:
X^B = Normal vision
X^b = Color blindness

The possible genotypes in both males and females are

$X^B X^B$ = Female with normal color vision
$X^B X^b$ = Carrier female with normal color vision
$X^b X^b$ = Female who is color blind
$X^B Y$ = Male with normal vision
$X^b Y$ = Male who is color blind

Note that the second genotype is a carrier female because although a female with this genotype appears normal, she is capable of passing on an allele for color blindness. Color-blind females are rare because they must receive the allele from both parents; because color-blind males are more common since they need only one recessive allele in order to be color blind. The allele for color blindness has to be inherited from their mother because it is on the X chromosome; males only inherit the Y chromosome from their father.

Now, let us consider a particular cross. If a heterozygous woman reproduces with a man with normal vision, what are the chances of their having a color-blind daughter? a color-blind son?

Parents $X^B X^b \times X^B Y$

Inspection indicates that all daughters will have normal color vision because they all will receive an X^B from their father. The sons, however, have a 50% chance of being color blind, depending on whether they receive an X^B or an X^b from their mother. The inheritance of a Y chromosome from their father cannot offset the inheritance of an X^b from their mother.

Figure 23.9 illustrates the use of the Punnett square when doing X-linked problems. Notice that when indicating the results of a cross involving an X-linked gene, you give the phenotypic ratios for males and females separately.

Figure 23.10 gives a pedigree chart for a recessive X-linked gene and lists ways to recognize this pattern of inheritance.

Recessive X-Linked Disorders

Color Blindness
In humans, there are 3 genes involved in distinguishing color because there are 3 different types of cones, the receptors for color vision (p. 376). Two of these are X-linked genes; one affects the green-sensitive cones, whereas the other affects the red-sensitive cones. About 6% of men in the United States are color blind due to a mutation involving green perception, and about 2% are color blind due to a mutation involving red perception.

Hemophilia
There are about 100,000 hemophiliacs in the United States. The most common type of hemophilia is hemophilia A, due to the absence or minimal presence of a particular clotting factor called factor VIII. *Hemophilia* is called the bleeder's disease because the affected person's blood is unable to clot. Although hemophiliacs do bleed externally after an injury, they also suffer from internal bleeding, particularly around joints. Hemorrhages can be checked with transfusions of fresh blood (or plasma) or concentrates of the clotting protein. Unfortunately, some hemophiliacs have contracted AIDS after using concentrated blood from untested donors, but this cannot occur if they use a purified form of the concentrate from donors who have been tested.

Figure 23.9
Cross involving X-linked genes. The male parent is normal, but the female parent is a carrier; an allele for color blindness is located on one of her chromosomes. Therefore, each son stands a 50-50 chance of being color blind.

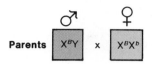

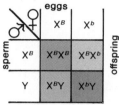

Results   females—all normal
          males—1 normal:1 color blind

Genotypes	Phenotypes
$X^B X^B$	female, normal vision
$X^B X^b$	female, carrier
$X^b X^b$	female, color blind
$X^B Y$	male, normal vision
$X^b Y$	male, color blind

Key:
X^B = Normal vision
X^b = Color blind

Figure 23.10

Sample pedigree chart for a recessive X-linked genetic disorder. Which female in the chart is a human carrier?

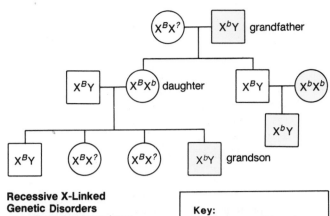

Recessive X-Linked Genetic Disorders
- An affected son can have parents who have the normal phenotype.
- In order for a female to have the characteristic, her father must also have it. Her mother must have it or be a carrier.
- The characteristic often skips a generation from the grandfather to the grandson.
- If a woman has the characteristic, all of her sons will have it.

Key:
$X^B X^B$ = Normal female
$X^B X^b$ = Carrier female
$X^b X^b$ = Color-blind female
$X^B Y$ = Normal male
$X^b Y$ = Color-blind male

23.1 Critical Thinking

1. Early in this century, geneticists performed this cross:

	♀		♂
P	red-eyed	×	white-eyed
F_1	red-eyed		red-eyed

   From these results, they knew which characteristic is dominant?

2. They went on to perform this cross:

	♀		♂
$F_1 \times F_1$	red-eyed	×	red-eyed
F_2	red-eyed		1 : 1 red- to white-eyed

   Are these results explainable if the allele for red/white eye color is on the Y chromosome but not on the X chromosome? on the X chromosome but not on the Y chromosome? Explain.

3. How do these results support the hypothesis that genes are on the chromosomes?

At the turn of the century, hemophilia was prevalent among the royal families of Europe. All of the afflicted males could trace their ancestry to Queen Victoria of England (fig. 23.11). Because none of Queen Victoria's forebearers or relatives was affected, it seems that the gene she carried arose by mutation either in Victoria or in one of her parents. Her carrier daughters, Alice and Beatrice, introduced the gene into the ruling houses of Russia and Spain. Alexis, the last heir to the Russian throne before the Russian Revolution, was a hemophiliac. The present British royal family has no hemophiliacs because Victoria's eldest son, King Edward VII, did not receive the gene and therefore could not pass it on to any of his descendants.

Muscular Dystrophy

Muscular dystrophy, as the name implies, is characterized by a wasting away of the muscles. The most common form, *Duchenne muscular dystrophy,* is X-linked and occurs in about one out of every 25,000 male births. Symptoms, such as waddling gait, toe walking, frequent falls, and difficulty in rising, may appear as soon as the child starts to walk. Muscle weakness intensifies until the individual is confined to a wheelchair. Death usually occurs during the teenage years; therefore, affected males are rarely fathers. The recessive allele remains in the population by passage from carrier mother to carrier daughter.

Recently, the gene for muscular dystrophy was isolated, and it was discovered that the absence of a protein, now called dystrophin, is the cause of the disorder. Much investigative work determined that dystrophin is involved in the release of calcium from the calcium-storage sacs (p. 363) in muscle fibers. The lack of dystrophin causes calcium to leak into the cell, which promotes the action of an enzyme that dissolves muscle fibers. When the body attempts to repair the tissue, the formation of fibrous tissue occurs, and this cuts off the blood supply so that more and more cells die.

A test now is available to detect carriers for Duchenne muscular dystrophy.

There are sex-linked genes on the X chromosome that have nothing to do with sex characteristics. Males have only one copy of these genes, and if they inherit a recessive allele, it is expressed.

Figure 23.11

A simplified pedigree showing the X-linked inheritance of hemophilia in European royal families. Because Queen Victoria was a carrier, each of her sons had a 50% chance of having the disease and each of her daughters had a 50% chance of being a carrier. This pedigree shows only the affected individuals. Many others are unaffected, such as the members of the present British royal family.

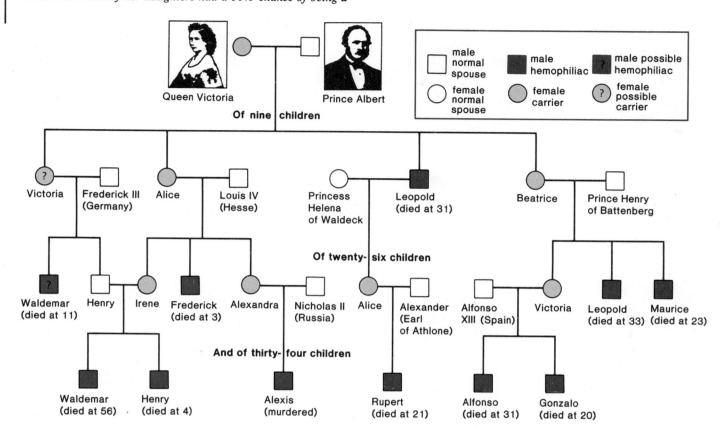

Practice Problems*

1. Both the mother and the father of a male hemophiliac appear to be normal. From whom did the son inherit the allele for hemophilia? What are the genotypes of the mother, the father, and the son?

2. A woman is color blind. What are the chances that her sons will be color blind? If she is married to a man with normal vision, what are the chances that her daughters will be color blind? will be carriers?

3. Both parents are right handed (*R* = right handed, *r* = left handed) and have normal vision. Their son is left handed and color blind. Give the genotype of all persons involved.

4. Both the husband and the wife have normal vision. The wife gives birth to a color-blind daughter. What can you deduce about the girl's father?

*Answers to problems are on page 505.

Fragile-X Syndrome

Fragile-X syndrome is one of the most common genetic causes of mental retardation, second only to Down syndrome. It is called fragile-X syndrome because under the right laboratory conditions, a lesion can be seen in the X chromosome. Nevertheless, the condition is believed to be caused by a particular gene on the X chromosomes.

Figure 23.12

Baldness, a sex-influenced characteristic. Due to hormonal influences, the presence of only one gene for baldness causes the condition in the male, whereas the condition does not occur in the female unless she possesses both genes for baldness.

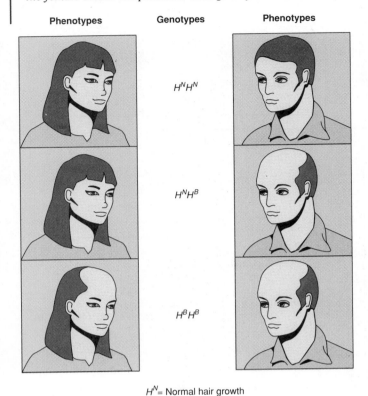

Phenotypes	Genotypes	Phenotypes
	$H^N H^N$	
	$H^N H^B$	
	$H^B H^B$	

H^N= Normal hair growth

H^B= Pattern baldness

Figure 23.13

Complete linkage (hypothetical). In this example, the genes for dimples and type of fingers are linked. Even though the parents are dihybrids, the offspring show only 2 possible phenotypes.

Parents

D d × D d
B b   B b

Gametes

D d × D d
B b   B b

Offspring

D D   D d   D d   d d
B B   B b   B b   b b

Key:
D = Dimples
d = No dimples
B = Blunt fingers
b = Pointed fingers

Results:
3 dimples and blunt fingers
1 no dimples and pointed fingers

The manner in which fragile-X syndrome is inherited does not follow the usual X-linked pattern of inheritance. For example, males with the defective gene can have normal intelligence, but their daughters can have fragile-X-positive sons. Also, investigators have found a number of girls with the fragile-X lesion who are subnormal in intelligence. It appears that the allele is turned off if inherited from the father but turned on if inherited from the mother. The hypothesis that the activity of a gene is dependent on the sex that passes the gene on is a new one in genetic inheritance that is discussed further in the next chapter.

Sex-Influenced Traits

Not all traits we associate with the sex of the individual are due to sex-linked genes. Some are simply sex-influenced traits. Sex-influenced traits are characteristics that often appear in one sex but only rarely appear in the other. It is believed that these traits are governed by genes that are turned on or off by hormones. For example, the secondary sex characteristics, such as the beard of a male and the developed breasts of a female, probably are controlled by the balance of hormones.

Baldness (fig. 23.12) is believed to be c
testosterone because males who take the hormone to increase masculinity begin to lose their hair. A more detailed explanation has been suggested by some investigators. It has been reasoned that due to the effect of hormones, males require only one gene for the trait to appear, whereas females require 2 genes. In other words, the gene acts as a dominant in males but as a recessive in females. This means that males born to a bald father and a mother with hair *at best* have a 50% chance of going bald. Females born to a bald father and a mother with hair *at worst* have a 25% chance of going bald.

Another sex-influenced trait of interest is the length of the index finger. In women, the index finger is at least equal to if not longer than the fourth finger. In males, the index finger is shorter than the fourth finger.

Mapping the Human Chromosomes

A chromosome map indicates the various gene loci on a particular chromosome. There are several techniques that can be used to determine the loci of genes, and several are explored here. Some of these techniques have been in existence for some time, and others only just now have been perfected.

Once we know the location of a gene, it can be excised, the sequence of its nucleotides can be determined, and most important, its effects on the phenotype can be studied.

Linkage Data

As illustrated in figure 23.13, a chromosome pair has a series of genes. Alleles on the same chromosome are said to form a **linkage group.** Mendel's law of independent assortment cannot hold for linked genes because they tend to appear together in the same gamete. Therefore, traits controlled by linked genes tend to be inherited together.

To take a hypothetical example, remember that dimples are dominant over no dimples and blunt fingers are dominant over pointed fingers. If a dihybrid reproduces with a dihybrid, you expect 4 possible phenotypes among the offspring. But as figure 23.13 shows, only 2 phenotypes appear if the genes are absolutely linked in the manner illustrated. When doing linkage problems, it is better to use the method illustrated in figure 23.13 than a Punnett square so that the genes can be shown on a single chromosome.

When a dihybrid is crossed with a recessive, you normally expect all possible phenotypes among the offspring. If linkage is present, however, the number of possible phenotypes could be reduced to 2 types. To take an actual example, it has been reported that the ABO blood group genes and the gene for a very unusual dominant condition called nail-patella syndrome (NPS) are on the same chromosome. A person with NPS has fingernails and toenails that are reduced or absent and a kneecap (patella) that is small. In one family, the female parent had the genotype *BO* for blood type and the genotype *Nn* for NPS; furthermore, it could be established that the allele *B* was on the same chromosome as *N* and that the allele *O* was on the same chromosome as *n*. Notice in figure 23.14*a* that if linkage holds, this individual would form only 2 possible gametes.

The male parent in this example had the recessive genotype for both traits and therefore could form only one type of gamete carrying the recessive alleles of each gene, as illustrated in figure 23.14*a*. Therefore, assuming linkage, the children of this couple should have only 2 possible phenotypes: blood type B with NPS and blood type O without NPS. However, at least 10% of all offspring showed a phenotype in which blood type B was found without NPS and blood type O was found with NPS. This indicates that crossing-over occurred (fig. 23.14*b*).

Figure 23.14

*Linkage (in practice). a. If linkage were complete,
then only 2 phenotypes would appear among the
offspring resulting from this cross. b. In practice,
linkage is not complete because of crossing-over.
In this actual example, 10% of the offspring had
recombinant characteristics because crossing-over
had occurred.*

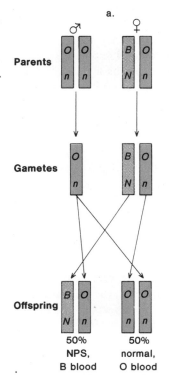

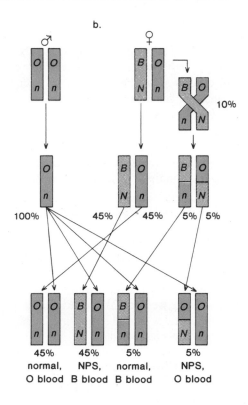

Key:
BO = Type B blood
OO = Type O blood
N = Nail-patella syndrome (NPS)
n = Normal

Crossing-Over

When tetrads form during meiosis, the nonsister chromatids of homologous
chromosomes can exchange portions by a process of breaking and then reas-
sociating (see fig. 4.15). The gametes that receive recombined chromosomes
are called recombinant gametes. Recombinant gametes indicate that the
linkage between 2 genes has been broken by crossing-over. Figure 23.14 shows
how recombinant gametes produced the unexpected phenotypes in our example.

It has been possible to begin to map chromosomes by studying the crossover frequency of linked genes. Alleles distant from one another are more likely
to be separated by crossing-over than alleles that are close together. Thus, the
crossover frequency indicates the distance between 2 alleles on a chromosome.
Each percentage of crossing-over is interpreted as a distance of one map unit.
Using these frequencies, then, it is possible to indicate the order of the alleles
on the chromosome.

*The presence of linkage groups changes the expected results of genetic crosses. The
frequency of recombinant gametes that occur due to the process of crossing-over
has been used to map chromosomes.*

Human-Mouse Cell Data

Human and mouse cells are mixed together in a laboratory dish, and in the
presence of inactivated virus of a special type, they fuse (fig. 23.15). As the
cells grow and divide, some of the human chromosomes are lost, and eventually the daughter cells contain only a few human chromosomes, each of which

Figure 23.15

Human-mouse cell hybrids. In the presence of a fusing agent, human fibroblast cells sometimes join with mouse tumor cells to give hybrid cells having nuclei that contain both types of chromosomes. Subsequent cell division of the hybrid cell produces clones that have lost most of their human chromosomes, allowing the investigator to study these chromosomes separate from all other human chromosomes.

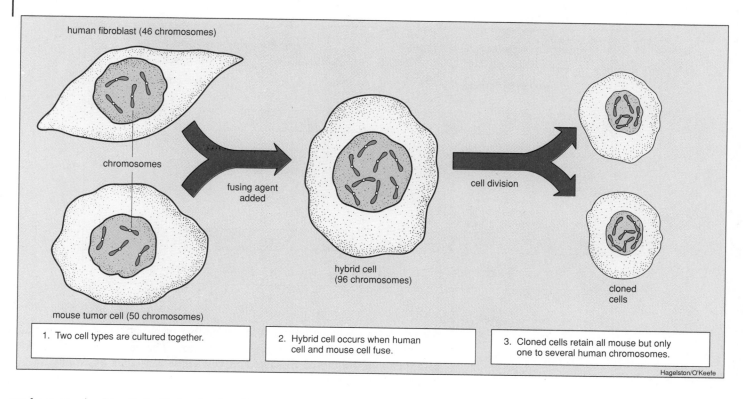

human fibroblast (46 chromosomes)

chromosomes

fusing agent added

hybrid cell (96 chromosomes)

cell division

cloned cells

mouse tumor cell (50 chromosomes)

1. Two cell types are cultured together.

2. Hybrid cell occurs when human cell and mouse cell fuse.

3. Cloned cells retain all mouse but only one to several human chromosomes.

Hagelston/O'Keefe

can be recognized by their distinctive banding pattern (fig. 23.5). Analysis of the proteins made by the various human-mouse cells enables scientists to determine which genes to associate with which human chromosomes.

Sometimes, it is possible to obtain a human-mouse cell that contains only one human chromosome or even just a portion of a chromosome. This technique has been very helpful to those researchers who have been studying the genes that are located on the number-21 chromosome (p. 492).

Genetic Marker Data

A new mapping technique works directly with DNA and uses genetic markers to tell if an individual has a defective allele. The exact location of the defective allele is not known, but the close proximity of the allele and the marker can be assumed because they almost always are inherited together. For a marker to be dependable, it should be inherited with the defective allele at least 98% of the time.

Detection of genetic markers makes use of special bacterial enzymes called **restriction enzymes,**[1] each of which cuts the DNA strand at a specific nucleotide sequence, producing a particular pattern of DNA fragments. A marker, which is really a DNA mutation, alters the normal pattern of DNA fragments resulting from restriction enzyme use. Notice in figure 23.16 the different sizes of the fragments—the polymorphism that exists—between the normal individual and the affected individual. Scientists refer to the differences in the observed fragment lengths as restriction fragment length polymorphisms (RFLPs). These polymorphisms can be discovered by comparing

[1]These enzymes are called restriction enzymes because bacteria use them to restrict the growth of viruses whenever viral DNA enters bacteria. Scientists extract the enzymes from bacterial cells and use them in order to cleave DNA in the laboratory.

Figure 23.16

Use of a genetic marker to detect the approximate location of a gene or to test for a genetic disorder when the exact location of the gene is unknown. a. DNA from the normal individual has certain restriction enzyme cleavage sites near the gene in question. b. DNA from another individual lacks one of the cleavage sites, and this loss indicates that they almost certainly have the genetic disorder. Experience has shown that this genetic marker is almost always present when an individual has the disorder. In other cases, a gain in a cleavage site is the genetic marker.

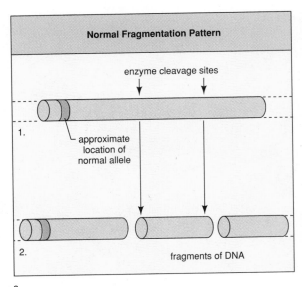

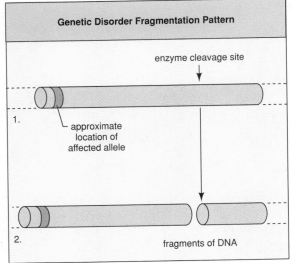

a.

b.

the DNA fragment pattern in a large number of closely related normal and abnormal individuals or by using DNA probes. The current tests for sickle-cell anemia, Huntington disease, and Duchenne muscular dystrophy all are based on the presence of a marker.

DNA Probe Data

An exact but radioactive sequence of nucleotides then can be prepared in the laboratory and used as a probe. A DNA probe seeks out the chromosome, from among any others, that shares the same sequence of nucleotides. It then binds to it. The chromosomes then can be exposed to a photographic film that shows this pair bound together.

As discussed in chapter 25, the U.S. government is now supporting the Human Genome Project for mapping the human chromosomes. DNA probes more efficiently detect genetic markers and also can test whether an individual is a carrier for or is affected by a genetic disorder. This type of test is now available for NF (neurofibromatosis, p. 482) and fragile-X syndrome, and others are expected soon.

Several new techniques—human-mouse cell preparations, and genetic markers and probes—now make it possible to map the human chromosomes at a faster rate than formerly was possible.

Summary

Humans inherit 22 autosomal and one sex chromosome from each parent. The father determines the sex of the offspring because the mother only gives an X chromosome while the father gives an X or a Y chromosome.

Nondisjunction of chromosomes during meiosis causes the inheritance of an abnormal number of chromosomes. For example, Down syndrome results when an individual inherits 3 number-21 chromosomes. Abnormal sex chromosome inheritance also occurs because of nondisjunction. Females who are XO have Turner syndrome, and those who are XXX are metafemales. Males with Klinefelter syndrome are XXY; there are also XYY males. Chromosome mutations that occur during meiosis give rise to abnormalities. Cri du chat syndrome is caused by the inheritance of a number-5 chromosome that has a portion deleted.

X-linked genes are genes on the X chromosome that have nothing to do with the sex of the individual. Because males receive only one X chromosome, they are more subject to disorders caused by recessive X-linked genes. There are numerous ways to detect an X-linked inheritance pattern, and these are listed in figure 23.10. Three well-known X-linked disorders are color blindness, hemophilia, and Duchenne muscular dystrophy.

All the genes on one chromosome form a linkage group that is broken only when crossing-over occurs. Genes that are linked do not follow Mendel's law of segregation because they tend to go together into the same gamete. If crossing-over occurs, a cross gives all possible phenotypes among the offspring, but the expected ratio is changed greatly. Linkage data contribute to our knowledge of the location of genes on chromosomes; crossing-over data help determine the order of genes since genes that are distant from one another tend to crossover more often than those that are close. Human-mouse cell data, genetic marker data, and DNA probes are new ways to help map the human chromosomes.

Study Questions

1. What is the normal chromosome inheritance of a human? (p. 490)
2. What is the most common autosomal abnormality seen in humans? What causes this abnormality? (p. 491)
3. Draw a diagram that shows how nondisjunction can occur during meiosis I and during meiosis II. (p. 492)
4. Name and describe 4 chromosome mutations. (p. 493)
5. Describe Turner syndrome, metafemales, Klinefelter syndrome, and XYY males. (pp. 495–96)
6. Name 4 ways to recognize an X-linked recessive disorder. Why do males exhibit such disorders more often than females? (p. 498)
7. Explain the occurrence of sex-influenced traits. How do they differ from sex-linked traits? (p. 500)
8. What is a linkage group, and how can the occurrence of linkage groups help to map the human chromosomes? (p. 501)
9. How do human-mouse cell data, genetic marker data, and DNA probe data help to map the human chromosomes? (pp. 502–504)

Objective Questions

1. An XXY individual has _____ syndrome.
2. A(n) _____ is the movement of a chromosome segment from one chromosome to another, nonhomologous chromosome.
3. Whereas females are XX, males are _____ .
4. A female who is a carrier for color blindness has the genotype _____ .
5. A male who is X^BY _____ (is, is not) color blind.
6. Genes ABCDE are all on the same chromosome. They are part of a(n) _____ group.
7. Among the genes listed in question 6, AE are _____ (more likely, less likely) to crossover than CD.
8. Genetic markers and DNA probes sometimes can be used to _____ an individual for a genetic disorder.

Additional Genetics Problems

1. John is the only member of his family with hemophilia. What are the chances that a newborn brother also will be a hemophiliac? (p. 496)
2. In fruit flies, X^R = red eye and X^r = white eye. a. If a white-eyed male reproduces with a homozygous red-eyed female, what phenotypic ratio is expected for males? for females? b. If a white-eyed female reproduces with a red-eyed male, what phenotypic ratio is expected for males? for females? (p. 496)
3. A woman who is homozygous dominant for widow's peak is a carrier for color blindness reproduces with a man who is heterozygous for widow's peak and has normal vision. What are the genotypes of these parents? What are the chances of their having a color-blind son with a widow's peak? (p. 499)
4. In fruit flies, gray body (G) is dominant over black body (g). A female fly heterozygous for both gray body and red eyes reproduces with a red-eyed male heterozygous for gray body. What phenotypic ratio is expected for males? for females? (p. 499)
5. Imagine that the ability to curl the tongue is dominant and that this characteristic is linked to a rare form of mental retardation, which is also dominant. The parents are both dihybrids, with the 2 dominant alleles on one chromosome and the 2 recessive alleles on the other. What phenotypic ratio is expected among the offspring if crossing-over does not occur? (p. 502)
6. Give 2 reasons for deciding that this is a pedigree chart for an X-linked trait. What is the genotype of the starred individual? (p. 498)

Selected Key Terms

Down syndrome (down sin′drōm) human congenital disorder associated with an extra number-21 chromosome. *491*

Klinefelter syndrome (klīn′fel-ter sin′drōm) a condition caused by the inheritance of a chromosome abnormality in number; an XXY individual. *496*

linkage group (lingk′ij grup) alleles on the same chromosome are linked in the sense that they tend to move together to the same gamete; crossing-over interferes with linkage. *501*

sex linked (seks lingkt) allele located on the sex chromosomes. *496*

metafemale (met″ah-fe′māl) a female who has 3 X chromosomes. *495*

Turner syndrome (tur′ner sin′drōm) a condition caused by the inheritance of an abnormality in chromosome number; an X chromosome lacks a homologous counterpart—XO. *495*

X-linked gene (eks lingkt jēn) an allele located on the X chromosome. *496*

XYY male (eks wi wi māl) a male who has an extra Y chromosome. *496*

Chapter 24

The Molecular Basis of Inheritance

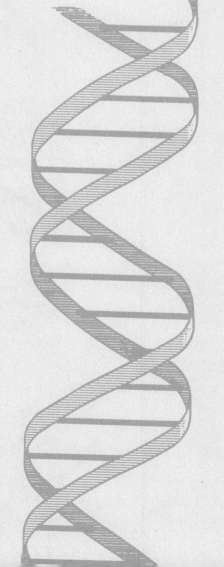

Chapter Concepts

1 DNA is the genetic material, and therefore its structure and function constitute the molecular basis of inheritance.

2 DNA is able to replicate, to mutate, and to control the phenotype of the cell and the organism.

3 DNA directs protein synthesis, a process that also requires the participation of RNA.

4 Regulatory genes control the activity of other genes involved in protein synthesis.

5 Gene mutations range from those that have little effect to those that have an extreme effect.

6 Cancer develops when there is a loss of genetic control due to mutations in particular genes called oncogenes.

Chapter Outline

DNA
    The Structure of DNA
    Functions of DNA
Protein Synthesis
    The Code of Heredity
    Transcription
    Translation
    Summary of Protein Synthesis
Control of Gene Expression
    The Control of Gene Expression in
        Prokaryotes
    The Control of Gene Expression in
        Eukaryotes
Gene Mutations
    Substitutions, Alterations, and
        Deletions of Bases
    Transposons
Cancer, a Failure in Genetic Control
    Causes of Cancer

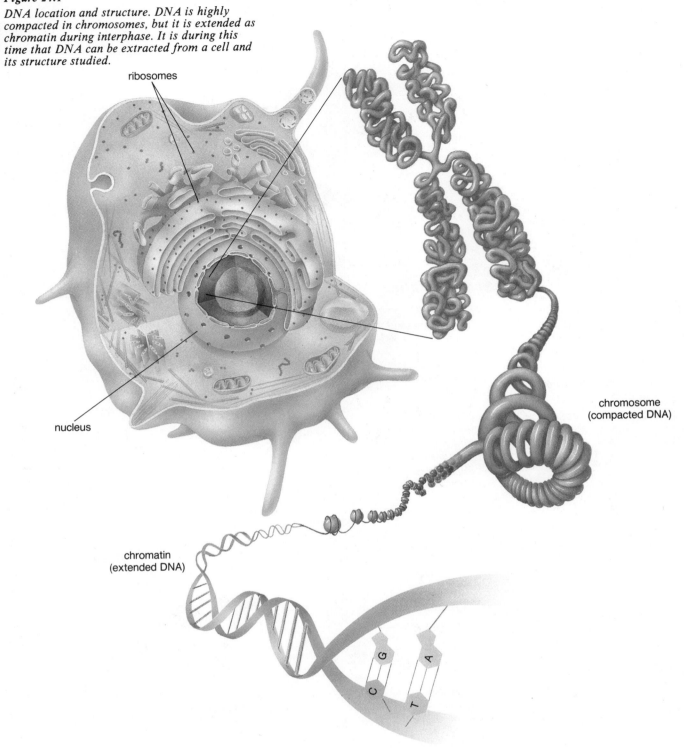

Figure 24.1

DNA location and structure. DNA is highly compacted in chromosomes, but it is extended as chromatin during interphase. It is during this time that DNA can be extracted from a cell and its structure studied.

ribosomes

nucleus

chromosome
(compacted DNA)

chromatin
(extended DNA)

C G

T A

When the sperm fertilizes the egg, a new individual comes into being. Each of the gametes contributes genes that direct the functioning of the individual from conception to death. The genes are on the chromosomes in the nucleus of a cell (fig. 24.1), but of what are the genes composed?

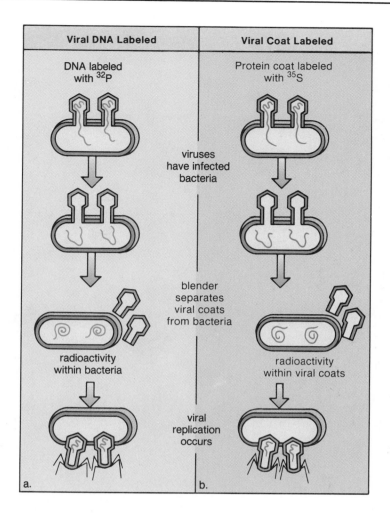

Figure 24.2

Proof of DNA function. T virus is composed of a DNA core and protein coat only. It was reasoned that whichever of these enters a bacterium and controls viral replication is the genetic material. a. In this experiment, ^{32}P was used to label viral DNA. The coats were removed by agitation in a blender, and the radioactively labeled DNA entered the cell. Because replication proceeded normally, DNA is the genetic material. b. In this experiment, ^{35}S was used to label the protein coat. When the cells were agitated in a blender, the radioactively labeled protein coats were removed. Because replication proceeded normally, protein is not the genetic material.

Labels within the figure:

Viral DNA Labeled	Viral Coat Labeled
DNA labeled with ^{32}P	Protein coat labeled with ^{35}S

viruses have infected bacteria

blender separates viral coats from bacteria

radioactivity within bacteria — radioactivity within viral coats

viral replication occurs

a.   b.

DNA

In the mid-1900s, it was known that the genes are on the chromosomes and that the chromosomes contain both DNA and protein, but it was uncertain which of these was the genetic material. Scientists turned to experiments with viruses to resolve this uncertainty because they knew that viruses are tiny particles having just 2 parts: an outer coat of protein and an inner core of nucleic acid, quite often DNA.

They chose to work with a virus called a T virus (the T simply means "type"), which infects bacteria. They wanted to determine which part of a T virus, the outer protein coat or the inner DNA core, enters a bacterium and takes over its machinery so that it produces more viruses. They began their search by culturing bacteria and viruses in radioactive sulfur (^{35}S) or radioactive phosphorus (^{32}P) until they had a batch with ^{35}S-labeled protein coats and ^{32}P-labeled DNA. Then, they allowed these viruses to attach to new bacteria (fig. 24.2) to determine whether the labeled protein or the labeled DNA enters the cell. They found that only DNA enters the cell and takes over its metabolism. More viral particles then are made. In other words, DNA is the genetic material.

The Structure of DNA

DNA is a nucleic acid that contains multiple copies of just 4 nucleotides. Each nucleotide is a complex of 3 subunits: phosphoric acid (phosphate), a pentose

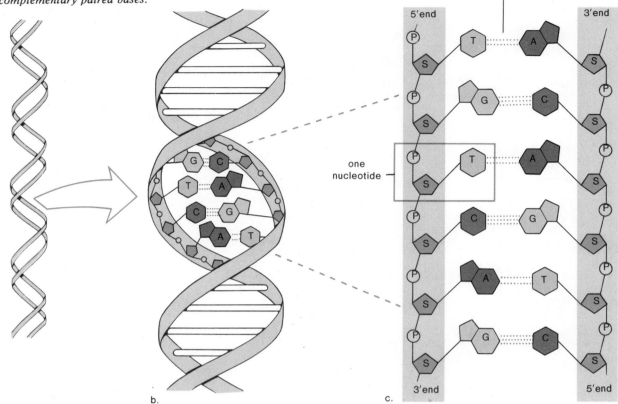

Figure 24.3

Overview of DNA structure. a. The double helix structure is a (b) twisted ladder. c. DNA unwound shows that the sides of the ladder are composed of sugar and phosphate molecules and the rungs are complementary paired bases.

sugar (deoxyribose), and a nitrogen-containing, organic base. There are 4 different bases: 2 are **purines,** which have a single ring, and 2 are **pyrimidines,** which have a double ring (fig. 1.31).

Purines	*Pyrimidines*
Adenine (A)	Thymine (T)
Guanine (G)	Cytosine (C)

When nucleotides join, they form a polymer, or a strand, in which the backbone is made up of an alternating pattern of phosphate-sugar-phosphate-sugar molecules, with the bases to one side of the backbone (fig. 1.33). DNA contains 2 such strands; therefore, it is double stranded. The 2 strands of DNA twist about one another in the form of a **double helix** (fig. 24.3*a* and *b*). The 2 strands are held together by hydrogen bonds between purine and pyrimidine bases. A is always paired with T, and G is always paired with C, or vice versa. This is called **complementary base pairing.** If we unwind the DNA helix, it resembles a ladder (fig. 24.3*c*). The sides of the ladder are made entirely of phosphate and sugar molecules, and the rungs of the ladder are made only of the complementary paired bases (fig. 24.4). The bases can be in any order, but regardless of the number of any particular base pair, the number of purine bases always equals the number of pyrimidine bases.

There are 2 important aspects of DNA structure:
1. *DNA is a double helix with sugar-phosphate backbones on the outside and paired bases on the inside.*
2. *Complementary base pairing occurs. Adenine (A) is paired with thymine (T) and guanine (G) is paired with cytosine (C). The number of purines (A + G) equals the number of pyrimidines (T + C).*

Figure 24.4

Complementary base pairing. Because a purine (A or G) is always paired with a pyrimidine (T or C), the number of purines is always equal to the number of pyrimidines in DNA. Also, the bases pair in such a way that the sugar-phosphate groups are oriented in different directions. This means that the strands of DNA end up running antiparallel to one another; that is, where one strand (fig. 24.3c) has a 5' end (phosphate is attached to the fifth carbon) the other has a 3' end (phosphate is attached to the third carbon).

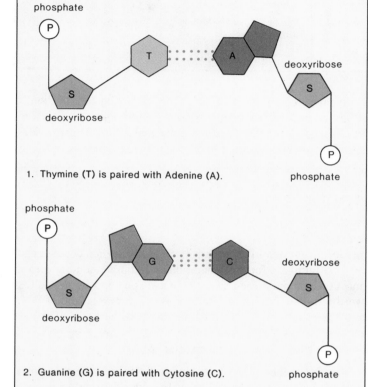

1. Thymine (T) is paired with Adenine (A).

2. Guanine (G) is paired with Cytosine (C).

Figure 24.5

Structure of RNA. RNA is single stranded. The backbone contains the sugar ribose instead of deoxyribose. The bases are guanine (G), uracil (U), cytosine (C), and adenine (A).

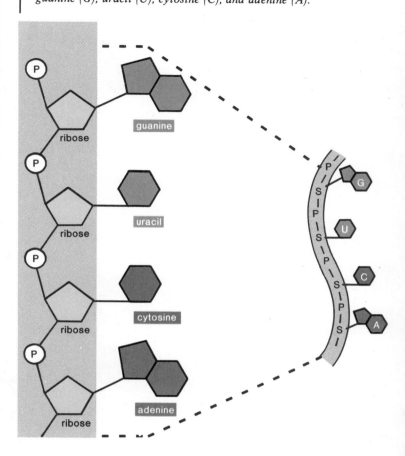

The structure of the DNA molecule first was determined by 2 young scientists, James Watson and Francis Crick, in the early 1950s. The data that were available to them and the way they used them to deduce DNA's structure is reviewed in the reading on page 514.

The Structure of RNA versus the Structure of DNA

RNA is also a nucleic acid composed of multiple copies of nucleotides (fig. 24.5). However, the pentose sugar in RNA is ribose, not deoxyribose. Also, the pyrimidine thymine does not appear in RNA; it is replaced by the pyrimidine uracil. Therefore, RNA contains 4 bases: the purines adenine (A) and guanine (G) and the pyrimidines cytosine (C) and uracil (U). Finally, RNA is single stranded and does not form a double helix in the same manner as DNA (table 24.1).

There are 3 different classes of RNA: messenger RNA (mRNA), ribosomal RNA (rRNA), and transfer RNA (tRNA). Each of these has specific functions in protein synthesis.

Functions of DNA

Any hereditary material has at least 3 functions. The hereditary material must

1. be able to *store information* that is used to control both the development and the metabolic activities of the cell and the organism;

Table 24.1 DNA Compared to RNA

	DNA	RNA
Function	Genetic material; controls protein synthesis	DNA helper; involved in protein synthesis
Sugar	Deoxyribose	Ribose
Bases	Adenine, guanine, thymine, cytosine	Adenine, guanine, uracil, cytosine
Strands	Double stranded with base pairing	Single stranded
Helix	Yes	No

2. be stable—it must *replicate* with high accuracy during cell division and must be transmitted from generation to generation;
3. be able to *undergo rare changes* called **mutations.** Mutations provide genetic variability, which is acted upon during evolution.

The genetic material must be able to store information, to replicate with few errors, and to undergo mutations.

The Replication of DNA

The double-stranded structure of DNA lends itself to replication because each strand can serve as a template for the formation of a complementary strand. A **template** is most often a mold used to produce a shape opposite to itself. In this case, the word template is appropriate because each new strand of DNA has a sequence of bases opposite, or complementary, to the bases of the old strand of DNA.

Replication requires the following steps (fig. 24.6):

1. The 2 strands that make up DNA unwind and "unzip" (i.e., the weak hydrogen bonds between the paired bases break). A special enzyme called helicase unwinds the molecule.
2. New complementary nucleotides, always present in the nucleus, move into place by the process of complementary base pairing.
3. The complementary nucleotides are joined so that DNA is again double stranded. This step is carried out by the enzyme **DNA polymerase.**

When the process is finished, 2 complete double-stranded DNA molecules are present, identical to each other and to the original molecule.

Although DNA replication can be easily explained, it is actually an extremely complicated process involving many more steps and enzymes than those discussed here.

Semiconservative Replication　The replication process is termed semiconservative because each new double helix has one parental (old) strand and one new strand. In other words, one of the parental strands is conserved, or present, in each daughter double helix.

During DNA replication, DNA is unwound and unzipped, and a new strand is formed complementary to each original strand. This is called semiconservative replication.

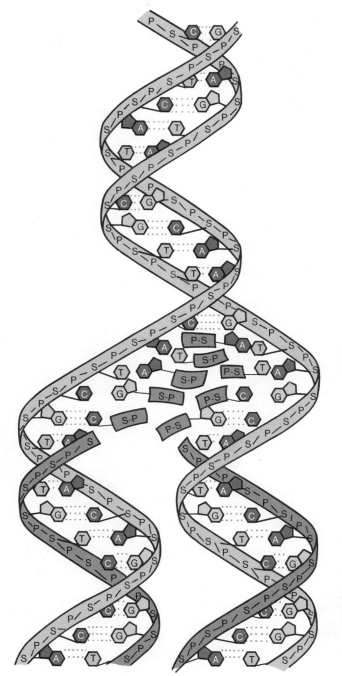

Figure 24.6

DNA replication. Replication is called semiconservative because each new double helix is composed of an old parental strand and a new daughter strand.

Region of parental DNA helix. (Both backbones are light.)

Region of replication (simplified). Parental DNA is unwound and unzipped. New nucleotides are pairing with those in parental strands.

Region of completed replication. Each double helix is composed of an old parental strand (light) and a new daughter strand (dark). Notice that each double helix is exactly like the other and also like the original parental double helix.

The Accuracy of Replication   It does not always happen that the bases pair correctly with one another as replication proceeds, but the error rate is minimized because DNA polymerase has a "proofreading" function. It checks each pairing as soon as it occurs, and if it finds that a mistake has been made, it removes the incorrect nucleotide and replaces it with a correct one. If an error is not corrected, then a gene mutation has occurred.

The Activity of DNA

The occurrence of gene mutations contributed to the conclusion that DNA controls the metabolism of the cell. For example, the metabolic pathway outlined in figure 24.7 was discovered in the early 1900s. In this pathway, 3 genetic disorders are known to occur. In the disorder known as *phenylketonuria*

In 1951, James Watson, an American biologist, began an internship at the University of Cambridge, England. There he met Francis Crick, an English physicist, who was interested in molecular structures. They began to try to build a model that would show the molecular structure of DNA, the known hereditary material. They knew that their model should explain the manner in which DNA can vary from species to species and even from individual to individual and also that it should show how it is possible for DNA to replicate (make a copy of itself) so that copies of the heredi-

Solving the Puzzle

tary material can be passed on to daughter cells and from the parent to the offspring.

Bits and pieces of data were available to Watson and Crick, and they undertook to solve the puzzle by putting the pieces together. This is what they knew from research done by others:

1. DNA is a polymer of nucleotides, each one having a phosphate group, the sugar deoxyribose, and a nitrogen-containing, organic base. There are 4 nucleotides that differ according to the base: adenine (A) and guanine (G) are purines, while cytosine (C) and thymine (T) are pyrimidines.
2. A chemist, Erwin Chargaff, had determined in the late 1940s that regardless of the species under consideration, the number of purines in DNA always equals the number of pyrimidines and that the amount of adenine equals the amount of thymine—that is, A = T—and the amount of guanine equals the amount of cytosine—that is, G = C. These findings came to be known as *Chargaff's rules*.

3. Rosalind Franklin and Maurice Wilkins, working at King's College, London, had just prepared an X-ray diffraction photograph (fig. *a*) of DNA. It showed that DNA is a double helix of constant diameter and that the bases are regularly stacked on top of one another.

Using this data, Watson and Crick deduced that DNA has a twisted, ladder-type structure: the sugar-phosphate molecules make up the sides of the ladder and the bases make up the rungs of the ladder. Further, they determined that if A is always hydrogen bonded with T and G is always hydrogen bonded with C (in keeping with Chargaff's rules), then the rungs always have a constant width (as required by the X-ray photograph).

Watson and Crick built an actual model of DNA out of wire and tin (fig. *b*). This double-helix model does indeed allow for differences in DNA structure between species because while A must always pair with T and G must always pair with C, there is no set order in the sequencing of these pairs. Also, the model provides a means by which DNA can replicate, as Watson and Crick pointed out in their original paper: "It has not escaped our notice that the specific pairing we have postulated immediately suggests a possible copying mechanism for the genetic material."

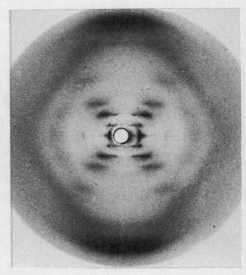

a.

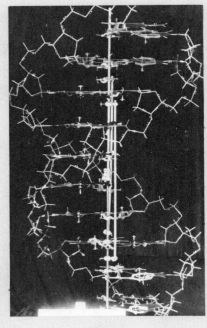

b.

DNA structure. a. *X-ray diffraction photograph of DNA taken by Rosalind Franklin. The crossing pattern of dark spots in the center of the picture indicated that DNA is helical. The dark regions at the top and bottom of the photograph showed that base pairs are stacked on top of one another.* b. *A portion of the actual wire and tin model constructed by Watson and Crick.*

(PKU), the phenylpyruvate accumulates in the body and spills over into the urine because the enzyme needed to convert phenylalanine to tyrosine is missing. If the condition is not treated, the continued accumulation of phenylpyruvate can cause mental retardation. **Albinism** results because tyrosine cannot be converted to melanin, the natural pigment in human skin. The genetic disorder *alkaptonuria* results if the enzyme needed to metabolize homogentisate is missing. Because homogentisate is excreted, the urine in these individuals turns a dark color.

Figure 24.7

Metabolic pathway by which phenylalanine is converted to other metabolites. a. If the enzyme that converts phenylalanine to tyrosine is defective, tyrosine is converted to phenylpyruvate instead, and the accumulation of this substance leads to PKU *(phenylketonuria). b. If the enzyme that converts tyrosine to melanin is defective, albinism results. c. If homogentisate cannot be metabolized, alkaptonuria, a condition characterized by dark urine, results.*

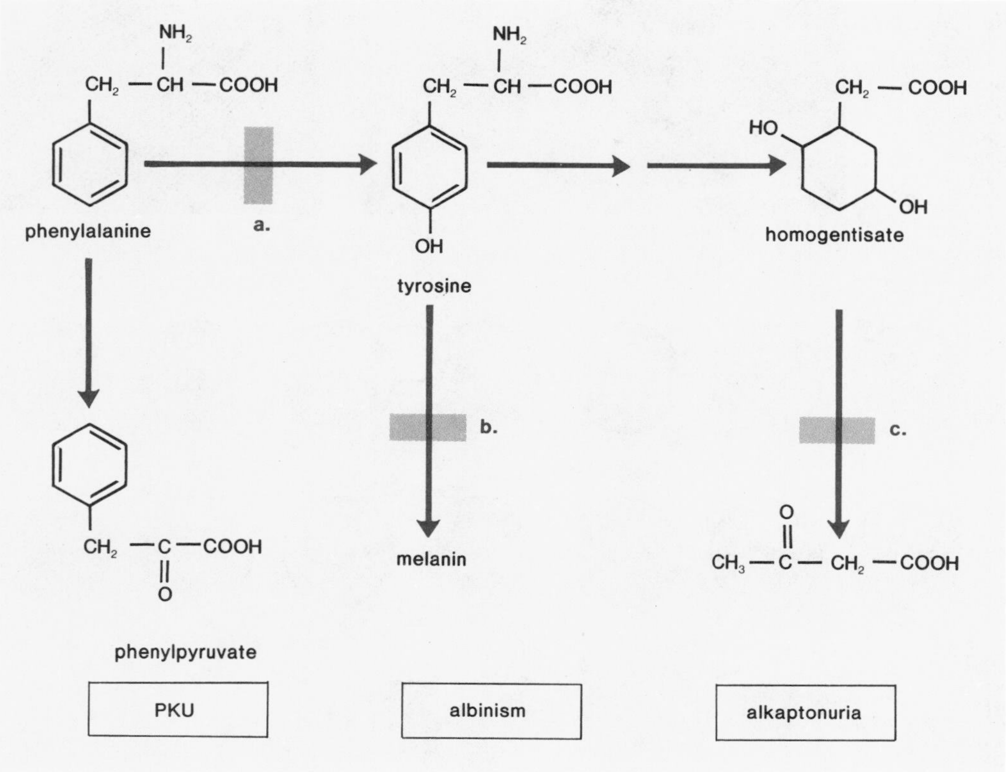

At first, these conditions were called simply **inborn errors of metabolism,** and only later was it confirmed that the genetic fault lay in the absence of particular enzymes within cells. This suggested that genes (DNA) control the metabolism of the cell by controlling the production of enzymes in the cell. This suggestion was called the one gene–one enzyme theory.

Because enzymes are proteins, the one gene–one enzyme theory soon was broadened to the one gene–one protein theory. Some proteins are concerned with the structure, rather than metabolism, of the cell. Therefore, in controlling protein production, DNA also controls the structure of the cell.

Later, it was pointed out that some proteins, like hemoglobin, have more than one type of polypeptide chain (see fig. 12.3). In persons with sickle-cell anemia, it is only the β polypeptide chain that has an altered sequence of amino acids (fig. 24.8) compared to the normal β chain. Therefore, it may be more appropriate to state that a gene (DNA) controls the sequence of amino acids in a polypeptide. Today, we define a gene as a section of a DNA molecule that determines the sequence of amino acids in a single polypeptide chain of a protein.

Genes (DNA) control the primary structure of proteins. In this way, genes control the structure and the metabolism of the cell.

24.1 Critical Thinking

1. DNA stores information. What is the information, and where is it stored? What is your evidence that DNA stores this information?
2. DNA replicates. What is there about DNA that allows it to replicate? What is your evidence that DNA replicates prior to cell division? What is your evidence that replication occurs in the nucleus?
3. DNA mutates. What part of DNA permits mutations to occur? What evidence in nature shows that DNA mutates?

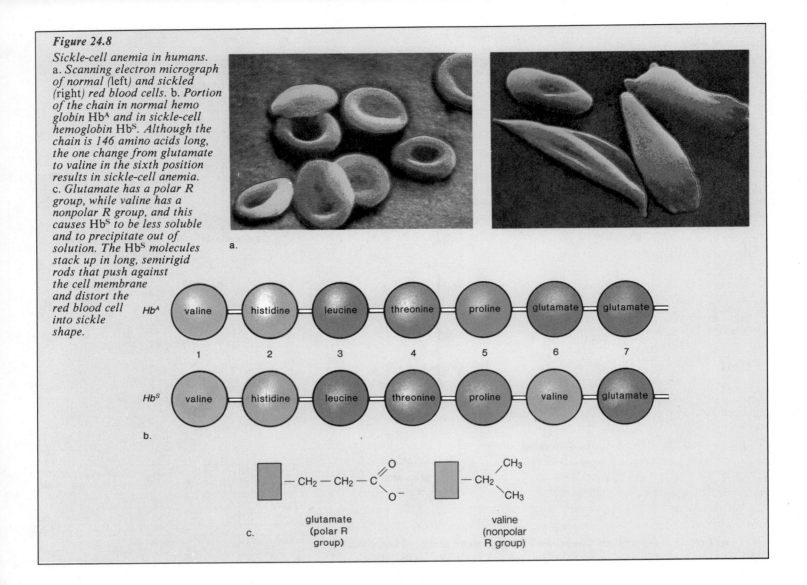

Figure 24.8

Sickle-cell anemia in humans.
a. Scanning electron micrograph of normal (left) and sickled (right) red blood cells. b. Portion of the chain in normal hemoglobin Hb^A and in sickle-cell hemoglobin Hb^S. Although the chain is 146 amino acids long, the one change from glutamate to valine in the sixth position results in sickle-cell anemia. c. Glutamate has a polar R group, while valine has a nonpolar R group, and this causes Hb^S to be less soluble and to precipitate out of solution. The Hb^S molecules stack up in long, semirigid rods that push against the cell membrane and distort the red blood cell into sickle shape.

Protein Synthesis

The fact that DNA controls the production of proteins at first may seem surprising when we consider that genes are located in the nucleus of higher cells but proteins are synthesized at the ribosomes in the cytoplasm. However, although DNA is found only in the nucleus (fig. 24.1), RNA exists in both the nucleus and the cytoplasm.

The central dogma of modern genetics is diagrammed in this manner:

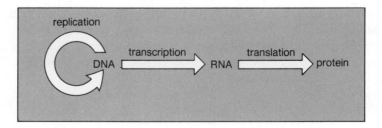

The diagram indicates that DNA not only serves as a template for its own replication, it is also a template for RNA formation. The arrow between DNA and RNA indicates that RNA is transcribed from the DNA template. **Transcription** is making an RNA molecule that is complementary to a portion of

DNA. Following transcription, RNA moves into the cytoplasm. Photographs are available that show that radioactively labeled RNA moves from the nucleus to the cytoplasm, where protein synthesis occurs.

The arrow between RNA and protein in the diagram indicates that proteins are synthesized according to RNA's instructions. The specific type of RNA that has this job is called **messenger RNA (mRNA)**. During **translation,** the information carried by RNA (in nucleotides) is used to produce the correct order of amino acids in a polypeptide (fig. 24.9). You can see that these terms are appropriate because transcribing a document means making a close copy of it, while translating a document means putting it in an entirely different language.

The Code of Heredity

During transcription, DNA provides mRNA with a message that directs the order of amino acids during translation, when protein synthesis occurs. The message cannot be contained in the sugar-phosphate backbone because it is constant in every DNA molecule. However, the order of the bases in DNA and mRNA can and does change. Therefore, it must be the bases that contain the message. The order of the bases in DNA must code for the order of the amino acids in a polypeptide. Can 4 bases provide enough combinations to code for 20 amino acids? If the code was a doublet (any 2 bases stand for one amino acid), it would not be possible to code for 20 amino acids (table 24.2), but if the code was a triplet, then the 4 bases could supply 64 different triplets, far more than needed to code for 20 different amino acids. It should come as no surprise, then, to learn that the code is a triplet code.

To crack the code, a cell-free experiment was done: artificial RNA was added to a medium containing bacterial ribosomes and a mixture of amino acids. Comparison of the bases in the RNA with the resulting polypeptide allowed investigators to decipher the code. Each three-letter unit of an mRNA molecule is called a **codon.** All 64 mRNA codons have been determined (fig. 24.10). Sixty-one triplets correspond to a particular amino acid; the remaining 3 are stop codons that signal polypeptide chain termination. The one codon that stands for the amino acid methionine is also a start codon that signals polypeptide initiation.

The Universal Genetic Code

Research indicates that the genetic code is essentially universal. The same codons stand for the same amino acids in all living things, including bacteria, plants, and animals. This illustrates the remarkable biochemical unity of living things and suggests that all living things have a common evolutionary ancestor.

DNA contains a code, and its message is passed to mRNA during transcription. Sixty-one of the 64 triplet codons stand for particular amino acids, and the other 3 codons are stop codons. During translation, the order of the codons in mRNA determines the order of the amino acids in a protein.

Transcription

During transcription, the DNA code is passed to mRNA. Therefore, the code is transcribed, or copied.

Messenger RNA

Following transcription (fig. 24.11), mRNA has a sequence of bases complementary to DNA; wherever A, T, G, or C is present in the DNA template, U, A, C, or G is incorporated into the mRNA molecule. A segment of the DNA helix unwinds and unzips, and complementary RNA nucleotides pair with DNA nucleotides of one strand. When these RNA nucleotides are joined by

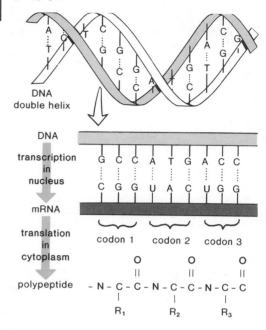

Figure 24.9

Transcription and translation in protein synthesis. Transcription occurs in the nucleus when DNA acts as a template for mRNA synthesis. Translation occurs in the cytoplasm when the sequence of the codons found in mRNA determines the sequence of the amino acids in a polypeptide.

Table 24.2 *Number of Bases in Code*

Number of Bases in Code	Number of Amino Acids Specified
1	4
2	16
3	64

mRNA codons. These mRNA codons are complementary to the code in DNA. The DNA code is degenerate; most of the 20 amino acids have more than one codon. In this chart, notice that each of the codons comprise a letter from the first, second, and third positions. For example, find the square where C (from the first position) and A (from the second position) come together, and then look across to the right. You will notice the letters for the third position of the codons for histidine and glutamine.

First Base	Second Base				Third Base
	U	C	A	G	
U	UUU phenylalanine	UCU serine	UAU tyrosine	UGU cysteine	U
	UUC phenylalanine	UCC serine	UAC tyrosine	UGC cysteine	C
	UUA leucine	UCA serine	UAA stop	UGA stop	A
	UUG leucine	UCG serine	UAG stop	UGG tryptophan	G
C	CUU leucine	CCU proline	CAU histidine	CGU arginine	U
	CUC leucine	CCC proline	CAC histidine	CGC arginine	C
	CUA leucine	CCA proline	CAA glutamine	CGA arginine	A
	CUG leucine	CCG proline	CAG glutamine	CGG arginine	G
A	AUU isoleucine	ACU threonine	AAU asparagine	AGU serine	U
	AUC isoleucine	ACC threonine	AAC asparagine	AGC serine	C
	AUA isoleucine	ACA threonine	AAA lysine	AGA arginine	A
	AUG (start) methionine	ACG threonine	AAG lysine	AGG arginine	G
G	GUU valine	GCU alanine	GAU aspartate	GGU glycine	U
	GUC valine	GCC alanine	GAC aspartate	GGC glycine	C
	GUA valine	GCA alanine	GAA glutamate	GGA glycine	A
	GUG valine	GCG alanine	GAG glutamate	GGG glycine	G

an enzyme called RNA polymerase, an mRNA molecule results. mRNA has a sequence of bases that are triplet codons complementary to the DNA triplet code.

After the mRNA strand is processed, it passes from the cell nucleus into the cytoplasm. There, it becomes associated with the ribosomes.

RNA Processing

Most genes in eukaryotes are interrupted by segments of DNA that are not part of the gene. These portions are called introns because they are *intra*gene segments. The other portions of the gene are called exons because they are ultimately *ex*pressed. A gene is expressed when a protein product results. When DNA is transcribed, the mRNA contains bases that are complementary to both exons and introns, but before the mRNA exits from the nucleus, it is *processed*—the nucleotides complementary to the introns are removed enzymatically.

There has been much speculation about the role of introns in the genes of eukaryotes. It is possible that introns allow crossing-over within a gene during meiosis. It is also possible that introns divide a gene into domains that can be joined in different combinations to give novel genes and protein products, a process that perhaps facilitates evolution.

During transcription, mRNA is made complementary to one of the DNA strands. It then contains a sequence of codons and moves into the cytoplasm, where it becomes associated with the ribosomes.

Human Reproduction, Development, and Inheritance

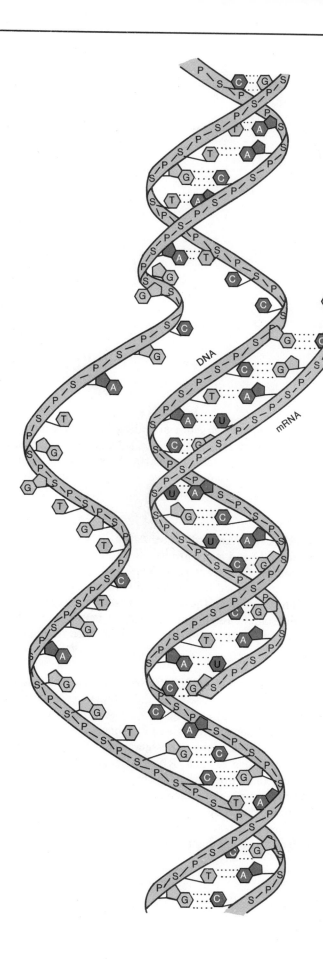

Figure 24.11

Transcription. A portion of DNA unzips, and by complementary base pairing, mRNA has a sequence of codons dictated by the sequence of bases in DNA. Transcription occurs in the nucleus.

This mRNA transcript is ready to move into the cytoplasm.

Transcription is going on here—the nucleotides of mRNA are joined in an order complementary to a strand of DNA.

One portion of DNA—a particular gene or genes—is transcribed at a time.

DNA

mRNA

The Molecular Basis of Inheritance

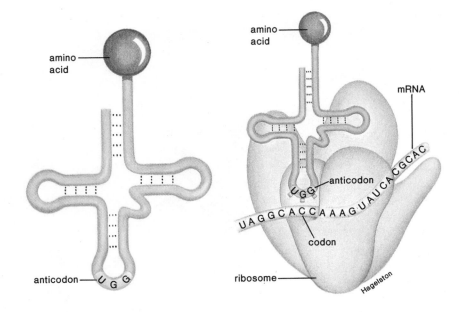

Figure 24.12
Anticodon-codon base pairing. a. tRNA molecules have an amino acid attached to one end and an anticodon at the other end. If the anticodon is UGG, the amino acid is threonine (see text). b. The anticodon of a tRNA molecule is complementary to a codon. The pairing between codon and anticodon ensures that the sequence of amino acids in a polypeptide is that directed originally by DNA.

a. Transfer RNA (tRNA)

b. Transfer RNA (tRNA) at ribosome

Translation

During translation, the sequence of codons in mRNA dictates the order of amino acids in a polypeptide. This is called translation because the sequence of bases in DNA is translated into a particular sequence of amino acids. Translation requires the involvement of several enzymes and 2 other types of RNA: ribosomal RNA and transfer RNA.

Ribosomal RNA

Ribosomal RNA (rRNA) is called structural RNA because it makes up the ribosomes (fig. 2.5) and is not involved in coding. Ribosomes are composed of 2 subunits, each with characteristic RNA and protein molecules. The rRNA molecules are transcribed from DNA in the region of the nucleolus. The proteins are manufactured in the cytoplasm but then migrate to the nucleolus, where the ribosomal subunits are assembled before they migrate to the cytoplasm. Ribosomes play an important role in coordinating protein synthesis.

Transfer RNA

Small molecules of **transfer RNA (tRNA)** bring the amino acids from the cytoplasm to the ribosomes located in the cytoplasm. A particular amino acid attaches to a tRNA at one end (fig. 24.12a). Attachment requires ATP energy, and the resulting bond is a high-energy bond represented by a wavy line. Therefore, the entire complex is designated as tRNA ~ amino acid.

At the other end of each tRNA molecule, there is a specific **anticodon** complementary to an mRNA codon (fig. 24.12b). When the tRNA molecule comes to the ribosome, the anticodon pairs with a codon. Let us consider an example: For the codon ACC, what will be the tRNA molecule's anticodon, and what amino acid will be attached to the tRNA molecule? Inspection of figure 24.10 allows us to determine this:

codon	anticodon	amino acid
ACC	UGG	threonine

Human Reproduction, Development, and Inheritance

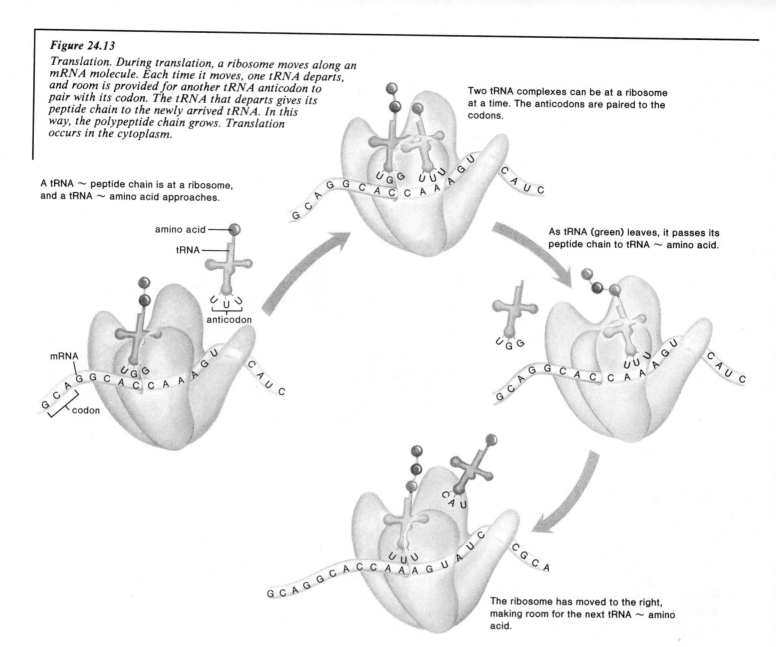

Figure 24.13

Translation. During translation, a ribosome moves along an mRNA molecule. Each time it moves, one tRNA departs, and room is provided for another tRNA anticodon to pair with its codon. The tRNA that departs gives its peptide chain to the newly arrived tRNA. In this way, the polypeptide chain grows. Translation occurs in the cytoplasm.

A tRNA ~ peptide chain is at a ribosome, and a tRNA ~ amino acid approaches.

amino acid

tRNA

anticodon

mRNA

codon

Two tRNA complexes can be at a ribosome at a time. The anticodons are paired to the codons.

As tRNA (green) leaves, it passes its peptide chain to tRNA ~ amino acid.

The ribosome has moved to the right, making room for the next tRNA ~ amino acid.

In this way, it is the order of the codons of the mRNA that determines the sequence of amino acids in the polypeptide.

During translation, tRNA molecules, each carrying a particular amino acid, travel to the mRNA, and through complementary base pairing between anticodon and codon, the tRNA molecules and therefore the amino acids in a polypeptide chain are sequenced in a predetermined order.

The Process of Translation

Protein synthesis requires 3 steps: initiation, elongation, and termination. During *initiation*, a ribosome binds to an mRNA molecule. Initiation always begins with a codon that stands for the amino acid methionine. First, the smaller ribosomal subunit binds to mRNA, and then the larger subunit joins to the smaller subunit, giving a complete ribosomal structure. *Elongation* occurs as the polypeptide chain grows in length (fig. 24.13). A ribosome is large enough to accommodate 2 tRNA molecules; the peptide chain attached to the tRNA molecule in the first position is transferred to the tRNA ~ amino acid complex in the second position. The ribosome then moves laterally so that the next

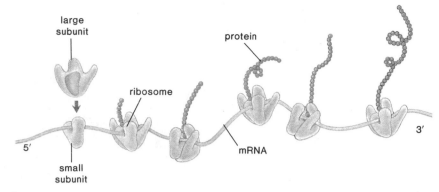

Figure 24.14

Polysome structure. a. Several ribosomes move along an mRNA molecule at a time. They function independently of each other; therefore, several polypeptides can be made at the same time. b. Electron micrograph of a polysome.

large subunit

protein

ribosome

5'

small subunit

mRNA

3'

a.

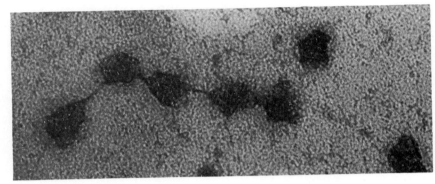

b.

mRNA codon is available to receive the next tRNA ~ amino acid complex. In this manner, the peptide chain grows and the primary structure of a protein comes about. The secondary and tertiary structures of a polypeptide appear after termination, as the amino acids, in a predetermined sequence within the polypeptide chain, interact with one another.

Several ribosomes, collectively called a **polysome,** can move along one mRNA at a time; therefore, several polypeptides of the same type can be synthesized at once (fig. 24.14). *Termination* per ribosome occurs at a stop codon on the mRNA. The ribosome dissociates into its 2 subunits and falls off the mRNA molecule. Protein synthesis continues until the very last ribosome has come to the stop codon.

Summary of Protein Synthesis

The following list, along with table 24.3 and figure 24.15, provides a brief summary of the steps involved in protein synthesis.

1. DNA, which always remains in the nucleus, contains a series of bases that serve as a *triplet code* (every 3 bases stands for an amino acid).
2. During transcription, one of the 2 strands of DNA serves as a template for the formation of mRNA, which contains *codons* (sequences of 3 bases) complementary to the DNA.
3. mRNA is processed (modified) before leaving the nucleus.
4. mRNA moves into the cytoplasm and becomes associated with the *ribosomes,* which are composed of rRNA and proteins.
5. tRNA molecules, each of which is bonded to a particular amino acid, have *anticodons* that are complementary to the codons in mRNA.
6. During translation, the linear sequence of mRNA codons determines the order in which the tRNA molecules and their attached amino acids arrive at the ribosomes, and this determines the primary structure (linear sequence of amino acids) of a protein.

Table 24.3 *Participants in Protein Synthesis*

Name of Molecule	Special Significance	Definition
DNA	Code	Sequence of DNA bases in threes
mRNA	Codon	Sequence of RNA bases complementary to DNA code
tRNA	Anticodon	Sequence of 3 bases complementary to codon
rRNA	Ribosome	Site of protein synthesis
Amino acid	Building block for protein	Transported to ribosome by tRNA
Protein	Enzyme	Amino acids joined in a predetermined order

Figure 24.15

Summary of protein synthesis. Transcription occurs in the nucleus, and translation occurs in the cytoplasm (blue). During translation, the codons borne by mRNA dictate the order of the amino acids in the polypeptide.

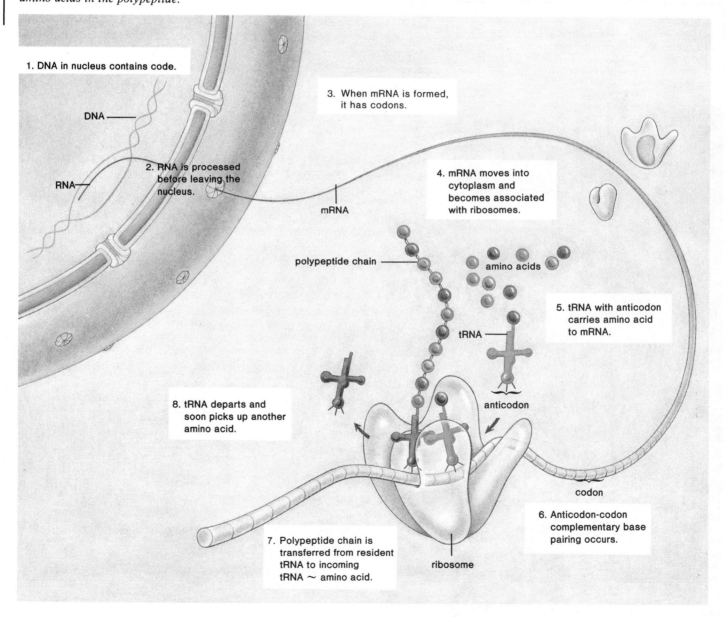

1. DNA in nucleus contains code.

DNA

RNA

2. RNA is processed before leaving the nucleus.

3. When mRNA is formed, it has codons.

mRNA

4. mRNA moves into cytoplasm and becomes associated with ribosomes.

polypeptide chain

amino acids

5. tRNA with anticodon carries amino acid to mRNA.

tRNA

anticodon

8. tRNA departs and soon picks up another amino acid.

codon

6. Anticodon-codon complementary base pairing occurs.

7. Polypeptide chain is transferred from resident tRNA to incoming tRNA ~ amino acid.

ribosome

This is a segment of a DNA molecule. What are (1) the RNA codons, (2) the tRNA anticodons, and (3) the sequence of amino acids in the polypeptide?

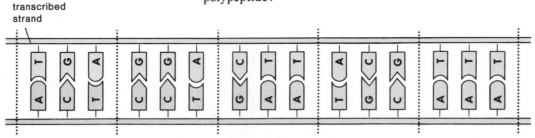

transcribed strand

*Answers to the problem are on page 532.

Figure 24.16

Levels at which control of gene expression occurs in eukaryotic cells. Transcriptional and posttranscriptional control occur in the nucleus; translational and posttranslational control occur in the cytoplasm.

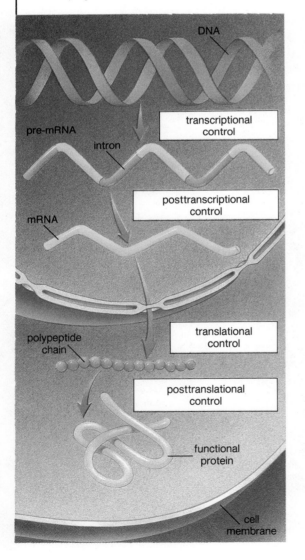

DNA

pre-mRNA

intron

transcriptional control

mRNA

posttranscriptional control

polypeptide chain

translational control

posttranslational control

functional protein

cell membrane

Definition of a Gene

Classical (Mendelian) geneticists thought of a gene as a particle on a chromosome. To molecular geneticists, however, a gene is a sequence of DNA nucleotide bases that code for a product. Most often the gene product is a polypeptide chain. However, there are exceptions because rRNA and tRNA, for example, are gene products themselves. They are involved in protein synthesis, but they do not code for protein.

Control of Gene Expression

You might think that genetic control occurs only at the level of the gene, but in eukaryotes, there are 4 levels at which regulation is possible (fig. 24.16):

1. **Transcriptional control**—mechanisms that control which genes are transcribed and/or the rate at which transcription occurs.
2. **Posttranscriptional control**—differential processing of mRNA and/or the rate at which mRNA leaves the nucleus.
3. **Translational control**—how soon and how long the mRNA is active in the cytoplasm.
4. **Posttranslational control**—how soon the protein just translated becomes functional. Also, some functional proteins are subject to feedback control in the manner described in figure 5.6.

Each of these affects the expression of a gene, that is, how much gene product (protein) there is in a cell at a given time.

Regulation of gene expression in eukaryotes involves at least 4 levels of control.

As a background to a more detailed discussion of eukaryotic control, let us consider how prokaryotes regulate the expression of their genes.

The Control of Gene Expression in Prokaryotes

Usually, regulation of gene expression in prokaryotes occurs at the level of gene transcription. The **operon** model explains how bacteria can regulate at one time the production of several enzymes in a particular metabolic pathway. An **operon** includes the following elements:

Regulator gene—a gene that codes for a repressor protein molecule. The repressor molecule binds to the operator and prevents RNA polymerase from binding to the promoter.

Promoter—a short sequence of DNA where RNA polymerase first attaches when a gene is to be transcribed.

Operator—a short sequence of DNA where the repressor binds, preventing RNA polymerase from attaching to the promoter. This often is called the on/off switch of transcription.

Structural genes—one to several genes of a metabolic pathway that are transcribed as a unit.

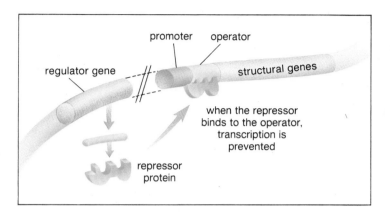

Notice in this list the distinction between a regulator gene and a structural gene. A structural gene codes for a protein, and a regulator gene regulates the activity of (a) structural gene(s). Therefore, although each cell contains a full complement of genes, only certain ones need to be active at any one time.

The Lac Operon

The *lac* operon was the first operon discovered. Ordinarily, the bacterium *E. coli* uses glucose as its energy source; however, if it is denied glucose and given the milk sugar lactose instead, it immediately begins to make 3 enzymes needed to metabolize lactose (fig. 24.17).

Notice that the structural genes in this operon are normally inactive; that is, the regulator gene codes for a repressor protein molecule that automatically attaches to the operator, preventing transcription from occurring. The operon becomes active when the repressor joins with an inducer molecule—lactose—and the complex is unable to bind with the operator. Therefore, this is called an *inducible operon* model.

In the lac *operon, the repressor ordinarily can bind to the operator, but it is unable to do so when lactose is present. This is an example of an inducible operon.*

The Tryp Operon

There are other operons in *E. coli* that ordinarily are turned on rather than off. For example, the prokaryotic cell ordinarily produces 5 enzymes needed for the synthesis of the amino acid tryptophan. However, if tryptophan is provided, these enzymes no longer are produced. In this operon, the regulator gene codes for a repressor that ordinarily is unable to attach to the operator. The repressor has a binding site for tryptophan, and if tryptophan is present, it binds to the repressor. Now the complex is able to bind to the operator. Tryptophan is called the corepressor, and the *tryp* operon is an example of a *repressible operon*.

In the tryp *operon, the repressor ordinarily cannot bind to the operator, but it is able to do so when tryptophan is present. This is an example of a repressible operon.*

24.3 Critical Thinking

1. Multicellular organisms, like humans, have different phenotypes. Do the phenotypes of various cells in an organism differ? How do you know?
2. Do all cells of an organism ordinarily possess a complete set of genes, including regulator genes and structural genes? How do you know?
3. Would you expect to find transcripts (mRNA) of all genes in all cells or only certain transcripts? Why?
4. How would you account for the fact that only certain structural genes are active in certain cells?

Figure 24.17
The lac *operon, a model for an inducible operon. a. The regulator gene codes for a repressor that is normally active. When active, the repressor binds to the operator and prevents RNA polymerase from attaching to the promoter. Therefore, transcription of the 3 structural genes does not occur. b. When*

lactose is present, it binds to the repressor, which changes the repressor's shape, preventing it from binding to the operator. Now RNA polymerase binds to the promoter; transcription and translation of the 3 structural genes follows.

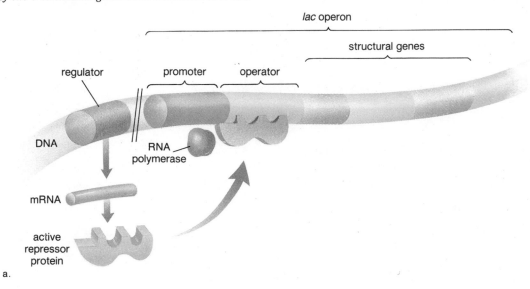

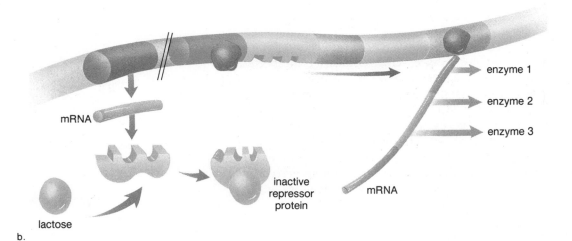

The Control of Gene Expression in Eukaryotes

Although there are 4 levels of genetic control in eukaryotic cells (fig. 24.16), more is known about transcriptional control; therefore, we will limit our remarks to this level. Transcriptional control recently has taken on even greater significance because it appears that certain genes are turned on or off depending on whether they are inherited from the mother or the father. It is said that various genes are believed to be "imprinted," that is, turned on or off depending on whether they go through the process of spermatogenesis or oogenesis.

Chromatin and the Regulation of Transcription

For a gene to be transcribed in eukaryotes, the chromosome in that region must first decompact (p. 82). For example, during the development of a *Drosophila* larva, first one and then another of the chromosome's bands bulge out, forming chromosome puffs (fig. 24.18). The use of radioactive uridine, a label

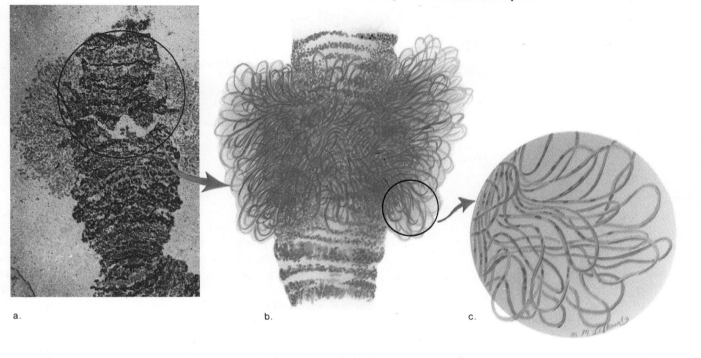

Figure 24.18
Chromosome puffs. a. Electron micrograph of a larval chromosome that has a puff in one region. b. Artist's interpretation of a puffed region. c. Puffs contain loops of DNA, where enzymes can transcribe mRNA from the DNA template. This interpretation is supported by the observation that varying regions of the chromosomes exhibit puffs as development of Drosophila and other insects proceeds.

a.

b.

c.

specific for RNA, indicates that DNA is being actively transcribed at chromosome puffs. It could be that genes ordinarily are inactive in eukaryotes, and they must be turned on to be active.

Regulator Proteins and Transcription
There is evidence of regulator proteins in eukaryotes that attach to promoter regions prior to RNA polymerase attachment. These proteins are believed to help RNA polymerase bind to the promoter. Therefore, it suggests that regulator proteins such as these are needed to turn eukaryotic genes on.

A unique feature in eukaryotes is the presence of regulatory DNA sequences called enhancers. An enhancer increases the frequency of structural gene transcription up to several hundredfold. Enhancers need not be located near the gene(s) they are affecting. Enhancers may be sites for attachment of regulator proteins that help to loosen chromatin structure so transcription can begin.

Gene Mutations

We previously studied chromosome mutations (fig. 23.6), but there are also gene mutations. As you can see in table 24.4, a **gene mutation** is any alteration in the code of a single gene or any change in its expression. Gene mutations do not necessarily have a deleterious effect; some have no effect at all, and some even have a beneficial effect.

Substitutions, Alterations, and Deletions of Bases
Mutations involving a change in the DNA sequence of bases are of 3 types (table 24.5). *Substitutions,* which involve a change in a single base, may have no effect at all if the new codon happens to stand for the same amino acid as

Table 24.4 Types of Mutations

Type of Mutation	Definition
Chromosome mutation	A rearrangement of chromosome parts, as described in figure 23.6, which may or may not result in a change in the phenotype.
Gene mutation	A change in the genetic code for a gene or in the expression of the gene. Usually results in a change in the phenotype.
Germinal mutation	A mutation that manifests itself in the gametes so that it is passed on to offspring.
Somatic mutation	A mutation that occurs in the body cells and that very likely is not passed on to offspring.

Table 24.5 Gene Mutations

Base Change		Worst Result
Normal	TAC'GGC'ATG	
Substitution	TAG'GGC'ATG	Change in one amino acid or change to stop signal
Deletion	ACG'GCA'TG	Polypeptide altered completely
Addition	ATA'CGG'CAT'G	Polypeptide altered completely

the old codon (fig. 24.10). Other base substitutions may lead to an amino acid substitution in the protein product. For example, a change from GAG to GUG causes glutamate to be replaced by valine, as it is in sickle-cell hemoglobin (fig. 24.8). This particular substitution causes a drastic deleterious effect in the phenotype. It is also possible for a base substitution to result in a new stop codon; this causes transcription to be terminated before the polypeptide is formed fully. This type of base change is being investigated as a possible cause of hemophilia, the X-linked clotting disorder (p. 497).

Additions and *deletions* of bases are expected to result in profound alterations in the DNA code. If the altered allele codes for an enzyme, the enzyme most likely would be nonfunctional because the sequence of amino acids would be so greatly affected.

Transposons

Transposons are specific DNA sequences that have the remarkable ability to move within and out of chromosomes. Their movement to a new location sometimes alters neighboring genes, particularly by increasing or decreasing their expression. Although "movable elements" in corn were described 40 years ago, their significance was realized only recently. So-called jumping genes now have been discovered in bacteria, fruit flies, and humans, and it is likely that all organisms have such elements.

Cancer, a Failure in Genetic Control

Cancer cells exhibit characteristics that indicate they have experienced a severe failure in the control of genetic expression. These characteristics are discussed in the following paragraphs.

Cancer cells exhibit uncontrolled and disorganized growth. Normal cells only divide about 50 times, but cancer cells enter the cell cycle (fig. 4.4) over and over again and never fully differentiate.

In tissue culture, normal cells grow in only one layer because they adhere to the glass. They stop dividing once they make contact with their neighbors, a phenomenon called contact inhibition. Cancer cells have lost contact inhibition and grow in multiple layers, most likely because of cell surface changes. In the body, a cancer cell divides to form a growth, or tumor, that invades and destroys neighboring tissue (fig. 24.19). The cells are disorganized because they do not differentiate into the tissue of the organ and therefore never can help to fulfill the function of the organ. To support their growth, cancer cells

Figure 24.19

Metastasis. In the body, cancer cells form a tumor, a disorganized mass of cells undergoing uncontrolled growth. As a tumor grows, it may invade underlying tissues. Some of the cells leave this primary tumor and move through layers of tissue into blood vessels or lymphatic vessels. After traveling through these vessels, the metastatic cells start new tumors elsewhere in the body. A carcinoma is a cancer that begins in epithelial tissue; a sarcoma is one that begins in connective tissue.

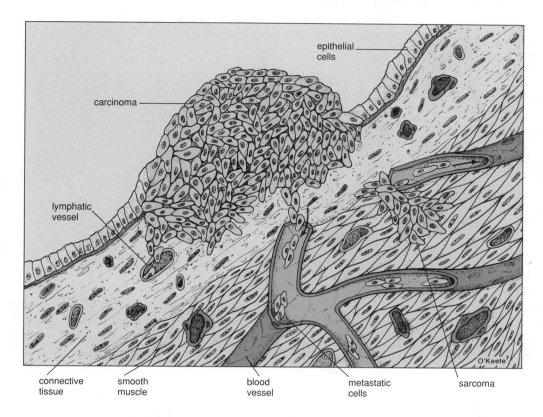

release a growth factor that causes neighboring blood vessels to branch into the cancerous tissue. This phenomenon has been termed vascularization, and some modes of cancer treatment are aimed at preventing vascularization.

Cancer cells detach from the tumor and spread around the body. Cancer cells produce hydrolytic enzymes that allow them to invade underlying tissues. After traveling through the blood vessels or the lymphatic vessels, cancer cells start new tumors elsewhere in the body. This process is called metastasis. If a tumor is found before metastasis has occurred, the chances of a cure are greatly increased. This is the rationale for early detection of cancer.

Benign tumors have a slower growth rate, contain more differentiated cells, become encapsulated, and do not invade or metastasize. A wart is a benign tumor. *Malignant tumors* are described as aggressive; they have a more rapid growth rate, contain more undifferentiated cells, and have a tendency to invade or metastasize.

Cancer cells grow and divide uncontrollably, and they often metastasize, forming new tumors wherever they locate.

Causes of Cancer

One theory about the development of cancer, or carcinogenesis, suggests that it is a two-step process involving (1) initiation and (2) promotion.

During initiation, a gene mutation or a chromosome mutation occurs that makes cancerous growth a possibility at a later date. Carcinogens, agents that can contribute to the development of cancer, are *initiators*. For example,

Figure 24.20

Summary of the development of cancer. A virus can pass an oncogene to a cell. A normal gene, called a proto-oncogene, can become an oncogene because of a mutation caused by a chemical or radiation. The oncogene either expresses itself to a greater degree than normal or else expresses itself inappropriately.

Thereafter, the cell is cancerous. Cancer cells usually are destroyed by the immune system, and the individual only develops cancer when the immune system fails to perform this function.

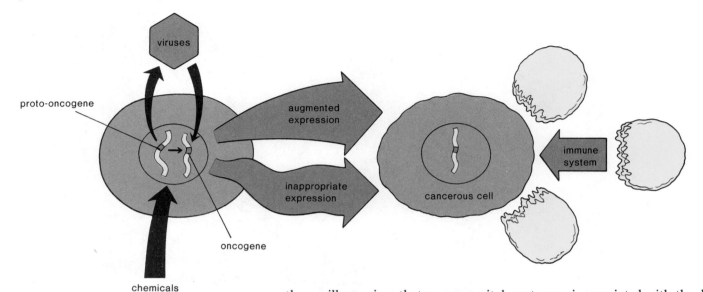

the papilloma virus that causes genital warts now is associated with the development of cervical cancer. Radiation, including ultraviolet (UV) radiation, radon (a radioactive gas made by the natural decay of radium), and X rays, damages the normal bonding patterns between DNA nucleotides. Chemical carcinogens, such as pesticides, are known to cause base sequence changes in DNA. Cigarette smoke plays a significant role in the development of lung cancer because it contains chemical carcinogens.

A cancer *promoter* is any influence that triggers a cell to start growing in an uncontrolled manner. It is possible that promotion simply involves a second change in the DNA brought about by another one of the same factors just discussed. In other words, cumulative DNA changes can result finally in uncontrolled growth. It also is possible that a promoter provides the environment that causes altered cells to form a tumor. For example, there is some evidence to suggest that a diet rich in saturated fats and cholesterol is a cancer promoter. Considerable time may elapse between initiation and promotion, and this is why cancer more often is seen in older rather than younger individuals.

One model suggests that carcinogenesis involves a gene mutation or a chromosome mutation (initiation) and a second influence that triggers cancerous growth (promotion).

Oncogenes

It is now clear that cells contain proto-oncogenes, genes that can be transformed into **oncogenes,** or cancer causing genes (fig. 24.20). These genes are not alien to the cell; they are normal, essential genes that have undergone a mutation that leads to augmented or inappropriate expression. An oncogene, known to cause both lung cancer and bladder cancer, differs from a normal gene by a change in only one nucleotide. It is believed that almost any type of mutation can convert a proto-oncogene into an oncogene. For instance, in addition to a gene mutation, a chromosome rearrangement can place a normally inactive structural gene next to an active promoter. If this structural gene is a proto-oncogene, it now may become an oncogene. An oncogene also can be introduced into a cell by a virus.

The Function of Oncogenes Several oncogenes have been located and studied. All of these are mutated forms of normal genes that regulate growth and cell division, but these normal genes all do not have the same type of function. There are numerous elements that affect the growth and the development of cells. These include growth factors and growth factor inhibitors that act on the surface of the cell, receptors for these substances, proteins that carry signals from the receptors, and nuclear functions that regulate the response to these influences. Oncogenes are genes that cause one of these elements to malfunction in such a way that the cell divides repeatedly to produce a tumor. Exactly how this comes about is not known yet.

Recently, investigators discovered genes they call anti-oncogenes that seem to suppress cancer. One anti-oncogene appears to be on the same chromosome as genes associated with some forms of colon, lung, bone, and breast cancers. The inactivation of the **anti-oncogene** can bring on one of these forms of cancer.

Genes that encode elements necessary to growth and division of cells sometimes become oncogenes due to a gene mutation or a chromosome mutation.

Summary

DNA is the genetic material; it can store information, it can replicate, and it can mutate. During replication, DNA becomes "unzipped," and then a complementary strand forms opposite to each original strand. DNA directs protein synthesis. During transcription, mRNA is made complementary to one of the DNA strands. It then contains codons and moves to the cytoplasm, where it becomes associated with the ribosomes. During translation, tRNA molecules, attached to their own particular amino acid, travel to the mRNA, and through complementary base pairing, the tRNAs and therefore the amino acids in a polypeptide chain are sequenced in a predetermined way. This sequence of events pertains to structural genes. The prokaryote operon model explains how one regulator gene controls the transcription of several structural genes. In eukaryotes, the chromosome has to decompact before transcription can begin.

A gene mutation is an alteration in the normal sequence of bases within a gene. Cancer is characterized by a lack of control: the cells grow uncontrollably and metastasize. Perhaps cancer requires a two-step process involving first an initiator and then a promoter. Oncogenes are normal genes that bring on cancer when they mutate because they control some aspect of cell growth and/or cell division.

Study Questions

1. Describe the experiment that designated DNA rather than protein as the genetic material. (p. 509)
2. Describe DNA structure and RNA structure. (pp. 509–10)
3. Explain how DNA replicates. (p. 512)
4. Various genetic disorders indicate that DNA controls the formation of proteins. Name and discuss some of these diseases. (pp. 513–15)
5. If the code is TTA'TGC'TCC'TAA, what are the codons and what is the sequence of amino acids? (p. 518)
6. List the 6 steps involved in protein synthesis. (p. 522)
7. What are the 4 levels of possible regulation of gene expression in eukaryotes? (p. 524)
8. What is the operon model of structural gene control? Explain the *lac* operon and the *tryp* operon. (p. 525)
9. The substitution of one base for another base in DNA can have what effect on the phenotype? Why would you expect additions and deletions of bases to have a major effect on the phenotype? (pp. 527–28)
10. What 2 characteristics of cancer cells show they no longer can control their metabolism? (p. 528) Describe a two-step process that explains the development of cancer. (p. 529) What is an oncogene? (p. 530)

Objective Questions

1. The backbone of DNA is made up of _____ and _____ molecules.
2. Replication of DNA is semiconservative, meaning that each new helix is composed of a(n) _____ strand and a(n) _____ strand.
3. The base _____ in DNA is replaced by the base uracil in RNA.
4. The DNA code is a(n) _____ code, meaning that every 3 bases stands for a(n) _____ .
5. The 3 types of RNA that are necessary to protein synthesis are _____ , _____ , and _____ .
6. When mRNA is processed within the eukaryotic nucleus, the portions complementary to _____ in DNA are removed.
7. Which of the 3 types of RNA carries amino acids to the ribosomes? _____
8. Another name for transposons is _____ .
9. The _____ model explains regulation of gene transcription in prokaryotes.

10. Cancer cells contain _____ ,
which actively direct the production of
_____ factors.

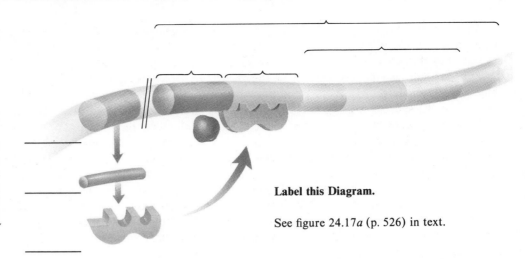

Label this Diagram.

See figure 24.17*a* (p. 526) in text.

Selected Key Terms

anticodon (an″tĭ-ko′don) a "triplet" of nucleotides in transfer RNA that pairs with a complementary triplet (codon) in messenger RNA. *520*

codon (ko′don) a "triplet" of nucleotides in messenger RNA that directs the placement of a particular amino acid into a polypeptide chain. *517*

complementary base pairing (kom″plĕ-men′tă-re bās pār′ing) pairing of bases between nucleic acid strands; adenine is always paired with either thymine (DNA) or uracil (RNA), and cytosine is always paired with guanine. *510*

double helix (dŭ′b′l he′liks) a double spiral often used to describe the three-dimensional shape of DNA. *510*

messenger RNA (mes′en-jer) mRNA; a nucleic acid (ribonucleic acid) complementary to genetic DNA and bearing a message to direct cell protein synthesis at the ribosome. *517*

oncogene (ong′ko-jēn) a gene that contributes to the transformation of a normal cell into a cancerous cell. *530*

operon (op′er-on) a group of structural and regulating genes that functions as a single unit. *524*

polysome (pol′e-sōm) a cluster of ribosomes attached to the same mRNA molecule and participating in the synthesis of the same polypeptide. *522*

purine (pu′rin) nitrogen-containing, organic base found in DNA and RNA that has 2 interlocking rings, as in adenine and guanine. *510*

pyrimidine (pi-rim′ĭ-din) nitrogen-containing, organic base found in DNA and RNA that has just one ring, as in cytosine and thymine. *510*

regulator gene (reg′u-lah-tor″ jēn) gene that codes for proteins involved in regulating the activity of structural genes. *524*

replication (re″pli-ka′shun) the duplication of DNA; occurs when the cell is not dividing. *512*

ribosomal RNA (ri′bo-sōm″al) rRNA; RNA occurring in ribosomes, structures involved in protein synthesis. *520*

structural gene (struk′tūr-al jēn) gene that directs the synthesis of enzymes and structural proteins in the cell. *525*

template (tem′plāt) a pattern that serves as a mold for the production of an oppositely shaped structure; one strand of DNA is a template for the complementary strand. *512*

transcription (trans-krip′shun) the process resulting in the production of a strand of mRNA that is complementary to a segment of DNA. *516*

transfer RNA (trans′fer) tRNA; molecule of RNA that carries an amino acid to a ribosome engaged in the process of protein synthesis *520*

translation (trans-la′shun) the process by which the sequence of codons in mRNA dictates the sequence of amino acids in a polypeptide. *517*

Chapter 25

Recombinant DNA and Biotechnology

Chapter Concepts

1 Modern biotechnology grew out of recombinant DNA technology.

2 Using recombinant DNA technology, it is possible to have bacteria (and other types of cells) mass-produce various products.

3 Naturally occurring bacteria can be bioengineered to perform useful services in agriculture and industry.

4 Plants lend themselves to genetic manipulation because cultured cells will grow into entire plants. Improved crops are expected soon.

5 Improved breeds of livestock are also expected, and laboratory research with bioengineered mice is expected to provide much needed genetic information.

6 Gene therapy in humans is largely experimental, but procedures are being developed to cure genetic disorders.

Chapter Outline

Basic Biotechnology Laboratory
  Techniques
    Recombinant DNA
    Other Biotechnology Techniques
Biotechnology Products
    Hormones and Similar Types of
      Proteins
    DNA Probes
    Vaccines
Genetically Engineered Bacteria
    Uses for Bacteria in Agriculture and
      Industry
    Ecological Considerations
Transgenic Organisms
    Transgenic Plants
    Transgenic Animals
Gene Therapy in Humans
    Modifying Stem Cells
    Developing Delivery Systems
    The Human Genome Project

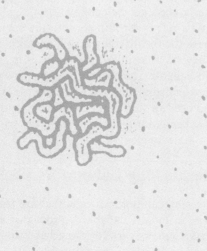

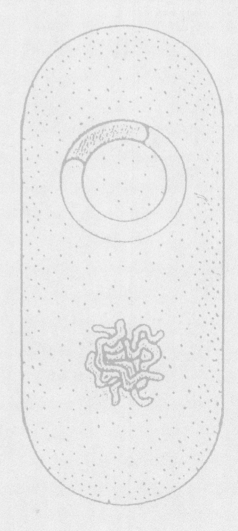

Figure 25.1

Biotechnology, an industrial endeavor.
a. Laboratory procedures are adapted to mass-
produce the product. b. Microbes are grown in
huge tanks called fermenters because they were
first used for yeast fermentation in the
production of wine. c. The product is purified
and (d) packaged.

a.

b.

c.

d.

Biotechnology, the use of a natural biological system to produce a product or to achieve an end desired by human beings, is not new. Plants and animals have been bred to give a particular phenotype since the dawn of civilization. The biochemical capabilities of microorganisms have been exploited for a very long time. For example, the baking of bread and the production of wine are dependent on yeast cells to carry out fermentation reactions.

Today, however, biotechnology is first and foremost an industrial process (fig. 25.1) that provides products because we are able to carry out **genetic engineering** of bacteria. The product can be an organic chemical of interest, a protein that is useful as a vaccine, or a drug to promote human health. Engineered bacteria need not be confined to a chemical plant or a laboratory; they soon may be released routinely into the environment to clean up pollutants, to increase the fertility of the soil, or to kill insect pests. Biotechnology even extends beyond unicellular organisms; it is now possible to alter the genotype and subsequently the phenotype of plants and animals. An agricultural revolution of unequaled magnitude is not inconceivable even in the near future, and there is no need to stop here. Gene therapy in humans may be just over the horizon. We begin our discussion, however, with the basic procedures and applications of biotechnology.

Basic Biotechnology Laboratory Techniques

Genetic engineering can produce cells that are **transformed.** These cells contain a foreign gene and are capable of producing a new and different protein. Often the foreign gene is carried into a bacterium, a yeast cell, a plant cell, or an animal cell as a part of recombinant DNA.

Figure 25.2

Gene cloning using bacteria and viruses. a. A plasmid is removed from a bacterium and is used to make recombinant DNA. After the recombined plasmid is taken up by a host cell, cloning is achieved when the host cell and the recombinant DNA of the *plasmid reproduce. b. Viral DNA is removed from a virus and is used to make recombinant DNA. Virus containing the recombinant DNA infects a host bacterium. Cloning is achieved when the virus reproduces and then leaves the host cell.*

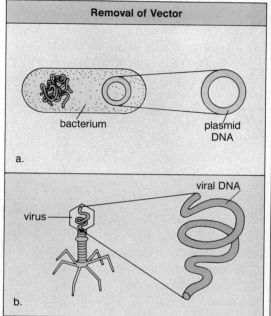
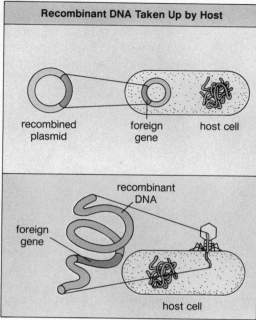
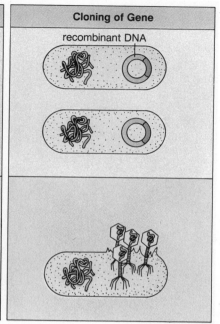

Removal of Vector	Recombinant DNA Taken Up by Host	Cloning of Gene

Recombinant DNA

Recombinant DNA contains DNA from 2 or more different sources. To make recombinant DNA, a technician often begins by selecting a **vector,** the means by which recombinant DNA is introduced into the host cell.

Vectors

The most common vector is a plasmid (fig. 25.2*a*). **Plasmids** are small accessory rings of DNA taken from bacteria. Plasmids used as vectors have been removed from bacteria and have had a foreign gene inserted into them. Treated cells take up a plasmid, and after it enters, the plasmid continues to reproduce as usual. Whenever the host reproduces, the plasmid, including the foreign gene, is copied. Eventually, there are many copies of the plasmid and therefore many copies of the foreign gene. The gene now is said to have been **cloned.**

Viral DNA also can be used as a vector to carry recombinant DNA into a cell (fig. 25.2*b*). When a virus containing recombinant DNA attacks a cell, the viral DNA enters the cell.[1] Here, it can direct the reproduction of many more viruses. Each virus derived from a viral vector contains a copy of the foreign gene. Therefore, viral vectors allow cloning of a particular gene, too. Viral vectors also are used to create genomic libraries. A **genomic library** is a collection of engineered viruses that together carry all the genes of a species. Since each virus carries only a short sequence of DNA, it takes about 10 million viruses to carry all the genes of a mouse.

Retroviruses sometimes are used to carry recombinant DNA into mammalian cells, including human cells. Retroviruses are RNA viruses containing an enzyme that can make a DNA copy (called cDNA) of their RNA genes. Thereafter, the cDNA is integrated into host DNA (see fig. 28.4). This is the preferred methodology when the permanent transformation of a particular cell line is desired.

[1]Viruses are discussed on pages 584–89.

Enzymes

The introduction of foreign DNA into a plasmid or into viral DNA to give recombinant DNA is a two-step process (fig. 25.3). First, the plasmid (or viral) DNA is cut open, and then the foreign DNA is inserted into this opening. Both of these steps require a specific type of enzyme.

One type of enzyme has the ability to cut a DNA molecule into discrete pieces. Such enzymes occur naturally in some bacteria, where they stop viral reproduction by cutting up viral DNA. These are called **restriction enzymes** because they *restrict* the growth of viruses. Each type of restriction enzyme— and over a hundred now are known—cleaves DNA at a specific location called a *restriction site*. For example, one restriction enzyme always cleaves double-stranded DNA when it has this sequence of bases:

$$. . . . .G A A T T C. . . . .$$
$$. . . . .C T T A A G. . . . .$$

Furthermore, this enzyme always cleaves each strand between a G and an A in this manner:

```
. . . . .G                     A A T T C. . . .
. . . . .C T T A A             G. . . .
```

Notice that there is now a gap into which a piece of foreign DNA can be placed if it ends in bases complementary to those exposed by the restriction enzyme. To ensure this, it is only necessary to cleave the foreign DNA with the same type of restriction enzyme. The single-stranded but complementary ends of the 2 DNA molecules are called "sticky ends" because they adhere by complementary base pairing. These ends facilitate insertion of foreign DNA into vector DNA.

The second enzyme needed for preparation of a vector is **DNA ligase,** a bacterial enzyme that seals any breaks in a DNA molecule. Genetic engineers use this enzyme to seal the foreign piece of DNA into the vector. Gene splicing now is complete, and a recombinant DNA molecule has been prepared.

The Result

Once the recombined plasmid has been taken up by the host cell, it is expected to function normally—reproducing along with the host DNA and producing the protein of the foreign gene. If the inserted plasmid is replicated and this gene is actively expressed, the investigator later can recover either the cloned gene (e.g., insulin gene) or a protein product (e.g., insulin) (fig. 25.3).

Other Biotechnology Techniques

Aside from recombinant DNA, other techniques are common in biotechnology, including the following:

1. **Gene sequencing,** or determining the order of nucleotides in a gene. There are automated DNA sequencers that make use of computers to accomplish gene sequencing at a fairly rapid rate.
2. Preparation of a gene in vitro (within laboratory glassware) using cellular components.
3. Manufacture of a gene; that is, the nucleotides are joined in the correct sequence, or if desired, a mutated gene is prepared by altering the sequence.
4. Joining the regulatory regions of a viral or bacterial gene to an isolated or prepared gene so that transcription is ensured. Genes spliced into vectors very often are accompanied by regulatory regions, and mammalian genes can be put under the control of bacterial promoters.
5. Insertion of an isolated or prepared gene directly into a host cell rather than first into a vector.

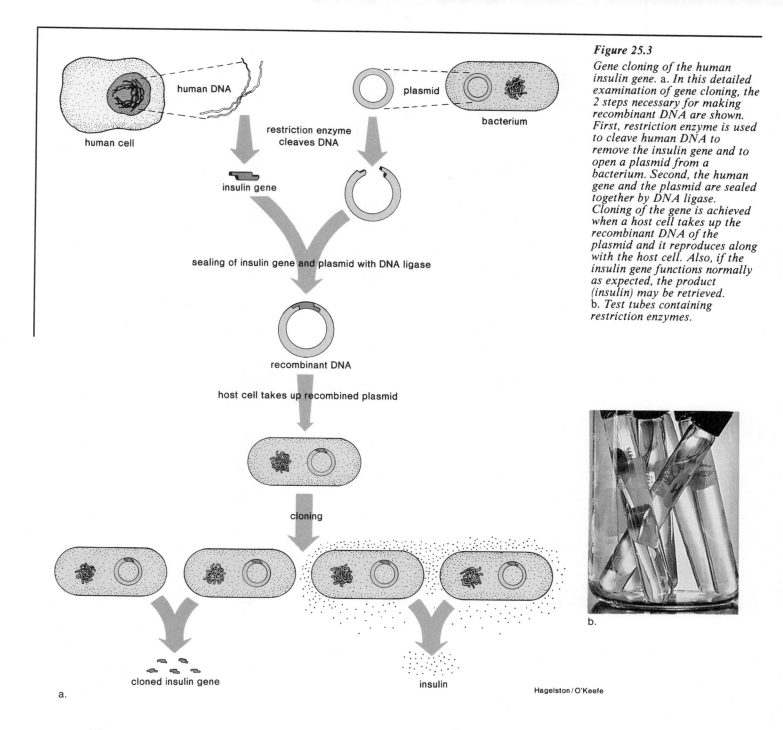

Hagelston/O'Keefe

Figure 25.3
Gene cloning of the human insulin gene. a. In this detailed examination of gene cloning, the 2 steps necessary for making recombinant DNA are shown. First, restriction enzyme is used to cleave human DNA to remove the insulin gene and to open a plasmid from a bacterium. Second, the human gene and the plasmid are sealed together by DNA ligase. Cloning of the gene is achieved when a host cell takes up the recombinant DNA of the plasmid and it reproduces along with the host cell. Also, if the insulin gene functions normally as expected, the product (insulin) may be retrieved. b. Test tubes containing restriction enzymes.

Labels in figure:
- human cell
- human DNA
- plasmid
- bacterium
- restriction enzyme cleaves DNA
- insulin gene
- sealing of insulin gene and plasmid with DNA ligase
- recombinant DNA
- host cell takes up recombined plasmid
- cloning
- cloned insulin gene
- insulin
- a.
- b.

Cells can be genetically engineered to produce a new protein product. Often, vectors are used to carry the necessary gene into the cell.

Biotechnology Products

Table 25.1 shows at a glance many of the types of biotechnology products that now are available. There are 3 categories in the table: hormones and similar types of proteins, DNA probes, and vaccines. Monoclonal antibodies (p. 272) are not now but may soon be produced by recombinant DNA technology.

Table 25.1 Representative Biotechnology Products

Hormones and Similar Types of Proteins	DNA Probes	Vaccines
Treatment of Humans	*Diagnostics in Humans*	*Use in Humans*

Insulin
Growth hormone
tPA (tissue plasminogen activator)
Interferon (alpha, beta*, gamma*)
Erythropoietin
Interleukin-2
Clotting factor VIII*
Human lung surfactant*
Atrial natriuretic factor*
Superoxide dismutase*
Alpha-I-antitrypsin*
Various growth factors*
	Detection of criminals
	Legionnaires' disease
	Walking pneumonia
	Sickle-cell anemia
	Cystic fibrosis
	Tendency for emphysema, thalassemia,
	  retinoblastoma
	Hemophilia B
	Sexually transmitted diseases
	Polycystic kidney disease
	Paternity testing
	Monitoring bone marrow transplants
	Susceptibility to atherosclerosis, diabetes, and
	  hypertension
	Periodontal disease
	Huntington disease*
	Duchenne muscular dystrophy*
	Cancers (chronic myelogenous leukemia, T- and
	  B-cell cancer, others*)
		Hepatitis B
		Herpes* (oral and genital)
		AIDS*
		Hepatitis A*
		Group B meningococci*
		Hepatitis C*
		Malaria*

Treatment of Animals		*Use in Animals*
Growth hormone (cows, pigs, chickens*) Lymphokines*		Hoof-and-mouth disease Scours Brucellosis Rift Valley fever Herpes Influenza Pseudorabies Feline leukemia virus* Rabies*

*Expected in the near future. These may be available now, but presently they are not made by recombinant DNA technology.

Hormones and Similar Types of Proteins

One impressive advantage of biotechnology is that it allows mass production of proteins that are very difficult to obtain otherwise. For example, human growth hormone previously was extracted from the pituitary gland of cadavers, and it took 50 glands to obtain enough for one dose. Insulin previously was extracted from the pancreas glands of slaughtered cattle and pigs; it was expensive and sometimes caused allergic reactions in recipients. Not long ago, few of us knew of tPA (tissue plasminogen activator), a protein present in the body in minute quantities that activates an enzyme to dissolve blood clots. Now tPA is a biotechnology product that is used to treat heart attack victims.

A study of the list of prospective protein products indicates that some of the most troublesome and serious afflictions in humans soon may be treatable: clotting factor VIII may be available for hemophilia; human lung surfactant may be available for premature infants with respiratory distress syndrome; and atrial natriuretic factor may be available for hypertension. This list will grow because bacteria (or other cells) can be engineered to produce virtually any protein.

Hormones also are being produced for use in animals. It is no longer necessary to feed steroids to farm animals; they can be given growth hormone, which produces a leaner meat that is more healthful for humans. Cows given bovine growth hormone (bGH) produce 25% more milk than usual, which should make it possible for dairy farmers to maintain fewer cows and to cut down on overhead expenses.

Human Reproduction, Development, and Inheritance

Figure 25.4

Possible biotechnology scenario. a. *A polypeptide is removed from a cell, and the amino acid sequence is determined. (This hormone is growth hormone releasing factor.) From this, the sequence of nucleotides in DNA can be deduced.* b. *The DNA synthesizer can be used to string nucleotides together in the* correct order. c. *A small section of the gene can be used as a DNA probe to test fetal cells for an inborn error of metabolism. (The manufactured gene could be placed in a bacterium to produce more of the protein, which then possibly could be used as treatment of the defect.)*

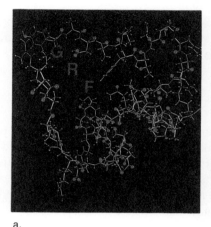

a.

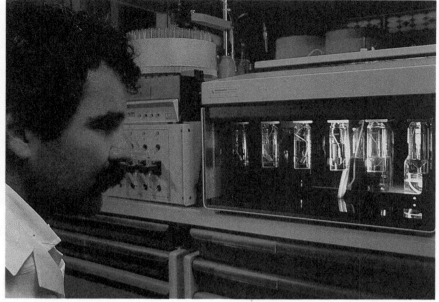

b.

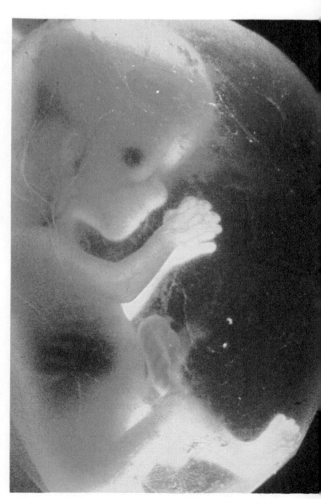

c.

DNA Probes

Biotechnology is contributing greatly to diagnostics by making DNA probes (table 25.1) and monoclonal antibodies. Recombinant DNA technology can be used to clone monoclonal antibodies, which also can be made by the method described on page 272. To construct a DNA probe it is only necessary to use a DNA synthesizer to make single stranded DNA with a particular sequence of bases. Each strand can be used as a **DNA probe** because it seeks out and binds to any complementary DNA strand present in body fluids or removed from body cells (fig. 25.4). DNA probes can be used presently in paternity suits (consider that one chromosome of each pair is inherited from the father) and sometimes in court cases to identify an individual whose body cells are found at the scene of a crime. They can be used also to diagnose an infection by indicating if the gene of an infectious organism is present. When available, DNA probes can tell us whether a gene coding for a hereditary defect or causing a cell to be cancerous is present.

Vaccines

Recombinant DNA technology can produce pure and therefore safe vaccines. Bacteria and viruses have surface proteins, and a gene for just one of these proteins can be placed in a plasmid. The host cell for the plasmid produces many copies of the surface protein, which then can be used as a vaccine. Right now, the only recombinant vaccine on the market is for hepatitis B, but you can see from table 25.1 that the ones expected within a few years are of great importance: a vaccine for malaria and another for AIDS. Malaria has been a scourge on humankind for many thousands of years, and AIDS is a deadly disease known to modern humans.

Vaccines have been produced for the inoculation of farm animals. Each of the illnesses listed—hoof-and-mouth disease, scours, and others—causes an untold number of illnesses and deaths each year. These were a severe drain on the time, the energy, and the resources of farmers.

Biotechnology products include hormones and similar types of proteins, DNA probes, and vaccines. These products are of enormous importance to the fields of medicine and animal husbandry.

Genetically Engineered Bacteria

Most often, bacteria are used to clone a gene or to mass-produce a product (fig. 25.3). However, in other instances, bacteria taken from the natural environment have been engineered to perform other types of services.

Uses for Bacteria in Agriculture and Industry

Agriculture

Genetically engineered bacteria can be used to promote the health of plants. For example, bacteria that normally live on plants have been changed from frost-plus to frost-minus bacteria. Field tests showed that these bioengineered bacteria protect the vegetative parts of plants from frost damage. Also, a bacterium that normally colonizes the roots of corn plants now has been endowed with genes (from another bacterium) that code for an insect toxin.

Many other recombinant DNA applications in agriculture are thought to be possible. For example, *Rhizobium* is a bacterium that forms nodules on the roots of leguminous plants. Here, the bacteria fix atmospheric nitrogen in a form that can be used by the plant (see fig. 34.12). It might be possible to transfer the necessary genes to other bacteria that then can infect nonleguminous plants. This would reduce the amount of fertilizer needed on agricultural fields.

Pollution Cleanup

There are naturally occurring bacteria that can degrade most any type of chemical or material. Bacteria can be selected for their ability to degrade a particular substance, and then this ability can be enhanced by genetic engineering. For example, microbes recently were used to help clean up the beaches of Alaska after a massive oil spill. These were naturally occurring bacteria that eat oil, but there are genetically engineered bacteria that could have done an even better job (fig. 25.5).

Bioremediation

Organic chemicals often are synthesized by having catalysts act on precursor molecules or by using bacteria to carry out the synthesis. Today, it is possible to go one step further and to manipulate the genes that code for these enzymes. For example, biochemists discovered a strain of bacteria that is especially good at producing phenylalanine, an organic chemical needed to make aspartame, the dipeptide sweetener better known as NutraSweet™. They isolated, altered, and formed a vector for the appropriate genes so that various bacteria could be geared to produce phenylalanine.

Mineral Processing

Many major mining companies already use bacteria to obtain various metals. Genetic engineering may enhance the ability of bacteria to extract copper, uranium, and gold from low-grade sources. At least 2 mining companies plan to test genetically engineered organisms with enhanced bioleaching capabilities.

Ecological Considerations

There are those who are very concerned about the deliberate release of genetically engineered microbes (GEMs) into the environment. Ecologists point out that these bacteria might displace those that normally reside in an ecosystem, and the effects could be deleterious. Others rely on past experience

---

25.1 Critical Thinking

1. What experiment would you suggest to support the hypothesis that insulin produced by biotechnology will have fewer side effects than insulin taken from the organs of livestock? Why do you expect your results to support the hypothesis?

2. What experiment would you suggest to support the hypothesis that meat from cattle bioengineered to contain extra growth hormone genes is safe for human consumption? Why do you expect your results to support the hypothesis? Your answer should consider the fact that growth hormone is a protein.

3. What experiment would you suggest to support the hypothesis that plants engineered to contain genes for an insect toxin against insects is safe for human consumption? Why might your results not support the hypothesis?

4. What experiment would you suggest to support the hypothesis that bacteria engineered to clean up a pollutant disappear when the pollutant is gone? What would make them disappear?

Human Reproduction, Development, and Inheritance

Figure 25.5

Oil-eating bacteria, engineered and patented by investigator Dr. Chakrabarty. In the inset, the flask toward the front contains oil and no bacteria; the flask toward the rear contains the bacteria and is almost clear of oil. Now that engineered organisms (e.g., bacteria and plants) can be patented, there is even more impetus to create them.

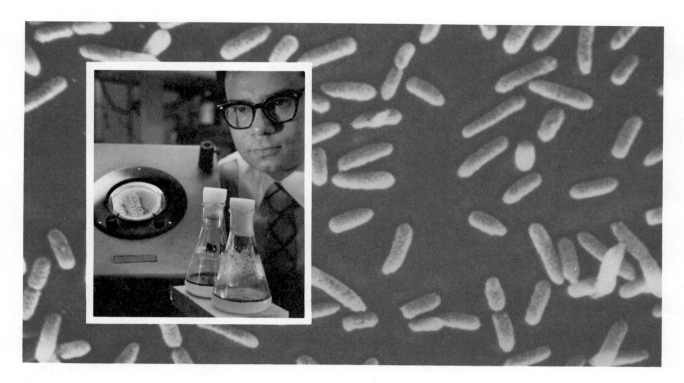

with GEMs, primarily in the laboratory, to suggest that these fears are unfounded. Tools now are available to detect, to measure, and even to disable cell activity in the natural environment. It is hoped that these eventually will allow GEMs to play a significant role in agriculture and environmental protection.

Transgenic Organisms

We have considered only the insertion of foreign genes into bacteria or viruses. If a foreign gene is inserted into an immature cell, will it be expressed in the cells of a multicellular organism? Indeed, this has been achieved. Plants and animals that have a foreign gene inserted into them are called **transgenic organisms.**

Transgenic Plants

Plants, in particular, lend themselves to genetic manipulation because it is possible to grow plant cells in tissue culture, and each cell can be stimulated to produce an entire plant (fig. 25.6). Previously, tissue culture was used to provide genetic carbon copies of strawberry, asparagus, and oil palm plants for commercial use. Tissue culture also was used to screen for cells that have a particular property, such as resistance to an herbicide. It is obviously less time consuming and less costly to screen millions of cells in a flask than to screen adult plants.

Methods for Introducing Genes into Plants

Beside these procedures, it also is possible to apply recombinant DNA technology to plant cells. The presence of the plant cell wall hinders plasmid uptake, but plant cell walls can be removed to give "naked" cells called **protoplasts**

Figure 25.6

Cloning of entire plants from tissue cells. a. Sections of carrot root are cored, and thin slices are placed in a nutrient medium. b. After a few days, the cells form a callus, a lump of undifferentiated cells. c. After several weeks, the callus begins sprouting cloned carrot plants. d. Eventually the carrot plants can be moved from culture medium to potting soil.

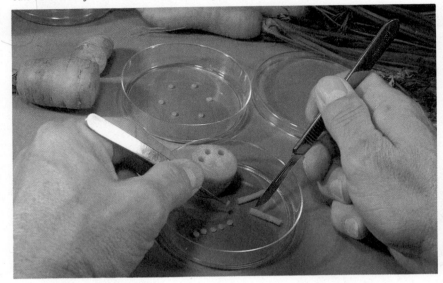

a.

b.

c.

d.

(fig. 25.7). Protoplasts take up plasmids. The vector of choice is the Ti (tumor-inducing) plasmid from the bacterium *Agrobacterium tumefaciens*. Ordinarily, the Ti plasmid invades plant cells and causes a cancerlike growth called crown gall disease, but when used as a vector, the plasmid is engineered to lack virulence. This particular method of gene engineering has been the most successful in dicots. Monocots (most cereal plants are monocots) usually do not grow from protoplasts, and only recently has it been possible to get corn and rice plants to grow from protoplasts. Also, monocot protoplasts ordinarily do not take up the Ti plasmid in tissue culture. One solution is to simply irradiate the protoplasts with a laser beam while they are suspended in a liquid containing foreign DNA. Laser beams make tiny self-sealing holes in the cell membrane through which genetic material can enter.

Other methods that bypass the protoplast stage entirely have been developed recently. Some investigators are using a "particle gun," or high-velocity microprojectile technology. In this system, DNA is carried through the cell wall and membrane into the cytoplasm on the surface of small (0.5–5μm) metal

Human Reproduction, Development, and Inheritance

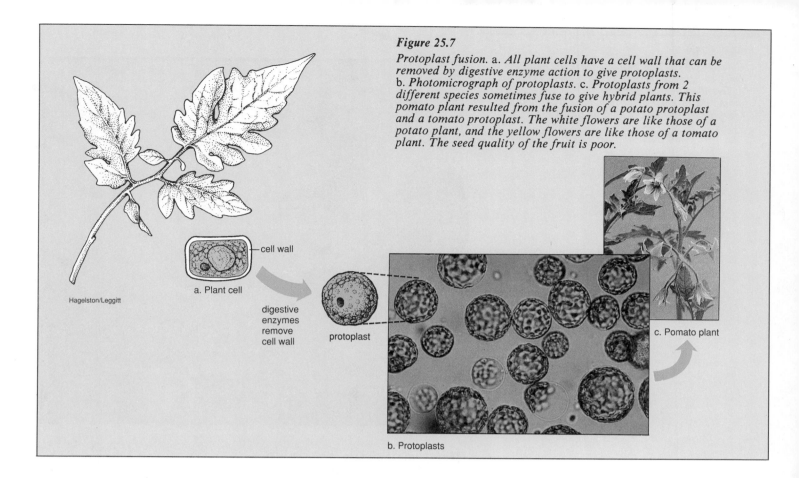

Figure 25.7

Protoplast fusion. a. *All plant cells have a cell wall that can be removed by digestive enzyme action to give protoplasts.* b. *Photomicrograph of protoplasts.* c. *Protoplasts from 2 different species sometimes fuse to give hybrid plants. This pomato plant resulted from the fusion of a potato protoplast and a tomato protoplast. The white flowers are like those of a potato plant, and the yellow flowers are like those of a tomato plant. The seed quality of the fruit is poor.*

Hagelston/Leggitt

a. Plant cell

cell wall

digestive
enzymes
remove
cell wall

protoplast

b. Protoplasts

c. Pomato plant

particles that have been accelerated to speeds of several hundred meters per second. With this method, any type of cell can receive foreign DNA and later can be cultured to give an entire plant (fig. 25.6).

Transgenic Plant Applications

About 50 types of genetically engineered plants that resist either insects, viruses, or herbicides (fig. 25.8) now have entered small-scale field trials. The major crops that currently can be improved in this way are soybean, cotton, alfalfa—dicots—and rice—a monocot. However, even genetically engineered corn is expected to reach the marketplace by the year 2000.

Other possible results of plant biotechnology are foreseen. As we previously discussed (p. 204), plant seeds are not complete protein sources because they often lack a particular amino acid needed by animals. Protein-enhanced beans, corn, soybeans, and wheat now are being developed. Plants such as cotton traditionally have been a source of polymers, but others, like rubber, now are synthetically made. This often leads to environmental pollution; therefore, it may be better to turn to genetically engineered plants to supply synthetic polymers. Plants also can be engineered to produce foreign proteins, such as neuropeptides, blood factors, and growth hormones. Therefore, they could be another source of the biotechnology products listed in table 25.1.

Biotechnology in plants is facilitated by the fact that entire plants can grow from cultured cells. Dicots take up a Ti plasmid, but other techniques are being developed to bioengineer monocots.

Figure 25.8

Genetically engineered cotton plants. The boll on the left is from a plant that was not engineered to resist cotton bollworm larvae and is heavily infested. The boll on the right is from a plant that was engineered to resist cotton bollworm larvae and will go on to give a normal yield of cotton.

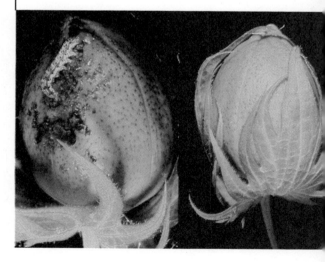

Figure 25.9

Chimeric mice. The coats of these mice are mottled because their cells are derived from 2 different sources of embryonic cells. Genetically manipulated embryonic stem cells from another source were introduced into an intact blastocyst.

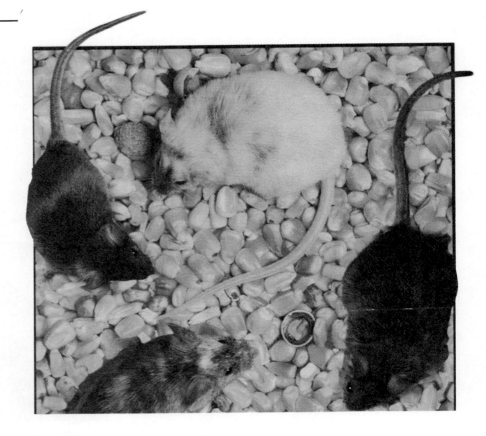

Transgenic Animals

Genetic engineering of animals has begun. The most common procedure is to microinject animal eggs with a certain gene. For example, researchers in Minnesota want to inject the bGH gene into lake walleye eggs in the hope the gene will establish itself and will be transmitted to all the cells of the organism, eventually resulting in a larger fish. This same procedure has been used experimentally in chickens, cows, pigs, rabbits, and sheep.

A systematic study of the results of one such experiment in pigs showed that after bGH gene was incorporated into the genome (genes of the organism), it was passed on to the offspring. In 2 successive generations, the pigs grew larger and they produced a leaner meat. However, the pigs were not healthy. They showed a high incidence of gastric ulcers, arthritis, dermatitis, and cardiovascular and renal disease. The researchers think that it might be possible to find a breed of pigs that would show only the beneficial effects and not the detrimental effects of having a bGH gene.

Most laboratory investigations of gene expression in animals have utilized the mouse as the experimental animal. For example, the sheep gene that codes for the milk protein betalactoglobulin (BLG) was microinjected into mice eggs. (Mice normally produce milk that has no BLG at all). Of 46 offspring successfully weaned, 16 carried the BLG sequence, and the females among them later produced BLG-rich milk. One produced BLG at 5 times the concentration found in sheep milk. Some of the transformed mice passed on the BLG gene to their offspring.

A new procedure for introducing a mutated gene into mouse embryos has been developed. The mutated gene is cloned and then microinjected into embryo-derived stem (ES) cells. They are called stem cells because they are totipotent embryonic cells and show no differentiation yet. The introduced DNA is a modification of a gene normally found in mice, and in a few cells, it replaces the resident gene at a particular locus by recombination. It is said that one of the mice genes has been "targeted" for replacement. The experimenter cultures these targeted ES cells and then microinjects them into blastocysts. After implantation into a foster mother, the resulting animals are

Figure 25.10

Retroviruses as vectors. a. A retrovirus has an RNA chromosome. When this chromosome enters a cell, reverse transcription occurs. In this way, it is possible to retrieve viral DNA. b. Now, a foreign human gene can be inserted into viral DNA. c. Transcription inside a host cell produces an RNA copy of the recombinant DNA. d. A virus now is used to carry the recombinant RNA into a human host cell. Following reverse transcription, recombinant DNA is inserted into the human genome.

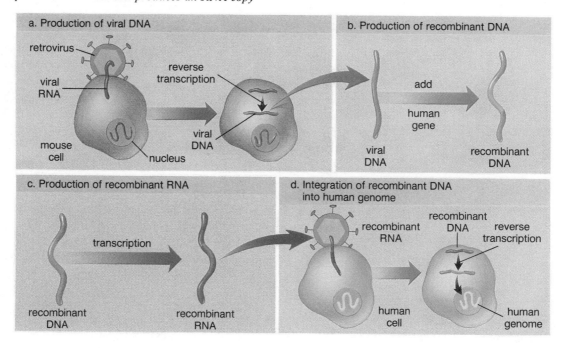

a. Production of viral DNA
b. Production of recombinant DNA
c. Production of recombinant RNA
d. Integration of recombinant DNA into human genome

chimeric—some of the cells of the organism carry the introduced gene and some do not (fig. 25.9). However, among the offspring of the chimeric mice, there are those that have the introduced gene in all their cells. It is believed that this experimental procedure will allow investigators to study the metabolic effects of many specific types of genes in a mammalian organism.

Biotechnology in animal husbandry eventually may produce livestock with improved characteristics. Biotechnology in laboratory animals will help us to study the phenotypic effects of particular mammalian genes in the living animal.

Gene Therapy in Humans

Investigators are striving to use biotechnology as a means to cure or to treat human genetic disorders and to treat other human ills as well. This is called **gene therapy.** As one investigator phrases it, "Gene therapy is not just for genetic disease anymore."

Modifying Stem Cells

Originally, investigators concentrated on the possibility of curing certain genetic disorders by introducing a normal gene into bone marrow stem cells. Bone marrow stem cells were chosen as possible recipients because they give rise to other blood cells. It was reasoned that if the stem cells and then many of the blood cells contained a normal functioning gene, they would supply the molecule needed by the patient. For example, some children have SCID (severe combined immune deficiency) because they are unable to produce a normal enzyme called adenosine deaminase (ADA), which is essential to the T and B lymphocytes. The experimental design was to introduce the gene coding for ADA into bone marrow stem cells by means of a retrovirus (fig. 25.10). The necessary stem cells would be removed from the patient, treated, and then returned to the patient by injection.

Recombinant DNA and Biotechnology

Investigators pursuing this line of attack against genetic disorders have run into difficulties. For example, the stem cells are hard to get a hold of and are not infected by a retrovirus easily. Therefore, bone marrow gene therapy is on hold until stem cells can be further studied. If a method for growing large numbers of stem cells in vitro can be discovered, then it might be possible to develop a way to perform the procedure.

Developing Delivery Systems

In the meantime, it has been found that a retrovirus can be used as a vector for lymphocytes. The only gene experiment conducted in humans so far used lymphocytes from a patient about to die from advanced melanoma. These cells were removed from the patient, treated with interleukin (p. 271), and engineered to contain an antibiotic (neomycin)-resistant gene that is easy to detect. The detected presence of the neomycin-resistant gene in the lymphocytes proved that it is possible to use lymphocytes to place new genes in the human body without ill effects. Researchers now want to put the ADA gene in lymphocytes of SCID patients.

Endothelial cells also can be genetically engineered to make a particular protein, but how can these cells be inserted into the body? One novel idea is to put them in "organoids," artificial organs that can be implanted in the abdominal cavity. Organoids have been made by coating angel-hair Gore-Tex fibers with collagen and adding a growth factor for blood vessels (fig. 25.11).

Polymerase Chain Reaction

The polymerase chain reaction (PCR) is a way to make multiple copies of a single gene, or any specific piece of DNA, in a test tube. Further, the process is very specific—the targeted DNA sequence can be less than one part in a million of the total DNA sample! This means then that a single gene among the entire human genome can be amplified (copied) using PCR.

The process takes its name from DNA polymerase, the enzyme that carries out replication in a cell. The reaction is a chain reaction because DNA polymerase is allowed to carry out replication over and over again until there exists a million or more copies of the targeted DNA sequence.

PCR is specific because it makes use of *primers*, sequences of about 20 nucleotides that are complementary to the nucleotides on either side of target DNA. DNA polymerase cannot start a replication process, it can only continue or extend it. Therefore, after the primers are in place, DNA polymerase copies only the targeted DNA sequence.

One cycle of PCR amplification requires these steps:

• DNA denaturation—at a high temperature of 95°C, the DNA strands unzip.

• Primer annealing—at reduced temperature of 54°C, the primers bind complementarily to either side of target DNA.

• DNA polymerase extension—at an increased temperature of 72°C, target DNA is copied.

Successful PCR requires careful attention to certain details. The primers have to be long enough so that their sequence is unique, assuring that they will bind only to either side of the target DNA. Primer concentration must be high enough so that the primers will bind to the unzipped DNA faster than the original strands of DNA bind together. Temperature and salt concentrations are also critical in this regard. A recent advance is the use of a newly discovered thermostable bacterial DNA polymerase, called Taq polymerase, that is not destroyed by the high temperature required for DNA denaturation at the beginning of each PCR cycle. Using this enzyme means that you do not have to add more DNA polymerase for each cycle and this has led to the introduction of automated PCR machines and the wide application of the PCR process.

In particular PCR has simplified the use of DNA probes, sequences of DNA that are used to detect the presence of a mutated gene, or pathogen DNA, for example. After PCR amplification it becomes a lot easier for a probe to detect that target DNA is present. The binding of the probe is detectable because the probe is labeled either radioactively, or with a fluorescent dye.

Specifically, PCR can be used in the following areas:

Molecular Biology Research

If a multigram quantity of DNA is needed rather than the microgram usually made available by PCR, it is much less expensive to clone the DNA using *E. coli*. But once a normal gene is cloned and the needed primer sequence is known, it is possible to use PCR to amplify a mutant gene to see how it differs from the normal gene.

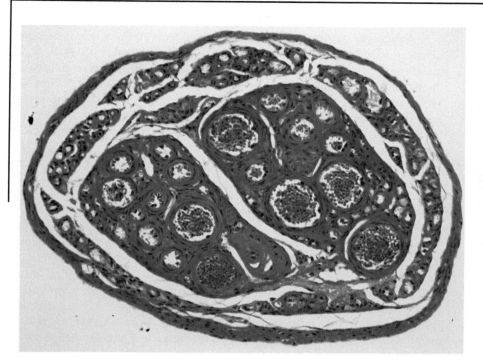

Figure 25.11

Cross section of an organoid. In order to make an organoid, a growth factor that stimulates blood vessel formation is added to Gore-Tex fibers coated with collagen. Once implanted in the abdominal cavity of a rat, blood vessels extend from the fibers to the animal's natural liver. In this cross section, note the presence of abundant vessels (red color) lined with endothelial cells and surrounded by layers of smooth muscle. These vessels could serve as a means to transfer proteins, made by genetically engineered endothelial cells, into the general circulation.

Or to take another example, PCR can help determine if a certain mRNA is present during a particular stage of development. First, the enzyme reverse transcriptase can be used to make DNA copies (called cDNA) of the RNAs from a particular embryonic cell; then the target DNA can be amplified using PCR; and finally, a DNA probe can be used to determine if the cDNA and therefore the mRNA of interest is present in the embryonic cell.

Genetic Disorders

PCR analysis, consisting of PCR amplification followed by the use of a DNA probe, can be used to determine if a mutant gene responsible for a genetic disorder is present in an individual; a single fetal or embryonic cell. Therefore, following in vitro fertilization, PCR analysis can be used to ensure the implantation of only normal embryos—ones that lack the faulty gene.

Infectious Diseases

Blood tests currently used to diagnose human illness often detect the presence of antibodies to the pathogen rather than the pathogen itself. For example, the blood test for AIDS does not detect HIV itself, but instead confirms the presence of antibodies to HIV in the blood. PCR analysis can be used to detect the presence of the pathogen before the immune system has even begun an antibody response. It then becomes possible to begin treatment of a disease like AIDS earlier than before.

Cancer Diagnosis

It has long been the custom for physicians to save the tissues of persons who have died from various illnesses. Some of these are cancer patients. PCR is making it possible to analyze the DNA of deceased patients along with living patients who currently have a particular type of cancer. Using this method, for example, investigators have found that 90% of pancreatic tumors contain a particular mutated gene.

As another example, a certain form of leukemia is characterized by a chromosomal translocation that can be detected by utilizing PCR analysis. Early detection of the presence of the chromosomal translocation can lead to early treatment, especially in patients who are believed to be in remission.

Human Genome Project

Investigators are beginning to use PCR to help them map the human chromosomes. The DNA of a single sperm cell can be analyzed to detect the presence of genetic markers, even ones that are quite close together. PCR also allows scientists to determine whether a particular DNA sequence in one chromosomal fragment is also found in another fragment so they can be placed in the correct order.

Evolutionary Relations of Organisms

Scientists used PCR to amplify DNA segments from a quagga, an extinct zebralike animal whose only remains consisted of bits of dried skin. This allowed them to determine that the quagga was a zebra rather than a horse. In other studies DNA sequences from a 7,000-year-old mummified human brain and from a 17 to 20 million-year-old plant fossil were analyzed. It is expected that PCR will contribute greatly to providing information about evolutionary relationships.

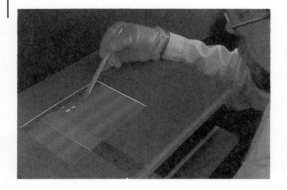

Figure 25.12

Sequencing DNA using gel electrophoresis. The DNA is cleaved into fragments that vary in length by only one nucleotide. Each fragment ends with an A, T, C, or G marked by one of 4 fluorescent dyes. The fragments are subjected to gel electrophoresis, during which they migrate on a gel in an electric field. Shorter fragments migrate further than longer fragments. Under proper lighting, the investigator (or a computer) simply can read off the sequence of the nucleotides.

Once in the body, the endothelial cells begin to line the newly developed blood vessels, which spread out to adjoining organs.

Organoids possibly could be used to help treat AIDs patients. As mentioned previously, the AIDS virus attaches to a receptor known as CD4. If there are extra soluble CD4 receptors in blood, they might serve as decoys; the virus might combine with them instead of a CD4 on a helper T lymphocyte. Endothelial cells or lymphocytes first could be engineered to contain the gene that produces CD4 and then could be placed in an organoid destined for the abdominal cavity. Once inside the patient's body, the cells would produce CD4 decoys continuously.

Endothelial cells within organoids might make a good delivery system for any type of "genetic drug" needed by a patient.

The Human Genome Project

Some feel that gene therapy in humans will be easier once the human chromosomes are completely mapped and sequenced. The U.S. government recently decided to provide the funds necessary for such a project. The first step will be to use DNA probes to find genetic markers about 10 million bases apart. This will narrow the location of a gene from 3 billion bases in the entire genome to 10 million bases between 2 markers. Eventually it is hoped that the entire genome will be base sequenced (fig. 25.12). Even with the use of automated DNA sequencers, it is estimated that sequencing the human genome will take about 30 years unless better technology is developed. The amount of paper needed to record the results will be staggering, requiring on the order of 200 books the size of an encyclopedia volume.

Some researchers believe that it will be of little benefit to know the sequence of the bases in the human genome. Others believe that this knowledge will take us one step closer to being able to manufacture genes (and their control elements) without having to splice them out from their neighbors in a hit-or-miss fashion. Then it will be possible to study the protein coded by the gene to determine how it functions normally in the cell. This knowledge will contribute greatly, for example, to our understanding of how the immune system functions and how defective genes wreak havoc on the body.

Summary

Biotechnology is being expanded greatly by DNA technology. To achieve a genetically engineered cell, a vector first is prepared. Plasmids (and viruses) are used as vectors to carry a foreign gene into a cell. First, a restriction enzyme is used to cleave plasmid DNA and foreign DNA. Then, DNA is sealed into the vector DNA by DNA ligase. When the recombinant plasmid replicates or the virus reproduces, the foreign gene is cloned. The protein produced by the foreign gene also can be collected. This is the basis for the production of biotechnology products, such as hormones, DNA probes, and vaccines.

Transgenic organisms also have been made. Plants lend themselves to genetic manipulation because whole plants will grow from cultured cells. Plants resistant to pests and herbicides and containing complete protein soon will be available commercially. The possibility also exists of developing improved livestock through biotechnology.

Human gene therapy is being investigated. Researchers are currently interested in using lymphocytes and endothelial cells as delivery systems for genetic drugs. The U.S. government is committed to mapping and sequencing the entire human genome. It is hoped this information will assist human gene therapy.

Study Questions

1. Explain the recombinant DNA technology for cloning a gene. (p. 535)
2. List and explain other types of laboratory procedures used in biotechnology. (p. 536)
3. Categorize and give examples of types of biotechnology products available today. (pp. 537–39)
4. Naturally occurring bacteria have been bioengineered to perform what services? What are the ecological concerns regarding their release into the environment? (p. 540)
5. Why are plants good candidates for genetic engineering, and what are the problems involved in using this method to achieve agriculturally significant plants? (pp. 541–44)
6. Describe the current situation in regard to transgenic animals. (p. 544)
7. What types of problems have to be solved before gene therapy in humans becomes a reality? (p. 545)
8. What is the human genome project, and why might the project be useful? (p. 546)

Objective Questions

1. Plasmids and viruses can be _____ for carrying foreign DNA into host cells.
2. DNA probes seek out _____ DNA and bind with it.
3. _____ produced by recombinant DNA are not expected to cause infection.
4. _____ (monocots or dicots) take up plasmids.
5. Plants can be bioengineered to be resistant to _____ and _____ .
6. The bovine _____ gene has been microinjected into various types of livestock eggs.
7. _____ , not plasmids, are used commonly as vectors in research with human cells.
8. The human _____ project is being backed by the U.S. government.

Selected Key Terms

biotechnology (bi″o-tek-nol′o-je) the use of a natural biological system to produce a commercial product. *534*

cloned (klōnd) genes from an external source that have been reproduced by bacteria. *535*

DNA ligase (li′gās) an enzyme that links DNA fragments; used in genetic engineering to join foreign DNA to the vector DNA. *536*

DNA probe (prōb) known sequences of DNA that are used to find complementary DNA strands; can be used diagnostically to determine the presence of particular genes. *539*

gene sequencing (jēn se′kwens-ing) determination of the base sequence of a gene. *536*

gene therapy (jēn thēr-ah-pe) use of transplanted genes to overcome an inborn error in metabolism. *545*

genetic engineering (je-net′ik en″ji-nēr′ing) alteration of the genome of an organism by technological processes. *534*

genomic library (je-nom′ik li′brer-e) a collection of engineered viruses that together carry all of the genes of the species. *535*

plasmid (plaz′mid) a circular DNA segment that is present in bacterial cells but is not part of the bacterial chromosome. *535*

protoplast (pro′to-plast) naked cells that have had their cell walls removed; used in genetic engineering to allow a plasmid to enter a plant cell. *541*

recombinant DNA (re-kom′bi-nant) DNA having genes from 2 different organisms often produced in the laboratory by introducing foreign genes into a bacterial plasmid. *535*

restriction enzyme (re-strik′shun en′zīm) enzyme that stops viral reproduction by cutting viral DNA; used in genetic engineering to cut DNA at specific points. *536*

transformed (trans-formd′) cell that has been altered by genetic engineering and is capable of producing new protein. *534*

transgenic organism (trans-jen′ik orgah-nizm) a multicellular organism that has a foreign gene inserted in it. *541*

vector (vek′-tor) a carrier, such as a plasmid or a virus, for recombinant DNA that introduces a foreign gene into a host cell. *535*

Further Readings for Part Four

Antebi, E., and D. Fishlock. 1986. *Biotechnology: Strategies for life.* Cambridge, Mass.: The MIT Press.

Bishop, J.M. March 1982. Oncogenes. *Scientific American.*

Chambon, P. May 1981. Split genes. *Scientific American.*

Chilton, M. June 1983. A vector for introducing new genes into plants. *Scientific American.*

Cohen, S. N., and J. A. Shapiro. February 1980. Transposable genetic elements. *Scientific American.*

Darnell, J. E. October 1983. The processing of RNA. *Scientific American.*

Dickerson, R. E. December 1983. The DNA helix and how it is read. *Scientific American.*

Drlica, K. 1984. *Understanding DNA and gene cloning.* New York: John Wiley and Sons.

Elkington, J. 1985. *The gene factory: Inside the science and business of biotechnology.* New York: Carroll and Graf Publishers.

Feldman, M., and L. Eisenback. November 1988. What makes a tumor cell metastatic? *Scientific American.*

Glover, D. M. 1984. *Gene cloning: The mechanics of DNA manipulation.* New York: Chapman and Hall.

Guttmacher, Alan F. 1986. *Pregnancy, birth, and family planning.* New York: New American Library.

Kieffer, G. H. 1987. *Biotechnology, genetic engineering, and society.* Reston, Va.: National Association of Biology Teachers.

Nilsson, L. 1977. *A child is born.* Rev. ed. New York: Delacorte Press.

Patterson, D. August 1987. The causes of Down syndrome. *Scientific American.*

Ptashne, M. November 1982. A genetic switch in a bacterial virus. *Scientific American.*

———. January 1989. How gene activators work. *Scientific American.*

Ross, J. April 1989. The turnover of messenger RNA. *Scientific American.*

Shepard, J. F. May 1982. The regeneration of potato plants from leaf-cell protoplasts. *Scientific American.*

Tompkins, J. S., and C. Rieser. June 1986. Special report: Biotechnology. *Science Digest.*

Torrey, J. G. July–August 1985. The development of plant biotechnology. *American Scientist.*

Volpe, E. P. 1983. *Biology and human concerns.* 3d ed. Dubuque, Ia.: Wm. C. Brown Publishers.

Weinberg, R. A. November 1983. A molecular basis of cancer. *Scientific American.*

———. September 1988. Finding the anti-oncogene. *Scientific American.*

White, R., and J. Lalouel. February 1988. Chromosome mapping with DNA markers. *Scientific American.*

Part Five

Evolution and Diversity

Evolution depends on the retention of genetic changes that have been tested by the environment. This process, termed natural selection, results in adaptation to both the abiotic and biotic environment.

A gradual increase in chemical complexity produced the first cell(s) and this (these) evolved into all the forms of life we see about us. Taxonomists try to classify living things according to their evolutionary relationship; therefore, when we study taxonomy we are also studying evolutionary history. This text recognized five kingdoms: Monera (bacteria), Protists (protozoans and unicellular algae), Fungi (molds and mushrooms), Plants (multicellular algae and terrestrial plants), and Animals.

Humans are primates, animals adapted to living in trees. They share a common ancestor with apes, some of whom still live in trees. The first human ancestor may have left the trees when grasslands replaced trees in Africa. Walking erect could have been an adaptation to this change of habitat. Later, tool use and intelligence evolved together, and allowed humans to eventually take up a hunting way of life.

Culture, which began with tool use, soon also included art, science, and religion. Unfortunately, twentieth-century culture tends to make humans unaware of their natural place in the biosphere.

The evolution of silk production is best exemplified by spiders. The silk threads of this web are outlined by fog-formed droplets. Not all spiders build webs but they all make silk. They use it variously; for example, in construction of egg sacs, nursery tents, and drag lines.

Chapter 26

The Origin of Life

Chapter Concepts

1 The first cell or cells most likely arose by a slow process of chemical evolution.

2 It is generally accepted that the first cell was an anaerobic heterotroph.

3 Autotrophic nutrition made life on land possible by releasing free oxygen (O_2) into the atmosphere.

4 Although life may have originated by chemical evolution, this process does not occur today. Today, life comes only from life.

Chapter Outline

Chemical Evolution
    The Primitive Atmosphere
    Simple Organic Molecules
    Macromolecules
Biological Evolution
    Protocells
    True Cells

Today we do not believe that life arises spontaneously from nonlife; we say life comes only from life. But if this is so, how did the first form of life come about? We can assume that the first form of life was very simple—a single cell (or cells) that could grow, reproduce, and mutate. Because it was the very first living thing, the cell had to have come from nonliving chemicals. Perhaps a slow progression of chemicals from the simple to the complex finally resulted in a living cell. In other words, a **chemical evolution** produced the first form of life. Table 26.1 lists the steps that could have occurred to produce life. The evidence for these steps is based on our knowledge of the primitive earth and on experiments performed in the laboratory.

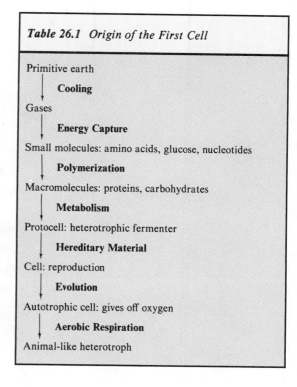

Table 26.1 Origin of the First Cell

Primitive earth
→ **Cooling**
Gases
→ **Energy Capture**
Small molecules: amino acids, glucose, nucleotides
→ **Polymerization**
Macromolecules: proteins, carbohydrates
→ **Metabolism**
Protocell: heterotrophic fermenter
→ **Hereditary Material**
Cell: reproduction
→ **Evolution**
Autotrophic cell: gives off oxygen
→ **Aerobic Respiration**
Animal-like heterotroph

Chemical Evolution

The sun and the planets probably formed from aggregates of dust particles and debris about 4.6 billion years ago. Intense heat produced by gravitational energy and radioactivity of some atoms caused the earth to become stratified into a core, a mantle, and a crust. Heavier atoms of iron and nickel became the molten liquid core, and dense silicate minerals became the semiliquid mantle. The lighter molecules of silicon, aluminum, and iron formed the crust.

The temperature was so hot that atoms could not permanently bind; whenever bonds formed, they were quickly broken. Cooling had to occur before an atmosphere could develop.

The Primitive Atmosphere

The gases of the **primitive atmosphere** were not the same as those of today's atmosphere. Originally, it was proposed that the earliest atmosphere contained a lot of hydrogen (H_2) because it is the most abundant element in our solar system. Later, it was suggested that lightweight atoms, including hydrogen, would have been lost as the earth formed because the earth's gravitational field was not strong enough to hold them. Now it is thought that the primitive atmosphere was produced by outgassing from the interior, particularly by volcanic action, after the earth formed. If this was the case, the atmosphere consisted mostly of water vapor (H_2O), nitrogen (N_2), and carbon dioxide (CO_2), with only small amounts of hydrogen and carbon monoxide (CO). The primitive atmosphere, with little if any free oxygen, was a **reducing atmosphere,** as opposed to the **oxidizing atmosphere** of today. This was fortuitous because oxygen (O_2) attaches to organic molecules, preventing them from joining to form larger molecules.

The primitive atmosphere was a reducing atmosphere that contained little if any oxygen.

At first, the earth was so hot that water was present only as a vapor that formed dense, thick clouds. Then, as the earth cooled, water vapor condensed to liquid water, and rain began to fall (fig. 26.1a). It rained in such quantity that the oceans of the world were produced.

Simple Organic Molecules

The atmospheric gases, dissolved in rain, were carried down into newly forming oceans. The remaining steps shown in table 26.1 took place in and about the sea, where life is believed to have originated. The dissolved gases, although relatively inert, are believed to have reacted with one another to form simple organic compounds when they were exposed to the strong outside **energy sources** present on the primitive earth. These energy sources included heat from volcanoes and meteorites, radioactivity from the earth's crust, powerful electric discharges in lightning, and solar radiation, especially ultraviolet radiation (fig. 26.1b). In a classic experiment (fig. 26.2), Stanley Miller showed that an atmosphere containing methane (CH_4), ammonia (NH_3), and hydrogen (H_2)

The Origin of Life

Figure 26.1
A model for the origin of life.

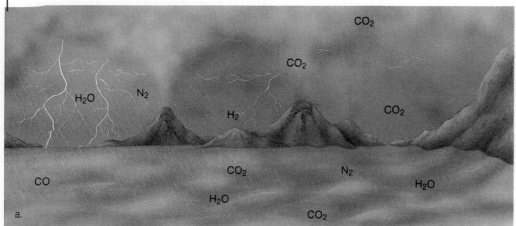

a. The primitive atmosphere contained gases, including water vapor that escaped from volcanoes; as the latter cooled, some gases were washed into the oceans by rain.

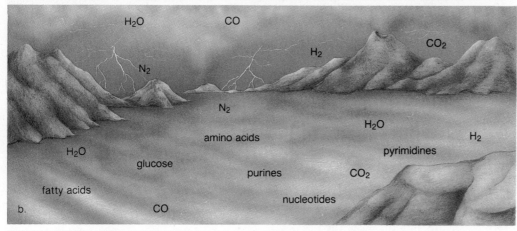

b. The availability of energy from volcanic eruption (shown here) and lightning allowed gases to form simple organic molecules.

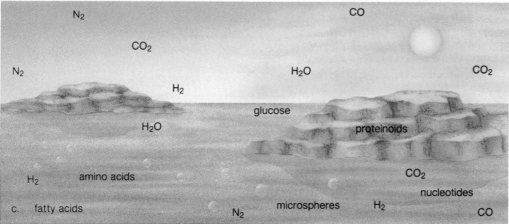

c. Amino acids splashed up onto rocky coasts could have polymerized to give proteinoids that would have become microspheres when they reentered the water.

d. Eventually various types of prokaryotes and then eukaryotes evolved. Some of the prokaryotes were oxygen-producing photosynthesizers. The presence of oxygen in the atmosphere was needed for aerobic respiration to evolve.

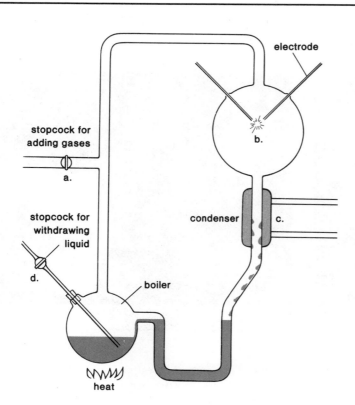

Figure 26.2
Origin of small organic molecules. In Miller's experiment, gases were (a) admitted to the apparatus, (b) circulated past an energy source (electric spark), and (c) cooled to produce (d) a liquid that could be withdrawn. Upon chemical analysis, the liquid was found to contain various simple organic molecules.

could have produced organic molecules. These gases were dissolved in water and were circulated in a closed system past an electric spark. After a week's run, Miller discovered that a variety of amino acids and other organic acids had been produced. Since that time, other investigators have achieved similar results utilizing other combinations of gases dissolved in water.

These experiments indicate that the primitive gases not only could but probably did react with one another to produce simple organic compounds that accumulated in the oceans for hundreds of millions of years. With the accumulation of these simple organic compounds, the oceans became a thick, hot, **organic soup** containing a variety of organic molecules.

Cooling caused water vapor to turn to rain that formed the oceans. Here, as Miller and others have shown, atmospheric gases could have reacted with one another under the influence of an outside energy source to produce simple organic molecules.

Macromolecules

The newly formed organic molecules combined to form still larger molecules and macromolecules (fig. 26.1*c*). Sidney Fox of the University of Miami has shown that amino acids polymerize abiotically when exposed to dry heat. Further, these so-called **proteinoids** contain amino acids that join in a preferred manner. He suggests that amino acids collected in shallow puddles along the rocky shore and the heat of the sun caused proteinoids to form as drying took place. The idea that proteins were the first cellular macromolecules to arise is attractive because proteins are enzymes that do the work of the cell and allow it to grow. When proteinoids are exposed to water, they form **microspheres.**

Other investigators believe that because nucleic acids store genetic information, they were necessary for cells to arise. Up until recently, this suggestion had little appeal because it was believed that nucleic acids had no enzymatic ability. However, it now has been shown that when introns are removed from mRNA in the nucleus (p. 518), a form of RNA called catalytic

26.1 Critical Thinking

1. Why did Fox show that polymerization of amino acids can occur without enzymes?
2. Why do investigators believe that proteins alone could not form a cell? What property do they lack?
3. By what mechanism would it be possible for RNA to replicate in the first cell?
4. Why was it necessary for this RNA to have enzymatic properties?

Figure 26.3

Protocells. Protocells may have been (a) proteinoid microspheres, which have a variety of cellular characteristics (table 26.2) or (b) liposomes, which automatically form when phospholipid molecules are placed in water.

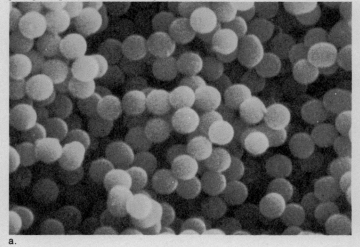
a.

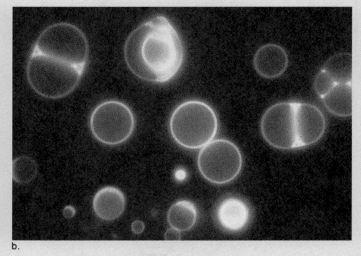
b.

Table 26.2 Similar Properties of Proteinoid Microspheres and Today's Cells

Stability (standing, centrifugation, sectioning)
Microscopic size
Variability in shape but uniformity in size
Numerousness
Stainability
Ultrastructure (electron microscope)
Double-layered boundary
Differential passage of molecules through boundary
Catalytic activities
Patterns of association
Propagation by "budding" and fission
Growth by acquiring materials
Motility
Propensity to form junctions and to communicate

From Sidney W. Fox, "Chemical Origins of Cells" in *Chemical & Engineering News,* 49, 50, December 6, 1971. Copyright © 1971 American Chemical Society. Reprinted by permission.

RNA does the splicing. Therefore, it now appears that RNA could have been the first cellular macromolecule to arise because RNA molecules could have allowed the first cell to both grow and reproduce without the help of either proteins or DNA.

Biological Evolution

Biological evolution began whenever cells began to compete with one another for limited resources. The most successful of these cells reproduced at the expense of the other types of cells.

Protocells

The first cells may have been **protocells,** structures that carry on certain cellular functions but are not true cells. For example, Fox has shown that proteinoid microspheres (fig. 26.3*a*), have properties similar to today's cells (table 26.2), including a membrane that is differentially permeable. If lipids are made available to microspheres, they tend to associate with them. The outer boundary, then, is a lipid-protein membrane. Even so, it is also possible that a lipid membrane formed first and that the protocell developed afterward. Phospholipid molecules automatically form droplets called **liposomes** in a liquid environment (fig. 26.3*b*). Perhaps the first membrane formed in this manner, rather than by way of microspheres.

Some researchers support the work of Soviet chemist A. I. Oparin, who was one of the first researchers on the origin of life. As early as 1938, Oparin showed that under appropriate conditions of temperature, ionic composition, and pH, concentrated mixtures of macromolecules tend to give rise to complex units called **coacervate droplets.** Coacervate droplets have a tendency to absorb and to incorporate various substances from the surrounding solution. Eventually, a semipermeable type of boundary may form about the droplet.

The Heterotroph Hypothesis

We might now ask how the protocell acquired nutrients for growth. Nutrition was not a problem because the protocell existed in the oceans, which at that time were an organic soup containing simple organic molecules that served as sources of building blocks and energy. Therefore, the protocell was a **heterotroph,** an organism that takes in preformed food. Notice that this theory suggests that heterotrophs preceded **autotrophs,** organisms that make their own organic food.

At first, the protocell may have used preformed ATP for its energy needs, but as this supply of ATP dwindled, natural selection favored any cells that could extract energy from carbohydrates in order to transform ADP to ATP. Glycolysis is a common metabolic pathway in living things, and this testifies to its early evolution in the history of life. Since there was no free oxygen, we can assume that the protocell was anaerobic and carried on a form of fermentation:

2 ADP   2 ATP

glucose → alcohol and/or acids + CO_2 + energy

It seems logical that the protocell at first had limited ability to break down organic molecules and that it took millions of years for glycolysis to evolve completely. It is of interest that Fox showed that microspheres have some catalytic activities, and others found that coacervate droplets incorporate enzymes if they are available in the medium.

The protocell could have been a simple sphere that contained polypeptides having some degree of enzymatic ability. The protocell most likely was a heterotrophic fermenter.

True Cells

No matter whether catalytic RNA or the protocell came first, eventually a true cell evolved. A true cell not only has cytoplasm, it also contains genetic material surrounded by a cell membrane. Not only is a true cell capable of growing, it also is capable of reproduction.

If the first genes were composed of RNA, how did DNA come into the picture? Viruses that have RNA genes use their RNA as a template to form DNA. Perhaps with time, reverse transcription of this type occurred within the protocell, and this is how DNA genes arose.

True cells have cytoplasm, genetic material, and a cell membrane. They are capable of growing and reproducing.

Fossils of Primitive Cells

There is no recorded history for evolution prior to the true cell; the earliest fossils (about 3.5 billion years old) are presumed to have been prokaryotic cells. They are found in fossilized stromatolites (*stroma* = bed and *lithos* = stone), pillarlike structures composed of sedimentary layers containing communities of photosynthetic microorganisms (fig. 26.4). Living stromatolites exist even today in shallow waters off the west coast of Australia.

Autotrophs

The autotrophs found in ancient stromatolites are believed to have been anaerobic photosynthesizers that did not give off oxygen (O_2). They most likely prospered because once the supply of organic molecules in the organic soup began to decline, cells that could use solar energy to produce carbohydrates had an advantage. A billion years later, oxygen-producing photosynthesizers (cyanobacteria) evolved. These autotrophs, which make their own food using carbon dioxide (CO_2) and water (H_2O), release oxygen as a by-product.

With the evolution of oxygen-producing photosynthesizers (fig. 26.1*d*), the atmosphere became an oxidizing one instead of a reducing one. This is an example of the profound effect that life has on the physical conditions of the earth. As discussed in the reading on page 558, there are those who have been

Figure 26.4
The oldest prokaryotic fossils. a. This filamentous prokaryotic fossil was found in (b) layers of an ancient stromatolite. c. Living stromatolites are located in shallow waters off the shores of western Australia.

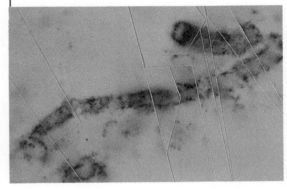
a.

b.

c.

The Origin of Life

The Gaia hypothesis (after the classical Greek word for our Mother Earth) maintains that life keeps the climate and the chemical composition of the earth suitable for life. For example, photosynthesizing organisms put oxygen (O_2) into the atmosphere, and its level has been maintained at about 20% for the last half-billion years. If, by chance, this level rises to 30%, there would be disastrous fires any time lightning occurred. What

The Gaia Hypothesis

keeps the level so constant? Those who support the Gaia hypothesis point to the presence of bacteria that produce methane (CH_4), which combines with oxygen to form carbon dioxide (CO_2). It is these bacteria, they say, that keep atmospheric levels of oxygen suitable for all other living things.

To take another example, the sun has been heating up since its formation, yet the temperature of the earth always has been about the same. About 4 billion years ago, when the sun was perhaps 25% less luminous, the primitive atmosphere contained more ammonia (NH_3) and methane, which acted like the glass of a greenhouse to trap the heat of the sun near the earth. Today, carbon dioxide put into the atmosphere by all living things plays a more prominent role in creating a greenhouse effect. As the sun heats up, less and less of a greenhouse effect is needed to maintain a suitable temperature for life. How has this been regulated up until now? Those who support the Gaia hypothesis point out that there are microorganisms in the oceans with calcium carbonate ($CaCO_3$) shells (see figure). When carbon dioxide enters an ocean, it become bicarbonate and is utilized by these organisms to form their shell. When these organisms die, their shell sinks to the bottom of the ocean; in this way, carbon dioxide in effect is removed from the atmosphere.

While many biologists agree that living things play a role in maintaining suitable conditions for life on earth, they doubt there is anything purposeful about it. Quite often those who support the Gaia hypothesis see the earth as a giant geophysiological organism that is purposefully maintaining itself. In this way, the Gaia hypothesis seems more like a religion than a science, and therefore it would be difficult to devise ways to test the hypothesis.

Those who refute the Gaia hypothesis point out that inorganic processes also play a role in maintaining the chemical composition and the temperature of the earth. For example, oxygen no doubt failed to quickly build up in the atmosphere because much of it was taken up as iron was oxidized—there is much iron in the earth's crust. Also, carbon dioxide is washed out of the atmosphere by rain and in the process becomes carbonic acid. Carbonic acid combines with calcium silicate in rocks to give sediments containing carbon compounds. In other words, inorganic and organic processes have contributed to keeping the earth suitable for life.

Another argument against the Gaia hypothesis is that while some life forms have benefited from changed conditions, other forms have not. For example, when oxygen was put into the atmosphere, it may have helped aerobic organisms to evolve, but the populations of anaerobic organisms could decline only. The history of life contains many species of organisms that became extinct due to changing conditions. New ones that are more suited to the present conditions then evolve.

The Gaia hypothesis challenges us to see the earth at a geophysiological level of organization. Life is not passively acted upon by physical processes; it has contributed to changing and to shaping the world we now live in.

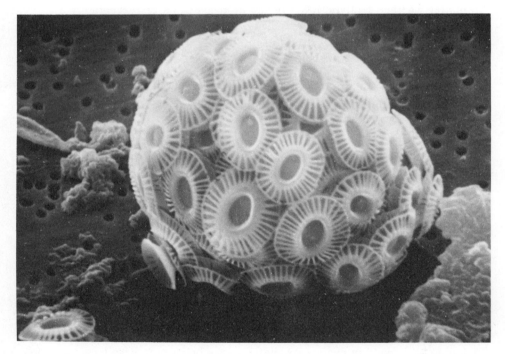

Coccolithophore. *This microorganism has calcium carbonate, ($CaCO_3$) plates embedded in its cell wall and exemplifies marine organisms that take calcium carbonate out of the sea. Calcium carbonate forms after carbon dioxide (CO_2) enters from the air; therefore, this action helps to decrease the amount of carbon dioxide in the atmosphere. The sea is full of these tiny organisms that are only 60 μm in diameter (a human hair is 100-200 μm thick).*

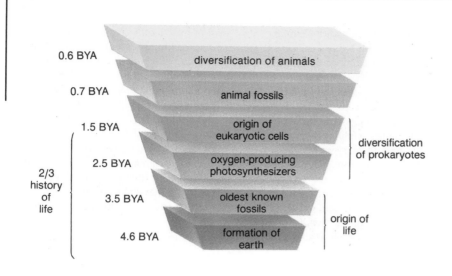

Figure 26.5
Some of the major events that occurred in the early history of the earth. Two-thirds of the history of the earth involved the origin and diversification of prokaryotic cells.
(BYA = billions of years ago)

so struck with this realization they have formulated a new hypothesis. The Gaia hypothesis states that living things function in such a way as to preserve the earth for life.

Aerobic Respiration

The presence of oxygen (O_2) in the atmosphere meant that most environments were no longer suitable for **anaerobic organisms,** and they began to decline in importance. The photosynthetic cyanobacteria proliferated, and eventually the atmosphere contained a stable amount of oxygen. Still, it was another billion years before **aerobic respiration** and the eukaryotic cell evolved. As figure 26.5 shows, a large portion of the history of life was devoted to prokaryotic evolution, during which time various metabolic pathways came into existence.

The presence of oxygen in the atmosphere eventually permitted life forms, such as plants and animals, to invade the land. Oxygen in the upper atmosphere forms ozone (O_3), which filters out the ultraviolet rays of the sun. Before the formation of this so-called **ozone shield,** the amount of radiation striking the earth would have destroyed land-dwelling organisms.

The evolution of autotrophs caused oxygen to enter the atmosphere. Oxygen allows aerobic respiration to occur and forms an ozone shield that protects the earth from ultraviolet radiation.

We have shown that life could have come into existence by means of chemical evolution. However, we do not believe that this same chemical evolution is occurring today for the following reasons.

1. Certain energy sources, particularly ultraviolet radiation, are unavailable. The ozone layer now acts as a shield to prevent these rays from reaching the earth in great quantity.
2. While the first atmosphere is believed to have been a reducing one that therefore promoted the buildup of organic molecules, today's atmosphere is an oxidizing atmosphere that tends to break down organic molecules.
3. Living organisms present today use any newly formed organic molecules for food.

Today, life is believed to come only from life because if any organic molecules did form, they would be oxidized by oxygen or utilized by preexisting life.

Summary

The gases of the primitive atmosphere probably included hydrogen (H_2), methane (CH_4), water (H_2O) vapor, and ammonia (NH_3). There was no free oxygen (O_2). As the newly formed earth cooled, water vapor-turned-to-rain produced the oceans in which the gases were dissolved. In the oceans, gases reacted with one another under the influence of an outside energy source (lightning or ultraviolet radiation) to give small organic molecules. These polymerized to produce large organic molecules similar to proteins or nucleic acids. In the laboratory, amino acid polymers form spheres resembling cell-like bodies called proteinoid microspheres, which have many features in common with today's cells. Although proteins can carry on enzymatic reactions, they cannot store genetic information. Recently, it was shown that some forms of RNA have enzymatic properties. This lends support to the idea that RNA was the first cellular molecule with the ability to both carry on metabolism and store genetic information.

A precellular stage called protocells was anaerobic and carried on heterotrophic nutrition. Eventually, a true cell, capable of both metabolism and reproduction, came into existence. Biological evolution began when primitive cells competed with one another for limited resources. Eventually, autotrophs, which make their own food, were anaerobic and did not release oxygen. Later, photosynthesizers that split water to release oxygen arose. This oxygen formed an ozone shield that permitted plants and animals to live on the land.

Today, life is believed to come only from life because an appropriate energy source is lacking to cause the formation of organic molecules. If they did form, they would be oxidized by oxygen or eaten by preexisting life.

Study Questions

1. What was the primitive earth like when it first formed? (p. 553)
2. What gases are believed to have been in the primitive atmosphere? (p. 553).
3. Under what conditions was it possible for gases to react with one another to produce small organic molecules? (p. 553)
4. Describe an experiment that shows that small organic molecules can form from primitive gases. (pp. 553–55)
5. What evidence is there that a protein may have been the first cellular molecule? What evidence is there that RNA may have been the first cellular molecule? (pp. 555–56)
6. What types of respiration and nutrition would a protocell have had? (pp. 556–57)
7. The first true cell must have had which structural and physiological properties? (p. 557)
8. Describe the earliest fossils (p. 557). Aerobic autotrophs put what gas into the atmosphere? (p. 557)
9. Why does life not arise by chemical evolution today? (p. 559)

Objective Questions

1. An _____ evolution is believed to have preceded an organic evolution.
2. Miller's experiment showed that the first gases could have reacted together to form _____ .
3. Amino acid polymerization could have produced cell-like bodies called _____ .
4. The protocell carried on _____ nutrition and fed on the organic material in the oceans.
5. The protocell was a(n) _____ organism because there was no free oxygen in the atmosphere.
6. Once the protocell could grow and _____ , it was a true cell.
7. The earliest fossils found so far have been _____ , organisms capable of making their own food.
8. The presence of oxygen in the atmosphere permitted _____ respiration to evolve.
9. Today, we believe that life comes only from _____ .

Answers to Objective Questions

1. chemical 2. small organic molecules 3. proteinoid microspheres 4. heterotrophic 5. anaerobic 6. reproduce 7. autotrophs 8. aerobic 9. life

aerobic respiration (a″er-ōb′ik res″pi-ra′shun) a cellular process by which molecules are oxidized with the release of energy. *559*

anaerobic organism (an-a″er-ob′ik or-gă-niz-ŭm) organism that acquires energy by processes that do not utilize oxygen. *559*

autotroph (aw′to-trōf) an organism that is capable of making its food (organic molecules) from inorganic molecules. *557*

biological evolution (bi-o-loj′ĕ-kal ev″o-lu′shun) changes that have occurred in life forms from the origination of the first cell or cells to the many diverse forms in existence today. *556*

chemical evolution (kem′i-kal ev″o-lu′shun) a gradual increase in the complexity of chemical compounds that is believed to have brought about the origination of the first cell or cells. *553*

coacervate droplet (ko-as′er-vāt drŏp-let) a mixture of polymers that may have preceded the origination of the first cell or cells. *556*

energy source (en′er-je sor′se) way by which energy can be made available to organisms from the environment. *553*

heterotroph (het′er-o-trof″) an organism that takes in preformed food. *557*

microsphere (mi-kro-sfēr) structure composed only of protein that looks like a cell and carries on many cellular functions; a possible early step in cell evolution. *555*

organic soup (or-gan′ik sōop) an expression used to refer to the oceans before the origin of life when they contained newly formed organic compounds. *555*

oxidizing atmosphere (ok′si-dīz-ing at′mos-fēr) an atmosphere that contains oxidizing molecules, such as O_2, rather than reducing molecules, such as H_2. *553*

ozone shield (o′zōn shēld) a layer of O_3 present in the upper atmosphere that protects the earth from damaging ultraviolet light. *559*

primitive atmosphere (prim′i-tiv at′mos-fēr) the gases that were in the atmosphere when the earth first arose. *553*

protocell (pro′to-sel) the structure that preceded the true cell in the history of life. *556*

reducing atmosphere (re-dūs′ing at′mos-fēr) an atmosphere that contains reducing molecules, such as H_2, rather than oxidizing molecules, such as O_2. *553*

Chapter 27

Evolution

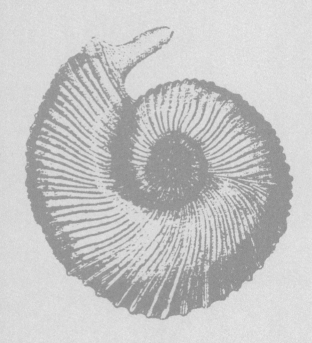

Chapter Concepts

1 Life evolved from the first cell(s) into all the forms of life, present or extinct.

2 Many fields of biology provide evidence that evolution has occurred.

3 Evolution explains both the unity (sameness) and the diversity of life.

4 Natural selection occurs when the better-adapted members of a population reproduce to a greater degree than the less well-adapted members.

5 New species come about when a population is first geographically isolated and later reproductively isolated from other similar populations.

Chapter Outline

Evidences for Evolution
    The Fossil Record
    Comparative Anatomy
    Comparative Embryology
    Vestigial Structures
    Comparative Biochemistry
    Biogeography
The Evolutionary Process
    Population Genetics
    The Synthetic Theory
    Examples of Natural Selection
Speciation
    The Process of Speciation
    Adaptive Radiation

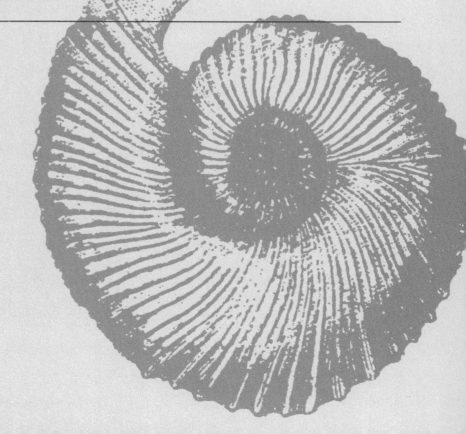

a.

b.

Figure 27.1
Fossils. a. *Impression of fern leaves on a rock.*
b. *Small dinosaur fossil bones being exposed by chipping away rocks holding them.*

 Evolution is the process that explains the unity and the diversity of life. Data from various fields of biology give us evidences that all living things are descended from a common source and evolution produced the myriad of organisms now present on earth.

Evidences for Evolution

The Fossil Record

Our knowledge of the history of life is based primarily on the fossil record. **Fossils** (fig. 27.1) are the remains or the evidences of some organism that lived long ago. Most fossils are formed when an organism is buried in mud or sand before the hard mineralized parts have decayed. A fossil can be the remains of this part or it can be the impression, or mold, that the part made in the rock developing about it. Sometimes, fossils are formed by a replacement of the original organic material by a durable mineral, such as silica.

The earth has recognizable layers, or strata, and each layer has its own mix of fossils. The boundaries between the strata, where one mix of fossils gives way to another, provide the basis for dividing geological time as shown in table 27.1. The strata alone cannot provide dates for fossil deposits; they only can tell us the sequence of the organisms over time. Dates can be calculated, however, from a study of the radioactive isotopes in the rocks of each

Table 27.1 The Geological Time Scale

Era	Period	Millions of Years Ago	Plant Life	Animal Life
Cenozoic	Quaternary	2.5	Increase in number of herbaceous plants	Age of Human Civilization
	Tertiary	65	Dominance of land by flowering plants	First hominids appear Dominance of land by mammals and insects
Mesozoic	Cretaceous	130	Flowering plants spread Cone-bearing trees decline	Dinosaurs become extinct
	Jurassic	180	First flowering plants appear	First mammals and birds appear Age of Dinosaurs
	Triassic	230	Dominance of land by ferns and cone-bearing trees	First dinosaurs appear
Paleozoic	Permian	280	First seed plants	Expansion of reptiles Decline of amphibians
	Carboniferous	350	Age of great coal-forming forests, including club mosses, horsetails, and ferns	First reptiles appear Age of Amphibians
	Devonian	400	Low-lying primitive, vascular plants appear on land	First amphibians move onto land First insects appear Age of Fishes
	Silurian	435		
	Ordovician	500	Unicellular marine algae abundant	First fishes (jawless) appear Age of Invertebrates
	Cambrian	600		
Precambrian (Proterozoic)		700 1,000 2,500	Origin of nonshelled invertebrates Origin of complex (eukaryotic) cells	
Archeozoic		3,500 4,600	Oldest fossils (prokaryotic cells)	
			Formation of earth	

stratum. Radioactive isotopes decay at a constant rate; therefore, by determining how much decay has taken place since the rocks first formed, a date can be assigned to the rocks.

The data gathered so far lead us to believe that the solar system is 4.6 billion years old. A large portion of the history of the earth was devoted to the evolution of prokaryotes; eukaryotic cells did not evolve until about 1.5 billion years ago. It was during the time of prokaryotic diversification—between 3.5 and 1.5 billion years ago—that most of the metabolic pathways evolved. Single-celled organisms remained in the oceans and increased in complexity, until multicellular organisms evolved—about 650 million years ago. Fossils of multicellular organisms are much easier to detect than those of simpler ones; therefore, the history of life is better documented from this time on (table 27.1).

Plants invaded the land environment about 400 million years ago, and they were followed by fungi, invertebrates, and finally vertebrates. The Age of Reptiles lasted from 300 million to 65 million years ago, when dinosaurs died out. Mammals originated about 150 million years ago, but it was not until the dinosaurs had vanished that they became abundant. Direct ancestors to humans do not appear until about 3 million years ago, which is very recent compared to how long life has been evolving. In fact, if the history of the earth is measured using a 24-hour time scale that starts at midnight, humans do not appear until one-half minute before the next midnight (fig. 27.2).

Evolution and Diversity

Figure 27.2

History of life on earth. The outer ring of this diagram shows the history of the earth as measured on a 24-hour time scale starting at midnight. (The inner ring shows the actual years starting at 4.6 billion years.) If the history of the earth is measured as if it had all happened in 24 hours, the Cambrian Period would not start until 8 P.M.! This means that a very large portion of life's history is devoted to the evolution of single-celled organisms. The first multicellular organisms do not appear until just after 8 P.M., and humans are not on the scene until less than a minute before midnight.

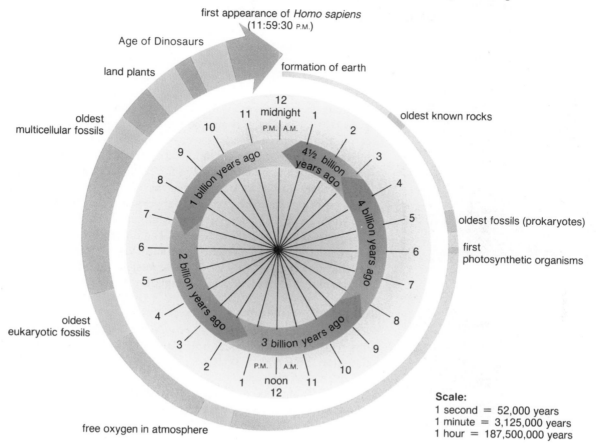

first appearance of *Homo sapiens* (11:59:30 P.M.)

Age of Dinosaurs

land plants

formation of earth

oldest multicellular fossils

oldest known rocks

oldest fossils (prokaryotes)

first photosynthetic organisms

oldest eukaryotic fossils

free oxygen in atmosphere

Scale:
1 second = 52,000 years
1 minute = 3,125,000 years
1 hour = 187,500,000 years

The Pace of Evolution

Most evolutionists assume that evolution occurs gradually and that a single species changes slowly—over millions of years—to give rise to new species. This *gradualistic model* of evolutionary change implies that the fossil record contains a plentiful supply of intermediate forms, which are fossil remains having characteristics of 2 different groups (fig. 27.3).

In contrast to the gradualistic model, a few paleontologists (those who study fossils) recently proposed a *punctuated equilibrium model* of evolutionary change. The fossil record shows instances in which species tend to remain the same for a long time, and then suddenly, new species appear. In other words, an *equilibrium* phase (long periods without change) is *punctuated* by a rapid burst of change during which some species become extinct and new species arise. According to the punctuated equilibrium model, there are few intermediate forms in the fossil record because such forms exist only during the period of rapid change before a quiet period returns. These paleontologists are in the process of gathering data to support their hypothesis.

The fossil record traces in broad terms the history of life and more specifically allows us to study the history of particular groups.

Figure 27.3

Intermediate forms. a. *An artist's conception of* Archaeopteryx. *This fossil had features common to both reptiles and birds. Note the indication of feathers, a birdlike feature, and the long bony tail, a reptilian feature.* b. Peripatus. *This animal has features common to both annelids and arthropods. It is obviously a segmented animal; its excretory, reproductive, and nervous systems are similar to those of the annelids, while its circulatory and respiratory systems are similar to those of the arthropods.*

a.

b.

Extinctions

The fossil record indicates that on the average, 2–5 families, each containing many species, became extinct every million years. These extinctions, termed *background extinctions,* are explained on the basis of an inability to adapt to changing environmental conditions. The fossil record also gives evidences of periods of *mass extinctions,* when about 20 families became extinct per million years.

The mass extinction that occurred at the end of the Cretaceous Period is well known as the time when the dinosaurs disappeared. Geologists found that there is an abnormally high level of iridium in clay of the late Cretaceous Period and speculated that this was caused by a worldwide fallout of radioactive material created by a comet or an asteroid impact at the time of the dinosaur disappearance. While this may seem to be an absurd idea, evidence suggests that it very well may be the case: (1) paleontologists found evidence of other such mass extinctions every 26 million years or so and (2) astronomers tell us that the movement of our solar system in the Milky Way possibly could cause the earth to be subjected to such regular but infrequent bombardments.

Others do not look to the heavens for the causes of mass extinctions. They point out that the continents always have not been as they are now. At one time, there was only one huge continent called Pangaea. Pangaea eventually broke into 2 major continents, and during the Cretaceous Period, these fragmented into the 7 present-day continents. Even now, these continents are drifting slowly apart (fig. 27.4). It is possible that changes in climate due to continental drift are sufficient to explain mass extinctions.

Comparative Anatomy

Related species share a *unity of plan.* For example, reproductive organs of all flowering plants are basically similar and all vertebrate animals have essentially the same type of skeleton.

Unity of plan allows organisms to be classified into taxonomic categories (table 27.2) and makes it possible to construct **evolutionary trees** (fig. 27.5). These trees tell us how the members of a group are believed to be related and how their history can be traced by way of common ancestors. **Common ancestors** are found at the points of divergence in an evolutionary tree. In the diversity chapters that follow, we have occasion to construct evolutionary trees showing the presumed relationships of major groups of organisms.

Organisms that share a unity of plan exhibit **homologous structures,** similarities in structure that have arisen through descent from a common ancestor.

Figure 27.4

Continental drift. The continents drift because they are on mammoth pieces of crust that are formed at midoceanic ridges and disappear at oceanic trenches. Continental drift possibly contributed to the mass extinctions seen in the fossil record.

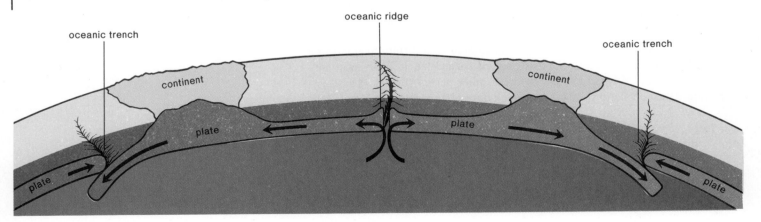

Figure 27.5

Evolutionary tree of certain primate species based on a
biochemical study of their genes. The length of the branches
indicates the DNA difference found between groups. With the
help of the fossil record, it is possible to suggest a date at which
each group diverged from the other.

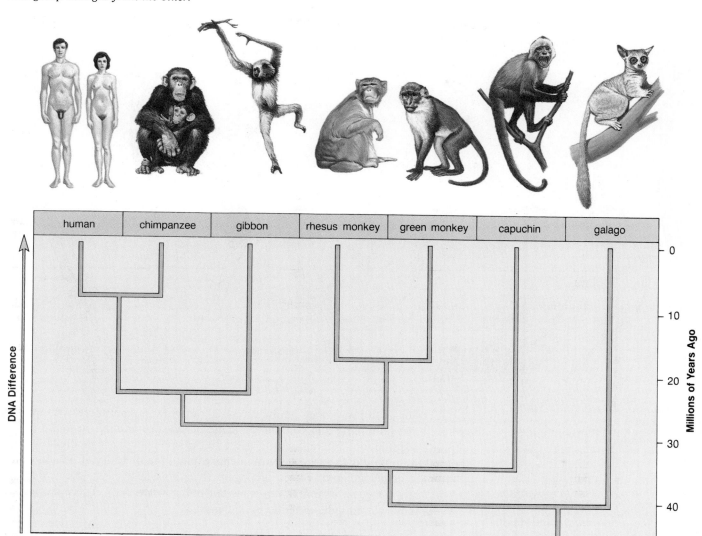

Table 27.2 *Hierarchy of Classification*

Category	Description		
Species	A type of organism distinguishable from all other types		
Genus	Contains related species		
Family	Contains related genera	more characteristics in common	characteristics in common
Order	Contains related families		
Class	Contains related orders		
Phylum (animals) Division (plants)	Contains related classes		are more distinctive
Kingdom	Contains related phyla (divisions)		

Figure 27.6

Homologous structures. The bones are color coded so that you can note the similarity in the bones of the forelimbs of these vertebrates. This similarity is to be expected; all vertebrates trace their ancestry to a common ancestor.

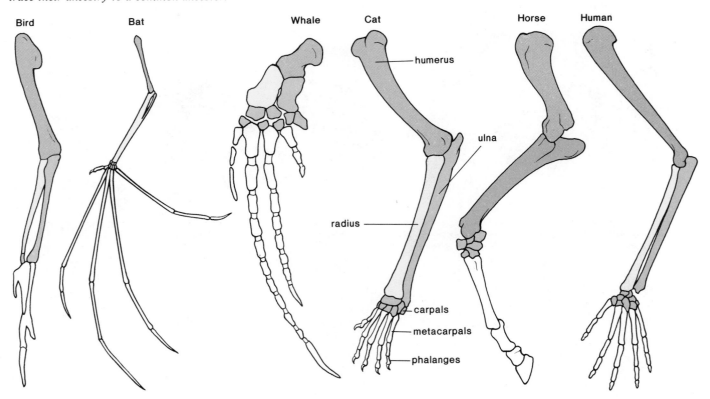

For example, among adult terrestrial vertebrates, their forelimbs are organized similarly and contain the same bones. The animals in figure 27.6 are adapted to a different way of life, and the same bones are modified variously. Adaptations to various environments explain the diversity of life. In contrast to homologous structures, **analogous structures,** such as an insect wing and a bird wing, have similar functions but differ in anatomy. Therefore we know they evolved independently of one another.

Organisms that are descended from a common ancestor share a unity of plan that indicates they are related closely. Comparative anatomy helps investigators to classify organisms and to construct evolutionary trees showing the presumed relationships of major groups of organisms.

Comparative Embryology

Some groups of organisms share the same types of embryonic stages. For example, the larva of certain lower chordates (the phylum containing vertebrates) is strikingly similar to that of certain echinoderms (e.g., the starfish). As expected, the embryonic stages of all vertebrates are also similar (fig. 27.7). Therefore, during development, a human embryo at one point has gill clefts, even though it never will breathe by means of gills as do fishes, and a rudimentary tail, even though it never will have a long tail as do some four-legged vertebrates. In this way, embryological observations indicate evolutionary relationships.

Vestigial Structures

An organism can have structures that are underdeveloped and seemingly useless, but these same structures can be fully developed and functional in related

Figure 27.7

Figure 27.7

A chick embryo and a pig embryo at comparable stages have many features in common. a. Chick embryo. b. Pig embryo.

a.

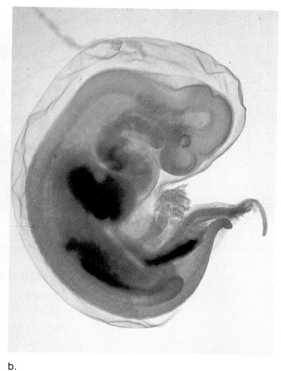

b.

organisms. These are called **vestigial structures.** Some vestigial structures in humans are remnants of a nictitating membrane, which in birds can be drawn over the eye, and the caudal vertebrae, which are more fully developed and functional in mammals that have tails. Figure 27.8 shows these and other vestigial structures in humans. The presence of these structures is understandable when we realize that related organisms share common genes.

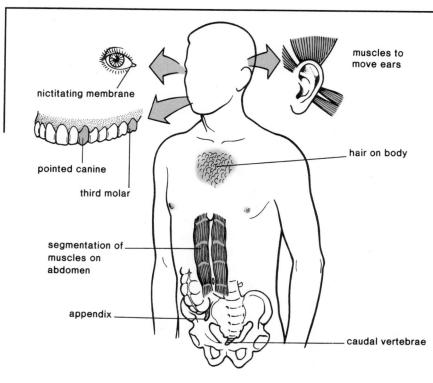

Figure 27.8

Vestigial structures. Human beings have various vestigial structures, such as those shown. These indicate our relationship to animals in which these structures are fully developed and functional.

muscles to
move ears

nictitating membrane

hair on body

pointed canine

third molar

segmentation of
muscles on
abdomen

appendix

caudal vertebrae

Evolution   **569**

Figure 27.9

Biogeography. Unrelated plants and animals are adapted similarly when the environments are similar. Cacti in North American deserts (a) and spurges in African deserts (b) both are adapted to arid environments. Coyotes in North American grasslands (c) and jackals in African grasslands (d) both are carnivorous predators that prey on small animals and eat carrion.

a.   b.   c.   d.

Comparative Biochemistry

Almost all living organisms use the same basic biochemical molecules, including DNA, ATP, and many identical or nearly identical enzymes. It seems that these molecules evolved very early in the history of life and have been passed on ever since.

Amino acid sequences in certain proteins, like hemoglobin and cytochrome c, have been determined for various animals, as have DNA nucleotide differences between animals. When 2 animals first diverge from a common ancestor, the genes and the proteins of the organisms involved are nearly identical, but as time passes, each accumulates individual changes in gene and protein structure. Therefore, the degree of difference in these molecules provides some indication of how long ago they diverged from a common ancestor. In other words, DNA and amino acid differences can be used as a kind of *molecular clock* to indicate evolutionary time. Figure 27.5 shows the results of one such study of DNA differences. Investigators found that evolutionary trees based on biochemical data are quite similar to those based on the fossil record. Whenever the same conclusions are drawn from independent data, they substantiate scientific theory—in this case, evolution.

Biochemistry provides evidences for the evolution of life from a common source; it also can be used to construct evolutionary trees. These trees agree in general with those based on the fossil record and on anatomical data.

Biogeography

In chapter 34, we have an opportunity to study the geographic distribution of plants and animals. It is observed that similar but geographically separate environments have different plants and animals that are adapted similarly. For example, plants of the cactus family are found in the deserts of southwestern North America, while members of the spurge family are found in the deserts of Africa. Both types of plants have made similar adaptations to arid habitats. There also are North American mammals and African mammals that are adapted similarly to living in a grassland environment (fig. 27.9).

On the other hand, plants and animals that live in different environments are variously adapted. Narrow-leaved trees are found in northern coniferous forests, but broad-leaved evergreen trees are found in tropical rain forests.

Evolution explains the history and the diversity of life. Evidences for evolution can be taken from the fossil record, comparative anatomy, comparative embryology, and comparative biochemistry. Vestigial structures and biogeography also support the occurrence of evolution.

The Evolutionary Process

Modern evolutionists emphasize that individuals are members of a population. A study of population genetics allows us to see when and if evolution has occurred. Therefore, a brief look at population genetics is in order.

Population Genetics

For the sake of our discussion, a **population** is defined as a group of interbreeding individuals living in a particular area. The various alleles and their frequency constitute the genetic makeup, or **gene pool,** of the population. If nothing upsets the equilibrium, we expect the gene pool to remain constant generation after generation. For example, suppose it is known that one-fourth of all persons in a human population are homozygous dominant for widow's peak, one-half are heterozygous, and one-fourth are homozygous recessive for continuous hairline. What will be the ratio of genotypes in the next generation?

Using the key given in the previous chapter—W = widow's peak and w = continuous hairline—we can describe the population in this manner:

$$\tfrac{1}{4}\ WW + \tfrac{1}{2}\ Ww + \tfrac{1}{4}\ ww$$

Necessarily, the homozygous dominant individuals will produce one-fourth of all the gametes of the population, and these gametes all will carry the dominant allele, W; the heterozygotes will produce one-half of all the gametes, but one-fourth will be W and one-fourth will be w; the homozygous recessive will produce one-fourth of all the gametes, and they will be w. Therefore, one-half of the gametes will be W and one-half will be w.

Assuming that all possible gametes have an equal chance to combine with one another, then as the Punnett square shows, the next generation will have exactly the same ratio of genotypes as the previous generation:

	$\tfrac{1}{2}\ W$	$\tfrac{1}{2}\ w$
$\tfrac{1}{2}\ W$	$\tfrac{1}{4}\ WW$	$\tfrac{1}{4}\ Ww$
$\tfrac{1}{2}\ w$	$\tfrac{1}{4}\ Ww$	$\tfrac{1}{4}\ ww$

Results: $\tfrac{1}{4}\ WW + \tfrac{1}{2}\ Ww + \tfrac{1}{4}\ ww$

This example indicates that dominant alleles necessarily do not take the place of recessive alleles; recessive alleles do not disappear; and sexual reproduction, in and of itself, cannot bring about a change in the allelic frequencies of the population.

The Binomial Equation

Two investigators, G. H. Hardy and W. Weinberg, recognized that it is possible to use the binomial equation to calculate the genotypic and the allelic frequencies of a population:

$$p^2 + 2pq + q^2 = 1.00$$

27.1 Critical Thinking

1. In reference to figure 27.9, why would you have expected to find coyotes in both North American grasslands and in African grasslands?
2. Explain, on the basis of evolution, why there are coyotes in North American grasslands but jackals in African grasslands.
3. What evidence would you gather to support your explanation?

In this case, p represents the frequency of the dominant allele and q represents the frequency of the recessive allele. Therefore,

$p^2 =$ homozygous dominant individuals
$q^2 =$ homozygous recessive individuals
$2pq =$ heterozygous individuals

The real value of this mathematical approach to population genetics is that by observation or inspection it is possible to determine the percentage of individuals who are recessive, and from this it is possible to calculate the frequencies of the alleles and the genotypes. Frequencies are given in decimals rather than fractions or percentages. For our example, then,

if $q^2 = 0.25$,	$q = 0.50$
$p^2 = 0.25$,	$p = \underline{0.50}$
$2pq = \underline{0.50}$	1.00
1.00	

Notice that $p + q$ (frequencies of the 2 alleles) must equal 1.00 and $q^2 + p^2 + 2pq$ (frequencies of the various genotypes) also must equal 1.00.

To take another example, suppose by inspection we determine that 1% of the population does not have freckles. Therefore, 99% of the population does have freckles. Of these, how many are homozygous dominant? How many are heterozygous?

To answer these questions, first convert 1% to a decimal. Then we know that $q^2 = 0.01$ and therefore $q = 0.1$. Since $p + q = 1.0$, then we know that $p = 0.9$ and therefore p^2 (frequency of the population that is homozygous dominant) $= 0.81$. To determine the frequency of the heterozygote, we simply realize that so far we have accounted for only 0.82 of the population, and therefore $0.18 =$ heterozygous. If you prefer, you can calculate that $2pq = 0.18$. In summary, we have found that

Homozygous recessive $= 0.01 = 1\%$ do not have freckles
Homozygous dominant $= 0.81$
Heterozygous $= 0.18$ $\Big\} = 99\%$ does have freckles

Practice Problems*

1. A student crosses fruit flies heterozygous for wing length with flies recessive for wing length ($L =$ long wings; $l =$ short wings). What genotypic frequencies are expected among the first generation offspring and each generation thereafter?
2. Four percent of the members of a population of pea plants are short. What are the frequencies of this recessive allele and the dominant allele? What are the genotypic frequencies in this population?
3. Twenty-one percent of a population is homozygous dominant, 49% is heterozygous, and 30% is recessive. What percentage of the next generation is predicted to be recessive?

*Answers to problems are on page 582.

The Hardy-Weinberg Law

Use of the binomial equation told Hardy and Weinberg that theoretically it is possible for the gene pool of a population to remain constant generation after generation. In nonmathematical terms, the Hardy-Weinberg law states that the frequencies of alleles in a population remains the same in each succeeding generation as long as certain conditions are met. These conditions are (1) the population is large and mating is random, (2) no mutations occur, (3) there is no gene flow, and (4) there is no natural selection.

Evolution and Diversity

Figure 27.10

*Evolution and the gene pool. Without evolution, (a) the gene
pool is constant generation after generation, but (b) if certain
gametes are selected for reproduction at the expense of others,
gene pool frequencies do not remain constant, and evolution
occurs.*

a.   25% = light coat   **First Generation**   75% = heavy coat

b.   37.5% = light coat   **Second Generation**   62.5% = heavy coat

Evolution   Most often, investigators find that these conditions rarely are met.
For example, suppose that 75% of a bear population has the dominant phenotype for a heavy coat of hair and 25% has the recessive phenotype for a light
coat. We predict that the second generation will have the exact same phenotypic frequencies as the first generation. When we sample the next generation,
we find a change in the phenotypic distribution (fig. 27.10). This indicates that
evolution has occurred in this population. *When the gene pool does not stay
constant, then evolution has occurred.*

*The concept of gene pool offers a way to recognize when evolution has occurred.
The frequency of genes in the gene pool of a population stays constant unless
evolution has occurred.*

The Synthetic Theory

The explanation of evolution in terms of modern genetic principles is a synthesis; it takes data and hypotheses from all sources and blends them into one

Table 27.3 *The Evolutionary Process*

Produce Variation	Reduce Variation
Mutation Gene flow Recombination	Genetic drift Natural selection

Figure 27.11
Variations among individuals of a population. It is easy for us to note that humans vary one from the other. However, the same is true for populations of any organism.

whole. According to the synthetic theory, the evolutionary process requires 2 steps (table 27.3):

1. Production of genotypic and therefore phenotypic variations.
2. Sorting out of these variations through successive generations.

The Production of Variations

It is readily apparent that members of a human population (fig. 27.11) vary, but just as humans differ one from the other, so do members of other populations. The daisies on the hill and the earthworms in your backyard are not genotypically or phenotypically the same. Metabolic, structural, and behavioral differences exist between them. Genetic variations in the gene pool of sexually reproducing diploid populations have 3 sources: mutation, gene flow, and recombination.

Mutation   There are chromosome mutations (p. 495) and there are gene mutations (p. 527) as discussed previously. Mutations are the only original source of allelic changes. Many times, observed mutations, such as those that cause human genetic disorders, seem to harm rather than to benefit an individual. This may be because members of a population are so adapted (suited) to an environment that only nonbeneficial changes are apparent. Nevertheless, recessive nonobserved mutations may be occurring that could be beneficial should the environment change.

In organisms that lack sexual reproduction, such as bacteria, genotypic variability is dependent entirely on mutations. This is sufficient because the short generation time of these organisms allows new mutations to be tested immediately by the environment.

Gene Flow   **Gene flow** occurs when individuals immigrate and emigrate between populations, bringing new genes into the pool of each population. The sharing of genes can cause the 2 populations to become adapted similarly but also can keep each one from becoming very closely adapted to a local environment.

Recombination   Recombination of alleles occurs during meiosis due to crossing-over between nonsister chromatids of homologous chromosomes and independent assortment of chromosomes (p. 89). It also occurs at fertilization when 2 different gametes join. Recombination gives the offspring a genotype and possibly a phenotype that differs from either parent. Mutation is the ultimate source of variation, but the recombined genotype supplied by sexual reproduction is usually of much greater importance.

Genetic variation in a population of sexually reproducing diploid organisms has 3 sources: mutation (both chromosome and gene), gene flow (emigration or immigration), and recombination (at the time of meiosis or fertilization).

The Reduction in Variations

Genetic drift and *natural selection* both act in such a way that variations in a population are sorted out and reduced, but only natural selection consistently results in adaptation. Adaptations can be structural (land animals breathe by means of lungs), physiological (desert animals make do with metabolic water), or behavioral (some animals forage at night and others forage in the daytime).

Genetic Drift   **Genetic drift,** as diagrammed in figure 27.12*a*, is a reduction in gene pool variation that occurs purely due to chance. It should be visualized

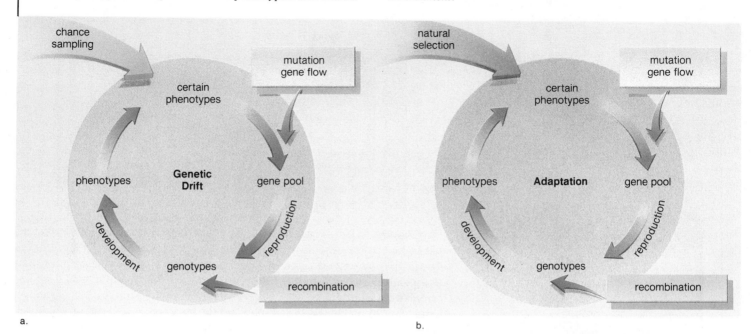

Figure 27.12

Genetic drift versus adaptation. In both cases, mutation, gene flow, and recombination are sources of genetic variation. The various genotypes develop into various phenotypes. a. A chance sampling of these phenotypes can lead to genetic drift. b. Selection of adapted phenotypes can lead to adaptation to the environment.

a.

b.

as a chance drifting toward certain genes so that others are eliminated. In the diagram, *chance sampling* means that only a few individuals among all the various phenotypes available produce offspring.

Genetic drift operates in both large and small populations, but it is significant only in small populations: Imagine that by chance a small number of individuals, representing a fraction of the gene pool, found a colony, and then during gamete formation, only certain of their genes are passed on to the next generation. As a result, a severe reduction in genetic variation compared to the original population takes place, and genetic drift occurs. This combination of circumstances has been observed historically and is called the *founder principle.* For example, an investigation of a small religious group called the Dunkers showed that the blood type was 60% blood type A. Since the frequency of this blood type is 40% in the United States and 45% in West Germany (the country from which the Dunkers emigrated), the high occurrence of blood type A among the Dunkers can be explained only by genetic drift.

Genetic drift is *not* expected to produce adaptation to the environment because phenotypes are not selected for reproduction; rather, chance alone determines who reproduces. We can imagine that a natural disaster, for example, would reduce a large population severely; only a few individuals would remain to reproduce. Only those genes that happened to be passed on to the next generation then would be available in the gene pool. As gene pool variation decreases, the possibility of fixation of a few genes and therefore genetic drift increases.

Natural Selection   **Natural selection** (fig. 27.12*b*) is the process by which populations become adapted to their environment. The reading on page 578 outlines how Charles Darwin, the father of evolution, explained evolution by

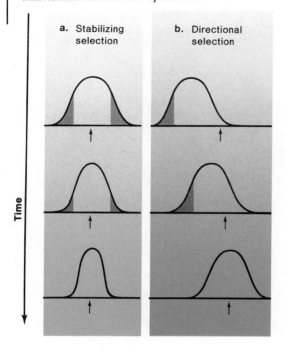

a. Stabilizing selection

b. Directional selection

Time

natural selection. Here, we restate these steps in the context of modern evolutionary theory. In evolution by natural selection, the fitness of an individual is measured by how reproductively successful its offspring are in the next generation.

Evolution by natural selection requires

1. variation. The members of a population differ from one another.

2. inheritance. Many of these differences are heritable genetic differences.

3. differential adaptedness. Some of these differences affect how well an organism is adapted to its environment.

4. differential reproduction. Individuals that are better adapted to their environment are more likely to reproduce, and their fertile offspring will make up a greater proportion of the next generation.

Natural selection operates by means of both biotic and abiotic factors that influence survival and reproduction. Biotic factors involve other organisms in the environment, and abiotic factors involve physical conditions. For example, an organism that is better able to escape a predator would be selected—that is, would be more likely to have offspring. Similarly, an organism better able to withstand the climate would be selected and again would be more likely to have offspring. By this gradual process, each generation of population contains more members that are adapted.

Figure 27.13 indicates that natural selection has 2 common effects: (1) to stabilize variations and (2) to direct variations. **Stabilizing selection** tends to eliminate atypical phenotypes and to enhance adaptation of the population to current environmental circumstances, but directional selection selects an extreme phenotype better adapted to a new environmental circumstance. **Directional selection** occurs during a time when the environment is changing rapidly or when members of a population are adapting to a new environmental situation.

The Maintenance of Variations

Even after evolution has occurred and the amount of variation has been reduced, some variations always are maintained. The more variations in a population, the greater its potential to adapt to new changes in its environment.

There are both genetic and ecological means for the maintenance of variations. *Diploidy* (organisms that have 2 of each kind of gene) helps to maintain variation because recessive genes can remain hidden in the gene pool, serving as a potential source of future phenotypic variations. *Pleiotropic genes* (genes that affect several different characteristics) can produce variations aside from those that adapt the organism to a current environment. When a characteristic is controlled by several genes *(polygeny)*, a change in one of these can produce an unexpected variation. Such phenotypic variations can be neutral or harmful to the organism.

The environment actually can promote the maintenance of 2 distinctly different phenotypes due to opposing selection pressures. In Africa, persons with normal red blood cells carry oxygen more efficiently but are more susceptible to malaria, while those with sickle-shaped cells escape malaria but die from sickle-cell anemia. For this reason, more heterozygotes than homozygotes survive. This **heterozygote superiority** maintains the frequency of the sickle-cell allele in the population.

The maintenance of all 3 phenotypes (normal cell, sickled cell, and sickle-cell anemia) in the population is an example of balanced polymorphism.

Polymorphism is the coexistence of 2 or more forms in a population, and *balanced polymorphism* occurs when the frequency of the varied forms remains the same in each generation.

The environment itself can vary and therefore can call for different adaptations at different times, such as seasonal changes within certain environments. Members of a population can become adapted generally, never specializing for any particular season, or they can become *polymorphic*, having a different phenotype for each season. Polymorphic adaptation is demonstrated by the arctic hare, which has a white coat in the winter and a brown coat in the summer.

Genetic drift and natural selection act in such a way that gene pool variations are reduced, but only natural selection consistently results in adaptation. Natural selection has 2 possible primary effects: (1) to stabilize population variations and (2) to direct population variations. Eventually, stabilization always occurs, except genetic variations are maintained because of genetic and ecological reasons.

Examples of Natural Selection

While natural selection usually takes many hundreds or even thousands of years to produce a noticeable change in the phenotype, there are a few examples of rapid adaptation.

Industrial Melanism

Before the industrial revolution in England, collectors of a moth called the peppered moth (fig. 27.14) noted that most moths were light colored, although occasionally a dark-colored moth was captured. Several decades after the industrial revolution, however, the black moths made up 99% of the moth population in polluted areas. An explanation for this rapid change can be found in natural selection. The color of the moths, dark or light, is caused by their genetic makeup; black is a mutation that occurs with some regularity. Moths rest on the trunks of trees during the day; if they are seen there by predatory birds, they are eaten. As long as the trees in the environment are light in color, the light-colored moths live to reproduce, but once the trees turn black due to industrial pollutants, natural selection enables the dark moths to avoid being eaten and to survive and to reproduce. Therefore, the black phenotype becomes the more frequent one in the population.

Resistance

Indiscriminate use of antibiotics and pesticides has caused bacteria to become resistant to these chemicals. When these chemicals are used, only the resistant organisms survive and reproduce. A good example is provided by malaria, a disease caused by an infection of the liver and the red blood cells (see fig. 28.14). The dread *Anopheles* mosquito transfers the disease-causing protozoan *Plasmodium* from person to person. In the early 1960s, international health authorities thought that malaria soon would be eradicated. The administration of a new drug, chloroquine, was more effective than quinine, and DDT spraying killed the mosquitoes. But in the mid-1960s, *Plasmodium* showed signs of chloroquine resistance, and worse yet, mosquitoes were becoming resistant to DDT. A few drug-resistant parasites and a few DDT-resistant mosquitoes had survived and multiplied, making the fight against malaria more difficult than ever. New tactics have to be devised. Recombinant DNA procedures have enabled researchers to identify proteins that it is hoped will be effective as vaccines in humans.

It has been possible to witness the process of natural selection in several recent instances: the adaptation of moths to polluted areas, the adaptation of bacteria to modern drugs, and the adaptation of insects to pesticides.

Figure 27.14
Industrial melanism. a. In nonpolluted areas, the tree trunks are white. A black moth (lower left) can be seen easily, but a white moth (upper right) cannot be seen on a tree trunk. Predatory birds therefore are likely to prey on dark moths in nonpolluted areas. b. In polluted areas, the tree trunks are dark. A white moth (lower left) can be seen easily, but a black moth (upper right) cannot be seen on a tree trunk. Predatory birds therefore are likely to prey on white moths in polluted areas.

a.

b.

At the age of 22, Charles Darwin signed on as a naturalist with the HMS Beagle, a ship that took a five-year trip around the world in the latter half of the nineteenth century. Because the ship sailed in the Southern Hemisphere where life is more abundant and varied, Darwin encountered forms of life very different from those in his native England.

Even though it was not his original intent, Darwin began to realize and to gather evidence that life forms change over time and from place to place. He read a book by

Charles Darwin and the Theory of Natural Selection

Charles Lyell, a geologist, who suggested the world is very old and has been undergoing gradual changes for many, many years. Darwin found the remains of a giant ground sloth and an armadillo on the east coast of South America and wondered if these extinct forms were related to the living forms of these animals. When he compared the animals of Africa to those of South America, he noted that the African ostrich and the South American rhea, although similar in appearance, were actually different animals. He reasoned that they had a different line of descent because they were on different continents. When Darwin arrived at the Galápagos Islands, he began to study the 13 species of finches (fig. 27.16), whose adaptations could best be explained by assuming they had diverged from a common ancestor. With this type of evidence, Darwin concluded that species evolve (change) with time.

When Darwin arrived home, he spent the next 20 years gathering data to support the principle of organic evolution. His most significant contribution to this principle was his theory of natural selection, which explains how a species becomes adapted to its environment. Before formulating the theory, he read an essay on human population growth written by Thomas Malthus. Malthus observed that although the reproductive potential of humans is great, there were many environmental factors,

such as availability of food and living space, that tend to keep the human population within bounds. Darwin applied these ideas to all populations of organisms. For example, he calculated that a single pair of elephants could have 19 million descendants in 750 years. He realized that other organisms have even greater reproductive potential than this pair of elephants; yet, usually the number of each type of organism remains about the same. Darwin decided there is a constant struggle for existence, and only a few members of a population survive to reproduce. The ones that survive and contribute to the evolutionary future of the species are by and large the better-adapted individuals. This so-called survival of the fittest causes the next generation to be better adapted than the previous generation.

Darwin's theory of natural selection was nonteleological. Organisms do not strive to adapt themselves to the environment, rather, the environment acts on them to select those individuals that are best adapted. These are the ones that have been "naturally selected" to pass on their characteristics to the next generation. In order to emphasize the nonteleological nature of Darwin's theory, it often is contrasted to Jean-Baptiste Lamarck's theory, another nineteenth-century naturalist (see figure). The Lamarckian explanation for the long neck of the giraffe was based on the assumption that the ancestors of the modern giraffe were trying to reach into the trees to browse on high-growing vegetation. Continual stretching of the neck caused it to become longer, and this acquired characteristic was passed on to the next generation. Lamarck's theory is teleological because according to him, a species shapes its own future. This type of explanation has not stood the test of time, but Darwin's theory of evolution by natural selection has been substantiated fully by later investigations.

These are the critical elements in Darwin's theory.

Variations. Individual members of a species vary in physical characteristics. Physical variations can be passed

from generation to generation. (Darwin was never aware of genes, but we know today that the inheritance of the genotype determines the phenotype.)

Struggle for existence. The members of all species compete with each other for limited resources. Certain members are able to capture these resources better than others.

Survival of the fittest. Just as humans have artificial breeding programs and select which plants and animals reproduce, so there is a natural selection by the environment of which organisms survive and reproduce. While Darwin emphasized the importance of survival, modern evolutionists emphasize the importance of unequal reproduction. In any case, however, the selection process is not teleological. Certain members of the population are selected to produce more offspring simply because they happen to have a variation that makes them more suited to the environment.

Adaptation. Natural selection causes a population of organisms and ultimately a species to become adapted to the environment. The process is slow, but each subsequent generation includes more individuals that are better adapted to the environment.

Can natural selection account for the origin of new species and for the great diversity of life? Yes, if we are aware that life has been evolving for a very long time and that variously adapted populations can arise from a common ancestor.

Darwin was prompted to publish his findings only after he received a letter from another naturalist, Alfred Russel Wallace, who had come to the exact same conclusions about evolution. Although both scientists subsequently presented their ideas at the same meeting of the famed Royal Society in London in 1858, only Darwin later gathered together detailed evidence in support of his ideas. He described his experiments and reasonings at great length in *The Origin of Species by Means of Natural Selection*, a book still studied by all biologists today.

Early giraffes probably had short necks that they stretched to reach food.

Early giraffes probably had necks of various lengths.

Their offspring had longer necks that they stretched to reach food.

Natural selection due to competition led to survival of the longer-necked giraffes and their offspring.

Eventually, the continued stretching of the neck resulted in today's giraffe.

Eventually, only long-necked giraffes survived the competition.

a. Lamarck's theory

b. Darwin's theory

Mechanism of evolution. *This diagram contrasts (a) Jean-Baptiste Lamarck's theory as to how evolution occurs, called the theory of acquired characteristics, with that of (b) Charles Darwin, called the theory of natural selection. Lamarck's theory is not supported by data, but Darwin's theory is supported by data.*

Figure 27.15

First stage of speciation is geographic isolation (a and b). Second stage of speciation is reproductive isolation (c and d).

a. A species has many subpopulations. Each rabbit represents one subpopulation.

b. Geographic isolation results in 2 separate gene pools.

c. Subpopulations of one gene pool evolve differently from those of the other gene pool. Two separate species arise.

d. Reproductive isolation has occurred. Rabbits of one species (gray) do not reproduce with rabbits of the other species (white).

27.3 Critical Thinking

1. Organisms generally are adapted to conserve energy. Which one—premating or postmating isolating mechanisms (see table 27.4)—represents a waste of energy? Why?

2. During the process of speciation, which types of mechanisms most likely evolve first? Why?

3. Which of the premating isolating mechanisms would you expect to be operating among Darwin's finches? What other type of premating mechanism might there be that is not listed in table 27.4?

Table 27.4 *Reproductive Isolating Mechanisms of Coexisting Species*

Isolating Mechanisms	Example
Premating	
Habitat	Species at same locale occupy different habitats
Temporal	Species mate at different seasons or different times of day
Behavioral	In animals, courtship behavior differs or they respond to different songs, calls, pheromones, or other signals
Mechanical	Genitalia unsuitable to one another
Postmating	
Gamete mortality	Sperm cannot reach or fertilize egg
Zygote mortality	Hybrid dies before maturity
Hybrid sterility	Hybrid survives but is sterile and cannot reproduce

---

Speciation

Usually, a species occupies a certain geographic range within which there are several subpopulations. For our present discussion, **species** is defined as a group of interbreeding subpopulations that share a gene pool and that are isolated reproductively from other species. The subpopulations of the same species exchange genes, but different species do not exchange genes. Reproductive isolation of the gene pools of similar species is accomplished by such mechanisms as those listed in table 27.4. If **premating isolating mechanisms** are in place, reproduction never is attempted. If **postmating isolating mechanisms** are in place, reproduction may take place, but it does not produce fertile offspring.

The Process of Speciation

Speciation has occurred when one species gives rise to 2 species. As an example, consider a species of rabbits that has several subpopulations. In figure 27.15a, each subpopulation is represented by a single rabbit. All the subpopulations share a common gene pool, and therefore all rabbits are the same color.

It generally is accepted that speciation is a two-stage process. Stage I is geographic isolation. Suppose that a canal is dug to divert water from a nearby source, and the canal separates the subpopulations of rabbits into 2 groups. Some subpopulations now are isolated geographically from the other subpopulations (fig. 27.15b). Stage II is reproductive isolation. In figure 27.15c, we see that the 2 sets of subpopulations are colored differently. This symbolizes that each set is now evolving separately. There are several reasons for this: (1) Certain variations may be present in only one set of subpopulations—because the gene pools are smaller, these variations have a better chance of being passed on; (2) each gene pool now experiences different mutations and recombination of alleles; and (3) the environment is different for each set of subpopulations, and each set is subject to different selective pressures.

Given enough time, we expect that even if the physical barrier is removed, the 2 sets of subpopulations will not be able to reproduce with one another (fig. 27.15d). The second stage of speciation is complete—what was formerly one species has become 2 species because they are isolated reproductively by the mechanisms listed in table 27.4.

Speciation is the origin of species. This usually requires geographic isolation followed by reproductive isolation.

Evolution and Diversity

Figure 27.16

Darwin's finches. The Galápagos Islands are close enough to South America that finches could have come there from the mainland. Once they arrived, the original population on one island is presumed to have spread out to the other islands, where varying selective pressures would have caused divergent evolution to occur. With time, the various populations of finches were unable to reproduce with one another. The ancestral form from which the Galápagos finches are descended most likely had a strong beak capable of crushing seeds. Each of the present-day species of finches has a bill adapted to a particular way of life. For example, (a) the large tree finch grinds fruit and insects with a parrotlike bill. The small ground finch (b) has a pointed bill and eats tiny seeds and ticks picked from iguanas. The woodpecker finch (c) has a stout, straight bill that chisels through tree bark to uncover insects, but because it lacks a woodpecker's long tongue, it uses a tool—usually a cactus spine or a small twig—to ferret out insects.

a. Large tree finch

b. Small ground finch

c. Woodpecker finch

Adaptive Radiation

One of the best examples of speciation is provided by the finches on the Galápagos Islands, which very often are called Darwin's finches because Darwin (p. 578) first realized their significance as an example of how evolution works. The Galápagos Islands, located 600 miles west of Ecuador, South America, are volcanic but do have forest regions at higher elevations. The 13 species of finches (fig. 27.16), placed in 3 genera, are believed to be descended from mainland finches that migrated to one of the islands some years ago. Therefore, Darwin's finches are an example of *adaptive radiation,* or the proliferation of a species by adaptation to different ways of life. We can imagine that after the original population of a single island increased, some individuals dispersed to other islands. The islands are ecologically different enough to have promoted divergent feeding habits. This is apparent because although the birds physically resemble each other in many respects, they have different bills, each of which is adapted to a particular food-gathering method. There are seed-eating ground finches with bills appropriate to cracking small-, medium-, or large-size seeds; insect-eating tree finches, also with different size bills; and a warbler-type finch with a bill adapted to nectar gathering. Among the tree finches, there is a woodpecker type, which lacks the long tongue of a true woodpecker but makes up for this by using a cactus spine or a twig to ferret out insects.

One frequently cited example of speciation is the evolution of several species of finches on the Galápagos Islands. This is also an example of adaptive radiation because the various species have different ways of life.

Summary

The fossil record; comparative anatomy, embryology and biochemistry; vestigial structures; and biogeography all give evidences of evolution. Evolution is a process that is believed to involve a gradual change in gene frequencies within the gene pool of a population. The Hardy-Weinberg law states that the gene pool remains constant only under certain prescribed conditions, and therefore evolution is expected to occur.

Evolution is presented as a two-step process requiring the production of genotypic variations (by mutation, gene flow, and recombination) and a sorting out of these variations (by genetic drift and natural selection). Several examples of natural selection at work are given.

Subpopulations that share the same gene pool all are part of the same species.

Speciation is the origin of species, and this usually requires geographic isolation followed by reproductive isolation. One frequently cited example of speciation is the evolution of several species of finches on the Galápagos Islands. This is an example of

adaptive radiation because the various species of finches became adapted to different ways of life.

Study Questions

1. Show that the fossil record, comparative anatomy, comparative embryology, comparative biochemistry, and biogeography all give evidences that evolution has occurred. (pp. 563–71)
2. What is an evolutionary tree, and how are evolutionary trees constructed? (p. 566)
3. What is a population and a gene pool? (p. 571)
4. What is the Hardy-Weinberg law? (p. 572) If genotype gg is found in 16% of a population, what is the frequency of the g allele? the G allele? What proportion of the next generation will be gg if the law holds? (p. 572)
5. Name and describe the sources of gene pool variations in a population made up of diploid sexually reproducing individuals. (p. 574)
6. Name and contrast 2 processes that reduce gene pool variations. (pp. 574–76)
7. Name several reasons for the maintenance of variation in a gene pool. (p. 576)
8. Give 2 modern examples of natural selection. (p. 577)
9. Define a species. How do new species originate? (p. 580)
10. When is adaptive radiation apt to take place? (p. 581)

Objective Questions

1. If the Hardy-Weinberg law holds, evolution _____ (always or does not) occur(s).
2. Twenty-one percent of a population is homozygous recessive. If the Hardy-Weinberg law holds, what percentage is expected to be homozygous recessive in the next generation? _____
3. Can sexual reproduction in and of itself cause evolution to occur? _____
4. Recombination of genes, mutation, and _____ are expected to increase the amount of variation among individuals of a population.
5. Natural selection and _____ are expected to reduce the amount of

variation among individuals of a population.
6. The fossil record shows that the brain size of humans has increased steadily since they evolved. This is an example of _____ selection.
7. There are genetic reasons for the maintenance of variation among individuals of a population, but _____ factors are also important.
8. During the first stage of speciation, populations become _____ .
9. During the second stage of speciation, populations become _____ .

10. Two species of butterflies have different courtship behavioral patterns. This is an example of a(n) _____ isolating mechanism of the _____ type.

Selected Key Terms

analogous structure (ah-nal′o-gus struk′-tūr) similar in function but not in structure; particularly in reference to similar adaptations. *568*

common ancestor (kŏ′mun an′ses-tor) an ancestor to 2 or more branches of evolution. *566*

evolutionary tree (ev″o-lu′shun-ar-e trē) diagram describing the evolutionary relationship of groups of organisms. *566*

fossil (fos″l) any remain of an organism that has been preserved in the earth's crust. *563*

gene flow (jēn flo) the movement of genes from one population to another via reproduction between members of the populations. *574*

gene pool (jēn pool) the total of all the genes of all the individuals in a population. *571*

genetic drift (jĕ-net′ik drift) evolution by chance processes alone. *574*

homologous structure (ho-mol′o-gus struk′tūr) similar in structure but not necessarily function; homologous structures in animals share a common ancestry. *566*

natural selection (nat′u-ral sĕ-lek′shun) the process by which better adapted

organisms are favored to reproduce to a greater degree and to pass on their genes to the next generation. *575*

population (pop″u-la′shun) all the organisms of the same species in one place. *571*

species (spe′shēz) a group of similarly constructed organisms capable of interbreeding and producing fertile offspring; organisms that share a common gene pool. *580*

vestigial structure (ves-tij′e-al struk′tūr) the remains of a structure that was functional in some ancestor but is no longer functional in the organism in question. *569*

Evolution and Diversity

Chapter 28

Viruses and Kingdoms Monera, Protista, and Fungi

Chapter Concepts

1 Viruses are acellular; whether they should be considered living organisms is questionable.

2 The monerans are prokaryotes, while the protistans and the fungi are eukaryotes.

3 The kingdom Monera includes bacteria, which are important organisms despite their small size.

4 The kingdom Protista contains algae, protozoans, and 2 types of molds (slime and water). Algae are the plantlike protists; protozoans are the animal-like protists; slime molds and water molds are the funguslike protists.

5 The kingdom Fungi contains the most complex organisms to rely on saprophytic nutrition.

Chapter Outline

Viruses
    Life Cycles
    Viroids and Prions
    Viral Infections
Kingdom Monera
    The Structure of Bacteria
    The Reproduction of Bacteria
    The Metabolism of Bacteria
    Types of Bacteria
Kingdom Protista
    Protozoans
    Algae
    Slime Molds and Water Molds
Kingdom Fungi
    Black Bread Molds
    Sac Fungi
    Club Fungi
    Other Types of Fungi

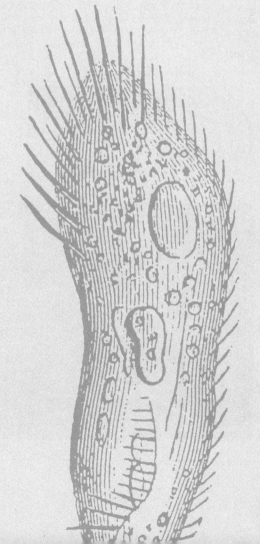

In this and the next 2 chapters, our aim is to discuss living organisms from the simple to the complex and from the primitive (earliest evolved) to the most advanced (most recently evolved). We can discuss how living organisms may be related only in the broadest of terms, since detailed information is often lacking. It is also important to remember that no living group of organisms is the direct ancestor of another living group of organisms, although it is possible for 2 living groups to have a common ancestor.

It is curious that we begin our discussion with viruses when they are not even included in the classification table found in appendix D. We begin with viruses only because they are on the borderline between living and nonliving things.

Viruses

Viruses are considered acellular because they do not have a cellular type of organization. The entire infectious unit (25–200 nm in size), known as a virion, is composed of at least 2 parts: *an outer coat of protein* and *an inner core of nucleic acid.* The coat, called a **capsid,** sometimes is surrounded by a membranous outer envelope.

In general, viruses are classified according to whether DNA or RNA serves as the genome and whether the nucleic acid is single or double stranded. The organization of the capsid is also important because it determines the overall shape of the virus. Most commonly, viruses have a helical, polyhedral, or complex symmetry (fig. 28.1). If the virus has an envelope, the capsid symmetry is not obvious.

Figure 28.1

Viruses: micrographs below and drawings above depicting symmetry. a. The tobacco mosaic virus is an RNA virus that attacks tobacco. The drawing shows how the nucleic acid core circles within a capsid composed of individual protein units.

b. The adenovirus is a DNA virus that causes respiratory and intestinal infections in humans. c. T₄ bacteriophage, a DNA virus that attacks bacteria, has a complex structure. d. The influenza virus is an RNA virus whose protein capsid is enclosed by an envelope.

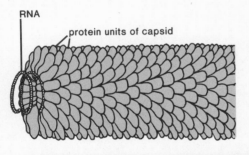

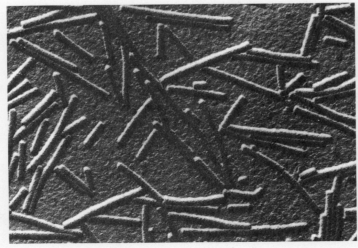

a.

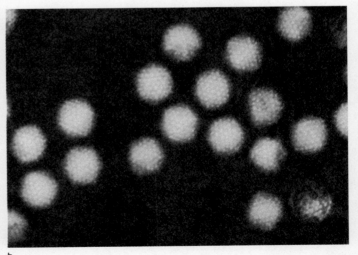

b.

The size of the virus particle and host specificity are also useful characteristics for classification purposes. Viruses of bacteria (called **bacteriophages**), plants, fungi, invertebrates, and vertebrates are known. Further, each virus has a particular host; for example, plant viruses attack only specific plants, and animal viruses attack only specific animals. The human disease-causing viruses even attack only certain types of cells because there must be a match between a protein in the outer coat or the envelope of the virus and a receptor in the cell's outer surface before a virus can gain entry.

Upon entry into a cell, viruses are capable of reproduction. Because viruses will reproduce only inside a living cell, they are called *obligate parasites*. In the laboratory, active animal viruses are replicated by injecting them into living cells, such as live chick embryos (fig. 28.2). Outside living cells, viruses are nonliving and are stored in liquid nitrogen. It is reasonable, therefore, to ask if viruses should be considered alive.

Viruses are acellular obligate parasites that always have a coat of protein and a nucleic acid core.

Figure 28.2
Inoculation of live chick eggs with virus particles. A virus only reproduces inside a living cell, not because it uses the cell for nutrients but rather because it takes over the machinery of the cell.

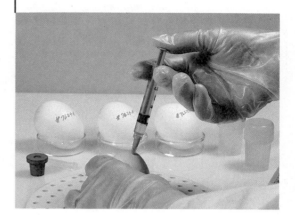

Life Cycles

Bacteriophages

Two types of bacteriophage life cycles, termed the lytic and the lysogenic cycles (fig. 28.3), have been studied carefully.

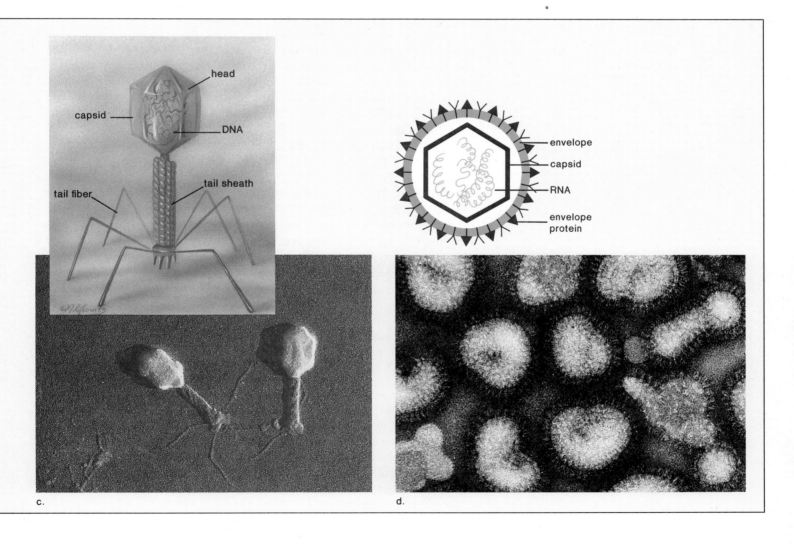

Figure 28.3

Lytic life cycle versus lysogenic life cycle. In the lytic life cycle, a bacteriophage reproduces within the host bacterial cell, and then the cell is lysed (broken open), allowing the virus particles to escape. In the lysogenic cycle, the DNA of a bacteriophage integrates into the host DNA and becomes a prophage. Thereafter, the prophage is replicated along with the bacterial chromosome and is passed to all the daughter cells. When and if the viral DNA leaves the chromosomes, the lysogenic cycle can be followed by the lytic cycle.

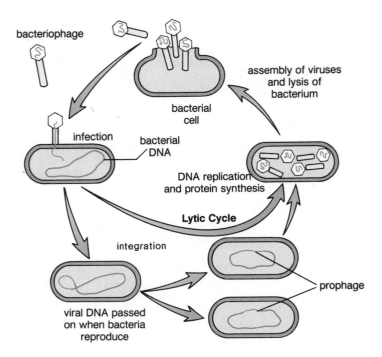

bacteriophage

assembly of viruses and lysis of bacterium

bacterial cell

infection

bacterial DNA

DNA replication and protein synthesis

Lytic Cycle

integration

prophage

viral DNA passed on when bacteria reproduce

Lysogenic Cycle

The Lytic Cycle When a bacteriophage collides with a bacterium, it attaches to a specific receptor. An enzyme digests away part of the bacterial cell wall, and the viral DNA enters the bacterial cell. Once inside, the viral DNA brings about disintegration of host DNA and takes over the operation of the cell. Viral DNA replication, utilizing the nucleotides within the host cell, produces many copies of viral DNA. Transcription occurs, and mRNA molecules utilize host ribosomes to bring about the production of multiple copies of coat protein. Viral DNA and capsids are *assembled* to produce about 100 virus particles. In the meantime, viral DNA has directed the synthesis of lysozyme, an enzyme that disrupts the cell wall, and the particles are released. The bacterial cell then dies.

The Lysogenic Cycle Some bacteriophages do not undergo a lytic life cycle immediately. Instead, the viral DNA is integrated into the bacterial DNA. In this stage, the bacteriophage is called a *prophage.* The prophage is replicated along with the host DNA, and all subsequent cells, called *lysogenic cells,* carry a copy of the prophage. Certain environmental factors, such as ultraviolet radiation, can induce the lytic cycle. The prophage leaves the bacterial chromosome; replication of viral DNA, production of capsids, assemblage, and cell lysis follow.

During the lytic cycle of a bacteriophage, the bacterial cell dies when the viral particles burst from the cell. During the lysogenic cycle, viral DNA integrates itself into bacterial DNA for an indefinite period of time.

Animal Viruses

Animal viruses may have a membranous outer envelope. Such viruses enter a cell by endocytosis, and *uncoating* releases viral nucleic acid from the capsid. These viruses leave a cell by exocytosis, or *budding,* and in this way, they acquire the membranous envelope (fig. 28.4). The envelope often contains glycoproteins that interact with the next host cell's membrane, permitting entry of the virus into this cell.

Evolution and Diversity

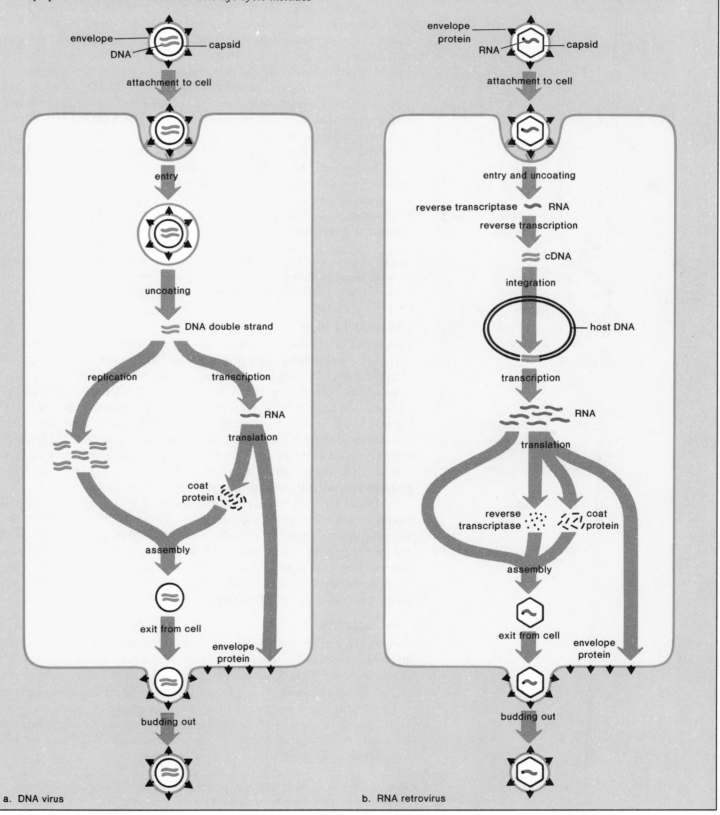

Figure 28.4

Life cycles of animal viruses. a. DNA virus. After entering by endocytosis, the virus becomes uncoated. The DNA then codes for proteins, some of which are capsid (coat) proteins and some of which are envelope proteins. Assembly follows replication of the DNA. When the virus exits by budding, it is enclosed by an envelope made up of host cell membrane lipids and viral envelope proteins. b. RNA retrovirus. The life cycle includes steps not seen in (a). The RNA genes are transcribed to cDNA (DNA copied off of RNA) that is integrated into the host DNA. Transcription produces many copies of the RNA genes, which also serve to direct the synthesis of 3 types of proteins: the enzyme reverse transcriptase, the capsid (coat) protein, and the envelope protein. Again, the virus buds from the host cell.

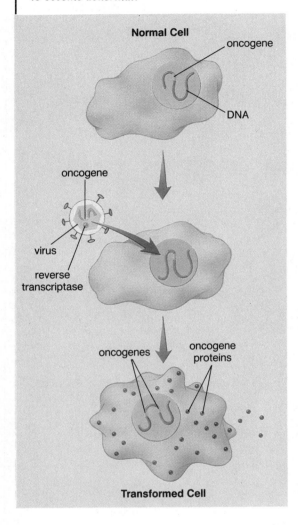

Figure 28.5

Retroviruses and the development of cancer. The normal cell contains one oncogene but is not yet cancerous. The virus also contains one oncogene that it passes to the cell. Now, the presence of 2 oncogenes causes the cell to be transformed. It becomes cancerous because it produces a protein, most likely a growth factor, that causes the cell to become abnormal.

Normal Cell

oncogene

DNA

oncogene

virus

reverse transcriptase

oncogenes

oncogene proteins

Transformed Cell

DNA Animal Viruses   After a DNA virus with an envelope has gained entry to the cell and uncoating has occurred, transcription, translation, and assembly occur as we have just described for bacteriophages. Then, as each virus buds from the cell, it acquires its envelope. The DNA codes for coat protein and also envelope glycoprotein (fig. 28.4*a*).

RNA Animal Viruses   Some animal viruses have RNA genomes; that is, only RNA is enclosed within the capsid. In most of these viruses, the single strand of RNA serves as a template for the production of double-stranded RNA. This unique molecule then serves as a template for multiple replicas of the genetic material and for the transcription of mRNA molecules. Special enzymes, termed *RNA replicase* and *RNA transcriptase,* perform these tasks.

Other RNA viruses, called **retroviruses** (fig. 28.4*b*), have an enzyme called *reverse transcriptase* that carries out RNA → DNA transcription. Following replication, the resulting double-stranded DNA (called *cDNA* because it contains a copy of the viral RNA) is integrated into the host chromosome for an indefinite period of time before reproduction of the virus and budding from the host cell occurs. Because integration occurred, RNA viruses sometimes carry host genes into a new host cell.

Retroviruses are of extreme interest because these are the viruses that bring cancer-causing oncogenes into a host cell (fig. 28.5). Also, the AIDS viruses are retroviruses.

Animal viruses often have a membranous envelope that they acquire by budding from the host cell. Retroviruses are RNA viruses that carry out RNA → DNA transcription and integrate this cDNA into the host cell.

Viroids and Prions

Viroids and prions are very unusual infectious particles that can be compared to viruses. **Viroids** differ from viruses in that they consist only of a short chain of naked RNA. The number of nucleotides present is not sufficient to code for a capsid, and there is no evidence that the viroid RNA ever is transcribed into a protein. We know of their presence only because they cause diseases in plants; the first viroid to be recognized is the cause of potato spindle tuber disease, and viroids also are known to have attacked coconut trees in the Philippines and chrysanthemums in the United States. The manner in which they cause disease is not known, but it is hypothesized that they are reproduced by the host cell and interfere with gene regulation in these cells.

Prions also differ markedly from viruses. They consist of a glycoprotein having only one polypeptide of about 250 amino acids. DNA that codes for the polypeptide is not found in the host cell; therefore, it is not known how they manage to be reproduced. Like viroids, these particles are known only because they have been associated with a disease. It is possible that they are the cause of nervous system diseases previously termed slow-virus diseases because it takes a long time for symptoms to worsen.

Viral Infections

Viruses are best known for causing infectious diseases in animals and plants. In plants, infectious diseases can be controlled only by destroying those plants that show symptoms of disease. In animals, especially humans, they are controlled by administering vaccines and only recently by the administration of antiviral drugs, as discussed in the reading on page 589. Some well-studied viral diseases in humans are flu, mumps, measles, polio, rabies, and infectious hepatitis. As mentioned previously, retroviruses cause AIDS and have been implicated in the development of some forms of cancer.

Viruses cause diseases of bacteria, plants, and animals. They also have been implicated in the development of some forms of cancer.

Antibiotics and Antiviral Drugs

An antibiotic is a chemical that selectively kills bacteria when it is taken into the body as a medicine. Since the introduction of the first antibiotics in the 1940s, there has been a dramatic decline in deaths due to pneumonia, tuberculosis, and other infections.

Most antibiotics are produced naturally by soil microorganisms. Penicillin is made by the fungus *Penicillium*; streptomycin, tetracycline, and erythromycin all are produced by the bacterium *Streptomyces*. Sulfa, a chemotherapeutic agent rather than an antibiotic, an analogue of a bacterial growth factor, can be produced in the laboratory.

A few antibiotics are metabolic inhibitors specific for bacterial enzymes. This means that they poison bacterial enzymes without harming host enzymes. Penicillin blocks the synthesis of the bacterial cell wall; streptomycin, tetracycline, and erythromycin block protein synthesis; and sulfa prevents the production of a coenzyme.

There are problems associated with antibiotic therapy. Some patients are allergic to antibiotics, and the reaction can be fatal. Antibiotics not only kill off disease-causing bacteria, they also reduce the number of beneficial bacteria in the intestinal tract. These beneficial bacteria may have checked the spread of a pathogen that now is free to multiply and to invade the body. The use of antibiotics sometimes prevents natural immunity from occurring, leading to the need for recurring antibiotic therapy. Most important, perhaps, is the growing resistance of certain strains of bacteria to antibiotics. While penicillin used to be 100% effective against hospital strains of *Staphylococcus aureus*, today it is far less effective. Tetracycline and penicillin, long used to cure gonorrhea, now have a failure rate of more than 20% against certain strains of gonococcus. Most physicians believe that antibiotics should be administered only when absolutely necessary. Some believe that if antibiotic use is not strictly limited then resistant strains of bacteria will replace present strains completely and antibiotic therapy no longer will be effective at all. They are very opposed to the current practice of adding antibiotics to livestock feed in order to make animals grow fatter because resistant bacteria are easily transferred from animals to humans.

The development of antiviral drugs has lagged far behind the development of antibiotics. Viruses lack most enzymes and instead utilize the metabolic machinery of the host cell. Rarely has it been possible to find a drug that successfully interferes with viral reproduction without also interfering with host metabolism. One such drug, however, call vidarabine, was approved in 1978 for treatment of viral encephalitis, an infection of the nervous system. Acyclovir (ACV) seems to be helpful in treating genital herpes, and the drugs AZT (zidovudine) and DDI (dideoxyinosine) are being used in AIDS patients.

Penicillium chrysogenum. *The antibiotic penicillin is prepared from this bacterium.*

Kingdom Monera

In the classification system used by this text, the kingdom Monera contains all the different types of bacteria.

The Structure of Bacteria

Prokaryotic cells (fig. 2.15) are very small (1–10 μm in length and 0.2–0.3 μm in width), and they do not have the cytoplasmic organelles found in eukaryotic cells, except ribosomes (table 2.2). They do have DNA, but it is not contained within a nuclear envelope; therefore, they are said to lack a nucleus. They have respiratory enzymes, but no mitochondria, and if they possess chlorophyll, it may be found within thylakoids, but there are no chloroplasts.

The eubacteria have a cell wall containing unique amino sugars cross-linked by peptide chains that may be surrounded by a capsule. Some bacteria can move by means of flagella, and some can adhere to surfaces by means of short, fine, hairlike appendages called pili.

Bacteria occur in 3 basic shapes (fig. 28.6): *rod* (bacillus), *spherical* or round (coccus), and *spiral* (e.g., helical shape called a spirillum). The bacilli and the cocci may form chains of a length typical of the particular bacterium.

Flagella of a bacterium. These threadlike locomotor organs are only about 20 nm wide and 15 or 10 m long. The flagella of bacteria are composed of a protein called flagellin and do not contain microtubules like the flagella of eukaryotes. They can be variously placed and only certain bacteria have a cluster of flagella at one end as seen here.

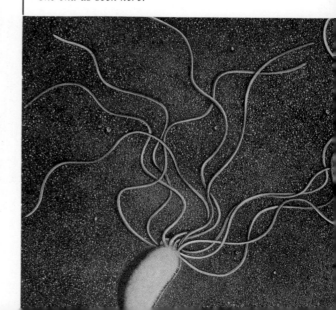

Figure 28.6
Scanning electron micrographs of bacteria. a. Spherical-shaped bacteria. b. Rod-shaped bacteria. c. Spiral-shaped bacteria with flagella used for locomotion. See figure 2.15 for generalized drawings of bacteria.

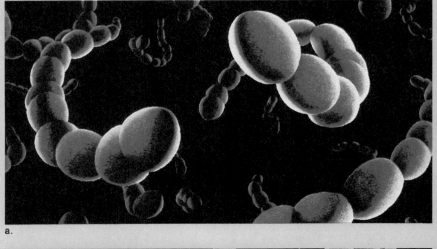

a.

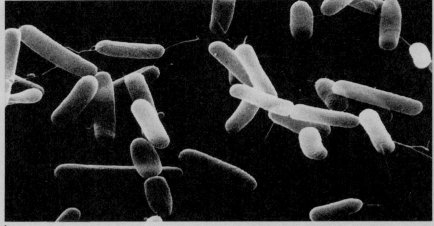

b.

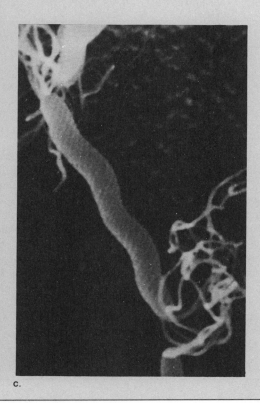

c.

Prokaryotic cells lack a nucleus and most of the other organelles found in eukaryotes. They occur in 3 basic shapes: rod, round, and spiral.

The Reproduction of Bacteria

Prokaryotes reproduce asexually by means of **binary fission.** First, the single circular chromosome duplicates; there then are 2 chromosomes attached to the inside of the cell membrane. The chromosomes are separated by an elongation of the cell that pushes the chromosomes apart. Then the cell membrane grows inward, and the cell wall forms, dividing the cell into 2 daughter cells, each of which now has its own chromosome (fig. 28.7).

Sexual reproduction does not occur among bacteria; however, genetic recombination is accomplished in 3 ways. *Conjugation* has been observed between bacteria when the so-called male cell passes DNA to the female cell by way of a sex pilus. *Transformation* occurs when a bacterium binds to and then takes up DNA released into the medium by dead bacteria. During *transduction,* bacteriophages carry portions of bacterial DNA from one cell to another.

When faced with unfavorable environmental conditions, some bacteria form *endospores.* During endospore formation, the cell shrinks, rounds up within the former cell membrane, and secretes a new, thicker wall inside the old one (fig. 28.8). Endospores are amazingly resistant to extreme temperatures, drying out, and harsh chemicals, including acids and bases. When conditions are again suitable for growth, the spore absorbs water, breaks out of the inner shell, and becomes a typical bacterial cell, reproducing once again by binary fission.

The Metabolism of Bacteria

Some bacteria are *obligate anaerobes* and are unable to grow in the presence of oxygen. A few serious illnesses, such as botulism, gas gangrene, and tetanus, are caused by anaerobic bacteria. Some other bacteria, called *facultative anaerobes,* are able to grow in either the presence or the absence of oxygen. Most bacteria, however, are aerobic and like animals require a constant supply of oxygen to carry out cellular respiration.

Every type of nutrition is found among bacteria except holozoism (eating whole food). Some autotrophic bacteria are photosynthetic; they use light as a source of energy to produce their own food. These *photosynthetic bacteria* can be divided into 2 types—those that are primitive (evolved first) and those that are advanced (evolved later). The characteristics of these 2 types of photosynthetic bacteria are as follows:

Primitive	*Advanced*
Photosystem I only	Photosystems I and II
Do not give off O_2	Do give off O_2
Unique type of chlorophyll	Type of chlorophyll found in plants

The green sulfur bacteria and the purple sulfur bacteria are primitive photosynthetic bacteria. They do not give off oxygen because they do not use water as an electron donor; instead, they use hydrogen (H_2) and hydrogen sulfide (H_2S). These bacteria usually live in anaerobic conditions, such as the muddy bottom of marshes, and they cannot photosynthesize in the presence of oxygen. The cyanobacteria (fig. 28.9) are advanced photosynthetic bacteria and carry on photosynthesis in the same manner as plants.

Some autotrophic bacteria are *chemosynthetic bacteria,* which oxidize inorganic compounds to obtain the necessary energy to produce their own food. Among the inorganic compounds oxidized by specific bacteria are ammonia,

Figure 28.7

Reproduction in bacteria. Above, the single chromosome is seen attached to the cell membrane, where it is replicating. As the cell membrane lengthens, the 2 chromosomes separate. Once fission has taken place, each bacterium has its own chromosome.

Figure 28.8

Endospore formation. This bacterium (Clostridium botulinum) *contains an endospore, the dark oval at the lower end of the cell. An endospore normally protects the organism's DNA from exposure to environmental conditions that could destroy it. However, sterilization in an autoclave, a container that maintains steam under pressure, can kill endospores.*

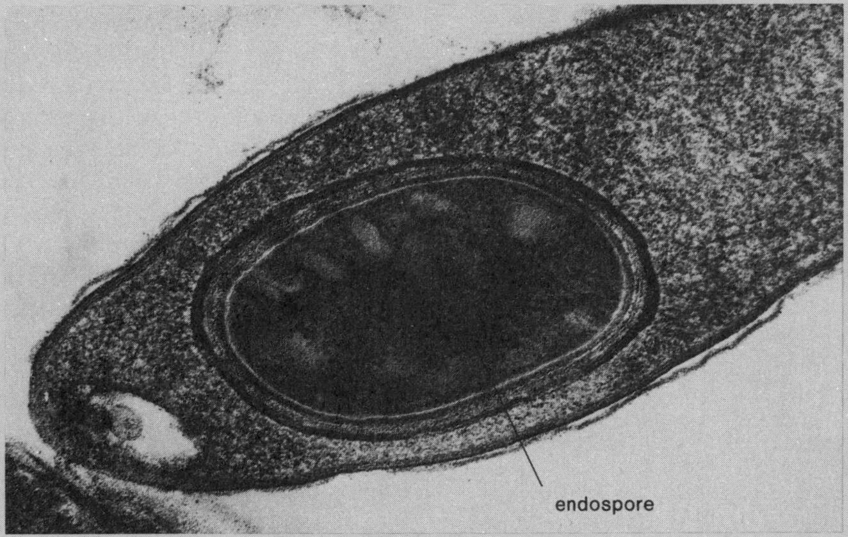

endospore

Viruses and Kingdoms Monera, Protista, and Fungi

Figure 28.9

Diversity among the cyanobacteria. a. In Chroacoccus, single cells are grouped together in a common gelatinous sheath. b. Filaments of cells occur in Oscillaturia.

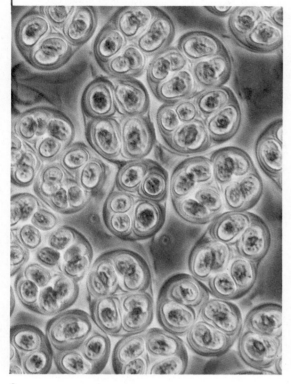

a.

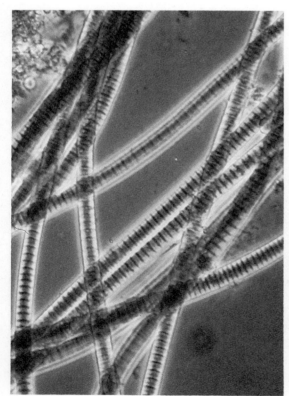

b.

nitrite, sulfur, hydrogen, and ferrous iron. In the nitrogen cycle (p. 728), nitrifying bacteria oxidize ammonia to nitrate, a nutrient for plants.

The majority of bacteria are free-living aerobic heterotrophs and feed on dead organic matter by secreting digestive enzymes and absorbing the products of digestion. The so-called saprophytic decomposers play a critical role in recycling matter in ecosystems and in making inorganic molecules available to photosynthesizers. The metabolic capabilities of various heterotrophic bacteria are exploited by human beings who use them to perform services ranging from the digestion of sewage and oil to the production of such products as alcohol, vitamins, and even antibiotics. By means of gene splicing (see fig. 25.3), bacteria now can be used to produce useful substances, such as human insulin and human growth hormone.

Bacteria are often symbiotic; they live in association with other organisms. The nitrogen-fixing bacteria in the nodules of legumes are mutualistic, as are the bacteria that live within our own intestinal tract. We provide the bacteria with a home, and they provide us with certain vitamins. Commensalistic bacteria reside on our skin, where they usually cause no problems. Parasitic bacteria are responsible for a wide variety of plant and animal diseases. Common human infections caused by parasitic bacteria include strep throat, diphtheria, typhoid fever, and gonorrhea.

The majority of bacteria are free-living heterotrophs (saprophytic decomposers) that contribute significantly to recycling matter through ecosystems. Many are also symbiotic heterotrophs, including those that cause disease.

Types of Bacteria

There are many different types of bacteria. For example, *Bergey's Manual,* a standard for classification of bacteria since the 1920s, has 4 separate volumes that list the different types of bacteria. We will divide the 33 major groups of bacteria into the archaebacteria and the eubacteria. It is interesting to observe that on the basis of molecular analysis of ribosomal RNA, it has been suggested that there should be only 3 kingdoms: the archaebacteria, the eubacteria, and the eukaryotes. This tells us that the archaebacteria and the eubacteria are probably not closely related.

Archaebacteria

Most likely, archaebacteria were the earliest prokaryotes. Their cell wall, cell membrane, and ribosomes do not have the same composition as the eubacteria.

The **archaebacteria** are able to live in the most extreme environments, perhaps representing the kinds of habitats that were available when the earth first formed. The methanogens are anaerobic and live in swamps and marshes, producing the methane known as marsh gas. They also live in the guts of organisms, including humans. The halophiles live where it is salty, such as the Great Salt Lake in Utah. Curiously, a type of rhodopsin pigment (related to the one found in our own eyes) allows them to carry on a primitive type of photophosphorylation for ATP production. The thermoacidophiles live where it is both hot and acidic. Those that live in the hot sulfur springs of Yellowstone National Park obtain energy by oxidizing sulfur.

Eubacteria

Most bacteria are **eubacteria,** which can be divided in many different ways. One way to classify eubacteria is to consider their reaction to the **Gram stain.** After applying this stain, the gram-positive bacteria appear purple and the gram-negative bacteria appear pink to red (fig. 28.10). This difference is dependent on the construction of the cell wall. The gram-positive cell wall is a single, thick layer, while the gram-negative wall has several thinner layers. Some gram-negative bacteria, such as the rickettsias and chlamydias, are much smaller than the average size of bacteria. The rickettsias, which cause Rocky

Mountain spotted fever and typhus, and the chlamydias, which cause a sexually transmitted urethritis, actually grow inside other cells.

Among the photosynthetic bacteria, the cyanobacteria are of special importance; therefore, these are discussed further here.

Cyanobacteria　Cyanobacteria (fig. 28.9), formerly called blue-green algae, are the most prevalent of the photosynthetic bacteria. The cyanobacteria carry on photosynthesis in a manner similar to that of plants; they possess chlorophyll *a* and evolve oxygen. They also have other pigments that can mask the color of chlorophyll, giving them, for example, not only blue-green but also red, yellow, brown, or black colors. We mentioned in chapter 26 that the cyanobacteria are believed to be responsible for first introducing oxygen into the primitive atmosphere.

Cyanobacteria can be unicellular, filamentous, or colonial. The filaments and colonies are not considered multicellular because each cell is independent of the others. Cyanobacteria lack any visible means of locomotion, although some glide when in contact with a solid surface and others oscillate (sway back and forth). Some cyanobacteria have a special advantage because they possess heterocysts, thick-walled cells without nuclei where nitrogen fixation occurs. The ability to photosynthesize and also to fix atmospheric nitrogen (N_2) means that their nutritional requirements are minimal.

Cyanobacteria are common in fresh water, in soil, and on moist surfaces but also are found in inhospitable habitats, such as hot springs. In fresh water, cyanobacteria sometimes are responsible for the bloom associated with cultural eutrophication. They also form symbiotic relationships with a number of organisms, such as ferns and even at times invertebrates, like corals. In association with fungi, they form **lichens** (p. 605), which can grow on rocks. Therefore, cyanobacteria may have been among the first organisms to colonize land.

Cyanobacteria are photosynthesizers that sometimes also fix atmospheric nitrogen. They first introduced oxygen into the atmosphere and probably were among the first organisms to colonize land.

Kingdom Protista

The protists are eukaryotes; their cells have all the organelles we studied in chapter 2. Unicellular organisms are predominant in kingdom Protista, and even the multicellular forms lack the tissue differentiation that is seen in more complex organisms. The protists are grouped according to their mode of nutrition and other characteristics into the following categories: (1) protozoans, which are the animal-like protists—they are motile and ingest their food like animals; (2) algae, which are the plantlike protists because they are photosynthetic like plants; and (3) slime molds and water molds, which are the funguslike protists. Slime molds are heterotrophic and produce windblown spores like fungi, but unlike fungi, they ingest their food. Water molds are heterotrophic and saprophytic like the fungi. A **saprophyte** carries on external digestion and absorbs the resulting nutrients across the cell membrane. Unlike the fungi, water molds produce swimming zoospores.

Protists, even those that are only one cell, should not be considered simple organisms. Each cell is organized to cope with the environment; alone it can carry out all the functions assumed by various tissues and organs in more complex organisms.

Protozoans

Protozoans are small (2 μm–1,000 μm), usually colorless unicellular organisms that lack a cell wall. Like animals, they tend to have special structures for food gathering and locomotion; excretion and respiration are carried out

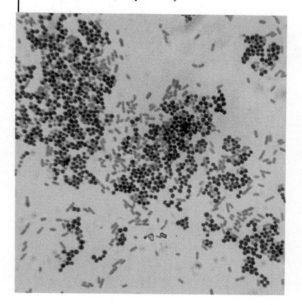

Figure 28.10
Photomicrograph of bacteria subjected to Gram staining. The gram-positive bacteria are purple, and the gram-negative bacteria are red. These bacteria are Staphylococcus aureus *and* Escherichia coli, *respectively.*

across the cell membrane. Although sexual exchange sometimes is observed, reproduction occurs by cell division. Protozoans are classified according to their locomotor organelles (table 28.1).

Amoeboids

Amoeboids (phylum Sarcodina), such as *Amoeba proteus* (fig. 28.11), are a small mass of cytoplasm without any definite shape. They move about and feed by means of cytoplasmic extensions called pseudopodia, or false feet. A pseudopodium forms when the cytoplasm streams forward in a particular direction.

The organelles within an amoeba include food or digestive vacuoles and contractile vacuoles. *Food vacuoles* are characteristic of holozoic protozoans and are formed within an amoeba when a morsel of food is surrounded by pseudopodia. This form of phagocytosis produces a vacuole that later becomes a digestive vacuole. *Contractile vacuoles* first collect excess water from the cytoplasm and then appear to "contract," releasing the water through a temporary opening in the cell membrane. Contractile vacuoles most often are seen in freshwater protozoans.

Table 28.1 *Types of Protozoans*

Protozoan	Locomotor Organelles	Example
Amoeboid	Pseudopodia	*Amoeba*
Ciliate	Cilia	*Paramecium*
Flagellate	Flagella	*Trypanosoma*
Sporozoa	No locomotor organelle	*Plasmodium*

Figure 28.11

Amoeba proteus, *a protozoan that moves by formation of pseudopodia. Note also the unique organelles, including the food vacuoles and the contractile vacuole. This artist's representation is based on electron micrographs of this organism. Arrows indicate the flow of cytoplasm as a newly formed pseudopodium takes shape.*

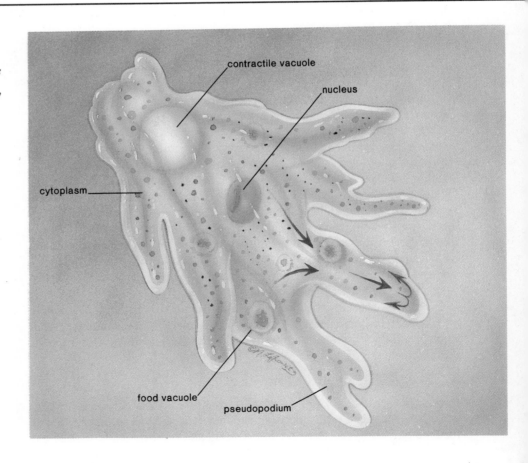

Evolution and Diversity

Two forms of marine amoeba have shells. *Foraminifera* have a chalky, sometimes many-chambered shell; these organisms were so numerous at one time that their remains built the White Cliffs of Dover. *Radiolaria* secrete a beautiful skeleton of silica that becomes the ooze of the ocean floor. One form of amoeba, *Entamoeba histolytica,* causes amoebic dysentery in humans.

Ciliates

The **ciliates** (phylum Ciliophora), such as those in the genus *Paramecium* (fig. 28.12), are the most complex of the protozoans. Hundreds of cilia project through tiny holes in the outer covering, or pellicle. Lying in the cytoplasm just beneath the pellicle are numerous oval capsules that contain **trichocysts.** The trichocysts can be discharged as long, barked threads, which are used for defense. When a paramecium feeds, food is swept down a gullet, below which food vacuoles form. Following digestion, the soluble nutrients are absorbed by the cytoplasm and the nondigestible residue is eliminated at the anal pore.

Ciliates have 2 types of nuclei: a large macronucleus and one or more small micronuclei. The macronucleus controls the normal metabolism of the cell, while the micronuclei are concerned with reproduction. Following meiotic division of the micronuclei, 2 paramecia may exchange these in a sexual process called **conjugation.** During conjugation, 2 paramecia join and exchange chromosome material.

Zooflagellates

Protozoans that move by means of flagella sometimes are called **zooflagellates** (phylum Zoomastigina) to distinguish them from unicellular algae that have flagellate.

Many zooflagellates enter symbiotic relationships. *Trichonympha collaris* lives in the gut of termites and enzymatically converts wood to soluble carbohydrates, which is digested easily by the insect. The trypanosomes (fig. 28.13) cause African sleeping sickness and are transmitted to vertebrates by the tsetse fly. The tsetse fly, which becomes infected when it takes a blood meal from a diseased animal, passes on the disease when it feeds on another victim. The white blood cells in an infected animal accumulate around the

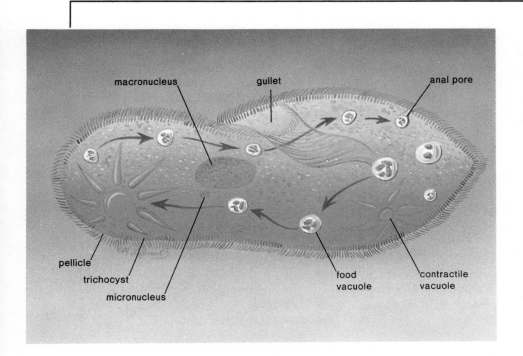

macronucleus   gullet   anal pore

pellicle
trichocyst
micronucleus   food vacuole   contractile vacuole

Figure 28.12
Diagram illustrating the principal structures found in Paramecium caudatum. *Despite its complexity, a paramecium is a single-celled organism. When discharged, trichocysts are poisonous, threadlike darts that are used for defense and capturing prey. Arrows show path of food vacuoles from formation to discharge of nondigestible particles.*

Figure 28.13

Figure 28.13

Trypanosome infection. A stained blood smear from a patient suffering with African sleeping sickness showing trypanosomes among the blood cells.

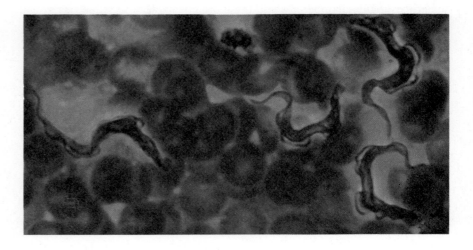

Figure 28.14

Life cycle of Plasmodium vivax. *Asexual reproduction occurs in humans, while the sexual life cycle takes place within the* Anopheles *mosquito.*

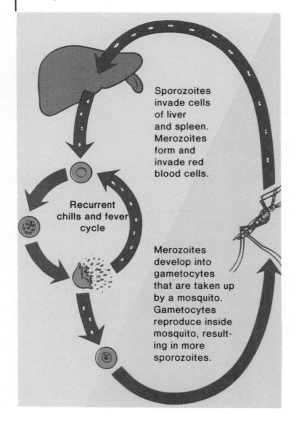

Sporozoites invade cells of liver and spleen. Merozoites form and invade red blood cells.

Recurrent chills and fever cycle

Merozoites develop into gametocytes that are taken up by a mosquito. Gametocytes reproduce inside mosquito, resulting in more sporozoites.

blood vessels leading to the brain and cut off the circulation. The lethargy characteristic of the disease is caused by an inadequate supply of oxygen for normal brain alertness.

Sporozoa

The **sporozoa** (phylum Sporozoa) are nonmotile parasites with a complicated life cycle that always involves the formation of infective spores. The most important human parasite among the sporozoa is *Plasmodium vivax* (fig. 28.14), the causative agent of one type of malaria. When a human being is bitten by an infected female *Anopheles* mosquito, the parasite eventually invades red blood cells. The chills and the fever of malaria occur when the infected cells burst and release toxic substances into blood.

The eradication of malaria has centered on the destruction of the mosquito, since without this host the disease cannot be transmitted from one human being to another. However, the use of pesticides has caused the development of resistant strains of mosquitoes. It is hoped that genetic engineering techniques soon will result in the production of a vaccine.

The protozoans are animal-like in that they are heterotrophic and motile. They are classified according to the type of locomotor organelle employed (table 28.1).

Algae

The term **algae** is used for aquatic organisms that photosynthesize as do terrestrial plants. Algae produce the food that maintains communities of organisms both in the oceans and in bodies of fresh water. They commonly are named for the type of pigment they contain; therefore, there are green, golden brown, brown, and red algae. All algae contain green chlorophyll, but they also can contain other pigments that mask the color of the chlorophyll. Algae are grouped according to their pigmentation and biochemical differences, such as the chemistry of the cell wall and the chemical compound used to store excess food.

Life Cycles

There are 3 types of life cycles exhibited by living organisms, and all these types are found among the algae: in the **haplontic cycle,** the adult is haploid; in **alternation of generations,** a haploid generation alternates with a diploid generation; in the **diplontic cycle,** the adult is always diploid. These cycles are diagrammed in figure 28.15, and are further contrasted in table 28.2.

Figure 28.15

Three types of life cycles in organisms. a. Haplontic life cycle is typical of algae and fungi. Notice that the adult is haploid, the gametes are sometimes isogametes (look alike), and the only diploid part of the cycle is the zygote, which undergoes meiosis to produce spores. b. Alternation of generations life cycle is typical of plants. Notice that there are 2 generations. The sporophyte (2N) produces the spores by meiosis, and the gametophyte (N) produces gametes. c. Diplontic life cycle typical of animals. Notice that the adult is always diploid and meiosis produces heterogametes (egg and sperm). The earliest cycle may have been the haplontic, which could have led to both the alternation of generations and the diplontic cycles.

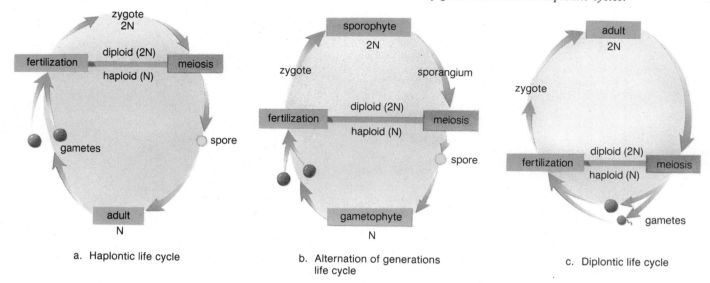

a. Haplontic life cycle

b. Alternation of generations life cycle

c. Diplontic life cycle

Table 28.2 Life Cycles

Name	Chromosome Number in Adult(s)		Spores
Haplontic	Haploid only		Yes
Alternation of generations	Haploid	Diploid	Yes
Diplontic	Diploid only		No

28.1 Critical Thinking

1. In reference to the life cycles (fig. 28.15), the timing of what one element determines whether or not a diploid adult results?
2. Hypothesize how the haplontic life cycle may have given rise to the alternation of generations life cycle.
3. Hypothesize how the alternation of generations life cycle may have given rise to the diplontic life cycle.

In the first 2 cycles, meiosis produces **spores,** structures that mature to a haploid generation. Plants, which are studied in the next chapter, use the alternation of generations life cycle. In the diplontic life cycle, meiosis produces gametes, the only haploid structures found within the cycle. Animals, including human beings, follow the diplontic cycle.

In general, an organism exhibits one of 3 life cycles (haplontic, alternation of generations, or diplontic). All 3 of these are seen among the algae.

Green Algae

There are single-celled, colonial, filamentous, and multicellular green algae (phylum Chlorophyta). It sometimes is suggested that the green algae are ancestral to the first plants because both of these groups possess chlorophylls *a* and *b,* both store reserve food as starch, and both have cell walls that contain cellulose.

Flagellated Green Algae   *Chlamydomonas* (fig. 28.16) is a single-celled green algae that has been studied in detail by means of the electron microscope. It has a definite cell wall and a single, large, cup-shaped chloroplast that contains a pyrenoid, a dense body where starch is stored. A red-pigmented eyespot is

Figure 28.16

Chlamydomonas, *a motile green alga.* a. *Structure of* Chlamydomonas. *b.* Chlamydomonas *has the haplontic life cycle. During sexual reproduction, mature adults produce gametes that fuse to give a zygote. The zygote becomes a thick-walled* zygospore *that enters a period of dormancy. Upon germination, meiosis produces zoospores that grow to become haploid adults. These individuals usually reproduce asexually, with each of the original zoospores giving rise to many individuals.*

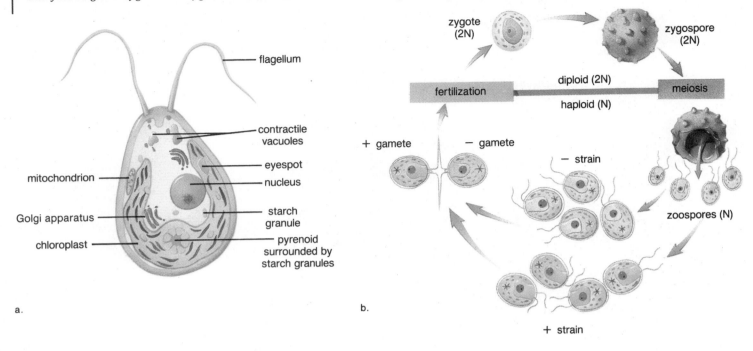

Figure 28.17

Volvox, *a green algal colony.* a. *The adult* Volvox *colony often contains daughter colonies, which are asexually produced by special cells.* b. *During sexual reproduction, colonies produce a definite sperm and egg. Some produce either sperm or egg, and others produce both sperm and eggs.*

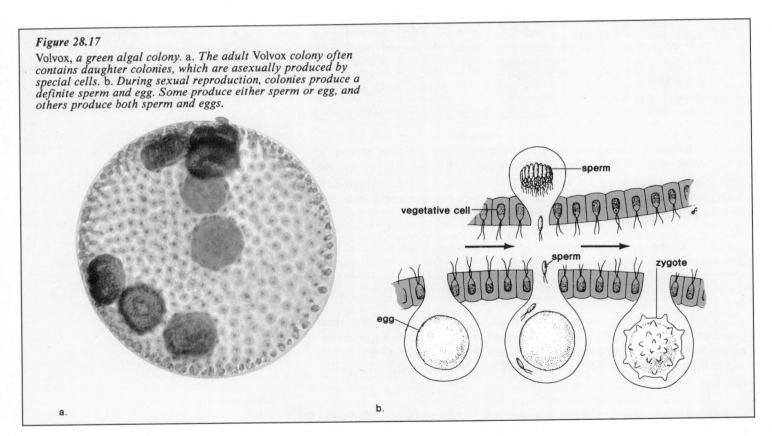

Evolution and Diversity

sensitive to light and helps to bring the organism into the light, where photosynthesis can occur. The 2 flagella project from the anterior end and move the cell freely toward the light. *Chlamydomonas* has both animal-like and plantlike characteristics in that it is motile and yet it makes its own food; it is believed that such an organism is a good example of the most primitive protist.

Chlamydomonas also is a good example of an organism that follows the haplontic life cycle (fig. 28.16*b*). Usually, this protist practices asexual reproduction, and the adult divides to give *zoospores* (flagellated spores) that resemble the parent cell. During sexual reproduction, gametes of 2 different strains come into contact and join to form a zygote. A heavy wall forms around the zygote, and it becomes a zygospore. The zygospore is able to survive until conditions are favorable for germination and subsequent production of 4 zoospores by meiosis. The gametes shown in figure 28.16*b* are **isogametes;** that is, they look exactly alike.

The life cycle of *Chlamydomonas* illustrates the primary differences between asexual and sexual reproduction. Sexual reproduction is simply reproduction that involves gametes. Distinct and separate sexes are not required, nor are heterogametes (dissimilar gametes, such as egg and sperm). Sexual reproduction aids the process of evolution because it offers means to produce variations in addition to mutations.

Colonial Green Algae   A number of colonial (loose association of cells) forms occur among the flagellated green algae. A *Volvox* **colony** is a hollow sphere with thousands of cells arranged in a single layer surrounding a watery interior. The cells of a *Volvox* colony, each one of which resembles a *Chlamydomonas* cell, cooperate in that the flagella beat in a coordinated fashion. Some cells are specialized for reproduction, and each of these can divide asexually to form a new daughter colony (fig. 28.17*a*). This daughter colony resides for a time within the parental colony. A daughter colony leaves the parental colony by releasing an enzyme that dissolves a portion of the matrix of the parental colony, allowing it to escape. During sexual reproduction, there are **heterogametes**—large nonmotile eggs and small flagellated sperm (fig. 28.17*b*).

Filamentous Green Algae   Filaments are end-to-end chains of cells that form after cell division occurs in only one plane. *Spirogyra* (fig. 28.18), found in green masses on the surface of ponds and streams, has chloroplasts that are ribbonlike and arranged in a spiral within the cell. These species practice a form of sexual reproduction called *conjugation,* during which the contents of the cells act as isogametes. The 2 filaments line up next to one another, and the contents of the cells of one filament move into the cells of the other filament, forming 2N zygotes. These zygotes survive the winter, and in the spring, they undergo meiosis to produce new haploid filaments.

Multicellular Sheets   Multicellular *Ulva* is commonly called sea lettuce because of its leafy appearance (fig. 28.19). *Ulva* shows the alternation of generations life cycle in the same way as terrestrial plants except that (1) the spores are flagellated, (2) there are isogametes, and (3) both generations look exactly alike. In plants, one generation is typically dominant over (longer lasting than) the other.

Green algae are a diverse group that have some of the same characteristics as plants. All 3 types of life cycles (fig. 28.15) are seen, but Ulva *has a life cycle that has 2 distinct generations like that of plants.*

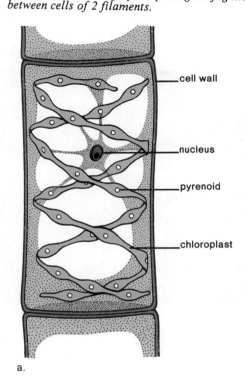

cell wall

nucleus

pyrenoid

chloroplast

a.

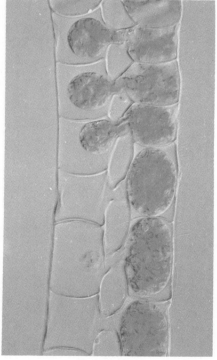

b.

Figure 28.19

Ulva, *a green alga.* a. *Photomicrograph of* Ulva, *commonly called sea lettuce.* b. *Members of the genus* Ulva *undergo alternation of generations life cycle in which the sporophyte and the gametophyte have the same appearance. The sporophyte produces spores by meiosis, and the gametophyte produces isogametes that give a zygote upon fertilization.*

a.

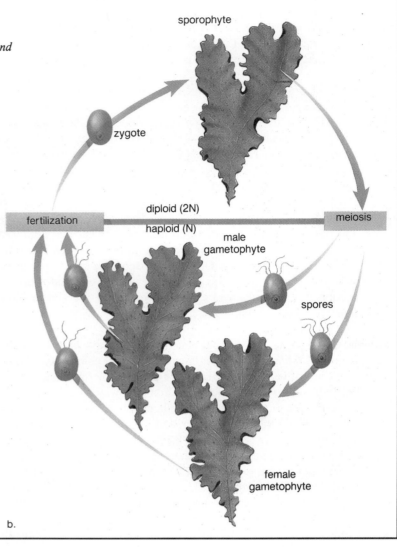

sporophyte

zygote

fertilization — diploid (2N) — meiosis

haploid (N)

male gametophyte

spores

female gametophyte

b.

28.2 Critical Thinking

1. For reasons given in the text (p. 597), it is believed that green algae are ancestral to plants. Give 2 reasons why a member of the genus *Ulva* is a good candidate for this ancestor.
2. Some believe that the euglenoids are ancestral to both plants and animals. How would a euglenoid have to change in order to become more plantlike? How would a euglenoid have to change in order to become more animal-like?
3. Suppose you used cellular structure to determine which protist is ancestral to animals. Which organelle would be of greatest interest to you? Why?

Seaweeds

Multicellular green algae (such as *Ulva*), red algae, and brown algae are all seaweeds. Although all seaweeds have chlorophyll, this green pigment sometimes is masked by red or brown pigments. While at one time it was believed that these accessory pigments determined the depth at which these algae are found, systematic studies have shown that there is no difference in the overall depth distribution of red, brown, and green seaweeds.

Although the seaweeds are multicellular as are plants, this does not mean that they are closely related to plants. Multicellularity most likely arose independently in several different lines.

Brown algae (phylum Phaeophyta) range from small forms with simple filaments to large forms between 50 and 100 m long (fig. 28.20). Large brown algae often are observed along the rocky shoreline in the north temperate zone, where they are pounded by waves as the tide comes in and are exposed to dry conditions at low tide. These plants are anchored firmly by holdfasts, and when the tide is in, their broad flattened blades are buoyed by air bladders. When the tide is out, they do not dry out because their cell walls contain a mucilaginous, water-retaining material.

Most brown algae have the alternation of generations life cycle, but some species of *Fucus* are unique in that they have the diplontic life cycle (fig. 28.15), in which meiosis produces gametes and the adult is always diploid, as in animals.

Figure 28.20

Diversification among the brown algae. Both Laminaria, *a type of kelp, and* Fucus, *known as rockweed, are examples of brown algae that grow along the shoreline. In deeper waters, giant kelps, such as* Nereocystis *and* Macrocystis *often form spectacular underwater forests. Individuals of* Sargassum *sometimes break off from their holdfasts and form floating masses, where life forms congregate in the ocean. Brown algae provide food and habitat for marine organisms and even have been harvested for human food and sources of fertilizer in several parts of the world. They are also a source of algin, a pectinlike material that is added to ice cream, sherbet, cream cheese, and other products to give them a stable, smooth consistency.*

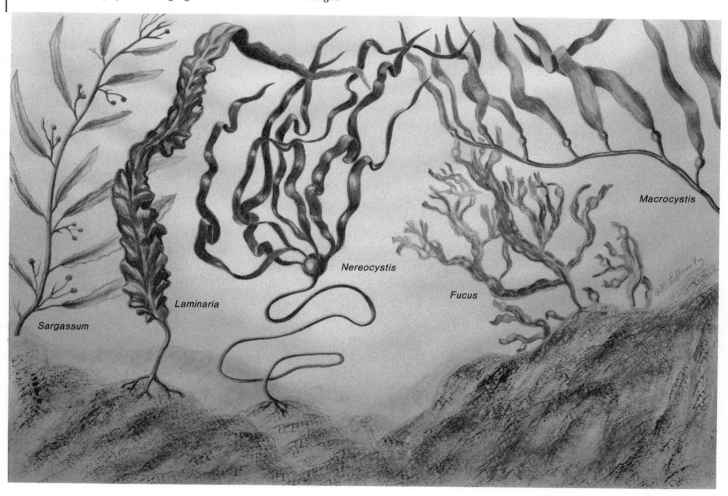

Like the brown algae, the red algae (phylum Rhodophyta) are multicellular, but they occur chiefly in warmer seawaters, growing both in shallow waters and as deep as light penetrates. Some forms of red algae are filamentous, but more often they are complexly branched, with the branches having a feathery, flat, and expanded or ribbonlike appearance. Coralline algae are red algae that have cell walls impregnated with calcium carbonate. In some instances, they contribute as much to the growth of coral reefs as coral animals do.

The seaweeds are multicellular protista that are grouped according to the type of accessory pigment they contain.

Other Types of Algae

There are several groups of exclusively unicellular algae in kingdom Protista. They also contain chloroplasts and carry out photosynthesis in a manner similar to plants.

Figure 28.21

Euglena anatomy. Euglena *is typical of those protozoans that have both animal-like and plantlike characteristics. A very long flagellum propels the body, which is enveloped by a flexible pellicle made of protein. A photoreceptor shaded by an eyespot allows* Euglena *to find light, after which photosynthesis can occur in the numerous chloroplasts. In addition to the pyrenoids, which store starch, there are starch granules in the cytoplasm.*

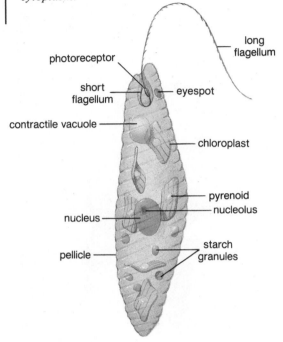

Euglenoids (phylum Euglenophyta) are freshwater organisms having both animal-like and plantlike characteristics (fig. 28.21). Their animal-like characteristics include motility and a flexible body wall. Euglenoids move by means of a long flagellum that projects to the anterior. Because they are bounded by a flexible pellicle instead of a rigid cell wall, they can assume different shapes as the underlying cytoplasm undulates or contracts.

The plantlike characteristics of euglenoids include chloroplasts. A photoreceptor shaded by an eyespot at the base of one flagellum allows euglenoids to judge the direction of light. After they move toward light, photosynthesis takes place. Carbohydrate is stored as starch.

Dinoflagellates (phylum Pyrrophyta) have 2 flagella; one is free, but the other is located in a transverse groove that encircles the animal. The beating of these flagella causes the organism to spin like a top. The cell wall, when present, frequently is divided into closely joined polygonal plates of cellulose. At times there are so many of these organisms in the ocean that they cause a condition called "red tide." The toxins given off in these red tides cause widespread fish kills and can cause paralysis in humans who eat shellfishes that have fed on the dinoflagellates.

Usually, the dinoflagellates are an important source of food for small animals in the ocean. They also live as symbiotes within the bodies of some invertebrates. For example, corals usually contain large numbers of these organisms, and this allows them to grow much faster than otherwise.

Diatoms (phylum Chrysophyta) (fig. 28.22) have a golden brown accessory pigment in their chloroplasts that can mask the color of chlorophyll. The structure of a diatom often is compared to a box because the cell wall has 2 halves, or valves, with the larger valve acting as a "lid" for the smaller valve. When diatoms reproduce, each receives only one old valve. The new valve fits inside the old one.

The cell wall of the diatom has an outer layer of silica, a common ingredient of glass. The valves are covered with a great variety of striations and markings that form beautiful patterns when observed under the microscope. Diatoms are among the most numerous of all unicellular algae in the oceans. As such, they serve as an important source of food for other organisms. Their remains, called diatomaceous earth, accumulate on the ocean floor and are mined for use as filtering agents, soundproofing materials, and scouring powders.

Figure 28.22

Diatom diversity. *These unicellular algae may be variously colored as shown in (a) and (b), but they have a unique golden-brown pigment in addition to chlorophyll within the chloroplasts. These beautiful patterns are markings on their silica-embedded walls.*

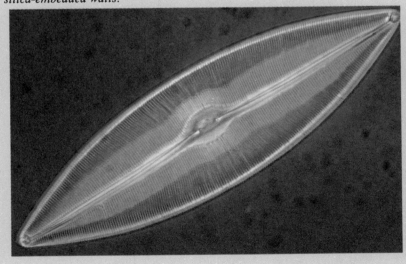
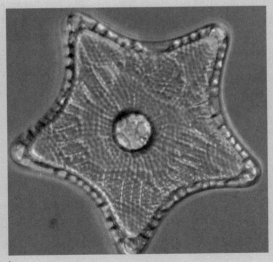

a.

b.

Slime Molds and Water Molds

Slime molds and water molds have 2 funguslike characteristics. These 2 types of protista live on dead organic matter and produce windblown spores during one phase of their life cycle.

Slime Molds

There are 2 types of **slime molds** (phylum Myxomycota): cellular and acellular. They both have an amoeboid stage that lives on rotting logs or dead agricultural crops. In cellular slime molds, the total mass is composed of individual amoeboid cells, but in acellular slime molds, the mass is multinucleated and is called a **plasmodium.** In another part of their life cycle, slime molds form fruiting bodies, stalks that produce windblown spores. The spores germinate to give cells that join to form a zygote. This zygote eventually begins the cycle again.

Water Molds

Some **water molds** (phylum Oomycota) live in the water, where they parasitize fishes, forming furry growths on the gills of the fishes. In spite of their common name, others live on land and parasitize insects and plants. A water mold was responsible for the potato famine in the 1840s that caused many Irish to come to the United States. Most water molds are saprophytic and live off dead organic matter, however.

Water molds have a threadlike body like that of fungi, but the cells are diploid, not haploid, and their life cycle is diplontic (fig. 28.15). Also, the cell walls are composed largely of cellulose, quite unlike the fungi.

Kingdom Fungi

Fungi (singular, fungus), like bacteria, are most commonly saprophytic decomposers that assist in the recycling of nutrients in ecosystems. Some fungi are parasitic, causing serious diseases in plants and animals. The bodies of all fungi, except unicellular yeast, are made up of filaments called **hyphae.** A hypha is an elongated cylinder containing a mass of cytoplasm and many haploid nuclei, which may or may not be separated by cross walls. A collection of hyphae is called a **mycelium.**

Fungi reproduce in accordance with the haplontic life cycle (fig. 28.15), and most are adapted to life on land in that they produce windblown spores. Most fungi reproduce both asexually and sexually; classification is largely based on the mode of sexual reproduction.

Fungi are saprophytic heterotrophic eukaryotes composed of hyphal filaments (a mycelium). Fungi produce spores during both sexual and asexual reproduction, and the major groups of fungi are distinguishable on the basis of sexual reproduction.

Black Bread Molds

Black bread molds (division Zygomycota) belonging to the genus *Rhizopus* often are used as an example of fungi. These molds exist as a whitish or grayish haploid mycelium on bread or fruit (fig. 28.23). During asexual reproduction, some hyphae grow upright and bear a spherical **sporangium** within which thousands of spores are formed.

During sexual reproduction, hyphae of 2 different strains (usually called plus and minus) reach out to one another and form a diploid zygote that darkens as it enlarges into a *zygospore.* After remaining dormant for several months, meiosis occurs and the zygospore germinates, producing a short haploid hypha and sporangium. The sporangium release windblown spores.

Figure 28.23

Life cycle of the black bread mold Rhizopus. *(lower left) Asexually, a mycelium gives rise to spore-producing sporangia. (above) Sexually, the tip ends of hyphae from opposite mating strains can fuse, giving a zygote, the only diploid portion of the cycle. After a period of dormancy, meiosis is followed by germination of the zygospore and production of a sporangium. Windblown spores, an adaptation to land, produce mycelia.*

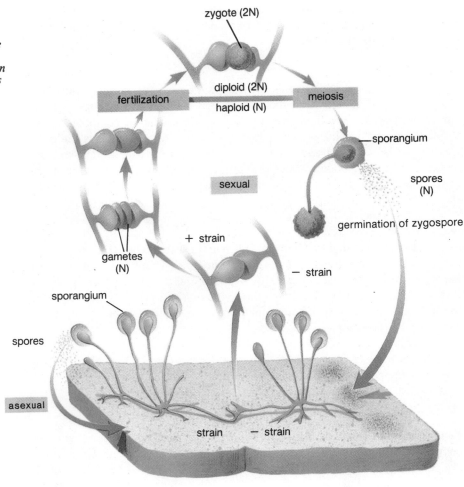

28.3 Critical Thinking

1. Members of the kingdom Fungi are adapted to living on land. What characteristic of fungi would make you think that other organisms (such as plants and animals) must have been present on land before fungi?

2. Both *Chlamydomonas* and fungi follow the haplontic life cycle. What elements in the cycle show that the former is adapted to living in the water and the latter is adapted to living on land?

3. The mycelia of sac fungi and club fungi are found in the ground, and the fruiting bodies usually appear after a rain. What does this tell you abut the adaptation of fungi to living on land?

Sac Fungi

There are many different types of sac fungi (division Ascomycota) (fig. 28.24*a*), most of which produce asexual spores called **conidia.** During sexual reproduction, sac fungi form spores called ascospores within saclike cells called asci. In most species, the asci are supported within **fruiting bodies,** a collection of specialized hyphae. In cup fungi, the fruiting body takes the shape of a cup.

Yeasts are sac fungi that do not form fruiting bodies. In fact, yeasts are different from all other fungi in that they are unicellular and most often reproduce asexually by budding. Yeasts, as you know, carry out fermentation as follows:

$$glucose \rightarrow carbon\ dioxide + alcohol$$

In baking, the carbon dioxide (CO_2) from this reaction makes bread rise. In the production of wines and beers, it is the alcohol that is desired, and the carbon dioxide escapes into the air.

Blue-green molds, notably *Penicillium,* are also sac fungi. These molds grow on many different organic substances, such as bread, fabric, leather, and wood. They are used by humans to provide the characteristic flavor of Camembert and Roquefort cheeses; more important, they produce the antibiotic penicillin. Another mold, the red bread mold *Neurospora,* was used in the experiments that helped to decipher the function of genes.

Unfortunately, sac fungi are also the cause of chestnut tree blight and Dutch elm disease, resulting in the death of most of these trees in the United States. Also, powdery mildew, apple scab, and ergot, a disease of cultivated cereals, are caused by sac fungi.

Figure 28.24

Sac and club fungi. a. Colorful cup fungi (a sac fungus). b. Bracket fungi (a club fungus) growing on a tree limb. c. Mushrooms (a club fungus) of the genus Mycena have bell-shaped caps.

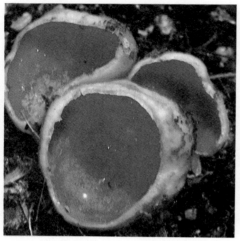

a.

b.

c.

Figure 28.25

Lichen structure. Lichens come in many shapes and sizes, but each is a symbiotic relationship between a fungus and an alga. Whether the fungus contributes to the relationship is debatable, and it even may be parasitic on the alga. a. Diagrammatic longitudinal section of a crustose (flat) lichen. b. Photo of a crustose lichen. Lichens are important soil formers.

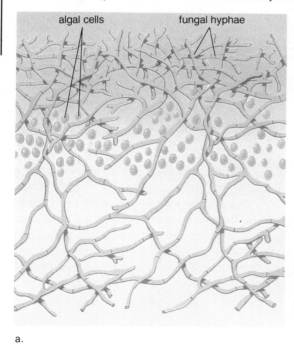

algal cells   fungal hyphae

a.

b.

A **lichen** is a symbiotic relationship between an alga and a fungus (fig. 28.25). The fungal portion of lichens is usually a sac fungus, while the algal part of this symbiosis is usually a green alga. Lichens can live on bare rock or in poor soil and are able to survive great extremes in temperature and moisture in all regions of the world. Reindeer moss is a lichen that is an important food source for arctic animals.

Figure 28.26

Life cycle of a mushroom. Beginning at lower right, hyphae from 2 different strains fuse and produce a mycelium in which there are haploid nuclei from both strains. The mycelium gives rise to a mushroom, consisting of a stalk and a cap. The gills on the underside of the cap are lined with basidia (club-shaped structures). After nuclei fuse to give a diploid nucleus, meiosis in each basidium produces basidiospores, which are windblown. If conditions are favorable, each basidiospore germinates into hyphae. (upper right) Scanning electron micrograph of a basidium and basidiospores.

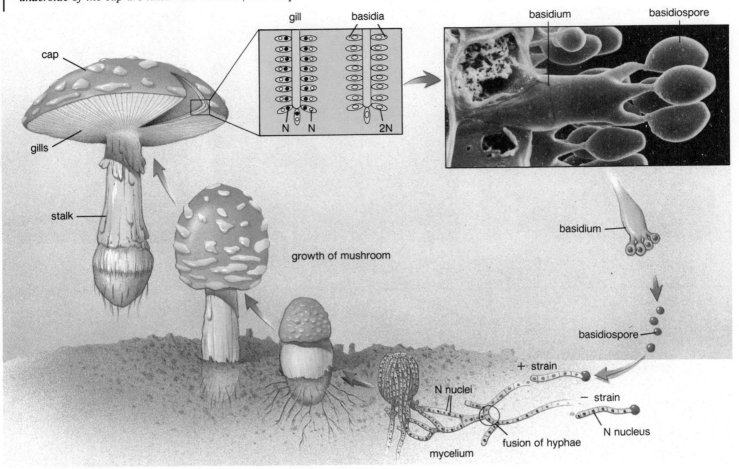

Club Fungi

Among the club fungi (division Basidiomycota) (fig. 28.24*b* and *c*), asexual reproduction is accomplished by formation of conidia. As a result of sexual reproduction, members of this group form club-shaped structures called *basidia,* often within fruiting bodies. The mushroom, the puffball, and the bracket (shelf) fungi are club fungi; the visible portions of these are actually fruiting bodies, and the mycelia lie beneath the surface. On the underside of a mushroom cap, the basidia project from the gills. Within each basidium, a diploid nucleus undergoes meiosis to produce basidiospores, which are windblown (fig. 28.26).

Club fungi are economically important. Mushrooms are raised and are sold commercially as a delicacy. Rusts and smuts are parasitic club fungi that attack grains, resulting in great economic loss and necessitating expensive control measures. They do not have conspicuous fruiting bodies, and generally occur as vegetative hyphae that produce spores of various kinds. On the other hand, the mycelia of club fungi that lie beneath the soil often form beneficial symbiotic relationships with plants, notably pine trees. These *mycorrhizae* (fungus roots) help the trees to garner nutrients from the soil and to grow at a faster rate. Therefore, foresters make sure the tree roots are exposed to club fungi before they are planted.

Other Types of Fungi

Some fungi (division Deuteromycota) cannot be assigned to a definite group because the sexual portion of the life cycle has not been observed. For this reason, these fungi sometimes are called "imperfect fungi." The fungi that cause ringworm and athlete's foot belong to this group, as does the yeast *Candida albicans,* which causes thrush, a mouth infection, and moniliasis, a fairly common vaginal infection in females who take the birth control pill.

During sexual reproduction, the black bread molds produce spores in sporangia; the sac fungi produce spores in saclike cells; and the club fungi produce spores in club-shaped structures. The sac and club fungi typically have fruiting bodies. The sexual life cycles for certain fungi that parasitize humans are unknown.

Summary

Viruses are acellular obligate parasites that have a protein coat and a nucleic acid core. A virus must enter a host cell before reproduction is possible. The kingdom Monera includes prokaryotic, usually single-celled organisms, the archaebacteria and the eubacteria. The archaebacteria are adapted to living in extreme habitats. Most information relates to the eubacteria. Reproduction is by binary fission, but sexual exchange occasionally takes place. Some bacteria form endospores that can survive the harshest of treatment except sterilization. Usually bacteria are aerobic, but some are facultative anaerobes or even obligate anaerobes. All types of nutrition are found except holozoism. Many bacteria are saprophytes, but the cyanobacteria are important photosynthesizers.

The protists are eukaryotes. The kingdom Protista contains the protozoans, which are heterotrophic and ingest food; the algae, which carry on photosynthesis; and the slime molds and the water molds, which have some characteristics of fungi. Protozoans are classified according to the type of locomotor organelle. Green algae are diverse; some are unicellular or colonial flagellates, some are filamentous, and some are multicellular sheets. The latter are seaweeds, as are brown and red algae. Three types of algae are exclusively unicellular: euglenoids, dinoflagellates, and diatoms. Slime molds have an amoeboid stage and then form fruiting bodies that produce spores. Water molds have threadlike bodies.

Fungi are typically saprophytic heterotrophic eukaryotes composed of hyphae filaments that form a mycelium. Along with heterotrophic bacteria, they are organisms of decay. The fungi produce spores during both sexual and asexual reproduction. The major groups of fungi are distinguishable by their mechanism of sexual reproduction.

Study Questions

1. Compare and contrast the bacteriophage lytic and lysogenic cycles. (p. 586)
2. Describe the life cycle of a DNA animal virus and an RNA retrovirus. (pp. 586–88)
3. Describe the prokaryotic characteristics shared by most bacteria. (p. 589)
4. What are the 3 shapes of bacteria? (p. 589) How do bacteria reproduce? (p. 589) What are endospores? (p. 590)
5. Discuss the importance of bacteria, including cyanobacteria. (pp. 592–93)
6. Give an example of each type of protozoan studied. Describe the anatomy of those that are free living and the life cycle of a parasitic one. (pp. 593–96)
7. What are the 3 types of life cycles found among living things? (p. 596) Show that *Chlamydomonas* and black bread mold have the haplontic life cycle and that *Ulva* undergoes the alternation of generations life cycle. (pp. 599, 600, 603) What type of life cycle does *Fucus* have? (p. 600)
8. List the distinguishing features of the euglenoids, the dinoflagellates, and the diatoms, mentioning any anatomical and ecological aspects that are of importance. (p. 603)
9. Why are the slime molds and the water molds sometimes called the funguslike protista? (p. 603)
10. Describe the anatomical features of the fungi, and tell how fungi are classified. (p. 603)
11. Describe the structure and the life cycle of black bread mold. (pp. 603–4)
12. Define a fruiting body, and name 2 groups of fungi that typically have fruiting bodies. (p. 604)

Objective Questions

1. Viruses always have a(n) _____ core and a(n) _____ coat called a capsid.
2. The _____ are RNA viruses that carry an enzyme for RNA → DNA transcription.
3. All different types of _____ are classified as monerans.
4. Most bacteria are saprophytic decomposers, meaning that they _____ .
5. In contrast, cyanobacteria are _____ , using the energy of the sun to make their own _____ .
6. Amoeba move by means of _____ , and ciliates move by means of _____ .
7. In the haplontic life cycle, the only diploid stage is the _____ .
8. In *Spirogyra*, zygotes form following the process of _____ .
9. *Ulva* has the life cycle _____ , as do land plants.
10. The body of a fungus is a(n) _____ that contains filamentous _____ .

11. The _____ fungi and the
    _____ fungi both have fruiting
    bodies.

12. A lichen is a symbiotic relationship
    between a _____ and an
    _____.

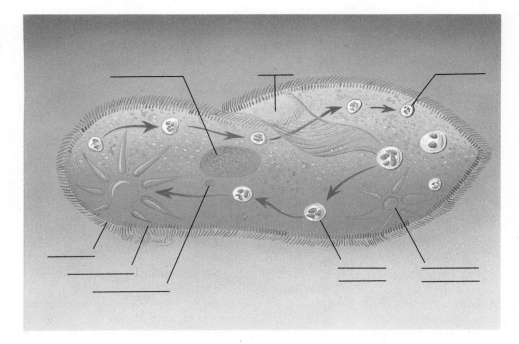

Label this Diagram.

See figure 28.12 (p. 595) in text.

Selected Key Terms

bacteriophage (bak-te′re-o-fāj′′) a virus that infects a bacterial cell. *585*

bacterium (bak-te′re-um) prokaryote that lacks the organelles of eukaryotic cells; archaebacteria and eubacteria. *589*

binary fission (bi′na-re fish′un) reproduction by division into 2 equal parts by a process that does not involve a mitotic spindle. *590*

colony (kol′o-ne) an organism that is a loose collection of cells that are specialized and cooperate to a degree. *599*

conjugation (kon′′ju-ga′shun) sexual union in which the nuclear material of one cell enters another. *595*

cyanobacterium (si′′ah-no-bak-te′re-um) photosynthetic prokaryote that contains chlorophyll and releases O$_2$; formerly called blue-green alga. *593*

diatom (di′ah-tom) one of a large group of fresh and marine unicellular algae having a cell wall consisting of 2 silica-impregnated valves that fit together as in a pill box. *602*

dinoflagellate (di′′no-flaj′e̅ lāt) one of a large group of marine unicellular algae that have 2 flagella; one circles the body, while the other projects posteriorly. *602*

euglenoid (u-gle′noid) one of a small group of unicellular algae that are bounded by a flexible pellicle and move by flagella. *602*

fungus (fung′gus) an organism, usually composed of strands called hyphae, that lives chiefly on dead organic matter; e.g., mushroom and mold. *603*

hypha (hi′fah) one filament of a mycelium that constitutes the body of a fungus. *603*

isogamete (i′′so-gam′et) one of 2 gametes whose union produces a zygote, but which have a similar appearance. *599*

lichen (li′ken) fungi and algae coexisting in a symbiotic relationship that is described as controlled parasitism of fungi on the algae. *605*

mycelium (mi-se′le-um) a mass of hyphae that make up the body of a fungus. *603*

protozoan (pro′′to-zo′an) animal-like protist that is classified according to means of locomotion: amoeba, flagellate, ciliate. *593*

retrovirus (ret′′ro-vi′rus) virus that contains only RNA and carries out RNA → DNA transcription. *588*

saprophyte (sap′ro-fit) heterotrophic organism such as bacteria and fungi that externally breaks down dead organic matter before absorbing the products. *593*

sporangium (spo-ran′je-um) a structure within which spores are produced. *603*

Chapter 29

The Plant Kingdom

Chapter Concepts

1 Plants, unlike algae, protect the embryo. This is an adaptation that facilitates a land existence.

2 Vascular plants have a system that not only transports water but also provides internal support.

3 In nonseed plants, spores disperse the species, and in seed plants, seeds disperse the species.

4 In seed plants, pollen transports the sperm nucleus to the egg nucleus.

Chapter Outline

Characteristics of Plants
    The Alternation of Generations Life Cycle
Nonvascular Plants
    Bryophytes
Vascular Plants
    Vascular Plants without Seeds
    Vascular Plants with Seeds
Seed Plants
    Gymnosperms
    Angiosperms
Comparisons between Plants

Figure 29.1

Representatives of the dominant types of land plants on earth today. a. Most plants today are flowering plants. This is a tiger lily. b. Mosses are low lying and usually found in moist locations. This is hairy-cap moss. c. Ferns were much more prominent in ancient times. This is a bracken fern. d. The gymnosperms are seed plants, as are flowering plants. This is a conifer in which the seeds are borne on cones.

a.

b.

c.

d.

Plants can be distinguished from algae even by a most cursory examination (fig. 29.1). Some plant characteristics are shared by algae, but some are not.

Characteristics of Plants

Plants are photosynthetic organisms with the following characteristics. Plants

1. contain chlorophylls *a* and *b* and cartenoids; they store reserve food as starch and have cellulose cell walls.
2. lack the power of motion or locomotion by means of contracting fibers.
3. are multicellular and have cells specialized to form tissues and organs.
4. have a life cycle that is identified as alternation of generations.

Table 29.1 *Comparison of Water Environment with Land Environment*

Water	Land
1. The surrounding water prevents the organism from drying out; that is, it prevents desiccation.	1. To prevent desiccation, the organism obtains water, provides it to all body parts, and possesses a covering that prevents evaporation.
2. The surrounding water buoys up the organism and keeps it afloat.	2. An internal structure helps a large body to oppose the pull of gravity.
3. The water prevents desiccation and allows easy transport of reproductive units, such as zoospores and swimming sperm.	3. In plants, the reproductive units may be adapted to transport by wind currents or by motile animals. Animals may provide a water environment for swimming sperm.
4. The surrounding water prevents the fertilized egg (zygote) from drying out.	4. The developing zygote is protected from possible desiccation.
5. The water maintains a relatively constant environment in regard to temperature, pressure, and moisture.	5. The organism may be capable of withstanding extreme external fluctuations in temperature, humidity, and wind.

5. have sex organs with an outer layer of nonreproductive cells that can prevent desiccation of gametes.
6. protect the developing diploid embryo from drying out by providing it with water and nutrients within the female reproductive structure.

All green algae fulfill the first and second characteristics listed here. In addition, members of the genus *Ulva* are multicellular (characteristic 3) and follow the alternation of generations life cycle (characteristic 4). *Fucus* is multicellular and has sex organs (characteristic 5). However, none of the green algae fulfill characteristic 6. It is clear that this characteristic is useful particularly to organisms that are adapted—as plants are—to living on the land.

A land existence offers some advantages to plants. One advantage is the greater availability of light for photosynthesis since water, even if clear, filters out light. Another advantage is that carbon dioxide (CO_2) and oxygen (O_2) are present in higher concentrations and diffuse more readily in air than in water. Water, however, provides many services that must be fulfilled in other ways when an organism lives on land (table 29.1).

Plants are multicellular photosynthesizers that are adapted to living on land. All plants protect the embryo from drying out.

The Alternation of Generations Life Cycle

All plants have a two-generation life cycle known as **alternation of generations.** This means that a plant exists in 2 forms: the haploid generation is the **gametophyte** that produces gametes; and the diploid generation is the **sporophyte** that produces spores by meiosis. Spores are haploid structures that develop or mature into a gametophyte plant.

In plants, one generation, either the gametophyte or the sporophyte, is *dominant* over the other generation: the dominant generation lasts longer, is larger and more conspicuous, and is actually the one that laypeople refer to as the plant because they do not realize that the other generation exists. Alternation of generations is diagrammed in figure 29.2. Notice the following in this life cycle:

Figure 29.2

Alternation of generations life cycle. In this life cycle, the zygote does not undergo meiosis and instead develops into the sporophyte. The sporophyte produces haploid spores that develop into the gametophyte, and the gametophyte produces gametes.

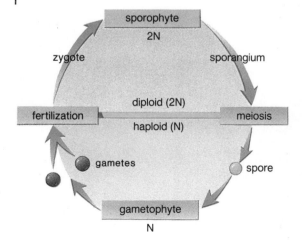

Figure 29.3

Marchantia, *a liverwort. Marchantia can reproduce asexually by means of gemmae—minute bodies that give rise to new plants. As shown here, the gemmae are located in cuplike structures called gemma cups.*

Meiosis produces **spores,** haploid structures that develop into the gametophyte (N generation). The **gametophyte produces the gametes** (egg and sperm).

A zygote (2N) results when the sperm fertilizes the egg and develops into the sporophyte. The **sporophyte produces spores** by meiosis in structures called sporangia.

All plants have a life cycle that shows an alternation of generations; some have a dominant gametophyte and some have a dominant sporophyte.

Nonvascular Plants

The nonvascular plants are restricted to the bryophytes; all other plants have vascular tissue.

Bryophytes

The **bryophytes** (division Bryophyta) include liverworts and mosses. Although liverworts (fig. 29.3), with flattened, lobed bodies, are widespread and well known, they represent but a small fraction of the total number of bryophyte species. Most species of liverworts are "leafy" and look somewhat like mosses, but examination shows that the body of a liverwort has distinct top and bottom surfaces, with numerous **rhizoids** (rootlike hairs) projecting into the soil. In contrast, a moss has a stemlike structure with radially arranged, leaflike structures (fig. 29.1*b*). Rhizoids anchor the plant and absorb minerals and water from the soil. Because bryophytes do not have vascular tissue, *they lack true roots, stems, and leaves.* Instead, they have rhizoids, stemlike structures, and leaflike structures.

The Moss Life Cycle

In mosses, the gametophyte is dominant—it is longer lasting. In some mosses, there are separate male and female gametophytes (fig. 29.4). At the tip of a male gametophyte are **antheridia** (singular, antheridium), in which swimming sperm are produced. After a rain or a heavy dew, the sperm swim to the tip of a female gametophyte, where eggs have been produced within the **archegonia** (singular, archegonium). Antheridia and archegonia are both multicellular structures, and each has an outer layer of jacket cells that protects the enclosed gametes from desiccation, or drying out. After an egg is fertilized, the developing sporophyte is retained within the archegonium. The sporophyte, which is dependent—indeed parasitic—on the gametophyte, consists of a *foot* that grows down into the gametophyte tissue, a *stalk,* and an upper *capsule,* or sporangium, where meiosis occurs and where haploid spores are produced. In some species of mosses, a hoodlike covering is carried upward by the growing sporophyte. When this covering and the capsule lid fall off, the spores are mature and ready to escape. The release of spores is controlled by one or 2 rings of "teeth" that project inward from the margin of the capsule. The teeth close the opening when the weather is wet but curl up and free the opening when the weather is dry. This appears to be the mechanism that allows spores to be released at times when they most likely will be dispersed by the wind.

When a spore lands on an appropriate site, it germinates. The single row of cells that first appears branches, giving an algalike structure called a *protonema.* After about 3 days of favorable growing conditions, new moss plants are seen at intervals along the protonema. Each of these consists of the rootlike rhizoids and the upright shoots of a moss gametophyte. The gametophytes produce gametes, and the moss life cycle begins again.

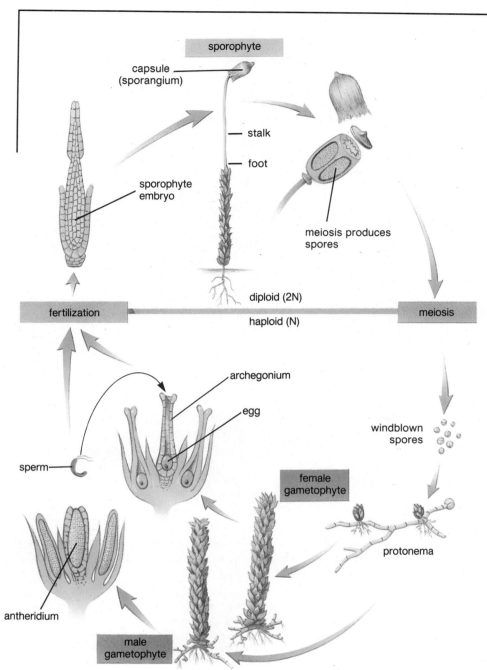

Figure 29.4

Moss life cycle. The male gametophyte (a leafy shoot) produces swimming sperm in antheridia. They use external moisture to reach egg-producing archegonia within the female shoot. Following fertilization, the diploid zygote develops into the sporophyte that produces haploid spores by meiosis. These spores develop into the separate male and female gametophytes.

Labels in figure: sporophyte; capsule (sporangium); stalk; foot; sporophyte embryo; meiosis produces spores; diploid (2N); haploid (N); fertilization; meiosis; archegonium; egg; sperm; windblown spores; female gametophyte; protonema; antheridium; male gametophyte

Adaptations of Bryophytes

Bryophytes are adapted to living in moist locations on land. Neither generation has vascular tissue for water transport, and the sperm swim in external moisture to reach the egg. However, the diploid embryo is protected from drying out by remaining within the archegonium for a time, and the organism is dispersed to new locations by windblown spores.

The Importance of Bryophytes

Certain bryophytes colonize rocks and slowly convert them to soil that can be used for the growth of other organisms.

Sphagnum, also called bog or peat moss, has commercial importance. This moss has special nonliving cells that can absorb moisture, which is why peat moss is often used in gardening to improve the water-holding capacity

Figure 29.5

Rhynia major, *the simplest and earliest known vascular plant. The leafless stem of this plant, which thrived about 350–400 million years ago, was green and carried on photosynthesis. The terminal sporangia apparently released their spores by splitting longitudinally.*

of the soil. In some areas, like bogs, where the ground is wet and acidic, dead mosses, especially sphagnum, accumulate and do not decay. This accumulated moss, called peat, can be used as fuel.

The bryophytes include the inconspicuous liverworts and the mosses, plants that have a dominant gametophyte. Bryophytes lack vascular tissue, and fertilization requires an outside source of moisture. Windblown spores disperse the species.

Vascular Plants

All the other plants studied here have vascular tissue and therefore are called vascular plants (also called **tracheophytes**), which are believed to have evolved sometime during the late Silurian Period (table 27.1).

You might suppose that the bryophytes evolved before the vascular plants, but actually the bryophytes appear later in the fossil record than the tracheophytes. Two suggestions to explain this have been proposed: (1) both the bryophytes and the tracheophytes evolved separately from a green algal ancestor, or (2) the bryophytes, which are nonvascular, evolved from a vascular ancestor. The second suggestion is not held in high regard because it would mean that the bryophytes lost a feature (vascular tissue) that they formerly had.

There is an extinct group of plants known from the fossil record, the **rhyniophytes** (fig. 29.5), that may have been the ancestral tracheophytes because they provide the first evidence of vascular tissue. The tracheophytes have 2 types of vascular tissue. **Xylem** conducts water and minerals up from the

Figure 29.6

Representative primitive tracheophytes. These plants were widespread when the tracheophytes first evolved. a. Whisk fern (division Psilophyta). b. Club moss (division Lycophyta). c. Horse tail (division Sphenophyta).

a.

b.

c.

soil, and **phloem** transports organic nutrients from one part of the body to another. Because they have vascular tissue, the specialized body parts of tracheophytes can be called properly roots, stems, and leaves.

Among the tracheophytes, the diploid generation, or the sporophyte, is dominant. One advantage in having a dominant diploid generation is that if a faulty gene is present, it can be masked by a functional gene. Among plants, this is also the generation that has vascular tissue. Vascular tissue in an important adaptation to the land environment because it transports water from the roots to the leaves. Also, xylem, with its strong-walled cells, supports the body of a plant against the pull of gravity. The tallest organisms in the world are vascular plants—the redwood trees of California.

The tracheophytes have vascular tissue, and this helps to explain why these are the most diverse and widely distributed plants.

Vascular Plants without Seeds

There are both nonseed plants and seed plants among the tracheophytes. The earliest vascular plants did not produce seeds.

Primitive Vascular Plants

The primitive vascular plants (fig. 29.6) include the whisk ferns (division Psilophyta), the club mosses (division Lycophyta), and the horsetails (division Sphenophyta). The psilophytes are of particular interest because they may be the most primitive. *Psilotum* (fig. 29.6a) has been classified as a psilophyte (whisk fern), though currently there is some question as to whether it is instead a true fern. Whatever its correct taxonomic designation, *Psilotum* bears considerable resemblance to the extinct rhyniophytes. The sporophyte consists of stems with scalelike structures but no leaves. There is a horizontal stem (lacking roots), from which rhizoids grow, and there are green, photosynthetic, upright branches with tiny, scalelike structures that grow upward. Sporangia are located on the branches. The gametophyte is separate from and smaller

Figure 29.7

Fern life cycle. The heart-shaped prothallus (the gametophyte, much enlarged here) produces swimming sperm in antheridia. They use external water in order to reach egg-producing archegonia. Following fertilization, the diploid zygote develops into a sporophyte—the familiar fern plant with large fronds. Spores produced by meiosis are released from sori located on the underside of the fronds. Each can germinate to give a gametophyte plant.

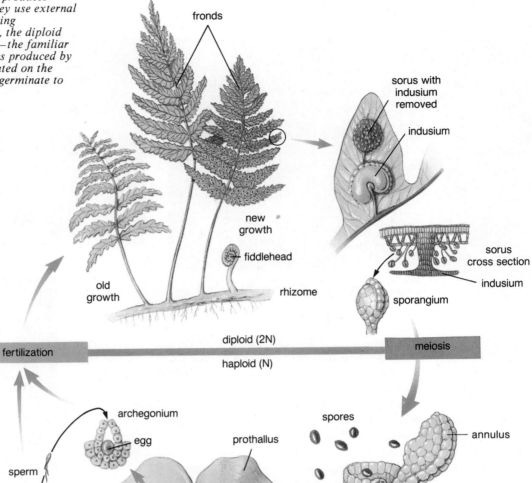

than the sporophyte; it also lacks vascular tissue. In fact, the life cycle of *Psilotum* is very close to that of the fern, which is discussed next.

Ferns

Ferns (division Pterophyta) vary in appearance. Many are low lying, but there are also tall tree ferns in the tropics. Figure 29.7 shows the life cycle of a common fern of the temperate zone, with a horizontal stem (rhizome) from which hairlike roots project downward and large **fronds** (leaves) project upward. Young fronds grow in a curled-up form called fiddleheads, which unroll as they grow. The fronds often are subdivided into a large number of leaflets.

The sporophyte fern plant represents the dominant generation. Sporangia develop in clusters called **sori** (singular, sorus) (fig. 29.8*a*), which are protected by a covering, the *indusium*. Within the sporangia, meiosis occurs and spores are produced. As a band of thickened cells on the rim of a sporangium (the *annulus*) (fig. 29.8*b*) dries out, it moves backward, and the spores are released. The gametophyte is a tiny (1–2 cm), heart-shaped structure called a **prothallus.** The antheridia and archegonia develop on the underside of a prothallus. Typically, the archegonia are at the notch and antheridia are toward the tip between the rhizoids. Fertilization takes place when moisture is present because the spiral-shaped sperm must swim from the antheridia to the archegonia. The resulting zygote soon develops into a sporophyte embryo consisting of a foot, a root, a stem, and a leaf. The root grows down into the soil, and the stem grows upward through the prothallus notch. As the sporophyte matures, the prothallus shrivels and disappears.

Adaptations of Ferns   For 2 reasons, ferns are likely to be found in habitats that are at least seasonally moist. First, the gametophyte lacks vascular tissue, and second, the gametophyte produces swimming sperm that use external moisture to swim to the eggs in the archegonia. Once established, the sporophyte of some ferns, like the bracken fern *Pteridium aquilinum,* can spread by vegetative reproduction into drier areas because this generation has vascular tissue. As the rhizomes grow horizontally in the soil, the fiddleheads grow up as new fronds.

The Importance of Nonseed Vascular Plants

During the Carboniferous Period (table 27.1), the horsetails, the club mosses, and the ferns were abundant, very large, and treelike (fig. 29.9). For some unknown reason, a large quantity of these plants died and did not decompose completely. Instead, they were compressed and compacted to form the coal that we still mine and burn today. (Oil was formed similarly but most likely formed in marine sedimentary rocks that included animal remains.)

In nonseed vascular tracheophyte plants, such as ferns, there is a dominant vascular sporophyte. These plants usually are found in moist environments because of an independent, nonvascular, gametophyte that produces swimming sperm.

Vascular Plants with Seeds

There are 2 groups of plants that produce seeds: the gymnosperms and the angiosperms. The gymnosperms produce naked seeds; that is, they are not

Figure 29.8
Fern sporophyte anatomy. a. *A photomicrograph of the underside of a leaflet showing sori (indusium removed).* b. *A scanning electron micrograph of sporangia within a sorus. When the rim (annulus) contracts, a sporangium breaks open and the spores are released (see fig. 29.7).*

a.

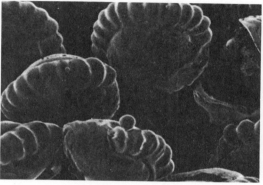

b.

Figure 29.9
Drawing of a carboniferous swamp. Carboniferous swamps are believed to have contained plants with fernlike foliage (left), treelike club mosses (left), and treelike horsetails (right).

Figure 29.10
Life cycle in seed plants. Notice there are
separate male and female gametophytes.

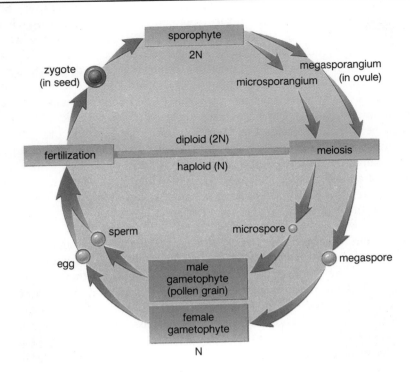

enclosed by fruit. Most people, for example, are familiar with the winged seed of pine trees. The angiosperms produce covered seeds; that is, they are enclosed by fruit. For example, most people are familiar with peaches, pears, and apples within which seeds are found. However, many vegetables are also fruits to a botanist. For example, a string bean is a fruit that contains many seeds in a row.

Seed Plants

Seed plants also undergo the alternation of generations life cycle; however, this cycle has been modified to allow them to reproduce without external water (fig. 29.10). Seed plants produce **heterospores,** called microspores and megaspores, instead of homospores, or identical spores. Microspores develop into immature male gametophytes while still retained within a microsporangium. After they are released, **pollen grains** develop into mature sperm-bearing male gametophytes. Therefore, pollen in seed plants replaces the external swimming sperm in nonseed plants. **Pollination,** the transfer of the male gametophyte to the vicinity of the female gametophyte, is dependent on wind or animals, rather than an external source of water.

A megaspore develops into an egg-bearing female gametophyte while still retained within the megasporangium. (The megasporangium is located within a structure called an **ovule.**) After fertilization, the zygote becomes an embryonic plant enclosed within a seed, the mature ovule. While nonseed plants are dispersed by means of spores, the seeds in seed plants serve as the dispersal package. A seed contains an embryonic sporophyte and stored food enclosed within a protective seedcoat. Seeds are resistant to adverse conditions, such as moisture or temperature extremes.

Figure 29.10 diagrams the life cycle of seed plants. In this life cycle, the following are true:

1. The dominant generation is the diploid sporophyte.
2. Meiosis produces microspores within microsporangia and megaspores within megasporangia. Each megasporangium is found within an ovule.

Figure 29.11

*Representatives of the lesser-known gymnosperm divisions.
a. Cycads resemble palm trees but are gymnosperms that
produce cones. Male plants have pollen cones, and female plants
have seed cones like the one shown here. b. Ginkgos exist only as
a single species, the maidenhair tree. This photograph of a
female plant features the seeds. Male plants have pollen cones.*

a.

b.

3. The mature male gametophyte is the pollen grain, and this structure
   contains a sperm that fertilizes the egg. The female gametophyte is
   retained within the ovule, and within the ovule an egg is produced.
4. Fertilization results in an embryo that lies inside the original ovule.
   Upon maturation, the ovule becomes the released seed.
5. The seed contains the new sporophyte and stored food, usually within
   several protective layers.

*While nonseed plants are dispersed by spores, the seeds in seed plants disperse the
species.*

Gymnosperms

The gymnosperms, which appear in the fossil record before angiosperms (table
27.1), produce naked **seeds;** that is, the seeds are not enclosed by fruit. There
are 4 divisions of gymnosperms (division Cycadophyta, division Ginkgophyta,
division Gnetophyta, and division Coniferophyta), but only 3 of these are considered here. **Cycads** are cone-bearing, palmlike plants found today mainly in
tropical and subtropical regions (fig. 29.11*a*). Only one species of **ginkgo,** the
maidenhair tree (fig. 29.11*b*), survives today. The maidenhair tree was restricted largely to ornamental gardens in China until it was discovered that
it does quite well in polluted areas. Because female trees produce rather smelly
seeds, it is the custom to use only male trees, propagated vegetatively, in city
parks.

The largest group of gymnosperms is the cone-bearing **conifers** (fig.
29.1*d*), which include pine, cedar, spruce, fir, and redwood trees. These trees
have needlelike leaves that are well adapted to not only hot summers but also
cold winters and high winds. Most gymnosperms are evergreen trees.

Figure 29.12

Life cycle of a pine tree. The mature sporophyte (pine tree) has female pine cones, which produce megaspores that develop into female gametophytes, and male pine cones, which produce microspores that develop into male gametophytes (mature pollen grains). Following fertilization, the immature sporophyte is present in seeds located on the female cones.

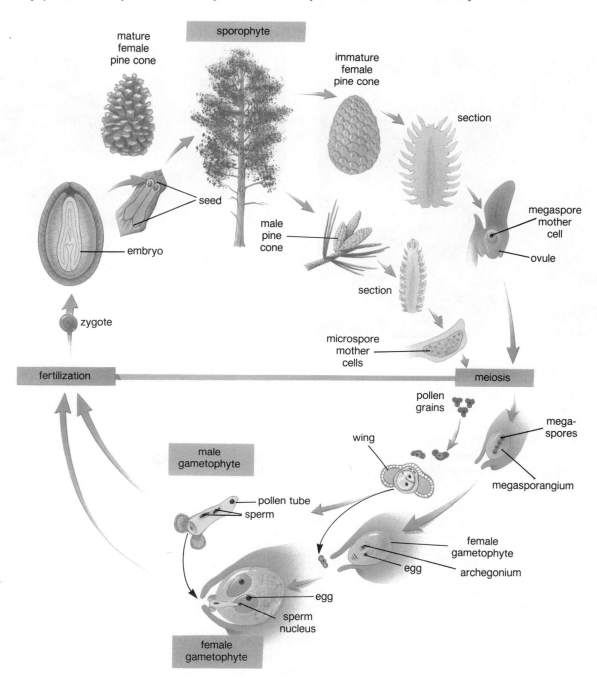

The Pine Life Cycle

The life cycle illustrated in figure 29.12 is a good example of a conifer's life history. The sporophyte is dominant, and its microsporangia and megasporangia are located on the scales of the cones. There are 2 types of cones—male and female.

The Male Pine Cone   Typically, the male pine cones are quite small and develop near the tips of lower branches. Each scale of the male cone has 2 or more microsporangia on the underside. Inside the microsporangia are microspore mother cells that undergo meiosis and produce 4 **microspores.** Each microspore undergoes meiosis and develops into a mature pollen grain, which is a sperm-bearing male gametophyte. The pollen grain has 2 lobular wings and is carried by the wind. Pine trees release so many pollen grains during pollen season that everything in the area may be covered by a dusting of yellow, powdery pine pollen.

The Female Pine Cone   The female pine cones are larger than the male pine cones and are located near the top of the tree. Each scale of the female cone has 2 ovules that lie on the upper surface. An ovule is surrounded by a thick, layered coat with an opening at one end. Within the ovule, a megaspore mother cell undergoes meiosis, producing 4 **megaspores.** Only one of these divides mitotically and goes on to develop into the mature female gametophyte, which has 2–6 archegonia, each containing a single, large egg lying near the ovule opening.

Pollination and Fertilization   During pollination, pollen grains are transferred from the male cone to the female cone. Once enclosed within the female cone, the pollen grain develops a pollen tube that slowly grows toward the ovule. The pollen tube discharges 2 nonflagellated sperm. One of these fertilizes an egg in the ovule, and the other degenerates. Fertilization takes place 15 months after pollination and is an entirely separate event from pollination, which is simply the transfer of pollen.

Seed Formation   After fertilization, the ovule matures and becomes the seed composed of the embryo, its stored food, and a seed coat. Finally, in the third season, the female cone, by now woody and hard, opens to release its seeds, whose wings are formed from a thin, membranous layer of the cone scale. When a seed germinates, the sporophyte embryo develops into a new pine tree, and the cycle is complete.

Adaptations of Gymnosperms

The reproductive pattern of conifers has several important innovations not found in the plants that have been considered so far. These differences make the conifers better adapted to reproduction in a dry environment. First, the gametophyte is not dependent on external moisture. The nonmotile sperm are enclosed within the pollen grain and are delivered to the egg by the growth of the pollen tube. Therefore, fertilization does not require external water for swimming sperm. Second, the female gametophyte is protected from drying out by remaining within the ovule, which still is attached to the cone. Third, the immature sporophyte is dispersed by the seed, within which it is enclosed. The seed need not germinate until conditions are favorable to support growth.

The Importance of Gymnosperms

Conifers grow on large areas of the earth's surface and are economically important. They supply much of the wood used for construction of buildings and production of paper. They also produce many valuable chemicals, such as those extracted from resin, a waxy substance that protects the conifers from attack by fungi and insects.

Perhaps the oldest and the largest trees in the world are conifers. Bristlecone pines in the Nevada mountains are known to be more than 4,500 years old (fig. 29.13), and a number of redwood trees in California are 2,000 years old and more than 90 m tall.

A conifer is the most typical example of a gymnosperm. In its life cycle, windblown pollen grains replace swimming sperm. Following fertilization, the seed develops from the ovule, a structure that has been protected within the body of the sporophyte plant. The seeds are uncovered and are dispersed by the wind.

Figure 29.13
Bristlecone pines, perhaps the oldest living plants in the world.

Angiosperms

Angiosperms (division Anthophyta) are the flowering plants. All hardwood trees, including all the deciduous trees of the temperate zone and the broad-leaved evergreen trees of the tropical zone, are angiosperms, although sometimes the flowers are inconspicuous. All herbaceous (nonwoody) plants common to our everyday experience, such as grasses and most garden plants (fig. 29.1), are flowering plants. Angiosperms are adapted to every type of habitat, including water (e.g., water lilies and duckweed).

The angiosperms are divided into 2 classes (fig. 7.2): the *monocots* (e.g., lily) and the *dicots* (e.g., buttercup). The monocots are almost always herbaceous, with flower parts in threes, parallel leaf veins, scattered vascular bundles in the stem, and one cotyledon, or seed leaf. The dicots are either woody or herbaceous and have flower parts usually in fours and fives, net leaf veins, vascular bundles arranged in a circle within the stem, and 2 cotyledons, or seed leaves.

The Flower

In angiosperms, the reproductive structures are located in a **flower** (fig. 29.14). The flower attracts insects and birds that aid in pollination (pollination by wind also occurs), and it produces seeds enclosed by **fruit.** There are many different types of fruits (table 29.2), some of which are fleshy (e.g., apples) and some of which are dry (e.g., peas enclosed by pods). The fleshy fruits sometimes are eaten by animals, which then may transport the seeds to a new location and deposit them during defecation. Both fleshy and dry fruits provide protection for the seeds. Many so-called vegetables are actually fruits—tomatoes, string beans, and squash, for example. Nuts, berries, and grains of wheat, rice, and oats are also fruits.

Angiosperms have well-developed vascular and supporting tissues that make them well adapted to terrestrial life. Their xylem tissue contains vessel elements, as well as tracheids. Other vascular plants, including virtually all gymnosperms, have only tracheids in their xylem. Whereas the gymnosperms are softwood trees, woody angiosperms are hardwood trees.

The Flowering Plant Life Cycle

The flowering plant life cycle is illustrated in figure 29.15.

The Development of the Gametophyte Within a flower, there is a diploid megaspore mother cell in each ovule of the ovary. The megaspore mother cell undergoes meiosis, producing 4 haploid megaspores. Three of these disintegrate, leaving one functional megaspore, whose nucleus divides mitotically until

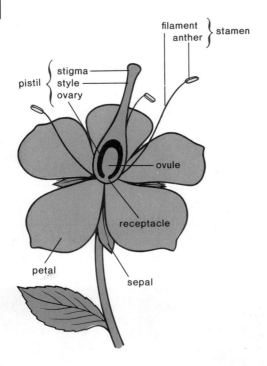

Figure 29.14
Anatomy of a flower. Flowers have an outer whorl of green, leaflike sepals and an inner whorl of usually colorful petals. At the very center of a flower is the pistil, composed of the stigma, the style, and the ovary. Grouped about the pistil are the stamens, each of which has an anther and a filament. Sometimes the pistil is called the female part of the flower and the stamens are called the male part of the flower. This is not strictly correct because they do not produce gametes; they produce megaspores and microspores respectively. A microspore goes on to become a male gametophyte that produces sperm, and a megaspore goes on to become a female gametophyte that produces an egg.

Table 29.2 Types	
Type	**Example**
Fleshy fruits	
Simple fruits	
Drupe	Peach, plum, olive
Berry	Grape, tomato
Pome	Apple, pear
Aggregate fruits	Strawberry, raspberry
Multiple fruits	Pineapple
Dry fruits	
Follicle	Milkweed, peony
Legume	Pea, bean, lentil
Capsule	Poppy
Achene	Sunflower "seed"
Nut	Acorn, hickory nut, chestnut
Grain	Rice, oat, barley

Figure 29.15

Life cycle of a flowering plant. Each ovule in the ovary of a pistil contains a megaspore mother cell that produces one functional haploid megaspore by meiosis. A megaspore divides mitotically 3 times, and the resulting structure with 8 nuclei and 7 cells is the female gametophyte (embryo sac). One of these cells is an egg. An anther contains many microspore mother cells, each of which produces 4 haploid microspores by meiosis. Each microspore divides mitotically, becoming a 2-celled pollen grain. When a pollen grain germinates, it contains a tube nucleus and 2 sperm nuclei. This is the mature male gametophyte. Following pollination, double fertilization occurs; one sperm nucleus unites with the egg nucleus, giving a zygote, and the other unites with the polar nuclei, giving a 3N endosperm nucleus. The zygote becomes an embryo, and the endosperm nucleus becomes stored food contained within a seed. Germination of the seed gives a sporophyte plant.

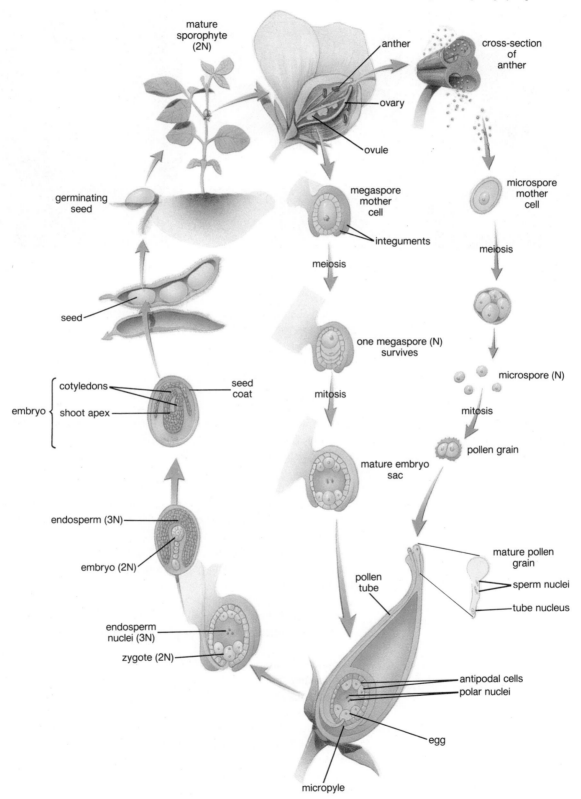

Figure 29.16

Flower adaptations that make them attractive to their pollinators. a. *A bee-pollinated flower is a color other than red (bees cannot detect this color), and has color guides leading to the source of nectar, a narrow flower tube that allows passage of the bee feeding apparatus, and a landing platform, where the reproductive structures of the flower brush up against the bee's body.* b. *A butterfly-pollinated flower is often a composite,* containing many individual flowers. The broad expanse provides room for the butterfly to land, after which it lowers it proboscis into each flower in turn. c. Hummingbird-pollinated flowers are curved back, allowing the bird to insert its beak to reach the rich supply of nectar. While doing this, the bird's head touches the reproductive structures.

a.

b.

c.

there are 8 haploid nuclei. This is the female gametophyte, which sometimes is called the embryo sac. At one end of the embryo sac there are 3 cells, one of which is the egg cell.

Male gametophytes are produced in the stamens. An anther contains 4 pollen sacs with many microspore mother cells, each of which undergoes meiosis to produce 4 haploid microspores. After a mitotic division, each microspore has 2 cells, one of which later divides again to give 2 sperm.

Pollination and Fertilization Pollination, which is simply the transfer of pollen from the anther to the stigma, is brought about by wind or with the assistance of a particular pollinator, an animal that carries the pollen from the anther to the stigma (fig. 29.16). A plant and its pollinator are adapted to one another. They have a mutualistic relationship. Each benefits—the plant uses the pollinator to ensure cross-pollination, and the pollinator uses the plant as a source of food in the form of nectar. Pollinators have mouth parts adapted for gathering nectar from flowers.

When a pollen grain lands on a stigma of the same species, it germinates, forming a pollen tube. The pollen tube grows as it passes between the cells of the stigma and the style to reach the female gametophyte. Now, *double fertilization* takes place. One sperm nucleus from the pollen tube unites with the egg nucleus, forming a zygote, and the other sperm nucleus unites with the polar nuclei, forming a triploid (3N) endosperm nucleus.

Seeds and Fruits With *double fertilization,* a number of processes begin. 1. The endosperm nucleus divides, forming the endosperm. **Endosperm** is a nutrient material for the developing embryo and sometimes for the young seedling as well. 2. The zygote develops into an embryo. 3. The outer layers (integuments) of the ovule harden and become the seed coat. A *seed* is a structure formed by the maturation of the ovule; it contains a sporophyte embryo plus stored food. The ovary and sometimes other floral parts develop into the fruit. A *fruit* is a mature ovary that usually contains seeds. Therefore, angiosperms are said to have covered seeds.

Figure 29.17

Angiosperms as food. Flowering plants are sources of food for the biosphere.

In the life cycle of flowering plants, the ovule becomes the seed and the ovary becomes the fruit.

Adaptations of Angiosperms

Like the gymnosperms, the angiosperms are well adapted to a land environment. They have innovations based on the production of flowers. Their flowers attract appropriate animal (e.g., insects) pollinators that increase the efficiency of pollination by their habit of visiting particular types of plants (p. 624). The fruits produced by flowers often are specialized to help with dispersal of the seeds.

It also should be mentioned that the vascular tissue of angiosperms contains vessel elements in addition to tracheids, and the improvement in water flow allows them to thrive in drier habitats. These are all selective advantages for the angiosperms.

The Importance of Angiosperms

The angiosperms are the major food producers in most terrestrial ecosystems (fig. 29.17). They provide food that sustains most of the animals on land, including humans. Still, it has been observed, as discussed in the reading on page 626, that humans use relatively few types of plants for food. Grasses, alfalfa, and clover also are used as food or forage for livestock.

Since we earlier stressed that all our food is ultimately derived from plants, it may come as somewhat of a surprise to learn that relatively few species of plants are involved. Of the 800,000 kinds of plants estimated to be in existence, only about 3,000 species have provided food, even in the form of nuts, berries, and other fleshy fruits.

Virtually all of these food plants are angiosperms, or flowering plants. This is not surprising when you recall that only the angiosperms have seeds enclosed in fruits, and it is fruits that are used by humans for food.

Of the 3,000 plants noted above, only 150 species have been cultivated extensively and have entered the commerce of the world. And of the 150, only 12 species are really important—indeed it can be said that these 12 plants stand between humanity and starvation. If all 12 or even if a few of these cultivated plants were eliminated from the earth, millions of people would starve.

Three of these all-important species are cereals—**wheat, corn,** and **rice**; the last alone supplies the energy required by 50%

of the people of the world. It is a remarkable fact that each of these cereals, or grains, is associated with a different major culture or civilization—wheat with Europe and the Middle East, corn or maize with the Americas, and rice with the Far East. Three of the 12 food plants are so-called root crops—**white,** or **Irish potato** (not a root but a **tuber,** an enlarged tip of a rhizome, or horizontal underground stem); **sweet potato**; and **cassava,** or **manioc**

or **tapioca,** from which millions of people in the tropics of both hemispheres derive their basic food. Two of the 12 are sugar-producing plants—**sugar cane** and **sugar beet.** Another pair of species are legumes—the **common bean** and the **soybean,** both important sources of vegetable protein and hence sometimes referred to as the "poor person's meat." The final 2 plants of this august company are tropical tree crops—**coconut** and **banana.** . . .

"Twelve Plants Standing between Man and Starvation." From Tippo/Stern, *Humanistic Botany.* © 1977 W.W. Norton & Company, New York, NY. Reprinted by permission.

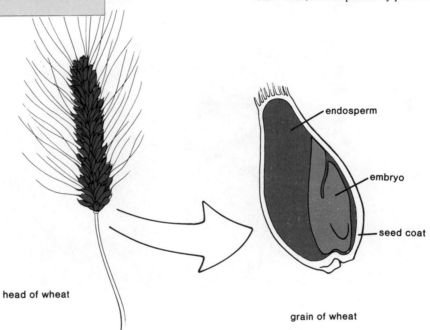

head of wheat

grain of wheat

Humans use angiosperms for many other functions. Their wood becomes the timber used for construction and furniture production. It also is used for fuel (firewood), particularly in poorer countries. Flax and cotton are sources of natural fiber for making cloth, and cellulose can be treated to yield rayon. Plant oils not only are used in cooking but also in making perfumes and medicines. Spices are from various parts of plants: peppercorns are small, berrylike structures from a vine; cinnamon comes from the bark of a tree; and cloves are dried flower buds. Various drugs come from angiosperm plants, including morphine and heroin from the juice of the poppy and marijuana from the leaves and flowers of the plant *Cannabis, sativa.*

Comparisons between Plants

We have seen that plants are adapted to a land existence. Some prefer a moist location and others can tolerate rather dry conditions. Table 29.3 compares the adaptations of the nonseed and seed plants in these regards.

The role of the haploid and diploid stages in the life cycle of various plants can be correlated with adaptation to moist versus dry environments (fig.

Table 29.3 *Adaptation Summary*

Plant	Generations	Reproduction	Representative
Nonseed Plants		**Windblown spores disperse the species.**	
Bryophytes	The sporophyte is dependent on the dominant gametophyte; both generations lack vascular tissue.	Swimming sperm require a source of outside moisture.	Moss
Primitive tracheophytes	The 2 generations are independent; the dominant sporophyte has vascular tissue.	Swimming sperm require a source of outside moisture.	Fern
Seed Plants		**Seeds disperse the species.**	
Gymnosperms (naked seeds)	The gametophyte is dependent on the dominant sporophyte; the sporophyte has vascular tissue.	Pollen grains replace swimming sperm; windblown seeds.	Pine
Angiosperms (covered seeds)	Adapted in the same manner as gymnosperms.	Insect pollination; fruits aid dispersal of seeds.	Flower

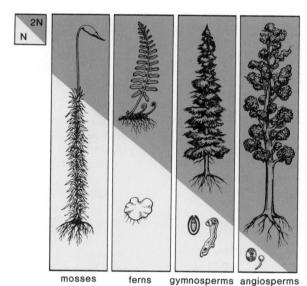

mosses   ferns   gymnosperms   angiosperms

Figure 29.18
The relative importance of the diploid (2N) and haploid (N) generations among plants. In mosses, the gametophyte is dominant and larger than the dependent sporophyte. In ferns, the sporophyte is dominant and larger than the gametophyte. Both generations are independent. In gymnosperms and angiosperms, the sporophyte is dominant and very much larger than the dependent and microscopic gametophyte.

29.18). Mosses, with a dominant gametophyte and a small, dependent sporophyte, are adapted to moist locations and have a limited distribution elsewhere. Ferns have well-developed sporophyte bodies and vascular tissue, but they still require very wet conditions for growth of a small, independent gametophyte and fertilization by swimming sperm.

The gymnosperms and the angiosperms are distributed widely on land because the large, dominant sporophyte is well adapted to relatively dry environments. Furthermore, the delicate spores, gametophytes, gametes, zygotes, and embryos are enclosed within protective coverings produced by the sporophyte plant.

29.2 Critical Thinking

1. Most terrestrial organisms protect the gametes and the embryo from drying out. Compare the reproductive adaptations of humans to that of trees.
2. Most terrestrial organisms have a transport system. Compare the transport adaptations of a tree to that of humans.
3. Many terrestrial organisms have an internal skeleton to oppose the force of gravity. Compare the skeletal adaptations of a tree to that of humans.
4. In the temperate zone, humans remain active in the winter; deciduous trees do not. Explain.

Summary

Plants are multicellular, photosynthetic organisms adapted to a land existence. Among the various adaptations, all plants protect the developing embryo from drying out. Plants have an alternation of generations life cycle, but some have a dominant gametophyte and others have a dominant sporophyte. The bryophytes, which include the liverworts and the mosses, are nonvascular plants and therefore lack true roots, stems, and leaves. In the moss life cycle, spores disperse the species, the gametophyte is dominant, and external water is required for sperm to reach the egg.

The vascular plants (tracheophytes) have a dominant diploid generation. In nonseed tracheophytes, spores disperse the species.

The extinct rhyniophytes may be the ancestral vascular plant. Today, the primitive vascular plants include the whisk ferns, the club mosses, and the horsetails. In these plants as well as in ferns, there is a separate and water-dependent gametophyte that produces swimming sperm.

Gymnosperms and angiosperms are vascular plants that produce seeds. They have a life cycle that includes heterospores and male and female gametophytes. The male gametophyte is the pollen grain, and this structure replaces the swimming sperm of nonseed tracheophytes. The female gametophyte is retained within the ovule, a

sporophyte structure. Gymnosperms (meaning naked seeds) are exemplified by the conifers, in which the sporangia are located on cones. Angiosperms (meaning covered seeds) are the flowering plants that have seeds enclosed by fruit. Angiosperms provide most of the food that sustains terrestrial animals, and they are the source of many products used by humans.

Study Questions

1. What characteristics define plants? (pp. 610–11)
2. Draw a diagram to describe the moss life cycle, and point out significant features of this cycle. (p. 612)
3. The primitive vascular plants comprise which plants? At what time in the history of the earth were they larger and more abundant than today? (pp. 615, 617)
4. Diagram the fern life cycle, and point out significant features of this cycle. (p. 616)

5. Describe the pine life cycle, and point out the significant features of this cycle. (p. 620)
6. List 3 innovations that are observed in the gymnosperm plant life cycle that are not seen in the life cycle of ferns. (p. 621)
7. Describe the life cycle of flowering plants, pointing out any features not observed in gymnosperms. (p. 623)
8. List all the ways that angiosperms are adapted to a land existence. (p. 625)

9. For each group of plants studied, list one feature of ecological relevance and one feature of human relevance. (pp. 613, 617, 621, 626)
10. Draw a diagram that shows the increasing dominance of the sporophyte generation among the 4 groups of plants studied. How is this feature related to adaptation to a relatively dry environment? (pp. 626–27)

Objective Questions

1. All plants protect the _____ from drying out.
2. In the alternation of generations life cycle, the sporophyte (2N) produces _____ .
3. In the moss life cycle, the dominant generation is the _____ .
4. Mosses and ferns are apt to be found in moist locations because the sperm must _____ in external water to the egg.

5. In the fern life cycle, the gametophyte is independent and _____ from the sporophyte.
6. In the seed plant life cycle, there are _____ and therefore separate male and female gametophytes.
7. In the seed plant life cycle, _____ replace swimming sperm.
8. In the life cycle of the pine tree, ovules are found on _____ pine cones.

9. Angiosperms practice double fertilization; one sperm nucleus unites with the _____ nucleus and the other unites with the _____ nuclei.
10. In angiosperms, the ovule becomes the _____ and the ovary becomes the _____ .

Selected Key Terms

alternation of generations (awl″ter-na′shun uv jen″ĕ-ra′shunz) a life cycle typical of plants in which a diploid sporophyte alternates with a haploid gametophyte. *611*

antheridium (an″ther-id′e-um) male organ in certain nonseed plants where swimming sperm are produced. *612*

archegonium (ar″kĕ-go′ne-um) female organ in certain nonseed plants where an egg is produced. *612*

bryophyte (bri′o-fīt) a plant group that includes mosses and liverworts. *612*

frond (frond) the large leaf of a fern plant containing many leaflets. *616*

fruit (frōōt) a mature ovary enclosing seed(s). *622*

gametophyte (gam′ĕ-to-fīt) the haploid generation that produces gametes in the life cycle of a plant. *611*

heterospore (het′er-o-spor) a nonidentical spore, such as a microspore and a megaspore, produced by the same plant. *618*

megaspore (meg′ah-spōr) in seed plants, spore that develops into the female gametophyte. *621*

microspore (mi′kro-spōr) in seed plants, spore that develops into a pollen grain. *620*

ovule (o′vūl) in seed plants, a structure that contains the megasporangium, where meiosis occurs and the female gametophyte is produced; develops into the seed. *618*

pollen grain (pol′en grān) male gametophyte of seed plants. *618*

prothallus (pro-thal′-us) a small, heart-shaped structure, which is the gametophyte of the fern. *617*

seed (sēd) a mature ovule that contains an embryo with stored food enclosed in a protective coat. *619*

sorus (so′rus) a cluster of sporangia found on the underside of fern leaves. *617*

spore (spōr) usually a haploid reproductive structure that develops into a haploid generation. *612*

sporophyte (spo′ro-fīt) spore-producing diploid generation of a plant. *611*

tracheophyte (trā′kē-a-fīt) vascular plant, including ferns and seed plants, that has a dominant sporophyte. *614*

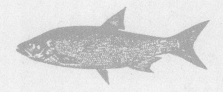

Chapter 30

Chapter Concepts

1 Animals are classified according to certain criteria, such as body plan, symmetry, number of germ layers, and level of organization.

2 There is an increase in complexity of organization when the animal groups are surveyed from those first evolved to those most recently evolved.

3 Animals are adapted to their way of life; for example, those adapted to an inactive life can be contrasted to those adapted to an active life, and those adapted to an aquatic existence can be contrasted to those adapted to a terrestrial existence.

Chapter Outline

Evolution and Classification
    The Evolution of Animals
    The Classification of Animals
Primitive Invertebrates
    Sponges
    Cnidarians
    Flatworms
    Roundworms
Advanced Invertebrates
    Mollusks
    Annelids
    Arthropods
    Echinoderms
Chordates
    Protochordates
    Vertebrate Chordates
    Comparisons between Vertebrates

The Animal Kingdom

Figure 30.1

The diverse animal kingdom. A hydra (a) and a bobcat (b) are both multicellular heterotrophic organisms that must take in food. Note the radial symmetry of the aquatic hydra that has tentacles for capturing small prey. Cats stalk their prey, and it is important not to be seen. The black-spotted brown coat of a bobcat blends in well with dense vegetation. Locating prey by sound may be required, and the bobcat has ear tufts that are thought to help hearing.

a.

b.

Animals (fig. 30.1) are heterotrophic and must take in food. In contrast to stationary green plants that absorb energy from the sun and make their own organic food, animals are nongreen and possess some means of locomotion that enables them to acquire food.

A predator that actively seeks and captures food best exemplifies the animal way of life. Predators have bilateral symmetry, good musculature, and a well-developed nervous system, including sense organs. All of these help the animal to seek prey and to escape enemies. Good predators also have a means of seizing and digesting their food.

All animals digest their food, carry on gas exchange, excrete waste, circulate nutrient and waste products to and from cells, coordinate their movements, protect themselves, and reproduce and disperse the species. The more complex animals have organ systems to carry out these functions; in simple animals, these functions sometimes are carried out by specialized tissues.

Evolution and Classification

The Evolution of Animals

All phyla of animals had evolved by the beginning of the Paleozoic Era some 600 million years ago (table 27.1). The evolutionary tree of animals (fig. 30.2) indicates that animals are believed to have arisen from protozoans—perhaps a colonial form whose cells differentiated into various types of cells. The evolutionary tree of animals resembles a tree with 2 main branches. The animal phyla located on the main trunk of the tree are referred to as the *primitive invertebrates* in this text, and the animals of the main 2 branches include the *advanced invertebrates* and the vertebrates. **Invertebrates** lack a dorsal backbone, while **vertebrates** have a backbone made up of vertebrae.

Animals are divided into 3 groups. The primitive invertebrate phyla form the trunk of an evolutionary tree. The advanced invertebrates and the vertebrates are distributed on 2 main branches of the tree.

The Classification of Animals

A study of the evolution of animals indicates that the most complex animals have the most advanced features listed in table 30.1. Classification of animals

Figure 30.2

Evolutionary tree of the animal kingdom. All animals are believed to be descended from protozoans; however, the sponges may have evolved separately from the rest of the animals.

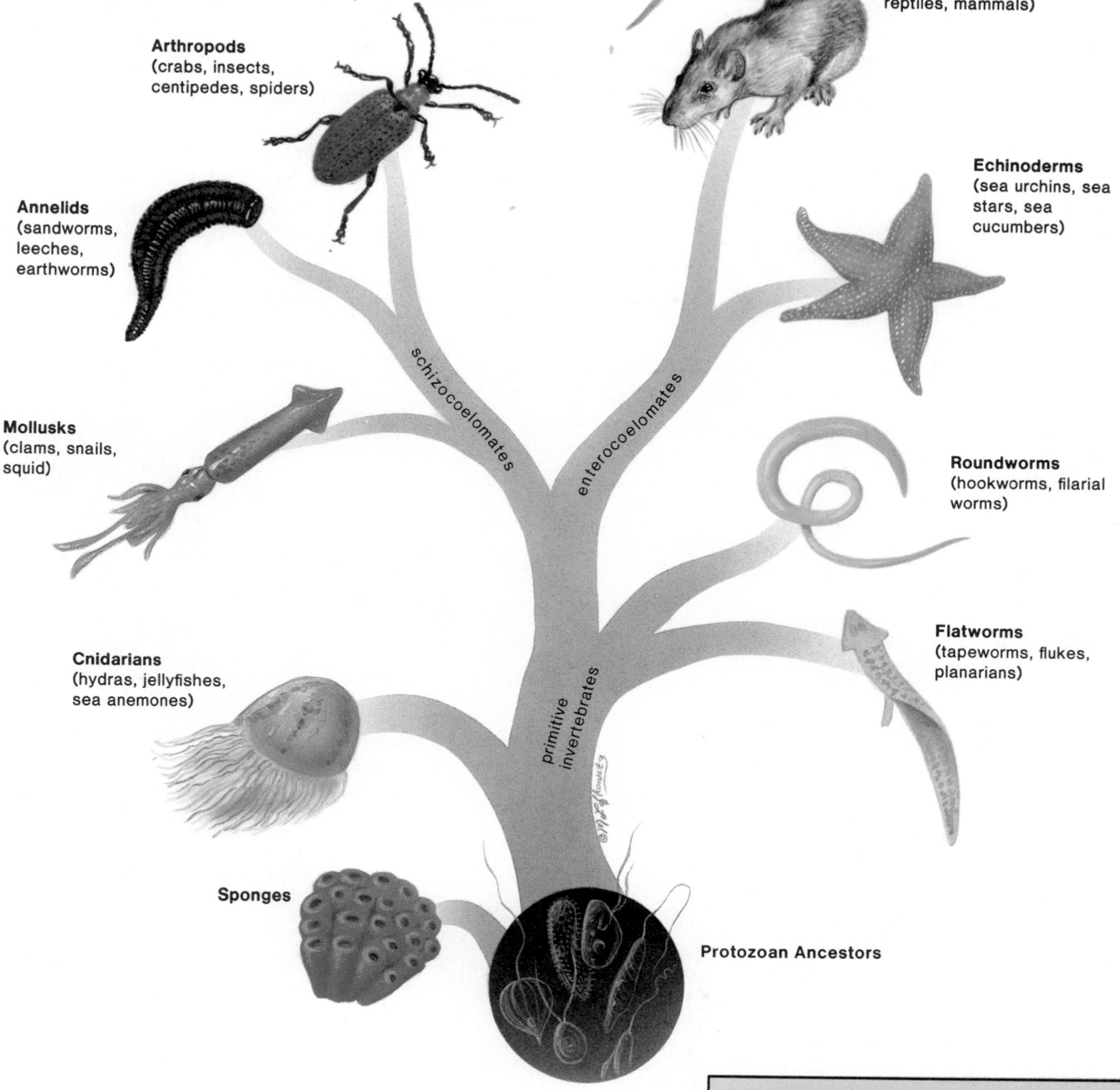

Chordates
(fishes, amphibians, reptiles, mammals)

Arthropods
(crabs, insects, centipedes, spiders)

Echinoderms
(sea urchins, sea stars, sea cucumbers)

Annelids
(sandworms, leeches, earthworms)

Mollusks
(clams, snails, squid)

Roundworms
(hookworms, filarial worms)

schizocoelomates

enterocoelomates

Cnidarians
(hydras, jellyfishes, sea anemones)

Flatworms
(tapeworms, flukes, planarians)

primitive invertebrates

Sponges

Protozoan Ancestors

therefore is based on type of body plan, symmetry, number of germ layers, level of organization, type of body cavity, and presence or absence of segmentation.

Two body plans (fig. 30.3) are observed in the animal kingdom: the **sac plan** and the **tube-within-a-tube plan.** Animals with the sac plan have only one opening, which is used both as an entrance for food and an exit for waste. Animals with the tube-within-a-tube plan have an entrance for food and an exit for waste. These 2 openings allow specialization of parts to occur along the length of the tube.

Asymmetry means that the animal has no particular symmetry. **Radial symmetry** means that the animal is organized circularly, and just as with a wheel, it is possible to obtain 2 identical halves no matter how the animal is

30.1 Critical Thinking

1. Using the listing on pages 610–11 for the characteristics of plants, list, if possible, a corresponding characteristic for animals.
2. What one animal characteristic can be associated with an animal's need to acquire food?
3. Why is it reasonable to assume that animals arose from protozoans?
4. Some animals can regenerate from a small part of the whole. Why would you expect only simple animals and not those with highly differentiated tissues to be able to do this?

sliced longitudinally. **Bilateral symmetry** means that the animal has definite left and right halves; only one longitudinal cut down the center of the animal produces 2 equal halves (fig. 30.4). Radially symmetrical animals tend to be attached to a substrate, or to be *sessile*. This type of symmetry is useful to these animals because it allows them to reach out in all directions from one center. Bilaterally symmetrical animals tend to be active and to move forward with one anterior end. This end develops a head region (called *cephalization*) that is acutely aware of the environment and aids the animal in its forward progress.

In most animals, there are 3 germ layers (ectoderm, endoderm, and mesoderm) during development (p. 439), but in a few animals, there are only 2 germ layers (ectoderm and endoderm). Such animals have the tissue level of organization. Animals with 3 germ layers have the organ level of organization.

Table 30.1 Primitive versus Advanced Features

	Most Primitive	Primitive	Advanced	Most Advanced
Body plan	None	Sac plan	Tube-within-tube plan	Tube-within-tube plan with specialization of parts
Symmetry	None	Radial	Bilateral	Bilateral with cephalization
Germ layers	None	2	3	3
Level of organization	None	Tissue	Organ	Organ system
Body cavity	Acoelomate	Acoelomate	Pseudocoelom	True coelom
Segmentation	Nonsegmented	Nonsegmented	Segmented	Segmented with specialization of parts

Figure 30.3
Body plans. a. *The sac plan with only one opening.* b. *The tube-within-a-tube plan with 2 openings.*

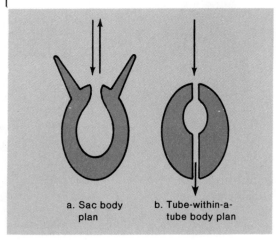

a. Sac body plan   b. Tube-within-a-tube body plan

Figure 30.4
Types of symmetry. a. *Bilateral symmetry. Notice that only the one longitudinal cut shown gives 2 identical halves of the animal.* b. *Radial symmetry. Notice that any longitudinal cut, such as those shown, gives 2 identical halves of the animal.*

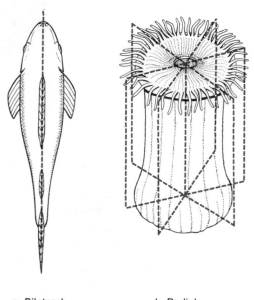

a. Bilateral symmetry   b. Radial symmetry

Some animals are acoelomate; they do not have a **coelom,** or space within which the internal organs are located (fig. 30.5). Animals that do have a coelom are either pseudocoelomates, having a pseudocoelom, or coelomates, having a true coelom. A pseudocoelom is lined with mesoderm beneath the body wall only, but a true coelom is lined with mesoderm both beneath the body wall and around the gut. Besides serving as a location for the internal organs, a coelom can have other functions. In animals that lack blood vessels, its fluids aid the movement of materials, like nutrients or metabolic wastes. In animals that lack a skeleton, a fluid-filled coelom acts as a *hydrostatic skeleton.* This coelom not only offers some resistance to the contraction of muscles, it also permits flexibility; the animal then can change shape and can perform a variety of movements.

Some animals are nonsegmented, and some have repeating units called segments. It is easy to tell, for example, that an earthworm (see fig. 30.19) is segmented because its body appears to be a series of rings. Segmentation leads to specialization of parts in that the various divisions of the body can differentiate for specific purposes.

Classification of animals considers type of body plan, symmetry, number of germ layers, presence or absence of a coelom, and segmentation. Among the invertebrates, it is possible to observe an increasing complexity of these features.

Primitive Invertebrates

The primitive invertebrates include sponges, cnidarians, flatworms, and roundworms. By studying the animals in this order, an increase in complexity that may reflect the order in which these animals evolved is observed. Nevertheless, it is difficult to determine their exact evolutionary relationships.

Sponges

Most **sponges** (phylum Porifera) are marine and are more abundant in warm ocean water, near the coast. Some sponges grow on rocks and are brightly colored, appearing almost lichenlike when seen at a distance. Sponges often are shaped like vases. They all have a central cavity, but the body wall may be simple or convoluted with canals. Regardless, the wall of a sponge is perforated by numerous *pores* surrounded by contractile cells capable of regulating pore size.

The wall of a sponge contains 3 types of cells (fig. 30.6). The outer cells are flattened *epidermal cells.* The inner cells are **collar cells** with flagella, whose constant movement produces water currents that flow through the pores into the central cavity and out through the upper opening of the body, called the **osculum.**

Sponges are **sessile filter feeders.** This means that they remain in one place as an adult, and the food they acquire filters through the pores. Microscopic food particles brought by the water are engulfed by the collar cells and are digested by them in food vacuoles or they are passed to the amoeboid cells for digestion.

The *amoeboid cells* within the wall of a sponge not only act as a circulatory device to transport nutrients from cell to cell, they also produce spicules and the sex cells, the egg and the sperm. **Spicules** are the needle-shaped structures that serve as the internal skeleton of sponges. Sponges are classified according to the type of spicule; some make spicules of calcium carbonate, others of glass, and still others use spongin, a proteinlike substance.

The eggs and the sperm are released into the water, where fertilization occurs. Fertilization results in a zygote, which develops into a ciliated larva that may swim to a new location. Sponges also reproduce by budding, and this

Figure 30.5

Comparison of mesoderm organization. a. In acoelomate animals, the mesoderm is packed solidly. b. Pseudocoelomate animals have mesodermal tissue inside the ectoderm, but not adjacent to the gut endoderm. c. In coelomate animals, there is mesodermal tissue both inside the ectoderm and adjacent to the gut endoderm. True coeloms are body cavities completely lined by mesodermal tissue. Mesenteries hold organs in place within the body cavity.

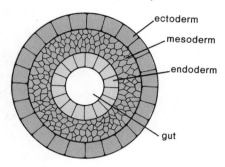

a. Acoelomate
(no coelom)

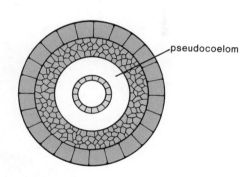

b. Pseudocoelomate
(coelom incompletely
lined by mesoderm)

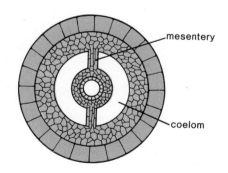

c. Coelomate
(coelom completely
lined by mesoderm)

Figure 30.6

Generalized sponge anatomy. Epidermal cells form an outer layer of cells, and collar cells line the central cavity and the canals. The collar cells (enlarged) have flagella, and as they beat, they move water through the pores as indicated by the arrows. Food particles in the water are trapped by the collar cells and are digested within their food vacuoles. Amoeboid cells transport nutrients from cell to cell; spicules comprise an internal skeleton of some sponges.

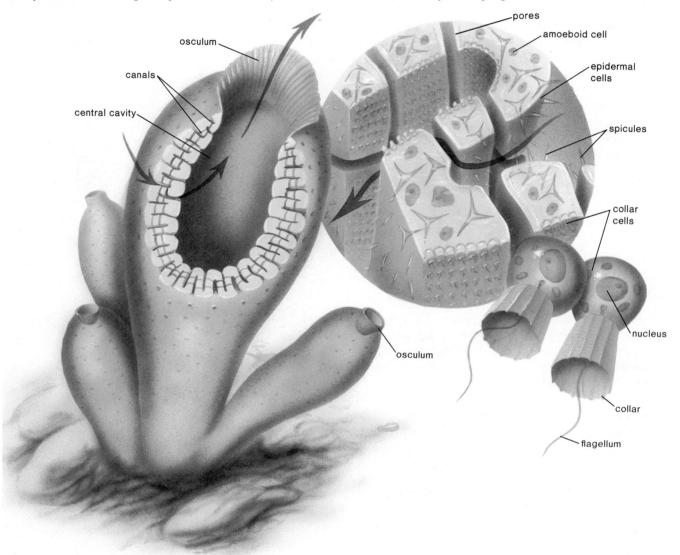

process produces whole colonies that can become quite large. Like many less-specialized organisms, sponges are capable of **regeneration,** or growth of a whole from a small part.

The cellular organization of sponges is different from that of other animals, and the main opening of a sponge is used only as an exit, not an entrance. Further, movement is limited to the beating of the flagella, constriction of the osculum, and larval-stage swimming. Sponges are classified according to type of spicule.

Cnidarians

The body of a **cnidarian** (phylum Cnidaria) is a hollow, 2-layered sac, which accounts for the former name of these organisms—*coelenterate* means hollow sac. The outer layer of the sac, the ectoderm, is separated from the inner layer, the endoderm, by a jellylike material called **mesoglea.** All cnidarians have specialized stinging cells now termed cnidocytes, from a Greek word meaning sea nettles. Within these cells are the **nematocysts,** which are long, spirally

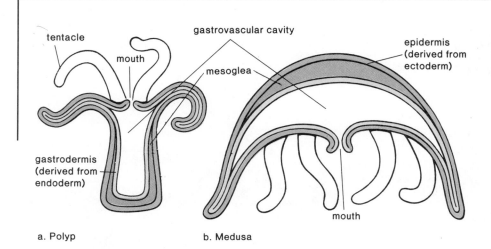

Figure 30.7

The 2 body forms of cnidarians. The drawings indicate the manner in which the 2 tissue layers (epidermis and gastrodermis), separated by a packing material, the mesoglea, surround the central gastrovascular cavity. a. The polyp form, with the mouth upward, is usually attached to surfaces. b. The medusa form, with the mouth downward, is free swimming.

coiled, hollow threads. When the trigger of a cnidocyte is touched, the discharged thread, which sometimes contains poison, stuns either prey or enemy (see fig. 30.9).

Cnidarians have radial symmetry (fig. 30.4), and there is typically a ring of *tentacles* surrounding the mouth region. Some cnidarians, referred to as **hydroids** or **polyps** (fig. 30.7), have a tubular shape, with the mouth region directed upward. Others, which have a bell shape with the mouth region directed downward, are called jellyfishes, or **medusae.** The polyp is adapted to a sessile life, while the medusa is adapted to a floating or free-swimming existence. At one time, both body forms may have been a part of the life cycle of all cnidarians, because today we see an alternation of generations[1] life cycle of these 2 forms in certain cnidarians, such as members of the genus *Obelia.* When the alternation of generations life cycle does exist, the polyp stage produces medusae, and the medusae, which produce eggs and sperm, disperse the species.

Cnidarians are quite diversified (fig. 30.8). The *Portuguese man-of-war,* whose nematocysts can cause serious or even fatal poisoning in humans, is a colony of polyps suspended from a large medusoid form that serves as a gas-filled float. Many species of jellyfishes, such as *Aurelia,* show the alternation

[1]This is not the same as the alternation of generations life cycle in plants because here both generations are diploid.

Figure 30.8

Cnidarian diversity. a. Medusae of Aurelia. *A thick layer of jelly (mesoglea) gives medusae buoyancy and accounts for their common name, jellyfishes. b. Living coral polyps. Each polyp is an individual but has a chalky skeleton that is joined to its neighbor's skeleton. The polyps feed on zooplankton but also contain symbiotic algae that contribute to their nutrition. c. Sea anemone. These animals are called the "flowers of the sea," but despite their appearance, they are carnivorous animals.*

a.

b.

c.

Figure 30.9

Generalized hydra anatomy. There are 2 tissue layers separated by mesoglea. The cells contain muscle fibers, and the mesoglea contains a nerve net; therefore, the animal can move. In fact, it is carnivorous and uses its nematocysts to stun and its tentacles to capture prey. Inset shows that before discharge, a nematocyst is tightly coiled within a cnidocyte (stinging cell) and after discharge, it is extended.

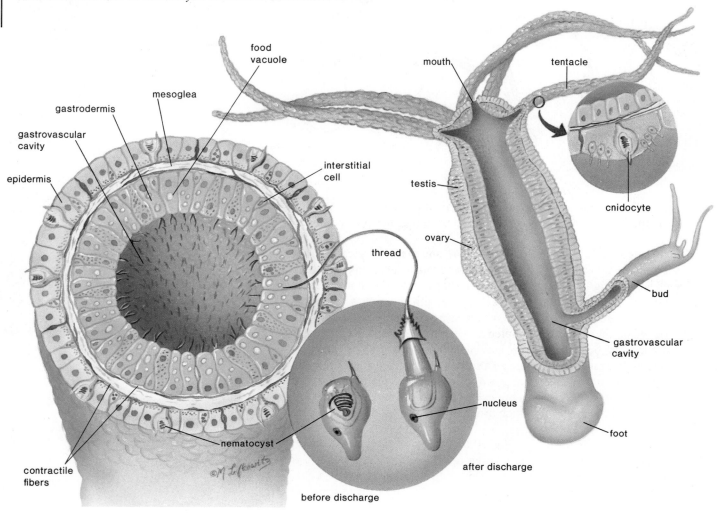

of generations life cycle, but the 2 generations are not equal—the medusa is the primary stage and the polyp remains quite small and insignificant. *Sea anemones* are solitary polyps with thick walls. They can be brightly colored and can look like beautiful flowers. **Corals** are similar to sea anemones, but they have calcium carbonate skeletons. Some corals are solitary, but most are colonial, with either flat, rounded, or upright and branching colonies. The slow accumulation of coral skeletons has formed reefs in the South Pacific, including the Great Barrier Reef along the eastern coast of Australia. An ancient coral reef that now lies beneath Texas is the source of petroleum for that state.

Hydras

Hydras (fig. 30.9) are likely to be found attached to underwater plants or rocks in most lakes and ponds. The hydra body is a small, tubular polyp about 7.5 mm in length. Although hydras usually remain in one place, they may glide along on their base, or foot, or even move rapidly by means of somersaulting. Hydras, like other animals capable of locomotion, possess both muscle and nerve cells. The nerve cells form a connecting network throughout the mesoglea known as the **nerve net.** The nerve net makes contact with the outer layer of cells, called the **epidermis,** and the inner layer of cells, called the **gastrodermis.** These cells contain contractile fibers.

The cells of the gastrodermis secrete digestive juices that pour into a central cavity. The enzymes begin the digestive process, which is completed within food vacuoles when small pieces of the prey are engulfed by the cells of the gastrodermis. Nutrient molecules are passed by diffusion to the rest of the cells of the body. The presence of the large inner cavity makes it possible for all cells to exchange gases directly with the surrounding medium. Because this function is carried out by a vascular system in more complex animals, the cavity is known as a **gastrovascular cavity.**

In *Hydra,* the gastrodermis contains interstitial or embryonic cells capable of becoming other types of cells. For example, they can produce the ovary and the testes and probably account for the animal's great regenerative powers. Like the sponge, a whole cnidarian can grow from a small piece. When conditions are favorable, small outgrowths, or *buds,* appear, pinch off, and begin to live independently.

The cnidarians are radially symmetrical *and have a* sac body plan. *There are 2* tissue layers, *the epidermis and the gastrodermis, derived from the embryonic germ layers, the ectoderm and the endoderm. The presence of* nematocysts *is a unique feature.*

Flatworms

Flatworms (phylum Platyhelminthes) have 3 germ layers. The presence of mesoderm not only gives bulk to the animal, it also allows for greater complexity of internal structure. Free-living forms have muscles and excretory, reproductive, and digestive organs. The worms lack respiratory and circulatory organs, but since the body is flattened, diffusion alone is adequate for the passage of oxygen (O_2) and other substances from cell to cell. In free-living forms, bilateral symmetry and good cephalization are very important, as is a well-developed nervous system, including sense organs. This combination makes them efficient predators.

Flatworms are nonsegmented, lack a coelom, and have the sac body plan with only one opening. Therefore, if we analyze them according to table 30.1, we see that they have a combination of primitive and advanced features.

There are 3 classes of flatworms: one is free living and 2 are parasitic. The free-living specimen, the planarian, best exemplifies the characteristics of the phylum.

Planarians

Freshwater planarians (e.g., *Dugesia*) (fig. 30.10) are small (several mm–several cm), literally flat worms. Some tend to be colorless; others have brown or black pigmentation. Planarians live in lakes, ponds, streams, and springs, where they feed on small living or dead organisms, such as worms and crustacea.

Because planarians live in fresh water, water tends to enter the body by osmosis. They have an excretory organ that largely rids the body of excess water. The organ consists of a series of interconnecting canals that run the length of the body on each side. The beating of cilia in the **flame cells** (so named because the beating of the cilia reminded some early investigator of the flickering of a flame) keeps the water moving toward the excretory pores.

The digestive organ is tripartite (having 3 branches) and ramifies throughout the body. It begins with a pharynx, which ejects from the mouth to suck food particles into the digestive organ.

Figure 30.10

Flatworm anatomy as exemplified by a planarian. a. Photomicrograph of planarian Dugesia. *This photo shows the pharynx leading to the digestive organ, which has been darkly stained to illustrate the manner in which it ramifies throughout the body. b. Details of organ anatomy. (1) Excretory organ is composed of canals that have flame cells (enlarged drawing), whose beating cilia draw in fluid that is excreted by way of* excretory pores. (2) The nervous system has a ladder appearance because cross branches stretch between longitudinal fibers that extend the length of the animal. (3) The digestive organ is tripartite (has 3 branches), and because the flatworms are hermaphroditic, there are both male and female reproductive organs.

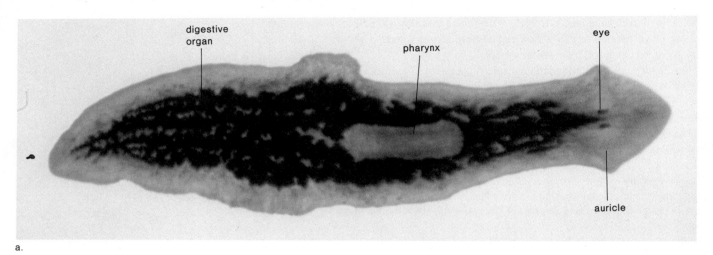

digestive organ

pharynx

eye

auricle

a.

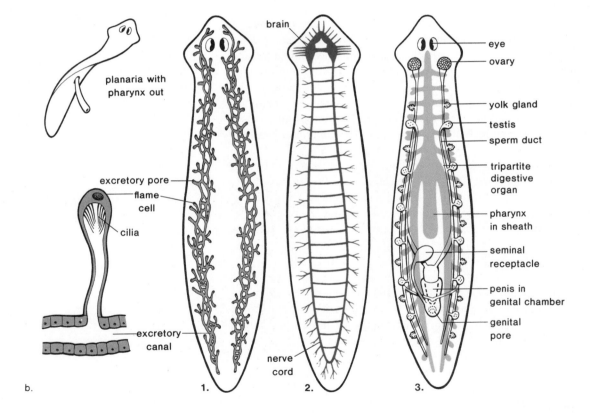

planaria with pharynx out

excretory pore

flame cell

cilia

excretory canal

brain

nerve cord

eye

ovary

yolk gland

testis

sperm duct

tripartite digestive organ

pharynx in sheath

seminal receptacle

penis in genital chamber

genital pore

b.

1.

2.

3.

The planarian has a **ladder-type nervous organ.** There is a small brain and 2 lateral nerve cords joined by cross branches. Planarians show good cephalization. Aside from the brain, there are light-sensitive organs (the eyes) and chemosense organs (the auricles). They have well-developed muscles, and their ciliated epidermis allows them to glide along a film of mucus.

Planarians are **hermaphroditic animals,** which means that they possess both male and female sex organs. The worms practice cross-fertilization; the

Table 30.2 *Free-living Worms versus Parasitic Worms*

	Planarians	Tapeworms	Flukes
Body wall	Ciliated epidermis	Glycocalyx covers tegument	Glycocalyx covers tegument
Cephalization	Yes—Eyes and auricles	No—Scolex with hooks and suckers	No—Oral sucker
Nervous connections	Nerves and brain	Reduced	Reduced
Digestive organ	Ramifies	Absent	Reduced
Reproductive organs	Hermaphroditic	Greatly increased in volume	Increased in volume
Larva	Absent	Present	Present

penis of one is inserted into the genital pore of the other. The fertilized eggs hatch in 2–3 weeks as tiny worms.

Planarians often are used in biology laboratories to illustrate regeneration. If a worm is cut crosswise, it usually grows a new head or a new tail as is appropriate. Planarians also have been used in so-called memory experiments. In these experiments, planarians were trained to swim mazes and then were cut up and fed to untrained planaria. When the cannibals subsequently were taught the same task, they learned faster than the first set. The exact significance of these experiments is debatable, but they have led to some interesting student speculations as to how best to acquire the knowledge of teachers.

Parasitic Flatworms

There are 2 types of parasitic flatworms: tapeworms (cestodes) and flukes (trematodes). The structure of both these worms illustrates the modifications that occur in parasitic animals (table 30.2). Concomitant with the loss of predation, there is an absence of cephalization; the anterior end notably carries hooks and/or suckers for attachment to the host. The parasite acquires nutrient molecules from the host, and the digestive system is reduced. The tegument, a specialized body wall resistant to host digestive juices, is covered by the glycocalyx, a mucopolysaccharide coating. The extensive development of the reproductive system, with the production of millions of eggs, may be associated with difficulties in dispersing the species. Both parasites utilize a *secondary host,* or intermediate host, to transport the species from primary host to primary host. The *primary host* contains the sexually mature adult; the secondary host(s) contain(s) the larval stage or stages.

A **tapeworm** has a head region (fig. 30.11*a*) containing hooks and suckers for attachment to the intestinal wall of the host. Behind the head region, called the **scolex,** there is a short neck and then a long series of proglottids. **Proglottids** are segments, each of which contains a full set of both male and female sex organs. Therefore, the tapeworm is little more than a reproductive factory. There are excretory canals but no digestive system and only the rudiments of nerves.

After fertilization, the proglottids become nothing but a bag filled with developing embryos (larvae). Mature proglottids such as these break off, and as they pass out with the host's feces, the larvae, enclosed by a protective covering, are released.

If feces-contaminated food is fed to pigs or cattle, the larvae escape when their covering is digested. They burrow through the intestinal wall and travel in the bloodstream to finally lodge and encyst in muscle. Here, a *cyst* means a small, hard-walled structure that contains an immature worm. When humans

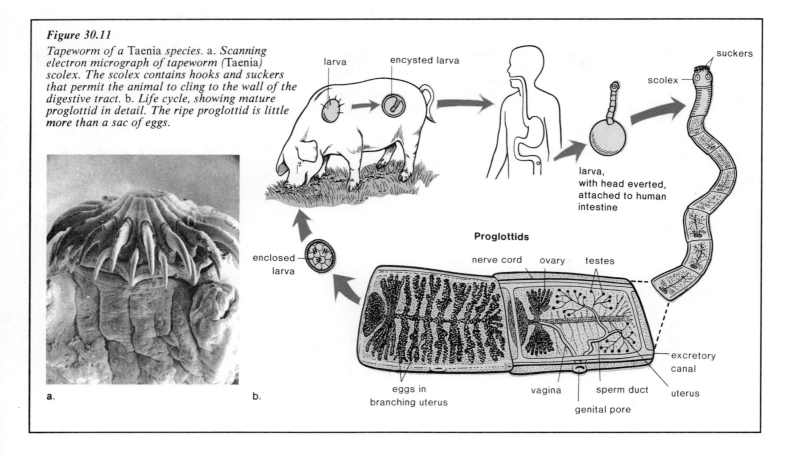

Figure 30.11

Tapeworm of a Taenia species. a. Scanning electron micrograph of tapeworm (Taenia) scolex. The scolex contains hooks and suckers that permit the animal to cling to the wall of the digestive tract. b. Life cycle, showing mature proglottid in detail. The ripe proglottid is little more than a sac of eggs.

a.

b.

Figure 30.12

Fluke anatomy. Scanning electron micrograph of fluke (Gorgoderina attenuata) oral sucker. Note upraised structures, which are believed to be sensory in nature.

eat raw or rare infected meat, the worms break out of the cyst, attach themselves to the intestinal wall, and grow to adulthood. Then the cycle begins again (fig. 30.11*b*).

There are many different types of **flukes,** usually designated by the type of vertebrate organ they inhabit; for example, there are blood, liver, and lung flukes. While the structure may vary slightly, in general the fluke body tends to be oval and to elongate with no definite head except that the oral sucker, surrounded by sensory papillae, is at the anterior end (fig. 30.12). Usually, there is at least one other sucker for attachment to the host. Internally, a fluke has reduced digestive, nervous, and excretory systems. There is a well-developed reproductive system, and the adult fluke is usually hermaphroditic, although there are exceptions.

A blood fluke causes **schistosomiasis** in Africa and South America. This disease is especially prevalent in areas with irrigation ditches because the secondary host is a freshwater snail. The disease is spread when egg-laden human feces get into the water and newly hatched larvae enter the snails. Asexual reproduction occurs within the snail, and the resulting larvae penetrate human skin to enter blood vessels, where they mature.

The Chinese liver fluke requires 2 hosts: the snail and the fish. Humans contract the disease when they eat uncooked fish. The adult flukes reside in the liver and deposit their eggs in the bile duct, which carries the eggs to the intestine.

Although flatworms have a sac body plan, they are more complex than cnidarians because they have 3 germ layers and possess true organs. The free-living forms exhibit cephalization and bilateral symmetry.

Roundworms

Roundworms (phylum Nematoda), as their name implies, are round rather than flat worms. They have a smooth outside body wall, indicating that they

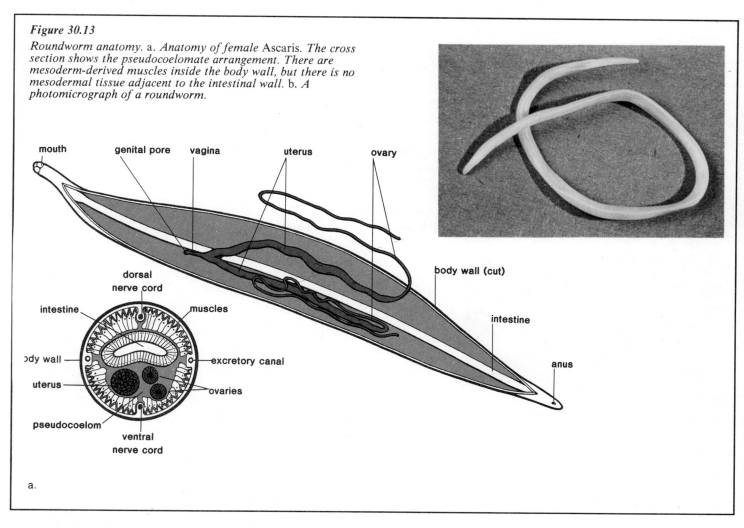

Figure 30.13

Roundworm anatomy. a. Anatomy of female Ascaris. *The cross section shows the pseudocoelomate arrangement. There are mesoderm-derived muscles inside the body wall, but there is no mesodermal tissue adjacent to the intestinal wall. b. A photomicrograph of a roundworm.*

are nonsegmented. These worms, which are generally colorless and less than 5 cm long, occur almost anywhere—in the sea, in fresh water, and in the soil—in such numbers that thousands of them can be found in a small area.

Roundworms possess 2 anatomical features not seen in more primitive animals: a tube-within-a-tube body plan (fig. 30.3) and a body cavity. The body cavity is a **pseudocoelom** (fig. 30.5*b*), or a cavity incompletely lined with mesoderm. This fluid-filled pseudocoelom provides space for the development of organs, substitutes for a circulatory system by allowing easy passage of molecules, and provides a type of skeleton. Worms in general do not have an internal or external skeleton, but they do have a *hydrostatic skeleton,* a fluid-filled interior that supports muscle contraction and enhances flexibility.

When roundworms are analyzed according to table 30.1, they are seen to have features associated with advanced animals except that they are non-segmented. Roundworms are thought to be a side branch to the main evolution of animals; they may have arisen from an ancestor that also produced coelomate animals.

Ascarids

Most roundworms are free living, but a few are parasitic. *Ascaris,* a large parasitic roundworm, often is studied as an example of this phylum.

Ascaris (fig. 30.13) females tend to be larger (20–35 cm in length) than males, which have an incurved tail. Both sexes move by means of a characteristic whiplike motion because only longitudinal muscles lie next to the body wall.

The ascarid internal organs, including the tubular reproductive organs, lie within the pseudocoelom. Because mating produces embryos that mature

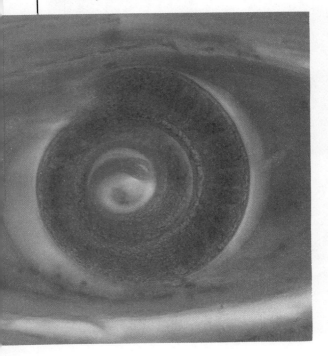

in the soil, the parasite is limited to warmer environments. When larvae within their protective covering are swallowed, they escape and burrow through the intestinal wall. Making their way through the organs of the host, they move from the intestine to the liver, the heart, and then the lungs. Within the lungs, growth takes place, and after about 10 days, the larvae migrate up the windpipe to the throat, where they are swallowed, allowing them once again to reach the intestine. Then the mature worms mate, and the female produces embryo-containing eggs that pass out with the feces. To complete this life cycle, as with other roundworms, feces must reach the mouth of the next host; therefore, proper sanitation is the best means to prevent infection with *Ascaris* and other parasitic roundworms.

Other Parasites

Trichinosis is a serious infection of humans caused by *Trichinella* roundworms. Humans contract the disease when they eat rare pork containing encysted larvae. After maturation, the female adult burrows into the wall of the small intestine and produces live offspring that are carried by the bloodstream to the skeletal muscles, where they encyst (fig. 30.14). Because humans normally are not eaten by any other animals, these larvae rarely reach another host. However, the cycle can be completed if pigs eat infected pig meat or infected rats.

Elephantiasis is caused by a roundworm called the filarial worm, which utilizes the mosquito as a secondary host. Because the adult worms reside in lymphatic vessels, collection of fluid is impeded, and the limbs of an infected human can swell to an enormous size (see fig. 13.2). When a mosquito bites an infected person, it transports larvae to new hosts.

Other roundworm infections are more common in the United States. Children frequently acquire a pinworm infection, and hookworm is seen in the southern states. A hookworm infection can be very debilitating because the worms attach to the intestinal wall and feed on blood.

Roundworms possess 2 advanced features: a body cavity and a tube-within-a-tube body plan. The body cavity is a pseudocoelom rather than a true coelom. The tube-within-a-tube plan, in contrast to the sac plan, has both a mouth and an anus.

Advanced Invertebrates

All the advanced invertebrate phyla have a true coelom (fig. 30.5c). Nevertheless, they can be divided into 2 groups on the basis of embryological evidence (fig. 30.15). In mollusks, annelids, and arthropods, the coelom forms by splitting of the mesoderm. Therefore, they are called the **schizocoelomates.** Also, marine mollusks and annelids have larvae of the **trochophore** type (top-shaped with a band of cilia at the midsection). Because the blastopore (the site of invagination of the endodermal germ layer during development, p. 438) becomes the mouth, they also are called the **protostomes.** In echinoderms and chordates, the coelom forms by outpocketing of the primitive gut. Therefore, they are called the **enterocoelomates.** Also, echinoderms and certain invertebrate chordates have larvae of the **dipleurula** type (bands of cilia are placed as shown in fig. 30.15). Because the blastopore became the anus, these animals also are called the **deuterostomes.**

All higher invertebrates have a true coelom. They are divided into 2 groups on the basis of embryological evidence: mollusks, annelids, and arthropods are schizocoelomates, while echinoderms and chordates are enterocoelomates.

Figure 30.15

Schizocoelomates versus enterocoelomates. In schizocoelomates, the coelom forms by splitting of the mesoderm, the trochophore larva is typical, and the blastopore becomes the mouth. In enterocoelomates, the coelom forms by outpocketing of the primitive gut, the dipleurula larva is typical, and the blastopore becomes the anus.

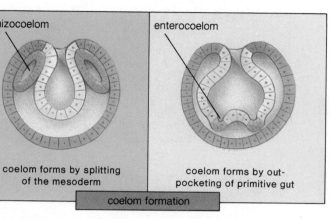

Schizocoelomates	Enterocoelomates
mollusks annelids arthropods	echinoderms chordates

schizocoelom / enterocoelom

coelom forms by splitting of the mesoderm / coelom forms by out-pocketing of primitive gut

coelom formation

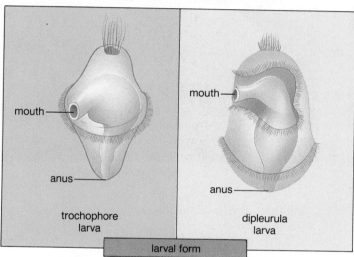

mouth / anus

trochophore larva / dipleurula larva

larval form

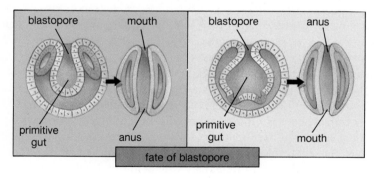

blastopore — mouth / blastopore — anus

primitive gut — anus / primitive gut — mouth

fate of blastopore

Mollusks

Mollusks (phylum Mollusca) are a very large and diversified group containing many thousands of living and extinct forms. However, all forms of mollusks have a body composed of at least 3 distinct parts:

1. **Visceral mass:** the soft-bodied portion that contains internal organs.
2. **Foot:** the strong, muscular portion used for locomotion.
3. **Mantle:** the membranous or sometimes muscular covering that envelops but does not completely enclose the visceral mass. The *mantle cavity* is the space between the 2 folds of the mantle. The mantle may secrete a shell.

Figure 30.16

Molluskan diversity. a. A scallop with sensory tentacles extended between the valves. Humans eat only the single, large muscle that holds the 2 halves of the shell together. b. An octopus moving over the surface of coral in the Pacific Ocean shows that it completely lacks a protective shell. c. While most mollusks are marine, snails are adapted to living on land, and their mantle tissue is capable of gas exchange with air. During copulation, each inserts a penis into the mantle cavity of its partner.

a.

b.

c.

In addition to these 3 parts, many mollusks show cephalization and have a head region with eyes and other sense organs.

The division of the body into distinct areas seems to have allowed diversification to occur because there are many different types of mollusks adapted to various ways of life (fig. 30.16). The molluskan groups can be distinguished by the modification of their foot. In the gastropods ("belly-footed"), including snails, conches, and nudibranchs, the foot is ventrally flattened, and the animal moves by muscle contractions that pass along the foot. While nudibranchs, also called sea slugs, lack a shell, most other gastropods have a coiled shell in which the visceral mass spirals. Snails are adapted to life on land. For example, their mantle is richly supplied with blood vessels and functions as a lung when air is moved in and out through respiratory pores.

In cephalopods ("head-footed"), including octopuses and squids, the foot has evolved into tentacles about the head. Aside from the tentacles that seize the prey, cephalopods have powerful jaws and a radula (toothy tongue) to tear prey apart. Cephalization aids these animals in recognizing prey and in escaping enemies. The eyes are superficially similar to those of vertebrates and have a lens and a retina with photoreceptors. The brain is formed from a fusion of ganglia, and nerves leaving the brain supply various parts of the body. An especially large pair of nerves controls the rapid contraction of the mantle, allowing these animals to move quickly by a jet propulsion of water. Rapid movement and the secretion of a brown or black pigment from an ink gland help cephalopods to escape their enemies. Octopuses have no shell, and squids have only a remnant of one concealed beneath the skin.

In bivalves, such as clams, oysters, and scallops, the foot is laterally compressed. Another name for these animals is pelecypods, which means "hatchet

Table 30.3 *Comparison of Clam and Squid*

	Clam	Squid
Food getting	Filter feeder	Active predator
Skeleton	Heavy shell for protection	No external skeleton
Circulation	Open	Closed
Cephalization	None	Marked
Locomotion	Hatchet foot	Jet propulsion
Nervous system	3 separate ganglia	Brain and nerves

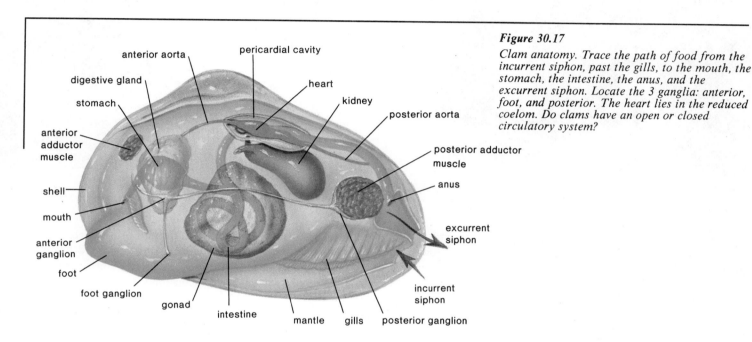

Figure 30.17
Clam anatomy. Trace the path of food from the incurrent siphon, past the gills, to the mouth, the stomach, the intestine, the anus, and the excurrent siphon. Locate the 3 ganglia: anterior, foot, and posterior. The heart lies in the reduced coelom. Do clams have an open or closed circulatory system?

foot." They are called bivalves because there are 2 parts to the shell. Because of the accessibility of clams, they often are studied as an example of this phylum. However, as outlined in table 30.3, a clam is adapted to an inactive life, and other mollusks, like a squid, are adapted to an active life.

Clams

In a clam, such as *Anodonta,* the shell is secreted by the mantle and is composed of calcium carbonate with an inner layer of *mother-of-pearl.* If a foreign body is placed between the mantle and the shell, pearls form as concentric layers of shell are deposited about the particle.

Within the mantle cavity, the gills (fig. 30.17) hang down on either side of the visceral mass, which lies above the foot. **Gills** are composed of vascularized, highly convoluted, thin-walled tissue specialized for gas exchange.

The heart of a clam lies just below the hump of the shell within the pericardial cavity, the only remains of the coelom. Therefore, the coelom of the clam is said to be *reduced.* The heart pumps blue blood, containing the pigment hemocyanin instead of red hemoglobin, into vessels that lead to the various organs of the body. Within the organs, however, blood flows through spaces, or *sinuses,* rather than through vessels. Such a circulatory system is called an *open circulatory system* because the blood is not contained within blood vessels all the time. This type of circulatory system can be associated

Figure 30.18

Polychaetes. a. *A sandworm, such as* Neanthes, *has fleshy lobes, the parapodia on each body segment. They are used for swimming and as respiratory organs. Numerous chitinous bristles grow out from the parapodia, and hence they are polychaetes, or "many-bristled."* b. *This worm is a predator; its small prey are captured by a pair of strong chitinous jaws that evert with a part of the pharynx when the worm is feeding.* c. *Compare (b) to the head region of a sedentary tubeworm, which extends long ciliated tentacles into the water. The cilia create currents, filter food particles, and move these particles toward the mouth.*

a.

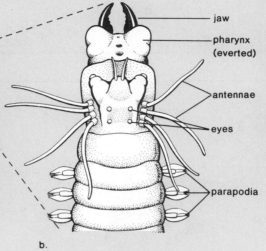

b.

c.

with an inactive animal because it is an inefficient means of transporting blood throughout the body. An active animal needs to have oxygen and nutrients transported quickly to rapidly working muscles, while an inactive animal is able to survive with a sluggish system for transporting these necessities.

The clam nervous system (fig. 30.17) is composed of *3 pairs of ganglia* (anterior, foot, and posterior), which all are connected by nerves. Clams lack cephalization. The foot projects anteriorly from the shell, and by expanding the tip of the foot and pulling the body after it, the clam moves forward.

The clam is a filter feeder, meaning that it feeds on small particles that have been filtered from the water environment. Food particles and water enter the mantle cavity by way of the *incurrent siphon,* a posterior opening between the 2 valves. Mucous secretions cause smaller particles to adhere to the gills, and cilia action sweeps them toward the mouth. Many inactive animals are filter feeders because this method of feeding does not require rapid movement.

The digestive system of the clam includes a mouth, a stomach, and an intestine, which coils about in the visceral mass and then goes right through the heart before ending in an anus. The anus empties at an *excurrent siphon,* which lies just above the incurrent siphon. There is also an accessory organ of digestion called a digestive gland. It is seen readily that the tube-within-a-tube plan leads to specialization of parts in the clam.

There are 2 excretory kidneys in the clam (fig. 30.17), which lie just below the heart and remove waste from the pericardial cavity for excretion into the mantle cavity. The clam excretes ammonia (NH_3), a poisonous substance that requires the concomitant excretion of water. Land-dwelling animals tend to excrete a less toxic substance in a more concentrated form.

The male or female gonad of a clam can be found about the coils of the intestine. While all clams have some type of larval stage, only marine clams have a trochophore larva. The presence of the trochophore larva (fig. 30.15) among some mollusks indicates a relationship to the annelids, some members of which also have this type of larval stage.

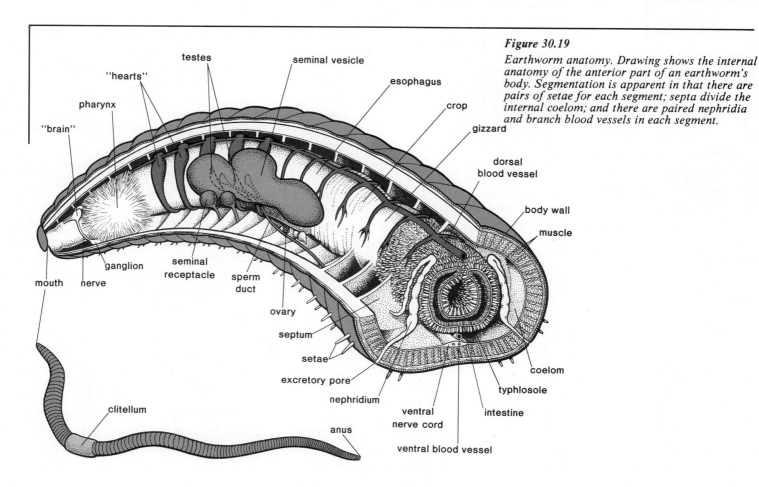

Figure 30.19
Earthworm anatomy. Drawing shows the internal anatomy of the anterior part of an earthworm's body. Segmentation is apparent in that there are pairs of setae for each segment; septa divide the internal coelom; and there are paired nephridia and branch blood vessels in each segment.

Mollusks are a very diverse group with a body plan always composed of 3 parts: the foot, the mantle, and the visceral mass. They possess all advanced features except segmentation (table 30.1).

Annelids

The primary characteristic of annelids (phylum Annelida) compared to the other groups studied is segmentation; obvious rings encircle the body, and even the well-developed coelom is partitioned by membranous septa. Both segmentation and an ample coelom are important advances, facilitating the development of specialization of parts as seen in later phyla.

While the earthworm is used here as an example of the annelids, marine worms, such as sandworms (*Neanthes*), may be more representative. Sandworms are distinguished by the presence of a pair of fleshy lobes, the **parapodia,** on each body segment. These are used not only in swimming but also as respiratory organs, where the expanded surface area allows for exchange of gases. Numerous chitinous bristles grow from the parapodia—hence the name polychaetes, or "many bristled." These worms are predators. They prey on crustaceans and other small animals, which are captured by a pair of strong chitinous *jaws* that evert with a part of the pharynx when *Neanthes* is feeding. Associated with its way of life, *Neanthes* shows cephalization and has a head region with sense organs, including eyes and antennae (fig. 30.18).

Earthworms

The earthworm (*Lumbricus*) (fig. 30.19) is adapted to living in damp soil, where a moist body wall can be used for gas exchange. Therefore, earthworms more likely are found above ground after a rain than when the weather is hot and dry.

Table 30.4 Segmentation in the Earthworm
1. Body rings
2. Coelom divided by septa
3. Setae on each segment
4. Ganglia and lateral nerves in each segment
5. Nephridia in each segment
6. Branch blood vessels in each segment

The earthworm lacks obvious cephalization and feeds on leaves or any other organic matter, living or dead, that can be taken conveniently into its mouth along with dirt. Food drawn into the mouth by the action of the muscular pharynx is stored in a crop and is ground up in a thick, muscular gizzard. Digestion and absorption occur in a long intestine, whose dorsal surface is expanded by a **typhlosole** that allows additional surface area for absorption. Notice that the tube-within-a-tube plan has allowed specialization of the digestive system to occur.

Locomotion in the earthworm is suitable to its way of life, and each segment of the body has 4 pairs of **setae,** or slender bristles. The setae are inserted into the dirt, and then the body is pulled forward. Both a circular layer and a longitudinal layer of muscle in the body wall make it possible for the worm to move and to change its shape. Muscle contraction is aided by the fluid-filled coelomic compartments that act as a hydrostatic skeleton.

The nervous system (fig. 30.19) consists of an anterior, dorsal, ganglionic mass, or a brain, and a long *ventral solid nerve cord* with ganglionic swellings and lateral nerves in each segment. When invertebrates are compared to vertebrates, it often is said that the former have a ventral solid nerve cord, while the latter have a dorsal hollow nerve cord.

The excretory system consists of paired **nephridia** (fig. 30.19), or coiled tubules, in each segment. Nephridia have 2 openings: one is a ciliated funnel that collects coelomic fluid, and the other is an exit in the body wall. Between the 2 openings is a convoluted region where waste material is removed from the blood vessels about the nephridium.

The earthworm has an extensive *closed circulatory system.* Hemoglobin-containing blood moves anteriorly in a dorsal blood vessel and then is pumped by 5 pairs of hearts into a ventral blood vessel. As the ventral blood vessel takes blood toward the posterior regions of the worm's body, it gives off branches in every segment.

The worms are *hermaphroditic,* with a complete set of organs for both sexes. The male organs of an earthworm are the testes, the seminal vesicles, and the sperm ducts; the female organs are the ovaries, the oviducts, and the seminal receptacles. Copulation occurs when 2 worms lie ventral surface to ventral surface with the heads pointing in opposite directions. The **clitellum,** a smooth girdle about each worm secretes mucus, after which the sperm leave the sperm ducts and travel to the seminal receptacles of the partner. The clitellum later produces a slime tube, which is moved along over the head of a worm by muscular contractions. Into this tube are deposited eggs from the oviducts and the sperm from the seminal receptacles. The slime tube forms a cocoon within which the miniature worms develop. There is no larval stage.

The annelids show the most obvious segmentation of any phylum of animals. Table 30.4 lists structures that have repeating units, illustrating segmentation in earthworms.

Annelids are segmented worms, and most organ systems show evidence of segmentation. These worms have a well-developed coelom divided by septa, a closed circulatory system, and a ventral solid nerve cord.

Arthropods

The arthropods (phylum Arthropoda) have more species (approximately 900,000) than any other group of animals and often are said to be the most successful of all the animals. The phylum includes animals adapted to living in water, such as crayfishes, lobsters, and shrimps, and animals adapted to living on land, such as spiders, insects, centipedes, and millipedes (fig. 30.20).

Arthropods have an external skeleton containing **chitin,** a strong, flexible polysaccharide. The skeleton serves many functions—protection, attachment

Figure 30.20

Arthropod diversity. There are 5 major classes of Arthropods.
a. Class Crustacea is represented by this crab with a large
carapace and 5 pairs of legs. The first pair are pinching claws.
b. Class Arachnida is represented by this land-dwelling scorpion,
with 4 pairs of legs, poisonous claws, and stinging tail. c. Class
Chilopoda is represented by this centipede, a carnivorous
animal, with a pair of appendages on every segment. d. Class
Diplopoda is represented by this millipede, a scavenger that
seems to have 2 pairs of appendages on each segment because
every 2 segments are fused. Class Insecta is represented by (e) a
swallowtail butterfly with wings and (f) an ant with clearly
defined head, thorax, and abdomen. Insects have 3 pairs of legs.

a.

b.

c.

d.

e.

f.

for muscles, and prevention of desiccation on land, for example. The appendages also are covered by the skeletal material, but they are jointed. The presence of **jointed appendages** is a great advance in the animal kingdom and aids locomotion on land.

An external skeleton is not without difficulties, however. Because this particular skeleton does not grow larger, arthropods **molt,** or shed the skeleton periodically.

Specialization of parts is seen readily in the arthropod body; it is not composed of a series of like segments but rather, due to fusion of segments, is composed of 3 parts—head, thorax, and abdomen. The head shows good cephalization with sense organs. The sense organs include **antennae** (or feelers) and eyes. The eyes are of 2 types: the **compound eye** and the **simple eye.** The compound eye is not seen in any other phylum. It is composed of many complete visual units grouped in a composite structure: each visual unit contains a separate lens and a light-sensitive cell. In the simple eye, a single lens covers many light-sensitive cells.

The coelom, so well developed in the annelids, is *reduced* in the arthropods and is composed chiefly of the space about the reproductive system. Instead of a coelomic cavity, there is a **hemocoel,** or blood cavity, consisting of vessels and sinuses (open spaces), where the blood flows about the organs. The dorsal heart keeps the blood moving around in the sinuses. Arthropods, like most mollusks, have an open circulatory system.

Crayfishes

Crayfishes are in the class Crustacea along with lobsters, shrimps, copepods, and crabs. Figure 30.21*a* gives a view of the external anatomy of the crayfish, and it can be seen that the head and the thorax are fused into a **cephalothorax,** which is covered on the top and the sides by a nonsegmented **carapace.** The abdominal segments, however, are marked off clearly.

A pair of stalked compound eyes and 2 pairs of antennae are on the head. Chitinous jaws and mouthparts also are present. The appendages in the thorax include accessory mouthparts, *pinching claws,* and 4 pairs of *walking legs;* the abdominal segments are equipped with *swimmerets,* small paddlelike structures. The first pair of swimmerets in the male are quite strong and are used to pass sperm to the female. The last 2 segments bear the *uropods* and the *telson,* which make up a fan-shaped tail used for swimming backwards.

Ordinarily, the crayfish lies in wait for prey. It faces out from an enclosed spot with the claws extended and the antennae moving about. If a small animal, dead or alive, happens by, it is quickly seized and carried to the mouth. When a crayfish does move about, it generally crawls slowly but may swim rapidly backwards by using the heavy abdominal muscles.

Respiration in the crayfish is by means of *gills* (fig. 30.21*a*), which lie above the walking legs and are protected by the carapace. Gills, as we have seen, are typical organs of respiration in water-dwelling animals. The crayfish, like the clam, has blue blood containing the pigment hemocyanin, which aids in the transport of oxygen.

Internally, the digestive system (fig. 30.21*b*) includes a stomach, which is divided into 2 main regions: an anterior portion has a *gastric mill*—chitinous teeth that grind coarse food, and a posterior region, which acts as a filter to prevent coarse particles from entering the digestive glands, where absorption takes place.

The crayfish nervous system is quite similar to that of the earthworm. There is a brain from which a solid *ventral nerve cord* passes posteriorly. Along the length of the nerve cord, periodic ganglia give off lateral nerves.

The excretory system (fig. 30.21*b*) consists of a pair of *green glands* lying in the head region anterior to the esophagus. Each organ possesses a glandular region for waste removal—a bladder and a duct that opens ventrally at the base of the antennae.

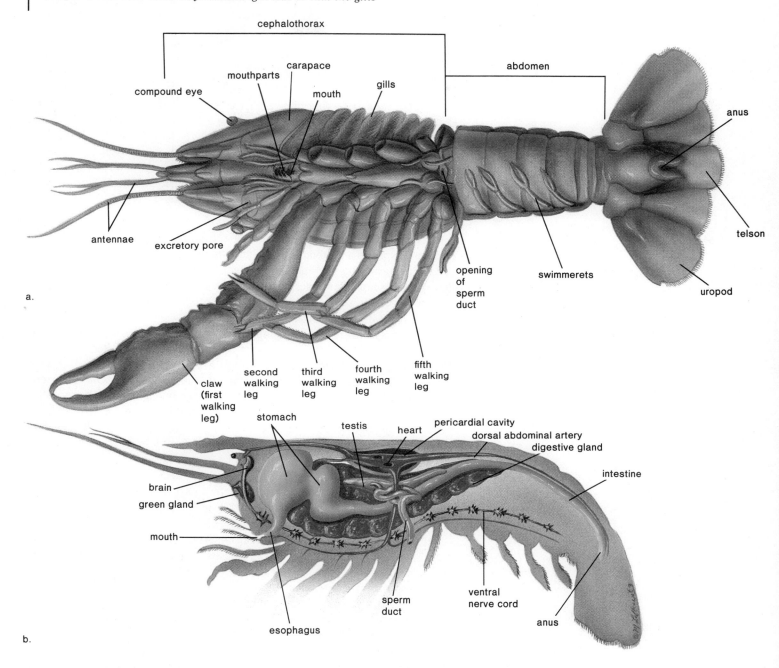

The sexes are separate in the crayfish. The white testes of the male are located just ventral to the pericardial cavity. From each side, a coiled sperm duct passes ventrally and opens to the outside at the base of the fifth walking leg. Sperm transfer is accomplished by the modified first 2 swimmerets of the abdomen. In the female, the ovaries are located in a position similar to that occupied by the testes, and the oviducts pass ventrally, opening near the bases of the third pair of walking legs. There is a cuticular fold between the bases of the fourth and fifth pair that serves as a seminal receptacle. Following fertilization, the eggs are attached to the swimmerets.

Table 30.5 compares the crayfish to the grasshopper to illustrate how one is adapted to the water and the other is adapted to the land.

Table 30.5 *Comparison of Crayfish and Grasshopper*

	Crayfish	Grasshopper
Locomotion	Legs and uropods	Hopping legs and wings
Respiration	Gills	Tracheae
Excretion	Liquid waste by way of green gland	Solid waste by way of Malpighian tubules
Circulation	Blue blood	Colorless blood
Nervous system	Cephalization	Cephalization with tympanum
Reproduction	In male, modified swimmerets pass sperm to female; in female, eggs are attached to swimmerets	Penis in male; ovipositor in female

Grasshoppers

Insects (class Insecta) comprise one of the largest animal groups—both in number of species and in number of individuals—perhaps because of the presence of *wings*. Wings enhance the insects' ability to survive by providing a new way of escaping enemies, finding food, facilitating mating, and dispersing the species. Representative examples of insects, of which the grasshopper is studied in detail, are given in figure 30.20.

Every system of the grasshopper (*Romalea*) (fig. 30.22) is adapted to life on land. There are *3 pairs* of legs, and one of these pairs is suited to jumping. There are 2 pairs of wings; the forewings are tough and leathery and when folded back at rest, they protect the broad, thin hindwings. On the lateral surface, the first abdominal segment bears a large **tympanum** for the reception of sound waves. The posterior region of the exoskeleton in the female has 2 pairs of projections that form an ovipositor for digging a hole in which eggs are layed.

The digestive system (fig. 30.22*b*) is suitable for a grass diet. The mouth has mouthparts to grind the food, and salivary secretions contain enzymes. Food is stored temporarily in the crop before passing into the gizzard, where it is finely ground. Digestion is completed in the stomach, and nutrients are absorbed in outpockets called gastric caeca.

Excretion is carried out by means of **Malpighian tubules,** which extend into the hemocoel and empty into the digestive tract. A solid nitrogenous waste (uric acid) is excreted, conserving water.

Respiration occurs when air enters small tubules called **tracheae** by way of openings in the exoskeleton called **spiracles.** The tracheae branch and rebranch, finally ending in moist areas, where the actual exchange of gases takes place. The movement of air through this complex of tubules is not a passive process; air is pumped through by a series of bladderlike structures (air sacs), which are attached to the tracheae near the spiracles. Air enters the anterior 4 spiracles and exits by the posterior 6 spiracles. This mechanism of breathing, found in insects and arachnids (e.g., spiders and scorpions), is an adaptation to breathing air.

The heart is a slender, tubular organ that lies against the dorsal wall of the abdominal exoskeleton. Blood passes into the hemocoel, where it circulates before finally returning to the heart. The blood is colorless because it lacks a respiratory pigment since the tracheal system transports gases.

Reproduction is adapted to life on land. The male has 2 testes and associated ducts that end in the penis. The female has ovaries that occupy the whole dorsal part of the animal and oviducts that end in the vagina. The sperm

Figure 30.22

Anatomy of a female grasshopper illustrating many adaptations to life on land. a. Externally, the hard skeleton prevents loss of water. There are spiracles, openings in the skeleton that admit air into tracheae (air tubes for respiration). The tympanum uses air waves for sound reception, and the hopping legs and the wings are for locomotion. The ovipositor deposits eggs in soil.

b. Internally, the digestive system is adapted to digesting grass. The Malpighian tubules excrete a solid nitrogenous waste (uric acid). A seminal receptacle receives sperm from the male, which has a penis. Internal fertilization prevents the gametes from drying out.

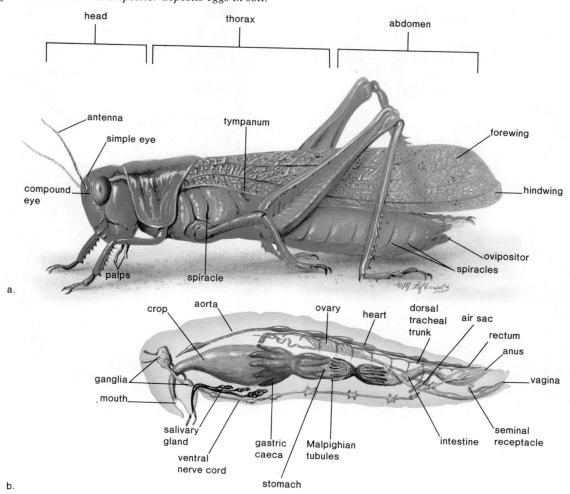

received during copulation are stored in the seminal receptacles for future use. Fertilization is internal, usually occurring during late summer or early fall. The female deposits the fertilized eggs in the ground using her ovipositor.

Insects are land animals that often have wormlike larval stages and undergo metamorphosis. **Metamorphosis** means a change, usually a drastic one, in form and shape. Some insects undergo what is called complete metamorphosis, in which case they have 3 stages of development: *larval stage, pupal stage,* and finally *adult stage*. Metamorphosis occurs during the pupal stage, when the animal is enclosed within a hard covering. The animal that is best known for metamorphosis is the butterfly, whose larval stage is the caterpillar and whose pupal stage is the cocoon; the adult is the butterfly. Grasshoppers undergo incomplete metamorphosis, which is a gradual rather than a drastic change in form. The immature stages of the grasshopper are called nymphs rather than larvae, and they are recognizable as grasshoppers even though they differ somewhat in shape and form. Metamorphosis is controlled by hormones.

Arthropods are the most numerous and varied of all the animal phyla. They have an external skeleton and jointed appendages. Segmentation has led to specialization of parts within the various organ systems.

Figure 30.23
Representative echinoderms. a. *A starfish uses the section of its many tube feet to open a clam.* b. *Sea cucumber, whose general overall shape resembles the vegetable of this name.*
c. *Sea urchins, demonstrating their many external spines.*

a.

b.

c.

Echinoderms

The echinoderms (phylum Echinodermata) include only marine animals—starfishes, sea urchins, sea cucumbers, feather stars, sea lilies, and sand dollars (fig. 30.23). The most familiar of these is the starfish, and we will study this representative. An echinoderm begins life as the bilateral dipleurula larva and then becomes a *radially symmetrical* adult, with a body plan based on *5 parts.* Their other unique feature is the **water vascular system,** which is used as a means of locomotion. They also have a carbonaceous **endoskeleton,** whose projecting spines give the phylum its name, Echinodermata, meaning "spiny skin."

Starfishes

The starfish (*Asterias*), sometimes called the sea star, is found commonly along rocky coasts. It has a five-rayed body plan with an *oral* (mouth) side and an *aboral* (anus) side (fig. 30.24). The oral side is actually the underside, and the aboral side is the upper side. On the aboral side there are various structures that project through the body wall: (1) spines that project from the endoskeletal plates; (2) pincerlike structures called *pedicellarie,* which keep the surface free of small particles; and (3) skin gills, which serve for respiratory exchange. The mouth is located on the oral surface, where each of the 5 arms has a groove lined by little **tube feet.**

Figure 30.24

Starfish anatomy. Like other echinoderms, starfishes have a water vascular system shown here in color. Water enters the sieve plate and eventually is sent into tube feet by the action of the ampullae. Each arm of a starfish contains digestive glands, gonads, and the water vascular system.

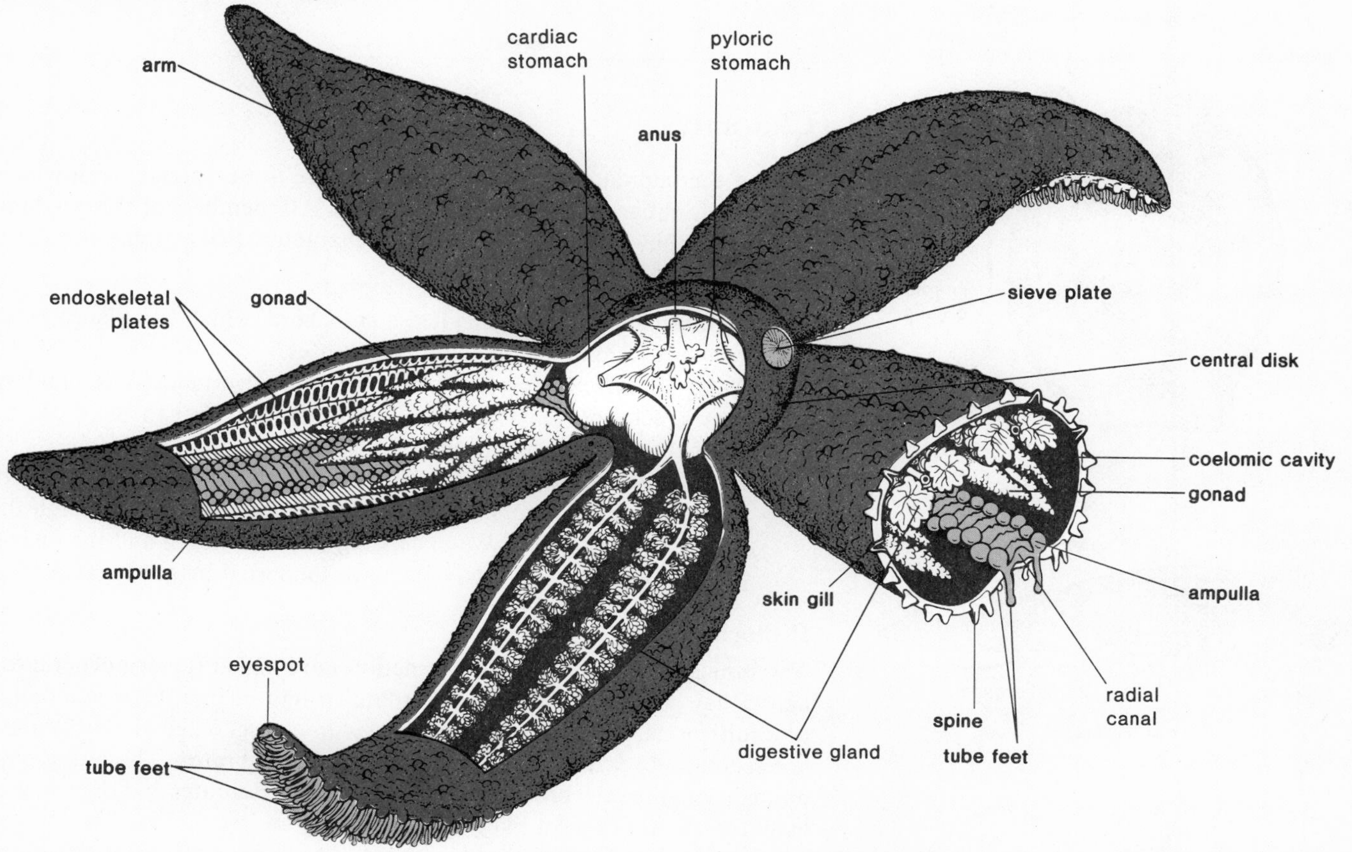

Starfishes feed on mollusks. When a starfish attacks a clam, it arches its body over the shell (fig. 30.23a), and by the concerted action of the tube feet, forces the clam to open. Then it everts a portion of its stomach to digest the contents of the clam.

The mouth of a starfish opens into a narrow esophagus, which in turn leads to an expanded stomach. The stomach has 2 portions: the saclike cardiac, which can be everted as described, and the narrower pyloric, which is connected to a short intestine. The anus opens on the aboral or upper side of the animal.

Each of the 5 arms contains a well-developed coelom, a pair of large digestive glands that secrete powerful enzymes into the pyloric portion of the stomach, and gonads, which open on the aboral surface by very small pores. The nervous system consists of a central nerve ring that supplies radial nerves to each arm. A light-sensitive eyespot is at the tip of each arm.

Coelomic fluid, circulated by ciliary action, performs many of the normal functions of a circulatory system; the water vascular system is purely for locomotion. Water enters this system through a structure on the aboral side called the **sieve plate,** or madreporite. From there, it passes through a short canal, called the *stone canal,* to a *ring canal,* which surrounds the mouth. From the ring canal, *5 radial canals* extend into the arms. From the radial canals, many lateral canals extend into the tube feet. One lateral canal goes to each tube foot, where it ends in the *ampulla.* When the ampulla contracts, the water is

Figure 30.25

Protochordates. a. Tunicate anatomy. Gill slits
are the only chordate feature retained by the
adult. b. Lancelet anatomy. This animal retains
all 3 chordate characteristics as an adult.

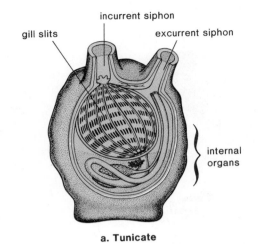

a. Tunicate

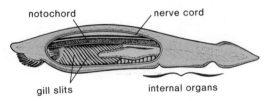

b. Lancelet

forced into the tube foot, expanding it and giving it suction. By alternating
the expansion and the contraction of the tube feet, the starfish moves along
slowly.

*Echinoderms are radially symmetrical and have a spiny skin. They move by tube
feet, which are part of their water vascular system. Their other body systems are
rather primitive.*

Chordates

Among the chordates (phylum Chordata) are those animals with which we
are most familiar, including human beings. All members of this phylum are
observed to have the following 3 basic characteristics at some time in their
life history.

1. A dorsal supporting rod called a **notochord,** which is replaced by the
   vertebral column in the adult vertebrates.
2. A *dorsal hollow nerve cord,* in contrast to invertebrates, which have a
   ventral solid nerve cord. By hollow, it is meant that the cord contains
   a canal that is filled with fluid.
3. *Pharangeal pouches or gill clefts (slits),* which are seen only during
   embryological development in most vertebrate groups, although they
   persist in adult fishes. Water passing into the mouth and the pharynx
   goes through the gill slits, which are supported by gill bars.

Protochordates

The tunicates and the lancelets sometimes are called the **protochordates** because they possess all 3 typical chordate structures in either the larval and/
or adult forms, as did the first chordates to evolve. They also are called *invertebrate chordates* because they are not vertebrates. These 2 groups of
animals link the vertebrates to the rest of the invertebrates and show how modestly the chordates most likely began.

A **tunicate** (subphylum Urochordata), or sea squirt (fig. 30.25a), appears to be a thick-walled, squat sac with 2 openings, an incurrent siphon and
an excurrent siphon. Inside the central cavity of the animal are numerous gill
slits, the only chordate feature retained by the adult. The larva of the tunicate,
however, has a tadpole shape and possesses the 3 chordate characteristics. It
has been suggested that such a larva may have become sexually mature without
developing the other adult tunicate characteristics. If so, it may have evolved
into a fishlike vertebrate.

A **lancelet** (subphylum Cephalochordata) (fig. 30.25b) is a chordate that
shows the *3 chordate characteristics as an adult.* In addition, segmentation
is present, as witnessed by the fact that the muscles are segmentally arranged
and the nerve cord gives off periodic branches.

*The invertebrate chordates include the tunicates and the lancelets. A lancelet is the
best example of a chordate that possesses the 3 chordate characteristics—
notochord, nerve cord, and gill slits—as an adult.*

Vertebrate Chordates

Vertebrates (subphylum Vertebrata) have all the most advanced characteristics listed in table 30.1. They are segmented chordates in which the notochord is replaced in the adult by a *vertebral column* composed of individual
vertebrae. The skeleton is internal, and in all the vertebrates, there is not only
a backbone but also a skull, or cranium, to enclose and to protect the brain.

In higher vertebrates, other parts of the skeleton serve as attachment for muscles and for protection of internal organs of the thoracic cavity and the abdomen. All vertebrates have a *closed circulatory system* in which red blood is contained entirely within blood vessels. They show good cephalization with sense organs; the eyes develop as outgrowths of the brain, and the ears serve as equilibrium devices in aquatic vertebrates plus sound wave receivers in land vertebrates. The kidneys are important excretory and water-regulating organs that conserve or rid the body of water as appropriate.

Two comparisons often are made between invertebrates and vertebrates: (1) invertebrates have a *ventral solid* nerve cord, while vertebrates have a dorsal hollow nerve cord called the spinal cord; and (2) invertebrates have an *external* skeleton, while vertebrates have an *internal* skeleton. There are, however, many invertebrates that do not have a nerve cord or an external skeleton.

As a group, the vertebrates are the dominant animals in the world today. They are found in every habitat, from the ocean floor to the mountaintop and in the forest and the desert. Included in the group are the 3 classes of fishes and one class each of amphibians, reptiles, birds, and mammals. Figure 30.26 represents an evolutionary tree of the vertebrates, and you can see that it is possible to trace the evolution of the vertebrates from fishes to amphibians to reptiles to both birds and mammals. All but the fishes are **tetrapods,** meaning that they have 4 limbs.

Vertebrates are animals in which the vertebral column has replaced the notochord. Vertebrates often are compared to the invertebrates, which lack a vertebral column.

Fishes

There are 3 classes of fishes: the jawless fishes, the cartilaginous fishes, and the bony fishes. Living representatives of the *jawless fishes* are cylindrical, up to a meter long, with smooth, scaleless skin and no jaws or paired fins. There are 2 families of jawless fishes: *hagfishes* and *lampreys.* The hagfishes are scavengers, feeding mainly on dead fishes, while some lampreys are parasitic. When parasitic, the round mouth of the lamprey serves as a sucker by which it attaches itself to another fish and taps into its circulatory system.

Cartilaginous fishes (fig. 30.27a) are the sharks, the rays, and the skates, which have skeletons of cartilage instead of bone. The dogfish shark is a small shark often dissected in biology laboratories to show the main features of the vertebrate body. Other sharks are well known to us as vicious predators that attack human swimmers. One of the most dangerous sharks inhabiting both tropical and temperate waters is the hammerhead shark. The largest sharks, the whale sharks, feed on small fishes and marine invertebrates and do not attack humans. Skates and rays are rather flat fishes that live partly buried in the sand and feed on mussels and clams.

Bony fishes (fig. 30.27b) are by far the most numerous and varied of the fishes. Most of the fishes we eat, such as perch, trout, flounder, and haddock, are a type of bony fish called *ray-finned fishes*. These fishes have a *swim bladder* that aids them in changing their depth in the water. By secreting gases into the bladder or by absorbing gases from it, these fishes can change their density and thus go up or down in the water. "Ray-finned" refers to the fact that the fins are thin and are supported by bony rays. Another type of bony fish, called the *lobe-finned fishes,* evolved into the amphibians. These fishes not only have fleshy appendages that could be adapted to land locomotion, they also have a lung[2] that is used for respiration. A type of lobe-finned fish called the coelacanth, which exists today, is the only "living fossil" among the fishes.

---

[2]The swim bladder of ray-finned fishes is believed to be derived from an ancient lung.

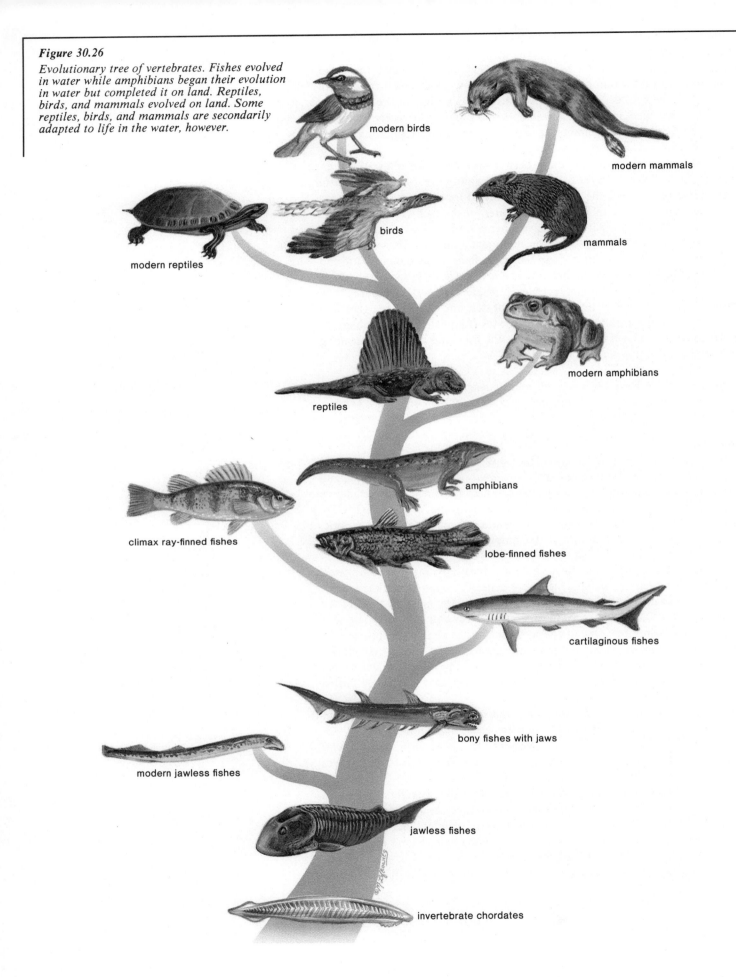

Figure 30.26

Evolutionary tree of vertebrates. Fishes evolved in water while amphibians began their evolution in water but completed it on land. Reptiles, birds, and mammals evolved on land. Some reptiles, birds, and mammals are secondarily adapted to life in the water, however.

modern birds

modern mammals

modern reptiles

birds

mammals

reptiles

modern amphibians

amphibians

climax ray-finned fishes

lobe-finned fishes

cartilaginous fishes

bony fishes with jaws

modern jawless fishes

jawless fishes

invertebrate chordates

Evolution and Diversity

Figure 30.27

Representative cartilaginous fish and bony fish. a. The bull shark is a cartilaginous fish. b. The freckled grouper is a bony fish.

a.

b.

Fishes are adapted to life in the water. Their streamlined shape, fins, and muscle action are all suitable to locomotion in the water. Their body is covered by *scales,* which protect the body but do not prevent water loss. Fishes breathe by means of *gills,* respiratory organs that are kept continuously moist by the passage of water through the mouth and out the gill slits. As the water passes over the gills, oxygen (O_2) is absorbed by blood and carbon dioxide (CO_2) is given off. The heart of a fish is a simple pump, and the blood flows through the chambers, including a nondivided atrium and ventricle, to the gills only (see fig. 30.33*a*). Oxygenated blood leaves the gills and goes to the body proper.

Generally speaking, reproduction in the fishes requires external water; sperm and eggs usually are shed into the water, where fertilization occurs. The zygote develops into a swimming larva that can fend for itself until it develops into the adult form.[3]

The most primitive fishes are jawless. Sharks, skates, and rays are cartilaginous, while the bony fishes include all other well-known fishes.

Amphibians

The living amphibians include *frogs, toads, newts,* and *salamanders* (fig. 30.28*a*). These animals have distinct walking legs, each with 5 or fewer toes. This represents an adaptation to land locomotion. Respiration is accomplished by the use of small, relatively *inefficient lungs,* supplemented by gaseous exchange through the skin. Therefore, the skin is smooth, moist, and glandular. This is a distinct disadvantage in a dry environment; therefore, frogs spend most of their time in or near freshwater. All amphibians possess 2 nostrils that

---

[3]Some fishes, such as sharks, practice internal fertilization and retain their eggs during development. Their young are born alive.

Figure 30.28

Representative amphibian and reptile. a. This male bullfrog is an amphibian. Note the thin, moist skin, an indication that adult amphibians cannot live in a dry environment. b. This rough green snake is a reptile. Note the scaly, dry skin, an indication that a snake can live in a dry environment.

a.

b.

are connected directly with the mouth cavity. Air enters the mouth by way of the nostrils, and when the floor of the mouth is raised, air is forced into the lungs.

With the development of lungs, there is a change in the circulatory system. The amphibian heart has a divided atrium but a single ventricle (see fig. 30.33*b*). The right atrium receives impure blood with little oxygen from the body proper, and the left atrium receives purified blood from the lungs that has just been oxygenated, but these 2 types of blood are mixed partially in the single ventricle. Mixed blood is then sent, in part, to the skin, where further oxygenation can occur.

Nearly all the members of this class lead an amphibious life—that is, the larval stage lives in the water and the adult stage lives on the land. The adults must return to the water, however, for reproduction. Just as with the fishes, the sperm and the eggs are discharged into the water and fertilization results in a zygote that develops into the familiar tadpole. The tadpole undergoes metamorphosis into the adult before taking up life on the land.

Amphibians are adapted to a dual existence. The appendages function on land and in the water, lungs and skin are used for respiration, and external water is utilized for reproduction.

Metamorphosis allows amphibians to switch from an aquatic to a land existence. However, they retain anatomical features suitable to water and they return to the water for reproduction.

Reptiles

The reptiles living today are *turtles, alligators, snakes,* and *lizards* (fig. 30.28). Reptiles with limbs, such as lizards, are able to lift their body off the ground, and the body is covered with hard, *horny scales* that protect the animal from desiccation and from predators. Both of these features are adaptations to life on land.

Reptiles have well-developed lungs enclosed in a protective rib cage. When the rib cage expands, the lungs expand, and air rushes in. The creation of a partial vacuum establishes a negative pressure that causes air to rush into the lungs. The atrium of the heart is always separated into right and left chambers, but division of the ventricle varies. There is always at least one interventricular septum, but it is incomplete in all but the crocodiles; therefore, exchange of oxygenated and deoxygenated blood between the ventricles occurs in all but the crocodile.

Perhaps the most outstanding adaptation of the reptiles is that they have a means of reproduction suitable to a land existence. There is usually no need for external water to accomplish fertilization because the penis of the male passes sperm directly to the female. After *internal fertilization* has occurred, the egg is covered by a protective, leathery shell and is laid in an appropriate location.

The *shelled egg* made development on land possible and eliminated the need for a swimming-larva stage during development. It provides the developing embryo with oxygen, food, and water; it removes nitrogenous wastes; and it protects the embryo from drying out and from mechanical injury. This is accomplished by the presence of *extraembryonic membranes* (see fig. 21.13).

Reptiles do not regulate their body temperature. Sometimes, animals that cannot maintain a constant temperature—that is, fishes, amphibians, and reptiles—are called *cold blooded*. Actually, however, they take on the temperature of the external environment. If it is cold externally, they are cold internally; and if it is hot externally, they are hot internally. Reptiles try to regulate body temperatures by exposing themselves to the sun if they need warmth or by hiding in the shadows if they need cooling off. This works reasonably well in most areas of the world.

Reptiles are well adapted to a land environment. They have a scaly skin that prevents loss of water, and they can reproduce on land because they lay a shelled egg.

Birds

Birds (fig. 30.29) are characterized by the presence of feathers, which are actually modified reptilian scales. There are many orders of birds, including birds that are flightless (ostrich), web footed (penguin), divers (loons), fish eaters (pelicans), waders (flamingos), broad billed (ducks), birds of prey (hawks), vegetarians (fowl), shorebirds (sandpipers), nocturnal (owl), small (hummingbirds), and songbirds, the most familiar of the birds.

Nearly every anatomical feature of a bird can be related to its *ability to fly*. The anterior pair of appendages (wings) is adapted for flight; the posterior is variously modified, depending on the type of bird. Some are adapted to swimming, some to running, and some to perching on limbs. The breastbone is enormous and has a ridge to which the flight muscles are attached. Respiration is efficient since the lobular lungs form *air sacs* throughout the body, including the bones. The presence of these sacs means that the air circulates one way through the lungs during both inspiration and expiration; "used" air is not trapped in the lungs. Another benefit of air sacs is that the air-filled, hollow bones lighten the body and aid flying. Birds have a four-chambered heart (see fig. 30.33c) that completely separates oxygenated blood from deoxygenated blood.

Birds have well-developed brains, but the enlarged portion seems to be the area responsible for instinctive behavior. Therefore, birds follow very definite patterns of migration and nesting.

30.2 Critical Thinking

Both insects and birds are adapted to life on land.

1. How does each carry on respiration?
2. How does each prevent drying out of the animal?
3. How does each prevent drying out of the gametes and the embryo?
4. What is the reproductive strategy of each animal? How do they differ?

Figure 30.29
Bird diversity. a. *Chinstrap penguin feeding chicks. Penguins are adapted to arctic conditions.* b. *Flamingos are exotic birds of the tropics.*

a.

b.

Birds are *warm blooded;* like mammals, they are able to maintain a constant internal temperature. This may be associated with their efficient nervous, respiratory, and circulatory systems. Also, their feathers provide insulation.

Birds can maintain a constant internal temperature. All organ systems are adapted for flight.

Mammals

The chief characteristics of mammals are *hair* and *mammary glands* that produce milk to nourish the young. Human mammary glands are called breasts.

Mammals are adapted to life on land and have limbs that allow them to move rapidly. In fact, an evaluation of mammalian features leads us to the obvious conclusion that they lead active lives. The brain is well developed; the lungs are expanded not only by the action of the rib cage but also by the contraction of the *diaphragm,* a horizontal muscle that divides the thoracic cavity from the abdominal cavity; and the heart has *4 chambers* (see fig. 30.33*c*). The internal temperature is constant, and hair, when abundant, helps to insulate the body.

The mammalian brain is enlarged due to the expansion of the foremost part—the cerebral hemispheres. These have become convoluted and have expanded to such a degree that they hide many other parts of the brain from view.

Mammals are classified according to their means of reproduction: there are *egg-laying* mammals, mammals with *pouches* for immature embryos, and **placental mammals.**

Figure 30.30

Monotreme and marsupial, 2 rare types of mammals. a. Spiny anteater is a monotreme that lays a shelled egg. b. Koala bear is a marsupial, whose young are born immature and complete their development within the mother's pouch.

a.

b.

Monotremes   Monotremes are egg-laying mammals, represented by the duck-billed platypus and the spiny anteater (fig. 30.30*a*). In the same manner as birds, the female monotreme incubates the eggs, but after hatching, the young are dependent upon the milk that seeps from glands on the abdomen of the female. Therefore, monotremes retained the reptilian mode of reproduction while evolving hair and mammary glands. The young are blind, helpless, and completely dependent on the parent for some months. The mouth is variously modified among the monotremes. The platypus has a horny, bill-like structure somewhat resembling that of a duck, while the anteater has an elongated, cylindrical snout.

Marsupials   Another primitive group of mammals is the *marsupials,* such as opossums, kangaroos, and koala bears. In marsupials, the young are born in a very immature state and finish their development in the mother's abdominal pouch, called the marsupium. For example, when an opossum is born after only 12–16 days of gestation, it is blind, naked, grublike, and no larger than a honeybee. Using clawed forelimbs, the newborn crawls toward the mother's fur-lined pouch. Once there, it attaches itself to a nipple. After 4 or 5 weeks in the pouch, an opossum spends an additional 8 or 9 weeks clinging to the mother's back.

Placental Mammals   The vast majority of living mammals are placental mammals (fig. 30.31). In these mammals, the extraembryonic membranes have been modified for internal development within the uterus of the female. The chorion contributes to the fetal portion of the placenta, while a portion of the uterine wall contributes to the maternal portion. Here, nutrients, oxygen, and waste are exchanged between fetal and maternal blood. These mammals not

Figure 30.31

Mammalian diversity. a. *White-tailed deer feeding fawn. The white of the tail is exposed when the tail is raised as a sign of imminent danger.* b. *Torpedo-shaped harbor seals are excellent swimmers and divers, having short paddlelike flippers and a thick layer of subcutaneous fat (blubber).*

a.

b.

only have a long embryonic period, they also are dependent on their parents until the nervous system is developed fully and they have learned to take care of themselves.

Placental mammals can be classified into 12 orders, 9 of which can be considered major (table 30.6). A study of these reveals that mammals have differentiated, or become specialized, largely according to mode of locomotion and how they get their food.

The anatomy and the physiology of human beings exemplifies vertebrate, especially mammalian, anatomy and physiology. Chapters 9–19 therefore can be used for detailed information regarding vertebrate and mammalian anatomy and physiology.

Table 30.6 *Some Major Orders of Placental Mammals*

Insectivora (moles, shrews)	Primitive; small, sharp-pointed teeth
Chiroptera (bats)	Digits support membranous wings
Carnivora (dogs, bears, cats, sea lions)	Long canine teeth; pointed teeth
Rodentia (mice, rats, squirrels, beavers, porcupines)	Incisor teeth grow continuously
Lagomorpha (rabbits, hares, pikas)	Chisel-like incisors; hind legs longer than front legs; herbivorous
Perissodactyla (horses, zebras, tapirs, rhinoceroses)	Large, long-legged, one or 3 toes, each with hoof; grinding teeth
Artiodactyla (pigs, cattle, camels, buffalos, giraffes)	Medium to large; 2 or 4 toes, each with hoof; many with antlers or horns
Cetacea (whales, porpoises)	Medium to very large; paddlelike forelimbs; hind limbs absent
Primates (lemurs, monkeys, gibbons, chimpanzees, gorillas, humans)	Mostly tree dwelling; head freely movable on neck; 5 digits, usually with nails; thumbs and/or large toes usually opposable

Adapted Table 18.5 from *Essentials of Biology,* Second Edition, by Willis H. Johnson, copyright © 1974 by Saunders College Publishing, a division of Holt, Rinehart and Winston, Inc., reprinted by permission of the publisher.

Placental mammals are far more numerous than the monotremes and marsupials (pouched) mammals. They are adapted to life in the air, in the water, and on the land. Humans are classified as primates, mammals that are adapted to living in trees.

Comparisons between Vertebrates

Vertebrates, like other animals, are adapted to their way of life. Figure 30.32 shows that fishes breathe by means of gills, respiratory organs appropriate to life in the water. Amphibians have small, ineffectual lungs that must be supplemented by use of the skin as a respiratory organ. Reptiles have more efficient lungs with a rib cage, which not only protects the lungs but helps to fill them with air. Birds have lungs expanded by air sacs that allow one-way flow of air, and mammals have highly subdivided lungs surrounded by a rib cage and separated from the abdominal cavity by a diaphragm. These anatomical features make breathing by negative pressure possible in reptiles, birds, and mammals.

Figure 30.33 compares the circulatory systems of the vertebrates. Fishes have a nondivided atrium and ventricle, and the heart pumps blood only to the gills. Amphibians have a heart in which there is a right and a left atrium but only a single ventricle, where oxygenated and deoxygenated blood are partially mixed before being sent, in part, to the skin for further oxygenation. The reptiles have a right and a left atrium and a ventricle that has 2 partial septa. Some mixing of oxygenated and deoxygenated blood is not a serious disadvantage for the cold-blooded amphibians and reptiles since their oxygen demands are relatively low. Birds and mammals have a four-chambered heart in which division of the atria and the ventricles is complete and there is no opportunity for mixing to occur. The right side of the heart pumps blood to the lungs, and the left side of the heart pumps blood to the rest of the body.

A comparison of the eggs of vertebrates shows that fish and amphibian eggs are generally small with little yolk; these eggs are deposited into the water, where they may develop into a swimming larva. Both reptiles and birds lay a

Figure 30.32
Breathing mechanisms of vertebrates. Fishes breathe by means of gills; amphibians have poorly developed lungs; reptiles have a rib cage and lungs; birds have lungs with air sacs; and humans have well-developed lungs plus a rib cage and a diaphragm.

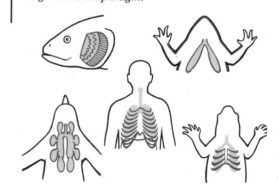

Figure 30.33

A comparison of circulatory paths in vertebrates. a. A fish heart with only one atrium and one ventricle pumps blood to the gills. b. In an amphibian there are pulmonary and systemic circuits, but the heart has 2 atria and only one ventricle. c. The heart of *birds and mammals is a double pump; one half of the heart pumps blood to the lungs and the other half pumps blood to the body.*

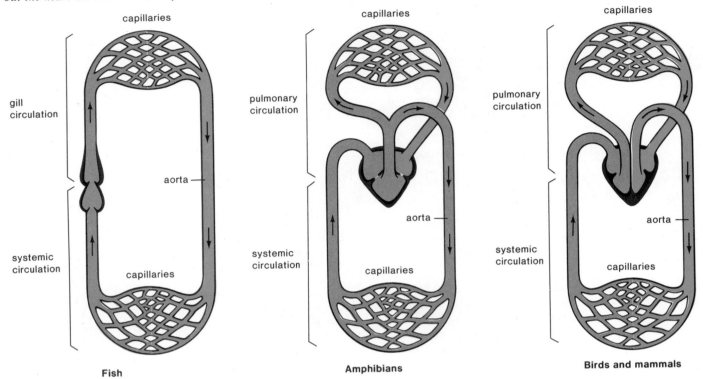

Fish

Amphibians

Birds and mammals

■ Oxygenated blood
■ Deoxygenated blood
■ Mixed blood

shelled egg with extraembryonic membranes to take over the functions previously performed by external water. The placental mammals have modified membranes that permit internal development, during which time the mother provides for the needs of the developing fetus (chap. 21).

A comparison of the vertebrates shows that the respiratory system in terrestrial forms is adapted to breathing air; the circulatory system is adapted to maintaining separate pulmonary and systemic circuits; and there is a development of the ability to reproduce on land (table 30.7).

Table 30.7 *Comparison of Vertebrates (in General)*

	Fishes	**Amphibians**	**Reptiles**	**Birds**	**Mammals**
Habitat	Water	Water when larvae; land when adults	Land	Land	Land
Heart	Nondivided atrium and ventricle	2 atria and nondivided ventricle	2 atria and partially divided ventricle	2 atria and 2 ventricles	2 atria and 2 ventricles
Respiration	Gills	Gills when larvae; lungs and skin when adults	Lungs and rib cage	Lungs, rib cage, and air sacs	Lungs, rib cage, and diaphragm
Fertilization	External	External	Internal	Internal	Internal
Egg	Small, no shell, develops externally	Small, no shell, develops externally	Large, with shell, develops externally	Large, with shell, develops externally	Small, no shell, develops internally

Summary

Classification of animals considers type of body plan, symmetry, number of germ layers, presence or absence of a coelom, and segmentation. Among the primitive invertebrates, it is possible to observe an increase in complexity regarding these features if the various groups are arranged as in table 30.8. While the flatworms and the roundworms have specialized organs, all of the primitive invertebrates carry out respiration and circulation by diffusion.

All of the advanced invertebrates have the tube-within-a-tube body plan, 3 germ layers, and a true coelom. Embryological evidence suggests that the echinoderms and the chordates are related. Echinoderms are unique among the advanced invertebrates because of their radial symmetry and locomotion by tube feet. At some time in their life history, all chordates have a dorsal notochord, dorsal hollow nerve cord, and pharyngeal pouches. Lancelets retain all 3 features as an adult. The vertebrates are chordates in which the notochord is replaced by a vertebral column in the adult. The evolutionary tree indicates that amphibians evolved from lobe-finned fishes; reptiles evolved from amphibians; and birds and mammals evolved from reptiles.

Adaptation to land life begins in the amphibians and continues in reptiles, which are able to reproduce on land due to a shelled egg with extraembryonic membranes. Only birds and mammals are warm blooded, and able to maintain a constant internal temperature; they also have a heart that separates oxygenated blood from deoxygenated blood.

Table 30.8 Classification Features

	Sponges	Cnidarians	Flatworms	Roundworms
Body plan	— — —	Sac plan	Sac plan	Tube-within-a-tube plan
Symmetry	Radial or none	Radial	Bilateral	Bilateral
Germ layers	— — —	2	3	3
Level of organization	— — —	Tissues	Organ	Organ
Type of coelom	— — —	— — —	— — —	Pseudocoelom
Segmentation	— — —	— — —	— — —	— — —

Study Questions

1. The primitive invertebrates comprise which group of animals? (p. 633) Compare the representatives of these animals in regard to body plan, symmetry, germ layers, level of organization, coelom, and segmentation. (pp. 633–42, 667)
2. Compare the representatives of the 4 primitive invertebrate phyla in regard to nervous conduction, musculature, digestion, excretion, and reproduction. (pp. 633–42)
3. Describe the life cycle and the structure of a tapeworm. Compare the anatomy of free-living flatworms with that of the fluke and the tapeworm. (pp. 637–40)
4. What biological data are used to divide advanced animals into 2 groups? (p. 642)
5. Compare the clam, the earthworm, the crayfish, the grasshopper, and the starfish with respect to nervous, digestive, skeletal, excretory, circulatory, and respiratory systems and means of reproduction and locomotion. (pp. 643–55)
6. Compare the adaptations of the clam to those of the squid to show that the clam is adapted to an inactive life and the squid is adapted to an active life. (p. 645)
7. Compare the adaptations of the crayfish to those of the grasshopper to show that the crayfish is adapted to an aquatic existence while the grasshopper is adapted to a terrestrial existence. (p. 652)
8. Name and describe unique features of echinoderm anatomy and physiology. (pp. 644–56)
9. Compare the adaptations of vertebrates to a land existence. (p. 665)

Objective Questions

1. The function of collar cells in a sponge is _____ .
2. Cnidarians have the _____ body plan and are _____ symmetrical.
3. Planarians have the _____ type of nervous system and a(n) _____ excretory system.
4. The intermediate host are for a tapeworm either _____ or _____ .
5. Pinworm, trichinosis, hookworm, and elephantiasis are all _____ worm infections.
6. In protostomes, the first embryonic opening becomes the _____ .
7. In today's mollusks, the coelom is much _____ and limited to the region around the _____ .
8. Earthworms have external rings signifying that they are _____ animals.
9. The water vascular system of echinoderms consists of canals and _____ feet.
10. The 3 chordate characteristics are a(n) _____ , _____ , and _____ .
11. The _____ and the _____ are primitive chordates.
12. The 3 classes of fishes, _____ , _____ , and _____ , indicate in general their order of evolution.

13. Amphibians evolved from _____ fishes that had primitive lungs.

14. Whereas amphibians must return to the _____ to reproduce, reptiles lay _____ that contain _____ membranes.

15. Both _____ and mammals maintain a constant internal _____ .

16. There are 3 types of mammals:

_____ ,

_____ , and

_____ .

17. Dogs, cats, horses, mice, rabbits, bats, whales, and humans are all _____ mammals.

Schizocoelomates

enterocoelomates

Protistan ancestors

Label this Diagram.
See figure 30.2 (p. 631) in text.

Answers to Objective Questions

1. to keep water moving through the central cavity 2. sac, radially 3. ladder, flame-cell 4. cattle or pigs 5. round 6. mouth 7. reduced, heart 8. segmented 9. tube 10. notochord, dorsal hollow nerve cord, pharyngeal pouches 11. tunicates, lancelets 12. jawless, cartilaginous, bony 13. lobe-finned 14. water, eggs, extraembryonic 15. birds, temperature 16. monotremes, marsupials, placental 17. placental

Selected Key Terms

bilateral symmetry (bi-lat′er-al sim′ĕ-tre) having a right and a left half so that only one vertical cut gives 2 equal halves. *632*

chitin (ki′tin) flexible, strong polysaccharide forming the exoskeleton of arthropods. *648*

compound eye (kom′pownd i) arthropod eye composed of multiple lenses. *650*

deuterostome (du′ter-o-stōm″) member of a group of animal phyla in which the anus develops from the blastopore and a second opening becomes the mouth. *642*

dipleurula (di-ploor′u-lah) a larval form unique to the deuterostomes that indicates that they are related. *642*

filter feeder (fil′ter fēd′er) an animal that obtains its food, usually in small particles, by filtering it from water. *633*

hermaphroditic animal (her′maf′ro-di-tik″ an′ĕ-mĕl) an animal having both male and female sex organs. *638*

invertebrate (in-ver′tĕ-brāt) an animal that lacks a vertebral column. *630*

jointed appendage (joint′ed ah-pen′dij) the flexible exoskeleton extension found in arthropods that is used as sense organ and mouthpart, and for locomotion. *650*

Malpighian tubule (mal-pig′i-an tu′būl) organ of excretion, notably in insects. *652*

mesoglea (mes″o-gle′ah) a jellylike packing material between the ectoderm and the endoderm of cnidarians. *634*

nematocyst (nem′ah-to-sist) a threadlike structure in stinging cells of cnidarians that can be expelled to numb and to capture prey. *634*

notochord (no′to-kord) dorsal supporting rod that exists in all chordates sometime in their life history; replaced by the vertebral column in vertebrates. *656*

protostome (pro′to-stōm) member of a group of animal phyla in which the mouth develops from the blastopore. *642*

radial symmetry (ra′de-al sim′ĕ-tre) regardless of the angle of a vertical cut made at the midline of an organism, 2 equal halves result. *631*

sessile (ses′il) organisms that lack locomotion and remain stationary in one place, such as plants or sponges. *633*

seta (se′ta) bristle, especially that of the segmented worms. *648*

trachea (tra′ke-ah) an air tube of insects. *652*

trochophore (tro′ko-fōr) a larval form unique to the protostomes that indicates they are related. *642*

vertebrate (ver′tĕ-brāt) animal possessing a backbone composed of vertebrae. *630*

visceral mass (vis′er-al mas) soft-bodied portion of a mollusk that includes internal organs. *643*

Chapter 31

Human Evolution

Chapter Concepts

1 Humans are primates, and many of their physical traits are the result of their ancestors' adaptations to an arboreal existence.

2 Humans share a common ancestor with apes. Our lineage did not diverge from the apes' lineage until about 5–10 million years ago.

3 All fossils placed in the same family as modern humans were bipedal tool users. Tool use and increased intelligence evolved together.

4 Hunting and language have been identified as characteristics that most likely led to the development of culture. Culture distinguishes us from the apes.

5 All human races are classified as *Homo sapiens sapiens*.

Chapter Outline

The Evolution of Humans
  Prosimians
  Anthropoids
  Hominoids
  Hominids
  Australopithecines
Humans
  Homo habilis
  Homo erectus
  Homo sapiens

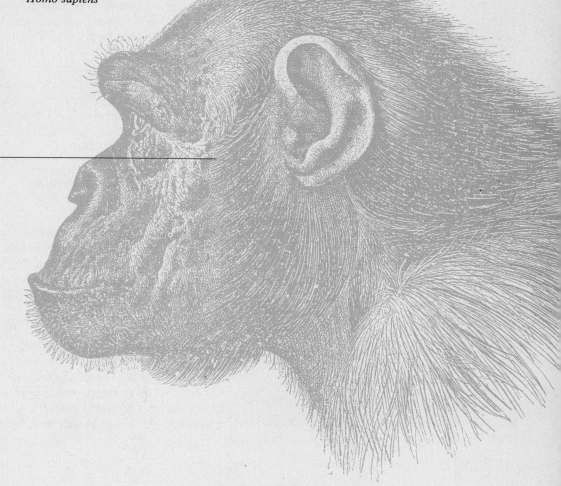

Table 31.1 *Classification of Primates*

Category	Animals	Category	Animals
Phylum Chordata		Superfamily	*Dryopithecus*
Subphylum Vertebrata		Hominoidea	Modern apes
Class Mammalia			Humans
Order	Prosimians		
Primates	Lemurs	Family	*Australopithecus*
	Tarsiers	Hominidae	*Homo habilis*
	Anthropoids		*Homo erectus*
	Monkeys		*Homo sapiens*
	Apes		
	Humans	Genus *Homo*	*Homo habilis*
		(Humans)	*Homo erectus*
			Homo sapiens

Figure 31.1

Tree shrew. A tree shrew is a small mammal with feet adapted for climbing in trees, where it feeds on fruits and insects. Notice that the ears are similar to the ears of the primates, being rounded with folded edges. A tree shrew's brain is relatively large, but the olfactory portion is surprisingly small. Tree shrews may resemble the ancient ancestors of modern primates.

31.1 Critical Thinking

How can you tell that humans are

1. animals?
2. chordates and vertebrates?
3. mammals?
4. primates?

Humans are mammals in the order Primates (table 31.1). The first **primates** may have resembled today's tree shrews, rat-sized animals with a snout, claws, and sharp front teeth (fig. 31.1). By 50 million years ago, however, primates had evolved characteristics suitable to moving freely through the trees. Their limbs became adapted to swinging and leaping from branch to branch. Their hands were especially dexterous and mobile because their thumbs were opposable; that is, they closed to meet the fingertips. Therefore, these animals easily could reach out and bring food, such as fruits, to the mouth. Claws were replaced by nails, which allowed a tree limb to be grasped and released freely.

A snout is common in animals in which a sense of smell is of primary importance. In primates, the sense of sight is more important, and the snout has shortened considerably, allowing the eyes to move to the front of the head. This resulted in binocular (three-dimensional) vision, permitting primates to make accurate judgments about the distance and the position of adjoining tree limbs.

One birth at a time became the norm with primates; it would have been difficult to care for several offspring as large as primates in trees. The period of postnatal maturation was prolonged, giving immature young an adequate length of time to learn complex behavior patterns.

Figure 31.2 and table 31.2 indicate more specifically when the various groups we will discuss evolved.

The Evolution of Humans

Prosimians

The first primates were **prosimians,** a term meaning "premonkeys." The prosimians are represented today by several types of animals, including the lemurs (fig. 31.2), which have a squirrel-like appearance, and the tarsiers, curious mouse-sized creatures with enormous eyes suitable for their nocturnal way of life. Tarsiers have a flattened face, and their digits terminate in nails.

Anthropoids

Monkeys, along with apes and humans, are **anthropoids.** Monkeys evolved from the prosimians about 38 million years ago, when the weather was warm and vegetation was like that of a tropical rain forest. There are 2 types of monkeys: the New World monkeys, which have long prehensile (grasping) tails and flat noses, and the Old World monkeys, which lack such tails and have protruding noses. Two of the well-known New World monkeys are the spider monkey and

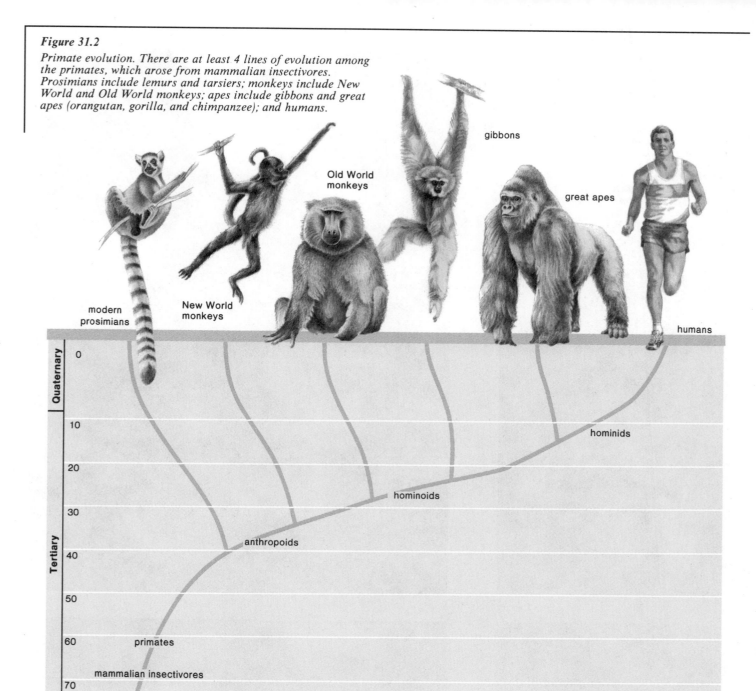

Figure 31.2

Primate evolution. There are at least 4 lines of evolution among the primates, which arose from mammalian insectivores. Prosimians include lemurs and tarsiers; monkeys include New World and Old World monkeys; apes include gibbons and great apes (orangutan, gorilla, and chimpanzee); and humans.

gibbons

Old World monkeys

great apes

modern prosimians

New World monkeys

humans

hominids

hominoids

anthropoids

primates

mammalian insectivores

Millions of Years Ago

Leggitt

Table 31.2 Periods and Epochs of the Cenozoic Era

Note: Compare this table to figure 31.2.

Era	Period	Epoch	Millions of Years before Present	Biological Events
Cenozoic	Quaternary	Recent	0.01–present	Modern humans
		Pleistocene	2.5	Early humans
	Tertiary	Pliocene	10–2.5	Hominids
		Miocene	25–10	Hominoids
		Oligocene	38	Anthropoids
		Eocene	54	Prosimians
		Paleocene	65	First placental mammals
Mesozoic			230	

Figure 31.3

Ape diversity. a. Of the apes, gibbons are the most distantly related to humans. They dislike coming down from trees, even at watering holes. They extend a long arm into the water and then drink collected moisture from the back of the hand. b. Orangutans are solitary except when they reproduce. Their name means "forest man"; early Malayans believed that they were intelligent and could speak but did not because they were afraid of being put to work. c. Gorillas are terrestrial and live in groups in which a silver-backed male such as this one is always dominant. d. Of the apes, chimpanzees sometimes seem the most humanlike.

a.

b.

c.

d.

the capuchin, the "organ grinder's monkey." Some of the better-known Old World monkeys are now ground dwellers, such as the baboon and the rhesus monkey, which has been used in medical research.

Apes and humans, the other anthropoids, evolved later.

Primates evolved from shrewlike mammals and became adapted to living in trees, as exemplified by skeletal features, good vision, and reproductive habits. Primates are represented by prosimians, monkeys, apes, and humans. Monkeys, apes, and humans are anthropoids.

Hominoids

Humans are more closely related to apes (fig. 31.3) than to monkeys. **Hominoids** include apes and humans. There are 4 types of apes: the **gibbon**, the **orangutan**, the **gorilla**, and the **chimpanzee**. The gibbon is the smallest of the apes, with a body weight ranging from 5 to 10 kg. Gibbons have extremely long arms that are specialized for swinging between tree limbs. The orangutan is large (75 kg) but nevertheless spends a great deal of time in trees. In contrast, the gorilla, the largest of the apes (185 kg), spends most of its time on the ground. Chimpanzees, which are at home both in the trees and on the ground, are the most humanlike of the apes in appearance and are used sometimes in psychological experiments.

Hominoid Ancestor

About 25 million years ago, the weather began to turn cooler and drier. In places, the tropical forests gave way to grasslands. At this time, apes became abundant and widely distributed in Africa, Europe, and Asia. Among these, members of the genus *Dryopithecus* are of particular interest because they are thought to be a possible hominoid ancestor—the last common ancestor to all other apes and humans. The dryopithecines were forest dwellers that probably spent most of the time in trees. The bones of their feet, however, indicate that if they did sometimes walk on the ground, they walked on all 4 limbs, using the knuckles of their hands to support part of their weight. Such "knuckle walking" is retained by modern apes but not by humans, who stand erect (fig. 31.4).

The skull of the dryopithecines had a sloping brow, heavy eyebrow ridges (called supraorbital ridges), and jaws that projected forward. These, too, are features retained by apes but not by humans. In contrast to apes, humans have a high brow and lack the supraorbital ridges. The human face is flat, and there is no snout, and the canine teeth are comparable in size to the other human teeth.

Hominids

It is not known when the human lineage split from that of the apes, but biochemical evidence suggests the division occurred about 5–10 million years ago. Biochemists compared the structure of proteins and DNA in modern apes and humans and used the molecular clock concept explained on page 570 to work backwards and to arrive at this date. Amazingly, biochemists found only a 2% difference between the sequence of DNA bases in humans and those in chimpanzees.

Fossil evidence indicates that hominids evolved in Africa, but we do not know what the very first **hominids**, humans and fossils in the human lineage, were like nor whether they lived on trees or walked on the ground. The cooling trend that began earlier continued. If the first hominids occasionally did come down out of the trees, perhaps they began to assume an upright **bipedal** posture (walking on 2 limbs). This would have enhanced survival on the savanna because it would have allowed hominids to see over tall grass and would have freed the hands for tool use.

Figure 31.4

Evolution of anatomical differences between apes and humans.
a. As outlined in the table, the dryopithecines had primitive
characteristics that could have led to both modern-day apes and
humans. Compare the modern human skeleton (b) to the modern
ape (gorilla) skeleton (c). Humans are bipedal, while modern-
day apes are knuckle walkers. In the ape, the pelvis is very long
and tilts forward, whereas in humans it is short and upright.
The ape shoulder girdle is more massive than that of humans,
and the head hangs forward because of the angle of attachment
of the vertebral column to the skull. In apes, the foramen
magnum (hole in skull through which the spinal cord passes) is
well to the rear of the skull; in humans it is almost directly in
the bottom center of the skull. Also, in humans the spine has an
S-shaped curve, allowing a better weight distribution and
improved balance when the body is upright.

(b) and (c) From "The Antiquity of Human Walking" by John Napier. Copyright
© 1967 by Scientific American, Inc. All rights reserved.

a. Comparison of *Dryopithecus*, Modern Apes, and Modern Humans

Feature	*Dryopithecus*	Modern Apes	Modern Humans
Brain size	Small brain and skull	Slightly enlarged	Very much enlarged
Face	Sloping brow, heavy eyebrow ridges, and projection of face	Same as primitive	High brow, reduced eyebrow ridges, and flat face
	Rectangular-shaped jaw with large molars and long canine teeth	Same as primitive	U-shaped jaw with small molars and shortened canine teeth
Locomotion	Quadrupedal locomotion	Same as primitive	Bipedal locomotion
	Limbs of equal length	Forelimbs elongated	Forelimbs shortened
	Opposable thumb and toe	Same as primitive	Opposable thumb retained

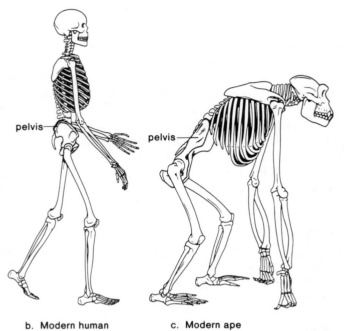

b. Modern human   c. Modern ape

The first hominid (ancestor leading to humans) arose at a time when a change in
weather reduced the size of the African forests, making it advantageous to move on
the ground.

New ideas regarding speciation have been used to interpret the hominid fossil record. Although at one time scientists attempted to place each hominid fossil in a straight line from the most primitive to the most advanced, it now is reasoned that several hominid species existed at the same time. In other words, hominids underwent adaptive radiation (p. 581). Figure 31.5 indicates a possible evolutionary tree for known hominids (see also table 31.1).

Australopithecines

The fossils classified in the genus *Australopithecus* (Southern Apeman) were found in southern Africa. These fossils date from about 4 million to 1.5 million years ago. The **australopithecines** (fig. 31.6) were about 1.5 m tall, with a brain that ranged in size from 300 to 500 cc (cubic centimeters).

Three species of *Australopithecus* have been identified; *A. afarensis, A. africanus,* and *A. robustus*. Because of the small brain, the large canine teeth, and the protruding face, it was suggested that *A. afarensis,* also called Lucy (from the song "Lucy in the Sky with Diamonds"), is the most primitive of the 3 fossils. Even so, the knee, the ankle, and the hipbones indicate that this hominid may have walked erect. This would mean that an enlarged brain was *not* needed for the evolution of bipedal locomotion.

The hominid *A. robustus,* as its name implies, was larger than *A. africanus*. Certain anatomical differences may be associated with diet. The massive cranium and face and the large cheek teeth of *A. robustus* indicate a

Figure 31.5

Human evolutionary tree. Adaptive radiation is evident because this tree indicates that various hominids coexisted. This is only one possible tree; others have been suggested.

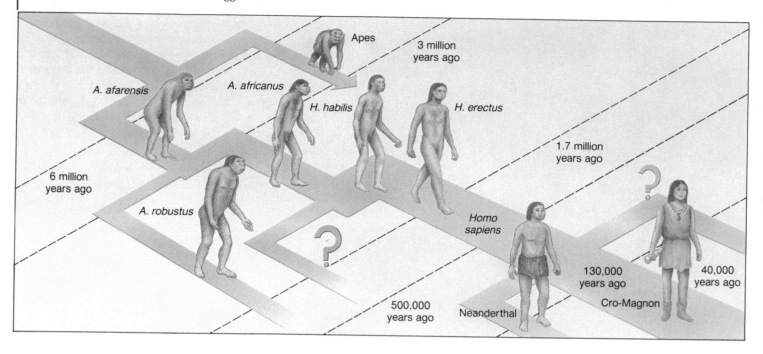

Apes

3 million years ago

A. afarensis

A. africanus

H. habilis

H. erectus

1.7 million years ago

6 million years ago

A. robustus

?

Homo sapiens

?

130,000 years ago

40,000 years ago

Cro-Magnon

500,000 years ago

Neanderthal

Figure 31.6

Australopithecus africanus. *While at one time these hominids were thought to have been hunters, a study of cut marks left on animal bones suggests that they most likely were scavengers that sought and confiscated the kill of coexisting predators and scavengers.*

vegetarian diet. By comparison, the gracile facial features of *A. africanus* indicate a more varied diet, possibly including meat, because facial bones—notably eyebrow ridges—facial muscles, and teeth need not be as large in meat eaters. It now is believed that the australopithecines were not hunters but scavenged the meat that they ate.

The earliest stone tools were found in East Africa and date from about 2 million years ago. Some of these very crude stone tools were found among the remains of *A. robustus*. Perhaps this australopithecine was a toolmaker just as *Homo habilis* people were. Supporting this contention is the broadness of the thumb, which probably could flex at the terminal knuckle. Even so, the manner in which the teeth developed and the final thickness of the enamel suggest that the hominid *A. robustus* is on a side branch of the human evolutionary tree and became extinct. Only the hominid *A. africanus* is believed to have evolved into *Homo habilis*.

One possible human evolutionary tree shows the bipedal Australopithecus afarensis *as a common ancestor to 2 other members of this genus.* A. africanus *is in the mainstream of human evolutionary history, but* A. robustus *is believed to have become extinct.*

Humans

Humans are distinguishable by certain traits. Among these are a bipedal gait, a large brain, the making of tools, the eating of meat and green plants (omnivorous), the pair-bonding between male and female, and the ability to create language, art, and music.

Paleontologists disagree on which fossils belong to the human genus. Some contend that even the 3 australopithecine fossils should be included in the genus, and others want to exclude *Homo habilis,* the first fossil given the designation *Homo.*

Homo habilis

Homo habilis is dated at about 3 million years ago, making these people contemporary with the hominid *A. robustus*. The skeletal remains indicate that *Homo habilis* people evolved from *A. africanus. Homo habilis* means "handyman"; the fossil was given this name because it was found with stone tools. At the time this fossil was named, it was thought that *Homo habilis* people were the first to make tools. Now, however, it appears the hominid *A. robustus* may have made tools also.

The skull of *Homo habilis* has a brain capacity of 600–800 cc, while modern humans have a brain size of 1,360 cc. At one time, it was thought that only people with a brain size of at least 1,000 cc could make tools. Because *Homo habilis* people had a smaller brain than this and yet made tools, are we to think that the making of tools was necessary to the evolution of an enlarged brain? This seems to be an unnecessary question. Increased brain capacity, no matter how slight, would have permitted better toolmaking. This combination would have been selected because toolmaking would have fostered survival in a grassland habitat. Therefore, as the brain became increasingly larger, tool use became more sophisticated.

Homo habilis, *people of at least 2 million years ago, may not have been highly intelligent but did make tools. Intelligence and the making of tools probably evolved together.*

Homo erectus

Homo erectus people (fig. 31.7) evolved in Africa about 1.5 million years ago and then later spread throughout Eurasia. Their fossils have an average brain

Figure 31.7

Homo erectus *people. This drawing shows these people had the use of fire, made stone tools, and may have been big-game hunters.*

size of 1,000 cc, but the shape of the skull indicates that the areas of the brain necessary for memory, intellect, and language probably were not well developed.

Homo erectus people had a posture and locomotion (i.e., a striding gait, fig. 31.8) similar to that seen in modern humans. They made tools of a better quality than the tools of predecessors—these tools had symmetrical bifaces. *Homo erectus* people used fire and ate meat. How much meat they ate is questionable, and some researchers do not believe that they were big-game hunters. However, even though concrete evidence is lacking, circumstantial evidence suggests that *Homo erectus* people could hunt big game:

- Social cooperation was present. Old and diseased skeletons were found, indicating that frail and sick people were allowed to live beyond the time of usefulness. Because humans are relatively small, they need to cooperate in order to kill large animals.
- They had sharp cutting devices. The stone tools found with fossil remains are of a better quality than those found previously.
- They had a wide geographic range. Hunters need to spread out in order to find sufficient game. *Homo erectus* people spread out into Europe and Asia.
- The newborn's brain was small. The adult pelvis of *Homo erectus* is too narrow to allow a large-brained baby to pass through. The brain of

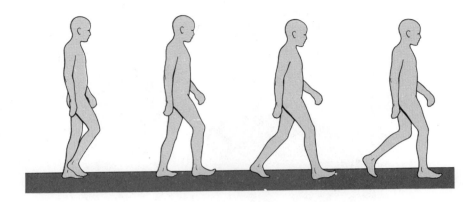

Figure 31.8

The striding gait of modern humans. Each limb alternately goes through a stance phase and a swing phase. In the first drawing, the left limb is in the stationary stance phase and the right limb is beginning the swing phase. First, the knee is bent and extended forward. Then as the knee straightens, the heel and the ball of the foot touch the ground. In the last 2 drawings, the left limb has entered the swing phase and the right limb is entering the stance phase.

the newborn most likely increased in size after birth, which could only happen if the lactating mother received good nutrition (such as hunting might provide).

Homo erectus *people had a large brain, walked like we do, and used fire. They even may have been big-game hunters.*

Homo sapiens

Many years (about 1.5 million) separate people from *Homo erectus* the Neanderthal people. During these years, *Homo sapiens* people must have evolved. This early version of our own species probably was not radically different from *Homo erectus,* however.

There is evidence that by 100,000 years ago, there were at least 3 distinct populations of humans. The Neanderthal people lived in Europe and western Asia; in Africa, there were people increasingly like modern humans in anatomy; and in eastern Asia, there were people unlike either of these other 2 populations. We only have detailed information about the Neanderthal people.

Neanderthals

The **Neanderthal** people (*Homo sapiens neanderthalis*) take their name from Germany's Neander Valley, where one of the first Neanderthal skeletons was discovered. Their earliest fossils are 130,000 years old, and their latest fossils are only 32,000 years old. Neanderthals evolved and lived in Eurasia during the last Ice Age.

The Neanderthals (fig. 31.9) had massive eyebrow ridges, and the nose, the jaws, and the teeth protruded far forward. The forehead was low and sloping, and the lower jaw sloped back without a chin. Surprisingly, the Neanderthal brain was, on the average, slightly larger than that of modern humans. The brain was 1,400 cc, whereas that of most modern humans is 1,360 cc.

The Neanderthals were heavily muscled, especially in the shoulders and the neck. The bones of the limbs were shorter and thicker than those of modern humans. It is hypothesized that a larger brain than that of modern humans was required for control of the extra musculature.

Were the Neanderthals adapted to a cold climate? Did they have a modern appearance? These questions have been answered variously. Recently, it was fashionable to stress how the Neanderthals were like us in appearance; currently, researchers once again are emphasizing their primitive features—the low forehead, the lack of a chin, and the squat skeleton with massive musculature, for example. Nevertheless, the Neanderthals seem to have been more culturally advanced than *Homo erectus* people. Although most remains were found in caves, there is evidence that those who lived in the open built houses. They buried their dead with flowers—an expression of grief perhaps—and with

Figure 31.9
The Neanderthals. This drawing shows that the nose and the
mouth of these people protruded from the face, and the muscles
were massive. They made stone tools and perhaps were hunters.

tools, as if they thought these would be needed in a future life. Some researchers even have suggested that Neanderthals had a religion and that because great piles of bear skulls were found in their caves, the bear played a
role in this religion.

*The Neanderthals were more massive than modern humans, and they had a larger
brain. Perhaps they were adapted to a cold climate because they lived in Eurasia
during the last Ice Age.*

Cro-Magnon

Everyone agrees that **Cro-Magnon** (*Homo sapiens sapiens*), a fossil named
for Cro-Magnon, France, where these remains were first found, has a modern
appearance. There is good evidence now that modern humans evolved in Africa
some 60,000 years ago, and they then spread into Europe and Asia. It seems
unlikely, then, that the Neanderthals, which have not been found in Africa,
evolved into Cro-Magnon people. While some interbreeding may have taken
place between these 2 types of people after Cro-Magnon arrived in Europe,
most likely Cro-Magnons simply replaced the Neanderthals, who died out.

Figure 31.10

Cro-Magnon people painting on cave walls. Some of these paintings still can be observed today. Of all the animals, only humans developed a culture that includes technology and the arts.

The Cro-Magnons had an advanced form of stone technology that included the making of compound tools; stone flakes were fitted to a wooden handle. They were the first to throw spears, enabling them to kill animals from a distance. They also were the first to make knifelike blades. They were such accomplished hunters that some researchers believe they were responsible for the extinction of many larger mammals, such as the giant sloth, the mammoth, the saber-toothed tiger, and the giant ox during the Upper Pleistocene Epoch.

Because language would have facilitated their ability to hunt such large animals, it is quite possible that meaningful speech began at this time. There are very few DNA base sequence differences between modern humans and modern apes. Is it possible that these different sequences are in genes controlling the ability to speak? Perhaps the ability to speak led to language and the development of modern culture. Some believe so. It is our culture and way of life that separate us from the apes. Others point out that cooperative hunting requires language; therefore, this would have led to socialization and the advancement of culture. Humans are believed to have lived in small groups, the men going out to hunt by day, while the women remained at home with the children. It is possible that a hunting way of life shaped the behavior patterns that still are seen today.

Cro-Magnon people lived during the Reindeer Age, a time when great reindeer herds spread across Europe. They used every part of the reindeer, including the bones and the antlers, from which they sculpted many small figurines. They also painted beautiful drawings of animals on cave walls in Spain and France (fig. 31.10). Perhaps they also had a religion, and these artistic endeavors were an important part of their form of worship.

If Cro-Magnon people did cause the extinction of many types of animals, it may account for the transition from a hunting existence to an agricultural existence about 12,000–15,000 years ago. This agricultural period extended from that time to about 200 years ago, when the Industrial Revolution began. At this time, many people began to live in cities, in large part divorced from nature and endowed with the philosophy of exploitation and control of nature. Only recently have we begun to realize that the human population, like all other organisms with which we share an evolutionary history, should work with, rather than against, nature.

Cro-Magnons were modern in appearance. They were expert hunters and developed an advanced culture. They could speak and had a language, which may account for both their hunting ability and their culture.

a.

b.

c.

d.

e.

Human Races

All human races of today also are classified as *Homo sapiens sapiens*. This is consistent with the biological definition of species because it is possible for all types of humans to interbreed and to bear fertile offspring. The close relationship between the races is supported by biochemical data showing that differences in amino acid sequence between 2 individuals of the same race are as great as those between 2 individuals of different races.

It is generally accepted that racial differences developed as adaptations to climate. Although it might seem as if dark skin is a protection against the hot rays of the sun, it has been suggested that it is actually a protection against

31.2 Critical Thinking

Refer to figure 31.5 to answer the following questions.

1. Why might the upright posture of *A. afarensis* be an adaptation to the environment?
2. How does the hominid evolutionary tree illustrate adaptive radiation (p. 581 and 676)?
3. How do the advancements of *Homo erectus* and Cro-Magnon support the concept of punctuated equilibrium (p. 565)?
4. How do you know that all human races should be considered the same species?

ultraviolet ray absorption. Dark-skinned persons living in southern regions and white-skinned persons living in northern regions absorb the same amount of radiation. (Some absorption is required for vitamin D production.) Other features that correlate with skin color, such as hair type and eye color, may simply be side effects of genes that control skin color.

Differences in body shape represent adaptations to temperature. A squat body with short limbs and a short nose retains more heat than an elongated body with long limbs and a long nose. Also, the "almond" eyes, flattened nose and forehead, and broad cheeks of the Oriental are believed to be adaptations to the extremely cold weather of the last Ice Age.

While it always has seemed to some that physical differences warrant assigning human races to different species, this contention is not borne out by the biochemical data mentioned previously.

Summary

Primates evolved from shrewlike mammalian insectivores and became adapted to living in trees. The first primates were prosimians, followed by monkeys, apes, and humans. Monkeys, apes, and humans are all anthropoids. Only humans and their immediate ancestors are hominids.

Humans and apes share a common ancestor, which may have been *Dryopithecus,* a fossil commonly regarded as the first hominoid. This fossil had generalized characteristics that could be ancestral to the specific characteristics of both apes and humans. The first hominid did not appear until about 5–10 million years ago and may have been bipedal. The

weather was drier and grasslands were replacing forests at this time.

One possible human evolutionary tree shows *Australopithecus afarensis* as a common ancestor to both *A. africanus* and *A. robustus*. These fossils had small brains but probably walked erect. The hominid *A. robustus* was a vegetarian that is believed to have become extinct. *Homo habilis* people most likely evolved from the hominid *A. africanus*. *Homo habilis* people had a larger brain than the australopithecines and made stone tools.

Homo erectus is the first fossil to have a brain size of 1,000 cc and to walk with a

striding gait. *Homo erectus* people used fire and may have been big-game hunters. *Homo sapiens neanderthalis* fossil remains are found in Europe and Asia. The Neanderthals did not have modern human physical traits but did have culture. *Homo sapiens sapiens* (Cro-Magnon) people evolved in Africa and then replaced the Neanderthals in Europe.

Cro-Magnons had language and were expert hunters. In a relatively short period of time, humans developed an advanced culture that tended to prevent them from integrating into the biosphere. All human races belong to the same species.

Study Questions

1. Name several primate characteristics still retained by humans. (p. 670)
2. What animals mentioned in this chapter, whether living or extinct, are anthropoids? hominoids? hominids? humans? (p. 670)
3. Draw an evolutionary tree that includes all primates. Discuss each member of the tree. (p. 671)
4. Which fossil may have been the hominoid ancestor? Contrast the

characteristics of this ancestor to those of apes and modern humans. (p. 673)
5. When did the first hominids probably evolve? What was the weather like? Why would bipedal locomotion have been advantageous? (p. 673)
6. Draw and discuss an evolutionary tree for hominids. (p. 675)
7. Contrast the 3 species of australopithecines. Why is it believed that the hominid *A. robustus* made stone tools? (p. 674)

8. Did hominids need a large brain to walk erect or to make tools? How do you know? (pp. 674–76)
9. How was *Homo erectus* different from the preceding hominids? Give evidence that these people were big-game hunters. (pp. 676–77)
10. Where are Neanderthal fossils found? Describe the appearance of these people and their culture. (p. 678)
11. Where did modern humans evolve? What may have caused Cro-Magnons to develop such an advanced culture? (pp. 679–80)

Objective Questions

Place each animal below in the highest category possible.

1. Neanderthals   a. prosimians
2. lemurs   b. anthropoids
3. australopithecines   c. hominoids
4. chimpanzees   d. hominids
           e. *Homo*
5. Primates are adapted to life in the _____ .

6. Among living animals, a(n) _____ best resembles the common ancestor to all primates.
7. The fossil _____ may have been the hominoid ancestor.
8. Based on biochemical evidence, the first hominid is believed to have evolved as late as _____ – _____ million years ago.

9. The fossil known as Lucy could probably walk _____ but had a(n) _____ brain.
10. _____ is the first fossil to be classified as *Homo sapiens sapiens.*

Selected Key Terms

anthropoid (an'thro-poid) higher primate, including only monkeys, apes, and humans. *670*

australopithecine (aw''strah-lo-pith'e-sīn) referring to one of the 3 species of *Australopithecus,* the first generally recognized hominids. *674*

chimpanzee (chim-pan'ze) a small great ape that is related closely to humans and is used sometimes in psychological studies. *673*

Cro-Magnon (kro-mag'non) the common name for the first fossils to be accepted as representative of modern humans. *679*

Dryopithecus (dri''o-pith'e-cus) a genus of extinct apes that may have included or resembled a common ancestor to both apes and humans. *673*

gibbon (gib'on) the smallest ape; well known for its arm-swinging form of locomotion. *673*

gorilla (go-ril'ah) the largest of the great apes, which is as closely related to humans as to other apes. *673*

hominid (hom'ĭ-nid) member of a family of upright, bipedal primates that includes australopithecines and modern humans. *673*

hominoid (hom'ĭ-noid) member of a superfamily that contains humans and the great apes. *673*

Homo erectus (ho'mo ĕ-rek'tus) the earliest nondisputed species of humans, named for their erect posture that allowed them to have a striding gait. *676*

Homo habilis (ho'mo hah'bĭ-lis) an extinct species that may include the earliest humans, having a small brain but quality tools. *676*

Neanderthal (ne-an'der-thawl) the common name for an extinct subspecies of humans whose remains are found in Europe, Asia, and Africa. *678*

orangutan (o-rang'oo-tan'') one of the great apes; large with long red hair. *673*

primate (pri'māt) animal that belongs to the order Primates, the order of mammals that includes prosimians, monkeys, apes, and humans. *670*

prosimian (pro-sim'e-an) primitive primate, such as lemurs, tarsiers, and tree shrews. *670*

Further Readings for Part Five

Bonatti, E. March 1987. The rifting of continents. *Scientific American.*

Brock, T. D., and M. T. Madigan. 1988. *Biology of microorganisms.* 5th ed. Englewood Cliffs, N.J.: Prentice Hall.

Cairns-Smith, A. G. June 1985. The first organisms. *Scientific American.*

Dickerson, R. E. September 1981. Chemical evolution and the origin of life. *Scientific American.*

Dodson, E. O., and P. Dodson. 1985. *Evolution: Process and product.* 3d ed. Boston: Prindle, Weber, and Schmidt.

Eckert, R., and D. Randall. 1983. *Animal physiology.* 2d ed. San Francisco: W. H. Freeman.

Feder, M. E., and W. W. Burggren. November 1985. Skin breathing in vertebrates. *Scientific American.*

Futuyma, E. J. 1986. *Evolutionary biology.* 2d ed. Sunderland, Mass.: Sinauer.

Gallo, R. C. December 1986. The first human retrovirus. *Scientific American.*

Gosline, J. M., and M. E. De Mont. January 1985. Jet-propelled swimming in squids. *Scientific American.*

Grant, P. R. November–December 1981. Speciation and the adaptive radiation of Darwin's finches. *American Scientist.*

Hadley, N. F. July 1986. The arthropod cuticle. *Scientific American.*

Hay, R. L., and M. D. Leaky. February 1982. The fossil footprints of Laetoli. *Scientific American.*

Hickman, Z. P., and L. S. Roberts. 1988. *Integrated principles of zoology.* 8th ed. St. Louis: C. V. Mosby Co.

Hirsch, M. S., and J. C. Kaplan. April 1987. Antiviral therapy. *Scientific American.*

Hogle, J. M. March 1987. The structure of poliovirus. *Scientific American.*

Koehl, M. A. R. December 1982. The interaction of moving water and sessile organisms. *Scientific American.*

Margulis, L. 1982. *Early life.* Boston: Science Books International.

Mossman, D. J., and W. A. S. Sarjeant. January 1983. The footprints of extinct animals. *Scientific American.*

Raven, P. H., et al. 1986. *Biology of plants.* 4th ed. New York: Worth Publishers, Inc.

Simons, K., et al. February 1982. How an animal virus gets into and out of its host cell. *Scientific American.*

Stebbins, G. L., and F. J. Ayala. July 1985. The evolution of Darwinism. *Scientific American.*

Vidal, G. February 1984. The oldest eukaryotic cells. *Scientific American.*

Volpe, E. P. 1981. *Understanding evolution.* 4th ed. Dubuque, Ia.: Wm. C. Brown Publishers.

Part Six

Behavior and Ecology

 The behavior patterns of organisms increase the chances of survival and allow them to interact with other species. These interactions are the framework for ecosystems, units of the biosphere in which energy flows and chemicals cycle. Mature natural ecosystems contain populations that remain constant in size and require the same amount of energy and chemicals each year.

In contrast, the human population constantly increases in size and uses more energy and raw materials each year. Because energy is used inefficiently and raw materials are not cycled properly, the human ecosystem is dependent on natural ecosystems to absorb pollutants. The natural ecosystems are no longer able to support the human ecosystem in this manner, and we must find ways to use energy more efficiently and to recycle materials so that sustainable growth is possible. Furthermore, the preservation of the natural communities, called biomes, is beneficial to all ecosystems. Preserving the biomes helps to ensure the continuance of the biosphere.

The countries of the world are divided into 2 groups. The developed countries enjoy a comparatively high standard of living, consume more resources, and create much of the pollution in the world. The less-developed countries have a comparatively low standard of living and consume fewer resources, but they are responsible for most of the world's population increase.

Red fox litter mates peek out of a burrow. Male and female foxes may be monogamous or one male fox may live with several females. Each group has its own territory, marked by scent, which is used for gathering food, preferably small rodents. Foxes communicate with each other by barks, howls, soft whimpers, and screams.

Chapter 32

Behavior Patterns

Chapter Concepts

1 Competition, predation, and symbiotic relationships are common behavior patterns.

2 Competition leads to diversity of species because no 2 species occupy the same niche.

3 Predator and prey coevolve; therefore, a predator rarely overkills its prey population.

4 Symbiotic relationships usually require a close relationship between 2 species that coevolve.

Chapter Outline

Competition
    An Example of Competition
    The Exclusion Principle
Predation
    Predation Adaptations
    Prey Defense Adaptations
Symbiosis
    Parasitism
    Commensalism
    Mutualism

Behavior patterns increase the chances of survival for the individual and therefore the species. A **population** is all the members of a species in one locale. Certain behavior patterns, such as competition and predation, are also important in controlling the size of populations. Only when populations remain at an appropriate size—one to the other—can a community of populations continue to sustain itself.

Competition

Similar types of species with the same needs are likely to compete with one another for resources, such as water, food, sunlight, and space. The outcome of **competition** is often the predominance of one species and the virtual elimination of the other.

An Example of Competition

A classic example of competition concerns 2 species of barnacles. Barnacles are attached to rocks in the intertidal zone, and an investigator noticed that one species (genus *Chthamalus*) occupies the upper part of the intertidal zone along a Scottish coast, while another species (genus *Balanus*) occupies the lower intertidal zone. It was found that if one species was removed, each species could live in at least a portion of the other's zone (fig. 32.1). Since *Balanus* could occupy a large portion of the zone occupied by *Chthamalus,* it must be competition that prevents it from doing so. In some manner, *Chthamalus* is better adapted to the upper intertidal zone than *Balanus.*

Human Intervention

The fact that successful competition can cause one species to increase in size at the expense of another inadvertently has been demonstrated by human intervention. The carp, a fish imported to the United States from the Orient, is able to tolerate polluted water. Therefore, this fish now is often more prevalent than our own native fishes. Melaleuca, an ornamental tree that thrives in wet habitats, became a pest in the Everglades National Park after being introduced into Florida. The burro, which originated in Ethiopia and Somalia and is adapted to a dry environment, now is threatening the existence of deer, pronghorn antelope, and desert bighorn sheep in the Grand Canyon.

The Exclusion Principle

Laboratory experiments have shown that one species can cause another to become extinct. For example, species X and species Y both vie for the same resource when grown together in the same container. Eventually, one replaces the other entirely. Which population is successful depends on the environmental conditions. In figure 32.2*a,* the environmental conditions are favorable for species X, while in figure 32.2*b,* the environmental conditions are favorable for species Y.

Such experiments have led biologists to formulate and to support a **competitive exclusion principle,** which states that no 2 species can occupy the same niche at the same time. **Niche** is the term used to refer to the role a species plays in a community of organisms. To describe the niche of a species, it is necessary to state all the requirements and the activities of the species. Table 34.2 lists factors to be included when describing the niche of a particular plant or animal species.

Diversity

Competition leads to diversity of species because similar species evolve to have different niches. While it may seem as if several species living in the same area are occupying the same niche, it is usually possible to find slight differences. For example, the 3 species of monkeys in figure 32.3 have no difficulty

Figure 32.1

Effects of competition. Competition prevents 2 species of barnacles from occupying as much of the intertidal zone as possible. Both exist in the area of competition between Chthamalus *and* Balanus. *Above this area, only* Chthamalus *survives, and below it, only* Balanus *survives.*

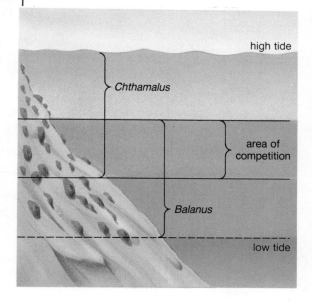

Figure 32.2

Extinction due to competition with a species better adapted to the environment. a. In this experiment, species X survives and species Y becomes extinct because the environmental conditions were favorable to species X. b. In this experiment, the environmental conditions changed, with the result that species Y survives and species X becomes extinct.

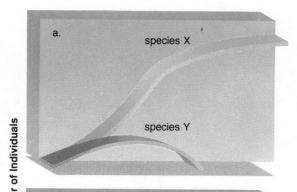

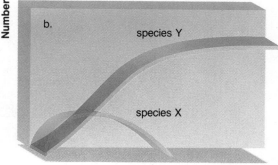

living in close proximity because they have different, although sometimes overlapping, habitats and food requirements.

When 2 species compete for the same resources, one usually eliminates or restricts the range of the other. This is in keeping with the exclusion principle, which states that no 2 species can occupy the same niche.

Predation

Predation, simply defined, is one organism feeding on another, called the **prey.** Examples include monkeys feeding on flower bulbs, a bear feeding on salmon, and a lion feeding on a gazelle. Through the evolutionary process, predators become specialized to capture their prey and prey become adapted to escaping their predators. In this manner, predators and prey **coevolve.**

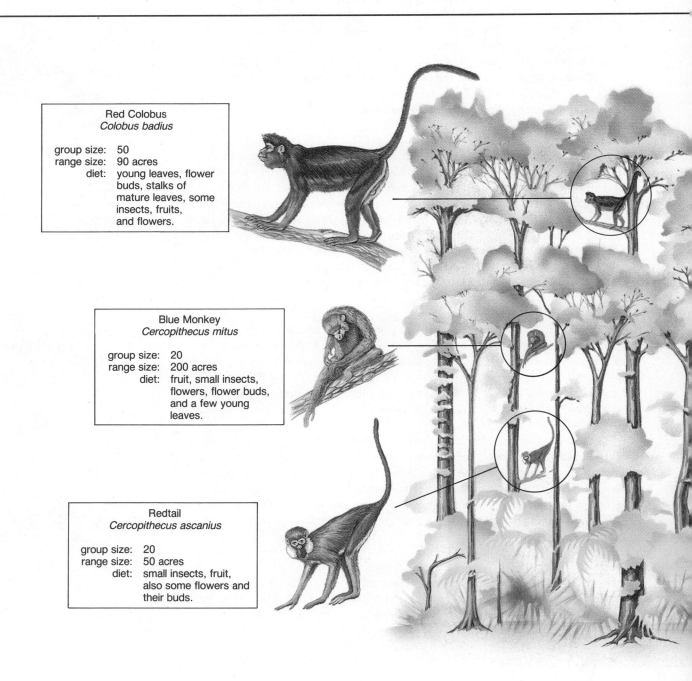

Red Colobus
Colobus badius

group size: 50
range size: 90 acres
diet: young leaves, flower buds, stalks of mature leaves, some insects, fruits, and flowers.

Blue Monkey
Cercopithecus mitus

group size: 20
range size: 200 acres
diet: fruit, small insects, flowers, flower buds, and a few young leaves.

Redtail
Cercopithecus ascanius

group size: 20
range size: 50 acres
diet: small insects, fruit, also some flowers and their buds.

Behavior and Ecology

Predation Adaptations

The grasslands of all continents support populations of **grazers** that feed on grasses and browsers that feed on shrubs and trees. The anatomy of **browsers** enables them to reach high into trees—giraffes have long necks and elephants have long snouts. The most prevalent terrestrial herbivores are the insects, which have evolved efficient and diverse means of eating plants.

Some carnivores, such as a lynx, go out alone and seek their prey (fig. 32.4). While most birds of prey are specialized for hunting, seizing, and killing small terrestrial animals, some, like the osprey and the kingfisher, are specialized for fishing. Instead of seeking prey, some solitary predators lie in wait. The octopus hides within a protective shelter until an unsuspecting prey, such as a crab, happens by. The octopus then quickly catches the prey with its arms and carries it home to be eaten.

Predators, such as lions and wolves, prey on animals larger than themselves; therefore, they often hunt their prey as a group. Lions typically prey

Gray-Cheeked Mangabey
Cercoebus albigena

group size: 15
range size: 1,000 acres
diet: inner bark of trees, fruits, insects, other small animals, also a few leaves and flowers.

Black-and-White Colobus
Colobus guereza

group size: 10
range size: 40 acres
diet: young leaves, mature leaves, also some buds, flowers, and fruit.

L'Hoest's Monkey
Cercopithecus l'Hoesti

group size: 20
range size: unknown
diet: fruits, shoots of herbs, mushrooms, and insects.

Figure 32.3
Diversity of monkey species in a tropical rain forest. All of these monkeys can coexist in a tropical rain forest because they have different niches. Each prefers to live at a different height above ground, and each feeds on slightly different foods.

Figure 32.4

Montana Lynx, a solitary predator. A long strong forelimb grabs its main prey, the snowshoe hair, with sharp claws. A lynx lives in northern forests, its brownish-gray coat blends in well against the trunk of trees and its long legs help walking through deep snow.

Figure 32.5

Example of predator-prey cycling. As the hare population increases in size so does the lynx population, but then the hare population, followed by the lynx population, suffers a dramatic *decrease in size. While it may seem as if this cycling is due to "overkill" by the lynx population, other factors could be involved.*

(These data are based on the number of lynx and snowshoe hare pelts received by the Hudson Bay Company in the years indicated.)

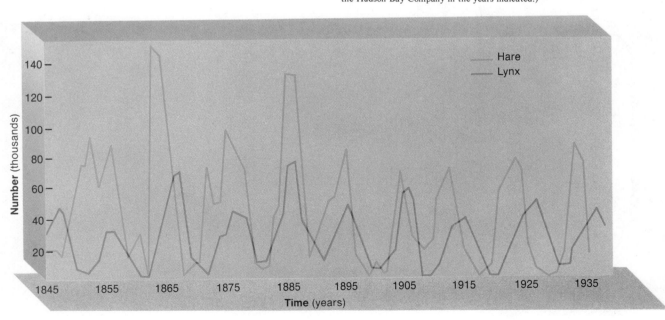

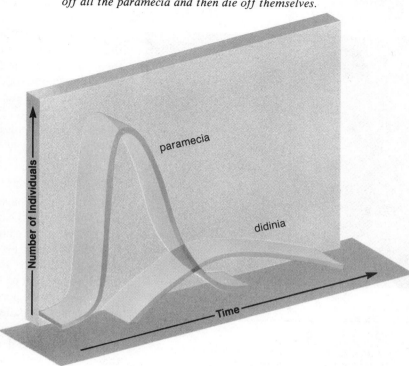

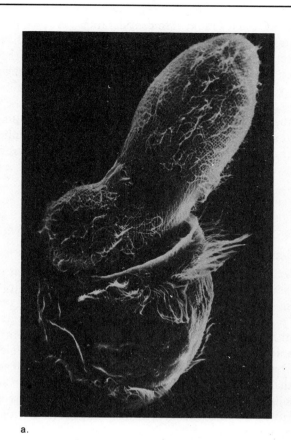

a.

b.

on animals that have been separated from their herd. Although only a small number of lions participate in the kill, the food is shared with the rest of the pride.

The Control of Prey Population Size

A carnivore helps to control the size of its prey population. Biologists reasoned that there would be an oscillation in the size of predator and prey populations as illustrated in figure 32.5. The oscillation was expected because it was reasoned that as the number of prey increased, predation also would increase until finally the prey population would suffer a decline. This would be followed by a reduction in the number of predators until the prey population eventually would begin to recover. Then the cycle would begin again. Rarely are such oscillations observed, however. Therefore, it is believed that factors other than predation caused the oscillation in the lynx and hare populations observed between 1845 and 1935 (fig. 32.5). For example, the prey population may have been reduced because of some other reason, such as lack of food due to the weather, and this, in turn, may have led to a decline in the predator population.

One important consideration in the predator-prey relationship is illustrated by this experiment. When didinia and paramecia are placed in the same test tube, the didinia capture and eat all the paramecia and then die off from lack of food (fig. 32.6). However, if refuge, such as sediment at the bottom of the test tube, is provided for the paramecia, part of the prey population survives and reproduces. This suggests that predator-prey relationships can stabilize when predator pressure eases as the prey population decreases in size. In nature, this can be the result of prey being able to hide or of predators switching to a more abundant prey species.

32.1 Critical Thinking

There are 2 types of competition: scramble competition, in which all organisms have equal access to resources, and contest competition, in which a contest decides which organism will have access to resources.

1. When a large population of blowfly larvae are provided with a limited amount of food in the laboratory, most larvae die from lack of food. Which type of competition most likely took place? Why?
2. In a baboon troop, certain males receive food and mate first. These males receive this treatment because they successfully fought other males. Which type of competition occurred? Why?
3. In nature, blowflies are not restricted to a particular territory, although baboons are. How does this information help to explain why scramble competition is seen in the one species and contest competition is seen in the other?

Figure 32.7

Startle display as an antipredator defense. The South American lantern fly has a large false head that resembles an alligator. This may frighten a predator into thinking it is facing a dangerous animal.

Figure 32.8

Warning coloration as an antipredator defense. The skin secretions of poison-arrow frogs are so poisonous that they were used by native South Americans to make their arrows instant lethal weapons. The coloration of these frogs warns others to beware.

Human Intervention

Sometimes humans have neglected to take into consideration that predators help to keep prey populations in check. For example, in the past, coyotes have been killed indiscriminately in the West without regard for the fact that they help to keep the prairie dog population under control. Similarly, when the dingo, a wild dog in Australia, was killed off because it attacked sheep, the rabbit and wallaby populations greatly increased. In contrast, humans formerly kept the burro population in the Grand Canyon within reasonable limits by killing them for meat. However, since a federal law was passed in 1971 forbidding the killing of burros, their numbers have increased until they now are destructive pests and threaten the existence of native animals, as previously discussed.

Predators help to control the size of prey populations. Whether oscillations in predator and prey populations occur still is being investigated.

Prey Defense Adaptations

While predators have evolved strategies to secure the maximum amount of food with minimal expenditure of energy, prey organisms have evolved strategies to escape predation.

Plants try to avoid predation. The sharp spines of the cactus, the pointed leaves of the holly, and the tough, leathery leaves of oak trees all discourage predation by insects. Plants even produce poisonous chemicals, some of which interfere with the normal metabolism of the adult insect and others which act as hormone analogues that interfere with the development of insect larvae. Some insects still can inhabit trees that produce toxins because the insects have evolved either detoxifying enzymes or a method of keeping the toxin within their body in a manner that is not harmful to them.

Animals have varied **antipredator defenses.** Some of the more effective defenses are concealment, fright, warning coloration, and vigilance.

Concealment. Some animals attempt to blend in with their background. Caterpillars look like twigs, katydids can look like sprouting green leaves, and moths can resemble the bark of trees. Larger animals, too, try to conceal themselves by looking like their environment. Decorator crabs cover themselves with debris, and green herons, even young ones, attempt to look like straight tree branches.

Fright. Other types of prey try to startle or to frighten predators (fig. 32.7). Some make themselves appear much larger than they are, such as the frilled lizard of Australia that can open up folds of skin around its neck. Many moths have eyelike spots on their underwings that they can flash to startle a bird long enough for escape.

Warning coloration. Poisonous animals tend to be brightly colored as a warning to those that may want to prey upon them. Yellow and black striped bees and wasps, yellow or red caterpillars or plant bugs, and ladybird beetles all are advertising that they are not good to eat. The skin secretions of the poison-arrow frog (fig. 32.8), found in tropical regions, are so poisonous that South American Indians long have used these secretions to poison their arrow tips. When the arrow strikes, it instantly paralyzes the victim.

Vigilance. Flocks of birds, schools of fishes, and herds of mammals stick together as protection against predators. Grazing herbivores are constantly on the alert; if one begins to dart away, they all run. Baboons that detect predators visually and antelope that detect predators by smell sometimes forage together, giving double protection against stealthy predators. The gazellelike springboks of southern Africa jump stiff legged straight up and down (as much as 4 m) a number of times as a warning. The jumble of shapes and motion may confuse an attacking lion, and the herd can escape.

Behavior and Ecology

Mimicry

Mimicry occurs when one species resembles another that possesses an anti-predator defense. For example, the brightly colored monarch butterfly (fig. 32.9a) causes predaceous birds to become sick and sometimes to vomit. Monarch butterfly larvae feed on milkweed plants. These plants contain digitalis-like compounds, one form of which becomes incorporated into the adult butterfly tissue. This compound activates nerve centers controlling vomiting. Birds that have eaten a poisonous monarch butterfly avoid monarch butterflies in the future.

The viceroy butterfly (fig. 32.9b) mimics the monarch butterfly's coloration but is not toxic. Birds eagerly eat viceroy butterflies unless they have had previous experience with a poisonous monarch. Because the butterflies closely resemble each other, birds then avoid both types of butterflies. The queen butterfly also mimics the monarch, but the queen is also poisonous.

A mimic that lacks the defense of the organism it resembles is called a Batesian mimic. The viceroy butterfly is a Batesian mimic of the monarch. Therefore, Batesian mimicry involves the resemblance of a harmless species (the mimic) to a harmful species (the model). A mimic that possesses the same defense as its model is called a Müllerian mimic. The queen butterfly and the monarch butterfly are mimics of one another—both animal species are harmful and resemble each other. In such cases, predators learn more rapidly to avoid contact with any animal that has that particular pattern.

Prey defenses include concealment, fright, warning coloration, and vigilance. Sometimes, an animal with a successful antipredator defense is mimicked by other animals that lack the defense (Batesian mimic) or that have the same defense (Müllerian mimic).

Symbiosis

Symbiosis (table 32.1) is an intimate relationship between 2 different species. As with predation, coevolution occurs, and the species become closely adapted to one another. In parasitism, the parasite benefits, but the host is harmed; in commensalism, one species benefits, but the other is unaffected; and in mutualism, both species benefit.

Parasitism

Parasitism is similar to predation in that the **parasite** derives nourishment from the host (just as the predator derives nourishment from its prey). Usually, however, the host is larger than the parasite, and an efficient parasite does not kill the host, at least until its own life cycle is complete. Viruses are always parasites, as are a number of bacteria, protists, plants, and animals. The smaller parasites tend to be endoparasites, which live in the bodies of the hosts. Larger parasites tend to be ectoparasites and remain attached to the exterior of the host by means of specialized organs and appendages.

Just as predators can reduce dramatically the size of a prey population that lacks a suitable defense, so parasites can reduce the size of a host population that has no defense. Thousands of elm trees have died in this country from the inadvertent introduction of Dutch elm disease, which is caused by a parasitic fungus. A new method of treating Dutch elm disease utilizes bacteria that produce a fungicide when injected into the tree. Another means of curbing this fungal infection of trees is to control the bark beetle that spreads the disease.

Many other parasites use a secondary host for dispersal or completion of the stages of development. For example, the deer tick has 3 different hosts in its life cycle. The adult tick feeds on white-tailed deer, the larva feeds primarily on white-footed mice, and the nymph occasionally feeds on humans.

Figure 32.9
The monarch butterfly (a) and its mimic, the viceroy butterfly (b). Birds avoid both types of butterflies even though only the monarch is distasteful to them.

a.

b.

Table 32.1 Symbiosis

	Species 1	Species 2
Parasitism	+	−
Commensalism	+	0
Mutualism	+	+

Key: + = benefit
− = harm
0 = no effect

With the exception of AIDS, Lyme disease is the most newly recognized serious infectious disease in the United States in terms of number of cases and potential for severe illness. Lyme disease has been reported in 43 states, but most cases occur in the Northeast (New York, New Jersey, Connecticut, and Massachusetts), the upper Midwest (Minnesota and Wisconsin), and the Pacific Northwest (California and Oregon). Lyme disease takes its name from Lyme, Connecticut, where the disease first was recognized in 1975.

Lyme Disease, a Parasitic Infection

The cause of Lyme disease is a spirochete (*Borrelia burgdorferi*), a coiled bacterium that is spread by the bite of a deer tick (*Ixodes dammini* in the East and *Ixodes ricinus* in the West). Adult deer ticks are only about two-thirds the size of dog ticks. They are so named because adult ticks feed and mate on white-tailed deer. Ticks feed on the blood of their host and triple in size following a meal (see figure). Ticks are arthropods that go through a number of stages (egg, larva, nymph, adult) as they develop; molting occurs between the stages.

After a female tick has fed and mated in the fall, she drops down onto the ground and lays her eggs. When these eggs hatch in the spring, they become larvae that feed primarily on white-footed mice. These mice are an important reservoir for the Lyme

disease bacterium; when mice are infected, the tick becomes infected. The fed larvae overwinter and molt the next spring to become nymphs that are not much larger than the head of a pin. It is during the nymph stage, between May and September, that ticks are most apt to feed on humans and to pass on to them the bacterium that causes Lyme disease. The nymph is dark brown and hard bodied and is mistaken easily for a scab or a piece of dirt on the skin. Other likely hosts for nymphs are white-footed mice, raccoons, and skunks. After feeding, the nymph develops into an adult, and the life cycle repeats itself.

Altogether, ticks remain on their hosts only about 2 1/2 weeks. The unfed adults, larvae, and nymphs spend most of their time attached to grasses or shrubs waiting for an appropriate host to come by. Infestation is greatest in wooded areas, but people also have been bitten in their own backyards, especially if the yards are adjacent to wooded areas.

The effects of Lyme disease vary from mild to severe. First, some victims, but not all, get a rash that is called a bull's-eye rash because it has a red center encircled by a light area and then a red area again. The rash can get as large as a dinner plate. At this time, infected persons possibly feel like they have a mild case of flu. In about a month or so, nausea and neck and joint pain similar to that caused by arthritis may develop. Finally, a few may experience heart blockage, sometimes even requiring

the installation of a pacemaker, and neurological disorders that mimic the symptoms of meningitis and encephalitis. Fierce headaches, facial paralysis, depression, and a temporary loss of memory also can occur. Loss of muscular coordination sometimes makes the disease seem like multiple sclerosis. Lyme disease is called "the great imitator" because it easily is misdiagnosed as a number of other illnesses.

It is never too late to be treated for Lyme disease. A blood test can detect antibodies to the bacterium in the system. If the blood test is positive, the administration of an antibiotic, like penicillin, tetracycline, or erythromycin, cures the condition. Antitick sprays, which discourage ticks from choosing a human host in the first place, are available. Some home owners are trying to control the tick population in their vicinity by killing off the mice and ticks with chemicals. A more natural solution can be attempted by local governments. It is possible to release into the environment tiny parasitic wasps that lay their eggs only in ticks. The developing wasp larvae kill the ticks.

The life cycle of a deer tick. (a) *Adult ticks feed and mate on white-tailed deer, which accounts for the common name of the ticks. Female adults lay their eggs in soil and then die.* (b) *In the spring and summer, larvae feed mainly on white-footed mice, they then overwinter.* (c) *During the next summer, nymphs feed on white-footed mice or other animals, including humans. If a nymph is infected with the bacterium, humans get Lyme disease.* (d) *Deer tick before feeding and after feeding. Actual size is shown along with enlarged size.*

As discussed in the reading above, the deer tick is a carrier for a bacterium that causes Lyme disease.

Social parasitism occurs when one species exploits another species. The cuckoo lays eggs in nests of songbirds, and the newly hatched cuckoo ejects its nest mates. Therefore, the songbird parents attend only to the cuckoo. Slave-making ants of the species *Polyergus rufescens* raid the ant colonies of the slave species *Formica fusca*. They destroy any resisting defenders with their mandibles, which are shaped like miniature sabers. *Polyergus* ants are so specialized that they only can groom themselves. To eat, they must beg slave workers for food. The slave workers not only provide food for the slave-making ants, they also care for the eggs, the larvae, and the pupae of their captors.

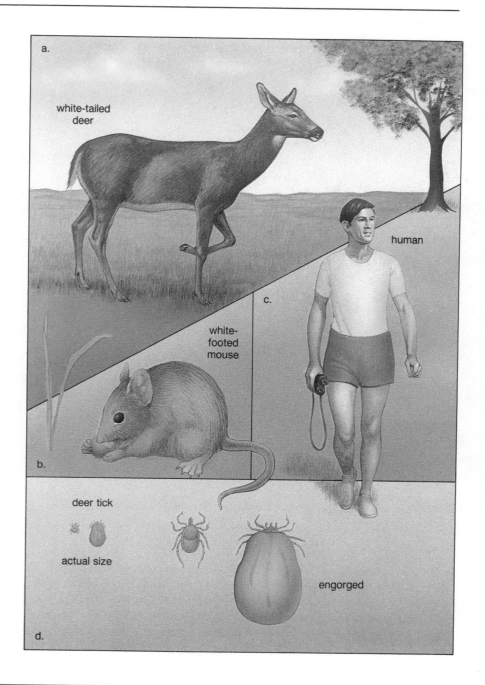

a. white-tailed deer

b. white-footed mouse

c. human

d. deer tick

actual size

engorged

In parasitism, the host species is harmed. Parasites, which sometimes help to control the size of their host population, may require a secondary host for dispersal.

Commensalism

Commensalism is a relationship between 2 species in which one species is benefited and the other is neither benefited nor harmed. Often, the host species provides a home and/or transportation for the other species. Barnacles that attach themselves to the backs of whales and the shells of horseshoe crabs are provided with both a home and transportation. Remoras are fishes that attach themselves to the bellies of sharks by means of a modified dorsal fin acting as a suction cup. The remoras obtain a free ride and also feed on the remains of

Figure 32.10

Example of commensalism. Clownfishes live among a sea anemone's tentacles and yet are not seized and eaten as prey. The reason this relationship is maintained is not known.

the shark's meals. Epiphytes grow in the branches of trees, where they receive light, but they take no nourishment from the trees. Instead, their roots obtain nutrients and water from the air. Clownfishes (fig. 32.10) live within the tentacles and the gut of sea anemones. Because most fishes avoid the poisonous tentacles of the anemones, the clownfishes are protected from predators. Perhaps this relationship borders on mutualism because the clownfishes actually may attract other fishes on which the anemone can feed. The sea anemone's tentacles quickly paralyze and seize other fishes as prey.

In commensalism, one species is benefited and the other is unaffected. Often the host simply provides a home and/or transportation.

Mutualism

Mutualism is a symbiotic relationship in which both members of the association benefit. Mutualistic relationships often allow organisms to obtain food or to avoid predation.

Bacteria that reside in the human intestinal tract are provided with food, but they also provide humans with vitamins, which are molecules we are unable to synthesize for ourselves. Termites would not even be able to digest wood if not for the protozoans that inhabit their intestinal tract. The bacteria in the protozoans digest cellulose, which termites cannot digest. Mycorrhizae, also called fungal roots, are symbiotic associations between the roots of plants and

Figure 32.11

Mutualistic relationship. The bullhorn acacia is adapted to provide a home for Pseudomyrmex ferruginea, *a species of ant that protects the acacia from other insects. a. Thorns are hollow and the ants live inside. b. Bases of leaves have nectaries (openings), where ants feed. c. Leaves have nodules at the tips that ants harvest for larvae food.*

fungal hyphae. Mycorrhizal hyphae improve the uptake of nutrients for the plant, protect the plant's roots against pathogens, and produce plant growth hormones. In return, the fungus obtains carbohydrates from the plant. As we discussed on page 624, flowers and their pollinators have coevolved and are dependent upon one another. The flower is benefited when the pollinator carries pollen to another flower, ensuring cross-fertilization, and the flower provides food for the pollinator.

In tropical America, the bullhorn acacia (fig. 32.11) is adapted to provide a home for ants of the species *Pseudomyrmex ferruginea*. Unlike other acacias, this species has swollen thorns with a hollow interior, where ant larvae can grow and develop. In addition to housing the ants, the acacias provide them with food. The ants feed from nectaries at the base of the leaves and eat fat- and protein-containing nodules called Beltian bodies, which are found at the tips of some of the leaves. Bullhorn acacias have leaves throughout the year, while related acacia species lose their leaves during the dry season. The ants constantly protect the plant from herbivorous insects because unlike other ants, they are active 24 hours a day.

Cleaning symbiosis (fig. 32.12) is a phenomenon believed to be quite common among marine organisms. There are species of small fishes and shrimps that specialize in removing parasites from larger fishes. The large fishes line up at the "cleaning stations" and wait their turn. The small fishes feel so secure they even clean the mouths of the larger fishes. Not everyone plays fair, however, because there are small fishes that mimic the cleaners and take a bite out of the larger fishes, and cleaner fishes sometimes are found in the stomachs of the fishes they clean.

In mutualism, both species benefit and the 2 species often are closely adapted to one another.

32.2 Critical Thinking

Sometimes it is difficult to determine what type of symbiotic relationship organisms have.

1. In lichens, fungal hyphae characteristically penetrate algal cells via specialized organs called haustoria. What type of symbiotic relationship is suggested by this information? Why?
2. In lichens, algae are surrounded and mechanically protected by the meshwork of fungal hyphae, which absorb water and minerals from the substrate. If the algae benefit from this arrangement, what type of relationship do the 2 organisms have? Why?
3. Both the fungi and the algae found in a lichen can exist separately. Is this counter to a parasitic relationship, a mutualistic relationship?
4. The fungi do not overgrow or kill off the algal portion of a lichen. Is this counter to a parasitic relationship but consistent with a mutualistic one?
5. Do you think relationships that do not fall neatly into one of the 3 types of symbioses are possible?

Figure 32.12

Cleaner wrasse in the mouth of a spotted sweetlip. This is a mutualistic relationship. The cleaner wrasse is feeding off parasites on the spotted sweetlip, and the spotted sweetlip is being cleaned by the cleaner wrasse.

Summary

Certain interactions are typical between species. When 2 species compete for the same resources, one usually eliminates or restricts the range of the other. This is in keeping with the competitive exclusion principle.

Predators and prey coevolve. Herbivores prey on plants, and carnivores prey on other animals. Even so, both predatory and prey population sizes are maintained at a constant level. Antipredator defense adaptations include concealment, fright,

warning coloration, and vigilance. Sometimes, an animal with a successful antipredator defense is mimicked by other animals that have the same defense or lack the defense.

There are 3 types of symbiotic relationships. In parasitism, the host species is harmed. Parasites, which sometimes help to control the size of their host population, may require a secondary host for dispersal. The anatomy and the life-style of an animal parasite can be contrasted to those of an

active predator. In commensalism, one species is benefited and the other is unaffected. Often the host simply provides a home and/or transportation. In mutualism, both species benefit. The 2 species often are adapted closely to one another, such as flowers and their pollinators. There are also examples of mutualistic species that live together in the same locale, such as ants that live in bullhorn acacia trees, but in cleaning symbiosis, the relationship is transitory.

Study Questions

1. Give an example to show that competition between species results in the elimination or the restriction of the range of the other. (p. 687)
2. Give examples to show that it is sometimes unwise to bring a new competitor into an area. (p. 687)
3. Define the competitive exclusion principle, and discuss the concept of an organism's niche. (p. 687)
4. Would it be correct to say that the greater the species variety in an area, the more niches there must be? (p. 687)
5. Give examples to show that predators have a beneficial effect, helping to stabilize population sizes in an area. (p. 691)
6. Give examples to show that plants have defenses against herbivores. (p. 692)
7. Give examples of animal antipredator defenses that prevent predator populations from overkilling prey populations. (p. 692)
8. Define a mimic, and give 2 examples involving the monarch butterfly. (p. 693)
9. What are the 3 types of symbiotic relationships? Give several examples of each. (pp. 693–97)
10. Give 2 examples of social parasitism. (p. 694)

Objective Questions

1. The competitive exclusion principle states that no 2 species can occupy the same _____ .
2. Predators are _____ to capturing prey, and prey are _____ to escaping predators.
3. Predators that eat vegetation are _____ , and those that eat other animals are _____ .
4. Both competition and predation are factors that _____ population sizes.
5. _____ is present when one species resembles another that possesses an overt antipredator defense.
6. _____ mechanisms prevent predators from capturing all the prey and prevent parasites from killing off their hosts.
7. The symbiotic relationship _____ is exemplified by the habit of barnacles to attach themselves to whales.
8. In the symbiotic relationship known as mutualism, _____ .
9. The adaptations of flowers to attract specific pollinators and the pollinators' adaptations to exploit the flower as a source of nutrients is termed

   _____ .

Selected Key Terms

antipredator defense (an″tǐ-pred′ah-tor de-fens′) a physiological activity or behavior or a structural modification that protects an organism from its predators. *692*

browser (browz′er) animal that feeds on high-growing vegetation, such as shrubs and trees. *689*

cleaning symbiosis (klēn′ing sim″bi-o′sis) a mutualistic relationship in which one type of organism benefits by cleaning another, which is benefited by being cleaned of debris and parasites. *697*

coevolve (ko-e-volv′) the interaction of 2 species such that each influences the evolution of the other species. *688*

commensalism (kŏ-men′sal-izm) the relationship of 2 species in which one lives on or with the other without conferring either benefit or harm. *695*

competition (kom″pě-tish′un) interaction between members of the same or different species for a mutually required resource. *687*

competitive exclusion principle (kom-pet′ĭ-tiv eks-kloo′zhun prin′sĭ-p′l) an observation that no 2 species can continue to compete for the same resources because one species eventually becomes extinct. *687*

grazer (gra′zer) animal that feeds on low-lying vegetation, such as grasses. *689*

mimicry (mim′ik-re) the resemblance of an organism to another that has a defense against a common predator. *693*

mutualism (mu′tu-al-izm″) a relationship between 2 organisms of different species that benefits both organisms. *696*

niche (nich) the functional role and position of an organism in the ecosystem. *687*

parasite (par′ah-sīt) an organism that resides externally on or internally within another organism and does harm to this organism. *693*

parasitism (par′ah-si″tizm) symbiotic relationship in which an organism derives nourishment from and does harm to a host. *693*

predation (pre-da′shun) the eating of one organism by another. *688*

prey (pra) organism that serves as food for a particular predator. *688*

symbiosis (sim″be-o′sis) an intimate association of 2 dissimilar species, including commensalism, mutualism, and parasitism. *693*

Chapter 33

The Biosphere

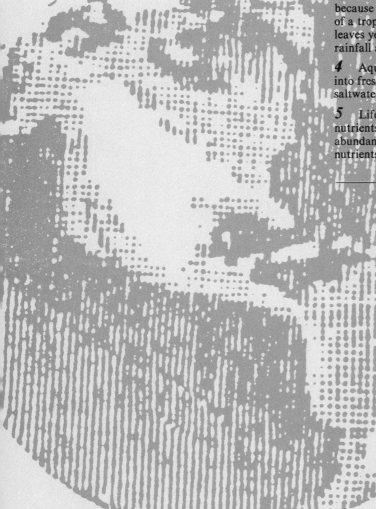

Chapter Concepts

1 Communities of organisms on land, called biomes, are adapted to climate.

2 Only forests have adequate rainfall to support large populations of trees.

3 The trees of a temperate deciduous forest lose their leaves in the fall because cold weather sets in; the trees of a tropical rain forest keep their leaves year-round because of adequate rainfall and high temperature.

4 Aquatic communities are divided into freshwater communities and saltwater communities.

5 Life in coastal communities, where nutrients are adequate, is much more abundant than in the open seas, where nutrients are scarce.

Chapter Outline

Biomes
    Treeless Biomes
    Forests
    Latitude versus Altitude
Aquatic Communities
    Freshwater Communities
    Saltwater Communities

Figure 33.1

Three major biomes in the United States, each containing its own mix of plants and animals. Temperature and amount of rainfall largely determine what type of biome is found where. a. A deciduous forest is typical of the eastern United States. b. A prairie biome is found in the Midwest. c. A desert is located in the Southwest.

b.

a.

c.

The earth is enveloped by the biosphere—a thin realm composed of water, land, and air—where organisms are found. The **biosphere** contains communities of populations whose interactions were discussed in the previous chapter. The largest communities on land are called **biomes** (fig. 33.1). There is no equivalent term for large aquatic communities, which is unfortunate because both the aquatic environment and the land environment contain large communities that have unique, defined characteristics.

Biomes

The location of the biomes studied in this text are indicated in figure 33.2. We will discuss treeless biomes (deserts, tundra, grasslands, and chaparral) and certain of the forests (taiga, temperate deciduous forests, and tropical rain forests).

The Biosphere

Treeless Biomes

Deserts

Deserts (fig. 33.1c) occur in regions where annual rainfall is less than 25 cm. The rain that does fall is subject to rapid runoff and evaporation. The days are hot because the lack of cloud cover allows the sun's rays to penetrate easily, but the nights are cold because the heat escapes easily into the atmosphere.

Some deserts, like the Sahara in Africa, have little or no vegetation, but most have a variety of plants. The best-known desert perennials in this country are the succulent, leafless cacti, which have stems that store water and also carry on photosynthesis. All cacti have extensive root systems that can absorb great quantities of water during brief periods of rainfall. Their spines provide a means of defense against desert herbivores. Desert vegetation also includes small-leaved, nonsucculent shrubs, such as sagebrush, a densely branched evergreen, and creosote, with leaves that turn brown and drop off. The deciduous

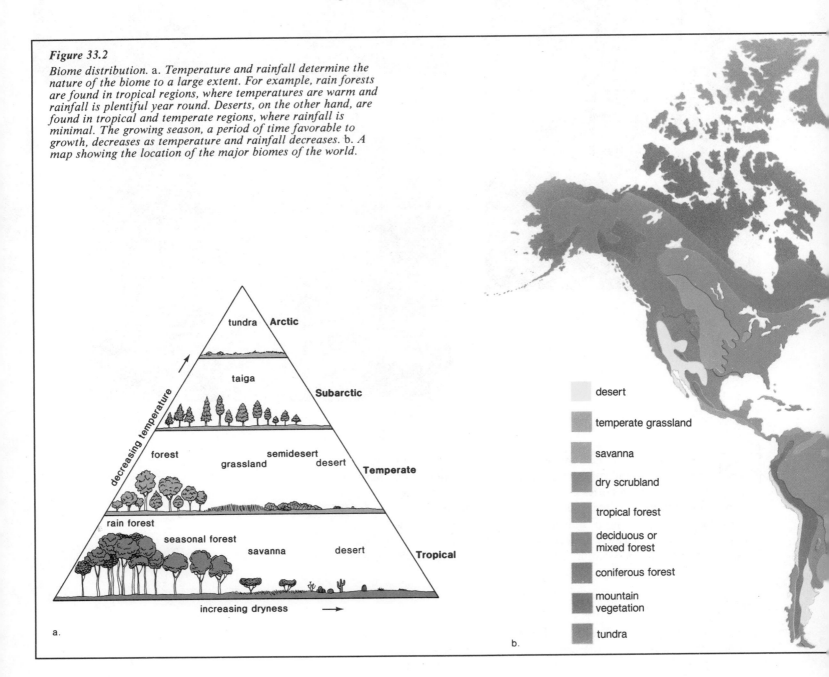

Figure 33.2

Biome distribution. a. Temperature and rainfall determine the nature of the biome to a large extent. For example, rain forests are found in tropical regions, where temperatures are warm and rainfall is plentiful year round. Deserts, on the other hand, are found in tropical and temperate regions, where rainfall is minimal. The growing season, a period of time favorable to growth, decreases as temperature and rainfall decreases. b. A map showing the location of the major biomes of the world.

Behavior and Ecology

mesquite tree, which has deep roots, also is common. Desert annuals exist most of the year as seeds; they burst into flower during the limited period of time when moisture and temperature are favorable.

Some animals are adapted to the desert environment. A desert has numerous insects, some of which, like the annual plants, have a compressed life cycle. They pass through the stages of development from pupa to pupa within a very short time because the pupa can remain inactive until it rains again. Reptiles, especially lizards and snakes, are perhaps the most characteristic group of vertebrates found in deserts, but running birds (e.g., the roadrunner) and rodents (e.g., the kangaroo rat) are also well known. Larger mammals, like the coyote, prey on the rodents, as do the hawks.

Most deserts are located in high-temperature areas that receive less than 25 cm of rainfall a year. Plants and animals living in deserts have adaptations that protect them from the sun's rays and the shortage of water.

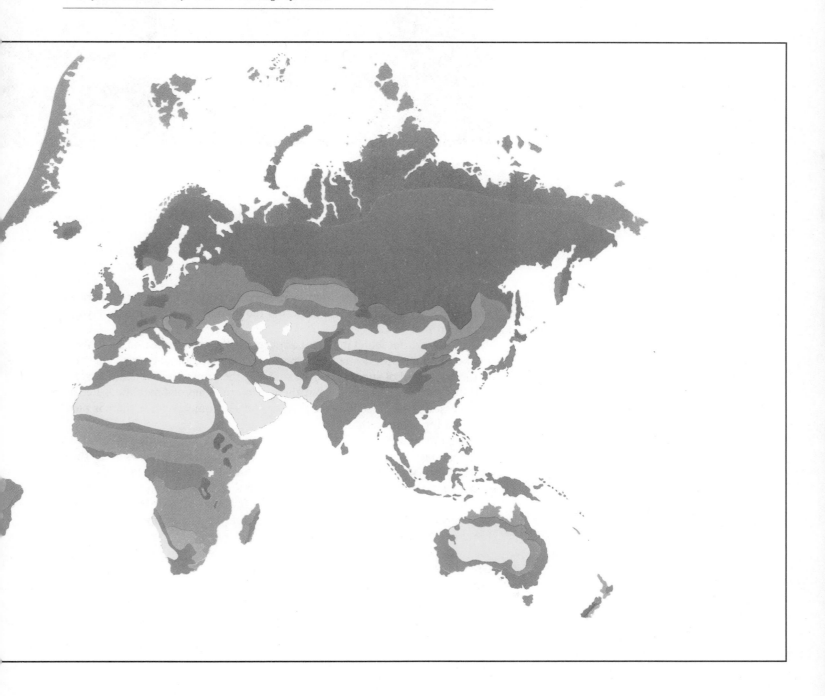

Figure 33.3
Tundra biome in the autumn. a. *Overall view
illustrates the lack of trees.* b. *Caribou bull
grazes on the low-lying vegetation.*

a.

b.

Tundra

The arctic **tundra** (fig. 33.3) biome encircles the earth just south of ice-covered polar seas in the Northern Hemisphere. (A similar community, called the alpine tundra, occurs above the timberline on mountain ranges.) The arctic tundra is cold and dark much of the year. Because rainfall amounts to only about 20 cm a year, the tundra possibly could be considered a desert, but melting snow makes water plentiful in the summer, especially because so little evaporates. Only the topmost layer of earth thaws; the **permafrost** beneath this layer is always frozen.

Trees are not found in the tundra because the growing season is too short, their roots cannot penetrate the permafrost, and they cannot become anchored in the boggy soil of summer. In the summer, the ground is just about covered with sedges and shortgrasses, but there also are numerous patches of lichens and mosses. Dwarf woody shrubs flower and seed quickly while there is plentiful sun for photosynthesis.

A few animals live in the tundra year-round. For example, the ratlike lemming stays beneath the snow; the ptarmigan, a grouse, burrows in the snow during storms; and the musk-ox conserves heat because of its thick coat and short, squat body. In the summer, the tundra is alive with numerous insects and birds, particularly shorebirds and waterfowl that migrate inland. Caribou and reindeer also come, along with the wolves that prey upon them. Polar bears are common near the coast.

Grasslands

Grasslands occur where rainfall is greater than 25 cm but is generally insufficient to support trees. The extensive root system of grasses allows them to recover quickly from drought and fire, which occur frequently in this biome. The matted roots also efficiently absorb surface water and prevent invasion by most trees. Over the years, organic matter builds up in the rich soil, which often is exploited for agriculture.

Figure 33.4
Savanna biome. a. The grasses provide food for many types of herbivores, such as these zebras. b. Male lions are carnivores that feed on zebras.

a.

b.

In the tropics, particularly in much of Africa, there is a tropical grassland called the **savanna** (fig. 33.4). The savanna has few trees because of a severe dry season. One tree that is found here is the flat-topped acacia, which sheds its leaves during a drought.

The African savanna supports the greatest variety and number of large herbivores of all the biomes. Elephants and giraffes are browsers that feed on tree vegetation. Antelopes, zebras, wildebeests, water buffalo, and rhinoceroses are grazers that feed on grasses. Any plant litter that is not consumed by grazers is attacked by a variety of small organisms, among them termites. Termites build towering nests in which they tend fungal gardens, their source of food. The herbivores support a large population of carnivores. Lions and hyenas hunt in packs, cheetahs hunt singly by day, and leopards hunt singly by night.

A **prairie** is a temperate grassland. When traveling east to west across the U.S. Midwest, the tall-grass prairie (fig. 33.1*b*) gradually gives way to the short-grass prairie. Although grasses dominate both types of prairie, they are interspersed by forbs, herbs other than grasses. These often catch the eye because they have colorful flowers, whereas grasses do not.

Because there are so few trees, only grazers are found in the prairie. Small mammals, such as mice, prairie dogs, and rabbits, typically burrow in the ground, although they usually feed above ground. Hawks, snakes, badgers, coyotes, and foxes feed on these mammals. Large herds of buffalo—estimated at hundreds of thousands—once roamed the prairies and plains, as did herds of pronghorn antelope.

Grasslands occur where rainfall is greater than 20 cm but is insufficient to support many trees. The savanna supports the greatest number of different types of herbivores, which serve as food for carnivores. Much of the U.S. Midwest was once a prairie.

Figure 33.5
*Chaparral biome. This biome contains drought-
and fire-resistant shrubs.*

Chaparral

In parts of South Africa, western Australia, central Chile, around the Med-
iterranean Sea, and in California, most of the rain falls in winter and the summers are very dry. Here, there is a dense scrubland known as **chaparral** (fig.
33.5) in this country. The shrubs have small but thick evergreen leaves that
often are coated with a waxy material that prevents loss of moisture from the
leaves. Their thick underground stems survive the dry summers and frequent
fires and can sprout new growth. Rodents and reptiles abound in this biome.

Forests

Taiga

The **taiga** is a coniferous forest extending in a broad belt across northern Eurasia and North America. The climate is characterized by cold winters and
cool summers with a growing season of about 130 days. Rainfall ranges between 40 and 100 cm per year, with much of it in the form of heavy snows.
The great stands of evergreen narrow-leaved trees, such as spruce, fir, and pine
(fig. 33.6), are interrupted by many lakes and swamps. ("Taiga" is Russian
for swampland.) Beneath the dense tree canopy (upper layer of leaves that is
the first to receive sunlight) there is little light, but a ground cover of lichens,
mosses, and ferns usually is found.

Compared to other forests, the taiga has relatively few consumer species.
In the summer, insects attack the trees and are eaten by warblers and fly-
catchers. Other birds, such as crossbills and grosbeaks, extract seeds from the
cones of the trees. Birds of prey and other carnivores—weasel, lynx, and wolf—

feed on small animals, such as the plentiful rodents. The large herbivores—moose and bear—are apt to be found in clearings or near the water's edge, where small trees and shrubs are found.

Temperate Deciduous Forests

A temperate deciduous forest (fig. 33.1*a*) is found south of the taiga in eastern North America, eastern Asia, and much of Europe. The climate here is moderate, with relatively high rainfall (75–150 cm per year). The seasons are well defined, and the growing season ranges between 140 and 300 days. The trees, such as oak, beech, and maple, are **deciduous;** they lose their leaves in the fall and grow them in the spring. Enough sunlight penetrates the canopy for the development of a well-developed understory: a layer of shrubs is followed by herbaceous plants and then often a ground cover of mosses and ferns. Animal life is plentiful. Birds and rodents provide food for bobcats, wolves, and foxes. The white-tailed deer has increased in number of late, while the black bear, an omnivore, has decreased in number but still is found commonly in some parts of the country.

Tropical Rain Forests

Tropical rain forests[1] in South America, Africa, and the Indo-Malayan region occur at the equator, where it is always warm (between 20 and 25° C) and rainfall is plentiful (in excess of 250 cm per year). This is the richest biome,

[1]In some tropical regions, there are tropical deciduous forests. Here, broad-leaved trees lose their leaves because of a dry season.

33.1 Critical Thinking

1. Using the data given in figure 33.2*a*, hypothesize what 2 environmental conditions are most influential in determining the type of biome.
2. If a deciduous forest is destroyed by fire, what less complex biome might take its place? If a grassland biome is overgrazed, what less complex biome might take its place?
3. The greater amount of life in tropical rain forest is related to a plentiful supply of sunlight. Explain why this might be so.
4. The great diversity of a tropical rain forest is explained by the competitive exclusion principle. Why?
5. How might the great diversity of a tropical rain forest be related to a plentiful supply of varied foods?

Sharing the planet Earth with us are at least 4–5 million species of animals, plants, and microorganisms, most of which are poorly known. About two-thirds of them have not even been given scientific names.

Of all the world's species, roughly 10% occur in the United States and Canada together. We can state with confidence that most of these plants, vertebrates, and larger invertebrates and fungi have been studied and classified. Furthermore, we now know that some 5–10% of North America's

The Urgency of Tropical Conservation

species are threatened or endangered, and we are making advances to protect them.

To think that by "taking care of our own" we are saving a large proportion of the world's organisms is to delude ourselves. It is to the tropics that we must turn. For instance, about one-third of the total number of Earth's plants, animals, and microorganisms occur in Latin America. Of these, only about one-sixth have been catalogued. Of the remainder, amounting to nearly 30% of all the planet's organisms, we know absolutely nothing. Some scientists, including Terry Erwin of the U.S. National Museum of Natural History, have estimated that the total number of species in Latin America might be much higher. His calculations indicate a regional total of

perhaps 15 million species. Whatever the final figure, the number of unknown species is staggering.

For example, South America's fresh waters are inhabited by an estimated 5,000 fish species, only about 3,000 of which have been named. (This amounts to about one-eighth of all the world's fish species.) On the eastern slopes of the Andes, 80 or more species of frogs and toads often exist within a single square mile—almost as many as in all of temperate North America. Approximately the size of the state of Colorado, Ecuador harbors more than 1,300 species of birds—roughly twice as many as those inhabiting the United States and Canada.

Moreover, hundreds of new species of plants—including many dozen tree species—are being discovered in Latin America every year, as are dozens of new species of terrestrial vertebrates and fishes. Over one-third of the nearly 800 species of reptiles and amphibians known to occur in Ecuador have been discovered since 1970, and many more still are being found there. In fact, the northern Andean countries of Colombia, Ecuador, and Peru combined are home to about one-sixth of all the Earth's biota—some 750,000 species that are also the most poorly known organisms on our planet. . . .

The actual and potential importance of species for human welfare cannot be overstated. About 85% of our food is

derived directly or indirectly from 20 species of plants, and some 60% comes from only 3: corn, wheat, and rice. It stands to reason that among the world's remaining estimated 235,000 known flowering plant species, there must be many more that could provide important sources of food—not to mention medicines, oils, chemicals, and renewable fuels. Our continued survival depends on our ability to use plants, animals, and microorganisms extensively and wisely. Yet, we know absolutely nothing about most tropical organisms, and we certainly have not examined them for their potential value to humans. By saving the tropical forests and their myriad plant and animal species, we also will be investing wisely in the future survival of our own species.

Despite its riches, tropical vegetation worldwide is being altered or eradicated at an alarming rate. In 1981, the Tropical Forest Resources Assessment Project of the United Nation's Food and Agriculture Organization (FAO) estimated that 44% of the tropical rain forests already had been degraded or destroyed by 1980. Recent estimates based on satellite imagery and field surveys show that most deforestation is now occurring in Brazil and India. India is losing 3.7 million acres per year and Brazil is losing about ten times that number of acres per year. Brazil's loss alone is more than 50 acres per minute. If present trends continue, the rich tropical moist forests of many developing countries will disappear in the next 2–3 decades. . . .

The greatest consequence of all, however, is that of biological extinction—the extermination of a major fraction of

both in the different kinds of species found and the total amount of living matter. The need to preserve tropical rain forests is discussed in the reading on page 708.

The rain forest has a complex structure, with many layers of canopy (fig. 33.7). Some of the broad-leaved evergreen trees grow to 50 m or taller, some to 35 m, and some to 15 m. These tall trees often have trunks buttressed at ground level to prevent their toppling over. Lianas, or woody vines, which encircle the tree as it grows, also help to strengthen the trunk. The rain forest understory becomes dense in open areas or clearings, where light contributes to the development of a thick jungle.

Although there is animal life on the ground (the paca, the agouti, the peccary, the armadillo, and the coati), most animals live in the trees. Insect life is so abundant that the majority of species have not been identified yet.

a.

Diversity of life in tropical rain forests.
*Destroying the tropical rain forests will
greatly reduce the variety of life on earth.
Already vulnerable are (a) the green
iguana, (b) South American ocelot, and
(c) the blue-yellow macaw.*

b.

c.

Earth's plants, animals, and microorganisms during the lifetimes of most people living today. Many tropical organisms have a very narrow geographical range and highly specific ecological requirements. At least 40% of the world's species occur in the very tropical forests that may not survive the next few decades. The loss of only half these organisms would amount to the permanent disappearance from our planet of at least 750,000 species—a far greater number than the total number of species found in the United States and Canada. If we wish to preserve the world's biota, can it be doubted that our major concern in the closing years of the 20th century and the early years of the 21st century must be with the tropics? . .

To find an extinction event of this magnitude in the geological record, we need to go back some 65 million years to the end of the Cretaceous Period, when—possibly spurred by a cloud thrown up by the collision of a gigantic meteorite with the Earth—a large proportion of the Earth's organisms, including the dinosaurs, became extinct. The extinction that is taking place now will occur during our life and that of our children. But we can ameliorate its effects by learning about tropical organisms and using our knowledge to save them. Our contribution to preserving natural diversity will be minute if we do not look beyond our own borders to the tropics. This task is one of extreme urgency. If we succeed, we shall play an important role in shaping the quality of life for all who come after us.

Reprinted from *The Nature Conservancy News* January-March 1986. Copyright 1986, The Nature Conservancy. Used with permission.

The various birds, such as the hummingbird, the parakeet, the parrot, and the toucan, are often beautifully colored. Amphibians and reptiles also are well represented by many types of snakes, lizards, and frogs. Monkeys are well-known primates that feed on the fruits of the trees. The largest carnivores are the big cats—the jaguars in South America and the leopards in the Old World.

Not only do many animals spend their entire lives in the canopy, some plants do also. **Epiphytes** are plants that grow on tall trees but do not parasitize them. Instead, some have roots that take moisture and minerals leached from the canopy and others catch daily rain and debris in special hollow leaves. The most common epiphytes belong to the pineapple, orchid, and fern families.

Whereas the soil of temperate deciduous forests is rich enough for agricultural purposes, the soil of a tropical rain forest is not. Numerous organisms quickly break down any litter, and nutrients are recycled immediately to

Figure 33.7

Tropical rain forest. This biome contains hardwood trees that are green the entire year. Epiphytes are plants that use the trees for support but do not parasitize them. They get their nutrients from water ladened with minerals and debris from the canopy above.

the plants. Of the minerals, only aluminum and iron sometimes remain near the surface, producing a red-colored soil known as laterite. When the trees are cleared, laterite bakes in the hot sun to a bricklike consistency and will not support crops. Slash-and-burn agriculture is the only type that thus far has been successful in the tropics. Trees are felled and burned, and the ashes provide enough nutrients for several harvests. Thereafter, the forest is allowed to regrow, and a new section is utilized for agriculture. These matters are discussed further on page 744.

Forests require adequate rainfall. The taiga has the least amount of rainfall. The temperate deciduous forest has trees that grow and shed their leaves with the seasons. The tropical rain forest is the least studied and the most complex of all the biomes.

Latitude versus Altitude

Latitude

If you travel from the Southern Hemisphere to the Northern Hemisphere, it is possible to observe first a tropical rain forest, followed by a temperate deciduous forest, and then the taiga and the tundra, in that order. This shows that the location of the biomes is influenced by temperature.

Altitude

It is also possible to observe a similar sequence of biomes by traveling from the bottom to the top of a mountain (fig. 33.8). These transitions are largely due to decreasing temperature as the altitude increases, but soil conditions and rainfall are also important.

In general, the same series of biomes can be observed according to both latitude and altitude.

Aquatic Communities

Aquatic communities can be divided into 2 types: freshwater (inland) communities and saltwater (oceanic or marine) communities.

Freshwater Communities

The freshwater communities are lakes, ponds, rivers, and streams. Here, we consider only the composition of a **lake,** a body of water that has some depths that are always dark. This study is sufficient because the other freshwater communities have a similar mix of organisms.

Lakes

Lakes have 3 life zones: the littoral zone, which is closest to the shore; the limnetic zone, which is the sunlit main body of the lake; and the profundal zone, which is the deep part where light does not penetrate.

Aquatic plants are rooted in the shallow littoral zone of a lake (fig. 33.9). Here, there are microscopic organisms that cling to plants and rocks. In the limnetic zone, there are a few types of organisms, like the water strider, that are found on the surface of the lake. In the water, there are microscopic, floating organisms known as **plankton.** Phytoplankton are photosynthesizing algae, and zooplankton are protozoans and tiny crustaceans. Small and large fishes also are found in the limnetic zone. In the profundal zone, there are mollusks, crustaceans, and worms that feed on debris that falls from above.

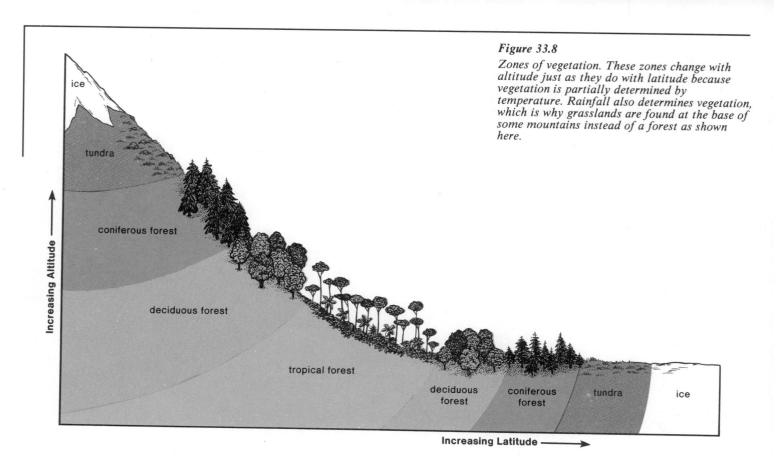

Figure 33.8

Zones of vegetation. These zones change with altitude just as they do with latitude because vegetation is partially determined by temperature. Rainfall also determines vegetation, which is why grasslands are found at the base of some mountains instead of a forest as shown here.

ice

tundra

coniferous forest

deciduous forest

tropical forest

deciduous forest

coniferous forest

tundra

ice

Increasing Altitude →

Increasing Latitude →

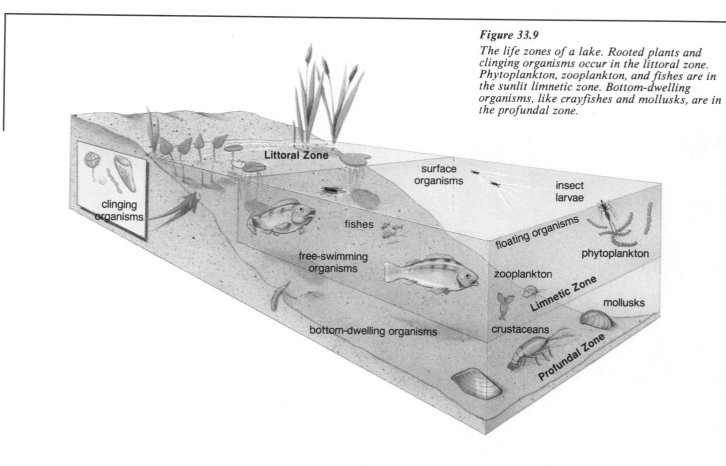

Figure 33.9

The life zones of a lake. Rooted plants and clinging organisms occur in the littoral zone. Phytoplankton, zooplankton, and fishes are in the sunlit limnetic zone. Bottom-dwelling organisms, like crayfishes and mollusks, are in the profundal zone.

clinging organisms

Littoral Zone

surface organisms

insect larvae

fishes

floating organisms

phytoplankton

free-swimming organisms

zooplankton

Limnetic Zone

mollusks

bottom-dwelling organisms

crustaceans

Profundal Zone

The Biosphere

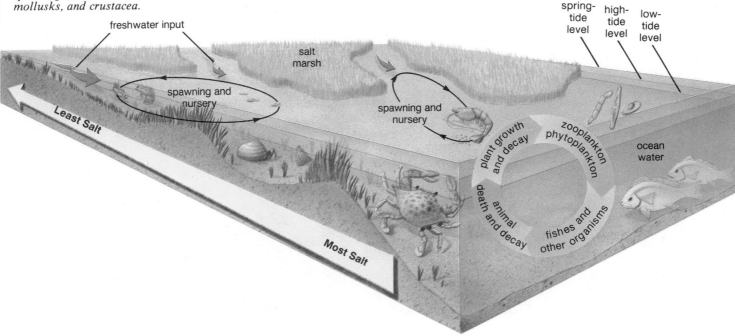

Figure 33.10

Estuary structure and function. Since an estuary is located where a river flows into the ocean, it receives nutrients from the land and also from the sea by way of the tides. Decaying grasses also provide nutrients. Estuaries serve as nurseries for the spawning and the rearing of the young of many species of fishes, mollusks, and crustacea.

Figure 33.11

A close-up of a coral reef. A coral reef is a unique community of marine organisms. The diversity and richness of form and color are the result of the optimal environmental conditions of tropical seas.

Saltwater Communities

Saltwater or marine communities begin along the shoreline. These coastal communities are discussed before the oceans themselves are taken up.

Coastal Communities

An **estuary** forms where a large river flows into the ocean (fig. 33.10). The river brings fresh water into the estuary. Because of the tides, the sea brings salt water into this same area. Therefore, an estuary is characterized by a mixture of fresh water and salt water; this is called brackish water. Most estuaries are rather shallow with good sunlight penetration. The temperature of estuarine waters is generally higher than that of the surrounding marine or freshwater environments. Further, an estuary acts as a nutrient trap; the tides bring nutrients from the sea, and at the same time, they prevent the seaward escape of nutrients brought by the river.

Only a few types of small fishes live permanently in an estuary, but many types develop there; therefore, both larval and immature fishes are present in large numbers. It has been estimated that well over half of all marine fishes develop in the protective environment of an estuary; this is why estuaries are called the nurseries of the sea. Shrimps and mollusks, too, use the estuary as a nursery.

Salt marshes dominated by salt-marsh cord grass are often adjacent to estuaries. Salt marshes contribute nutrients and detritus to estuaries, in addition to being an important habitat for wildlife.

The **rocky shore** offers a firm substratum to which organisms can attach. Large macroscopic brown algae cling to the rocks and offer support for other algae, protists, and small crustacea. Barnacles, mussels, and snails can cling firmly to rocks or seaweeds. These close up tightly during the periods the tide is out.

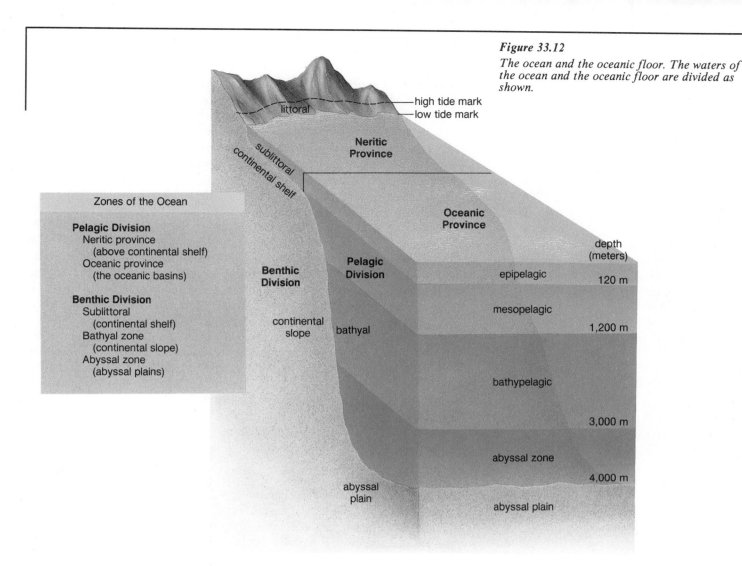

Figure 33.12
The ocean and the oceanic floor. The waters of the ocean and the oceanic floor are divided as shown.

high tide mark
low tide mark

littoral

sublittoral
continental shelf

Neritic
Province

Oceanic
Province

Pelagic
Division

Benthic
Division

Zones of the Ocean

Pelagic Division
Neritic province
(above continental shelf)
Oceanic province
(the oceanic basins)

Benthic Division
Sublittoral
(continental shelf)
Bathyal zone
(continental slope)
Abyssal zone
(abyssal plains)

continental
slope

bathyal

depth
(meters)

epipelagic

120 m

mesopelagic

1,200 m

bathypelagic

3,000 m

abyssal zone

4,000 m

abyssal
plain

abyssal plain

The shifting, unstable sands of a **sandy beach** do not provide a suitable substratum for the attachment of organisms. Many organisms, like ghost crabs and sand hoppers, burrow above high tide and feed at night whenever the tide is out. Sandworms and sand (ghost) shrimps remain within their burrows and feed on detritus. Shorebirds that feed on the burrowers are links with land communities.

Coral reefs (fig. 33.11) are found in warm tropical seas, where the water is still shallow. Their chief constituents are stony corals (phylum Cnidaria), which have a calcium carbonate (limestone) exoskeleton, and carbonaceous red and green algae. Most of the solid part of a coral reef is composed of the skeletons of dead coral; the outer layer contains living organisms.

A reef is densely populated with animal life. There are many types of small fishes, all of which are beautifully colored. In addition, the numerous crevices and caves provide shelter for filter feeders (sponges, sea squirts, fan worms) and for scavengers (crabs and sea urchins). The barracuda and the moray eel prey on these animals.

Oceans

The oceans cover approximately three-quarters of our planet. The geographic areas and zones of an ocean are shown in figure 33.12. It is customary to place the organisms of the oceans into either the pelagic division (open waters) or the benthic division (ocean floor).

The Pelagic Division   The pelagic division includes the neritic province and the oceanic province. There is a greater concentration of organisms in the neritic province than in the oceanic province because sunlight penetrates the waters of the neritic province and this is where you find the most nutrients. Phytoplankton is food not only for zooplankton but also for small fishes. These small fishes in turn are food for commercial fishes—herring, cod, and flounder.

The oceanic province (fig. 33.13) is divided into the epipelagic, mesopelagic, and bathypelagic zones. Only the epipelagic zone is brightly lit, or euphotic; the mesopelagic zone is in semidarkness, and the bathypelagic zone is in complete darkness.

The epipelagic zone does not have a high concentration of phytoplankton because of the lack of nutrients. These photosynthesizers, however, still support a large assembly of zooplankton because the oceans are so large. The zooplankton are food for herrings and bluefishes, which in turn are eaten by larger mackerels, tunas, and sharks. Flying fishes, which glide above the surface, are preyed upon by dolphins, not to be confused with mammalian porpoises, which also are present. Whales are other mammals found in the epipelagic zone. Baleen whales strain krill (small crustacea) from the water, and the toothed sperm whales feed primarily on the common squid.

Animals in the mesopelagic zone are carnivores, are adapted to the absence of light, and tend to be translucent, red colored, or even luminescent. There are luminescent shrimps, squids, and fishes, such as lantern and hatchet fishes.

The bathypelagic zone is in complete darkness except for an occasional flash of bioluminescent light. Carnivores and scavengers are found in this zone. Strange-looking fishes with distensible mouths and abdomens and small, tubular eyes feed on infrequent prey.

The Benthic Division   The benthic division includes organisms that live on the continental shelf, the continental slope, and the abyssal plain (fig. 33.12). These are the organisms of the sublittoral, bathyal, and abyssal zones. Note that the sublittoral zone lies beneath the waters of the neritic province and the bathyal and abyssal zones lie beneath the waters of the oceanic province. The sublittoral zone takes its name from the fact that it occurs after the littoral zone. The coastal communities we already discussed (p. 712) are in the littoral zone.

Seaweed grows in the first part of the sublittoral zone, and it can be found in batches on outcroppings as the water gets deeper. However, nearly all benthic organisms are dependent on the slow rain of plankton and detritus from the sunlit waters above. There is more diversity of life in the sublittoral and bathyal zones than in the abyssal zone. In these first 2 zones, clams, worms, and sea urchins are preyed upon by starfishes, lobsters, crabs, and brittle stars.

The abyssal life zone is inhabited by animals that live just above and in the dark abyssal plain. It once was thought that few animals exist in this zone because of the intense pressure and the extreme cold. Yet, many invertebrates live here by feeding on debris floating down from the mesopelagic zone. Sea lilies rise about the seafloor; sea cucumbers and sea urchins crawl around on the sea bottom; and tubeworms burrow in the mud.

The flat abyssal plain is interrupted by enormous underwater mountain chains called oceanic ridges. Along the axes of the ridges, crustal plates are spreading apart and molten magma is rising to fill the gap. At specific sites of these tectonic spreading zones, seawater percolates through cracks and is heated to about 350° C, causing sulfate to react with water and form hydrogen sulfide (H_2S). The water temperature–hydrogen sulfide combination has been found to support communities that contain huge tubeworms and clams (see fig. 6.11). Chemosynthetic bacteria that obtain energy from oxidizing hydrogen sulfide live within these organisms.

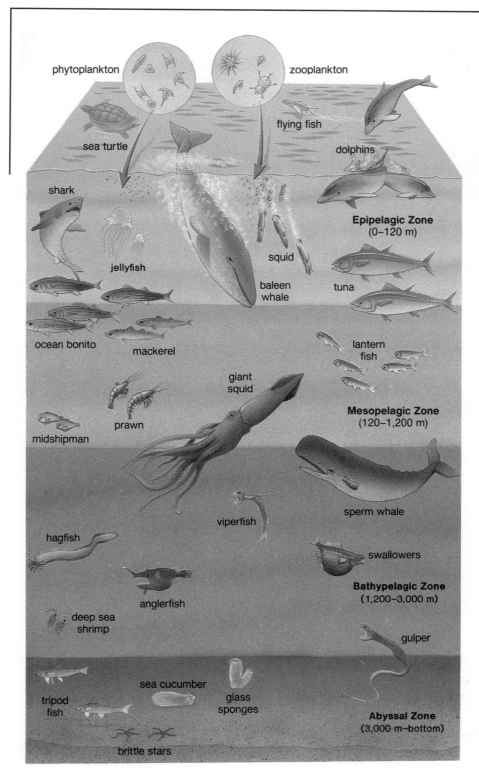

phytoplankton

zooplankton

sea turtle

flying fish

dolphins

shark

Epipelagic Zone
(0–120 m)

jellyfish

squid

baleen
whale

tuna

ocean bonito

mackerel

lantern
fish

giant
squid

Mesopelagic Zone
(120–1,200 m)

midshipman

prawn

sperm whale

viperfish

hagfish

swallowers

anglerfish

Bathypelagic Zone
(1,200–3,000 m)

deep sea
shrimp

gulper

tripod
fish

sea cucumber

glass
sponges

Abyssal Zone
(3,000 m–bottom)

brittle stars

Figure 33.13

The zones of the oceanic province (pelagic division). The epipelagic zone contains the most organisms because this is where you find the phytoplankton and the zooplankton. Only carnivores are found in the mesopelagic zone. Both carnivores and scavengers are found in the bathypelagic zone. The abyssal zone is a zone of the benthic division, but it is shown here because it lies beneath the bathypelagic zone. The abyssal zone is characterized especially by certain echinoderms (sea cucumbers and sea lilies) and tubeworms. A few kinds of fishes, such as the tripod fish, also are found here.

In the pelagic division of the ocean, only the neritic province and the epipelagic zone of the oceanic province receive sunlight and contain the organisms with which we are most familiar. The organisms of the benthic division are dependent upon debris that floats down from above.

Summary

The biosphere contains major communities of organisms. On land, these are called biomes. Biomes can be divided into those that are treeless and those that have trees. Deserts are high-temperature areas receiving less than 25 cm of rainfall a year. They contain organisms that are adapted to a dry environment. The arctic tundra is the northernmost biome and contains organisms that are adapted to a cold climate. Grasslands occur where rainfall is greater than 20 cm but is insufficient to support trees. Forests require adequate rainfall. The taiga, the northernmost of the forests, contains narrow-leaved evergreen coniferous trees, and a temperate deciduous forest contains broad-leaved trees that shed their leaves in the fall. Tropical rain forests, with a warm, wet climate all year, contain the greatest diversity of life forms. The series of biomes on mountain slopes mirrors the sequence of biomes according to latitude because temperature also decreases with altitude.

Aquatic communities are divided into freshwater and saltwater communities. Freshwater lakes have various life zones, each with typical organisms. Coastal communities include estuaries, salt marshes, rocky shores, and sandy beaches. Coral reefs occur in shallow tropical seas. The open ocean can be divided into divisions, provinces, and zones. Only the neritic province and the epipelagic zone of the oceanic province receive adequate sunlight to support photosynthesis, and this limits the concentration of life in the ocean.

Study Questions

1. Describe the climate and the populations of a desert in North America. (p. 702)
2. Describe the location, the climate, and the populations of the arctic tundra. (p. 704)
3. Describe the climate and the populations of the African savanna. (p. 705)
4. Describe the location and the climate of the North American prairie. What are its populations of organisms? (p. 705)
5. Describe the location, the climate, and the populations of the taiga. (p. 706)
6. Describe the location, the climate, and the populations of temperate deciduous forests in North America. (p. 707)
7. Describe the location, the climate, and the populations of tropical rain forests. (p. 707)
8. Name the terrestrial biomes you would expect to find when going from the base to the top of a mountain. (p. 710)
9. Describe the life zones of a lake and the organisms you would expect to find in each zone. (p. 710)
10. Describe the coastal communities, including coral reefs, and discuss the importance of estuaries to the productivity of the ocean. (p. 712)
11. Describe the zones of the open ocean and the organisms you would expect to find in each zone. (p. 714)

Objective Questions

1. The major terrestrial communities are called _____ .
2. Trees are not plentiful in a desert or a grassland because of the reduced amount of _____ .
3. The tropical grassland of Africa is called a(n) _____ .
4. Broad-leaved evergreen trees are found in a(n) _____ forest.
5. Narrow-leaved evergreen trees are found in a biome called the _____ .
6. In the U.S. Midwest, the _____ biome is found.
7. At the highest altitudes and latitudes, a _____ biome is found.
8. An estuary is very productive because it acquires _____ brought by both river flow and tidal action.
9. The chief builders of coral reefs are _____ and carbonaceous _____ .
10. The pelagic division is the open ocean, and the benthic division is the _____ .

Label this Diagram.
See figure 33.8 (p. 711) in text.

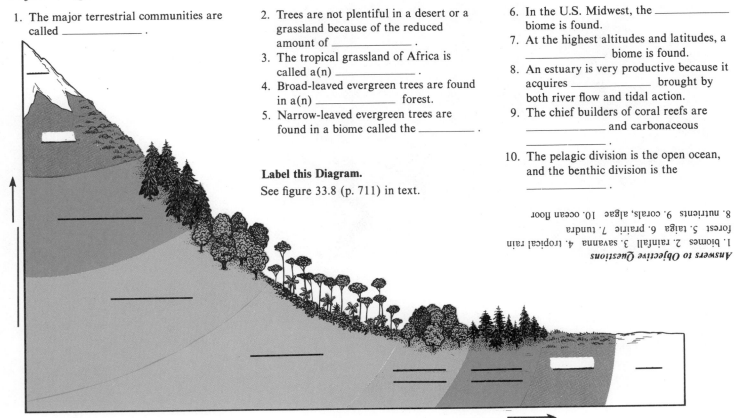

Selected Key Terms

biome (bi'ōm) one of the major climax communities present in the biosphere, characterized by a particular mix of plants and animals. *701*

chaparral (shap-ĕ-ral') a biome of broad-leaved evergreen shrubs forming dense thickets. *706*

coral reef (kor'al rēf) a structure found in tropical waters formed by the buildup of coral skeletons, where many and various types of organisms reside. *713*

deciduous (de-sid'u-us) plant that sheds leaves at certain seasons. *707*

desert (dez'ert) an arid biome characterized especially by plants, such as cacti, that are adapted to receiving less than 25 cm of rain per year. *702*

epiphyte (ep'i-fīt) nonparasitic plant that grows on the surface of other plants, usually above the ground, such as arboreal orchids and Spanish moss. *709*

estuary (es'tu-a-re) an area where fresh water meets the sea; thus, an area with salinity intermediate between fresh water and saltwater. *712*

permafrost (perm'ah-frost) earth beneath surface in tundra that remains permanently frozen. *704*

plankton (plank'ton) floating microscopic organisms found in most bodies of water. *710*

prairie (prar'e) a grassland biome of the temperate zone that occurs when rainfall is greater than 25 cm but less than 40 cm. *705*

salt marsh (sawlt marsh) coastal grassland exposed to seasonal flooding. *712*

savanna (sah-van'ah) a grassland biome that has occasional trees and is associated particularly with Africa. *705*

taiga (ti'gah) a biome that forms a worldwide northern belt of coniferous trees. *706*

tropical rain forest (trŏp'i-kĕl rān for'est) a biome of equatorial forests that remain warm year-round and receive abundant rain. *707*

tundra (tun'drah) a biome characterized by lack of trees due to cold temperatures and the presence of permafrost year-round. *704*

Chapter 34

Ecosystems

Chapter Concepts

1 Climax communities, containing many and varied populations, arise by succession, a series of developmental steps.

2 Ecosystems are units of the biosphere in which populations interact with each other and with the physical environment.

3 Energy is dissipated and does not cycle in an ecosystem; therefore, there is a need for a continual supply of solar energy.

4 The same chemical elements do cycle through an ecosystem, and there is no need for a continual outside supply.

5 In contrast to mature natural ecosystems, the human ecosystem is characterized by an ever-greater input of energy and materials each year.

Chapter Outline

The Nature of Ecosystems
    Habitat and Niche
    Ecosystem Composition
Energy Flow
    Food Chains and Food Webs
    Ecological Pyramids
Chemical Cycling
    The Carbon Cycle
    The Nitrogen Cycle
The Human Ecosystem
    The Country
    The City
    The Solution

a.

b.

Figure 34.1

Primary succession. These photos show a possible early sequence of events by which bare rock becomes a climax community. a. Lichens growing on bare rock. b. Individual plants taking hold. c. Perennial herbs spreading out over the area.

c.

When life first arose, it was confined to the oceans, but about 450 million years ago organisms began to colonize the bare land. Eventually, the land supported many complex communities of living things. Similarly, today we can observe a series of sequential events by which bare rock becomes capable of sustaining many organisms. We call this primary succession (fig. 34.1). During **succession,** a sequence of communities replaces one another in an orderly and predictable way, until finally there is a climax community, a mix of plants and animals that is typical of that area. Secondary succession also is observed when a climax community that has been disturbed once again takes on its former state. Figure 34.2 shows a possible secondary succession for abandoned farmland in the eastern United States.

In the previous chapter, we studied some of the climax communities found in the biosphere. For example, in the United States not too long ago, most of the Northeast was a deciduous forest, a prairie was common to the Midwest, and a desert covered the Southwest (see fig. 33.1).

The process of succession leads to a climax community that contains plants and animals characteristic of the area.

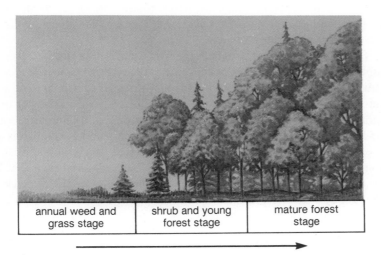

annual weed and grass stage	shrub and young forest stage	mature forest stage

Figure 34.2

Secondary succession. This drawing shows a possible sequence of events by which abandoned farmland becomes a climax community again. During secondary succession in the Northeast, grass and weeds are followed by shrubs and trees and finally by a mature climax forest.

Table 34.1 Ecological Terms

Term	Definition
Ecology	Study of the interactions of organisms with each other and with the physical environment.
Population	All the members of the same species that inhabit a particular area.
Community	All the populations that are found in a particular area.
Ecosystem	A community and its physical environment; has living (biotic) components and nonliving (abiotic) components.
Biosphere	The portion of the surface of the earth (air, water, and land) where living things exist.

The Nature of Ecosystems

When we study a community, we are considering only the populations of organisms that make up that community, but when we study an **ecosystem,** we are concerned with the community plus its physical environment. **Ecology** is the study of the interactions of organisms with each other and with the physical environment. A review of ecological terms is given in table 34.1.

An ecosystem possesses both living (biotic) and nonliving (abiotic) components. The abiotic components include soil, water, light, inorganic nutrients, and weather variables. The biotic components of the ecosystem have a habitat and a niche.

Habitat and Niche

Each population in an ecosystem has a habitat and a niche. The **habitat** of an organism is its place of residence, that is, where it can be found, such as under a log or at the bottom of the pond. The **niche** of an organism is its profession or total role in the community. A description of an organism's niche (table 34.2) includes its interactions with the physical environment and with the other organisms in the community. One important aspect of niche is how the organism acquires its food.

Producers are autotrophic organisms with the ability to carry on photosynthesis and to make food for themselves (and indirectly for the other populations as well). In terrestrial ecosystems, the producers are predominantly green plants, while in freshwater and saltwater ecosystems, the dominant producers are various species of algae.

Consumers are heterotrophic organisms that use preformed food. It is possible to distinguish 4 types of consumers, depending on their food source. **Herbivores** feed directly on green plants; they are termed primary consumers. **Carnivores** feed only on other animals and are therefore secondary or tertiary consumers. **Omnivores** feed on both plants and animals. Therefore, a caterpillar feeding on leaf is a herbivore; a green heron feeding on a fish is a carnivore; a human eating both leafy green vegetables and beef is an omnivore. **Decomposers** are organisms of decay, such as bacteria and fungi, that break down **detritus,** nonliving organic matter, to inorganic matter, which can be used again by producers. In this way, the same chemical elements can be used over and over again in an ecosystem.

Ecosystem Composition

When we diagram the components of an ecosystem, as in figure 34.3, it is possible to illustrate that every ecosystem is characterized by 2 fundamental phenomena: energy flow and chemical cycling. Energy flow begins when pro-

Table 34.2 Aspects of Niche

Plants	Animals
Season of year for growth and reproduction	Time of day for feeding and season of year for reproduction
Sunlight, water, and soil requirements	Habitat and food requirements
Relationships with other organisms	Relationships with other organisms
Effect on abiotic environment	Effect on abiotic environment

Behavior and Ecology

Figure 34.3

Ecosystem composition. A diagram illustrating energy flow and chemical cycling through an ecosystem. Energy does not cycle because all the energy that is derived from the sun eventually dissipates as heat.

ducers absorb solar energy, and chemical cycling begins when producers take in inorganic nutrients from the physical environment. Thereafter, producers make food for themselves and indirectly for the other populations of the ecosystem. Energy flow occurs because all the energy content of organic food eventually is lost to the environment as heat. Therefore, most ecosystems cannot exist without a continual supply of solar energy. However, the original inorganic elements are cycled back to the producers, and no new input is required.

Within an ecosystem, energy flows and chemicals cycle.

Energy Flow

Energy flows through an ecosystem because when one form of energy is transformed into another form, there is always a loss of some usable energy as heat. For example, the conversion of energy in one molecule of glucose to 38 molecules of ATP represents only 50% of the available energy in a glucose molecule. The rest is lost as heat. This means that as one population feeds on another and as decomposers work on detritus, all of the captured solar energy that was converted to chemical-bond energy by algae and plants is returned to the atmosphere as heat. Therefore, energy flows through an ecosystem and does not cycle.

Figure 34.4

Examples of food chains. a. *Terrestrial.*
b. *Aquatic.*

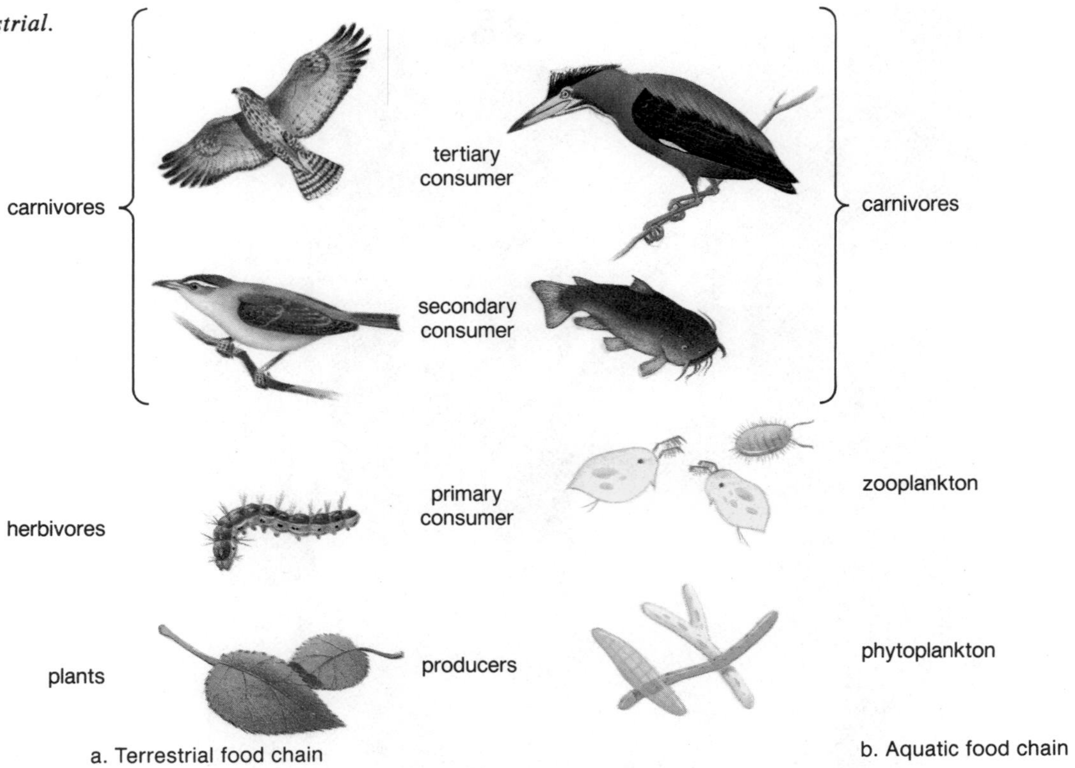

carnivores

tertiary
consumer

carnivores

secondary
consumer

herbivores

primary
consumer

zooplankton

plants

producers

phytoplankton

a. Terrestrial food chain

b. Aquatic food chain

Food Chains and Food Webs

Energy flows through an ecosystem as the individuals of one population feed on those of another. A **food chain** indicates who eats whom in an ecosystem. Figure 34.4 depicts examples of a terrestrial food chain and an aquatic food chain. It is important to realize that each represents just one path of energy flow through an ecosystem. Natural ecosystems have numerous food chains, each linked to others to form a complex **food web.** For example, figure 34.5 shows a deciduous forest ecosystem in which plants are eaten by a variety of insects, and in turn, these are eaten by several different birds, while any one of the latter may be eaten by a larger bird, such as a hawk. Therefore, energy flow is better described in terms of **trophic** (feeding) **levels,** each one further removed from the producer population, the first (photosynthetic) trophic level. All animals acting as primary consumers are part of a second trophic level, and all animals acting as secondary consumers are part of the third level, and so on.

The populations in an ecosystem form food chains in which the producers produce food for the other populations, which are consumers. While it is convenient to study food chains, the populations in an ecosystem actually form a food web, in which food chains join with and overlap one another.

One of the food chains depicted in figure 34.4 is part of the forest food web shown in figure 34.5 and the other food chain is part of the aquatic food web from the freshwater pond ecosystem shown in figure 34.6. Both of these food chains are called grazing food chains because the primary consumer feeds on a photosynthesizer. In some ecosystems (forests, rivers, and marshes), the primary consumer feeds mostly on detritus (dead organisms). This *detritus*

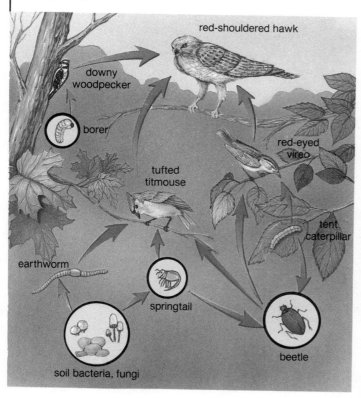

Figure 34.5

A deciduous forest ecosystem. The arrows indicate the flow of energy in a food web.

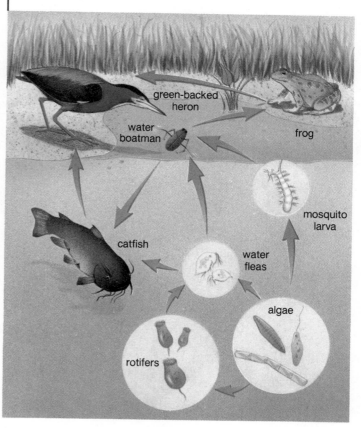

Figure 34.6

A freshwater pond ecosystem. The arrows indicate the flow of energy in a food web.

food chain accounts for more energy flow than the grazing food chain whenever most organisms die without having been eaten. In the forest, an example of a detritus food chain is

detritus ——→ soil bacteria ——→ earthworms

A detritus food chain often is connected to a grazing food chain, as when earthworms are eaten by a robin. Eventually, however, as dead organisms decompose, all the solar energy that was taken up by the producer populations is dissipated as heat. Therefore, energy does not cycle.

Ecological Pyramids

The trophic structure of an ecosystem can be summarized in the form of an **ecological pyramid.** The base of the pyramid represents the producer trophic level, and the apex is the highest-level consumer, called the top predator. The other consumer trophic levels are in between the producer and the top predator levels. There are 3 kinds of pyramids. One is a *pyramid of numbers,* based on the number of organisms at each trophic level. A second is the *pyramid of biomass.* Biomass is the weight of living material at some particular time. To calculate the biomass for each trophic level, an average weight for the organisms at each level is determined and then the number of organisms at each level is estimated. Multiplying the average weight by the estimated number gives the approximate biomass for each trophic level. A third pyramid, the

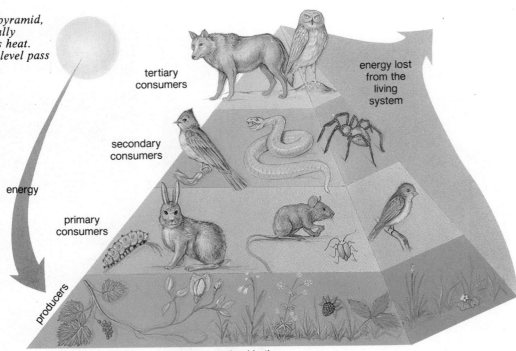

tertiary consumers

energy lost from the living system

secondary consumers

energy

primary consumers

producers

energy retained in the living system

pyramid of energy (fig. 34.7), illustrates that each succeeding trophic level is smaller than the previous level. Less energy is found in each succeeding trophic level for the following reasons.

1. Of the food available, only a certain amount is captured and eaten by the next trophic level. After all, prey are adapted in many ways to escape their predators, as discussed on page 692.
2. Some of the food that is eaten cannot be digested and exits the digestive tract as waste.
3. Only a portion of the food that is digested becomes part of the organism's body. The rest is used as a source of energy.

In regard to the last point, we have to realize that a significant portion of food molecules is used as an energy source for ATP buildup in mitochondria. This ATP is needed to build the proteins, carbohydrates, and lipids that compose the body. ATP also is needed for such activities as muscle contraction and nerve conduction. Figure 34.7 indicates the amount of energy in one trophic level that is unavailable to the next trophic level.

The energy considerations associated with ecological pyramids have implications for the human population. It generally is stated that only about 10% of the energy available at a particular trophic level is incorporated into the tissues at the next level. This being the case, it can be estimated that 100 kg of grain could, if consumed directly, result in 10 human kg; however, if fed to cattle, the 100 kg of grain would result in only 1 human kg. Therefore, a larger human population can be sustained by eating grain than by eating grain-fed animals. Humans generally need some meat in their diet, however, because this is the most common source of the essential amino acids, as discussed in chapter 10.

In a food web, each successive trophic level has less total energy content. This is because some energy is lost to the environment as heat when energy is transferred from one trophic level to the next.

34.1 Critical Thinking

1. Using the data given in figure 34.7, explain why you would expect mice (herbivores) to be more common than weasels, foxes, or hawks (carnivores) in the environment.
2. Explain why you would expect food chains to be short—4 or 5 links at most.
3. The population size of a top predator is not held in check by another predator population. Again with reference to figure 34.7, why does a top predator population not increase constantly in size?
4. What would you expect to happen to an ecosystem if one of the secondary consumer populations suffered a collapse?

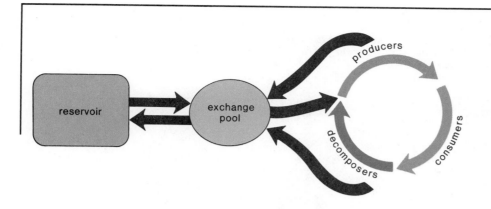

Figure 34.8

Components of a chemical cycle. The reservoir stores the chemical and the exchange pool makes it available to producers. The chemical then cycles through food chains. Decomposition returns the chemical to the exchange pool once again if it has not already returned by another process.

Chemical Cycling

In contrast to energy, inorganic nutrients do cycle through large natural ecosystems. Because there is minimal input from the outside, the various elements essential for life are used over and over. For each element, the cycling process (fig. 34.8) involves (1) a reservoir—that portion of the earth that acts as a storehouse for the element; (2) an exchange pool—that portion of the environment from which the producers take their nutrients; and (3) the biotic community—through which elements move along food chains to and from the exchange pool.

The Carbon Cycle

The relationship between photosynthesis and respiration should be kept in mind when discussing the carbon cycle. Recall that for simplicity's sake, this equation in the forward direction represents respiration, and in the other direction, it is used to represent photosynthesis:

$$C_6H_{12}O_6 + 6O_2 \rightleftharpoons 6CO_2 + 6H_2O$$

The equation tells us that respiration releases carbon dioxide (CO_2), the molecule needed for photosynthesis. However, photosynthesis releases oxygen (O_2), the molecule needed for respiration. From figure 34.9, it is obvious that animals are dependent on green organisms, not only to produce organic food and energy but also to supply the biosphere with oxygen.

In the carbon cycle, organisms in both terrestrial and aquatic ecosystems (fig. 34.10) exchange carbon dioxide with the atmosphere. On land, plants take up carbon dioxide from the air, and through photosynthesis, they incorporate carbon into food that is used by autotrophs and heterotrophs alike. When organisms respire, a portion of this carbon is returned to the atmosphere as carbon dioxide.

In aquatic ecosystems, the exchange of carbon dioxide with the atmosphere is indirect. Carbon dioxide from the air combines with water to give bicarbonate (HCO_3^-), a source of carbon for algae that produce food for themselves and for heterotrophs. Similarly, when aquatic organisms respire, the carbon dioxide they give off becomes bicarbonate. The amount of bicarbonate in the water is in equilibrium with the amount of carbon dioxide in the air.

Carbon Reservoirs

Living and dead organisms contain organic carbon and serve as one of the reservoirs for the carbon cycle. The world's biota, particularly trees, contain

Figure 34.9

Relationship between photosynthesis and respiration. Animals are dependent on plants for a supply of oxygen (O_2) and organic food and plants are dependent on animals for a supply of carbon dioxide (CO_2).

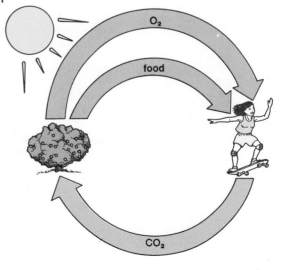

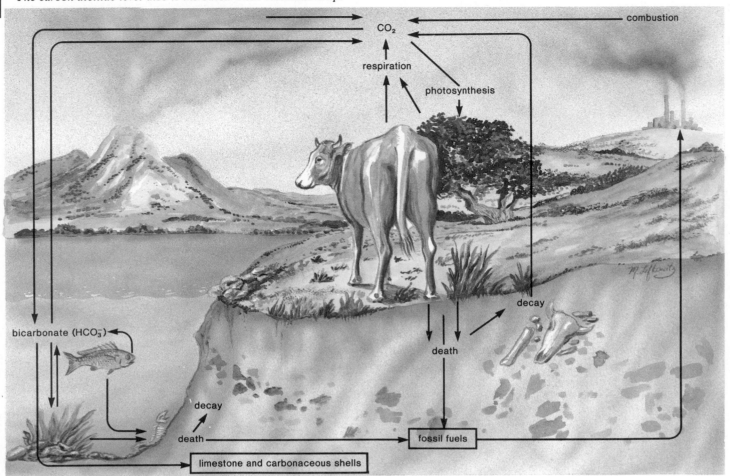

Figure 34.10

Carbon cycle. Photosynthesizers take up carbon dioxide (CO_2) from the air or bicarbonate (HCO_3^-) from the water. They and all other organisms return carbon dioxide to the environment. The carbon dioxide level also is increased when volcanoes erupt and fossil fuels are burned. Presently, the oceans are a primary reservoir for carbon in the form of limestone and carbonaceous shells.

800 billion metric tons of organic carbon, and an additional 1,000–3,000 billion metric tons are estimated to be held in the remains of plants and animals in the soil. Before decomposition can occur, some of these remains are subjected to physical processes that transform them into coal, oil, and natural gas. We call these materials the fossil fuels. Most of the fossil fuels were formed during the Carboniferous Period, 280–350 million years ago, when an exceptionally large amount of organic matter was buried before decomposing. Another reservoir is the inorganic carbonate that accumulates in limestone and in carbonaceous shells. The oceans abound in organisms, some microscopic, that are composed of carbonaceous shells and accumulate in ocean bottom sediments. Limestone is formed from these sediments by geological transformation.

Human Influence on the Carbon Cycle

The activities of human beings have increased the amount of carbon dioxide (CO_2), and other gases in the atmosphere. Data from monitoring stations record an increase of 20 ppm (parts per million) in carbon dioxide in only 22 years. (This is equivalent to 42 billion metric tons of carbon.) This buildup is attributed primarily to the burning of fossil fuels and the destruction of the

world's forests (fig. 34.11). When we do away with forests, we reduce a reservoir that takes up excess carbon dioxide. At this time, the oceans are believed to be taking up most of the excess carbon dioxide; the burning of fossil fuels in the last 22 years has probably released 78 billion metric tons of carbon, yet the atmosphere registers an increase of "only" 42 billion metric tons.

As discussed on page 753, there is much concern that an increased amount of carbon dioxide (and other gases) in the atmosphere is causing global warming. These gases allow the sun's rays to pass through, but they absorb and reradiate heat back to the earth, a phenomenon called the *greenhouse effect.*

In the carbon cycle, carbon dioxide is removed from the atmosphere by photosynthesis but is returned by respiration. Living things and dead matter are carbon reservoirs. The oceans, because they abound with carbonaceous shells and accumulate limestone, are also major carbon reservoirs.

The Nitrogen Cycle

Nitrogen is an abundant element in the atmosphere. Nitrogen gas (N_2) makes up about 78% of the atmosphere by volume, yet nitrogen deficiency commonly limits plant growth. Plants cannot incorporate nitrogen gas into organic compounds and therefore depend on various types of bacteria to make nitrogen available to them (fig. 34.12).

Nitrogen Fixation

The reduction of nitrogen gas (N_2) prior to the incorporation of nitrogen into organic compounds is called **nitrogen fixation.** For example, some cyanobacteria in aquatic ecosystems and some free-living bacteria in the soil reduce nitrogen gas to ammonium (NH_4^+), which then is subject to processes leading to organic compounds. Other *nitrogen-fixing bacteria* infect and live in nodules on the roots of legumes (fig. 34.12). They fix atmospheric nitrogen gas and incorporate nitrogen into organic compounds that not only they themselves use, but the host plant as well.

Figure 34.12

Nitrogen cycle. Several types of bacteria are at work: nitrogen-fixing bacteria reduce nitrogen gas (N_2); nitrifying bacteria, which include both nitrite-producing and nitrate-producing bacteria, convert ammonium (NH_4^+) to nitrate; and the denitrifying bacteria convert nitrate back to nitrogen gas. Humans contribute to the cycle by using nitrogen gas to produce nitrate for fertilizers.

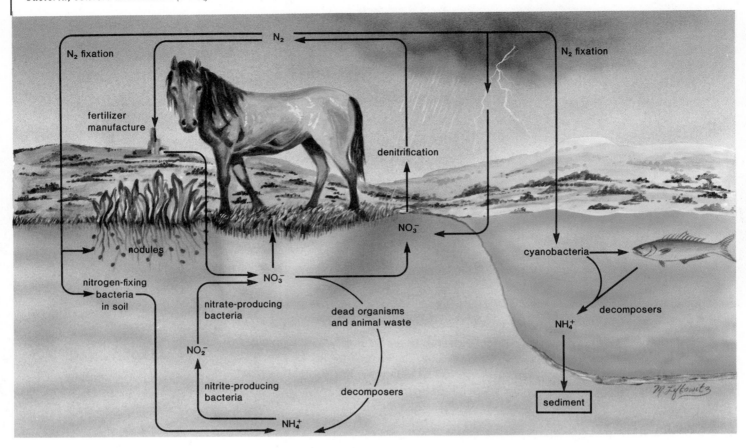

Nitrogen fixation also occurs after plants take up nitrates (NO_3^-) from the soil. Plants use nitrates as their source of nitrogen when they make amino acids and nucleic acids, for example.

Nitrification and Denitrification

Nitrogen gas (N_2) is converted to nitrate (NO_3^-) in the atmosphere when cosmic radiation, meteor trails, and lightning provide the high energy needed for nitrogen to react with oxygen. Also, humans make a most significant contribution to the nitrogen cycle when they convert nitrogen gas to nitrate for use in fertilizers.

Ammonium (NH_4^+) in the soil also is converted to nitrate by bacteria in a process called nitrification. Ammonium gets into the soil in 2 ways. As mentioned previously, some free-living bacteria in the soil reduce nitrogen gas to ammonium. Decomposers also produce ammonium when they decompose waste and dead organic matter. Other types of bacteria carry out nitrification. First, nitrite-producing bacteria convert ammonium to nitrite (NO_2^-), and then nitrate-producing bacteria convert nitrite to nitrate. These 2 groups of bacteria are called the *nitrifying bacteria*. Notice that there is a subcycle in the nitrogen cycle that involves only nitrates, ammonium, and nitrites. This subcycle does not depend on the presence of nitrogen gas at all (fig. 34.12).

Denitrification is the conversion of nitrate to nitrogen gas. There are *denitrifying bacteria* in both aquatic and terrestrial ecosystems. Denitrification counterbalances nitrogen fixation but not completely. There is more nitrogen fixation, especially due to fertilizer production.

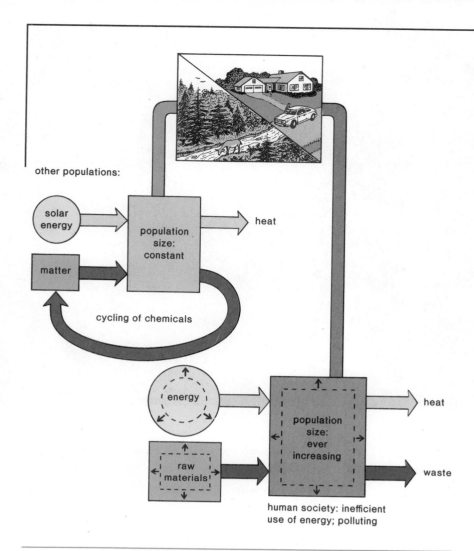

Figure 34.13

Human ecosystem versus a natural ecosystem. In natural ecosystems, population sizes remain about the same year after year and materials cycle and energy is used efficiently. In the human ecosystem, the population size consistently increases, resulting in much pollution because of inadequate cycling of materials and inefficient use of supplemental energy.

other populations:

solar energy

heat

matter

population size: constant

cycling of chemicals

energy

heat

raw materials

population size: ever increasing

waste

human society: inefficient use of energy; polluting

In the nitrogen cycle: nitrogen-fixing bacteria (in nodules and in the soil) reduce nitrogen gas, and thereafter nitrogen can be incorporated into organic compounds; nitrifying bacteria convert ammonium to nitrate; denitrifying bacteria convert nitrate back to nitrogen gas.

The Human Ecosystem

Mature natural ecosystems tend to be stable and to exhibit the characteristics listed in table 34.3. The sizes of the many and varied populations are held in check by the interactions we discussed in chapter 32; the energy that enters and the amount of matter that cycles is appropriate to support these populations. **Pollution,** defined as any undesirable change in the environment that can be harmful to humans and other life, and excessive waste do not occur normally. The human ecosystem that replaces natural ecosystems is quite different, however.

Human beings have replaced natural ecosystems with one of their own making, as is depicted in figure 34.13. This ecosystem essentially has 2 parts: the *country,* where agriculture and animal husbandry are found, and the *city,* where most people live and where industry is carried on. This representation of the human ecosystem, although simplified, allows us to see that the system requires 2 major inputs: *fuel energy* and *raw materials* (e.g., metals, wood, synthetic materials). The use of these necessarily results in *pollution* and *waste* as outputs.

Table 34.3 Ecosystems

Natural	Human
Independent	Dependent
Cyclical (except energy)	Noncyclical
Nonpolluting	Polluting
Renewable solar energy	Nonrenewable fossil fuel energy
Conserves resources	Uses up resources

Figure 34.14

The country. Crops are tended with heavy farming equipment that operates on fossil fuel. High yields are dependent upon a generous supply of fertilizers, pesticides, herbicides, and water.

The Country

Modern U.S. agriculture produces exceptionally high yields per acre, but this bounty is dependent on a combination of the 5 variables given here.

1. **Planting of a few genetic varieties.** The majority of farmers specialize in growing one of these. Wheat farmers plant the same type of wheat, and corn farmers plant the same type of corn (fig. 34.14). This so-called **monoculture agriculture** is subject to attack by a single type of parasite. For example, a single parasitic mold reduced the 1970 corn crop by 15%, and the results could have been much worse because 80% of the nation's corn acreage was susceptible.

2. **Heavy use of fertilizers, pesticides, and herbicides.** *Fertilizer* production requires a large energy input, and fertilizer runoff contributes to water pollution. *Pesticides* reduce soil fertility because they kill off beneficial soil organisms as well as pests, and some pesticides, for example alar, have been accused of increasing the long-term risk of cancer, particularly in children. *Herbicides,* especially those containing the contaminant dioxin, have been charged with causing adverse reproductive effects and cancer.

3. **Generous irrigation.** River waters sometimes are redirected for the purpose of irrigation, in which case "used water" returns to the river carrying a heavy concentration of salt. The salt content of the Rio Grande River in the Southwest is so high that the government has built a treatment plant to remove the salt. Water also is taken sometimes from aquifers (underground rivers), whose water content can be so reduced that it becomes too expensive to pump out more water. Farmers in Texas already are facing this situation.

4. **Excessive fuel consumption.** Energy is consumed on the farm for many purposes. Irrigation pumps already have been mentioned, but large farming machines also are used to spread fertilizers, pesticides, and herbicides and to sow and to harvest the crops. It is not incorrect to suggest that modern farming methods transform fossil fuel energy into food energy.

Behavior and Ecology

Supplemental fossil fuel energy also contributes to animal husbandry yields. At least 50% of all cattle are kept in *feedlots*, where they are fed grain. Chickens are raised in a completely artificial environment, where the climate is controlled and each bird has its own cage to which food is delivered on a conveyor belt. Animals raised under these conditions often have antibiotics and hormones added to their feed to increase yield.

5. **Loss of land quality.** Evaporation of excess water on irrigated lands can result in a residue of salt. This process, termed salinization, makes the land unsuitable for the growth of crops. Between 25 and 35% of the irrigated western croplands are thought to have excessive salinity. Soil erosion is also a serious problem. It is said that we are *mining the soil* because farmers are not taking measures to prevent the loss of topsoil. The Department of Agriculture estimates that erosion is causing a steady drop in the productivity of land equivalent to the loss of 1.25 million acres per year. Even more fertilizers, pesticides, and energy supplements will be required to maintain yield.

Organic Farming

Some farmers have given up this modern means of farming and instead have adopted organic farming methods. This means that they do not use applications of fertilizers, pesticides, or herbicides. They use cultivation of row crops to control weeds, crop rotation to combat major pests, and the growth of legumes to supply nitrogen fertility to the soil. Some farmers use natural predators and parasites instead of pesticides to control insects (fig. 34.15). Most of these farmers switched farming methods because they were concerned about the health of their family and livestock and had found that the chemicals were sometimes ineffective.

A study of about 40 farms showed that organic farming for the most part was just as profitable as conventional farming. Crop yields were lower, but so were operating costs. Organic farms required about two-fifths as much fossil energy to produce one dollar's worth of crop. The method of plowing and the utilization of crop rotation resulted in one-third less soil erosion. The researchers concluded it would be well to determine how far farmers can move in the direction of reduced agricultural chemical use and still maintain the quality of the product. They noted that a modest application of fertilizer would have improved the protein content of the crop.

If biotechnology techniques eventually are able to endow plants with an innate resistance to pests and the ability to fix nitrogen, organic farming most likely will be practiced by most farmers.

The City

The city (fig. 34.16) is dependent on the country to meet its needs. For example, each person in the city requires several acres of land for food production. Overcrowding in cities does not mean that less land is needed; each person still requires a certain amount of land to ensure survival. Unfortunately, however, as the population increases, the suburbs and the cities tend to encroach on agricultural areas and rangeland.

The city houses workers for both commercial businesses and industrial plants. Solar and other renewable types of energy rarely are used; cities currently rely mainly on fossil fuel in the form of oil, gas, electricity, and gasoline. The city does not conserve resources. An office building, with continuously burning lights and windows that cannot be opened, is an example of energy waste. Another example is people who drive cars long distances instead of carpooling or taking public transportation and who drive short distances instead of walking or bicycling. Materials are not recycled, and products are designed for rapid replacement.

Figure 34.15
Biological control of pests. Here, ladybugs are being used to control the cottony-cushion scale insect on citrus trees. If a pesticide is used, the ladybugs will be killed.

Figure 34.16
The city. The city is dependent upon the country to supply it with food and other resources. The larger the city, the more resources required to support it.

The burning of fossil fuels for transportation, commercial needs, and industrial processes causes air and water pollution. This pollution is compounded by the chemical and solid waste pollution that results from the manufacture of many products. Consider that any product used by the average consumer (house, car, washing machine) causes pollution and waste, both during its production and when it is disposed. Humans themselves produce much sewage that is discharged into bodies of water, often after only minimal treatment.

The Solution

Table 34.3 lists the characteristics of the human ecosystem as it now exists. Just as the city is not self-sufficient and depends on the country to supply it with food, so the whole human ecosystem is dependent on the natural ecosystems to provide resources and to absorb waste. Fuel combustion by-products, sewage, fertilizers, pesticides, and solid wastes all are added to natural ecosystems in the hope that these systems will cleanse the biosphere of these pollutants. But we have replaced natural ecosystems with our human ecosystem and have exploited natural ecosystems for resources, adding even more pollutants, to the extent that the remaining natural ecosystems have become overloaded.

Natural ecosystems have been destroyed and overtaxed because the human ecosystem is noncyclical and because an ever-increasing number of people want to maintain a standard of living that requires many goods and services. But we can call a halt to this spiraling process if we achieve zero population growth and if we conserve energy and raw materials. Conservation can be achieved in 3 ways: (1) wise use of only what is actually needed; (2) recycling of nonfuel minerals, such as iron, copper, lead, and aluminum; and (3) use of renewable energy resources and development of more efficient ways to utilize all forms of energy. Figure 34.17*a* presents a diagrammatic representation of what is needed to maintain the delicate balances of the human and natural ecosystems. As a practical example, consider a plant that was built in Lamar, Colorado, which produces methane from feedlot animals' wastes

Behavior and Ecology

Figure 34.17

Figure 34.17

Modified human ecosystem. a. In order to cut down on the amount of lost heat and waste matter, heat could be used more efficiently and discarded materials could be recycled. b. For example, instead of allowing cattle waste to enter a water supply, it could be sent to a conversion plant that produces methane gas. (The remaining residue could be converted into feed for cattle.) Excess heat, which arises from burning of methane gas to produce electricity, could be cycled back to the conversion plant.

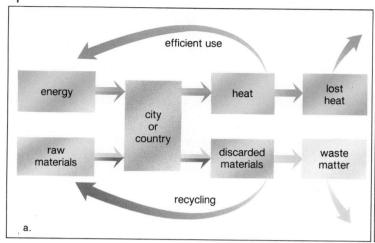

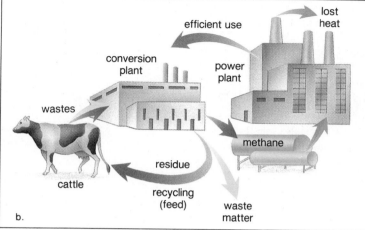

(fig. 34.17b). The methane is burned in the city's electrical power plant, and the heat given off is used to incubate the anaerobic digestion process that produces the methane. In addition, a protein feed supplement is produced from the residue of the digestion process. This system represents a cyclical use of material and an efficient use of energy similar to that found in nature. Many other such processes for achieving this end have been and will be devised. However, as long as the human ecosystem on the whole remains inefficient and noncyclical, it will continue to cause pollution.

Summary

The process of succession from either bare rock or disturbed land results in climax communities. An ecosystem is a community of organisms plus the physical environment. Each population in an ecosystem has a habitat and a niche. Some populations are producers and some are consumers. Consumers can be herbivores, carnivores, omnivores, or decomposers.

Energy flow and chemical cycling are important aspects of ecosystems. Food chains are paths of energy flow through an ecosystem. Grazing food chains always begin with a producer population that is capable of producing organic food, followed by a series of consumer populations. Detritus food chains begin with dead organic matter that is consumed by decomposers. Eventually, all members of a food chain die and decompose. Then the very same chemical elements again are made available to the producer population, although the energy has been dissipated as heat. Therefore, energy does not cycle through an ecosystem.

The food chains form an intricate food web in which there are various trophic (feeding) levels. All producers are on the first level, all primary consumers are on the second level, and so forth. To illustrate that energy does not cycle, it is customary to arrange the various trophic levels in the form of an energy pyramid. A pyramid results because each trophic level contains less energy than the previous level.

Each chemical cycle involves a reservoir, where the element is stored; an exchange pool, from which the populations take and return nutrients; and the populations themselves. In the carbon cycle, the reservoir is organic matter, carbonaceous shells, and limestone. The exchange pool is the atmosphere; photosynthesis removes carbon dioxide (CO_2), and respiration and combustion add carbon dioxide. In the nitrogen cycle, the reservoir is the atmosphere, but nitrogen gas (N_2) must be converted to nitrate (NO_3^-) for use by producers. Nitrogen-fixing bacteria, particularly in root nodules, make organic nitrogen available to plants. Other bacteria active in the nitrogen cycle are the nitrifying bacteria, which convert ammonium to nitrate, and denitrifying bacteria, which convert nitrate to nitrogen gas again.

In mature natural ecosystems, the populations usually remain the same size and need the same amount of energy each year. Additional material inputs are minimal because matter cycles. In the human ecosystem, the population size constantly increases; more energy is needed each year, and additional material inputs also increase. In the country, farmers plant only certain high-yield varieties of plants, which require such supplements as fertilizers, pesticides, and water. In the city, the populace is wasteful of energy and materials. Therefore, there is much pollution. It would be beneficial for us and future generations to find ways to use excess heat and to recycle materials.

Study Questions

1. What is succession, and how does it result in a climax community? (p. 719)
2. Define habitat and niche. (p. 720)
3. Name 4 different types of consumers found in natural ecosystems. (p. 720)
4. What is the difference between a food chain and a food web? Define a trophic level. (p. 722)
5. Give an example of a grazing food chain and a detritus food chain for a terrestrial and for an aquatic ecosystem. (pp. 722–23)
6. Draw an energy pyramid, and explain why such a pyramid can be used to verify that energy does not cycle. (p. 724)
7. What are the reservoir and the exchange pool of a chemical cycle? (p. 725)
8. Describe the carbon cycle. How do humans contribute to this cycle? (pp. 725–26)
9. Describe the nitrogen cycle. How do humans contribute to this cycle? (pp. 727–28)
10. Contrast the characteristics of mature natural ecosystems with those of the human ecosystem. (pp. 729–32)

Objective Questions

1. Chemicals cycle through the populations of an ecosystem, but energy is said to _____ because all of it eventually is dissipated as heat.
2. When organisms die and decay, chemical elements are made available to _____ populations once again.
3. Organisms that feed on plants are called _____ .
4. A pyramid of energy illustrates that there is a loss of energy from one _____ level to the next.
5. There is a loss of energy because one form of energy can never be _____ completely into another form.
6. Forests are a(n) _____ for carbon in the carbon cycle.
7. In the carbon cycle, when organisms _____ , carbon dioxide (CO_2) is returned to the exchange pool.
8. Humans make a significant contribution to the nitrogen cycle when they convert nitrogen gas (N_2) to _____ for use in fertilizers.
9. During the process of denitrification, nitrate is converted to _____ .
10. Natural ecosystems utilize the same amount of energy per year, but the human ecosystem utilizes a(n) _____ .

Label this Diagram.
See figure 34.3 (p. 721) in text.

Answers to Objective Questions
1. flow 2. producer 3. herbivores 4. trophic 5. transformed 6. reservoir 7. respire 8. nitrate 9. nitrogen gas 10. increasing amount

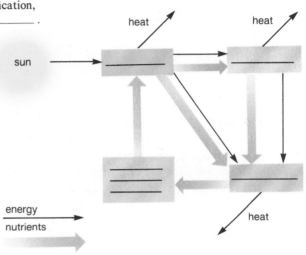

Selected Key Terms

consumer (kon-su′mer) a member of a population that feeds on members of other populations in an ecosystem. *720*

decomposer (de-kom-po′zer) organism of decay (fungi and bacteria) in an ecosystem. *720*

denitrification (de-ni″tri-fi-ka′shun) the process of converting nitrate to nitrogen gas; is a part of the nitrogen cycle. *728*

detritus (de-tri′tus) nonliving organic matter. *720*

ecological pyramid (e″ko-log′i-kal pir′ah-mid) pictorial graph representing the biomass, organism number, or energy content of each trophic level in a food web from the producer to the final consumer populations. *723*

ecology (e-kol′o-je) the study of the interactions of organisms between each other and the physical environment. *720*

ecosystem (ek″o-sis′tem) a biological community, together with the associated abiotic environment. *720*

food chain (fōod chān) a succession of organisms in an ecosystem that are linked by an energy flow and the order of who eats whom. *722*

food web (fōod web) the complete set of food links between populations in a community. *722*

habitat (hab′i-tat) the natural abode of an animal or plant species. *720*

niche (nich) total description of an organism's functional role in an ecosystem, from activities to reproduction. *720*

nitrogen fixation (ni′tro-jen fik-sa′shun) a process whereby nitrogen gas is reduced prior to the incorporation of nitrogen into organic compounds. *727*

pollution (pŏ-lu′shun) detrimental alteration of the normal constituents of air, land, and water due to human activities. *729*

producer (pro-du′ser) organism that produces food and is capable of synthesizing organic compounds from inorganic constituents of the environment; usually the green plants and algae in an ecosystem. *720*

succession (suk-se′shun) a series of ecological stages by which the community in a particular area gradually changes until there is a climax community that can maintain itself. *719*

trophic level (trof′ik lev′el) a categorization of species in a food web according to their feeding relationships from the first level autotrophs through succeeding levels of herbivores and carnivores. *722*

Chapter 35

Human Population Concerns

Chapter Concepts

1 A population undergoing exponential growth has an ever-greater increase in numbers and a shorter doubling time, and it may outstrip the carrying capacity of the environment.

2 The world is divided into the developed countries and the less-developed countries; mainly the less-developed countries presently are undergoing exponential population growth.

3 Human activities cause land, water, and air pollution and threaten the integrity of the biosphere.

4 A sustainable world is possible if economic growth is accompanied by ecological preservation.

Chapter Outline

Exponential Population Growth
    The Growth Rate
    The Doubling Time
    The Carrying Capacity
Human Population Growth
    Developed Countries
    Less-Developed Countries
The Human Population and Pollution
    Land Degradation
    Water Pollution
    Air Pollution
A Sustainable World
    The Steady State

Figure 35.1

Developed countries versus less-developed countries. a. *In the
developed countries, most people enjoy a high standard of living.*
b. *In the less-developed countries, the majority of people are
poor and have few amenities.*

a.

b.

The countries of the world today are divided into 2 groups (fig. 35.1).
The developed countries, typified by countries in North America
and Europe, are those in which population growth is under control
and the people enjoy a good standard of living. The less-developed countries,
typified by countries in Latin America, Africa, and Asia, are those in which
population growth is out of control and the majority of people live in poverty.
(Sometimes the term *third-world countries* is used to mean the less-developed
countries. This term was introduced by those who thought of the United States
and Europe as the first world and the USSR as the second world.)

Before we explore why the world now is divided into the developed and
less-developed countries, it is necessary to study exponential population growth
in general.

Exponential Population Growth

The human growth curve is an *exponential curve* (fig. 35.2). In the beginning,
growth of the human population was relatively slow, but as more reproducing
individuals were added, growth increased, until the curve began to slope steeply
upward. It is apparent from the position of 1990 on the growth curve in figure
35.2 that growth is now quite rapid. The world population increases at least
the equivalent of a medium-sized city every day (200,000), and the combined
populations of the United Kingdom, Norway, Ireland, Iceland, Finland, and
Denmark every year. These startling figures are a reflection of the fact that a
very large world population is undergoing exponential growth.

Mathematically speaking, **exponential growth,** or geometric increase,
occurs in the same manner as compound interest; that is, the percentage increase is added to the principal before the next increase is calculated. Referring specifically to populations, consider the hypothetical population sizes in
table 35.1. This table illustrates the circumstances of world population growth

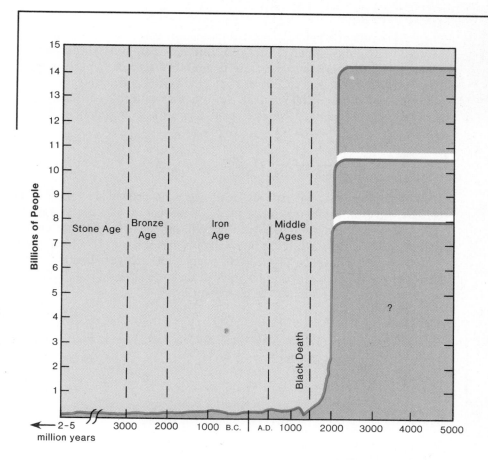

Figure 35.2
Growth curve for human population. The human population now is undergoing rapid exponential growth. Since the growth rate is declining, it is predicted that the population size will level off at 8, 10.5, or 14.2 billion, depending upon the speed with which the growth rate declines.

Table 35.1 *Exponential Growth of Hypothetical Populations*

Population Size	Increase (%)	Actual Increase in Numbers	Population Size	Increase (%)	Actual Increase in Numbers	Population Size
500,000,000	2.00	10,000,000	510,000,000	1.99	10,149,000	520,149,000
3,000,000,000	2.00	60,000,000	3,060,000,000	1.99	60,894,000	3,120,894,000
5,000,000,000	2.00	100,000,000	5,100,000,000	1.99	101,490,000	5,201,490,000

at the moment: the percentage increase has decreased and yet the size of the population grows by a greater amount each year. The increase in size is dramatically large because the world population is very large.

In our hypothetical examples (table 35.1), an initial increase of 2% added to the original population size followed by a 1.99% increase results in the third-generation size listed in the last column. Notice that

1. in each instance, the second generation has a larger increase than the first generation because the second generation's population is larger than the first;
2. because of exponential growth, the lower percentage increase (i.e., 1.99% compared to 2%) still brings about larger population growth;
3. the larger the population, the larger the increase for each generation.

The percentage increase is termed the **growth rate,** which is calculated per year.

The Growth Rate

The growth rate of a population is determined by considering the difference between the number of persons born (birthrate, or natality) and the number of persons who die per year (death rate, or mortality). It is customary to record these rates per 1,000 persons. For example, the USSR at the present time has a birthrate of 20 per 1,000 per year, but it has a death rate of 10 per 1,000 per year. This means that Russia's population growth, or simply its growth rate, is

$$\frac{20 - 10}{1,000} = \frac{10}{1,000} = \frac{1.0}{100} = 1.0\%$$

Notice that while birthrate and death rate are expressed in terms of 1,000 persons, the growth rate is expressed per 100 persons, or as a percentage.

After 1750, the world population growth rate steadily increased until it peaked at 2% in 1965, but it has fallen slightly since then to 1.7%. Yet, there is an ever-greater increase in the world population each year because of exponential growth. The explosive potential of the present world population can be appreciated by considering the doubling time.

The Doubling Time

Table 35.2 shows that the **doubling time** (d) for a population can be calculated by dividing 70 by the growth rate (gr):

$$d = \frac{70}{gr}$$

d = Doubling time
gr = Growth rate
70 = Demographic constant

If the present world growth rate of 1.8% continues, the world population will double in 39 years.

$$d = \frac{70}{1.7} = 39 \text{ years}$$

This means that in 39 years, the world would need double the amount of food, jobs, water, energy, and so on to maintain the same standard of living.

It is of grave concern to many individuals that the amount of time needed to add each additional billion persons to the world population has taken less and less time (table 35.3). The world reached its first billion around 1800—some 2 million years after the evolution of humans. Adding the second billion took about 130 years, the third billion took about 30 years, and the fourth billion took only about 15 years. However, if the growth rate continues to decline, this trend would reverse itself, and eventually there would be zero population growth. Then population size would remain steady. Therefore, figure

Table 35.2 *Relationship between Growth Rate and the Doubling Time of a Population*

Growth Rate (%)	Doubling Time (years)
0.25	280
0.50	140
1.00	70
2.00	35
3.00	23

Table 35.3 *World Population Increase*

Billions of People	Time Needed (years)	Year of Increase
First	2–5 million	1800
Second	130	1930
Third	30	1960
Fourth	15	1975
Fifth	12	1987
Sixth (projected)	11	1998

Source: Elaine M. Murphy, *World Population: Toward the Next Century*, (Washington, DC: Population Reference Bureau, November 1981), page 3.

Behavior and Ecology

35.2 shows 3 possible logistic curves: the population may level off at 8, 10.5, or 14.2 billion, depending on the speed with which the growth rate declines.

The Carrying Capacity

Examining the growth curves for nonhuman populations reveals that the populations tend to level off at a certain size. For example, figure 35.3 gives the actual data for the growth of a fruit fly population reared in a culture bottle. At the beginning, the fruit flies were adjusting to their new environment and growth was slow. Then, because food and space were plentiful, they began to multiply rapidly. Notice that the curve begins to rise dramatically just as the human population curve does now. At this time, it can be said that the population is demonstrating its **biotic potential.** Biotic potential is the maximum growth rate under ideal conditions. Biotic potential usually is not demonstrated for long because of an opposing force called **environmental resistance.** Environmental resistance includes all the factors that cause early death of organisms and therefore prevents the population from producing as many offspring as it might otherwise do. As far as the fruit flies are concerned, we can speculate that environmental resistance included the limiting factors of food and space. The waste given off by the fruit flies also may have contributed to keeping the population size down. When environmental resistance sets in, biotic potential no longer is possible, and the slope of the growth curve begins to decline. This is called the inflection point of the curve.

The eventual size of any population represents a compromise between the biotic potential and the environmental resistance. This compromise occurs

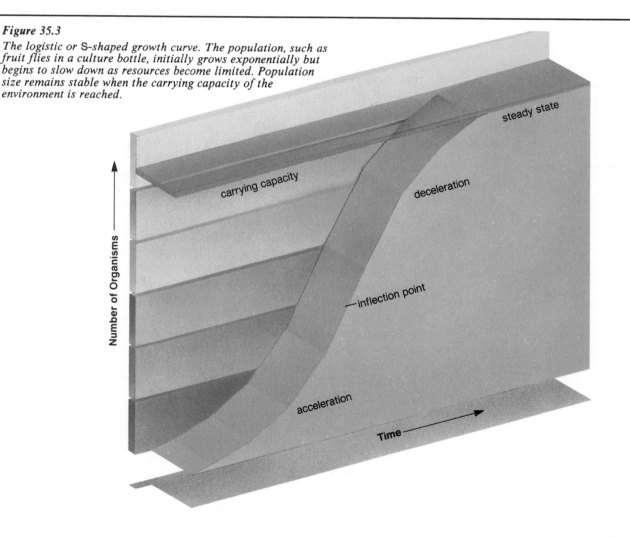

Figure 35.3

The logistic or S-shaped growth curve. The population, such as fruit flies in a culture bottle, initially grows exponentially but begins to slow down as resources become limited. Population size remains stable when the carrying capacity of the environment is reached.

at the **carrying capacity** of the environment. The carrying capacity is the maximum population that the environment can support—for an indefinite period.

The carrying capacity of the earth for humans is not certain. Some authorities think the earth potentially is capable of supporting 50–100 billion people. Others think we already have more humans than the earth can support adequately.

Human Population Growth

The human population has undergone 3 periods of exponential growth. *Toolmaking* may have been the first technological advance that allowed the human population to enter a period of exponential growth. *Cultivation of plants* and *animal husbandry* may have allowed a second period of growth. The *Industrial Revolution,* which occurred about 1850, promoted the third phase. At first, only certain countries of the world became industrialized, and these countries now are called the developed countries.

Developed Countries

The Industrial Revolution, which also was accompanied by a medical revolution, took place in the Western world. In addition to European and North American countries, Russia and Japan also became industrialized. Collectively, these countries often are referred to as the **developed countries.** The developed countries doubled their size between 1850 and 1950 (fig. 35.4), largely due to a decline in the death rate. This decline is attributed to the influence of modern medicine and improved socioeconomic conditions. Industrialization raised personal incomes, and better housing permitted improved hygiene and sanitation. Numerous infectious diseases, such as cholera, typhus, and diphtheria, were brought under control.

The decline in the death rate in the developed countries was followed shortly by a decline in the birthrate. Between 1950 and 1975, populations in the developed countries showed only modest growth (fig. 35.4) because the growth rate fell from an average of 1.1% in 1950 to 0.8% in 1975.

The Demographic Transition

Overall, the growth rate in developed countries has gone through 3 phases (table 35.4). In phase I, prior to 1850, the growth rate was low because a high death rate canceled out the effects of a high birthrate; in phase II, the growth rate was high because of a lowered death rate; and in phase III, the growth rate was again low because the birthrate had declined. These phases now are known as the **demographic transition.** In seeking a reason for the transition, it has been suggested that as industrialization occurred, the population became concentrated in the cities. Urbanization may have contributed to the decline in the growth rate because in the city, children were no longer the boon they were in the country. Instead of contributing to the yearly income of the family, they represented a severe drain on its resources. It also could be that urban living made people acutely aware of the problems of crowding, and for this reason, the birthrate declined. Also, some investigators believe that there was a direct relationship between improvement in socioeconomic conditions and the birthrate. They point out that as the developed nations became wealthier, as infant mortality was reduced, and as educational levels increased, the birthrate declined.

Regardless of the reasons for the demographic transition, it caused the rate of growth to decline in the developed countries. The growth rate for the developed countries is now about 0.6%, and their overall population size is about one-third that of the less-developed countries. A few developed countries—Austria, Denmark, East Germany, Hungary, Sweden, West Germany—are not growing or actually are losing population.

Behavior and Ecology

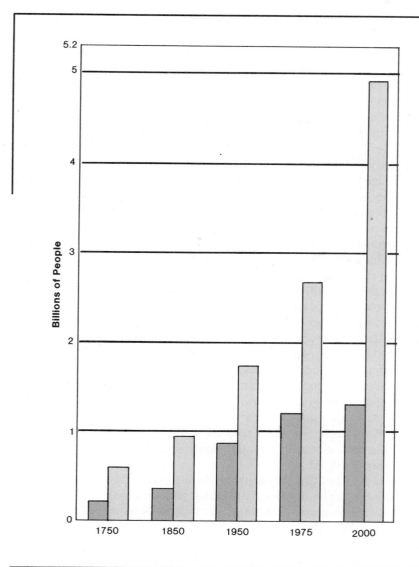

Figure 35.4

Size of human population in developed versus less-developed countries. The population of the developed countries (blue) increased between 1850 and 1950, but the size is expected to increase little between 1975 and 2000. In contrast, the population size of the less-developed countries (yellow) increased in the past and is expected to increase dramatically in the future also.

From "The Populations of the Underdeveloped Countries" by Paul Demeny. Copyright © 1974 by Scientific American, Inc. All rights reserved.

Table 35.4 *Analysis of Annual Growth Rates in Developed Countries*

Phase	Birthrate	Death Rate	Annual Growth Rate
I	High	High	Low
II	High	Low	High
III	Low	Low	Low

At this time, the United States has a population of over 249 million and still is growing. It is projected that the population could level off at about 300 million around 2080 and then start to decline. This projection assumes that at that time, the average number of children per woman will be 1.8, life expectancy will be age 81, and immigration will have declined to 500,000 newcomers a year. If immigration was not occurring, the U.S. population would decline to 220 million by 2080.

Less-Developed Countries

Countries such as those in Latin America, Asia, and Africa collectively are called the **less-developed countries** because they either are nonindustrialized or newly industrialized. (Some researchers are in favor of placing the nonindustrialized countries in a third category, called the least-developed countries.) The mortality rate of the less-developed countries began to decline steeply following World War II. This decline was prompted not by socioeconomic development, but by the importation of modern medicine from the developed countries. Various illnesses were brought under control due to immunization, use of antibiotics, improved sanitation, and use of insecticides. Although the death rate declined, the birthrate did not decline to the same extent; therefore, the populations of the less-developed countries began—and still are today—increasing dramatically (fig. 35.4). The less-developed countries were unable to cope adequately with such rapid population expansion. Today, many people in these countries are underfed, ill housed, unschooled, and living in abject poverty. Many of these poor have fled to the cities, where they live in makeshift shanties on the outskirts.

The growth rate of the less-developed countries as a whole finally peaked at 2.4% during 1960–1965. Since that time, the decrease in the death rate has slowed, and the birthrate is falling slowly. It is hoped that the overall growth rate will decline to 1.8% by the end of the century. At that time, about two-thirds of the world population will be in the less-developed countries.

Some investigators believe that the demographic transition occurs when a less-developed country begins to enjoy the benefits of economic development. Yet, during the 1980s, population outstripped economic growth in most less-developed countries, and two-thirds experienced a fall in per capita income. The least-developed countries, located largely in Africa, do not show any evidence yet of the demographic transition, and in these countries, the environmental resources are being depleted. Just 40 years ago, Ethiopia, for example, had a 30% forest cover; 12 years ago, it was down to 4%; and today, it may be 1%. Under such circumstances, the economic growth needed to check the birthrate may not occur.

Others point out that the countries with the greatest decline in the birthrate are those with the best family-planning programs. From this, it can be argued that such programs indeed can help to bring about a stable population size in the less-developed countries. Nevertheless, it has been found that certain socioeconomic factors also have contributed to a decline in the developing countries' growth rate. A relatively high gross national product (GNP), urbanization, low infant mortality, increased life expectancy, literacy, and education all had a dampening effect on the growth rate.

Age Structure Comparison

Laypeople are sometimes under the impression that if each couple had 2 children, zero population growth would take place immediately. However, **replacement reproduction,** as it is called, still would cause most countries today to continue growth due to the age structure of the population. If more young women are entering the reproductive years than there are older women leaving them behind, then replacement reproduction will give a positive growth rate.

Reproduction is at or below replacement level in some 20 developed countries, including the United States. Even so, some of these countries will continue to grow modestly, in part because there was a baby boom after World War II. Young women born in these baby boom years are now in their reproductive years, and even if each one has less than 2 children, the population still will grow. Also keep in mind that even the smallest of growth rates can add a considerable number of individuals to a large country. For example, a growth rate of 0.7% added about 1.5 million people to the U.S. population in 1990.

Many developed countries have a stabilized age structure diagram (fig. 35.5), but most less-developed countries have a youthful profile—a large proportion of the population is below the age of 15. Since there are so many young women entering the reproductive years, the population still will expand greatly

Behavior and Ecology

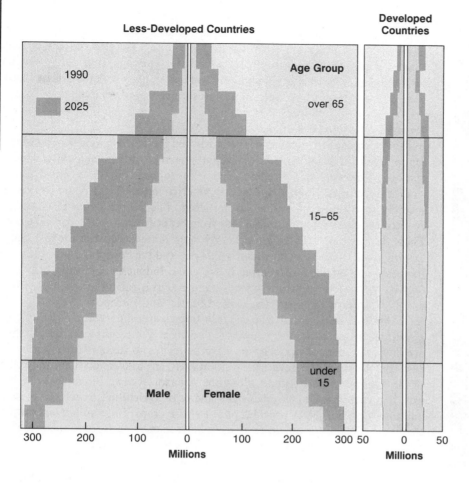

Less-Developed Countries

1990

2025

Age Group

over 65

15–65

under 15

Male   Female

300 200 100 0 100 200 300
Millions

Developed Countries

50 0 50
Millions

Figure 35.5

Age structure diagram for less-developed and developed countries. The less-developed countries contain 75% of the world population versus 25% in the developed countries. The less-developed countries will continue to expand for many years because of their youthful profile, but the developed countries are approaching stabilization.

From "The Growing Human Population" by Nathan Keyfitz. Copyright © 1989 by Scientific American, Inc. All rights reserved.

even after replacement reproduction is attained. The more quickly replacement reproduction is achieved, however, the sooner zero population growth will result.

The Human Population and Pollution

As the human population increases in size, more energy and materials are consumed. Because the human population does not use energy efficiently and does not recycle materials (see fig. 34.17), pollutants are added to all parts of the biosphere—land, water, and air.

Land Degradation

The land has been degraded in many ways. Here, we will discuss only those of the greatest concern.

Soil Erosion and Desertification

Soil erosion causes the productivity of agricultural lands to decline. It occurs when wind blows and rain washes away the topsoil to leave the land exposed without adequate cover. The U.S. Department of Agriculture estimates that erosion is causing a steady drop in the productivity of farmland equivalent to the loss of 1.25 million acres per year. To maintain the productivity of eroding land, more fertilizers, more pesticides, and more energy must be used.

One answer to the problem of erosion is to adopt soil conservation measures. For example, farmers could use strip-cropping and contour farming (fig. 35.6).

Figure 35.6
Contour farming. Crops are planted according to the lay of the land to reduce soil erosion. This farmer has planted alfalfa in between the strips of corn to replenish the nitrogen content of the soil. Alfalfa, a legume, has root nodules that contain nitrogen-fixing bacteria.

Desertification leads to the transformation of marginal lands to desert conditions. Desertification has been particularly evident along the southern edge of the Sahara Desert in Africa, where it is estimated that 350,000 square miles of once-productive grazing land has become desert in the last 50 years. However, desertification also occurs in this country. The U.S. Bureau of Land Management, which opens up federal lands for grazing, reports that much of the rangeland it manages is in poor or bad condition, with much of its topsoil gone and with greatly reduced ability to support forage plants.

Tropical Rain Forest Destruction

In developed countries, much of the hardwood forests long ago were converted to rapidly growing softwood forest plantations to provide humans with a source of wood. However, there are still virgin rain forests (fig. 35.7a) in Southeast Asia and Oceania, Central and South America, and Africa. These forests are severely threatened by human exploitation. The people living in developed countries want all sorts of things made from beautiful and costly tropical woods. These desires have created a market for such lumber. Most of the loss due to logging is in Southeast Asia, and in Malaysia and the Philippines, for example, the best commercial forests already are gone. Indonesia is cutting heavily to feed its new wood-exporting business, and Brazil is racing to catch up. Much of this wood goes to Japan, the United States, and Europe.

Another reason that tropical rain forests are undergoing destruction is a result of the needs of the people who live there. For example, in Brazil, there are large numbers of persons who have no means to support their family. To ease social unrest, the government allows citizens to own any land they clear in the Amazon Forest (occurs along the Amazon River). In tropical rain forests, it is customary to practice **slash-and-burn agriculture,** in which trees are cut down and burned to provide space to raise crops. Unfortunately, however, the land is fertile for only a few years. In tropical rain forests, the inorganic nutrients immediately cycle back to producers and do not accumulate in the soil. Therefore, despite the massive amount of growth above ground, the soil itself is nutrient-poor. Once the cleared land is unable to sustain crops, the farmer moves on to another part of the rain forest to slash and burn again.

Cattle ranchers are the greatest beneficiaries of deforestation, and increased ranching is therefore another reason for tropical rain forest destruction. The ranchers in Brazil are so aggressive that they actually force colonists to sell them newly cleared land at gunpoint. The cattle that are raised provide beef for export, and much of it is bought by U.S. fast-food chains. Because the imported beef is cheaper, hamburgers can be sold for 5 cents less than if the beef was purchased in the United States. The destruction of tropical forests on this account has been termed the "hamburger connection."

A newly begun pig-iron industry also indirectly results in further exploitation of the rain forest. The pig iron must be processed before it is exported, and smelting the pig iron requires the use of charcoal. The largest pig-iron company acknowledges having paid for construction of 1,500 small makeshift ovens used by peasants who burn trees from the rain forests to produce the charcoal.

There are currently 3 primary reasons for tropical rain forest destruction: (1) logging to provide hardwoods for export, (2) slash-and-burn agriculture, and (3) cattle ranching. Industrialization in countries having extensive tropical rain forests no doubt will become an important future reason also.

The Loss of Biological Diversity   As is discussed in the reading on page 708, tropical rain forests are much more biologically diverse than temperate forests. For example, temperate forests across the entire United States contain about 400 tree species. In the rain forest, a typical four-square-mile area holds

Behavior and Ecology

a.

Figure 35.7

Tropical rain forest sustainability. a. In its natural state, a tropical rain forest is immensely rich in vegetation and breathtakingly beautiful. b. Slash-and-burn agriculture is the first step toward destruction of a portion of the rain forest. With this method, agriculture cannot be sustained for more than a few years because the soil is infertile and does not hold moisture. c. If the trees are not cut down, rubber tappers can earn a sustainable living by tapping the same trees year after year.

c.

b.

as many as 750 types of trees. The fresh waters of South America are inhabited by an estimated 5,000 fish species; on the eastern slopes of the Andes, there are 80 or more species of frogs and toads; and in Ecuador, there are more than 1,200 species of birds—roughly twice as many as those inhabiting all of the United States and Canada. Therefore, a very serious side effect of deforestation in tropical countries is the loss of biological diversity.

Altogether, about half of the world's species are believed to live in tropical forests. A National Academy of Sciences study estimated that a million species of plants and animals are in danger of disappearing within 20 years as a result of deforestation in tropical countries. Many of these life forms never have been studied, and yet many possibly could be useful. At present, our entire domesticated crop production around the world relies on fewer than 30 species of plants and animals that have been domesticated during the last 10,000 years! It is quite possible that many additional species of wild plants and animals now living in tropical forests could be domesticated. As matters now stand, the clearing of tropical forests very likely will prevent humans from ever having the opportunity to utilize more than a tiny fraction of the earth's biological diversity.

While it may seem like an either-or situation—either biological diversity or human survival—there is a growing recognition that this is not the case. Natives who harvest rubber from the trees (fig. 35.7c) can have a sustainable income from the same trees year after year. A recent study calculated the market value of rubber and exotic produce, like the aguaje palm fruit, that can be harvested continually from the Amazon forest. It concluded that selling these products would yield more than twice the income of either lumbering or cattle ranching.

There is much worldwide concern about the loss of biological diversity due to the destruction of tropical rain forests. The myriad of plants and animals that live there possibly could benefit human beings.

Waste Disposal

Every year, the U.S. population discards billions of metric tons of solid wastes, much of it on land. Solid wastes include not only household trash (fig. 35.8) but also sewage sludge, agricultural residues, mining refuse, and industrial wastes. Some of these solid wastes contain substances that cause human illness and sometimes even death; they are called **hazardous wastes.**

Hazardous wastes, such as heavy metals, chlorinated hydrocarbons, or organochlorides, and nuclear wastes, are subject to biological magnification (fig. 35.9). Decomposers are unable to break down these wastes, and further they remain in the body and are not excreted. Therefore, they become more concentrated as they pass along a food chain. Notice in figure 35.9 that the number of dots representing DDT become more concentrated as it passes from producer to tertiary consumer. Biological magnification is most apt to occur in aquatic food chains; there are more links in aquatic food chains than there are in terrestrial food chains. Humans are the final consumers in both types of food chains, and in some areas, human milk contains detectable amounts of DDT and PCBs, which are organochlorides.

The dumping of hazardous wastes directly endangers public health. Chemical wastes buried over a quarter of a century ago in Love Canal, near Niagara Falls, have seriously damaged the health of some residents there. Similarly, the town of Times Beach, Missouri, is abandoned because workers spread an organochloride (dioxin)-laced oil on the city streets, leading to a myriad of illnesses among its citizens. In other places, such as in Holbrook, Massachusetts, manufacturers have left thousands of drums in abandoned or

Figure 35.8

Dump sites. These not only cause land and air pollution, they also allow chemicals to enter groundwater, which later may become part of human drinking water.

Behavior and Ecology

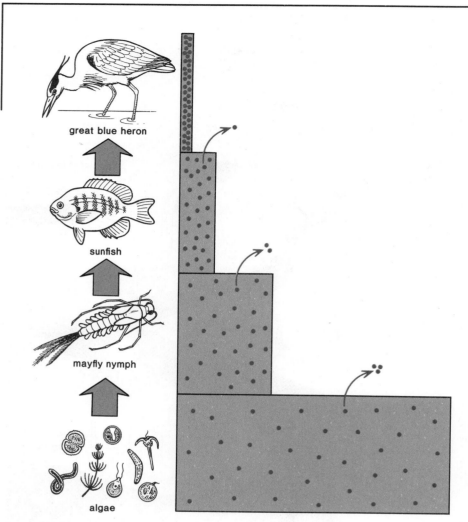

Figure 35.9
Biological magnification. A poison (dots), such as DDT, that is minimally excreted (arrows) becomes maximally concentrated as it passes along a food chain due to the reduced size of the trophic levels.

great blue heron

sunfish

mayfly nymph

algae

uncontrolled sites, where toxic chemicals are oozing out into the ground and are contaminating the water supply. Illnesses, especially forms of cancer, are quite common not only in Holbrook but also in adjoining towns.

Water Pollution

Pollution of surface water, groundwater, and the oceans is of major concern today.

Surface Water Pollution

All sorts of pollutants from various sources enter surface waters, as depicted in figure 35.10. Sewage treatment plants can be built to help degrade organic wastes, which otherwise can cause oxygen depletion in lakes and rivers. As the oxygen level decreases, the diversity of life is greatly reduced. Also, human feces can contain pathogenic microorganisms that cause cholera, typhoid fever, and dysentery. In less-developed countries, where the population is growing and where waste treatment is practically nonexistent, many children die each year from these diseases.

Typically, sewage treatment plants use bacteria to break down organic matter to inorganic nutrients, like nitrates and phosphates, which then enter surface waters. These types of nutrients, which also can enter waters by fertilizer runoff and soil erosion, lead to **cultural eutrophication,** an acceleration of the natural process by which bodies of water fill in and disappear. First, the

Figure 35.10

Sources of water pollution. Many bodies of water are dying due to the introduction of sediments and surplus nutrients.

Source: Adapted from U.S. Environmental Protection Agency, Office of Water Supply and Solid Waste Management Programs, "Waste Disposal Practices and Their Effects on Ground Water" *Executive Summary.* (Washington, DC, U.S. Government Printing Office, 1977).

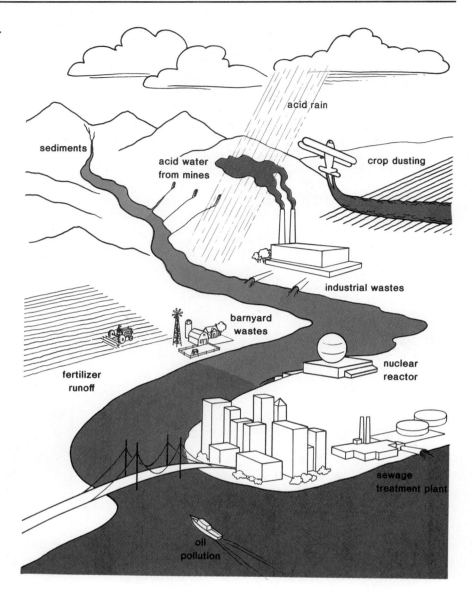

nutrients cause overgrowth of algae. Then, when the algae die, oxygen is used up by the decomposers, and the water's capacity to support life is reduced. Massive fish kills are sometimes the result of cultural eutrophication.

Industrial wastes can include heavy metals and organochlorides, such as pesticides. These materials are not degraded readily under natural conditions nor in conventional sewage treatment plants. Sometimes, they accumulate in the bottom mud of deltas, and estuaries of highly polluted rivers cause environmental problems if they are disturbed. Industrial pollution is being addressed in many industrialized countries but usually has low priority in less-developed countries.

Some pollutants enter bodies of water from the atmosphere. Acid deposition has caused many lakes to become sterile in the industrialized world because acid leaches aluminum and iron out of the soil. A high concentration of these ions kills fishes and other forms of aquatic life. Adding lime is sometimes helpful against acidification of a lake.

Groundwater Pollution

Figure 35.11 shows the ways—both intentionally and unintentionally—that pollutants can reach underground rivers called aquifers. In areas of intensive

Behavior and Ecology

Figure 35.11

Sources of groundwater pollution. Discontinuance of these means of industrial waste disposal has been difficult to achieve because citizens do not want to have waste disposal plants located near them.

Source: Adapted from U.S. Environmental Protection Agency, Office of Water Supply and Solid Waste Management Programs, "Waste Disposal Practices and Their Effects on Ground Water" *Executive Summary.* (Washington, DC, U.S. Government Printing Office, 1977).

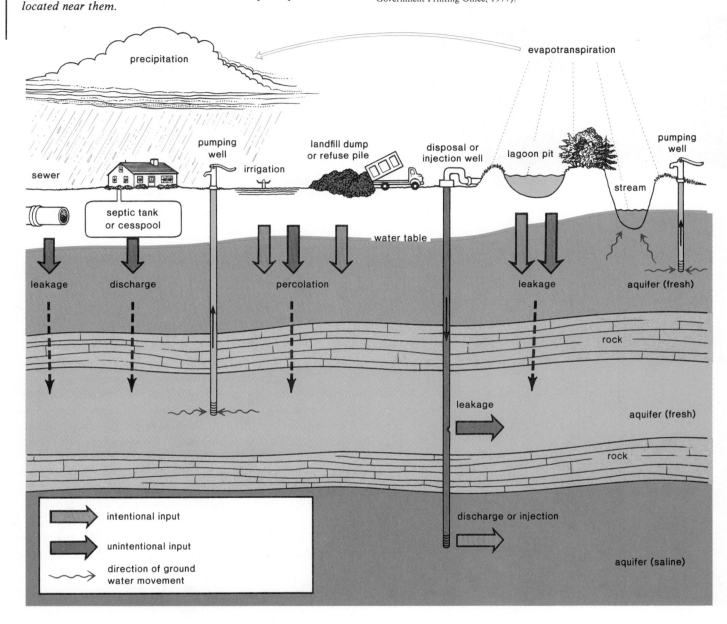

animal farming or where there are many septic tanks, ammonium (NH_4^+) released from animal and human waste is converted by soil bacteria to soluble nitrate that moves down through the soil (percolates) into underground water supplies. Between 5 and 10% of all wells examined in the United States have nitrate levels higher than the recommended maximum.

Industry also pollutes aquifers. Previously, industry was accustomed to running wastewater into a pit. The pollutants then could seep into the ground. Wastewater and chemical wastes also have been injected into deep wells from which the pollutants constantly discharge. Both of these customs have been or are in the process of being phased out. However, it is very difficult for industry to find other ways to dispose of wastes. More adequately managed and controlled waste treatment plants are needed, but because citizens do not wish to live near such plants, towns often are successful in preventing their construction.

Figure 35.12

Oceanic Pollution

Coastal regions are not only the immediate receptors for local pollutants, they are also the final receptors for pollutants carried by rivers that empty at the coast. Waste dumping also occurs at sea, and ocean currents sometimes transport both trash and pollutants back to shore. Examples are the nonbiodegradable plastic bottles, pellets, and containers that now commonly litter beaches and the oceans' surfaces (fig. 35.12). Some of these, such as the plastic that holds a six-pack of beer, cause the death of birds, fishes, and marine mammals that mistake them for food and get entangled in them.

Offshore mining and shipping add pollutants to the oceans. Some 5 million metric tons of oil a year, or more than one gram per 100 square meters of the oceans' surfaces, end up in the oceans. Large oil spills kill plankton, fish larvae, and shellfishes, as well as birds and marine mammals. The largest spill may have occurred on March 24, 1989, when the tanker *Exxon Valdez* struck a reef in Alaska's Prince William Sound and leaked 44 million liters of crude oil. Although petroleum is biodegradable, the process takes a long time because the low-nutrient content of seawater does not support a large bacterial population. Once the oil washes up onto beaches, it takes many hours of work and millions of dollars to clean it up.

Adequate sewage treatment and waste disposal are necessary to prevent the pollution of rivers and oceans. New methods also are needed to prevent pollution of underground water supplies.

Air Pollution

The atmosphere has 2 layers, the stratosphere and the troposphere. The stratosphere is a layer that lies 15–50 km above the surface of the earth. Here, the energy of the sun splits oxygen (O_2) molecules. These individual oxygen (O) atoms then combine with molecular oxygen to give ozone (O_3). This ozone layer is called a shield because it absorbs the ultraviolet rays of the sun, preventing them from striking the earth. If these rays did penetrate the atmosphere, life on earth would not be possible because living things cannot tolerate heavy doses of ultraviolet radiation. The troposphere is the atmospheric layer closest to the earth's surface; it ordinarily contains the gases nitrogen (N_2)—78%, oxygen (O_2)—21%, and carbon dioxide (CO_2)—0.3%.

Four major concerns—photochemical smog, acid deposition, the greenhouse effect, and the destruction of the ozone shield—are associated with the air pollutants listed in figure 35.13. You can see that fossil fuel burning and vehicle exhaust are primary sources of gases associated with air pollution. These 2 sources are related because gasoline is derived from petroleum, a fossil fuel.

Photochemical Smog

Photochemical smog contains 2 air pollutants—nitrogen oxides (NO_x) and hydrocarbons (HC)—that react with one another in the presence of sunlight to produce ozone (O_3) and PAN (peroxylacetyl nitrate). Both nitrogen oxides and hydrocarbons come from fossil fuel combustion, but additional hydrocarbons come from various other sources as well, including industrial solvents.

Ozone and PAN commonly are referred to as oxidants. Breathing ozone affects the respiratory and nervous systems, resulting in respiratory distress, headache, and exhaustion. These symptoms are particularly apt to appear in young people; therefore, in Los Angeles, where ozone levels are often high, schoolchildren must remain inside the school building whenever the ozone level reaches 0.35 ppm (parts per million by weight). Ozone is damaging especially to plants, resulting in leaf mottling and reduced growth (fig. 35.14).

Normally, warm air near the ground is able to escape into the atmosphere. Sometimes, however, air pollutants, including smog and soot, are trapped near the earth due to a long-lasting thermal inversion. During a **thermal**

Behavior and Ecology

Figure 35.13

Air pollutants. These are the gases, along with their sources, that contribute to 4 environmental effects of major concern: photochemical smog, acid deposition, the greenhouse effect, and the destruction of the ozone shield. An examination of the sources of these gases shows that vehicle exhaust and fossil fuel burning are the chief contributors.

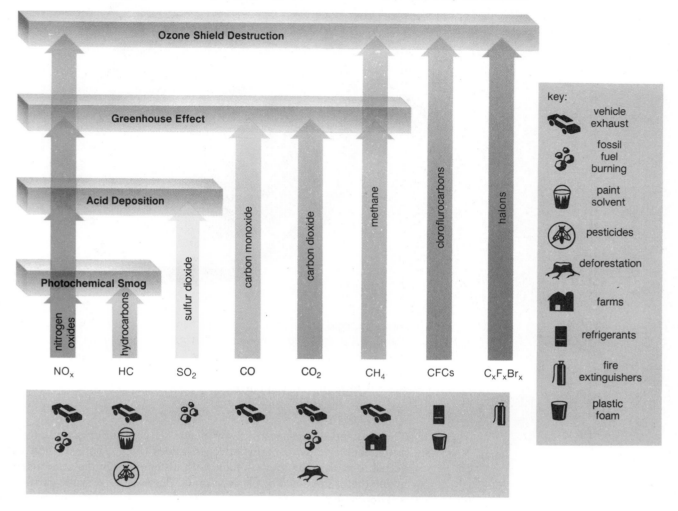

Ozone Shield Destruction

Greenhouse Effect

Acid Deposition

Photochemical Smog

nitrogen oxides — NO$_x$

hydrocarbons — HC

sulfur dioxide — SO$_2$

carbon monoxide — CO

carbon dioxide — CO$_2$

methane — CH$_4$

clorofluorocarbons — CFCs

halons — C$_x$F$_x$Br$_x$

key:
vehicle exhaust
fossil fuel burning
paint solvent
pesticides
deforestation
farms
refrigerants
fire extinguishers
plastic foam

Figure 35.14

Effect of ozone on plants. The milkweed in (a) was exposed to ozone and appears unhealthy; the milkweed in (b) was grown in an enclosure with filtered air and appears healthy.

a.

b.

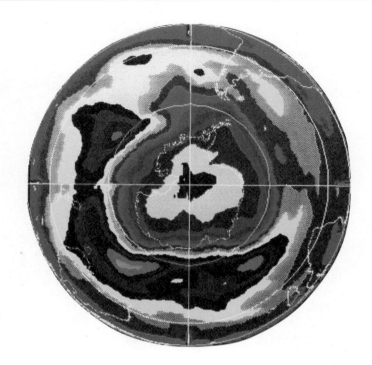

Figure 35.15

Ozone depletion. This has been confirmed by observations of a drastic decrease in ozone at the South Pole of Antarctica each spring during the late 1970s and the 1980s. The so-called ozone hole appears as a large white area at the center of this picture, taken by the total ozone mapping spectrometer aboard NASA's Nimbus 7 satellite on 5 October 1987, when the ozone loss reached nearly 60%.

inversion, cold air is at ground level beneath a layer of warm, stagnant air above. This often occurs at sunset, but turbulence usually mixes these layers during the day. Some areas surrounded by hills are particularly susceptible to the effects of a temperature inversion because the air tends to stagnate and there is little turbulent mixing.

Acid Deposition

The coal and oil burned by power plants releases sulfur dioxide (SO_2), and automobile exhaust contains nitrogen oxides (NO_x); both of these are converted to acids when they combine with water vapor in the atmosphere, a reaction that is promoted by ozone in smog. These acids return to earth as either wet deposition (acid rain or snow) or dry deposition (sulfate and nitrate salts).

As discussed earlier in this text in the reading on page 26, **acid deposition** now is associated with dead or dying lakes and forests, particularly in North America and Europe. Acid deposition also corrodes marble, metal, and stonework, an effect that is noticeable in cities. It also can degrade our water supply by leaching heavy metals from the soil into drinking-water supplies. Similarly, acid water dissolves copper from pipes and from lead solder that is used to join pipes.

Ozone Shield Destruction

Chlorofluorocarbons, or CFCs, of which the most important is Freon, are heat-transfer agents used in refrigerators and air conditioners. They also are used as foaming agents in such products as styrofoam cups and egg cartons. Formerly, they were used as propellants in spray cans, but this application now is banned in the United States.

It was known that CFCs would drift up into the stratosphere, but it was believed that they would be nonreactive there during their 150-year life span. However, it now is apparent that when the temperature drops, these compounds react chemically on frozen particle surfaces within stratospheric clouds and release active chlorine that then can react with the ozone in the **ozone shield.** Once freed, a single atom of chlorine destroys about 100,000 molecules of ozone before finally settling to the earth's surface as chloride years later. Measurements suggest that 3% and perhaps up to 5% of the global ozone layer

Figure 35.16

Greenhouse effect. a. The greenhouse effect is caused by the accumulation in the atmosphere of certain gases, such as carbon dioxide (CO₂), that allow the rays of the sun to pass through but absorb and reradiate heat back to the earth. b. The greenhouse gases. This graph shows the fraction of warming caused by the gases carbon dioxide, CFCs, methane, and nitrous oxide for the *decades from 1950 to 2020. There was no accumulation of CFCs in the 1950s because they were not being manufactured to any degree as yet. By 2020, carbon dioxide and all of the other gases taken together each will contribute about 50% to the projected global warming.*

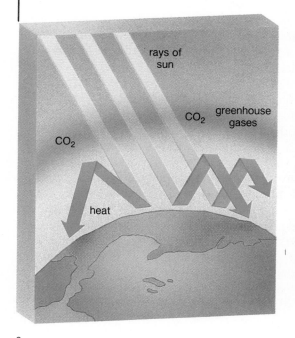

a.

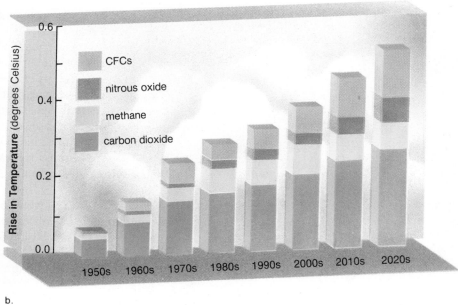

b.

already has been destroyed by CFCs. A more drastic effect has been found in the Antarctic and the Arctic. At the South Pole, up to 50% of the ozone is destroyed each spring over an area the size of North America (fig. 35.15). There is a similar reduction, called an ozone hole, but not as severe, over the Arctic as well.

Depletion of the ozone layer will allow more ultraviolet rays to enter the troposphere. The incidence of human cancer, especially skin cancer, can be expected to increase, and plants and animals living in the top microlayer of the oceans will begin to die. Increased ultraviolet radiation also will hasten the rate at which smog is formed.

The Greenhouse Effect

It is predicted that certain air pollutants will cause the temperature of the earth to rise, perhaps by as much as 0.8° C per decade (1.36° F/decade). These gases let the sun's rays pass through, but they absorb and reradiate their heat back toward the earth (fig. 35.16*a*). This is called the **greenhouse effect** because the glass of a greenhouse allows sunlight to pass through, then traps the resulting heat inside the structure. The pollutants that cause global warming include

- carbon dioxide (CO_2) (from fossil fuel and wood burning)
- nitrous oxide (NO_2) (primarily from fertilizer use and animal wastes)
- methane (from biogas, bacterial decomposition, particularly in the guts of animals, sediments, and flooded rice paddies)
- CFCs (from Freon mentioned earlier)

It so happens that together, nitrous oxide, methane, and CFCs cause a more severe greenhouse effect than carbon dioxide, and their combined effect is expected to equal the effect from carbon dioxide by the year 2020 (fig.

35.16*b*). Just now, carbon dioxide levels are 25% higher than they were in 1860. Most of this excess is due to industrialization, but tropical **deforestation** is now a major contributor as well. Burning one acre of primary forest puts 200,000 kg of carbon dioxide into the air; moreover, the trees are no longer available to act as a sink to take up carbon dioxide during photosynthesis.

It appears that global warming has already begun: the 4 hottest years on record have occurred during the 1980s; there has been greater warming in the winters than summers; there has been greater warming at high altitudes than near the equator; the stratosphere is cooler and the lower atmosphere is warmer than formerly. All of these effects had been predicted by computer models of the greenhouse effect.

The ecological effects of global warming are expected to be severe. First of all, the sea level will rise—melting of the polar ice caps will add more water to the sea and water expands when it heats as well. This will cause flooding of many coastal regions and the possible loss of many cities, like New York, Boston, Miami, and Galveston in the United States. Coastal ecosystems, such as marshes, swamps, and bayous, would normally move inland to higher ground as the sea level rises, but many of these ecosystems are blocked in by artificial structures and may be unable to move inland. If so, the loss of fertility will be immense.

There also may be food loss because of regional changes in climate. Not only will there be greater heat, there will be drought in the midwestern United States, and the suitable climate for growing wheat and corn will shift to Canada, where the soil is not as suitable.

Drastic measures are recommended to hold the quantity of greenhouse gases to their present level. A 50% decrease in consumption of fossil fuels is recommended, and we must find more efficient ways to acquire energy from cleaner fuels, such as natural gas. We should use alternative energy sources, such as solar and geothermal energy and even perhaps nuclear power, more aggressively.

Not only should tropical rain forest deforestation be halted, extensive reforesting all over the globe should take place. These forests could be used as a source of wood for fuel, and at the same time, the carbon dioxide given off by the burning of the fuel would be absorbed by the new trees coming along.

Manufacture and use of CFCs should be eliminated completely. The United States and the European countries already have agreed to reduce CFC production by 85% as soon as possible and to try to ban the chemicals altogether by the end of the century.

Outdoor air pollutants are involved in causing 4 major environmental effects: photochemical smog, acid deposition, ozone shield destruction, and the greenhouse effect. Each pollutant may be involved in more than one of these.

While each of these environmental effects is bad enough when considered separately, they actually feed on one another, making the total effect much worse than is predicted for the total of each separately.

A Sustainable World

Economic growth often is accompanied by environmental degradation, and there is great concern that as the less-developed countries become more developed, environmental degradation will increase to the point that the human population will outstrip the carrying capacity of the planet. Without economic growth, however, the demographic transition may not occur, and the sheer number of people in the less-developed countries will bring about environmental degradation to such a degree that the effects will be felt worldwide, not just in the immediate area.

[1]From George Morris, *Overpopulation: Everyone's Baby.* London: Priory Press Limited, 1973, p. 24.

The answer to this dilemma is economic growth without the side effect of environmental degradation. This is called sustainable growth. Certain developed countries, such as the Scandinavian countries and the United States to a degree, are beginning to learn to protect the environment. Energy consumption is decreasing even as economic growth continues. Industries are beginning to recycle their waste to prevent environmental pollution. Citizens are learning to recycle their trash. More should be done, and ecologically sound practices must be exported to the less-developed countries very quickly. Only sustainable economic development will ensure the continuance of the world's human population.

Once zero population growth has been achieved, then we can begin to consider the possibility of a steady state in which population and resource consumption remain constant.

The Steady State

A stable population size would be a new experience for humans. Figure 35.17 shows the age structure for a hypothetical stable population. Such a population has many benefits, as discussed here.

1. Over 40% of the people would be fairly youthful, with only about 15% in the senior citizen category. There would be proportionately fewer children and teenagers than in a rapidly expanding population.
2. The quality of life for children might increase substantially since fewer unwanted babies might be born and the opportunity would exist for children to receive the loving attention each needs.
3. There might be increased employment opportunities for women and a generally less competitive workplace because newly qualified workers would enter the job market at a more moderate rate.
4. Creativity need not be impaired. A study of Nobel Prize winners showed that the average age at which the prize-winning work was done was over 30.

Environmental preservation would be the most important consideration in a steady-state world. Renewable energy sources, such as solar energy, would play a greater role in meeting energy needs. Pollution would be minimized. Ecological diversity would be maintained, and overexploitation would cease (fig. 35.18). Ecological principles would serve as guidelines for specialists in

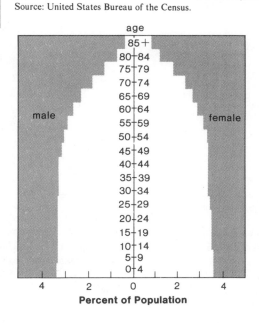

Figure 35.17
Age and sex structure of a hypothetical human population that remains the same size each year.
Source: United States Bureau of the Census.

Figure 35.18
In the steady state, environmental preservation will be an important consideration.

all fields, creating a unified approach to the environment. In a steady-state world, all people would strive to be aware of the environmental consequences of their actions, consciously working toward achieving balance in the ecosystems of their planet.

In a **steady-state society,** there would be no yearly increase in population nor resource consumption. It is forecast that under these circumstances, the quality of life would improve.

What would our culture be like if we had steady-state manufacturing and a steady-state population? Perhaps it would be greatly improved. Certainly, there are no limits to growth in knowledge, education, art, music, scientific research, human rights, justice, and cooperative human interactions. In a steady-state world, the general sense of fearful competition among peoples might diminish, allowing human compassion and creativity to prosper as never before.

Summary

The human population is expanding exponentially, and even though the growth rate has declined, there is a large increase each year—the doubling time is now about 40 years. The developed countries underwent a demographic transition between 1950 and 1975, but the less-developed countries just now are undergoing demographic transition. In these countries, where the average age is less than 15, it will be many years before reproduction replacement will mean zero population growth.

On land, soil erosion sometimes reduces soil quality and leads to desertification. The tropical rain forests are being reduced in size, and the loss of biological diversity will be immense. Solid wastes, including hazardous wastes, are deposited on land. These are not biodegradable and are subject to biological magnification.

Surface waters, groundwater, and the oceans all are being polluted. Organic materials can be broken down in sewage treatment plants, but the nutrients made available to algae can lead to cultural eutrophication. It is difficult to rid surface waters and particularly groundwater of hazardous wastes. The oceans are the final recipients of all the pollutants that enter water. In addition, some materials are dumped purposefully or accidentally directly into the oceans.

In the air, hydrocarbons and nitrogen oxides (NO_x) react to form smog, which contains ozone and PAN. Sulfur dioxide (SO_2) and nitrogen oxides react with water vapor to form acids that contribute to acid deposition. Ozone destruction is associated particularly with CFCs, which rise into the stratosphere and react with frozen particles within clouds to release chlorine. Chlorine causes ozone to break down. Several gases—carbon dioxide (CO_2), nitrous oxide (NO_2), methane, and CFCs—are called the greenhouse gases because they trap heat.

Economic growth is required particularly in the less-developed countries. All countries of the world should try to have sustainable growth, and this necessitates preservation of the environment. Eventually, it may be possible to have a steady state in which neither population nor resource consumption increases.

Study Questions

1. Define exponential growth. (p. 736) Draw a growth curve to represent exponential growth, and explain why a curve representing population growth usually levels off. (p. 739)
2. Calculate the growth rate and the doubling time for a population in which the birthrate is 20 per 1,000 and the death rate is 2 per 1,000. (p. 738)
3. Define demographic transition. When did the developed countries undergo demographic transition? When did the less-developed countries undergo demographic transition? (pp. 740–42)
4. Give at least 3 differences between the developed countries and the less-developed countries. (pp. 740–42)
5. Give 2 reasons why the equality of the land is being degraded today. What is desertification? (p. 743)
6. Give 3 reasons why tropical rain forests are being destroyed. What is another possible reason in the future? (p. 744)
7. What is the primary ecological concern associated with the destruction of rain forests? (p. 744)
8. What are the 3 types of hazardous wastes that contribute to pollution on land? (p. 746)
9. What are several ways in which underground water supplies can be polluted? (p. 748)
10. What substances contribute to air pollution? What are their sources? Which ones are associated with photochemical smog, acid deposition, destruction of the ozone shield, or the greenhouse effect? (pp. 750–54)

Objective Questions

1. After a country has undergone the demographic transition, the death rate and the birthrate both are _____ (high or low).
2. If a country has a pyramid-shaped age structure diagram, most individuals are _____ (prereproductive, reproductive, or postreproductive).
3. Less-developed countries are not as _____ as developed countries.
4. When a population is undergoing exponential growth, the increase in number of people each year is _____ than the year before. (higher or lower)
5. The chemicals best associated with ozone depletion are _____ .
6. The gas best associated with the greenhouse effect is _____ .
7. Sewage is biodegradable, but the nutrients released can lead to _____ of surface water.
8. Pesticides and radioactive wastes both are subject to biological _____ .

Selected Key Terms

acid deposition (as'id dep''o-zish'un) the return to earth as rain or snow of the sulfate or nitrate salts of acids produced by commercial and industrial activities on earth. *752*

biotic potential (bi-ot'ik po-ten'shal) the maximum population growth rate under ideal conditions. *739*

carrying capacity (kar'e-ing kah-pas'ĭ-te) the largest number of organisms of a particular species that can be maintained indefinitely in an ecosystem. *740*

cultural eutrophication (kul'tu-ral u''tro-fĭ-ka'shun) enrichment of a body of water, causing excessive growth of producers and then death of these and other inhabitants. *747*

demographic transition (dem-o-graf'ik tran-zi'shun) the change from a high birthrate to a low birthrate so that the growth rate is lowered. *740*

desertification (dez-ert''ĭ-fĭ-ka'shun) desert conditions caused by human misuse of land. *744*

developed country (de-vel'opt kun'trē) industrialized nation that typically has a strong economy and a low rate of population growth. *740*

doubling time (dŭ'b'l-ing tĭm) the number of years it takes for a population to double in size. *738*

environmental resistance (en-vi''ron-men'tal re-zis'tans) sum total of factors in the environment that limit the numerical increase of a population in a particular region. *739*

exponential growth (eks''po-nen'shal grōth) growth, particularly of a population, in which the increase occurs in the same manner as compound interest. *736*

greenhouse effect (grēn'hows ĕ-fekt) carbon dioxide (CO₂) buildup in the atmosphere as a result of fossil fuel combustion; retains and reradiates heat, effecting an abnormal rise in the earth's average temperature. *753*

hazardous waste (haz'er-dus wāst) waste containing chemicals hazardous to life. *746*

photochemical smog (fo''to-kem'ĭ-kal smog) air pollution that contains nitrogen oxides (NOₓ) and hydrocarbons that react to produce ozone and peroxylacetyl nitrate (PAN). *750*

ozone shield (o'zōn shēld) a layer of O₃ present in the upper atmosphere that protects the earth from damaging ultraviolet light. Nearer the earth, ozone is a pollutant. *752*

replacement reproduction (re-plās'ment re''pro-duk'shun) a population in which each person is replaced by only one child. *742*

Further Readings for Part Six

Alcock, J. 1984. *Animal behavior, an evolutionary approach.* 3rd ed. Sunderland, Mass.: Sinauer, Associates.

Batie, S. S., and R. G. Healy. February 1983. The future of American agriculture. *Scientific American.*

Begon, J. J., Harper, and C. Townsend. 1986. *Ecology: Individuals, populations, and communities.* Sunderland, Mass.: Sinauer, Associates.

Berner, R. A., and A. C. Lasaga. March 1989. Modeling the geochemical carbon cycle. *Scientific American.*

Blaustein, A. R., and R. K. O'Hara. January 1986. Kin recognition in tadpoles. *Scientific American.*

Brown, L. R., et al. 1990. *State of the world: 1990.* New York: W. W. Norton and Co.

Bushbacher, R. J. January 1986. Tropical deforestation and pasture development. *BioScience.*

Environment. All issues of this journal contain articles covering modern ecological problems.

Hamakawa, Y. April 1987. Photovoltaic power. *Scientific American.*

Horn, M. H., and R.N. Gibson. January 1988. Intertidal fishes. *Scientific American.*

Houghton, R. A., and G. M. Woodwell. April 1989. Global climatic change. *Scientific American.*

Miller, J. T. 1988. *Living in the Environment.* 5th ed. Belmont, Calif.: Wadsworth.

Mohnen, V. A. August 1988. The challenge of acid rain. *Scientific American.*

Nebel, B. J. 1987. *Environmental science: The way the world works.* 2d ed. Englewood Cliffs, N.J.: Prentice-Hall.

Odum, H. T. 1983. *Systems ecology: An introduction.* New York: John Wiley and Sons.

O'Leary, P. R., P. W. Walsh, and R. H. Ham. December 1988. Managing solid waste. *Scientific American.*

Power, J. F., and R. F. Follett, March 1987. Monoculture. *Scientific American.*

Ricklefs, R. E. 1986. *Ecology.* 3rd ed. New York: Chiron Press.

Schneider, S. H. May 1987. Climate modeling. *Scientific American.*

Shaw, R. W. August 1987. Air pollution by particles. *Scientific American.*

Smith, R. L. 1985. *Ecology and field biology.* 3rd ed. New York: Harper & Row, Publishers, Inc.

Sumich, J. 1988. *Biology of marine life.* 4th ed. Dubuque, Ia.: Wm. C. Brown Publishers.

Wursig, B. April 1988. The behavior of baleen whales. *Scientific American.*

Appendix A

Periodic Table of the Elements

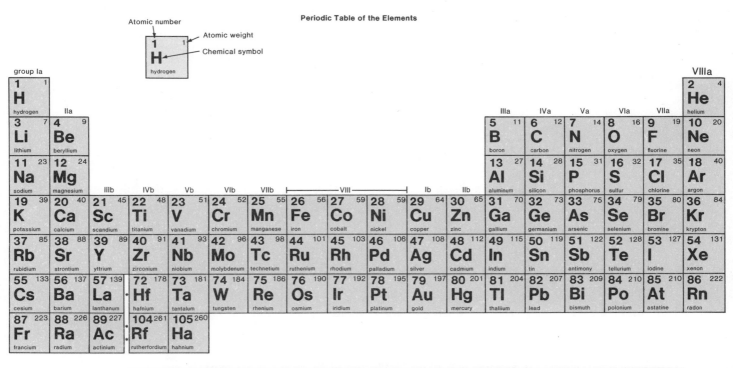

Periodic Table of the Elements

Appendix B

Drugs of Abuse

	Drugs	Often Prescribed Brand Names	Medical Uses	Potential Physical Dependence	Potential Psychological Dependence	Tolerance
Narcotics	Opium	Dover's Powder, Paregoric	Analgesic, antidiarrheal	High	High	Yes
	Morphine	Morphine	Analgesic	High	High	Yes
	Codeine	Codeine	Analgesic, antitussive	Moderate	Moderate	Yes
	Heroin	None	None	High	High	Yes
	Meperidine (Pethidine)	Demerol, Pethadol	Analgesic	High	High	Yes
	Methadone	Dolophine, Methadone, Methadose	Analgesic, heroin substitute	High	High	Yes
	Other Narcotics	Dilaudid, Leritine, Numorphan, Percodan	Analgesic, antidiarrheal, antitussive	High	High	Yes
Depressants	Chloral Hydrate	Noctec, Somnos	Hypnotic	Moderate	Moderate	Probable
	Barbiturates	Amytal, Butisol, Nembutal, Phenobarbital, Seconal, Tuinal	Anesthetic, anticonvulsant, sedation, sleep	High	High	Yes
	Glutethimide	Doriden	Sedation, sleep	High	High	Yes
	Methaqualone	Optimil, Parest, Quaalude, Somnafac, Sopor	Sedation, sleep	High	High	Yes
	Tranquilizers	Equanil, Librium, Miltown Serax, Tranxene, Valium	Antianxiety, muscle relaxant, sedation	Moderate	Moderate	Yes
	Other Depressants	Clonopin, Dalmane, Dormate, Noludar, Placydil, Valmid	Antianxiety, sedation, sleep	Possible	Possible	Yes
Stimulants	Cocaine*	Cocaine	Local anesthetic	Possible	High	Yes
	Amphetamines	Benzedrine, Biphetamine, Desoxyn, Dexedrine	Hyperkinesis, narcolepsy, weight control	Possible	High	Yes
	Phenmetrazine	Preludin	Weight control	Possible	High	Yes
	Methylphenidate	Ritalin	Hyperkinesis	Possible	High	Yes
	Other Stimulants	Bacarate, Cylert, Didrex, Ionamin, Plegine, Pondimin, Pre-Sate, Sanorex, Voranil	Weight control	Possible	Possible	Yes
Hallucinogens	LSD	None	None	None	Degree unknown	Yes
	Mescaline	None	None	None	Degree unknown	Yes
	Psilocybin-Psilocyn	None	None	None	Degree unknown	Yes
	MDA	None	None	None	Degree unknown	Yes
	PCP†	Sernylan	Veterinary anesthetic	None	Degree unknown	Yes
	Other Hallucinogens	None	None	None	Degree unknown	Yes
Cannabis	Marijuana Hashish Hashish Oil	None	Glaucoma	Degree unknown	Moderate	Yes

From: *Drugs of Abuse* Produced by the Affairs in Cooperation with the Office of Public Science and Technology.
*Designated a narcotic under the Controlled Substances Act.
†Designated a depressant under the Controlled Substances Act.

Duration of Effects (in hours)	Usual Methods of Administration	Possible Effects	Effects of Overdose	Withdrawal Syndrome
3–6	Oral, smoked	Euphoria, drowsiness, respiratory depression, constricted pupils, nausea	Slow and shallow breathing, clammy skin, convulsions, coma, possible death	Watery eyes, runny nose, yawning, loss of appetite, irritability, tremors, panic, chills and sweating, cramps, nausea
3–6	Injected, smoked			
3–6	Oral, injected			
3–6	Injected, sniffed			
3–6	Oral, injected			
12–24	Oral, injected			
3–6	Oral, injected			
5–8	Oral			
1–16	Oral, injected	Slurred speech, disorientation, drunken behavior without odor of alcohol	Shallow respiration, cold and clammy skin, dilated pupils, weak and rapid pulse, coma, possible death	Anxiety, insomnia, tremors, delirium, convulsions, possible death
4–8	Oral			
4–8	Oral			
4–8	Oral			
4–8	Oral			
2	Injected, sniffed	Increased alertness, excitation, euphoria, dilated pupils, increased pulse rate and blood pressure, insomnia, loss of appetite	Agitation, increased body temperature, hallucinations, convulsions, possible death	Apathy, long periods of sleep, irritability, depression, disorientation
2–4	Oral, injected			
2–4	Oral			
2–4	Oral			
2–4	Oral			
Variable	Oral	Illusions and hallucinations (with exception of MDA), poor perception of time and distance	Longer, more intense "trip" episodes, psychosis, possible death	Withdrawal syndrome not reported
Variable	Oral, injected			
Variable	Oral			
Variable	Oral, injected, sniffed			
Variable	Oral, injected, smoked			
Variable	Oral, injected, sniffed			
2–4	Oral, smoked	Euphoria, relaxed inhibitions, increased appetite, disoriented behavior	Fatigue, paranoia, possible psychosis	Insomnia, hyperactivity, and decreased appetite reported in a limited number of individuals

Appendix C

The Metric System

The Metric System

Standard Metric Units		Abbreviations
Standard unit of mass	gram	g
Standard unit of length	meter	m
Standard unit of volume	liter	l

Common Prefixes		Examples
kilo	1,000	a kilogram is 1,000 grams
centi	0.01	a centimeter is 0.01 meter
milli	0.001	a milliliter is 0.001 liter
micro (μ)	one-millionth	a micrometer is 0.000001 (one-millionth) of a meter
nano (n)	one-billionth	a nanogram is 10^{-9} (one-billionth) of a gram
pico (p)	one-trillionth	a picogram is 10^{-12} (one-trillionth) of a gram

Think Metric Length

1. The speed of a car is 60 miles/hr or 100 km/hr.
2. A man who is 6 feet tall is 180 cm.
3. A 6-inch ruler is 15 cm.
4. One yard is almost a meter (0.9 m).

Units of Length

Unit	Abbreviation	Equivalent
meter	m	approximately 39 in
centimeter	cm	10^{-2} m
millimeter	mm	10^{-3} m
micrometer	μm	10^{-6} m
nanometer	nm	10^{-9} m
angstrom	Å	10^{-10} m

Length Conversions

1 in = 2.5 cm	1 mm = 0.039 in
1 ft = 30 cm	1 cm = 0.39 in
1 yd = 0.9 m	1 m = 39 in
1 mi = 1.6 km	1 m = 1.094 yd
	1 km = 0.6 mi

To Convert	Multiply by	To Obtain
inches	2.54	centimeters
feet	30	centimeters
centimeters	0.39	inches
millimeters	0.039	inches

Think Metric Volume

1. One can of beer (12 oz) contains 360 ml.
2. The average human body contains between 10 and 12 pints of blood or between 4.7 and 5.6 liters.
3. One cubic foot of water (7.48 gallons) is 28.426 liters.
4. If a gallon of unleaded gasoline costs $1.00, a liter costs 38¢.

Units of Volume

Unit	Abbreviation	Equivalent
liter	l	approximately 1.06 qt
milliliter	ml	10^{-3} l (1 ml = 1 cm^3 = 1 cc)
microliter	μl	10^{-6} l

Volume Conversions

1 tsp = 5 ml	1 pt = 0.47 l	1 ml = 0.03 fl oz
1 tbsp = 15 ml	1 qt = 0.95 l	1 l = 2.1 pt
1 fl oz = 30 ml	1 gal = 3.8 l	1 l = 1.06 qt
1 cup = 0.24 l		1 l = 0.26 gal

To Convert	Multiply by	To Obtain
fluid ounces	30	milliliters
quarts	0.95	liters
milliliters	0.03	fluid ounces
liters	1.06	quarts

Think Metric Weight

1. One pound of hamburger is 448 grams.
2. The average human male brain weighs 1.4 kg (3 lb 1.7 oz).
3. A person who weighs 154 lbs weighs 70 kg.
4. Lucia Zarate weighed 5.85 kg (13 lbs) at age 20.

Units of Weight

Unit	Abbreviation	Equivalent
kilogram	kg	10^3 g (approximately 2.2 lb)
gram	g	approximately 0.035 oz
milligram	mg	10^{-3} g
microgram	μg	10^{-6} g
nanogram	ng	10^{-9} g
picogram	pg	10^{-12} g

Weight Conversions

1 oz = 28.3 g	
1 lb = 453.6 g	1 g = 0.035 oz
1 lb = 0.45 kg	1 kg = 2.2 lb

To Convert	Multiply by	To Obtain
ounces	28.3	grams
pounds	453.6	grams
pounds	0.45	kilograms
grams	0.035	ounces
kilograms	2.2	pounds

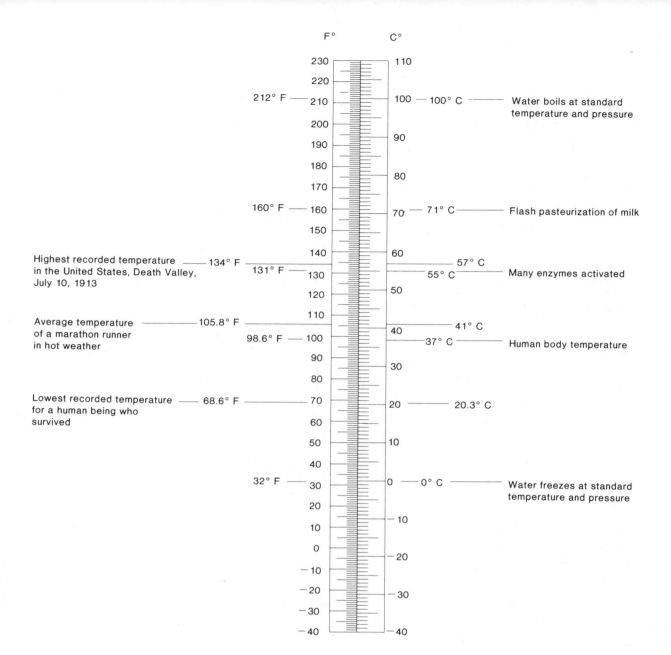

F°   C°

- 212° F — 210 — 100 — 100° C ——— Water boils at standard temperature and pressure
- 160° F — 160 — 70 — 71° C ——— Flash pasteurization of milk
- Highest recorded temperature in the United States, Death Valley, July 10, 1913 — 134° F
- 131° F — 130 — 57° C
- 55° C ——— Many enzymes activated
- Average temperature of a marathon runner in hot weather — 105.8° F — 41° C
- 98.6° F — 100 — 37° C ——— Human body temperature
- Lowest recorded temperature for a human being who survived — 68.6° F — 70 — 20 — 20.3° C
- 32° F — 30 — 0 — 0° C ——— Water freezes at standard temperature and pressure

To convert temperature scales:

Fahrenheit to Celsius $°C = \dfrac{5}{9} (°F - 32)$

Celsius to Fahrenheit $°F = \dfrac{9}{5} (°C + 32)$

The classification system given here is a simplified one, containing all the major kingdoms as well as the major divisions (called phyla in the kingdom Prostista and the kingdom Animalia). The text does not discuss all divisions and phyla listed here.

Kingdom Monera

Prokaryotic, unicellular organisms. Nutrition principally by absorption, but some are photosynthetic or chemosynthetic.

> Division Archaebacteria: methanogens, halophiles, and thermoacidophiles
> Division Eubacteria: all other bacteria, including cyanobacteria (formerly called blue-green algae)

Kingdom Protista

Eukaryotic, unicellular organisms (and the most closely related multicellular forms). Nutrition by photosynthesis, absorption, or ingestion.

> Phylum Sarcodina: amoeboid protozoans
> Phylum Ciliophora: ciliated protozoans
> Phylum Zoomastigina: flagellated protozoans
> Phylum Sporozoa: parasitic protozoans
> Phylum Chlorophyta: green algae
> Phylum Pyrrophyta: dinoflagellates
> Phylum Euglenophyta: *Euglena* and relatives
> Phylum Chrysophyta: diatoms
> Phylum Rhodophyta: red algae
> Phylum Phaeophyta: brown algae
> Phylum Myxomycota: slime molds
> Phylum Oomycota: water molds

Kingdom Fungi

Eukaryotic organisms, usually having haploid or multinucleated hyphal filaments. Spore formation during both asexual and sexual reproduction. Nutrition principally by absorption.

> Division Zygomycota: black bread molds
> Division Ascomycota: sac fungi
> Division Basidiomycota: club fungi
> Division Deuteromycota: imperfect fungi (i.e., means of sexual reproduction not known).

Kingdom Plantae

Eukaryotic, terrestrial, multicellular organisms with rigid cellulose cell walls and chlorophyll *a* and *b*. Nutrition principally by photosynthesis. Starch is the food reserve.

> Division Bryophyta: mosses and liverworts
> Division Psilophyta: whisk ferns
> Division Lycophyta: club mosses
> Division Sphenophyta: horsetails
> Division Pterophyta: ferns
> Division Cycadophyta: cycads
> Division Ginkgophyta: ginkgoes
> Division Gnetophyta: gnetae
> Division Coniferophyta: conifers
> Division Anthophyta: flowering plants
> > Class Dicotyledonae: dicots
> > Class Monocotyledonae: monocots

Kingdom Animalia

Eukaryotic, usually motile, multicellular organisms without cell walls or chlorophyll. Nutrition principally ingestive, with digestion in an internal cavity.

> Phylum Porifera: sponges
> Phylum Cnidaria: radially symmetrical marine animals
> > Class Hydrozoa: *Hydra,* Portuguese man-of-war
> > Class Scyphozoa: jellyfishes
> > Class Anthozoa: sea anemones and corals

Phylum Platyhelminthes: flatworms
    Class Turbellaria: free-living flatworms
    Class Trematoda: parasitic flukes
    Class Cestoda: parasitic tapeworms
Phylum Nematoda: roundworms
Phylum Rotifera: rotifers
Phylum Mollusca: soft-bodied, unsegmented animals
    Class Polyplacophora: chitons
    CLass Monoplacophora: *Neopilina*
    Class Gastropoda: snails and slugs
    Class Bivalvia: clams and mussels
    Class Cephalopoda: squids and octopuses
Phylum Annelida: segmented worms
    Class Polychaeta: sandworms
    Class Oligochaeta: earthworms
    Class Hirudinea: leeches
Phylum Arthropoda: joint-legged animals; exoskeleton
    Class Crustacea: lobsters, crabs, barnacles
    Class Arachnida: spiders, scorpions, ticks
    Class Chilopoda: centipedes
    Class Diplopoda: millipedes
    Class Insecta: grasshoppers, termites, beetles
Phylum Echinodermata: marine; spiny, radially symmetrical animals
    Class Crinoidea: sea lilies and feather stars
    Class Asteroidea: sea stars
    Class Ophiuroidea: brittle stars
    Class Echinoidea: sea urchins and sand dollars
    Class Holothuroidea: sea cucumbers
Phylum Chordata: dorsal supporting rod (notochord) at some stage; dorsal hollow nerve
    cord; pharyngeal gill pouches or slits
    Subphylum Urochordata: tunicates
    Subphylum Cephalochordata: lancelets
    Subphylum Vertebrata: vertebrates
        Class jawless fishes (lampreys, hagfishes)
        Class Chondrichthyes: cartilaginous fishes (sharks, rays)
        Class Osteichthyes: bony fishes
            Subclass Sarcopterygii: lobe-finned fishes
            Subclass Actinopterygii: ray-finned fishes
        Class Amphibia: frogs, toads, salamanders
        Class Reptilia: snakes, lizards, turtles
        Class Aves: birds
        Class Mammalia: mammals
            Subclass Prototheria: egg-laying mammals
                Order Monotremata: duckbilled platypus, spiny anteater
            Subclass Metatheria: marsupial mammals
                Order Marsupialia: opossums, kangaroos
            Subclass Eutheria: placental mammals
                Order Insectivora: shrews, moles
                Order Chiroptera: bats
                Order Edentata: anteaters, armadillos
                Order Rodentia: rats, mice, squirrels
                Order Lagomorpha: rabbits and hares
                Order Cetacea: whales, dolphins, porpoises
                Order Carnivora: dogs, bears, weasels, cats, skunks
                Order Proboscidea: elephants
                Order Sirenia: manatees
                Order Perissodactyla: horses, hippopotamuses, zebras
                Order Artiodactyla: pigs, deer, cattle
                Order Primates: lemurs, monkeys, apes, humans
                    Suborder Prosimii: lemurs, tree shrews, tarsiers, lorises, pottos
                    Suborder Anthropoidea: monkeys, apes, humans
                        Superfamily Ceboidea: New World monkeys
                        Superfamily Cercopithecoidea: Old World monkeys
                        Superfamily Hominoidea: apes and humans
                            Family Hylobatidae: gibbons
                            Family Pongidae: chimpanzees, gorillas, orangutans
                            Family Hominidae: *Australopithecus,* Homo erectus,* Homo
                                sapiens sapiens*
                            *extinct

Answers to I.1 Critical Thinking

1. Sweetener S is being varied in order to determine its effects on the body.
2. The control group is subjected to all conditions (i.e., living in a cage kept at a certain temperature and eating the same food) except it is not receiving sweetener S.
3. Chance of bladder cancer development is dependent upon quantity of sweetener S consumed.
4. Yes, the conditions mentioned in answer 2 are constant.
5. A control group increases confidence in results. For example, if there is no bladder cancer in control mice, it is more likely that development of bladder cancer is caused by sweetener S.

Answers to I.2 Critical Thinking

1. There is no observation or experiment that can be performed to prove the hypothesis false.
2. You can perform a controlled experiment. The experimental group is fed biotin-free food. The control group is expected to remain healthy, while the experimental group is expected to get sick.
3. Religious beliefs are not subjected to the process described in figure I.8; scientific beliefs arise from this process.

Answers to 1.1 Critical Thinking

1. Water contains a small number of atoms.
2. Water has covalent bonding.
3. Water (a) absorbs heat and gives off heat—this helps to keep the body warm; (b) is cohesive—this helps in fluid transport; and (c) is a solvent in bodies—this helps in chemical reactions.
4. Water does not contain carbon (C).

Answers to 1.2 Critical Thinking

1. Actin and myosin differ in the sequence of their amino acids.
2. Starch is not as branched as glycogen.
3. The length of the chain and the placement of unsaturated bonds in the 2 acids may be different.

Answers to 2.1 Critical Thinking

1. 1/10,000,000
2. 1,000
3. 100
4. Yes, it is much smaller.

Answer to 2.2 Critical Thinking

1. Carbon (C) is found in carbohydrates and fats, while sulfur is unique to the amino acids cysteine and methionine and therefore proteins.

2. Radiation will appear first at the ribosomes (polysomes) and subsequently at the region of the nuclear pore and then at the nucleolus.
3. Radiation will appear first at the rough ER and subsequently at the Golgi apparatus and then at the cell membrane.

Answers to 2.3 Critical Thinking

1. There are no centrioles in plant cells.
2. Centrioles are present in the microtubule organizing region.
3. Centrioles give rise to basal bodies that organize the microtubules in cilia and flagella. Cilia and flagella are associated with animal cells.

Answers to 3.1 Critical Thinking

1. Proteins are made at the ribosomes. Carbohydrate chains might be added in the Golgi apparatus.
2. Proteins travel to the cell membrane via vesicles formed by the Golgi apparatus.
3. It is not likely that this person will contract AIDS because the CD4 receptor is missing.
4. An example of structural proteins is the proteins of the cell membrane.

Answer to 3.2 Critical Thinking

1. Alcohol can cross the lipid bilayer by simple diffusion. Water enters by way of a channel protein.
2. Na^+ is pumped across by the sodium-potassium pump. Cl^- diffuses out by way of a channel protein.
3. Amino acids enter by facilitated diffusion. Proteins enter by endocytosis.
4. The proteins would be digested after the endocytic vesicle fuses with a lysosome containing hydrolytic enzymes.

Answers to 3.3 Critical Thinking

1. The number of exocytic and endocytic cells must be about the same, assuming equal sizes of vacuoles.
2. The exocytic vesicles must be adding membrane at the forward end (right), and the endocytic vesicles must be forming at the hind end (left).
3. The virus must somehow get out of the vesicle in order to enter the cytoplasm.

Answers to 4.1 Critical Thinking

1. You can tell that chromatin is metabolically active because it is (a) the extended form of the genetic material (b) seen in metabolically active cells, such as those shown in figure 2.3 (pp. 46–47).
2. Yes, they might differ. It is not necessary to inherit exactly the same form of a gene from each parent.

Appendix E

Answers to the Critical Thinking Questions

3. Various disorders result because there are too many proteins (enzymes) of the same kind.
4. If a particular gene is not needed for the maturation of the sperm and the egg, a defective form cannot have an effect.

Answers to 4.2 Critical Thinking
1. This is useful because nutrients enter and wastes exit cytoplasm at the cell membrane.
2. DNA replication (chromosome duplication) provides chromosomes for the daughter cells.
3. The increase is useful because newly formed cells must receive organelles from the dividing cell.
4. Examples of such specialized cells are nerve and muscle cells.

Answers to 4.3 Critical Thinking
1. Colchicine disrupts the spindle apparatus. Specifically, it prevents microtubule assembly.
2. Asexual reproduction, requiring only mitosis, produces cells (offspring) that have the same kinds of chromosomes as the mother cell (parent).
3. Meiosis is a part of sexual reproduction. Because of meiosis, the daughter cells can have any combination of the haploid number chromosomes, and the zygote has a different combination of the haploid number of chromosomes than either parent.
4. Yes, the production of variation allows new types of organisms to evolve.

Answers to 5.1 Critical Thinking
1. Correct pH and a warm temperature are the recommended conditions.
2. The yield could be increased if more pepsin or more egg white, whichever is in short supply, was added.
3. With irreversible inhibition, the reaction stops; with reversible inhibition, the reaction continues at a reduced rate.

Answers to 5.2 Critical Thinking
1. Oxygen (O_2) is the final acceptor for hydrogen (H) atoms at the end of the respiratory chain. The chain can continue to produce ATP only if oxygen is present.
2. Carbon dioxide (CO_2) is produced as molecules are broken down by the transition reaction and the Krebs cycle.
3. Carbohydrates are sources of glucose. The body realizes 38 ATP molecules for each molecule of glucose that is broken down completely to carbon dioxide and water.

Answers to 6.1 Critical Thinking
1. A plant takes (CO_2) and (H_2O) from the environment and gives (O_2) to the environment.
2. Oxygen comes from the thylakoids and the breakdown of water. Carbon dioxide goes into the stroma and becomes carbohydrate.
3. Oxygen is used for cellular respiration, and carbon dioxide comes from cellular respiration.
4. Some glucose molecules are used in cellular respiration; some become starch; and some become cellulose and other molecules needed by plants.

Answers to 6.2 Critical Thinking
1. This energy is used to make ATP molecules for cellular metabolism.
2. The energy required is 686 Kcal/mole. This energy comes from the sun.
3. The energy captured by photosynthesis produces glucose; the energy released by cellular respiration comes from glucose breakdown.
4. Glucose is an energy source within organisms.

Answers to 6.3 Critical Thinking
1. The matrix of a mitochondrion compares to the stroma of a chloroplast.
2. Opposite in the matrix of a mitochondrion, glucose products are being oxidized (CO_2 is being given off); in the stroma of a chloroplast, CO_2 is being reduced to carbohydrate.
3. The cristae of the mitochondrion compares to the thylakoid membrane of the chloroplast. The intermembrane space of the mitochondrion compares to the thylakoid space of the chloroplast.
4. Same. Both cristae and thylakoid membrane contain a cytochrome system for ATP production. Both the intermembrane space and the thylakoid space are (H^+) ion reservoirs.

Answers to 7.1 Critical Thinking
1. Leaf epidermis prevents drying out; it is covered by a waxy cuticle. Leaf epidermis allows gas exchange; it contains stomata.
2. The spongy layer carries on gas exchange; there are air spaces next to these cells. The spongy layer carries on photosynthesis; the cells contain chloroplasts.
3. Leaf veins transport water and minerals and organic substances; they contain xylem and phloem.
4. C_3 leaves have a palisade layer and a spongy layer; C_4 leaves have mesophyll cells in a ring around bundle sheath cells. In C_4 leaves, but not in C_3 leaves, mesophyll cells pass CO_2 to bundle sheath cells.

Answers to 8.1 Critical Thinking
1. Atmospheric pressure is the pressure of the air. This shows that atmospheric pressure cannot raise water to the height of a tall tree.
2. Transpiration occurs and pulls on water.
3. This suggests that transpiration could raise water to the top of trees.

Answers to 9.1 Critical Thinking
1. Epithelial cells have flat surfaces; therefore, they can be placed easily next to one another. This makes them suitable for covering a surface.
2. Casparian strip. Both tight junctions and the Casparian strip prevent materials from moving between the cells.
3. The long, tubular cells contain actin filaments and myosin filaments, and these account for their ability to contract. Because the cells run the length of a muscle, when they contract, the muscle contracts. The object would move left.
4. Nerve cells conduct nerve impulses sometimes over long distances. The long, skinny process, or fiber, of a nerve cell makes this possible.

Answers to 9.2 Critical Thinking
1. It fluctuates.
2. Internal conditions do not stay exactly the same—they fluctuate about a mean point.
3. Hot and cold body temperatures activate the receptor and the regulator center.
4. Yes, because conditions have to become either hot or cold in order to stimulate the receptor and the regulator center. Then, there is an adaptive response in the opposite direction.

Answers to 10.1 Critical Thinking
1. The epithelial portion of mucous membrane produces the enzymes.
2. The digestive system provides nutrients needed by cells.
3. The liver regulates the output of glucose to keep the amount in blood fairly constant.
4. Glucose is used for energy; amino acids are used for protein synthesis.

Answers to 10.2 Critical Thinking
1. No. Experiment is as follows: tube 1— pepsin, water, and egg white so pH is neutral; tube 2—pepsin, water, egg white, $NaHCO_3$ so pH is basic; tube 3— same as tube 4 in figure 10.14, which is expected to show the best digestion.

2. No. Experiment is as follows: tube 1—pepsin, HCl water (control); tube 2—pepsin, HCl, water, starch; tube 3—same as tube 4 in figure 10.14, which is expected to be the only tube that will show digestion.
3. No. Experiment is as follows: 3 tubes, all having the contents of test tube 4 in figure 10.14. Place one in the cold, one at room temperature, and one in the incubator at body temperature. The last tube is expected to show the best digestion.
4. The contents are bile, fat, water, lipase, and $NaHCO_3$.

Answers to 10.3 Critical Thinking
1. The cell is lacking a particular enzyme.
2. A mutation must have occurred.
3. The plant would die. Plants take in only inorganic nutrients and must make all their organic nutrients.
4. Animals take in preformed food so they can depend on their diet to supply amino acids they cannot make.

Answers to 11.1 Critical Thinking
1. a. upper right side of heart; b. lower left side of heart; c. upper left side of heart; d. lower right side of heart.
2. The pulmonary circuit is not functioning. No, in the embryo and the fetus, blood is oxygenated at the placenta.

Answers to 11.2 Critical Thinking
Smoking
1. increases the heartbeat
2. decreases the bore of arteries
3. closes capillary beds in fingers and toes
4. increases resistance of blood flow

Answers to 11.3 Critical Thinking
1. lungs
2. brain
3. liver
4. Coronary arteries are the first blood vessels off the aorta; most likely, any clots in the coronary arteries or the capillaries formed right there.

Answers to 12.1 Critical Thinking
1. Proteins cannot cross cell membranes so there must be a way by which they can get into blood from the liver.
2. Blood volume decreases because blood proteins exert osmotic pressure.
3. The pH of blood would fluctuate more.
4. Each cell makes its own proteins from amino acids at the ribosomes.

Answers to 12.2 Critical Thinking
1. When CO_2 is added in the tissues, the reaction is driven to the right.

2. When CO_2 diffuses out of the lungs, the reaction is driven to the left.
3. When H^+ is added, more H_2CO_3 forms and since it dissociates little, the H^+ is taken up.
4. When OH^- is added, it combines with the H^+, forming water. Since water dissociates little, the OH^- is taken up.

Answers to 12.3 Critical Thinking
1. Type O has neither antigen A nor antigen B on the red blood cells and theoretically would not clump in a recipient's blood. It now is known that matching blood is more involved than just this consideration.
2. Type AB has neither anti-A antibody nor anti-B antibody in the plasma and theoretically would not cause clumping of red blood cells. It now is known that matching blood is more involved than just this consideration.
3. It is inherited. Genes exist for blood type.
4. Most likely there are other antigens aside from A and B on the red blood cells.

Answers to 13.1 Critical Thinking
1. When blood pressure rises, more fluid is pushed out of the capillary; when blood pressure decreases, less fluid is pushed out of the capillary.
2. When osmotic pressure decreases, less fluid is retrieved by the blood; when osmotic pressure rises, more fluid is retrieved by the blood.
3. Osmotic pressure is much lower because it is largely dependent on a protein concentration gradient between blood and tissue fluid. Blood pressure will decrease because less fluid is being retrieved by the blood. Tissue fluid will continue to form as long as blood pressure is higher than osmotic pressure.

Answers to 13.2 Critical Thinking
1. B cells produce antibodies, when stimulated by helper T cells. T cell maturation occurs in the thymus.
2. No, all B cells do not bind because each B cell is specific for only one type of antigen.
3. T cells do not recognize an antigen unless it is presented by an APC.
4. They communicate by surface-to-surface interaction and by chemical signals (i.e., lymphokine molecules).

Answers to 13.3 Critical Thinking
1. An antigen is a foreign substance that induces an immune response by B or T cells.

2. First, the process described in figure 13.15 has to occur 4 times—once for each type of T cell. Tag each prepared monoclonal antibody with a different dye. Apply all 4 types of antibodies to the blood sample, and view the sample with a microscope, looking for the different types of dyes.
3. Prepare a monoclonal antibody against HSV-2. Apply it to a sample of HSV-2 and HSV-1. The antibody should combine only with HSV-2.
4. Only the desired substance combines with the antibody, and therefore it is separated out. (There is a chemical process that later releases the described substance from the tube.)

Answers to 14.1 Critical Thinking
1. With negative pressure, air is drawn in, and with positive pressure, air is pushed in. Frogs force air into the lungs by gulping it.
2. There is no mixing of used air with new air coming in; therefore, more O_2 enters the blood.
3. Frogs practice skin breathing. Reptiles have better-developed lungs.
4. Movement of the diaphragm assists in creating the negative pressure that draws air into the lungs.

Answers to Critical Thinking 14.2
1. CO_2 stimulates breathing; O_2 does not.
2. The buildup of CO_2 stimulates breathing.
3. CO_2 in blood raises the pH of blood.
4. Sense receptors usually are stimulated by the presence of something.

Answers to 15.1 Critical Thinking
1. Urea is a single molecule, page 000. Urine is a mixture of molecules and ions (table 15.2).
2. The force is osmotic pressure.
3. This increases blood pressure in the glomerulus.
4. The rate increases because osmotic pressure would decrease in blood and increase in Bowman's capsule.

Answers to 15.2 Critical Thinking
1. To say that urine is 95% water only indicates how much water per solutes there is.
2. Carriers can only work so fast. The fluid is moving in the proximal convoluted tubule, and in the meantime, glucose has gone by.

3. Refer to this equation:

$$CO_2 + H_2O \rightleftharpoons H_2CO_3 \rightleftharpoons H^+ + HCO_3^-$$

When lungs excrete CO_2, the equation is driven to the left and blood becomes more basic. When kidneys excrete the HCO_3^-, the equation is driven to the right and blood becomes more acidic.
4. The pH affects enzymes, causing a change in shape so that they do not function as well.

Answers to 15.3 Critical Thinking
1. Filtration, to a degree, and excretion are part of hemodialysis. Reabsorption and tubular excretion are absent.
2. The membrane is semipermeable, and proteins are too large to pass through.
3. Reabsorption into blood does not occur during hemodialysis. The dialysate is always the area of lesser urea concentration.
4. Glucose in the same concentration as is normal for blood should be added.

Answers to 16.1 Critical Thinking
1. The nerve impulse travels along a membrane and is dependent on the Na^+ and K^+ across the membrane.
2. A reading lower than -65 mV is expected. The resting potential is -65 mV, and inhibitory neurotransmitters increase the polarity.
3. Synaptic vesicles occur only at one end of an axon.
4. The degree of contraction depends on the number of neurons that are stimulating muscle fibers.

Answers to 16.2 Critical Thinking
1. Interneurons can take nerve impulses across the spinal cord from one side to the other.
2. Neither leg would respond because nerve impulses would never reach the cord.
3. The right leg would still be able to respond.
4. Neither leg would respond because interneurons would be destroyed.

Answers to 17.1 Critical Thinking
1. Bone is living tissue: it grows and heals; it is supplied with blood and nerves; and it contains cells.
2. Bone strength has to equal muscle strength or movement of muscles can cause bones to crack or to break.
3. These are for attachment of muscles.
4. The wide pelvis might be associated with childbirth.

Answers to 17.2 Critical Thinking
1. The tendons of muscles extend across joints. When muscles contract, they pull on the bone to which a tendon is inserted.
2. When muscles contract, they can only shorten—they do not get longer.
3. Nerves cause muscles to contract.
4. The legs support the weight of the body and are larger.

Answers to 17.3 Critical Thinking
1. Yes, all myofibrils contract because a muscle fiber does not have degrees of contraction.
2. Yes, it gets closer to the center.
3. Myoglobin has the higher affinity or else it could never receive O_2 from hemoglobin.
4. Mitochondria in muscle fibers use the oxygen to receive H at the electron transport system.

Answers to 18.1 Critical Thinking
1. Possible pro argument: Sense organs are sensitive to outside stimuli. Possible con argument: Sense organs do not imbibe anything; outside stimuli cause them to generate nerve impulses.
2. The subject would report sensation only in the appropriate body part.
3. A mechanistic view is supported because the stimulus causes generation of nerve impulses that go to the brain, where integration occurs. Only mechanical and biochemical mechanisms are involved.

Answers to 18.2 Critical Thinking
1. One possible categorization: focusing—lens, cornea, humors, ciliary body; vision—retina (rods, cones, fovea), optic nerve; other—iris, pupil, choroid, sclera (except for cornea). Justification: Some parts of the eye are concerned with focusing the light, some with bringing about vision, and some have neither of these functions. Glasses usually correct focusing.
2. Pigments are all colored molecules that usually are capable of absorbing energy. These 3 pigments absorb solar energy.
3. There must be a neural pathway between the eyes and the pineal gland.

Answers to 18.3 Critical Thinking
1. The evolution of the ear in a sequence of animals is needed to support the hypothesis. You would expect to find stages by which changes led to the mammalian ear.
2. Most likely, the inner ear evolved from the lateral line. Most likely, the outer ear and the middle ear evolved

otherwise. The human inner ear has mechanoreceptors sensitive to fluid pressure waves. The outer ear receives sound waves in the air, and the middle ear transmits and amplifies these.

Answers to 19.1 Critical Thinking
1. You would expect to find sugar in the urine because the body would contain no insulin; blood sugar would rise and spill over into the urine.
2. No, your findings only prove that blood sugar rises when the pancreas is missing.
3. You have to get a supply of pure insulin, inject it in an animal, and show that the blood sugar lowers.
4. Yes, you now know that the presence of both the pancreas and insulin lowers blood sugar. The logical conclusion is that the pancreas is the source of insulin.
5. You actually could extract insulin from the pancreas.

Answers to 19.2 Critical Thinking
1. You might want to change the ending of the definition to "target cell, organ, or organism." Figure 19.19 shows the liver as a target organ for insulin and a moth as a target organism for a pheromone.
2. A target cell must have receptors for the environmental signal.
3. Behavior is a good term because it encompasses a wide variety of responses. For example, a target cell changes its metabolism and might begin to produce a product; the liver begins to store glucose as glycogen in response to insulin; and the male moth begins to fly toward the female.
4. Low blood sugar is the stimulus for the secretion of insulin, and negative feedback shuts down its production. When a nerve impulse is crossing a synapse. The neurotransmitter then either is broken down or is taken up by the presynaptic membrane.

Answers to 20.1 Critical Thinking
1. Due to negative feedback, the intake of anabolic steroids causes the anterior pituitary to stop producing gonadotropic hormones, leading to atrophy of the interstitial cells of the testes.
2. To test the hypothesis, administer anabolic steroids to mice and collect data on resulting blood levels of gonadotropic hormones. Remove the testes and using a microscope, look for atrophy of tissues.

3. Anabolic steroids raise the level of LDL in the blood. This could lead to increased risk of heart disease.
4. To test the hypothesis, administer anabolic steroids to mice and collect data on resulting blood levels of LDL. Remove coronary blood vessels and look for the presence of plaque.

Answers to 20.2 Critical Thinking

1. Due to negative feedback, the administration of estrogen and progesterone causes the anterior pituitary to stop producing FSH and LH, and no follicles or oocytes mature. Without egg production, there can be no pregnancy.
2. Administer this birth control pill to mice. Collect data on blood levels of FSH and LH. Remove the ovaries, and examine them for the presence of mature follicles.
3. Postmenopausal women who take birth control pills should have minor or no levels of FSH and LH in the blood.
4. Administer birth control pills to postmenopausal women and collect data on blood levels of FSH and LH.

Answers to 21.1 Critical Thinking

1. When a cell inherits a certain cytoplasmic composition, only certain genes are activated. These activated genes begin to direct the synthesis of particular proteins. Some of these proteins may be secreted to act as specific signals for other cells.
2. Tissue A gives off certain signals that influence the morphogenesis of tissue B. Because of this, tissue B gives off certain signals that influence the morphogenesis of tissue C. Tissue C then gives off signals and so forth.

Answers to 22.1 Critical Thinking

1. Alternative hypotheses (a) factors do not segregate; therefore, all parental gametes would be the same—that is, Yy. (b) factors do segregate; therefore, 2 parental gametes are possible—that is, Y and y.
2. If the gametes were always Yy, then the phenotype green would not have appeared. Since green does appear, then the second hypothesis is supported.
3. Alternative hypotheses for figure 22.6: (a) factors do not assort independently; therefore, the gametes always will be, for example, either WS or ws; (b) factors do assort independently; therefore, 4 gametes are possible—WS, Ws, wS, ws.

4. If the factors do not assort, then there would be fewer phenotypes among the offspring. Since there are all 4 possible phenotypes among the offspring, the factors had to assort independently of one another.

Answers to 22.2 Critical Thinking

1. The fault is an inability to produce melanin because of an enzyme defect.
2. An inability to produce a normal enzyme is most likely recessive.
3. The possible crosses are $aa \times aa$; $Aa \times aa$; $Aa \times Aa$. Yes, Aa individuals are carriers.
4. A functioning gene would ensure that all skin cells are capable of producing melanin.

Answers to 23.1 Critical Thinking

1. Red eye is dominant.
2. No, because females do not have a Y chromosome, and yet they have red eye color. Yes, because males have only one X chromosome, and this explains why only males have white eyes in the F_2 generation.
3. The results are explainable only on the basis that the red/white allele is on the X chromosome.

Answers to 24.1 Critical Thinking

1. The information is the DNA's code for protein synthesis, and it is stored in the sequence of bases. The evidence is that the sequence of amino acids in a protein parallels the code in DNA.
2. DNA can replicate because complementary base pairing occurs between DNA strands. Your evidence is the existence of duplicated chromosomes. Chromosomes are found only in the nucleus.
3. The sequence of bases in DNA can change, and this allows mutation. Living things differ from one another. Also, inborn errors of metabolism occur.

Answers to 24.2 Critical Thinking

1. It would show up first in the nucleus and then in the cytoplasm.
2. Put your sample mRNA and rRNA in a test tube with the cellular elements needed for protein synthesis. Analyze the sequence of amino acids in the resulting polypeptide, and see which of these apparently is directing the sequence.
3. Figure 24.14b shows ribosomes moving along the mRNA.
4. Yes, because there are nucleotides, not amino acids, in RNA. If amino acids are present, they must have combined with tRNA.

Answers to 24.3 Critical Thinking

1. Yes, the phenotypes of cells differ. For example, muscle cells look quite different from nerve cells.
2. Yes, all cells contain all genes. The process of mitosis ensures that all cells receive a full complement of genes.
3. Only certain mRNA transcripts will be found in different cell types. Cells that look different must be constructed differently and must contain different proteins. For example, only muscle cells contain actin and myosin.
4. Regulator genes control which structural genes are active.

Answers to 25.1 Critical Thinking

1. Inject a large number of diabetics with both types of insulin (at different times), and observe any effects. The biotechnology insulin is expected to show fewer side effects because it is human insulin, not cattle or pig insulin—the sequence of amino acids is expected to be closer to that of the individual receiving the insulin. Also, it might be purer—it does not contain any substances other than insulin.
2. Feed the meat to 2 groups of human volunteers, and observe any effects. Since growth hormone is a protein, any present in the meat is denatured upon cooking or digested upon eating.
3. Same experiment as described in answer 2. First, feed the plants to animals, and if no effects are observed, then feed the plants to humans. The toxin might be harmful to humans.
4. Keep testing for the presence of the pollutant and the bacteria. See if the bacteria disappears. They run out of food, that is, the pollutant.

Answers to 26.1 Critical Thinking

1. There would have been no enzymes present before the first protein formed.
2. Proteins are unable to store genetic information or to replicate.
3. By the same mechanism as DNA—complementary base pairing.
4. Enzymes are needed for replication to occur and for other metabolic processes.

Answers to 27.1 Critical Thinking

1. Coyotes are adapted to a grassland environment.
2. Coyotes have their own line of descent in North America, and jackals have their own line of descent in Africa.
3. Look for fossils that show the different lines of descent.

Answers to 27.2 Critical Thinking

1. The plant species contains variations that are inheritable. On the mountaintop, plants that were shorter tended to survive and to reproduce, until only shorter plants were observed there.
2. You expect the plants still to be short because only these genes now are present in the gene pool.
3. Directional selection (toward shorter plants) was followed by stabilizing selection (most plants, then, tend to have genes for shortness).

Answers to 27.3 Critical Thinking

1. Postmating is wasted energy because animals have put energy into mating, and the offspring are not viable/fertile.
2. Premating evolves first because it represents incomplete reproductive isolation, while postmating represents complete reproductive isolation.
3. Habitat and behavioral mechanisms are good candidates. The visual mechanism is not listed, and most likely the birds are able to visually recognize their own species.

Answers to 28.1 Critical Thinking

1. If the zygote undergoes meiosis, there is no diploid adult.
2. It meiosis is delayed until after there is a diploid adult, the alternation of generations life cycle occurs.
3. If haploid spores join (then they act like gametes), the haploid generation is eliminated and the diplontic life cycle occurs.

Answers to 28.2 Critical Thinking

1. Members of the genus are multicellular, and they have the alternation of generations life cycle.
2. The organism would have to lose the flagella and gain a cell wall to be more plantlike. To be more animal-like, the organism would have to lose the chloroplasts.
3. You would look for centrioles because they give rise to cilia and flagella, structures associated with animal cells.

Answers to 28.3 Critical Thinking

1. Fungi live on dead organic matter. Without other living things, there is no dead organic matter.
2. In the *Chlamydomonas* cycle, the adult, the spores, and the gametes are flagellated. In the fungal cycle, the adult, the spores, and the gametes are not flagellated.
3. The body of a fungus may not have the ability to withstand dryness.

Answers to 29.1 Critical Thinking

1. No, vascular tissue would not be found because bryophytes lack vascular tissue.
2. Try to grow them in a dry location. Observe results.
3. The gametophyte (N) is dominant, and it bears the burden of adaptation to the environment.
4. If vascular tissue evolved, it is possible the bryophytes might spread into more habitats on land, but only if they also lost their dependence on water for reproduction.

Answers to 29.2 Critical Thinking

1. In humans, the male passes sperm directly to the female, who retains the egg within her body. In trees, pollen, which can resist drying out, carries the sperm to the vicinity of the egg. Then the pollen germinates to give a pollen tube, through which the sperm passes.
2. Humans have a blood vascular system; trees have xylem.
3. Humans have an internal skeleton of bone; trees are supported by xylem.
4. Humans can maintain a warm internal temperature; deciduous trees lose their leaves and become inactive.

Answers to 30.1 Critical Thinking

1. Animals (a) are heterotrophic, (b) locomote by means of contracting fibers, (c) are multicellular with specialized tissues, (d) have a diplontic life cycle, (e) have sex organs, and (f) do not always protect the zygote and the embryo.
2. An animal locomotes by means of contracting fibers.
3. Protozoans locomote and are heterotrophic.
4. Complex animals have tissues and organs that are highly differentiated.

Answers to 30.2 Critical Thinking

1. Insects have tracheae, and birds have lungs.
2. Insects have an external skeleton, and birds have feathers.
3. The male passes sperm to the female in both. Insects deposit eggs in water or in soil or provide them with a hard covering. Birds lay hard-shelled eggs.
4. Insects typically lay thousands of eggs and do not tend the young. Birds lay a few eggs and care for each offspring for a few months.

Answers to 31.1 Critical Thinking

1. Humans take in preformed food and have the power of locomotion by means of contracting fibers.

2. As embryos, humans have the 3 chordates characteristics: notochord, pharyngeal pouches, and dorsal hollow nerve cord. In adults, the vertebral column replaces the notochord.
3. Humans have hair and mammary glands.
4. Humans have long arms and legs, 5 fingers and 5 toes, opposable thumbs, nails and not claws, no snout and binocular vision with a poor sense of smell, and a large brain.

Answers to 31.2 Critical Thinking

1. Perhaps the upright posture allowed *A. afarensis* to see over tall grasses.
2. Branching occurs where the hominids are adapted variously. For example, *A. africanus* and *A. robustus* had different diets, and Neanderthals in Europe and *H. sapiens* in Africa were built differently.
3. *H. erectus* is markedly different from the previous fossils; many advances are evident. Similarly, Cro-Magnon is much more advanced than previous fossils.
4. All races can interbreed.

Answers to 32.1 Critical Thinking

1. This is scramble competition. The larvae all compete for the same resource, and each receives minimal food. Few, then, survive.
2. This is contest competition. With this type of competition, the unsuccessful may not survive and reproduce but the successful will survive and reproduce.
3. In nature, female blowflies lay many eggs in various locations. The larvae are likely to survive in some of these locations. Scramble competition is not a disadvantage under these circumstances. In nature, baboons stay within a territory. Under these circumstances, contest competition ensures that some members of the population survive.

Answers to 32.2 Critical Thinking

1. This suggests that the fungi are parasitic on the algae in a lichen because organic molecules can pass from algae to fungi via these organs.
2. This suggests that the fungi and the algae are mutualistic in a lichen because each member of the relationship benefits.
3. Yes, both parasitism and mutualism involve some sort of dependency.
4. Yes, because parasites usually harm their host; in a mutualistic relationship both members benefit.

5. Not all relationships fall neatly into one of the 3 types. Perhaps the best description of this relationship is controlled parasitism of the fungi on the algae.

Answers to 33.1 Critical Thinking

1. The conditions are temperature and amount of rainfall.
2. A grassland might replace a deciduous forest and a desert might replace a grassland.
3. The more sunlight; the more photosynthesis; the more life supported.
4. Similar species cannot play the same role (occupy the same niche) in a community; therefore, diversification occurs.
5. Varied and plentiful food sources provide different ways of getting food.

Answers to 33.2 Critical Thinking

1. A squid has adaptations for a water environment, not a land environment. It is streamlined and moves by jet propulsion—both of which are adaptations for locomotion in water. A squid breathes by means of gills.
2. A monkey has adaptations for a forest environment, not a grassland environment. It swings from limb to limb of a tree and mainly eats fruits.

3. A zebra has adaptations for a grassland environment, not a tropical rain forest environment. It has long legs for running, and it eats grass.
4. A polar bear has adaptations for the arctic tundra and not a desert. Its coat is thick for warmth and white for camouflage, and it feeds on fish.
5. From these answers, it appears that animals are adapted to their environment in terms of locomotion and food sources.

Answers to 34.1 Critical Thinking

1. Secondary consumers have more energy available to them (in the form of food) than tertiary consumers.
2. The pyramid indicates that there is less energy available at each trophic level. Eventually, there is not enough energy to support another population.
3. The size of a top predator population is controlled by the amount of food energy available to it.
4. The other secondary consumer populations would increase in size due to less competition, and the ecosystem would remain about the same.

Answers to 35.1 Critical Thinking

1. There can be 4 future generations. At that time, there will be 62 rabbits—32 newly born and 30 previous parents. If each new pair of rabbits now produces 4, there will be 126 rabbits, too many for the watering hole.
2. First generation: 2 rabbits; second generation: 6 rabbits; third generation: 14 rabbits; fourth generation: 30 rabbits; fifth generation: 62 rabbits. Yes. This is exponential growth.
3. It is a J-shape curve. The curve would fall dramatically.

Answers to 35.2 Critical Thinking

1. Formerly, it was assumed that each country was responsible for its own standard of living. This attitude is changing because planners now think in global terms.
2. Increasingly, people in the developed countries believe that health care for all people should be the same.
3. Most people now believe that private citizens and industry should find ways to ensure an ecologically fit world for future generations. If we do not, the standard of living will decrease dramatically.
4. Increasingly, people are beginning to think in global terms.

A

accommodation (ah-kom''o-da'shun)   lens adjustment in order to see close objects.   *373*

acetylcholine (as''ĕ-til-ko'lēn)   ACh; a neurotransmitter substance secreted at the ends of many neurons; responsible for the transmission of a nerve impulse across a synaptic cleft.   *328*

acetylcholinesterase (as''ĕ-til-ko''lin-es'ter-ās)   AChE; an enzyme that breaks down acetylcholine.   *328*

acid (as'id)   a solution in which pH is less than 7; a substance that contributes or liberates hydrogen ions (protons) in a solution.   *25*

acid deposition (as'id de'po-si-''shun)   an accumulation of acids from the atmosphere into lakes and forests, particularly in North America and Europe.   *752*

acromegaly (ak''ro-meg'ah-le)   condition resulting from an increase in growth hormone production after adult height has been achieved.   *391*

acrosome (ak'ro-sōm)   covering on the tip of a sperm that contains enzymes necessary for fertilization.   *412*

ACTH (adrenocorticotropic hormone);   hormone secreted by the anterior lobe of the pituitary gland that stimulates activity in the adrenal cortex.   *391*

actin (ak'tin)   one of the two major proteins of muscle; makes up thin filaments in myofibrils of muscle fibers. *See* myosin.   *36*

action potential (ak'shun po-ten'shal)   the change in potential propagated along the membrane of a neuron; the nerve impulse.   *325*

active acetate (ak'tiv as'ĕ-tāt)   an acetyl group attached to coenzyme A; a product of the transition reaction that links glycolysis to the Krebs cycle.   *107*

active site (ak'tiv sīt)   the region on the surface of an enzyme where the substrate binds and where the reaction occurs.   *101*

active transport (ak'tiv trans'port)   transfer of a substance into or out of a cell from a lesser to a greater concentration by a process that requires a carrier and expenditure of energy.   *73*

adaptation (ad''ap-ta'shun)   the fitness of an organism for its environment, including the process by which it becomes fit, in order that it may survive and reproduce; also the adjustment of sense receptors to a stimulus so that the stimulus no longer excites them.   *368*

adenosine triphosphate (ah-den'o-sēn tri-fos'fat) *See* ATP.

adrenocorticotropic hormone (ah-dre''no-kor''te-ko-trop'ik hor'mon)   *See* ACTH.

aerobic (a''er-ōb'ik)   growing or metabolizing only in the presence of oxygen, as in aerobic respiration.   *105*

aerobic respiration (a''er-ōb'ik res''pir-a'shun)   respiration in the presence of oxygen.   *559*

agglutination (ah-gloo''ti-na'shun)   clumping of cells, particularly in reference to red blood cells involved in an antigen-antibody reaction.   *254*

aging (āj'ing)   progressive changes in the human over time, leading to loss of physiological function and eventual death.   *461*

agranular leukocytes (ah-gran'u-lar lu'ko-sīt)   white blood cells that do not contain distinctive granules.   *250*

albinism (al'bi-nizm)   a genetic defect identified by the lack of melanin in the skin.   *514*

albumin (al-bu'min)   plasma protein of the blood with transport and osmotic functions.   *241*

aldosterone (al''do-ster'ōn)   a hormone, secreted by the adrenal cortex that functions in regulating sodium and potassium concentrations of the blood.   *313, 396*

algae (al'-jā)   aquatic organisms that carry on photosynthesis.   *596*

allantois (ah-lan'to-is)   one of the extra-embryonic membranes; in reptiles and birds, a pouch serving as a repository for nitrogenous waste; in mammals a source of blood vessels to and from the placenta.   *446*

allele (ah-lēl')   an alternative form of a gene that occurs at a given chromosome site (locus).   *469*

all-or-none response (al'or-nun' re-spons')   phenomenon in which a muscle fiber contracts completely when it is exposed to a stimulus of threshold strength.   *354*

alternation of generations (awl''ter-na'shun uv jen''ĕ-ra'shunz)   a life cycle typical of plants in which a diploid sporophyte alternates with a haploid gametophyte.   *165, 596, 611*

alveoli (al-ve'o-lī)   saclike structures that are the air sacs of a lung.   *283*

amino acid (ah-me'no as'id)   a unit of protein that takes its name from the fact that it contains an amino group (NH_2) and an acid group (COOH).   *30*

ammonia (ah-mo'ne-ah) NH_3, a nitrogenous waste product resulting from deamination of amino acids.   *300*

amnion (am'ne-on)   an extraembryonic membrane, a sac around the embryo that contains fluid.   *446*

amoeboids (ah-me'boyds)   protozoans that move by means of pseudopodia.   *594*

ampulla (am-pul'lah)   base of a semicircular canal in the inner ear.   *379*

amylase (am'i-lās)   an enzyme that catalyzes chemical breakdown of starch to maltose; salivary amylase works in the mouth, and pancreatic amylase works in the small intestine.   *200*

anabolic steroid (an''ah-bol'ik ster'oyd)   a synthetic steroid that mimics the effect of testosterone.   *400*

anaerobic organism (an-a''er-ob'ik or-gă-niz-ŭm)   organism that acquires energy by processes that do not utilize oxygen.   *559*

anaerobic respiration (an-a''er-ōb'ik res''pi-ra'shun)   growing or metabolizing in the absence of molecular oxygen.   *559*

Glossary

analogous structure (ah-nal'o-gus struk'-tūr) similar in function but not in structure; particularly in reference to similar adaptations. *568*

anaphase (an'ah-fāz) stage in mitosis during which chromatids separate, forming chromosomes. *86*

anemia (ah-nē'mē-ah) inefficient oxygen-carrying ability by the blood due to hemoglobin shortage. *245*

antennae (an-ten'e) sensory organs located on arthropod head. *650*

anterior pituitary (an-tē're-or pi-tu'ĭ-tār'e) a front lobe of the pituitary gland that produces 6 hormones and is controlled by the hypothalamus. *388*

anther (an'ther) that portion of a stamen in which pollen is formed. *165*

antheridium (an''ther-id'e-um) male organ in certain nonseed plants where swimming sperm are produced. *612*

anthropoids (an'thro-poidz) advanced primates, including only monkeys, apes, and humans. *670*

antibody (an'ti-bod''e) a protein produced in response to the presence of some foreign substance in the blood or tissues. *250, 264*

antibody-mediated immunity (an''ti-bod''e me'de-ā-tid i-mu'ni-te) body line of resistance with B cells producing antibodies. *264*

anticodon (an''ti-ko'don) a "triplet" of 3 nucleotides in transfer RNA that pairs with a complementary triplet (codon) in messenger RNA. *520*

antidiuretic hormone (an''ti-di''u-ret'ik hōr'mōn) (ADH); sometimes called vasopressin, a hormone secreted by the posterior pituitary that controls the rate at which water is reabsorbed by the kidneys. *313, 388*

antigen (an'ti-jen) a foreign substance, usually a protein, that stimulates the immune system to produce antibodies. *251, 263*

anti-oncogene (an''ti-on'ko-jēn'') a gene that apparently suppresses cancer. *531*

antipredator defense (an''ti-pred'ah-tor de-fens') a physiological or behavioral activity or a structural modification that protects an organism from its predators. *692*

anus (a'nus) inferior outlet of the digestive tube. *196*

anvil (an'vil) the middle bone of the 3 ossicles of the middle ear. *378*

aorta (a-or'tah) major systemic artery that receives blood from the left ventricle. *228*

appendicular skeleton (ap''en-dik'u-lar skel'ĕ-ton) portion of the skeleton forming the upper extremities, pectoral girdle, lower extremities, and pelvic girdle. *348*

appendix (ah-pen'diks) a small, tubular appendage that extends outward from the cecum of the large intestine. *195*

aqueous humor (a'kwe-us hu'mor) watery fluid that fills the anterior chamber of the eye. *371*

archaebacteria (ar'ke-bak-te're-ah) monerans that are able to live under adverse circumstances and represent an early branch of living organisms. *592*

archegonium (ar''kĕ-go'ne-um) female organ in certain nonseed plants where an egg is produced. *612*

archenteron (ar-ken'ter-on) central cavity or primitive gut in the animal embryo. *438*

arterial duct (ar-te're-al dukt) ductus arteriosus; fetal connection between the pulmonary artery and the aorta. *456*

arteriole (ar-te're-ōl) a vessel that takes blood from an artery to a capillary. *219*

artery (ar'ter-e) a vessel that takes blood away from the heart; characteristically possessing thick elastic walls. *219*

aster (as'ter) short rays of microtubules that appear at the ends of the spindle apparatus in animal cells during cell division. *85*

asymmetry (a-sim'ĕ-tre) lacking symmetry. *631*

atom (at'om) smallest unit of matter non-divisible by chemical means. *17*

atomic number (ah-tom'ik num'ber) the number of protons within the nucleus of an atom. *18*

atomic weight (ah-tom'ik wāt) the number of protons plus the number of neutrons within the nucleus of an atom. *18*

ATP (adenosine triphosphate); a compound containing adenine, ribose, and 3 phosphates, 2 of which are high-energy phosphates; the "common currency" of energy for most cellular processes. *41*

atria (a'tre-ah) chambers; particularly the upper chambers of the heart that lie above the ventricles. (*sing.* atrium) *222*

atrioventricular (a''tre-o-ven-trik'u-lar) a structure in the heart that pertains to both the atria and ventricles; for example, an atrioventricular valve is located between an atrium and a ventricle. *222*

atrioventricular node (a''tre-o-ven-trik'u-lar nōd) *See* AV node.

auditory canal (aw'dĭ-to''re kah-nal') a tube in the external ear that lies between the pinna and the tympanic membrane. *378*

auditory nerve (aw'di-to''re nurv) a nerve sending the signal for sound from the inner ear to the temporal lobe of the brain. *381*

australopithecines (aw''strah-lo-pith'e-sĭnz) referring to 3 species of *Australopithecus,* the first generally recognized hominids. *674*

autosome (aw'to-sōm) a chromosome other than a sex chromosome. *80*

autotroph (aw'to-trōf) an organism that is capable of making its food (organic molecules) from inorganic molecules. *123*

AV node (atrioventricular node); a small region of neuromuscular tissue that transmits impulses received from the SA node to the ventricular walls. *225*

axial skeleton (ak'se-al skel'ĕ-ton) portion of the skeleton that supports and protects the organs of the head, neck, and trunk. *348*

axon (ak'son) process of a neuron that conducts nerve impulses away from the cell body. *320*

B

bacteria (bak-te're ah) prokaryotes that lack the organelles of eukaryotic cells; archaebacteria and eubacteria, including cyanobacteria. *589*

bacteriophage (bak-te're-o-fāj'') a virus that infects a bacterial cell. *585*

basal bodies (ba'sal bod'ēz) short cylinders having a circular arrangement of 9 microtubule triplets (9 + 0 pattern) located within the cytoplasm at the bases of cilia and flagella. *57*

base (bās) a solution in which pH is more than 7; a substance that contributes or liberates hydroxide ions in a solution; alkaline; opposite of acidic. Also, in genetics the chemicals adenine, guanine, cytosine, thymine, and uracil that are found in DNA and RNA. *25*

bilateral symmetry (bi-lat'er-al sim'ĕ-tre) the condition of having a right and left half so that only one vertical cut gives two equal halves. *632*

bile (bīl) a secretion of the liver that is temporarily stored in the gallbladder before being released into the small intestine where it emulsifies fat. *198*

binary fission (bi'na-re fish'un) reproduction by division into two equal parts by a process that does not involve a mitotic spindle. *590*

biological evolution (bi-o-loj'ĕ-kal ev''o-lu'shun) changes that have occurred in life forms from the origination of the first cell to the many diverse forms in existence today. *556*

biome (bi'ōm) one of the major climax communities present in the biosphere characterized by a particular mix of plants and animals. *701*

biosphere (bi'o-sfer) that part of the earth's surface and atmosphere where living organisms exist. *7, 701*

biotechnology (bi''o-tek-nol'o-je) use of a natural biological system to make a new product or achieve a desired end. *534*

biotic potential (bi-ot'ik po-ten'shal) the maximum population growth rate under ideal conditions. *739*

biped (bi'ped) an organism walking on 2 limbs. *673*

black bread mold (blak bred mōld) a fungi of genus *Rhizopus* forming a whitish or grayish mycelium on bread or fruit. *603*

blade (blād) the main portion of a leaf. *150*

blind spot (blīnd spot) area containing no rods and cones where the optic nerve passes through the retina. *372*

blood (blud) connective tissue, composed of cells separated by plasma, which transports substances in the cardiovascular system. *178*

blood pressure (blud presh'ur) the pressure of blood against the wall of a blood vessel. *229*

B lymphocyte (lim'fo-sīt) a lymphocyte that matures in the bone marrow and differentiates into antibody-producing plasma cells when stimulated by the presence of a specific antigen. *264*

bone (bōn) connective tissue having a hard matrix of calcium salts deposited around protein fibers. *178*

Bowman's capsule (bo'manz kap'sūl) a double-walled cup that surrounds the glomerulus at the beginning of the kidney tubule. *305*

bradykinin (brad''e-ki'nin) a substance found in damaged tissue that initiates nerve impulses resulting in the sensation of pain. *252*

breathing (brēth'ing) entrance and exit of air into and out of the lungs. *280*

bronchi (brong'ki) the 2 major divisions of the trachea leading to the lungs. *282*

bronchiole (brong'ke-ōl) the smaller air passages in the lungs of mammals. *283*

browser (browz'er) an animal that feeds on higher-growing vegetation such as shrubs and trees. *689*

bryophyte (bri'o-fīt) a nonvascular plant including liverworts and mosses. *612*

buffer (buf'er) a substance or compound that prevents large changes in the pH of a solution. *28*

C

C₃ photosynthesis (se thre fo''to-sin'thē-sis) photosynthesis that utilizes the Calvin cycle to take up and then fix carbon dioxide; so named because the first molecule detected after CO_2 uptake is a C_3 molecule. *131*

C₄ photosynthesis (se for fo''to-sin'thē-sis) photosynthesis in which the first detected molecule following CO_2 uptake is a C_4 molecule. Later, this same CO_2 is made available to the Calvin cycle. *136*

calcitonin (kal''si-to'nin) hormone secreted by the thyroid gland that helps to regulate the level of blood calcium. *395*

Calvin-Benson cycle (kal'vin ben'sun si'kl) a circular series of reactions by which CO_2 fixation occurs within chloroplasts. *131*

capillaries (kap'i-lar''ēz) microscopic vessels connecting arterioles to venules through whose thin walls molecules either exit or enter the blood. *219*

capsid (kap'sid) the outer coat of a virus composed of protein subunits. *584*

carapace (kar'ah-pās) upper covering (shell) of some animals. *650*

cardiac muscle (kar'de-ak mus'l) specialized type of muscle tissue found only in the heart. *180*

carnivore (kar ni-vōr) an animal that feeds only on animals. *720*

carpal (kar'pal) a bone that is located in the human wrist. *349*

carrier (kar'e-er) a molecule that combines with a substance and actively transports it through the cell membrane; an individual that unknowingly transmits an infectious or genetic disease. *72, 479*

carrying capacity (kar'e-ing kah-pas'i-te) the largest number of organisms of a particular species that can be maintained indefinitely in an ecosystem. *740*

cartilage (kar'ti-lij) a connective tissue in which the cells lie within lacunae embedded in a flexible matrix. *178*

Casparian strip (kas-par'e-an strip) band of waxy material found on endodermal cells of plants; prevents passage of molecules outside of cells. *147*

CCK (cholecystokinin); hormone produced by duodenum that stimulates release of bile from gallbladder. *198*

cell (sel) the structural and functional unit of an organism; the smallest structure capable of performing all the functions necessary for life. *44*

cell body (sel bod'e) portion of a nerve cell that includes a cytoplasmic mass and a nucleus and from which the nerve fibers extend. *320*

cell cycle (sel si'kl) a cyclical series of phases that includes cellular events before, during, and after mitosis. *83*

cell-mediated immunity (sel me'de-ā-tid i-mu'ni-te) body line of resistance with T cells destroying antigen-bearing cells. *266*

cell membrane (sel mem'brān) a membrane that surrounds the cytoplasm of cells and regulates the passage of molecules into and out of the cell. *65*

cell plate (sel plāt) a structure that forms between two plant cells during telophase and marks the location of the cell membrane and cell wall. *88*

cell wall (sel wal) a protective barrier outside the cell membrane of a bacterial, fungal, algal, or plant cell. *67*

cellular respiration (sel'u-lar res''pi-ra'shun) the metabolic reactions that provide ATP energy to a cell. *54, 280*

cellulose (sel'u-lōs) a polysaccharide composed of glucose molecules; the chief constituent of a plant's cell wall. *35*

central canal (sen'tral kah-nal') tube within the spinal cord that is continuous with the ventricles of the brain and contains cerebrospinal fluid. *334*

central nervous system (CNS) (sen'tral ner'vus sis'tem) the brain and spinal cord in vertebrate animals. *320*

centriole (sen'tre-ōl) a short, cylindrical organelle in animal cells that contains microtubules in a 9 + 0 pattern and is associated with the formation of the spindle during cell division. *57*

centromere (sen'tro-mēr) a region of attachment for a chromosome to a spindle fiber that is generally seen as a constricted area. *80*

cephalothorax (sef''ah-lo-tho'raks) fusion of head and thoracic regions displayed by some arthropods. *650*

cerebellum (ser''ē-bel'um) the part of the vertebrate brain that controls muscular coordination. *336*

cerebral hemisphere (ser'ē-bral hem'i-sfēr) one of the large, paired structures that together constitute the cerebrum of the brain. *336*

cerebrospinal fluid (ser''ē-bro-spi'nal floo'id) fluid present in ventricles of brain and in central canal of spinal cord. *334*

chaparral (shap-e-ral') a biome of broad-leaved evergreen shrubs forming dense thickets. *706*

chemical evolution (kem'i-kal ev''o-lu'shun) a gradual increase in the complexity of chemical compounds that is believed to have brought about the origination of the first cell or cells. *553*

chemosynthesis (ke''mo-sin'the-sis) the process of making food by using energy derived from the oxidation of reduced molecules in the environment. *137*

chimpanzee (chim-pan'ze) a small great ape that is closely related to humans and is frequently used in psychological studies. *673*

chitin (ki'tin) flexible, strong polysaccharide forming the exoskeleton of arthropods. *648*

chlorophyll (klo'ro-fil) the green pigment found in photosynthesizing organisms that is capable of absorbing energy from the sun's rays. *55, 125*

chloroplast (klo'ro-plast) a membrane-bounded organelle that contains chlorophyll and where photosynthesis takes place. *54*

cholecystokinin (ko''le-sis''to-ki'nin) *See* CCK.

chorion (ko're-on) an extraembryonic membrane that forms an outer covering around the embryo; in reptiles and birds, it functions in gas exchange; in mammals it contributes to the formation of the placenta. *446*

chorionic villi (ko''re-on'ik vil'i) treelike extensions of the chorion of the mammalian embryo, projecting into the maternal tissues. *452*

choroid (ko'roid) the vascular, pigmented middle layer of the wall of the eye. *371*

chromatids (kro'mah-tidz) the two identical parts of a chromosome following replication of DNA. *80*

chromatin (kro'mah-tin) threadlike network in the nucleus that is made up of DNA and proteins. *49, 80*

chromosomes (kro'mo-sōmz) rod-shaped bodies in the nucleus, particularly during cell division, that contain the hereditary units or genes. *49, 80*

cilia (sil'e-ah) hairlike projections that are used for locomotion by many unicellular organisms and have various purposes in higher organisms. *57*

ciliary muscle (sil'e-er''e mus'el) a muscle that controls the curvature of the lens of the eye. *371*

ciliates (sil'e-āts) protozoans that move by means of cilia. *595*

circadian rhythm (ser''kah-de'an rith'm) a regular physiological or behavioral event that occurs on an approximately 24-hour cycle. *403*

circumcision (ser''kum-sizh'un) removal of the foreskin of the penis. *415*

citrate (sit'rāt) citric acid, the beginning substrate of the Krebs cycle. *112*

citric acid cycle (sit'rik as'id si'kl) *See* Krebs cycle.

clavicle (klav'i-k'l) a slender, rodlike bone located at the base of the neck that runs between the sternum and the shoulders. *349*

cleaning symbiosis (klēn'ing sim''bi-ō'sis) a mutualistic relationship in which one type of organism gains benefit by cleaning another that is benefited by having been cleaned of debris and parasites. *697*

clitellum (kli-tel'um) the smooth girdle on the body of an earthworm, secreting mucus. *648*

clone (klōn) asexually produced organisms having the same genetic makeup; also DNA fragments from an external source that have been reproduced by *E. coli*. *535*

clotting (klot'ing) process of blood coagulation, usually when injury occurs. *248*

cnidarian (ni-dah're-an) small aquatic animals having radial symmetry and bearing stinging cells with nematocysts. *634*

coacervate droplet (ko-as'er-vāt drŏp-let) a mixture of polymers that may have preceded the origination of the first cell or cells. *556*

cochlea (kok'le-ah) that portion of the inner ear that resembles a snail's shell and contains the organ of Corti, the sense organ for hearing. *379*

cochlear canal (kok'le-ar kah-nal') canal within the cochlea bearing small hair cells which function as hearing receptors. *379*

codon (ko'don) a "triplet" of 3 nucleotides in messenger RNA that directs the placement of a particular amino acid into a polypeptide chain. *517*

coelom (se'lom) body cavity of higher animals that contains internal organs such as those of the digestive system. *184, 633*

coenzyme (ko-en'zīm) a nonprotein molecule that aids the action of an enzyme, to which it is loosely bound. *103*

coenzyme A (ko-en'zīm a) coenzyme which participates in the transition reaction and carries the organic product to the Kreb's cycle. *111*

coevolve (ko-e-volv') the interaction of two species such that each determines the evolution of the other species. *688*

cohesion-tension theory (ko-he'zhun ten'shun the'o-re) explanation for upward transportation of water in xylem based upon transpiration created tension and the cohesive properties of water molecules. *156*

collar cells (kol'ler selz) flagella-bearing cells found in inner layer of the wall of sponges. *633*

collecting duct (kŏ-lekt'ing dukt) a tube that receives urine from several nephrons within a kidney. *305, 310*

colon (ko'lon) the large intestine of vertebrates. *196*

colony (kol'o-ne) an organism that is a loose collection of cells that are specialized and cooperate to a degree. *599*

colostrum (kŏ-los'trum) watery, yellowish-white fluid produced by the breasts. *424*

columnar epithelium (ko-lum'nar ep''ĭ-the le-um) pillar-shaped cells usually having the nuclei near the bottom of each cell and found lining the digestive tract, for example. *173*

commensalism (kŏ-men'sal-izm) the relationship of two species in which one lives on or with the other without conferring either benefit or harm. *695*

common ancestor (kŏ'mun an'ses-tor) an ancestor to two or more branches of evolution. *566*

compact bone (kom-pakt'bōn) hard bone consisting of Haversian systems cemented together. *178, 350*

companion cell (kom-pan'yun sel) a specialized cell that lies adjacent to a sieve-tube cell. *157*

competition (kom''pē-tish'un) interaction between members of the same or different species for a mutually required resource. *687*

competitive exclusion principle (kom pet'ĭ-tiv eks-kloo'zhun prin'sĭ-p'l) an observation that no two species can continue to compete for the same exact resources since one species will eventually become extinct. *687*

complement system (kom'plē-ment sis'tem) a group of proteins in plasma that produce a variety of effects once an antigen-antibody reaction has occurred. *263*

complementary base pairing (kom''plē-men'tă-re bās pār'ing) pairing of bases found in DNA and RNA; adenine is always paired with either thymine (DNA) or uracil (RNA) and cytosine is always paired with guanine. *510*

compound (kom'pownd) 2 or more atoms of different elements that are chemically combined. *19*

compound eye (kom'pownd i) arthropod eyes composed of multiple lenses. *650*

concentration gradient (kon sen trā'shun grā'di-ent) the difference in the solute concentration between 2 regions. *69*

cones (kōnz) bright-light receptors in the retina of the eye that detect color and provide visual acuity; specialized structures composed of scale-shaped leaves in conifers. *371*

conidia (ko-nid'e-ah) spores produced by sac and club fungi during asexual reproduction. *604*

conifer (kon'ĭ-fer) a cone-bearing seed plant, mostly trees, such as pines. *619*

conjugation (kon''ju-ga'shun) a sexual union in which the nuclear material of one cell enters another. *595*

connective tissue (kŏ-nek'tiv tish'u) a type of tissue, characterized by cells separated by a matrix that often contains fibers. *176*

consumers (kon-su'merz) organisms of one population that feed on members of other populations in an ecosystem. *720*

control (kon-trōl') in experimentation a sample that undergoes all the steps in the experiment except the one being tested. *10*

coral (kor'al) a cnidarian that has a calcium carbonate skeleton whose remains accumulate to form reefs. *636*

coral reef (kor'al rēf) a structure found in tropical waters formed by the buildup of coral skeletons where many and various types of organisms reside. *713*

cork cambium (kork kam'be-um) meristem tissue that produces cork. *148*

coronary artery (kor'ŏ-na-re ar'ter-e) an artery that supplies blood to the wall of the heart. *228*

corpus callosum (kor'pus kah-lo'sum) a mass of white matter within the brain, composed of nerve fibers connecting the right and left cerebral hemispheres. *338*

corpus luteum (kor'pus lut'e-um) a body, yellow in color, that forms in the ovary from a follicle that has discharged its egg. *418*

cortex (kor'teks) in animals, the outer layer of an organ; in plants, the tissue beneath the epidermis in certain stems. *304*

cortisol (kor'ti-sol) a glucocorticoid secreted by the adrenal cortex. *396*

cotyledon (kot''ĭ-le'don) the seed leaf of the embryo of a plant. *142, 167*

covalent bond (ko'va-lent bond) a chemical bond between atoms that results from the sharing of a pair of electrons. *21*

covalent reaction (ko'va-lent rē ak'shun) a chemical change that involves the formation of a covalent bond. *21*

Cowper's glands (kow'perz glandz) 2 small structures located below the prostate gland in males. *414*

cranial nerve (kra'ne-al nerv) nerve that arises from the brain. *329*

creatine phosphate (kre'ah-tin fos'făt) a compound unique to muscles that contains a high-energy phosphate bond. *361*

creatinine (kre-at'ĭ-nin) excretion product from creatine phosphate breakdown. *301*

cretinism (kre'tin-izm) a condition resulting from a lack of thyroid hormone in an infant. *394*

cristae (kris'tah) shelflike projections that extend into the matrix of the mitochondrion. *107*

cri du chat syndrome (krē'du-kat sin'drŏm) a group of body malfunctions caused by a deletion of chromosome 5. *495*

Cro-Magnon (kro-mag'non) the common name for the first fossils to be accepted as representative of modern humans. *679*

crossing-over (kros'ing o'ver) the exchange of corresponding segments of genetic material between nonsister chromatids of homologous chromosomes during synapsis of meiosis I. *89*

cuboidal epithelium (ku-boid'al ep''ĭ-the'le-um) cube-shaped cells found lining, for example, the kidney tubules. *173*

cultural eutrophication (kul'tu-ral u''tro-fi-ka'shun) enrichment of a body of water causing excessive growth of producers and then death of these and other inhabitants. *747*

cyanobacteria (si''ah-no-bak-te're-ah) photosynthetic prokaryotes that contain chlorophyll and release O_2; formerly called blue-green algae. *593*

cycad (si'kad) a cone-bearing, palmlike plant found in the tropics and subtropics. *619*

cyclic photophosphorylation (sik'lik fo''to-fos''for-i-la'shun) the synthesis of ATP by a cytochrome system within chloroplasts utilizing electrons that are repeatedly solar energized. *130*

cystic fibrosis (sis'tik fi-bro'sis) a lethal genetic disease involving problems with the functions of the mucous membranes in the respiratory and digestive tracts. *480*

cytochrome system (si''to-krōm sis'tem)   *See* respiratory chain.   *129*

cytokinesis (si''to-ki-ne'sis)   division of the cytoplasm of a cell.   *87*

cytoplasm (si'to-plazm)   the ground substance of cells located between the nucleus and the cell membrane.   *44*

cytoskeleton (si''to-skel'ĕ-ton)   filamentous protein structures found throughout the cytoplasm that help maintain the shape of the cell.   *55*

cytotoxic T cells (si''to-tok'sik te selz)   T lymphocytes that attack cells bearing foreign antigens.   *266*

D

data (da'tah)   experimentally derived facts.   *9*

daughter cells (daw'ter selz)   cells formed by division of a mother cell.   *84*

deamination (de-am''ĭ-na'shun)   removal of an amino group ($-NH_2$) from an amino acid or other organic compound.   *109, 199*

deciduous (de-sid'u-us)   plants that shed their leaves at certain seasons.   *707*

decomposers (de-kom-po'zerz)   organisms of decay (fungi and bacteria) in an ecosystem.   *720*

dehydration synthesis (de''hi-dra'shun sin'thĕ-sis)   the joining together of molecules to form macromolecules by removing components that form water.   *29*

dehydrogenase enzyme (de-hi'dro-jen-ās en'zim)   an enzyme that accepts hydrogen atoms, speeding up the process of dehydrogenation.   *104*

deletion (dē lē'shun)   a chromosome mutation caused by breakage and loss of a fragment of chromosome.   *494*

demographic transition (dem-o-graf'ik tran-zi'shun)   a population change from a high birthrate to a low birthrate so that the growth rate is lowered.   *740*

dendrite (den'drīt)   process of a neuron, typically branched, that conducts nerve impulses toward the cell body.   *320*

denitrification (de-ni''tri-fi-ka'shun)   the process of converting nitrate to nitrogen gas; is a part of the nitrogen cycle.   *728*

deoxyribonucleic acid (de-ok''se-ri''bo-nu-kle'ik as'id)   *See* DNA.

dermis (der'mis)   the thick skin layer that lies beneath the epidermis.   *182*

desert (de'zert)   an arid biome characterized especially by plants such as cacti that are adapted to receiving less than 25 cm rain per year.   *702*

desertification (dez-ert''i-fi-ka'shun)   desert conditions caused by human misuse of land.   *744*

detritus (de-tri'tus)   nonliving organic matter.   *720*

developed countries (de-vel'opt kun'trēz)   industrialized nations that typically have a strong economy and a low rate of population growth.   *740*

diabetes insipidus (di''ah-be'tēz in-sip'i-dus)   condition characterized by an abnormally large production of urine due to a deficiency of antidiuretic hormone.   *390*

diabetes mellitus (di''ah-be'tez mĕ-li'tus)   condition characterized by a high blood glucose level and the appearance of glucose in the urine due to a deficiency of insulin production or uptake by cells.   *398*

diaphragm (di'ah-fram)   a sheet of muscle that separates the chest cavity from the abdominal cavity in higher animals. Also, a birth control device inserted in front of the cervix in females.   *285*

diastole (di-as'to-le)   relaxation of heart chambers.   *223*

diastolic blood pressure (di''ah-stol'ik blud presh'ur)   arterial blood pressure during the diastolic phase of the cardiac cycle.   *230*

diatoms (di'ah-tomz)   a large group of fresh and marine unicellular algae having a cell wall consisting of two silica impregnated valves that fit together as in a pill box.   *602*

dicot (di'kot)   (dicotyledon); a type of angiosperm distinguished particularly by the presence of two cotyledons in the seed.   *142*

differentiation (dif''er-en''she-a'shun)   the process and developmental stages by which a cell becomes specialized for a particular function.   *436*

diffusion (di-fu'zhun)   the movement of molecules from an area of greater concentration to an area of lesser concentration.   *68*

digit (dij'it)   a finger or toe.   *349*

dihybrid (di-hi'brid)   the offspring of parents who differ in two ways; shows the phenotype governed by the dominant alleles but carries the recessive alleles.   *475*

dinoflagellates (di''no-flaj'e-lāts)   a large group of marine unicellular algae that have two flagella; one circles the body while the other projects posteriorly.   *602*

dipeptide (di-pep'tīd)   a molecule consisting of only two amino acids joined by a peptide bond.   *31*

dipleurula (di-ploor'u-lah)   a larval form unique to the deuterostomes that indicates that they are related.   *642*

diplontic life cycle (dip-lon'tik līf' si'k'l)   life cycle typical of animals in which the adult is always diploid because meiosis occurs after maturity is reached.   *596*

diploid (dip'loid)   the 2N number of chromosomes; twice the number of chromosomes found in gametes.   *81*

directional selection (di-rek'shun-al sĕ-lek'shun)   natural selection that favors an atypical phenotype.   *576*

disaccharide (di-sak'ah-rid)   a sugar such as maltose that contains 2 units of a monosaccharide.   *33*

dissociate (dis-so'she-āt)   the breakdown of a compound into its ionic or elemental components.   *25*

distal convoluted tubule (dis'tal kon'vo-lūt-ed tu'būl)   a highly coiled region of a nephron that is distant from Bowman's capsule.   *305, 310*

DNA (deoxyribonucleic acid);   a nucleic acid, the genetic material that directs protein synthesis in cells.   *40*

DNA ligase (de'en-ā li'gās)   a bacterial enzyme that seals breaks in the DNA molecule after cleavage, for example, by a restriction enzyme.   *536*

DNA polymerase (de'en-ā pol-im'er-ās)   an enzyme catalyzing the union of complementary base pairs in the formation of a DNA strand.   *512*

DNA probe (de'en-ā prōb)   a single strand of DNA used to locate a complementary DNA strand in a cell or body fluid by its binding activity.   *539*

dominant allele (dom'i-nant ah-lēl')   hereditary factor that expresses itself when the genotype is heterozygous.   *471*

dorsal root ganglion (dor'sal rōot gang'gle-on)   a mass of sensory neuron cell bodies located in the dorsal root of a spinal nerve.   *331*

double helix (dŭ'b'l he'liks)   a double spiral often used to describe the three-dimensional shape of DNA.   *510*

doubling time (dŭ-b'ling tim)   the number of years it takes for a population to double in size.   *738*

Down syndrome (down sin'drōm)   a human congenital disorder associated with an extra 23rd chromosome.   *491*

Dryopithecus (dri''o-pith'e-cus)   a genus of extinct apes that may have included or resembled a common ancestor to both apes and humans.   *673*

duodenum (du''o-de'num)   the first portion of the small intestine in vertebrates into which ducts from the gallbladder and pancreas enter.   *195*

duplication (du'pli-kā-shun)   chromosome mutation in which the chromosome segment occurs more than once on the same chromosome.   *494*

dyad (di'ad)   a chromosome having two chromatids held together at a centromere.   *89*

E

ecological pyramid (e''ko-log'i-kal pir'ah-mid)   a pictoral representation of the trophic structure of an ecosystem with producers at the base and the highest-level consumers at the top.   *723*

ecology (e-kol'o-je)   the study of the interactions of species between themselves and the physical environment.   *720*

ecosystem (ek''o-sis'tem)   a biological community together with the associated abiotic environment.   *7, 720*

ectoderm (ek'to-derm)   the outer germ layer of the embryonic gastrula; it gives rise to the skin and nervous system.   *437*

edema (ĕ-de'mah)   swelling due to tissue fluid accumulation in the intercellular spaces.   *259*

effector (ĕ-fek'tor)   a structure that allows an organism to respond to environmental stimuli such as the muscles and glands.   *320*

elastic cartilage (e-las'tik kar'ti-lij)   cartilage composed of elastic fibers allowing greater flexibility.   *178*

electrocardiogram (e-lek″tro-kar′de-o-gram″) ECG or EKG; a recording of the electrical activity associated with the heartbeat. 225

electroencephalogram (e-lek″tro-en-sef′ah-lo-gram″) EEG; a graphic recording of the brain's electrical activity. 338

electron (e-lek′tron) a subatomic particle that has almost no weight and carries a negative charge; travels in an orbital, called a shell, about the nucleus of an atom. 17

element (el′ĕ-ment) the simplest of substances consisting of only one type of atom; i.e., carbon, hydrogen, oxygen. 18

elephantiasis (el″ĕ-fan-ti′ah-sis) a disease caused by a parasitic nematode that blocks a lymphatic vessel and characterized by extreme swelling of a limb. 642

emulsification (e-mul″si-fi′ka′shun) the act of dispersing one liquid in another. 38

endocrine gland (en′do-krin gland) a gland that secretes hormones directly into the blood or body fluids. 173, 386

endocytosis (en″do-si-to′sis) a process in which a vesicle is formed at the cell membrane to bring a substance into the cell. 75

endoderm (en′do-derm) an inner layer of cells that line the primitive gut of the gastrula. It becomes the lining of the digestive tract and associated organs. 437

endodermis (en″do-der′mis) a plant tissue consisting of a single layer of cells that surrounds and regulates the entrance of materials into particularly the vascular cylinder of roots. 144

endometrium (en″do-me′tre-um) the lining of the uterus that becomes thickened and vascular during the uterine cycle. 418

endoplasmic reticulum (en-do-plaz′mic rĕ-tik′u-lum) a complex system of tubules, vesicles, and sacs in cells; sometimes having attached ribosomes. 50

endoskeleton (en″do-skel′ĕ-ton) calcareous supportive internal tissue of echinoderms and vertebrates. 654

endosperm (en′do-sperm) a nutrient material for developing the plant embryo. 624

energy of activation (en′er-je uv ak″ti-va′shun) the amount of energy that a molecule must gain to become sufficiently "excited" to enter into a chemical reaction. 100

energy sources (en′er-je sorsz) ways by which energy can be made available to organisms from the environment. 553

enterocoelomate an animal in which the coelom forms as an outpocketing of the primitive gut. 642

environmental resistance (en-vi″ron-men′tal re-zis′tans) sum total of factors in the environment that limit the numerical increase of a population in a particular region. 739

enzyme (en′zim) a protein catalyst that speeds up a specific reaction or a specific type of reaction. 29

epicotyl (ep″i-kot′il) plant embryo portion above the cotyledon(s) that contributes to stem development. 167

epidermis (ep″i-der′mis) the outer layer of cells of organisms including plants. Also the outer layer of skin composed of stratified squamous epithelium. 144, 181, 637

epididymis (ep″i-did′i-mis) coiled tubules next to the testes where sperm mature and may be stored for a short time. 412

epiglottis (ep″i-glot′is) a structure that covers the glottis during the process of swallowing. 193, 282

epinephrine (ep″i-nef′ron) a hormone produced by the adrenal medulla that stimulates "fight or flight" reactions. 396

epiphyte (ep′i-fit) nonparasitic plant that grows on the surface of other plants, usually above the ground, such as arboreal orchids and Spanish moss. 709

epithelial tissue (ep″i-the′le-al tish′u) a type of tissue that lines cavities and covers the external surface of the body. 173

equilibrium (e″kwi-lib′re-um) a state of balance; a steady state where forces are equalized. 380

erection (ĕ-rek′shun) referring to a structure such as the penis when it is turgid and erect as opposed to flaccid and lacking turgidity. 415

erythrocyte (ĕ-rith′ro-sit) a red blood cell that contains hemoglobin and carries oxygen from the lungs to the tissues in vertebrates. 178

esophagus (ĕ-sof′ah-gus) a tube that transports food from the mouth to the stomach. 193

essential amino acid (ĕ-sen′shal ah-me′no as′id) 9 different amino acids required in the human diet because the body cannot make them. 204

estuary (es′tu-a-re) an area where fresh water meets the sea; thus, an area with salinity intermediate between fresh water and seawater. 712

eubacteria (u″bak-te′re-ah) group containing most species of bacteria except for the archaebacteria. 592

euglenoids (u-gle′noidz) a small group of unicellular algae that are bounded by a flexible pellicle and move by flagella. 602

eukaryotic (u″kar-e-ot′ik) possessing the membranous organelles characteristic of complex cells. 44

eustachian tube (u-sta′ke-an tub) extension from the middle ear to the nasopharynx for equalization of air pressure on the eardrum. 378

evolutionary tree (ev″o-lu′shun-ar-e tre) a diagram describing the evolutionary relationship of groups of organisms. 566

excretion (eks-kre′shun) removal of metabolic wastes. 299

exocrine gland (ek′so-krin gland) secreting externally; particular glands with ducts whose secretions are deposited into cavities, such as salivary glands. 173

exocytosis (eks″o-si-to′sis) a process in which an intracellular vesicle fuses with the cell membrane so that the vesicle's contents are released outside the cell. 77

exophthalmic goiter (ek″sof-thal′mik goi′ter) an enlargement of the thyroid gland accompanied by an abnormal protrusion of the eyes. 395

expiration (eks″pi-ra′shun) a process of expelling air from the lungs; exhalation. 280

exponential growth (eks″po-nen′shal groth) growth, particularly of a population, in which the total number increases in the same manner as compound interest. 736

external respiration (eks-ter′nal res″pi-ra′shun) exchange between air and blood of oxygen and carbon dioxide. 280

extraembryonic membranes (eks″trah-em″bre-on′ik mem′branz) in embryology, membranes that are not a part of the embryo but are necessary to the continued existence and health of the embryo. 445

F

facilitated diffusion (fah-sil′i-tat-ed di-fu′zhun) passive transfer of a substance into or out of a cell along a concentration gradient by a process that requires a carrier. 73

FAD a coenzyme of oxidation; a dehydrogenase that participates in hydrogen (electron) transport within the mitochondria. 112

fatigue (fah-teg′) muscle relaxation in the presence of stimulation due to energy reserve depletion. 355

fatty acid (fat′e as′id) an organic molecule having a long chain of carbon atoms and ending in an acidic group. 37

femur (fe′mur) the thighbone found in the upper leg. 349

fermentation (fer″men-ta′shun) anaerobic breakdown of carbohydrates that results in organic end products such as alcohol and lactic acid. 114

fibrin (fi′brin) insoluble, fibrous protein formed from fibrinogen during blood clotting. 248

fibrinogen (fi-brin′o-jen) plasma protein that is converted into fibrin threads during blood clotting. 248

fibroblasts (fi′bro-blasts) cells that form fibers in connective tissues. 176

fibrocartilage (fi″-bro-kar′ti-lij) cartilage with a matrix of strong collagenous fibers. 178

fibrous connective tissue (fi′brus ko-nek′tiv tish′u) tissue composed mainly of closely packed collagenous fibers and found in tendons and ligaments. 176

fibula (fib′u-lah) a long slender bone located on the lateral side of the tibia. 349

filament (fil′ah-ment) a threadlike structure such as the thick (myosin) and thin (actin) filaments found in myofibrils of muscle fibers; in flowering plants, the stalk that supports the anther within a stamen. 165

filter feeder (fil′ter fed′er) an animal that obtains its food, usually in small particles, by filtering it from water. 633

fimbriae (fim′bre-a) fingerlike extensions from the oviduct near the ovary. 418

flagella (flah-jel′ah) slender, long processes used for locomotion by the flagellate protozoans, bacteria, and sperm. 57

flame cell (flām sel) excretory organ of flatworms. 637

flower (flow′er) the blossom of a plant that contains the reproductive organs of angiosperms. 622

fluid-mosaic model (floo′id mo-za′ik mod′el) proteins form a mosaic pattern within a bilayer of lipid molecules having a fluid consistency. 46

fluke (flook) a parasitic flatworm; member of class Trematoda. 640

focusing (fo-kus-ing) manner by which light rays are bent by the cornea and lens, creating an image on the retina. 372

follicle (fol′i-kl) a structure in the ovary that produces the egg and particularly the female sex hormone, estrogen. 417

follicle-stimulating hormone (fol′li-k′l stim′u-lāt″ing hor′mon) See FSH.

food chain (food chān) a series of organisms in a feeding chain, starting with a producer and ranging through succeeding levels of consumers. 722

food web (food web) the complete set of food links between populations in a community. 722

foot (foot) a strong organ of locomotion in mollusks. 643

foreskin (for′skin) skin covering the glans penis in uncircumcised males. 414

formed element (form′d el′ĕ-ment) those elements that make up the solid portion of blood; they are either cells (erythrocytes and leukocytes) or derived from a cell (platelets). 239

fossils (fos″lz) any remains of an organism that have been preserved in the earth's crust. 563

fovea centralis (fo′ve-ah sen-tral′is) region of the retina, consisting of densely packed cones, which is responsible for the greatest visual acuity. 372

frond (frond) the large leaf of a fern plant containing many leaflets. 616

frontal lobe (fron′tal lōb) area of the cerebrum responsible for voluntary movements and higher intellectual processes. 336

fruit (froot) a mature ovary enclosing seed(s). 622

fruiting bodies (froot′ing bod′es) a spore-bearing structure found in certain types of fungi, such as mushrooms. 604

FSH (follicle-stimulating hormone); a hormone secreted by the anterior pituitary gland that stimulates the development of an ovarian follicle in a female or the production of sperm in a male. 415

furrowing (fur′o-ing) a constriction of the cell membrane that accompanies cytokinesis in animal cells. 87

G

gallbladder (gawl′blad-er) a saclike organ associated with the liver that stores and concentrates bile. 198

gallstone (gawl′ston) a precipitated crystal of cholesterol or calcium carbonate formed from bile within the gallbladder or bile duct. 199

gamete (gam′et) a reproductive cell that joins with another in fertilization to form a zygote; most often an egg or sperm. 81

gametophyte (gam′ĕ-to-fīt) the haploid generation that produces gametes in the life cycle of a plant. 611

ganglion (gang′gle-on) a collection of neuron cell bodies outside the central nervous system. 329

gap junction (gap junk′shun) cell junction in which protein molecules form cell-to-cell channels allowing molecule exchange. 176

gastric gland (gas′trik gland) the gland within the stomach wall that secretes gastric juice. 194

gastrin (gas′trin) a hormone secreted by stomach cells to regulate the release of pepsin by the stomach wall. 198

gastrodermis (gas″tro-der′mis) layer of cells found lining the body cavity in cnidarians. 637

gastrovascular cavity (gas″tro-vas′ku-lar kav′i-te) a central cavity, with only one opening, of a lower animal in which digestion takes place and where nutrients are distributed to the cells lining the cavity. 637

gene flow (jēn flo) the movement of genes from one population to another via reproduction between members of the populations. 574

gene mutation (jēn mu-tā′shun) an alternation in the code of a single gene with a subsequent change in its expression. 527

gene pool (jēn pool) the total of all the genes of all the individuals in a population. 571

gene sequencing (jēn se′kwens-ing) determining the order of nucleotides in a gene. 536

gene therapy (jēn ther′ah-pe) the use of biotechnology to treat genetic disorders and illnesses. 545

genetic drift (jĕ-net′ik drift) evolution by chance processes alone. 574

genetic engineering (je net′ik en″ji-nēr′ing) the use of bacteria by humans to synthesize a desired product. 534

genomic library (je-nom′ik li′brer e) a collection of engineered viruses that together carry all of the genes of a species. 535

genotype (je′no-tīp) the genetic makeup of any individual. 471

gerontology (jer-on-tol′o ji) the study of aging. 461

gibbon (gib′on) the smallest ape; well known for its arm-swinging form of locomotion. 673

gills (gilz) gas exchange organs found in fishes and other types of marine and freshwater animals. 645

ginkgo (gink′go) a division of gymnosperms with only one current species. 619

glial cell (gli-′al sel) a nervous system cell that supports and protects neurons. 180

globulin (glob′u-lin) plasma protein of the blood; beta and alpha globulins are transport molecules; gamma globulins are antibodies. 241

glomerular filtrate (glo-mer′u-lar fil′trat) a solution formed in the nephron by the passage of filterable components from the glomerulus to the Bowman's capsule. 308

glomerulus (glo-mer′u-lus) a cluster; for example, the cluster of capillaries surrounded by Bowman's capsule in a kidney tubule. 307

glottis (glot′is) slitlike opening in the larynx between the vocal cords. 193, 282

glucagon (gloo′kah-gon) hormone secreted by the pancreatic islets of Langerhans that causes the release of glucose from glycogen. 398

glycerol (glis′er-ol) an organic compound that serves as a building block for fat molecules. 37

glycogen (gli′ko-jen) the storage polysaccharide found in animals that is composed of glucose molecules joined in a linear-type fashion but having numerous branches. 35

glycolysis (gli-kol′i-sis) the metabolic pathway that converts glucose to pyruvate. 107

Golgi apparatus (gol′ge ap″ah-ra′tus) an organelle that consists of concentrically folded membranes and functions in the packaging and secretion of cellular products. 51

gonadotropic hormone (go-nad″o-trop′ik hor′mon) a type of hormone that regulates the activity of the ovaries and testes; principally FSH and LH (ICSH). 391

gorilla (go-ril′ah) the largest of the great apes, which are as closely related to humans as to other apes. 673

Graafian follicle (graf′e-an fol′li-k′l) a mature follicle within the ovaries that houses a developing egg. 417

Gram stain (gram stān) a staining test done in the biology lab to differentiate between gram-positive and gram-negative bacterial species. 592

grana (gra′nah) stacks of flattened membranous vesicles in a chloroplast where chlorophyll is located and potosynthesis begins. 125

granular leukocytes (gran′u-lar lu′ko-sit) white blood cells that contain distinctive granules. 249

grazer (gra′zer) an animal that feeds on low-lying vegetation such as grasses. 689

greenhouse effect (gren′hows ĕ-fekt) a carbon dioxide buildup in the atmosphere as a result of fossil fuel combustion; retains and re-radiates heat, creating an abnormal rise in the earth's average temperature. 753

growth (groth) an increase in the number of cells and/or the size of these cells. 436

growth hormone (groth hor′mon) GH or somatotropin; hormone released by the anterior lobe of the pituitary gland that promotes the growth of the organism. 391

growth rate (groth rāt) the yearly percentage of increase or decrease in the size of a population. 737

guard cell (gahrd sel) a bean-shaped, epidermal cell; one found on each side of a leaf stoma; their activity controls stoma size. 151

H

habitat (hab'i-tat)   the natural abode of an animal or plant species.   *720*

hammer (ham'er)   the middle ear ossicle adhering to the tympanic membrane.   *378*

haploid (hap'loid)   the N number of chromosomes; half the diploid number; the number characteristic of gametes that contain only one set of chromosomes.   *81*

haplontic life cycle (hap-lon'tik līf si'kl) life cycle typical of protists in which the adult is always haploid because meiosis occurs after zygote formation and before maturity is reached.   *596*

hard palate (hard pal'at)   anterior portion of the roof of the mouth that contains several bones.   *192*

hazardous waste (haz'er-dus wāst)   waste containing chemicals hazardous to life.   *746*

HCG (human chorionic gonadotropic);   a gonadotropic hormone produced by the chorion that functions to maintain the uterine lining.   *423*

heart (hart)   a muscular organ located in the thoracic cavity responsible for maintenance of blood circulation.   *221*

helper T cell (hel'per te sel)   a T lymphocyte that stimulates certain other T and B lymphocytes to perform their respective functions.   *266*

heme (hēm)   the iron-containing portion of a hemoglobin molecule.   *241*

hemocoel (he'mo-sēl)   residual coelom found in anthropods that is filled with blood.   *650*

hemoglobin (he''mo-glo'bin)   a red, iron-containing pigment in blood that combines with and transports oxygen.   *241, 290*

hepatic portal vein (he-pat'ik por'tal vān)   vein leading to the liver formed by the merging of blood vessels from the villi of the small intestine.   *198*

herbaceous stem (her-ba'shus stem)   nonwoody stem.   *147*

herbivore (her'bi vōr)   an animal that feeds directly on plants.   *720*

hermaphroditic animal (her'maf'ro-di-tik'' an'ē-mēl)   an animal having both male and female sex organs.   *638*

hernia (her'ni-ah)   an opening and separation of the abdominal wall, through which part of an organ protrudes.   *414*

heterogamete (het''er-o-gam'et)   a different kind of sex cell, a large and nonmotile egg or a small and flagellated sperm.   *599*

heterospore (het'er-o-spor)   a nonidentical spore such as a microspore and a megaspore produced by the same plant.   *618*

heterotroph (het'er-o-trof'')   an organism that takes in preformed food.   *123, 557*

heterotroph hypothesis (het'er-o-trof'' hi-poth'e-sis)   the suggestion that the protocell and first cell(s) were heterotrophs.   *557*

heterozygote superiority (het''er-o-zi'gōt su-per''ē-or'i-tē)   a phenomenon in which the heterozygote of a population survives better than the homozygotes, thus maintaining its population frequency.   *576*

heterozygous (het''er-o-zi'gus)   having two different alleles (as *Aa*) for a given trait.   *471*

hexose (hek'sōs)   a 6-carbon monosaccharide.   *33*

histamine (his'tah-min)   a substance produced by basophil-derived mast cells in connective tissue that causes capillaries to dilate and release immune and other substances.   *253*

homeostasis (ho''me-o-sta'sis)   the maintenance of conditions, particularly the internal environment of birds and mammals: maintenance of temperature, blood pressure, pH, and other body conditions within narrow limits.   *185*

hominid (hom'i-nid)   a member of the family of upright, bipedal primates (family Hominidae) that includes modern humans.   *673*

hominoid (hom'i-noid)   a member of a superfamily containing humans and the great apes.   *673*

Homo erectus (ho'mo ē-rek'tus)   the earliest nondisputed species of humans, named for their erect posture that allowed them to have a striding gait.   *676*

Homo habilis (ho'mo hah'bi-lis)   an extinct species that may include the earliest humans, having a small brain but making quality tools.   *676*

homologous (ho-mol'o-gus)   similarly constructed; homologous chromosomes have the same shape and contain genes for the same traits; homologous structures in animals share a common ancestry.   *89*

homozygous (ho''mo-zi'gus)   having identical alleles (as *AA* or *aa*) for a given trait; pure breeding.   *471*

hormone (hor'mōn)   a chemical messenger produced in small amounts in one body region that is transported to another body region.   *159, 386*

humerus (hu'mer-us)   a heavy bone that extends from the scapula to the elbow.   *349*

Huntington disease (hunt'ing-tun di zēz)   a fatal genetic disease marked by neurological disturbances and failure of brain regions.   *482*

hyaline cartilage (hi'ah-līn kar'ti-lij)   cartilage composed of very fine collagenous fibers and a matrix of a clear milk-glass appearance.   *178*

hydrogen bond (hi'dro-jen bond)   a weak attraction between a hydrogen atom carrying a partial positive charge and an atom of another molecule carrying a partial negative charge.   *24*

hydroid (hi'droid)   a tubular-shaped polyp displayed by some cnidarians.   *635*

hydrolysis (hi-drol'i-sis)   the splitting of a bond within a larger molecule by the addition of the components of water.   *29*

hydrolytic enzyme (hi-dro-lit'ik en'zīm)   an enzyme that catalyzes a reaction in which the substrate is broken down with the addition of water.   *200*

hypertonic solution (hi''per-ton'ik so-lu'shun) one that has a greater concentration of solute, a lesser concentration of water than the cell.   *72*

hypha (hi'fah)   one filament of a mycelium that constitutes the body of a fungus.   *603*

hypocotyl (hi''po-kot'il)   a plant embryo portion below the cotyledon(s) that contributes to stem development.   *167*

hypothalamus (hi''po-thal'ah-mus)   a region of the brain, the floor of the third ventricle, that helps maintain homeostasis.   *336*

hypothesis (hi-poth'ē-sis)   a scientific theory that is capable of explaining present data and that may be used to predict the outcome of future experimentation.   *9*

hypotonic solution (hi''po-ton'ik so-lu'shun)   one that has a greater concentration of water, a lesser concentration of solute than the cell.   *72*

I

implantation (im''plan-ta'shun)   the attachment to and penetration of the embryo into the lining (endometrium) of the uterus.   *423*

impotency (im'po-ten''se)   failure of the penis to achieve erection.   *415*

induction (in-duk'shun)   a process by which one tissue controls the development of another, as when the embryonic notochord induces the formation of the neural tube.   *444*

inflammatory reaction (in-flam'ah-to''re re-ak'shun)   a tissue response to injury that is characterized by dilation of blood vessels and an accumulation of fluid in the affected region.   *252*

inner ear (in'er ēr)   the portion of the ear consisting of a vestibule, semicircular canals, and the cochlea where balance is maintained and sound is transmitted.   *378*

innervate (in'er-vāt)   to activate an organ, muscle, or gland by motor neuron stimulation.   *320*

innominate (i-nom'i-nāt)   one of 2 hipbones that form the pelvis.   *349*

inorganic (in''or-gan'ik)   refers to compounds and reactions between compounds that do not contain both carbon and hydrogen.   *24*

insertion (in-ser'shun)   the end of a muscle that is attached to a movable part.   *350*

inspiration (in''spi-ra'shun)   the act of breathing in.   *280*

insulin (in'su-lin)   a hormone produced by the pancreas that regulates carbohydrate storage.   *398*

interferon (in''ter-fēr'on)   a protein formed by a cell infected with a virus that can increase the resistance of other cells to the virus.   *263*

internal respiration (in-ter'nal res''pi-ra'shun) exchange between blood and tissue fluid of oxygen and carbon dioxide.   *280*

interneuron (in''ter-nu'ron)   a neuron that is found within the central nervous system and takes nerve impulses from one portion of the system to another.   *321*

interphase (in'ter-fāz)   the interval between successive cell divisions; during this time the chromosomes are in an extended state and are active in directing protein synthesis.   *83*

inversion (in-ver'zhun)   a chromosome mutation caused by a 180 degree turn-around by the fragment of a chromosome.   *494*

invertebrate (in-ver'tĕ-brāt) an animal that lacks a vertebral column. *630*

ion (i'on) an atom or group of atoms carrying a positive or negative charge. *19*

ionic bond (i-on'ik bond) a chemical union between oppositely-charged ions. *20*

ionic reaction (i-on'ik re-ak'shun) a chemical reaction in which atoms acquire or lose electrons. *19*

iris (ī'ris) a muscular ring that surrounds the pupil and regulates the passage of light through this opening. *371*

islets of Langerhans (i'lets uv lahng'er-hanz) distinctive groups of cells within the pancreas that secrete insulin and glucagon. *398*

isogametes (i''so-gam'ēts) gametes whose union produces a zygote, but which have a similar appearance. *599*

isotonic solution (i''so-ton'ik so-lu'shun) one that contains the same concentration of solute and water as does the cell. *71*

isotope (i'so-tōp); an atom with the same number of protons and electrons but differing in the number of neutrons and therefore in weight. *19*

J

jointed appendage (joint'ed ah-pen'dij) the flexible exoskeleton extension found in arthropods that is used as sense organs, mouthparts, and locomotion. *650*

K

karyotype (kar'e-o-tīp) the arrangement of all the chromosomes within a nucleus by pairs in a fixed order. *80*

kidney (kid'nē) an organ in the urinary system that produces and excretes urine. *303*

Klinefelter syndrome (klīn'fel-ter sin'drōm) a condition caused by the inheritance of a chromosome abnormality in number; an XXY individual. *496*

Krebs cycle (krebz si'kl) a series of reactions found within the matrix of mitochondria that give off carbon dioxide. Also called the citric acid cycle because the reactions begin and end with citric acid. *107*

L

labium (la'be-um) a fleshy border or liplike fold of skin, as in the labia majora and labia minora of the female genitalia. *419*

lacteal (lak'te-al) a lymph vessel in a villus of the intestinal wall of mammals. *195*

lactogenic hormone (lak''to-jen'ik hor'mōn) LTH; a hormone secreted by the anterior pituitary that stimulates the production of milk from the mammary glands. *391*

lacuna (lah-ku'nah) a small pit or hollow cavity, as in bone or cartilage where a cell or cells are located. *178*

ladder-type (lad'er tīp) planarian nervous system with a small brain and two lateral nerve cords joined by cross-bridges. *638*

lake (lāk) a freshwater body with some depths that are always dark. *710*

lancelet (lans'let) a type of protochordate that has the 3 chordate characteristics as an adult; formerly called amphioxus. *656*

lanugo (lah-nu'go) downy hair with which a fetus is born; fetal hair. *456*

larynx (lar'ingks) the structure that contains the vocal cords; voice box. *282*

leaf vein (lēf vān) the structure that contains vascular tissue in a leaf. *150*

lens (lenz) a clear membranelike structure found in the eye behind the iris. The lens brings objects into focus. *371*

lenticel (len'ti-sel) a pocket of loosely arranged cells in cork that permit gas exchange. *148*

less-developed country (les'devel'opd kun'trē) nonindustrialized or newly-industrialized country. *741*

leukemia (lu-ke'me-ah) a form of cancer characterized by uncontrolled production of leukocytes in red bone marrow or lymphoid tissue. *249*

leukocyte (lu'ko-sit) white blood cell of which there are several types each having a specific function in protecting the body from invasion by foreign substances and organisms. *178*

lichen (li'ken) fungi and algae coexisting in a symbiotic relationship that is described as controlled parasitism of fungi on the algae. *593*

ligament (lig'ah-ment) a dense connective tissue that joins bone to bone. *177, 351*

light-harvesting antennae (līt har'vest-ing an-ten'ah) pigment molecules that collect solar energy in the chloroplast. *127*

limbic system (lim'bik sis'tem) an area of the forebrain implicated in visceral functioning and emotional responses; involves many different centers of the brain. *338*

linkage group (lingk'ij grup) alleles on the same chromosome are linked in the sense that they tend to move together to the same gamete; crossing over interferes with linkage. *501*

lipase (li'pās) a fat-digesting enzyme secreted by the pancreas. *201*

lipids (lip'idz) a group of organic compounds that are insoluble in water; notably fats, oils, and steroids. *35*

liposome (lip'o-sōm) droplets formed by phospholipids in a liquid environment. *556*

liver (liv'er) a large organ in the abdominal cavity that has many functions vital to continued existence such as production of blood proteins and detoxification of harmful substances. *198*

loop of Henle (loop uv hen'lē) U-shaped portion of the nephron tubule. *305, 310*

loose connective tissue (loos kō-nek'tiv tish'u) tissue composed mainly of fibroblasts that are widely separated by a matrix containing collagen and elastin fibers and found beneath epithelium. *176*

lumen (lu'men) the cavity inside any tubular structure, such as the lumen of the gut. *193*

luteinizing hormone (lu'te-in-iz''ing hor'mōn) hormone produced by the anterior pituitary gland that stimulates the development of the corpus luteum in females and the production of testosterone in males. *415*

lymph (limf) fluid having the same composition as tissue fluid and carried in lymph vessels. *259*

lymphatic system (lim-fat'ik sis'tem) vascular system that takes up excess tissue fluid and transports it to the bloodstream. *258*

lysosome (li'so-sōm) an organelle in which digestion takes place due to the action of hydrolytic enzymes. *52*

M

macrophage (mak'ro-fāj) a large cell derived from a monocyte that ingests foreign material and cellular debris. *253*

Malpighian tubule (mal-pig'i-an tu'būl) an organ of excretion, notably in insects. *652*

maltose (mawl'tōs) a disaccharide composed of 2 glucose units. *33*

mantle (man't'l) fleshy fold that envelops the visceral mass of mollusks. *643*

matrix (ma'triks) the secreted basic material or medium of biological structures, such as the matrix of cartilage or bone. *107, 176*

medulla (mĕ-dul'ah) the inner portion of an organ; for example, the adrenal medulla. *304*

medulla oblongata (mĕ-dul'ah ob''long-gah'tah) the lowest portion of the brain that is concerned with the control of internal organs. *336*

medusa (mĕ-du'sah) a bell-shaped, free-swimming stage resembling a jellyfish that is capable of sexual reproduction in the life cycle of some sessile cnidarians. *635*

megaspore (meg'ah-spōr) in seed plants, a spore that develops into the female gametophyte. *621*

meiosis (mi-o'sis) a type of cell division that occurs during the production of gametes or spores by means of which the 4 daughter cells receive the haploid number of chromosomes. *81*

meiosis I (mi-o'sis wun) that portion of meiosis during which homologous chromosomes synapse and then later separate. Includes prophage I, metaphase I, anaphase I, and telophase I. *89*

meiosis II (mi-o'sis too) that portion of meiosis during which sister chromatids separate. Includes prophase II, metaphase II, anaphase II, and telophase II. *89*

memory B cells (mem'o-re be selz) persistent population of B cells that produce a specific antibody and account for the development of active immunity. *264*

memory T cells (mem'o-re te sels) persistent population of cells capable of secreting lymphokines to stimulate macrophages and B cells. *266*

meninges (mĕ-nin'jēz) protective membranous coverings about the central nervous system. *334*

meniscus (mĕ-nis'kus) a piece of fibrocartilage that separates the surfaces of bones in the knee. *351*

menopause (men′o-pawz) termination of the ovarian and uterine cycle in older women. *424*

menstruation (men″stroo-a′shun) loss of blood and tissue from the uterus at the end of a female uterine cycle. *420*

meristem tissue (mer′i-stem tish′u) plant tissue that always remains undifferentiated and capable of dividing to produce new cells. *143*

mesoderm (mes′o-derm) the middle germ layer of an animal embryo that gives rise to the muscles, connective tissue, and circulatory system. *438*

mesoglea (mes″o-gle′ah) a jellylike packing material between the ectoderm and endoderm of cnidarians. *634*

mesophyll (mes′o-fil) the middle tissue of a leaf made up of parenchyma cells that carries on photosynthesis and gas exchange. *151*

messenger RNA (mes′en jer) mRNA; a nucleic acid (ribonucleic acid) complementary to genetic DNA and bearing a message to direct cell protein synthesis at the ribosome. *517*

metabolism (mĕ-tab′o-lizm) all of the chemical changes that occur within cells. *98*

metacarpal (met″ah-kar′pal) a bone found in the palm of the hand. *349*

metabolic pool (met″ah-bol′ik pool) substrates in a cell used for biosynthesis. *109*

metafemale (met′ah-fe-māl) a female with 3 X chromosomes per body cell. *495*

metal (met′al) class of elements, which in reactions, characteristically lose electrons and become positively charged ions. *20*

metamorphosis (met″ah-mor′fo-sis) change in form as when a tadpole becomes an adult frog or as when an insect larva develops into the adult. *653*

metaphase (met′ah-fāz) stage in mitosis during which chromosomes are at the equator of the mitotic spindle. *85*

metatarsal (met″ah-tar′sal) a bone found in the foot between the ankle and the toes. *349*

MHC protein (major histocompatibility protein); a surface molecule that serves as a genetic marker. *268*

microfilament (mi″kro-fil′ah-ment) an extremely thin fiber found within the cytoplasm that is involved in the maintenance of cell shape and movement of cell contents. *55*

microsphere (mi-kro-sfēr) structure composed only of protein that looks like a cell and carries on many cellular functions; a possible early step in cell evolution. *555*

microspore (mi′kro-spor) in seed plants, a spore that develops into a pollen grain. *620*

microtubule (mi″kro-tu′bul) an organelle composed of 13 rows of globular proteins; found in multiple units in several other organelles such as the centriole, cilia, and flagella. *56*

middle ear (mid′l ēr) a portion of the ear consisting of the tympanic membrane, the oval and round windows, and the ossicles where sound is amplified. *378*

mimicry (mim′ik-re) the resemblance of an organism to another that has a defense against a common predator. *693*

mineral (min′er-al) an inorganic, homogeneous substance. *210*

mitochondrion (mi″to-kon′dre-on) an organelle in which aerobic respiration produces the energy molecule, ATP. *54*

mitosis (mi-to′sis) type of cell division in which the daughter cells receive the exact chromosome and genetic makeup of the mother cell; occurs during growth and repair. *83*

molting (mōlt′ing) shedding all or part of an outer covering; in arthropods, periodic shedding of parts of the exoskeleton to allow increase in size. *650*

monoclonal antibodies (mon″o-klon′al an″ti-bod″ēz) antibodies of one type that are produced by cells that are derived from a lymphocyte that has fused with a cancer cell. *272*

monocot (mon′o-kot) (monocotyledon); a type of angiosperm in which the seed has only one cotyledon, such as corn and lily. *142*

monoculture agriculture (mon′o-kul-chur ag′ri-kul-chur) the planting and growing of only one species of crop. *730*

monohybrid (mon″o-hi′brid) the offspring of parents who differ in one way only; shows the phenotype of the dominant allele but carries the recessive allele. *473*

monosaccharide (mon″o-sak′ah-rīd) a simple sugar; a carbohydrate that cannot be decomposed by hydrolysis. *33*

morphogenesis (mor″fo-jen′i-sis) the movement of cells and tissues to establish the shape and structure of an organism. *436*

mother cell (muth′er sel) a cell that divides producing daughter cells. *84*

motor neuron (mo′tor nu′ron) a neuron that takes nerve impulses from the central nervous system to the effectors. *320*

multicellular (mul″ti-sel′u-lar) composed of many cells. *44*

muscle action potential (mus′el ak′shun po-ten′shal) an electrochemical change due to increased sarcolemma permeability that is propagated down the T system and results in muscle contraction. *363*

muscle fiber (mus′el fi′ber) skeletal muscle cell. *358*

muscle spindle (mus′el spin′dul) modified skeletal muscle fiber that can respond to changes in muscle length. *353*

muscular tissue (mus′ky-lar tish′u) a type of tissue that contains cells capable of contracting; skeletal muscles are attached to the skeleton, smooth muscle is found within walls of internal organs, and cardiac muscle comprises the heart. *179*

mutation (mu-ta′shun) a change in the genetic material. *512*

mutualism (mu′tu-al-izm″) a relationship between two organisms of different species that benefits both organisms. *696*

mycelium (mi-se′le-um) a mass of hyphae that make up the body of a fungus. *603*

myelin sheath (mi′ĕ-lin shēth) the fatty cell membranes that cover long neuron fibers and give them a white, glistening appearance. *322*

myocardium (mi″o-kar′de-um) heart muscle. *221*

myofibrils (mi″o-fi′brilz) the contractile portions of muscle fibers. *359*

myogram (mi′o-gram) a recording of a muscular contraction. *354*

myosin (mi′o-sin) one of two major proteins of muscle; makes up thick filaments in myofibrils and is capable of breaking down ATP. *See* actin. *360*

myxedema (mik″sĕ-de′mah) a condition resulting from a deficiency of thyroid hormone in an adult. *394*

N

NAD a coenzyme of oxidation; a dehydrogenase that frequently participates in hydrogen transport. *104*

NADP a coenzyme of reduction; a hydrogenase that frequently donates hydrogen atoms to metabolites. *126*

natural selection (nat′u-ral sĕ-lek′shun) the process by which better adapted organisms are favored to reproduce to a greater degree and pass on their genes to the next generation. *575*

Neanderthal (ne-an′der-thawl) the common name for an extinct subspecies of humans whose remains are found in Europe, Asia, and Africa. *678*

negative feedback (neg′ah-tiv fēd′bak) mechanism that is activated by a surplus imbalance and acts to correct it by stopping the process that brought about the surplus. *186*

nematocyst (nem′ah-to-sist) a threadlike structure in stinging cells of cnidarians that can be expelled to numb and capture prey. *634*

nephridia (nĕ-frid′e-ah) excretory tubules found in invertebrates; notably the segmented worms. *648*

nephron (nef′ron) the anatomical and functional unit of the vertebrate kidney; kidney tubule. *305*

nerve (nerv) a bundle of long nerve fibers that run to and/or from the central nervous system. *180, 327*

nerve impulse (nerv im′puls) an electrochemical change due to increased membrane permeability that is propagated along a neuron from the dendrite to the axon following excitation. *322*

neurilemma (nu″ro-lem′ah) the outermost wrapping of a nerve fiber, promoting regeneration. *322*

neurofibromatosis (nu″ro-fi′bro mah to′sis) a genetic disease marked by development of neurofibromas under the skin and muscles. *482*

neuromuscular junction (nu″ro-mus′ku-lar jungk′shun) the point of contact between a nerve cell and a muscle cell. *363*

neuron (nu′ron) nerve cell that characteristically has 3 parts: dendrite, cell body, axon. *180, 320*

neurotransmitter substance (nu″ro-trans-mit′er sub′stans) a chemical made at the ends of axons that is responsible for transmission across a synapse. *327*

neurula (nu'roo-lah) the early embryonic stage during which the primitive nervous system forms. *439*

neutron (nu'tron) a subatomic particle that has a weight of one atomic mass unit, carries no charge, and is found in the nucleus of an atom. *17*

niche (nich) the functional role and position of an organism in the ecosystem. *687, 720*

nitrogen fixation (ni'tro-jen fik-sa'shun) a process whereby free atmospheric nitrogen is converted into compounds, such as ammonia and nitrates, usually by soil bacteria. *727*

node of Ranvier (nōd uv Ran' vē ā) gap in the myelin sheath around a nerve fiber. *325*

noncyclic photophosphorylation (non-sik'lik fo''to-fos''for-i-la'shun) the passage of solar-energized electrons within chloroplasts from water to NADP; results in the generation of ATP and NADPH₂. *129*

nondisjunction (non''dis-jungk'shun) the failure of homologous chromosomes or sister chromatids to separate during the formation of gametes. *491*

norepinephrine (nor''ep'ĭ-nef'ron) NA; excitatory neurotransmitter active in the peripheral and central nervous systems. *328, 396*

notochord (no'to-kord) a dorsal supporting rod that exists in all chordates sometime in their life history; replaced by the vertebral column in vertebrates. *656*

nuclear envelope (nu'kle-ar en'vē-lōp) the double membrane that surrounds the nucleus and is continuous with the endoplasmic reticulum. *48*

nucleic acid (nu-kle'ik as'id) a large organic molecule made up of nucleotides joined together; for example, DNA and RNA. *40*

nucleolus (nu-kle'o-lus) an organelle found inside the nucleus; composed largely of RNA for ribosome formation. *49*

nucleotide (nu'kle-o-tīd) a molecule consisting of 3 subunits: phosphoric acid, a 5-carbon sugar, and a nitrogenous base; a building block of a nucleic acid. *40*

nucleus (nu'kle-us) a large organelle containing the chromosomes and acting as a control center for the cell; center of an atom. *17, 48*

nutrient (nu'trē-ent) a substance found in food and used by the body to maintain health. *202*

O

obesity (o-bēs'ĭ-te) condition in which body weight is more than 120% of the ideal weight. *216*

occipital lobe (ok-sip'ĭ-tal lōb) area of the cerebrum responsible for vision, visual images and other sensory experiences. *336*

olfactory cells (ol-fak'to-re selz) sense organs whose stimulation results in the sensation of smelling. *369*

omnivore (om' ni vor) an animal that feeds on both plants and animals. *720*

oncogene (ong'ko-jēn) a gene that contributes to the transformation of a cell into a cancerous cell. *530*

oogenesis (o''o-jen'ĕ-sis) production of egg in females by the process of meiosis and maturation. *92*

operator (op'er-a-tor) the sequence of DNA in an operon to which the repressor protein binds. *525*

operon (op'er-on) an operator gene and the adjacent group of genes it controls. *524*

optic nerve (op'tik nerv) a nerve that carries nerve impulses from the retina of the eye to the brain. *371*

orangutan (o-rang'oo-tan'') one of the great apes; large with long red hair. *673*

organ (or'gan) a structure composed of two or more tissues functioning as a unit. *180*

organ system (or'gan sis'tem) a group of related organs working together. *180*

organelle (or''gah-nel') specialized structures within cells such as the nucleus, mitochondria, and endoplasmic reticulum. *46*

organic (or-gan'ik) pertaining to any aspect of living matter; organic compounds contain both carbon and hydrogen atoms. *24*

organic soup (or-gan'ik soop) an expression used to refer to the ocean before the origin of life when it contained newly formed organic compounds. *555*

organ of Corti (or'gan uv kor'ti) a portion of the inner ear that contains the receptors for hearing. *379*

orgasm (or'gazm) physical and emotional climax during sexual intercourse; results in ejaculation in the male. *415*

origin (or'ĭ-jin) end of a muscle that is attached to a relatively immovable bone. *352*

osculum (os'ku-lum) opening to the exterior of a sponge central cavity. *633*

osmosis (oz-mo'sis) the movement of water from an area of greater concentration of water to an area of lesser concentration of water across a semipermeable membrane. *69*

osmotic pressure (oz-mot'ik presh'ur) pressure generated by the osmotic flow of water. *70*

ossicles (os'-ĭ-k'lz) the tiny bones found in the middle ear: hammer, anvil, and stirrup. *378*

osteoblast (os'te-o-blast'') a bone-forming cell. *350*

osteoclast (os'te-o-klast'') a cell that causes the erosion of bone. *350*

osteocyte (os'te-o-sīt) a mature bone cell. *350*

otoliths (o'to-liths) granules associated with ciliated cells in the utricle and saccule. *379*

outer ear (out'er ēr) portion of the ear consisting of the pinna and the auditory canal. *378*

oval opening (o'val o'pen-ing) foramen ovale; an opening between the 2 atria in the fetal heart. *456*

oval window (o'val win'do) opening between the stapes and the inner ear. *378*

ovarian cycle (o-va're-an si'k'l) monthly occurring changes in the ovary that effects the level of sex hormones in the blood. *420*

ovaries (o'var-ez) female gonads, the organs that produce eggs, and estrogen and progesterone; the base of the pistil in angiosperms. *164, 417*

ovulation (o''vu-la'shun) the discharge of a mature egg from the follicle within the ovary. *418*

ovule (o'vŭl) structure that contains megasporangium in seed plants where meiosis occurs and the female gametophyte is produced. *165, 618*

oxidation (ok''sĭ-da'shun) the loss of electrons (inorganic) or the removal of hydrogen atoms (organic). *23*

oxidative decarboxylation (ok''sĭ-da'tiv de''kar-bok''sĭ-la'shun) a reaction that involves the release of carbon dioxide as oxidation occurs. *112*

oxidizing atmosphere (ok'sĭ-dīz-ing at'mos-fēr) an atmosphere that contains oxidizing molecules such as O_2 rather than reducing molecules such as H_2. *553*

oxygen debt (ok'sĭ-jen det) oxygen that is needed to metabolize lactic acid, a compound that accumulates during vigorous exercise. *116, 362*

oxytocin (ok''se-to'sin) a hormone released by the posterior pituitary that causes contraction of uterus and milk letdown. *390*

ozone shield (o'zōn shēld) a layer of O_3 present in the upper atmosphere that protects the earth from damaging ultraviolet light. Nearer the earth, ozone is a pollutant. *559, 752*

P

pacemaker (pās'māk-er) *See* SA node.

palisade mesophyll (pal'ĭ-sād mes'o-fil) the upper layer of the mesophyll of a leaf. *151*

pancreas (pan'kre-as) an elongate, flattened organ in the abdominal cavity that secretes enzymes into the small intestine (exocrine function) and hormones controlling blood sugar (endocrine function). *197*

Pap test (pap' test) an analysis done on cervical cells for detection of cancer. *419*

parapodia (par''ah-po'de-ah) footlike fleshy lobes found on the segments of marine annelids. *647*

parasite (par'ah-sīt) an organism that resides externally on or internally within another organism and does harm to this organism. *693*

parasitism (par'ah-si''tizm) symbiotic relationship in which an organism derives nourishment from and does harm to a host. *693*

parasympathetic nervous system (par''ah-sim''pah-thet'ik ner'vus sis'tem) that part of the autonomic nervous system that usually promotes those activities associated with a normal state. *333*

parathyroid hormone (par''ah-thi'roid hor'mōn) (PTH); a hormone secreted by the parathyroid glands that affect the level of calcium and phosphate in the blood. *395*

parenchyma (pah-reng'kĭ-mah) relatively unspecialized cells that make up the fundamental tissue of plants. *144*

parietal lobe (pah-ri'ĕ-tal lōb) area of the cerebrum responsible for sensations involving temperature, touch, pressure, and pain, as well as speech. *336*

parturition (par''tu-rish'un) the processes that lead to and include the birth of a mammal and the expulsion of the extraembryonic membranes through the terminal portion of the female reproductive tract. *458*

pectoral girdle (pek'to-ral ger'd'l) portion of the skeleton that provides support and attachment for the arms. *349*

pelvic girdle (pel'vik ger'd'l) portion of the skeleton to which the legs are attached. *349*

pelvic inflammatory disease (pel'vik in-flam'ah-to''re dĭ-zēz) PID; a disease state of the reproductive organs caused by an organism that is sexually transmitted. *432*

pelvis (pel'vis) a bony ring formed by the innominate bones. Also a hollow chamber in the kidney that lies inside the medulla and receives freshly prepared urine from the collecting ducts. *304*

penis (pe'nis) male copulatory organ. *414*

pentose (pen'tōs) a 5-carbon sugar; deoxyribose is a pentose found in DNA; ribose is a pentose found in RNA. *33*

pepsin (pep'sin) a protein-digesting enzyme secreted by gastric glands. *200*

peptide bond (pep'tĭd bond) the bond that joins 2 amino acids. *31*

pericycle (per''ĭ-si'kl) a single layer of tissue interior to the endodermis that produces secondary roots. *147*

periodontitis (per''e-o-don-tĭ'tis) inflammation of the gums. *191*

peripheral nervous system (pĕ-rif'er-al ner'vus sis'tem) PNS; nerves and ganglia that lie outside the central nervous system. *320*

peristalsis (per''ĭ-stal'sis) a rhythmical contraction that serves to move the contents along in tubular organs such as the digestive tract. *193*

peritubular capillary (per''ĭ-tu'bu-lar kap'i-lar''e) capillary that surrounds a nephron and functions in reabsorption during urine formation. *307*

permafrost (perm'ah-frost) earth beneath surface in tundra that remains permanently frozen. *704*

peroxisome (pĕ-roks'ĭ-sōm) an organelle involved in oxidation of molecules and other metabolic reactions. *51*

petals (pet'alz) leaves of a flower that are often colored. *164*

petiole (pet'e-ōl) a structure connecting a leaf to a stem. *150*

PG *See* prostaglandins.

pH a measure of the hydrogen ion concentration; any pH below 7 is acid and any pH above 7 is basic. *28*

phagocytosis (fag''o-si-to'sis) the taking in of bacteria and/or debris by engulfing; cell eating. *75, 253*

phalanges (fah-lan'jēz) bones of the finger and thumb. *349*

pharynx (far'ingks) a common passageway (throat) for both food intake and air movement. *192, 282*

phenotype (fe'no-tīp) the outward appearance of an organism caused by the genotype and environmental influences. *472*

phenylketonuria (fen''il-ke''to-nu're-ah) a genetic disease stemming from the lack of an enzyme to metabolize the amino acid phenylalanine. *482*

pheromone (fer'o-mōn) a chemical substance secreted by one organism that influences the behavior of another. *404*

phloem (flo'em) the vascular tissue in plants that transports organic nutrients. *144, 615*

phospholipid (fos''fo-lip'id) lipids containing phosphorus that are particularly important in the formation of cell membranes. *65*

photochemical smog (fo''to-kem'i-kal smog) a combination of nitrogen oxides and hydrocarbons reacting in sunlight to produce ozone and PAN. *750*

photoperiodism (fo''to-pe're-od-izm) a response to light and dark; particularly in reference to flowering in plants. *161*

photosynthesis (fo''to-sin'thĕ-sis) the process of making carbohydrate from carbon dioxide and water by using the energy of the sun. *54, 136*

Photosystem I and II (fo'to-sis''tem) molecular units located within the membrane of a thylakoid that capture solar energy, making photophosphorylation possible. *127*

physiograph (fiz'e-o-graf) instrument used to record a myogram. *354*

phytochrome (fi'to-krōm) a plant pigment that is involved in photoperiodism in plants. *162*

pineal gland (pin'ē-al gland) a gland either at the skin surface (fish, amphibians) or in the third ventricle of the brain, producing melatonin. *400*

pinna (pin'nah) outer, funnellike structure of the ear that picks up sound waves. *378*

pinocytosis (pin''o-si-to'sis) the taking in of fluid along with dissolved solutes by engulfing; cell drinking. *77*

pistil (pis't'l) part of the flower that contains a stigma, style, and ovary. *164*

pituitary gland (pi-tu'i-tār''e gland) a small gland that lies just below the hypothalamus and is important for its hormone storage and production activities. *388*

placental mammal (plah-sen'tal mam'al) a mammal having internal fetal development supported by the presence of a placenta. *662*

plankton (plank'ton) free-floating microscopic organisms found in most bodies of water. *710*

plasma cell (plaz'mah sel) a cell derived from a B cell lymphocyte that is specialized to mass produce antibodies. *264*

plasmid (plaz'mid) a circular DNA segment that is present in bacterial cells but is not part of the bacterial chromosome. *535*

plasmodium (plaz-mo'di-um) multinucleated, acellular mass in slime molds. *603*

plasmolysis (plaz-mol'ĭ-sis) contraction of the cell contents due to the loss of water. *72*

platelet (plāt'let) a formed element that is necessary to blood clotting. *248*

pleural membrane (ploo'ral mem'brān) a serous membrane that encloses the lungs. *285*

plumule (ploo'mūl) the shoot tip and first 2 leaves of a plant. *167*

polar bodies (po'lar bod'ēz) nonfunctioning daughter cells that have little cytoplasm and are formed during oogenesis. *92*

pollen grain (pol'en grān) male gametophyte generation of seed plants. *166, 618*

pollination (pol''i-na'shun) the delivery by wind or animals of pollen to the stigma of a pistil in flowering plants and leading to fertilization. *166, 618*

pollution (po̽-lu'shun) detrimental alteration of the normal constituents of air, land, and water due to human activities. *729*

polymorphism (pol''ē-mor'fizm) the coexistence of 2 or more genetic forms of a species in a population. *577*

polyp (pol'ip) the sedentary stage in the life cycle of cnidarians; a benign growth. *635*

polypeptide (pol''e-pep'tĭd) a molecule composed of many amino acids linked together by peptide bonds. *31*

polysaccharide (pol''e-sak'ah-rīd) a macromolecule composed of many units of sugar. *34*

polysome (pol'e-sōm) a cluster of ribosomes all attached to the same mRNA molecule and thus all participating in the synthesis of the same polypeptide. *522*

population (pop''u-la'shun) all the organisms of the same species in one place. *7, 571, 687*

posterior pituitary (pos-tēr'e-or pi-tu'i-tār''e) back lobe of the pituitary gland that secretes ADH and oxytocin produced by the hypothalamus. *388*

postganglionic axon (pōst''gang-gle-on'ik ak'son) axon that is located after an autonomic ganglion. *332*

postmating isolating mechanisms (pōst'māt-ing i'so-lāt-ing mek'ah-nizmz) an anatomical or physiological difference between two species that prevents successful reproduction after mating has taken place. *580*

postsynaptic membrane (pōst''si-nap'tik mem'brān) in a synapse the membrane of the neuron opposite the presynaptic membrane. *327*

prairie (praré) a grassland biome of the temperate zone that occurs when rainfall is greater than 25 cm but less than 40 cm. *705*

predation (pre-da'shun) the eating of one organism on another. *688*

preganglionic axon (pre''gang-gle-on'ik ak'son) axon that is located before an autonomic ganglion. *332*

premating isolating mechanisms (pre'māt-ing i'so-lāt-ing mek'ah-nizmz) an anatomical or behavioral difference between two species that prevents the possibility of mating. *580*

pressure filtration (presh'ur fil'tra'shun) the movement of small molecules from the glomerulus into Bowman's capsule due to action of blood pressure. *308*

pressure-flow hypothesis (presh'ur flo hi-poth'ē-sis) theory explaining phloem transport; osmotic pressure following active transport of sugar into phloem brings about a flow of sap from a source to a sink. *159*

presynaptic membrane (pre''si-nap'tik mem'brān) in a synapse the membrane of the neuron opposite the postsynaptic membrane. *327*

prey (pra) organisms that serve as food for a particular predator. *688*

primate (pri'māt) an animal that belongs to the order Primates, the order of mammals that includes prosimians, monkeys, apes, and humans. *670*

primitive atmosphere (prim'ī-tiv at'mos-fēr) the gases that were found in the atmosphere when the earth first arose. *553*

producers (pro-du'serz) organisms that produce food because they are capable of synthesizing organic compounds from inorganic constituents of the environment; usually the green plants and algae in an ecosystem. *720*

proglottids (pro-glot'idz) the body sections of a tapeworm. *639*

prokaryotic (pro''kar-e-ot'ik) lacking the organelles found in complex cells; bacteria including cyanobacteria are prokaryotes. *44*

prolactin (pro-lak'tin) *See* lactogenic hormone. *391*

promoter (pro-mo'ter) a sequence of DNA in an operon where DNA polymerase begins transcription. *525*

prophase (pro'fāz) early stage in mitosis during which chromatin condenses so that chromosomes appear. *84*

proprioceptor (pro''pre-o-sep'tor) sensory receptor that assists the brain in knowing the position of the limbs. *369*

prosimians (pro-sim'e-anz) primitive primates such as lemurs, tarsiers, and tree shrews. *670*

prostaglandins (pros''tah-glan'dinz) hormones that have various and powerful local effects. *405*

prostate gland (pros'tāt gland) a gland in males, located about the urethra at the base of the bladder, whose secretions contribute to seminal fluid. *414*

protein (pro'te-in) a macromolecule composed of one or several long polypeptides. *31*

proteinoid (pro'te-in-oid) a phase, consisting of polypeptides only, during the chemical evolution of the first cell or cells. *555*

prothallus (pro'thal-us) a small, heart-shaped structure, which is the gametophyte of the fern. *617*

prothrombin (pro-throm'bin) plasma protein that functions in the formation of blood clots. *248*

protocell (pro'to-sel) a structure that precedes the evolution of the true cell in the history of life. *556*

protochordates (pro''to-kor'dāts) the invertebrate chordates that possess the 3 chordate characteristics in either the larval or adult form; tunicates and lancelets. *656*

proton (pro'ton) a subatomic particle found in the nucleus of an atom that has a weight of one atomic mass unit and carries a positive charge; a hydrogen ion. *17*

protoplast (pro'to-plast) a plant cell with the cell wall removed. *541*

protostome (pro'to-stōm) member of a large group of animal phyla in which the mouth develops from the blastopore. *642*

protozoans (pro''to-zo'anz) animal-like protists that are classified according to means of locomotion: amoebas, flagellates, ciliates. *593*

proximal convoluted tubule (prok'si-mal kon'vo-lūt-ed tu'būl) highly coiled region of a nephron near Bowman's capsule. *305, 308*

pseudocoelom (su''do-se'lom) a coelom incompletely lined by mesoderm. *641*

pseudostratified (su''-do strat'e-fīd) the appearance of layering in some epithelial cells when actually each cell touches a base line and true layers do not exist. *173*

pulmonary circuit (pul'mo-ner''e ser'kit) that part of the circulatory system that takes deoxygenated blood to, and oxygenated blood away from, the lungs. *226*

pulse (puls) vibration felt in arterial walls due to expansion of the aorta following ventricle contraction. *228*

Punnett square (pun'et skwār) a gridlike device that enables one to calculate the expected results of simple genetic crosses. *473, 490*

pupil (pu'pil) an opening in the center of the iris of the eye. *371*

pure (pūr) *See* homozygous.

purines (pu'rinz) nitrogenous bases found in DNA and RNA that have 2 interlocking rings, as in adenine and guanine. *510*

pus (pus) thick yellowish fluid composed of dead phagocytes, dead tissue, and bacteria. *253*

pyrimidine (pi-rim'ī-din) a nitrogenous base found in DNA and RNA that has just one ring, as in cytosine and thymine. *510*

pyruvate (pi'roo-vāt) the end product of glycolysis; pyruvic acid. *107*

R

radial symmetry (ra'de-al sim'ē-tre) regardless of the angle of a vertical cut made at the midline of an organism, 2 equal halves result. *631*

radicle (rad'ī-k'l) the embryonic root of a plant. *167*

radioactive (ra''de-o-ak'tiv) a property of certain elements or isotopes by which the nucleus emits particles and/or rays in order to stabilize itself. *19*

radius (ra'de us) an elongated bone located on the thumb side of the lower arm. *349*

receptor (re-sep'tor) a sense organ specialized to receive information from the environment. Also a structure found in the membrane of cells that combines with a specific chemical in a lock and key manner. *320, 367*

recessive allele (re-ses'iv ah-lēl') hereditary factor that expresses itself only when the genotype is homozygous. *471*

recombinant DNA (re-kom'bi-nant) DNA having genes from 2 different organisms, often produced in the laboratory by introducing foreign genes into a bacterial plasmid. *535*

red bone marrow (red bōn mar'o) blood-cell-forming tissue located in spaces within bones. *350*

reducing atmosphere (re-dūs'ing at'mos-fer) an atmosphere that contains reducing molecules such as H_2 rather than oxidizing ones such as O_2. *553*

reduction (re-duk'shun) the gain of electrons (inorganic); the addition of hydrogen atoms (organic). *23*

regeneration (re-jen''er-a'shun) regrowth of tissue; formation of a complete organism from a small portion. *634*

regulator gene (reg'u-lah-tor'' jēn) genes that code for proteins that are involved in regulating the activity of structural genes. *524*

REM (rapid eye movement); a stage in sleep that is characterized by eye movements and dreaming. *338*

replacement reproduction (re-plās'ment re''pro-duk'shun) a population in which each person is replaced by only one child. *742*

replication (re''pli-ka'shun) the duplication of DNA; occurs when the cell is not dividing. *512*

residual volume (res ij'u-al vol'ūm) the amount of air remaining in the lungs after a forceful expiration. *288*

resolving power (re-solv'ing pow'er) in microscopy, the capacity to distinguish between 2 points. *44*

respiratory chain (re-spi'rah-to''re chan) a series of carriers within the inner mitochondrial membrane that pass electrons one to the other from a higher level to a lower energy level; the energy released is used to build ATP; also called the electron transport system; the cytochrome system. *107*

resting potential (rest'ing po-ten'shal) the voltage recorded from inside a neuron when it is not conducting nerve impulses. *323*

restriction enzyme (re-strik'shun en'zīm) a bacterial enzyme capable of cutting DNA at a specific nucleotide sequence, producing a pattern of fragments. *503, 536*

reticular activating system (rē tik'u-lar ak'ti-vā-ting sis'tem) a complex network of cell bodies and fibers in the brain, extending from the medulla to the cerebrum. *336*

retina (ret'ī-nah) the innermost layer of the eyeball that contains the rods and cones. *371*

retroviruses (ret''ro-vi'rus-ez) viruses that contain only RNA and carry out RNA to DNA transcription. *588*

Rh factor (ar'ach fak'tor)   a type of antigen on the red blood cells.   *254*

rhizoid (ri'zoid)   hairlike extensions anchoring a liverwort into the ground.   *612*

rhodopsin (ro-dop'sin)   visual purple, a pigment found in the rods of a type of receptor in the retina of the eye.   *375*

rhyniophytes (rī'nē-o-fītz)   an extinct plant group that may have been the ancestral tracheophytes.   *614*

ribonucleic acid (ri''bo-nu-kle'ik as'id)   *See* RNA.

ribosomal RNA (ri'bo-sōm''al)   rRNA; RNA occurring in ribosome structures involved in protein synthesis.   *520*

ribosome (ri'bo-sōm)   a minute particle, found attached to endoplasmic reticulum or loose in the cytoplasm, that is the site of protein synthesis.   *51*

ribs (ribz)   bones hinged to the vertebral column and sternum which, with muscle, define the top and sides of the chest cavity.   *285, 348*

ribulose bisphosphate (ri'bu-lōs bis-fos'fat)   RuBP; the molecule that acts as the acceptor for carbon dioxide within the Calvin-Benson cycle.   *132*

RNA (ribonucleic acid);   a nucleic acid important in the synthesis of proteins that contains the sugar ribose; the bases uracil, adenine, guanine, cytosine; and phosphoric acid.   *40*

rocky shore (rok'ē shor)   a coastal community with a firm and rocky substrate.   *712*

rods (rodz)   dim-light receptors in the retina of the eye that detect motion but no color.   *371*

root cap (root kap)   thimble-shaped mass of parenchyma cells which protects the apical meristem of a root.   *145*

rough endoplasmic reticulum (ruf en''do-plas'mik rē-tik'u-lum)   RER; endoplasmic reticulum having attached ribosomes.   *50*

round window (rownd win'do)   a membrane-covered opening between the inner ear and the middle ear.   *378*

RuBP *See* ribulose bisphosphate.

S

SA node (sinoatrial node);   a small region of neuromuscular tissue that initiates the heartbeat. Also called the pacemaker.   *225*

saccule (sak'ūl)   a saclike cavity that makes up part of the membranous labyrinth of the inner ear; receptor for static equilibrium.   *379*

sac plan (sak plan)   body plan possessed by animals having a single opening.   *631*

salivary amylase (sal'i-ver-e am'i-lās)   an enzyme in the saliva that initiates the digestion of starch.   *200*

salivary gland (sal'i-ver-e gland)   a gland associated with the mouth that secretes saliva.   *192*

salt marsh (sawlt marsh)   coastal grassland exposed to seasonal flooding.   *712*

sandy beach (san'dē bēch)   a coastal community with a sand substrate along the shoreline.   *713*

saprophyte (sap'ro-fit)   a heterotrophic organism such as bacteria and fungi that externally breaks down dead organic matter before absorbing the products.   *593*

sarcolemma (sar''ko-lem'ah)   the membrane that surrounds striated muscle cells.   *358*

sarcomere (sar''ko-mir)   a contractile unit of the myofibril in a skeletal muscle cell.   *360*

sarcoplasmic reticulum (sar''ko-plaz'mik rē-tik'u-lum)   membranous network of channels and tubules within a muscle fiber, corresponding to the endoplasmic reticulum of other cells.   *359*

savanna (sah-van'ah)   a grassland biome that has occasional trees and is particularly associated with Africa.   *705*

scapula (skap'u-lah)   a broad somewhat triangular bone located on either side of the back.   *349*

schizocoelomate (skiz'o-sel-o-māt)   an animal in which the coelom forms by the splitting of the mesoderm.   *642*

scientific method (si''en-tif'ik meth'ud)   characteristic process by which scientists test their conclusions. Consists of hypothesis generation, observation and experimentation, and results in testable theories.   *8*

sclera (skle'rah)   the white fibrous outer layer of the eyeball.   *371*

sclerenchyma (skle-reng'ki-mah)   a support tissue in plants made of hollow cells with thickened walls.   *144*

scolex (sko'leks)   the head region of a tapeworm.   *639*

secretin (se-kre'tin)   hormone secreted by the small intestine that stimulates the release of pancreatic juice.   *198*

seed (sēd)   a mature ovule that contains an embryo with stored food enclosed in a protective coat.   *619*

selective reabsorption (sē-lek'tiv re''ab-sorp' shun)   movement of nutrient molecules, as opposed to waste molecules, from the contents of the nephron into the blood at the proximal convoluted tubule.   *309*

semen (se'men)   the sperm-containing secretion of males; seminal fluid plus sperm.   *415*

semicircular canals (sem''e-ser'ku-lar kah-nal'z)   tubular structures within the inner ear that contain the receptors responsible for the sense of dynamic equilibrium.   *379*

seminal fluid (sem'i-nal floo'id)   a fluid produced by various glands situated along the male reproductive tract.   *414*

seminal vesicle (sem'i-nal ves'i-k'l)   a convoluted saclike structure attached to vas deferens near the base of the bladder in males.   *414*

seminiferous tubules (sem''i-nif'er-us tu'būlz)   highly coiled ducts within the male testes that produce and transport sperm.   *412*

sensory neuron (sen'so-re nu'ron)   a neuron that takes nerve impulses to the central nervous system; afferent neuron.   *320*

sepals (se'palz)   the leafy divisions of the calyx found in a whorl at the base of petals.   *164*

septum (sep'tum)   a partition or wall such as the septum in the heart, which divides the right half from the left half.   *222*

serum (se'rum)   light-yellow liquid left after clotting of the blood.   *248*

sessile (ses'il)   organisms that lack locomotion and remain stationary in one place, such as plants or sponges.   *633*

setae (se'te)   bristles, especially those of the segmented worms.   *648*

sex chromosome (seks kro'mo-sōm)   a chromosome responsible for the development of characteristics associated with maleness or femaleness; an X or Y chromosome.   *80*

sex linked (seks lingkt)   alleles located on sex chromosomes.   *496*

schistosomiasis (skis'tō-so-mi''a-sis)   a disease caused by the blood fluke in Africa and South America.   *640*

sickle-cell anemia (sik'l sel ah-ne'me-ah)   a genetic disorder due to the homozygous genotype of the sickle-cell gene, producing sickle-shaped cells and loss of oxygen-carrying power in the blood.   *485*

sieve plate (siv plāt)   madreporite; a large pore through which water enters echinoderms.   *655*

sieve-tube cell (siv tūb sel)   specialized cells that form a linear array running vertically through phloem that functions in transport of organic nutrients.   *157*

simple eye (sim'pl I)   a sensory organ in which a single lens covers light-sensitive cells.   *650*

simple goiter (sim'p'l goi'ter)   condition in which an enlarged thyroid produces low levels of thyroxin.   *394*

sinoatrial node (si''no-a'tre-al nōd)   *See* SA node.

sinus (si'nus)   a cavity, as the sinuses in the human skull and the blood sinuses of some animals with open circulatory systems.   *348*

skeletal muscle (skel'ē-tal mus'el)   the contractile tissue that comprises the muscles attached to the skeleton; also called striated muscle.   *179*

slash-and-burn agriculture (slash and burn ag'ri-kul''tūr)   the cutting down and burning of trees to provide space to raise crops.   *744*

sliding filament theory (slīd'ing fil'ah-ment the'o-re)   the movement of actin in relation to myosin that accounts for muscle contraction.   *360*

slime mold (slīm' mold)   fungal-like protist with cellular and acellular types.   *603*

small intestine (smal in-tes'tin)   long, tubelike chamber of the digestive tract between the stomach and large intestine.   *195*

smooth endoplasmic reticulum (smooth en''do-plas'mik rē-tik'u-lum)   SER; endoplasmic reticulum without attached ribosomes.   *50*

smooth muscle (smooth mus'el)   the contractile tissue that comprises the muscles found in the walls of internal organs.   *180*

soft palate (sŏft pal'at)   entirely muscular posterior portion of the roof of the mouth.   *192*

solute (sol'ūt)   a substance dissolved in a solvent to form a solution.   *70*

solvent (sol'vent)   a fluid such as water that dissolves solutes.   *70*

somatic nervous system (so-mat′ik ner′vas sis′tem)   that portion of the PNS containing motor neurons that control skeletal muscles.   331

sorus (so′rus)   a cluster of sporangia found on the underside of fern leaves (plural: sori).   617

species (spe′shēz)   a group of similarly constructed organisms that are capable of interbreeding and producing fertile offspring; organisms that share a common gene pool.   5, 580

spermatogenesis (sper″mah-to-jen′ĕ-sis)   the production of sperm in males by the process of meiosis and maturation.   92

sphincter (sfingk′ter)   a muscle that surrounds a tube and closes or opens the tube by contracting and relaxing.   194

spicules (spik′ūlz)   needle-shaped structures produced by some sponges that function as supportive inner skeleton.   633

spinal nerve (spi′nal nerv)   a nerve that arises from the spinal cord.   329

spindle fibers (spin′d′l fi′berz)   microtubule bundles in eukaryotic cells that are involved in the movement of chromosomes during mitosis and meiosis.   84

spiracles (spir′ah-k′lz)   respiratory openings in arthropods.   652

sponge (spunj)   a member of the phylum Porifera.   633

spongy bone (spun′je bōn)   porous bone found at the ends of long bones.   178, 350

spongy mesophyll (spun′je mes′o-fil)   the lower layer of the mesophyll of a leaf.   151

sporangium (spo-ran′je-um)   a structure within which spores are produced.   603

spore (spōr)   a haploid reproductive cell, produced by the diploid sporophyte of a plant, which asexually gives rise to the haploid gametophyte.   165, 597, 612

sporophyte (spo′ro-fīt)   spore-producing diploid generation of a plant.   165, 611

sporozoa (spo″ro-zo′ah)   nonmotile parasitic protozoans.   596

spot desmosome (spot des′mo-sōm)   joining of 2 cells by intercellular filaments that penetrate cytoplasmic plaques, one in each cell.   176

squamous epithelium (skwa′mus ep″i-the′le-um)   flat cells found lining the lungs and blood vessels, for example.   173

stabilizing selection (sta′bil-īz″ing sĕ-lek′shun)   effect of natural selection eliminating atypical phenotypes.   576

stamen (sta′men)   part of flower, composed of filament and anther, where pollen grains are produced.   165

starch (starch)   the storage polysaccharide found in plants that is composed of glucose molecules joined in a linear-type fashion.   34

steady-state society (ste′dē stāt so-si′e-te)   population without a yearly increase in the number of organisms or the consumption of resources.   756

stereoscopic vision (ste″re-o-skop′ik vizh′un)   the product of 2 eyes and both cerebral hemispheres functioning together allowing depth perception.   374

sternum (ster′num)   the breastbone to which the ribs are ventrally attached.   348

steroid (ste′roid)   a lipid soluble, biologically active molecules having 4 interlocking rings; examples are cholesterol, progesterone, testosterone.   39

stigma (stig′mah)   the uppermost part of a pistil.   164

stirrup (stŭr′up)   middle ear ossicle adhering to the oval window.   378

stoma (sto′mah)   an opening in the leaves of plants through which gas exchange takes place (pl. stomata).   151

stratified (strat′i-fīd)   layered, as in stratified epithelium, which contains several layers of cells.   173

stretch receptors (strech re-sep′torz)   muscle fibers, which upon stimulation, cause muscle spindles to increase the rate at which they fire.   369

striated (stri′āt-ed)   having bands; cardiac and skeletal muscle are striated with bands of light and dark.   179

stroma (stro′mah)   the interior portion of a chloroplast.   125

stroma lamellae (stro′mah lah-mel′e)   membranous connections between adjacent thylakoids of the grana.   125

structural genes (struk′tūr-al jēnz)   genes that code for enzymes or proteins otherwise necessary to the structure and function of the cell.   525

style (stīl)   the long slender part of the pistil.   164

subcutaneous layer (sub″ku-ta′ne-us la′er)   a tissue layer found in vertebrate skin that lies just beneath the dermis and tends to contain adipose tissue.   182

succession (suk-se′shun)   a series of ecological stages by which the community in a particular area gradually changes until there is a climax community that can maintain itself.   719

summation (sum-ma′shun)   ever greater contraction of a muscle due to constant stimulation that does not allow complete relaxation to occur.   327, 355

supressor T cell (su-pres′or T sel)   T lymphocyte that suppresses certain other T and B lymphocytes from continuing to divide and perform their respective functions.   266

symbiosis (sim″bi-o′sis)   an intimate association of two dissimilar species including commensalism, mutualism, and parasitism.   693

sympathetic nervous system (sim″pah-thet′ik ner′vus sis′tem)   that part of the autonomic nervous system that usually causes effects associated with emergency situations.   332

synapse (sin′aps)   the region between 2 nerve cells where the nerve impulse is transmitted from one to the other; usually from axon to dendrite.   327

synapsis (si-nap′sis)   the attracting and pairing of homologous chromosomes during prophase I of meiosis.   89

synaptic cleft (si-nap′tik kleft)   small gap between presynaptic and postsynaptic membranes.   327

synaptic ending (si-nap′tik end′ing)   the knob at the end of an axon at a synapse.   327

synovial joint (si-no′ve-al joint)   a freely movable joint.   351

systemic circuit (sis-tem′ik ser′kit)   that part of the circulatory system that serves body parts other than the gas-exchanging surfaces in the lungs.   226

systole (sis′to-le)   contraction of the heart chambers.   223

systolic blood pressure (sis-tol′ik blud presh′ur)   arterial blood pressure during the systolic phase of the cardiac cycle.   230

T

taiga (ti′gah)   a biome that forms a worldwide northern belt of coniferous trees.   706

tapeworm (tāp werm)   parasitic flatworm; member of class cestoda.   639

tarsal (tahr′sal)   a bone of the ankle in humans.   349

taste bud (tāst bud)   organ containing the receptors associated with the sense of taste.   369

taxonomy (tak-son′o-me)   the science of naming and classifying organisms.   6

Tay-Sachs disease (tā saks′ di zēz)   a lysosomal storage disease that is inherited and causes neurological impairment and death.   481

tectorial membrane (tek-te′re-al mem′brān)   membrane within the organ of Corti that transmits nerve impulses to the brain.   379

telophase (tel′o-fāz)   stage of mitosis during which diploid number of daughter chromosomes are located at each pole.   87

template (tem′plāt)   a pattern that serves as a mold for the production of an oppositely shaped structure; one strand of DNA is a template for the complementary strand.   512

temporal lobe (tem′po-ral lōb)   area of the cerebrum responsible for hearing and smelling, also the interpretation of sensory experience and memory.   336

tendon (ten′don)   dense connective tissue that joins muscle to bone.   176, 352

testcross (test kros)   the backcross of a heterozygote with the recessive in order to determine its genotype.   474

testes (tes′tēz)   the male gonads, the organs that produce sperm and testosterone.   412

testosterone (tes-tos′tĕ-rōn)   the most potent androgen.   416

tetanus (tet′ah-nus)   sustained muscle contraction without relaxation.   355

tetany (tet′ah-ne)   severe twitching caused by involuntary contraction of the skeletal muscles due to a lack of calcium.   395

tetrad (tet′rad)   a set of 4 chromatids resulting from the pairing of homologous chromosomes during prophase I of meiosis.   89

tetrapod (tet′rah-pod)   having 4 limbs.   657

thalamus (thal′ah-mus) a mass of gray matter, located at the base of the cerebrum in the wall of the third ventricle, that receives sensory input. *336*

theory (the′o-rē) a concept supported by a large number of observations by the scientific method. *12*

thermal inversion (ther′mal in-ver′zhun) temperature inversion such that warm air traps cold air and its pollutants near the earth. *750*

thrombin (throm′bin) enzyme that converts fibrinogen to fibrin threads during blood clotting. *248*

thylakoid (thi′lah-koid) an individual flattened vesicle found within a granum (pl. grana). *125*

thymus (thi′mus) an organ that lies in the neck and chest area and is absolutely necessary to the development of immunity. *400*

thyroid-stimulating hormone (thi′roid stim′u-lāt″ing hor′mon) *See* TSH.

thyroxin (thi-rok′sin) the hormone produced by the thyroid that speeds up the metabolic rate. *394*

tibia (tib′e-ah) the shinbone found in the lower leg. *349*

tidal volume (ti′dal vol′um) amount of air normally moved in the human body during an inspiration or expiration. *287*

tight junction (tīt junk′shun) cell membranes from 2 cells interlock in a zipperlike fashion preventing leakage into or out of a tissue. *176*

T lymphocyte (lim′fo-sīt) a lymphocyte that matures in the thymus and occurs in 4 varieties, one of which kills antigen-bearing cells outright. *264*

tone (tōn) the continuous partial contraction of muscle; also the quality of a sound. *353*

trachea (tra′ke-ah) in vertebrates, the windpipe; in insects, the tracheae are the air tubes. *652*

tracheophyte (trā′ke-a-fīt) vascular plant, including ferns and seed plants, that has a dominant sporophyte. *614*

tracheids (tra′ke-idz) a component of xylem made of long, tapered nonliving cells. *155*

trait (trāt) a specific term for a distinguishing feature studied in heredity. *469*

transcription (trans-krip′shun) the process that results in the production of a strand of mRNA that is complementary to a segment of DNA. *516*

transfer RNA (trans′fer) tRNA; molecule of RNA that carries an amino acid to a ribosome engaged in the process of protein synthesis. *520*

transformed (trans-formed′) cell that has been altered by genetic engineering and is capable of producing new protein. *534*

transgenic organism (trans jen′ik or′gah-nizm) plant or animal that has received a foreign gene. *541*

transition reaction (tran-zish′un re-ak′shun) a reaction within aerobic cellular respiration during which hydrogen and carbon dioxide are removed from pyruvic acid; connects glycolysis to Krebs cycle. *107*

translation (trans-la′shun) the process by which the sequence of codons in mRNA dictates the sequence of amino acids in a polypeptide. *517*

translocation (trans-lō-ka′shun) chromosome mutation caused by the movement of a chromosome fragment from one chromosome to a nonhomologous chromosome. *494*

transpiration (tran″spi-ra′shun) the evaporation of water from a leaf; pulls water from the roots through a stem to leaves. *156*

trichinosis (trik″i-no′sis) disease caused by a roundworm in which the larva are found in cysts within muscle cells. *642*

trichocyst (trik′o-sist) threadlike darts released by some ciliates that may help in defense against predators or in capturing prey. *595*

trophic level (tro′fic levl) a categorization of species in a food web according to their feeding relationships from the first level autotrophs through succeeding levels of herbivores and carnivores. *722*

trochophore (tro′ko-fōr) a larval form unique to the protostomes that indicates they are related. *642*

trophoblast (trof′o-blast) the outer membrane that surrounds the human embryo and, when thickened by a layer of mesoderm, becomes the chorion, an extraembryonic membrane. *448*

tropical rain forest (trŏp′i-kĕl rān for′est) a biome of equatorial forests that remain warm year-round and receive abundant rain. *707*

trypsin (trip′sin) a protein-digesting enzyme secreted by the pancreas. *201*

TSH (thyroid-stimulating hormone); hormone that causes the thyroid to produce thyroxin. *391*

tube feet (tūb fēt) rows of small tube-shaped appendages in echinoderms that are used in locomotion. *654*

tube-within-a-tube plan (tūb with-in′ ah tūb plan) body plan of animals having two openings. *631*

tubular excretion (tu′bu-lar eks-kre′shun) the movement of certain molecules from the blood into the distal convoluted tubule so that they are added to urine. *310*

tundra (tun′drah) a biome characterized by lack of trees, due to cold temperatures and the presence of permafrost year round. *704*

tunicate (tu′ni-kāt) a type of protochordate in which only the larval stage has the 3 chordate characteristics; and of these only gills are found in the adult. *656*

turgor pressure (tur′gor presh′ur) osmotic pressure that adds to the strength of the cell. *72*

Turner syndrome (tur′ner sin′drōm) a condition caused by the inheritance of an abnormality in chromosome number; an X chromosome lacks a homologous counterpart-XO. *495*

twitch (twich) a brief muscular contraction followed by relaxation. *355*

tympanic membrane (tim-pan′ik mem′brān) membrane located between outer and middle ear that receives sound waves; the eardrum. *378*

tympanum (tim′pah-num) sound receptor found in terrestrial animals, for example, the grasshopper. *652*

typhlosole (tif′lo-sōl) a longitudinal fold in the intestine of annelids that enhances absorption capacity. *648*

U

ulcer (ul′ser) an open sore in the wall of the digestive tract. *194*

ulna (ul′nah) an elongated bone found within the lower arm. *349*

umbilical arteries and vein (um-bil′i-kal ar′ter-ez and vān) fetal blood vessels that travel to and from the placenta. *456*

umbilical cord (um-bil′i-kal kord) cord connecting the fetus to the placenta through which blood vessels pass. *451*

urea (u-re′ah) primary nitrogenous waste of mammals derived from amino acid breakdown. *300*

ureters (u-re′terz) tubes that take urine from the kidneys to the bladder. *303*

urethra (u-re′thrah) the tube that takes urine from bladder to outside. *303*

uric acid (u′rik as′id) waste product of nucleotide breakdown. *300*

urinalysis (u″ri-nal′i-sis) a medical procedure in which the composition of a patient's urine is determined. *315*

urinary bladder (u′ri-ner″e blad′der) an organ where urine is stored before being discharged by way of the urethra. *303*

uterine cycle (u′ter-in si′k′l) monthly occurring changes in the characteristics of the uterine lining. *420*

utricle (u′tre-k′l) saclike cavity that makes up part of the membranous labyrinth of the inner ear receptor for static equilibrium. *379*

V

vaccine (vak′sēn) antigens prepared in such a way that they can promote active immunity without causing disease. *270*

vacuole (vak′u-ōl) a membrane-bounded cavity, usually fluid filled. *52*

vagina (vah-ji′nah) the copulatory organ and birth canal in females. *419*

valve (valv) an opening that opens and closes, insuring one-way flow only; common to the systemic veins, the lymphatic veins, and to the heart. *221*

vas deferens (vas def′er-ens) that part of the male reproductive tract leading from the epididymis to the urethra. *412*

vascular bundle (vas′ku-lar bundle) tissues that include xylem and phloem enclosed by a sheath and typically found in herbaceous plant stems. *148*

vascular cambium (vas′ku-lar kam′be-um) a cylindrical sheath of meristematic tissue that produces secondary xylem and phloem. *148*

vascular cylinder (vas′ku-lar sil′in-der) a central region of dicot roots that contains the vascular tissues, xylem, and phloem. *147*

a virus, for recombinant DNA that introduces a foreign gene into a host cell. *535*

vein (vān) a vessel that takes blood to the heart. *219*

venae cavae (ve′nah ka′vah) large systemic veins that return blood to the right atrium of the heart. Inferior vena cava collects blood from lower body regions. Superior vena cava collects blood from upper body regions. *223*

venous duct (ve′nus dukt) ductus venosus; fetal connection between the umbilical vein and the inferior vena cava. *456*

ventilation (ven″ti-la′shun) breathing; the process of moving air into and out of the lungs. *284*

ventricle (ven′tri-k′l) a cavity in an organ such as the ventricles of the heart or the ventricles of the brain. *222, 334*

venule (ven′ūl) a vessel that takes blood from capillaries to veins. *219*

vernix caseosa (ver′niks ka″se-o′sah) cheese-like substance covering the skin of the fetus. *456*

vertebrae (ver′tĕ brā) the bones of the vertebral column. *348*

vertebral column (ver′te-bral kol′um) the backbone of vertebrates through which the spinal cord passes. *348*

vertebrate (ver′tĕ-brāt) animal possessing a backbone composed of vertebrae. *630*

vessel cell (ves′el sel) an individual vessel element in xylem. During development, vessel cells lose their contents and end walls so that they form a continuous pipeline in xylem. *155*

vestigial structure (ves-tij′e-al struk′tūr) the remains of a structure that was functional in some ancestor but is no longer functional in the organism in question. *569*

villi (vil′i) fingerlike projections that line the small intestine and function in absorption. *195*

visceral mass (vis′er-al mas) soft-bodied portion of a mollusk that includes internal organs. *643*

vital capacity (vi′tal kah-pas′i-tē) maximum amount of air moved in or out of the human body with each breathing cycle. *287*

vitamin (vi′tah-min) essential requirement in the diet. Needed in small amounts, that are often part of coenzymes. *209*

vitreous humor (vit′re-us hu′mor) the substance that fills the posterior chamber of the eye. *371*

vocal cords (vo′kal kordz) folds of tissue within the larynx that create vocal sounds when they vibrate. *282*

vulva (vul′vah) the external genitalia of the female that surround the opening of the vagina. *419*

W

water mold (wah′ter mold) protist, usually saprophytic, with a threadlike body similar to fungi. *603*

water vascular system (wah′ter vas′ku-lar sis′tem) a series of canals that take water to the tube feet of echinoderms allowing them to expand. *654*

X

X-linked gene (eks lingkt jēn) an allele located on the X chromosome. *496*

xylem (zi′lem) the vascular tissue in plants that transports water and minerals. *144, 614*

XYY male (eks wi wi māl) a male that has an extra Y chromosome. *496*

Y

yellow bone marrow (yel′o bōn mar′o) fat-storage tissue found in the cavities within certain bones. *350*

yolk (yōk) a rich nutrient material in the egg of certain vertebrate embryos. *437*

yolk sac (yōk sak) one of the extraembryonic membranes within which yolk is often found. *446*

Z

zooflagellates (zo″o-flaj′ĕ-lāts) protozoans that move by means of flagella. *595*

zygote (zi′gōt) diploid cell formed by the union of 2 gametes, the product of fertilization. *81*

Credits

Photographs

History of Biology
Leeuwenhoek: Bettmann Newsphotos; **Darwin:** © 78 John Moss/Black Star; **Pasteur:** The Bettmann Archive; **Koch:** Bettmann Newsphotos; **Pavlov:** UPI/Bettmann; **Lorenz:** UPI/Bettmann; **McClintock:** AP/Wide World Photos, Inc.; **Jarvik:** UPI/Bettmann Newsphotos; **Gallo:** Reuters/Bettmann Newsphotos

Part Openers
Part One: © Francis Leroy; **Part Two:** © Hal Clason/Tom Stack & Associates; **Part Three:** © Francis Leroy

Introduction
I.1: Gail Shumway; I.3 (all): © Dwight Kuhn; I.4: Eliot Porter; I.5: Davis C. Fritts; I.7 (left): © Roy Whitehead/Photo Researchers, Inc.; I.7 (right): Michael Melford/The Image Bank; I.9: © Leonard Lessin/Peter Arnold, Inc.

Chapter 1
1.1: Warren Garst/Tom Stack & Associates; 1.5b: Journalism Services; 1.11: Brett Froomer/The Image Bank; 1.17: Photo Researchers, Inc.; page 33: © Francis Leroy; 1.23b: © Farrell Grehan/Photo Researchers, Inc.; 1.24b: © Don Fawcett/Photo Researchers, Inc.; 1.25b: © Werner H. Muller/Peter Arnold, Inc.

Chapter 2
2.1a: © Biophoto Associates/Photo Researchers, Inc.; 2.1b: Stephen L. Wolfe; 2.1c: CNRI/Science Photo Library/Photo Researchers, Inc.; 2.1d: © Doug Wechsler/Animals, Animals/Earth Scenes; 2.4a: © Dr. Stephen Wolfe; 2.4b: © Francis Leroy; 2.5a: © W. Rosenberg/Biological Photo Service, © Dr. James A. Lake, *Journal of Molecular Biology,* 105, 131–59, 1976 © Academic Press, Inc. (London) Ltd; 2.6a: © David M. Phillips/Visuals Unlimited; 2.8: © K. G. Murti/Visuals Unlimited; 2.10a: © Dr. Keith Porter; 2.11a: © Dr. Herbert Israel/Cornell University; 2.12a: © M. Schliwa/Visuals Unlimited; 2.14c: © Dr. Keith Porter; page 61: © Francis Leroy

Chapter 3
3.3b: © Biophoto Associates/Photo Researchers, Inc.; page 76: © Francis Leroy

Chapter 4
4.1: © Donald Yeager/Camera M. D. Studios; 4.7: © Edwin A. Reschke; 4.8a: © Francis Leroy; 4.9–4.11: © Edwin A. Reschke; 4.12 (both): © R. G. Kessel and C. Y. Shih; *Scanning Electron Microscopy* © 1976, Springer Verlag, Berlin; 4.13b: © John D. Cunningham/Visuals Unlimited

Chapter 5
5.1: © Tim Davis/Photo Researchers, Inc.; 5.4: © Courtesy of Dr. Alfonso Tramontano; page 115 (all): © Bob Coyle

Chapter 6
6.3c: © Gordon Leedale/Biophoto Associates; 6.5c: © Dr. Kenneth Miller; page 134: © Francis Leroy; 6.11: R. R. Hessler/Scripps Institution of Oceanography

Chapter 7
7.4b: Carolina Biological Supply; 7.5: © John D. Cunningham/Visuals Unlimited; 7.6: © J. R. Waaland, Univ. of Washington/Biological Photo Service; 7.7a, 7.8a: Carolina Biological Supply; 7.8a,b: © Runk Schoenberger/Grant Heilman Photography; 7.10b: Carolina Biological Supply; 7.12b: © J. H. Troughton and L. Donaldson

Chapter 8
8.2b: © Biological Photo Service; 8.4a: © David M. Phillips/Visuals Unlimited; 8.5b: © George Wilder/Visuals Unlimited; page 161: Courtesy of R. J. Weaver; 8.10: © Frank B. Salisbury; 8.11: © Michael Wotton/Weyerhaeuser Company; page 166: © Francis Leroy; 8.14: © Michael Godfrey

Chapter 9
9.2a,b: © Edwin A. Reschke; 9.2c: © Manfred Kage/Peter Arnold, Inc.; 9.3b–9.11: © Edwin A. Reschke; page 183: © Margaret C. Berg/Berg and Associates

Chapter 10
10.5b: Kessel, R. G. & Kardon, R. H., *Tissues and Organs: A Text Atlas of Scanning Electron Microscopy,* 1979 by W. H. Freemann & Company. Reprinted with permission; 10.7b: © Edwin A. Reschke; 10.8: © Martin M. Rotker/Taurus Photos; 10.9c,d: Kessel, R. G. & Kardon, R. H., *Tissues and Organs: A Text Atlas of Scanning Electron Microscopy,* © 1979 W. H. Freeman & Co.; 10.13: © Dr. Sheril D. Burton; 10.17: © Robert Frerck Productions; 10.20a,b: WHO; 10.20c,d: Centers for Disease Control, Atlanta GA; 10.21: © James Blank/FPG; 10.23: © 1988, Alan Carey/The Image Works; page 215: © Michael Phillip Manhelm/Marilyn Gartman Agency

Chapter 11
11.7a: © Igaku Shoin Ltd.; page 235: © Lewis Lainey

Chapter 12
12.3a: © Lennart Nilsson: *Behold Man* Little Brown & Company, Boston; 12.8: © Manfred Kage/Peter Arnold, Inc.; 12.10a: L. W. Diggs, M.D. *The Morphology of Human Blood Cells;* 12.10b: Dr. Etienne de Harven and Ms. Nina Lampen, Sloan-Kettering Institute; page 253: © Francis Leroy; 12.13a: Stuart I. Fox

Chapter 13
13.2: Keetin, *Biological Science,* Norton Publ.; 13.4: © Photo Courtesy of Boehringer, Ingelheim International GMBH, photo courtesy Lennart Nilsson; 13.7a: © R. Feldman—Dan McCoy/Rainbow; 13.8a,b: © Boehringer Ingelheim International GMBH, photo courtesy Lennart Nilsson; 13.11a: Guy Gillette/Photo Researchers, Inc.; 13.13: © Bill Bachman/Photo Researchers, Inc.

Chapter 14
14.1: © Bob Coyle; 14.5: © John Watney Photo Library; 14.13a: © Lennart Nilsson; 14.13b: Richard Feldman, National Institute of Health, Dept. of Health & Human Services; 14.14: American Lung Association/Patricia Delaney; 14.15a: © Manfred Kage/Peter Arnold, Inc.; 14.15b: © CNRI/Science Photo Library/Photo Researchers, Inc.; page 295 (both): © Martin Rotker/Taurus Photos

Chapter 15
15.2: © Phil Degginger; 15.3: © Stephen J. Kraseman, The National Audubon Society Collection/Photo Researchers, Inc.; 15.10a: © Gennars/Photo Researchers, Inc.

Chapter 16
16.1a: © David Madison/Bruce Coleman, Inc.; 16.1b: Kent M. Van De Graaff, *Human Anatomy,* 2d ed. Copyright © 1988 Wm. C. Brown Publishers, Dubuque, Iowa. All Rights Reserved. Reprinted by permission. 16.3b: © J. D. Robertson; 16.4: © Linda Bartlett, 1981; page 329: Francis Leroy; 16.15: © W. S. Ormerod, Jr./Visuals Unlimited; 16.18: © Dan McCoy/Rainbow; 16.20: © Marilyn Gartmen Agency; 16.22: © Bunde/Unicorn Stock Photos; 16.23a: © Frerck/Odyssey Productions, Inc.; 16.23b: © Gigli/Photo Reseachers, Inc.

Chapter 17
17.9a: International Biomedical, Inc.; page 357: © Alan D. Levenson Photography, 1985; 17.11b: © Dr. H. E. Huxley, Cambridge University; 17.13a: © Victor B. Eichler

Chapter 18
18.1: © Kaiser Porclein, Ltd.; 18.6d: © Dr. Frank Werblin, University of CA at Berkley, Zoology Dept.; 18.12c: © Ken Touchton; page 383: © Kaiser Porclein, Ltd.

Chapter 19

19.6: © Bettina Cirone/Photo Researchers, Inc.; **19.7a,b:** © Dr. Charles A. Blake/The Pituitary Gland/Carolina Reader #118; **19.9:** © Lester V. Bergman & Associates, Inc.; **19.10:** Courtesy of F. A. Davis Company, Philadelphia, PA and Dr. R. H. Kampmeier; **19.11:** from *Clinical Endocrinology & Its Physiological Basis* by Arthur Grollman, 1964. Used by permission of J. B. Lippincott Company; **19.12:** © Lester V. Bergman & Associates, Inc.; **19.15:** Arthur Grollman, 1964. Used by permission of J. B. Lippincott Company; **19.16:** © F. A. Davis Co., Philadelphia, PA and Dr. R. H. Kampmeier

Chapter 20

20.1: Dr. G. Schatten; **20.3b:** © BioPhoto Associates/Photo Researchers, Inc.; **20.7b:** © B. Baganvandoss/Photo Researchers, Inc.; **20.12:** © Hank Morgan/Science Source/Photo Researchers, Inc.; **20.15a:** © Lightdale/Photo Researchers, Inc.; **20.15b:** © Robert Settineri/Sierra Productions; **20.16:** George J. Wilder/Visuals Unlimited

Chapter 21

21.1a,b: © Edelmann/ *First Days of Life*/Black Star; **page 448:** © Francis Leroy; **21.8a:** © Lennart Nilsson; **21.23:** Dan Rubin/for Mutual of America; **21.25:** © Martin R. Jones/Unicorn Stock Photos

Chapter 22

22.1: © John Hicks/Australasian Nature Trans.; **22.3a,b:** © Bob Coyle; **22.11c:** © '89 J. Callahan, M.D./IMS/University of Toronto/Custom Medical Stock Photo; **22.12a:** © Daemmrich/The Image Works; **22.13:** London Express; **22.15b:** © Bill Longcore/Photo Researchers, Inc.

Chapter 23

23.1: Courtesy Lennard Nilsson © Boehringer Ingeheim GMBH; **23.4a:** © Jill Cannafax/EKM-Nepenthe; **page 493:** © Robinson/National Jewish Hospital & Research; **23.5:** © John Cunningham/Visuals Unlimited; **23.8a:** © F. A. Davis Company, Philadelphia and R. H. Kampmeier; **23.8b:** From M. Bartalos & T. A. Baramski: Medical Cytogenetics. © 1967 Williams and Wilkins Company; **page 513(a):** Biophysics Dept. Kings College, London; **page 513(b):** J. D. Watson

Chapter 24

24.8a: © Bill Longcore/Photo Researchers, Inc.; **24.14b:** A. Rich; **24.18a:** © A. L. Olins, University of Tennessee/Biological Photo Service

Chapter 25

25.1a–d: Courtesy of Genetech, Inc.; **25.3b:** © E. Hartmann/Magnum, Inc.; **25.4a:** © Sanofi Recherche; **25.4b:** © Cetus Research Place; **25.4c:** Petit Format/Nestle/Science Source/Photo Researchers, Inc.; **25.5 (both):** General Electric Research and Development Center; **25.6 (all):** © Runk/Schoenberger Grant Heilman Photography; **25.7b,c:** © Biophoto Associates/Photo Researchers, Inc., © Laboratories Carlsberg; **25.8:** Monsanto Company; **25.9:** Photograph by Vergil Sweeden, *Science* Nov., 1989, Vol. 246, pp. 725–856; **25.11:** John A. Thompson, *Science* Nov, 1989, p. 747; **25.12:** © Hoffmann-La Roche and Roche Institute

Chapter 26

26.3a: © Science Source/Sidney Fox/Visuals Unlimited; **26.3b:** © David W. Dreamer; **26.4a:** © S. M. Awramik/University of California/Biological Photo Service; **26.4b:** © Doug Sokell/Visuals Unlimited; **26.4c:** © Francois Gohier/Photo Researchers, Inc.; **page 558:** Scripps Institute of Oceanography

Chapter 27

27.1a: © Thomas Taylor/Photo Researchers, Inc.; **27.1b:** © William E. Fergusen; **27.3a:** American Museum of Natural History, Dept. of Library Services; **27.3b:** © Edward S. Ross; **27.7a,b:** Carolina Biological Supply; **27.9a:** © John D. Cunningham/Visuals Unlimited; **27.9b:** © John M. Trager/Visuals Unlimited; **27.9c:** © Dominique Brand/Tom Stack and Associates; **27.9d:** © Charles G. Summers, Jr./Tom Stack and Associates;

27.11: © D. P. Hershkowitz/Bruce Coleman, Inc.; **27.14a,b:** © J. A. Bishop & L. M. Cook; **27.16a:** © F. B. Gill/VIREO; **27.16b:** © Miguel Castro/Photo Researchers, Inc.; **27.16c:** © Alan Root/Bruce Coleman, Inc.

Chapter 28

28.1a: © Carl Zeiss, Inc.; **28.1b (top):** © Richard Feldman, National Institute of Health; **28.1b (bottom):** © Robert Caughey/Visuals Unlimited; **28.1c:** © Michael M. Wurtz, Biozentrum der Universitat Basel Abt Mikrobiologie; **28.1d:** © Omikron/Photo Researchers, Inc.; **28.2:** © Dr. E. R. Degginger; **page 589:** © Pfizer, Inc.; **28.6a:** © Francis Leroy; **28.6b:** D. M. Phillips/Visuals Unlimited; **28.8:** © TJ Beveridge/ University of Guelph/Biological Photo Service; **28.9a:** © R. Knauft/Photo Researchers, Inc.; **28.9b:** © Eric Grave/Science Source/Photo Researchers, Inc.; **28.10:** © Leon J. Le Beau/Biological Photo Service; **28.13:** © Eric V. Grave/Photo Researchers, Inc.; **28.17a:** Carolina Biological Supply; **28.18b:** © M. I. Walker/Photo Researchers, Inc.; **28.19a:** © John D. Cunningham/Visuals Unlimited; **28.22a:** Bio Photo Associates/Photo Researchers, Inc.; **28.22b:** © Eric Grave/Photo Researcher, Inc.; **28.24a:** © Bob Coyle; **28.24b:** © Kitty Kohout/Root Resources; **28.24c:** © Bob Coyle; **28.26b:** © Gordon Leedale/Bio Photo Associates

Chapter 29

29.1a: © Leonard Lee Rue III/Bruce Coleman, Inc.; **29.1b:** © John Shaw/Tom Stack and Associates; **29.1c:** © William F. Ferguson; **29.1d:** © Kent Danner/Photo Researchers, Inc.; **29.3:** © P. Gates University of Durham/Biological Photo Service; **29.6a:** Carolina Biological Supply; **29.6b:** © Bio Photo Associates; **29.6c:** Carolina Biological Supply; **29.8a:** © Albert Kunigh/Valan Photos; **29.8b:** © John N. A. Lott/McMaster University/ Biological Photo Service; **29.9:** © Field Museum of Natural History, Chicago; **29.11a:** © Jack D. Swenson/Tom Stack and Associates; **29.11b:** © John D. Cunningham/Visuals Unlimited; **29.13:** USDA, Forest Service; **29.16a:** © N. Smythe/Photo Researchers, Inc.; **29.16b:** © H. Eisenbeiss/Photo Researchers, Inc.; **29.16c:** © Antheny Mercieca/Photo Reseachers, Inc.; **29.17a:** © Edward S. Ross; **29.17b:** © Bob Coyle; **29.17c:** © Leonard Lee Rue, III

Chapter 30

30.1a: © Bruce Russell/Bio Media Associates; **30.1b:** © Stephen J. Krasemann/Peter Arnold, Inc.; **30.8a:** © Bill Cartsinger/Photo Researchers, Inc.; **30.8b:** © Ron Taylor/Bruce Coleman, Inc.; **30.8c:** Carolina Biological Supply; **30.10a:** © Michael DiSpezio/Images; **30.11a:** © Dr. Fred Whittaker; **30.12:** © Paul Nollen and Matthe Nadakavukaren; **30.13b:** © Fred Marski/Visuals Unlimited; **30.14:** © Jim Solliday/Biological Photo Service; **30.16a,b:** © Biophoto Associates; **30.16c:** © Rick Poley; **30.18a:** © Michael Di Spezio/Images; **30.18c:** © Gary Milburn/Tom Stack and Associates; **30.20a:** © Robert Evans/ Peter Arnold, Inc.; **30.20b:** © Rob and Melissa Simpson/Valan Photos; **30.20c:** © Dwight R. Kuhn; **30.20d:** © John Mac Gregor/Peter Arnold, Inc.; **30.20e,f:** © John Fowler/Valan Photos; **30.23a:** © Michael Di Spezio/Images; **30.23b:** © Bud Higdon; **30.23c:** © Robert A. Ross; **30.27a,b:** © Douglas Faulkner/Sally Faulkner Collection; **30.28a:** © Leonard Lee Rue, III; **30.28b:** © Andrew Odum/Peter Arnold, Inc.; **30.29b:** © Paul R. Erlich, Stanford University/Biological Photo Service; **30.29c:** © Bob Coyle; **30.30a:** © Oxford Scientific Films/Animals, Animals/Earth Scenes; **30.30b:** © Dallas Heaton/Uniphoto; **30.31a:** © Bob Coyle; **30.31b:** © E. S. Ross

Chapter 31

31.1: © Stouffer Enterprises, Inc./ Animals, Animals; **31.3a:** © Martha Reeves/Photo Researchers, Inc.; **31.3b:** © Tom McHugh/Photo Researchers, Inc.; **31.3c:** © George Holton/Photo Researchers, Inc.; **31.3d:** © Tom McHugh/Photo Researchers, Inc.; **31.10:** © American Museum of Natural History; **31.11a:** © Linda Bartlett/

Photo Researchers, Inc.; **31.11b:** © Delta Willis/Bruce Coleman, Inc.; **31.11c:** © Jocelyn Burt/Bruce Coleman, Inc.; **31.11d:** © Larry Voight; **31.11e:** © Stephen Trimble

Chapter 32

32.4a,b: © Fritz Polking; **32.5:** Fig 7.6 from *Fundamentals of Ecology,* Third Edition, by Eugene P. Odum, copyright © 1971 by Saunders College Publishing, a division of Holt, Rinehart and Winston, Inc., reprinted by permission of the publisher. **32.6a:** © Dr. Gregory Antipa and H. S. Wesenbergand; **32.7:** © National Audubon Society/Photo Researchers, Inc.; **32.8:** © Z. Leszczynski/Animals, Animals; **32.9a,b:** © J. A. L. Cooke/Oxford Scientific Picture Library; **32.10:** © Cliff B. Frith/Bruce Coleman, Inc.; **32.11a–c:** © Dr. Daniel Janzen; **32.12:** © Bill Wood/Bruce Coleman, Inc.

Chapter 33

33.1a: © Peter Kaplan/Photo Researchers, Inc.; **33.1b:** © Frank Miller/Photo Researchers, Inc.; **33.1c:** © Pat O'Hara; **33.3a:** W. E. Ruth/Bruce Coleman, Inc.; **33.3b:** © John Shae/Bruce Coleman, Inc.; **33.4a:** © Joe McDonald/Tom Stack & Associates; **33.4b:** William Johnson/Stock Boston; **33.5:** Brian Parker/Tom Stack & Associates; **33.6:** © Bob Pool/Tom Stack & Associates; **33.7:** © Brian Parker/Tom Stack & Associates; **page 709(a):** © Stephen Dalton/Photo Researchers, Inc.; **page 709(b):** © Erwin & Peggy Bauer/Bruce Coleman, Inc.; **page 709(c):** © Bruce Coleman/Bruce Coleman, Inc.; **33.10:** From D. Correll, "Estuarine Productivity" in *BioScience,* 28:649, 1978. Copyright 1978 by the American Institute of Biological Sciences, Washington, DC. Reprinted by permission of the publisher and author. **33.11:** © Douglas Faulkner/Sally Faulkner Collection

Chapter 34

34.1a: © Stephen Kraseman/Peter Arnold, Inc.; **34.1b:** © Richard Ferguson/William Ferguson; **34.1c:** © Mary Thatcher/Photo Researchers, Inc.; **34.3a:** © Michael Galbridge/Visuals Unlimited; **34.11:** © Jacques Jangoux/Peter Arnold, Inc. **34.15:** © Jane Windsor, Division of Plant Industry, Florida Dept. of Agriculture, Gainesville, FL

Chapter 35

35.1a: © Dr. E. R. Degginger; **35.1b:** © Ingelborg Lippman/Peter Arnold, Inc.; **35.6:** © Bob Coyle; **35.7a:** © Sydeny Thomson/Animals, Animals/Earth Scenes; **35.7b:** © G. Prance/Visuals Unlimited; **35.8:** © Gary Milburn/Tom Stack and Associates; **35.12:** © Don Riepe/Peter Arnold, Inc.; **35.14a,b:** © Dr. John Skelly; **35.15:** NASA

Line Art

Chapter 10

10.3: From Kent M. Van De Graaff, *Human Anatomy,* 2d ed. Copyright © 1988 Wm. C. Brown Publishers, Dubuque, Iowa. All Rights Reserved. Reprinted by permission.

Chapter 11

11.2: From Kent M. Van De Graaff and Stuart Ira Fox, *Concepts of Human Anatomy and Physiology,* 2d ed. Copyright © 1989 Wm. C. Brown Publishers, Dubuque, Iowa. All Rights Reserved. Reprinted by permission; **11.6:** From James E. Crouch, *Functional Human Anatomy,* 2d ed. Copyright © 1972 Lea & Febiger, Philadelphia, PA. Reprinted by permission; **11.7b:** From John W. Hole, Jr., *Human Anatomy and Physiology,* 5th ed. Copyright © 1990 Wm. C. Brown Publishers, Dubuque, Iowa. All Rights Reserved. Reprinted by permission; **11.12:** From Stuart Ira Fox, *A Laboratory Guide to Human Physiology: Concepts and Clinical Applications,* 5th ed. Copyright © 1990 Wm. C. Brown Publishers, Dubuque, Iowa. All Rights Reserved. Reprinted by permission; **11.14:** From Stuart Ira Fox,

Human Physiology, 3d ed. Copyright © 1990 Wm. C. Brown Publishers, Dubuque, Iowa. All Rights Reserved. Reprinted by permission.

Chapter 12

Page 243: © c. Anthony Hunt, University of California, San Francisco.

Chapter 13

13.6: From Stuart Ira Fox, *Human Physiology,* 3d ed. Copyright © 1990 Wm. C. Brown Publishers, Dubuque, Iowa. All Rights Reserved. Reprinted by permission.

Chapter 14

14.11: From John W. Hole, Jr., *Human Anatomy and Physiology,* 5th ed. Copyright © 1990 Wm. C. Brown Publishers, Dubuque, Iowa. All Rights Reserved. Reprinted by permission.

Chapter 15

15.4: From Kent M. Van De Graaff, *Human Anatomy,* 2d ed. Copyright © 1988 Wm. C. Brown Publishers, Dubuque, Iowa. All Rights Reserved. Reprinted by permission; **15.9a:** From Kent M. Van De Graaff and Stuart Ira Fox, *Concepts of Human Anatomy and Physiology,* 2d ed. Copyright © 1989 Wm. C. Brown Publishers, Dubuque, Iowa. All Rights Reserved. Reprinted by permission; **page 317:** From Kent M. Van De Graaff and Stuart Ira Fox, *Concepts of Human Anatomy and Physiology,* 2d ed. Copyright © 1989 Wm. C. Brown Publishers, Dubuque, Iowa. All Rights Reserved. Reprinted by permission.

Chapter 16

16.1b: From Kent M. Van De Graaff, *Human Anatomy and Physiology,* 2d ed. Copyright © 1988 Wm. C. Brown Publishers, Dubuque, Iowa. All Rights Reserved. Reprinted by permission; **16.14:** From Kent M. Van De Graaff, *Human Anatomy,* 2d ed. Copyright © 1988 Wm. C. Brown Publishers, Dubuque, Iowa. All Rights Reserved. Reprinted by permission.

Chapter 17

17.10: From John W. Hole, Jr., *Human Anatomy and Physiology,* 5th ed. Copyright © 1990 Wm. C. Brown Publishers, Dubuque, Iowa. All Rights Reserved. Reprinted by permission; **17.13b,c:** From John W. Hole, Jr., *Human Anatomy and Physiology,* 5th ed. Copyright © 1990 Wm. C. Brown Publishers, Dubuque, Iowa. All Rights Reserved. Reprinted by permission.

Chapter 19

19.17: From Kent M. Van De Graaff and Stuart Ira Fox, *Concepts of Human Anatomy and Physiology,* 2d ed. Copyright © 1989 Wm. C. Brown Publishers, Dubuque, Iowa. All Rights Reserved. Reprinted by permission.

Chapter 20

20.2: From John W. Hole, Jr., *Human Anatomy and Physiology,* 5th ed. Copyright © 1990 Wm. C. Brown Publishers, Dubuque, Iowa. All Rights Reserved. Reprinted by permission; **20.3c:** From Kent M. Van De Graaff and Stuart Ira Fox, *Concepts of Human Anatomy and Physiology,* 2d ed. Copyright © 1989 Wm. C. Brown Publishers, Dubuque, Iowa. All Rights Reserved. Reprinted by permission; **20.3d:** From John W. Hole, Jr., *Human Anatomy and Physiology,* 5th ed. Copyright © 1990 Wm. C. Brown Publishers, Dubuque, Iowa. All Rights Reserved. Reprinted by permission; **20.4:** From Kent M. Van De Graaff and Stuart Ira Fox, *Concepts of Human Anatomy and Physiology,* 2d ed. Copyright © 1989 Wm. C. Brown Publishers, Dubuque, Iowa. All Rights Reserved. Reprinted by permission; **20.6:** From John W. Hole, Jr., *Human Anatomy and Physiology,* 5th ed. Copyright © 1990 Wm. C. Brown Publishers, Dubuque, Iowa. All Rights Reserved. Reprinted by permission; **20.7a:** From John W. Hole, Jr., *Human Anatomy and Physiology,* 5th ed. Copyright © 1990 Wm. C. Brown Publishers, Dubuque, Iowa. All Rights Reserved. Reprinted by permission; **20.10:** From John W. Hole, Jr., *Human Anatomy and Physiology,* 5th ed. Copyright © 1990 Wm. C. Brown Publishers,

Dubuque, Iowa. All Rights Reserved. Reprinted by permission; **20.11:** From Kent M. Van De Graaff and Stuart Ira Fox, *Concepts of Human Anatomy and Physiology,* 2d ed. Copyright © 1989 Wm. C. Brown Publishers, Dubuque, Iowa. All Rights Reserved. Reprinted by permission; **page 434:** From John W. Hole, Jr., *Human Anatomy and Physiology,* 5th ed. Copyright © 1990 Wm. C. Brown Publishers, Dubuque, Iowa. All Rights Reserved. Reprinted by permission.

Chapter 21

21.16: From Kent M. Van De Graaff and Stuart Ira Fox, *Concepts of Human Anatomy and Physiology,* 2d ed. Copyright © 1989 Wm. C. Brown Publishers, Dubuque, Iowa. All Rights Reserved. Reprinted by permission; **21.19:** From John W. Hole, Jr., *Human Anatomy and Physiology,* 5th ed. Copyright © 1990 Wm. C. Brown Publishers, Dubuque, Iowa. All Rights Reserved. Reprinted by permission; **21.21:** From Kent M. Van De Graaff and Stuart Ira Fox, *Concepts of Human Anatomy and Physiology,* 2d ed. Copyright © 1989 Wm. C. Brown Publishers, Dubuque, Iowa. All Rights Reserved. Reprinted by permission.

Chapter 22

22.12b: From E. Peter Volpe, *Biology and Human Concerns,* 3d ed. Copyright © 1983 Wm. C. Brown Publishers, Dubuque, Iowa. All Rights Reserved. Reprinted by permission.

Chapter 23

23.3: From Robert F. Weaver and Philip Hedrick, *Genetics.* Copyright © 1989 Wm. C. Brown Publishers, Dubuque, Iowa. All Rights Reserved. Reprinted by permission. **23.12:** From E. Peter Volpe, *Biology and Human Concerns,* 3d ed. Copyright © 1983 Wm. C. Brown Publishers, Dubuque, Iowa. All Rights Reserved. Reprinted by permission.

Chapter 27

27.8: From Storer, et al., *General Zoology,* 6th ed. Copyright © McGraw-Hill Book Company, New York, NY. Reprinted by permission.

Chapter 28

28.17b: Copyright © 1974 Kendall/Hunt Publishing Company.

Chapter 32

32.5: From Eugene P. Odum, *Fundamentals of Ecology,* 3d ed. Copyright © 1971 Holt, Rinehart & Winston.

Chapter 33

33.10: From D. Correll, "Estuarine Productivity" in *BioScience,* 28:649, 1978. Copyright 1978 by the American Institute of Biological Sciences, Washington, DC.

Appendix B

Source: *Drugs of Abuse,* produced by the Affairs in cooperation with the Office of Public Science and Technology.

Illustrators

Laurel Antler
5.3

Chris Creek
10.1; Objective Questions art, page 217; 13.3, 14.2; Objective Questions art, page 297; 15.1, 15.9a; Objective Questions art, page 316; 19.1.

Anne Greene
15.5, 18.5; Objective Questions art, page 384; 21.15.

Kathleen Hagelston
Line work on 2.3a,b; 3.3a, 3.5, 3.10; Chapter 4 reading art, page 82; 4.4, 6.5, 6.6, 6.12, 7.10a, 8.2a, 8.5a, 9.4, 16.3a, 18.10, 24.12, 24.13, 24.14a, 24.15.
Also provided sketches for the following illustrators and figures: **Marjorie C. Leggitt** 10.18, 14.3b, 15.11, 18.10, 23.4, 25.7a. **Laurie O'Keefe** 13.5, 13.8c, 23.15, 25.3a.

Anthony Hunt
Chapter 12 reading art, page 243.

Carlyn Iverson
I.2, 1.5, 4.2, 5.6, 5.10, 6.1, 6.2, 6.3; Objective Questions art, page 140; 7.2, 7.3, 9.12, 9.14, 9.16, 10.2, 10.4, 10.5a, 10.7a, 10.12, 11.1, 11.16, 12.1, 12.4, 14.4, 15.7, 15.12, 15.13, 17.5, 17.7, 18.13, 18.14, 18.15, 18.16, 20.8, 21.4, 21.5, 21.8b, 32.3, 34.7.

Also provided sketches for most of the art new to this editon.

George V. Kelvin
Chapter 13 reading art, page 275.

Mark Lefkowitz
The following figures copyright © 1989 Mark Lefkowitz. All Rights Reserved. I.10, 4.8b, 8.12; Objective Questions art, page 169; 16.8, 16.10, 16.11, 16.13, 16.16, 18.4; Chapter 22 reading art, page 470; 24.18b,c, 28.1c, 28.11, 28.12; Objective Questions art, page 608; 28.20, 30.2, 30.6, 30.9, 30.17, 30.22, 31.6, 31.7, 31.9, 34.10, 34.12. The following figures copyright © 1990 Mark Lefkowitz. All Rights Reserved. 11.3; Objective Questions art, page 237; 11.4, 15.6, 30.21, 30.26.

Marjorie C. Leggitt
31.2.

Ron Mclean
20.7a.

Steve Moon
11.2, 20.11.

Diane Nelson
11.7b; Chapter 11 reading art, figure b, page 235; 14.11, 21.16.

Laurie O'Keefe
2.7, 9.9, 12.5, 22.11a,b, 24.19.

Precision Graphics
10.16, 10.19, 10.22, 13.7b, 14.9, 14.12, 16.21, 19.14, 20.14; text art, page 462; 479, 22.12, 23.12, 23.16a,b, 34.9.

Mike Schenk
10.3, 16.1b, 16.14, 19.17, 21.21.

Tom Waldrop
9.1, 13.1, 16.12, 17.10, 17.13b,c; Chapter 19 reading art, page 402; 20.2, 20.4, 20.6, 20.10; Objective Questions art, page 434.

John Walters & Associates
11.14.

Rolin Graphics
Introduction: I.2; text art, page 6; I.6, I.8, I.11. **Chapter 1:** 1.2, 1.3, 1.4, 1.5a, 1.6, 1.7, 1.8, 1.9, 1.10; reading art, page 27; 1.12, 1.13, 1.14, 1.15, 1.16; text art, pages 30, 31; 1.18, 1.19, 1.20, 1.21, 1.22, 1.23a, 1.24a, 1.25a, 1.26, 1.27; text art, page 38; 1.28, 1.29, 1.30, 1.31, 1.32, 1.33; Objective Questions art, page 42. **Chapter 2:** 2.2, 2.3a,b, 2.5, 2.6b, 2.9; text art, page 54; 2.10b, 2.11b, 2.13a, 2.15; Objective Questions art, page 62. **Chapter 3:** 3.1, 3.2, 3.4, 3.6, 3.7, 3.8, 3.9, 3.11; Objective Questions art, page 78. **Chapter 4:** 4.5; text art, page 84; 4.6, 4.14, 4.15, 4.16, 4.17, 4.18. **Chapter 5:** 5.2; text art, pages 99, 101; 5.5, 5.7, 5.8, 5.9; text art, page 105; 5.11, 5.12, 5.13, 5.14; text art, page 111; 5.15, 5.16, 5.17; text art, page 114. **Chapter 6:** Text art, pages 125, 126, 129; 6.4, 6.8, 6.9, 6.10; text art, pages 133, 135; reading art, page 136; text art, pages 138, 139. **Chapter 7:** 7.1, 7.2, 7.4c, 7.7b, 7.9, 7.11, 7.12. **Chapter 8:** 8.1, 8.3; text art, page 156; 8.4b,c, 8.6, 8.7, 8.8, 8.9, 8.13. **Chapter 9:** 9.10, 9.15. **Chapter 10:** Reading art, page 198; 10.14, 10.15. **Chapter 11:** 11.5, 11.8, 11.9, 11.10, 11.15. **Chapter 12:** 12.2, 12.3b, 12.6, 12.7; text art, page 248; 12.14. **Chapter 13:** 13.6, 13.9, 13.10, 13.14. **Chapter 14:** Text art, page 279; 14.7, 14.8, 14.10; text art, pages 290 & 291. **Chapter 15:** 15.8, 15.10b, 15.14. **Chapter 16:** 16.5, 16.6; text art, page 327. **Chapter 17:** 17.1, 17.2, 17.3, 17.4, 17.6, 17.8; text art, page 361. **Chapter 18:** 18.2, 18.3, 18.6. **Chapter 19:** 19.2, 19.3, 19.4, 19.5, 19.8, 19.13, 19.18, 19.19; reading art, page 403. **Chapter 20:** 20.3a, 20.5, 20.9, 20.13. **Chapter 21:** 21.9, 21.10; text art, pages

442 and 445; 21.14; text art, page 451; reading art, page 455; 21.24. **Chapter 22**: 22.5, text art, pages 473, 478; 22.9, 22.10. **Chapter 23**: 23.2, 23.6, text art, page 495; 23.10, Critical Thinking text art, page 498. **Chapter 24**: 24.1, 24.2, 24.8*c*; text art, page 516; 24.10, 24.16; text art, pages 525, 532. **Chapter 25**: 25.2, 25.10. **Chapter 26**: 26.1, 26.5. **Chapter 27**: 27.2, 27.5; text art, page 571; 27,10, 27.12; text art, page 576; reading art, page 579. **Chapter 28**: 28.3, 28.5, 28.7, 28.15, 28.16, 28.17*b*, 28.21, 28.23, 28.25*a*, 28.26. **Chapter 29**: 29.2, 29.4, 29.5, 29.7, 29.10, 29.12, 29.15. **Chapter 30**: 30.15. **Chapter 31**: 31.5. **Chapter 32**: 32.1, 32.2, 32.5, reading art, page 695. **Chapter 33**: 33.2*b*, 33.9, 33.10, 33.12, 33.13. **Chapter 34**: 34.2, art for figure 34.3, 34.4, 34.5, 34.6; text art, page 725; 34.17; Objective Questions art, page 734. **Chapter 35**: 35.3, 35.5, 35.13, 35.16. Other artwork throughout the text were rendered by Laurel Antler, Anne Greene, Ruth Krabach, and Mildred Rinehart.

A

ABA. *See* Abscisic acid (ABA)
A band, 360
Abdomen, and vestigial structures, 569
ABO blood type, and ABO system, 253–54, 484–85, 501
Abortion, 426
Abscisic acid (ABA), 160
Absorption spectrum, 124
Abuse, drug, 340–44. *See also* Intravenous drug abusers (IVDA)
Abyssal zones, 713, 714, 715
Accessory glands, 190
Accessory organs, 197
Accommodation, and eye, 373–74
Acetate, active. *See* Active acetate (AA)
Acetylcholine (ACh), 328, 339–40, 362–63
Acetylcholinesterase (AChE), 328
ACh. *See* Acetylcholine (ACh)
AChE. *See* Acetylcholinesterase (AChE)
Achondroplasia, 482
Acid chyme, 194
Acid deposition, 26–27, 752
Acid group, 31
Acid rain, 26
Acids
    abscisic (ABA), 160
    and acid deposition, 752
    and acidic solutions, 27
    amino, 30–31, 32, 199
    and bases, 25–29
    citric, 112
    deposition, 26–27
    DNA. *See* Deoxyribonucleic acid (DNA)
    essential amino, 204
    fatty, 37–38
    GABA. *See* Gamma-aminobutyrate (GABA)
    gibberellic (GA), 160, 161
    glycine, 339
    hydrochloric, 25, 194, 198, 200
    indolacetic (IAA), 160
    LSD. *See* Lysergic acid diethylamide (LSD)
    mRNA. *See* Messenger RNA (mRNA)
    nucleic, 40–41
    pantothenic, 209
    RNA. *See* Ribonucleic acid (RNA)
    rRNA. *See* Ribosomal RNA (rRNA)
    tRNA. *See* Transfer RNA (tRNA)
    uric, 300–301, 310
Acoelomate animals, 633
Acquired characteristics, theory of, 579
Acquired immune deficiency syndrome (AIDS), 274–75, 429–30, 694
    and AIDS-related complex (ARC), 429

and artificial blood, 243
and AZT and DDI, 589
and biotechnology products, 539
and birth defects, 454–55
and hemophilia, 497
and infections, 243
and lymphokines, 271
and newborns, 429
and retroviruses, 588
and T cells, 266
and thymus, 400
transmission, 429–30
treatment and transmission, 274–75
and viruses, 429, 588
Acromegaly, 391, 392, 393
Acrosome, 412, 413
ACTH. *See* Adrenocorticotropic hormone (ACTH)
Actin, 360, 361, 363
Action potential, 325
Activation, energy of, 99, 100
Active acetate (AA), 107, 109, 111, 200
Active immunity, 270, 271
Active reabsorption, 309
Active site, 101
Active transport, 68, 73
Actomyosin, 361–62
ACV. *See* Acyclovir (ACV)
Acyclovir (ACV), 589
ADA. *See* Adenosine deaminase (ADA)
Adam's apple, 193, 282
Adaptation
    and Darwin's theory, 578
    defense, 692–93
    and genetic drift, 575
    polymorphic, 577
    predation, 689–92
    and skin, 368
Adaptive radiation, 581
Addison's disease, 393, 397
Adenine, 510, 511, 512
Adenoids, 192
Adenosine deaminase (ADA), 545, 546
Adenosine diphosphate (ADP)
    and aerobic cellular respiration, 106, 108, 138
    and ATP cycle, 105
    and carbon dioxide regeneration, 135
    and heterotrophs, 557
    and muscle contraction, 361
    and photosynthesis, 133, 134, 135, 138
Adenosine triphosphate (ATP), 54, 55, 59, 73, 74
    and ADP, 105
    and aerobic cellular respiration, 105, 106, 107, 108, 109, 110, 111, 138–39
    and ATP cycle, 105
    and bacteria, 592
    and breathing, 279
    and cAMP, 387, 388
    and carbon dioxide regeneration, 135

and chemiosmotic ATP synthesis, 108, 113, 131
and comparative biochemistry, 570
and ecosystem energy flow, 721–24
and electron transport system, 112–14
energy reaction, 104–5, 114
and glucose molecule, 114
and heterotrophs, 557
and innervation, 363
and minerals, 210
and muscle contraction, 361
and muscular exercise, 355
and oxygen debt, 361–62
and photophosphorylation, 129–31
and photosynthesis, 127, 128, 129, 130, 131, 132, 133, 134, 135, 138–39
structure, 41
ADH. *See* Antidiuretic hormone (ADH)
Adipose cells, 177
Adipose tissue, 177
Adolescence, 461
ADP. *See* Adenosine diphosphate (ADP)
Adrenal cortex, 386, 392, 395–98
Adrenal glands, 395–98
Adrenalin, 392
Adrenal medulla, 392, 395–98
Adrenocorticotropic hormone (ACTH), 391, 392, 393, 396
Adulthood, 461–65
Adult stage, 653
Advanced, and primitive, 632
Advanced glycosylation end product (AGE), 462
Advanced invertebrates, 630, 642–56
Aerobic cellular respiration. *See* Cellular respiration
Aerobic respiration, 553, 559
Afferent arteriole, 307
Afferent neuron, 320–22
African sleeping sickness, 596
Afterbirth, 459, 460
Age
    and age structure comparison, 742–43
    and height, 460
    and steady state, 755
    See also Aging
AGE. *See* Advanced glycosylation end product (AGE)
Age structure comparison, 742–43
Agglutination, 251, 254
Aging, 461–65
    and body systems, 463–65
    and genes, 461
    and organs, 463
    theories of, 461–63
    See also Age
Agranular leukocytes. *See* White blood cells (leukocytes)
Agriculture
    Department of. *See* United States Department of Agriculture (USDA)

and genetically engineered
  bacteria, 540–41
monoculture, 730
slash-and-burn, 744, 745
Agrobacterium tumefaciens, 542
AID. *See* Artificial insemination by
  donor (AID)
AIDS. *See* Acquired immune
  deficiency syndrome (AIDS)
AIDS-related complex (ARC). *See*
  Acquired immune deficiency
  syndrome (AIDS)
Air
  inspired and expired, 284
  passage of, 280–84
  path, 281
  pollution, 750–54
  structure and function, 281
  See also Breathing
Air sacs, 69, 661, 665
Albinism, 514
Albumin, 240, 241
Alcohol, and abuse, 340, 341–42
Alcoholic beverages, and
  fermentation, 115
Aldosterone. *See* Mineralocorticoids
Alfalfa, 148, 744
Algae, 596–602
  green, 597–600
  life cycles, 596–97
Algin, 601
Alkalinity, 26
Alkaptonuria, 514
Allantois, 446
Alleles, 469
  and autosomal disorders, 479
  multiple, 484–85
Allen, Sandy, 391
Allergens, 274–75
Allergies, 274–75
Alligators, 660
All-or-none law, 354–55
Alpha globulins. *See* Globulins
Alpha Therapeutic, 243
Alpha waves, 338
Alterations, of bases, 527–28
Alternation of generations, 165,
  596–97, 600, 611–12
Altitude, and latitude, 710
Alveoli, 69, 281, 283, 423, 424
Alzheimer disease, 340, 465
Ameboids, 594–95
American Academy of Dermatology,
  183
American Indian, 681
Amino acids, 30–31, 32, 339
  and antibodies, 264–65, 266
  and blood plasma, 240
  and capillary exchange, 246, 247
  essential, 204
  and liver, 199
  and protein synthesis
    participants, 523
Amino group, 31
Ammonia, 310
  and blood plasma, 240, 241
  and capillary exchange, 246, 247
  and chemical evolution, 553
  and excretion, 300

Amniocentesis, 454, 455
Amnion, 446, 449
Amoeba proteus, 594
Amoeboid cells, 633
Amphibians, 659–60, 666
Ampulla, 379, 380, 381, 655
Amylase, 200, 201, 202
Anabolic steroids, 400, 401–2, 416
Anaerobes, facultative and obligate,
  591
Anaerobic organisms, 559
Analogous structures, 568
Anaphase, 85, 86–87, 90, 91, 92
Anatomy
  breasts, 424
  capillary bed, 220
  chloroplasts, 125
  cilia, 58
  clam, 645
  comparative. *See* Comparative
    anatomy
  crayfish, 651
  dicot herbaceous stem, 148
  ear, 378–80
  earthworm, 647
  Euglena, 602
  eye, 372
  female breasts, 424
  fern sporophyte, 617
  flagella, 58
  flatworm, 628
  flower, 164–67, 622
  flukes, 640
  grana, 128
  grasshopper, 653
  heart, 221, 222
  human, 170–407
  human skin, 181
  hydra, 636
  intestinal lining, 196
  kidneys, 304
  lancelet, 656
  leaf, 155
  long bone, 351
  muscle fiber, 358–60
  nephrons, 305, 307
  ovary and follicle, 417
  pancreas, 398
  pea flower, 470
  placenta, 453
  root, 155
  roundworm, 641
  skeletal muscle, 358
  skeletal muscle fiber, 364
  skin, 181
  spinal cord, 330
  spindle fibers, 86
  Spirogyra, 599
  sponge, 634
  terminal bud, 147
  tunicate, 656
  whole muscle, 352–54
Ancestors, common, 566
Androgens, 392, 415, 416
Anemia, 245, 485–86, 504
Anemone, 635
Angina pectoris, 234
Angiosperms, 622–26, 627
Angiotensin, 313–14, 396, 397

Animal cells, 46, 48, 60, 61
  and meiosis, 89
  and mitosis, 84
  and osmosis, 71
Animal kingdom, 629–68
Animals
  acoelomate, 633
  classification, 567, 630–33
  coelomate, 633
  evolution, 630
  and geological time scale, 564
  hermaphroditic, 638–39
  life cycle of, 82, 83
  and niches, 720
  organization of, 4
  pseudocoelomate, 633
  transgenic, 544–45
  and viruses, 586–88
Annelids, 647–48
Annulus, 617
Anopheles mosquito, 577, 596
Anorexia nervosa, 216, 402
Antagonistic muscles, 353
Anteater, 663
Antennae, 650
Anterior pituitary. *See* Pituitary
Anther, 165
Antheridia, 612
Anthophyta, 622–26
Anthropoids, 670–73
Antibiotics, 293, 430, 589
Antibodies, 250, 251–52, 264–65
  monoclonal, 272–73
  structure, 266
Antibody-mediated immunity,
  264–65
Antibody titer, 270
Anticodon-codon base pairing,
  520–21, 522
Antidiuretic hormone (ADH), 392,
  403, 405
  and diabetes insipidus, 390
  and kidneys regulatory functions,
    313
  and posterior pituitary, 388–90,
    392
Antigen, 263
Antigen-antibody reaction, 251–52
Antigen-presenting cell (APC),
  268–69
Antihypertensive drugs, 233
Anti-oncogenes, 531
Antipredator defenses, 692
Antiviral drugs, 589
Anus, 190, 191, 196
Anvil, 378
Aorta, 228
Aortic bodies, 285–86
APC. *See* Antigen-presenting cell
  (APC)
Apes, 670, 674
Appendages, jointed, 650
Appendicular skeleton, 347, 348,
  349
Appendix, 195, 197, 569
Aqueous humor, 371, 372
Aquatic communities, 710–15
Aquatic food chain, 722
ARC. *See* Acquired immune
  deficiency syndrome (AIDS)

Archaebacteria, 592
Archaeopteryx, 566
Archegonia, 612
Archenteron, 438
Archeozoic, 564
Arrector pili muscle, 182
Arteries, 219, 220
  and capillary exchange with
    tissues, 246–47
  common carotid, 230
  coronary, 228
  ducts, 456
  and human circulatory system,
    229
  plaque, 233, 235–36
  pulmonary, 227
  umbilical, 456
Arterioles, 219, 220, 314
  afferent and efferent, 307
  and blood flow velocity, 231
Arteriosclerosis, 233
Arthritis, 352
Arthropods, 648–53
Artificial blood, 242, 243
Artificial heart implants, 236
Artificial insemination by donor
  (AID), 427
Artificial kidney, 315–16
Artiodactyla, 665
Ascaris, 641–42
Ascending colon, 196
Ascomycota, 604
Asexual reproduction, 164
Aspirin, and heart disease, 235
Assortment, independent. *See*
  Independent assortment
Asterias, 654
Asters, 85
Astigmatism, 377
Asymmetry, 631
Atherosclerosis, 233–36
Atmospheres, oxidizing, primitive,
  and reducing, 553
Atomic number, 18
Atomic weight, 18
Atoms, 17–24
ATP. *See* Adenosine triphosphate
  (ATP)
Atria, 222
Atrial natriuretic hormone, 403
Atrioventricular (AV) nodes, 225,
  226
Atrioventricular (AV) valves, 222
Auditory canal, 378
Auditory nerve, 381
Aurelia, 635
Australian, 681
Australopithecines, 674–76
Australopithecus, 670, 674–76
Australopithecus africanus, 675
Autoimmune diseases, 276
Automobile accidents, and alcohol
  abuse, 341
Autonomic nervous system, 225,
  328, 332–34
Autosomal disorders, dominant and
  recessive, 479, 480–83,
  491–95
Autosomes, 80

Autotrophic bacteria, 137–38
Autotrophs, 123, 557–59
Auxins, 160, 161
AV nodes. *See* Atrioventricular (AV) nodes
Axial skeleton, 347, 348
Axillary buds, 147
Axons, 180, 320–22, 323, 326, 332
AZT. *See* Zidovudine (AZT)

B

Backcross, 479
Background extinctions, 566
Bacteria, 262
    autotrophic, 137–38
    chemosynthetic, 591
    denitrifying, 728
    genetically engineered, 540–41
    gonnorheal, 432
    metabolism of, 591–92
    nitrifying, 728
    nitrogen-fixing, 727
    photosynthetic, 137, 591
    reproduction of, 590–91
    structure of, 589–90
    types of, 592–93
Bacteriophages, life cycle, 585–86
Balance, 367, 379, 380
Balanced diet, 202
Balanced polymorphism, 577. *See also* Restriction fragment length polymorphism (RFLP)
Balanus, 687
Baldness, 416, 500, 501
Ball-and-socket joints, 351
Banana, 626
Barley, 115
Barnacles, 694
Barr body, 493
Basal bodies, 57
Basal-cell carcinoma, 183
Basal metabolic rate (BMR), 212
Basal metabolism, 212–13
Basal nuclei, 338
Basement membrane, 176
Base pairing, complementary. *See* Complementary base pairing
Bases
    and acids, 25–29
    alterations, substitutions, deletions, 527–28
    and basic solutions, 27
    and codes, 517, 518, 519, 520, 522, 527–28
Basidiomycota, 606
Basidiospores, 606
Basidium, 606
Basophils, 239, 244, 249–51
Bates, Cheri, 428
Bathyal zone, 713, 714, 715
Bathypelagic zone, 713, 714, 715
Beach, 713
Beagle. See HMS *Beagle*
Beans
    common, 626
    nutrient and energy content, 214
    seedling, 4
Beer, 115

Bees, and pollination, 624
Behavior
    and circadian rhythms, 403
    and ecology, 684–757
    feeding, 5
    and food, 213–15
    patterns, 686–99
Benign tumors, 529
Benthic division, 713, 714, 715
Bergey's Manual, 592
Beta globulins. *See* Globulins
Betalactoglobulin (BLG), 544
Beta waves, 338
bGH. *See* Bovine growth hormone (bGH)
Bicarbonate ion, 243
Biceps, 353
Bifocals, 378
Bilateral symmetry, 632
Bile, 197, 198–200
    and excretion, 299–300
    pigments, 301
Binary fission, 590
Binding sites, 363
Binomial
    equation, 571–72
    names, 6
Bioavailable, 211
Biochemistry
    comparative, 570
    and eye, 375–76
    of vision, 376
Biogenesis theory of biology, 12
Biogeography, 570–71
Biogerontology Laboratory, 356
Biological clocks, 163, 403
Biological diversity, loss of, 744
Biological evolution, 556–59
Biological magnification, 747–50
Biology
    photo-. *See* Photobiology
    plant, 120–69
    unifying theories of, 12
Biomass, pyramid of. *See* Pyramid of biomass
Biomes, 701–10
Bioremediation, and genetically engineered bacteria, 540–41
Biosphere, 7, 700–717
Biosystem, 720
Biotechnology
    and hormones, 538
    laboratory techniques, 534–37
    products, 537–40
    and recombinant DNA, 533–49
Biotic potential, 739
Biotin, 209
Bipedal posture, 673
Birds, 5, 661–62, 666
Birth, 458–60
    and birth control, 425–29
    and birth defects, 454–55
    and birthrate, 738, 741
    human development after, 460–65
Bitter, 369
Black bread molds, 603–4
Blackheads, 182
Bladder, urinary, 303
Blade, 150

Blastocoel, 437
Blastocyst, 437, 448
Blastula, 436, 437
BLG. *See* Betalactoglobulin (BLG)
Blindness, color, 376, 497
Blind spot, 372
Blood, 178–79, 238–56
    ABO system and ABO types, 253–54, 484–85, 501
    and alcohol, 341
    artificial, 242, 243
    and artificial kidney, 315–16
    and blood volume, 313
    clotting function, 248–49
    and cold blooded, 661
    components of, 240
    composition of, 239
    erythrocytes. *See* Red blood cells (erythrocytes)
    and fetal circulation, 456–58
    flow velocity, 231
    globulins, 240, 241
    and homeostasis, 240
    and infection fighting, 249–53
    and kidneys, 304, 312–14
    leukocytes. *See* White blood cells (leukocytes)
    and nephrons, 307
    path in heart, 223
    pH, 185, 312–14
    plasma. *See* Blood plasma
    platelets. *See* Blood platelets (thrombocytes)
    pressure. *See* Blood pressure
    and red bone marrow, 244
    red cells. *See* Red blood cells (erythrocytes)
    and respiration, 289, 291
    Rh system, 254–55, 485
    thrombocytes. *See* Blood platelets (thrombocytes)
    and tissue fluid formation, 185
    typing, 253–55
    vessels, 219–21
    volume, 241
    and warm blooded, 662
    white cells. *See* White blood cells (leukocytes)
Blood cells
    red. *See* Red blood cells (erythrocytes)
    white. *See* White blood cells (leukocytes)
Blood clotting. *See* Blood
Blood fluke, 640
Blood plasma, 178, 240–41
    and blood components, 240
    proteins, 240–41
    See also Blood
Blood platelets (thrombocytes), 178, 239, 240
    and blood clotting, 248–49
    See also Blood
Blood pressure, 229–31
    measurement of, 229–30
    systolic and diastolic, 230
    See also Blood
Blood typing. *See* Blood
Blood vessels, 239. *See also* Blood
Blood volume. *See* Blood

Bloody show, 458
Blubber, 664
Blue baby, 458
B lymphocyte. *See* Lymphocytes
BMR. *See* Basal metabolic rate
Bobcat, 630
Body fluids, 248–49
Body membranes, 184–85
Body plans, 631, 632
Body systems, and aging, 463–65
Body temperature control, 186–87
Bonds
    covalent, 21, 32
    double, 23
    hydrogen, 24, 32
    ionic, 20, 32
    peptide, 31, 101
Bone(s), 178
    compact and spongy, 350
    fetal, 456
    growth and development, 350
    and homologous structures, 568
    long, 349–50, 351
    marrow. *See* Bone marrow *and* Red bone marrow *and* Yellow bone marrow
    of skeleton, 347, 348
Bone marrow, and lymphoid organs, 259–60
    red. *See* Red bone marrow
    yellow. *See* Yellow bone marrow
    See also Bone(s)
Bony fish, 657, 659
Booster, 270
Borrelia burgdorfei, 10, 694
Bovine growth hormone (bGH), 538, 544
Bowman's capsule, 305, 308, 314
Brachydactyly, 482
Bradykinin, 252–53
Brain, 335–40
    and cerebral lobes, 336–38
    conscious, 336–38
    and eye, 375
    and nervous system, 328
    and neurotransmitters, 339–40
    unconscious, 336
    ventricles, 334
Breads, nutrient and energy content, 214
Breast
    anatomy, 424
    cancer, 206
    feeding, 271
    female, 423–24
Breastbone. *See* Sternum
Breathing, 280–84
    and ATP, 279
    mechanisms of, 284–88, 665
    nervous control of, 286
    and oxygen, 279
    See also Air
Bristlecone pines, 621
British Petroleum, 546
Bronchi, 281, 282–83
Bronchioles, 281, 283
Bronchitis, 293, 294
Brown algae, 601
Browsers, 689

Bryophytes, 612, 613, 627
Bucci, Enrico, 243
Buds
    and animal viruses, 586
    axillary, 147
    hydra, 637
    limb, 441
    taste, 367, 369, 370
    terminal, 147
Buffers, 28
Bulimia, 216
Bullfrog, 660
Burgdorfer, Willy, 10
Burning, 727, 744, 745
Burns, 183
Buttercup, 622
Butterfly, 624, 693
Bypass surgery, coronary, 236

C

Cacti, 570
Calcitonin, 392, 395
Calcium, 210, 211, 350
    and blood clotting, 248–49
    and diet, 211
    and excretion, 300
    ions, 103, 359, 361, 363
Calories, empty, 205
Calvin-Benson cycle, 131–35
Cambium, 148
Cambrian, 564
cAMP. See Cyclic adenosine
        monophosphate (cAMP)
Camplobacter pyloridis, 195
Canals
    Haversian. See Haversian canals
    membranous. See Membranous
        canals
Cancer
    and AIDS, 274
    breast, 206
    causes, 529
    cells, 84
    development, 530
    and genetic control, 528–31
    and initiation and promotion,
        529–31
    lung, 294–96
    and lymphokines, 271
    and retroviruses, 588
    skin, 182
    and tanning, 183
    and thymus, 400
    types of, 295
Candida albicans, 607
Canines, 190, 191, 192, 569
Cannabis delirium, 342
Cannabis psychosis, 342
Cannabis sativa, 342, 626
Capacity
    carrying, 739–40
    lung. See Lungs
    vital. See Vital capacity
CAPD. See Continuous ambulatory
        peritoneal dialysis (CAPD)
Capillaries, 219, 220
    and capillary bed anatomy, 220

and exchanges within tissues,
        246–47
    lymphatic, 247
    and nephrons, 307
    peritubular, 307
    and respiration, 290
Capsid, 584
Capuchin, 673
Carabaminohemoglobin, 243
Carapace, 650
Carbohydrates, 33–35, 65
    complex, 205, 206
    and nutrition, 204–6
    and photosynthesis, 136–37
Carbon, hydrates of, 33
Carbon atom, 18
Carbon cycle, 725–27
Carbon dioxide, 107, 112, 301
    and blood plasma, 240, 241
    and capillary exchange, 246, 247
    and chemical evolution, 553
    and excretion, 299–300
    fixation, 132–33
    and greenhouse effect, 753–54
    and homeostasis, 185
    and oxygen debt, 362
    reduction of, 131–37
    and respiration, 290, 291
    and tree burning, 727
Carbonic anhydrase, 245
Carboniferous, 564, 726
Carboniferous swamp, 617
Carbon monoxide
    and chemical evolution, 553
    molecule and electron-dot
        formula, 242
Carbon reservoirs, 725–26
Carboxyl group, 31
Carcinoma, 182–83
Cardiac conduction system, 225
Cardiac cycle, 223–24
Cardiac muscle, 179, 180
Cardiovascular disease, 231, 234–35
Caribou, 704
Caries, 191
Carnivora, 665
Carnivores, 123, 570, 665, 705, 720
Carnivorous predators, 570
Carotid artery, common, 230
Carotid bodies, 285
Carpals, 349
Carrier proteins, 72–75
Carriers, 67
    and genetic disorders, 479
    transport by, 72–75
Carrion, 570
Carrying capacity, 739–40
Cartilage, 178, 456
Cartilaginous disk, 350
Cartiliginous fish, 657, 659
Casparian strip, 146, 147
Cassava, 626
Catalytic RNA. See Ribonucleic
        acid (RNA)
Cataracts, 374
Cat's cry syndrome. See Cri du chat
        syndrome
Caucasian, 681
Caudal vertebrae, and vestigial
        structures, 569

CCK. See Cholecystokinin (CCK)
cDNA. See Deoxyribonucleic acid
        (DNA)
Cecum, 197
Cell body, 320–22
Cell division, 79–96
Cell-mediated immunity, 266–69
Cell membrane, 46, 47, 48, 59,
        64–78
Cell plate, 88
Cells, 4, 14–118
    adipose, 177
    amoeboid, 633
    animal, 46, 48, 60, 61, 71, 84, 89
    APC. See Antigen-presenting
        cell (APC)
    and cancer, 84, 182–83, 528–31
    and cell body, 320–22
    and cell division, 79–96
    and cell membrane, 46, 47, 48,
        59, 64–78
    and cell plate, 88
    and cell theory, 44–47
    and cell wall, 47, 48, 59, 64–78
    collar, 633
    companion, 157
    and cytoplasm, 44
    cytotoxic T, 266–69
    daughter, 84, 85, 88, 95
    epidermal, 633
    ES. See Embryo-derived stem
        (ES) cells
    eukaryotic, 44–45, 47–59, 60, 61
    flame, 637
    glial, 180, 322
    guard, 151, 157
    hair, 381, 382, 383
    Helper T, 266–69, 429
    human-mouse, 502–3
    interstitial, 412, 413, 415
    junctions between, 176
    of lumen, 309
    lysogenic, 586
    mast, 274–75
    memory B, 264–65
    memory T, 266–69
    and microspheres, 555, 556
    and molecule passage, 68
    mother, 84, 85
    muscle, 358
    neurosecretory, 388
    nurse, 412, 413
    olfactory, 367, 370
    origin of, 553
    plant, 46, 47, 48, 60, 61, 71, 88,
        93, 143–45
    plasma, 264–65
    prokaryotic, 44–45, 59–60, 61
    and proteinoids, 555, 556
    and protocells, 556
    red blood. See Red blood cells
        (erythrocytes)
    Schwann. See Schwann cells
    sertoli, 412, 413
    sieve-tube, 157
    stem, 244, 545–46
    striated, 179, 180
    structure and function, 43–63
    suppressor T, 266–69
    T, 266–69

true, 557–59
    types of, 44–47
    white blood. See White blood
        cells (leukocytes)
    and zone of cell division, 145
Cell theory, of biology, 12, 44–47
Cellular metabolism, 97–118
Cellular respiration, 54, 98, 280
    aerobic, 105–14
    and photosynthesis compared,
        138–39
Cellulose
    molecule, 67
    structure and function, 35
Cell wall, 47, 48, 59, 64–78
Cenozoic, 564, 671
Centipedes, 648
Central canal, of spinal cord, 334
Central nervous system (CNS), 320,
        321, 328, 330, 332, 334–40
    and brain. See Brain
    and muscles, 353, 354
    and spinal cord. See Spinal cord
    and STD, 431
Centrioles, 46, 48, 57, 59, 85
Centromere, 80
Centromeric fibers, 87
Cephalization, 632, 644
Cephalochordata, 656
Cephalopods, 644
Cephalothorax, 650
Cerebellum, 336
Cerebral hemispheres, 336
Cerebrospinal fluid, 334
Cerebrum, 336, 337
Cervical cap, 425
Cervix, 459
Cestodes, 639
Cetaea, 665
CFC. See Chlorofluorocarbons
        (CFC)
Chains, food. See Food chains
Chakrabarty, 541
Chance sampling, 575
Channels, 67, 69
Chaparral, 706
Characteristics, acquired. See
        Acquired characteristics
Chargaff, Erwin, 514
Chargaff's rules, 514
Chatman, Donald, 428
CHD. See Coronary heart disease
        (CHD)
Cheese, 601
Cheilosis, 208
Chemical cycling, and ecosystems,
        725–28
Chemical evolution, 553–56
Chemicals, and photochemical smog,
        750–52
Chemical senses, 369–71
Chemiosmotic ATP synthesis, 108,
        113, 131
Chemistry
    bio-. See Biochemistry
    inorganic and organic, 24
    and life, 16–42
Chemoreceptors, 367, 369–71
Chemosynthesis, 137–38

Chemosynthetic bacteria, 591
Chiasma, optic, 375
Chick, 446, 569, 585
Chickens, 731
Chick extraembryonic membranes, 446
Childhood, 461
Chimeric mice, 544
Chimpanzee, 673
China, 546
Chinstrap penguin, 662
Chipping sparrow, 5
Chiroptera, 665
Chitin, 648
Chlamydia, 430, 432, 454
Chlamydia trachomatis, 432
Chlamydomonas, 597, 598, 599
Chlorine, 210
Chlorine gas, formation of, 22
Chlorofluorocarbons (CFC), 752–53
Chlorophyll, 55, 124, 125
Chlorophyta, 597
Chloroplasts, 47, 48, 53, 54–55, 59, 125–27
    anatomy, 125
    function, 132
    interior, 134
    structure, 55
Cholecystokinin (CCK), 198
Cholesterol, 39–40, 46, 67, 207–9
Chordata, 670
Chordates, invertebrate and vertebrate, 656–66
Chorion, 446, 448
Chorionic villi, and chorionic villi sampling, 452–53, 454, 455
Choroid, 371, 372
Chroacoccus, 592
Chromatids, 80
Chromatin, 49, 80, 82
Chromosome inheritance
    and abnormal autosomes, 491–95
    patterns of, 489–506
Chromosome mutations, 493–95, 527–28
Chromosomes, 49, 80
    abnormal sex, 495–96
    and autosomal disorders, 479
    defined, 82
    diploid and haploid, 81–82
    DNA location and structure, 508
    and DNA and RNA synthesis, 82
    and genes, 469
    homologous, 89
    homologous pair, 469
    inheritance. *See* Chromosome inheritance
    mapping, 501–4
    movement of, 86–87
    mutations. *See* Chromosome mutations
    sex, 80
    structure, 82
Chrysophyta, 602
Chthamalus, 687
Chyme, 194
Cilia, 46, 48, 57–59
Ciliary muscle, 371, 372

Ciliates, 594, 595
Ciliophora, 595
Circadian rhythms, 403
Circuits, pulmonary and systemic, 223, 227, 228
Circulation, 218–37
    and circulatory disorders. *See* Circulatory disorders
    fetal, 456–58
    and nephrons, 307
    paths. *See* Circulatory paths
    system. *See* Circulatory system
Circulatory disorders, 231–33
Circulatory paths, of vertebrates, 666
Circulatory system, 184, 219–28
    closed, 648, 657
    features of, 228–31
    human, 229
    open, 645
Circumcision, 415
Cirrhosis, 200, 341–42
Citrate, 112
Citric acid cycle, 112
City, and ecosystems, 731–32
Clam, 137, 645–47, 655
Clark, Barney, 236
Class, 6, 567
Classification
    of animals, 630–33
    and evolution, 630–33
    hierarchy of, 567
    of organisms, 6
    of primates, 670
    of vertebrates, 667
Clavicle, 349
Claws, 650
Cleaner wrasse, 698
Cleaning symbiosis, 697
Cleavage, 437, 447, 504
Cleft, synaptic, 327, 329
Climax community, 719
Clinical Microbiology Review, 10
Clitellum, 648
Clitoris, 419
Clocks, biological. *See* Biological clocks
Cloning, gene, 535, 537, 542
Closed circulatory system, 648, 657
Clotting, blood. *See* Blood
Clover, 162
Clownfish, 696
Club fungi, 605, 606
Club moss, 615, 617
Cnidarians, 634–37, 667, 713
CNS. *See* Central nervous system (CNS)
Coacervate droplets, 556
Coastal communities, 712–13
Coat, seed, 167
Cobalt, 210
Cocaine, 342–43
    abuse, 340
    and cocaine babies, 454
Coccolithophore, 558
Cochlea, 379, 380
Cochlear canal, 379
Cocklebur, 162
Coconut, 626

Codes, and bases, 517, 518, 519, 520, 522, 527–28
Codons, 517, 518, 519, 520, 522
Coelenterate, 634
Coelom, 184, 437, 438, 440, 633
Coelomate animals, 633
Coenzyme, 103–4, 111
Coenzyme A (CoA), 111
Coevolve, 688
Cohesion-tension theory, of water transport, 156, 157
Coitus interruptus, 425
Cold blooded, 661
Colds, 292
Collar cells, 633
Collecting duct, 305, 310
Colon, 196, 197
Colonial green algae, 599
Colony, 599
Coloration, warning, 692
Color blindness, 376, 497
Colostrum, 424
Columbia University, 183
Columnar epithelium, 173–76
Coma
    alcoholic, 341
    diabetic, 399
Commensalism, 693, 695–96
Common ancestors, 566
Common bean, 626
Common carotid artery, 230
Common cold, 292
Community, 720
    aquatic, 710–15
    climax, 719
    coastal, 712–13
    freshwater, 710–11
    saltwater, 712–15
Compact bone, 178, 350
Companion cells, 157
Comparative anatomy, and evolution, 566–68
Comparative biochemistry, 570
Comparative embryology, 568
Competition, 687–88
Competitive exclusion principle, 687
Competitive inhibition, 102–3
Complementary base pairing, 510
Complement system, and immunity, 263
Complete linkage, 500
Complete proteins, 204
Complex carbohydrates, 205, 206
Components, of hemoglobin, 244
Composition
    of eukaryotic structures, 48
    of inspired air and expired air, 284
    of urine, 302
Compound eye, 650
Compounds, inorganic and organic, 19, 24
Concealment, 692
Concentration
    and concentration gradient, 69
    and diffusion, 68
    substrate, 101–2
Conches, 644

Conclusion, and experimentation, 9–10
Condom, 425
Conduction, 180, 325
Conduction deafness, 381
Cones, 371, 372, 376, 377
Congenital syphilis, 433
Congestive heart failure, 236
Conidia, 604
Conifers, 619–21
Conjugation, 590
Connective tissue, 173, 174–75, 176–79
Conscious brain, 336–38
Conservation, tropical, 708–9
Constant regions, 264
Constipation, 197
Consumers, 720
Consumption, fuel, 730–31
Continental drift, 566
Continuous ambulatory peritoneal dialysis (CAPD), 315–16
Contour farming, 744
Contractile elements, 360
Contractile vacuoles, 594
Contraction, muscle, 360–61
Contraction period, 355
Contragestation, 426
Control
    biological, 731
    body temperature, 186–87
    of breathing, 286
    and cancer, 528–31
    of digestive gland secretion, 198
    experiments, 10–12
    and gene expression, 524–27
    and heartbeat, 225
    hypothalamus-pituitary-ovaries, 421, 423
    hypothalamus-pituitary-testes, 415
    posttranscriptional, 524
    posttranslational, 524
    of reproductive system, 425–29
    sample, 10
    transcriptional, 524
    translational, 524
Controlled experiments, 10–12
Control sample, 10
Convoluted tubules, proximal and distal, 305, 308, 310
Coordination, and aging, 464–65
Copper, 210
Coral reef, 712, 713
Corals, 636
Cork cambium, 148
Corn, 115, 147, 626
Cornea, 374
Coronary arteries, 228
Coronary bypass surgery, 236
Coronary heart disease (CHD), 295
Corpus callosum, 338
Corpus luteum, 418, 420, 453
Corrective lenses, 377–78
Cortex, adrenal, 304, 386, 392, 395–98
Corti, organ of. *See* Organ of Corti
Cortisol. *See* Glucocorticoids
Costa Rica, 3

Cotton plants, 543
Cotyledons, 142, 167
Countercurrent mechanism, 310
Country
    developed, 740–41
    and ecosystems, 730–31
    less-developed, 741–43
    third-world, 736
Covalent bonds, 21, 32
Covalent reactions, 21–23, 32
Cowper's gland, 411, 414
Coyotes, 570
C photosyntheses. See
    Photosynthesis
Crack, 343
Cranial nerves, and nervous system,
    328, 329
Cranium, 348
Crash, 343
Crash diets, 215
Crayfish, 648, 650–51, 652
Cream cheese, 601
Creams, and birth control, 425
Creatine phosphate, 361–62
Creatinine, 300, 310
Creighton, Matthew, 401
Crenation, 72
Cretaceous, 564, 708
Cretinism, 393, 394
Crick, Francis, 511, 514
Cri du chat syndrome, 495
Cristae, 48, 54, 107
Crocodiles, 661
Cro-Magnon, 679–80
Crops, 730
Crop yields, 136
Cross bridges, 361–62, 363
Crosses
    and backcross, 479
    and inheritance, 471
    one-trait, 472–74, 475
    two-trait, 474–78
Crossing-over, 89, 502
Crustose lichen, 605
Cuboidal epithelium, 173–76
Cuckoo, 694
Cultural eutrophication, 747
Currents, 401
Curve, exponential, 736
Cushing's syndrome, 393, 397
Cuticle, 151
Cuticular plate, 383
Cyanobacteria, 592, 593
Cycads, 619
Cyclic adenosine monophosphate
    (cAMP), 387, 388
Cyclic AMP. See Cyclic adenosine
    monophosphate (cAMP)
Cyclic electron pathway, 128–31
Cyclic GMP, 375, 376
Cyclic photophosphorylation, 130,
    131
Cyst, 639
Cystic fibrosis, 480–81
Cystitis, 315
Cytochrome, 113, 129
Cytochrome system, 129
Cytokinesis, 87
Cytokinins, 160, 161

Cytoplasm, 44, 358
Cytosine, 510, 511, 512
Cytoskeleton, 46, 47, 48, 55–56, 59
Cytotoxic T cells, 266–69

D

Dark reaction, 127
Darwin, Charles, 575, 578–79, 581
Darwin's finches, 581
Data, presenting and reporting, 9,
    10, 12
Daughter cell, 84, 85, 88, 95
Day length, 161–63
DDI. See Dideoxyinosine (DDI)
DDT, 577, 746, 747
Dead space, 288
Deafness, 381–83
Deamination, 109, 199
Death, 738, 741
    and alcohol, 341
    and cancer, 182
    and EEG, 338
Death-causing cancer, 182
Death rate, 738, 741
Debt, oxygen. See Oxygen debt
Decarboxylation, oxidative, 112
Deciduous forests, 707
Decompacting, 83
Decomposers, 720
Decomposition, 720, 725
Deductive reasoning, 9
Deep-sea ecosystem, 137
Deer, 664
Deer tick, 10, 694–95
Defecation, 197, 299–300, 302
Defecation reflex, 197
Defects, birth. See Birth defects
Defense adaptations, 692–93
Defenses, antipredator, 692. See also
    Immunity
Deforestation, 754
Dehydrogenase enzymes, 104
Del Amo Hospital, 402
Deletion, 494, 495, 527–28
Delirium, 342
Demographic transition, 740–41
Denaturation, and proteins, 102
Dendrites, 180, 320–22, 331, 332
Denitrification, 728
Denitrifying bacteria, 728
Dentin, 191, 192
Deoxyribonucleic acid (DNA),
    40–41, 49, 53, 80, 509–16
    and AGE, 462
    and AIDS, 274
    and autoimmune diseases, 276
    and bacteria, 590
    and biotechnology, 533–49
    and cDNA, 535, 588
    and chromosomes, 82, 494,
        503–4, 508
    and comparative biochemistry,
        570
    and complement system of
        immunity, 263
    and DNA copy, 535
    and DNA ligase, 536, 537
    and DNA polymerase, 512

and DNA probes, 504, 537, 538,
    539
and evolutionary trees, 567
functions, 509, 511–16
and gene expression, 524–27
and gene sequencing, 536, 548
and gene therapy, 545–48
and genetic markers, 503–4
and genital warts, 431
and Human Genome Project,
    504, 548
and immunotherapy, 270
and inborn errors of metabolism,
    515
and information storage, 511
location, 508
and lymphokines, 271
and macromolecules, 556
and mutations, 494, 503–4, 512,
    527–28
and natural selection, 577
and pneumonia, 293
and protein synthesis, 516–24
recombinant, 533–49
replication, 83, 512–13
and retroviruses, 545
structure, 508, 509–11, 512
and Tay-Sachs disease, 481
and transcription, 516–22,
    526–27
and transgenic organisms,
    541–45
and translation, 516–22, 526–27
and true cells, 557–59
and vaccines, 537, 538, 539
as vector, 545
and viral DNA, 535, 545, 585
and viruses, 584, 585, 588
and vitamins, 209
and Watson and Crick, 514
Deoxyribose, 34
Department of Agriculture. See
    United States Department of
    Agriculture (USDA)
Department of Health and Human
    Services. See United States
    Department of Health and
    Human Services
Department of Natural Resources.
    See United States
    Department of Natural
    Resources
Dependence, 340
Dependent variable, 10
Depolarization, 325
Depo-Provera, 426
Deposition, acid. See Acid deposition
Depression, 340, 465, 481
Dermatitis, 208
Dermis, 181, 182
DES (synthetic estrogen), 426, 454
Descending colon, 196
Desertification, 743–44
Deserts, 702–3
Designer drugs, 344
Desmosome, spot, 176, 177
Detrius, 720, 722–23
Detrius food chain, 722–23
Deuteromycota, 607

Deuterostomes, 642
Developed countries, 740–41
Development, 435–67
    after birth, 460–65
    and birth defects, 454–55
    of bones, 350
    of cancer, 530
    early stages, 436–41
    of embryo, 445–60
    of fetus, 445–60
    human, 408–549
    of lancelet, 437, 438
Devonian, 564
Diabetes, 276, 390, 393, 398–400
    and diabetic coma, 399
    insipidus, 390, 393
    mellitus, 393, 398–400
Dialysis, CAPD. See Continuous
    ambulatory peritoneal
    dialysis (CAPD)
Diaphragm, 285, 286, 287, 425, 662
Diarrhea, 197
Diastole, 223–24
Diastolic blood pressure, 230
Diatoms, 602
Dicot herbaceous stem, anatomy,
    148
Dicots, 142–43, 622
    anatomy, 148
    roots, 145–47
    seed and germination, 168
Dideoxyinosine (DDI), 274, 430,
    589
Didinium, 691
Diet, 211–16
    balanced, 202
    and behavior, 213–15
    and calcium, 211
    crash, 215
    and daily energy requirement,
        212–13
    and hypertension, 232, 233
    and sodium, 211
    and sugar, 205
Differential permeability, 68
Differentiation, 436, 441–45
Diffusion
    and respiration, 288, 291
    simple and facilitated, 68–69, 73
Digestion, 189–217
    accessory organs, 197
    and enzymes, 200–202
    and gland secretion control, 198
    and hormones, 198
    and nutrition, 202–11
Digestive juices, 398
Digestive system, 184, 190–202
Digits, 349
Dihybrid, 475
Dilation, of cervix, 459
Dinoflagellates, 602
Dinosaurs, 563
Dioxins, 746
Dipeptide, 31
Diphtheria, 592
Dipleurula, 642, 643
Diploid, 81–82, 83, 576
Diploidy, 576
Diplontic cycle, 596–97

Directional selection, 576
Disaccharides, 33–34
Disease, 740
  autoimmune, 276
  cardiovascular. *See*
    Cardiovascular disease
  CHD. *See* Coronary heart
    disease (CHD)
  HDN. *See* Hemolytic disease of
    the newborn (HDN)
  and hypotheses, 9–10
  lung, 293
  Lyme, 694–95
  PID. *See* Pelvic inflammatory
    disease (PID)
  STD. *See* Sexually transmitted
    diseases (STD)
  Tay-Sachs. *See* Tay-Sachs
    disease
  tickborne, 10
  and vitamin deficiency, 208
  See also Disorders
Disks
  cartilaginous, 350
  embryonic, 449
Disorders
  adrenal cortex, 397
  autosomal, dominant and
    recessive, 479, 480–83,
    491–95
  circulatory, 231–33
  eating, 215–16
  genetic, 479–83
  liver, 199–200
  lung, 293
  neurotransmitter, 339–40
  polygenic genetic, 484
  skin, 182
  See also Disease
Disposal, waste, 746–47
Dissociate, 25
Dissociation, 25
Distal convoluted tubule, 305, 310
Distribution, biome, 702
Diuretic, anti-. *See* Antidiuretic
    hormone (ADH)
Diversity, 687–88
  biological, 744
  and cyanobacteria, 592
  and evolution, 550–683
  and origin of life, 552–61
  of steroids, 39
  in tropical rain forest, 709
Division, 6, 567. *See also* Cell
    division
DNA. *See* Deoxyribonucleic acid
    (DNA)
DNA copy. *See* Deoxyribonucleic
    acid (DNA)
DNA ligase. *See* Deoxyribonucleic
    acid (DNA)
DNA polymerase. *See*
    Deoxyribonucleic acid
    (DNA)
DNR. *See* United States
    Department of Natural
    Resources
Dominance, 485–86

Dominant disorders, autosomal,
    482–83
Dominant generations, 612
Dominant traits, 471
Donor heart transplants, 236
Dopamine, 339
Dorsal hollow nerve cord, 656
Dorsal root ganglia, 331–32
Double bonds, 23
Double fertilization, 167, 624
Double helix, 510
Double pump, 223
Douche, 425
Douglas fir, 164
Dover, 595
Down syndrome, 491–93, 499
Droplets, coacervate, 556
Drosophila melanogaster, 494, 526
Drugs
  abuse, 340–44
  and AIDS, 274–75
  antihypertensive, 233
  antiviral, 589
  and birth defects, 454–55
  designer, 344
  gateway, 342
  IVDA. *See* Intravenous drug
    abusers (IVDA)
  and synapses, 340
  and thrombolytic therapy,
    234–35
  and tolerance, 340
Dryopithecus, 670, 673
Duchenne muscular dystrophy, 498,
    504
Duckbilled platypus, 663
Ducks, 661
Ducts
  arterial, 456
  collecting, 305, 310
  mammary, 423, 424
  thoracic, 259
  venous, 456
Ductus arteriosus, 456
Ductus deferens. *See* Vas deferens
Ductus venosus, 456
Dugesia, 628, 637
Dunkers, 575
Duodenum, 195, 197
Duplication, 494, 495
Dutch elm disease, 604
Dwarfism, 393, 482
Dyad, 89
Dynamic equilibrium, 380

E

Ear, 367, 378–83
  anatomy of, 378–80
  and eardrum, 378
  function, 379
  and hearing, 381
  physiology of, 380–83
  and vestigial structures, 569
Eardrum, 378
Earthworm, 647–48
Eating disorders, 215–16
ECG. *See* Electrocardiogram (ECG)
Echinodermata, 654–56

Echinoderms, 654–56
Ecological pyramids, 723–24
Ecological terms, 720
Ecology, 720
  and behavior, 684–757
  and ecological pyramids, 723–24
  and genetically engineered
    bacteria, 540–41
Ecosystems, 7–8, 718–34
  and chemical cycling, 725–28
  city, 731–32
  composition, 720–21
  country, 730–31
  deep-sea, 137
  and ecological pyramids, 723–24
  and energy flow, 721–24
  human, 729–33
  natural, 729
  nature of, 720–21
Ectoderm, 437, 438, 439, 440, 449,
    451, 632
Ectopic pregnancy, 418, 432
Edema, 259, 313, 394
Edward VII, King, 498
EEG. *See* Electroencephalogram
    (EEG)
Effacement, 459
Effector, 320–22, 332
Efferent arteriole, 307
Efferent neuron, 320–22
Egg, 165, 411, 416, 442, 448, 454,
    455
Ejaculation, 415
Elastic cartilage, 178
Elbow, 349
Electrocardiogram (ECG), 225, 226
Electroencephalogram (EEG), 338
Electromagnetic spectrum, 124
Electron, 17
  and electron-dot formula, 23, 242
  and electron transport system,
    113
  pathways, 128–31
Elements, 18
  contractile, 360
  formed, 239, 240
  periodic table of, 18
  vessel, 155
Elephantiasis, 259, 642
Elongation, 146, 521
Embolism, pulmonary, 236
Embolus, 233
Embryo
  chick, 569
  and embryonic disk, 449
  frog, 440
  human, 80, 436, 438, 439, 441,
    445–60
  and stem cells. *See* Embryo-
    derived stem (ES) cells
  pig, 569
Embryo-derived stem (ES) cells, 544
Embryology, comparative, 568
Embryonic disk, 449
Emphysema, 294
Empty calories, 205
Emulsification, 38
Endocardium, 222

Endocrine glands, 173, 386–88,
    400–404
Endocrine system, 386
Endocytosis, 68, 75–77
Endoderm, 437, 438, 439, 440, 449,
    632
Endodermis, 144
Endometriosis, 428
Endometrium, 418
Endoplasmic reticulum (ER), 46,
    47, 48, 50–51, 52, 59, 200,
    358, 359
Endorphins, 339
Endoskeleton, 654
Endosperm, 624
Endospores, 590–91
Endosymbiotic theory, 53
End products, nitrogenous, 300
Energy, 104–5
  of activation, 99, 100
  and ATP, 104–5
  and cellular respiration and
    photosynthesis compared,
    138–39
  and diet, 212–13
  and energy-related organelles,
    53–55
  and energy sources, 553
  flow, 721–24
  fuel, 729
  and glucose molecule, 114
  pyramid of, 724
  radiant, 123–24
  solar, 124
  and sunlight, 127–31
Energy-related organelles, 53–55
Engineering, genetic, 534, 540–41
Entamoeba histolytica, 595
Enterocoelomates, 642, 643
Environment
  and carrying capacity, 739–40
  and environmental resistance,
    739
  and environmental signals, 404–5
  and homeostasis, 185–87
  hypotonic and hypertonic, 71
  and plants responses to stimuli,
    159–63
  and steady state, 755–56
Environmental resistance, 739
Environmental signals, 404–5
Enzymes, 29, 49
  and coenzymes, 103–4
  dehydrogenase, 104
  and digestion, 198, 200–202
  and DNA, 536
  and energy of activation, 99, 100
  and enzyme-substrate complex,
    101
  and fermentation, 115
  and food path, 192
  function, 67
  hydrolytic, 52
  and metabolism, 99–101
  and pancreas, 197
  reactions, 101–3
  restriction, 503, 536
  and substrates, 99–100, 101
Enzyme-substrate complex, 101

Eocene, 671
Eosinophils, 239, 244, 249–51
Epicotyl, 167
Epidermal cells, 633
Epidermis, 144, 151, 181, 635, 637
Epididymis, 411, 412
Epiglottis, 192, 193, 282
Epilepsy, 338
Epinephrine, 396
Epipelagic zone, 713, 714, 715
Epiphytes, 709
Epithelial tissue, 173–76
Epstein-Barr virus, 250
Equations, for photosynthesis, 125–27
Equilibrium, 379, 380
ER. *See* Endoplasmic reticulum (ER)
Era, and geological time scale, 564
Erection, 415
Erosion, 743–44
Erwin, Terry, 708
Erythrocytes. *See* Red blood cells (erythrocytes)
Erythroxylum cocoa, 342
ES cells. *See* Embryo-derived stem (ES) cells
Escherichia coli, 196, 593
Esophagus, 190, 191, 192, 193–94
Essential amino acids, 204
Estrogen, 350, 392, 420, 421, 424, 426, 453, 454
Estuary, structure and function, 712
Ethylene, 160
Eubacteria, 592
Euglena, anatomy of, 602
Euglenoids, 602
Euglenophyta, 602
Eukaryotes, and gene expression control, 526–27
Eukaryotic cells, 44–45, 47–59, 60, 61
Eunuch, 393
Eurasian kingfisher, 690
Eustachian tubes, 282, 378
Eutrophication, cultural, 747
Everglades National Park, 687
Evolution, 562–82
    animals, 630
    biological, 556–59
    chemical, 553–56
    and classification, 630–33
    and diversity, 550–683
    evidences for, 563–71
    and evolutionary process, 571–79
    and extinctions, 566
    and gene pool, 571, 572
    gradualistic model, 565
    and hierarchy of classification, 567
    human, 669–83
    and origin of life, 552–61
    pace of, 565
    primates, 671
    punctuated equilibrium model, 565
    and speciation, 580–81
    synthetic theory of, 573–77
    theory, of biology, 12
    trees. *See* Evolutionary trees

Evolutionary process. *See* Evolution
Evolutionary trees, 566, 567, 631
    human, 675
    of vertebrates, 658
Evolution theory, of biology, 12
Excitatory, 340
Exclusion principle, 687
Excretion, 298–318
    organs of, 299–303
    substances of, 299–303
    tubular, 307, 310
Excretory system, 184
Excurrent siphon, 646
Exercise
    and breathing, 279
    effect of, 355–58
Exhalation, 69
Exocrine glands, 173
Exocytosis, 68, 75–77
Exophthalmic goiter, 393, 394, 395
Experimental variable, 10
Experimentation, and conclusion, 9–10
Experiments, 10–12
Expiration, 280, 286, 287
Expiratory reserve volume, 287
Exponential curve, 736
Exponential growth, 736, 737
External respiration, 280, 288–91
External skeleton, 657
Extinction
    and competition, 687
    and evolution, 566
Extraembryonic membranes, 445, 446, 449, 661
Exxon Valdez, 750
Eye, 367, 371–78
    and accommodation, 373–74
    anatomy of, 372
    and biochemistry, 375–76
    and brain, 375
    and color blindness, 376
    compound, 650
    and corrective lenses, 377–78
    and focusing, 372–73
    functions and parts, 372
    physiology of, 372–77
    simple, 650

F

Facilitated diffusion. *See* Diffusion
Facultative anaerobes, 591
FAD, 112
    and electron transport system, 113
    and NAD, 112, 114
    and respiratory chain, 112–14
    and vitamins, 209
Fad diets, 213–15
Falk, Stephen, 383
Fallopian tubes, 418
Family, 6, 567, 670
FAO. *See* United Nation's Food and Agriculture Organization (FAO)
Faras, Anthony, 546
Farming
    contour, 744

organic, 731
Farsightedness, 377
FAS. *See* Fetal alcohol syndrome (FAS)
Fatigue, muscle, 355
Fats
    and blood plasma, 240
    and breast cancer, 206
    hydrolysis, 37
    nutrient and energy content, 214
    and nutrition, 206–7, 208
    and oils, 37
    and proteins compared, 205
    synthesis, 37
Fat-soluble vitamins, 209
Fatty acids, 37–38
FDA. *See* United States Food and Drug Administration (FDA)
Feces, 197
Feedback
    and competitive inhibition, 102, 103
    negative, 186
Feeding behavior, 5
Feedlots, 731
Feelers, 650
Feet. *See* Foot
Female gametophyte, 166
Female grasshopper, 653
Female pine cone, 621
Female reproductive system, 416–25
    and birth control, 425–29
    and external genitalia, 419
    functions of, 416
    genital tract, 418–19
    and hormones, 419–25
    organs of, 416
    and orgasm, 419
Females, height, 460
Femur, 349
Fentanyl (MMPP), 344
Fermentation, 114–16, 534
Fermenters, 534
Ferns, 615, 616–17, 627
Fertilization, 92–93, 411, 447, 448, 621, 624
    double, 167, 624
    internal, 661
    IVF. *See* In vitro fertilization (IVF)
Fertilizers, 301, 601, 728, 730
Fetal alcohol syndrome (FAS), 342, 454
Fetal cannabis syndrome, 342
Fetus, human, 436, 453, 456–58
Fiber
    anatomy of, 364
    asbestos, 294
    centromeric, 87
    Gore-Tex, 547
    muscle, 179, 331, 358–60, 360–62
    nerve, 180
    Purkinje, 225, 226
    spindel, 84–85, 86
Fibrillation, ventricular, 226
Fibrin, 248–49
Fibrinogen, 240, 241, 248–49
Fibroblasts, 176

Fibrocartilage, 178
Fibrosis, pulmonary, 294
Fibrous connective tissue, 176–77
Fibula, 349
Fight or flight, 332, 334
Filamentous green algae, 599
Filaments, 165
    micro-. *See* Microfilaments
    thick and thin, 360, 363
    See also Sliding filament theory
Filter feeders, 633
Filtration, pressure, 307
Fimbriae, 418
Finches, 581
Fingers, 349, 501
Fir tree, 164
Fish, 657–59, 666
    and genetic engineering, 546–47
    marine, 300
Fission, binary, 590
Fitness, checklist, 357
Fixation
    carbon dioxide, 132–33
    nitrogen, 727–28
Flagella, 46, 48, 57–59, 594
    anatomy, 58
    of bacterium, 589
    of green algae, 597–99
Flagellates, 594
Flame cells, 637
Flamingos, 661, 662
Flatworms, 637–40, 667
    anatomy, 628
    parasitic, 639–40
Flight, 661–62
Floating ribs, 348
Floor, oceanic. *See* Oceanic floor
Flow. *See* Pressure-flow theory of phloem transport
Flowers, 622, 627
    anatomy, 164–67, 622
    and flowering plants, 142–45, 159–63, 622–25
    and fruit, 167
Flu. *See* Influenza
Fluid-mosaic model of intracellular membrane, 46, 65
Fluids
    body, 248–49
    cerebrospinal, 334
    seminal, 414
    synovial, 351
    tissue, 239
Flukes, 639, 640
Fluosol-DA, 243
Foams, and birth control, 425
Focusing, of eye, 372–73
Folacin, 209
Follicles, 417, 418
    anatomy, 417
    hair, 182
Follicle-stimulating hormone (FSH), 391, 392, 415, 420, 421, 422, 426, 454, 461
Fontanels, 348, 456
Food
    angiosperms as, 625, 626
    and behavior, 213–15
    and breast feeding, 271

and eating disorders, 215–16
and food chains, 722–23
and food groups, 203
and minerals, 210
nutrient and energy content, 214
path of, 190, 191, 192
plants as, 145
and starvation, 626
and food vacuoles, 594
and vitamins, 209
and food webs, 722–23
Food and Drug Administration. *See* United States Food and Drug Administration (FDA)
Food chains, 722–23
Food groups, 203
Food vacuoles, 594
Food webs, 722–23
Foot, 350, 643
Foramen ovale, 456
Foraminifera, 595
Foreskin, 414
Forests, 706–10
    deforestation, 754
    destruction, 744, 745
Formed elements, 239, 240
Formica fusca, 694
Formulas, electron-dot, structural, molecular, 23, 242
Fossils, 557, 563–66
Founder principle, 575
Fovea centralis, 372
Fowls, 661
Fragile-X syndrome, 499–500, 504
Franklin, Rosalind, 514
Freckled grouper, 659
Free-living worms, 639
Freinhar, Jack, 402
Freshwater communities, 710–11
Fright, 692
Frogs, 3, 440, 442, 443, 659, 660, 692
Fronds, 616
Frontal lobe, of brain, 336–38
Frost, 704
Fructose, 34
Fruit, 160, 161, 622, 624
    and flowers, 167
    nutrient and energy content, 214
Fruit fly. *See Drosophila melanogaster*
Fruiting bodies, 604
FSH. *See* Follicle-stimulating hormone (FSH)
Fucus, 600, 601, 611
Fuel, consumption and energy, 729, 730–31
Functions
    air, 281
    autonomic nervous system, 333
    B cell, 269
    blood-clotting, 248–49
    cell membrane, 64–78
    cells, 43–63
    cellulose, 35
    cell wall, 64–78
    centrioles, 85
    cerebral lobes, 337
    chloroplast, 132

DNA, 509, 511–16
ear, 379
enzymes, 67
estuary, 712
eukaryotic structures, 48
eye, 372
female reproductive system, 416
glycogen, 34, 35
Golgi apparatus, 52
hemoglobin, 244
hormones, 393
kidneys, 312–14, 315–16
male reproductive system, 411
minerals, 210
mitochondria, 108
muscle spindle, 354
nerves, 330
phloem, 155
phytochrome, 163
proteins, 66
sacromere contraction, 363
skeleton, 347
spindle fibers, 86
starch, 34, 35
and T cell, 269
urinary system, 303
vitamins, 209
xylem, 155
Fungi, 583, 603–7
    club, 605, 606
    imperfect, 607
    sac, 604, 605
Funny bone, 349
Furrowing, 87

G

GA. *See* Gibberellic acid (GA)
GABA. *See* Gamma-aminobutyrate (GABA)
Gaia hypothesis, 558, 559
Gait, striding, 678
Galapagos Islands, 578, 581
Gallbladder, 198, 199, 301
Galleria Mall GoGetters, 357
Gallstones, 199
Gamete intrafallopian transfer (GIFT), 427
Gametes, 81, 472–74, 599, 600, 611, 612
Gametophyte, 600, 611–12, 622–24
    female, 166
    generations, 165
Gamma-aminobutyrate (GABA), 339, 340, 341
Gamma globulins. *See* Globulins
Gamma rays, 124
Ganglia, 329, 331–32, 646
Gap junction, 176, 177
Gart gene, 493
Gases
    and blood plasma, 178, 240
    formation, 22
    and gas exchange, 69
    methane, 733
    and pollution, 751
Gas exchange, 69
Gastric glands, 194
Gastric mill, 650

Gastrin, 198
Gastrodermis, 635, 637
Gastropods, 644
Gastrovascular cavity, 637
Gastrula, 436, 437–39, 449
Gastrulation, 437, 449
Gates, sodium and potassium, 325
Gateway drug, 342
GEM. *See* Genetically engineered microbe (GEM)
Gemmae, 612
Gene expression, control of, 524–27
Gene flow, and evolution, 574
Gene inheritance, patterns, 468–88
Gene mutation, 527–28
Gene pool, 571, 572
General receptors, 367–69
Generations
    alternation of, 165, 596–97, 600, 611–12
    dominant, 612
    gametophyte, 165
    and Mendel's results, 470
Generative nucleus, 166
Genes, 469
    and aging, 461
    and cancer, 528–31
    cloning, 535, 537, 542
    definition of, 524
    Gart, 493
    and gene expression, 524–27
    and genetic disorders, 479–83
    pleiotropic, 576
    regulator, 524
    structural, 525
    X-linked, 496–500
Gene sequencing, 536, 548
Gene theory of biology, 12
Gene therapy, in humans, 545–48
Genetically engineered microbe (GEM), 540–41. *See also* Microbes
Genetic disorders, 479–83
Genetic drift
    and adaptation, 575
    and evolution, 574–76
Genetic engineering, 534, 540–41
Genetic markers, 503–4
Genetics
    and crosses, 471
    and gene mutations, 527–28
    and gene sequencing, 536, 548
    and gene therapy, 545–48
    and Mendel's results, 470
    polygenic disorders, 484
    population, 571–73
Genital herpes, 430–31
Genital tract
    female, 418–19
    male, 412
Genital warts, 430, 431
Genomic library, 535
Genotype, 471–72, 479
Genus, 6, 567, 670
Geographic isolation, 580
Geography, bio-. *See* Biogeography
Geological time scale, 564
Germinal mutations, 528
Germination, seed, 167–68

Germ layers, 438, 439, 440, 450
Germ layer theory, 450
Gerontology, 461, 465
Getchell, Bud, 356
GH. *See* Growth hormone (GH)
Giantism, 391, 393
Gibberellic acid (GA), 160, 161
Gibbon, 673
Giberellins, 160, 161
GIFT. *See* Gamete intrafallopian transfer (GIFT)
Gill, 645, 650, 656, 659, 665
Gill clefts, 656
Gill slits, 656
Ginkgos, 619
Giraffes, 578, 579
Glands
    accessory, 190
    adrenal, 395–98
    Cowper's, 411, 414
    digestive, 198
    endocrine, 173, 386–88, 400–404
    exocrine, 173
    gastric, 194
    green, 650
    lacrimal, 281
    mammary, 662
    master, 391
    oil, 182
    pancreas, 398–400
    parathyroid, 392, 394–95
    parotid, 192
    pineal, 392, 400–403
    pituitary, 313, 386, 388–93
    prostate, 411, 414
    salivary, 192
    sebaceous, 182
    submandibular, 192
    sudoriferous, 182
    sweat, 182
    thymus, 392, 400
    thyroid, 19, 386, 392, 393, 394–95
Glial cells, 180, 322
Globin, 241, 244
Globulins, alpha, beta, and gamma, 240, 241
Glomerular filtrate, 308
Glomerule, 307
Glomerulus, 307
Glottis, 192, 193, 281, 282, 283
Glucagon, 392, 398, 399
Glucocorticoids, 392, 393, 396
Glucose, 34, 105, 106, 107, 201, 202
    and ATP production, 114
    and blood plasma, 240, 241
    and capillary exchange, 246, 247
    and homeostasis, 185
    and liver, 199
    molecule, 114
    and nephrons, 308
    and oxygen debt, 362
Glycerol, 37, 201, 202
Glycine, 339
Glycogen
    and homeostasis, 185
    and liver, 199
    structure and function, 35
Glycolipids, 65, 67, 253
Glycolysis, 105, 106, 107, 108

as metabolic pathway, 109, 110
summary, 111
Glycoproteins, 65, 67
GM-CSF. See
Granulocytemacrophage
colony-stimulating factor
(GM-CSF)
GMP. See Guanosine
monophosphate (GMP)
GnRH. See Gonadotropic-releasing
hormone (GnRH)
Goat, 5
Goiter, exophthalmic and simple,
393, 394, 395
Golgi apparatus, 46, 47, 48, 51–52,
59, 61, 77, 88
Gonadotropic hormones, 391
Gonadotropic-releasing hormone
(GnRH), 415, 420, 428
Gonads, 386, 392, 393, 400
Gonorrhea, 430, 432, 454, 592
Gore-Tex fibers, 547
Gorilla, 673
Graafian follicle, 417, 418
Gradualistic model, of evolution, 565
Grafting, 164
Gram stain, 592, 593
Grana, 54, 125, 128
Grand Canyon, 692
Granular leukocytes. See White
blood cells (leukocytes)
Granulocytemacrophage colony-
stimulating factor (GM-
CSF), 271
Grapes, 115, 161
Grasshoppers, 652–53
Grasslands, 704–5
Gray crescent, 443
Grazers, 689
Great Spruce Head Island, 5
Green algae, 597–600
Green glands, 650
Greenhouse effect, 727, 753–54
Green snake, 660
Groundwater, pollution, 747–48
Grouper, 659
Groups, acid and carboxyl, 31
Groveman, George, 243
Growth, 436
of bones, 350
curve. See Growth curve
exponential, 736, 737
factors. See Growth factors
and growth rate, 737, 738
hormone. See Growth hormone
(GH)
human population, 740–43
and lymphokines, 271–72
plants, 141–53
population, 736–40
rate. See Growth rate
regulators in plants, 160
secondary, 148
of stems, 147–48
sustainable, 754–55
Growth curve, logistic or S-shaped,
739
Growth factors, 271–72, 405
Growth hormone (GH), 391, 392

Growth rate, 737, 738
Guanine, 510, 511, 512
Guanosine monophosphate (GMP),
375, 376
Guard cells, 151, 157
Guise, Kevin, 546, 547
Gymnosperms, 619–21, 627

H

Habitat, 720
Hackett, Perry, 546
Hagfish, 657
Hair, 182, 662
Hair cells, and ear, 381, 382, 383
Hair follicle, 182
Hammer, 378
Haploid, 81–82, 83
Haplontic cycle, 596–97
Harber, Leonard C., 183
Hard palate, 192
Hardy, G. H., 571–72
Hardy-Weinberg law, 572–73
Harvard University, 356
Harvesting, and light-harvesting
antenna, 127
Haversian canals, 178
Haversian systems, 350
Hawks, 661
Hazardous waste, 746
HCG. See Human chorionic
gonadotropic hormone
(HCG)
HD. See Huntington disease (HD)
HDL. See High-density lipoprotein
(HDL)
HDN. See Hemolytic disease of the
newborn (HDN)
Health, and respiration, 292–96
Health and Human Services. See
United States Department of
Health and Human Services
Hearing, 367, 381, 382
Heart, 221–23
anatomy of, 221, 222
artificial, 236
and aspirin, 235
blood path, 223
CHD. See Coronary heart
disease (CHD)
and coronary bypass surgery, 236
external, 221
and heart attack, 233–34
and heartbeat, 223–26
internal, 222
and medical and surgical
treatment, 234–36
pulmonary circuit and systemic
circuit, 223
and pulmonary edema, 259
sounds, 225
and thrombolytic therapy,
234–35
valves, 225
Heart attack. See Heart
Heartbeat. See Heart
Height, males and females, 460
Heimlich maneuver, 282
Helix, double. See Double helix

Helper, 103
Helper T cells, 266–69, 429
Heme, 241, 244
Hemispheres, cerebral, 336
Hemocoel, 650
Hemodialysis, 315–16
Hemoglobin, 211, 241–45
components, structure, function,
244
and excretion, 299–300
and hemoglobin saturation curve,
291
and life cycle of red blood cells,
245
molecule, 33
and respiration, 290, 291
Hemolytic disease of the newborn
(HDN), 255
Hemolytic jaundice, 199
Hemophilia, 497–98, 499
Hemorrhoids, 197, 236
Henle, loops of. See Loops of Henle
Hepatic portal vein, 198–99, 228
Hepatic vein, 228
Hepatitis, 200, 270
Herb, 719
Herbaceous stems, 147–48
Herbicides, 730
Herbivores, 123, 705, 720
Hereditary material, and origin of
cells, 553
Heredity, code of, 517–20
Hermaphroditic, 638–39, 648
Hermaphroditic animals, 638–39
Hermissenda, 339
Hernia, 414
Heroin, 340, 343–44
Herpes, 454
genital, 430–31
HSV. See Herpes simplex virus
(HSV)
Herpes simplex virus (HSV), 273
Heterogametes, 599
Heterospores, 618
Heterotroph hypothesis, and
biological evolution, 556–57
Heterotrophs, 123, 557
Heterozygote superiority, 576
Heterozygous, 471–72
Hex A, 481–82
Hexose sugars, 33
Hierarchy of classification, 567
High
alcohol, 341
runner's, 339
High-density lipoprotein (HDL),
207, 234–35
Hinge joints, 351
Hipbones, 349
Histamine, 252–53
Histones, 82
HIV. See Human immunodeficiency
virus (HIV)
HMS Beagle, 578
Homeostasis, 185–87
and blood, 240
and kidneys, 314
Hominidae, 670
Hominids, 673–74

Hominoidea, 670
Hominoids, 673
Homo, 670, 676–78
Homo erectus, 670, 676–78
Homogentisate, 515
Homo habilis, 670, 676
Homologous chromosomes, 89
Homologous pair, of chromosomes,
469
Homologous structures, 566, 568
Homo sapiens, 6, 670, 678, 681
Homo sapiens sapiens, 681
Homozygous, 471–72
Hormonal system. See Hormones
Hormones, 385–407
ACTH. See Adrenocorticotropic
hormone (ACTH)
ADH. See Antidiuretic hormone
(ADH)
aldosterone. See
Mineralocorticoids
atrial natriuretic hormone, 403
bGH. See Bovine growth
hormone (bGH)
and biotechnology products, 538
and blood plasma, 240
and cellular activity, 388
cortisol. See Glucocorticoids
and digestion, 198
and female reproductive system,
419–25
FSH. See Follicle-stimulating
hormone (FSH)
functions, 393
GH. See Growth hormone (GH)
GnRH. See Gonadotropic-
releasing hormone (GnRH)
gonadotropic, 391
HCG. See Human chorionic
gonadotropic hormone
(HCG)
and hormonal system, 184
ICSH. See Interstitial cell-
stimulating hormone (ICSH)
lactogenic. See Lactogenic
hormone
LH. See Luteinizing hormone
(LH)
LTH. See Lactogenic hormone
(LTH)
mechanism of hormonal action,
387–88
MSH. See Melanocyte-
stimulating hormone (MSH)
in plants, 159, 160
and pregnancy, 453
prolactin. See Lactogenic
hormone
PTH. See Parathyroid hormone
(PTH)
sex, 392, 393, 397
somatotropin. See Growth
hormone (GH)
steroid, 387–88
testosterone. See Androgens
thyrotropic. See Thyroid-
stimulating hormone (TSH)
TRH. See Thyroid-releasing
hormone (TRH)

TSH. *See* Thyroid-stimulating hormone (TSH)
vasopressin. *See* Antidiuretic hormone (ADH)
Horny scales, 660
Horse tail, 615, 616
Hosts, primary and secondary, 639
Hot flashes, 424
Housekeeper system, 334
HPV. *See* Human papillomaviruses (HPV)
HSV. *See* Herpes simplex virus (HSV)
Human, 670, 676–82
    anatomy, 170–407
    and ape compared, 674
    and carbon cycle, 726–27
    cells. *See* Human-mouse cell data
    chromosomes, mapping, 501–4
    circulatory system, 229
    development, 408–549
    ear. *See* Ear
    ecosystem, and natural ecosystem, 729
    embryo, 80, 436, 438, 439, 441, 445–60
    evolution, 669–83
    evolutionary tree, 675
    extraembryonic membranes. *See* Human extraembryonic membranes
    eye. *See* Eye
    fetus, 436, 453, 456–58
    and gene therapy, 545–48
    Genome Project. *See* Human Genome Project
    HCG. *See* Human chorionic gonadotropic hormone (HCG)
    HIV. *See* Human immunodeficiency virus (HIV)
    HPV. *See* Human papillomavirus (HPV)
    inheritance, 408–549
    intervention, 687, 692
    karyotype, preparation, 81
    and mouse cells. *See* Human-mouse cell data
    organization, 172–88
    organ systems, 184–85
    papillomavirus. *See* Human papillomavirus (HPV)
    physiology, 170–407
    and plants, 626
    population. *See* Human population
    races, 681–82
    reproduction, 408–549
    respiratory tract, 281
    skin, anatomy, 181
    species, 681–82
Human chorionic gonadotropic hormone (HCG), 423, 449, 453
Human extraembryonic membranes, 445, 446, 449
Human Genome Project, 504, 548

Human immunodeficiency virus (HIV), 274–75, 429
Human-mouse cell data, 502–3
Human papillomavirus (HPV), 431
Human population, 8, 735–57
    and carrying capacity, 739–40
    and doubling time, 738–39
    growth, 740–43
    and growth rate, 738
    and pollution, 743–54
    and steady state, 755
Humans. *See* Human
Humerus, 349
Hummingbird, 624, 661
Humors, vitreous and aqueous, 371, 372
Hunt, Anthony, 243
Huntington disease (HD), 339–40, 482–83, 504
Hyaline cartilage, 178
Hybridomas, 273
Hybrids, 473, 475, 502–3
Hydra, 630, 636–37
Hydrates of carbon, 33
Hydrocarbons, 750
Hydrochloric acid, 25, 194, 198, 200
Hydrogen
    bonds, 24, 32
    and chemical evolution, 553
    and hydrogen sulfide, 137
    ions, 28, 310
Hydroids, 635
Hydrolysis
    of maltose, 34
    of neutral fat, 37
    of organic polymer, 29
    and peptides, 31
Hydrolytic enzymes, 52
Hydrostatic skeleton, 633, 641, 648
Hydroxide ions, 28
Hygiene, 740
Hypercholesterolemia, 482
Hypertension, 231–33, 314
Hypertonic environment, 71
Hypertonic solutions, 72
Hyphae, 603, 606
Hypocotyl, 167
Hypothalamus, 336, 337, 386, 388–93, 415, 421, 422
    and ADH, 388–90, 392
    and body temperature control, 186–87
Hypotheses, formulation of, 8, 9–10
Hypothyroidism, 394
Hypotonic environment, 71
Hypotonic solutions, 72
H zone, 360

I

IAA. *See* Indolacetic acid (IAA)
I band, 360
Ice, and hydrogen bonding, 25
Ice cream, 601
ICSH. *See* Interstitial cell-stimulating hormone (ICSH)
I. dammini, 10
Ig. *See* Immunoglobins (Ig)
Image, 372, 374

Immune system, 184
    and AIDS. *See* Acquired immune deficiency syndrome (AIDS)
    and allergies, 274–75
    and autoimmune diseases, 276
    and illnesses and side effects, 273–76
    and immunotherapy. *See* Immunotherapy
    and tissue rejection, 275–76
    therapy. *See* Immunotherapy
    See also Immunity
Immunity, 252, 261–69
    active, 270, 271
    and AIDS. *See* Acquired immune deficiency syndrome (AIDS)
    antibody-mediated, 264–65
    cell-mediated, 266–69
    and complement system, 263
    and general defense, 261–63
    and immunotherapy. *See* Immunotherapy
    induced, 270, 271
    and lymphatic system, 257–77
    passive, 270, 271
    SCID. *See* Severe combined immune deficiency (SCID)
    and specific defense, 263–69
    therapy. *See* Immunotherapy
    See also Immune system
Immunizations, 270
Immunoglobins (Ig), 265, 266, 274–75
Immunotherapy, 270–73
Imperfect fungi, 607
Implant, 236, 426
Implantation, 423, 447, 449
Impotency, 415
Impulse, nerve, 322–25
Inborn errors of metabolism, 515
Incisors, 190, 191
Incomplete dominance, 485
Incomplete proteins, 204
Incurrent siphon, 646
Incus, 378
Independent assortment, Mendel's laws of, 476
Index finger, 501
Indolacetic acid (IAA), 160
Inducible operon model, 525, 526
Induction, 444, 445
Inductive reasoning, 9
Industrial melanism, 577
Inefficient lungs, 659
Infancy, 461
Infections
    and blood, 249–53
    and respiration, 292–93
    trypanosome, 596
    viral, 588
Infectious mononucleosis, 250
Inferior vena cava. *See* Venae cavae
Infertility, 426–29
Inflammatory reaction, 252–53
Influenza, 293
Information storage, and DNA, 511
Inguinal canals, 413

Inhalation, 69
Inheritance
    abnormal, 491–95
    abnormal sex chromosome, 495–96
    chromosome, 489–506
    and crosses, 471
    gene, 468–88
    human, 408–549
    molecular basis of, 507–32
    and multitraits, 474–79
    normal, 490–91
    patterns of, 468–88
    polygenic, 483–84
    of sex, 491
    sex-linked, 496–501
    and sickle-cell anemia, 485–86
    and traits, 469–74
Inhibin, 415, 426
Inhibition, competitive and noncompetitive, 102–3
Inhibitory, 340
Initiation, 521, 529–31
Inner cell mass, 437, 448
Inner ear, 378, 380
Innervate, 320–22, 362–64
Innervation, 320–22, 362–64
Innominate hipbones, 349
Inoculation, 585
Inorganic chemistry, 24
Inorganic compounds, 24
Inorganic molecules, 24–29
Inorganic nutrients, 145
Insecta, 652
Insectivora, 665
Insects, 648, 652, 665
Insertion, of muscle, 352, 353
Inspiration, 280, 285–86, 287
Inspiratory reserve volume, 287
Institute of Aerobics Research, 356
Institute of Hydrobiology, 546
Insulin, 392, 398, 399
Insulin shock, 399
Integration
    and aging, 464–65
    and summation, 327
Integumentary system, 180
Integuments, 624
Interferon, 263, 271
Interleukin, 271
Internal fertilization, 661
Internal respiration, 280, 289, 291–92
Internal skeleton, 657
Interneuron, 321, 332
Internode, 147
Interphase, 83
Interstitial cells, 412, 413, 415
Interstitial cell-stimulating hormone (ICSH), 415
Intervention, human, 687, 692
Intestines, 190, 191
    anatomy, 196
    large, 190, 191, 195–97, 302
    small, 190, 191, 195, 197
Intolerance, lactose, 202
Intracellular membrane, fluid-mosaic model, 46, 65
Intragene segments, 518

Intrauterine device (IUD), 425
Intravenous drug abusers (IVDA), 429
Introduction, 2–13
Invagination, 437
Inversion, 494, 495, 750, 752
Invertebrate chordates, 656
Invertebrates, advanced and primitive, 630, 633–56
Inverted image, 374
In vitro fertilization (IVF), 427
Involuntary smoking, 296
Iodine, 19, 210
Ionic bond, 20, 32
Ionic reactions, 19–21, 32
Ions, 19
    balance within kidneys, 312–14
    bicarbonate, 243
    and blood plasma, 178
    calcium, 359, 361, 363
    and coenzymes, 103–4
    and excretion, 301
    hydrogen and hydroxide, 28, 310
    potassium, 324
    and potential, 324
    sodium, 20, 324
Iris, 371, 372
Irish potato, 626
Iron, 210, 300
Irrigation, 730
Islets of Langerhans, 398
Isogametes, 599, 600
Isolating mechanisms, 580
Isolation, geographic and reproductive, 580
Isotonic solutions, 71
Isotopes, 19, 273, 564
IUD. See Intrauterine device (IUD)
IVDA. See Intravenous drug abusers (IVDA)
IVF. See In vitro fertilization (IVF)
Ixodes dammini, 694
Ixodes ricinus, 694

J

Jackals, 570
Jaffe, Robert, 428
Jaundice, 199–200
Jawless fish, 657
Jellies, and birth control, 425
Jellyfish, 635
Johnson, Ben, 401
Joints, 350–52, 369, 650
Journal of the American Medical Association, 356
Judd, Howard, 428
Junctions
    between cells, 176, 177
    neuromuscular, 363
    tight and gap, 176, 177
Jurassic, 564
Juxtaglomerular apparatus, 314

K

Kaposi's sarcoma, and AIDS, 274
Kapuscinski, Anne, 546, 547
Karyotype, 80, 81, 492, 493

Kcalories, 211, 212, 213, 214
Kelp, 601
Keratin, 181
Kidneys, 304–8
    and ADH, 313
    anatomy, 304
    artificial, 315–16
    and excretion, 299–300, 302
    functions, 312–14
    and homeostasis, 314
    and hypertension, 232, 233
    problems, 315–16
    and renal pyramids, 304
    replacement, 315–16
    transplant, 315
    and urine, 303
Kingdom, 6, 567
    animal, 629–68
    plant, 609–28
Kingdom Fungi, 583, 603–7
Kingdom Monera, 583, 589–93
Kingdom Protista, 583, 593–603
Kingfisher, 690
King's College, 514
Klinefelter syndrome, 493, 496
Knee jerk, 369
Koala bear, 663
Krebs cycle, 105, 106, 107, 108, 362
    as metabolic pathway, 109
    summary, 112

L

Labia majora, 419
Labia minora, 419
Lac operon, 525, 526
Lacrimal glands, 281
Lactase, 100
Lactate, 362
Lactation, 424
Lacteal, 195, 258
Lactogenic hormone (LTH), 391, 392, 424
Lactose, and lactose intolerance, 100, 202
Lacunae, 178
Ladder-type nervous organ, 638
Ladybugs, 731
Lagomorpha, 665
Lakes, and life zones, 710, 711
Lamarck, Jean-Baptiste, 578, 579
Lamarckian, 578, 579
Lamellae, 54, 67, 125
Laminaria, 601
Lampreys, 657
Lancelet
    anatomy, 656
    development, 437, 438
Land
    degradation, 743–47
    environment compared to water environment, 611
    quality loss, 731
Langerhans, islets of. See Islets of Langerhans
Lantern fly, 692
Lanugo, 456
Laparoscope, 428
Large intestine. See Intestines

Larval stage, 653
Larynx, 281, 282, 283
Latitude, and altitude, 710
Laws, 8
    Hardy-Weinberg. See Hardy-Weinberg law
    independent assortment, 476
    Mendel's. See Mendel's laws
    probability, 473, 477–78
    segregation, 476
Layers, skin, 181–82
LDL. See Low-density lipoprotein (LDL)
Leaf. See Leaves
Learning, and memory, 339
Leaves, 147, 150–52
    anatomy, 155
    and fronds, 616
    and leaf veins, 150, 151–52
Leg, 350
Legume, nutrient and energy content, 214
Lemur, 670
Lens, 371, 372, 374, 377–78
Leon, Arthur, 356
Less-developed countries, 741–43
Leukemia, 249
Leukocytes. See White blood cells (leukocytes)
LH. See Luteinizing hormone (LH)
Lianas, 708
Library, genomic, 535
Lichens, 593, 605, 719
Life
    characteristics, 4–6
    and chemistry, 16–42
    and geological time scale, 564
    and organic compounds, 41
    origin, 552–61
    See also Life cycle
Life cycle
    algae, 596–97
    alternation of generations, 611–12
    animals, 82, 83
    bacteriophages, 585–86
    deer tick, 694–95
    ferns, 616
    flowering plant, 165, 622–25
    lysogenic, 586
    lytic, 586
    moss, 612–14
    mushroom, 606
    pine tree, 620–21
    Plasmodium vivax, 596
    red blood cells, 245
    Rhizopus, 604
    seed plants, 618
    tapeworm, 640
    viruses, 585–88
    See also Life
Life zones, of lake, 711
Ligament, 177, 351
Ligase, DNA. See Deoxyribonucleic acid (DNA)
Light
    and eye, 375–76
    and light-harvesting antenna, 127
    and light microscope, 44, 45

    and light reaction, 127
    white or visible, and sunlight, 124, 127–31
Lily, 622
Limb buds, 441
Limbic system, 337, 338–39
Limnetic zone, 711
Linkage, 500, 501–2
Linkage group, 501
Linticels, 148
Lipase, 100, 201, 202
Lipids, 35–40, 100
    and lipid bilayer, 65–67
    and nutrition, 206–9
Lipoprotein
    HDL. See High-density lipoprotein (HDL)
    LDL. See Low-density lipoprotein (LDL)
Liposomes, 556
Lipscomb, David, 383
Littoral zone, 711
Liver, 197, 198–200, 245
    cirrhosis, 341–42
    disorders, 199–200
    and excretion, 299–300, 301, 302
Liver fluke, 640
Liverwort, 612
Lizards, 660
Lobe-finned fish, 657
Lobes, of brain, 336
Lobsters, 648
Logistic growth curve, 739
Long bones, 349–50, 351
Loons, 661
Loop of Henle, 305, 310–12
Loose connective tissue, 176, 177
Love Canal, 746
Low-density lipoprotein (LDL), 207, 234–35
LSD. See Lysergic acid diethylamide (LSD)
LTH. See Lactogenic hormone (LTH)
Lumbar puncture, 334
Lumbricus, 647
Lumen, 193, 309
Lung cancer. See Lungs
Lung capacities. See Lungs
Lungs, 665
    and cancer, 294–96
    and capacities, 287–88
    disorders, 293
    and excretion, 299–300, 302
    gas exchange in, 69
    and human respiratory system, 281, 283–84, 285, 286
    inefficient, 659
    smoking and nonsmoking, 295
Lupus. See Systemic lupus erythematosus (SLE)
Luteinizing hormone (LH), 391, 392, 415, 420, 421, 422
Lycophyta, 615
Lyell, Charles, 578
Lyme disease, 9–10, 694–95
Lymph, 239, 259
Lymphatic capillaries, 247

Lymphatic system, 258–61
  and immunity, 257–77
  and lymphoid organs, 259–61
Lymphatic vessels, 247, 258–59
Lymph nodes, 259, 260–61
Lymphocytes, 239, 244, 249–51
  B, 244, 250, 251, 260, 264–65,
    268, 269, 272
  and blood marrow, 244
  T, 244, 250, 260, 264, 265,
    266–69
Lymphoid organs, 259–60
Lymphoid stem cells, 244
Lymphokines, 271–72
Lynx, 690
Lysergic acid diethylamide (LSD),
  342
Lysogenic cells, 586
Lysogenic life cycle, 586
Lysosomes, 46, 48, 52–53, 59
Lytic life cycle, 586

M

McLean Hospital, 401
McMillan, Tom, 26
Macrocystis, 601
Macromolecules, 555–56
Macronutrients, 145
Macrophages, 76, 253, 260, 262
Magasporangium, 618
Magnesium, 210
Magnesium ions, 103
Magnification, biological, 747–50
Maidenhair tree, 619
Major histocompatibility complex
  (MHC) protein, 268–69, 276
Male pine cone, 620–21
Male reproductive system, 411–16
  and birth control, 425–29
  functions, 411
  and genital tract, 412–14
  and orgasm, 414–15
Males, height, 460
Malignant melanoma, 182
Malignant tumors, 529
Malleus, 378
Malpighian tubules, 652, 653
Maltase, 100, 201, 202
Malthus, Thomas, 578
Maltose, 33, 34, 100, 200, 201, 202
Mammalia, 670
Mammals, 662–65, 666
Mammary duct, 423, 424
Mammary glands, 662
Manganese, 210
Manioc, 626
Man-of-war, 635
Mantle, 643
Mantle cavity, 643
Mapping, human chromosomes,
  501–4
Marchantia, 612
Marfan syndrome, 482
Marijuana, and abuse, 340, 342
Marine fish, 300
Markers, genetic, 503–4

Marrow, bone. *See* Bone(s) *and*
  Bone marrow *and* Red bone
  marrow *and* Yellow bone
  marrow
Marshes, 712
Marsupials, 663
Masculinization, 393, 397
Mass extinctions, 566
Mast cells, 274–75
Master gland, 391
Materials, raw, 729
Matrix, 54, 107, 176, 178
Maturation, zone of, 146
Maximum stimulus, 355
Measurement, of blood pressure,
  229–30
Meats, nutrient and energy content,
  214
Mechanoreceptors, 367, 378–83
Medulla, adrenal, 304, 392, 395–98
Medulla oblongata, 336
Medusae, 635
Megakaryocyte, 244, 248–49
Megakaryocytoblast, 244
Megaspore, 165, 621
Meiosis, 81, 89–93, 476, 492, 598,
    611, 612, 618
  and animal cells, 89
  and life cycle of animals, 82, 83
  and mitosis compared, 93–95
  overview, 89
  and plant cells, 93
  stage of, 89–92
Melanin, 181
Melanism, industrial, 577
Melanocytes, 181
Melanocyte-stimulating hormone
  (MSH), 391, 392
Melanoma, 182
Melatonin, 392
Membrane potential, 322, 324
Membranes
  basement, 176
  body, 184–85
  canals. *See* Membranous canals
  cell. *See* Cell membrane
  extraembryonic, 445, 446, 449,
    661
  intracellular, 46, 65
  pleural, 285
  postsynaptic, 327
  potential. *See* Membrane
    potential
  presynaptic, 327
  structure, 46–47
  tectorial, 379
  tympanic, 378
  and vestigial structures, 569
Membranous canals, 50–53
Memory, and learning, 339
Memory B cells, 264–65
Memory T cells, 266–69
Mendel, Gregor, 469
Mendel's laws, 469–79
Meninges, 334
Menisci, 351
Menopause, 424–25, 461
Menstruation, 420
Meristem tissue, 143–44

Mesenteries, 633
Mesoderm, 438, 439, 440, 451, 632,
  633
Mesoglea, 634, 635
Mesopelagic zone, 713, 714, 715
Mesophyll tissue, 151
Mesozoic, 564, 671
Messenger RNA (mRNA), 49, 511,
  517–18, 519, 520, 521, 522,
  523, 555–56, 588
Metabolic pathways, 98–99
  and glycolysis, 110
  interrelationships, 109
  and phenylalanine, 515
  and subpathways, 107
  See also Metabolism
Metabolic pool, 109. *See also*
  Metabolism
Metabolism, 98–105
  of bacteria, 591–92
  basal, 212–13
  cellular, 97–118
  and hemoglobin, 244
  inborn errors of, 515
  and origin of cells, 553
  pathways. *See* Metabolic
    pathways
  pool. *See* Metabolic pool
Metabolites, 108–10
Metacarpals, 349
Metafemale syndrome, 493, 495
Metals, 18, 20, 300
Metamorphosis, 653
Metaphase, 85, 87, 90, 91, 92
Metastasis, 529
Metatarsals, 349
Methane, and chemical evolution,
  553, 733
MHC protein. *See* Major
  histocompatibility complex
  (MHC) protein
Mice, chimeric, 544. *See also* Mouse
Michael Reese Hospital, 428
Microbes, 534. *See also* Genetically
  engineered microbe (GEM)
Microfilaments, 55, 56, 196
Micrographs, 44, 45
Microminerals, 210
Micronutrients, 145
Microscopy, 44, 45
Microspheres, 555, 556
Microspore, 166, 620–21
Microtubules, 56, 58
Microvilli, 196, 309
Midbrain, 336
Middle ear, 378, 379
Midoceanic ridges, 137
Milk, nutrient and energy content,
  214
Milkweed, 751
Miller, Stanley, 553–55
Millipedes, 648
Mimicry, 693
Mineralocorticoids, 313–14, 392,
  396–97, 403. *See also*
  Minerals
Minerals
  functions and sources, 210
  and genetically engineered

bacteria, 540–41
  and vitamins, 209–11
  See also Mineralocorticoids
Mining, 731
Minnesota Sea Grant College
  Program, 546
Miocene, 671
Mitochondria, 46, 47, 48, 53, 54, 55,
  59, 107–8, 309
Mitosis, 83–88
  and animal cells, 84
  and animal life cycle, 82, 83
  and cell cycle, 83–84
  and meiosis compared, 93–95
  overview, 84
  and plant cells, 88
  stages of, 84–87
Mitral stenosis, 226
Mixed nerve, 330
M line, 360
MMPP. *See* Fentanyl (MMPP)
Modern apes, 670
Molars, 190, 191, 569
Molds, black bread, slime, and
  water, 603–4
Molecular basis, of inheritance,
  507–32
Molecular formula, 23
Molecules, 4, 19
  ATP, 106, 107, 108, 109, 110,
    113
  carbon monoxide, 242
  cellulose, 67
  cholesterol, 46
  and endorphins, 339
  and energy of activation, 100
  glucose, 114
  hemoglobin, 33
  histone, 82
  inorganic, 24–29
  and monomers, 29
  organic, 29–41
  and passage through cells, 68
  PGAL, 109
  phospholipid, 46
  polar, 24
  protein, 46
  simple organic, 553–55
  steroid, 39–40
  and urine formation, 306
Mollusks, 643–47, 655
Molting, 650
Monarch butterfly, 693
Monera, 583, 589–93
Mongoloid, 681
Moniliasis, 607
Monkeys, 670, 689
Monoclonal antibodies, 272–73
Monocot, 142–43, 622
  root, 147
  seed and germination, 168
Monoculture agriculture, 730
Monocytes, 239, 244, 249–51, 260,
  262
Monohybrids, 473
Monomers, 29
Mononucleosis, infectious. *See*
  Infecious mononucleosis
Monosaccharides, 33, 34

Monotremes, 663
Mons pubis, 419
Morphine, 343
Morphogenesis, 436, 441–45
Mortality, 738
Morula, 436, 437, 447
Mosquito, 577, 596
Moss, 627
    club, 615, 617
    life cycle, 612–14
    reindeer, 605
Mother, surrogate, 427
Mother cell, 84, 85
Mother Earth, 558
Mother-of-pearl, 645
Motor neuron, 320–22, 330, 331, 332
Mountain goat, 5
Mouse, cells, 502–3. See also Mice
Mouth, 190–92
Movement, of chromosomes, 86–87
mRNA. See Messenger RNA (mRNA)
MS. See Multiple sclerosis (MS)
MSH. See Melanocyte-stimulating hormone (MSH)
Mucous membrane layer, 193, 194
Multicellular organisms, 44
Multicellular sheet algae, 599–600
Multiple alleles, 484–85
Multiple Risk Factor Intervention Trial, 356
Multiple sclerosis (MS), 276
Multipotent stem cells, 244
Multitraits, and inheritance, 474–79
Muscle action potential, 363
Muscle cells, 358
Muscle fibers, 179, 331
    anatomy of, 358–60
    physiology of, 360–62
Muscles
    anatomy of, 358
    antagonistic, 353
    arrector pili, 182
    cardiac, 179, 180
    ciliary, 371, 372
    contraction, 360–61
    and exercise, 355–58
    and innervation, 362–63
    and muscle action potential, 363
    physiology, 354–55
    and senses, 369
    skeletal, 179, 352–64
    of skeleton, 347
    smooth, 179, 180
    and vestigial structures, 569
Muscle spindles, 353, 354, 369
Muscle twitch, 355
Muscular dystrophy, 498–99
Muscular tissue, 173, 174–75, 179–80
Musculoskeletal system, 184, 346–65
Mushroom, life cycle, 606
Mutations
    chromosome, 493–95, 527–28
    and evolution, 574
    gene, 527–28
    RFLP. See Restriction fragment

    length polymorphism (RFLP)
    types of, 528
    See also Polymorphism
Mutualism, 693, 696–98
Mycelium, 603, 604
Mycorrhizae, 606, 696–97
Myelin sheath, 322
Myeloblast, 244
Myeloid stem cells, 244
Myocardium, 221
Myofibrils, 358, 359–60, 364
Myogram, 354, 355
Myosin, 360, 361, 363, 364
Myxedema, 393, 394
Myxomycota, 603

N

NAD cycle, 104, 105, 107, 108, 111, 138–39
    and electron transport system, 113
    and FAD, 112, 114
    and fermentation, 114–16
    and respiratory chain, 112–14
    and vitamins, 209
NADP, 126–27, 135, 138–39
NADPH, 126–27, 128, 135
Nail-patella syndrome (NPS), 501
Naked seeds, 619
Nasal cavities, 281–82
Nasopharyngeal openings, 192, 193
Nasopharynx, 281
Natality, 738
National Academy of Sciences, 204, 210
National Heart, Lung, and Blood Institute, 207
National Institute for Fitness and Sport, 356
National Institutes of Health (NIH), 211
National Research Council, 204
Natural ecosystem, and human ecosystem, 729
Natural selection
    and evolution, 574–76
    examples, 577–79
    theory of, 578
NE. See Norepinephrine (NE)
Neanderthals, 678–79
Neanthes, 647
Nearsightedness, 377
Negative feedback, 186
Negative pressure, 286
Negroid, 681
Neisseria gonorrheae, 432
Nematocysts, 634
Nematoda, 640–41
Neohemocyte (NHC), 243
Nephridia, 647, 648
Nephrons, 305–7
    anatomy, 305, 307
    and blood and circulation, 307
    and reabsorption, 308
Nereocystis, 601
Nerve(s), 180
    auditory, 381
    and breathing, 286

    cranial, 328, 329
    optic, 371, 372
    spinal, 328, 329, 331
    and summation and integration, 327
    types of, 330
    vagus, 329
    See also Nervous system
Nerve cord
    dorsal hollow, 656
    solid ventral, 648, 650, 657
Nerve deafness, 381
Nerve fibers, 180
Nerve impulse, 322–25
Nerve net, 637
Nervous control, and heartbeat, 225
Nervous system, 184, 319–45
    autonomic. See Autonomic nervous system
    central. See Central nervous system (CNS)
    of clam, 646
    and nerve types, 330
    and neurons, 320–25
    and neurotransmitter substances, 327, 328, 339–40
    parasympathetic. See Parasympathetic nervous system
    peripheral. See Peripheral nervous system (PNS)
    somatic. See Somatic nervous system
    and summation and integration, 327
    sympathetic. See Sympathetic nervous system
    and transmission across a synapse, 325–28
    See also Nerve(s)
Nervous tissue, 173, 174–75, 180
Neural fold, 438, 451
Neural tube, 440, 451
Neurilemma, 322
Neurofibromatosis (NF), 482, 504
Neuromuscular junction, 363
Neurons, 180, 320–25
    afferent and efferent, 320–22
    structure, 320–22
    types of, 321
Neurosecretory cells, 388
Neurospora, 604
Neurotransmitter substances, 327, 328, 334, 339–40, 404–5
Neurula, 436, 437, 439–41
Neurulation, 437, 444
Neutral fat, synthesis and hydrolysis, 37
Neutron, 17
Neutrophils, 239, 244, 249–51
Newborn
    and AIDS, 429
    and birth defects, 454–55
    and genital herpes, 431
    HDN. See Hemolytic disease of the newborn (HDN)
Newt, 659
New World monkeys, 670
Nezhat, Camran, 428

NF. See Neurofibromatosis (NF)
NGU. See Nongonococcal urethritis (NGU)
NHC. See Neohemocyte (NHC)
Niacin, 209
Niche, 687, 720
Nictitating membrane, and vestigial structures, 569
Night length, 161–63
NIH. See National Institutes of Health (NIH)
Nitrification, 728
Nitrifying bacteria, 728
Nitrogen, and chemical evolution, 553
Nitrogen cycle, 727–28
Nitrogen fixation, 727–28
Nitrogen-fixing bacteria, 727
Nitrogen gas, formation of, 22
Nitrogenous end products, 300
Nitrogen oxides, 750
Nodes, 147
    AV. See Atrioventricular (AV) node
    lymph, 259, 260–61
    Ranvier, nodes of. See Nodes of Ranvier
    SA. See Sinoatrial (SA) node
Nodes of Ranvier, 325, 326
Noise, 383
Noncompetitive inhibition, 103
Noncyclic electron pathway, 128–31
Noncyclic photophosphorylation, 129, 130, 131
Nondisjunction, 491–93, 495, 496
Nongonococcal urethritis (NGU), 432
Nonmetals, 18, 20
Nonprotein, 103
Nonseed plants, 627
Nonvascular plants, 612–14
Noradrenalin, 392
Norepinephrine (NE), 328, 334, 339–40, 396, 405
Northside Hospital, 428
Northwestern University, 356
Nose, 281–82
Notochord, 440, 444, 451, 656
NPS. See Nail-patella syndrome (NPS)
Nuclear envelope, 48, 59
Nucleic acids, 40–41
    DNA. See Deoxyribonucleic acid (DNA)
    mRNA. See Messenger RNA (mRNA)
    RNA. See Ribonucleic acid (RNA)
    rRNA. See Ribosomal RNA (rRNA)
    tRNA. See Transfer RNA (tRNA)
    and viruses, 584
Nucleoid region, 59
Nucleolus, 46, 47, 48, 49, 59
Nucleosomes, 82
Nucleotides, 40
    and DNA probe, 504
    polymers of, 40

Nucleus, 17, 46, 47, 48–50, 166
Nudibranchs, 644
Numbers, pyramid of. *See* Pyramid
    of numbers
Nurse cells, 412, 413
Nutrients
    and blood plasma, 178, 240
    inorganic, 145
    macro- and micro-, 145
    transport in plants, 157–59
    See also Nutrition
Nutrition, 202–11
    and carbohydrates, 204–6
    and cholesterol, 207–9
    and eating disorders, 215–16
    and energy content of foods, 214
    and fats, 206–7, 208
    and food groups, 203
    and lipids, 206–9
    and proteins, 204
    See also Nutrients

O

Obelia, 635
Obesity, 216
Obligate anaerobes, 591
Obligate parasites, 585
Obstructive jaundice, 199
Occipital lobe, of brain, 336–38
Oceans, 713–15
    and pollution, 750
    and oceanic floor, 713
    and oceanic province, 713, 714,
        715
Octet rule, 19
Octopus, 644
Oil
    and fats, 37
    glands, 182
Olfactory cells, 367, 370
Oligocene, 671
Olympics, 401
Omnivores, 123, 720
Oncogenes, 530–31
One-trait crosses, 472–74, 475
Ontogeny, 441
Oocyte, 417
Oogenesis, 92–93
Oomycota, 603
Ooplasmic segregation, 443
Oparin, A. I., 556
Open circulatory system, 645
Operator, 525
Operon
    inducible model, 525, 526
    lac, 525, 526
    and operon model, 524–26
    repressible model, 525
    tryp, 525–26
Opium, 343
Opsin, 375
Optic chiasma, 375
Optic nerve, 371, 372
Orangutan, 673
Orbits, 17
Order, 6, 567
Ordovician, 564

Organelles, 46
    energy-related, 53–55
    eukaryotic cell, 47–59
Organic chemistry, 24
Organic compounds, 24, 41
Organic farming, 731
Organic molecules, 29–41, 553–55
Organic polymer, synthesis and
    hydrolysis, 29
Organic soup, 555
Organisms, 4
    anaerobic, 559
    classification of, 6
    multicellular, 44
    transgenic, 541–45
Organization
    animal, 4
    human, 172–88
    plant, 4, 141–53
Organ of Corti, 379, 380, 381, 382,
    383
Organs, 4, 180–85
    and aging, 463
    Corti. *See* Organ of Corti
    of excretion, 299–303
    female reproductive system, 416
    ladder-type nervous, 638
    lymphoid, 259–60
    and male reproductive system,
        411
    and organ systems, 180–85
    sense, 182
    sex, 386, 392, 393, 400
    target, 387, 392
    and urinary system, 303
    vegetative, 143
Organ systems, 180–85
Orgasm
    in females, 419
    in males, 414–15
*Origin of Species by Means of
    Natural Selection,* 578
Origins
    of muscle, 352, 353
    of life, 552–61
Oscillatoria, 592
Oscilloscope, 322, 323
Osculum, 633
Osmosis, 69–72, 73
    and animal cells, 71
    ATP. *See* Chemiosmotic ATP
        synthesis
    and osmotic pressure, 70, 246
    and plant cells, 71
Osmotic pressure, 70, 246
Osteoclasts, 350
Osteocytes, 350
Osteoporosis, 211, 350, 462
Ostrich, 661
Otoliths, 379, 380
Outer ear, 378, 379
Oval opening, 456
Oval window, 378
Ovarian cycle, 420
Ovaries, 164, 386, 392, 393, 400,
    412, 416, 417–18, 420
Overdosing, 343
Overkill, 690
Oviduct, 418, 447

Ovipositor, 653
Ovulation, 418, 420, 447
Ovule, 165, 618
Owls, 661
Oxidation, 105, 111
    and hydrogen sulfide, 137
    and oxidation-reduction
        reactions, 23–24
    and oxidative decarboxylation,
        112
    and oxidizing atmosphere, 553
    See also Oxygen
Oxidation-reduction reactions, 23–24
Oxidative decarboxylation, 112
Oxidizing atmosphere, 553
Oxygen, 107
    and blood plasma, 240, 241
    and capillary exchange, 246, 247
    and chemical evolution, 553
    and gas exchange in lungs, 69
    and hemoglobin, 242
    and oxygen debt, 116, 361–62
    and oxyhemoglobin, 242, 291
    and respiration, 279, 290, 291
    and stroke and heart attack, 233
    See also Oxidation
Oxygen debt, 116, 361–62
Oxyhemoglobin, 242, 291
Oxytocin, 389, 390, 392, 405
Ozone, and ozone shield, 559, 750,
    751, 752

P

Pacemaker, 225
Paffenbarger, Ralph, 356
Pain, 252, 367
Palates, 192
Paleocene, 671
Paleozoic, 564
Palisade mesophyll, 151
PAN, 750
Pancreas, 197, 392, 398–400
    anatomy of, 398
    and pancreatic amylase, 201, 202
    and pancreatic juice, 197
Pancreatic amylase, 201, 202
Pancreatic juice, 197
Pantothenic acid, 209
Pap test, 419
Paramecium, 594, 691
Paramecium caudatum, 595
Parapodia, 647
Parasite, 693–95
    flatworms, 639–40
    obligate, 585
    worms, 639
Parasitic flatworms, 639–40
Parasitic worms, 639
Parasitism, 693–95
Parasympathetic nervous system,
    225, 328, 333–34
Parathyroid glands, 392, 394–95
Parathyroid hormone (PTH), 392,
    394–95
Parenchyma, 144
Parietal lobe, of brain, 336–38
Parkinson disease, 339–40, 465
Parotid gland, 192

Particle, subatomic, 17
Particle gun, 542
Parturition, 458–60
Passive immunity, 270, 271
Passive reabsorption, 308
Pathways
    and air, 281
    electron, 128–31
    metabolic. *See* Metabolic
        pathways
    phenylalanine, 515
    reflexes, 332
    urine, 303–4
    vascular, 226–28
Patterns, behavior, 686–99
Pavlov, Ivan, 198
PCB, 746
Pea flower, anatomy, 470
Peanut butter, nutrient and energy
    content, 214
Peas, nutrient and energy content,
    214
Pecticides, 730
Pectin, 601
Pectoral girdle, 349
Pedicellarie, 654
Pelagic division, 713, 714, 715
Pelecypods, 644
Pelicans, 661
Pellagra, 208
Pelvic girdle, 349, 350
Pelvic inflammatory disease (PID),
    432
Pelvis, kidneys, 304
Penguin, 661, 662
Penicillin, 310, 433, 589
Penicillium, 604, 589
Penicillium chrysogenum, 589
Penis, 303, 411
Pentose sugars, 33
People's Republic of China, 546
Pepsin, 200, 201, 202
Peptic ulcer, 194
Peptidases, 201, 202
Peptide bond, 31, 101
Peptides, 200, 201, 202
    formation of, 31
    and hormones, 387, 388
Perfluorocarbon (PFC), 243
Pericardium, 222
Pericycle, 147
Period, and geological time scale,
    564
Periodic table of the elements, 18
Periodism, photo-. *See*
    Photoperiodism
Periodontitis, 191
Peripatus, 566
Peripheral nervous system (PNS),
    320, 328–34
Perissodactyla, 665
Peristalsis, 193
Peritubular capillary network, 307
Permafrost, 704
Permeability, 68
Permian, 564
Peroxisomes, 51
Petals, 164
Petiole, 150

Peyer's patches, 261
PFC. *See* Perfluorocarbon (PFC)
PGA. *See* Phosphoglycerate (PGA)
PGAL. *See* Phosphoglyceraldehyde (PGAL)
pH, 26, 32
    and acid deposition, 26–27
    and blood, 185, 245
    and digestive enzymes, 200, 201, 203
    and enzymatic reaction, 102
    and kidneys, 312–14
    and protocells, 556
    scale, 28–29
    and seminal fluid, 414
Phaeophyta, 600
Phagocytes, 68, 75, 253, 262
Phagocytosis, 68, 75, 253
Phalanges, 349
Pharangeal pouches, 656
Pharynx, 192–93, 281, 282
Phenotypes, 471–72, 479, 575
Phenylalanine, metabolic pathway of, 515
Phenylketonuria (PKU), 482, 513–14, 515
Pheromones, 404–5
Phlebitis, 236
Phloem, 144, 151, 615
    function, 155
    pressure-flow theory of transport, 158–59, 160
Phosphocreatine, 361–62
Phosphoglyceraldehyde (PGAL), 109, 133, 134, 135, 136
Phosphoglycerate (PGA), 132–33
Phospholipids, 65–67
    bilayer, 46, 65
    structure, 38–39
Phosphorus, 210
Phosphorylation, photo-. *See* Photophosphorylation
Photobiology, 183
Photochemical smog, 750–52
Photoperiodism, 161
Photophosphorylation, cyclic and noncyclic, 129, 130, 131
Photoreceptors, 371–78
Photosynthesis, 54–55, 60, 122–40, 602
    bacterial, 137, 591
    C, 131–36
    and carbohydrates, 136–37
    and carbon cycle, 726
    and cellular respiration compared, 138–39
    and chloroplasts, 125–27, 132
    and crop yields, 136
    equations for, 125–27
    overview, 127
    participants in, 139
    and PGAL, 133, 134, 135, 136
    and respiration, 725
    and subpathways, 126–27
Photosystems, 127, 128–31, 591
Phylogeny, 441
Phylum, 6, 567
Physical dependence, 340
Physiograph, 354, 355

Physiology
    ear, 380–83
    eye, 372–77
    human, 170–407
    muscle, 354–55
    muscle fibers, 360–62
    plants, 154–69
    red blood cells, 242
Phytochrome, 162–63
PID. *See* Pelvic inflammatory disease (PID)
Pig, embryo, 569
Pigments, 181, 301
Pill, 425, 426
Pineal gland, 392, 400–403
Pine cones, male and female, 620–21
Pines, 620–21, 627
Pinna, 378
Pinocytosis, 68, 77
Pistil, 164
Pith, 147
Pituitary gland, 386, 388–93, 394, 415, 421, 422
    and ADH, 388–90, 392
    anterior, 386, 388, 390, 391–93, 394, 415
    and blood volume, 313
    posterior, 386, 388–90, 392
PKU. *See* Phenylketonuria (PKU)
Placenta, 423, 452–53
Placental mammals, 662, 663–65
Planarians, 637–39
Plankton, 710
Plant biology, 120–69
Plant cells, 46, 47, 48, 60, 61
    and meiosis, 93
    and mitosis, 88
    and osmosis, 71
    types, 143–45
Plant kingdom, 609–28
Plants
    cells. *See* Plant cells
    characteristics of, 610–12
    classification, 567
    and cohesion-tension theory of water transport, 156, 157
    comparisons among, 626–27
    cotton, 543
    and environmental stimuli, 159
    flowering, 142–45, 622–25
    as food, 145
    and geological time scale, 564
    growth, 141–53, 160
    hormones in, 159, 160
    and humans, 626
    and life cycle of flowering plant, 165
    and niches, 720
    nonseed, 627
    nonvascular, 612–14
    nutrient transport, 157–59
    organization, 4, 141–53
    and ozone, 751
    physiology, 154–69
    reproduction, 154–69
    root system, 142, 143, 145–47
    seed, 618–26, 627
    shoot system, 142, 143, 147–50
    transgenic, 541–43

transport, 155–59
    vascular, 614–18
    vegetative organs and tissues, 143
Plaque, arterial, 233, 235–36
Plasma, blood. *See* Blood plasma
Plasma cells, 264–65
Plasmids, Ti, 542
Plasmodium, 577, 594, 596, 603
Plasmodium vivax, 596
Plasmolysis, 72
Plate, cell, 88
Platelets, blood. *See* Blood platelets (thrombocytes)
Platypus, 663
Pleiotropic genes, 576
Pleistocene, 671
Pleural membranes, 285
Pliocene, 671
Plumule, 167
Pneumocystis carinii, 293
Pneumonectomy, 296
Pneumonia, 293
PNS. *See* Peripheral nervous system (PNS)
Polar body, 92–93
Polarity, and water, 24
Polarization, 325
Polar molecule, 24
Pollination, 166, 618, 621, 624
    and pollen cones, 619
    and pollen grains, 166, 618
Pollution, 26, 729
    air, 750–54
    city, 731–32
    country, 730–31
    and genetically engineered bacteria, 540–41
    human population, 743–54
    ocean, 750
    water, 747–50
Polyergus rufescens, 694
Polygenic genetic disorders, 484
Polygenic inheritance, 483–84
Polygeny, 576
Polymerase chain reactions, 546–47
Polymerization, and origin of cells, 553
Polymers, 553
    nucleotides, 40
    organic, 29
Polymorphic adaptation, 577
Polymorphism, balanced, 577. *See also* Restriction fragment length polymorphism (RFLP)
Polypeptides, 31, 33, 387, 388
Polyps, 196, 635
Polysaccharides, 34–35
Polysomes, 51, 522
Pons, 336
Pool, metabolic. *See* Metabolic pool
Pope, Harrison, 401
Population(s), 7, 571, 580, 687, 691, 720
    and carrying capacity, 739–40
    and doubling time, 738–39
    and genetics, 571–73
    growth, 736–43
    human, 8, 735–57
    and pollution, 743–54

and steady state, 755
Pores, 633
Porifera, 633
Pork, 642
Porphyria, 482
Portuguese man-of-war, 635
Posterior pituitary. *See* Pituitary
Postganglionic axon, 332
Postmating isolating mechanisms, 580
Postsynaptic membrane, 327
Posttranscriptional control, 524
Posttranslational control, 524
Posture, bipedal, 673
Potassium, 210
    gates, 325
    ions, 103, 324
    and sodium-potassium pump, 74
Potato, 626
Potentials
    action, 325
    biotic, 739
    membrane, 322, 324
    muscle action, 363
    resting, 322, 323, 324
Power, resolving, 44
Prairie, 705
Precambrian, 564
Predation, and adaptation, 570, 688–93
Predators, 570
Preganglionic axon, 332
Pregnancy, 423
    ectopic, 418, 432
    hormones, 453
    tests, 423
Premating isolating mechanisms, 580
Premolars, 190, 191
Prepuce, 414
Presenting, data, 12
Pressure filtration, 307
Pressure-flow theory, of phloem transport, 158–59, 160
Pressures, 367
    and filtration, 307
    -flow theory, 158–59, 160
    negative, 286
    osmotic, 70, 246
    turgor, 72
Presumptive notochord, 444
Presynaptic membrane, 327
Prey, and defense adaptations, 688–93
Primary growth, of stems, 147–48
Primary host, 639
Primary succession, 719
Primates, classification and evolution, 665, 670, 671
Primitive, and advanced, 632
Primitive atmosphere, 553
Primitive invertebrates, 630, 633–42
Primitive streak, 451
Primitive tracheophytes, 627
Primitive vascular plants, 615
Prince William Sound, 750
Principle, and theory, 8, 12
Prion, 588
Prism, 124

Probability, laws of, 473, 477–78
Probe, DNA. *See* Deoxyribonucleic acid (DNA)
Processing, and aging, 464
Producers, 720
Product(s), 98–99, 101
    biotechnology, 537–40
    nitrogenous end, 300
Profundal zone, 711
Progesterone, 392, 420, 421, 424, 453
Proglotids, 639
Prokaryotes, and gene expression control, 524–26
Prokaryotic cells, 44–45, 59–60, 61
Prolactin. *See* Lactogenic hormone (LTH)
Promoter, 525
Promotion, and cancer, 529–31
Propagation, tissue culture, 164
Prophage, 586
Prophase, 84–85, 87, 90, 91, 92
Proprioceptors, 367, 369
Prosimians, 670
Prostaglandins, 405
Prostate gland, 411, 414
Proteinoids, 555, 556
Proteins, 29–33, 46, 65, 66
    and blood plasma, 178, 240–41
    carrier, 72–75
    and coenzymes, 103
    and complement system, 263
    complete and incomplete, 204
    and denaturation, 102
    and fat compared, 205
    functions, 66
    and hormones, 387, 388
    MHC. *See* Major histocompatibility complex (MHC) protein
    and myofibrils, 358
    and nutrition, 204
    and pressure filtration, 308
    structure levels, 31–33
    synthesis and DNA and RNA, 516–24
    and viruses, 584
Proterzoic, 564
Prothallus, 617
Prothrombin, 248–49
Protista, 583, 593–603
Protocells, 556–57
Protochordates, 656
Protonema, 612
Protons, 17
Protoplasts, 161, 541–42
Protostomes, 642
Protozoans, 593–96, 631
Proximal convoluted tubule, 305, 308
Pseudocoelom, 641
Pseudocoelomate animals, 633
Pseudomyrmex ferruginea, 697
Pseudostratified ciliated columnar epithelium, 173, 176
Pseudostratified epithelium, 173–76
Psilophyta, 615
Psilotum, 615, 616
Psychosis, 342

Pteridium aquilinum, 617
PTH. *See* Parathyroid hormone (PTH)
Puberty, 461
Pulmonary arteries, 227
Pulmonary circuit, 223, 226–27, 228
Pulmonary edema, 259
Pulmonary embolism, 236
Pulmonary fibrosis, 294
Pulmonary veins, 227
Pulp, of tooth, 192
Pulse, 228, 229, 230
Pumps, 74, 223
Punctuated equilibrium model, of evolution, 565
Punnett square, 473, 478, 490
Pupal stage, 653
Pupil, 371, 372
Purina Mills, 546
Purines, 510, 511
Purkinje fibers, 225, 226
Pus, 253
P wave, 225, 226
Pyelonephritis, 315
PYR. *See* Pyruvate (PYR)
Pyramids, 304
    biomass, 723–24
    ecological, 723–24
    energy, 724
    numbers, 723–24
Pyrimidines, 510, 511
Pyrrophyta, 602
Pyruvate (PYR), 107, 109, 110, 111, 114

Q

QRS wave, 225, 226
Quality, of land, 731
Quaternary, 564, 671

R

Races, human, 681–82
Radial symmetry, 631, 654
Radiant energy, 123–24
Radiation, adaptive, 581. *See also* Ultraviolet (UV) radiation
Radicle, 167
Radioactive iodine, 19
Radioactive isotopes, 19, 273, 564
Radiolaria, 595
Radioreceptors, 367
Radio waves, 124
Radius, 349
Radula, 644
Rain, 26
Rain forests, 689, 707, 708, 709, 710, 744, 745
Ranvier, nodes of. *See* Nodes of Ranvier
Rapid eye movement (REM) sleep, 338
RAS. *See* Reticular activating system (RAS)
Raw materials, 729
Ray-finned fish, 657
RDA. *See* Required dietary (daily) allowance (RDA)

Reabsorption
    active and passive, 308, 309
    and nephrons, 308
    selective, 307, 308–10
    of water, 310–12
Reactant, 98–99
Reactions
    among atoms, 19–24
    antigen-antibody, 251–52
    covalent, 21–23, 32
    dark, 127
    and enzymes, 101–3
    inflammatory, 252–53
    ionic, 19–21, 32
    light, 127
    oxidation-reduction, 23–24
    synthetic, 109–10
    transition, 105, 106, 107, 111
Reasoning, deductive and inductive, 9
Recapitulation, 441
Receptors, 67, 320–22, 331, 332
    general, 367–69
    and skin, 367–68
    and stimulus, 367–69
    stretch, 369
Recessive disorders, autosomal, 480–82
Recessive traits, 471
Recessive X-linked disorders, 497–500
Recombination, and evolution, 574
Rectum, 196, 197
Red blood cells (erythrocytes), 45, 178, 240, 241–45
    and blood clotting, 248–49
    composition of, 239
    life cycle of, 245
    and lymph system, 260
    pH of, 245
    physiology, 242
    and red bone marrow, 244
    and vitamins, 209
    See also Blood
Red bone marrow, 244, 248–49, 350. *See also* Bone(s) *and* Bone marrow *and* Yellow bone marrow
Red-eyed tree frogs, 3
Red tide, 602
Reduced hemoglobin, 242
Reducing atmosphere, 553
Reduction
    of carbon dioxide, 131–37
    and oxidation-reduction reactions, 23–24
Reefs, 712, 713
Reflex, 197, 331–32
Reflex action, 192–93
Reflex arc, 331–32
Regeneration, 132, 133, 134, 135, 634
Regions, constant and variable, 264
Regulator gene, 524
Regulators, growth in plants, 160
Regulatory substances, and blood plasma, 178
Reindeer Age, 680
Reindeer moss, 605

Rejection, tissue, 275–76
Relaxation period, 355
REM sleep. *See* Rapid eye movement (REM) sleep
Renal pelvis, 304
Renal pyramids, 304
Renin, 396
Renin-angiotensin-aldosterone system, 313–14, 396
Replacement reproduction, 742–43
Replication, DNA, 83, 512–13
Reporting, of data, 10
Repressible operon, 525
Reproduction
    asexual, 164
    of bacteria, 590–91
    control of, 425–29
    human, 408–549
    plants, 154–69
    replacement, 742–43
    sexual, 164–67
Reproductive isolation, 580
Reproductive system, 184, 410–34
    and aging, 465
    control of, 425–29
    female. *See* Female reproductive system
    male. *See* Male reproductive system
Reptiles, 660–61, 666
Required dietary (daily) allowance (RDA), 204, 209, 210, 211
Reservoirs, carbon, 725–26
Residual volume, 288
Resistance
    environmental, 739
    and natural selection, 577
Resolving power, 44
Respiration, 278–97
    aerobic. *See* Aerobic respiration
    cellular. *See* Cellular respiration
    components of, 280
    external. *See* External respiration
    and health, 292–96
    and infections, 292–93
    internal. *See* Internal respiration
    and origin of cells, 553
    and photosynthesis, 725
Respiratory center, 285
Respiratory chain, 105, 106, 107, 112–14
Respiratory system, 184
Respiratory tract, human, 281
Responses
    all-or-none, 354–55
    plants to environmental stimuli, 159–63
Responsibility, social. *See* Social responsibility
Resting potential, 322, 323, 324
Restriction enzymes, 503, 536
Restriction fragment length polymorphism (RFLP), 503–4. *See also* Polymorphism
Restriction site, 536
Results, of experiments, 11–12
Reticular activating system (RAS), 336, 338, 340

Retina, 371–72, 374
Retinal, 375
Retroviruses
    and cancer, 588
    as vectors, 545
Reverse anorexia, 402
Reverse transcriptase, 588
RFLP. See Restriction fragment
    length polymorphism (RFLP)
R group, and amino acids, 30, 32
Rh blood factor system, 254–55,
    454, 485
Rhizobium, 540
Rhizoids, 612
Rhizopus, 603, 604
Rhodopsin, 375
RhoGam, 454
Rhynia major, 614
Rhyniophytes, 614
Rhythm method, 425
Rhythms, circadian, 403
Rib cage, 286
Riboflavin, 209
Ribonuclease, 100
Ribonucleic acid (RNA), 40–41, 49
    and AIDS, 274
    and catalytic RNA, 555–56
    and chromosomes, 82
    and gene expression, 524–27
    and gene therapy, 545–48
    and macromolecules, 555–56
    messenger. See Messenger RNA
        (mRNA)
    processing, 518–19
    and protein synthesis, 516–17
    and retroviruses as vectors, 545
    ribosomal. See Ribosomal RNA
        (rRNA)
    and RNA polymerase, 518, 526,
        527
    and RNA replicase, 588
    and RNA transcriptase, 588
    structure, 511, 512
    and substrates, 100
    and transcription, 516–22,
        526–27
    transfer. See Transfer RNA
        (tRNA)
    and translation, 516–22, 526–27
    and true cells, 557–59
    and viruses, 584, 588
    and vitamins, 209
Ribose, 34
Ribosomal RNA (rRNA), 49, 51,
    511, 520, 522, 523, 524, 592
Ribosomes, 46, 47, 48, 51, 59
Ribs, 285, 348
Ribulose bisphosphate (RuBP), 132,
    133, 134, 135
Rice, 626
Rickets, 208
Ridges, midoceanic, 137
Ring canal, 655
Rio Grande, 730
Risk factors, 356
RNA. See Ribonucleic acid (RNA)
RNA polymerase. See Ribonucleic
    acid (RNA)

RNA replicase. See Ribonucleic acid
    (RNA)
RNA transcriptase. See Ribonucleic
    acid (RNA)
Rockweed, 601
Rocky shore, 712
Rodentia, 665
Rods, 371, 372, 376, 377
Romalea, 652
Romoras, 694
Roots
    anatomy, 155
    dicots, 145–47
    monocots, 147
    and root cap, 145
    and root system, 142, 143,
        145–47
    of tooth, 192
    woody, 151
Rough endoplasmic reticulum. See
    Endoplasmic reticulum (ER)
Round window, 378
Roundworms, 640–41, 667
Royal Society of London, 578
rRNA. See Ribosomal RNA
    (rRNA)
Rubber tappers, 745
RuBP. See Ribulose bisphosphate
    (RuBP)
Runner's high, 339
Rush, 343
Rusts, 606

S

Saccule, 379, 380
Sac fungi, 604, 605
Sac plan, 631, 632
Sacromeres, contraction structure
    and function, 363
Salamanders, 659
Saliva, 201, 202
Salivary amylase, 200, 201, 202
Salivary glands, 192
Salk, Jonas, 430
Salt(s), 28–29, 369
    and blood plasma, 178, 240
    and diet, 211
    and excretion, 301
    and nephrons, 308
Salt marshes, 712
Saltwater communities, 712–15
Sample, control, 10
Sampling, chance, 575. See also
    Chorionic villi sampling
Sandpipers, 661
Sandy beach, 713
Sanitation, 740
SA node. See Sinoatrial (SA) node
Saprophyte, 593
Sarcodina, 594
Sarcolemma, 358, 363
Sarcoma, and AIDS, 274
Sarcomeres, 360
Sarcoplasm, 358
Sarcoplasmic reticulum, 358, 359,
    364
Sargassum, 601
Saturated fatty acids, 37

Saturation, hemoglobin. See
    Hemoglobin
Savanna, 705
Scale, pH. See pH
Scales, 659, 660
Scallop, 644
Scanning electron micrograph
    (SEM), 44, 45
Scapula, 349
Schistosomiasis, 640
Schizocoelomates, 642, 643
Schizophrenia, 481
Schwann cells, 321–22
SCID. See Severe combined immune
    deficiency (SCID)
Science
    process of, 8–12
    and scientific method, 8–10
    and social responsibility, 12–13
Sclera, 371, 372
Sclerenchyma, 144
Scolex, 640
Screening, and eggs, 454, 455
Scrotum, 411, 412
Scurvy, 208
Sea anemone, 635
Seabirds, 301
Sea cucumber, 654
Sea lettuce, 600
Seals, 664
Sea urchins, 654
Seaweed, 600–601, 714
Sebaceous glands, 182
Secondary growth, 148
Secondary host, 639
Secondary succession, 719
Secretin, 198
Secretion, 52, 77
Seed(s), 165, 167–68, 624
    formation, 621
    naked, 619
    and seed coat, 167
    vascular plants without, 615–17
Seed coat, 167
Seedless grapes, 161
Seedling, 4
Seed plants, 618–26, 627
Segmentation, in earthworm, 648
Segregation
    Mendel's laws of, 476
    ooplasmic, 443
Selection
    directional and stabilizing, 576
    natural, 574–76, 577–79
Selective reabsorption, 307, 308–10
SEM. See Scanning electron
    micrograph (SEM)
Semen, 415
Semicircular canals, 379, 380, 381
Semiconservative replication, 513
Semilunar valves, 223
Seminal fluid, 414
Seminal vesicle, 411, 414
Seminiferous tubules, 412, 413, 415
Sense organs, 182
Senses, 366–84
    chemical, 369–71
    and muscles and joints, 369
    and receptors, 367–69
    special, 369–83

Sensory neurons, 320–22, 330, 331
Sensory system, 184
Seoul Olympics, 401
Sepals, 164
Septum, 222
·Sequencing, gene, 536, 548
Serotonin, 339
Serous membrane layer, 193
Sertoli cells, 412, 413
Serum, 248, 270
Sessile filter-feeding animals, 632,
    633
Setae, 648
Severe combined immune deficiency
    (SCID), 545, 546
Sex
    and height, 460
    inheritance of, 491
    ratios in U.S., 491
    and steady state, 755
Sex chromosomes, 80, 495–96
Sex hormones, 392, 393, 397
Sex-influenced traits, 500–501
Sex-linked inheritance, 496–501
Sex organs, 386, 392, 393, 400
Sexually transmitted diseases
    (STD), 429–33
Sexual reproduction, in plants,
    164–67
Sharks, 657, 694
Shells, of atom, 17, 20, 21, 22, 23
Sherbet, 601
Shettles, 409
Shock, 393, 399
Shoot system, 142, 143, 147–50
Shore, 712
Shorebirds, 713
Shrew, 670
Shrimp, 648
Sickle-cell anemia, 485–86, 504
Sickle-cell trait, 485–86
Sieve plate, 655
Sieve-tube cells, 157
Sight, 367, 377
Sightedness, 367, 377
Signals, environmental. See
    Environmental signals
Silurian, 564
Simple diffusion. See Diffusion
Simple epithelium, 173–76
Simple eye, 650
Simple goiter, 393, 394
Sinoatrial (SA) node, 225, 226
Sinuses, 348
Sinus tachycardia, 226
Siphilis, 430
Siphons, incurrent and excurrent,
    646
Skateboarding, 336
Skeletal muscle, 179, 352–64
Skeletal muscle fiber, 364
Skeleton, 347–58
    appendicular, 347, 348, 349
    axial, 347, 348
    bones of, 347, 348
    cyto-. See Cytoskeleton
    endo-. See Endoskeleton
    external, 657
    functions, 347

hydrostatic, 633, 641, 648
internal, 657
and long bones, 349–50
musculo-. *See* Musculoskeletal
system
and skeletal muscles, 352–64
structure of, 348–52
Skin, 180–84, 367–68
and aging, 464
anatomy, 181
and burns, 183–84
and cancer, 182
disorders, 182
and excretion, 302
and inheritance, 483–84
layers, 181–82
and tanning, 183
Skull, 348
Slash-and-burn agriculture, 744, 745
SLE. *See* Systemic lupus
erythematosus (SLE)
Sleep, REM. *See* Rapid eye
movement (REM) sleep
Sleeping sickness, 596
Sliding filament theory, of muscle
contraction, 360–61
Slime molds, 603
Slit lamp, 377
Small intestine. *See* Intestines
Smell, 367
Smith, Everett L., 356
Smog, photochemical, 750–52
Smoking
and cancers, 295
involuntary, 296
marijuana. *See* Marijuana
Smooth endoplasmic reticulum. *See*
Endoplasmic reticulum (ER)
Smooth muscle, 179, 180, 193
Smooth muscle layer, 193
Smuts, 606
Snail, 339, 644
Snakes, 660
Soaps, 38
Social parasitism, 694
Social responsibility, 12–13
Society, steady-state, 756
Sodium, 210
and diet, 211
ion, 20, 324
and nephrons, 308
and sodium gates, 325
Sodium hydroxide, 25
Sodium-potassium pump, 74, 324,
325
Soft palate, 192
Soil
erosion, 743–44
mining, 731
Solar energy, 124
Solute, 70
Solutions
acidic and basic, 27
hypertonic and hypotonic, 72
isotonic, 71
Solvent, 70
Somatic mutations, 528
Somatic nervous system, 328,
331–32

Somatotropin. *See* Growth hormone
(GH)
Somites, 439
Songbirds, 661
Sori, 617
Sounds, heart. *See* Heart
Soup, organic, 555
Sour, 369
Sources, energy. *See* Energy sources
Soybean, 626
Sparrow, 5
Special senses, 369–83
Speciation, 580–81
Species, 5, 6, 567
human, 681–82
isolating mechanisms, 580
Specificity, 103
Spectra, absorption and
electromagnetic, 124
Sperm, 165, 411, 412, 413, 448
Spermatic cords, 414
Spermatids, 413
Spermatocytes, 412, 413
Spermatogenesis, 92–93
Spermatogonia, 412
Spermatozoa, 412
Spermatozoan, 412
Sphagnum, 613
Sphenophyta, 615
Sphincters, 194
Sphygmomanometer, 230
Spicules, 633
Spiders, 648
Spinal cord, 334–35
anatomy, 330
craniosacral portion, 333–34
and nervous system, 328, 329,
331
thoracic-lumbar portion, 332–33
Spinal nerves, and nervous system,
328, 329, 331
Spinal tap, 334
Spindle fibers, 84–85, 86
Spindles, muscle, 353, 354, 369
Spiny anteater, 663
Spiracles, 652
Spirochete, 694
Spirogyra, 599
Spirometer, 287
Spleen, 245, 259, 261
Sponges, 631, 633–34, 667
Spongy bone, 178, 350
Spongy mesophyll, 151
Sporangium, 603, 617
Sporazoan, 594
Spores, 165, 166, 597, 611, 612
Sporophyte, 164, 165, 600, 611–12,
617
Sporozoa, 596
Spot desmosome, 176, 177
Spotted sweetlip, 698
Spurges, 570
Squamous-cell carcinoma, 182–83
Squamous epithelium, 173–76
Squid, 323, 645
S-shaped growth curve, 739
Stabilizing selection, 576
Stain, Gram, 592, 593
Stamens, 165

Stanford University, 356
Stapes, 378
Staphylococcus aureus, 589, 593
Starch, 34, 35, 200, 202
Starfish, 654–56
Starvation, and plants for food, 626
Static equilibrium, 380
STD. *See* Sexually transmitted
diseases (STD)
Steady state, 755–56
Steady-state society, 756
Steere, Allen C., 9–10
Stem(s)
alfalfa, 148
dicot herbaceous anatomy, 148
herbaceous, 147–48
primary growth, 147–48
secondary growth, 148
woody, 148, 151
Stem cells, modifying, 545–46. *See
also* Embryo-derived stem
(ES) cells
Stenosis, mitral, 226
Stereoscopic vision, 374–75
Sterility, 425, 426, 432
Sternum, 285, 348
Steroids, 29
anabolic, 400, 401–2, 416
diversity, 39
molecules, 39–40
and steroid hormones, 387–88
Stigma, 164
Stimulation, 340
Stimuli
maximum, 355
plants responses to, 159–63
and receptors, 367–69
threshold, 354–55
Stirrup, 378
Stomach, 190, 191, 194–95
Stomata, 151, 156, 157
Stone canal, 655
Strep throat, 293, 592
Streptococcus pyogenes, 293
Streptokinase, 234
Streptomyces, 589
Stress, and hypertension, 232, 233
Stretch receptors, 369
Striated cells, 179, 180
Striding gait, 678
Stroke, 233–34
Stroma, 54, 125
Stroma lamellae, 125
Stromatolites, 557
Structural formula, 23
Structural genes, 525
Structure(s)
and age comparison, 742–43
air, 281
analogous, 568
antibody, 266
ATP, 41
autonomic nervous system, 333
bacteria, 589–90
cells, 43–63
cellulose, 35
Chlamydomonas, 598
chloroplasts, 55

chromatin, 82
chromosome, 82
DNA, 508, 509–11
estuary, 712
glycogen, 34, 35
hemoglobin, 244
homologous, 566, 568
lichens, 605
membrane, 46–47
mitochondria, 55, 108
muscle spindle, 354
neurons, 320–22
phospholipids, 38–39
polysome, 522
protein, 31–33
RNA, 511, 512
sacromere contraction, 363
skeleton, 348–52
sporophyte, 164
starch, 34, 35
steady state, 755
vestigial, 568–69
Style, 164
Subatomic particles, 17
Subcutaneous skin layer, 181, 182
Subjucosal layer, 193
Sublittoral zone, 713, 714, 715
Submandibular glands, 192
Subpathways
and aerobic cellular respiration,
107
and photosynthesis, 126–27
Subpopulations, 580
Substance P, 343–44
Substitutions, of bases, 527–28
Substrate concentration, 101–2
Substrates, 99, 100
and enzyme-substrate complex,
101
and substrate concentration,
101–2
Succession, primary and secondary,
719
Sucrose, 34
Sudoriferous glands, 182
Sugar(s)
dietary, 205
pentose and hexose, 33
and plants, 159, 160
Sugar beet, 626
Sugarcane, 161, 626
Sulfa, 589
Sulfur, 210
Summation
and integration, 327
and tetanus, 355
Sunlight, and energy, 124, 127–31
Superfamily, 670
Superiority, heterozygote, 576
Superior vena cava. *See* Venae cavae
Suppressor T cells, 266–69
Surface water, pollution, 747–48
Surgery, coronary bypass, 236
Surrogate mothers, 427
Survival of the fittest, 578
Sustainable growth, 754–55
Swallowing, 192–93
Swamp, carboniferous, 617

Sweat glands, 182
Sweet, 369
Sweetlip, 698
Sweet potato, 626
Swim bladder, 657
Swimmerets, 650
Symbiosis, 605, 693–98
Symmetry
    radial, 654
    types of, 631–32
Sympathetic nervous system, 225,
    328, 332–33, 334
Synapses, 89
    and drugs, 340
    transmission across, 325–28
Synaptic cleft, 327, 329
Synaptic ending, 327
Synovial fluid, 351
Synovial joints, 351
Synthesis
    ATP. See Chemiosmotic ATP
        synthesis
    chemo-. See Chemosynthesis
    DNA and RNA, 82
    of maltose, 34
    of neutral fat, 37
    of organic polymer, 29
    proteins and DNA and RNA,
        516–24
Synthetic estrogen (DES), 426, 454
Synthetic reactions, 109–10
Synthetic theory, of evolution,
    573–77
Syphilis, 433
Systemic circuit, 223, 226–27, 228
Systemic lupus erythematosus
    (SLE), 276
Systole, 223–24
Systolic blood pressure, 230

T

Tachycardia, sinus, 226
Taenia, 640
Taiga, 706–7
Tanning, and skin cancer, 183
Tapeworms, 639, 640
Tapioca, 626
Tappers, 745
Target organ, 387, 392
Target tissue, 392
Tarsals, 349
Tarsiers, 670
Taste, and taste buds, 367, 369, 370
Taxonomists, 6
Tay-Sachs disease, 481–82
T cells, 266–69
TDF. See Testis determining factor
    (TDF)
Tear glands. See Lacrimal glands
Tectorial membrane, 379
Teeth, 190–92
Telophase, 85, 87, 90, 91, 92
Telson, 650
TEM. See Transmission electron
    micrograph (TEM)
Temperature, 367
    body control, 186–87
    enzymatic reaction, 102

Template, 512
Temporal lobe, of brain, 336–38
Temporate deciduous forests, 707
Tendons, 176–77, 352
Tension. See Cohesion-tension theory
Tentacles, 635
Terminal bud, 147
Terminal web, 196
Termination, 522
Terms, ecological, 720
Terrestrial food chain, 722
Tertiary, 564, 671
Testcross, 474, 475, 478
Testes, 386, 392, 393, 400, 411, 412,
    413, 415. See also Testis
    determining factor (TDF)
Testis determining factor (TDF),
    456, 491
Testosterone. See Androgens
Tetanus, and summation, 355
Tetany, 393
Tetrad, 89
Tetrahydrocannabinol (THC), 342
Tetrapods, 657
Thalamus, 336, 337
Thalidomide, 454
THC. See Tetrahydrocannabinol
    (THC)
Theory, 8
    of acquired characteristics, 579
    of aging, 461–63
    of biology, 12
    cell, 44–47
    cohesion-tension theory of water
        transport, 156, 157
    endosymbiotic, 53
    germ layer, 450
    natural selection, 578
    pressure-flow theory of phloem
        transport, 158–59, 160
    and principle, 12
    sliding filament theory of muscle
        contraction, 360–61
    synthetic, 573–77
Therapy
    gene, 545–48
    immuno-. See Immunotherapy
    thrombolytic, 234–35
Thermal inversion, 750, 752
Thiamin, 209
Thick filaments, 360, 363
Thigh bone, 349
Thin filaments, 360, 363
Third-world country, 736
Thistle tube, 70
Thompson seedless grapes, 161
Thoracic duct, 259
Threshold stimulus, 354–55
Throat, 281, 282, 283
Thrombin, 248–49
Thrombocytes. See Blood platelets
    (thrombocytes)
Thromboembolism, 233
Thrombolytic therapy, 234–35
Thrombus, 233
Thrush, 607
Thylakoids, 54, 125
Thymine, 510, 512
Thymosins, 392, 400

Thymus gland, 259, 260, 261, 392,
    400
Thyroid gland, 19, 386, 392, 393,
    394–95
Thyroid-releasing hormone (TRH),
    393
Thyroid-stimulating hormone
    (TSH), 391, 392, 393,
    394–95
Thyroptropic hormone. See Thyroid-
    stimulating hormone (TSH)
Thyroxin, 392, 393, 394–95
Tibia, 349
Tick, 10, 694–95
Tidal volume, 287
Tight junction, 176, 177
Times Beach, 746
Time scale, geological. See
    Geological time scale
Ti plasmid, 542
Tissue(s), 4
    adipose, 177
    and capillaries, 246
    connective, 173, 174–75, 176–79
    epithelial, 173–76
    meristem, 143
    mesophyll, 151
    muscular, 173, 174–75, 179–80
    nervous, 173, 174–75, 180
    plasminogen. See Tissue
        plasminogen activator (tPA)
    target, 392
    and tissue culture propagation,
        164
    and tissue fluid, 239
    and tissue rejection, 275–76
    tPA. See Tissue plasminogen
        activator (tPA)
    types, 173–80
    vegetative, 143–44
Tissue plasminogen activator (tPA),
    234, 538
Titer, antibody, 270
T lymphocyte. See Lymphocytes
Toads, 659
Tobacco. See Smoking
Tolerance, 340, 343, 344
Tone, muscle, 353
Tongue, and taste buds, 367, 369,
    370
Tonicity, 71–72
Tonsils, 192, 261
Tooth. See Teeth
Totipotency experiment, 442
Touch, 367
Toxins, 602
tPA. See Tissue plasminogen
    activator (tPA)
Trachea, 19, 173, 176, 193, 281,
    282, 652
Tracheids, 155
Tracheophytes, 614, 615, 627
Tracheostomy, 282
Traits
    and inheritance, 469–74
    multi-, 474–79
    sex-influenced, 500–501
    sickle-cell, 485–86

Transcription
    control, 524
    and DNA and RNA, 516–19,
        526–27
Transduction, 590
Transfer RNA (tRNA), 49, 511,
    520–23, 524
Transformation, and genetic
    engineering, 534, 590
Transgenic animals, 544–45
Transgenic organisms, 541–45
Transgenic plants, 541–43
Transition
    demographic, 740–41
    reaction, 105, 106, 107, 111
Translation
    control, 524
    and DNA and RNA, 517,
        520–22
Translocation, 494, 495
Transmission, and synapse, 325–28
Transmission electron micrograph
    (TEM), 44, 45
Transpiration, 156, 157
Transplant, kidneys, 315
Transport
    active, 68, 73
    and aging, 464
    by carriers, 72–75
    cohesion-tension theory of water
        transport, 156, 157
    nutrients in plants, 157–59
    in plants, 155–59
    pressure-flow theory of phloem
        transport, 158–59, 160
    water in plants, 155–57
Transport system, electron. See
    Electron transport system
Transposons, 528
Transverse colon, 196
Tree(s), 164
    burning of, 727
    evolutionary, 631, 658, 675
    life cycle, 620–21
Tree frogs, 3
Treeless biomes, 702–6
Tree shrew, 670
Trematodes, 639
Treponema pallidum, 433
TRH. See Thyroid-releasing
    hormone (TRH)
Triassic, 564
Triceps, 353
Trichinella, 642
Trichinosis, 642
Trichocysts, 595
Trichonympha collaris, 595
Triglycerides, 37
tRNA. See Transfer RNA (tRNA)
Trochophores, 642
Troendle, Gloria, 401
Trophic levels, 722
Trophoblast, 448, 451, 453
Tropical conservation, 708–9
Tropical Forest Resources
    Assessment Project, 708
Tropical rain forests, 689, 707, 708,
    709, 710, 744, 745
Tropomyosin, 363

Troponin, 363
True cells, 557–59
Trypanosoma, 594
Trypanosome infection, 596
Tryp operon, 525–26
Trypsin, 101, 102, 201, 202
TSH. *See* Thyroid-stimulating hormone (TSH)
Tubal ligation, 425
Tube feet, 654
Tube nucleus, 166
Tuber, 626
Tubercle, 293
Tuberculosis, 293–94
Tube-within-a-tube plan, 631, 632
Tubeworms, 137
Tubular excretion, 307, 310
Tubules
   convoluted, 305, 308
   Malpighian, 652, 653
   micro-. *See* Microtubules
   and muscle fiber, 358–59
   seminiferous, 412, 413, 415
Tumors, benign and malignant, 529
Tundra, 704
Tunicate, anatomy, 656
Turgor pressure, 72
Turner syndrome, 493, 495
Turtles, 660
T virus, 509
T wave, 225, 226
Twins, 448
Twitch, muscle, 355
Two-trait crosses, 474–78
Tympanic membrane, 378
Tympanum, 652
Typhlosole, 648
Typhoid fever, 592
Typing, blood. *See* Blood

U

Ulcers, 194–95
Ulna, 349
Ultraviolet (UV) radiation, 181
Ulva, 599–600, 611
Umbilical arteries, 456
Umbilical cord, 451
Umbilical veins, 456
Uncoating, and animal viruses, 586
Unconscious brain, 336
Unifying theories of biology, 12
United Nation's Food and Agriculture Organization (FAO), 708
United States Army Medical Corps, 486
United States Department of Agriculture (USDA), 205, 731, 743
United States Department of Health and Human Services, 205
United States Department of Natural Resources, 546
United States Food and Drug Administration (FDA), 243, 401
United States National Museum of Natural History, 708

Unity of plan, 566
University of California at Los Angeles Medical Center, 428
University of California at San Francisco, 243, 428
University of Cambridge, 514
University of Maryland School of Medicine, 243
University of Minnesota, 356, 546
University of Tennessee Noise Laboratory, 383
University of Texas, 402
University of Wisconsin-Madison, 356
Unsaturated fatty acids, 37
Uracil, 511, 512
Urea, 100
   and blood plasma, 240
   and excretion, 300–301
   and nephrons, 308
Urease, 100
Uremia, 315
Ureters, 303
Urethra, 303, 315, 411, 414
Urethritis, 315. *See also* Nongonococcal urethritis (NGU)
Uric acid, 300–301, 310
Urinalysis, 315
Urinary bladder, 303
Urinary system, 303–12
Urination, 304, 414
Urine, 303–7
   and ADH, 313
   composition, 302
   and excretion, 299–300
   formation, 306, 307–12
   path, 303–4
Urochordata, 656
Uropods, 650
USDA. *See* United States Department of Agriculture (USDA)
Uterine cycle, 418, 420–21
Uterine tubes, 418
Uterus, 418–19
Utricle, 379, 380
UV. *See* Ultraviolet (UV) radiation
Uvula, 192

V

Vaccines, 270, 271
   and DNA, 537, 538
   and respiratory infections, 292
Vacuoles, 46, 47, 48, 50–53, 59, 61, 594
Vagina, 416, 419
Vagus nerve, 329
Valves, 221
   atrioventricular, 222
   and blood flow velocity, 231
   heart, 225
   semilunar, 223
Variable regions, 264
Variables, dependent and experimental, 10
Variations
   and Darwin's theory, 578

maintenance of, 576–77
   production of, 574
Varicose veins, 236
Vascular bundles, 148
Vascular cambium, 148
Vascular cylinder, 147
Vascular pathways, 226–28
Vascular plants, 614–18
Vas deferens, 411, 412
Vasectomy, 425
Vasopressin. *See* Antidiuretic hormone (ADH)
Vectors
   and DNA, 535
   retroviruses as, 545
Vegetables, and nutrition and energy content, 204, 214
Vegetation, zones of, 711
Vegetative organs, 143
Vegetative tissues, 143
Veins, 219, 221
   and capillary exchange with tissues, 247
   hepatic, 228
   hepatic portal, 198–99, 228
   and human circulatory system, 229
   leaf. *See* Leaves
   pulmonary, 227
   umbilical, 456
   varicose, 236
Velocity, blood flow, 231
Venae cavae, superior and inferior, 223, 226, 227
Venous duct, 456
Ventilation, 284
Ventral solid nerve cord, 648, 650, 657
Ventricles, 222, 334
Ventricular fibrillation, 226
Venules, 219, 221, 231, 307
Vermiform appendix, 195
Vernix caseosa, 456
Vertebrae, 348, 569
Vertebral column, 348, 656
Vertebrata, 656, 670
Vertebrate chordates, 656–65
Vertebrates, 630
   breathing mechanisms, 665
   circulatory paths, 666
   classification features, 667
   evolutionary tree of, 658
Vesicle, 46, 47, 48, 61
Vessel elements, 155
Vessels
   blood, 219–21, 239
   lymphatic, 247, 258–59
Vestibule, 380, 419
Vestigial structures, 568–69
Viceroy butterfly, 693
Victoria, Queen, 498
Vigilance, 692
Villi, 195, 309. *See also* Chorionic villi
Viral DNA. *See* Deoxyribonucleic acid (DNA)
Viral infections, 588
Viral RNA. *See* Ribonucleic acid (RNA)

Viroids, 588
Viruses, 243, 250, 262, 429, 583, 584–89
   AIDS. *See* Acquired immune deficiency syndrome (AIDS)
   and animals, 586–88
   and antibiotics and antiviral drugs, 589
   and birth defects, 454
   and cancer, 530, 588
   and common cold, 292
   herpes (HSV). *See* Herpes simplex virus (HSV)
   HPV. *See* Human papillomaviruse (HPV)
   life cycles, 585–88
   and retroviruses, 545, 588
   T, 509
   as vectors, 545
Visceral mass, 643
Viscosity, blood, 241
Visible light, 124
Vision
   biochemistry of, 376
   stereoscopic, 374–75
Vital capacity, 287
Vitamins
   and blood clotting, 248–49
   and blood plasma, 240
   deficiency, 208
   fat-soluble and water-soluble, 209
   functions and sources, 209
   and minerals, 209–11
Vitis vinifera, 161
Vitreous humor, 371, 372
Vocal cords, 282
Voice box. *See* Larynx
Volume
   blood. *See* Blood
   lung. *See* Lungs
Volvox, 599
von Recklinghausen disease, 482
Vulva, 419

W

Walking, and fitness, 357
Wall, cell. *See* Cell wall
Wallace, Alfred Russel, 578
Warm blooded, 662
Warning coloration, 692
Warts, genital, 430, 431
Wastes, 729
   and blood plasma, 178, 240
   disposal, 746–47
   and excretion, 299–300
   hazardous, 746
   nitrogenous, 300
Water, 107
   and agriculture, 730
   and blood plasma, 178, 240
   and capillary exchange, 246, 247
   characteristics of, 24–25
   and cohesion-tension theory, 156, 157
   dissociation of, 25
   environment compared to land environment, 611

and excretion, 301
formation of, 22
and groundwater pollution, 747–48
and homeostasis, 185
and hydrolysis, 29
and nephrons, 308
and oxygen debt, 362
pollution, 747–50
reabsorption of, 310–12
and surface water pollution, 747–48
and synthesis, 29
transport in plants, 155–57
Water molds, 603
Water-soluble vitamins, 209
Water vascular system, 654
Watson, James, 511, 514
Wavelength, 124
Waves
and EEG, 338
and heartbeat, 225, 226
Webs, food. *See* Food webs
Weinberg, W., 571–72
Wheat, 626
Whiskey, 115
Whisk fern, 615
White blood cells (leukocytes), 178, 240, 249–51
agranular and granular, 239, 244, 249–51
composition of, 239
and immunity, 262
See also Blood
White Cliffs of Dover, 595
Whiteheads, 182
White light, 124

White potato, 626
Widow's peak, introduced, 471
Wilkins, Maurice, 514
Windows, oval and round, 378
Windpipe. *See* Trachea
Wine, 115
Wings, 661
Wisdom teeth, 190, 191
Withdrawal, 343, 344, 425
Woody root, 151
Woody stem, 148, 151
Worms
earth-. *See* Earthworm
free-living and parasitic, 639
round-. *See* Roundworms
Wrasse, 698

X

X-linked genes, 496–500
Xylem, 144, 146, 151, 155, 614–15
XYY syndrome, 493, 496

Y

Yeast, 115, 534
Yellow bone marrow, 350, 351. *See also* Bone(s) *and* Bone marrow *and* Red bone marrow
Yellowstone National Park, 592
Yield, crop, 136
Yolk, and yolk sac, 437, 438, 446, 449
York Barbell Club, 401

Z

Zea mays, 6
Zeatin, 160
Zebra, 705
Zidovudine (AZT), 274, 430, 589
Zinc, 210
Z lines, 360
Zone(s)
abyssal, 713, 714, 715
bathyal, 713, 714, 715
bathypelagic, 713, 714, 715
cell division, 145
elongation, 146
epipelagic, 713, 714, 715
life, 711
limnetic, 711
littoral, 711
maturation, 146
mesopelagic, 713, 714, 715
produndal, 711
sublittoral, 713, 714, 715
vegetation, 711
Zonula occludens, 196
Zonulysin, 374
Zooflagellates, 595
Zoomastigina, 595
Zoospores, 598
Zuo-Yan Zhu, 546
Zygospore, 603, 604
Zygote, 81, 447, 448